Pearson Australia

(a division of Pearson Australia Group Pty Ltd)
707 Collins Street, Melbourne, Victoria 3008
PO Box 23360, Melbourne, Victoria 8012
www.pearson.com.au

The *Pearson Mathematics 7–10* second edition series is an adaptation of the *Heinemann Mathematics 7–10* series.

First published 2017 by Pearson Australia
2024 2023 2022 2021
10 9 8 7 6 5 4 3 2 1

Series Publisher: Tim Carruthers
Content and Learning Specialists: Catherine McKenzie, Aynur Bulut
Project Managers: Tamara D'Mello-Pirois, Emma-Jane McCarroll, Jennifer Boyce
Production Managers: Diane Leyman, Shahara Ahmed
Editor: Liz Waud
Lead Editor: Daniel Hernández
Designers: Glen McClay, Nikki M Group, Anne Donald
Illustrators: Sunset Publishing Services, diacriTech, Nikki M Group
Typesetters: Nikki M Group
Rights & Permissions Editor: Justin Lim
Series Cover Design: Anne Donald

Printed in Malaysia (CTP-VVP)

National Library of Australia Cataloguing-in-Publication entry

Creator:	Nolan, Jennifer, author.
Title:	Pearson mathematics 10 student book / Jennifer Nolan [and twenty-seven others]
ISBN:	9781488656873 (paperback)
Target Audience:	For secondary school age.
Subjects:	Mathematics—Textbooks.
	Mathematics—Problems, exercises, etc.
	Mathematics—Study and teaching (Secondary)
Dewey Number:	510.712

Pearson Australia Group Pty Ltd ABN 40 004 245 943

Acknowledgements

All material identified by AC Australian CURRICULUM is material subject to copyright under the *Copyright Act 1968* and is owned by the Australian Curriculum, Assessment and Reporting Authority 2017.

ACARA neither endorses nor verifies the accuracy of the information provided and accepts no responsibility for incomplete or inaccurate information. In particular, ACARA does not endorse or verify that:

- The content descriptions are solely for a particular year and subject;
- All the content descriptions for that year and subject have been used; and
- The author's material aligns with the Australian Curriculum content descriptions for the relevant year and subject.

You can find the unaltered and most up to date version of this material at http://www.australiancurriculum.edu.au/ This material is reproduced with the permission of ACARA.

We thank the following for permission to reproduce copyright material. The following abbreviations are used in this list: t = top, b = bottom, l = left, r = right, c = centre.

Cover image: 123RF/ sakkmesterke

123RF: anmbph, p. 502b; Rafael Ben-Ari, p. 490t; bizoon, p. 306; bluekaizer, p. 424t; Carolina Jaramillo Castro, p. 636; Dean Drobot, p. 502t; Phiseksit Inthip, p. 378t; Chatchai Issariya, p. 247; jabiru, p. 430c; Rafael Ramirez Lee, p. 432; Chad McDermott, p. 216; mreco99, p. 640; nightman1965, p. 358; Adrienn Orbánhegyi, p. 341c; PhotosIndia.com LLC, p. 310tr; Tatiana Popova, p. 206b; rangizzz, p. 285; sakkmesterke, i; Richard Semik, p. 425b; Darryl Smith, p. 215b; Vlad Teodor, p. 128tl; Maksim Toome, p. 755t; Shane White, p. 254; Natallia Yeumenenka, p. 422t.

Australian Bureau of Statistics (ABS): © Commonwealth of Australia, 1370.0 - Measures of Australia's Progress, 2010, p. 113tl; Australian Social Trends, cat. no. 4102.0, p. 127.

Alamy Stock Photo: Allstar Picture Library, p. 532l; Tom Bean, p. 350b; Martin Berry, p. 62; Mark Burnett, p. 596; Classic Image, p. 57; Daisy Images, p. 583; epa european pressphoto agency b.v., p. 280; Rob Francis, p. 95; John Fryer, p. 622; Dennis Hallinan, p. 425c; Image Source Plus, p. 616; Justin Kase z11z, p. 314tr; Leslie Garland Picture Library, p. 366c; Chris Laurens, p. 28; PACIFIC PRESS, p. 562b; PhotoBliss, p. 348br; SciMed images, pp. 315b, 315t; speedpix, p. 125cl; Doug Steley B, p. 309; Trains and Planes, p. 302c; Travelscape Images, p. 560b; Ian Trower, p. 346; Lisa Werner, p. 423t; Wildlight Photo Agency, p. 113tr; Tengku Mohd Yusof, p. 11.

Australian Government Department of Defence: p. 368t.

Bureau of Meteorology (BOM): © Commonwealth of Australia, p. 145b.

Casio Computer Co., Ltd.: Casio ClassPad CAS screenshots and elements.

Department for Communities and Social Inclusion, Government of South Australia: Fact Sheet – 'The Pokies: The Odds', sourced on 14 June 2016, http://www.problemgambling.sa.gov.au/__data/assets/pdf_file/0013/11416/opg-fact-sheet-6.pdf, p. 131;

Fotolia: iofoto, p. 337t.

Getty Images: James Braund, p. 44; CBS Photo Archive, p. 382b; Bill Frakes, p. 120; Chris Graythen/Staff, pp. 236-7t; Paul Kane/Stringer, p. 215t; Matt King/Stringer, p. 88; David Madison, pp. 381t, 655b; Alfred Pasieka, p. 627; Science Photo Library, p. 626cr.

ImageFolk (was Amana/Corbis): Nano Calvo/ZUMA Press, p. 349; Sean Justice, p. 732.

Medical Xpress: 'Parents' top fears about teens' cellphone use—are they justified?', 18 November 2015, p. 128.

Microsoft Corporation: Used with permission from Microsoft., pp. 135t, 238r, 239bl, 239cl, 256bl, 256tr, 722bl, 722cr, 723bl, 723cr, 723t.

NASA Images: p. 291; SA/courtesy of nasaimages.org, p. 255.

Newspix: Angelo Soulas, p. 140.

Pearson Australia: p. 2; Malcolm Cross, pp. 342t, 357b, 656; Amanda Marasco, pp. 265, 666; Alice McBroom, pp. 63, 148, 372, 439, 492, 570, 706; Bronwyn Quint, p. 635.

Pearson Education Inc: Gareth Boden, p. 259; Trevor Clifford, p. 56 (kiwi fruit); Imagemore Co., Ltd, p. 456c; John Foxx Collection/Imagestate, p. 621; MIXA Co., Ltd, p. 81cr; Sozaijiten, p. 263br.

Press Association: Martyn Ziegler, p. 125.

Roy Morgan Research: "The brew crew: Australia's heaviest coffee drinkers", 6 March 2014, Roy Morgan Research, http://www.roymorgan.com/findings/5470-brew-crew-australias-heaviest-coffee-drinkers-201403060304, p. 126.

Shutterstock: Allies Interactive, p. 624-5 (background); Andy Dean Photography, p. 730tr; Subbotina Anna, pp. 58-9; Apples Eyes Studio, p. 490br; arka38, p. 490bl; Ashwin, p. 623; BarryTuck, p. 448; B Calkins, p. 263bl; Narudom Chaisuwon, p. 341b; clearviewstock, p. 111; cloki, p. 56 (pineapple); Stephen Coburn, p. 415; Corepics VOF, p. 758; EPSTOCK, p. 652t; Excellent backgrounds, p. 3; Gilmanshin, p. 588; Andrii Gorulko, p. 335 (plane); guentermanaus, p. 491tl; Hitdelight, p. 626bl; HomeStudio, p. 161tr; Nataliya Hora, p. 294br; Kaesler Media, pp. 192 (seat belt); John Kasawa, p. 728; Nickolay Khoroshkov, p. 491tr; Kichigin, p. 686tl; Robert Kneschke, p. 490bc; Konstantinos Kokkinis, p. 491b; koya979, p. 290; Simon Krzic, p. 295c; kuzmichs, p. 685; Reinhold Leitner, p. 534t; logoboom, pp. 192-3t; Steve Lovegrove, p. 425t; MaraZe, p. 319tr; Bob Mawby, p. 302t; Lorelyn Medina, p. 264t; MP_P, p. 366b; mtlapcevic, p. 429c; Pedro Nogueira, p. 625 (dna helix); Odua Images, p. 45; OksanaAriskina, p. 264t; O.V.D., p. 166; Ioan Panaite, p. 720; Andrew Park, p. 534b; Curly Pat, p. 149; BORTEL Pavel - Pavelmidi, p. 562t; Edyta Pawlowska, pp. 32-3; Photosani, p. 82; Portfolio, p. 490c; Nikolai Pozdniakov, p. 424c; rayjunk, p. 308tr; Federico Rostagno, p. 582c; Sergey Ryzhov, p. 194; Juan Sanchez, p. 109cr; Alexey V Smirnov, p. 641t; Aleksey Stemmer, p. 310c; StudioSmart, p. 428b; Cristian Zamfir, p. 702.

Texas Instruments Australia Pty Ltd: TI-Nspire CAS screenshots and elements.

The Australian: 'Can cabbages to Korea cover for coal's comedown?', by Rowan Callick, 17 Nov 2015, The Australian, p. 129.

Thinkstock: p. 609; 3bugsmom, p. 647c; BenGoode, p. 422b; Brand X Pictures, p. 132; Comstock, p. 582b; coryz, p. 449; cyda101, p. 430t; fstockfoto, p. 295t; GeorgiosArt, p. 694; ia_64, p. 595; jd_field, p. 380c; Jupiterimages, p. 456t; Dmitry Kalinovsky, p. 746; lisafx, p. 320; Ryan McVay, p. 589; perkmeup, p. 607; PhotoObjects.net, p. 275br; Photos.com, p. 122bl; Mara Radeva, p. 119; scol22, p. 755b; Sandra Van der steen, p. 715; Alexandr Stepanov, p. 20cr; TomFullum, p. 142; tomislz, p. 356c; Vinicius Tupinamba, p. 737; upthebanner, p. 729; wysokiniski, p. 294cr.

Wikimedia Commons: Maciej Jaros, p. 532r.

Every effort has been made to trace and acknowledge copyright. However, should any infringement have occurred, the publishers tender their apologies and invite copyright owners to contact them.

PEARSON Mathematics 7–10

Writing and Development Team

Evelyn Ashcroft
Canberra Mathematical Association President
Mathematics Consultant
Teacher—ACT

Dr Maria Athanassenas
Academic Consultant
Senior Lecturer—Monash University

Bob Aus
Contributing Author
Teacher—NSW

Rosetta Batsakis
Contributing Author
Teacher—Victoria

Deborah Bridge
Contributing Author
Teacher—NSW

Jennie Bucco
Contributing Author
Teacher—NSW

Aynur Bulut
Content and Learning Specialist
Pearson Australia

Terry Byers
Contributing Author
Teacher—Queensland

Greg Carroll
Contributing Author
Campus Principal—Victoria

Tim Carruthers
Lead Publisher (2nd Edition)
Content and Learning Specialist
Pearson Australia

Karen Chin
Contributing Author
Teacher—Victoria

David Coffey
Contributing Author
Teacher—Victoria

Evan Curnow
Development Editor (First Edition)
Pearson Australia

Mark Darrell
Mathematical Association of South Australia President
Assistant Principal—South Australia

Noel Davies
Author

Tony Daws
Contributing Author
Teacher—Victoria

Kiran Dhot
Contributing Author
Teacher—Tasmania, Victoria, Singapore

George Dimitriadis
Contributing Author
Teacher—Victoria

Mark Dusting
Contributing Author
Teacher—Victoria

Sia Evans
Contributing Author
Teacher—Victoria

Vebica Evans
Publisher (First Edition)
Pearson Australia

Patricia Fraser
Consultant
Teacher—Victoria

Belinda Frew
Contributing Author
Teacher—Victoria, NSW

Kelly Gallivan
Mathematics Education Consultant

Catherine Gatt
Consultant
Teacher—Victoria

Patricia Goss
Consultant

Dr John Gough
Deakin University
Mathematics Education Consultant

Bozenna Graham
Contributing Author & Technical Consultant
Teacher—Victoria

Joshua Harnwell
Contributing Author
Teacher—NSW

Daniel Hernández
Lead Editor (Mathematics)
Pearson Australia

Kate Hillery
Contributing Author
Teacher—NSW, Victoria

Gregory Hine
Consultant
Teacher—Western Australia

Cindy Hogan
Author
(Homework Program)

Damien Igoe
Contributing Author
Teacher—Queensland

Margaret Junor
Contributing Author
Teacher—Victoria

Helen Keating
Consultant

Leah Kelly
Consultant

Ingrid Kemp
Contributing Author
Teacher—Victoria

Dana Killmister
Contributing Author
Teacher—Victoria

Jana Kohout
Contributing Author
Teacher—Victoria

Richard Korbosky
Mathematical Association of WA President
Edith Cowan University
DAPMA Education Consultancy

Fiona Latrobe
Contributing Author & Technical Consultant
Teacher—Victoria

Donna Lannan
Contributing Author
Teacher—Victoria

Antje Leigh-Lancaster
Head of K–12 Learning Services (Mathematics)
Pearson Australia

Martin Lindsay
Contributing Author
Teacher—Victoria

Julian Lumb
Content and Learning Specialist
Pearson Australia

Amanda Marasco
Contributing Author
Teacher—Victoria

Anne Matheson
Contributing Author
Teacher—Victoria

Catherine McKenzie
Content and Learning Specialist
Pearson Australia

John McLaverty
Contributing Author
Teacher—Victoria

Gael McLeod
Content Developer (First Edition)
Pearson Australia

John McMillan
Contributing Author
Teacher—Victoria

Andrew Mentlikowski
Author

Rob Money
Consultant

Marian Nicolazzo
Literacy and ESL Consultant

Jennifer Nolan
Author

James O'Connor
Contributing Author
Teacher—Victoria, Western Australia

Gary Passmore
Aboriginal Cultural Studies Consultant—South Australia

Dr Anne Paterson
Mathematical Association of Western Australia
Teacher—Western Australia

Karen Perkins
Consultant
Teacher—Victoria

Trang Pham
Contributing Author
Teacher—Victoria

Geoff Phillips
Author

Annette Psereckis
Mathematics Consultant—Tasmania

Katherine Quane
Consultant
Teacher—NSW

Shanna Rankin
Consultant

Li Richardson
Consultant

Soumya Saini
Contributing Author
Teacher—Victoria

Aimee Shackleton
Contributing Author
Teacher—Victoria

Lindy Sharkey
Contributing Author
Teacher—Victoria

Neetu Sharma
Contributing Author
Teacher—Victoria

Nicola Silva
Contributing Author
Teacher—Queensland

Tanya Smith
Contributing Author
Teacher—Victoria

Chris Snell
Consultant
Teacher—Victoria

Jena Sous
Contributing Author
Teacher—Victoria

Professor Kaye Stacey
University of Melbourne
Mathematics Education Consultant

Ann Stephens
Contributing Author
Teacher—Victoria

Dirk Strasser
Author

Sofia Thapa
Contributing Author
Teacher—Victoria

Arthur Tsingoidas
Contributing Author
Teacher—Victoria

Peter Walsh
Consultant
Teacher—Victoria

Sahooda Walters
Contributing Author
Teacher—Victoria

Terry White
Author

Leanne Wilson
Contributing Author
Teacher—Victoria

Contents

PEARSON mathematics

Student Book

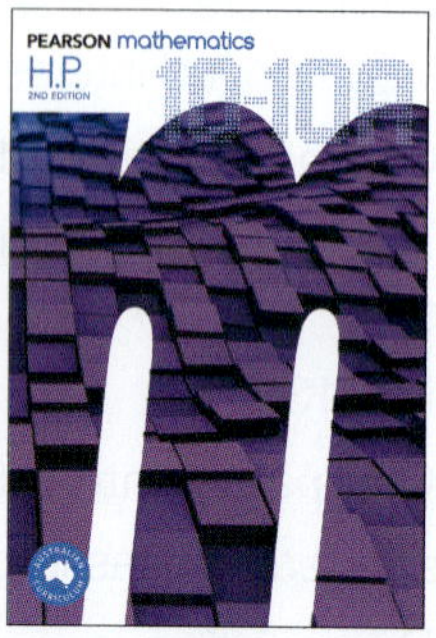

Homework Program

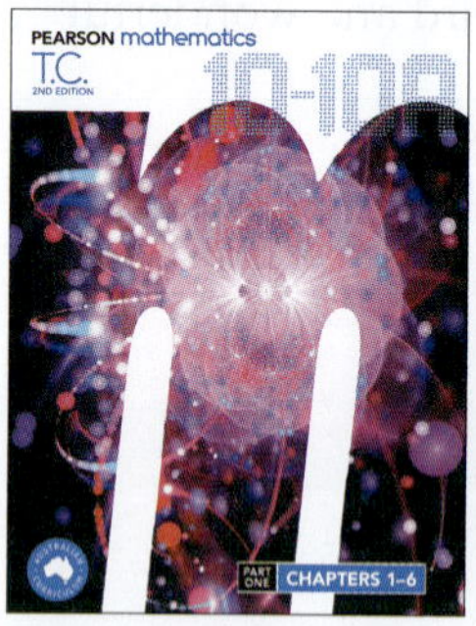

Teacher Companion 1

Teacher Companion 2

LS LightbookStarter

Lightbook Starter

eBook

Student Book

The Student Book for 10–10A provides maximum flexibility for teachers by including both 10 and 10A material in the same book. All 10A material is clearly labelled, enabling teachers to cover 10 or 10A course topics as appropriate for different classes.

The Second Edition Student Book includes updated questions, activities and design, with full coverage of the Australian Curriculum: Mathematics as well as the Victorian Curriculum: Mathematics.

It incorporates the latest research as well as feedback from teachers and learners across Australia.

Content caters for students of all abilities, with improved differentiation of all exercise questions and more questions for students consolidating their skills.

Homework Program

The Homework Program provides a collection of tear-out worksheets for students to practise and revise mathematical concepts.

Teacher Companion

The Teacher Companion makes lesson preparation easy by combining full-colour Student Book pages with teacher support including improved contextual teaching suggestions and strategies, class activities, extra questions, worked solutions and answers for every question in the Student Book.

AUSTRALIAN CURRICULUM

Pearson

Pearson Lightbook Starter

Lightbook Starter is an innovative digital resource powered by Pearson's award-winning Lightbook technology. It has been developed to help students learn key mathematical concepts, evaluate their understanding and track their progress. 'Before you begin' sections assess learner readiness before each chapter topic, while 'Check-in' questions can be used to evaluate learner understanding and practice after every chapter section.

Auto-correcting questions are linked to the Progress Tracker dashboard for easy analysis and viewing of results, which are mapped to progression through the Student Book as well as to Australian Curriculum: Mathematics and Victorian Curriculum: Mathematics content descriptions.

Pearson eBook

Much more than just pages on a screen, Pearson eBook is an online or offline version of your Student Book linked to interactive content, rich media resources and other useful content specifically developed for Mathematics. It supports you with appropriate online resources and tools for every section of the Student Book, including videos, eWorked Examples, interactive lessons, worksheets and more. Teacher resources include chapter tests, full teaching programs and curriculum mapping for the Australian Curriculum: Mathematics and for the Victorian Curriculum: Mathematics.

Pearson Places is the gateway to digital learning material for teachers and students across Australia. Access your content at **www.pearsonplaces.com.au**.

Professional Learning, Training and Development

Did you know that Pearson also offers teachers a diverse range of training and development product-linked learning programs? We are dedicated to supporting your implementation of **Pearson Mathematics**, but it doesn't stop there.

Our courses align closely with Pearson Mathematics Second Edition and offer an in-depth learning experience, combining both practical and theoretical elements, enabling you to implement the resource effectively in your classroom.

Find out more about our product-linked learning, workshops, courses and conferences at Pearson Academy **www.pearsonacademy.com.au**.

Using PEARSON mathematics

2ND EDITION

The Australian Curriculum: Mathematics aims to ensure that students:

- *are confident, creative users and communicators of mathematics, able to investigate, represent and interpret situations in their personal and work lives and as active citizens*
- *develop an increasingly sophisticated understanding of mathematical concepts and fluency with processes, and are able to pose and solve problems and reason in number and algebra, measurement and geometry, and statistics and probability*
- *recognise connections between the areas of mathematics and other disciplines and appreciate mathematics as an accessible and enjoyable discipline to study.*

Australian CURRICULUM

Pearson Mathematics shares these aims.

Pearson Mathematics Second Edition is a compelling research-based series, written by experienced and practising Australian teachers with the support of Australia's leading mathematics education experts. It has been specifically designed to scaffold students' fluency development, conceptual understanding, reasoning and problem-solving skills.

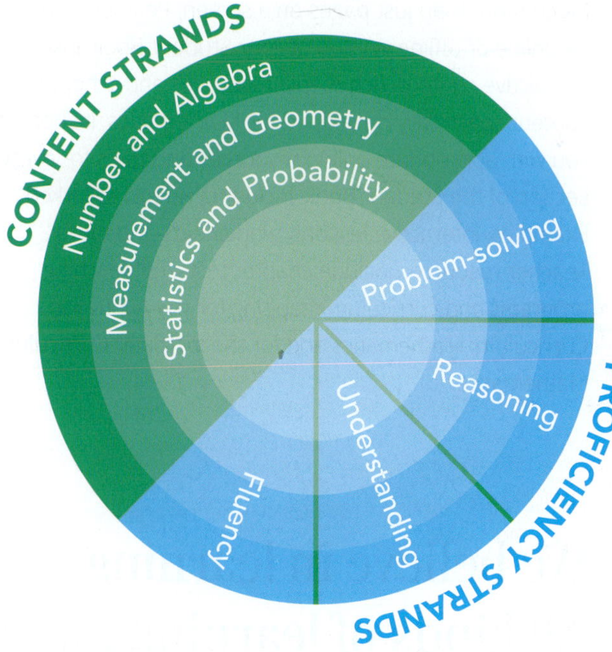

Supporting all learners

Differentiation

Pearson Mathematics Second Edition has been written and designed for the needs of the full ability spectrum of students in Australian classrooms.

1 Recall

Recall—Each chapter begins with a review of assumed and necessary knowledge for the chapter. For students needing extra revision, Recall Worksheets are available for each Recall question, with explanations to refresh understanding and exercises to practise skills.

3.1 Expanding brackets

Exercises—All Exercises include questions according to the Australian Curriculum Proficiency Strands: Fluency, Understanding and Reasoning. They have been carefully paced to help students build skills, develop deep conceptual understanding and apply learning. Every Exercise also has 'Open-ended' questions to encourage students' creative thinking and ability to communicate mathematics effectively.

Navigator

1, 2, 3, 4, 5, 6, 7, 10, 12, 13, 16, 19, 22	2 (a, b), 4, 5, 6, 7, 8, 9, 10, 12, 13, 14, 16, 19, 20, 21, 22	2 (a, b), 4 (a), 5 (a), 6 (a), 7 (a), 8, 9, 10, 11, 14, 15, 16, 17, 18, 19, 20, 21, 22, 23

Navigator—Three optional graded pathways through *every* Exercise, with every question rated as Foundation, Standard or Advanced level.

Challenge A

Challenge—With every Mixed Review in this Student Book there is a Challenge section for early finishers and advanced students.

Regular revision and reinforcement

Pearson Mathematics Second Edition has a broad range of cumulative and chapter-based revision.

Half-time 3

Half-time—A mid-chapter review of chapter content so far.

Chapter review 3

Chapter review—An end-of-chapter thorough review of chapter content.

Mixed review A

Mixed review—Cumulative revision that mixes content from previous chapters.

Encouraging inquiry and problem solving

Investigation

These are scaffolded to be accessible to all learners and structured following a 5e+ format.

Investigation

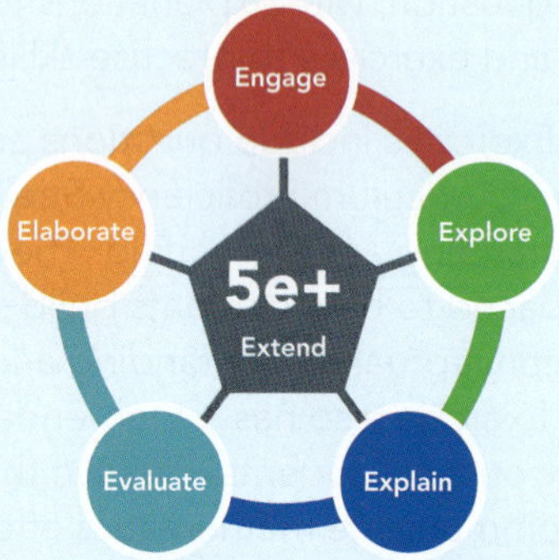

Exploration Spreadsheet and CAS

Pearson Mathematics Second Edition understands the importance places of students developing technological literacy. Explorations with CAS technology and spreadsheet software allow students to learn and practise their technical proficiency, while also helping to develop students' deeper understanding of the mathematical concepts covered.

Exploration Spreadsheet

Exploration CAS

Problem solving, Puzzles and Games

These are located throughout every chapter and give students fun opportunities to develop their problem-solving skills and logical reasoning.

Problem solving

Puzzle

Game

Exploration Coding

Algorithmic thinking is an increasingly relevant component of mathematics education for the 21st century. Algorithmic thinking and coding tasks encourage learners to understand and develop reasoning skills with computational procedures, algorithms and logical problem-solving, exploring a coding environment within a mathematical context.

Exploration Coding

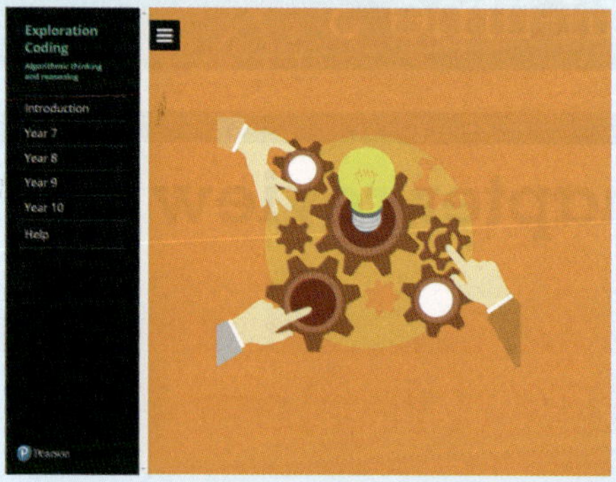

Real-life contexts and engagement

Exploration Task and STEM

Practical contexts provide useful learning opportunities and natural pathways into interesting, relevant mathematics.

Exploration tasks are rich tasks that encourage student engagement, questioning and creative thinking. STEM activities allow students to explore topics related to Science, Technology, Engineering and Design with a Mathematics perspective.

Exploration Task

Exploration STEM

Home Page

Anticipate the 'Why learn this?' question and provide a motivating entry into every new chapter topic, featuring an interview about mathematics in the workplace.

Pearson Mathematics Second Edition has been designed to capture students' interest, with the incorporation of Maths 4 Real, Gamespace and more.

Maths 4 Real

Scenarios that help students make connections with mathematics in the real world.

Gamespace

For year 7 to year 9, skill consolidation and reinforcement wrapped up in fun and quirky scenarios, including multi-player maths board games and solve-the-riddle, find-the-clue tasks and games.

1

Linear relationships

Maths saves the planet!

Name: Zoe Ryan
Job: Forest Carbon Specialist
Qualifications: Bachelor of Forest Science

Developing countries with tropical rainforests have had a high rate of deforestation in the past. However, a change in attitude over the past few years and a growing interest in the environment is halting the devastation. Zoe Ryan is a Forest Carbon Specialist. Her job is to measure the amount of carbon stored in a forest to calculate the amount of greenhouse gases that would be emitted if they were cut down.

About a third of Zoe's time is spent abroad, travelling to tropical regions in countries such as Borneo, Papua New Guinea, Peru, Ecuador and Mozambique. She has to travel on canoes through croc-infested waters to get to some of the forests. 'It's really wild; you wouldn't find these places in a Lonely Planet travel guide … there's always something totally unexpected, you just have to let go and go with it.'

In the forest, Zoe takes measurements of the trees and uses sampling in order to measure the carbon stock in the forest. After the quantity of avoided emissions is calculated, it is then audited and verified by an independent third party. Carbon credits are then sold in the carbon market to companies who want, or need, to reduce their carbon emissions debit. Deforestation contributes about 15% of the world's greenhouse gas emissions. Getting companies to buy carbon credits means that the forest stays protected.

Knowing how important it is to manage the forest sustainably, it is crucial for Zoe to be accurate with her maths. 'I can see that the more samples we put in, the lower our error, so the more carbon we're able to sell because our precision is greater. I feel like I'm genuinely doing something that's good for society, and it's also an adventure.'

Why learn this?

A variety of jobs use linear relationships and systems of linear equations to represent, analyse and solve a variety of problems, and to manipulate numbers to find quantities. Linear equations are used extensively in all branches of science and technology, including computer science, biology, chemistry, physics, anthropology and all the engineering professions. Other professionals such as economists, bankers, architects and photographers also use linear equations to understand information and to predict future results.

After completing this chapter you will be able to:

- solve linear equations
- find the gradient of a line joining two points or from an equation
- sketch linear graphs
- find the equations of parallel and perpendicular lines
- find the solutions for linear inequalities
- solve simultaneous linear equations.

1 Recall

Prepare for this chapter by attempting the following questions. If you have difficulty with a question, you can download a Recall Worksheet from the eBook or the Pearson Places website.

1 If $P = 2t - 7$, find P when:

(a) $t = 0$ (b) $t = 5$ (c) $t = -4$ (d) $t = 2$

2 Solve the following linear equations.

(a) $2a - 3 = 11$ (b) $\frac{b}{4} + 7 = 13$ (c) $3(2c - 5) = 45$ (d) $\frac{2d + 6}{5} = 4$

3 Plot these coordinate points on a Cartesian plane.

(a) $A(4, 3)$ (b) $B(0, 2)$ (c) $C(-1, 2)$ (d) $D(-2, -3)$

4 Copy and complete the table of values below for each of the following equations, then plot each graph for $-2 \le x \le 2$ on a Cartesian plane.

x	-2	-1	0	1	2
y					

(a) $y = x + 3$ (b) $y = x - 2$ (c) $y = 2 - x$ (d) $y = 3x$

5 Sketch the graphs of:

(a) $y = 3$ (b) $x = 2$

6 Write **(i)** the gradient and **(ii)** the y-intercept for the graphs of the following equations.

(a) $y = 3x + 5$ (b) $y = 3 - 2x$ (c) $2x + 3y = 4$

7 Find the gradient of the line joining the points with coordinates:

(a) (1, 3) and (4, -2) (b) (-3, -6) and (2, -1)

8 (a) Sketch the graphs of $x = 1$ and $y = 3$ on the same set of axes.

(b) Find the point of intersection of the two lines.

Exploration Task

You can download this activity from the eBook or the Pearson Places website.

Straight lines

In this activity you will explore how to match equations with different kinds of straight lines.

1.1 Linear equations

A **linear relationship** exists between two variables if the graph of the relationship is a **linear graph** (a straight line). In a linear relationship, a uniform change in one variable, the **independent variable**, produces a uniform change in the other, the **dependent variable**. The variables x and y are commonly used because they represent the axes of the Cartesian plane on which linear relationships are graphed.

A linear relationship is described by a **linear equation**. If you know the value of one of the variables, you can solve the equation to find the corresponding value of the other variable. You can solve a linear equation using inverse operations. Linear equations have only one solution.

Worked example 1 — W.E. 1

Solve each of the following equations to find the value of the unknown.

(a) $5x - 6 = 34$

(b) $\frac{3a}{2} + 4 = -1$

Thinking	Working
(a) 1 Use inverse operations in the opposite order to find the value of the unknown. (Here, add 6 to both sides and then divide both sides by 5.)	(a) $5x - 6 = 34$ $5x = 40$ $x = 8$
2 Check your answer by substituting the value of the pronumeral into the equation.	$\text{LHS} = 5x - 6$ $= 5 \times 8 - 6$ $= 34$ $= \text{RHS}$
(b) 1 Use inverse operations in the opposite order to find the value of the unknown. (Here, subtract 4 from both sides, multiply both sides by 2 and then divide both sides by 3.)	(b) $\frac{3a}{2} + 4 = -1$ $\frac{3a}{2} = -5$ $3a = -10$ $a = \frac{-10}{3}$ $= -3\frac{1}{3}$
2 Check your answer by substituting the value of the pronumeral into the equation.	$\text{LHS} = \frac{3a}{2} + 4$ $= \frac{3 \times -3\frac{1}{3}}{2} + 4$ $= \frac{3 \times -\frac{10}{3}}{2} + 4$ $= -\frac{10}{2} + 4$ $= -5 + 4$ $= -1$ $= \text{RHS}$

Equations where the unknown appears more than once

To solve equations of the type $5a - 3 = 2a + 6$ you must first add or subtract a **term** containing the pronumeral on both sides of the equation, so that the terms with pronumerals appear on one side only (usually the LHS). If the equations involve brackets, you can expand the brackets using the distributive law $a(b + c) = ab + ac$.

Worked example 2

W.E. 2

Solve each of the following equations to find the value of the unknown.

(a) $5a - 3 = 2a + 6$ **(b)** $2a - 5 = 6a + 1$ **(c)** $4(k + 1) = 3(6 - k)$

Thinking	Working
(a) 1 Add or subtract a term so that the terms with pronumerals appear on one side of the equation only. (Here, subtract $2a$ from both sides.)	(a) $5a - 3 = 2a + 6$ $3a - 3 = 6$
2 Use inverse operations in the opposite order to find the value of the unknown. (Here, add 3 to both sides and then divide both sides by 3.)	$3a = 9$ $a = 3$
3 Check your answer by substituting the value of the pronumeral into each side of the equation.	$\text{LHS} = 5a - 3$ $= 5 \times 3 - 3$ $= 15 - 3$ $= 12$ $\text{RHS} = 2a + 6$ $= 2 \times 3 + 6$ $= 6 + 6$ $= 12$ LHS = RHS
(b) 1 If the coefficient of the pronumeral term on the RHS is greater than the coefficient of the pronumeral on the LHS, then swap the sides of the equation. (Here, 6 > 2 so swap sides.)	(b) $2a - 5 = 6a + 1$ $6a + 1 = 2a - 5$
2 Add or subtract a term so that the terms with pronumerals appear on one side of the equation only.	$4a + 1 = -5$
3 Use inverse operations in the opposite order to find the value of the unknown.	$4a = -6$ $a = -\frac{6}{4}$
4 Simplify your solution if necessary.	$= -\frac{3}{2}$
5 Check your answer by substituting the value of the pronumeral into each side of the equation.	$\text{LHS} = 2a - 5$ $= 2 \times -\frac{3}{2} - 5$ $= -3 - 5$ $= -8$ $\text{RHS} = 6a + 1$ $= 6 \times -\frac{3}{2} + 1$ $= -9 + 1$ $= -8$ LHS = RHS

(c) 1 Expand brackets.	(c) $4(k+1)=3(6-k)$ $4k+4=18-3k$
2 Add or subtract a term so that the terms with pronumerals appear on one side of the equation only. (Here, add $3k$ to both sides.)	$7k+4=18$
3 Use inverse operations in the opposite order to find the value of the unknown. (Here, subtract 4 from both sides and then divide both sides by 7.)	$7k=14$ $k=2$
4 Check your answer by substituting the value of the pronumeral into both sides of the equation.	$\text{LHS}=4(k+1)$ $=4(2+1)$ $=4\times3$ $=12$ $\text{RHS}=3(6-k)$ $=3(6-2)$ $=3\times4$ $=12$ $\text{LHS}=\text{RHS}$

To solve an equation containing fractions, find the lowest common multiple (LCM) of the denominators, called the lowest common denominator (LCD). You then multiply both sides of the equation by the LCD. This simplifies the equation by eliminating the fractions. The equation can now be solved as shown previously.

Worked example 3

W.E. 3

Solve each of the following equations to find the value of the unknown.

(a) $\dfrac{2x+1}{3}=\dfrac{x-3}{5}$ (b) $\dfrac{x}{3}-2=\dfrac{x-1}{4}$

Method 1: Multiplying by the LCD

Thinking	Working
(a) 1 Find the LCD of all fractions in the equation.	(a) $\text{LCD}=15$
2 Multiply every term in the equation by the LCD. Insert brackets if required.	$\dfrac{2x+1}{3}=\dfrac{x-3}{5}$ $\dfrac{15(2x+1)}{3}=\dfrac{15(x-3)}{5}$
3 Divide the denominator of each term into the LCD.	$5(2x+1)=3(x-3)$
4 Expand any brackets.	$10x+5=3x-9$
5 Add or subtract a term so that the terms with the pronumeral is on one side of the equation only. (Here, subtract $3x$ from both sides.)	$7x+5=-9$
6 Use inverse operations to find the value of the unknown.	$7x=-14$ $x=-2$

7 Check the value of your solution by substituting the value of the pronumeral into both sides of the original equation.	$\text{LHS} = \frac{2 \times -2 + 1}{3} = \frac{-3}{3} = -1$ $\text{RHS} = \frac{-2 - 3}{5} = \frac{-5}{5} = -1$ LHS = RHS
(b) 1 Find the LCD of all fractions in the equation.	(b) LCD = 12 $\frac{x}{3} - 2 = \frac{x-1}{4}$
2 Multiply every term in the equation by the LCD. Insert brackets if required.	$\frac{12x}{3} - 12 \times 2 = \frac{12(x-1)}{4}$
3 Divide the denominator of each term into the LCD.	$4x - 24 = 3(x - 1)$
4 Expand any brackets.	$4x - 24 = 3x - 3$
5 Add or subtract a term so that the terms with the pronumerals are is on one side of the equation only. (Here, subtract $3x$ from both sides.)	$x - 24 = -3$
6 Use inverse operations to find the value of the unknown.	$x = 21$
7 Check the value of your solution by substituting the value of the pronumeral into both sides of the original equation.	$\text{LHS} = \frac{21}{3} - 2 = 7 - 2 = 5$ $\text{RHS} = \frac{21 - 1}{4} = \frac{20}{4} = 5$ LHS = RHS

Alternatively, you can write all fractions with a common denominator (the LCD) and then multiply both sides of the equation by the LCD.

Method 2: Writing all terms as fractions using the LCD

Thinking	**Working**
(a) 1 Find the LCD of all fractions in the equation.	(a) LCD = 15
2 Write all terms in the equation as fractions using the LCD as the denominator. Use brackets if required. (Here, $\frac{1}{3} = \frac{5}{15}$ and $\frac{1}{5} = \frac{3}{15}$.)	$\frac{2x+1}{3} = \frac{x-3}{5}$ $\frac{5(2x+1)}{15} = \frac{3(x-3)}{15}$
3 Multiply every term on both sides of the equation by the denominator (LCD) to eliminate the fraction.	$5(2x + 1) = 3(x - 3)$
4 Expand any brackets.	$10x + 5 = 3x - 9$

5 Add or subtract a term so that the terms with the pronumeral are on one side of the equation only. (Here, subtract $3x$ from both sides.)	$7x + 5 = -9$
6 Use inverse operations to find the value of the unknown.	$7x = -14$ $x = -2$
7 Check the value of your solution by substituting the value of the pronumeral into both sides of the original equation.	$\text{LHS} = \frac{2 \times -2 + 1}{3} = \frac{-3}{3} = -1$ $\text{RHS} = \frac{-2 - 3}{5} = \frac{-5}{5} = -1$ LHS = RHS
(b) 1 Find the LCD of all fractions in the equation.	(b) LCD = 12
2 Write all terms in the equation as fractions using the LCD as the denominator. Use brackets if required. (Here, $\frac{1}{3} = \frac{4}{12}$, $2 = \frac{24}{12}$ and $\frac{1}{4} = \frac{3}{12}$.)	$\frac{x}{3} - 2 = \frac{x-1}{4}$ $\frac{4x}{12} - \frac{24}{12} = \frac{3(x-1)}{12}$
3 Multiply every term on both sides of the equation by the denominator (LCD) to eliminate the fraction.	$4x - 24 = 3(x - 1)$
4 Expand any brackets.	$4x - 24 = 3x - 3$
5 Use inverse operations to find the value of the unknown.	$x = 21$
6 Check the value of your solution by substituting the value of the pronumeral into both sides of the original equation.	$\text{LHS} = \frac{21}{3} - 2 = 7 - 2 = 5$ $\text{RHS} = \frac{21 - 1}{4} = \frac{20}{4} = 5$ LHS = RHS

Writing worded information as equations

When mathematical information is given in words, you need to write it in the form of an equation before you solve it.

Worked example 4

W.E. 4

Write an equation for the following and solve it to find the unknown numbers.

The sum of three consecutive numbers is 21.

Thinking	Working
1 Use a variable to define one of the unknowns.	Let the first number be x.
2 'Consecutive' means 'following' so add 1 to x for the second number and add 2 to x for the third number.	Second number is $x + 1$. Third number is $x + 2$.
3 'Sum' tells us to add the three unknown numbers to give the total. (Here, it is 21.)	$x + (x + 1) + (x + 2) = 21$
4 Add like terms to simplify.	$3x + 3 = 21$
5 Use inverse operations in the opposite order to find the value of the unknown.	$3x = 18$ $x = 6$
6 Check your solution by substituting the value of the pronumeral back into the equation.	$\text{LHS} = 6 + (6 + 1) + (6 + 2)$ $= 6 + 7 + 8$ $= 21$ $= \text{RHS}$
7 State the answer in words.	The three consecutive numbers are 6, 7 and 8.

In the example above, you might find it simpler to let the middle number equal x. The three numbers would then be $x - 1$, x and $x + 1$. Try this for yourself.

1.1 Linear equations

Navigator

1 (columns 1–2), 2 (columns 1–2), 3, 4, 5 (columns 1–2), 6, 7, 8, 9, 10, 11, 12, 15, 17, 18, 20	1 (column 2), 2 (columns 2–3), 3, 4, 5 (columns 2–3), 6, 7, 8, 9, 10, 11, 12, 14, 15, 16, 17, 18, 20	1 (column 3), 2 (columns 2–3), 4, 5 (columns 2–3), 6, 7, 9, 10, 11, 13, 14, 15, 16, 17, 18, 19, 20, 21

Answers p.762

Fluency

1 Solve the following equations to find the value of the given variable. W.E. 1

(a) $4x + 3 = 15$ (b) $4x + 1 = 33$ (c) $3c + 7 = -2$

(d) $5a - 3 = 17$ (e) $7x - 2 = -44$ (f) $7d - 5 = -26$

(g) $6 - 5m = -4$ (h) $9 - 4n = 5$ (i) $-2 - 3x = 1$

(j) $19 = 6f + 7$ (k) $-9 = 4e - 1$ (l) $-3 = 5p - 4$

(m) $\frac{x}{8} - 5 = -5$ (n) $\frac{g}{7} + 2 = 5$ (o) $\frac{h}{4} - 7 = -3$

(p) $\frac{2k}{3}+1=3$ (q) $\frac{5x}{3}-2=8$ (r) $\frac{7y}{2}+3=3$

(s) $3-\frac{4x}{5}=2$ (t) $6-\frac{2x}{3}=4$ (u) $4-\frac{5x}{7}=-1$

W.E. 2

2 Solve the following equations to find the value of the unknown.

(a) $3a+7=a+9$ (b) $9d+4=2d-3$ (c) $5e+2=7-e$

(d) $7c-1=4c+5$ (e) $4b-3=2b+5$ (f) $4f-5=8-2f$

(g) $g-11=5-g$ (h) $2h+3=7-3h$ (i) $4k+13=7-2k$

(j) $2-3x=2x+7$ (k) $3-4x=2x-3$ (l) $-6-x=2x+1$

(m) $2(x+3)=6x$ (n) $3(2x+1)=18x$ (o) $5(7x+2)=65x$

(p) $3(m+2)=m-5$ (q) $2(3d-5)=2d+7$ (r) $4(2d-3)=6d+2$

(s) $5(a-2)=3(a+2)$ (t) $7(a+2)=2(2a-1)$ (u) $4(2g-3)=5(3g-4)$

(v) $7(2t+4)=4(7-2t)$ (w) $5(2h+3)=4(3h+1)$ (x) $2(4j+3)=3(1-j)$

3 (a) The equation $3a+5=-1$ has the solution:

A $a=\frac{4}{3}$ B $a=-\frac{1}{2}$ C $a=-2$ D $a=-18$

(b) The equation $\frac{d}{7}+3=-2$ has the solution:

A $d=7$ B $d=-\frac{5}{7}$ C $d=-7$ D $d=-35$

(c) The equation $\frac{2x}{3}-4=4$ has the solution:

A $x=-48$ B $x=0$ C $x=3$ D $x=12$

4 Find the lowest common multiple for each of the following sets of numbers.

(a) 5 and 7 (b) 8 and 16 (c) 5, 6 and 15

W.E. 3

5 Solve the following equations to find the value of the unknown.

(a) $\frac{a+7}{3}=\frac{a-3}{5}$ (b) $\frac{b-2}{2}=\frac{b+1}{3}$ (c) $\frac{2c-6}{3}=\frac{c+7}{4}$

(d) $\frac{3d+1}{2}=\frac{d-3}{4}$ (e) $\frac{2x+1}{3}=\frac{x-3}{5}$ (f) $\frac{5y-2}{3}=\frac{y+3}{4}$

(g) $\frac{3-2m}{3}=\frac{5-4m}{4}$ (h) $\frac{1-4k}{4}=\frac{3k-7}{2}$ (i) $\frac{2-3t}{4}=\frac{t-5}{3}$

(j) $\frac{a+7}{3}+2=-4$ (k) $\frac{p-3}{4}-1=2$ (l) $\frac{m-8}{5}-2=-9$

(m) $\frac{3p+1}{2}-\frac{2p}{5}=6$ (n) $\frac{3n-1}{5}+\frac{n}{2}=2$ (o) $\frac{3t+2}{4}-\frac{2t}{3}=-1$

(p) $\frac{4m-3}{7}+\frac{2m+5}{2}=3$ (q) $\frac{2k+3}{2}+\frac{5-k}{3}=2$ (r) $\frac{2x+3}{4}-\frac{3x-5}{3}=2$

W.E. 4

6 Write an equation for each of the following and solve it to find the unknown numbers.

(a) 6 is added to x to give a result of 13. (b) 7 is subtracted from y to give -2.

(c) Twice a certain number is 17. (d) One-third of a certain number is 12.

(e) Two more than three times a number is 23.

(f) Five less than four times a number is -3.

7 (a) $3x + 8 = 2 - x$ has the solution:

A $x = -3$ B $x = -\frac{3}{2}$ C $x = \frac{2}{3}$ D $x = 3$

(b) $7(y - 3) = 5(15 - y)$ has the solution:

A $y = 9$ B $y = 8$ C $y = 7$ D $y = 0$

(c) $\frac{3m - 2}{2} = \frac{m + 11}{3}$ has the solution:

A $m = 0$ B $m = 2$ C $m = 4$ D $m = 7$

Understanding

8 (a) x, y and z are consecutive whole numbers. If $x = 4$, what are the values of y and z?

(b) x, y and z are consecutive whole numbers. If $y = 12$, what are the values of x and z?

(c) x, y and z are consecutive even numbers. If $x = 4$, what are the values of y and z?

9 Find three consecutive even numbers whose sum is 126.

Hint
Add 2 to get the next even number.

10 Sam divided his bag of jubes so that he and his friends Nikki and Minh could share them. He gave himself twice as many as he gave Nikki, and Minh received three times as many as Nikki. If the bag contained 42 jubes, how many did each person receive?

11 A rectangular block of land has a length of x m and a width of $(x - 50)$ m. The perimeter of the block of land is 300 m. Find the value of x, and hence find the length and width of the block.

12 Suppose that p represents Penny's age.

(a) How old will she be in 7 years' time?

(b) Write an expression to represent twice her age in 7 years' time.

(c) What is Penny's age now if twice her age in 7 years' time is 48?

13 The width of a rectangular block of land is $\frac{2}{7}$ of its length. If its perimeter is 252 m, find its length.

14 To travel to a concert, Olivia and her friends decided to walk to the train station. The trip on the train covered 6 times the distance they had walked. They then rode in a taxi for the final 2 km to the concert venue. Altogether they travelled 12.5 km. How far did Olivia and her friends walk, and how far did they travel by train?

15 Ross is 20 years older than his niece Krystal. Krystal has worked out that in 5 years' time her uncle will be twice as old as she will be then. Find Krystal's age.

16 If $\frac{3}{5}$ of a number is subtracted from $\frac{2}{3}$ of the same number, the result is 3. Find the number.

Reasoning

17 **(a)** Expand $3(x - 3) = 6$.

(b) Solve $x - 3 = 2$.

(c) If you were now asked to solve $3(x - 3) = 6$, how could you find the solution using your answer to **(b)**?

(d) If you were now asked to solve $3x - 9 = 6$, how would you know the solution immediately?

18 Amy is twice as old as Lucy. Three years from now the sum of their ages will be 42. How old is Amy?

Hint

Distance = speed × time

19 Two cars start at the same time from the same point and travel in opposite directions. One car is travelling 10 km/h faster than the other. In 3 hours they are 288 km apart. Find the speed of each car.

Open-ended

20 Write three different equations, each with a solution of -2, where at least three mathematical operations are used.

21 Ms Grey wants Tim the tiler to lay square tiles to make a bordered rectangle with white tiles on the interior and black tiles on the border. For his first attempt Tim comes up with the following design, which contains 18 black and 12 white tiles.

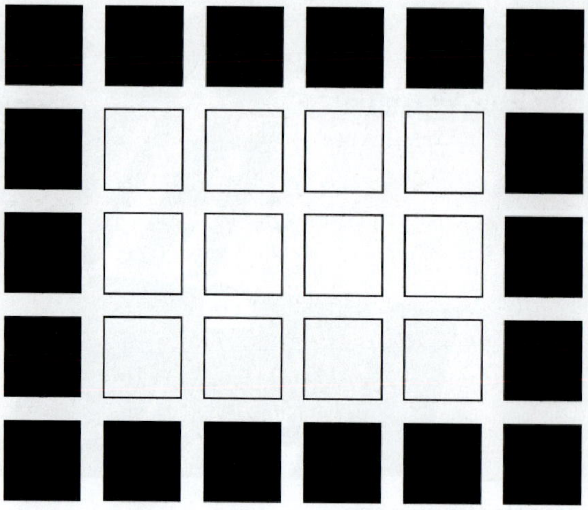

(a) If the rectangle remains 3 white tiles wide, but the number of white tiles in the length increases to l, how many black tiles will be required for the border?

(b) **(i)** Draw a different rectangular design that uses 12 white tiles.

(ii) How many black tiles would now be needed?

(iii) If the number of white tiles in the width remains constant, but the number of white tiles in the length increases to t, how many black tiles will be required for the border?

(c) Choose a different number of white tiles to make a rectangular pattern and find the number of black tiles (n) needed to make a border.

Exploration CAS

Equipment required: TI-Nspire CAS or Casio ClassPad CAS

Solving equations

Equations such as $\frac{2c-6}{3}=\frac{c+7}{4}$ can be slow or difficult to solve by hand. However, you can use your CAS to solve these quite easily.

Using TI-Nspire CAS

1 Add a **Calculator** page to your document by selecting the Calculator icon on the bottom row as shown.

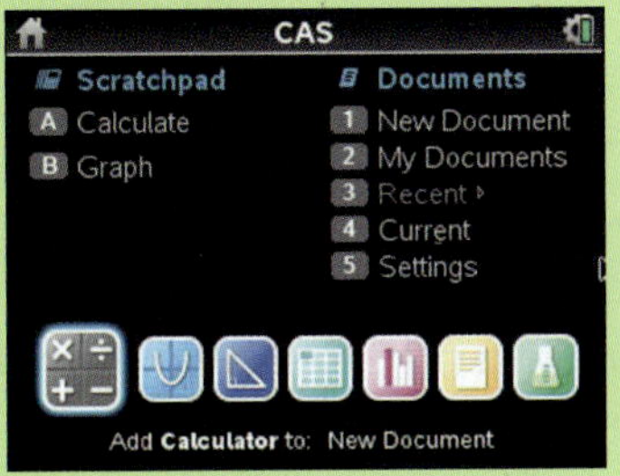

2 Select **menu** > **Algebra** > **Solve** and enter an equation, followed by a comma and the variable to be solved for (c). To use a fraction template press **ctrl** > ÷.

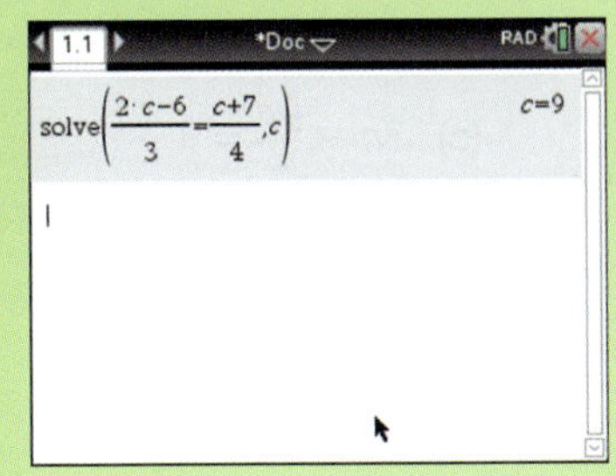

Using Casio ClassPad CAS

From the menu select **Main** and then make sure you are in Standard mode, with the word **Standard** shown at the bottom of the screen. If this is **Decimal** instead, then select the word to change it to **Standard**.

You should also select **Edit** > **Clear All Variables**, to make sure that no variables are set.

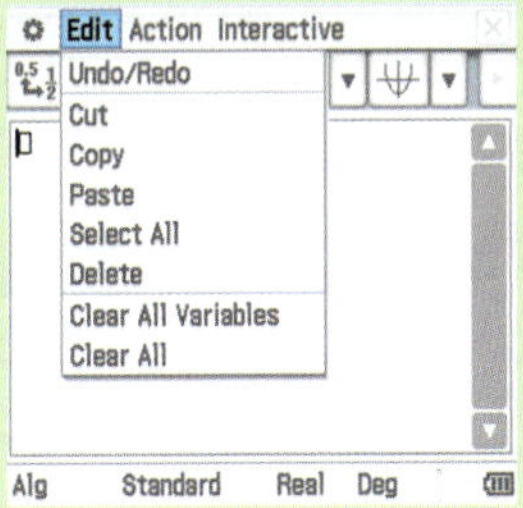

2 Select **Interactive** > **Equation/Inequality** > **solve** and then enter an equation, followed by a comma and the variable to be solved for (c). You can use the fraction template and the **Var** tab from the **Keyboard** to enter the equation in the main window.

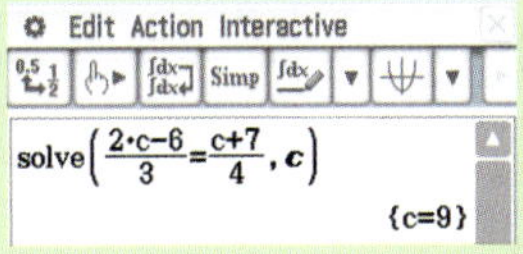

It is important to include the variable that you want to solve, even if there is only one variable.

Equations with different variables that represent known values, sometimes called literal equations, will often require you to find one variable in terms of another variable.

Using TI-Nspire CAS	Using Casio ClassPad CAS
Repeat the previous step to solve the equation $\frac{a+7b}{2}+2=4$, solving for a. 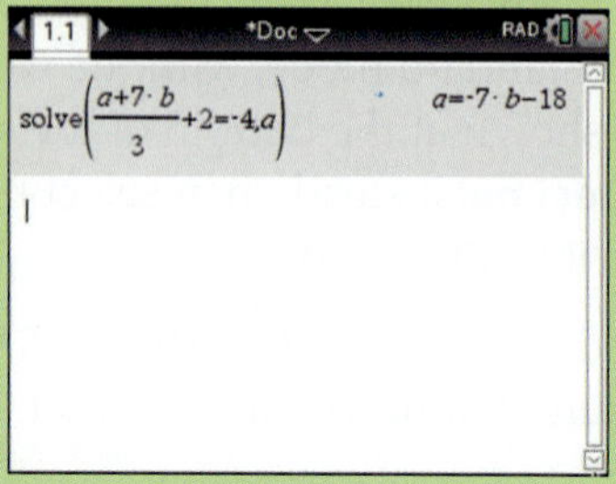	Repeat the previous step to solve the equation $\frac{a+7b}{2}+2=4$, solving for a. 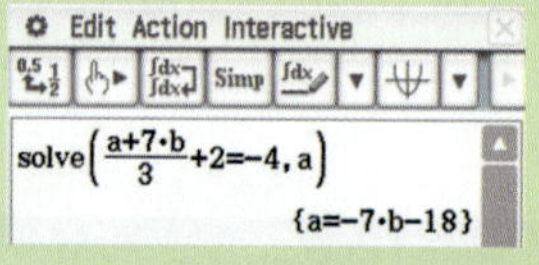
Repeat the previous step to solve the same equation again, but now solve for b. 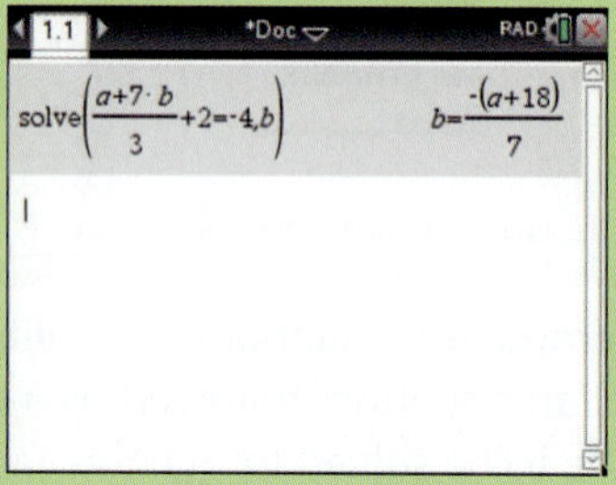	Repeat the previous step to solve the same equation again, but now solve for b. 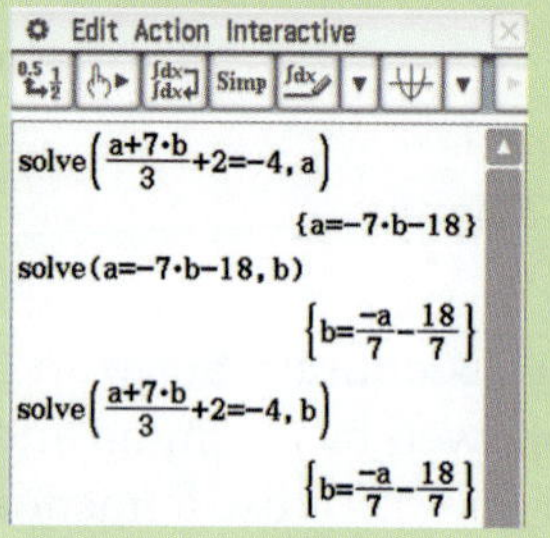

1 Solve the following equations for the given variable using your CAS.

(a) $\frac{3t+5}{4}=\frac{4t-5}{3}$ **(b)** $\frac{8t-7}{3}=4t+7$ **(c)** $3a-\frac{4a}{5}=9$

(d) $\frac{6p}{5}-4=2p-8$ **(e)** $\frac{2k-3}{4}=\frac{3k-2}{2}$ **(f)** $\frac{p}{2}+\frac{p}{10}=\frac{2p}{5}+4$

2 Solve the following literal equations for both variables shown.

(a) $\frac{a}{x-3}=\frac{2a-1}{2x+1}$ **(b)** $\frac{m+6n}{3}=\frac{2m+5n}{4}$ **(c)** $\frac{t+3}{2v}+\frac{t+5}{5v}=8$

(d) $\frac{a+1}{3b}+\frac{a+1}{4b}=9$ **(e)** $f-g=\frac{6f+g}{10}$ **(f)** $\frac{1}{p}(q-2)-\frac{1}{p}(2q-1)=5$

Gradient

1.2

The **gradient** of a line connecting two points is the rate at which the lines rises or falls. The gradient of a line is sometimes referred to as the slope of the line. Consider the distance–time graph of three different modes of transport to the right.

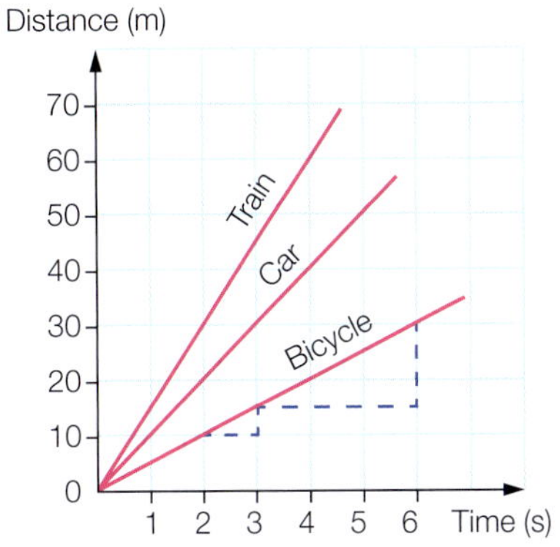

From 0 to 2 seconds, the distance travelled by the bicycle is 10 m. Another way of expressing this is that the bicycle travels 10 m in 2 seconds or is travelling at an average speed of 5 m/s. What distance is travelled by the bicycle from 2 to 3 seconds? From 3 to 6 seconds? Does the average speed change?

$$\text{Average speed} = \frac{\text{distance travelled}}{\text{time taken}}$$

From 2 to 3 seconds:

$$\begin{aligned}\text{Average speed} &= \frac{15 - 10 \text{ m}}{3 - 2 \text{ s}} \\ &= \frac{5 \text{ m}}{1 \text{ s}} \\ &= 5 \text{ m/s}\end{aligned}$$

From 3 to 6 seconds:

$$\begin{aligned}\text{Average speed} &= \frac{30 - 15 \text{ m}}{6 - 3 \text{ s}} \\ &= \frac{15 \text{ m}}{3 \text{ s}} \\ &= 5 \text{ m/s}\end{aligned}$$

The car travels 10 metres every second. Its average speed is 10 m/s. The train travels 15 m every second, so its average speed is 15 m/s.

A constant speed produces a linear graph. The steeper the line, the greater the speed.

The measure of a line's steepness is called the gradient or slope and is often represented by m. The gradient of a distance–time graph gives the speed.

The gradient is the same at any point along a straight line.

The steeper the line, the greater the magnitude of the gradient.

The gradient of a line can be found by considering any two points (x_1, y_1) and (x_2, y_2) on the line.

$$\begin{aligned}\text{Gradient of line } AB &= \frac{\text{vertical rise}}{\text{horizontal run}} \\ &= \frac{\text{difference in } y\text{-coordinates}}{\text{difference in } x\text{-coordinates}} \\ &= \frac{y_2 - y_1}{x_2 - x_1}\end{aligned}$$

where (x_1, y_1) is one point and (x_2, y_2) is another point.

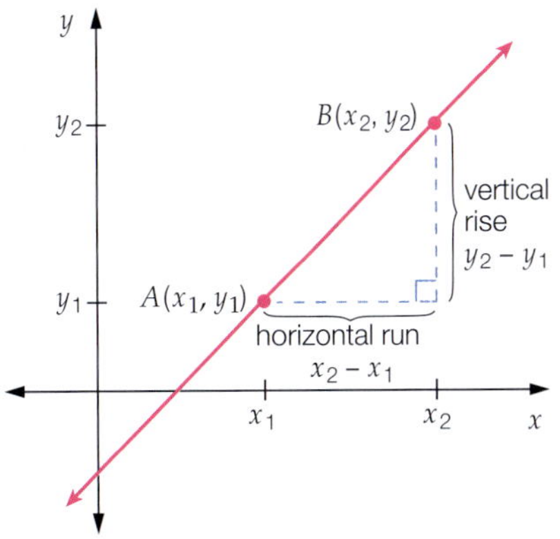

Usually (x_1, y_1) is the point to the left of (x_2, y_2) so that $x_2 - x_1$ is positive. If (x_2, y_2) is lower than (x_1, y_1), then $y_2 - y_1$ will be negative and the gradient will therefore be negative.

However, as $\frac{y_1 - y_2}{x_1 - x_2} = \frac{y_2 - y_1}{x_2 - x_1}$ it makes no difference which point you label (x_1, y_1) or (x_2, y_2).

The gradient of a line is the gradient of any interval within that line. The gradient is constant at any point on the line. The gradient of a line joining $A(x_1, y_1)$ and $B(x_2, y_2)$ is given by:

$$m = \frac{y_2 - y_1}{x_2 - x_1}$$

- The gradient is positive if the y-values increase as the x-values increase.
- The gradient is negative if the y-values decrease as the x-values increase.
- A horizontal line has zero gradient.
- A vertical line has an undefined gradient.
- Lines with the same gradient are parallel, $m_A = m_B$.
- Two lines are perpendicular if the product of their gradients is -1. One gradient is the negative reciprocal of the other $m_A = -\frac{1}{m_B}$. Horizontal and vertical lines are perpendicular.

Interactive

Gradient

Explore how the gradient of a line graph rises or falls as the values change.

Go to the eBook or the Pearson Places website to access this interactive.

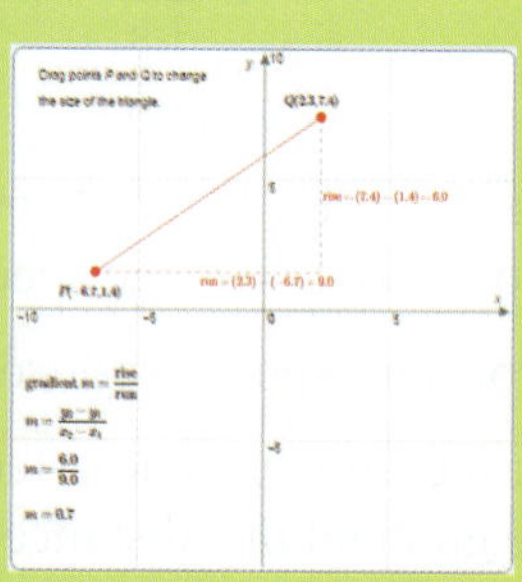

Worked example 5

W.E. 5

Find the gradient of each of the following lines through the given points.

(a) (2, 3) and (6, 9)

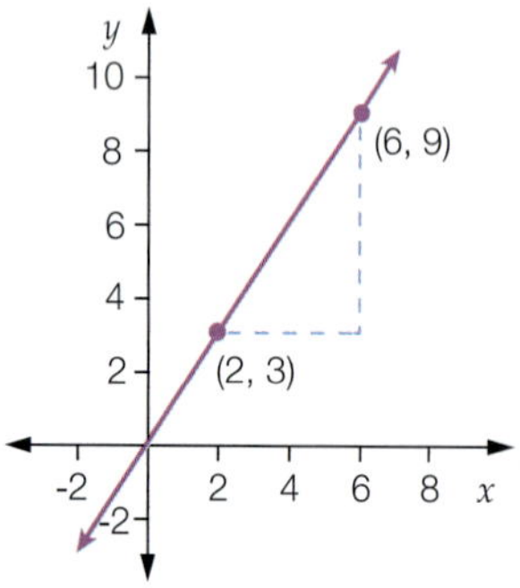

(b) (-4, 4) and (2, 1)

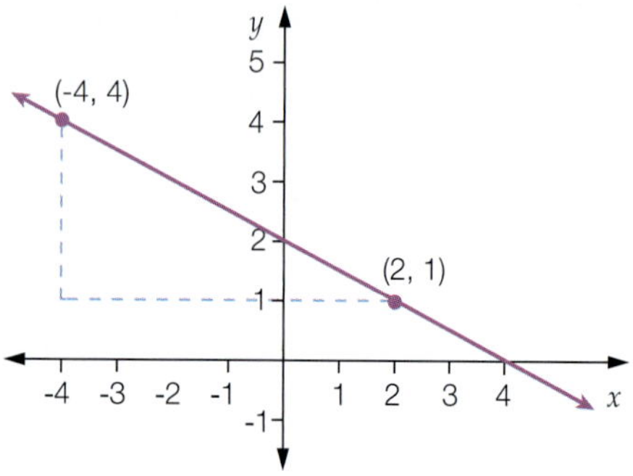

Thinking	Working
(a) 1 Identify (x_1, y_1) and (x_2, y_2).	(a) Let $(x_1, y_1) = (2, 3)$ and $(x_2, y_2) = (6, 9)$.
2 Find the gradient of the line using: $m = \frac{\text{vertical rise}}{\text{horizontal run}} = \frac{y_2 - y_1}{x_2 - x_1}$	$m = \frac{\text{rise}}{\text{run}} = \frac{y_2 - y_1}{x_2 - x_1} = \frac{9-3}{6-2} = \frac{6}{4} = \frac{3}{2}$
(b) 1 Identify (x_1, y_1) and (x_2, y_2).	(b) Let $(x_1, y_1) = (-4, 4)$ and $(x_2, y_2) = (2, 1)$.
2 $m = \frac{\text{vertical rise}}{\text{horizontal run}}$ (Here, there is a decrease of 3 units in y for an increase of 6 units in x, so the vertical 'rise' is -3.)	$m = \frac{\text{rise}}{\text{run}} = \frac{y_2 - y_1}{x_2 - x_1} = \frac{1-4}{2-(-4)} = \frac{-3}{6} = -\frac{1}{2}$

Finding the equation of a line

If you know the gradient and a point on the line, you can find the equation of the line using the gradient formula:

$$m = \frac{y_2 - y_1}{x_2 - x_1}$$

You can find the equation of any line if given m and a point on the line (x, y).

To find the equation, substitute the values of m and (x, y) into $m = \frac{y - y_1}{x - x_1}$.

Worked example 6 **W.E. 6**

Find the equation of a line with:

(a) a gradient of 2 that passes through the point (3, -5)

(b) a gradient of -4 that passes through the point (-4, 2).

Thinking	Working
(a) **1** Write the formula for equating gradients.	(a) $m = \frac{y - y_1}{x - x_1}$
2 Substitute values for the gradient and the point into the formula.	$2 = \frac{y - (-5)}{x - 3}$
3 Rearrange the equation and simplify.	$2(x - 3) = y + 5$ $y = 2x - 6 - 5$
4 Write an equation in $y = mx + b$ form.	$y = 2x - 11$
(b) **1** Write the formula for equating gradients.	(b) $m = \frac{y - y_1}{x - x_1}$
2 Substitute values for the gradient and the point into the formula.	$-4 = \frac{y - 2}{x - (-4)}$
3 Rearrange the equation and simplify.	$-4(x + 4) = y - 2$ $y - 2 = -4x - 16$
4 Write an equation in $y = mx + b$ form.	$y = -4x - 14$

1.2 Gradient

Navigator

Answers p. 763

1, 2, 3, 4 (columns 1–2), 5, 6, 7, 8, 10, 12, 13, 16 (d)	1, 2, 3, 4 (columns 2–3), 6, 7, 8, 9, 11, 12, 13, 15, 16 (b, e), 17	1, 2, 4 (column 3), 6, 7, 8, 9, 11, 12, 14, 15, 16, 17

Fluency

W.E. 5

1 Find the gradient of each of the following lines through the given points.

(a) (-2, -1) and (2, 1)

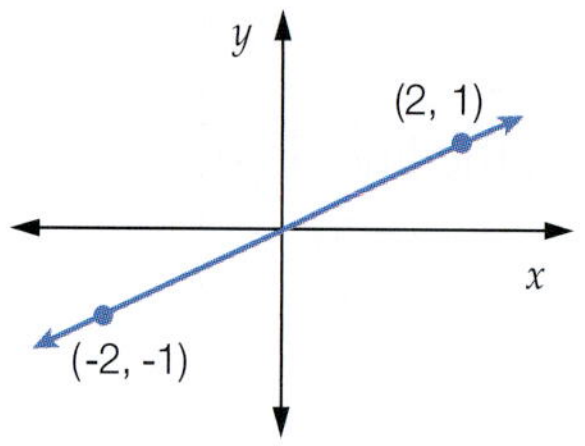

(b) (0, 2) and (1, 0)

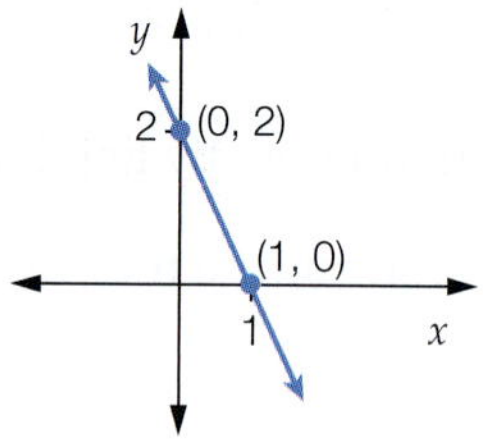

(c) (-2, 3) and (3, 3)

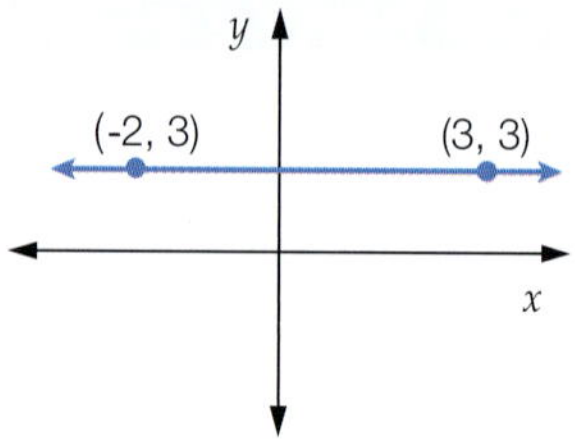

(d) (2, -1) and (2, 2)

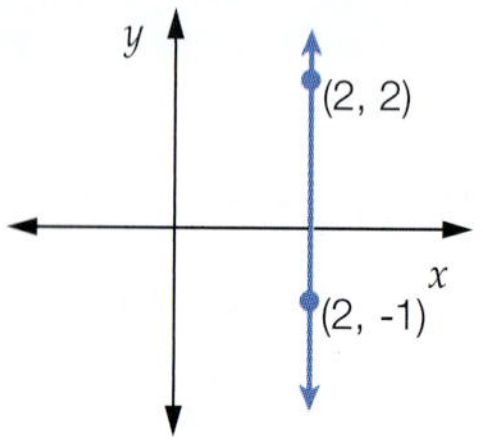

2 Find the equation of a line with:

W.E. 6

(a) a gradient of 6 that passes through the point (1, 3)

(b) a gradient of -2 that passes through the point (4, -1)

(c) a gradient of 3 that passes through the point (-4, -3)

(d) a gradient of -5 that passes through the point (3, -7).

3 Use the results from Question **1** to answer the following questions.

(a) If a line is sloping up to the right, does it have a positive or a negative gradient?

(b) If a line is sloping down to the right, does it have a positive or a negative gradient?

(c) What is the gradient of a horizontal line?

(d) What is the gradient of a vertical line?

4 Find the gradient of the line joining each of the following pairs of points.

(a) (2, 4) and (5, 5) (b) (0, 3) and (2, 7) (c) (4, 10) and (5, 6)

(d) (7, 5) and (9, 2) (e) (-2, 1) and (2, -4) (f) (-1, -3) and (0, -5)

(g) (0, 0) and (2, 6) (h) (2, 1) and (-2, -7) (i) (-2, 5) and (5, -2)

(j) (-4, 4) and (2, 0) (k) (-3, 2) and (2, 2) (l) (5, -4) and (-7, -4)

(m) (-8, 5) and (2, 1) (n) (-1, -1) and (6, 0) (o) (2, 5) and (2, 7)

5 (a) The gradient of the line shown in the graph is:

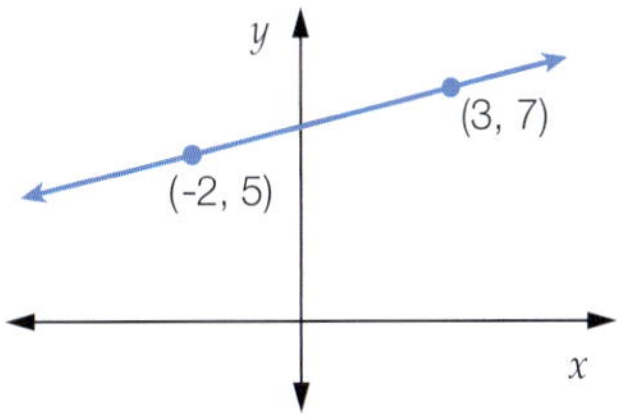

A $\frac{5}{2}$ B 2

C $\frac{2}{5}$ D $-\frac{2}{5}$

(b) The gradient of the line shown in the graph is:

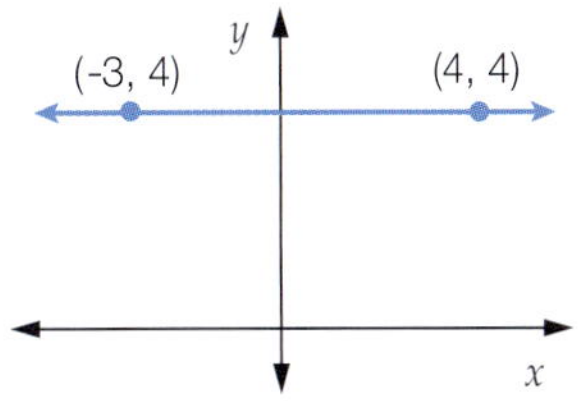

A 8 B 0

C -7 D undefined

(c) The gradient of the line shown in the graph is:

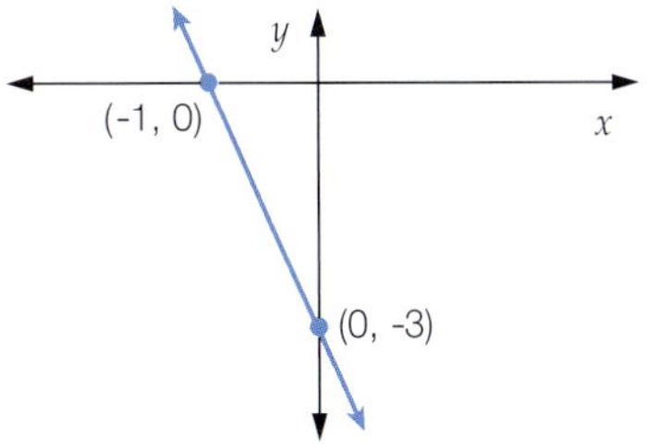

A $\frac{1}{3}$ B $-\frac{1}{3}$

C 3 D -3

Understanding

6 (a) A horizontal line is drawn through the points A (5, 7) and B (x, y). The coordinates for B could be:

A (7, 5) B (5, 9) C (9, 7) D (10, 14)

(b) A vertical line is drawn through the points A (5, 7) and B (x, y). The coordinates for B could be:

A (7, 5) B (5, 9) C (9, 7) D (10, 14)

(c) If the points A (-3, -7) and B (x, y) are on the same horizontal line, what could you say about the possible values for x and y?

(d) If the points A (-3, -7) and B (x, y) are on the same vertical line, what could you say about the possible values for x and y?

7 **(a)** Use the results from Question **4** to find a pair of points that are on:

(i) a horizontal line

(ii) a vertical line

(iii) the line with the steepest negative gradient

(iv) the line with the smallest positive gradient.

(b) Use the results from Question **4** to find two pairs of points that would give two parallel non-horizontal lines.

8 Alex is 200 m from home and decides to walk to the local shop. This graph shows Alex's distance from home (d) at time (t) on his 5-minute walk to the shop.

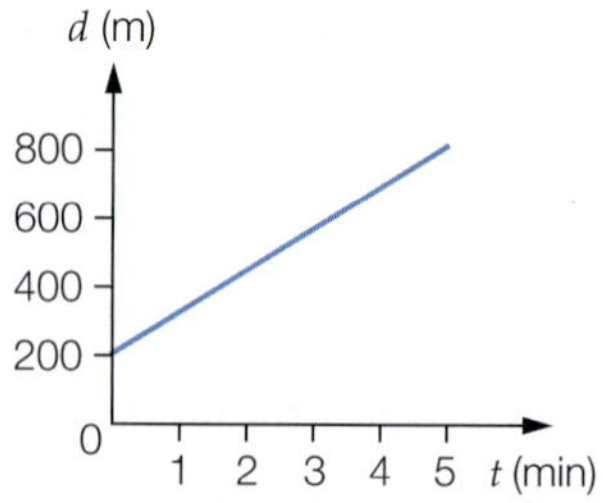

(a) How far was the shop from Alex's home?

(b) How far did Alex walk in 5 minutes?

(c) Find the gradient of this graph by first identifying two points that the line passes through.

(d) Find Alex's speed on the walk.

(e) How far was Alex from home after 2 minutes of walking?

(f) How far was Alex from home after 3 minutes of walking?

(g) Write the relationship or rule to find the distance from home (d) after t minutes.

(h) The graph of Alex's trip is a straight line. What does this tell you about Alex's walking speed?

9 An egg-timer contains grains of sand that pour through the neck of the timer, from the top bulb to the bottom, in 3 minutes.

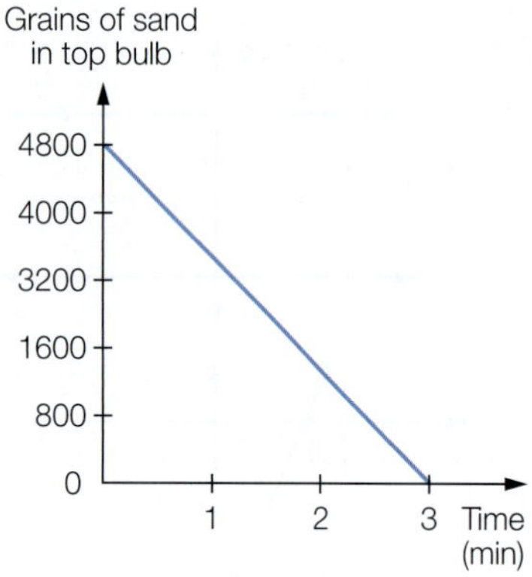

(a) How many grains of sand are in the egg-timer?

(b) How many grains of sand are left in the top bulb after 2 minutes?

(c) Find the gradient of this line by identifying two points that the line passes through.

(d) Find the rate of change of sand grains with respect to time. In other words, what is the change in the number of sand grains left in the top bulb each minute?

(e) What does the negative represent in your answer to part **(c)**?

(f) Write the relationship or rule to find the number of sand grains left in the top bulb (S) after t minutes.

(g) Use your rule to confirm your answer to **(b)**.

(h) What are the restrictions on the values that t can take?

Reasoning

10 **(a)** Find the equation of the line that passes through (2, 4) and (4, 9):

(i) using the point (2, 4) as (x_1, y_1) **(ii)** using the point (4, 9) as (x_1, y_1)

(b) Did you get the same equation for parts **(i)** and **(ii)**? What conclusion can you make from your answers to parts **(i)** and **(ii)**?

11 The gradient of the line joining the points (2, 1) and (6, a) is 4. Find the value of a.

12 A passenger jet is landing on the flight path shown. What is the gradient of the flight path?

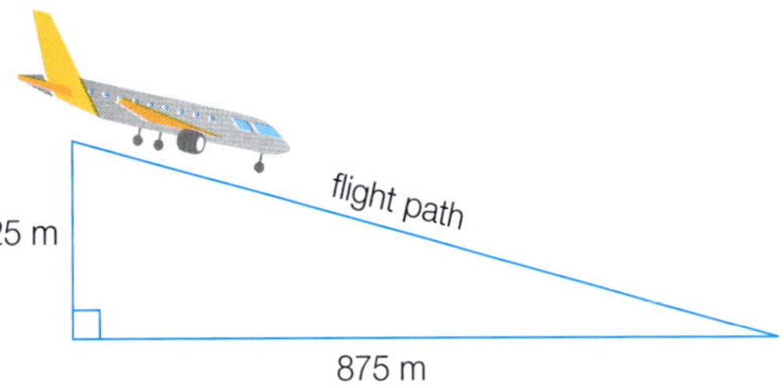

Hint

Use the vertex of the right angle of the triangle as (0, 0).

13 Consider the graph shown at right.

(a) Find the gradient of the line segment joining A and B.

(b) Find the gradient of the line segment joining B and C.

(c) Find the gradient of the line segment joining A and C.

(d) Explain why the answers found in **(a)**, **(b)** and **(c)** are all the same.

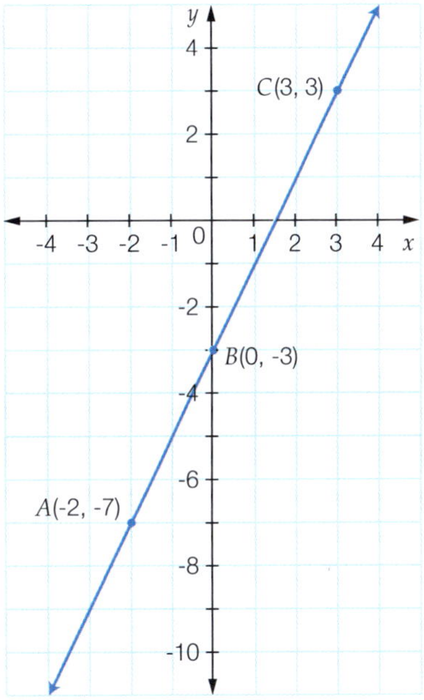

14 Show that the points (-1, -5), (2, 4) and (5, 13) are on the same straight line. This means that the points are collinear.

15 The points (1, 3), (3, 8) and (7, y) are on the same straight line (they are collinear).

(a) Find the gradient of the line that connects (1, 3) and (3, 8).

(b) What is the gradient of the line that connects (3, 8) and (7, y)?

(c) Use the answer from **(b)** to find the value of y.

(d) The point (x, -2) is on the same straight line. Find the value of x.

Open-ended

16 Identify three different sets of coordinates that when connected by a straight line produce each of the following gradients:

(a) -3 **(b)** -1 **(c)** $-\frac{1}{2}$ **(d)** 2 **(e)** 5

17 Write the coordinates of two points such that a line joining them has the same gradient as the line joining (3, 5) and (7, -7).

1.3 Sketching linear graphs

The graph of a linear relationship is a linear graph. You need to locate only two coordinate points in order to sketch a linear graph.

The equation of a linear graph can be written in the form:

$$y = mx + b$$

where m is the gradient and b is the y-coordinate of the y-intercept $(0, b)$.

Sketching linear graphs using the y-intercept and gradient

This method involves locating the point at which the graph crosses the vertical y-axis and then using the gradient to locate a second point. When two points are known, the straight line can be drawn through them.

Write the gradient as a fraction using the rule gradient $= \frac{\text{rise}}{\text{run}}$, and then identify the rise (units up or down) and the run (units left or right) to reach the next point.

Sometimes, the equation needs to be rearranged into the form $y = mx + b$.

Worked example 7 — W.E. 7

Find the y-intercept and the gradient and then sketch the graph of the line $y = 4x + 1$.

Thinking

1 Use the equation to identify the gradient (m) and the y-intercept $(0, b)$. Use gradient $= \frac{\text{rise}}{\text{run}}$ to identify the rise and the run. (Here, the rise is 4 and the run is 1.)

2 Rule axes for the Cartesian plane, label the x-axis and y-axis, then mark the y-intercept.

3 Starting at the y-intercept, move horizontally the distance of the run and move vertically the distance of the rise. (Here, move 1 unit to the right and 4 units up to find the second point (1, 5).)

4 Rule a line through the two points.

Working

$y = 4x + 1$ has a y-intercept at (0, 1).

Gradient $= \frac{4}{1}$ so rise = 4, and run = 1.

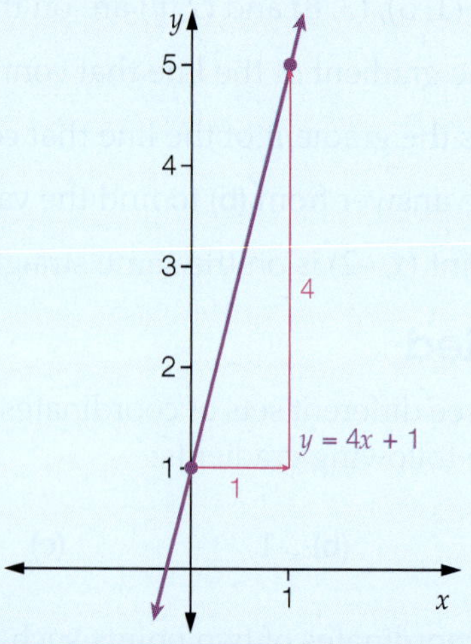

Worked example 8

W.E. 8

Find the y-intercept and gradient, and then sketch the graph of the line $3x + 2y = 4$.

Thinking	Working
1 Rearrange into the form $y = mx + b$.	$3x + 2y = 4$ $2y = -3x + 4$ $y = \frac{-3x}{2} + 2$
2 Use the equation to identify the y-intercept $(0, b)$ and gradient (m).	y-intercept is (0, 2). Gradient $= -\frac{3}{2}$, so rise = -3 and run = 2.
3 Draw and label a set of axes. Starting at the y-intercept, move horizontally the distance of the run and move vertically the distance of the rise. A negative rise means a move down. (Here, move 2 units to the right and 3 units down to find the second point (2, -1).)	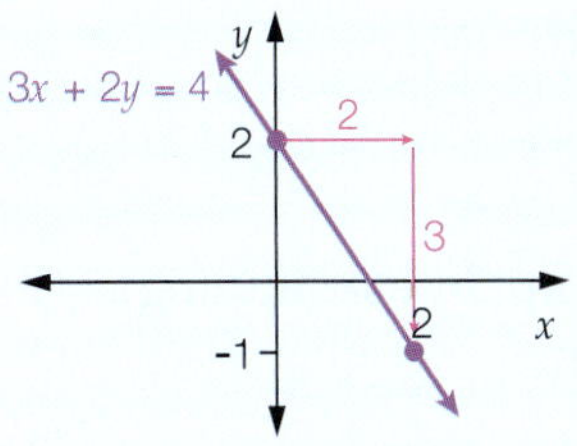
4 Rule a line through the two points.	

Interactive

Sketching linear graphs

Explore how the y-intercept and gradient affect the position and shape of linear graphs.

Go to the eBook or the Pearson Places website to access this interactive.

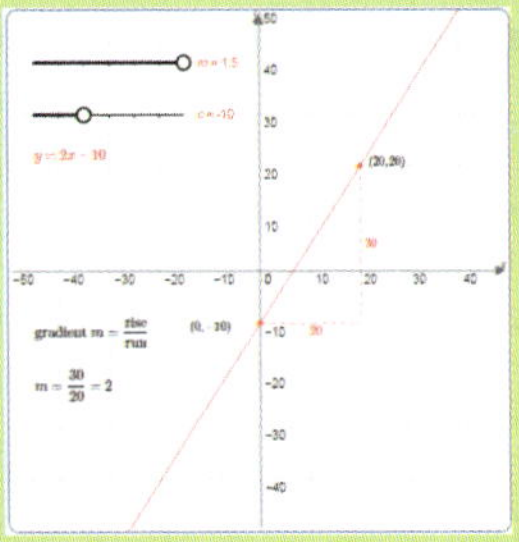

Sketching linear graphs using x- and y-intercepts

If a linear equation is not in the form of $y = mx + b$, it is often useful to find the x- and y-intercepts, and use them as the major features of a linear graph.

For example, consider the graph of $-3x + 2y = 6$.

The graph crosses the x-axis when $y = 0$.
Here, the x-intercept is (-2, 0).

The graph crosses the y-axis when $x = 0$.
Here, the y-intercept is (0, 3).

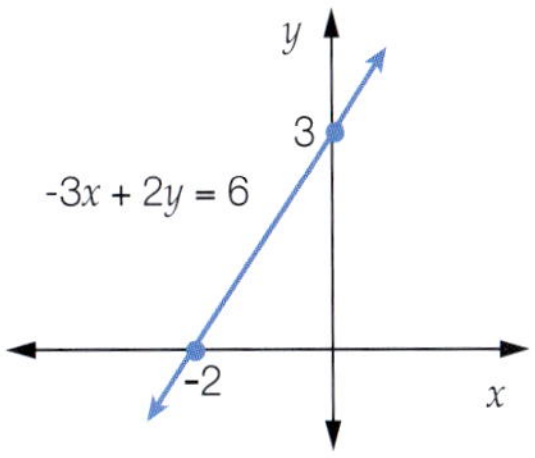

You can sketch a linear graph using the two intercepts.

- Substitute $y = 0$ and find the x-intercept.
- Substitute $x = 0$ and find the y-intercept.
- Mark these two points on a set of axes and then join using a straight line.
- Label the two intercepts and the equation of the line.

Worked example 9

W.E. 9

Find the x- and y-intercepts for $2x + 3y = 12$ and sketch the graph.

Thinking	Working
1 Find the x-intercept by substituting $y = 0$ into the equation.	$y = 0,\ 2x + 3 \times 0 = 12$ $2x = 12$ $x = 6$ x-intercept is $(6, 0)$.
2 Find the y-intercept by substituting $x = 0$ into the equation.	$x = 0,\ 2 \times 0 + 3y = 12$ $3y = 12$ $y = 4$ y-intercept is $(0, 4)$.
3 Mark in the x-intercept and y-intercept and rule a straight line through them.	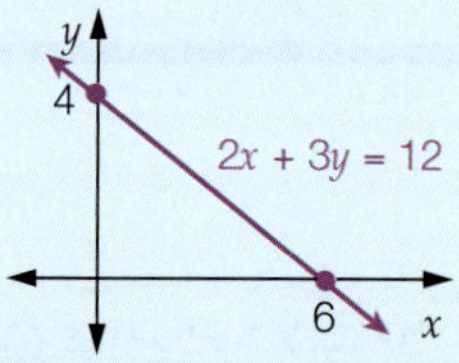

Note that for a sketch graph, only the x- and y-intercepts need to be shown on the axes; no other values for x and y need to be shown.

- To find the x-intercept, substitute $y = 0$ into the equation.
- To find the y-intercept, substitute $x = 0$ into the equation.

Sketching linear graphs that pass through the origin

If a line passes through the origin, then the x- and y-intercepts do not give sufficient information to sketch the graph. The origin gives only one coordinate point (0, 0), so a second point is needed to sketch the graph.

You can use the gradient and y-intercept method to sketch these graphs. Start at the origin (y-intercept) and then use the gradient to find the second point.

Worked example 10

W.E. 10

Sketch the graph of $y = 3x$, showing the y-intercept and one other point.

Thinking	Working
1 Use the gradient to identify the rise and run.	Gradient $= \frac{3}{1}$ so rise = 3 and run = 1.
2 Because the value of b in the equation is zero, plot the first point at the origin (0, 0).	
3 Find a second point by using the gradient. (Here, the rise is 3 and the run is 1.)	
4 Sketch the graph.	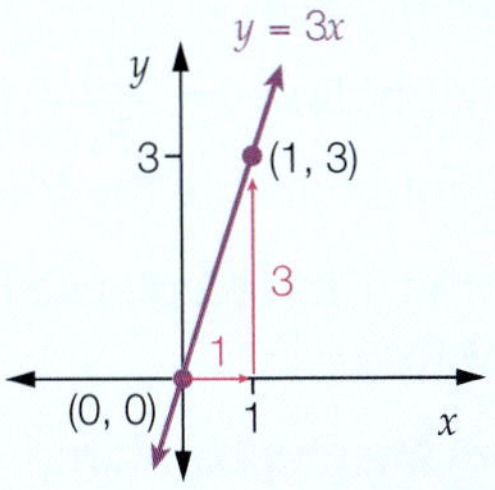

Lines parallel to an axis

Other linear relationships involving x and y may be parallel to an axis, as follows:

- $x = a$, where a is a constant, is the equation of a vertical line parallel to the y-axis. The gradient of these lines is undefined.
- $y = b$, where b is a constant, is the equation of a horizontal line parallel to the x-axis. The gradient of these lines is zero.

For example, $y = 5$ means that y always has the value 5, no matter what value x is given. Some coordinate points on this line are (2, 5), (0, 5) and (-3, 5).

The graph of $y = 5$ is a straight line parallel to the x-axis.

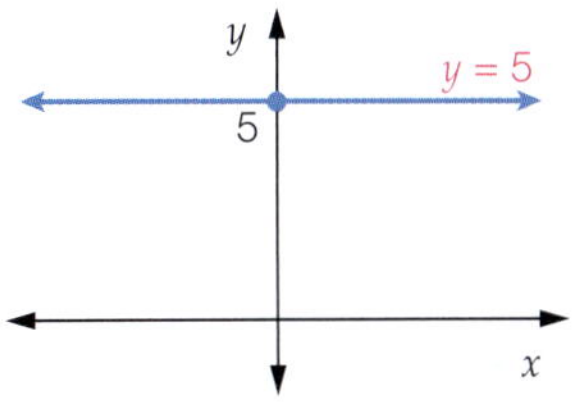

Similarly, for $x = -2$, x will always have the value -2, no matter what value of y is chosen. Some coordinate points on this line are (-2, 4), (-2, -1) and (-2, 0).

The graph of $x = -2$ is a straight line parallel to the y-axis.

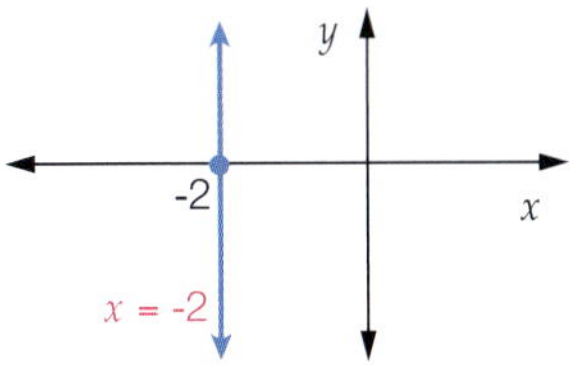

1.3 Sketching linear graphs

Navigator

1, 2, 3, 4, 5, 6, 8, 9 (columns 1–2), 12, 15, 16, 17, 18, 20	1, 2, 3, 4, 5, 6, 7, 8, 9 (columns 2–3), 10, 11, 14, 15, 16, 17, 18, 20, 21	1, 2, 4, 5, 6, 7, 9 (columns 3–4), 10, 11, 13, 14, 15, 16, 17, 18, 19, 20, 21

Answers p.764

Fluency

1 (a) Consider the straight line $y = 2x - 3$.

(i) Write the coordinates of the y-intercept.

(ii) Starting at the y-intercept, use the gradient $\left(m = \frac{\text{rise}}{\text{run}}\right)$ to find the coordinates of the point for $x = 1$.

(b) Consider the straight line $y = -3x + 4$.

(i) Write the coordinates of the y-intercept.

(ii) Starting at the y-intercept, use the gradient $\left(m = \frac{\text{rise}}{\text{run}}\right)$ to find the coordinates of the point for $x = 1$.

(c) Consider the straight line $y = \frac{3x}{4} + 3$.

(i) Write the coordinates of the y-intercept.

(ii) Starting at the y-intercept, use the gradient $\left(m = \frac{\text{rise}}{\text{run}}\right)$ to find the coordinates of the point for $x = 4$.

(d) Consider the straight line $y = \frac{-2x}{5} - 4$.

(i) Write the coordinates of the y-intercept.

(ii) Starting at the y-intercept, use the gradient $\left(m = \frac{\text{rise}}{\text{run}}\right)$ to find the coordinates of the point for $x = 5$.

2 For each of the following graphs, state:

(i) the coordinates of the x-intercept

(ii) the coordinates of the y-intercept

(iii) the gradient of the line

(iv) the equation of the line.

(a)

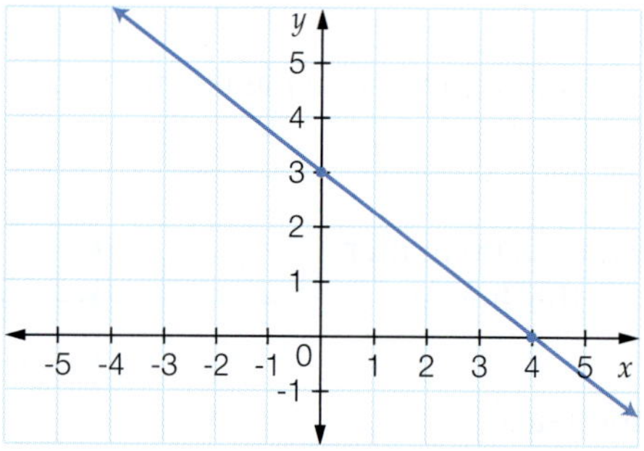

(b)

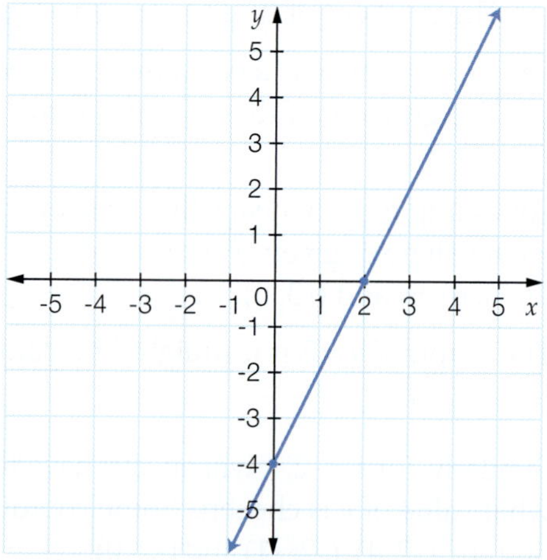

W.E. 7

3 State the y-intercept and the gradient, then sketch a graph of each line.

(a) $y = 3x + 2$ (b) $y = 4x - 3$ (c) $y = -2x + 1$ (d) $y = -3x + 2$

W.E. 8

4 Find the y-intercept and the gradient, then sketch a graph of each line.

(a) $x + 2y = -4$ (b) $x + 3y = 3$ (c) $3x + 4y = 8$

(d) $2x + 5y = -5$ (e) $2x - 3y = -1$ (f) $4x - 2y = 3$

W.E. 9

5 Find the x- and y-intercepts, then sketch a graph of each line.

(a) $2x + 3y = 6$ (b) $4x + y = 4$ (c) $x - 2y = 4$

(d) $x - 3y = 9$ (e) $4x + y = 2$ (f) $3x + y = 1$

W.E. 10

6 Sketch graphs of the following, showing the intercept on the axes and one other point for each.

(a) $y = 4x$ (b) $y = x$ (c) $y = -2x$ (d) $y = -3x$

7 (a) The equation of a linear graph with a y-intercept of 2 and a gradient of -3 is:

A $y = 3x + 2$ B $x + 3y = 2$ C $x - 3y = 2$ D $3x + y = 2$

(b) The equation of a linear graph with a gradient of $\frac{1}{2}$ and a y-intercept of -3 is:

A $2x - y = 6$ B $x - 2y = 6$ C $x - 2y = -6$ D $2x - y = -6$

(c) The equation of a linear graph with a y-intercept of $\frac{2}{5}$ and a gradient of $-\frac{1}{4}$ is:

A $5x + 20y = 8$ B $20x + 5y = 8$ C $y = \frac{2x}{5} - \frac{1}{4}$ D $\frac{2x}{5} - y + \frac{1}{4} = 0$

8 When using the y-intercept and gradient to sketch a graph, first mark in the y-intercept and then use the gradient to locate a second coordinate point.

(a) A line has a gradient of $\frac{2}{3}$. From the y-intercept, you move:

A 2 units to the right and 3 units up
B 2 units to the right and 3 units down
C 3 units to the right and 2 units up
D 3 units to the right and 2 units down

(b) A line has a gradient of $-\frac{3}{4}$. From the y-intercept, you move:

A 4 units right and 3 units up
B 3 units right and 4 units up
C 4 units right and 3 units down
D 3 units left and 4 units down

(c) A line has a gradient of 4. From the y-intercept, you move:

A 4 units right and 4 units up
B 1 unit right and 4 units up
C 4 units right and 4 units down
D 1 unit right and 4 units down

9 Sketch graphs of the following, showing the intercepts on the axes.

(a) $y = 1$ (b) $y = 4$ (c) $y = -5$ (d) $y = -2$
(e) $y = 0$ (f) $y = \frac{1}{2}$ (g) $x = 2$ (h) $x = 5$
(i) $x = -3$ (j) $x = -4$ (k) $x = \frac{5}{2}$ (l) $x = 0$

10 (a) Two coordinate points on the line $2x + y = 4$ are:

A (4, 0) and (0, 3)
B (-4, 0) and (0, -2)
C (-2, 0) and (0, -4)
D (2, 0) and (0, 4)

(b) Two coordinate points on the line $y = 6x$ are:

A (0, 6) and (6, 0)
B (0, 0) and (6, 6)
C (2, 12) and (0, 0)
D (2, 0) and (0, 3)

(c) Two coordinate points on the line $x = -3$ are:

A (0, 0) and (1, 3)
B (-3, 1) and (1, -3)
C (-3, 0) and (0, 0)
D (-3, -5) and (-3, 0)

Understanding

11 (a) Consider the graph shown at right. Write the value of the intercept with the vertical axis.

(b) Find the gradient of this graph.

(c) Write an equation linking the distance from Melbourne in kilometres (D) and the time in hours (T) in the form $y = mx + b$.

(d) What was the constant speed throughout the journey?

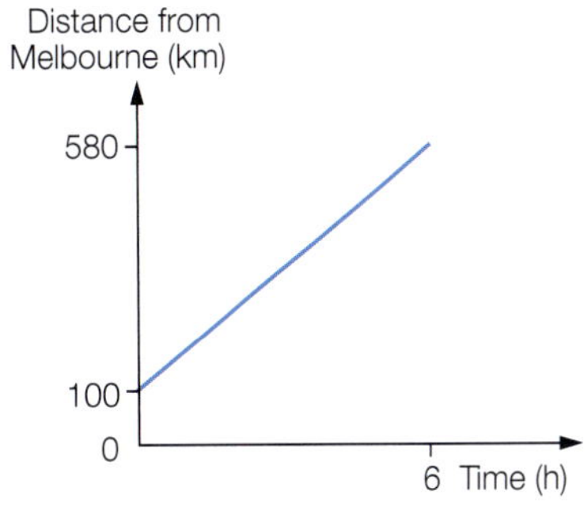

(e) If a journey started 250 km from Melbourne and a constant speed of 100 km/h was maintained, what equation could be written to represent the relationship between D and T in the form $y = mx + b$?

(f) A particular trip from Melbourne is represented by the equation $4D - 372T = 30$. What was the constant speed throughout the journey?

12 Darrio has borrowed $200 from his parents and has agreed to repay $40 each week.

(a) After 3 weeks, how much does Darrio still owe his parents?

(b) Write an equation linking the amount Darrio owes his parents (A) with the number of weeks (n).

(c) Find the amount owed when $n = 0$.

(d) Find the number of weeks when $A = 0$.

(e) Use your answers from **(c)** and **(d)** to sketch the graph of this relationship. (Put A on the vertical axis and n on the horizontal axis.)

Reasoning

13 A circular puddle has a radius of 15 cm. As the water evaporates at a constant rate of k cm^2/min, the puddle retains its circular shape.

(a) State the initial area of the puddle in terms of π.

(b) Write an equation relating the area A of the puddle after it has been evaporating for t minutes at the constant rate k cm^2/min.

(c) Write an equation involving k for the area of the puddle after 5 minutes.

(d) If the area of the puddle after 5 minutes is 175π cm^2 find the value of k.

(e) How long does it take the puddle to completely evaporate?

(f) What does the value $t = 22.5$ min represent on the graph of the linear equation found in **(c)**?

14 The annual yachting regatta of the Mathematical Club takes place on Lake Tetrahedra. It starts and finishes at the marina. Assume this is the point (0, 0) on a coordinate plane, where the x-coordinates run from west to east and the y-coordinates run from south to north. The yachts travel around six buoys that are located at various positions on the lake.

- The first buoy is located at the point (2, 4).
- The second buoy is located at the point (-8, 8).
- The third buoy is located 10 units to the south of the second buoy.
- The fourth buoy is located at the point (-3, 11).
- The fifth buoy is located 5 units to the east of the fourth buoy.
- The sixth buoy is located 4 units to the east of the first buoy.

(a) Plot the points on a Cartesian plane.

(b) Sketch the paths the yachts will take, travelling to each buoy and back to the marina.

(c) Determine the equation of each path taken by the yachts between the:

(i) start of the race at the marina and the first buoy

(ii) first and second buoys

(iii) second and third buoys

(iv) third and fourth buoys

(v) fourth and fifth buoys

(vi) fifth and sixth buoys

(vii) sixth buoy and the finish of the race at the marina.

(d) The race referee's speedboat breaks down at the point (4, 7.5). Will this interfere with the paths taken by the yachts?

15 After her bath, Anne removes the plug to empty it. The water flows out of the bathtub at a rate of 0.75 L/s. There was originally 225 L of water in the bath.

(a) Write an equation relating the volume of water left in the bath (V) with the time in seconds (T).

(b) How long does it take to empty the bath?

16 From the beginning of August, a local reservoir has been losing water at a constant rate. The Water Resources Department estimates that on 12 August the reservoir held 200 million kL of water, and on 21 August it held 164 million kL.

(a) Develop a linear equation relating V, the volume of water remaining (millions of kL), and n, the number of days since the beginning of August (when the reservoir was full).

(b) Draw a graph corresponding to the linear equation developed above to find the maximum capacity of the reservoir.

(c) Why should the graph only be drawn in the first quadrant?

17 A cubic block of ice has an edge of 10 cm. It melts so that its volume decreases at a constant rate (k) and the block remains a cube.

(a) What is the volume of the cube before it starts to melt?

(b) Write an equation relating (V) the volume of the cube after it has been melting for (t) minutes at a constant rate (k).

Hint
The change in volume = kt.

(c) What is the volume of the cube if it has an edge of 5 cm after 50 minutes?

(d) Find the value of k.

(e) What does the sign of k tell you about what is happening to the ice cube?

(f) Substitute the value of k found in part **(d)** to rewrite the equation found in part **(b)**.

(g) Draw a graph corresponding to the linear equation developed above to find how long it would take for the block of ice to completely melt.

18 At the start of the year, Ringo had \$600 in the bank. He has been earning nothing and spending \$20 each week. At the same time, Sophia started with \$100 in the bank. She has been earning \$30 each week and spending nothing.

(a) Develop two linear equations relating the number of weeks since the start of the year (t) to the amounts of money that Ringo and Sophia each have in their bank accounts (R and S, respectively).

(b) Draw graphs of the linear equations to find how many weeks it will take for Sophia to have twice the amount of money in the bank as Ringo.

Open-ended

19 Write three different linear equations that each have the same gradient, but with a different y-intercept. What property of the three lines indicates that they have the same gradient? Explain.

20 Linear equations are used every day to model real-life situations. The graph shows five lines that model the costs of five different mobile phone plans. Match each graph with its description below.

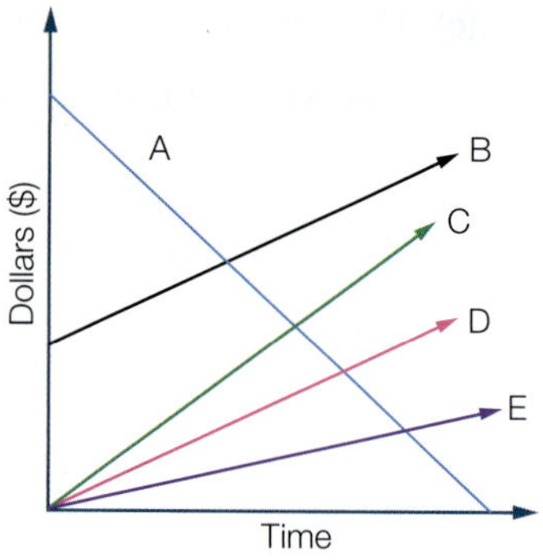

Note:

- The specific values of the gradients and y-intercepts are not needed for this question.
- The connection fee (also called 'flagfall') is the fixed cost when a call is made.
- The 'call rate' is the number of cents per minute charged for the time of a phone call.

(a) This graph models the cost of a phone call. There are no connection fees. The cost of a call is dependent on the time of the call.

(b) This graph models the cost of a phone call. There are no connection fees. Calls are charged at half the rate of the phone plan in **(a)**.

(c) This graph models the amount of credit remaining on a pre-paid monthly mobile phone plan.

(d) This graph models the cost of a phone call. There is a connection fee for every call made and the call rate is the same as the phone plan in **(a)**.

(e) This graph models the cost of a phone call. There are no connection fees, but calls are charged at one-and-a-half times the rate of the phone plan in **(a)**.

21 Sketch three different lines that have gradients between -1 and -2, and y-intercepts between 1 and 4.

Problem solving

Odd one out

Four of these points are on a single straight line.
Which point is the odd one out?

(2, 5) (-5, -7) (2, 4) (5, 13) (9, 21)

Strategy options

- Draw a diagram.
- Look for a pattern.
- Test all possible combinations.

Half-time 1

1 Sketch the graphs of the following, showing the intercepts on the axes. **1.3**

(a) $y = 3x + 2$ (b) $y = 5 - x$

(c) $y = 2x - 3$ (d) $y = 6 - 3x$

2 Calculate the gradient of the line passing through each pair of points. **1.2**

(a) (1, 3) and (5, 5) (b) (2, 4) and (6, -4)

(c) (-5, -3) and (5, 2) (d) (-4, 8) and (-2, -8)

3 Sketch the graphs of the following, showing the intercepts on the axes. **1.3**

(a) $2x - 3y = 6$ (b) $5x + 2y = 10$

(c) $3x - y = 6$ (d) $3x + 4y = 12$

4 The Celsius temperature scale is used in Australia, while the Fahrenheit temperature scale is used in the USA. The freezing point of water is 0 °C and the boiling point is 100 °C using the Celsius scale. The freezing point of water is 32 °F and the boiling point is 212 °F using the Fahrenheit scale. Show that the formula to convert a Fahrenheit temperature to a Celsius temperature can be expressed as $C = \frac{5}{9}(F - 32)$. **1.1**

5 Sketch the graph of each of the following, showing the intercept with the axes and one other point. **1.3**

(a) $y = -3x$ (b) $y = 4x$

(c) $x + 2y = 0$ (d) $3y = 5x$

6 Solve each of the following equations. **1.1**

(a) $6x + 2 = 14$ (b) $3(x + 5) = 21$

(c) $\frac{2x - 3}{7} = 5$ (d) $3x + 4 = 16 - x$

(e) $\frac{2x + 4}{5} = \frac{3x - 3}{3}$ (f) $\frac{3x + 2}{2} - \frac{x + 10}{7} = 5$

7 Sketch the graph of each of the following. **1.3**

(a) $y = 4$ (b) $x = -2$

(c) $y = -3$ (d) $x = 8$

8 Find the equation of the line with a: **1.2**

(a) gradient of 4 passing through the point (-1, 5)

(b) gradient of -1 passing through the point (3, -7).

Maths 4 Real

Mobile phone plans

There are just as many mobile phones as people in Australia and there are many different telecommunication companies that are after your dollar.

There are two main types of mobile phone plans:

1 **Pre-paid credit**—You don't have a regular monthly plan, but purchase credit in advance of your use. The credit is generally for a defined period of time, after which it expires. The credit is used for particular services such as calls and text messages. When you have spent your credit, you are unable to make additional calls or text messages.

2 **Post-paid plan or cap**—You enter a formal contract for a defined period (usually 12, 24 or 36 months) in which you pay the same amount of money each month for a defined quota of specific services such as calls, texts and internet data. If you don't use your defined quota, you still pay the monthly contract amount. However, if you exceed this quota you are charged on top of your monthly contract amount.

Each type of mobile plan has its pros and cons. A wide range of services are available within these two types of plans such as 'off-peak' and 'peak' call rates, which depend on the time of day. Generally, the peak rate applies from 8 am to 8 pm and the off-peak rate applies from 8 pm to 8 am

There is also a 'flagfall', or connection fee for most calls. To make it more complex, some plans have 'free' calls and/or text messages to other phones on their network. Choosing the best plan is tricky because of the huge number of products available.

Company	Access fee per month	Call rate (per 30 seconds)	Off-peak calls (per 30 seconds)	Text message	Flagfall	Bonus
Terla	$0	45c	30c	30c	0c	Nil
Big Phone	$10	30c	20c	20c	70c	$5 of free calls per month
4 Mobile	$20	20c	10c	10c	73c	$10 of free calls per month
Crazy Joe's	$45	10c	5c	5c	75c	$20 of free calls per month

Which is the best deal?

Your parents have agreed to buy you a mobile phone on the condition that you select a plan that costs no more than $50 per month. You do some research on the internet and obtain the above post-paid mobile phone plans from the major telecommunication companies in your area.

1 You estimate that you would make 180 minutes of calls during peak time, make 30 minutes of calls during the off-peak time and send 100 text messages during a month. Use this estimate to calculate the cost of each plan.

2 Do any of the plans cost $50 or less? Which is the best plan for your estimated usage? Explain your answer.

3 Upon further consideration you decide that you will make all calls during peak time. Create a linear equation that models the cost for each plan if all the calls are made during peak time, taking into account the access fee.

4 Use each of the equations to determine the cost of each plan if 4 hours of peak-time calls were made in the month.

5 Graph the four linear equations on the same set of axes (with time t along the horizontal axis and cost c along the vertical axis).

6 Compare and contrast each of the graphs to identify which plan would be good for a low user (e.g. 60 minutes of peak time calls per month), a moderate user (e.g. 180 minutes of peak time calls per month) and a heavy user (e.g. more than 240 minutes of peak-time calls per month).

1.4 Parallel and perpendicular lines

Parallel and perpendicular lines have specific properties associated with their gradients.

Parallel lines

Lines that are parallel to each other will never intersect because they have the same slope or gradient.

In the graph shown, you can see that the rise and run of both lines is identical. The lines have the same gradient, which in this case is 1.

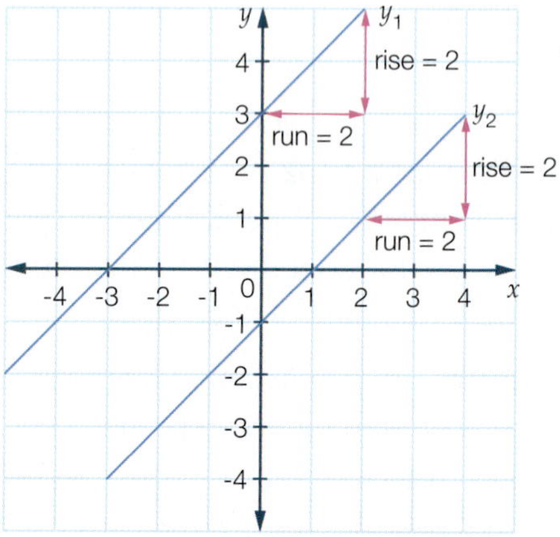

For line y_1: $m = \frac{\text{rise}}{\text{run}} = \frac{2}{2} = 1$ and y-intercept $= 3$

Therefore, the equation is:

$y_1 = x + 3$

For line y_2: $m = \frac{\text{rise}}{\text{run}} = \frac{2}{2} = 1$ and y-intercept $= -1$

Therefore, the equation is:

$y_2 = x - 1$

Two lines that are parallel have the same gradient: $m_1 = m_2$.

Interactive

Parallel lines

Compare the features of parallel lines.

Go to the eBook or the Pearson Places website to access this interactive.

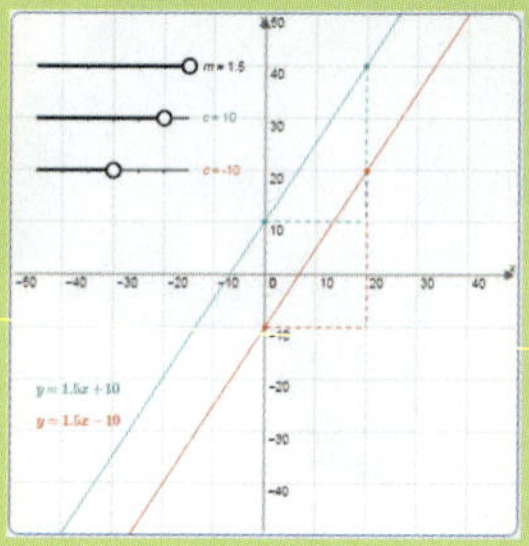

Worked example 11

W.E. 11

Show that the line with equation $2y = 4x - 9$ is parallel to the line with equation $3y - 6x + 5 = 0$.

Thinking	Working
1 Rearrange both equations using inverse operations so that both are in the form $y = mx + b$.	$2y = 4x - 9$ $y = 2x - \frac{9}{2}$ $m_1 = 2$ $3y - 6x + 5 = 0$ $3y = 6x - 5$ $y = 2x - \frac{5}{3}$ $m_2 = 2$
2 Compare the gradient values to show that $m_1 = m_2$.	$m_1 = m_2$ Therefore, the lines are parallel.

Perpendicular lines

Lines that are perpendicular intersect at right angles.

The line $y = -2x + 4$ has an x-intercept of (2, 0) and a y-intercept of (0, 4). If these points are rotated 90° anticlockwise about the origin they will be positioned at (0, 2) and (-4, 0). Joining these points gives us a line perpendicular to $y = -2x + 4$.

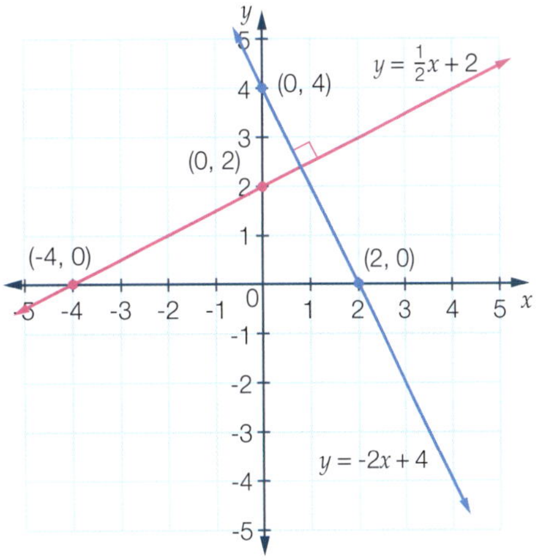

The gradient of this line is $\frac{1}{2}$ because it has a rise of 2 and a run of 4.

The y-intercept is (2, 0) so the equation of this line is $y = \frac{1}{2}x + 2$.

In the equation $y = -2x + 4$, the gradient $m_1 = -2$.

In the equation $y = \frac{1}{2}x + 2$, the gradient $m_2 = \frac{1}{2}$.

When you multiply m_1 by m_2 you obtain -1.

$$\begin{aligned} m_1m_2 &= -2 \times \tfrac{1}{2} \\ &= -1 \end{aligned}$$

When lines are perpendicular, the product of their gradients is -1.

$$m_1m_2 = -1$$

Therefore, the gradients m_1 and m_2 of perpendicular lines are the negative reciprocal of each other.

$$m_1 = -\frac{1}{m_2} \text{ and } m_2 = -\frac{1}{m_1}$$

Worked example 12

W.E. 12

Show that the line with equation $y = 3x - 10$ is perpendicular to the line with equation $6y + 2x + 12 = 0$.

Thinking	Working
1 Rearrange the equations using inverse operations so that both are in the form $y = mx + b$.	$y = 3x - 10$, $m_1 = 3$; $6y + 2x + 12 = 0$, $6y = -2x - 12$, $y = -\frac{1}{3}x - 2$, $m_2 = -\frac{1}{3}$
2 Identify the gradients m_1 and m_2, and multiply to determine whether $m_1 \times m_2 = -1$.	$m_1m_2 = 3 \times -\frac{1}{3} = -1$ Therefore, the lines are perpendicular.

1.4 Parallel and perpendicular lines

Navigator

Answers p.767

1, 2, 3, 4 (a–h), 5 (column 1), 6, 7, 8, 10, 12, 13, 15, 17	1, 2, 3, 4 (column 1), 5 (column 2), 6, 7, 8, 9, 10, 11, 12, 13, 14, 16, 17, 19	2, 3, 4 (e–j), 5 (column 2), 6, 7, 9, 10, 11, 13, 14, 16, 18, 19

Fluency

1 Write the gradient of a line that is **(i)** parallel to and **(ii)** perpendicular to the line with the equation:

(a) $y = 3x - 5$

(b) $y = -\frac{2}{3}x + 6$

W.E. 11

2 **(a)** Show that the line with equation $5x - 2y = 7$ is parallel to the line $2y = 5x + 11$.

(b) Show that the line with equation $10y = 5x + 3$ is parallel to the line $3x - 6y = 2$.

W.E. 12

3 **(a)** Show that the line with equation $4x + 3y = 7$ is perpendicular to the line $8y = 6x - 2$.

(b) Show that the line with equation $2x - 5y = 1$ is perpendicular to the line $4y + 10x = 3$.

4 Find the gradients of the following pairs of lines and determine whether they are parallel.

(a) $y = 3x - 10$ and $y = 3x + 10$

(b) $y = 2x + 5$ and $y = -2x + 6$

(c) $y = \frac{1}{2}x - 4$ and $y = \frac{1}{2}x + 5$

(d) $y + 4x = 4$ and $y = 4x - 8$

(e) $y - 5x = 9$ and $y = 9x + 5$

(f) $y - 2x = 9$ and $y - 2x - 4 = 0$

(g) $2y + 6x - 4 = 0$ and $y + 3x - 8 = 0$

(h) $4y + 12x - 4 = 0$ and $3y + 12x - 9 = 0$

(i) $-2y + 8x - 4 = 0$ and $4y - 16x - 8 = 0$

(j) $5x + 3y = 15$ and $3x - 15y = 10$

5 Find the gradients of the following pairs of lines and determine whether they are perpendicular.

(a) $y = 2x - 10$ and $2y + x + 12 = 0$

(b) $y - 4x = 10$ and $4y + 2x + 12 = 0$

(c) $y = \frac{1}{3}x$ and $6y + 2x + 12 = 0$

(d) $6y - 3x = 24$ and $2y + 4x + 12 = 0$

(e) $-4y + 3x = 16$ and $8y + 12x + 12 = 0$

(f) $5y - 4x = 25$ and $8y + 5x + 12 = 0$

6 Which of the following lines is parallel to $y = 4x + 12$?

A $y = -\frac{1}{4}x + 12$ B $4y = 4x + 12$ C $2y - 8x + 6 = 0$ D $2y + 8x + 12 = 0$

7 Which of the following lines is perpendicular to $y = 3x - 9$?

A $y = 3x + 9$ B $6y = 9x + 12$ C $y = -3x - 9$ D $9y + 3x = 27$

Understanding

8 Show that the line joining (-2, 2) and (4, 8) is parallel to $y = x + 10$.

9 Show that the line joining (-2, 5) and (4, -7) is parallel to $2x + y - 10 = 0$.

10 Show that the line joining (-4, 1) to (0, 4) is perpendicular to $y = -\frac{4}{3}x - 2$.

11 Show that the line joining (4, 5) to (9, 8) is perpendicular to $y = -\frac{5}{3}x + 10$.

Reasoning

12 (a) Which two of the following lines **A–D** are parallel?

(b) Which two of the following lines **A–D** are perpendicular?

A $y = 3x + 3$ B $3x + y = 3$ C $3x - y = 3$ D $3y = x + 3$

13 Find the equation of the line through (2, -3) that is perpendicular to $y = -\frac{1}{3}x + 11$.

14 Find the equations of the lines through (0, 5) that are parallel and perpendicular to $2x + y - 10 = 0$.

15 Consider the diagram at right. Find the gradients of the line segments:

(a) AB (b) BC (c) AC

(d) Dakota thinks that the triangle ABC is right-angled. Is she correct?

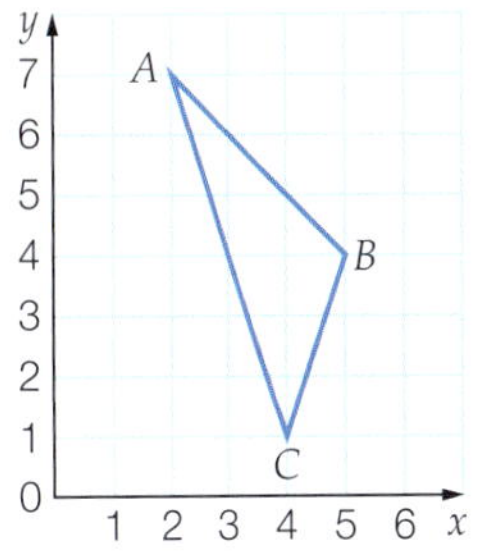

16 Show that the lines joining the points $A(-1, -1)$, $B(1, -4)$ and $C(2, 1)$ form a right-angled triangle ABC.

17 Show that the points $A(-3, 1)$, $B(-1, 3)$, $C(3, 1)$ and $D(1, -1)$ form the vertices of a parallelogram.

18 Show that the points $A(-4, 3)$, $B(-2, -2)$, $C(3, 0)$ and $D(1, 5)$ form the vertices of a square.

Open-ended

19 Identify three different straight-line equations that are:

(a) parallel to $3x - 6y + 9 = 0$ (b) perpendicular to $4x - 12y - 18 = 0$.

Problem solving

Area of a triangle

Find the area (in square units) of the triangle formed by the three lines whose equations are:

1 $y - x - 6 = 0$ 2 $x - 2y = 3$ 3 $x = -y + 6$

Strategy options

- Draw a diagram.
- Break problem into manageable parts.

Exploration CAS

Equipment required: TI-Nspire CAS or Casio ClassPad CAS

The art of linear relationships

The following designs were all created by graphing linear relationships. Your task is to work out the equations that can be graphed to create these designs. You may work alone or in small groups and can attempt the designs in any order. Write all the equations needed to create each design.

After you have finished these designs, make up your own. Challenge a friend to reproduce your new designs.

Using TI-Nspire CAS

From the **Home** screen, select **Scratchpad** > **Graph**. Then select **menu** > **Graph Entry/Edit** > **Function** and enter the equation to be graphed.

For each group of designs below, adjust the graph window settings by selecting **menu** > **Window/Zoom** > **Window Settings...** and then change the values as shown.

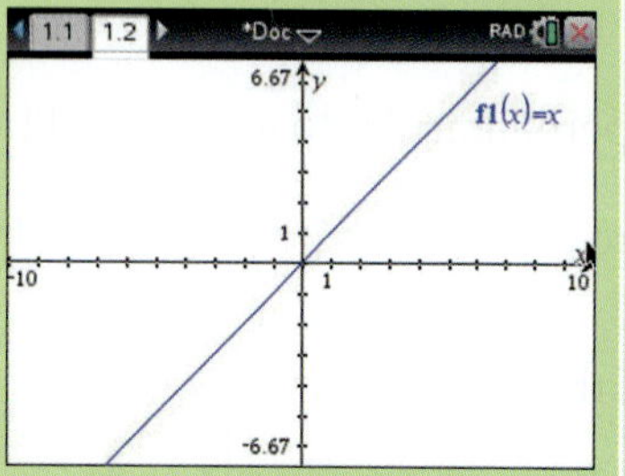

Using Casio ClassPad CAS

From the menu select **Graph & Table**. Enter your equations and tick the box next to each one.

For each group of designs below, adjust the graph window settings by selecting the view icon and then change the values as shown. Select the graph icon to view the graphs.

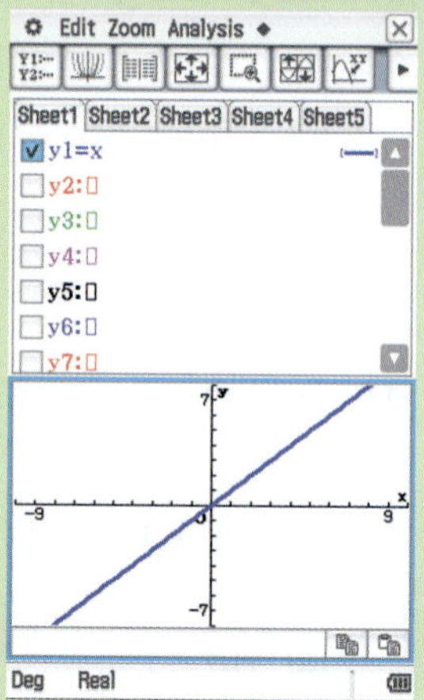

Use the given window settings for each group of designs.

Group A designs

Using TI-Nspire CAS

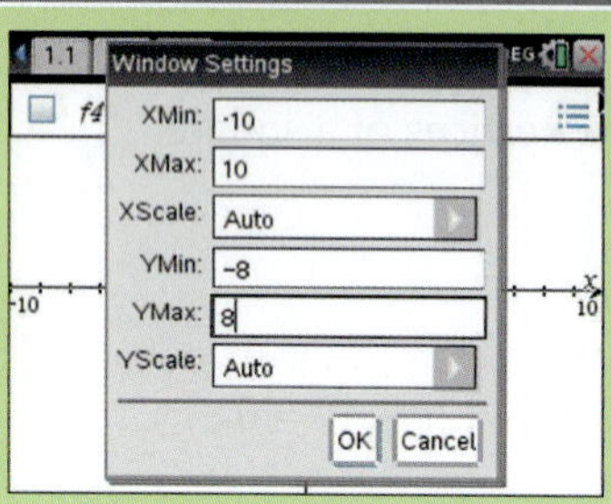

Using Casio ClassPad CAS

Use a scale of 1 for both axes.

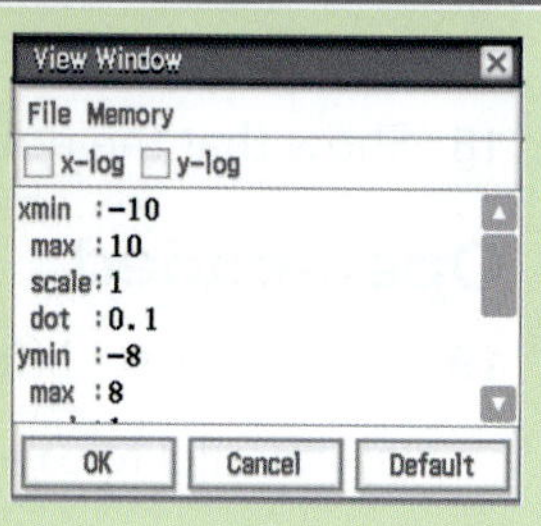

(a)

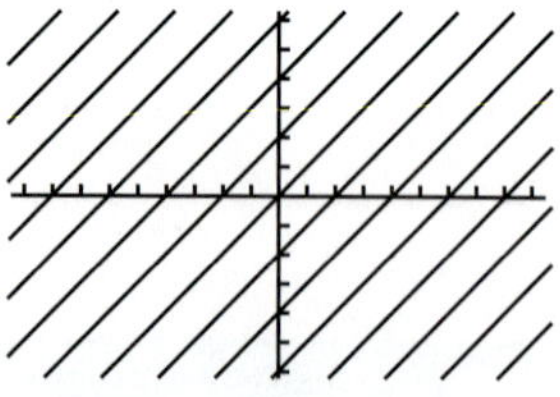

(b)

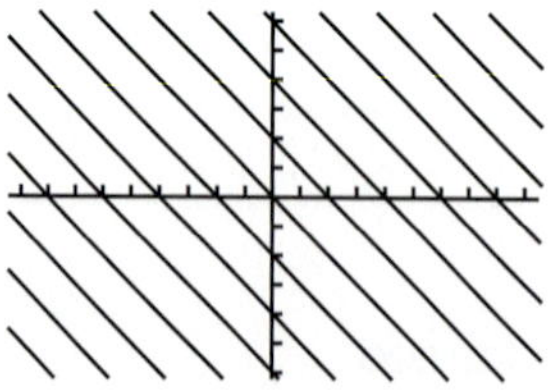

(c)

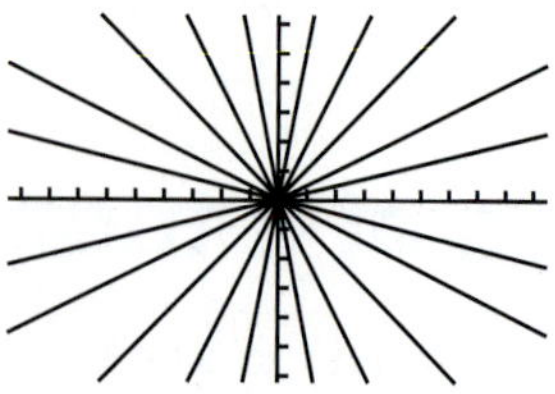

(d)

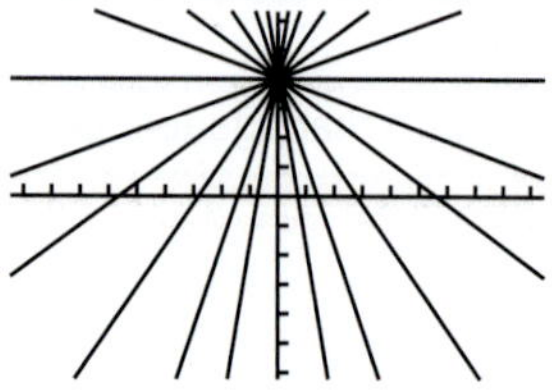

(e)

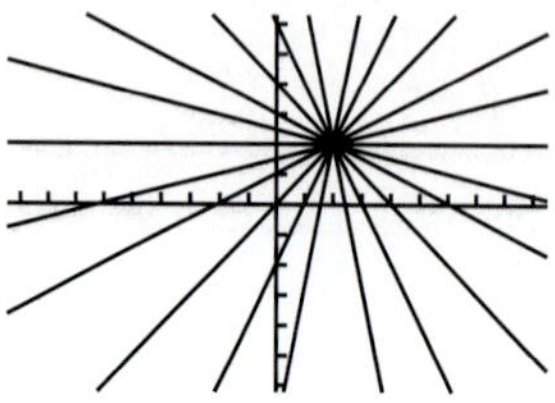

(f)

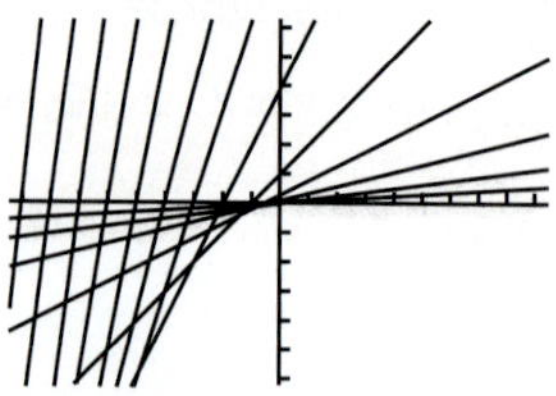

Group B designs

Using TI-Nspire CAS	Using Casio ClassPad CAS
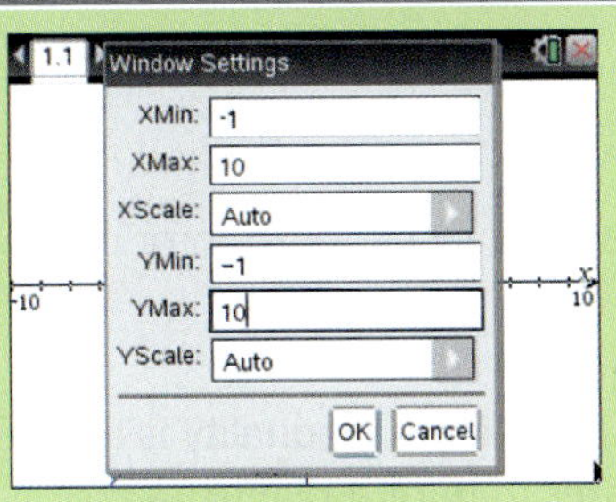	Use a scale of 1 for both axes. 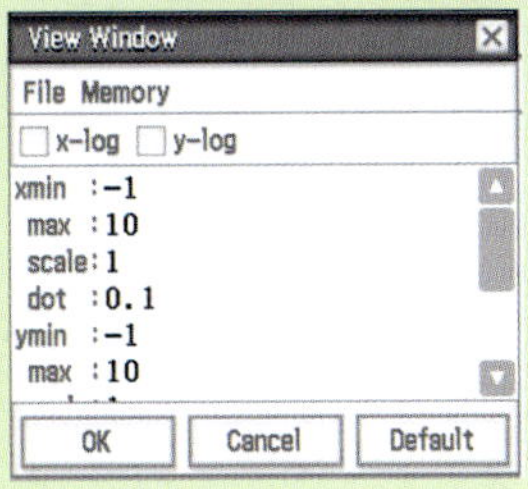

(a) 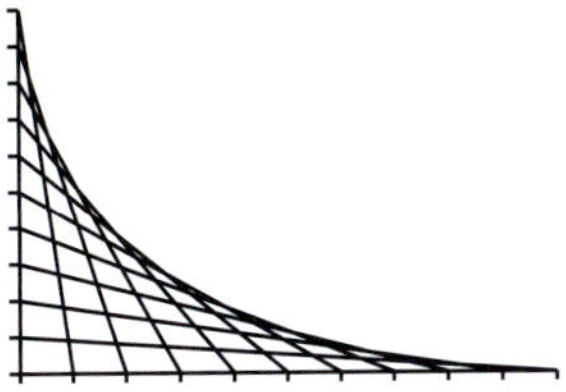

(b) 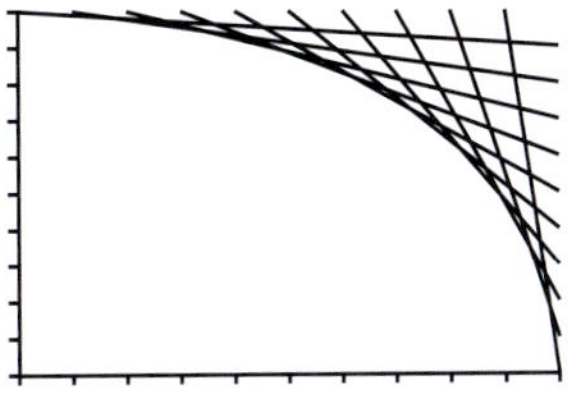

(c) 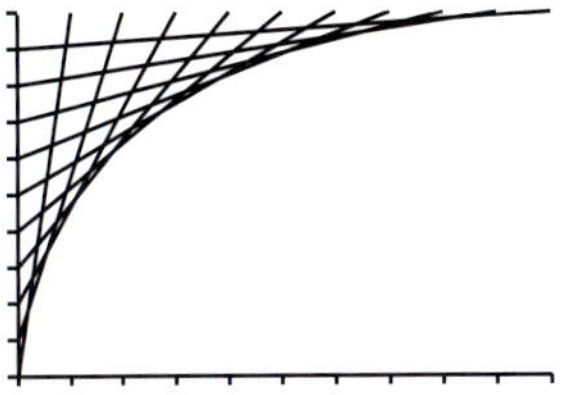

(d) 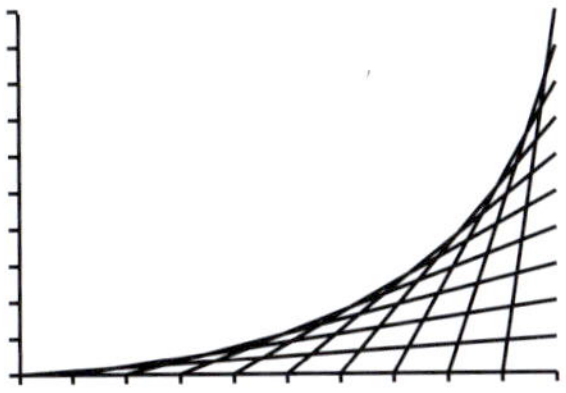

(e) 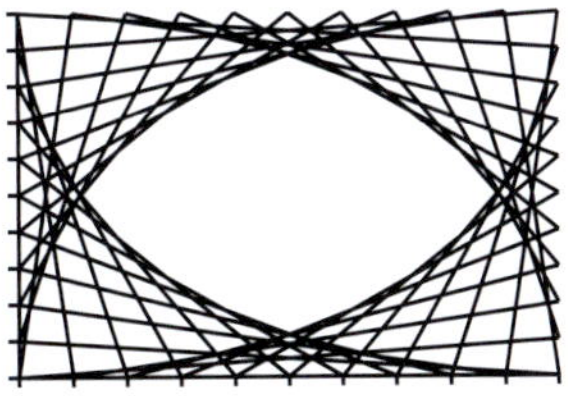

(f) 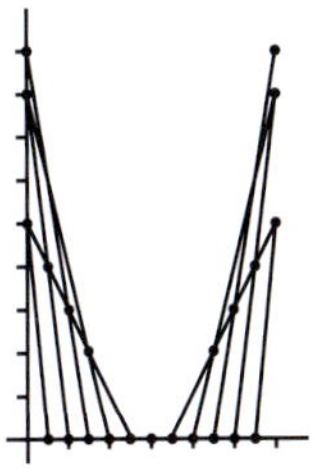

Group C designs

Using TI-Nspire CAS	Using Casio ClassPad CAS
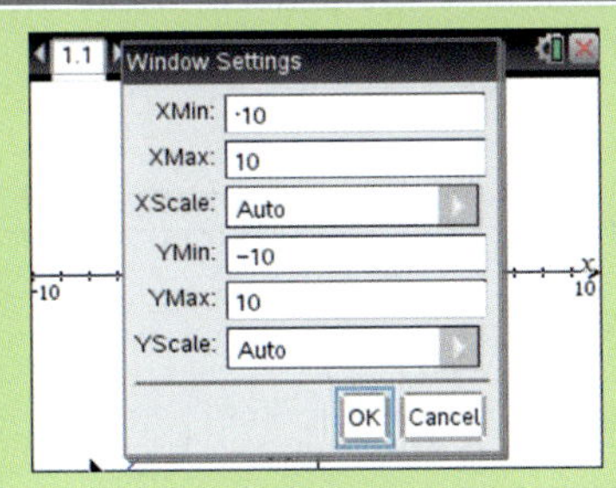	Use a scale of 1 for both axes. 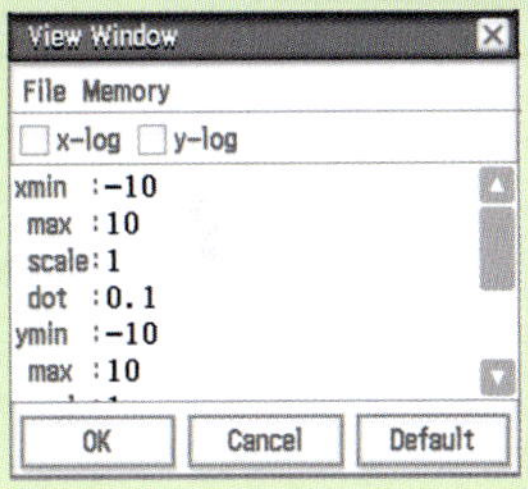

(a) 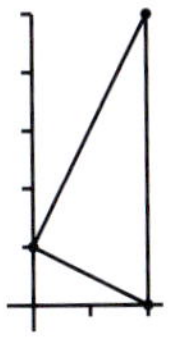

(b) 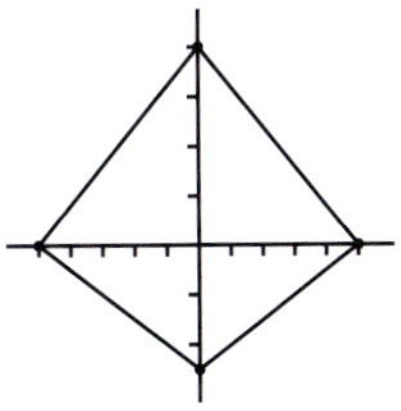

(c) 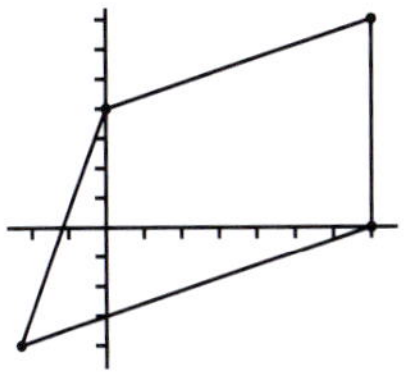

(d) 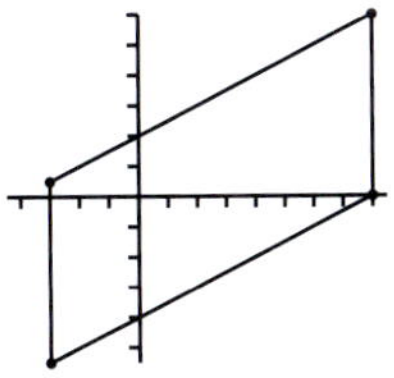

(e) 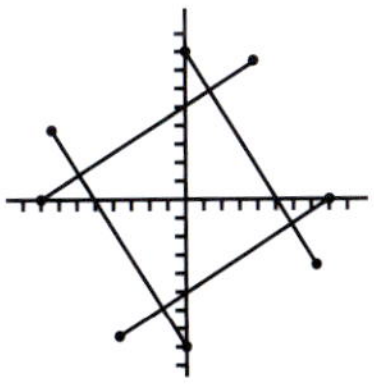

(f) 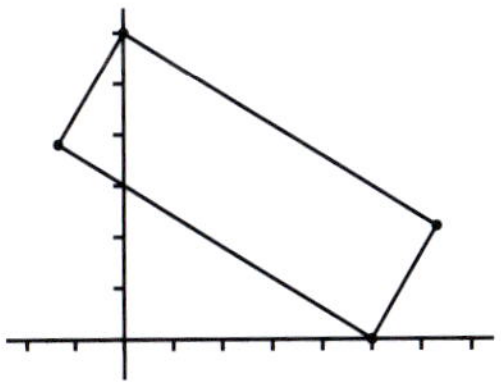

1.5 Linear inequalities

Inequality symbols

$<$ means less than

$\leq$ means less than or equal to

$>$ means greater than

$\geq$ means greater than or equal to

What is an inequality?

An equation is a statement of equality, such as $x = 17$. A **linear inequality** is a statement such as $x \geq 17$, which means that x can be any value greater than or equal to 17.

On a number line, $x \geq 17$ can be represented as:

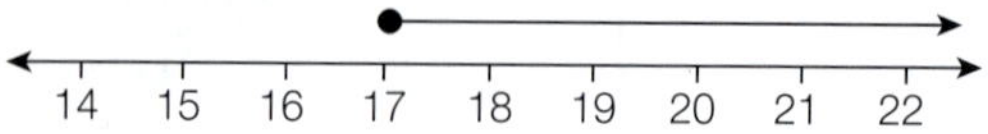

An arrow is used to indicate that the inequality represents all the numbers beyond this point. A closed circle at 17 indicates that 17 is included.

Similarly, on a number line $x < 2$ can be represented as:

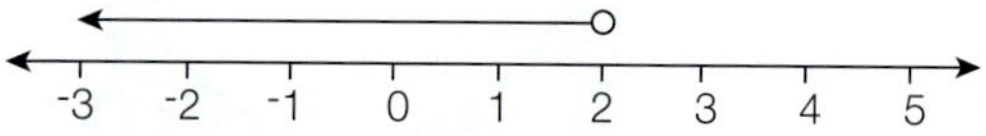

An open circle at 2 indicates that 2 is not included.

> If the inequality includes the end point, a closed circle is used; these inequalities use $\geq$ or $\leq$.
>
> If the inequality does not include the end point, an open circle is used; these inequalities use $>$ or $<$.

$-3 \leq x \leq 2$ represents all values of x greater than or equal to -3 and less than or equal to 2. The closed circles at the endpoints indicate that both -3 and 2 are included.

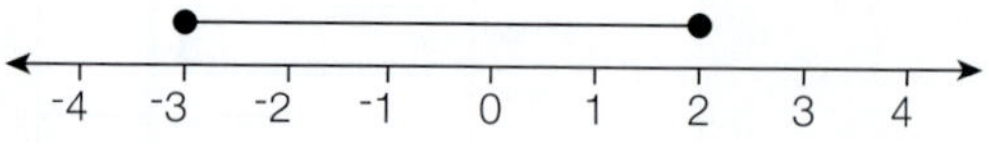

$-3 < x < 2$ represents all values of x greater than -3 and less than 2. The open circles at the endpoints indicate that -3 and 2 are not included.

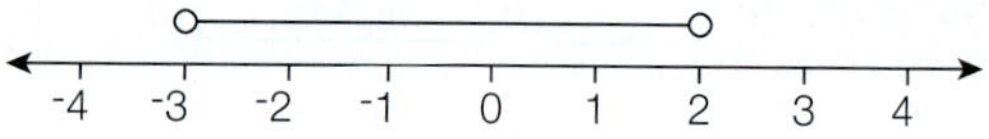

$-3 \leq x < 2$ represents all values of x greater than or equal to -3 and less than 2. The closed circle indicates that -3 is included. The open circle indicates that 2 is not.

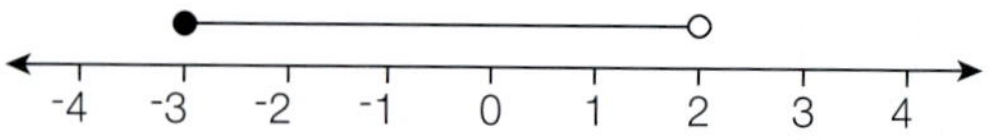

$-3 < x \leq 2$ represents all values of x greater than -3 and less than or equal to 2. The closed circle indicates that 2 is included. The open circle indicates that -3 is not.

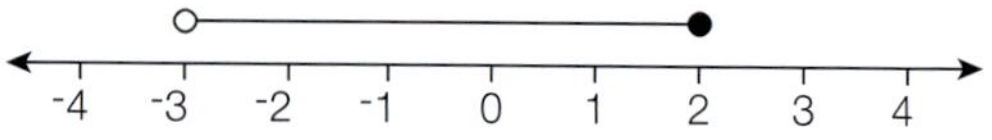

Worked example 13 — W.E. 13

Represent the values of x on a number line for each of the following.

(a) $x \leq 5$ **(b)** $x > -2$ **(c)** $3 < x \leq 7$

Thinking	Working
(a) x is less than or equal to 5, so draw a closed circle at 5 and an arrow to the left.	(a) $x \leq 5$ 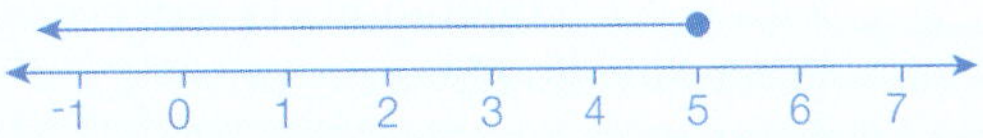
(b) x is greater than -2, so draw an open circle at -2 and an arrow to the right.	(b) $x > -2$ 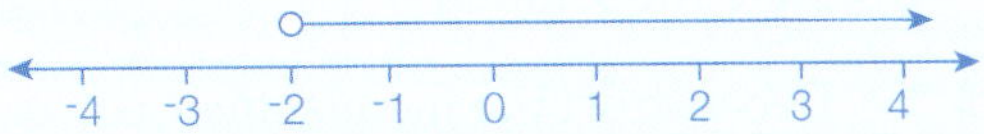
(c) x is greater than 3 and less than or equal to 7, so draw an open circle at 3 and a closed circle at 7. Join these with a line.	(c) $3 < x \leq 7$ 

To solve an inequality

You solve linear inequalities using the same inverse operations in the opposite order as you use to solve linear equations. Unlike linear equations, inequalities have a range of solutions, sometimes called the 'solution set'.

You also need to be careful when multiplying or dividing by a negative number, because this will affect the inequality symbol.

Consider $4 < 8$. If you multiply both sides by -1, what happens to the inequality symbol?

To keep the inequality correct, the symbol needs to be reversed:

$-4 > -8$

When you multiply or divide both sides of an inequality by a negative number, the inequality symbol is reversed.

$-2x < 8$

$x > \frac{8}{-2}$ (inequality is reversed)

$x > -4$

Worked example 14 — W.E. 14

Solve for x in each of the following inequalities.

(a) $x + 3 \geq 12$ **(b)** $4x - 3 < -7$

Thinking	Working
(a) 1 Deconstruct the inequality with an inverse operation (-3 from both sides).	(a) $x + 3 \geq 12$
2 State the solution.	$x \geq 9$
(b) 1 Deconstruct the inequality with inverse operations (+ 3, ÷ 4).	(b) $4x - 3 < -7$ $4x < -4$
2 State the solution.	$x < -1$

Worked example 15

W.E. 15

Solve for x in each of the following.

(a) $-4x \leq 20$

(b) $9 - 2x \geq 5$

Thinking	Working
(a) 1 Deconstruct the inequality with inverse operations (÷ -4). When dividing by a negative number, reverse the inequality.	(a) $-4x \leq 20$
2 State the solution.	$x \geq -5$
(b) 1 Deconstruct the inequalities with inverse operations (− 9, ÷ -2). When dividing by a negative number, reverse the inequality.	(b) $9 - 2x \geq 5$ $-2x \geq -4$
2 State the solution.	$x \leq 2$

1.5 Linear inequalities

Navigator

Answers p.768

1 (columns 1–2), 2 (columns 1–2), 3 (columns 1–2), 4, 5, 6, 7, 8, 9, 11 (columns 1–2), 12, 13 (columns 1–2), 15, 16	1 (columns 2–3), 2 (columns 2–3), 3 (columns 2, 3), 4, 5, 6, 7, 8, 9, 10, 11 (columns 2–3), 12, 13 (columns 2, 3), 15, 16, 17	1 (column 3), 2 (column 3), 3 (column 3), 5, 6, 7, 8, 9, 10, 11, 12, 13, 14, 16, 17, 18

Fluency

W.E. 13

1 Represent the values of x on a number line for each of the following.

(a) $x \geq 5$ (b) $x \leq 8$ (c) $x \geq -2$

(d) $x < -4$ (e) $x > -5$ (f) $x > 7$

(g) $x \leq 0$ (h) $x \geq 2$ (i) $x \leq -5$

(j) $x > 1$ (k) $x > 0$ (l) $x < -2$

(m) $-2 \leq x \leq 4$ (n) $1 \leq x \leq 6$ (o) $0 < x < 5$

(p) $-3 < x < -1$ (q) $-4 \leq x < -2$ (r) $-4 < x \leq 0$

W.E. 14

2 Solve for x in each of the following.

(a) $x + 6 > -4$ (b) $x - 3 < -8$ (c) $x + 12 \leq 2$

(d) $\frac{x}{5} \leq 7$ (e) $\frac{x}{9} \geq -1$ (f) $\frac{x}{7} > -4$

(g) $3x < -16$ (h) $5x > -27$ (i) $9x \leq -81$

(j) $7x - 3 \geq 53$ (k) $4x + 11 \leq 3$ (l) $3x + 24 > 0$

(m) $\frac{4x-1}{5} > 3$ (n) $\frac{2x+5}{3} < 9$ (o) $\frac{3x+2}{7} \leq 2$

(p) $\frac{3x}{4} + 1 \leq -5$ (q) $\frac{7x}{11} - 2 > 5$ (r) $\frac{4x}{5} - 10 > -2$

W.E. 15

3 Solve for y in each of the following.

(a) $-3y \geq 15$ (b) $-8y \leq 48$ (c) $-5y \leq 25$

(d) $-y < -12$ (e) $-2y > -42$ (f) $-3y > -36$

(g) $1 - 2y \leq 17$ (h) $8 - 5y \geq -2$ (i) $5 - 3y \leq 17$

(j) $5(2 - 3y) \geq 11$ (k) $4(3 - 7y) \leq -2$ (l) $3(5 - 2y) \geq -4$

(m) $\dfrac{8 - 3y}{5} > -1$ (n) $\dfrac{14 - 6y}{3} < -3$ (o) $\dfrac{12 - 5y}{4} > -2$

4 (a)

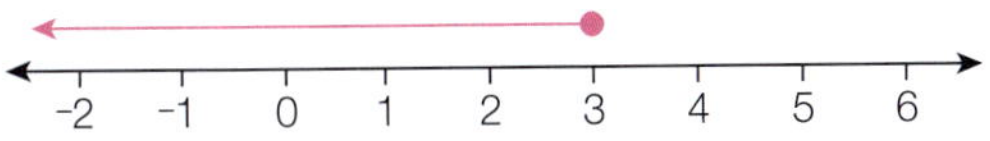

This number line represents:

A $x < 3$ B $x \leq 3$ C $x \geq 3$ D $x > 3$

(b)

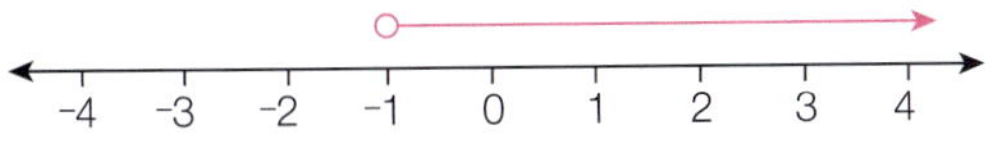

This number line represents:

A $x < -1$ B $x \leq -1$ C $x > -1$ D $x \geq -1$

(c)

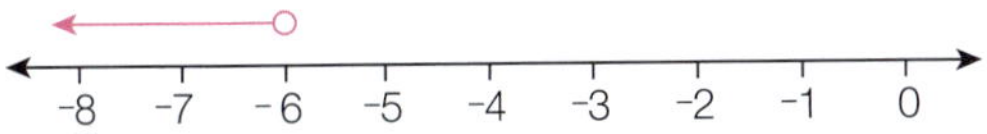

This number line represents:

A $x \leq -6$ B $x < -6$ C $x \geq -6$ D $x > -6$

(d)

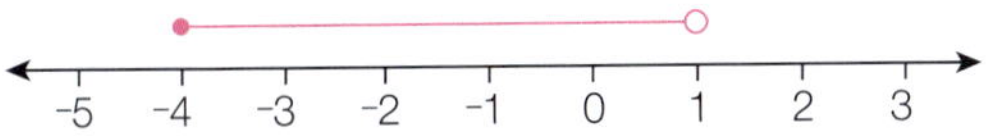

This number line represents:

A $-4 < x \leq 1$ B $-4 \geq x > 1$ C $-4 \leq x \leq 1$ D $-4 \leq x < 1$

5 Which alternative is *not* a possible solution for each of the following inequalities?

(a) $x \geq 4$

A $x = 0$ B $x = 4$ C $x = 5$ D $x = 7$

(b) $x - 3 < 4$

A $x = 0$ B $x = 3$ C $x = 4$ D $x = 7$

(c) $6 - x \leq 4$

A $x = -1$ B $x = 2$ C $x = 4$ D $x = 6$

6 Which alternative is *not* a possible solution to the inequality given?

(a) $2x + 3 \geq 7$

A $x = 1$ B $x = 2.1$ C $x = 3$ D $x = 5.2$

(b) $5x - 2 < -9$

A $x = -5$ B $x = 0$ C $x = -2$ D $x = -3$

(c) $11 - 4x < -1$

A $x = 10$ B $x = 7$ C $x = 4$ D $x = 2$

(d) $\dfrac{6x + 1}{2} > 5x - 2$

A $x = -2$ B $x = -1$ C $x = 1$ D $x = 2$

Understanding

Hint

The reserve price is the smallest amount that the house could be sold for.

7 The reserve price at an auction of a house is \$270 000. The house is sold at auction.

(a) Represent the selling price (P) on a number line.

(b) Write the linear inequality for the selling price (P).

(c) Write the inequality for the selling price (P) given that the house sold for less than \$320 000.

8 The speed limit on a particular freeway is 100 km/h. Write an inequality that represents all possible legal speeds on that freeway.

9 A sales representative needs to sell at least five vacuum cleaners a week to receive a bonus commission. Write an inequality for the number of vacuum cleaners (n) that the sales representative needs to sell to earn a bonus.

10 For the wool and other materials to make seven hand-knitted jumpers, Lucy and Lindsay spent \$385. Write an inequality to show the selling price (j) for each jumper if they want to make a profit.

11 Find the solutions for each of the following.

(a) $5x - 3 \geq 3 - 4x$ (b) $2x + 7 \leq 4 - 9x$ (c) $6x + 2 < 3x + 2$

(d) $2(x + 10) > 10 - 3x$ (e) $15x - 4 \leq 4(3x + 2)$ (f) $5(2x - 3) \geq 8x + 9$

(g) $\frac{2x+1}{7} > \frac{x-3}{4}$ (h) $\frac{3x-4}{2} < \frac{4x+7}{5}$ (i) $\frac{2-3x}{2} \leq \frac{5x-16}{3}$

12 Find the solution for each of the following.

(a) $2a + 3 < a - 4$ (b) $\frac{a}{3} - 2 \geq 4$ (c) $\frac{2a-2}{3} < \frac{a+4}{5}$

Hint

Reverse inequalities when multiplying or dividing by negative numbers.

13 Find the solution for each of the following.

(a) $y + 6 \leq 2y + 7$ (b) $20 - 3y > 5 + 2y$ (c) $7y - 4 \geq 3 + 10y$

(d) $13 - y < 5y + 11$ (e) $4 - \frac{y}{3} \leq -6$ (f) $7 - \frac{y}{6} \geq 1$

(g) $-3 - \frac{5y}{12} < 2$ (h) $\frac{-7y}{3} + 2 > 0$ (i) $\frac{3y+2}{11} > \frac{2y-5}{3}$

Reasoning

14 A shopkeeper can fit a total of only 120 2-litre cartons of milk (both full cream and low fat) into the refrigerator. He knows that he needs at least 50 2-litre cartons of full cream milk and that no more than 40 2-litre cartons of low fat milk will be needed.

(a) Let f represent the number of cartons of full cream milk and l represent the number of cartons of low fat milk. Write the two inequalities to show the number of cartons of each type of milk required by the shopkeeper and show them on a number line.

(b) Write an inequality to show the number of cartons of milk in the refrigerator.

15 Tony has spent $428 to buy 20 calculators wholesale.

(a) Write an inequality to represent the price (p) he could sell each calculator for and make a profit.

(b) Solve the inequality to find the minimum price that he could sell the calculators for and still make a profit.

16 At the local milk bar, Sue spends $3.45 on milk and bread. With the change from her $5 note, she buys some liquorice straps at 40c each.

(a) Write an inequality to represent the number of liquorice straps (n) she could buy.

(b) Solve the inequality to find the maximum number of liquorice straps she could buy.

Open-ended

17 Draw a number line showing numbers that could be greater than or equal to a negative number, but less than a positive number.

18 (a) Make up an inequality, with the variable on both sides of the inequality, that will produce solutions that are all negative.

(b) Make up an inequality, with the variable on both sides of the inequality, that will produce solutions that are all positive.

Problem solving

Win ratio

A tennis player computes her win ratio by dividing the number of matches she has won by the total number of matches she has played. At the start of a weekend, her win ratio is exactly 0.5 or $\frac{1}{2}$. Over the course of the weekend, she plays four matches, winning three and losing one. At the end of the weekend, her overall win ratio is greater than 0.503. What is the greatest number of matches she could have won before the weekend began?

Hint

Let the number of wins at the start of the weekend be w. Given the win ratio was 0.5 or $\frac{1}{2}$, what is the relationship between the number of wins, w, and the total number of games, n?

Strategy options

- Break problem into manageable parts.
- Test all possible combinations.

1.6 Simultaneous equations

You have learnt how to solve an equation to find an unknown quantity. But many problems involve more than one unknown quantity.

For example, suppose the McKillop and Morris families go to the cinema. The McKillop family pays \$48 for their tickets (two adults and three children). The Morris family pays \$44 for theirs (one adult and four children). The eldest son in the Morris family wants to pay for his own ticket, but no one can remember the individual ticket prices for adults and children.
This problem can be represented using two linear equations.

Let a = cost of an adult's ticket and c = cost of a child's ticket.

For the McKillop family: 2 adult's tickets + 3 child's tickets = \$48

so $2a + 3c = 48$ [1]

For the Morris family: 1 adult's ticket + 4 child's tickets = \$44

so $a + 4c = 44$ [2]

You now have two equations involving the unknowns a and c.
These are called **simultaneous equations**. To solve this problem, you need to find the values of a and c that make *both* equations true.

Solving simultaneous equations graphically

If you sketch the graphs of two linear equations, the point of intersection is on both graphs and will satisfy both equations. There will only be one point of intersection and therefore the equations will have only one solution. Therefore, graphs of linear equations can be used to solve simultaneous equations.

However, there are two important exceptions to this: parallel and coincident lines.

Parallel lines

Parallel lines do not intersect. As there is no point of intersection, the equations of lines that are parallel have no solution. Any equations that have the same gradient, but are not multiples of each other, will be parallel and will therefore have no solution. For example, $y = 2x + 4$ and $y = 2x + 6$ are parallel. No point is on both lines and therefore no point satisfies both equations. Thus, the simultaneous equations have no solution.

Coincident lines

If one equation is a multiple of the other, the gradients will be the same, but the equations do not represent parallel lines. The graph of one line will be identical to the graph of the other, because the equations represent the same line. These linear graphs are called **coincident lines**. Therefore, there are an infinite number of solutions to these type of equations because every point on either line satisfies both equations.

For example, $2x - y = 5$ and $4x - 2y = 10$ are coincident lines and have an infinite number of solutions.

Solving unique simultaneous equations

Equations that give non-parallel and non-coincident graphs are called **unique equations** because they have only one point of intersection and therefore only one solution.

Because the point of intersection satisfies both equations, graphs can be used to find the solution to simultaneous equations.

Method 1: Solving simultaneous equations graphically

Simultaneous equations involve two or more variables and are solved at the same time, either graphically or algebraically. The solution of simultaneous equations gives values for each variable that satisfy all equations. You can solve for any number of variables as long as you have the same number of unique equations. In this section, you will deal with two variables only.

Consider the cinema problem: $2a + 3c = 48$ [1]
$a + 4c = 44$ [2]

Both these equations can be graphed on the same set of axes.

For $2a + 3c = 48$:
$c = 0, 2a = 48$
$\therefore a = 24$
$a = 0, 3c = 48$
$\therefore c = 16$
x-intercept is (24, 0)
y-intercept is (0, 16)

For $a + 4c = 44$:
$c = 0, a = 44$
$a = 0, 4c = 44$
$\therefore c = 11$
x-intercept is (44, 0)
y-intercept is (0, 11)

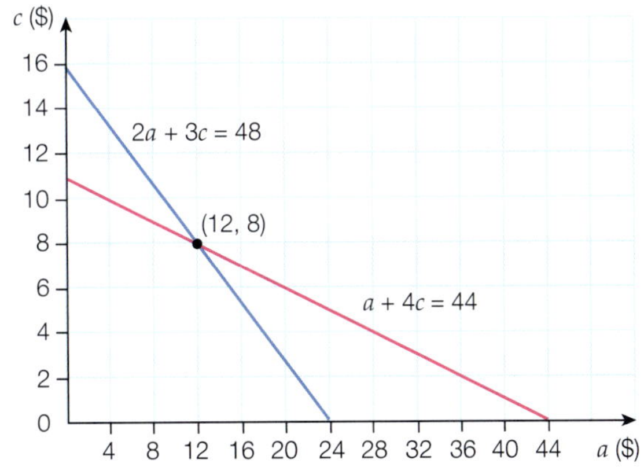

From the graph, it can be seen that the two lines cross at the point (12, 8). This means that $a = 12$, $c = 8$ satisfies both equations. You can check this by substituting these values for a and c back into the original equations.

For $2a + 3c = 48$:

$$\begin{aligned} \text{LHS} &= 2 \times 12 + 3 \times 8 \\ &= 24 + 24 \\ &= 48 \\ &= \text{RHS} \end{aligned}$$

For $a + 4c = 44$:

$$\begin{aligned} \text{LHS} &= 12 + 4 \times 8 \\ &= 12 + 32 \\ &= 44 \\ &= \text{RHS} \end{aligned}$$

So, an adult's ticket to the cinema costs \$12 and a child's ticket costs \$8.

Worked example 16

W.E. 16

(a) Sketch the pair of linear equations $y = x + 2$ and $y = 6 - x$ and find the point of intersection.

(b) Hence, find the solution to $y = x + 2$ and $y = 6 - x$.

Thinking	Working
1 Sketch the two lines clearly showing the point of intersection. You may have to extend your graphs to see the point where the lines cross. **2** Find the coordinates of the point of intersection on your graph.	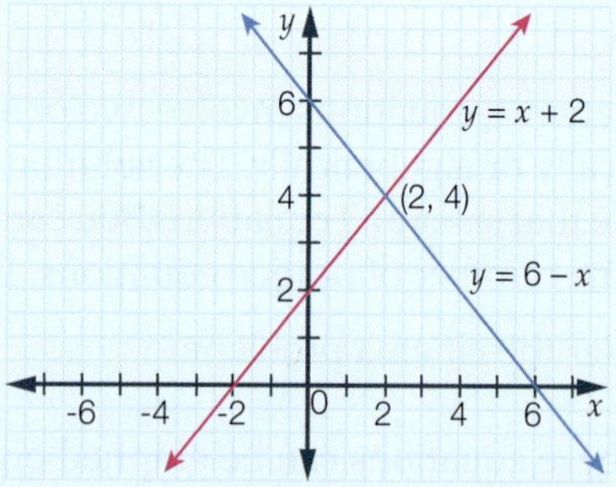
3 Write the coordinates of the point of intersection.	Point of intersection = (2, 4)
(b) The point of intersection gives the solution to the equations	(b) $x = 2$, $y = 4$ is the solution to the equations $y = x + 2$ and $y = 6 - x$

Notice also that the point of intersection (2, 4) satisfies both equations.

$$\begin{aligned} y &= x + 2 \\ &= 2 + 2 \ (x = 2) \\ &= 4 \end{aligned} \qquad \begin{aligned} y &= 6 - x \\ &= 6 - 2 \ (x = 2) \\ &= 4 \end{aligned}$$

You can also use CAS technology to find intersection points and solve simultaneous equations.

Using TI-Nspire CAS

From the Home screen, select **Scratchpad** > **Graph** and enter the equations to be graphed. Then select **menu** > **Analyze Graph** > **Intersection**. Set the lower bound to the left of the intersection point and set the upper bound to the right of the intersection point. The intersection coordinates will be displayed.

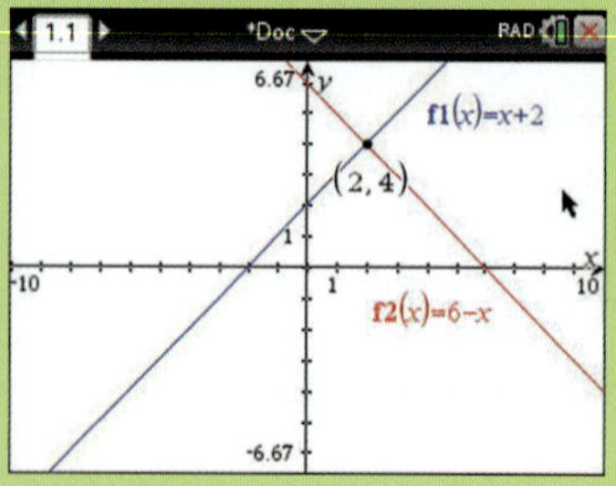

Using Casio ClassPad CAS

From the menu select **Graph & Table**. Enter the equations to be graphed, tick the box next to each one and select the Graph icon. Then select **Analysis** > **G-Solve** > **Intersection**. The intersection coordinates will be displayed.

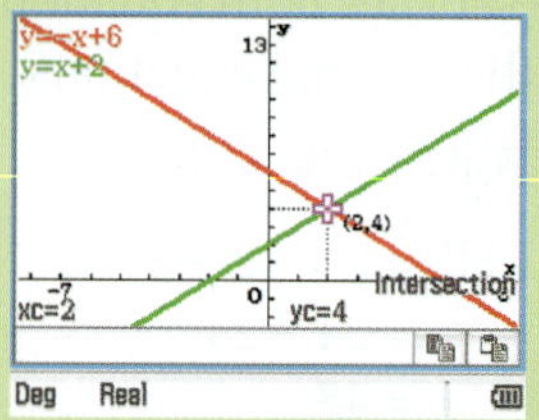

Method 2: Solving simultaneous equations algebraically

Using algebra can be a more accurate and efficient way of solving simultaneous equations than the graphical method. The graphical method does not work for more complex simultaneous equations with more than three unknowns and it is difficult to get exact solutions if one or both of the values at the point of intersection is not a whole number.

Simultaneous equations can be solved algebraically using either elimination or substitution.

Solving simultaneous equations by elimination

Use the elimination method when both the variables are on the same side of the equation.

You eliminate one of the variables by adding or subtracting the two equations or multiples of the two equations. When one of the variables has a coefficient of the same magnitude you can add or subtract the equations.

Worked example 17

W.E. 17

Solve the following simultaneous equations algebraically using the elimination method.

(a) $3x + 5y = -3$
$x - 5y = 19$

(b) $4x + 5y = -6$
$4x - 3y = 10$

Thinking	Working
(a) 1 Write the equations and label them.	(a) $3x + 5y = -3$ [1] $x - 5y = 19$ [2]
2 Eliminate the variable with the same coefficient by either adding or subtracting. (Here, you use equations [1] and [2] to eliminate y.) $3x + 5y = -3$ $+ \quad x - 5y = 19$ $4x + 0 = 16$	$4x = 16$ [1] + [2]
3 Solve the equation for the remaining variable.	$x = 4$
4 Find the value of the eliminated variable by substituting the value of the variable into one of the original equations. (Here, you substitute $x = 4$ into equation [1].)	$3x + 5y = -3$ [1] $3 \times 4 + 5y = -3$ $12 + 5y = -3$ $5y = -15$ $y = -3$
5 Check the solution by substituting the values into the other original equation.	$x - 5y = 19$ [2] LHS $= x - 5y$ $= 4 - 5 \times -3$ $= 4 + 15$ $= 19$ $=$ RHS
6 Write the solution for both variables.	$x = 4$, $y = -3$ or $(x, y) = (4, -3)$

Thinking	Working	
(b) 1 Write the equations and label them.	(b) $4x + 5y = -6$ $4x - 3y = 10$	[1] [2]
2 Eliminate the variable with the same coefficient by either adding or subtracting. (Here, you eliminate x by subtracting equation [2] from [1].)	$8y = -16$	[1] – [2]
3 Solve the equation for the remaining variable.	$y = -2$	
4 Find the value of the eliminated variable by substituting the value of the variable into one of the original equations. (Here, you substitute $y = -2$ into equation [1].)	$4x + 5y = -6$ $4x + 5 \times -2 = -6$ $4x - 10 = -6$ $4x = 4$ $x = 1$	[1]
5 Check the solution by substituting the values into the LHS of the other original equation. (Here, substitute $x = 1$ and $y = -2$ into $4x - 3y = 10$.) The value of the LHS should equal the value of the RHS.	LHS $= 4x - 3y$ $= 4 \times 1 - 3 \times -2$ $= 4 + 6$ $= 10$ = RHS	
6 Write the solution for both variables.	$x = 1, y = -2$ or $(x, y) = (1, -2)$	

However, not all pairs of simultaneous equations can be solved by simply adding or subtracting the two equations to eliminate one of the variables. If the coefficients of neither variable are equal in magnitude, you need to make the coefficients of one of the variables equal by multiplying by the LCM.

Worked example 18

W.E. 18

Solve the following simultaneous equations algebraically using the elimination method.

(a) $3x + 2y = 5$
$4x - 3y = -16$

(b) $2x + 5y = 9$
$x + 3y - 5 = 0$

Thinking	Working	
(a) 1 Write the equations and label them.	(a) $3x + 2y = 5$ $4x - 3y = -16$	[1] [2]
2 Choose the variable to eliminate. Label your new equations. (Here, multiply [1] by 3 and [2] by 2 so that the coefficient of y is 6 in both equations.)	$9x + 6y = 15$ $8x - 6y = -32$	[3] [1] × 3 [4] [2] × 2
3 Eliminate the variable with the same coefficient by either adding or subtracting. (Here, add [3] and [4] to eliminate y.)	$17x = -17$	[3] + [4]
4 Solve the equation for the remaining variable.	$x = -1$	

	5	Find the value of the eliminated variable by substituting the value of the variable into one of the original equations. (Here, substitute $x = -1$ into [1].)	$3x + 2y = 5$ [1] $3 \times -1 + 2y = 5$ $-3 + 2y = 5$ $2y = 8$ $y = 4$
	6	Check the solution by substituting the values into the LHS of the other original equation. (Here, substitute $x = -1$ and $y = 4$ into $4x - 3y = -16$.)	$\text{LHS} = 4x - 3y$ $= 4 \times -1 - 3 \times 4$ $= -4 - 12$ $= -16$ $= \text{RHS}$
	7	Write the solution for both variables.	$x = -1$, $y = 4$ *or* $(x, y) = (-1, 4)$
(b)	1	Write the equations and label them.	(b) $2x + 5y = 9$ [1] $x + 3y - 5 = 0$ [2]
	2	Rearrange one or both equations so that the variables are on the left-hand side of the equation and the numbers are on the right.	$2x + 5y = 9$ [1] $x + 3y = 5$ [3]
	3	Choose the variable to eliminate. Label your new equations. (Here, multiply [3] by 2 to obtain $2x$ in both equations.)	$2x + 6y = 10$ [4] [3] × 2 $2x + 5y = 9$ [1]
	4	Eliminate the variable with the same coefficient by either adding or subtracting. (Here, subtract [1] from [4].)	$y = 1$ [4] − [1]
	5	Find the value of the eliminated variable by substituting the value of the variable into one of the original equations. (Here, substitute $y = 1$ into [2].)	$x + 3y - 5 = 0$ [2] $x + 3 \times 1 - 5 = 0$ $x - 2 = 0$ $x = 2$
	6	Check the solution by substituting the values into the LHS of the other original equation. (Here, substitute $x = 2$ and $y = 1$ into $2x + 5y = 9$.)	$\text{LHS} = 2x + 5y$ $= 2 \times 2 + 5 \times 1$ $= 4 + 5$ $= 9$ $= \text{RHS}$
	7	Write the solution for both variables.	$x = 2$, $y = 1$ *or* $(x, y) = (2, 1)$

Useful strategies

It doesn't matter which variable you choose to eliminate when solving simultaneous equations. You will get the same solution for the variables.

- If you only need to change one equation to get the same coefficient of one of the variables, eliminate that variable.
- If the signs in front of one of the variables are opposite, eliminate that variable.
- If the signs in front of each of the variables are the same, choose the variable to eliminate where the coefficients are smallest in magnitude.
- If the signs in front of each of the variables are the same then multiplying one equation by a negative number and adding the equations is often easier than multiplying both by a positive number and subtracting the equations.

Solving simultaneous equations by substitution

This method is commonly used when one or both equations are written with one of the variables as the subject of the equation (e.g. $y = 3x + 5$ as in gradient intercept form or $x = 4 - 3y$).

Worked example 19

W.E. 19

Solve the following pairs of simultaneous equations algebraically using the substitution method.

(a) $y = 2x$
$x + y = 18$

(b) $2x + 3y = -4$
$x = y + 3$

Thinking	Working
(a) 1 Write the equations and label them.	(a) $y = 2x$ [1] $x + y = 18$ [2]
2 Find the equation that has one variable as the subject of the equation and substitute for this variable into the other equation. (Here, substitute $2x$ for y in [2].)	Substituting $2x$ for y in [2]: $x + (2x) = 18$ [3]
3 Simplify the equation and solve the equation for the remaining variable.	$3x = 18$ $x = 6$
4 Find the value of the eliminated variable by substituting the value of the variable into one of the original equations. (Here, substitute $x = 6$ into [1].)	$y = 2x$ [1] $= 2 \times 6$ $= 12$
5 Check the solution by substituting the values into the LHS of the other original equation. (Here, substitute $x = 6$ and $y = 12$ into $x + y = 18$.)	LHS $= x + y$ $= 6 + 12$ $= 18$ $=$ RHS
6 Write the solution for both variables.	$x = 6, y = 12$ or $(x, y) = (6, 12)$

(b) 1	Label the equations.	(b) $2x + 3y = -4$ [1] $x = y + 3$ [2]
2	Find the equation that has one variable as the subject of the equation and substitute for this variable into the other equation.	Substituting $y + 3$ for x in [1]: $2(y + 3) + 3y = -4$ [3]
3	Simplify the equation. (Here, substitute $(y + 3)$ for x in [1].)	$2y + 6 + 3y = -4$ $5y + 6 = -4$ $5y = -10$
4	Solve the equation for the remaining variable.	$y = -2$
5	Find the value of the eliminated variable by substituting the value of the variable into one of the original equations. (Here, substitute $y = -2$ into [2].)	$x = y + 3$ [2] $= -2 + 3$ $= 1$
6	Check the solution by substituting the values into the LHS of the other original equation. (Here, substitute $x = 1$ and $y = -2$ into $2x + 3y = -4$.)	LHS $= 2x + 3y$ $= 2 \times 1 + 3 \times -2$ $= 2 - 6$ $= -4$ = RHS
7	Write the solution for both variables.	$x = 1, y = -2$ or $(x, y) = (1, -2)$

Using simultaneous equations

Simultaneous equations often involve worded problems. You need to identify the unknowns, define them with a variable, write the information as equations and then solve them simultaneously. Consider this problem:

A furniture store sells tables and chairs. If a table costs \$60 and a chair costs \$25, and a total of 15 tables and chairs were sold for \$480, then how many tables and chairs were sold?

The number of tables and the number of chairs sold are the two variables. Let T = number of tables sold and C = number of chairs sold.

You can then write the equations:

$T + C = 15$ [1]

$60T + 25C = 480$ [2]

Solve these simultaneous equations using the elimination method.

$60T + 25C = 480$ [2]

$25T + 25C = 15 \times 25$ [3] [1] × 25
$= 375$

$35T = 480 - 375$ [4] [2] − [3]
$= 105$

$T = 3$

$3 + C = 15$
$C = 12$

So 3 tables and 12 chairs were sold.

In general, when solving problems using simultaneous equations, these steps need to be followed:

1 Identify the unknowns. Define each with a variable.
2 Write the information given as equations involving the unknowns. (If there are two unknowns, i.e. two pieces of information required, there must be two equations.)
3 Solve the equations simultaneously either using elimination, substitution or graphically.
4 Check your solution by substitution.
5 Write the final answer to the problem in words.

1.6 Simultaneous equations

Navigator

Answers p.769

1, 2 (columns 1–2), 3, 4 (a–h), 5 (a–f), 6, 7, 8, 9, 10, 11, 12, 14, 16, 19	2 (columns 2–3), 3, 4 (columns 1–2), 5 (columns 1–2), 6, 7, 8, 9, 10, 11, 12, 13, 14, 15, 16, 17, 18, 19, 21, 22	2 (columns 2–3), 3, 4 (column 3), 5 (column 3), 6, 9, 11, 12, 13, 14, 15, 16, 17, 18, 19, 20, 21, 22, 23

Equipment required: graph paper for Question **2**

Fluency

1 State the point of intersection for each of the following pairs of linear equations.

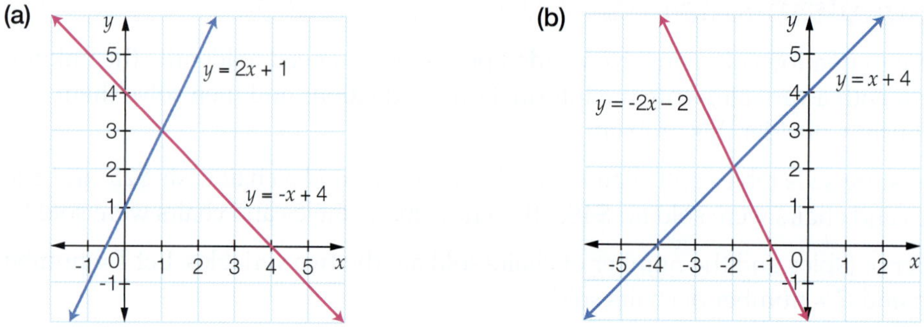

W.E. 16

2 (i) Sketch each pair of linear equations on graph paper and find the point of intersection.
(ii) Hence, find the solution for each pair of simultaneous equations.

(a) $y = x + 3$, $y = 5 - x$
(b) $y = x - 7$, $y = 3 - x$
(c) $x - y = 6$, $2x - y = 8$
(d) $3x - y = 6$, $x - 2y = 2$
(e) $y = 2x$, $y = 3x - 4$
(f) $y = 6 - 2x$, $y = 4x$
(g) $x = -2$, $y = 3x$
(h) $y = -x$, $y = 3$
(i) $y = x + 2$, $y = x + 4$

W.E. 17

3 Solve the following pairs of simultaneous equations using the elimination method.

(a) $x + y = 5$, $x - y = 3$
(b) $x + y = 1$, $x - y = 9$
(c) $2x + 3y = 9$, $5x - 3y = 12$
(d) $3x + 2y = 13$, $9x - 2y = 47$
(e) $3x + 2y = 9$, $-3x + y = 0$
(f) $7x + y = 8$, $-x + y = -8$

(g) $-2x + y = 7$
$2x + 3y = 5$

(h) $2x - 3y = -2$
$5x - 3y = 4$

(i) $-3x + y = -3$
$-3x - 2y = -12$

4 Solve the following pairs of simultaneous equations using the elimination method. **W.E. 18**

(a) $3x + 2y = 1$
$-2x + y = -3$

(b) $5x + 2y = -7$
$x + 7y = 25$

(c) $2x + 6y = 4$
$4x + 3y = -1$

(d) $2x + 5y = -1$
$4x + 3y = 5$

(e) $3x + 2y - 1 = 0$
$-2x + y + 3 = 0$

(f) $4x + 3y + 9 = 0$
$2x - 5y + 11 = 0$

(g) $3x - 2y = 7$
$2x - 3y = 3$

(h) $5x - 2y = -23$
$2x + 7y = 22$

(i) $4x + 2y - 6 = 0$
$-3x - 3y + 6 = 0$

(j) $\frac{x}{2} + y = 0$
$x - 3y = 10$

(k) $x + \frac{y}{4} = 8$
$2x - y = -2$

(l) $5x - 6y = 21$
$\frac{x}{3} + \frac{y}{2} = 5$

(m) $7x - y = -25$
$\frac{2x}{5} - \frac{y}{2} = 3$

(n) $\frac{2x}{3} + y + 10 = 0$
$4x - \frac{y}{4} + 10 = 0$

(o) $\frac{5x}{6} + \frac{3y}{5} + 2 = 0$
$7x + 3y + 27 = 0$

5 Solve the following pairs of simultaneous equations using the substitution method. **W.E. 19**

(a) $y = 2x$
$x + 5y = 22$

(b) $x + 3y = -30$
$y = 3x$

(c) $x = -2y$
$y = 3x - 14$

(d) $2x + y = 7$
$x = 6 - y$

(e) $x = y + 1$
$3x - y = 7$

(f) $y = 7 - 3x$
$4x - 3y = 18$

(g) $y = 1 - 2x$
$5x + 2y = 0$

(h) $x = 2 - 7y$
$2y = x - 11$

(i) $y = 6 - 3x$
$x = 2y - 19$

6 (a) For simultaneous equations $y = 2x - 5$ and $y = 3 - 2x$, the solution is:

A (0, -5) B (0, 3) C (2, -1) D (2, 1)

(b) For simultaneous equations $x + 2y = 8$ and $3x - y = 3$, the solution is:

A (6, 1) B (3, 6) C (4, 2) D (2, 3)

7 (a) Adding which pair of equations below will eliminate y?

A $x + y = 4$
$2x + y = 3$

B $3x + 2y = 6$
$-3x + y = 0$

C $x - y = 7$
$3x - y = 7$

D $2x - 3y = 4$
$x + 3y = 2$

(b) Adding which pair of equations in part **(a)** will eliminate x?

8 (a) $5x - 6y = 4$ [1]
$-3x + 8y = 2$ [2]

To eliminate x in the pair of equations above, you need to:

A multiply [1] by 3 and add to 5 times [2]

B multiply [2] by 3 and add to [1]

C multiply [1] by -3 and add to 5 times [2]

D multiply [2] by -3 and add to 5 times [1]

(b) $9x + 7y = 3$ [1]
$5x + 3y = -1$ [2]

To eliminate y in the pair of equations above, you need a new equation formed by:

A $3 \times [1] + 7 \times [2]$

B $5 \times [1] + 9 \times [2]$

C $3 \times [1] - 7 \times [2]$

D $5 \times [1] - 9 \times [2]$

9 (a) Simultaneous equations $y = x + 3$ and $2x - y = -11$ have the solution:

A $x = 3, y = 6$ B $x = -5, y = 1$ C $x = -5, y = -8$ D $x = -8, y = -5$

(b) Simultaneous equations $x = 2y$ and $4x - 7y - 3 = 0$ have the solution:

A (-3, -6) B (2, 4) C (3, 6) D (6, 3)

Understanding

10 During school holidays a cinema sells tickets for adults and children at discount prices. Let a represent the cost of an adult's ticket and c represent the cost of a child's ticket.

(a) Write an equation to show that tickets for 5 adults and 3 children cost \$65.

(b) Write an equation to show that tickets for 7 adults and 6 children cost \$100.

(c) Multiply both sides of the equation from **(a)** by 2.

(d) Now solve the equations from **(b)** and **(c)** simultaneously to find the cost of each type of ticket.

11 The fruit below cost \$14.05. The fruit below cost \$8.50.

How much did each pineapple and each kiwifruit cost? Assume that all pineapples are the same price and all kiwifruit are the same price.

12 Three salad rolls and four pies cost \$20.56 altogether. A salad roll costs \$1.16 more than a pie. How much does each cost?

13 The sum of x and y is 12. When y is subtracted from $3x$ the result is 16. Find x and y.

14 The sum of two numbers is 38 and their difference is 24. What are the numbers?

Reasoning

15 Two books have a total of 500 pages. If the number of pages in one book is 12 pages less than 3 times the number of pages in the other, how many pages are in each book?

16 A farmer uses 140 m of fencing wire to enclose a rectangular paddock. If the length of the paddock is 10 m longer than the width, find the dimensions of the paddock.

17 After buying a book, Libby received 18 coins totalling \$1.65 in change. If she was given only 5-cent and 10-cent coins, how many of each did she receive?

18 Five less than twice Michelle's present age is equal to Yesmin's present age. Two years ago, twice Yesmin's age was the same as 10 years more than triple Michelle's age. Find the present ages of the two girls.

19 Willow and Tony live in two cities 360 km apart. They leave their houses at the same time by car and travel towards each other at 80 km/h and 100 km/h, respectively.

(a) Write an equation for each journey representing the distance from Willow's house (D) at any time (t).

(b) How much time will pass before they meet on the highway?

20 It takes a barge 3 hours to travel 27 km upstream against the current. On its return journey, with the current, the barge can travel the same distance downstream and an additional 3 km in only 2 hours. Find the speed of the boat and the current.

21 The pirate Captain Greybeard left careful instructions for his crew to locate his buried treasure. Using the old oak tree as the origin, north–south as the y-axis, east–west as the x-axis and each unit on the number plane as 20 paces, the solution of four different pairs of simultaneous equations would lead to the burial site.

The first point, A, is at the intersection of the two lines $x = -3$ and $y = -7$.
The second point, B, is at the intersection of $2x - y = 7$ and $2x - 3y = -3$.
The third point, C, is at the intersection of $2y + x = 0$ and $2y = x - 8$.
The final point, D, is at the intersection of $y = 7x$ and $x + y = 8$.

(a) Find the coordinates of each of the points A, B, C and D.

(b) Plot the points A, B, C and D on the same number plane and join AB and CD.

(c) Find the coordinates of the intersection of the two lines, which will give the location of the buried treasure.

(d) Determine the equations of the lines AB and CD, and solve algebraically to show that the coordinates found in **(c)** are the solution of these equations.

Open-ended

22 Make up three different pairs of simultaneous equations, each having the same solution, where $x = 3$ and $y = 4$.

23 Emmanuel has three numbers. When he adds them together in pairs he gets three different totals: 11, 17 and 22. What are the three numbers Emmanuel had to start with? Can you describe a method that would enable you to work out the three numbers given any three totals?

Investigation

Breaking even

Equipment required: graph paper

The Big Question

Angelo's Ice-Cream Parlour sells ice-cream in a plain cone for \$3.50 and in a waffle cone for \$4. Angelo orders 200 plain cones and 120 waffle cones each day. He usually sells 220 ice-cream cones per day.

Some days he earns a lot more money than on other days and he wonders if there is a certain number of each type of cone that will maximise the amount of money he earns. (The amount he earns is called revenue.) What advice should Angelo be given to solve his problem?

Engage

1 **(a)** On a set of axes, sketch the graph of $y = 10 - x$.

(b) Plot the following points and decide which are:

(i) on the line

(ii) below the line

(iii) above the line.

(1, 9) (1, 6) (3, 5) (3, 7) (6, 4)
(6, 8) (4, 7) (4, 2) (2, 8) (2, 3)

(c) Shade the region above the line. Check by substitution that all points given in **(b)** that are in this shaded region satisfy the inequality $y > 10 - x$.

(d) Shade differently the region below the line. Check by substitution that all points given in **(b)** that are in this shaded region satisfy the inequality $y < 10 - x$.

2 On a new set of axes, sketch the graphs of $x + 2y = 6$ and $4x + 3y = 12$.

(a) Shade $x + 2y < 6$ and shade differently $4x + 3y > 12$.

(b) Show by substitution into the inequalities that (4, 0) is in the region where the shadings overlap.

Explore

3 To understand Angelo's problem and find a solution, complete the following.

(a) On the same set of axes, sketch the graphs of:

(i) $x + y = 220$ **(ii)** $x = 200$ **(iii)** $y = 120$

(b) Shade the region that is common to the graphs of:

(i) $x + y \leq 220$ **(ii)** $x \leq 200$ **(iii)** $y \leq 120$

(c) This shaded region is called the feasible region. Find the coordinates of all the vertices of this region and label these points on your graph.

(d) Substitute the coordinates of each vertex into the expression $3.5x + 4y$ and determine which vertex gives the maximum value of this expression.

Strategy options

- Look for a pattern.
- Draw a diagram.
- Test all possible combinations.

Explain

4 (a) Explain the meaning of $x + y \leq 220$ in relation to Angelo's ice-creams.

(b) Why did Angelo graph $x \leq 200$ and $y \leq 120$?

(c) Explain why the shaded region is called the feasible region.

(d) Explain what was calculated when the vertices of the feasible region were substituted into the expression $3.5x + 4y$.

Elaborate

5 Use the graphs you sketched and your calculations to advise Angelo about the numbers of each type of cone he should sell to maximise his revenue.

Evaluate

6 (a) How confident are you that you have given Angelo the correct advice? Justify your answer.

(b) What did you learn by completing **1**, **2** and **3**? How did this help you to solve the problem?

(c) How did drawing graphs help you to advise Angelo?

(d) Do you think you could have solved this problem using another method? If so, explain how.

Extend

Angelo finds that the days he earns the most revenue are not the days he makes the most profit, as the waffle cones are more expensive for him to buy than the plain cones.

7 If a plain cone with ice-cream costs him \$0.80, a waffle cone with ice-cream costs him \$1.50 and he has overheads (other costs such as wages, electricity and rent) of \$580 per day:

(a) write an equation to show the total cost (C) of producing x plain cones and y waffle cones

(b) write an equation to show the profit, P, of producing x plain cones and y waffle cones.

Hint

Profit = Revenue – Cost

8 (a) If $P = 0$, sketch the graph of the equation in **7 (b)**.

(b) Sketch the graph of $x + y = 220$ on the same axis.

(c) Write the coordinates of the point of intersection of these two graphs.

(d) What information does this point give Angelo?

(e) Advise Angelo about the number of plain cones and waffle cones he should sell to make the largest profit.

Chapter review 1

Maths literacy

coincident lines	independent variable	linear inequality	sketch
dependent variable	linear equation	linear relationships	unique equations
gradient	linear graph	simultaneous equations	

Copy and complete the following using the words and phrases from the list, where appropriate. A word or phrase may be used more than once.

1 A __________ indicates how two quantities that change at a constant rate are related to one another.

2 A __________ can be solved by performing inverse operations on both sides of the equation.

3 The measure of a line's steepness is called the __________.

4 __________ show that the relationship between two variables is a straight line.

5 For a __________, only two points need to be shown.

6 Two linear equations that contain the same variables and are not parallel or coincident can be solved using __________.

7 $3x + 4y = 7$ and $8y = 14 - 6x$ are examples of __________.

Fluency

1 Solve the following equations. **1.1**

(a) $2c - 7 = 9$ (b) $\frac{2x-3}{4} + 6 = -1$

(c) $4v - 5 = v + 4$ (d) $\frac{3a+1}{2} - \frac{2a+3}{5} = 1$

2 The gradient of the line joining the points (2, 9) and (5, -6) is: **1.2**

A 1 B 5 C $-\frac{1}{5}$ D -5

3 Find the gradient and y-intercept for each of the following. **1.2**

(a) $y = 5x + 2$ (b) $y = -2x$ (c) $2x - 5y = 10$ (d) $4x + y = 8$

4 Sketch graphs of the following. **1.3**

(a) $y = 2x - 5$ (b) $y = -2x$ (c) $x + 5y = 10$ (d) $y = 4x$

5 Sketch graphs of the following, showing the intercepts on the axes. **1.3**

(a) $4x - 5y = 20$ (b) $3x + 7y = 42$ (c) $y = 3 - 5x$ (d) $y = -2x + 6$

6 Show whether or not each pair of lines is parallel, using their gradients. **1.4**

(a) $x + 2y = 7$ and $x + 2y = 3$ (b) $x - y + 6 = 0$ and $3x - 2y + 4 = 0$

(c) $4x + y = 6$ and $4x + y + 6 = 0$ (d) $y = 4x - 3$ and $8x - 2y = 0$

7 Show whether or not each pair of lines is perpendicular, using their gradients. **1.4**

(a) $2x + 2y = 4$ and $2x + y = 8$

(b) $x - y + 9 = 0$ and $2x + 2y - 3 = 0$

(c) $2x - 4y = 9$ and $4x + 2y + 6 = 7$

(d) $3x + 2y = 8$ and $3y = 2x + 4$

8 Find the solutions for the following linear inequalities. **1.5**

(a) $9x \geq 21$

(b) $\frac{3x}{5} + 2 > -5$

(c) $7 - 6x < 2$

9 Solve the following simultaneous equations using the elimination method. **1.6**

$5x - 2y = 8$ $\quad$ $3x - 4y = -12$

10 Solve the following simultaneous equations using the substitution method. **1.6**

$6x - 5y = -21$ $\quad$ $y = 2x + 5$

Understanding

11 In an enclosure at the zoo, there are twice as many wombats as emus. Altogether, they have a total of 90 legs. How many wombats are there? **1.1**

12 To earn extra money, Martin works part-time in a butcher's shop. In a particular week during the school holidays he worked twice as many hours on Tuesday as he did on Monday but only $\frac{3}{4}$ the hours on Wednesday compared to Monday. For these 3 days he received \$127.50. If he is paid \$8.50 per hour, how many hours did Martin work on Monday? **1.1**

13 A swimming pool has been filled with water. **1.2**

The graph shows the volume of water (in litres) in the pool after a number of hours as it filled.

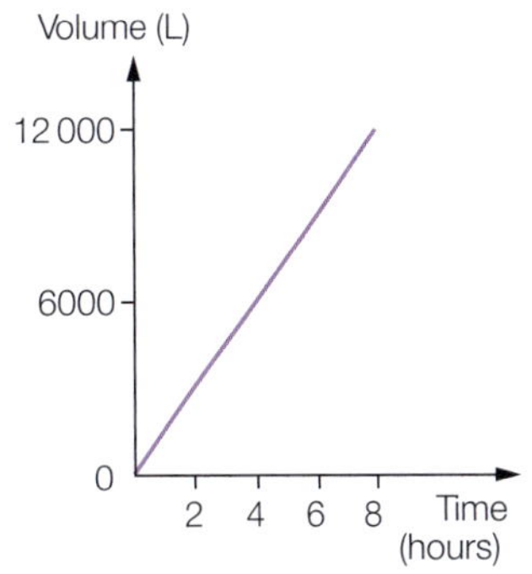

(a) How long did it take for the pool to contain 12 000 L of water?

(b) How many litres of water flowed into the pool each hour?

(c) What is the gradient of the graph?

(d) Write the relationship for the volume (V) in terms of the time (T).

14 Manny, a motorcycle courier, notes his distance from the centre of Melbourne on a delivery trip. **1.3**

Time (min)	0	10	20	30	40	50	60
Distance from Melbourne (km)	20	26	32	38	44	50	56

(a) Find an equation describing Manny's distance from Melbourne (d km) in terms of the time spent travelling (t minutes).

(b) Draw a graph of the equation found in part **(a)** for the hour of travel.

(c) What does the gradient of the graph represent?

(d) What does the intercept with the vertical d-axis represent?

(e) Find Manny's distance from Melbourne after he has been travelling for 14 minutes.

(f) How long does it take for Manny to be 48 km from Melbourne?

15 What is the equation of the line that passes through (6, 3) and is perpendicular to $y - x - 1 = 0$? **1.4**

16 What is the equation of the line that passes through (8, -2) and is parallel to $y - 5x = -1$? **1.4**

17 Find the solutions for the following linear inequalities. **1.5**

(a) $\frac{3x-4}{2} < \frac{2x+1}{3}$ (b) $3(2x-4) \geq 4(3x+1)$

18 Twice one number added to three times another number equals -3. If the first number is tripled and subtracted from 8 and the result is divided by 2, then the second number is obtained. Find the two numbers. **1.6**

19 Find the values of a and b in the following equations, given that the lines are perpendicular and intersect on the y-axis. **1.6**

$ax + 3y = 1$

$by = -x - 1$

Reasoning

20 Houses for sale at auction often have an EPR (Expected Price Range) in the advertisement. This means that it is expected that the house will be sold for a price somewhere between the values given. **1.5**

(a) If the EPR for a house is advertised as \$570 000–\$620 000, write this as an inequality for the selling price of the house (p).

(b) If the deposit must be 10% of the price paid, write an inequality to show how much deposit is needed if the price falls within the EPR.

(c) If the house sold for more than \$50 000 above the highest value in the EPR, write an inequality to show how much the house sold for.

(d) If the house sold for less than \$50 000 above the highest value in the EPR, write an inequality to show how much the house sold for.

21 Three times the distance (d) between two towns subtracted from 823 km would take 4 hours to travel at an average speed (s). However, five times the same distance subtracted from 1071 km would only take 3 hours to travel at the same average speed. How far is the distance (d) between the two towns and what is the average speed (s)? **1.6**

Statistics

2

Maths boosts sales

Name: Adam Driussi
Job: Director of The Quantium Group, an actuarial consulting firm
Qualifications: Bachelor of Economics (Actuarial Studies), Fellow of the Institute of Actuaries of Australia and Graduate Diploma in Applied Finance and Investment

What do you do with a sharp analytical mind and great communication skills? You become an actuary!

Actuaries analyse large quantities of data and provide strategic advice to help businesses solve a wide range of practical problems. Adam Driussi is director of an actuarial consulting firm. He says, 'I love coming to work every day and having to solve a different problem to the day before. The fact that you tend to get paid very well to do it is a very nice bonus.'

As an actuary, Adam and his team might analyse a client's customer database. Based on what they discover they will make a suggestion as to how a business can market to their customers more effectively. Adam explains: 'For example, eBay communicates with most of its customers via email. So we analysed the data to help eBay develop email campaigns that promoted new items that are relevant to each of eBay's millions of customers—as opposed to promoting the same items to every customer. The result was more than double the number of people responding to the campaigns and buying new items from eBay.'

Adam loved maths at school. 'My maths skills form the foundation for the problem solving and logic that I need to do my job every day. Without that foundation you could never solve the problems you get posed as an actuary.'

Why learn this?

We live in an age of 'Big Data', but we need to organise and analyse the data to draw conclusions or make decisions. Statistics are used by all sorts of organisations to understand complicated situations, make decisions and convince others of what to do. A knowledge of the processes used to gather and present data will help you to distinguish credible facts from unlikely fiction.

After completing this chapter you will be able to:

- find five-number summaries and use them to understand, analyse and compare data sets
- understand and construct data representations such as frequency curves, tables, box plots, dot plots, column graphs, histograms and scatter plots
- understand and construct data representations involving time-dependent bivariate data
- investigate and analyse the use of statistics in the media

10A

- draw and interpret lines of best fit
- calculate the mean and standard deviation and use these values to compare data sets.

Recall 2

Prepare for this chapter by attempting the following questions. If you have difficulty with a question, you can download a Recall Worksheet from the eBook or the Pearson Places website.

1 Find the mean, median (middle number), mode (most common number) and range (difference between largest and smallest numbers) for each of these sets of data.

(a) The number of siblings for your five best friends:
2 4 5 1 4

(b) The shoe size of your nine young cousins:
3.5 4.5 5.0 5.5 5.0 4.5 5.5 5.0 3.5

2 Classify each of the following as being nominal, ordinal, discrete or continuous data.

(a) Number of goals scored by a hockey team each week.

(b) Heights, in metres, of the members of your maths class.

(c) Sizes of shirts (S, M, L) on a rack.

3 The heights (in centimetres) of 20 students are as follows:

179.7 167.3 168.6 164.7 182.8 160.5 179.2 163.0 177.7 171.3
169.8 173.2 164.7 172.1 165.6 171.9 164.8 166.7 169.1 166.7

Round the heights to the nearest cm and then:

(a) draw a stem-and-leaf plot using intervals of 5 cm **(b)** find the median.

4 The distance jumped (in metres) by each competitor in the long jump was recorded and is shown in the frequency table.

(a) Find an estimate for the value of the mean.

(b) In which class interval is the median?

Class interval (m)	Frequency
5.75–<6.00	4
6.00–<6.25	6
6.25–<6.50	9
6.50–<6.75	13
6.75–<7.00	5
7.00–<7.25	3

Exploration Task

You can download this activity from the eBook or the Pearson Places website.

Problems with problems

In this activity you will explore how to work with statistics problems that have more than one possible answer. How can the answers be checked? What might all the answers have in common?

Cumulative frequency curves

2.1

Measures of centre for ungrouped data

Various **measures of centre** (sometimes called **measures of location**) can be used to represent the 'centre' of a set of numerical data. Each measure gives a useful piece of information about the data. Three measures of centre are the **mean**, the **median** and the **mode**.

The **mode** is the most common data value, so it is the data value with the highest frequency. Showing all the data values in a table can make it easier to identify the mode. If two (or more) data values have the same highest frequency, then the data is called **bimodal** (or **multimodal**) and all data values with the highest frequency are modal values.

The **median** is the middle value when the data set is written in order. When the number of data values n is odd or even:

- odd—the median is the middle data value, which is the $\left(\frac{n+1}{2}\right)$th value.
- even—the median is the value between the two middle values, which is the mean of the two middle values. You find this by adding together the $\left(\frac{n}{2}\right)$th value and the $\left(\frac{n}{2}+1\right)$th value and dividing by 2.

Showing all the data values in a table together with their **cumulative frequency** values is a useful way to identify the median. The cumulative frequency is the sum of the frequencies up to each data value.

The **mean** is the sum of all the data divided by the number of data values n.

$$\bar{x} = \frac{\Sigma x}{n}$$

Showing all the data values in a table can make it easier to calculate the mean.

- Each data value x can be multiplied by its frequency f to find the product xf.
- The column of frequency values f is added to find the sum Σf, which is the number of data values n.
- The column of xf values is added to find the sum Σxf, which is the sum of all the data
- The mean is the sum of the xf values (sum of all the data) divided by the sum of frequency values (number of data values).

  $$\bar{x} = \frac{\Sigma xf}{\Sigma f}$$

The mean is often called the **average**, which is used in everyday language to describe something that is typical or unexceptional.

Measures of centre for ungrouped data

Mode: most frequently occurring data value

Median: middle data value

- n is odd: $\left(\frac{n+1}{2}\right)$th data value
- n is even: mean of the $\left(\frac{n}{2}\right)$th and $\left(\frac{n}{2}+1\right)$th data values.

Mean: $\bar{x} = \frac{\Sigma x}{n} = \frac{\Sigma xf}{\Sigma f}$

Worked example 1

W.E. 1

Find the mode, median and mean for each set of data correct to 1 decimal place, if necessary.

(a) 6.2, 5.7, 2.1, 8.7, 1.3

(b)

x	16	17	18	19	20
f	8	20	15	4	1

Thinking	Working
(a) 1 Check for repeated values.	(a) There is no mode.
2 State the value of the mode.	The mode does not exist.
3 Order the data and locate the middle.	1.3, 2.1, 5.7, 6.2, 8.7 $n = 5$ (odd) $\frac{n+1}{2} = \frac{5+1}{2}$ $= 3$ The 3rd data value is 5.7.
4 State the value of the median.	The median is 5.7.
5 Write the formula for the mean.	$\bar{x} = \frac{\Sigma x}{n}$
6 Substitute the values and evaluate the mean.	$\text{mean} = \frac{1.3 + 2.1 + 5.7 + 6.2 + 8.7}{5}$ $= \frac{24}{5}$ $= 4.8$
(b) 1 Find the highest frequency and matching data value.	(b) $f = 20$, $x = 17$
2 State the value of the mode.	The mode is 17.
3 Extend the original table to include a cumulative frequency column.	(see table below)

x	f	Cumulative frequency
16	8	8
17	20	8 + 20 = 28
18	15	28 + 15 = 43
19	4	43 + 4 = 47
20	1	47 + 1 = 48

4 Decide on the middle position.

$n = 48$ (even), $\frac{n}{2} = 24$

The middle is between the 24th and 25th data values.

5 Find the value of the median.

The 9th–28th data values are all 17.

The median is 17.

6 Extend the original table to include an xf column.

7 Find the total of the f and xf columns.

x	f	xf
16	8	128
17	20	340
18	15	270
19	4	76
20	1	20
Total	48	834

8 Calculate the mean and round to the specified number of decimal places.

$$\bar{x} = \frac{\Sigma xf}{\Sigma f}$$
$$= \frac{834}{48}$$
$$= 17.4 \text{ (1 d.p.)}$$

Measures of centre for grouped data

Large numerical data sets may be presented in a frequency table with the data grouped into class intervals. Although this makes the data set easier to deal with, it means that the original data values are lost.

When dealing with grouped data, the mode, median and mean can be found in terms of class intervals rather than individuals values.

- A modal class interval can be found rather than a modal value. The modal class interval is the most common class interval with the highest frequency.
- For the mean, the midpoint of each class interval is used as its representative data value. The mean can then be estimated by finding the mean of these representative values. However, this estimate will not be as accurate as a calculation from the raw data values.
- The median class interval is the 'middle' of the class intervals, but it is more useful to find an estimated median by using a **cumulative frequency curve** (sometimes called an *ogive*).

A cumulative frequency curve, graphed against the data values, will show the sum of frequencies 'up to' each data value. This means it shows the number of data values that are 'less than' each particular data value.

By definition, half of the data values will be less than the median, so you can estimate the median by finding the half-way point on a cumulative frequency curve, as in the following example.

Note that when the number of data values is large, then the cumulative frequency values associated with the data values $\frac{n}{2}$ and $\frac{n+1}{2}$ will each give very similar values for the median.

Worked example 2

W.E. 2

The table shows the weights (kg) of 50 crates of equipment.

(a) Find the modal class.

(b) Calculate an estimate for the mean weight.

(c) Use a cumulative frequency curve to find an estimate for the median weight.

Class interval (kg)	Frequency
45–<50	3
50–<55	13
55–<60	8
60–<65	10
65–<70	4
70–<75	12

Thinking	Working
(a) 1 Find the highest frequency and matching interval.	(a) $f = 13$, 50–<55
2 State the modal class.	The modal class is 50–<55 kg.
(b) 1 Find the midpoint (x) of each interval. **2** Include an xf column. **3** Find the total of the f and xf columns.	(b) see table below
4 Calculate the mean.	$\bar{x} = \frac{\Sigma xf}{\Sigma f}$ $= \frac{3050}{50}$ $= 61$ The estimated mean is 61 kg.

Class interval (kg)	x	f	xf
45–<50	47.5	3	142.5
50–<55	52.5	13	682.5
55–<60	57.5	8	460
60–<65	62.5	10	625
65–<70	67.5	4	270
70–<75	72.5	12	870
	Total	50	3050

(c)

1 Extend the original table to include a cumulative frequency column and an individual data (x) column with an extra row at the top under the headings.

2 Let the x values be the same as the upper values of each interval with the extra place at the top matching the lower end of the first interval.

3 Fill in the cumulative frequency column with the number of data 'less than' each x value (Here, 0 then 3, then 3 + 13 = 16 and so on.)

4 Graph the cumulative frequency against x values with data values on the horizontal axis.

5 Join the points with straight lines (although sometimes a smooth curve is used).

6 Locate the half-way position on the cumulative frequency axis ($\frac{n+1}{2}$ gives 25.5 but $\frac{n}{2} = 25$ is simpler and sufficiently accurate).

7 Draw a horizontal line to the graph, then a vertical line to the data axis.

8 State the estimated median value.

(c)

Class interval (kg)	x	f	Cumulative frequency
<45	45	–	0
45–<50	50	3	3
50–<55	55	13	16
55–<60	60	8	24
60–<65	65	10	34
65–<70	70	4	38
70–<75	75	12	50

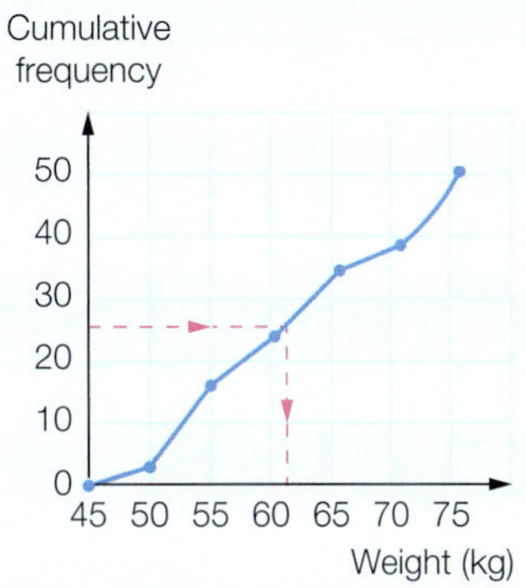

The median is about 61 kg.

Measures of centre for grouped data

The modal class is the interval with the highest frequency.

The mean can be estimated by using the midpoint of each class interval as its representative value. Mean: $\bar{x} = \frac{\Sigma x}{n} = \frac{\Sigma xf}{\Sigma f}$

The median can be estimated from the cumulative frequency curve:

- Use the endpoints of the intervals as the data values.
- Write the number of data 'less than' each of the endpoints, which is the cumulative frequency.
- Plot the points and join with straight lines.
- Read across from half-way up the cumulative frequency axis to ($\frac{n+1}{2}$, or $\frac{n}{2}$ is usually close enough), then down to the data axis.

Grouped discrete data

For discrete data that has been grouped into intervals, the cumulative frequency still represents the number of data values 'less than' each data value.

Consider the following grouped discrete frequency table and cumulative frequency table. There are no data values less than 20, there are five data values less than 30, etc. All 34 values are less than 70.

Class interval	Frequency
20–29	5
30–39	8
40–49	17
50–59	3
60–69	1

Class interval	Cumulative frequency
<20	0
<30	5
<40	13
<50	30
<60	33
<70	34

2.1 Cumulative frequency curves

Navigator

Answers p. 771

1, 2, 3, 4, 5, 6, 7 (a–b), 8 | 1, 2, 3, 4, 5, 6, 7 (a–b), 8 | 1, 2, 3, 4, 6, 7, 8

Fluency

W.E. 1

1 For each of the following data sets, calculate the **(i)** mode, **(ii)** median and **(iii)** mean. Round the mean to 1 decimal place where necessary.

(a) 5, 12, 2, 7, 8, 11, 4

(b) 22, 27, 35, 42, 15, 18, 30, 44, 28, 17

(c)

x	25	26	27	28	29
f	1	5	12	18	10

(d)

x	0	1	2	3	4	5	6
f	8	10	14	2	6	0	1

W.E. 2

2 For each set of grouped data **(i)** find the modal class, **(ii)** find an estimate for the mean and **(iii)** use the cumulative frequency curve to find an estimate for the median. Round the mean value correct to 1 decimal place where necessary.

(a)

Class interval	Frequency
0–<10	4
10–<20	5
20–<30	8
30–<40	11
40–<50	7

(b)

Class interval	Frequency
250–<300	37
300–<350	58
350–<400	26
400–<450	16
450–<500	9
500–<550	4

3 For each set of discrete grouped data find **(i)** the modal class, **(ii)** an estimate for the mean and **(iii)** use the cumulative frequency curve to find an estimate for the median. Round the mean value correct to 1 decimal place where necessary.

(a) The number of students present each day.

Class interval	Frequency
450–459	2
460–469	7
470–479	8
480–489	15
490–499	10
500–509	8

(b) The number of points scored by a rugby team.

Class interval	Frequency
0–9	2
10–19	3
20–29	3
30–39	4
40–49	1
50–59	1

4 The frequency table on the right lists the lengths of fish caught, measured in centimetres. Find the median length by constructing a cumulative frequency curve.

Class interval (cm)	Frequency
30–<35	7
35–<40	10
40–<45	5
45–<50	12
50–<55	13
55–<60	4

5 The following questions relate to the cumulative frequency graph shown on the right.

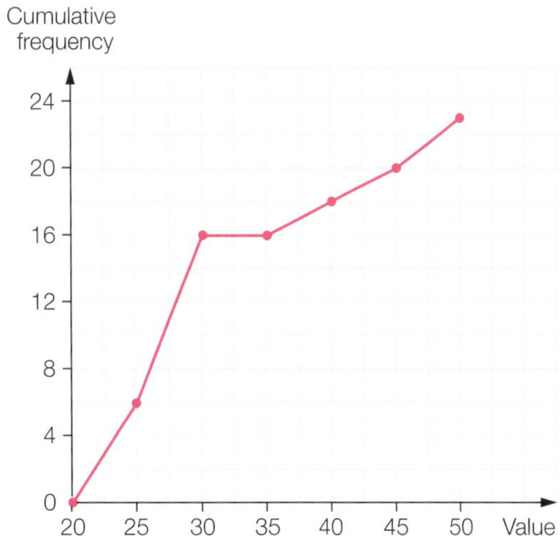

(a) The number of values in the data set is:

A 23 B 24

C 26 D 50

(b) The median is closest to:

A 12 B 27

C 28 D 29

(c) The class interval that has a frequency of zero is:

A 20–<25 B 30–<35

C 30–<36 D 31–<36

Understanding

6 The number of people boarding each of 25 successive trains at a station was as follows:

6	16	21	29	39	46	42	29	21	29	35	21	17
10	8	18	17	26	7	9	12	19	26	20	21	

(a) Find the median of this raw data.

(b) How many trains have fewer than 18 people?

(c) Group the data into the intervals 0–9, 10–19 etc. to draw a cumulative frequency curve.

(d) Use the graph to estimate the median number of people boarding each train and the number of times there were fewer than 18 people.

(e) Compare your answers in **(d)** with your answers from the raw data in **(a)** and **(b)**.

Reasoning

7 There is another way to estimate the median for grouped data, using the class interval that contains the median. If you know that the median is the third of eight results in a class interval 20–<25 (or 20–24 for discrete values), then the median is approximated by the value $20 + \frac{3}{8} \times 5 \approx 21.9$. Round to 1 decimal place in a calculation such as this.

(a) Use this method to find an approximation for the median for Question **2(a)**.

(b) Compare this with the value obtained using the cumulative frequency curve.

(c) Use this method to find an approximation for the median for Question **3(a)**.

(d) Compare this with the value obtained using the cumulative frequency curve.

Open-ended

8 Use the following information to construct a discrete data set with at least 19 results and then calculate the mean.

(a) median = 39, odd number of results

(b) median = 72.5, even number of results

Problem solving

Product codes

Most products that you buy have a barcode attached to them. This barcode represents the International Article Number (EAN-13) and is a 13-digit unique code for that product. As part of the system to ensure the correct numbers are scanned by a barcode reader, the 13th digit is a 'check' digit which is not actually part of the product identification.

The following are all legal product codes:

9316090060604	9337090060609
9326090060603	9338090060606
9336090060602	9339090060603

1 In the first column of codes, apart from the check digit, which digit changes? What do you notice about the sum of the two digits that are changing?

2 Predict the 'check' digit in the code 934609006060*.

3 In the second column of codes, apart from the check digit, which digit changes? Do these digits follow the same pattern as the first column of codes?

Not every digit acts in the same way. In fact the even-placed digits all act in the same way and the odd-placed digits all act in the same way, as shown:

×1	×3	×1	×3	×1	×3	×1	×3	×1	×3	×1	×3	×1
9	3	1	6	0	9	0	0	6	0	6	0	4

4 Calculate the sum of these multiplied values.

5 Do the same for each of the EAN-13 codes given. What conclusion can you reach?

6 Now calculate the check digit in each of the following EAN-13 codes:

(a) 933919006060* (b) 933929006060*

(c) 933928006060* (d) 933927006060*

Strategy options

- Work backwards.
- Test all possible combinations.

Box plots

2.2

Median, quartiles and range

The median is the 'middle' of a set of ordered data, so it divides the data into two halves.

It is sometimes useful to divide an ordered data set into quarters, which you can do by finding the 'middle' of each half. These 'middle' values of each half are called **quartiles**.

The lower quartile Q_L (or first quartile Q_1) is the 'middle' of the lower half of the data.

The upper quartile Q_U (or third quartile Q_3) is the 'middle' of the upper half of the data.

For example, consider the following sets of data.

$n = 8$ 6 8 | 9 9 | 13 14 | 14 17 Lower quartile: $Q_L = 8.5$ Median = 11 Upper quartile: $Q_U = 14$	$n = 9$ 21 26 | 28 30 [30] 34 36 | 39 39 $Q_L = 27$ Median = 30 $Q_U = 37.5$
$n = 10$ 5.6 5.9 [6.2] 6.7 6.7 | 6.9 7.2 [7.5] 7.5 7.9 $Q_L = 6.2$ Median = 6.8 $Q_U = 7.5$	$n = 11$ 14 18 [19] 25 25 [25] 27 27 [27] 28 29 $Q_L = 19$ Median = 25 $Q_U = 27$

Measures of spread

Aside from the statistical measures of centre (mean, median, mode), there are also 'measures of spread' that can be used to measure the spread of the set of data.

The **range** is the 'distance' across the data. It is the difference between the largest value and the smallest value.

The **interquartile range** (IQR) is the distance across the 'middle' of the data. It is the difference between the upper and lower quartiles.

Range = largest data value − smallest data value

Interquartile range = upper quartile − lower quartile

or

$\text{IQR} = Q_U - Q_L$

Worked example 3

W.E. 3

The following data set represents the heights (in cm) of the 38 members of the Collingwood AFL 2016 players list.

186	181	182	174	184	195	196	211	190	187
178	181	195	199	193	185	181	203	190	189
187	193	186	199	192	191	183	195	191	180
183	185	189	182	180	196	181	209		

Find the median, range and interquartile range of the data.

Thinking	Working
1 Order the data.	174 178 180 180 181 181 181 181 182 [182]
2 Find the middle of the data.	183 183 184 185 185 186 186 187 187 | 189
3 Find the middle of each half of the data.	189 190 190 191 191 192 193 193 [195] 195
	195 196 196 199 199 203 209 211
4 State the median.	Median = 188 cm
5 Calculate the range.	Range = largest − smallest data value = 211 − 174 = 37 cm
6 Calculate the interquartile range.	$IQR = Q_U - Q_L$ = 195 − 182 = 13 cm

Drawing box plots

When frequencies are large, or you are using grouped data, the interquartile range can be found using a cumulative frequency curve to find Q_U and Q_L. Q_U marks the upper quarter of results, while Q_L marks the lower quarter of results.

A **box plot**, also known as a box-and-whisker plot, is a graphical summary of the five key data values known as the **five-number summary** of a set of data:

- minimum value (smallest)
- lower quartile Q_L
- median
- upper quartile Q_U
- maximum value (largest)

The central part of the box plot, between the two quartiles, is the 'box'. This box contains the middle 50% of the data values. The lines extending to the minimum and maximum values are the 'whiskers', which contain the remaining upper and lower 25% of the data values.

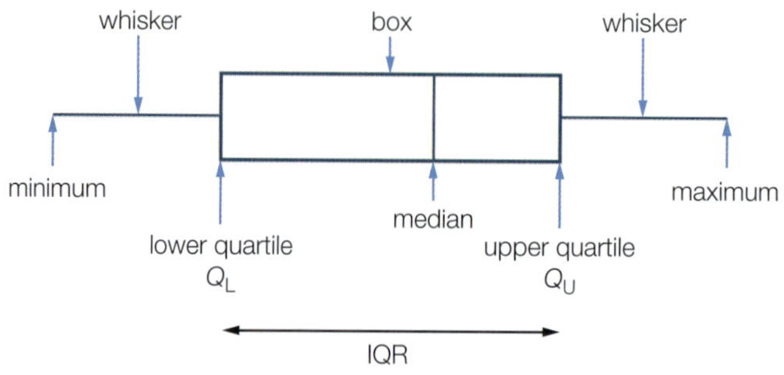

Worked example 4

W.E. 4

The weights of 20 packing crates (in kg) are shown in the frequency table.

(a) Calculate the range.

(b) Draw a cumulative frequency curve and use it to estimate the interquartile range.

(c) Draw a box plot to represent the data.

Weight (kg)	Frequency
45–<50	2
50–<55	3
55–<60	3
60–<65	6
65–<70	4
70–<75	2

Thinking / Working

(a) Range can be calculated by taking the smallest result from the largest. (Estimate these values by using the end points.)

(a) Range = largest result − smallest result
= 75 − 45
= 30

The range of weights is 30 kg.

(b) 1 Construct a cumulative frequency table.

(b)

Weight(kg)	Cumulative frequency
45	0
50	2
55	5
60	8
65	14
70	18
75	20

2 Draw the cumulative frequency curve.

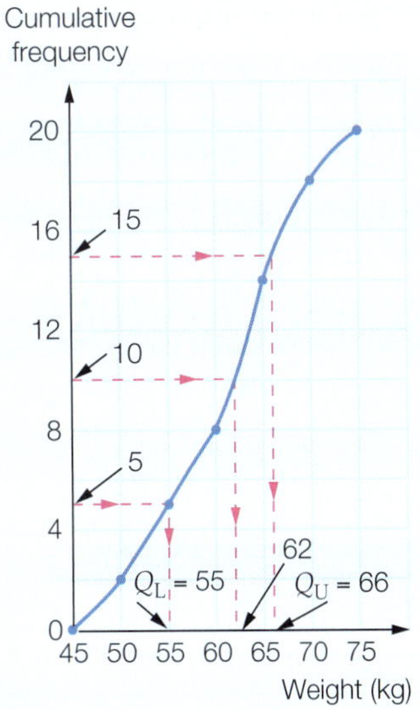

3 Read Q_U from $\frac{3n}{4}$ on the cumulative frequency axis and read Q_L from $\frac{n}{4}$.

$n = 20$, so read the 5th and 15th values.

Q_U: 66 (15th data value)

Q_L: 55 (5th data value)

4 Calculate the interquartile range (IQR).

$IQR = Q_U - Q_L$
$= 66 - 55$
$= 11$

The interquartile range is 11 kg.

(c) 1 Read the median from $\frac{n}{2}$ on the cumulative frequency axis.

(c) $n = 20$, so read the 10th value.
Median: 62 (10th data value)
Median = 62 kg

2 Draw a box plot using the median and the information found in part **(a)**.

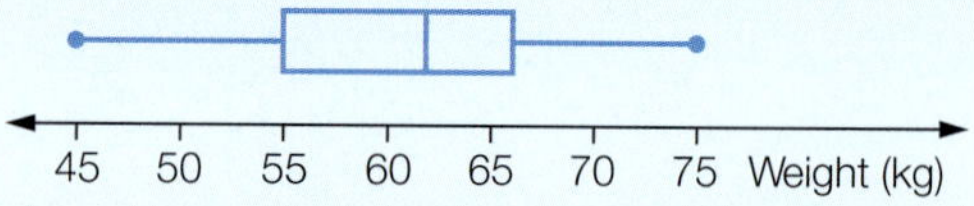

Notice that the box plot uses the data axis from the cumulative frequency curve. The descending lines from the median and quartiles produce the spacing for the 'box'.

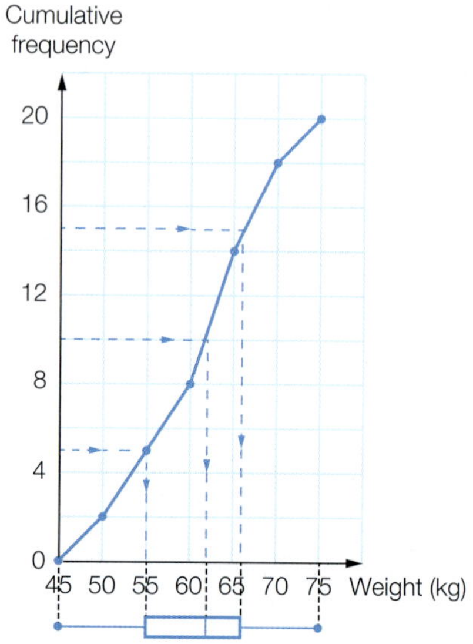

Using technology to find the five-number summary

You can use technology, such as a scientific calculator, CAS or spreadsheet, to find the five-number summary from a raw data list or a grouped data table.

Using TI-Nspire CAS

Add **Lists & Spreadsheet** to your document. Enter the data from the example above, with column **A** as 'weight' and column **B** as 'freq', using the midpoint of each class interval for the weight values, as shown. (The midpoint of 45–<50 is: $\frac{45+50}{2} = 47.5$)

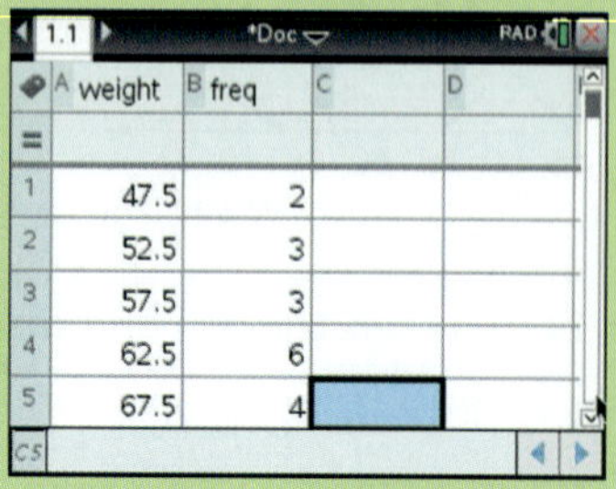

Using Casio ClassPad CAS

From the menu select **Statistics**. Enter the data from the example above, with **list1** as the weight data and **list2** as the frequencies, using the midpoint of each class interval for the weight values.

(The midpoint of 45–<50 is: $\frac{45+50}{2} = 47.5$)

Using TI-Nspire CAS	Using Casio ClassPad CAS
Select **menu** > **Statistics** > **Stat Calculations** > **One-Variable Statistics...** and confirm that **Num of Lists** is 1. Then set the **One-Variable Statistics** settings as shown. 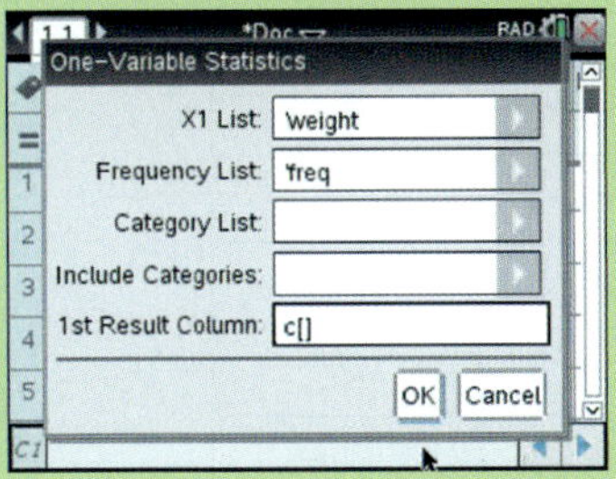	Select **Calc** > **One-Variable** and set the **XList** as **list1** and the **Freq** as **list2** as shown. 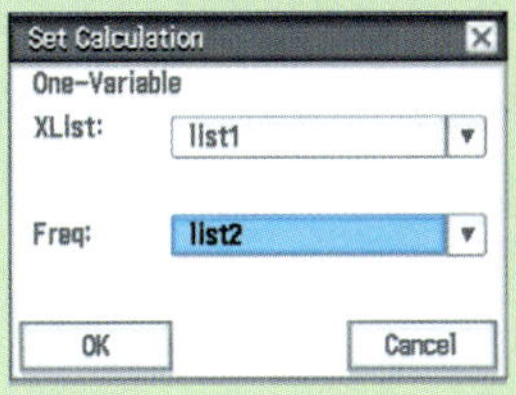
The statistics data should appear as shown. 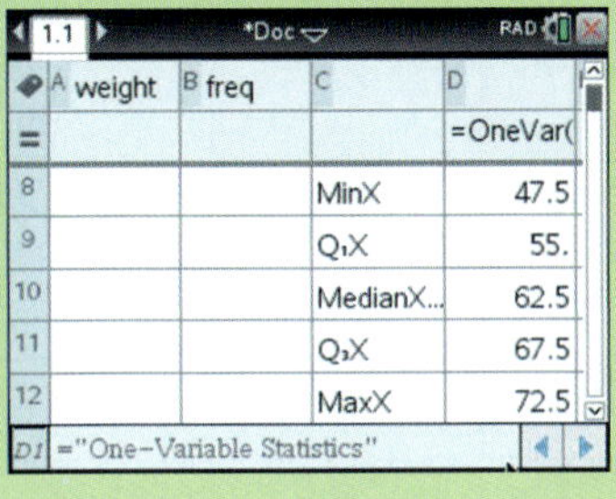	The statistics data should appear as shown. 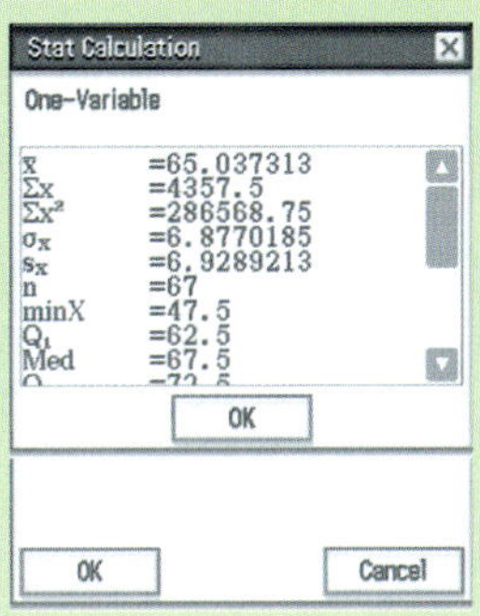

Note: If you had the raw data you could just enter each value individually in the first column, instead of using the class midpoints.

Percentiles and quantiles

To create a **percentage cumulative frequency curve**, you convert the cumulative frequencies for grouped data to proportions (percentages). These curves can be used to find a percentile, which is the value below which a percentage of the results occur. For example, the value below which 70% of the results occur is the 70th **percentile** or the 0.7 **quantile** (written as a decimal).

- The median is the 50th percentile or the 0.5 quantile.
- The lower quartile Q_L is the 25th percentile or the 0.25 quantile.
- The upper quartile Q_U is the 75th percentile or the 0.75 quantile.

Worked example 5 W.E. 5

The following table shows the data in class intervals from a survey of the weights of packing crates. Use the data to:

(a) draw a cumulative frequency table and then a percentage cumulative frequency table

(b) draw a percentage cumulative frequency curve

(c) evaluate the 80th percentile and the 0.3 quantile.

Weight (kg)	Frequency
45–<50	2
50–<55	3
55–<60	3
60–<65	6
65–<70	4
70–<75	2

Thinking	Working

(a) 1 Complete a cumulative frequency table.

(a)

Weight (kg)	Cumulative frequency
<45	0
<50	2
<55	5
<60	8
<65	14
<70	18
<75	20

2 Draw a table of percentage cumulative frequencies. Each percentage cumulative frequency is obtained by dividing the cumulative frequency by the total number of results (20) and multiplying by 100.

Weight (kg)	Cumulative frequency	Percentage cumulative frequency
<45	0	$\frac{0}{20} \times 100 = 0$
<50	2	$\frac{2}{20} \times 100 = 10$
<55	5	$\frac{5}{20} \times 100 = 25$
<60	8	$\frac{8}{20} \times 100 = 40$
<65	14	$\frac{14}{20} \times 100 = 70$
<70	18	$\frac{18}{20} \times 100 = 90$
<75	20	$\frac{20}{20} \times 100 = 100$

(b) Plot the percentage cumulative frequency against weight. The percentage cumulative frequency curve has a similar shape to that of the cumulative frequency curve, but it has a different vertical scale.

(b)

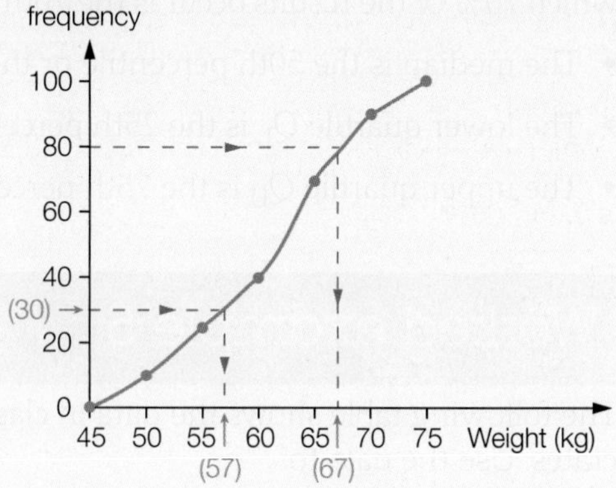

(c) Find the 80% point on the vertical axis. Draw a line horizontally from this point until it meets the graph. Then draw a line vertically down and read off the result. (See the dashed lines drawn on the graph.)
Similarly, convert the 0.3 quantile to a percentage ($0.3 \times 100 = 30\%$) and then find the 30% point on the vertical axis.

(c) The 80th percentile is 67 kg.
The 0.3 quantile is the 30% mark and is 57 kg.

A cumulative frequency is the number of results below a given value.

$$\text{Percentage cumulative frequency} = \frac{\text{cumulative frequency}}{\text{total number of results}} \times 100\%$$

For any percentage cumulative frequency graph, n% of results lie below the nth percentile.

Percentile and quantile values can be read from a percentage cumulative frequency graph. The nth percentile (or $0.01n$ quantile) is the value on the horizontal axis corresponding to n% on the vertical axis, where $0 \leq n \leq 100$.

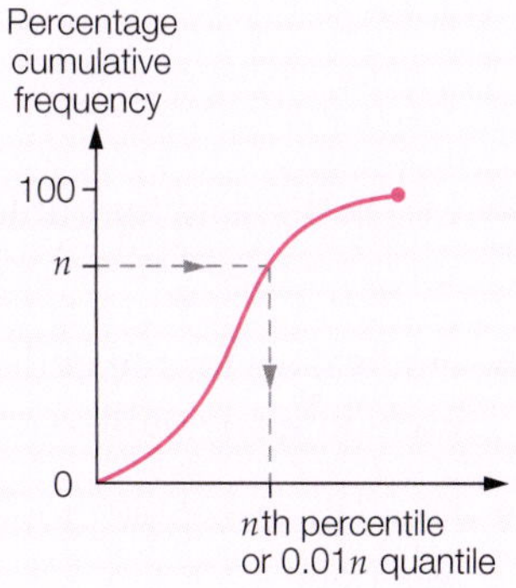

Outliers

When a data value is significantly different to most of the data, it may be an **outlier**. In any particular data set an outlier might be smaller than most of the other data values, or it might be bigger than most of the other data values, or there might be outliers at both ends of the data range.

A data value is considered an outlier if it is more than $1.5 \times \text{IQR}$ above Q_U, or if it is less than $1.5 \times \text{IQR}$ below Q_L.

When drawing a box plot that includes any outliers, indicate the outlier values by drawing them as dots. Do not stretch the whisker line to the outlier values.

A data value x is an outlier if:

$x < Q_L - 1.5 \times \text{IQR}$
$x > Q_U + 1.5 \times \text{IQR}$

Worked example 6 W.E.6

Construct a box plot for the following set of ordered data and identify any outliers.

3, 5, 5, 6, 6, 7, 7, 12, 20

Thinking	Working
1 Identify the median and quartiles.	3, 5 | 5, 6, [6] 7, 7 | 12, 20 Median = 6 $Q_L = 5$ $Q_U = 9.5$
2 Calculate the interquartile range.	$\text{IQR} = 9.5 - 5$ $= 4.5$

3 Find 1.5 × IQR above and below the interquartile region.

$1.5 \times IQR = 1.5 \times 4.5$
$= 6.75$

$Q_L - 1.5 \times IQR = 5 - 6.75$
$= -1.75$

$Q_U + 1.5 \times IQR = 9.5 + 6.75$
$= 16.25$

4 Check the data for outliers.

The only data value outside the range -1.75 to 16.25 is 20.

5 Find values for the ends of the whiskers, excluding any outliers.

Minimum = 3
Maximum = 12 (excluding 20)

6 Draw the box plot showing the outlier as a dot.

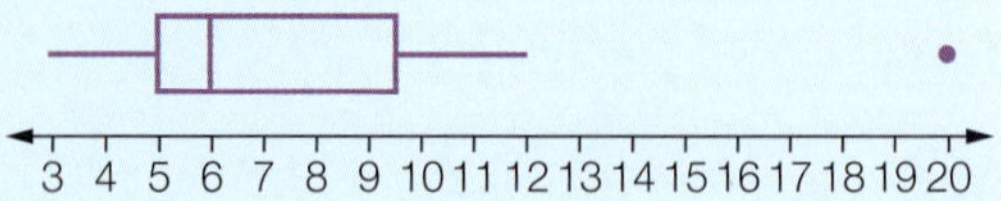

2.2 Box plots

Navigator

Answers p. 772

1, 2, 3, 4, 5, 6, 7, 8, 10, 11, 12 (a)	1, 2, 3, 4, 5, 6, 7, 8, 10, 11, 12 (a), 13, 14 (a)	1, 2, 3, 4, 6, 7, 8, 9, 11, 12, 13, 14

Fluency

W.E. 3

1 The following data represents the weight (in kg) of the 38 members of the Collingwood AFL 2016 player list.

92	84	75	81	77	101	108	106	89	82
79	84	89	96	92	77	87	101	86	85
89	87	88	93	78	91	77	102	87	85
80	93	89	84	82	102	83	113		

Find the median, range and interquartile range of the data.

W.E. 4

2 For each given data set, **(i)** calculate the range and **(ii)** draw a cumulative frequency curve and use it to estimate the interquartile range, then **(iii)** draw a box plot.

(a)

Class interval	Frequency
0–<10	4
10–<20	5
20–<30	8
30–<40	11
40–<50	7

(b)

Class interval	Frequency
60–69	9
70–79	12
80–89	19
90–99	17
100–109	13

3 Draw cumulative frequency and percentage cumulative frequency curves to represent the following data, and find the 60th percentile for each data set. W.E. 5

(a)

Class interval	Frequency
15–<16	4
16–<17	5
17–<18	7
18–<19	3
19–<20	4

(b)

Class interval	Frequency
150–<155	1
155–<160	2
160–<165	3
165–<170	4
170–<175	3

4 The number of points scored by a basketballer in a season of games is: W.E. 6

32	30	16	18	24	16
28	26	19	18	8	26
34	36	41	18	16	14
17	21	26	36	54	8

Construct a box plot that identifies any outliers in the data.

5 The following is a list of the heights, measured to the nearest centimetre, of the players at the local football club.

155	167	189	156	178	179	190	200
206	196	187	156	166	178	168	178
168	170	167	159	167	160	163	181
191	157	156	187	157	159	156	161
153	173	174	168	177	164	175	185
186	164	171	170	180	141	167	159

(a) Find the range of the heights.

(b) Calculate the interquartile range by finding the upper and lower quartiles.

(c) Calculate the values that are the cut-off values at either end of the data set after which the results would be considered outliers.

(d) Are there any outliers in the data set?

6 Each question below refers to the percentage cumulative frequency curve shown at right, which represents the heights of players in a football club.

(a) The 30th percentile is closest to:

A 170 B 172.5

C 175 D 184

(b) The 0.8 quantile is closest to:

A 170 B 188

C 191 D 194

(c) The interquartile range is closest to:

A 16 B 18

C 22 D 50

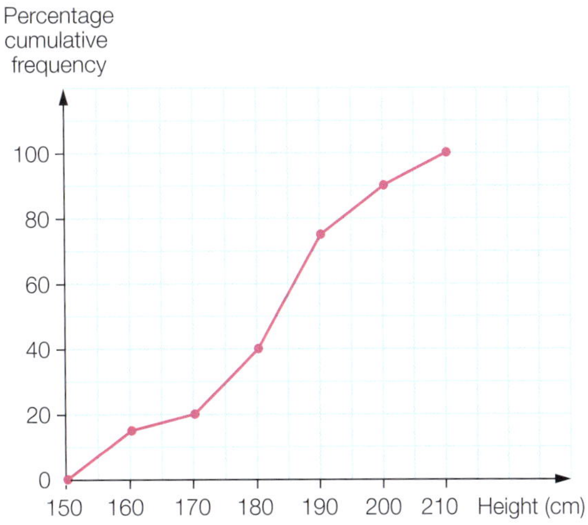

7 For the percentage cumulative frequency curve shown at right, outliers would be:

A results less than 32 cm

B results greater than 63 cm

C results less than 16.5 cm and greater than 63 cm

D none

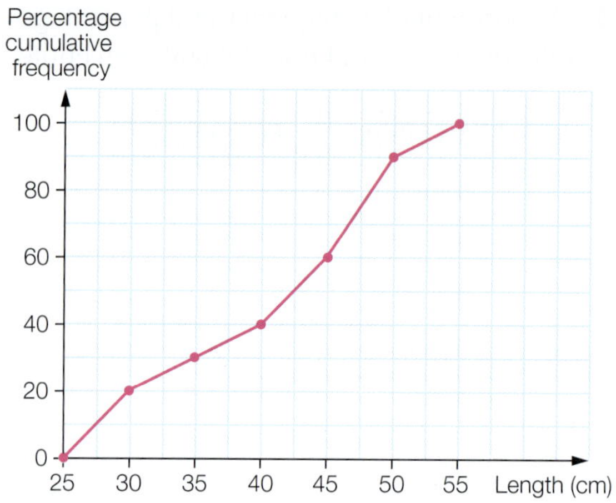

Understanding

8 The distances run (km) by athletes training for a marathon are summarised in this frequency table.

Class interval	Frequency
10–<15	4
15–<20	6
20–<25	13
25–<30	16
30–<35	11
35–<40	9

(a) Draw a cumulative frequency curve for this data.

(b) How many of the distances run were less than 17 km?

(c) Draw a percentage cumulative frequency curve.

(d) Find the 20th percentile and the 0.8 quantile.

9 The data below shows the distances travelled (km) by car in a week for 30 employees.

895	245	697	1098	573	1569	1296	1880	972	843
1108	1037	986	1265	1398	1107	1275	1117	1023	1062
1197	1047	1284	1402	438	569	974	1328	1194	1073

(a) Calculate the range of the data.

(b) Calculate the interquartile range by finding the upper and lower quartiles.

(c) Draw a box plot that shows the outliers.

10 For each of the following box plots:

(i) indicate the values below which, or above which, outliers would lie

(ii) decide whether there are any outliers in the data set displayed.

(a)

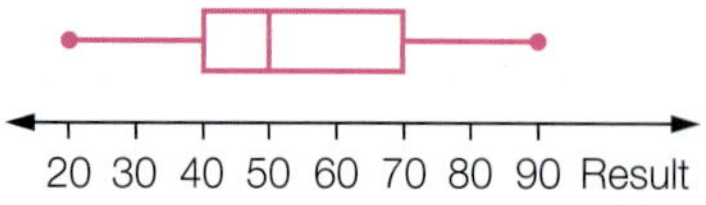

(b)

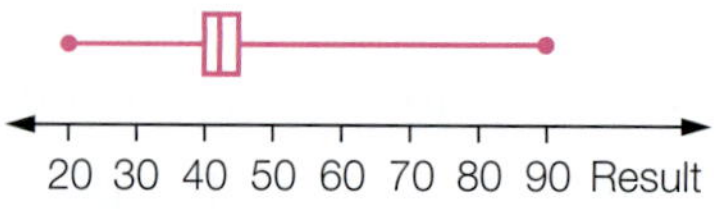

Reasoning

11 Consider this table, which shows the annual earnings for all the employees of a company.

(a) Calculate estimates for the mean and median annual incomes.

(b) Calculate estimates for the range and interquartile range.

(c) Are there any outliers in the data set? If so, list them.

(d) Remove any outlier(s) from the results and recalculate the mean, median, range and interquartile range.

(e) Comment on your results.

Annual income	Frequency
\$35 000–<\$40 000	17
\$40 000–<\$45 000	24
\$45 000–<\$50 000	18
\$50 000–<\$55 000	16
\$55 000–<\$60 000	1
\$60 000–<\$65 000	1
\$65 000–<\$70 000	2
\$70 000–<\$75 000	1
\$75 000–<\$80 000	1
\$80 000–<\$85 000	1

Open-ended

12 For each of the following box plots:

(i) construct a data set of at least 20 values that matches it

(ii) explain the reasoning you used to do this, including any assumptions made and alternative actions that might have been taken.

(a)

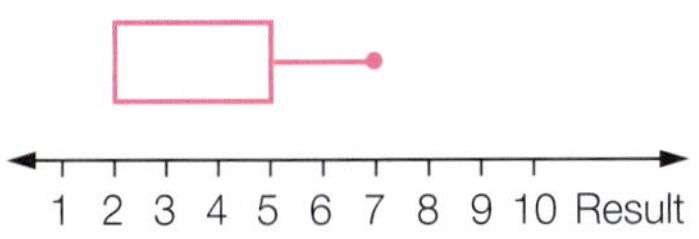

(b)

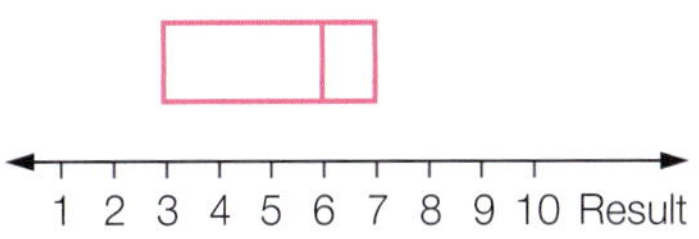

13 (a) Draw a cumulative frequency curve with an interquartile range of 35.

(b) Draw a percentage cumulative frequency curve with an interquartile range of 16.

14 (a) Construct a data set with at least 20 values that has an outlier at the upper end of the data set. The outlier is to be less than $Q_U + 2 \times \text{IQR}$.

(b) Construct a data set with at least 20 values that has an outlier at the lower end of the data set. The outlier is to be greater than $Q_L - 2 \times \text{IQR}$.

Problem solving

Draw that plot

A data set has the following characteristics:

- 50th percentile is 56
- 75th percentile is 80
- IQR is 50
- Range is 80

Draw a box plot to represent this information. Which of the five key values are you certain about? Give the maximum and minimum values possible for the values that you are unsure of.

Strategy options

- Work backwards.
- Draw a diagram.

Exploration CAS

Equipment required: TI-Nspire CAS or Casio ClassPad CAS, 1 large sheet of cardboard and thick marker pens for the whole class

Investigating data with box plots

1 Copy the table, and in the column 'Estimate of age', write your best guess for the age in years of each famous person listed. You can choose your own famous people if you prefer.

Name	Estimate of age	Actual age	Error
Taylor Swift			
Harrison Ford			
Malala Yousafzai			
Chris Hemsworth			
Cate Blanchett			
Barack Obama			
Jennifer Lawrence			
Will Ferrell			
Adam Goodes			
Prince George of Cambridge			
Justin Bieber			

2 In the next column, write the actual ages of these people after doing some research. Subtract the actual ages in column 3 from your estimated ages in column 2 and write the answer in column 4 ('Error'). Below is an example.

Name	Estimate of age	Actual age	Error
Susie Lee	58	60	-2
Joe Bloggs	38	35	3

3 Use your CAS to draw a box plot that identifies any outliers in the error data.

Using TI-Nspire CAS

Add **Lists & Spreadsheet** to your document. Label column **A** as 'error' and enter the data.

(The work that follows uses the example data -2, 3, 2, -4, 2, 0, 11.)

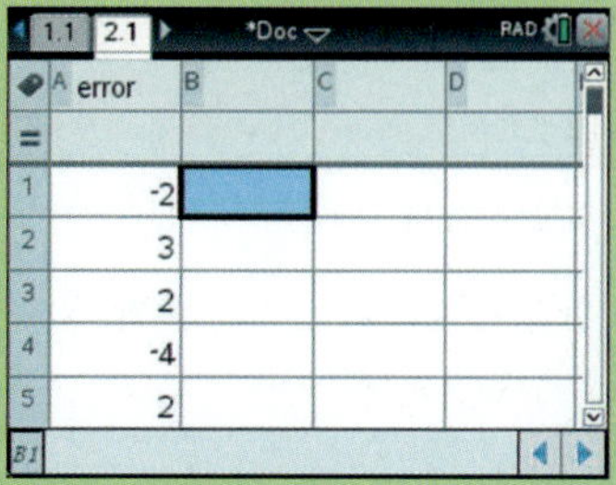

Using Casio ClassPad CAS

From the menu select **Statistics**. Enter the data in the first column.

(The work that follows uses the example data -2, 3, 2, -4, 2, 0, 11.)

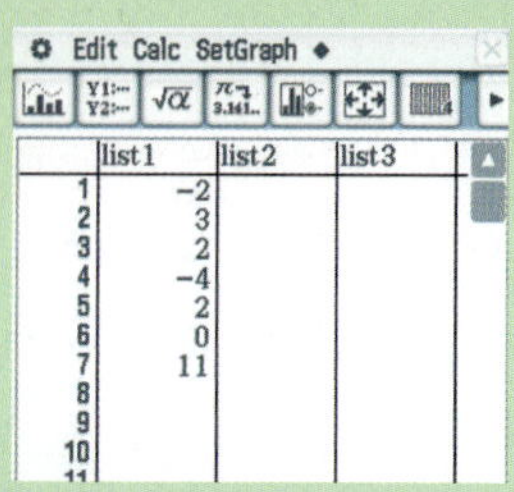

Using TI-Nspire CAS

Now go to the Home screen and add **Data & Statistics** to your document. At the bottom of the window, select 'error' as the variable. Then select **menu** > **Plot Type** > **Box Plot**. The box plot will be displayed, with the outlier shown as a dot.

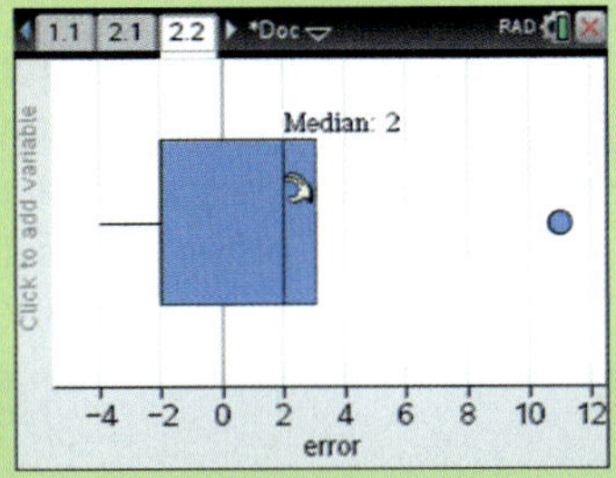

You can use the arrow keys to move the box plot and to show the key values.

Using Casio ClassPad CAS

Now select **SetGraph** > **Setting...** and set the graph settings as shown.

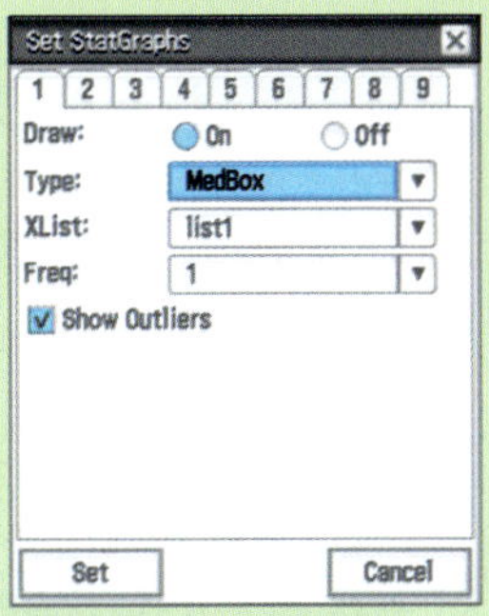

Select the draw graph icon and the box plot will be displayed, with the outlier shown as a dot.

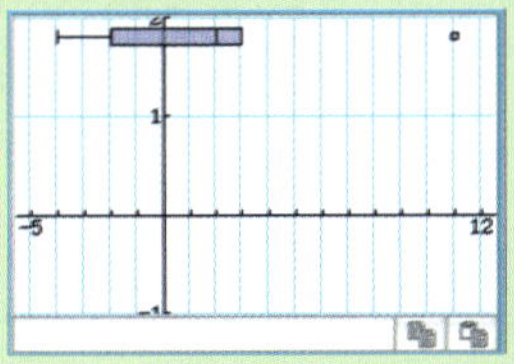

Select **Analysis** > **Trace** and then use the arrow keys to show the key values.

4 Make a neat sketch of your box plot in your exercise book together with values for **Xmin**, **Q_1**, **Median**, **Q_3** and **Xmax**. Use a ruler to create an appropriate scale marked in one-unit increments. Your plot should look something like the diagram shown at right.

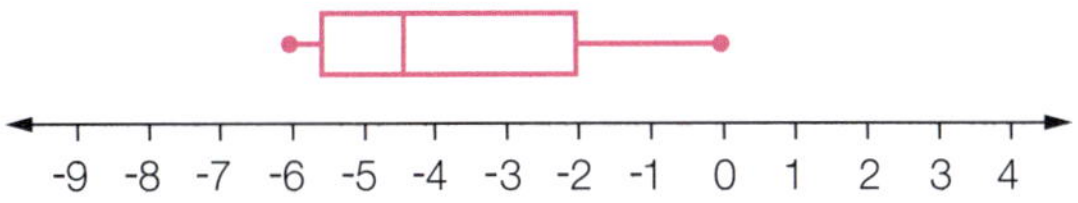

Interpreting box plots

1 Use your box plot to find the range and the IQR of your data.

2 Does your box plot indicate that your guesses were, on the whole, underestimated or overestimated? Why?

3 Your teacher will coordinate the following class activity. On a single large sheet of cardboard (i.e. one sheet for the whole class), draw a scale marked in units of 1 with 0 at the centre. The scale should be placed at the top of the sheet of cardboard. Using a thick marker pen, neatly draw your box plot on the cardboard. Make sure that all of the five key values of the box plot are aligned properly with the scale. Print your name next to your box plot.

After all of the box plots have been drawn, answer these questions.

(a) Who was the best estimator of the ages of the famous people? How do you decide?

(b) Who was the worst?

(c) Which plots had median values closest to zero?

(d) Who was the most consistent estimator? How can you tell? Is a consistent estimator always a good estimator?

(e) Who were inconsistent estimators? Why?

(f) Were there any box plots with extreme results? Discuss.

2.3 Comparing data sets

When faced with a lot of data, you need some skills to interpret the data correctly and make reasonable conclusions based on that data. So far you have made use of three measures of centre or location (mean, median and mode) and two measures of spread (range and interquartile range).

Statisticians combine these measures to analyse and compare sets of data. The mean is often linked with the range, because they are usually both affected by any small change to the data. The median is often linked with the interquartile range, because they are not always affected by any small change, unless the change is in the middle section of the data set.

Worked example 7

W.E. 7

For each of the following sets of data:

Set A:	1	1	1	2	3	3	3	4	4	4	5	5	5	5	6
	6	6	6	7	7	7	8	8	8	8	8	9	9	9	
Set B:	1	1	1	2	3	3	3	4	4	4	5	5	5	5	6
	6	6	6	7	7	7	8	8	8	8	8	9	9	19	

(a) find the mean

(b) find the median

(c) find the range

(d) calculate the interquartile range

(e) use your findings to compare the two sets of data.

Thinking	Working
(a) Add all the data values in each set and divide by the total number of data values to find the mean.	(a) Set A mean $= \frac{158}{29}$ $= 5.48$ Set B mean $= \frac{168}{29}$ $= 5.79$
(b) Find the middle value of each set.	(b) Set A median = 6 Set B median = 6
(c) Find the difference between the maximum and minimum values.	(c) Set A range = 9 − 1 = 8 Set B range = 19 − 1 = 18

(d) Find Q_L and Q_U for each set and find the difference to find the IQR.	(d) Set A $Q_L = 3.5$ Set A $Q_U = 8$ Set A IQR $= 8 - 3.5$ $= 4.5$ Set B $Q_L = 3.5$ Set B $Q_U = 8$ Set B IQR $= 8 - 3.5$ $= 4.5$
(e) Compare the calculations for each statistic.	(e) The two sets have the same median, but the mean of Set B is slightly higher than Set A. The two sets have the same interquartile range, but the range of Set B is greater than Set A. This is because Set B has the score of 19, which affects the mean and the range, but not the median (middle score) nor the interquartile range (spread of the middle half of the data).

Box plots are useful when comparing data sets. In particular, you can use a **parallel box plot**, which is two or more box plots drawn on the same set of axes. From the parallel box plot you can easily compare the median, range and IQR of the data sets, as well as decide which data set is more consistent. It is important that the scale used covers all values in both sets.

In the same way, a **parallel dot plot** can also be useful. When drawing a parallel dot plot it is common to put the scale in twice; it effectively serves as the dividing line between the two dot plots. (See Question 9 in Exercise 2.3.)

Worked example 8

W.E. 8

The heights (cm) of the students in two Year 8 classes were recorded.

8A:	148	149	150	150	153	154	154	154	155	155	156	156	156
	157	157	158	158	158	158	158	160	161	163	165	168	
8B:	150	151	152	152	152	153	153	154	155	155	155	156	156
	156	156	157	157	158	160	160	160	161	161	162	163	

(a) Find the median height for each class.

(b) Find the range for each class.

(c) Find the interquartile range for each class.

(d) Draw a parallel box plot for both classes.

(e) Compare the statistics for the two classes.

Thinking	Working
(a) Find the middle value of each class.	(a) 8A median = 156 8B median = 156
(b) Find the difference between the maximum and minimum values.	(b) 8A range $= 168 - 148$ $= 20$ 8B range $= 163 - 150$ $= 13$

(c) Find Q_L and Q_U for each set and find the difference to find the IQR.	(c) 8A $Q_L = 154$ 8A $Q_U = 158$ 8A IQR $= 158 - 154$ $= 4$ 8B $Q_L = 153$ 8B $Q_U = 160$ 8B IQR $= 160 - 153$ $= 7$
(d) Use the statistics to draw a parallel box plot.	(d) 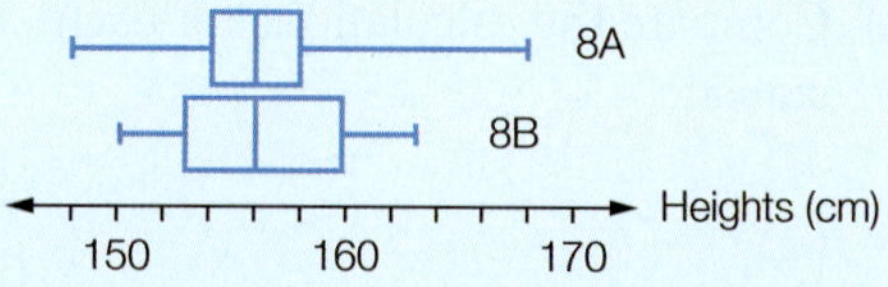
(e) Compare the calculations for each statistic.	(e) The medians for the two classes are the same but the range for 8A is much larger than for 8B. However, the IQR for 8A is much smaller than the IQR for 8B.

Comparing different data representations

When investigating a data set, presenting the data in different forms can give you a deeper understanding of the data.

Worked example 9

W.E. 9

The data set gives the height (cm) of each of the players on the 2016 playing list for Melbourne Storm in the NRL.

200 193 177 183 194 190 178 192 188 185
188 179 178 179 195 182 191 189 185 180
194 183 196 185 181 189 185 178 185 182
187 192 195 192

(a) Draw a histogram of the heights. Use class intervals of 5, with the first interval being 175–<180.

(b) Calculate the five-number summary for the data and use it to construct a box plot.

(c) Draw a dot plot of the data.

(d) Make some comments about what the three displays tell you about the data set.

Thinking	Working
(a) 1 Set up the class intervals. Count the number of data for each interval.	(a) see table below

Height (cm)	Frequency
175–<180	6
180–<185	6
185–<190	10
190–<195	8
195–<200	3
200–<205	1

2 Construct the histogram of the data in the frequency table.

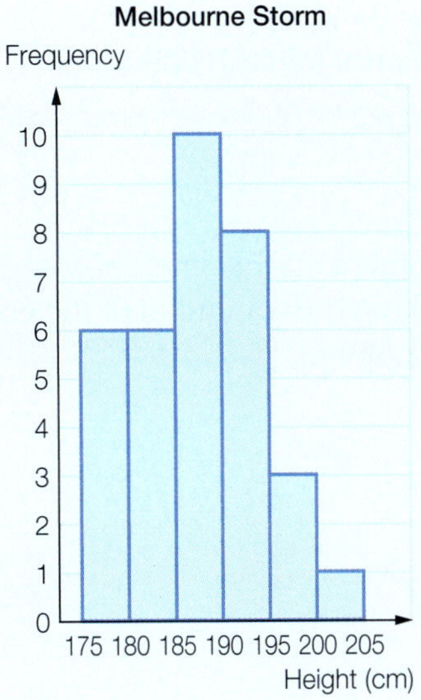

(b) 1 Find the five-number summary using a CAS or by hand assisted by a number sorter.

(b)

Min	Q_L	Medium	Q_U	Max
177	182	186	192	200

2 Draw the box plot using the five data values.

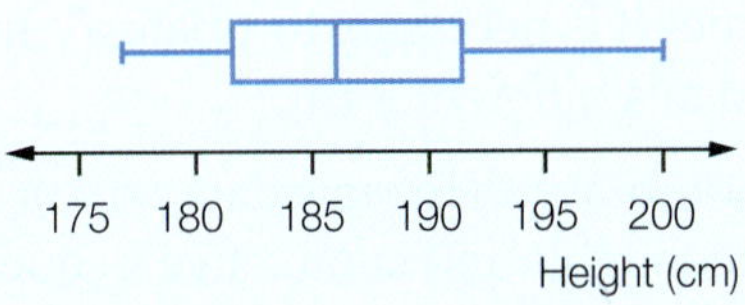

(c) 1 The dot plot does not need grouped data as the range is not too big.

Create a frequency table for individual data.

(c)

Height (cm)	Frequency
177	1
178	3
179	2
180	1
181	1
182	2
183	2
185	5
187	1
188	2
189	2
190	1
191	1
192	3
193	1
194	2
195	2
196	1
200	1

2 Draw the dot plot using the frequency for each data value.

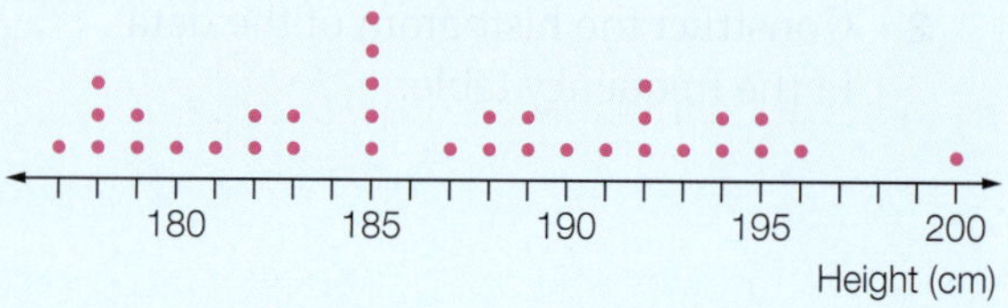

(d) 1 Comment on the differences in the displays, being mindful of the effect of grouping.

(d) Some accuracy has been lost in the histogram by grouping the data, making it appear that the heights range from 175 cm to 205 cm. The box plot displays the median, range and IQR accurately. It can be seen that the heights of the taller half of the players are more spread out than the heights of the shorter half. The dot plot has retained all raw data for us. It can be clearly seen that only one player was taller than 196 cm.

Vic Cumulative frequency distributions from dot plots, box plots and histograms

Sometimes it is necessary to create a cumulative frequency distribution from data that is supplied in a different form.

A dot plot shows all the raw data, so it is straightforward to use a dot plot to create a frequency distribution table and cumulative frequency curve.

A histogram of grouped data will only show data values for the endpoints of the class intervals, but this can still be used to form a cumulative frequency distribution.

Box plots show only five data values (the five-number summary). It is unusual to give the value of n, the number of data values, so a percentage cumulative frequency table with percentiles 0, 25, 50, 75 and 100 is the best cumulative frequency distribution that can be found from a box plot.

Worked example 10

W.E. 10

Use the graphical displays of heights for Melbourne Storm players (shown in the previous example) to create a cumulative frequency distribution.

(a) Form a cumulative frequency table from the histogram.

(b) Form a percentage cumulative frequency curve from the box plot.

(c) Form a cumulative frequency table from the dot plot.

Thinking / Working

(a) 1 Create a frequency table to match the histogram.

(a)

Height (cm)	Frequency
175–<180	6
180–<185	6
185–<190	10
190–<195	8
195–<200	3
200–<205	1

2 Write values from the ends of the class intervals as the data values and accumulate the frequencies.

Height (cm)	Cumulative frequency
<175	0
<180	6
<185	12
<190	22
<195	30
<200	33
<205	34

(b) 1 Find the five-number summary values.

(b)

Min	Q_L	Medium	Q_U	Max
177	182	186	192	200

2 Match the data values to their corresponding percentiles.

Height (cm)	Percentage cumulative frequency
177	0
182	25
186	50
192	75
200	100

3 Draw the percentage cumulative frequency curve.

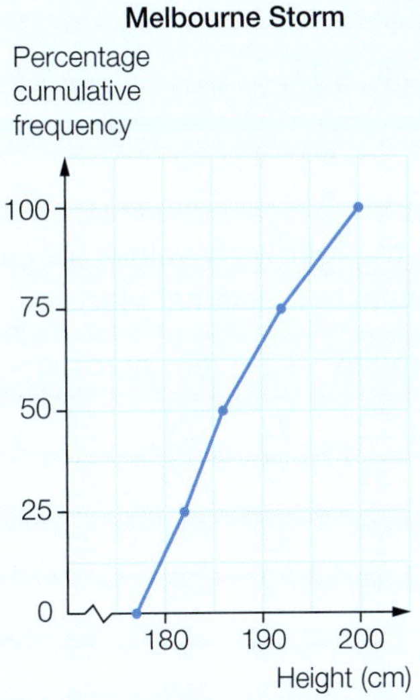

(c) 1 Create a frequency table for the individual data values.

2 Accumulate the frequencies.

(c)

Height (cm)	Frequency	Cumulative frequency
177	1	1
178	3	4
179	2	6
180	1	7
181	1	8
182	2	10
183	2	12
185	5	17
187	1	18
188	2	20
189	2	22
190	1	23
191	1	24
192	3	27
193	1	28
194	2	30
195	2	32
196	1	33
200	1	34

2.3 Comparing data sets

Navigator

Answers p.774

1, 2, 3, 5, 6, 7, 8, 9, 11	1, 2, 3, 4, 5, 6, 7, 8, 9, 10, 11, 12	1, 2, 3, 4, 6, 7, 8, 9, 10, 11, 12

Fluency

W.E. 7

1 Calculate the following values for each pair of data sets below. Where necessary round your answers to 2 decimal places.

(i) mean (ii) median (iii) range (iv) interquartile range

(a) Set A: 22 22 24 25 26 27 28 28 29 30
31 32 35 35 36 37 38 39 40

Set B: 12 22 24 25 26 27 28 28 29 30
31 32 35 35 36 37 38 39 40

(b) Set A: 44 44 46 47 47 48 48 48 51
52 52 52 54 55 58 58 58 60

Set B: 24 44 46 47 47 48 48 48 51
52 52 52 54 55 58 58 58 80

(c) Set A: 11 12 12 12 13 13 13 13 14 14 15
16 17 18 18 18 21 22 24 24 24 26

Set B: 11 12 12 12 13 13 13 13 13 13 13
16 17 18 18 18 21 22 24 24 24 26

(d) Make a few comments about what you have found by completing these calculations.

2 The heights (cm) of the students in two Year 8 classes were recorded. W.E. 8

8C:	145	146	147	147	149	150	150	151	152	153	154	155	155
	156	157	158	158	159	160	162	163	163	164	166	169	
8D:	147	148	149	150	150	150	152	153	154	154	155	155	155
	156	156	156	157	157	158	160	160	160	161	161	162	

(a) Find the median height for each class.

(b) Find the range for each class.

(c) Find the interquartile range for each class.

(d) Draw a parallel box plot for both classes.

(e) Compare the statistics for each class.

3 The following data list shows the weight (kg) of each player on the 2016 playing list for the Brisbane Broncos in the NRL. W.E. 9

95	108	94	97	95	102	101	103	89	94	98	92	107
92	115	85	89	107	112	100	102	98	106	110	91	108

(a) Draw a histogram of the heights. Use a class interval of 5, with the first interval being 85–<90.

(b) Calculate the five-number summary for the data and use it to construct a box plot.

(c) Draw a dot plot of the data.

(d) Make some comments about what the three displays tell you about the data set.

Vic 4 Use the graphical displays of weights of Brisbane Broncos players from the previous question to create cumulative frequency distributions, as follows: W.E. 10

(a) Form a cumulative frequency table from the histogram.

(b) Form a percentage cumulative frequency curve from the box plot.

(c) Form a cumulative frequency table from the dot plot.

5 (a) For each data set in the following parallel box plot, find the:

(i) median (ii) range (iii) IQR.

(b) Compare the statistics for the two data sets.

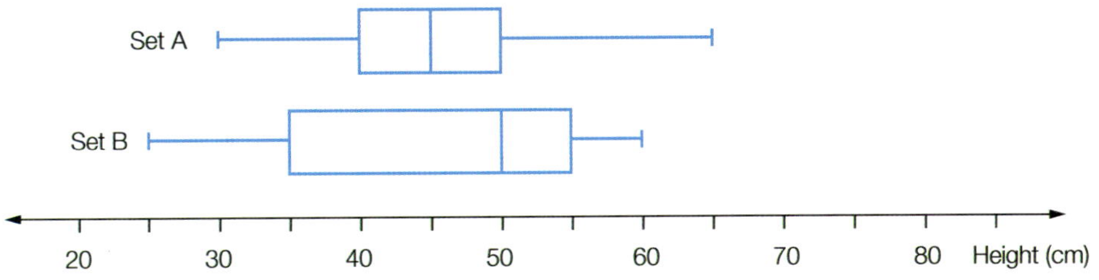

6 (a) The students in a class had their heights (cm) recorded. The following dot plot shows the results.

The interquartile range for this data set is closest to:

A 5 B 5.5

C 7.5 D 11

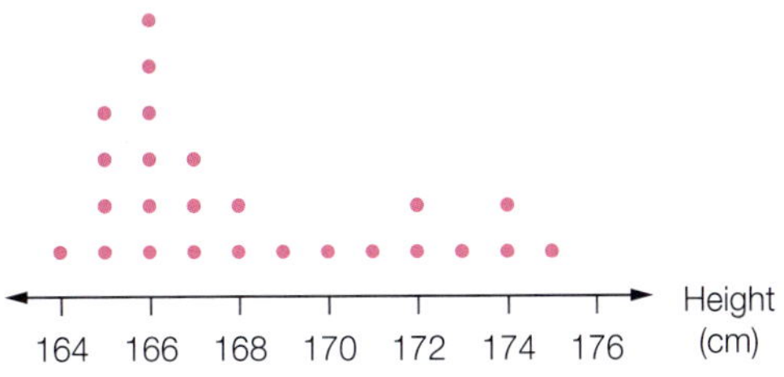

(b) The histogram below shows the weights, in kg, of the students in another class.

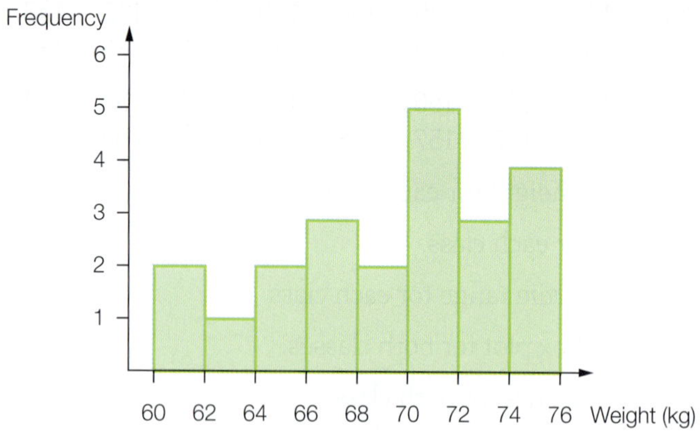

The median of this data set is likely to be closest to:

A 66 **B** 70 **C** 74 **D** 75

Understanding

7 Three classes attempted the same mathematics test, marked out of 100. The results for the three classes are summarised in the following table.

Class	Mean	Median	Range	IQR
A	64.32	68	72	27.5
B	64.12	63	31	19
C	60.92	63	55	38.5

Which class do you think performed best on the test?

8 The following dot plot shows the percentage of the world's total emissions of carbon dioxide produced by the 30 countries that contribute the most, rounded to the nearest whole number.

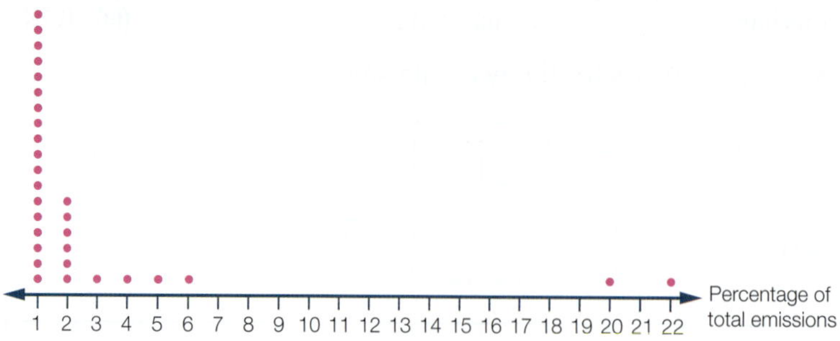

(a) Calculate the **(i)** mean and **(ii)** median for the data.

(b) Calculate the **(i)** range and **(ii)** interquartile range for the data.

(c) Which pair of values better describe this data set: mean and range, or median and interquartile range? Give a reason to support your answer.

9 The following parallel dot plot shows the heights of players at two netball clubs who play in the same competition. The heights have been rounded to the nearest centimetre.

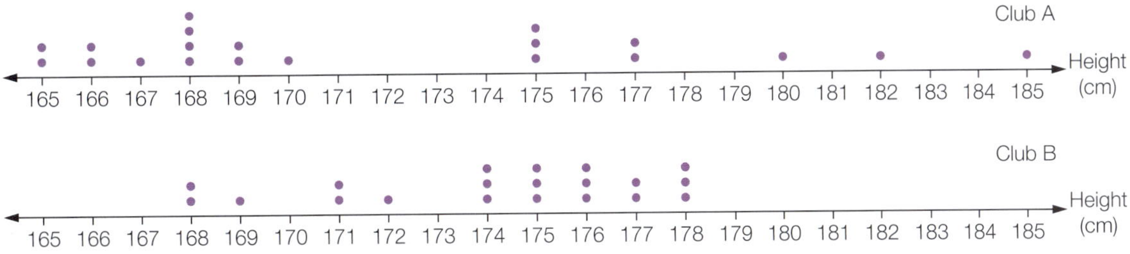

(a) Which club appears to have the taller players?

(b) Calculate the mean and median of the given data for the two clubs.

(c) Calculate the range and IQR for each of the two clubs.

(d) Draw a parallel box plot for the data.

(e) Do you still agree with your answer to part **(a)**? Use statistics to justify your answer.

(f) Compare the shape of the parallel dot plot with that of the parallel box plot.

Reasoning

10 The following table gives the height (cm) and weight (kg) of each of the players on the 2016 playing lists of the Brisbane and Essendon AFL teams.

Brisbane

Height:	200	184	182	186	191	187	184	177	195	190	196	190	188	187	198
	191	176	197	184	185	184	192	190	197	187	180	189	194	195	192
	194	183	188	185	185	201	173	196	194	199	174				
Weight:	92	84	83	85	86	98	83	76	93	80	91	85	84	87	95
	85	80	88	73	85	90	93	85	100	84	73	72	89	103	95
	87	84	90	88	85	96	77	99	74	103	78				

Essendon

Height:	185	187	180	202	179	174	187	201	185	190	177	186	189	188	189
	186	184	196	180	193	185	191	190	203	186	182	196	183	179	191
	191	182	190	182											
Weight:	77	77	83	106	83	72	88	90	80	83	66	69	93	96	81
	86	87	97	78	92	78	80	82	100	75	82	99	85	75	88
	92	84	94	79											

(a) Find the mean heights and weights for each club. Give your answer correct to 1 decimal place.

(b) Find the median heights and weights for each club. Give your answer correct to 1 decimal place.

(c) Find the range and IQR for the heights and weights for each club.

(d) Draw a parallel box plot for the heights of players in each club.

(e) Draw a parallel dot plot for the weights of players in each club.

(f) Which team is taller overall? Use the statistics to justify your decision.

(g) Which team is heavier overall? Use the statistics to justify your decision.

11 The following table shows the percentages of particular age ranges (in years) in the total Australian population and in the Aboriginal and Torres Strait Islander (ATSI) population, rounded to the nearest 0.1%.

	0–4	5–9	10–14	15–19	20–24	25–29	30–34	35–39	40–44	45–49	50–54	55–59	60–64	65–69	70–74	75–79	80–85	85+
Australia	6.7	6.3	6.5	6.9	7.6	7.6	6.9	7.2	6.9	7.0	6.5	5.9	5.4	4.0	3.1	2.3	1.7	1.2
ATSI	12.3	11.2	11.3	11.2	9.5	7.7	6.3	6.2	5.9	5.1	4.2	3.3	2.3	1.5	1.0	0.6	0.3	0.2

(a) Draw cumulative frequency curves for the two sets of data.

(b) Find the five-number summary for each set (use 90 as the maximum value).

(c) Draw a parallel box plot showing the two populations.

(d) Write a couple of sentences comparing the two populations.

Open-ended

12 A school collected information about the distances (km) that 160 students travel to school. The results were published only in box plot form, as shown below.

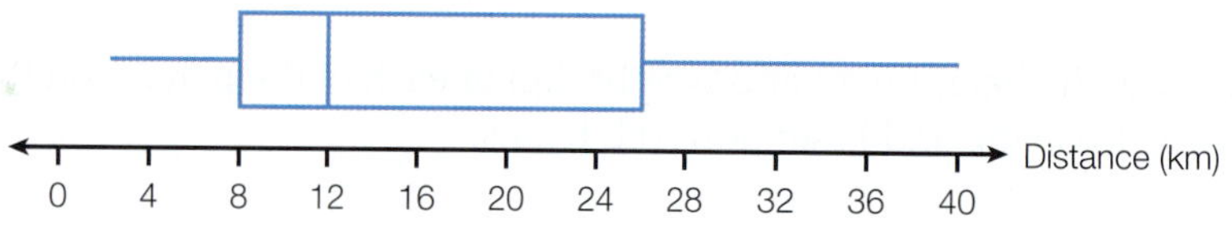

(a) Write a sentence or two describing what information the box plot gives you about the travel distances for the students at this school.

(b) What would you estimate to be the mean distance travelled by the students?

(c) **(i)** Draw a histogram that might represent the data collected. Use a class size of 4 with a first class interval of 0–<4, and so on.

(ii) Show that your histogram supports the values shown in the box plot.

(iii) Use your histogram to estimate the mean distance. How does this value compare to your estimate in **(b)**?

Problem solving

What do you mean?

1 It was reported that most Australians have more than the average number of legs. How can this be true?

2 Each week, for 4 weeks in a row, a market research company interviewed people asking questions about the performance of the Prime Minister. The results are as follows:

Week 1:	47 approved	42 disapproved	11 were unsure
Week 2:	68 approved	57 disapproved	7 were unsure
Week 3:	112 approved	106 disapproved	8 were unsure
Week 4:	77 approved	68 disapproved	7 were unsure

What would you say is the percentage approval for the Prime Minister? Explain why you think your answer is justified.

Strategy options

- Act it out.
- Make a table.

Exploration CAS

Equipment required: TI-Nspire CAS or Casio ClassPad CAS

Exploring grouped data

A data set is *discrete* if its values are distinct and separate (for example, values that are counted such as numbers rolled on a die: 1, 2, 3, 4, 5, 6).

A data set is *continuous* if its values can take on any value within a given interval (for example, values that are measured such as race times: 1.38 s, 1.41 s, 2.53 s, 4.10 s).

In this task, you will produce a frequency table for continuous data. The frequencies will be generated randomly by your CAS so that you will have a unique data set.

This task involves continuous data, so use the class centre to represent each interval of data values. Use the class centres 25, 35, 45, … 115 for the intervals 20–30, 30–40, … 10–120.

Using TI-Nspire CAS

Add a new **Lists & Spreadsheet** to your document. Enter the name of column **A** as 'centre' and the name of column **B** as 'freq'. In column **A** enter the numbers 25, 35, 45, … 115. In cell **B1** enter the formula =randint(2,12), which will find a random whole number value between 2 and 12. Then select cell **B1**, select **menu** > **Data** > **Fill** and use the down arrow to fill the same formula down the column to cell **B10**.

(Your screen should look similar to the image below, but the random values in column **B** will probably be different.)

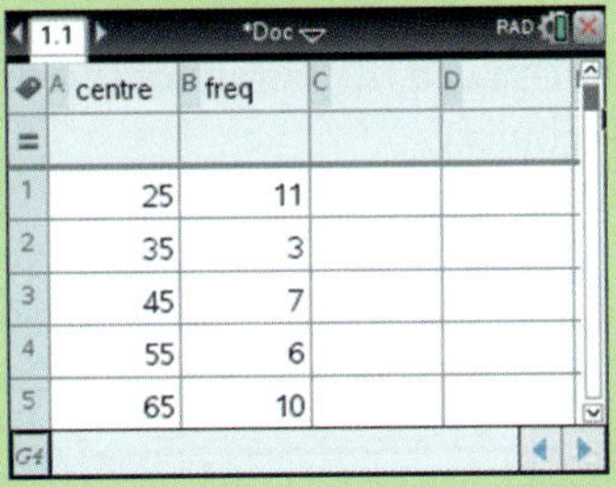

To draw a scatter plot of this data, add **Data & Statistics**. Use the cursor to select the bottom of the screen, where the text shows **Click to add variable**, and select the variable **centre**. Then select the left of the screen similarly, to select the variable **freq**. This will then display a scatter plot of the data, similar to the one shown below.

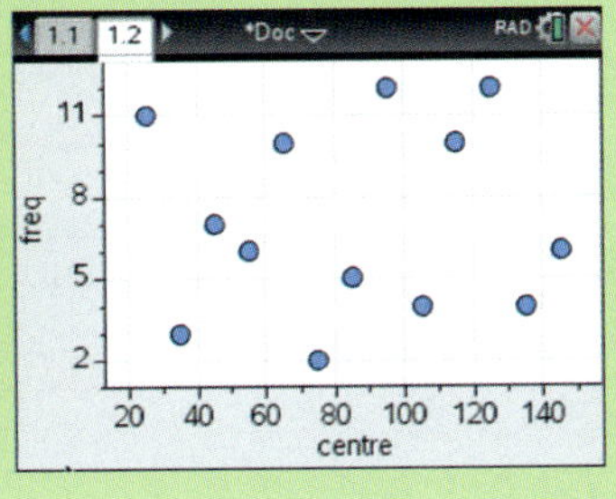

Using Casio ClassPad CAS

From the menu select **Statistics**. Into **list1**, enter the class centre values 25, 35, 45, … 115. In the **Cal** box at the bottom of **list2**, enter the formula **randList(10,2,12)** which will find 10 random whole number values between 2 and 12. (You can find **randList** in the **Catalog**.)

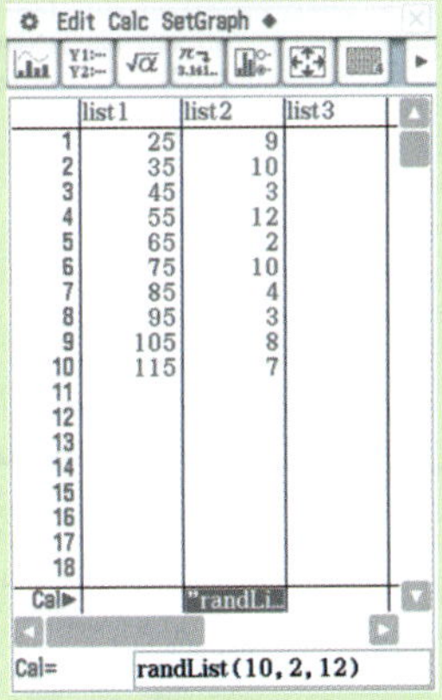

(Your screen should look similar to the image shown, but the random values in **list2** will probably be different.)

To draw a scatter plot of this data, select **SetGraph** > **Setting…** and set the **Scatter** plot settings as shown.

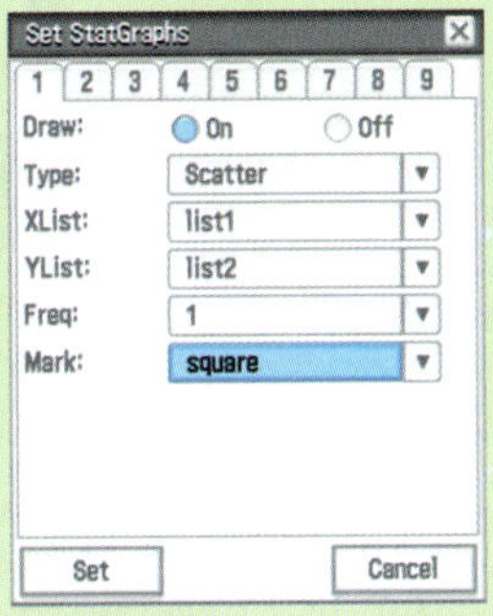

Then select [icon] and the scatter plot will appear, similar to the one shown.

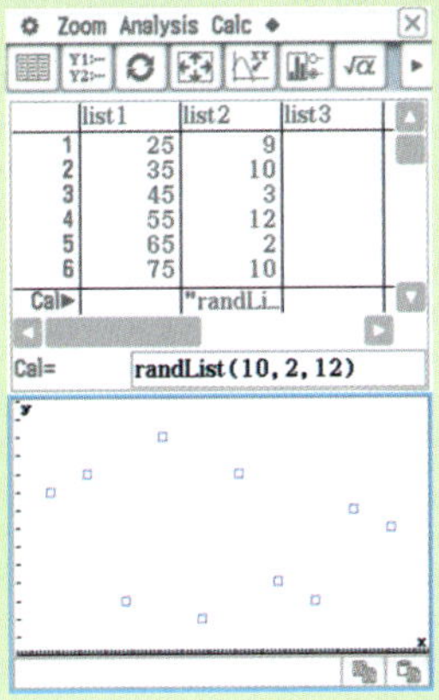

1 From your graph:

(a) Is it possible to give an estimate for the median? What is your estimate?

(b) What do you think is the range of the data?

Now construct a frequency histogram for your data.

Using TI-Nspire CAS

Add **Data & Statistics** to your document. Select **ctrl** > **menu** > **Add X Variable with Summary List** and set the **X List** as **centre** and the **Summary List** as **freq** as shown. Add XVariable with Summary List. Enter centre in X List and freq in Summary List as shown below.

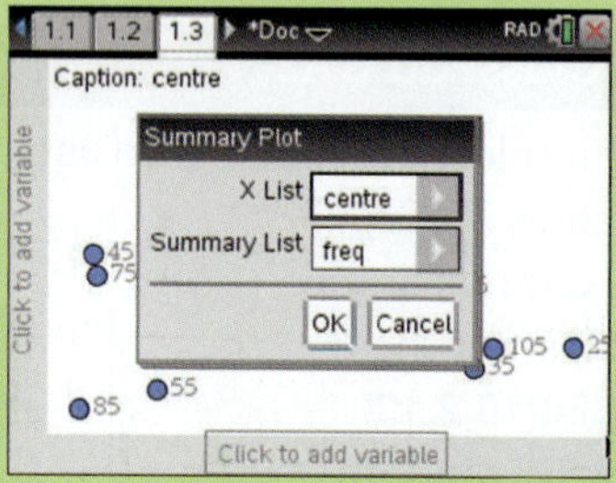

This will then display a histogram of the data, similar to the one shown below. You may also need to select **Bin Settings** > **Equal Bin Width** to set the **Width** to 10 and the **Alignment** to 25.

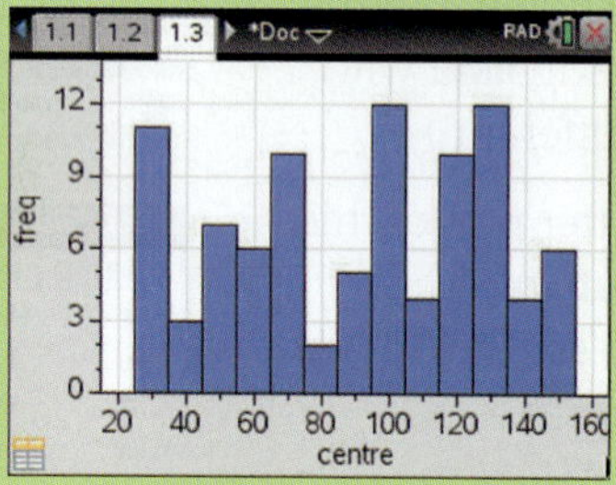

Using Casio ClassPad CAS

In the data section of the screen (the top half), select **SetGraph** > **Setting...** and then select tab **2**. Set the **Histogram** plot settings as shown.

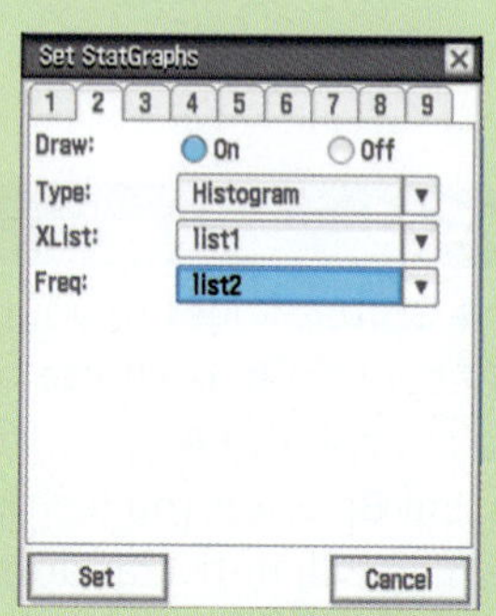

Then select the icon and set the histogram start point **HStart** at 20 and the step value **HStep** at 10 as shown.

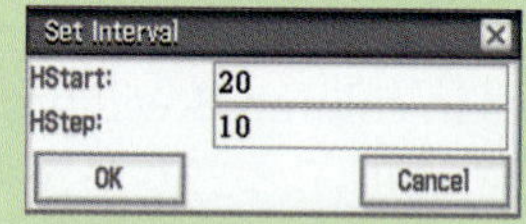

This will then display a histogram of the data, similar to the one shown below.

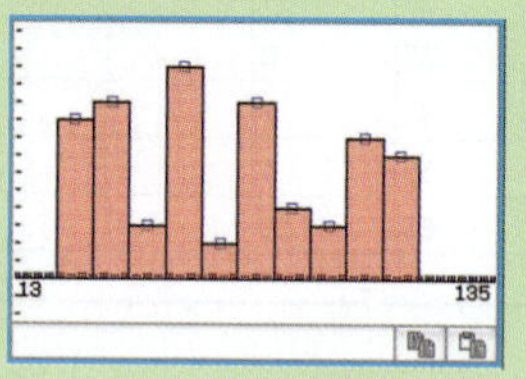

2 From your graph:

(a) Is it possible to give an estimate for the median? What is your estimate?

(b) What do you think is the range of the data?

(c) Did any of your answers change? If so, what might be a reason for any differences?

Summary statistics

Find the summary statistics for your data set.

Using TI-Nspire CAS

Return to your original lists of data. Select **menu** > **Statistics** > **Stat Calculations** > **One-Variable Statistics...** and confirm that **Num of Lists** is 1. Set the **One-Variable Statistics** settings as shown.

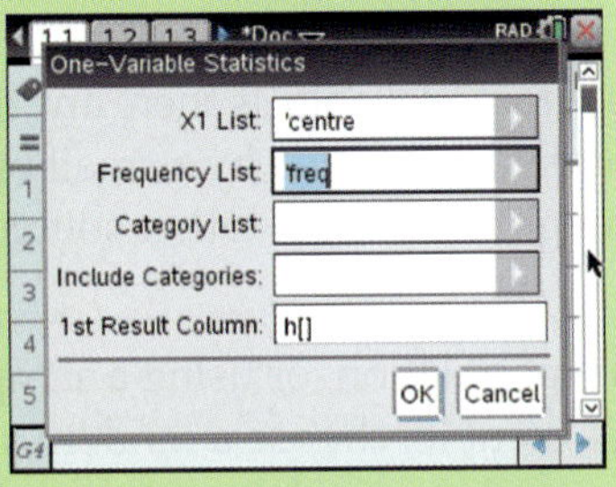

The statistics data should appear. The key statistics values may be labelled as follows:

$\bar{x}$	mean
n	number of data values
MinX	minimum value
Q_1X	lower quartile value
MedianX	median
Q_3X	upper quartile value
MaxX	maximum value

Using Casio ClassPad CAS

Select **Calc** > **One-Variable** and set the **XList1** as **list1** and the **Freq** as **list2**, as shown.

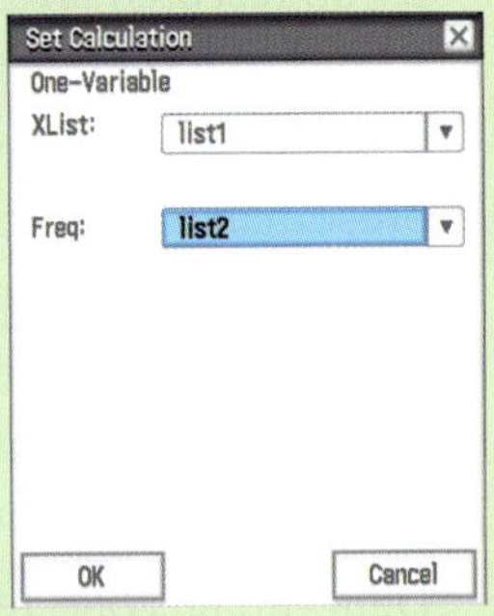

The statistics data should appear. The key statistics values may be labelled as follows:

$\bar{x}$	mean
n	number of data values
minX	minimum value
Q_1	lower quartile value
Med	median
Q_3	upper quartile value
maxX	maximum value

3 Record the key statistics values for your data set.

4 For **1** and **2** above you estimated the median based on graphical displays of the data. How accurate were your estimates compared to the calculated statistics? Do you think you would be more accurate with this sort of task in the future?

Box plot

You have collected the key values necessary to draw a box plot. You could draw it by hand or use your CAS.

Using TI-Nspire CAS

Return to your histogram. Select **menu** > **Plot Type** > **Box Plot**. This will then display a box plot of the data, similar to the one shown.

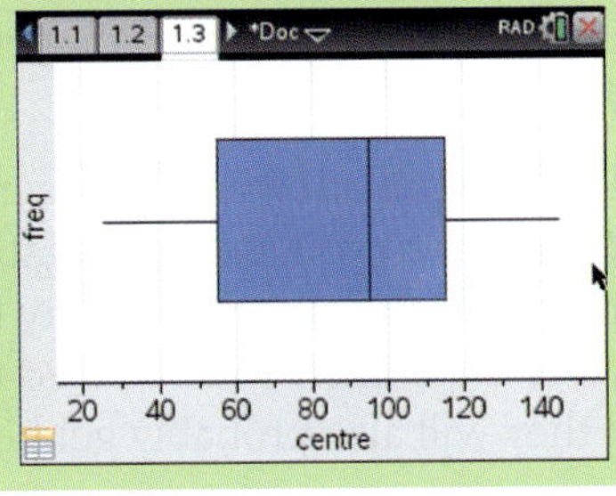

Using Casio ClassPad CAS

Select **SetGraph** > **Setting...** and set the box plot settings as shown. Also make sure that under the **SetGraph** menu, only **StatGraph1** is selected.

Then select [icon]. This will display a box plot of the data.

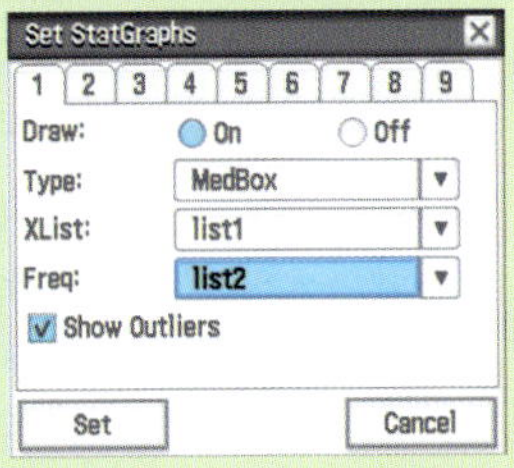

5 Were there any outliers with your box plot? How do outliers appear on your CAS?

2.4 Scatter plots and data investigations

A large-scale data investigation can be quite a challenge. You first need to decide whether to use **primary data** (data you collect yourself) or **secondary data** (data already collected by somebody else). This decision will depend on the type of investigation.

If you research a current issue such as the public transport needs in your suburb, then you would collect your own data.

If you research weather patterns over the last 10 years, then you would use secondary data, presumably from the Bureau of Meteorology.

When using secondary data, establish whether the source is reliable. Is there any likelihood of bias in the data? For example, you might expect that data published by a small business association on the need for small business tax cuts might be biased. Other sources of data would be needed to ensure the data was reliable.

Data can be collected using a census, from the whole population, or using a sample, from a small proportion of the population. When selecting a sample, choose a random sample to avoid a biased result. To select a random sample, use random numbers generated by technology, pick numbers out of a hat or roll dice.

Worked example 11

W.E. 11

Choose whether the following examples represent primary data or secondary data.

(a) You are doing a project on water storages, so you collect data from the newspaper on the levels of the dams in your state over the past 12 months.

(b) You survey youth in your area to find out what facilities they would like the council to provide.

Thinking	Working
(a) Has this material been collected by someone else? If yes, it is secondary data.	(a) This is secondary data.
(b) Has this material been collected by someone else? If no, it is primary data.	(b) This is primary data.

Identifying relationships

In this section you will deal with data that involves two variables, which is called **bivariate data**. When interpreting relationships that involve bivariate data, be careful not to jump to conclusions. Just because there appears to be a strong connection between variables does not mean that one was caused by the other.

For example, there might appear to be a strong relationship between the variables *head circumference* and *ability to recall times tables* when you are investigating primary school students. It would be dangerous to conclude that a student's ability to recall times tables is caused by their head circumference. It is more likely to be related to age—as a student gets older their head grows bigger and they will also probably get better at their times tables.

Scatter plots

A **scatter plot** can be used to display bivariate data and show any connections that might exist between the variables. A scatter plot uses the data to create a series of points drawn on a pair of axes, with each axis representing the values of one of the two variables.

Worked example 12

W.E. 12

The following table represents the heights (cm) and weights (kg) of 10 students.

Height (cm)	156	167	157	155	161	173	163	162	161	172
Weight (kg)	45	55	61	49	67	75	71	75	72	77

Draw a scatter plot of the data with height on the horizontal axis.

Thinking	Working
1 Draw a set of axes, labelled appropriately, with scales stretching far enough to fit all the data. (If you leave out some of the axis values, indicate this on the axis.)	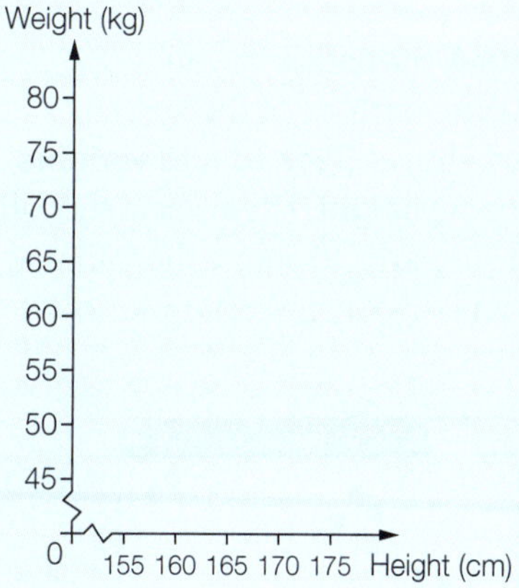
2 Mark each of the points on the graph.	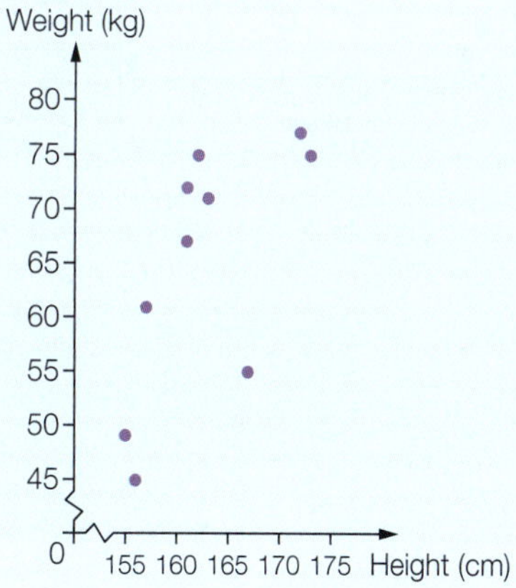

A scatter plot can be used to show a relationship between two variables. You can see what sort of trend there is in the positions of the points.

In the previous example, the scatter plot shows that taller students tend to weigh more. So, there appears to be a relationship between the two variables height and weight for the students who were measured. In this case, a tentative conclusion might be that as height increases, so does weight. However, you would need to examine more data before making a stronger statement.

It is important to note that bivariate data can include different points that have the same value on one or both of the axes. In graphical terms this means you might plot points in a vertical or a horizontal line.

Technology makes it much easier to create a scatter plot and to see any apparent trends. Using the same data as above, you can use your CAS to produce the scatter plot.

Using TI-Nspire CAS

Add **Lists & Spreadsheet** to your document. Enter the data from the previous example, with column **A** as 'height' and column **B** as 'weight', as shown.

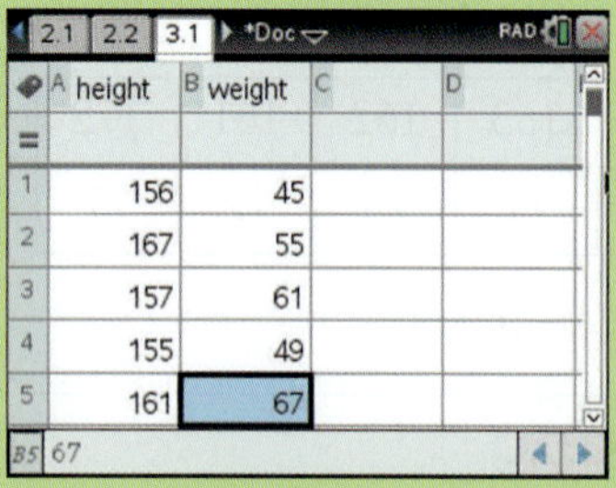

Now add **Data & Statistics** to your document. At the bottom of the screen, where the text shows **Click to add variable**, select the variable **height**. Then select the left of the screen similarly, to select the variable **weight**. This will display a scatter plot of the data.

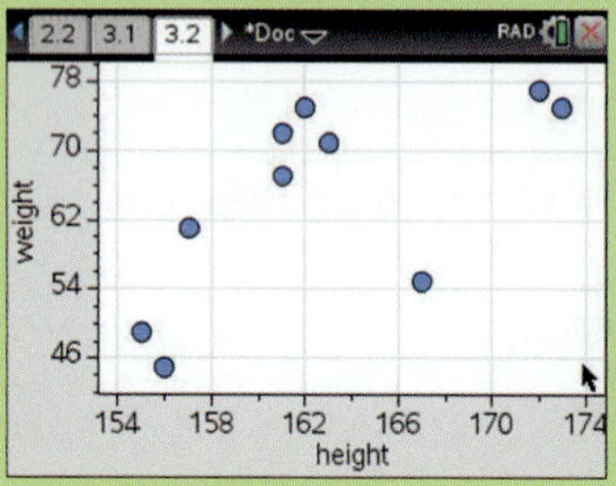

Using Casio ClassPad CAS

From the menu select **Statistics**. Enter the data from the previous example, with the height data in **list1** and the weight data in **list2**.

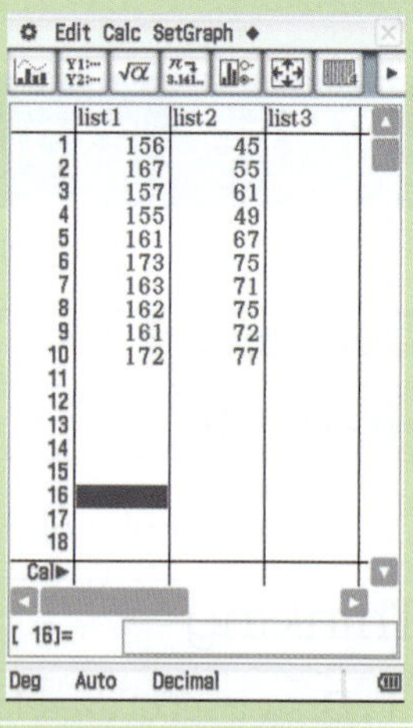

Now select **SetGraph** > **Setting...** and set the graph settings as shown.

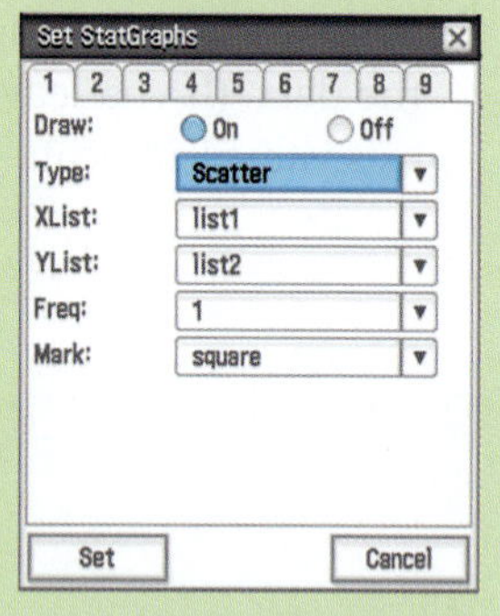

Select the draw graph icon and the scatter plot will be displayed.

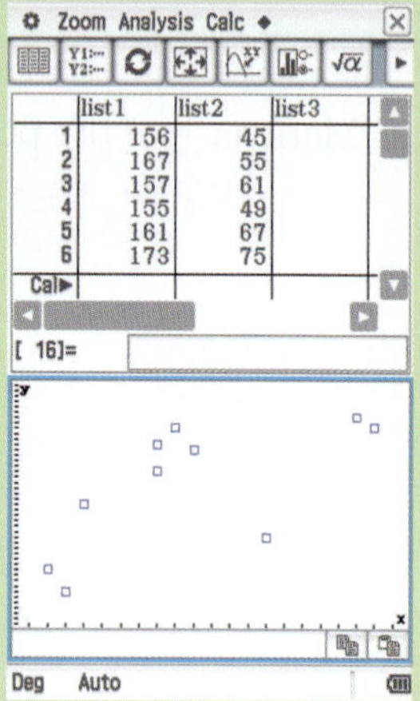

Describing relationships

A scatter plot makes it easier to see any relationships that might exist between the variables you are examining. In some cases, the points on the scatter plot will look as though they are close to a straight line. If one variable increases as the other decreases, then the linear relationship is negative. If both variables increase or decrease together, then the linear relationship is positive.

The following diagrams give some examples of possible linear relationships with different strengths.

Perfect negative linear relationship

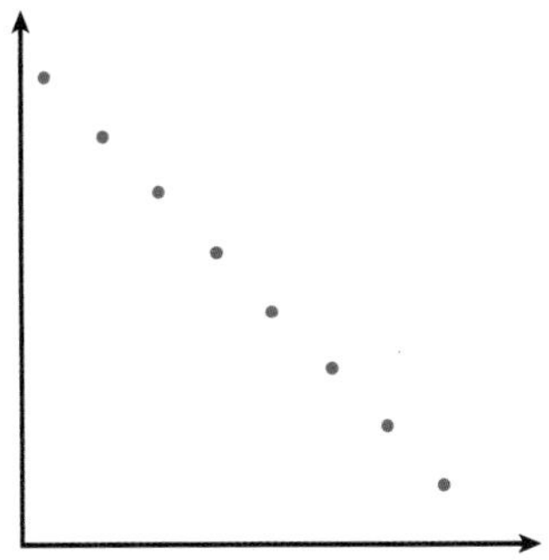

Description: As one variable increases the other decreases. There is a high degree of order in the placement of the points.

Example: Length and width of a rectangle for a fixed perimeter

Strong negative linear relationship

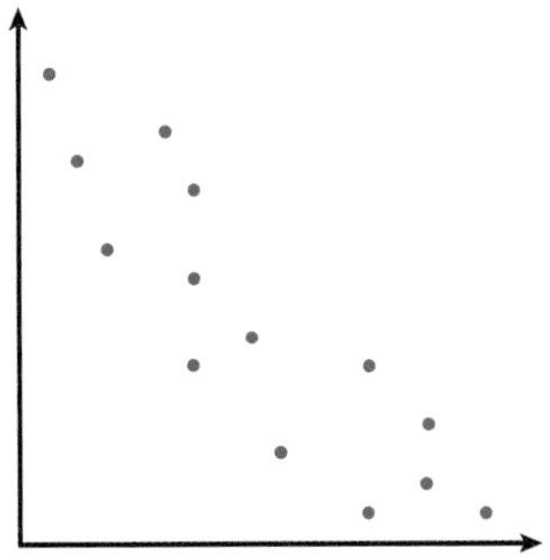

Description: As one variable increases the other generally decreases. There may be a few points that don't fit this pattern exactly.

Example: Age of a second-hand car and its value (excluding vintage cars)

Weak negative linear relationship

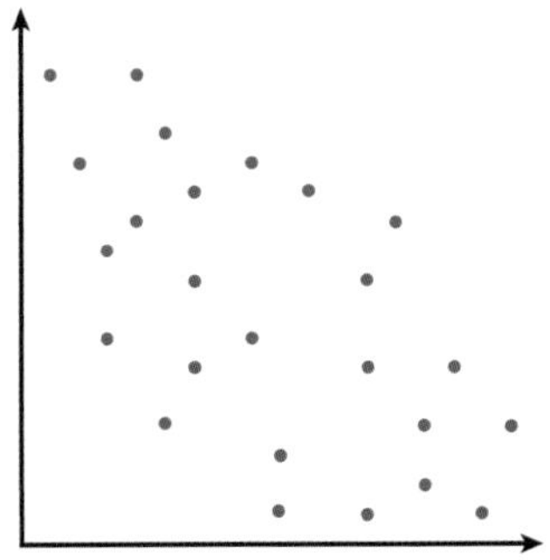

Description: There is some trend evident of one variable increasing as the other decreases, although there would certainly be a number of points that don't fit this pattern.

Example: Average number of hours spent keeping fit each week and the number of annual visits to a doctor

Weak positive linear relationship

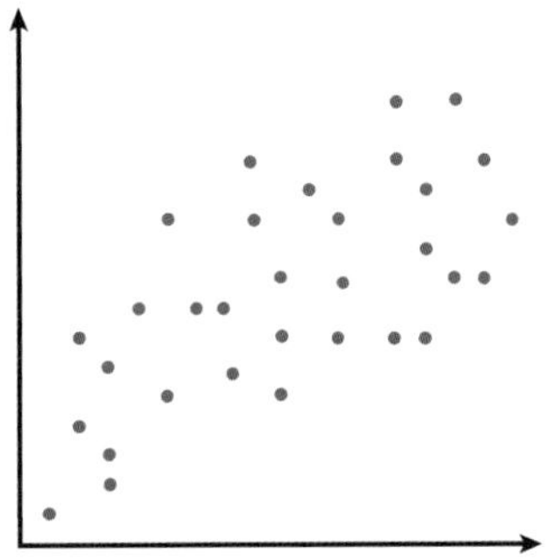

Description: There is some trend evident of either both variables increasing or both variables decreasing, although there would certainly be a number of points that don't fit this pattern.

Example: Money spent on a holiday and money spent buying a car

Strong positive linear relationship

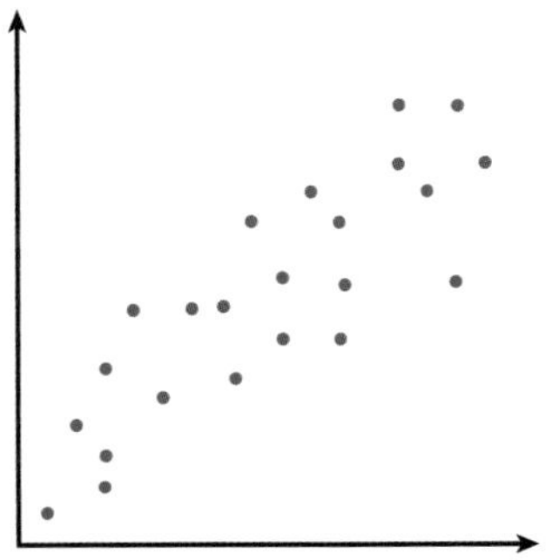

Description: As one variable increases the other generally increases. There may be a few points that don't fit this pattern exactly.

Example: Smoking and lung cancer

Perfect positive linear relationship

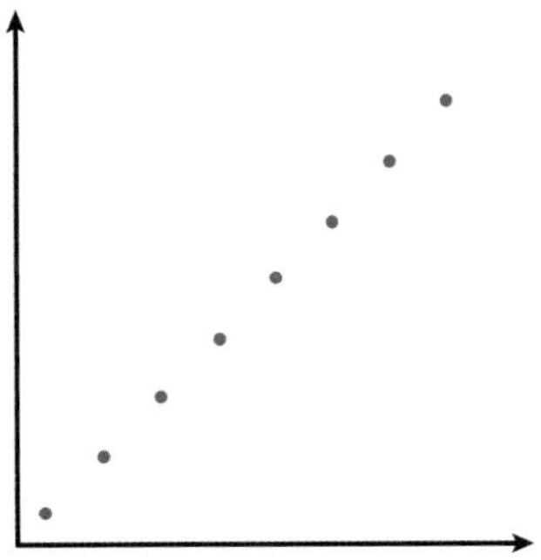

Description: As one variable increases so does the other. There is a high degree of order in the placement of the points.

Example: Volume and height of water in a cylindrical tank

No relationship

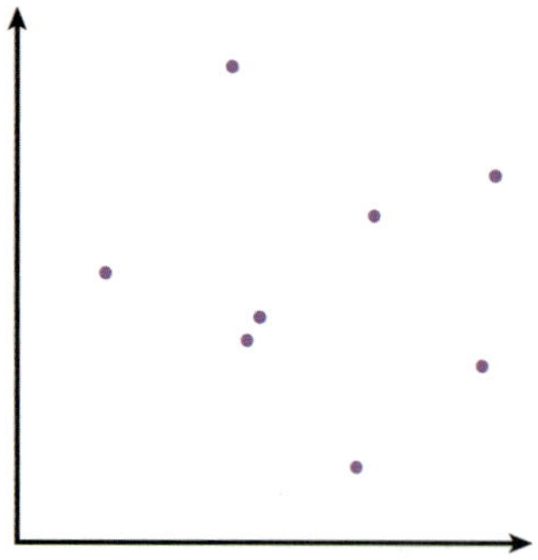

Description: There is no linear pattern evident in the placement of the points.

Example: Hours in a week spent cycling and playing the piano

Non-linear trend

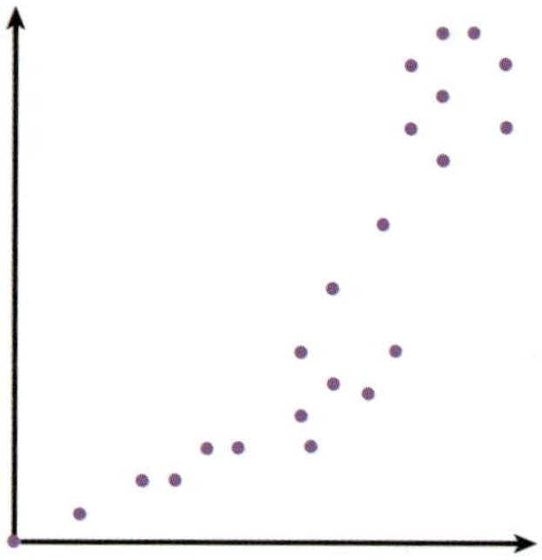

An apparent trend will not always be linear. In this scatter plot, a curve might be a more appropriate choice.

Scatter plots show whether the trend is linear or non-linear. If the trend is linear, then decide whether the strength of the relationship between the variables is strong, moderate or weak.

2.4 Scatter plots and data investigations

Navigator

Answers p.777

1, 2, 3, 4, 5, 6, 7, 9, 12, 13	1, 3, 4, 5, 6, 7, 8, 9, 11, 12, 13	1, 2, 3, 4, 5, 7, 8, 9, 10, 11, 12, 13

Fluency

W.E. 11

1 Choose whether the following examples represent primary data or secondary data.

(a) You sit on a street corner and record the colour of the cars passing by.

(b) You send a text message to 20 friends asking them whether or not they are watching a particular show on television.

(c) You write the scores for all the weekend's football matches from Monday's newspaper.

(d) Your teacher asks the class how many brothers and sisters they have and records the information on the board.

(e) You go to your favourite football team's website and write the height of each player.

W.E. 12

2 The following table represents the heights (cm) and weights (kg) of 10 football players. Draw a scatter plot of the data. Put height on the horizontal axis. Make a comment about the type of relationship, if any, that is displayed by the data.

Height (cm)	203	186	184	189	186	197	188	181	182	185
Weight (kg)	103	84	84	86	89	97	87	82	80	86

3 Are the following examples of a census or a sample?

(a) You ask every student in Year 10 their favourite movie.

(b) You ask the first thirty people who leave the canteen what they purchased.

(c) You send a questionnaire to every family in your school, but only 35% of the forms are returned.

(d) You ask the members of your cricket club how long they have been playing cricket.

4 You are conducting a survey on the topic 'The best team sport played in Australia' and are considering using the following questions. Decide whether each question is appropriate or inappropriate, and give a reason.

(a) How often do you participate in team sport?

(b) How often do you watch team sport on television?

(c) Which of the following team sports do you think is the best? Australian Rules Football, Rugby League, Rugby Union, Cricket, Basketball

(d) Do you think too much government money is spent supporting Olympic athletes?

5 Decide whether or not the following sampling methods would be likely to return biased results. Give a reason for each answer.

(a) You stand in a shopping centre and ask people who smile at you.

(b) You go to the local petrol station and ask people who are filling their car with diesel their opinions about environmental matters.

(c) You put a survey form in the letterbox of one hundred houses in your neighbourhood asking about support for the renovation of the local hall. You include a freepost envelope for the return of the survey.

(d) You find numbers in a telephone directory and call two hundred numbers to find out voting intentions at the next State election.

6 The data below shows the marks for 10 students taking physics and mathematics tests.

Student	A	B	C	D	E	F	G	H	I	J
Physics mark (out of 50)	42	38	44	41	46	31	25	30	33	36
Mathematics mark (out of 70)	66	60	65	59	65	50	38	58	60	60

(a) Draw a scatter plot of the data with the physics mark on the horizontal axis.

(b) Comment on any relationship evident between the marks.

7 For each of the following scatterplots state whether the trend is linear or non-linear. If linear, also state whether the trend is positive or negative and give an indication of strength.

(a)
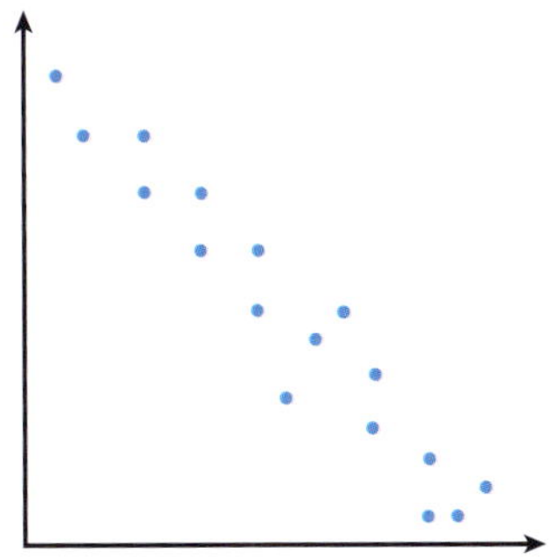

(b)

(c)
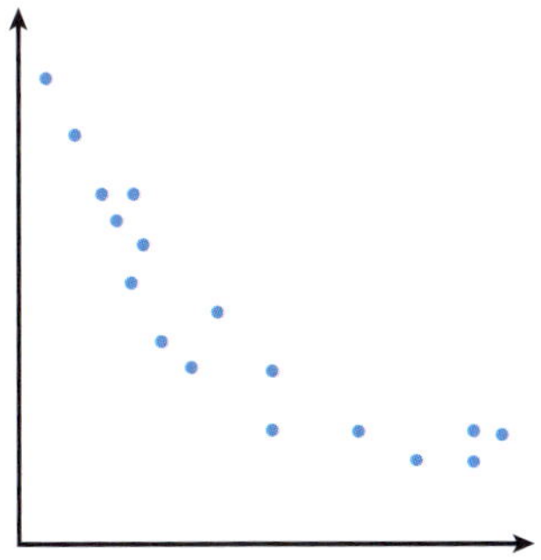

(d)
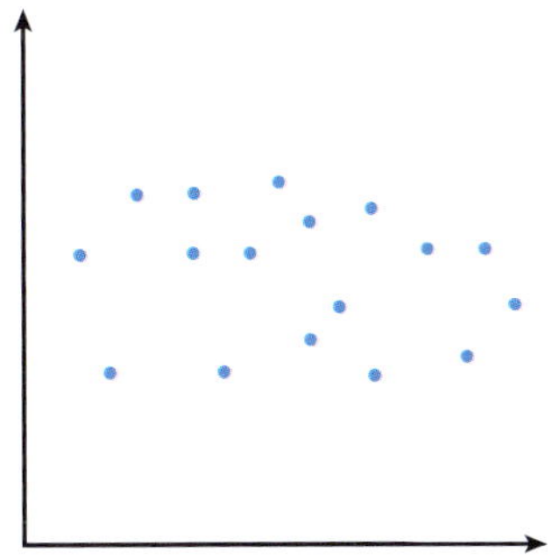

(e)
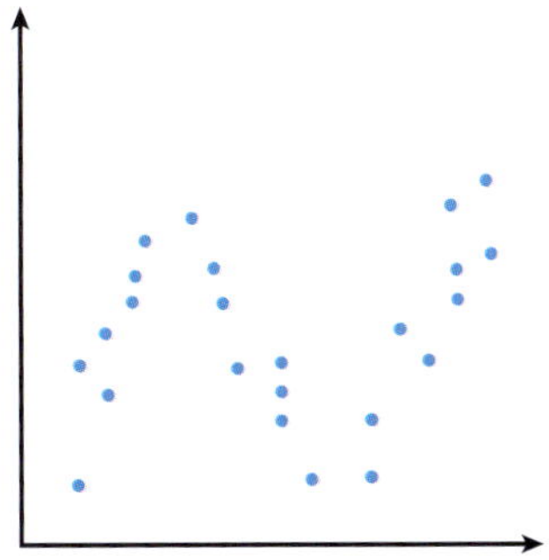

(f)
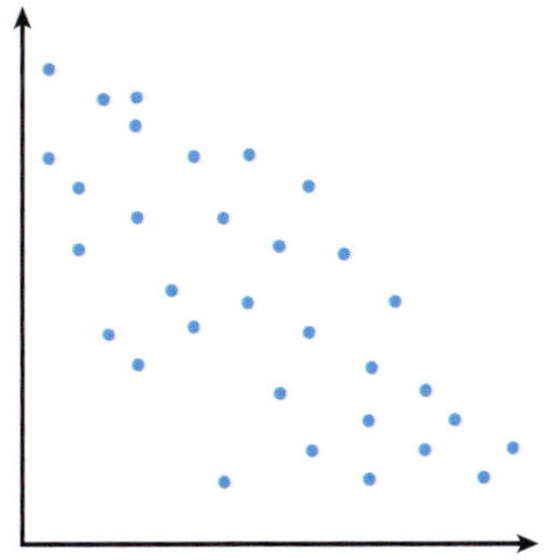

Understanding

8 You are a member of the Student Representative Council (SRC) at a school which has 1230 students. A suggestion has been made that a good quality coffee machine should be installed in the school canteen. You must conduct an investigation and report back to the SRC within 4 weeks.

(a) You decide that students and teachers are the stakeholders and parents need not be involved in the decision. How will this decision make your job easier?

(b) Why should you conduct a sample of the students?

(c) Why should you conduct a sample of the teachers even though there are only 110 of them?

(d) Your friend, who is very good at mathematics, suggests you should conduct a stratified random sample. She says that this means that the number of students sampled would be allocated in proportion to some criteria such as year level or gender or both and the proportion of teachers to students sampled would be in proportion to the number of teachers and students in the school.

Explain why this is a good suggestion.

(e) Make up at least three of the questions that would form part of your survey instrument. Test them on a few of your friends to see if there are any flaws in them.

9 The following table shows the width W and length L, both measured in centimetres, of 88 bivalves of the same species collected on a beach.

W (cm)	1.3	1.4	1.4	1.5	1.9	2.1	2.1	2.1	2.2	2.2	2.3	2.4	2.4	2.5
L (cm)	1.7	1.9	1.8	1.9	2.3	3.5	2.8	2.7	3.0	2.7	3.0	3.2	2.9	3.5

W (cm)	2.5	2.6	2.6	2.6	2.7	2.8	2.8	2.9	2.9	2.9	2.9	3.0	3.0	3.0
L (cm)	3.3	3.5	3.4	3.4	3.2	3.8	3.5	3.8	3.9	3.8	3.7	3.8	4.0	3.7

W (cm)	3.1	3.4	3.4	3.5	3.6	3.6	3.6	3.6	3.7	3.9	3.9	4.1	4.1	4.1
L (cm)	4.1	4.4	4.3	4.5	4.6	4.8	4.9	4.5	4.9	5.4	5.2	5.3	5.4	5.2

W (cm)	4.2	4.2	4.2	4.2	4.3	4.7	4.8	4.8	4.8	4.9	5.0	5.0	5.1	5.2
L (cm)	5.4	5.5	5.6	5.2	5.6	6.1	6.6	6.1	6.0	6.1	6.6	6.2	6.6	6.4

W (cm)	5.2	5.4	5.6	5.6	5.8	5.8	5.8	5.9	6.0	6.0	6.0	6.2	6.2	6.3
L (cm)	6.8	6.3	7.0	7.9	7.4	7.4	6.9	7.9	7.7	7.8	7.3	8.5	7.5	7.6

W (cm)	6.5	6.6	6.6	6.7	6.7	6.7	6.7	6.9	6.9	7.0	7.1	7.1	7.1	7.2
L (cm)	9.3	8.1	8.0	8.2	9.2	8.0	7.9	8.6	8.1	9.5	9.1	9.1	9.4	8.4

W (cm)	7.3	7.3	7.4	7.5
L (cm)	9.3	8.9	8.9	8.6

(a) Use a CAS, spreadsheet or some other technology to draw a scatter plot.

(b) Comment on any relationship between the measurements.

Reasoning

10 Countries with literacy rates regarded as low or medium were researched for a report. The following graph was produced as part of the report.

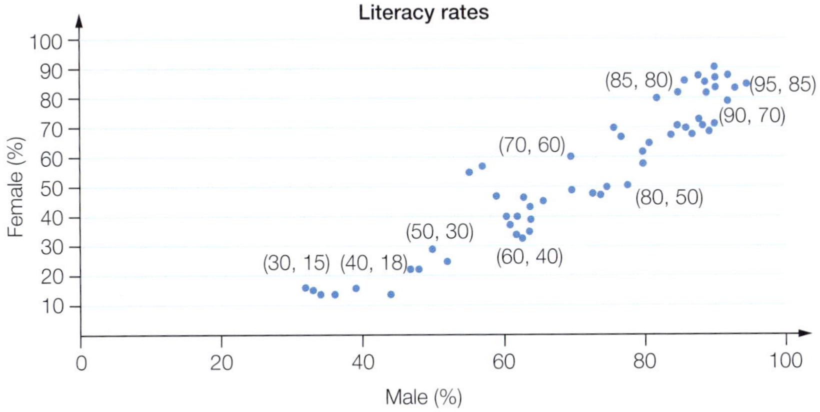

(a) Is the data primary or secondary?

(b) Describe any relationship you can see between male and female literacy rates. As part of this answer state whether you think, in general, male or female literacy rates are higher.

(c) Could your conclusions in **(b)** be applied to countries regarded as having high literacy rates?

11 An internet search produced data regarding the number of radios per 1000 people in the population. From countries with a literacy rate listed as low (litgrp of 1) or medium (litgrp of 2) the graph at right was produced.

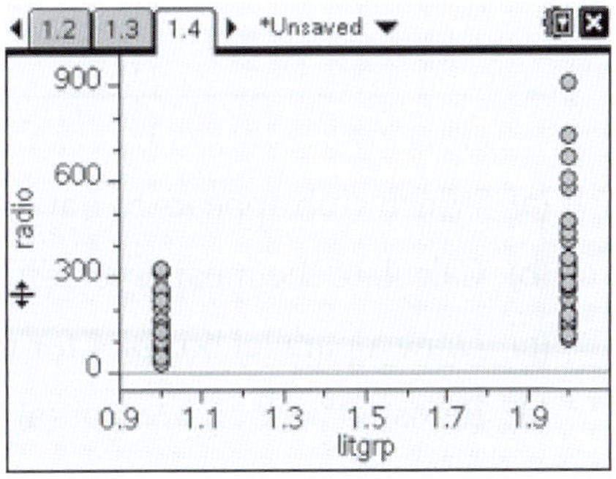

(a) What information can you gather from this graph?

(b) Given the following statistics, construct a parallel box plot for the data.

	Min	Q_L	Median	Q_U	Maximum
Low literacy	34	79	130	226	329
Medium literacy	107	178	306	434	907

(c) Use your parallel box plot to help you make a few statements about the two data sets.

(d) There is one data pair that looks like an outlier. Which pair is this?

Open-ended

12 Suggest a topic for which a census would be appropriate and which you could realistically undertake. Construct the survey instrument that you would use.

13 Suggest a topic for which a sample would be appropriate and which you could realistically undertake. Construct the survey instrument that you would use.

Half-time 2

2.2

1 Complete the following tasks for the data shown in the table.

(a) Find the range.

(b) Draw a cumulative frequency curve and use it to estimate the interquartile range.

(c) Draw a box plot.

Class interval	Frequency
10–<15	9
15–<20	15
20–<25	12
25–<30	17
30–<35	13
35–<40	11

2.3

2 The following parallel dot plot shows the heights of players at two hockey clubs who play in the same competition. The heights have been rounded to the nearest centimetre.

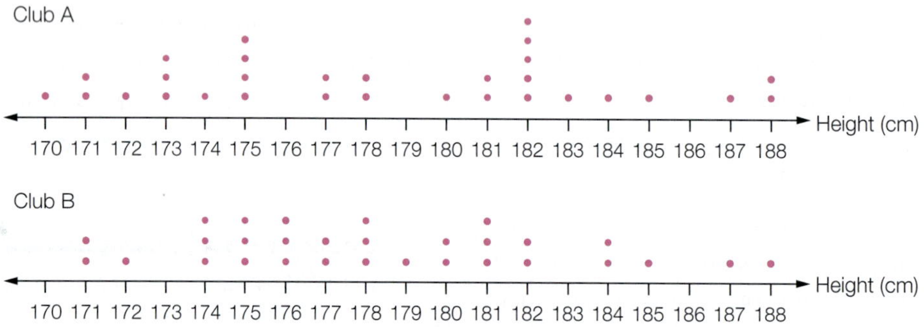

(a) Which club appears to have the taller players?

(b) Calculate the mean, median, range and IQR for each club.

(c) Draw a parallel box plot for the data.

(d) Reassess your answer for **(a)**. Justify your assessment using the calculated statistics.

(e) Which statistic is the most helpful in making your assessment?

2.1

3 Construct a cumulative frequency curve and use it to find an estimate for the median of the following data set.

Class interval	Frequency
0–<20	12
20–<40	19
40–<60	21
60–<80	32
80–<100	19

2.4

4 The following table gives the head circumferences and the lengths, both in cm, of 12 babies at birth. Draw a scatter plot of the data. Show head circumference on the horizontal axis. Make a comment about the type of relationship, if any, that is displayed by the data.

Head circumference (cm)	38	35	39	37	34	32
Length (cm)	51	50	54	53.5	50.5	47.5

Head circumference (cm)	33	35	35	37	39.5	36
Length (cm)	50	53	49	51	51	50.5

Time-related data

2.5

Bivariate data involves two variables. If you can choose the values of one of the variables before any trials begin, that variable will be the **independent variable**. The data values that pair up with the values you choose belong to the **dependent variable**.

In some cases, it is difficult to choose between the two variables being studied, to know or assume which variable might depend on the other. In cases like these you can take the first variable to be the variable for the horizontal axis, which is where the independent variable is shown.

If time is one of the variables, then it is almost always the independent variable. For example, if you were measuring the height of water on a tidal beach or river, you would graph height against time, with time on the horizontal axis.

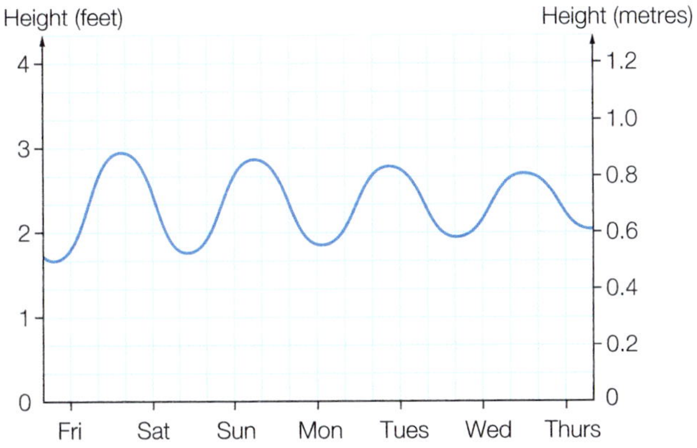

Scatter plots can be used to display time-related data. When you have large data sets it may be appropriate to use technology such as CAS or spreadsheet software to draw the scatter plot. However, for smaller data sets you should draw the scatter plot by hand.

Worked example 13

W.E. 13

The following table shows the temperatures, in °C, for Jerusalem over a 24-hour period from 8 am one morning to 7 am the next morning.

Time	8 am	9 am	10 am	11 am	noon	1 pm	2 pm	3 pm	4 pm	5 pm	6 pm	7 pm
Temperature (°C)	18.4	21	21.5	22.3	23.3	24.7	24.9	25.3	24.2	23.2	22	20.4

Time	8 pm	9 pm	10 pm	11 pm	mid-night	1 am	2 am	3 am	4 am	5 am	6 am	7 am
Temperature (°C)	18.9	18.2	18.2	17.9	17.1	16.5	16.36	15.9	15.6	15.6	15.4	16.5

(a) Draw a scatter plot to show these temperatures.

(b) What is the maximum temperature shown on the graph?

(c) Is this necessarily the maximum daily temperature? If not, what might that maximum temperature have been?

(d) Do the temperatures follow a similar pattern to what you would expect for where you live?

Thinking	Working
(a) Draw the scatter plot in the usual way. To be able to observe the changes it will be necessary to leave out some of the vertical axis.	(a) 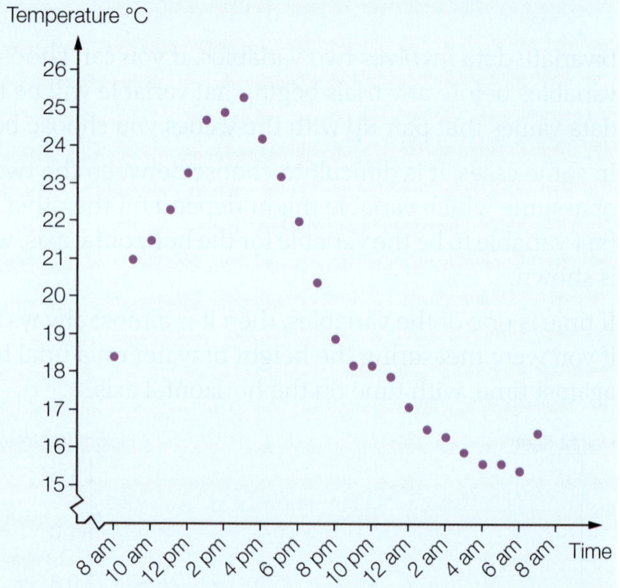
(b) Read the maximum value from the table.	(b) The maximum temperature recorded is 25.3 °C.
(c) Where else might the maximum have occurred?	(c) If this is not the absolute maximum it is likely to have been in the preceding hour, as the temperature falls away more quickly after 3 pm than it was rising before.
(d) Interpret the graph in relation to your own place of residence.	(d) It doesn't really matter where you live, it is hotter during the day than at night.

2.5 Time-related data

Navigator

Answers p. 779

1, 2, 3, 4, 5, 7, 8, 9	1, 2, 3, 4, 5, 6, 8, 9	1, 2, 3, 4, 5, 6, 9

Equipment required: CAS or spreadsheet would be useful for Questions **3–5**

Fluency

W.E. 13

1 The table below gives the wind speeds, in km/h, for Quebec in Canada over a 24-hour period.

Time	mid-night	1 am	2 am	3 am	4 am	5 am	6 am	7 am	8 am	9 am	10 am	11 am
Wind speed (km/h)	9	9	13	7	6	6	4	4	6	4	7	6

Time	noon	1 pm	2 pm	3 pm	4 pm	5 pm	6 pm	7 pm	8 pm	9 pm	10 pm	11 pm
Wind speed (km/h)	6	7	9	7	11	13	11	13	11	9	9	11

(a) Draw a scatter plot to show the wind speed.

(b) What seems unusual about the 2 am recording?

(c) Describe what is happening to the wind speed between 4 pm and 8 pm.

2 For the following pairs of variables, state which one you think is the independent variable. Give a brief reason for your answer.

(a) Test score and number of hours spent studying.

(b) Height, measured in cm, and age, measured in months, for children less than 3 years of age.

(c) Age at death and average number of cigarettes smoked per day.

(d) Amount of rubbish collected, measured in kg, and the number of people in a household.

3 The following data set indicates the number of bicycles that crossed northbound on the Sydney Harbour Bridge for 29 consecutive Fridays.

332	372	367	370	361	398	387	363	447	331	530	454	460	428	283
515	444	498	481	500	396	440	402	383	339	325	402	517	416	

Draw a scatter plot to represent the data. Use your CAS, if you have one, or draw the graph by hand. Comment on the graph produced.

Understanding

4 The following table gives the number of 15-year-old Australians in June of each year from 2000 to 2015.

Year	15-year-olds	Year	15-year-olds
2000	266 234	2008	281 873
2001	268 017	2009	283 307
2002	267 813	2010	286 499
2003	268 642	2011	282 465
2004	271 781	2012	283 757
2005	278 740	2013	282 767
2006	281 043	2014	284 970
2007	281 700	2015	287 772

(a) Use spreadsheet software, or some other suitable form of technology, to draw a scatter plot displaying the data.

(b) Comment on the trend seen in the number of 15-year-old Australians over this time period.

5 The following table gives the number of road deaths for Australian drivers and passengers in the period 2005–2014.

Year	Total	Drivers	Passengers
2005	1627	775	347
2006	1598	757	336
2007	1603	785	336
2008	1437	670	303
2009	1491	707	333
2010	1353	636	284
2011	1277	568	286
2012	1300	610	260
2013	1189	557	204
2014	1156	535	229

(a) Use technology to draw a composite scatter plot to represent the data. Use three different colours to represent the data sets for Total, Drivers and Passengers.

(b) Describe any trends you see in the data.

Reasoning

6 The graph below shows the recorded Australian population from 1788 until 2015.

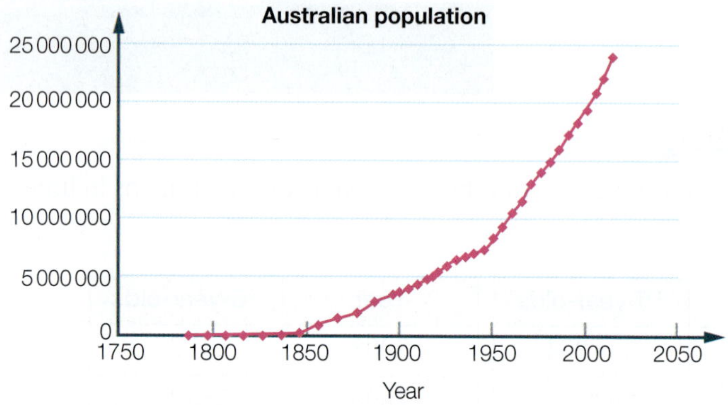

(a) Describe the trend you see in the graph.

The figures prior to 1961 do not include the Aboriginal and Torres Strait Islander populations. Estimates of the size of these populations before European settlement range from 300 000 to 1 000 000.

(b) How do you think the inclusion of the Aboriginal and Torres Strait Islander populations prior to 1961 would affect the graph?

(c) Why does the graph seem to go flat along the horizontal axis for the first 50 years?

(d) How do you think the early figures were collected?

(e) Which figures do you think are more reliable, the early ones or the later ones?

(f) How would you now estimate the Australian population?

7 The graph below shows the number of 90-year-old Australians from 1971 through to 2016. Comment on the trend you see in the graph and try to explain why this is the case. For what other ages would you expect to see this type of trend?

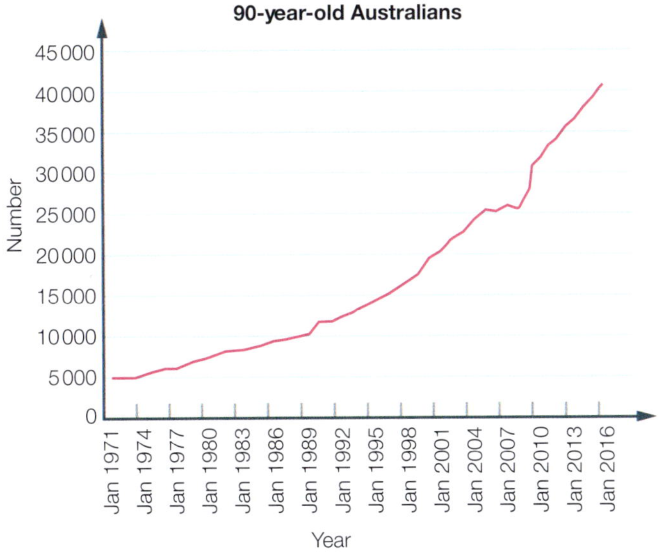

Open-ended

8 Sonika produced the graph below for the information given in the table. Suggest how she could improve her graph.

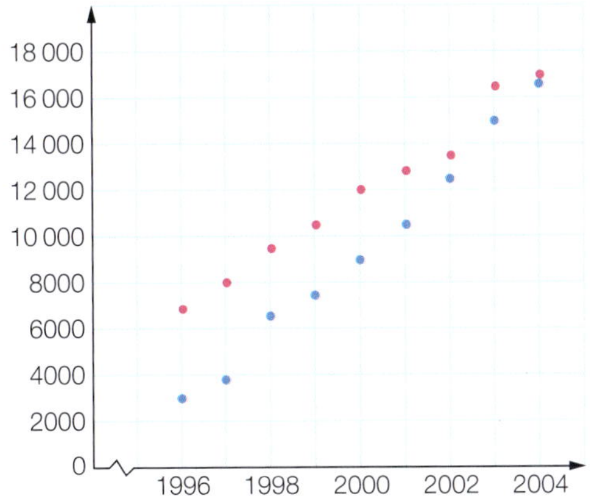

Mercedes–Benz E320 used car prices (November 2015)

Car year	Elegance	Avantgarde
1996	3000	7000
1997	3800	8000
1998	6500	9500
1999	7500	10 500
2000	9000	12 000
2001	10 500	12 800
2002	12 500	13 500
2003	15 000	16 500
2004	16 400	17 000

9 The table below gives some information about extinct and endangered species in Australia. These figures relate to the period after European settlement of Australia.

	Extinct	Critically endangered	Endangered	Vulnerable	Total
Total plant species and subspecies	49	81	523	665	1318
Mammals	27	4	33	55	119
Birds	23	6	41	61	131
Reptiles	0	2	14	37	53
Frogs	4	2	14	12	32
Fish	1	3	16	25	45
Invertebrates	0	16	12	6	34
Total animal species and subspecies	55	33	130	196	414

Of the 27 mammal species that are extinct, 10 of these have become extinct since 1900. Some other information about Australia's biodiversity is as follows:

- 82% of our mammals, 45% of our land birds, 89% of our reptiles and 93% of our frogs live nowhere else
- estimates of probable global losses of species are as high as 25% over the next 30 years
- 90% of native vegetation in the eastern temperate zone has been removed or replaced by introduced crops
- 50% of rainforests have been cleared
- the proportion of Australia covered by woodland or forest has been reduced by more than one-third since European settlement.

Your task is to produce some graphs to indicate how you think Australia's biodiversity has changed since European settlement. You could extend the investigation and research more information.

Lines of best fit

2.6

Scatter plots of bivariate data can be used to make predictions based on the data collected.

Where the relationship between the two numerical variables appears to be linear, it is useful to draw a **line of best fit**, find the equation of this line, and then use this equation to predict new data points.

The most simple (but least accurate) method is to draw the line of best fit approximately by eye.

Not all numerical bivariate data will have a linear relationship, so not all scatter plots will have a line of best fit. If the relationship does not appear to be linear, then you should not try to draw a straight line of best fit.

A line of best fit should follow the direction or trend of the data points and be positioned so that the distances of each point from the line are minimised. One way to achieve this is to try to have the same number of points above and below the line.

The line shown in the first diagram at right does not follow the trend of the data and does not minimise the distance of the points from the line. It is not a good line of best fit for the data.

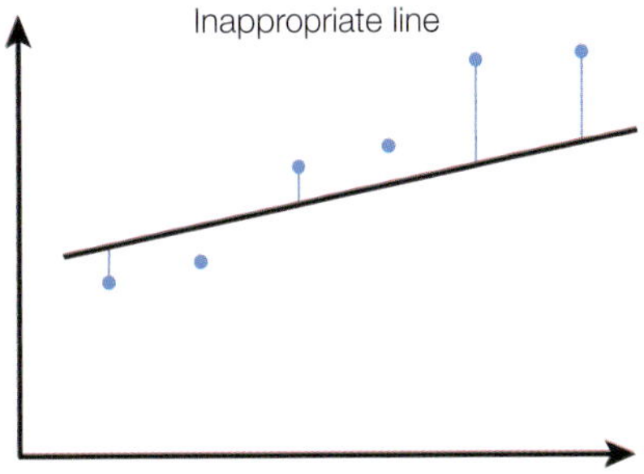

The line drawn for the same data points in the second diagram at right is a much better fit. Can you see why? The line follows the trend of the data, there are the same number of points above and below the line, and the distance between the points and the line is minimised.

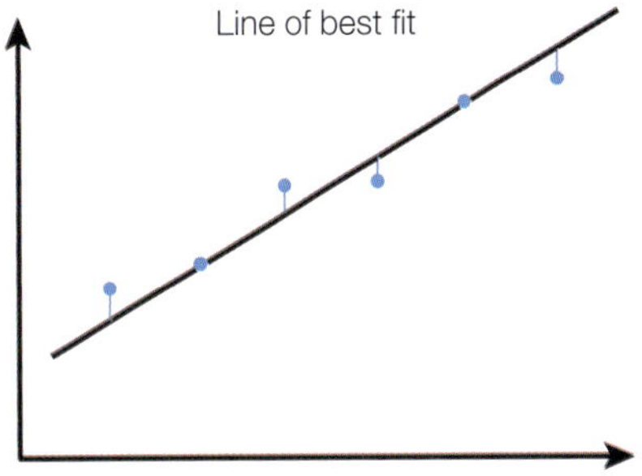

Technology can be used to draw a line of best fit, which mathematically minimises squares of the vertical distances between each point and the line. The values are squared so that positive distances for points above the line do not cancel out negative distances for points below the line. This is why it is called a 'least squares' **regression line**.

Technology can also be used to find the equation of the regression line.

Consider the following table of data that shows the marks for 10 students taking science and mathematics tests.

Student	A	B	C	D	E	F	G	H	I	J
Science mark (out of 30)	20	23	10	27	14	13	12	20	17	19
Mathematics mark (out of 40)	31	36	25	34	29	27	25	30	31	33

You can use technology to draw the scatter plot for this data and then to find the equation of the least squares regression line (line of best fit).

10A

Using TI-Nspire CAS	Using Casio ClassPad CAS
Add **Lists & Spreadsheet** to your document. Label column **A** as 'science' and label column **B** as 'maths'. Enter the data from above. Now go to the Home screen and add **Data & Statistics** to your document. At the bottom of the window, select **science** as the variable for the horizontal axis. At the side of the window, select **maths** as the variable for the vertical axis. The scatter plot will be displayed. 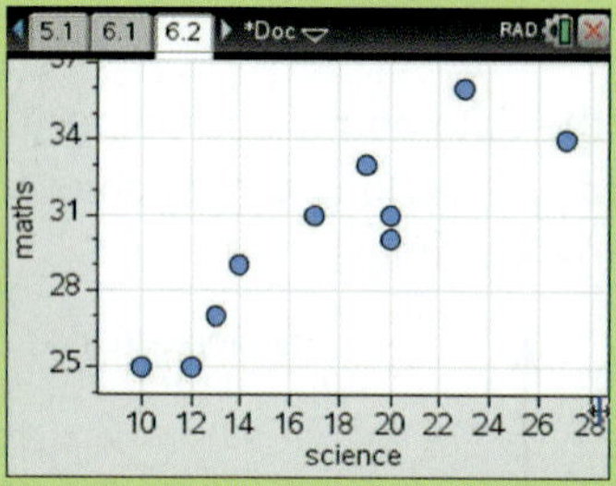	From the menu select **Statistics**. Enter the data from above, with the 'science' data in **list1** and the 'maths' data in **list2**. Select the draw graph icon and the scatter plot will be displayed. Then select **Calc** > **Regression** > **Linear Reg**. Set the **XList** as **list1** and **YList** as **list2**, as shown. 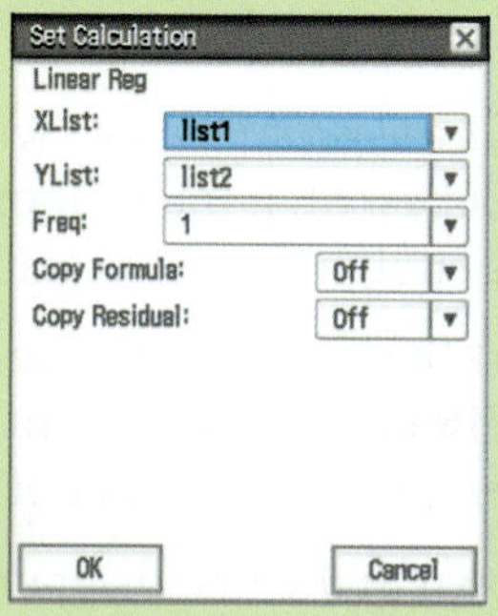
Now return to your **Lists & Spreadsheet**. Select **menu** > **Statistics** > **Stat Calculations** > **Linear Regression (*mx*+*b*)...** and set the **X List** as the **science** data and the **Y List** as the **maths** data. The values for the linear regression are displayed. 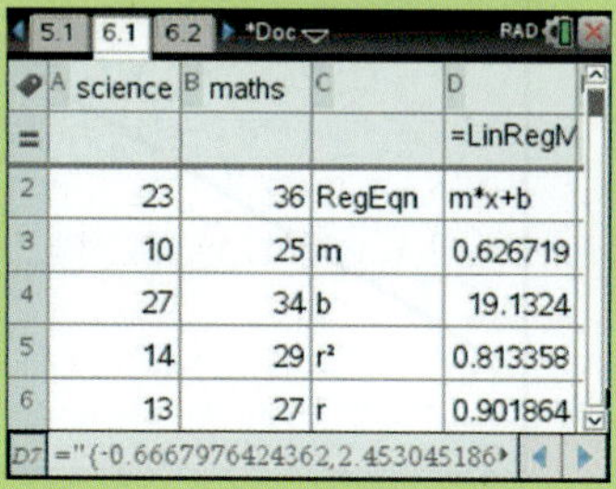 This shows that the linear regression (line of best fit) equation is $y = 0.627x + 19.132$. This means that the mathematics mark = 0.627 × science mark + 19.132 (approximately).	The values for the linear regression are displayed. 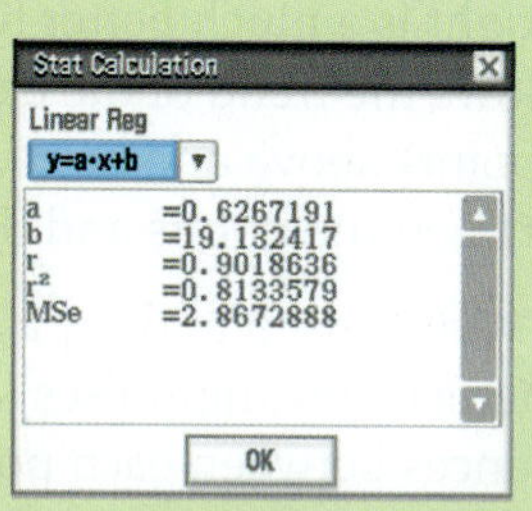 This shows that the linear regression (line of best fit) equation is $y = 0.627x + 19.132$. This means that the mathematics mark = 0.627 × science mark + 19.132 (approximately).

10A

Using TI-Nspire CAS	Using Casio ClassPad CAS
The linear equation can also be displayed on the scatter plot. Return to the scatter plot screen, then select **menu > Analyze > Regression > Show Linear (mx+b)**.	The linear equation will now be displayed on the scatter plot.

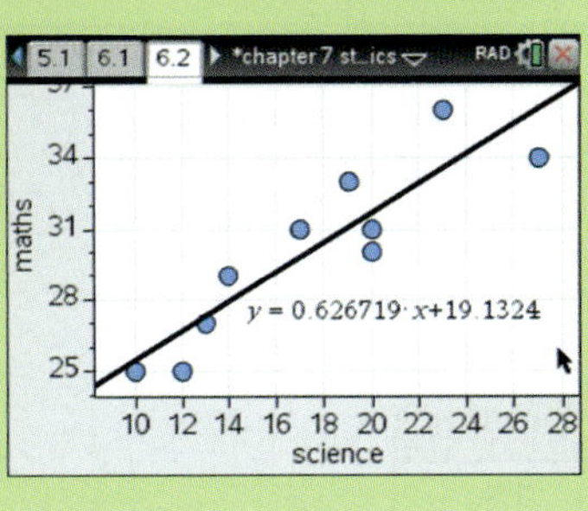

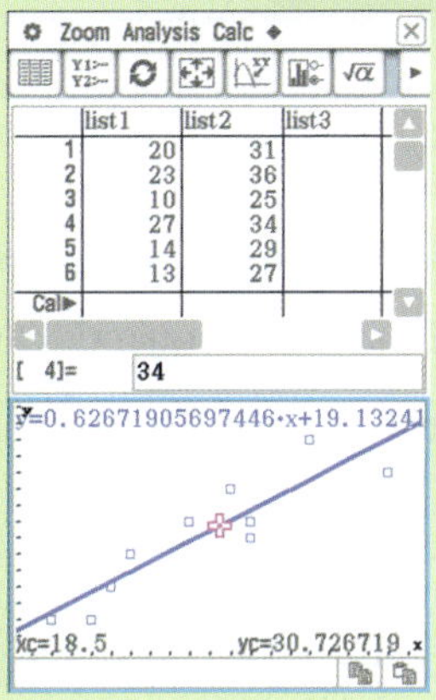

You can similarly use a spreadsheet to find the equation of the line of best fit. Using spreadsheet software such as Excel, select the data and create a scatter plot. You should then be able to add a 'Trendline' to the scatter plot to create the line of best fit.

Trendline is another name for 'line of best fit'.

Making predictions

The equation of the line of best fit can be used to make predictions.

- Making predictions for a value from within the given range of data is called **interpolation**.
- Making predictions for a value from outside the given range of data is called **extrapolation**.

In general, values predicted by interpolation will be more reliable than those predicted by extrapolation. This is because you cannot be sure that the same trend will continue outside the given range.

The equation for the line of best fit found in the previous example can be used to predict the mathematics mark for a student if you know the science mark.

$$\text{mathematics mark} = 0.627 \times \text{science mark} + 19.132$$

If Trev scored 22 for science, then his mathematics mark could be predicted by substituting the known value into the equation.

$$\begin{aligned}\text{mathematics mark} &= 0.627\\ &= 0.627 \times 22 + 19.132\\ &= 32.926\end{aligned}$$

If Trev's science mark is 22, then his predicted mathematics mark is 33.

If Sonya scored 29 for mathematics, then her science mark could be predicted by substituting the known value into the equation. Alternatively, you could use technology and switch the variables so that the mathematics mark is on the horizontal axis.

$$\begin{aligned}\text{mathematics mark} &= 0.627 \times \text{science mark} + 19.132\\ 29 &= 0.627 \times \text{science mark} + 19.132\\ \text{science mark} &= \frac{29 - 19.132}{0.627}\\ \text{science mark} &= 15.738\end{aligned}$$

If Sonya's mathematics mark is 29, then her predicted science mark is 16.

In both examples, the predicted value is within the range of known data values for that variable. The predicted mathematics mark of 33 is within the mark range of 25–36. The predicted science mark of 16 is within the mark range of 10–27. The predictions are both interpolations and would therefore be considered reliable.

2.6 Lines of best fit

Navigator

Answers p.780

1, 2, 3, 4, 5, 8, 9	1, 2, 3, 4, 5, 6, 8, 9	1, 2, 3, 4, 5, 6, 7, 8, 9

Equipment required: CAS or spreadsheet software

Fluency

1 The table gives the assignment and test results for a Year 10 class.

Assignment	Test
32	19
12	6
30	14
34	20
30	14
37	19
22	10
34	10
36	17
35	18

Assignment	Test
32	15
27	15
34	16
34	17
32	16
14	5
35	20
32	13
29	12
24	9

10A

(a) Plot the points and draw a line of best fit by eye.

(b) Find an equation for your line of best fit in the form $y = mx + b$.

Fiona's assignment mark was 25 but she was absent on the day of the test.

Enrico's test mark was 9 but he was absent when the assignment was completed.

(c) Use your graph to estimate the missing marks for Fiona and Enrico. Check your answers using your equations and using technology, if possible.

2 Find the equation of the line of best fit, using technology, for each of the following sets of data. Round your answers to 3 decimal places, where necessary. In each case, the required rule is of the form $y = \ldots$

(a)

x	0	1	2	3	4	5	6
y	4	16	29	42	54	65	76

(b)

x	0	2	5	7	10	12	14
y	-7	8	60	92	150	170	200

(c)

x	1	3	7	11	13	16	20
y	45	28	18	0	-5	-15	-45

(d)

x	3	6	8	10	13	15	18
y	26	12	1	-12	-24	-30	-48

3 (a) Using the line of best fit equation $d = 23.2t - 4.6$, what is the interpolated value for d when $t = 40$?

A 59 B 88 C 923 D 928

(b) (i) For the following table of values, use technology to find the equation of the line of best fit. Give your answer in the form $q = \ldots$

p	0	3	5	12	15
q	2	7	9	15	21

(ii) Assuming the relationship continues to be linear, what is the extrapolated value for q where $p = 25$?

A 32 B 30 C 28 D 25

Understanding

4 It is believed that the height of a male human at 30 months will be about half his full adult height. The table below gives the height (cm) for 15 boys measured at 30 months of age and again at 19 years of age (taken to be their full adult height).

30-month height (cm)	89.0	89.9	90.3	90.8	90.9	91.0	91.1	91.2
19-year height (cm)	178.0	177.1	179.6	181.8	184.0	180.5	182.0	183.1

30-month height (cm)	91.2	91.9	92.9	93.3	94.7	95.4	96.1
19-year height (cm)	180.1	185.1	182.0	186.3	187.4	187.9	189.4

(a) Draw a scatter plot of the data, putting the 30-month height on the horizontal axis.

(b) Draw, by eye, a line of best fit over your scatter plot.

(c) Do you think that the 30-month height is a good predictor of full adult height?

(d) It is predicted that a 94 cm 30-month-old boy will grow to a full height of 188 cm. What height does your line of best fit predict? Give your answer correct to the nearest cm.

10A

5 A chemist conducted an experiment in which the solubility (S units) of sodium chloride in water was tested for different temperatures (T °C). The results of the experiment are recorded in the following table.

Temperature T (°C)	0	10	20	30	40	50	60	70	80	90	100
Solubility S (units)	35.0	35.4	36.1	36.5	37.1	37.7	38.2	38.6	38.9	39.4	40.0

(a) Use technology to find the equation of the line of best fit when temperature is on the horizontal axis. Round your answer to 2 decimal places.

(b) Find the solubility of sodium chloride in water at:

(i) 25 °C (ii) 78 °C (iii) 120 °C

(c) Use technology to find the equation of the line of best fit when solubility is on the horizontal axis.

(d) Find the temperature at which sodium chloride has the following solubilities in water:

(i) 36.0 units (ii) 38.5 units (iii) 45.0 units

(e) What is the potential problem with the answers to **(b) (iii)** and **(d) (iii)**?

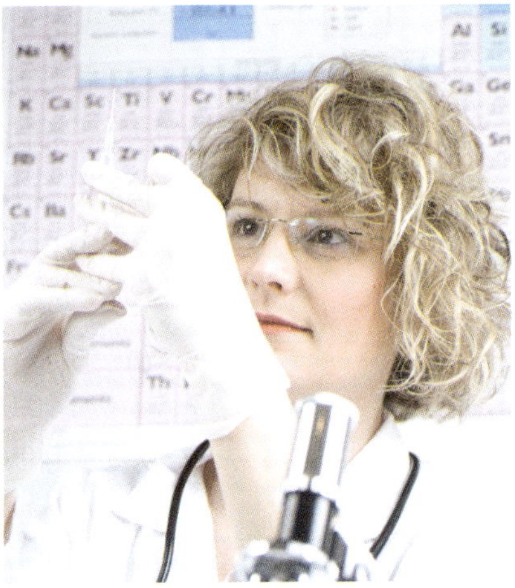

Reasoning

6 The following table gives the national running records for various distances for men for 10 different countries.

100 m (s)	10.39	10.31	10.17	12.18	10.35	10.11	10.98	10.60	10.61	10.46
200 m (s)	20.81	20.06	20.22	23.20	20.65	20.38	21.82	21.42	20.96	20.66
5000 m (min)	14.04	13.28	13.55	16.70	13.42	13.34	14.16	13.77	13.32	13.10

(a) Draw a scatter plot with the 100 m on the horizontal axis and the 200 m on the vertical axis.

(b) Does there appear to be a connection between the times for the 100 m and the 200 m?

(c) Draw a scatter plot with the 100 m on the horizontal axis and the 5000 m on the vertical axis.

(d) Does there appear to be a connection between the times for the 100 m and the 5000 m?

(e) Can you explain what you have found?

7 A complex spring system is tested under compression and the results are shown in the table below.

Compression Δx (mm)	0	20	40	60	80	100	120	140
Applied force F (N)	0	1.8	3.9	6.1	8.2	9.9	11.9	14.1

Compression Δx (mm)	160	180	200	220	240	260	280	300
Applied force F (N)	16.1	17.9	20.0	26.7	32.9	38.4	44.6	50.2

(a) Use technology to draw a scatter plot of the data, with compression on the horizontal axis. You should be able to see that the data splits into two groups. At which compression value does this occur?

(b) Find an equation for each of the two sets of data. Round your answers to 3 decimal places.

(c) Find the applied force required, to 1 decimal place, to produce a compression of:

(i) 75 mm (ii) 150 mm (iii) 225 mm

(d) Breaking the data into the same two groups, find the equations of the lines of best fit if applied force was on the horizontal axis.

(e) Find the compression, to the nearest whole number, if the spring system has an applied force of:

(i) 5 N (ii) 30 N (iii) 45 N

(f) The spring system has a maximum compression of 500 mm with an applied force of 90.0 N. Have you found an appropriate equation for this part of the graph? Explain why, listing any assumptions you have made. (Determine a new equation if necessary.)

10A

Open-ended

8 (a) What mistake has Emily made when drawing her line of best fit (shown at right)?

(b) Redraw the graph (note that coordinate pairs are all whole number pairs) and draw the line of best fit correctly.

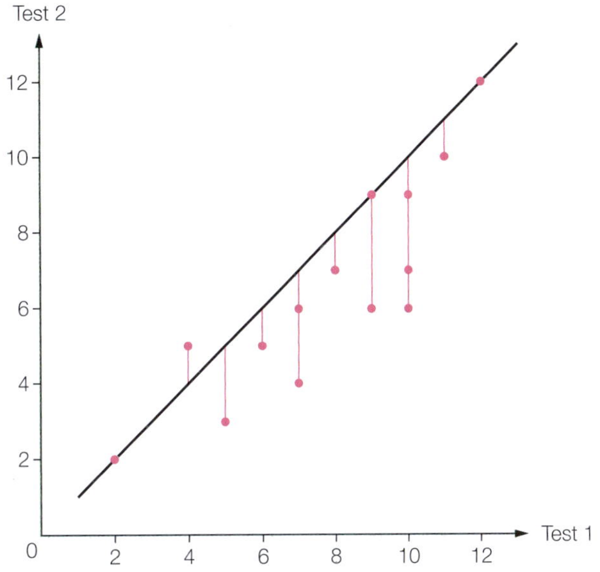

Class test results

9 A scatter plot has been drawn with its line of best fit drawn over it. The equation of the line of best fit has been given. Make up three questions you could ask using this information. The data refers to results on two mathematics tests.

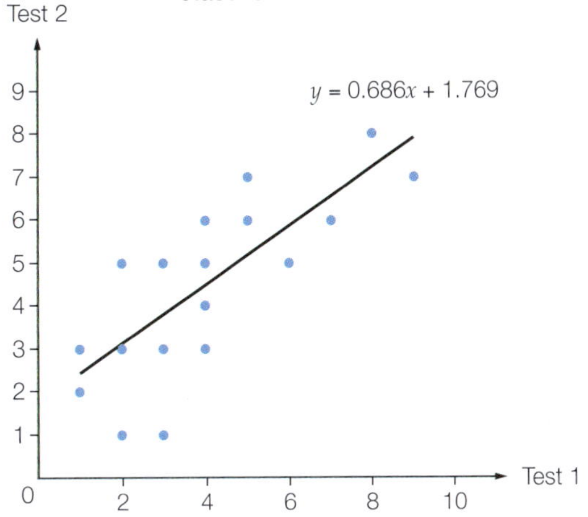

10A

Problem solving

Making the data fit

Equipment required: spreadsheet software may be useful

You have drawn a scatter plot and found that the equation of the line of best fit is $y = 3x - 4$, with the values rounded to the nearest whole number.

Can you find a data set of at least 10 number pairs that would produce this as the line of best fit?

Don't try to make all the data fit the line perfectly!

Strategy options

- Look for a pattern
- Work backwards.

Maths 4 Real

John Graunt –father of statistics

John Graunt is regarded as one of the first people to make use of statistical information and to draw conclusions from his analysis of the data collected. He collected information about a range of situations from the weekly Bills of Mortality produced in the city of London.

During the time that John Graunt collected and analysed his statistics, London was suffering an outbreak of the bubonic plague every few years. It is thought that this disease was carried to Europe by rats on merchants' ships as trade with Asia expanded towards the end of the Middle Ages.

The bubonic plague was sometimes called the Black Death, referring to the way the skin of a sufferer turned black. People were employed to go through the streets collecting dead bodies on carts and taking them away for disposal. It was soon observed that contact with the sufferer greatly increased the chances of falling victim to the disease yourself.

LONDON'S Dreadful Visitation:
Or, A COLLECTION of All the
Bills of Mortality
For this Present Year:
Beginning the 27th of December 1664. and ending the 19th. of December following:
As also, The GENERAL or whole years BILL:
According to the Report made to the KING's Most Excellent Majesty,
By the Company of Parish-Clerks of London. &c
LONDON:
Printed and are to be sold by E. Cotes living in Aldersgate-street. Printer to the said Company 1665.

The table that follows is a reproduction of part of Graunt's data regarding the plague.

'Anno 1592' means in the year 1592.
'Whereof of the Plague' means died as a result of the plague.
'Christned in the said year' gives the number of babies christened in the particular year.
For that time this gives a relatively accurate estimate of the number of births, as most babies were christened.

There have been in London, within this Age, four Times of great Mortality, that is to say, the years 1592, and 1593, 1603, 1625, and 1636.

There died *Anno* 1592 from *March* to *December*	25 886	Anno 1603 within the same space of time were Buried	37 294
Whereof of the *Plague*	11 503	Whereof of the *Plague*	30 561
Anno 1593	17 844	*Anno* 1625 within the same space	51 758
Whereof of the *Plague*	10 662	Whereof of the *Plague*	35 417
Christned in the said year	4021	Anno 1636 from *April* to *December*	23 359
		Whereof of the Plague	10 400

1 Draw a graph to represent the total deaths and deaths from plague for each year given. Make sure that the graph type you choose enables you to compare the years.

2 Which year was the worst for deaths from the plague as compared with deaths from all other causes? (This is not necessarily the year with the greatest number of plague deaths.)

The following extract shows some further data collected by Graunt. This time we are told about the number of burials not associated with the plague ('Buried in all'), the number of deaths associated with the plague ('Besides of the *Plague*') and the approximate number of children born (Christned) for each of the years 1604–1661.

Anno Dom.	Buried in all	Besides of the Plague	Christned	Anno Dom.	Buried in all	Besides of the Plague	Christned
1604	4323	896	5458	1628	7740	3	8564
1605	5948	444	6504	1629	8771	0	9901
1606	5796	2124	6614	1630	9237	1317	9315
1607	5670	2352	6582	1631	8288	274	8524
1608	6748	2262	6845	1632	9527	8	9584
1609	7545	4240	6388	1633	8392	0	9997
1610	7486	1803	6785	1634	10 899	1	9855
1611	6716	627	7014	1635	10 651	0	1034
1612	7778	64	6986	1636	12 959	10 400	9522
1613	7503	16	6846	1637	8681	3082	9160
1614	7367	22	7208	1638	13 261	363	10 311
1615	7850	37	7682	1639	9548	314	10 150
1616	8063	9	7985	1640	11 321	1450	10 850
1617	8280	6	7747	1641	11 767	1375	10 670
1618	9596	18	7735	1642	11 999	1274	10 370
1619	7999	9	8127	1643	12 216	996	9410
1620	9691	21	7845	1644	9441	1492	8104
1621	8112	11	8039	1645	9608	1871	7966
1622	8943	16	7894	1646	10 415	2365	7163
1623	11 095	17	7945	1647	10 462	3597	7332
1624	12 199	11	8299	1648	9283	611	6544
1625	18 848	35 417	6983	1649	10 499	67	5825
1626	7401	134	6701	1650	8749	15	5612
1627	7711	4	8408	1651	10 804	23	6071

3 Draw a graph that shows three pieces of information for each year: the number of deaths in total, the number of deaths associated with the plague, and the number of deaths not associated with the plague.

4 Calculate the difference between the number of births, as approximated by the Christned figures, and the total number of deaths for each of the years. Display this in a table.

5 Draw a graph to represent these figures; call it *Change in population*.

6 The number of births per year, as approximated by the figures in the Christned column, seems to fluctuate greatly. Can you offer some reasons for this? You may need to do some further research to answer this question.

Taking it further

Complete some research on the life and work of John Graunt. What conclusions did he draw from his data? How was his data used? Use your research to make a poster, slideshow or movie.

2.7 Statistics in the media

Discussions in the media often involve statistics, especially when someone wants to justify an opinion. You need to think carefully about what you hear and read, because statistics are not always as they seem. Some of the issues that might arise include:

- insufficient evidence being provided to support the conclusions made
- percentages used without an indication of the actual size of the survey
- raw figures used that are based on very different populations
- oversimplification of the results.

This is not to say everything you read or hear is incorrect, but it is a good idea to think critically about the key statistical issues involved. As an example, consider the two tables that follow, which accompanied a newspaper article.

Dressing with danger			
% of clothes that cause injuries		**% of injuries caused by clothes treated at Victorian hospitals**	
Shoes	48.1	Fracture, excluding tooth	18.8
Jumper, jersey, cardigan, shirt	13.0	Sprain/strain	17.2
Buttons	9.2	Open wounds	13.3
Jeans/pants	8.3	Foreign body	9.6
Socks	7.4	Burn	9.0
Clothes, not specified	5.8	Superficial	8.4
Zippers	4.7	Muscle/tendon injury	5.2
Nightwear	3.5	Dislocation	4.9
		Eye	2.9
		Bites	1.2
		Other	9.5

Source: Victorian Injury Surveillance and Applied Research System

At first glance, it appears that clothing injuries could be quite a serious problem. However, when you consider how many injuries in total get treated at Victorian hospitals, very few of them would be caused by clothing.

2.7 Statistics in the media

Navigator

Answers p.781

1, 2, 3, 4, 5, 6, 8	1, 2, 3, 4, 5, 6, 7, 8	1, 3, 4, 5, 6, 7, 8

Fluency

1 A section of an article from the London *Daily Mail* is shown on the following page.

(a) True or false: The article rates sports stars and managers in terms of their earnings in sport. Explain your answer.

(b) Who of the top 10 rich list, if any, experienced a drop in wealth in the year shown?

(c) Apart from the new people on the list, which sportsperson experienced the biggest percentage increase in their wealth over the year? State the percentage increase to the nearest whole percentage.

Lewis Hamilton remains Britain's richest sports star with £88m fortune as Wayne Rooney overtakes Jenson Button into second

- **Lewis Hamilton's fortune has spiralled by £20m to £88m in the past year**
- **Wayne Rooney is now second in the Sunday Times Sport Rich List**
- **The Manchester United captain's fortune is up £12m to £72m**
- **F1 driver Jenson Button is £1m behind Rooney in third place**

By Martyn Ziegler, Press Association
Published: 09:01 EST, 25 April 2015
Updated: 10:51 EST, 25 April 2015

Lewis Hamilton remains the richest sportsman in Britain with a fortune of £88 million but Wayne Rooney has overtaken Jenson Button to be in second place in the 2015 Sunday Times Sport Rich List.

Formula One world champion Hamilton, who lives as a tax exile in Monaco, remains on pole position in the list but his fortune has spiralled by £20 million on last year.

Hamilton has his sights on retaining his world championship title after winning three of the first four races of the season, and is set to sign a new contract with Mercedes worth £27 million a year.

The list was released just hours after Hamilton was pictured on the Rome set for the filming of comedy sequel Zoolander 2 on Friday.

Hamilton was seated front row with Olivia Munn and busily chatting in between takes for the film, which will be released in February 2016.

Hamilton is making his first live-action cameo on the big screen having previously lent his voice to Disney animation Cars 2.

The Sunday Times Rich List 2015—Britain's wealthiest sports stars and managers

Position	Sportsperson	Sport	2015	2014
1	Lewis Hamilton	Motor racing	£88m	£68m
2	Wayne Rooney	Football	£72m	£60m
3	Jenson Button	Motor racing	£71m	£63m
4	Andy Murray	Tennis	£48m	£40m
5	Rio Ferdinand	Football	£44m	£44m
6	Steven Gerrard	Football	£42m	£37m
7=	Luoi Deng	Basketball	£40m	£36m
7=	Jose Mourinho	Football	£40m	New
9	Frank Lampard	Football	£39m	£37m
10=	Sir Nick Faldo	Golf	£38m	New
10=	Ryan Giggs	Football	£38m	£36m
10=	Rory McIlroy	Golf	£38m	£28m
10=	Arsene Wenger	Football	£38m	New

2 (a) In the early stages of the nomination process for the Republican Party leading up to the 2012 election for the US President, an American news network produced a graph similar to the one shown here.

What problems can you see with this graph? Why do you think the graph has been drawn this way? Could you draw an accurate graph from the information provided?

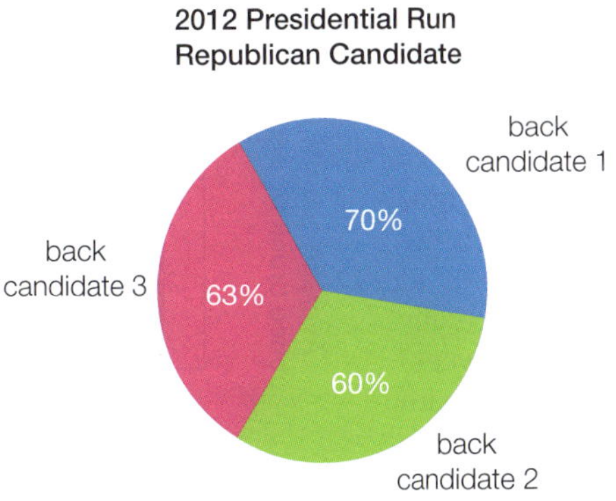

(b) Some graphs provide a simple and accurate visual representation of information. This box plot shows the winnings ($) in the New Jersey Pick-It Lottery. In this lottery you pick any three-digit number you like and the total prize pool is divided by the number of winning entries.

What do you think this graph is telling you and why do you think that is the case?

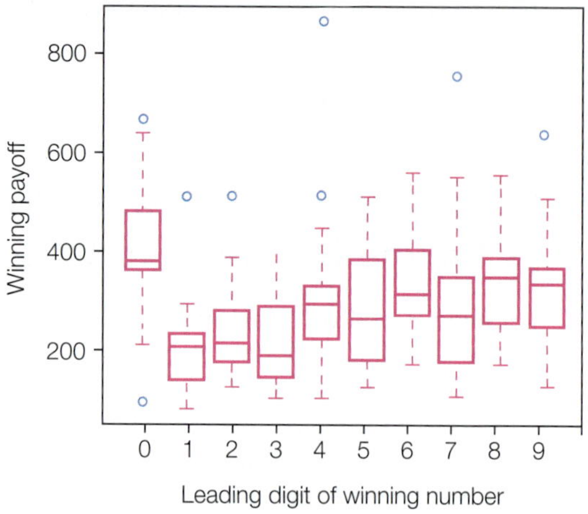

Understanding

3 Consider the following article and graph from Roy Morgan Research.

The brew crew: Australia's heaviest coffee drinkers

Is Australia kicking the caffeine habit? In the last decade, the average coffee consumption by Australian adults has declined slowly but steadily, from 10.5 cups to 9.2 cups per week. Or could we just be drinking a stronger brew, thus reducing the need for so many cups? After all, café visitation is on the rise, as is ownership of coffee-making machines.

The proportion of Australians aged 18+ who go to a café for a coffee or tea in an average 3-month period has grown gradually from 54% in the year to December 2009 to 56% in the year to December 2013. Meanwhile, the increase in people who own coffee makers has shot up, from 28% in 2009 to 36% in 2013.

Among Australia's biggest coffee drinkers are people who work long hours. In the year to December 2013, those who worked 60+ hours in any given week consumed an average 10.1 cups weekly, compared to 8.6 for non-workers or 8.8 for those who worked 35–39 hours.

In news that probably won't surprise the parents out there, it seems that having children also increases our need for caffeine. Whereas the average weekly coffee consumption for people who don't have kids is 7.2 cups, it rises to 9.6 cups for parents.

Perhaps a little more surprising is that consumption increases with the age of the kids. So while parents of infants under 2 years old actually drink less coffee (8.8 cups per week) than the national average, those with kids aged between 12 and 15 drink an average of 10.3 cups.

The rise of the home coffee maker

Even though it's hard to imagine when they'd get time to use it, because they work such long hours, people who work 60+ hours per week are significantly more likely than the average Aussie to have a coffee maker at home, with 44% of them owning one (up from 38% in the year to December 2009). People with kids also over-index on coffee-maker ownership at 39%, up from 32% in 2009.

Angela Smith, Group Account Manager—Consumer Products, Roy Morgan Research, says:

'The increased presence of coffee makers—which can be anything from stovetop cafetières to 'pod' machines such as Nespresso and Expressi—in Australian households is good news for the manufacturers of these items. Interestingly, it hasn't impacted adversely on café visitation: possibly because people are developing a taste for 'real' coffee over instant.

'While it makes sense that people who work long hours would consume more coffee, their need for caffeine goes beyond this, to the point where they also drink more Cola and energy drinks than people who work fewer hours.

'The news that parents of older children drink more coffee in an average week than those of infants may seem surprising, considering the stereotype of the sleep-deprived new parent, but this is simply a function of age. Our data shows that older people drink more coffee, and parents of older children are typically older than those of infants. Mind you, their extra caffeine intake could also be linked to the sleep they lose through lying awake at night, worrying about where their kids are or what they're up to on Snapchat…!'

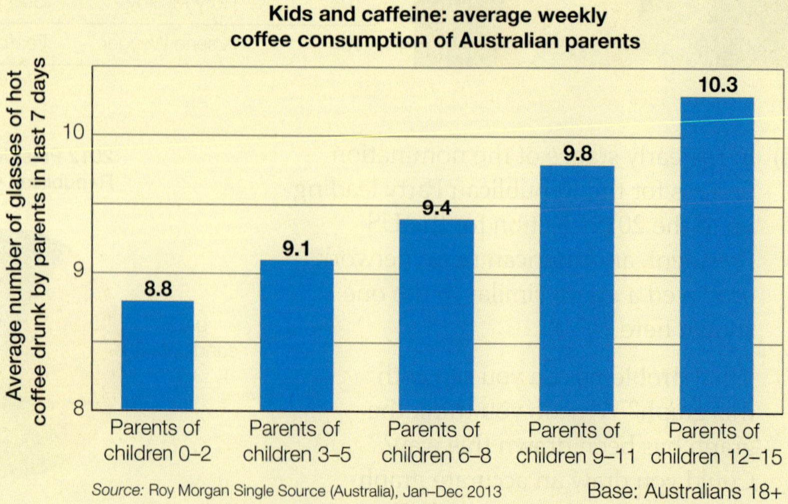

Margin of Error

The margin of error to be allowed for in any estimate depends mainly on the number of interviews on which it is based. Margin of error gives indications of the likely range within which estimates would be 95% likely to fall, expressed as the number of percentage points above or below the actual estimate. Allowance for design effects (such as stratification and weighting) should be made as appropriate.

Sample Size	Percentage Estimate			
	40%–60%	25% or 75%	10% or 90%	5% or 95%
5000	±1.4	±1.2	±0.8	±0.6
7500	±1.1	±1.0	±0.7	±0.5
10 000	±1.0	±0.9	±0.6	±0.4
20 000	±0.7	±0.6	±0.4	±0.3
50 000	±0.4	±0.4	±0.3	±0.2

(a) Considering the graph, what is the trend of coffee consumption when comparing parents with children in different age groups?

(b) According to the article, why is this trend surprising?

(c) What alternative reason was given to explain the trend?

(d) When parents are compared to non-parents, which group tends to drink more coffee? What is the difference in the number of cups consumed per week?

(e) Which group of people drink the greatest amounts of coffee? What else, according to the article, does this group of people have more than the average Australian? Why do you think this is so?

(f) According to the article, coffee consumption has decreased but visiting cafés and owning coffee makers has increased. What reasons can you think of to explain this?

4 The following article appeared on the Australian Bureau of Statistics website in 2015. (*Note:* 'LGA' stands for 'local government area'.)

The graph on the right shows the number of older Australians who moved to selected 'sea change' areas of the Queensland coast and their place of origin.

Around 70% of the sea-changers moved into separate houses. Almost half of these people moved into a house with 3 bedrooms. The sea-change moves were generally made by those in the younger end of the age range, with those between 55 to 64 in 2006 making up over 14,000 (or 67%) of the 22,000 movers.

Sea change moves are often associated with changes in people's employment circumstances, for example following retirement. Amongst the 14 000 55–64 year old movers to selected coastal LGAs in Queensland, 37% were employed in 2006 but were not in the labour force in 2011. This was higher than when compared to all 55–64 year olds in Australia, where 19% were employed in 2006 but were not in the labour force in 2011.

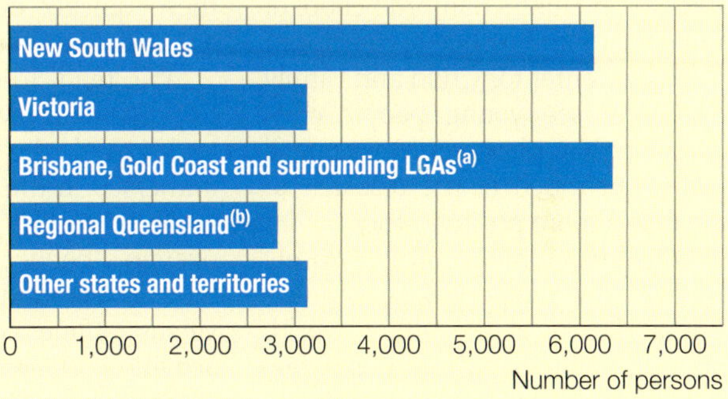

Footnote(s): (a) Includes LGAs of Brisbane, Gold Coast, Ipswich, Morton Bay, Redland and Logan. (b) Includes all other LGAs in Queensland outside of Brisbane, Gold Coast, Ipswich, Morton Bay, Redland and Logan and the selected coastal local government areas.

Source(s): Microdata: Australian Census Longitudinal Dataset, 2006-2011

(a) Explain why this article only appeared in 2015, when the initial data dates from 2006.

(b) How many people aged 55 and over shifted to the coast from somewhere else in Queensland in 2006? Estimate how many of these people retired during the period 2006–2011. Comment on the reliability of any assumptions you made.

(c) Did the article suggest that a person was more likely or less likely to retire after a sea change, compared to over 55s generally? Explain your reasoning.

5 The following article appeared on-line in *Medical Xpress* in November 2015.

Parents' top fears about teens' cellphone use—are they justified?

Parents' fears about their teenagers' heavy use of cell phones and social media may be exaggerated, according to a new report from Duke University researchers. However, there are important exceptions in the areas of cyberbullying and sleep disruption.

'Each generation worries about how young people are using their time,' said Candice Odgers, associate professor in Duke's Sanford School of Public Policy and associate director of the Duke Center for Child and Family Policy. 'We see young people constantly on their phones and assume ill effects, but much of the research to date tells a more positive story.'

Teenagers' online lives closely resemble their experiences, connections and risks in the offline world, and cellphone use alone poses few entirely new dangers, Odgers said.

The article by Odgers and Duke Ph.D. candidate Madeleine J. George, titled 'Seven Fears and the Science of How Mobile Technologies May Be Influencing Adolescents in the Digital Age,' appears online 18 Nov in the journal *Perspectives on Psychological Science*:

The review weighs commonly expressed fears regarding teenagers' use of mobile devices against existing research evidence. It calls for more rigorous research to evaluate how these quickly evolving technologies are impacting young people's lives.

'We tend to count hours spent using technology, rather than seek to understand the reasons teens are immersed in the digital world. When we look closely, we see considerable overlap between the underlying motivations and content of online versus offline communications and activities,' Odgers said.

There is little question that American adolescents are constantly connected. Almost 90% of adolescents own, or have access to, a mobile phone. They spend an average of 1.5 hours a day text messaging and the vast majority access the internet from their phones. They devote an average of 7.5 hours a day to digital media of all kinds.

But, contrary to the early internet age—when a small minority of teens were online and heavy internet use was a sign of offline problems—now, teens' online worlds mirror their offline lives.

Teens with strong offline social networks tend to reinforce and strengthen their relationships through online interactions, the review found. Rather than connecting with strangers, most adolescents use digital media to interact with friends and acquaintances already in their face-to-face social networks.

'The overlap between offline and online connections is so striking that viewing what happens online as somehow separate from teens 'real lives' is a false distinction,' said George.

For adolescents struggling with existing relationships, though, more time spent online can predict declines in well-being.

'If parents have concerns about their teen's face-to-face social interactions or activities, they probably have more reason to be concerned about online activities,' said George.

Parent–adolescent relationships online also appear to mirror offline relationships, the authors found. Although cellphone use may take away time spent with parents, if the existing relationship is strong, the new technology can allow more frequent, positive parent–child contact.

(a) The article explores the idea that adolescents are 'constantly connected'. Find three pieces of information used in the article to back this up.

(b) Is there evidence of how the data was collected? How would you imagine the data to be collected? Do you have any reason to suspect bias in the findings?

Reasoning

6 The following article appeared in the business section of the *Australian* in November 2015.

Can cabbages to Korea cover for coal's comedown?

THE AUSTRALIAN
12:00AM NOVEMBER 17, 2015

Rowan Callick
Asia-Pacific Editor

The South Korea story offers a salutary warning against expectations that Australia's bright new free-trade agreements will turn around our slumping export figures any time soon.

But it also showcases the fearlessness of Australian exporters—whose sales of cabbages and cauliflowers to the famous home of kimchi soared by 7944 per cent in the first half of this year.

Early data from the South Korea FTA, the first of the succession of deals concluded by Trade Minister Andrew Robb and his team of trade negotiators, shows that while producers have rapidly seized their new opportunities, especially in agribusiness, the bottom line for Australian exports remains in decline. For the value of iron ore and coal, especially, within the trade profile with South Korea—half the value of all exports—is such that it will take years for other sales to bulk up sufficiently to compensate.

The same story applies to our other big export markets except the US. Those commodities accounted last year for 42 per cent of our exports to Japan, 58 per cent to India, and an astounding 65 per cent of our sales to China. In the first half of this year, iron ore receipts from sales to South Korea were 34 per cent down, to $2.25 billion—a fall of $1.2bn.

The good news is that the big rises in exports following the South Korea FTA, representing the most impressive increases across the board, comprise $1bn.

Some of the agricultural successes are attributable not just to the FTA but at least as much to a shift in South Korea's import risk assessments, essentially the quarantine process, which formerly delayed accepting products for years.

This provides a huge boost, which surely is also driving a rise in jobs, especially in regional Australia.

Yet those combined revenues—not just the levels of increase but the entire sales—still fall $200m short of the drop in value of Australia's iron ore exports alone to South Korea in the same six months.

Food boom

Australian exports to Korea in the six months to July 1 post free trade agreement

Fish exports: 2838% to $852,000	**Cabbages and cauliflowers:** 7944% to $724,000	**Straw:** 482% to $2.3m
Wine: 39% to $7.5m	**Fresh or chilled beef:** 26% to $195m	**Nuts:** 234% to $10.8m
Olive oil: 1067% to $70,000	**Other liquor:** 5000% to $2.65m	**Frozen beef:** 37% to $476m
Citrus: 887% to $1.16m	**Other oils:** 167% to $917,000	**Dentists' preparations:** 1520% to $243,000
Sheep and goat meat: 30% to $35m	**Grapes:** 1357% to $1.95m	**Sausages:** 1389% to $804,000
Blankets: 4571% to $1.3m	**Butter sales:** 93% to $8m	**Dates, figs and tropical fruit:** 175% to $250,000
Bread, cakes and biscuits: 930% to $2.8m	**Gold:** 153% to $240m	**Potatoes:** 64% to $6.4m
Other fruit: 4616% to $3.5m	**Confectionery:** 2500% to $104,000	**Furniture:** 84% to $991,000

(a) An increase that 'soared by 7944 per cent' means it increased by how many times? Round your answer to the nearest whole number. How is it possible that such a huge percentage increase will not be enough to make up for the fall in value of other types of export?

(b) Potatoes were up '64% to $6.4m' and cabbages and cauliflowers were up '7944% to $724 000'. What were the values of these exports 6 months previously?

(c) What was the value of iron ore exports to South Korea 6 months previously?

7 The following fictional article is of a type often seen around election time. Read it and answer the questions that follow.

Poll – Voters back Smith

Mr John Smith's return as Prime Minister appears to have buoyed the government's electoral chances, with one in four voters saying they are at least 'a bit more likely' to vote for the government at the next federal election. Mr Smith was clearly preferred by female voters (47% support) over the leader of the opposition, Ms Rosa Casey (41%).

Men preferred Ms Casey (54%) to Mr Smith (45%). Support for the opposition among men was 47%, compared to the government's (45%), while from women the government attracted 47% support compared to the opposition's 44%. The survey also asked voters to identify those characteristics they believed Ms Casey or Mr Smith possessed. The results of the survey are summarised below.

SUNDAY BUGLE POLL

Voting intentions	%	Preferred PM	%
ALP	46	Casey	48
Coalition	46	Smith	46
Other/Don't know	8	Neither/Don't know	6

SUNDAY BUGLE POLL: RATING THE LEADERS

	Casey (%)	Smith (%)		Casey (%)	Smith (%)
Decisive	68	53	Sincere	22	55
Forward thinking	68	48	Too arrogant	75	15
Good for Australia	43	55	Honest	20	62
Weak	10	30	Capable	72	72
A good leader	66	43	None of the above	1	2

Source: Corio Market Research 1030 voters from Vic., NSW. Poll done by telephone.

(a) Do the survey results agree with the heading 'Voters back Smith'? Discuss.

(b) The survey was done by telephone—is this the best way to do such a survey? Discuss.

(c) Why have the results for choice of preferred PM been separated into male and female opinions in the article?

(d) Why don't the totals of male and female voters' opinions add up to 100%?

(e) Discuss the categories that describe Rosa Casey and John Smith, such as 'decisive' and 'forward thinking'. Comment on the appropriateness of these categories. What other categories might be appropriate?

(f) Is this survey just about the preferred Prime Minister or is it about something else?

Open-ended

8 The following advertisement was produced by the South Australian Government in 2014.

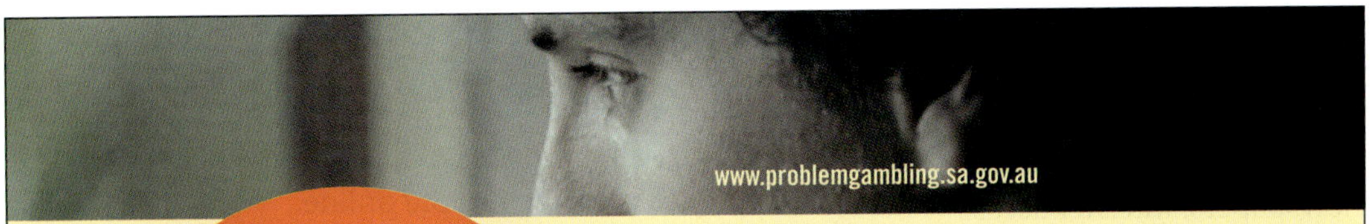

FACT SHEET

THE POKIES: THE ODDS

What are the odds? Odds are they are against you

If you gamble for the chance to win money, do you know what the odds are of actually winning?

Every form of gambling has an element of chance that helps determine the outcome. How much a person will win or lose is determined by the odds of the game and how much money is staked. It is important to remember that the 'house' (eg the casino, hotel or club) always has the advantage and that the odds are against the gambler.

THE ODDS ARE CLEARLY STACKED AGAINST THE GAMBLER - MEANING YOU SHOULD ALWAYS EXPECT TO LOSE.

THIS TABLE SHOWS THE ODDS OF WINNING ON SOME FORMS OF GAMBLING IN SOUTH AUSTRALIA

Bet	Odds of Winning
Poker Machines - Getting 5 Black Rhinos on Black Rhinos Machine (Top Prize) ($1 Bet per line)	1 in 9,765,625
Lotto - Winning First Division (playing 1 game)	1 in 8,145,060
Oz Lotto - Winning First Division (playing 1 game)	1 in 45,379,620
Powerball - Winning First Division (playing 1 game)	1 in 54,979,155
Keno - Winning First Division (Spot 10, playing 1 game)	1 in 8,911,711
Keno - Winning First Division (Spot 9, playing 1 game)	1 in 1,380,687
Super 66 - Winning First Division (playing 1 game)	1 in 1,000,000
The Pools - Winning First Division (playing 1 game)	1 in 2,760,681
Instant Scratchies - Winning First Division with $1 High Tier (playing 1 game)	1 in 1,500,000
Trackside - picking, at random, a trifecta in a 13 horse race	1 in 1,715

NOW SEE HOW YOUR ODDS OF WINNING COMPARE WITH NON-GAMBLING RELATED ACTIVITIES

Odds of non-gambling activities	Odds of Occurrence
Marriage ending in divorce	1 in 2.3 marriages
Chances of going bald - if you're a man	3 in 4 men
Dying from heart disease	1 in 4 people
Having your car stolen	1 in 142 cars
Dying from a venomous bite or sting	1 in 1,000,000 people
Being killed by lightning	1 in 1,603,250 people

GAMBLINGHELPLINE
1800 858 858
24/7 • **FREE** • CONFIDENTIAL

* Some figures taken from Centre for Gambling Research Fact Sheet - Gambling Odds 2003

The South Australian Government developed this advertisement in 2014 with funds from the Gamblers Rehabilitation Fund.

What do you think about this advertisement? What do you think about the statistics that are quoted in the advertisement? How would they have been calculated? Do you think this would be effective in reducing problem gambling?

Investigation

Newspaper polls

Equipment required: newspapers, internet access

Some newspapers conduct a daily poll on an issue of current interest. A couple of examples are shown below, taken from the *Herald Sun*.

Should politicians be banned from shopping centres?		Should Australia abandon the Union Jack and create a new flag?	
Total: 1695 votes		Total: 16 404 votes	
YES	NO	YES	NO
1266 votes	429 votes	5561 votes	10 843 votes
74.6%	25.4%	33.9%	66.1%

The Big Question

Are these polls reliable? To answer this you will have to consider what reliable means. This is not always a simple matter and you need to address the following issues.

- Is the question a fair question?
- Is the polling method reasonable?
- If the question was asked again the next day, would a similar result be expected?

Sometimes questions are phrased in such a way as to lead to a particular answer. In the polling industry this is known as a push poll.

Engage

To participate in these polls you can either call a telephone vote line or vote online.

1 What type of poll, of those you have studied, are these closest to in their design?

2 Why do you think there is such a difference in the total number of votes for the two examples shown? Is this likely to influence you in how reliable you think the results might be?

Explore

See if you can find some similar examples online or in another newspaper.

Explain

3 Is there a problem with these polls from the point of view of how the sample is obtained?

4 Do a statistical analysis of the results obtained in these two polls, or on a couple of other similar polls that you have found for yourself. Do the results appear to be reasonable?

Evaluate

5 Do you think the wording of the questions should be changed? If so, write the question the way you would like it to be asked.

6 Why do you think newspapers conduct these polls on a daily basis?

Extend

7 Find out more about 'push polling'.

8 Make up some examples of these one-question polls for yourself, based on some issues of local or current interest. Decide how you would go about conducting your poll so that the results are reliable.

Strategy options

- Break the problem into manageable parts.
- Have I seen a similar problem?

Standard deviation

2.8

One of the most useful statistical measures of spread is the **standard deviation**. The symbol σ or the letters SD are often used to represent the standard deviation.

Consider the following data set:

5, 6, 7, 10, 10, 12, 14, 16

The mean of the data is 10.

You can now calculate how far each of these values is away from the mean.

This is referred to as the score's *deviation* from the mean. The deviation values are:

-5, -4, -3, 0, 0, 2, 4, 6

If you add these together you get 0, which is not very helpful. In fact, you will *always* get 0. (You might like to think about why this is true.) To avoid this zero, the deviations can all be squared (to give positive values) before they are added.

The squared deviation values are:

25, 16, 9, 0, 0, 4, 16, 36

These add to 106.

You can then divide this by the number of values, in this case 8, to find the average deviation from the mean, which is called the **variance**. Here, the variance is 13.25.

10A

To find the standard deviation you then take the square root of the variance, which reverses the effect of the squaring. The standard deviation here is approximately 3.64.

Note that previously you used two measures of spread: range and interquartile range. The standard deviation is a more accurate measure of spread because it uses all data points, not just the difference between the extreme points.

$$\text{Variance } \sigma^2 = \frac{\text{sum of (deviations from mean)}^2}{\text{number of values}}$$

$$\begin{aligned}\text{Standard deviation } \sigma &= \sqrt{\text{variance}}\\ &= \sqrt{\frac{\text{sum of (deviations from mean)}^2}{\text{number of values}}}\\ &= \sqrt{\frac{\Sigma(x-\bar{x})^2}{n}}\end{aligned}$$

Note: σ is the lower-case Greek letter sigma and Σ is the capital Greek letter sigma.

Worked example 14

W.E. 14

Find the standard deviation, rounded to 2 decimal places, for the following data.

2, 3, 3, 3, 5, 7, 8, 11, 12

Thinking	Working
1 Find the mean of the data set.	Mean: $\frac{2+3+3+3+5+7+8+11+12}{9} = 6$
2 Find the deviations from the mean and then square them.	Deviations: -4, -3, -3, -3, -1, 1, 2, 5, 6 Square of deviations: 16, 9, 9, 9, 1, 1, 4, 25, 36
3 Add the squares of the deviations and divide by the number of values to find the variance.	Variance = $\frac{110}{9}$
4 Take the square root of the variance.	Standard deviation = $\sqrt{\frac{110}{9}} = 3.50$ (2 d.p.)

This calculation is not practical if there are too many data values, but in that case technology can be used. Standard deviation can be calculated using your CAS as shown below.

Using TI-Nspire CAS

Add **Lists & Spreadsheet** to your document and enter the data in column **A**. Select **menu** > **Statistics** > **Stat Calculations** > **One-Variable Statistics...** and set **Num of Lists** to 1. Then set the **One-Variable Statistics** settings as shown.

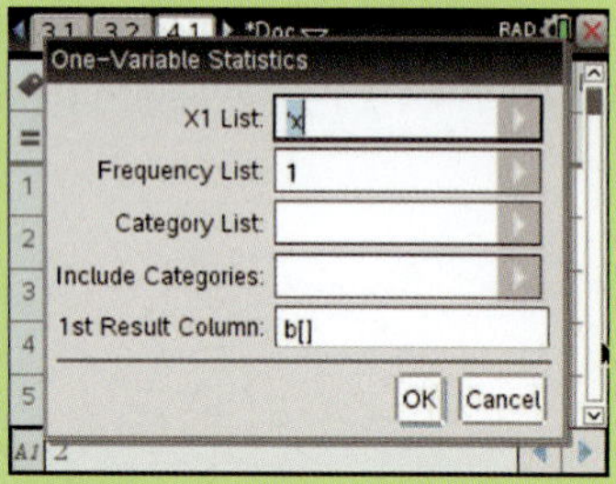

The statistics data should appear as shown. Scroll through the list to find the σx standard deviation value.

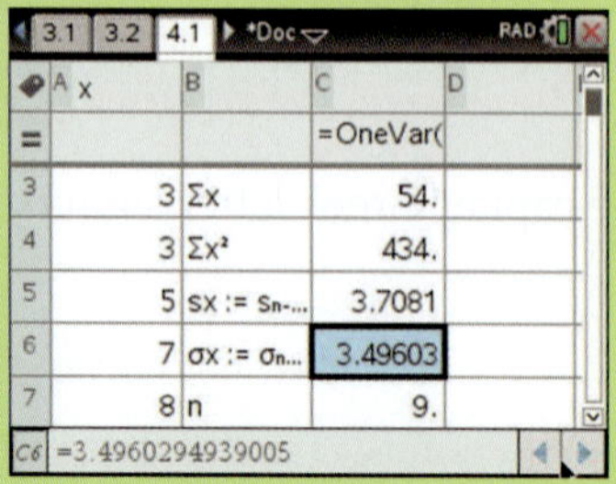

This is 3.50, rounded to 2 decimal places, just as found when calculating by hand.

Using Casio ClassPad CAS

From the menu select **Statistics** and enter the data in **list1**. Select **Calc** > **One-Variable** and set the **XList** as **list1** and **Freq** as 1 as shown.

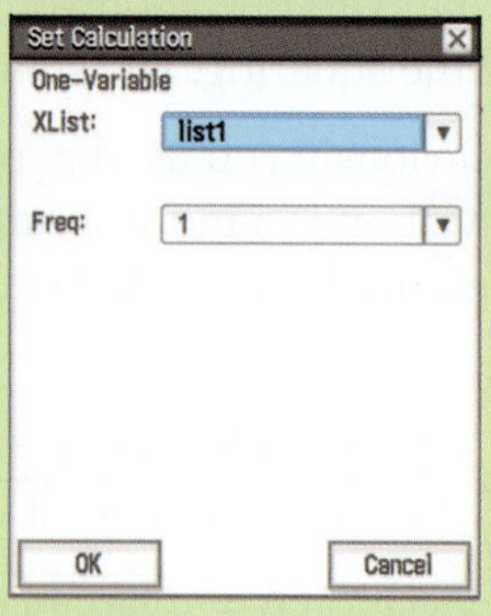

The statistics data should appear as shown. Scroll down to find the σ_x standard deviation value.

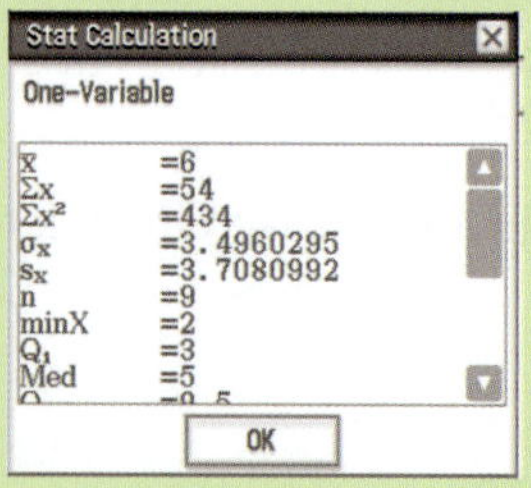

This is 3.50, rounded to 2 decimal places, just as found when calculating by hand.

Many scientific calculators can be used to calculate the mean and standard deviation. You can also use spreadsheet software.

In Excel you use **average** to calculate the mean and **stdevp** to calculate the standard deviation for a population.

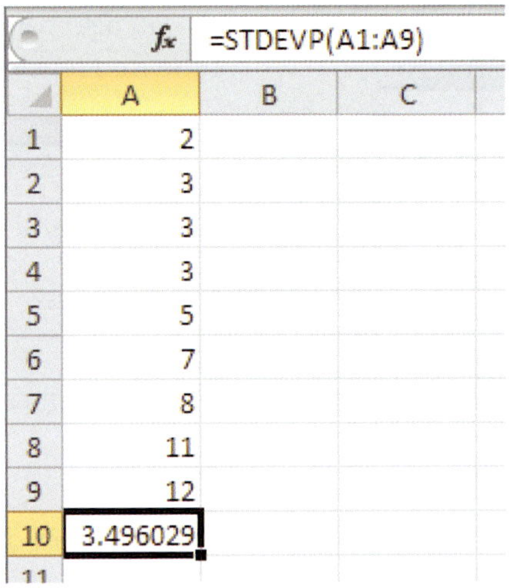

If you have a large data set in a frequency table you can still use your CAS to simplify the calculations. If the data is grouped, then use the class centre as the value in the first column.

The following example shows how to find the standard deviation for the following list of times taken by students to complete the 100 m sprint at the School Sports Carnival.

Time (s)	Frequency
10.50–<11.75	2
10.75–<11.00	4
11.00–<11.25	7
11.25–<11.50	3
11.50–<11.75	5
11.75–<12.00	10

Using TI-Nspire CAS

Enter the class centre values in column **A** and the frequencies in column **B**. Then select **menu** > **Statistics** > **Stat Calculations** > **One-Variable Statistics...** and set **Num of Lists** to 1. Then set the **One-Variable Statistics** settings as shown.

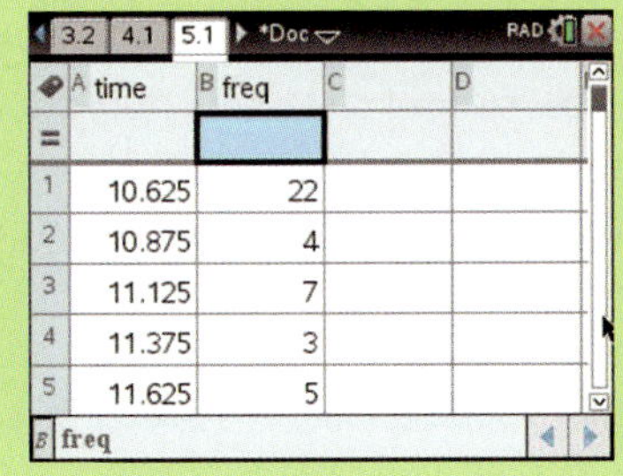

Using Casio ClassPad CAS

Enter the class centre values in **list1** and the frequencies in **list2**. Select **Calc** > **One-Variable** and set the **XList** as **list1** and **Freq** as **list2** as shown.

Using TI-Nspire CAS	Using Casio ClassPad CAS
The statistics data should appear as shown. Scroll through the list to find the σx standard deviation value. You can also confirm that **n** = 31, the sum of the frequencies.	The statistics data should appear as shown. Scroll through the list to find the σ_x standard deviation value. You can also confirm that **n** = 31, the sum of the frequencies.

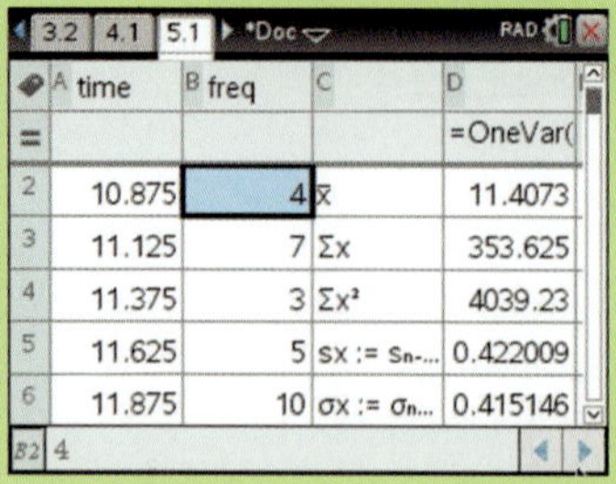

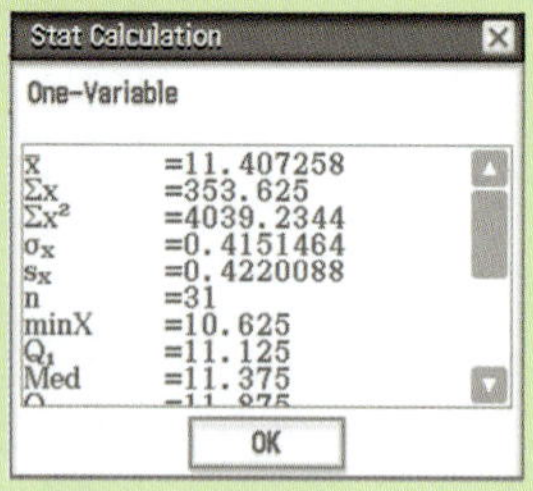

Interpreting the standard deviation

For two data sets that have the same mean, a smaller standard deviation value means that the data set is more closely packed. The data set with the smaller standard deviation is not necessarily bigger, stronger, longer (or whatever the variable represents), but just more consistently located near the mean value.

The mean and the standard deviation both use all of the data values, so they are often linked together. In a similar way, the median and the IQR are usually not affected by the extreme data values, so they are often linked together.

Worked example 15

W.E. 15

The table below shows the test results for 15 students in English and Mathematics. Both tests were scored out of 50. Find the mean and standard deviation for each data set and use these values to compare them. Round the values to 2 decimal places where necessary.

English	42	44	31	20	25	34	45	50	44	12	13	23	44	47	42
Maths	38	34	42	39	40	44	43	27	34	27	40	32	31	22	23

Thinking	Working
1 Calculate the mean and standard deviation.	English: mean = 34.4 SD = 12.39 Maths: mean = 34.4 SD = 6.98
2 Comment on the statistical values calculated.	*The two tests had the same mean score but the standard deviation was much less for the Mathematics test. This means the scores for the Mathematics test were more consistent. However, you can see that in the English test both the best and worst results, overall, were recorded.*

To analyse data, you need to decide which statistical values will be important to consider. Remember that the mean and standard deviation use all the data values. The median and IQR effectively use only the middle 50% of the data values. If there are outliers, it may be better to use the median and IQR. If the data set seems relatively uniformly distributed, then it may be better to use the mean and standard deviation.

Constructing distributions for the mean and standard deviation of simple random samples

It is often impractical to collect data from a whole population, so a random sample is taken that represents the population. The mean and standard deviation of the sample is calculated, and these statistics can be used to estimate the mean and standard deviation of the population. Small samples can be unreliable, but larger samples should give values close to the population.

The formula for the mean of a sample is the same as for the mean of the population.

Mean: $\bar{x} = \frac{\Sigma x}{n} = \frac{\Sigma xf}{\Sigma f}$

The formula for the standard deviation of a sample is slightly different from the population formula, using $n - 1$ instead of n, to help correct for mathematical sampling bias.

Population standard deviation: $\sigma_n = \sqrt{\frac{\Sigma(x - \bar{x})^2}{n}}$

Sample standard deviation: s or $\sigma_{n-1} = \sqrt{\frac{\Sigma(x - \bar{x})^2}{n - 1}}$

Note that your CAS offers both of these values each time you select 'calculate' for a set of data. Simply choose the '$n - 1$' or 's' alternative for sample standard deviation. The sample standard deviation value will always be larger than the population standard deviation value.

Worked example 16 W.E. 16

The data below gives the mass in grams of a population of 50 river rocks to be used in an ornamental garden. In this question, round mean values to 1 decimal place and standard deviations to 2 decimal places.

13 37 44 48 31 17 63 67 37 32 35 13 91 65 23 28 18 32 39 42 54 23 21 29 26
22 18 34 36 38 24 18 17 18 17 30 18 18 30 25 29 26 29 34 20 22 27 29 23 17

(a) Find the mean and standard deviation of the mass of the river rocks.

(b) Use random sampling methods to create 12 samples of 5 values each. Label the samples A, B, C, … L. Find the mean and standard deviation of each sample.

(c) Combine the samples of 5 in pairs to create 6 samples of 10: A and B; C and D; etc. Find the mean and standard deviation of each sample of 10.

(d) Combine the samples of 10 in pairs to create 3 samples of 20: A, B, C, D; E, F, G, H; etc. Find the mean and standard deviation of each sample of 20.

(e) Arrange the values of the sample means and standard deviations around the population values for each of the three different sample sizes. Comment on the distributions.

Thinking	Working
(a) 1 Enter the data into a CAS.	(a) Mean = 30.9 g
2 Read the mean value. Round correct to 1 decimal place, if necessary.	SD = 15.25 g
3 Read the population standard deviation. Round correct to 2 decimal places, if necessary.	

10A

(b) 1 Use a random number generator to produce 12 sets of 5 whole numbers from 1 to 50. Go through the set of 50 numbers picking out the selected data. For example, in Set A you pick the 26th, 39th, 22nd, 3rd and 2nd score.	(b) This is a sample answer only. A: Random numbers: 26, 39, 22, 3, 2 Data: 36, 30, 23, 44, 37 B: Random numbers: 25, 28, 12, 36, 21 Data: 26, 24, 13, 30, 54 C: Random numbers: 1, 29, 5, 13, 8 Data: 13, 34, 31, 91, 67 D: Random numbers: 48, 49, 38, 18, 44 Data: 29, 23, 18, 32, 34 E: Random numbers: 45, 17, 25, 19, 21 Data: 20, 18, 26, 39, 54 F: Random numbers: 14, 34, 8, 9, 12 Data: 65, 18, 67, 37, 13 G: Random numbers: 22, 37, 7, 10, 36 Data: 23, 18, 63, 32, 30 H: Random numbers: 36, 15, 23, 17, 2 Data: 30, 23, 21, 18, 37 I: Random numbers: 34, 1, 22, 28, 25 Data: 18, 13, 23, 24, 26 J: Random numbers: 40, 11, 48, 24, 19 Data: 25, 35, 29, 29, 39 K: Random numbers: 12, 29, 22, 25, 18 Data: 13, 34, 23, 26, 32 L: Random numbers: 13, 10, 30, 33, 32 Data: 91, 32, 18, 17, 18
2 Enter each set of data into a CAS. **3** Read the mean value. Round correct to 1 decimal place, if necessary. **4** Read the sample standard deviation. Round correct to 2 decimal places, if necessary	This is a sample answer based on the samples drawn in part (b). A: Mean = 34 g, SD = 7.91 g B: Mean = 29.4 g, SD = 15.13 g C: Mean = 47.2 g, SD = 31.29 g D: Mean = 27.2 g, SD = 6.61 g E: Mean = 31.4 g, SD = 15.06 g F: Mean = 40 g, SD = 25.38 g G: Mean = 33.2 g, SD = 17.57 g H: Mean = 25.8 g, SD = 7.66 g I: Mean = 20.8 g, SD = 5.26 g J: Mean = 31.4 g, SD = 5.55 g K: Mean = 25.6 g, SD = 8.32 g L: Mean = 35.2 g, SD = 31.81 g

10A

(c) 1 Enter each set of data into a CAS.	(c) This is a sample answer based on the samples drawn in part (b).
2 Read the mean value. Round correct to 1 decimal place, if necessary.	A, B: Mean = 31.7 g, SD = 11.63 g
3 Read the sample standard deviation. Round correct to 2 decimal places, if necessary.	C, D: Mean = 37.2 g, SD = 23.79 g E, F: Mean = 35.7 g, SD = 20.19 g G, H: Mean = 29.5 g, SD = 13.36 g I, J: Mean = 26.1 g, SD = 7.56 g K, L: Mean = 30.4 g, SD = 22.50 g
(d) 1 Enter each set of data into a CAS.	A, B, C, D: Mean = 34.5 g, SD = 18.44 g
2 Read the mean value. Round correct to 1 decimal place, if necessary.	E, F, G, H: Mean = 32.6 g, SD = 16.96 g I, J, K, L: Mean = 28.3 g, SD = 16.48 g
3 Read the sample standard deviation. Round correct to 2 decimal places, if necessary.	

(e) 1 Draw an axis for each set of mean values and mark the population mean.

2 Mark each sample mean on the axis, like a dot plot. Mark the position of the population mean.

(e)

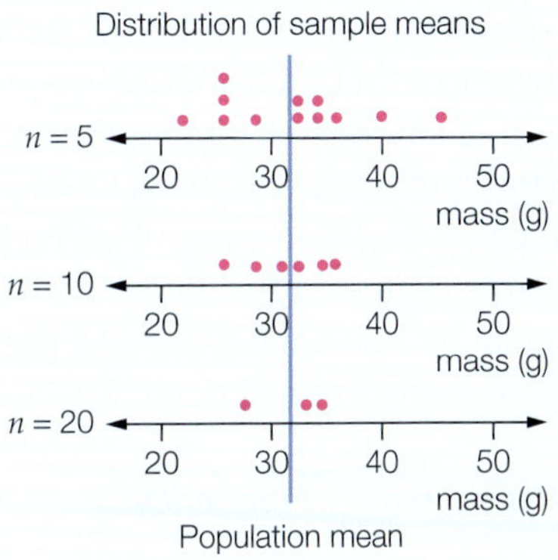

3 Repeat the process for the standard deviations.

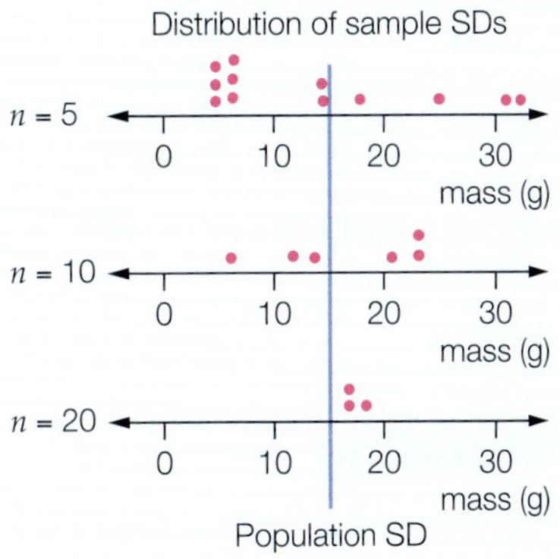

4 Comment on the distributions saying how they are similar and how they are different depending on sample size.

The mean and SD values for the samples were distributed either side of the population values. For the larger samples the sample values clustered closer to the population values. The larger the sample size, the closer the sample statistics are to the population.

2.8 Standard deviation

Navigator

Answers p.782

1, 2, 3, 4, 5, 6, 8, 10	1, 2, 3, 4, 5, 6, 7, 8, 10	2, 3, 4, 6, 7, 8, 9, 10

Fluency

W.E. 14

1 Find the standard deviation, rounded to 2 decimal places, for the following data.

11 12 15 15 17 19 21 24 26 29

W.E. 15

2 The table below shows the test results for 15 students in English and in Mathematics. Both tests were scored out of 50. Find the mean and standard deviation for each data set and use these values to compare them. Round the means to 1 decimal place and the standard deviations to 2 decimal places.

English	40	40	34	23	24	39	46	47	41	19	23	23	34	37	45
Maths	33	44	32	49	36	38	41	29	37	29	41	35	30	22	24

3 The grouped data shown below represents the distances, in km, travelled to school by a group of students. Estimate the mean and standard deviation (both in km) of the distances, using the centre of each class interval to represent the data points. Round the mean to 1 decimal place and the standard deviation to 2 decimal places.

Distance (km)	Frequency
0–<2	22
2–<4	28
4–<6	15
6–<8	7
8–<10	4
10–<12	2

10A

Understanding

4 The following table shows the heights, in m, of the ten highest mountains in each of Australia, New Zealand, Brazil and Sri Lanka.

Australia	2220	2185	2178	2166	2132	2048	2047	2045	2050	1979
New Zealand	3754	3500	3440	3279	3201	3183	3176	3163	3160	3157
Brazil	2973	2890	2882	2861	2849	2798	2791	2770	2734	2033
Sri Lanka	2524	2243	2240	2100	2100	2076	2036	2016	2010	1896

(a) Calculate the range of the data for each country.

(b) Calculate the mean and standard deviation of the data for each country.

(c) Write a sentence or two about what you have found.

5 (a) Find the mean, rounded to 1 decimal place, and standard deviation, rounded to 2 decimal places, for each of the following data sets:

(i) 13 14 14 15 15 15 16 16 19 20 22 25 26 28

(ii) 13 14 14 15 15 15 16 16 19 20 22 25 26 48

(iii) 3 14 14 15 15 15 16 16 19 20 22 25 26 28

(iv) 3 14 14 15 15 15 16 16 19 20 22 25 26 48

(b) Write a sentence or two about what this tells you.

6 You have recorded a data set as: 23 45 46 47 55 57 58 59 62 65 and calculated the mean and standard deviation. When checking your work you discover that the '23' should have been '43'. Changing this will:

A increase both the mean and standard deviation

B increase the mean and decrease the standard deviation

C decrease the mean and increase the standard deviation

D decrease both the mean and the standard deviation.

Reasoning

7 The screenshot shows some information taken from a music playlist. You will need to be careful how you deal with this data. Give your answers correct to the nearest second.

(a) Find the mean and standard deviation for songs listed as *Rock* (including *Rock*, *Classic Rock* and *RockNRoll*).

(b) Find the mean and standard deviation for songs listed as *Pop*.

(c) Find the mean and standard deviation for songs listed as neither *Rock* nor *Pop*.

(d) Draw a parallel box plot showing the three categories *Rock*, *Pop* and *Other*. Use decimal time values here.

(e) Write a couple of sentences comparing the three categories. Is there any evidence that *Rock* songs are a different length to *Pop* songs, for example?

Genre	Time	Genre	Time
Rock	2:07	Pop	2:25
Blues	3:25	Soundtrack	2:12
RnB	2:26	Rock	4:15
Blues	5:27	R&B	3:59
Rock	3:35	Blues	2:33
Country & F...	2:16	Blues	15:30
Rock	5:28	Rock	2:47
Rock	4:02	Rock	5:30
Blues	3:31	Rock	3:32
Blues	2:51	Rock	12:28
Rock	3:26	R&B	4:36
Rock	2:34	Rock	3:30
Pop	3:46	Pop	2:15
Classic Rock	4:01	Blues	3:17
Grunge	3:50	Country & F...	2:48
Pop	3:13	RockNRoll	3:05
Pop	2:25	Pop	3:00
Pop	2:30	Rock	3:51
Blues	2:09	Rock	3:43
Rock	3:31	R&B	3:56
Rock	5:51	Soundtrack	5:56
RockNRoll	1:59	Rock	2:52
Rock	3:32	R&B	4:10
Pop	3:49	Pop	3:52
Blues	4:20	Pop	2:40
Blues	2:46	Rock	3:34
Rock	3:20	Blues	6:00
Rock	3:32	R&B	4:47
Rock	3:32	Rock	5:55
Blues	4:07	Classical	3:57
Rock	8:33	Rock	3:42
Rock	8:53	Pop	4:14

Hint

Change the time in seconds to decimal parts of a minute.

10A

Vic W.E. 16

8 The data below gives the mass in grams of a population of 35 pebbles to be used in a flower arrangement. In this question, round mean values to 1 decimal place and standard deviations to 2 decimal places.

3	6	7	7	5	5	3	8	7	6	6	6	6	5	5	6	5	5
6	5	5	5	5	5	3	5	3	5	7	7	4	6	6	5	7	

(a) Find the mean and standard deviation of the mass of the pebbles.

(b) Use technology to generate random numbers, and use these to randomly select 12 samples of 5 values each from the data. Label the samples A, B, C, … L. Find the mean and standard deviation of each sample.

(c) Combine the samples of 5 in pairs to create 6 samples of 10: A and B; C and D; etc. Find the mean and standard deviation of each sample of 10.

(d) Combine the samples of 10 in pairs to create 3 samples of 20: A, B, C, D; E, F, G, H; etc. Find the mean and standard deviation of each sample of 20.

(e) Use a parallel dot plot to compare the distribution of sample means around the population mean for each of the three different sample sizes. Repeat the process for the distribution of sample standard deviations around the population standard deviation. Comment on the distributions.

Open-ended

9 (a) Construct a data set with at least 20 values that has a mean of between 20 and 22, a range of at least 15 and a standard deviation of between 2 and 4.

(b) Write a couple of sentences about how you went about this task.

(c) What effect did the condition about the range have on this task?

10A

10 Construct a data set, of at least 20 values, that would fit the following box plot and then find the mean and standard deviation of your data set.

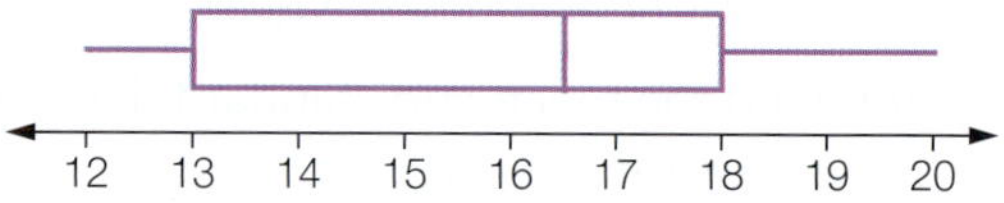

Problem solving

Keeping score

From her attempts to shoot 40 points in the season's first basketball game, Angela scored only 10 points. This gave her a scoring average of 25%. In the second game she had the chance to score 30 points but only scored enough to raise her scoring average over the two games to 50%.

(a) How many points did she score in the second game?

(b) What was her scoring average for the second game?

(c) What is the greatest scoring average Angela could have had after the second game, assuming her maximum in the second game was still 30 points?

Strategy options

- Guess and check.
- Work backwards.

Chapter review 2

Maths literacy

bivariate data	interquartile range	primary data
box plot	outlier	quantile
cumulative frequency curve	parallel box plot	quartile
dependent variable	parallel dot plot	range
five-number summary	percentage cumulative frequency curve	scatter plot
independent variable	percentile	secondary data
10A extrapolation	line of best fit	standard deviation
interpolation	regression line	variance

Copy and complete the following using the words and phrases from the list, where appropriate. A word or phrase may be used more than once.

1 Data taken from other people's research is __________.

2 The difference between the upper quartile and the lower quartile is called the __________.

3 A __________ is a graph that shows the sum of frequencies 'up to' each data value.

4 Data collected from your own observations is __________.

5 A__________ is a graph that shows the minimum, lower quartile, median, upper quartile and maximum values for a data set.

6 The value below which 80% of values are on a cumulative frequency curve can also be referred to as the 0.8 __________.

7 The difference between the largest observed value and the smallest is called the __________.

8 If a value is more than $1.5 \times \text{IQR}$ above Q_U or more than $1.5 \times \text{IQR}$ below Q_L then it is an __________.

10A 9 When you use a regression line to predict a value within the known range you are using __________. If the value is outside the known range you are using __________.

10A 10 The__________ is the square root of the __________.

Fluency

1 For each of the following cumulative frequency curves: **2.1, 2.2**

(i) What is the median?

(ii) What is the interquartile range?

(iii) Draw a box plot.

(a)

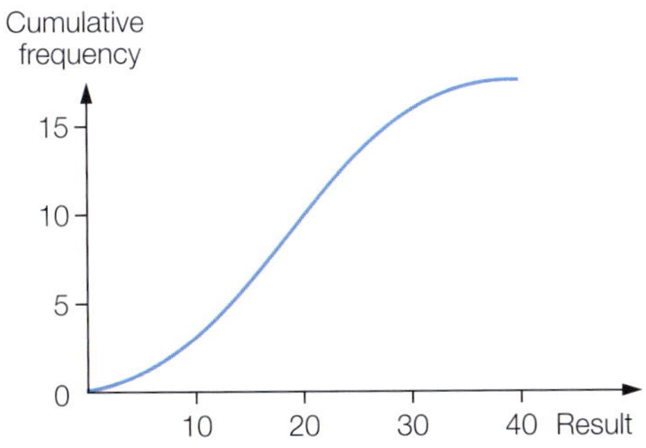

(b)

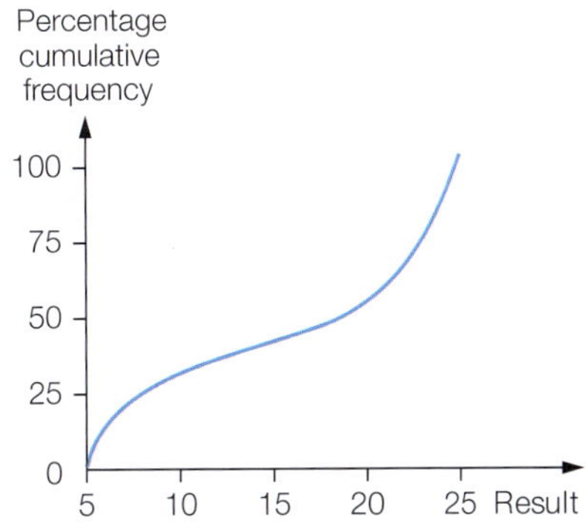

2 For the set of data shown at right: **2.2**

(a) find the median by drawing a percentage cumulative frequency curve

(b) find the interquartile range

(c) draw a box plot.

Class interval	Frequency
120–<125	5
125–<130	17
130–<135	32
135–<140	54
140–<145	24
145–<150	13

3 A frog in training achieves the following jump distances (in centimetres). On the basis of these jumps, how long would a jump need to be, for it to be considered an outlier? **2.2**

19 25 17 22 20 19 16 15 17 21

4 You have been working in a group on an analysis task. One member of the group has produced the following parallel box plot for the data that has been collected. Your task is to write the first draft of the analysis. This should include a comment about which team you think has been more successful. **2.3**

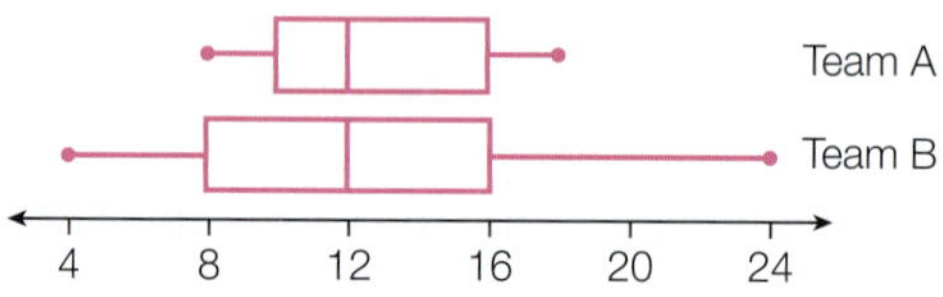

5 For each of the following occupations, state whether you think they deal with primary data or secondary data. **2.4**

(a) Meteorologist

(b) Television weather forecaster

(c) Insurance broker

(d) Cancer research scientist

10A 6 Find, rounded to 2 decimal places, the standard deviation for the following data sets. **2.8**

(a) 3 4 4 5 6 7 7 7 7 8 9 10 11 11 12 12 12 14 15 16

(b)

Number of flaws	Frequency
0	2
1	19
2	16
3	15
4	7

(c)

Weight (kg)	Frequency
0–<5	6
5–<10	11
10–<15	15
15–<20	8
20–<25	2

10A 7 Find the equation of the line of best fit, using a spreadsheet or some other form of technology, for each of the following sets of data. Round your answers to 3 decimal places, where necessary. In each case, the required rule is of the form $y = \ldots$ **2.6**

(a)

x	0	1	2	3	4	5	6
y	3	7	13	20	26	30	35

(b)

x	0	2	5	7	10	12	14
y	18	15	11	5	0	-6	-11

Understanding

8 Consider the following data: **2.1**

72.8 64.8 55.4 44.3 61.8 52.9 62.0 73.6 34.5 72.3
65.9 57.7 64.4 55.3 45.2 71.2 69.6 54.2 59.9 51.1

(a) Group the data into intervals of 10 (30–<40, 40–<50 etc.) and draw a cumulative frequency curve.

(b) Calculate the five-number summary for the data.

(c) Find the median from the cumulative frequency curve and compare it with the median calculated from the raw data.

9 The rainfall (in millimetres) recorded on each day of a month in a particular location was: **2.2**

12	15	17	16	25	2	44	3	2	0
4	6	8	10	0	11	3	6	3	8
6	15	15	17	0	22	17	21	30	33

(a) Calculate the range for the rainfall figures.

(b) Calculate the five-number summary for the data.

(c) Draw a box plot to represent the data.

10 For each of the following box plots: **2.2**

(i) indicate what values would be considered as outliers

(ii) decide whether there are any outliers in the data set displayed.

(a)

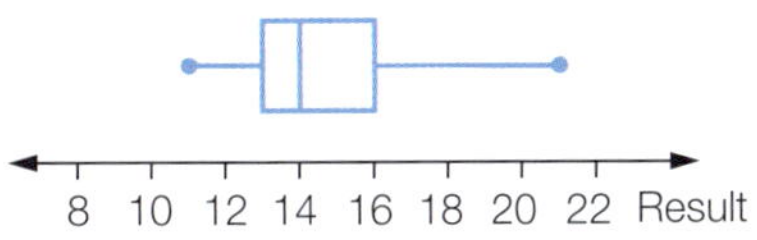

(b)

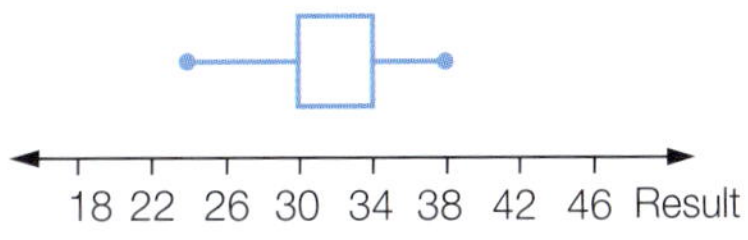

Reasoning

11 Some of the data available from the Bureau of Meteorology website for Perth (Metro) is shown below. **2.4**

Temperature (°C)	Jan	Feb	Mar	Apr	May	Jun	Jul	Aug	Sep	Oct	Nov	Dec	Annual
Mean maximum	30.9	31.3	29.5	25.6	22.4	19.3	18.3	18.8	20.0	22.9	26.2	28.8	24
Mean minimum	17.9	18.0	16.5	13.5	10.5	8.4	7.8	8.1	9.5	11.2	14.1	16.2	12

(a) Find the mean difference in temperature (°C) for each month.

(b) Draw a scatter plot showing mean maximum temperature, mean minimum temperature and mean difference in temperature.

(c) Describe the temperature pattern for Perth (Metro) using the graph as an aid.

(d) Could you use this graph to describe the temperature pattern for Kalgoorlie, in inland Western Australia? Give a reason for your answer.

(e) The daily maximum and minimum temperatures in Perth for June 2010 are given.

Minimum:	10.9	6.8	5.6	4.8	9.9	5.7	6.5	6.6	8.7	7.3	9.6
	6.9	4.3	2.2	2.3	6.1	12.0	9.7	10.8	13.1	6.4	9.4
	8.2	10.8	11.1	6.4	3.1	6.3	8.4	5.2			
Maximum:	16.2	17.9	17.6	17.0	16.9	14.3	15.5	15.9	15.1	16.3	16.1
	16.0	13.4	13.8	14.0	16.9	16.3	16.5	17.0	17.3	16.4	17.3
	18.2	18.2	15.2	14.2	14.5	14.6	11.9	14.0			

How did June 2010 compare to the overall data?

10A **12** The number of colony forming units per 100 mL of water of two different bacteria taken from a particular lake at weekly periods is recorded in the following table. The table also shows the maximum temperature (°C) for the day of recording.

2.6

Fecal coliform (CFU/100 mL)	350	73	180	150	200	130	2200	6300
E. coli (CFU/100 mL)	350	90	120	330	200	120	1600	5800
Temperature (°C)	15.3	14.6	16.8	16.8	18	24	19.6	21.7

Fecal coliform (CFU/100 mL)	360	1200	560	1700	19	430	320	87
E. coli (CFU/100 mL)	300	980	630	1800	16	270	270	160
Temperature (°C)	20.1	19.4	22.1	24.1	22	23	19	18

(a) In how many different ways can the data be grouped to compare a pair of variables?

Use technology to:

(b) draw a scatter plot with fecal coliform on the x-axis and *E. coli* bacteria on the y-axis

(c) find the equation of the line of best fit, assuming the data is linear.

13 The following fictional article contains information about the opinions of voters on the introduction of tolls on certain roads around Melbourne.

2.7

Toll roads – poll reaction

The State Government is facing a big electoral backlash over its new toll roads according to a *Sunday Bugle* poll conducted by Corio Market Research Centre in marginal seats along the route.

The poll found that at least half the voters surveyed avoid paying the tolls on the Tullamarine Freeway and Monash Freeway.

Residents of the affected areas have increased worries about traffic on their local suburban roads growing massively. The poll shows that the Government could lose three of the four seats in which the survey was conducted.

SUNDAY BUGLE POLL: TOLL ROADS

How marginal electorates would vote . . .

	Government %	Opposition %
Tullamarine	33	47
Essendon	43	43
Oakleigh	38	38
Knox	46	32

. . . and who avoid tolls

Tullamarine	59%
Essendon	50%
Oakleigh	52%
Knox	49%

Source: Telephone poll of 1200 voters conducted by Corio Market Research Centre

(a) Do the survey results support the article's suggestion of a 'big electoral backlash'?

(b) Add up the percentages for Government and Opposition voters in the Tullamarine electorate. Why don't they add up to 100%? Why is this important to the judgement of the poll?

(c) The poll was done by telephone. Which sampling methods would you use to get the best indication of how voters in each electorate are reacting to the toll roads?

(d) What is the significance of the four different locations?

14 The owner of a company holds a record of the salaries paid to all the employees. Annual salaries 3 years ago and today are given in the following table.

2.3

Salary range (× \$10 000)	Number of employees (3 years ago)	Number of employees (currently)
1–<3	12	9
3–<5	15	10
5–<7	9	9
7–<9	5	8
9–<11	3	7
11–<13	0	5

(a) Convert the data for employees to percentages, and then draw a graph that emphasises the difference between salaries paid 3 years ago and those paid currently.

(b) Contrast the means, medians and interquartile ranges for the two sets of data.

(c) Draw a box plot for each set of data.

(d) Describe, in your own words, the change in salaries over the past 3 years.

(e) How much more is the company paying in salaries today than it was 3 years ago?

(f) By what percentage has the average salary risen?

15 The grouped frequency table below shows the age distribution of the employees of a local company. The age given is the nearest whole number of years on 1 January of the current year. (So, if a worker was 22 years and 5 months old they would be recorded as 22, whereas a worker 22 years and 7 months old would be recorded as 23.)

2.1, 2.2

Age at 1 Jan (years)	16–20	21–25	26–30	31–35	36–40	41–45	46–50	51–55
Frequency	25	30	32	28	33	42	55	10

(a) Estimate the mean age, to 1 decimal place, giving a reason for your estimate.

10A (b) Calculate the mean and standard deviation for the ages of the employees, giving your answers rounded to 2 decimal places, where appropriate.

2.8

(c) Draw a cumulative frequency curve and use it to find the five-number summary.

(d) Draw a box plot of the data.

(e) Use your cumulative frequency curve to estimate the number of employees who are:

(i) under 32 (ii) over 48.

3 Algebra and quadratics, part 1

What would you do if you had control of the traffic lights?

Name: Miles Brown
Job: Communication Technical Services Officer
Qualifications: Advanced Diploma of Electronic Engineering

When 000 FIRE is called, there is an extensive network of equipment that needs to be activated and personnel that need to be alerted. Who makes sure that all the technology is in place so that an emergency can be handled as smoothly as possible?

Miles Brown works as a Communication and Technical Services Officer at the CFA (Country Fire Authority) and does exactly that.

'We're in charge of all the maintenance and installation of the communication and signalling systems for the CFA, from repairing radios and telephone equipment, to when there's a fire call; all the things that make the lights go on, sirens go on, doors open, and traffic lights change.

There's a lot of rural travel, working outdoors and physical tasks. Sometimes you can be running cabling, spending a day up in a roof or crawling around under a building. It's very hands on. That's one of the reasons that I chose it.'

Miles uses algebra to calculate the resistance of electronic circuits and measure the signal strength of radios.

Why learn this?

Algebra is a convenient way to describe and generalise patterns that occur in the real world. Algebraic techniques allow you to manipulate algebraic expressions into different forms to understand and compare these patterns. In this chapter, you will learn and practise skills to help you solve more complex equations and use variables to model a greater variety of situations.

After completing this chapter you will be able to:

- expand binomial products, expressing answers in simplest form
- factorise quadratic trinomials and other algebraic expressions by using various methods
- recognise and use special products to factorise expressions
- identify, describe, construct and transform algebraic and graphical representations of parabolas
- simplify expressions involving algebraic fractions.

3 Recall

Prepare for this chapter by attempting the following questions. If you have difficulty with a question, you can download a Recall Worksheet from the eBook or the Pearson Places website.

1 How many terms are in each expression?

(a) $x^2 + 6x + 9$ (b) $16x^2 - y^2$ (c) $xy + 2y + 4x + 8$

2 What is the coefficient of each of these terms?

(a) $4x$ (b) $-6ab$ (c) y (d) $-12x^2y$

3 Simplify: (a) $3 \times a + 5 \times b$ (b) $8 \times 3x - 8 \times y$

4 Expand and simplify if possible:

(a) $3(x + 2)$ (b) $5(x - 3)$ (c) $3x + 2(x - 4)$

5 Expand and simplify if possible:

(a) $-2(x + 3)$ (b) $-8(x - 5)$ (c) $-(x + 2)$

6 Simplify: (a) $(3x)^2$ (b) $(2y)^2$ (c) $(7k)^2$

7 Simplify: (a) $(4x)^2 - 2x^2$ (b) $p^2 - (3p)^2$ (c) $(-y)^2 + y^2$

8 Rearrange to make x the subject.

(a) $y = 2x + 3$ (b) $2a = \dfrac{x - 4b}{c}$ (c) $s = \dfrac{5x - at^2}{3}$

9 (a) If $p = 2m^2 - 3m + 5$, find the value of p when:

(i) $m = 1$ (ii) $m = 0$ (iii) $m = -1$

(b) Find the value of $x(x - 3)$ when:

(i) $x = 5$ (ii) $x = -3$ (iii) $x = 0$

10 Simplify the following.

(a) $\dfrac{2}{3} + \dfrac{1}{4}$ (b) $\dfrac{5}{6} - \dfrac{3}{8}$ (c) $\dfrac{a}{3} + \dfrac{4a}{3}$

11 Simplify the following.

(a) $\dfrac{3}{8} \times \dfrac{4}{9}$ (b) $\dfrac{x}{10} \times \dfrac{5}{x}$ (c) $\dfrac{5}{6} \div \dfrac{10}{9}$

Exploration Task

You can download this activity from the eBook or the Pearson Places website.

How can rectangles help with algebra?

In this activity you will explore how the area or grid method of multiplication can help when multiplying algebraic expressions.

3.1 Expanding brackets

Expanding brackets

The distributive law is used to multiply terms inside brackets by terms outside brackets, to remove the brackets. This is often called 'expanding the brackets'.

$a(b + c) = ab + ac$
$a(b - c) = ab - ac$

Worked example 1 W.E. 1

Expand and simplify each of the following.

(a) $3(4x - 2)$ **(b)** $x(x + 4) - 2(x - 1)$

Thinking	Working
(a) Multiply all terms inside the brackets by the term in front of the brackets.	(a) $3(4x - 2)$ $= 12x - 6$
(b) 1 Expand both sets of brackets by multiplying every term inside the brackets by the term in front of the brackets. Change all signs when multiplying by a negative number.	(b) $x(x + 4) - 2(x - 1)$ $= x^2 + 4x - 2x + 2$
2 Simplify by collecting like terms.	$= x^2 + 2x + 2$

Expanding binomial products

A **binomial product** is a product that has two terms in each expression being multiplied, such as $(x + 3)(x + 7)$. To understand how to find the expansion of binomial products like this, consider the rectangle below.

If the length of a rectangle is $(x + 7)$ units, and the width is $(x + 3)$ units then an expression for the area is:

$A = lw$
$A = (x + 7)(x + 3)$ square units (Note that 'square units' can be written as 'units^2'.)

You can separate the $x + 7$ into a length of x units and a length of 7 units. You can also separate $(x + 3)$ in a similar way.

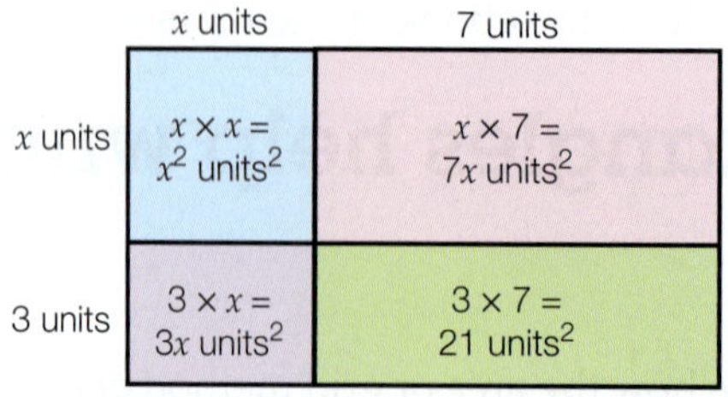

You can now add the areas of the four smaller rectangles and simplify the like terms.

$A = x^2 + 7x + 3x + 21$
$= x^2 + 10x + 21$ units^2

Both expressions represent the area of the same big rectangle, so the expressions are equal.

$(x + 7)(x + 3) = x^2 + 10x + 21$

This is the extended distributive law.

To help you remember how to multiply the terms together, think of the word FOIL:

First terms in each bracket
Outside terms in each bracket
Inside terms in each bracket
Last terms in each bracket

and then add them all together.

To expand binomial products, multiply each term in the second bracket by each term in the first bracket and add them together.

$$(a + b)(c + d) = \overset{F}{ac} + \overset{O}{ad} + \overset{I}{bc} + \overset{L}{bd}$$

Worked example 2

W.E. 2

Expand and simplify $(x + 5)(2x - 3)$.

Thinking	Working
1 Multiply each term in the second bracket by each term in the first bracket. $(x + 5)(2x - 3)$ (F, O, I, L)	$(x + 5)(2x - 3)$ $= 2x^2 - 3x + 10x - 15$
2 Simplify by collecting like terms.	$= 2x^2 + 7x - 15$

Expanding special binomial products

The following rules are useful when expanding special products.

The **difference of two squares** rule:

$$\begin{aligned}(a + b)(a - b) &= a^2 - ab + ba - b^2 \\ &= a^2 - b^2\end{aligned}$$

The **perfect square** rules:

$$\begin{aligned}(a + b)^2 &= (a + b)(a + b) \\ &= a^2 + ab + ba + b^2 \\ &= a^2 + 2ab + b^2\end{aligned}$$

$$\begin{aligned}(a - b)^2 &= (a - b)(a - b) \\ &= a^2 - ab - ba + b^2 \\ &= a^2 - 2ab + b^2\end{aligned}$$

Together, these rules can be used to expand special binomial products.

$(a - b)(a + b) = a^2 - b^2$ (difference of two squares)
$(a + b)^2 = a^2 + 2ab + b^2$ (perfect square)
$(a - b)^2 = a^2 - 2ab + b^2$ (perfect square)

Worked example 3

W.E. 3

Expand and simplify:

(a) $(x-3)(x+3)$ (b) $(3x+1)^2$

Thinking	Working
(a) Identify a special product and apply the difference of two squares rule.	(a) $(x-3)(x+3) = (x)^2 - (3)^2$ $= x^2 - 9$
(b) Identify a special product and apply the perfect square rule.	(b) $(3x+1)^2 = (3x)^2 + 2 \times 3x \times 1 + (1)^2$ $= 9x^2 + 6x + 1$

Expanding three factors

To multiply three or more factors, first multiply any two factors together and simplify the result. Then repeat this process until the expression is fully expanded. Look for special product expansions to expand more easily.

Worked example 4

W.E. 4

Expand and simplify.

(a) $x(x-2)(x+3)$ (b) $(x+3)(x+1)(x-4)$ (c) $(x-1)(x-5)(x+1)$

Thinking	Working
(a) 1 Expand and simplify the binomial product.	(a) $x(x-2)(x+3)$ $= x(x^2 + 3x - 2x - 6)$ $= x(x^2 + x - 6)$
2 Multiply by the single factor.	$= x^3 + x^2 - 6x$
(b) 1 Look for special product expansions. (Here, there are none so expand the first two brackets and simplify their product.)	(b) $(x+3)(x+1)(x-4)$ $= (x^2 + x + 3x + 3)(x-4)$ $= (x^2 + 4x + 3)(x-4)$
2 Expand the two remaining brackets.	$= x^3 - 4x^2 + 4x^2 - 16x + 3x - 12$
3 Simplify the expansion.	$= x^3 - 13x - 12$
(c) 1 Look for special product expansions. Rearrange the expansion if necessary. (Here, $(x-1)(x+1) = x^2 - 1$.)	(c) $(x-1)(x-5)(x+1)$ $= (x-1)(x+1)(x-5)$ $= (x^2 - 1)(x-5)$
2 Expand the two remaining brackets.	$= x^3 - 5x^2 - x + 5$

Using CAS to check expansions

CAS technology makes checking algebraic expansions easy.

You should already be able to work out by hand that the expansion of $x(x-4)(x+6)$ is x^3+2x^2-24x. On your CAS you can use an 'expand' command to find this expansion.

Using TI-Nspire CAS	Using Casio ClassPad CAS
Select **menu** > **Algebra** > **Expand** and then enter the expression, being careful to include the correct number of brackets and a multiplication symbol between each set of brackets as shown. 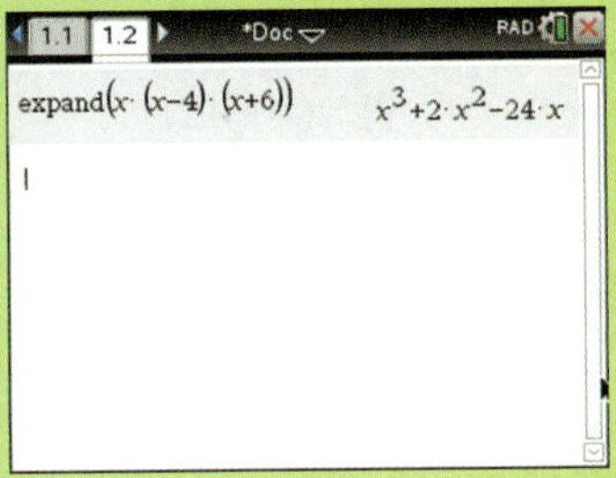	From the menu select **Main** and then enter the expression, including brackets around each part. Then highlight the expression and select **Interactive** > **Transformation** > **expand**. 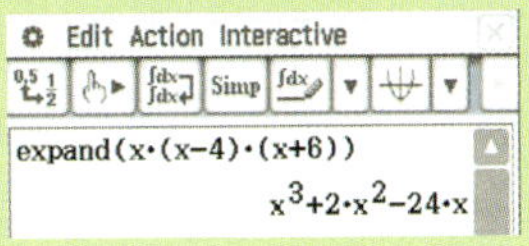

Expand the following and check the expansions using your CAS.

1 $3x(x-2)(x+7)$ **2** $(x-2)(x+8)(x+2)$ **3** $(x+1)(x-4)(x+11)$

If you have used the correct CAS instruction to expand the expressions, you should see these answers:

1 $3x^3+15x^2-42x$ **2** $x^3+8x^2-4x-32$ **3** $x^3+8x^2-37x-44$

3.1 Expanding brackets

Navigator

1 (columns 1–2), 2 (columns 1–2), 3 (columns 1–2), 4, 5 (column 1), 6, 7 (column 1), 8 (a–e), 9, 10, 13, 14, 16, 17	1 (columns 2–3), 2 (columns 2–3), 3 (columns 2–3), 4, 5 (column 2), 6, 7 (column 2), 8, 9, 10, 12, 13, 14, 15, 16, 17	1 (column 3), 2 (column 3), 3 (column 3), 5 (column 3), 6, 7 (column 3), 8, 10, 11, 12, 13, 15, 16, 17, 18

Answers p.786

Fluency

1 Expand and simplify each of the following. W.E.1

(a) $4(x+3)$ (b) $8(x+2)$ (c) $12(x+11)$

(d) $2(6x-1)$ (e) $7(2x-5)$ (f) $x(5x-4)$

(g) $-2(x+1)$ (h) $-x(x-7)$ (i) $-x(3x-6)$

(j) $x(x+2)+3(x+2)$ (k) $x(2x+1)+2(2x+1)$ (l) $x(5x+2)+3(5x+2)$

(m) $2x(x-3)-5(x+6)$ (n) $5x(2x+5)-4(3x-8)$ (o) $4x(3x-1)-6(2x+1)$

(p) $2x(x-4)+3x(x-2)$ (q) $3x(2x-5)-3x(3x-1)$ (r) $5x(x^2-1)-2x(4x-3)$

W.E. 2

2 Expand and simplify each of the following.

(a) $(x+1)(x+3)$ (b) $(x+4)(x+2)$ (c) $(x+5)(x+4)$

(d) $(x+8)(x-3)$ (e) $(x+7)(x-9)$ (f) $(x-3)(x+7)$

(g) $(x-1)(x-7)$ (h) $(x-4)(x-5)$ (i) $(x-9)(x-3)$

(j) $(2x+3)(x-2)$ (k) $(4x-5)(x+3)$ (l) $(7x+1)(x-4)$

(m) $(3x-1)(2x-5)$ (n) $(5x-3)(7-x)$ (o) $(6x-5)(2-3x)$

W.E. 3

3 Expand and simplify each of the following, using the rules for special products.

(a) $(2-x)(2+x)$ (b) $(7-x)(7+x)$ (c) $(6+x)(6-x)$

(d) $(2x+1)(2x-1)$ (e) $(3x+4)(3x-4)$ (f) $(5-3x)(5+3x)$

(g) $(x+6)^2$ (h) $(x+11)^2$ (i) $(8+x)^2$

(j) $(x-3)^2$ (k) $(x-7)^2$ (l) $(x-12)^2$

(m) $(2x+5)^2$ (n) $(3x+2)^2$ (o) $(7x-2)^2$

4 Choose the option that would correctly complete the equation.

(a) $7(x + \_\_) = 7x + 21$

A 3 B 14 C 21 D x

(b) $\_\_\_(x-3) = 5x - 15$

A 15 B 5 C 2 D 12

(c) $\_\_\_(10-8x) = 20x - 16x^2$

A $2x$ B 2 C x D 10

(d) $-6(\_\_\_\_\_) = 6x - 18$

A $x+3$ B $x-3$ C $-x+3$ D $x-18$

(e) $4(2x + \_\_) = 8x + 2$

A $\frac{1}{2}$ B 2 C 2 D 4

(f) $\frac{1}{2}(\_\_ + 20) = x + 10$

A x B 2 C 1 D $2x$

(g) $(x-3)(x + \_\_) = x^2 - 9$

A 9 B -9 C 3 D -3

(h) $(\_\_ + 1)(2x-1) = 4x^2 - 1$

A $4x$ B 2 C $2x$ D x

W.E. 4

5 Expand and simplify each of the following.

(a) $x(x+6)(x+4)$ (b) $x(x-1)(x+5)$ (c) $x(x+1)(x-4)$

(d) $-x(x-3)(x+2)$ (e) $-3x(x-2)(x-1)$ (f) $-4x(x-8)(x+3)$

(g) $(x+4)(x+2)(x+5)$ (h) $(x+2)(x-5)(x-1)$ (i) $(x-3)(x-6)(x-2)$

(j) $(x-3)(x+3)(x+5)$ (k) $(x+7)(x-2)(x+2)$ (l) $(2x+1)(x+4)(x-4)$

(m) $(x-3)(x+4)(x+4)$ (n) $(x+2)(x+2)(x-1)$ (o) $(x-6)(x+5)(x-6)$

Understanding

6 Jax did not get full marks for his solution to the following question:

Expand and simplify $(3x - 2)(2x - 3)$.

$(3x - 2)(2x - 3) = 3x \times 2x + 3x \times -3 + -2 \times 2x + -2 \times -3$ Line 1

$= 6x - 9x + 4x + 6$ Line 2

$= x + 6$ Line 3

(a) In which line are the errors?

(b) Write the correct solution.

7 Expand and simplify the following.

(a) $(4x - 3y)(2x + 5y)$ (b) $(5x + 2y)(7x + 6y)$ (c) $(3x - 4y)(5x - 8y)$

(d) $(2x + 7y)(2x + 7y)$ (e) $(7x + 6y)(7x + 6y)$ (f) $(5x + 8y)(5x + 8y)$

(g) $(4x - y)(4x - y)$ (h) $(6x - 5y)(6x - 5y)$ (i) $(9x - 7y)(9x - 7y)$

(j) $(5y - x)(5y + x)$ (k) $(3x - y)(3x + y)$ (l) $(2x + 3y)(2x - 3y)$

(m) $x(2x - 1)(3x + 5)$ (n) $x(3x + 2)(5x - 6)$ (o) $3x(4x + 7)(2x + 3)$

(p) $-x(3x - 4)(5x + 4)$ (q) $-2x(5x + 3)(3x + 4)$ (r) $-4x(2x - 3)(3x - 5)$

8 Find the shaded area, in terms of x, of each of these shapes.

(a)

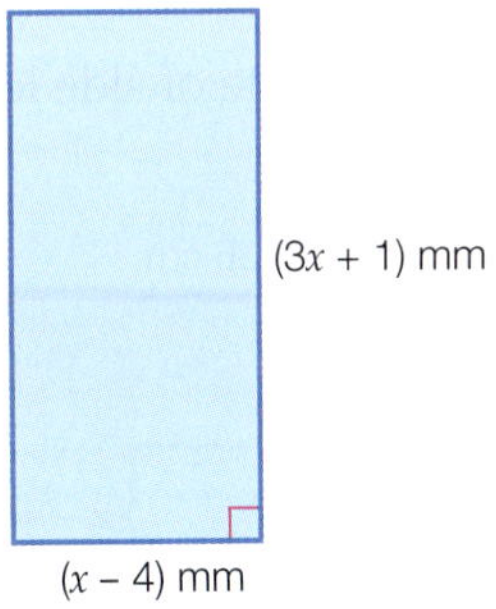

(b)

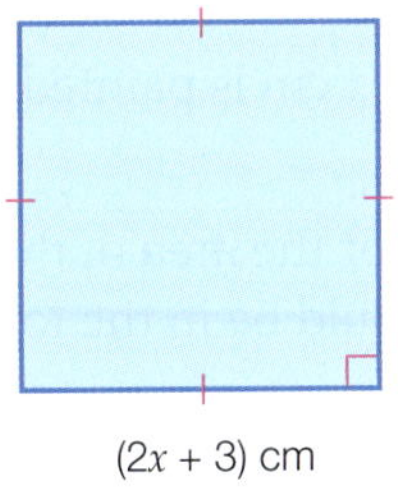

(c)

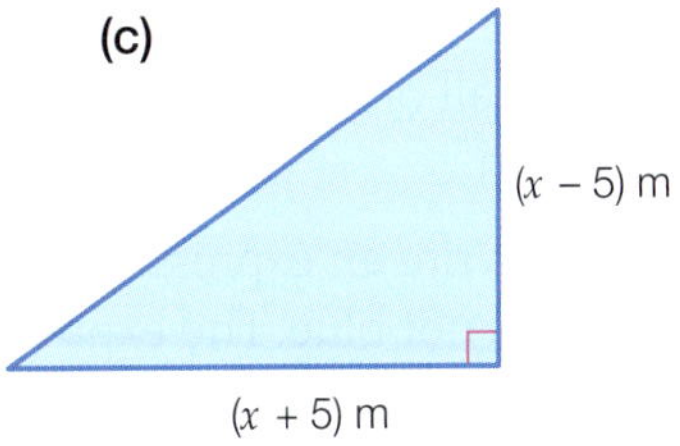

(d)

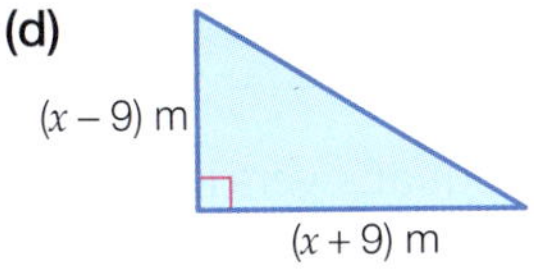

(e)

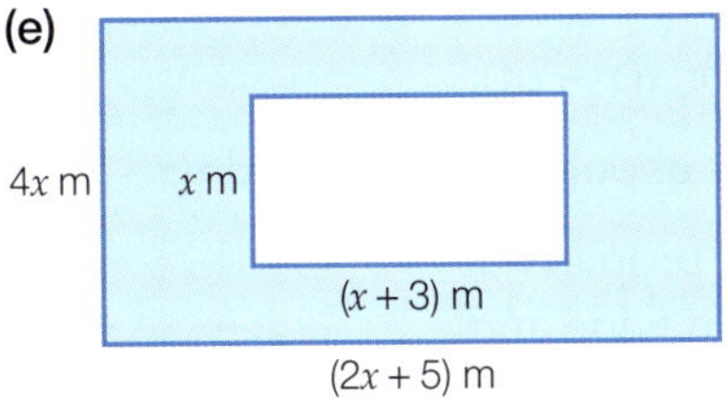

(f)

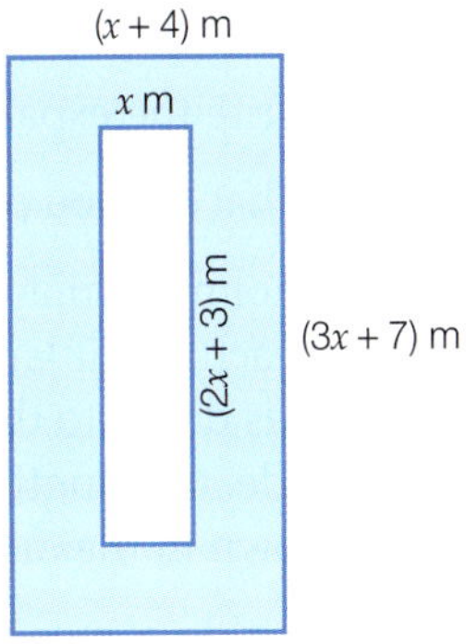

9 The expression $(x + 1)(x + 3)(x - 2)$ is *not* the same as:

A $(x + 3)(x + 1)(x - 2)$ B $(x - 3)(x + 1)(x + 2)$

C $(x + 1)(x - 2)(x + 3)$ D $(x - 2)(x + 1)(x + 3)$

10 (a) If x is the second of three consecutive even numbers, write an expression for the product of the three numbers.

(b) Expand the expression in **(a)**.

(c) Why is it simpler to make x the second and not the first number?

11 (a) Write the dimensions of the shaded rectangle.

(b) Write the binomial product used to find the area of the shaded region, then expand and simplify.

(c) Find the area of the square of side x cm and the rectangle of length 5 cm and width x cm.

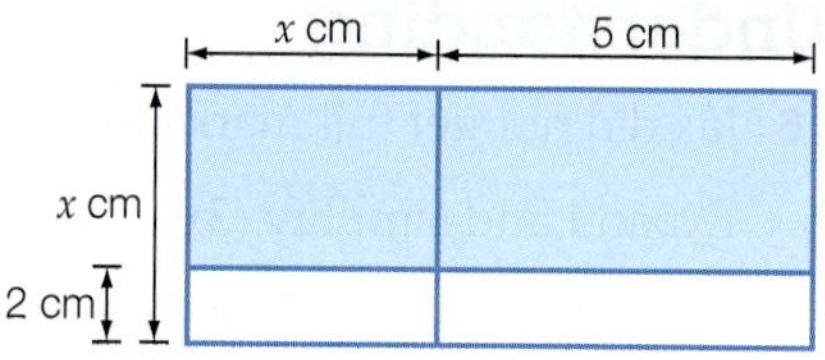

(d) Find the unshaded area, writing it as the sum of the areas of the two rectangles.

(e) Use your answers in **(c)** and **(d)** to find the area of the shaded region in simplest form.

(f) Compare your answers in **(b)** and **(e)**.

12 (a) Expand $x(x + a)(x + b)$.

(b) Use your answer to **(a)** to:

(i) write the coefficient of x^2

(ii) write the coefficient of x

(iii) show that $x(x + a)(x + b) = x^3 - 4x^2 - 12x$, when $a = -6$ and $b = 2$.

(c) For the expression $x(x + 25)(x - 4)$:

(i) write the values of a and b

(ii) use the expansion found in **(a)** to expand $x(x + 25)(x - 4)$.

Reasoning

13 A small blue square of side length x cm is painted on a square green tile of side length 36 cm.

(a) Write an expression (in cm^2) for the area of the tile that is not blue. The expression should be in the form of a product of factors.

(b) Use this product to find the area of the green part if the blue square has a side length of 6 cm.

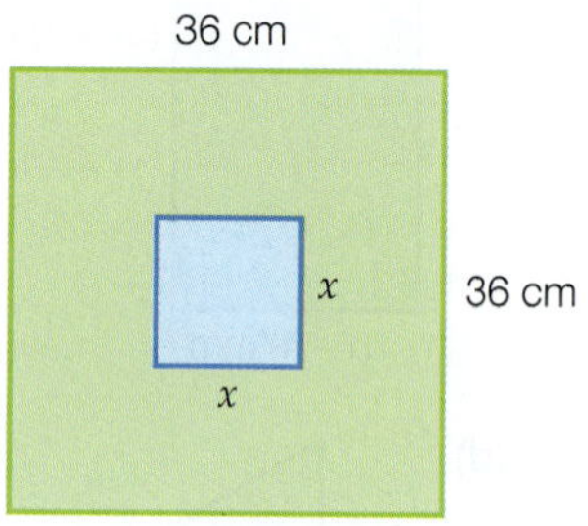

14 (a) Expand and simplify $(x - a)^2 - a^2$.

(b) Write the expansion for $x(x - 2a)$.

(c) What can you conclude about $(x - a)^2 - a^2$ and $x(x - 2a)$ from your answers in **(a)** and **(b)**?

15 The Holmes family own a property at the beach. Their land is x m wide and extends down to the water's edge. At low tide, the length of the land is 55 m, as shown in the diagram at right.

As the tide comes in, the length is reduced by y m.

(a) Write an expression for the length of the land at any time of day.

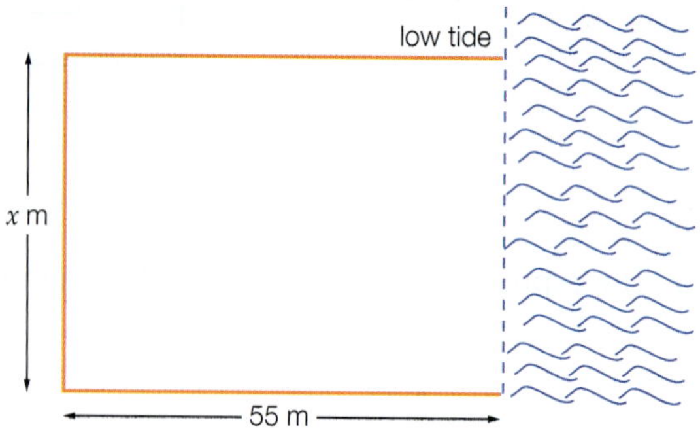

(b) Write an expression for the area of the land at any time of day.

(c) What is the area of the land at low tide?

(d) The family have decided to subdivide the land by reducing their land width by 30 m. Write an expression for the area of the remaining land at any time of the day.

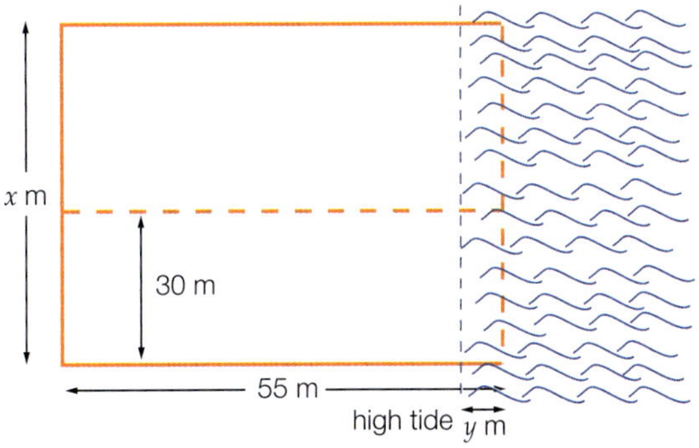

(e) Expand the expression.

(f) If the tide comes in 10 m at high tide, what is the area of land at high tide?

(g) What is the difference between the area of land at low tide before the subdivision and at high tide after the subdivision?

16 Consider the rectangular box shown.

(a) Write an expression (in cm^3) for the volume of the box in factorised form.

(b) Find the volume when x is 4 cm.

(c) Show that the volume of the box is 72 cm^3 when $x = 5$.

(d) Write the expression (in cm^3) for the volume in expanded form. Use this to calculate the volume when $x = 5$.

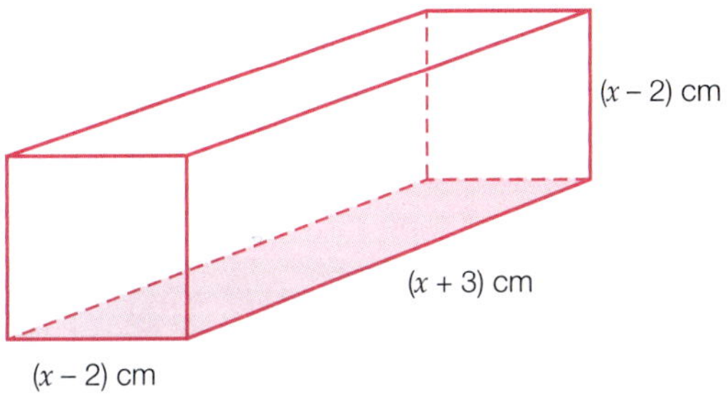

(e) State the dimensions (length, width and height) of the box when the volume is 72 cm^3.

(f) What is the smallest possible integer value of x so that the box exists?

(g) Find the volume of the box when x is the value found in part **(f)**.

Open-ended

17 Find two different sets of values for a, b, c and d so that the expanded, simplified form of $(ax + b)(cx + d)$ is a binomial expression.

Hint
A binomial expression contains two terms.

18 Find two different whole numbers representing x so that the value of the expression $(4 - 2x)(x + 3)(5x - 10)$ is:

(a) positive (b) negative (c) zero.

3.2 Factorising using common factors

Factorising is the opposite or reverse process to expanding.

In the previous section, expressions were converted from factorised form (with brackets) to expanded form (without brackets).

For example:

$$5(x+3) = 5 \times x + 5 \times 3$$
$$= 5x + 15$$

factorised form → expanded form

In other words, $5(x+3)$ can be expanded to $5x + 15$. In reverse, $5x + 15$ can be factorised to $5(x+3)$.

To factorise an expression, find the **highest common factor (HCF)** of the terms and write it outside a set of brackets that contains the remaining terms.

A common factor of -1 may also be taken out.

Worked example 5

W.E. 5

Factorise:

(a) $15xy - 9y$ **(b)** $8x^2 + 12x$ **(c)** $-12a^2b - 9ab^2$

Thinking	Working
(a) Find the HCF of the terms ($3y$). Write it in front of the brackets. Divide each term by the HCF and write the quotient inside the brackets.	(a) $15xy - 9y$ $= 3y \times 5x + 3y \times (-3)$ $= 3y(5x - 3)$
(b) Find the HCF of the terms ($4x$). Write it in front of the brackets. Divide each term by the HCF and write the quotient inside the brackets.	(b) $8x^2 + 12x$ $= 4x \times 2x + 4x \times 3$ $= 4x(2x + 3)$
(c) Find the HCF of the terms ($3ab$). Write it in front of the brackets. Take out the factor of -1 also. Divide each term by the negative of the HCF and write the quotient inside the brackets. Remember to change signs inside the brackets.	(c) $-12a^2b - 9ab^2$ $= -3ab \times 4a + -3ab \times 3b$ $= -3ab(4a + 3b)$

You can check your answer by expanding the brackets in your answer to get back to the original expression.

Sometimes the HCF is an expression of two or more terms.

Worked example 6

W.E. 6

Factorise $k(m+3)+5(m+3)$.

Thinking	Working
Find the HCF of the terms. (Here, it is $(m+3)$.) Write it in front of the brackets. Divide each term by the HCF and write this quotient inside the brackets.	$k(m+3)+5(m+3)$ $=(m+3)(k+5)$

Factorising four terms by grouping in pairs

For expressions with four terms , you can sometimes factorise by **grouping in pairs**, which is also referred to as 'two plus two' factorisation.

To factorise by grouping in pairs:

- Check that the expression has four terms.
- Group as two pairs of terms, with a common factor in each pair.
- Take out the common factor from each pair of terms, then write the expression as a binomial product.

Worked example 7

W.E. 7

Factorise each of the following expressions by grouping in pairs.

(a) $2ax-8ay-cx+4cy$ **(b)** $uw+vx+xu+vw$

Thinking	Working
(a) 1 Factorise the first two terms by finding the HCF. (Here, it is $2a$.) Factorise the next two terms by finding the HCF. (Here, it is $-c$.) Don't forget to change signs when required.	(a) $2ax-8ay-cx+4cy$ $=2a(x-4y)-c(x-4y)$
2 Find the new HCF and write it outside the brackets. (Here, it is $(x-4y)$.) Divide each term by the HCF and write the quotient inside the brackets.	$=(x-4y)(2a-c)$
(b) 1 Rearrange the pairs so there is a common factor in each pair.	(b) $uw+vx+xu+vw$ $=uw+xu+vw+vx$
2 Find and take out the HCF from the first two terms (here, it is u) and the HCF from the next two terms (here, it is v).	$=u(w+x)+v(w+x)$
3 Find the new HCF and write it outside the brackets. (Here, it is $(w+x)$.) Divide each term by the HCF and write the quotient inside the brackets.	$=(w+x)(u+v)$

3.2 Factorising using common factors

Navigator

Answers p. 787

1 (columns 1–2), 2 (column 1), 3 (column 1), 4 (column 1), 5, 6, 7, 9, 10, 13	1 (column 2), 2 (column 2), 3 (column 2), 4 (column 2), 5 (column 1), 6 (columns 2–3), 7, 8, 9, 10, 11, 13	2 (column 3), 3 (column 2), 4 (column 3), 6, 8, 9, 10, 11, 12, 13

Fluency

W.E. 5

1 Factorise the following expressions.

(a) $9x - 3$ (b) $-7x - 14$ (c) $3x + x^2$

(d) $8ab - b$ (e) $16mn - 4m^2$ (f) $25x^2y + 5xy$

(g) $-x^2 - 17x^2y$ (h) $2a^2b + 9ab^2$ (i) $10x^2 - 100x^3$

2 Factorise each expression.

(a) $4x + 8$ (b) $18x - 15$ (c) $24x - 9$

(d) $xy + 7x$ (e) $12kx + 21k$ (f) $32ab + 20b$

(g) $7x^2y + 7y^2$ (h) $9x^2 + 3xy$ (i) $48m^2 + 8m$

(j) $-6x - x^2$ (k) $-4x + x^2$ (l) $-24a^2 + 10a$

(m) $6ab - 18a^2b$ (n) $40x^2y + 16xy^2$ (o) $10cd^2 - 2cd$

W.E. 6

3 Factorise each expression.

(a) $x(k + 1) + 4(k + 1)$ (b) $a(x - 4) + 6(x - 4)$

(c) $m(n - 7) + 2(n - 7)$ (d) $x(p + 3) - 9(p + 3)$

(e) $n(q - 2) - 7(q - 2)$ (f) $x(u - 4) - 5(u - 4)$

(g) $d(c + 5) + (c + 5)$ (h) $a(d + 2) - (d + 2)$

(i) $k(m + 7) - (m + 7)$ (j) $x(x - 3) + 5(x - 3)$

W.E. 7

4 Factorise each of the following expressions by grouping in pairs.

(a) $xy + 3y + 5x + 15$ (b) $ab + 7b + 2a + 14$ (c) $mn + kn + 4m + 4k$

(d) $xy - y + 3x - 3$ (e) $3a - 3b + ac - bc$ (f) $km - 4m + 2k - 8$

(g) $15ab - 10b - 12a + 8$ (h) $6ab - 4b - 15a + 10$ (i) $8mn - 6n - 12m + 9$

(j) $vx - vy - ax + ay$ (k) $pq - rq - np + nr$ (l) $gh - fh - dg + df$

(m) $m^2n + np + m^2t + pt$ (n) $a^2b + ab^2 - ac - bc$ (o) $d^2k - dk - dr + r$

(p) $27xy - 10 + 18y - 15x$ (q) $36a + 5c - 4ac - 45$ (r) $25vw + 5v - 15w - 3$

(s) $32a^3c + 15b - 12abc - 40a^2$ (t) $21d^2 + 20fh - 12dh - 35df$ (u) $m^2nr - 20p + 4mpr - 5mn$

5 Answer true or false for each of these statements.

(a) $8a + 4ab = 4a(2 + b)$ (b) $2x - 6xy = 2x(1 - y)$

(c) $mn + 3n - 2m - 6 = (m - 3)(n + 2)$ (d) $x^2 - 4x + x - 4 = (x - 4)(x + 1)$

(e) $3x^2 + 15x - 2x - 10 = (x - 2)(3x + 5)$ (f) $4x^2 + 20x - 3x - 15 = (x - 5)(4x + 3)$

6 Factorise the following expressions using an appropriate strategy.

(a) $6s + 60 + st + 10t$ (b) $2 - 8a$ (c) $7x(x - 0.5) + 3(x - 0.5)$

(d) $p(q + 6) - 3(q + 6)$ (e) $25x^2 + 15x$ (f) $12ab + 21b - 4a - 7$

7 When factorised, $16f^2g - 12gf$ becomes:

A $2fg(8f - 6)$ B $4f^2g(4g - 3f)$ C $16fg(f - 12)$ D $4fg(4f - 3)$

8 The total area of the two shapes shown is:

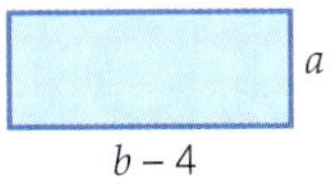

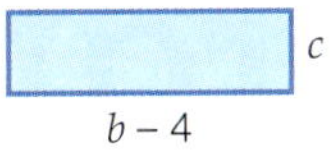

A $(a + 4)(b - c)$ B $(a + c)(b - 4)$ C $(a + b)(c - 4)$ D $(a - c)(b + 4)$

9 $10dk + 25d - 2k - 5$ factorises to:

A $(2k + 5)(5d - 1)$ B $5d(2k + 5)$ C $(2k - 5)(5d + 1)$ D $(10k + 1)(d - 5)$

Understanding

10 Five small bulldog clips are added to a box containing p small bulldog clips. Five large bulldog clips are added to another box that contains p large bulldog clips. Each small clip costs a cents and each large clip costs b cents.

(a) Give the expression (in cents) for the combined cost of the clips.

(b) Factorise the expression in part **(a)**.

(c) If the cost of one large clip is twice the cost of one small clip write a factorised expression for the total cost of the clips in terms of a and p.

11 (a) Draw and label a rectangle that has a length of p metres and a width of r metres.

(b) The length is reduced by a metres. Show this on your diagram and find the new area as the difference of two areas.

(c) The width of this shortened rectangle is now increased by b metres. Show this on your diagram. Label the dimensions and find an expression for the added area.

(d) Show that the area of the newly formed rectangle is $(pr - ar + pb - ab)$ m^2.

(e) Find the area as a product of the new length and width.

(f) Comment on your answers to parts **(d)** and **(e)**.

Reasoning

12 (a) A cone of height h and radius $2r$ is filled with ice-cream as shown at right. Give an expression for the total volume of ice-cream (m^3), assuming the scoop on top is hemispherical (half a sphere).

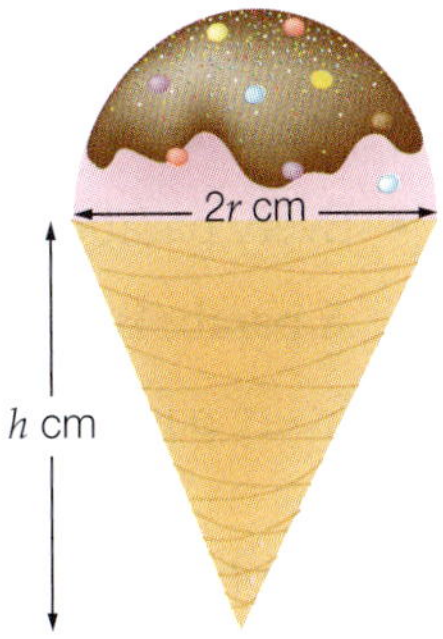

Hint

The volume of a sphere of radius r is $\frac{4\pi r^3}{3}$ and the volume of a cone of base radius r and height h is $\frac{\pi r^2 h}{3}$.

(b) Factorise the expression in part **(a)**.

(c) If $r = 3$ cm and $h = 10$ cm, find the total volume of ice-cream to the nearest cm^3.

(d) Another cone is empty except for the same-sized hemispherical scoop on top, which falls down into the cone. If the height of this cone is $2r$, use your expression in part **(b)** to show that the total volume is twice the volume of the cone, and so all the ice-cream will fill the cone exactly.

Open-ended

13 A paving tile has the shape and dimensions shown.

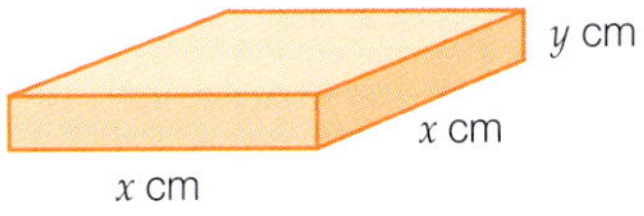

(a) Write an expression (in cm^2) for the total surface area.

(b) Factorise the expression in part **(a)**.

(c) Find two sets of values for x and y so that the volume (in cm^3) has numerically the same value as the surface area (in cm^2). The values do not have to be integers.

3.3 Identifying quadratic graphs and their equations

Quadratic relationships occur often in everyday life. The path of a ball in flight, the area of land, or the height of a ball thrown vertically in the air are all modelled by quadratic relationships. A quadratic relationship can be expressed as an equation in the general form of $y = ax^2 + bx + c$, where x is the independent variable and y is the dependent variable. The highest power of the independent variable is 2 and the highest power of the dependent variable is 1. All powers are integers.

The graph of a quadratic equation is called a **parabola**. A parabola has a minimum or maximum **turning point** or **vertex**, which is where the graph changes direction.

Every parabola is symmetrical about an **axis of symmetry**, a vertical line with an equation in the form $x = h$, where h is a constant equal to the x-value of the turning point. The shape of a parabola is given below.

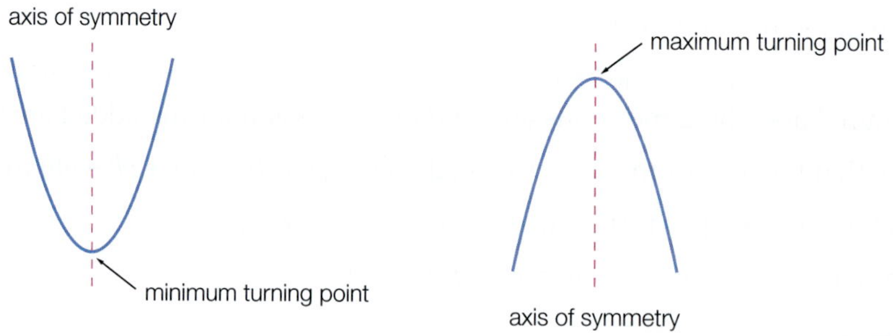

Features of a parabola

- A parabola is a graph of a quadratic relationship $y = ax^2 + bx + c$.
- Every parabola has a vertex or turning point.
- A parabola has a minimum turning point or a maximum turning point.
- Every parabola has an axis of symmetry.

The graph of a quadratic equation can be plotted using the following steps:

1. Complete a table of values.
2. Draw a set of appropriately sized, scaled and labelled axes.
3. Plot the points from the table.
4. Find the axis of symmetry or the turning point and mark it to help shape the graph.
5. Join the points with a smooth curve.
6. Label the graph with its equation.

Worked example 8

W.E. 8

Plot the graphs of the following for values of x between -3 and 3 inclusive:

(a) $y = x^2$ **(b)** $y = 2x^2 - 1$ **(c)** $y = 5 - x^2$

Thinking	Working

(a) 1 Write the equation and a table of values.

(a) $y = x^2$

x	-3	-2	-1	0	1	2	3
y	9	4	1	0	1	4	9

2 Draw a set of labelled Cartesian axes, plot the points and join them with a smooth curve. The scale does not have to be the same for both axes. Label the graph with its equation. You can also label the turning point.

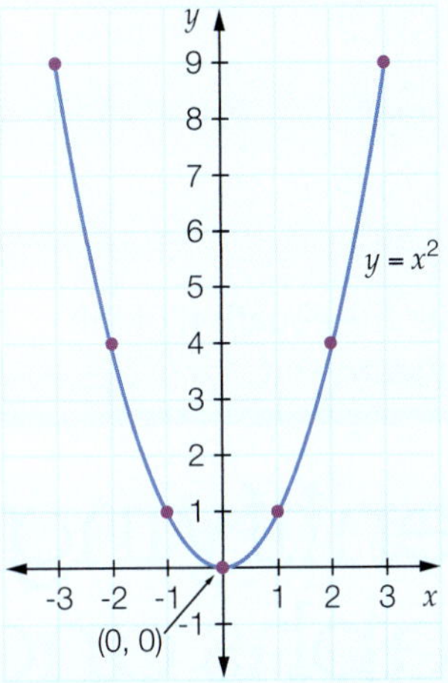

3 You can write the equation of the axis of symmetry.

The equation of the axis of symmetry is $x = 0$.

(b) 1 Write the equation and a table of values.

(b) $y = 2x^2 - 1$

x	-3	-2	-1	0	1	2	3
y	17	7	1	-1	1	7	17

2 Draw a set of labelled Cartesian axes, plot the points and join them with a smooth curve. The scale does not have to be the same for both axes. Label the graph with its equation. You can also label the turning point.

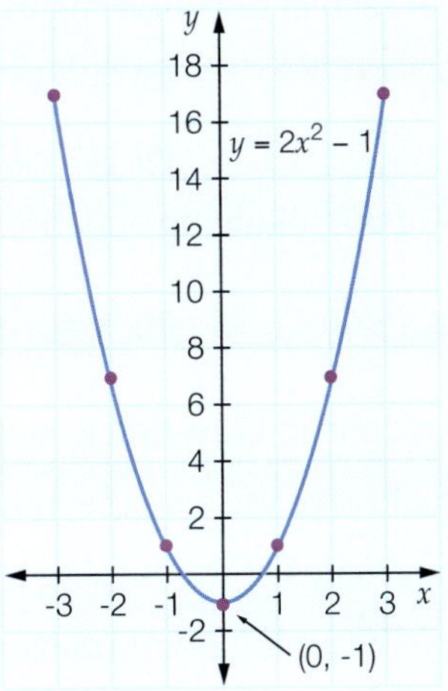

3 You can write the equation of the axis of symmetry.

The equation of the axis of symmetry is $x = 0$.

(c) 1 Write the equation and a table of values.

(c) $y = 5 - x^2$

x	-3	-2	-1	0	1	2	3
y	-4	1	4	5	4	1	-4

2 Draw a set of labelled Cartesian axes, plot the points and join them with a smooth curve. The scale does not have to be the same for both axes. Label the graph with its equation. You can also label the turning point.

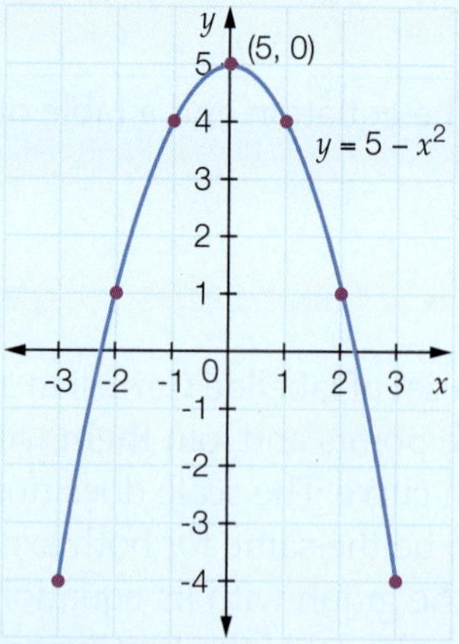

3 You can write the equation of the axis of symmetry.

The equation of the axis of symmetry is $x = 0$.

3.3 Identifying quadratic graphs and their equations

Navigator

Answers p.788

1, 2, 3, 4, 5, 6, 8, 9	1, 2, 3, 4, 5, 6, 7, 8, 9	1 (c–e), 2, 3, 4, 5, 6, 7, 9

Equipment required: graph paper for Questions **5**, **7**

Fluency

W.E. 8

1 Plot the graphs of each of the following using integer values of x between -3 and 3 inclusive.

(a) $y = x^2 + 1$ **(b)** $y = x^2 - 3$ **(c)** $y = x^2 + 5$

(d) $y = 2x^2$ **(e)** $y = -2x^2$ **(f)** $y = -2x^2 + 1$

2 Plot the graphs of the following equations for the given x-values.

(a) $y = \frac{1}{4}x^2$, $-3 \le x \le 3$ **(b)** $y = x^2 - 4x$, $-1 \le x \le 5$ **(c)** $y = -x^2 - 1$, $-2 \le x \le 2$

3 For each of the graphs in Question **2**, state:

(i) the equation of the axis of symmetry

(ii) the coordinates of the turning point

(iii) the type of turning point (maximum or minimum).

Understanding

4 (a) For each of the following, complete a table of values using integer values for x from -4 to 4 and then plot the graph.

(i) $y = (x - 1)(x + 3)$ (ii) $y = x(x + 3)$ (iii) $y = (4 + x)(2 - x)$

(b) Write the coordinates of the turning point for each graph and state whether it is a maximum or minimum.

5 The volume of a circular swimming pool, with a depth of $\frac{3}{\pi}$ (≈ 0.96 m), depends on the radius of the pool. The volume of the pool is given by $V = \pi r^2 h$, where V is the volume of the pool in cubic metres, r is the radius (in metres) and h is the height (in metres).

(a) Show that $V = 3r^2$.

(b) Explain why this relationship is quadratic.

(c) Complete the following table of values for $0 \leq r \leq 5$.

r	0	1	2	3	4	5
r^2						
V						

(d) Plot a graph of the relationship $V = 3r^2$. (Represent r on the horizontal axis.)

(e) Find the volume of the pool when the radius is:

(i) 2.5 m (ii) 3.8 m

(f) Find the radius of the pool if its volume is:

(i) 30 m^3 (ii) 65 m^3

(g) What is the difference in volume between pools of radius 1.5 m and 3.5 m?

Reasoning

6 A tablecloth has a length that is 0.4 m longer than its width.

(a) Write an equation for the area of the tablecloth in terms of its length l.

(b) Plot a graph of the area against the length.

(c) Use this graph to find the dimensions of the tablecloth if its area is 3.2 m^2.

7 A water fountain forms a parabolic shape. If p is the horizontal distance from the water outlet, then the height h is: $h = -\frac{1}{16}(p - 40)^2 + 100$

(h and p are in cm).

(a) Plot the graph of the relationship between h and p, with p on the horizontal axis with values from 0 to 80.

(b) How high does the water go up in the air?

(c) How far across from the water outlet does the water from the fountain reach?

(d) If the fountain starts in the centre of a circular pool and reaches to the edge of the pool, then what is the surface area of the pool?

Open-ended

8 Write an equation for a parabola that has a maximum turning point and plot the graph.

9 When Sian threw a ball up in the air and then caught it, the path of the ball traced a parabolic shape.

(a) Sketch what you think might have happened on a set of axes.

(b) Describe the main features of the graph and explain why it has a parabolic shape.

Problem solving

How much is enough?

Bob needs to build a picket fence from his house to the garage. The curve formed by the top of the pickets follows the path of a parabola with this equation:

$$y = \frac{3}{2.56}(x^2 - 3.2x + 2.56) + 1.2$$

where y is the height above the ground and x is the distance from the wall of the house.

The pickets are 50 mm wide and have a gap of 50 mm between them. The length of the fence is 3.2 m. Each picket is shaped to a point at the top.

What length of full width wood is used for the pickets? (A half-width picket is used at each end.)

Strategy options

- Draw a diagram.
- Make a model.
- Break problem into manageable parts.

Quadratic transformations

3.4

It is useful to explore the **transformations** of **dilation**, **reflection** and **translation** of quadratics. Transformations can be used to sketch simple quadratic graphs instead of plotting them point-by-point.

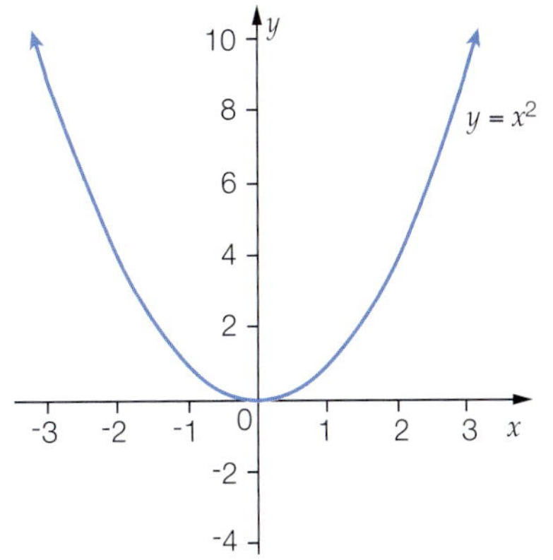

The basic quadratic relationship is $y = x^2$.

The parabola $y = x^2$ has a turning point, or vertex, at (0, 0). The turning point is a minimum.

The x-intercept and y-intercept are both at the point (0, 0).

Dilations $y = ax^2$

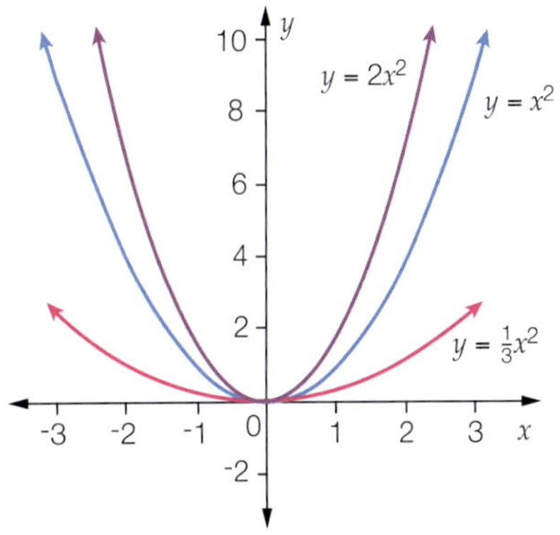

Graphing $y = x^2$, $y = \frac{1}{3}x^2$ and $y = 2x^2$ on the same set of axes shows that the coefficient of x^2 determines the shape of the graph, but the turning point is still the same. This effect is called dilation. The size of the coefficient of x^2 is called the **dilation factor**.

- A factor size greater than 1 ($a > 1$ or $a < -1$) will make the graph narrower.
- A factor size between 0 and 1 ($0 < a < 1$ or $-1 < a < 0$) will make the graph wider.

Note that if the coefficient is negative, the negative does not affect the size of the dilation factor. For example, the dilation factor of $y = -4x^2$ is 4.

Reflections $y = -x^2$

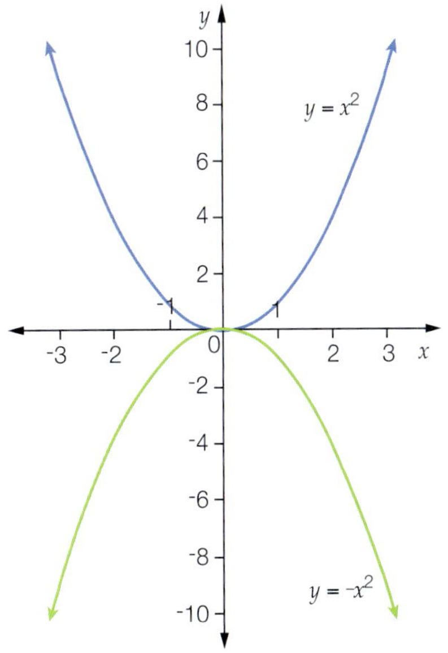

The graph of $y = -x^2$ is the same shape as the graph of $y = x^2$ except that it is upside down. This is called *inverted* and it is a reflection in the x-axis. The turning point is still at the point (0,0), but it is now a maximum instead of a minimum.

All graphs that have a negative coefficient for x^2 will have this same basic shape (an inverted parabola).

The turning point of a reflected (inverted) parabola is always a maximum.

Worked example 9

W.E. 9

For each of the following equations:

(a) $y = 4x^2$ **(b)** $y = \frac{1}{4}x^2$ **(c)** $y = -3x^2$

(i) state the dilation factor

(ii) determine whether the graph is narrower or wider than the parabola $y = x^2$

(iii) sketch the graph of $y = x^2$ on a set of Cartesian axes and then use transformations to sketch the graph of the given equation on the same set of axes.

Thinking	Working
(a) (i) The size of the coefficient of x^2 gives the dilation factor. Write the dilation factor.	(a) $a = 4$ (i) The dilation factor is 4.
(ii) If the dilation factor is greater than 1, the graph is narrower than $y = x^2$.	(ii) The graph is narrower than $y = x^2$.
(iii) Sketch the graphs of $y = x^2$ and $y = 4x^2$ on the same set of axes. Each point on $y = 4x^2$ has been dilated by the dilation factor (here, $a = 4$). Label the equation of this graph and its turning point. (Here, the turning point is the same as for $y = x^2$.)	(iii) 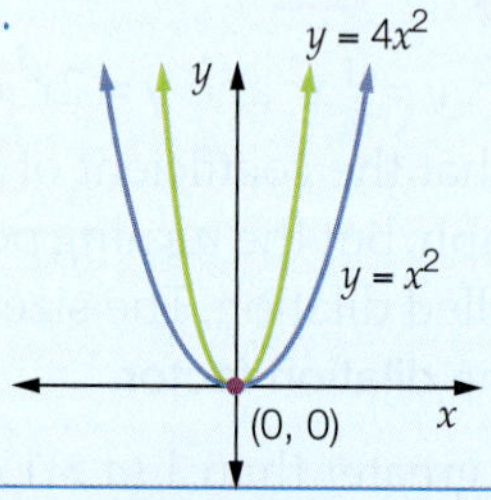
(b) (i) The size of the coefficient of x^2 gives the dilation factor. Write the dilation factor.	(b) $a = \frac{1}{4}$ (i) The dilation factor is $\frac{1}{4}$.
(ii) If the dilation factor is between 0 and 1, the graph is wider than $y = x^2$.	(ii) The graph is wider than $y = x^2$.
(iii) Sketch the graphs of $y = x^2$ and $y = \frac{1}{4}x^2$ on the same set of axes. Each point on $y = \frac{1}{4}x^2$ has been dilated by the dilation factor (here, $a = \frac{1}{4}$). Label the equation of this graph and its turning point. (Here, the turning point is the same as for $y = x^2$.)	(iii) 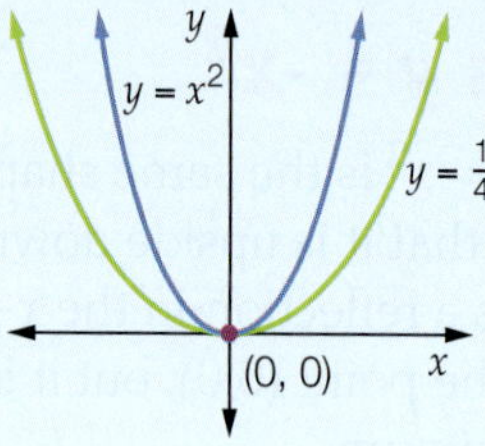
(c) (i) The size of the coefficient of x^2 gives the dilation factor. Write the dilation factor.	(c) $a = -3$ (i) The dilation factor is 3.
(ii) If the dilation factor is greater than 1, the graph is narrower than $y = x^2$.	(ii) The graph is narrower than $y = x^2$.

(iii) Sketch the graphs of $y = x^2$ and $y = -3x^2$ on the same set of axes. Each point on $y = -3x^2$ has been dilated by the dilation factor (here, $a = -3$). If the coefficient a is negative, then the graph is reflected in the x-axis. Label the equation of this graph and its turning point. (Here, the turning point is the same as for $y = x^2$.)

(iii)

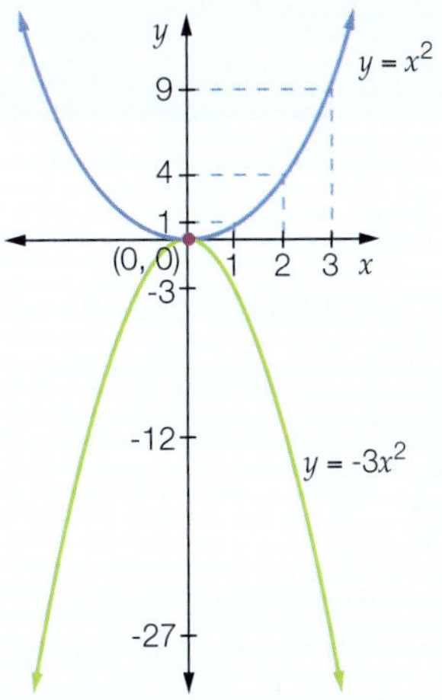

Interactive

Dilation of a quadratic

Explore how the coefficient of x^2 determines the dilation of a quadratic graph.

Go to the eBook or the Pearson Places website to access this interactive.

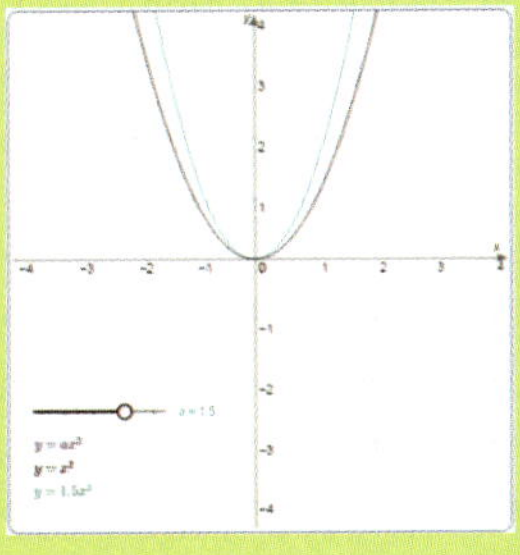

Horizontal translations $y = (x - h)^2$

Graphing $y = x^2$ with $y = (x - 1)^2$ and $y = (x + 2)^2$ shows that when a number is added or subtracted to x in the brackets, the basic graph $y = x^2$ is moved left or right. This movement is called a horizontal translation.

If $y = (x - h)^2$, then h indicates the size and direction of the horizontal translation.

- When h is positive, the graph is moved right h units.
- When h is negative, the graph is moved left h units.

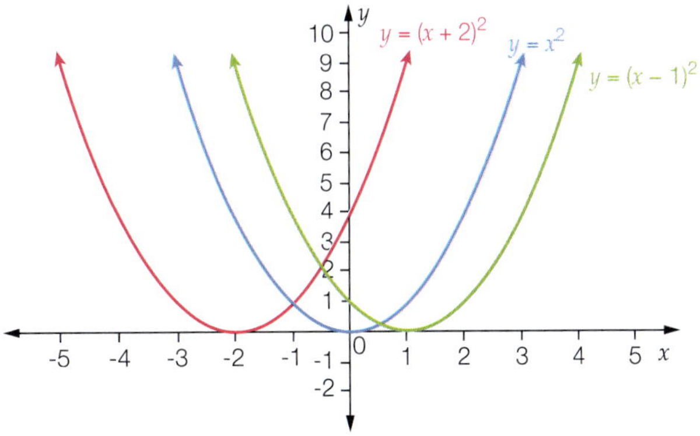

In the graph shown above:

$y = (x - 1)^2$ is the graph of $y = x^2$ translated 1 unit to the right.

$y = (x + 2)^2$ is the graph of $y = x^2$ translated 2 units to the left.

Worked example 10

W.E. 10

For each of the following equations:

(a) $y = (x + 1.3)^2$ (b) $y = (x - 2)^2$

(i) state the translation of the graph of $y = x^2$

(ii) sketch the graph of $y = x^2$ on a set of Cartesian axes and then use transformations to sketch the graph of the given equation on the same set of axes.

Thinking	Working
(a) (i) Compare the graph's equation to $y = (x - h)^2$ to identify the value of h (here, $h = -1.3$). If h is negative, $y = x^2$ is moved h units to the left.	(a) (i) $y = x^2$ is translated 1.3 units to the left.
(ii) Sketch the graphs. They are identical in shape, as every point on the graph of $y = x^2$ has been translated h units horizontally (here, 1.3 units to the left). Label the graphs and the turning points.	(ii) 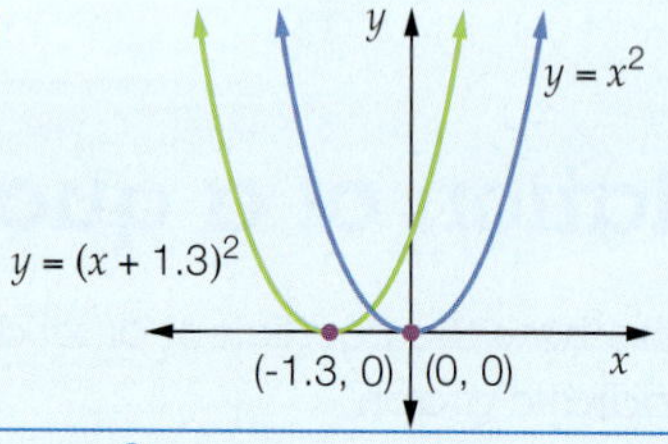
(b) (i) Compare the graph's equation to $y = (x - h)^2$ to identify the value of h (here, $h = 2$). If h is positive, $y = x^2$ is moved h units to the right.	(b) (i) $y = x^2$ is translated 2 units to the right.
(ii) Sketch the graphs. They are identical in shape, as every point on the graph of $y = x^2$ has been translated h units horizontally (here, 2 units to the right). Label the graphs and the turning points.	(ii) 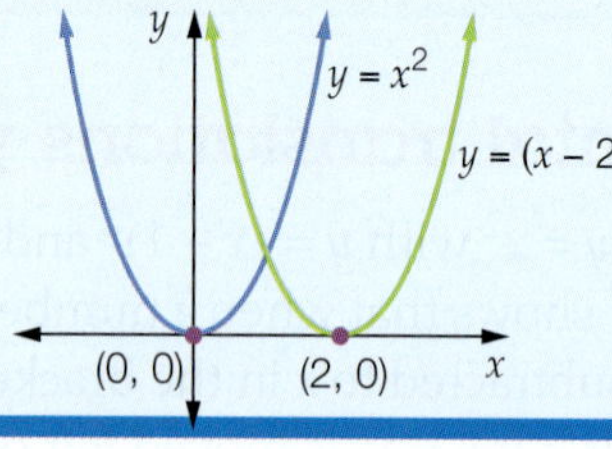

Interactive

Horizontal translation of a quadratic

Explore how the graph of a quadratic function can be translated to the left or right.

Go to the eBook or the Pearson Places website to access this interactive.

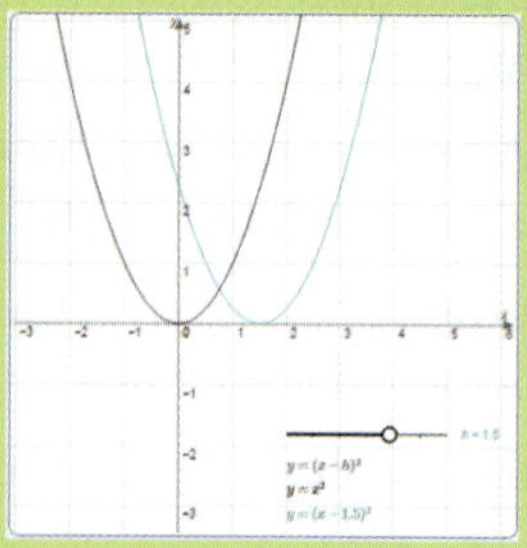

Vertical translations $y = x^2 + k$

Graphing $y = x^2$, $y = x^2 + 3$ and $y = x^2 - 4$ shows that the basic graph $y = x^2$ will move up when adding a number and down when subtracting a number. This movement is called a vertical translation.

If $y = x^2 + k$, the constant term k indicates the size and direction of the vertical translation.

- When k is positive, the graph is moved up k units.
- When k is negative, the graph is moved down k units.

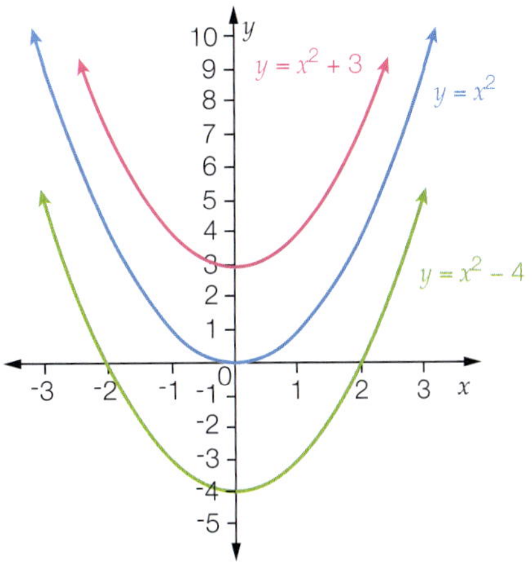

Worked example 11

W.E. 11

For each of the following equations:

(a) $y = x^2 - 3$ **(b)** $y = x^2 + 2.5$

(i) state the translation of the graph of $y = x^2$

(ii) sketch the graph of $y = x^2$ on a set of Cartesian axes and then use transformations to sketch the graph of the given equation on the same set of axes.

Thinking	Working
(a) (i) Compare the graph's equation to $y = x^2 + k$ to identify the value of k (here, $k = -3$). If k is negative, $y = x^2$ is moved k units down.	(a) (i) $y = x^2$ is translated 3 units down.
(ii) Sketch the graphs. They are identical in shape, as every point on the graph of $y = x^2$ has been translated k units vertically (here, the graph of $y = x^2$ is translated 3 units down). Label the graphs and the turning points.	(ii) 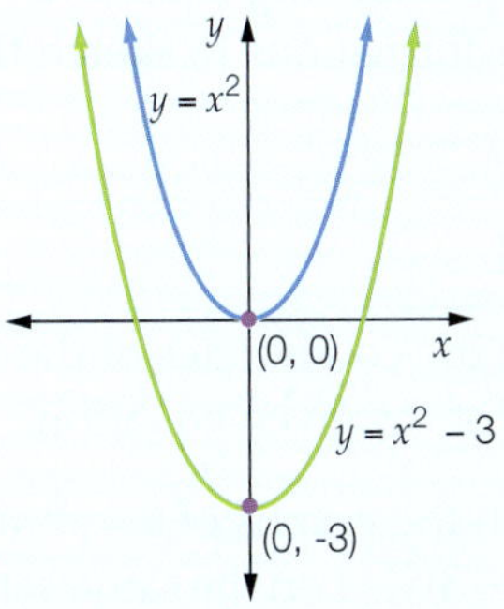
(b) (i) Compare the graph's equation to $y = x^2 + k$ to identify the value of k (here, $k = 2.5$). If h is positive, $y = x^2$ is moved k units up.	(b) (i) $y = x^2$ is translated 2.5 units up.
(ii) Sketch the graphs. They are identical in shape, as every point on the graph of $y = x^2$ has been translated k units vertically (here, the graph of $y = x^2$ is translated 2.5 units up). Label the graphs and the turning points.	(ii) 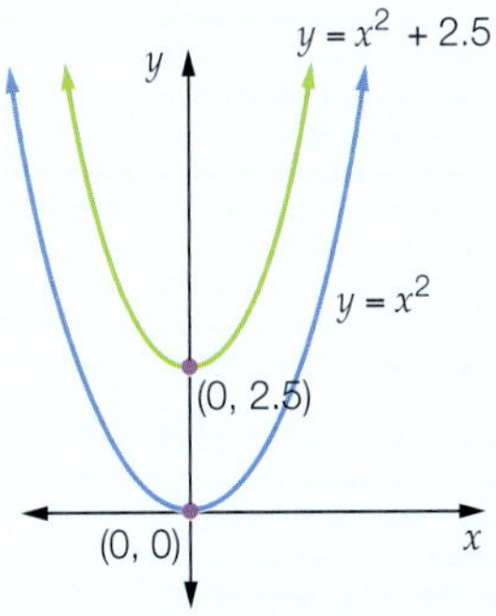

Interactive

Vertical translation of a quadratic

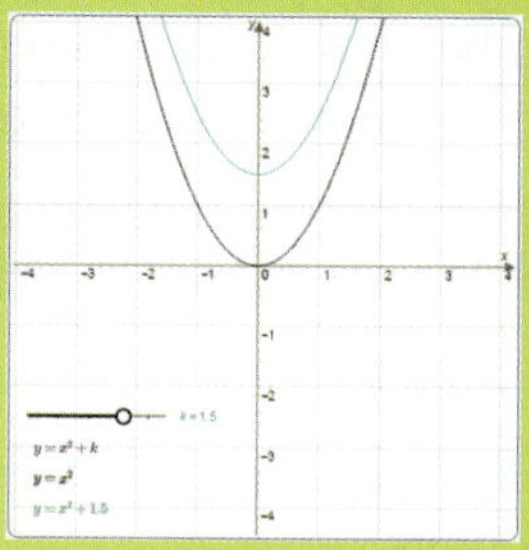

Explore how the graph of a quadratic function can be translated up or down.

Go to the eBook or the Pearson Places website to access this interactive.

Combined transformations $y = a(x - h)^2 + k$

You can combine horizontal and vertical translations.

These equations are in the form of $y = (x - h)^2 + k$, where h is the horizontal translation and k is the vertical translation.

This equation is said to be written in **turning point form**, because the turning point is translated from the origin (0, 0) to the point (h, k).

You can also combine dilations with translations by including the dilation factor as well. These equations are in the form of $y = a(x - h)^2 + k$. This dilates the parabola by the factor a while also translating the turning point from the origin to the point (h, k).

Worked example 12 — W.E. 12

For each of the following equations, sketch the graph of $y = x^2$ on a set of Cartesian axes and then use transformations to sketch the graph of the given equation on the same set of axes.

(a) $y = (x - 3)^2 - 4$ **(b)** $y = (x + 2)^2 + 3$

Thinking	Working
(a) 1 Write the coordinates of the turning point of the given equation.	(a) The turning point is at (3, -4).
2 Sketch the graphs of $y = x^2$ and $y = (x - 3)^2 - 4$ on the same set of axes. Each point on $y = (x - 3)^2 - 4$ has been translated h units horizontally and k units vertically (here, 3 units right and 4 units down). Label the graphs and the turning points.	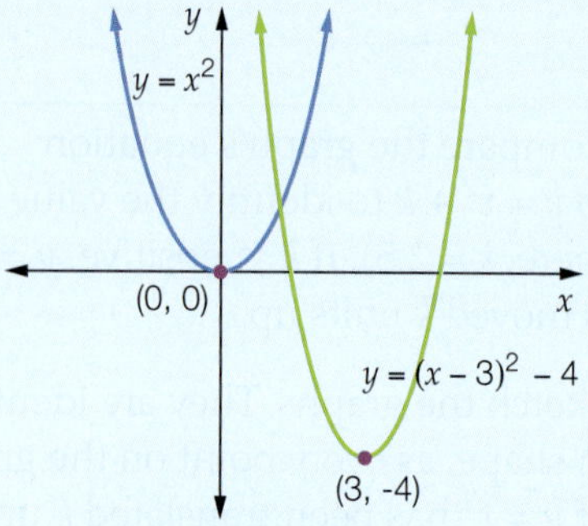

(b) 1 Write the coordinates of the turning point of the given equation.

(b) The turning point is at (-2, 3).

2 Sketch the graphs of $y = x^2$ and $y = (x + 2)^2 + 3$ on the same set of axes. Each point on $y = (x + 2)^2 + 3$ has been translated h units horizontally and k units vertically (here, 2 units left and 3 units up). Label the graphs and the turning points.

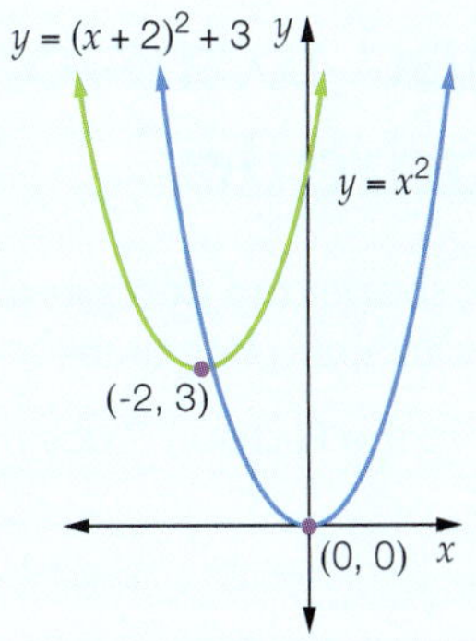

The graph of $y = x^2$ can be transformed to produce parabolas in the form of $y = a(x - h)^2 + k$, by:

- dilating it by a factor of a, making it narrower ($a > 1$ or $a < -1$) or wider ($-1 < a < 1$)
- reflecting it in the x-axis ($a < 0$)
- translating it h units horizontally to the right ($h > 0$) or to the left ($h < 0$)
- translating it k units vertically up ($k > 0$) or down ($k < 0$).

The order of transformations is important, so always dilate and reflect a graph before you translate it.

Worked example 13

W.E. 13

Using the graph shown, find the equation of the curve, assuming there is no dilation.

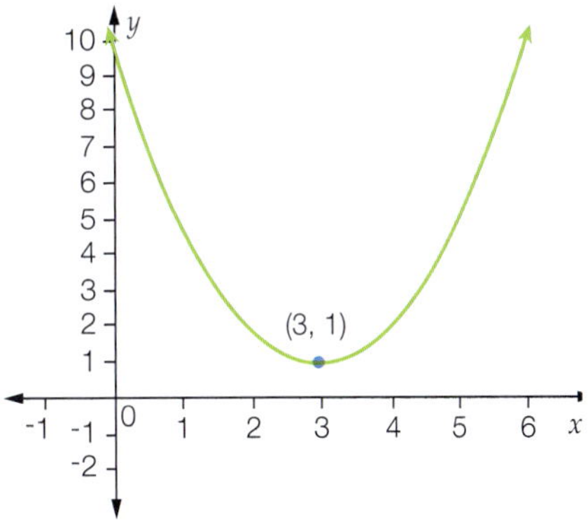

Thinking	Working
1 There is no dilation so use the equation $y = (x - h)^2 + k$.	$y = (x - h)^2 + k$
2 Identify the turning point from the graph to find the values of h and k.	The turning point is at (3, 1) so $h = 3$ and $k = 1$.
3 Write the equation.	$y = (x - 3)^2 + 1$

Interactive

Combined transformations of a quadratic

Explore how the shape and position of a quadratic graph can be modified by changing its values.

Go to the eBook or the Pearson Places website to access this interactive.

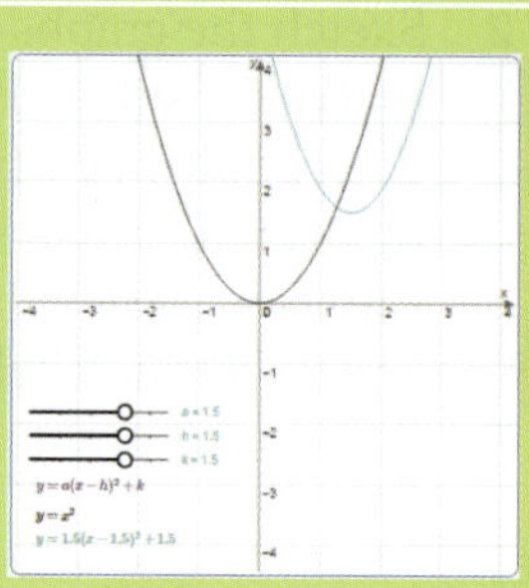

3.4 Quadratic transformations

Navigator

Answers p.790

1, 2, 3, 4, 5, 6, 7, 8, 9, 10, 11, 12, 13, 14, 16, 17	1, 2, 3, 4, 5, 6, 7, 8, 9, 10, 11, 12, 14, 16, 17	1, 2, 3, 4, 5, 7, 8, 10, 11, 12, 14, 15, 17

Fluency

W.E. 9

1 For each of the following equations:

(a) $y = 2x^2$ (b) $y = \frac{x^2}{2}$ (c) $y = -4x^2$

(i) state the dilation factor

(ii) determine whether the graph is narrower or wider than the parabola $y = x^2$

(iii) sketch the graph on the same set of axes as the graph of $y = x^2$.

W.E. 10

2 For each of the following equations:

(a) $y = (x - 1.3)^2$ (b) $y = (x + 2)^2$

(i) sketch the graph of $y = x^2$ on a set of Cartesian axes, then use transformations to sketch the graph of the given equation

(ii) state the translation of the graph of $y = x^2$.

W.E. 11

3 For each of the following equations:

(a) $y = x^2 + 6$ (b) $y = x^2 - 1.2$

(i) sketch the graph of $y = x^2$ on a set of Cartesian axes, then use transformations to sketch the graph of the given equation

(ii) state the translation of the graph of $y = x^2$.

W.E. 12

4 For each of the following equations, sketch the graph of $y = x^2$ on a set of Cartesian axes, then use transformations to sketch the graph of the given equation on the same set of axes.

(a) $y = (x + 3)^2 - 4$ (b) $y = (x - 2)^2 - 3$

5 Assuming there is no dilation, find the equation of each of the following curves.

W.E. 13

(a)

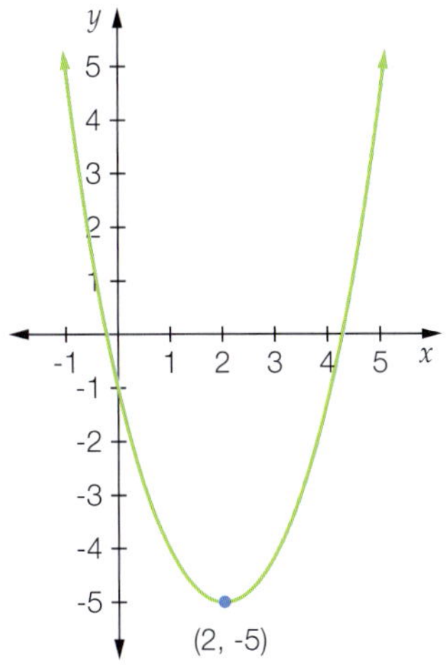

(b)

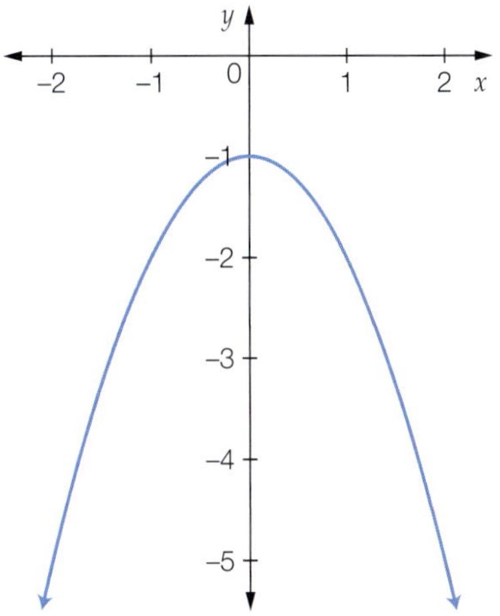

6 To sketch the graph of $y = (x + 5)^2 - 2$, the graph of $y = x^2$ will need to be translated:

A 5 units right and 2 units down
B 5 units left and 2 units down
C 5 units left and 2 units up
D 5 units right and 2 units up

7 On the same set of axes, sketch and label the graphs of:

(a) $y = x^2$ **(b)** $y = x^2 - 5$ **(c)** $y = (x + 4)^2$ **(d)** $y = (x - 1)^2 + 5$

8 On the same set of axes, sketch and label the graphs of:

(a) $y = x^2$ **(b)** $y = 3x^2$ **(c)** $y = \frac{x^2}{3}$ **(d)** $y = -x^2$

9 The equation of the translated parabola shown is:

A $y = (x - 3)^2 + 1$
B $y = (x + 3)^2 + 1$
C $y = (x + 1)^2 + 3$
D $y = (x + 3)^2 - 1$

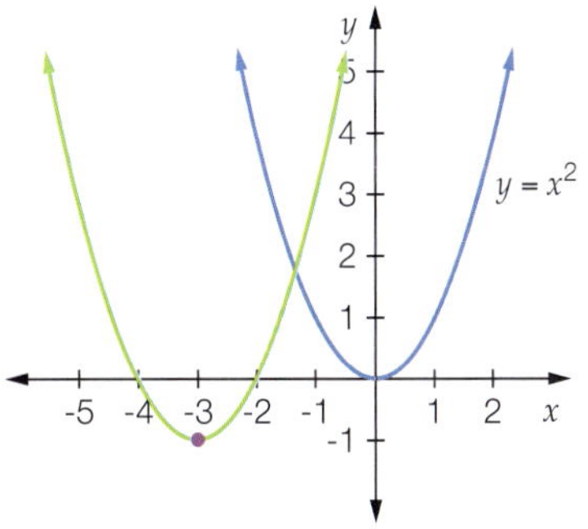

Understanding

10 **(a)** Sketch and label the graph of $y = (x + 2)^2$.

(b) Use the graph from part **(a)** to sketch and label the graph of $y = -(x + 2)^2$ on the same set of axes.

11 **(a)** Sketch the graph of $y = x^2 + 4$.

(b) On the same set of axes, sketch and label the graph formed by dilating the graph in part **(a)** by a factor of 2.

(c) Write the equation for this graph.

Hint

It will have the same turning point as the graph in part **(a)**.

12 **(a)** Sketch the graph of $y = (x - 3)^2$.

(b) On the same set of axes, sketch and label the graph formed by dilating the graph in part **(a)** by a factor of 2.

(c) Write the equation for this graph.

Reasoning

13 A quadratic graph has a turning point at (-3, 2) and a dilation factor of 1.

(a) Find the equation of this graph. (b) Sketch the graph of this equation.

14 A rope tied to each end of a horizontal bar hangs in the shape of a parabola $y = x^2$. The bottom of the rope is 2 m to the left of a person and 0.5 m above the flat ground. If the person is standing at the point with coordinates (0, 0), what is the equation of the rope?

15 An arch over a road is built in the shape of a parabola. It is modelled by the equation $y = -(x - h)^2 + k$ where the origin (0, 0) is the bottom left base of the arch. The road is 6 m wide and the highest point of the arch is 9 m above the flat ground. Find the coordinates of lights to be placed on the arch horizontally 1 m apart from the left base of the arch to the right base of the arch.

Open-ended

16 (a) Write an equation of a parabola that is a vertical translation of $y = x^2$. Sketch a graph of both $y = x^2$ and your equation on the same set of axes.

(b) Write an equation of a parabola that is a horizontal translation of $y = x^2$. Sketch a graph of both $y = x^2$ and your equation on the same set of axes.

17 Freya and Syd are asked to sketch the graph of $y = (x + 6)^2 - 5$. Syd gets the wrong answer. Freya's graph is correct. Syd thinks that the + sign in front of the 6 means that the graph could be moved 6 units to the right and the negative sign in front of the 5 means that the graph should be moved 5 units down. Where did Syd go wrong? What explanation could Freya give so that Syd doesn't make the same mistake again?

Problem solving

Finding curved areas

Erica wants to paint a wall of her architect-designed house. The top of the wall can be described by the equation $y = \frac{x^2}{50}$, where x is the horizontal distance and y is the vertical distance from point A, as shown below.

Strategy options

- Look for a pattern.
- Break problem into manageable parts.

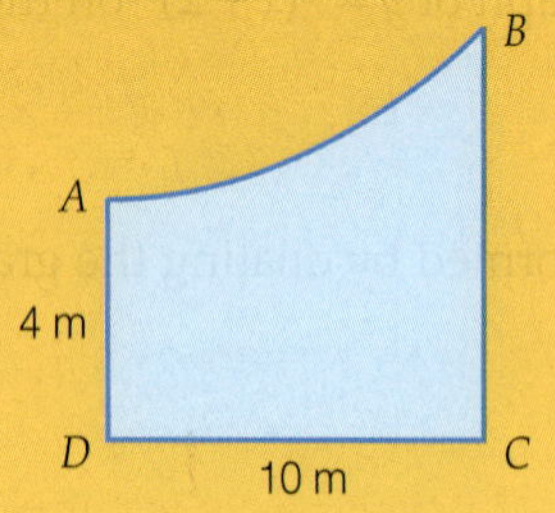

Some wall dimensions are shown on the diagram.

1 How high is the highest point on the wall?

2 Estimate the area by considering the shape to be a trapezium (*ABCD*).

Erica finds that the actual area of the wall is 46.6 m^2.

3 How accurate was your estimate? Explain the method you used.

Half-time 3

1 Expand and simplify. 3.1

(a) $(x + 2)(x + 6)$

(b) $(x - 5)(x + 7)$

(c) $(3x - 2)(x + 5)$

(d) $(2x - 1)(x + 3)$

2 When expanded, which of the following gives the expression $acx^2 + adx + bcx + bd$? 3.1

A $(bx + a)(cx + d)$

B $(ax + c)(bx + d)$

C $(dx + b)(ax + c)$

D $(ax + b)(cx + d)$

3 Factorise each of the following by taking out a common factor. 3.2

(a) $7x + 14y$

(b) $18x^2 + 24xy$

(c) $3(x - 2) + y(x - 2)$

(d) $3(x + 5) + 2y(x + 5)$

4 Expand and simplify. 3.1

(a) $(x - 3)(x + 3)$

(b) $(x + 8)(x - 8)$

(c) $(3x + y)(3x - y)$

(d) $(6x - 5y)(6x + 5y)$

5 Expand and simplify. 3.1

(a) $x(x + 2)(x + 3)$

(b) $3(x - 2)(x + 2)$

(c) $(3x - 4)(7x + 19)(3x + 3)$

(d) $6x(2x - 6)(4x - 1)$

6 Consider the equation $y = -x^2 + 1$. 3.3

(a) Set up a table of values using integer values of x from -3 to 3 inclusive.

(b) Plot the points on a number plane and join them with a smooth curve.

(c) Find the coordinates of the turning point from your graph.

(d) Find the equation of the axis of symmetry.

7 The graph $y = x^2$ is translated 3 units right and 2 units up. What is the equation of the new graph? 3.4

8 A house has a rectangular extension, which has a length of $(x + 6)$ m and a width of $(x + 4)$ m. An area 2 m wide and 4 m long is an untiled verandah. The remainder is a tiled family room. The cost of tiling is \$18 per square metre. 3.1

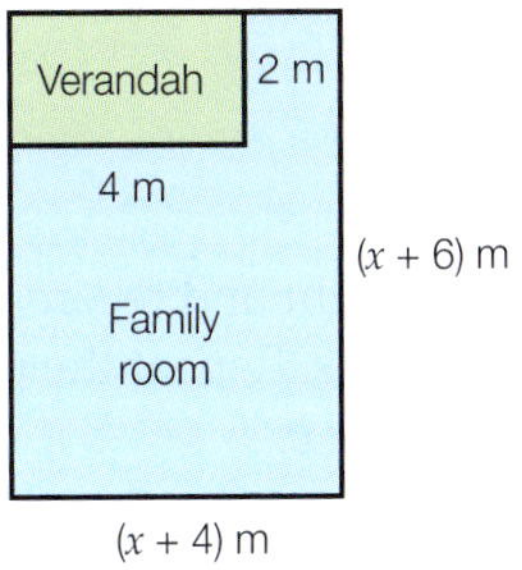

(a) Write an expression for the area of the family room in factorised form.

(b) Write an expression for the cost of tiling the family room.

(c) Find the cost of tiling the family room if x is 2 m.

9 Find, in expanded form, the square of the sum of $4x$ and $3y$. 3.1

Exploration CAS

Equipment required: TI-Nspire CAS or Casio ClassPad CAS

Investigating quadratic graphs and turning points

For a quadratic equation $y = ax^2 + bx + c$, how does varying b affect the turning point of the graph $y = ax^2 + bx + c$, for given values of a and c?

The position of the turning point and the axis of symmetry of the graph of $y = ax^2 + bx + c$ will change as b changes. To investigate systematically, you can set the value of a to 1 and c to 2, while varying b from -4 to 4 in steps of 1 unit.

1 First, use technology to find the coordinates of the turning points of $x^2 + bx + 2$ for integer values of b from -4 to 4.

Using TI-Nspire CAS

Add **Graphs** to your document and enter the equation **f1(x)=x²–4x+2** to be graphed.

To find the coordinates of the turning point, select **menu** > **Analyze Graph** > **Minimum**. Use the cursor to set the lower bound (a point to the left of the turning point) and then the upper bound (a point to the right of the turning point).

The coordinates of the turning point will be displayed. Record the coordinates in the table (see below).

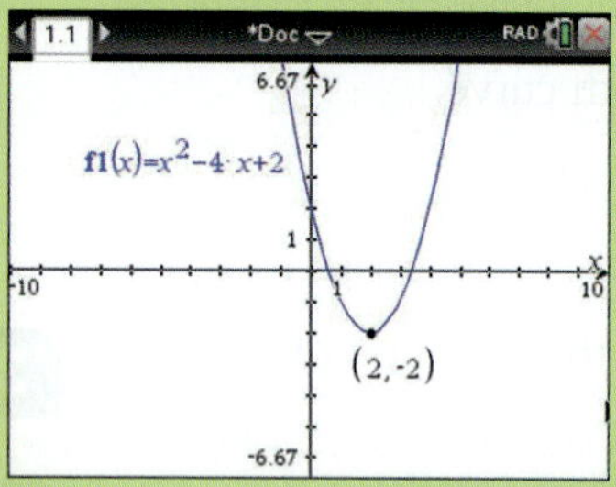

Using Casio ClassPad CAS

From the menu select **Graph & Table**. Enter **y1=x^2–4^x+2**, tick the box, and then select the Graph icon to display the graph.

To find the coordinates of the turning point, select **Analysis** > **G-Solve** > **Min**. The coordinates of the turning point will be displayed. Record the coordinates in the table (see below).

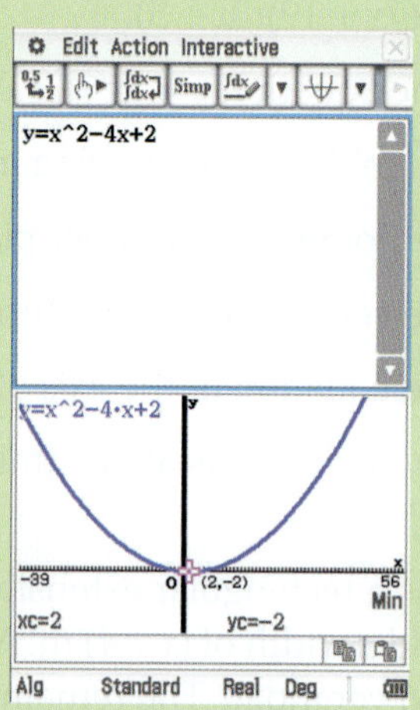

2 Copy the table on the right and record the turning point coordinates. Repeat the process for all the given values of b.

3 Calculate $x = \frac{-b}{2a}$ for each equation. Is this value the same as the x-coordinate of the turning point, for the corresponding value of b in the table?

4 Plot these coordinate pairs as points on the Cartesian plane.

b	x-coordinate of turning point	y-coordinate of turning point
-4	2	-2
-3		
-2		
-1		
0		
1		
2		
3		
4		

Using TI-Nspire CAS	Using Casio ClassPad CAS
Add **Lists & Spreadsheet** to your document. Enter the coordinates of turning points, with column **A** 'x' and column **B** 'y' as shown. To graph these coordinates, highlight both columns and then select **menu** > **Data** > **Quick Graph**.	From the menu select **Spreadsheet**. Enter the coordinates of turning points, with column **A**'x' and column **B** 'y'. To graph these coordinates, highlight both columns and then select **Graph** > **Scatter**.

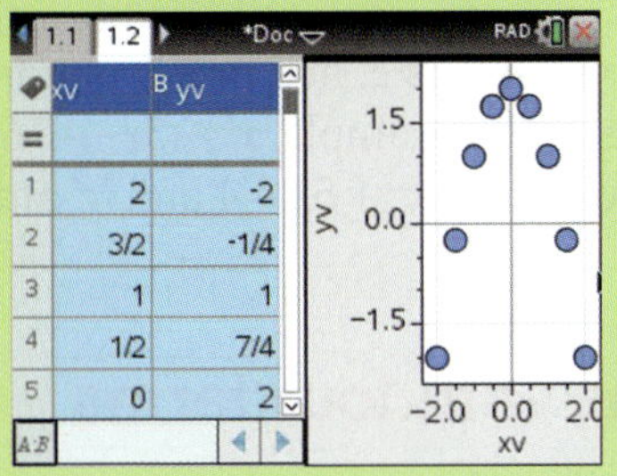

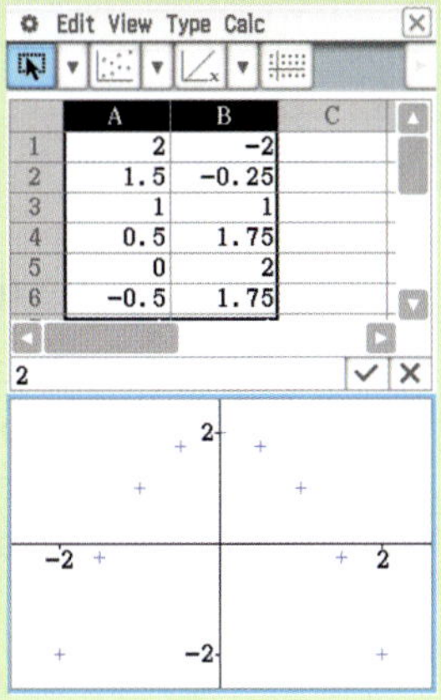

5 What kind of relationship would include this set of points? Try to guess the equation for the relationship. Record your guess, then use technology to find the equation.

Using TI-Nspire CAS	Using Casio ClassPad CAS
With the graph highlighted, select **menu** > **Analyze** > **Regression** and choose the type of relationship that you think fits the points displayed. If correct, the curve that appears will pass through the centre of each point. This will also give the rule for the relationship.	With the columns highlighted, select **Calc** > **Regression** and choose the type of relationship that you think fits the points displayed. If correct, the curve that appears will pass through the centre of each point. The display should also show that $r^2 = 1$, which means it is a perfect fit. This will also give the rule for the relationship.

6 Repeat steps **1**–**3** for the following quadratic equations.

(a) $y = 2x^2 + bx - 5$ (b) $y = -x^2 + bx + 4$

Can you see a pattern connecting the original equation and the equation of the path formed by the turning points as the value of b changes?

7 Find the x-coordinate of the turning point of $y = x^2 + bx + 2$. Substitute this x-coordinate into $y = x^2 + bx + 2$ to find the y-coordinate of the turning point. Verify that these coordinates are satisfied by the relationship in **5**.

Taking it further

8 Find the turning point of the equation $y = ax^2 + bx + c$. Verify that these coordinates satisfy the equation $y = -ax^2 + c$ to show that this is the path (locus) traced out by the vertices as b varies and a and c are held constant.

9 Investigate the effect on the graph of $y = ax^2 + bx + c$ when:

- c is varied while a and b are held constant
- a is varied while b and c are held constant.

Write a brief report on your investigation and state your conclusions.

3.5 Factorising monic quadratic expressions

A **quadratic expression** has the form of $ax^2 + bx + c$.

In a quadratic expression the highest power of the variable is 2. In a **monic quadratic** expression, the coefficient of the squared variable is 1. For example, $x^2 + 5x - 4$, $6 - 3a^2$, h^2 and $4n^2 + 7n$ are all quadratic expressions. Of these, only $x^2 + 5x - 4$ and h^2 are monic quadratic expressions.

Factorising quadratic expressions with four terms

In section 3.2 you factorised four terms by grouping in pairs.

Use the following checklist to see whether you can 'group in pairs' to factorise expressions into binomial products.

- The first term is a squared variable.
- The two middle terms contain the same variable.
- The last term is a constant.

Worked example 14

W.E. 14

Factorise $x^2 - 4x + 5x - 20$.

Thinking	Working
1 Write the expression. Check whether there is a factor common to all terms. (None.)	$x^2 - 4x + 5x - 20$
2 Group in pairs that have common factors ($x^2 - 4x$ has a common factor of x: $5x - 20$ has a common factor of 5). Take out the factor from each pair.	$= x(x - 4) + 5(x - 4)$
3 Take out the factor common to both groups $(x - 4)$.	$= (x - 4)(x + 5)$

Factorising quadratic trinomials

A **quadratic trinomial** consists of three terms. For example, $x^2 + 5x - 4$ is a quadratic trinomial. The monic quadratic trinomials that you will factorise in this section will all be in the form of $ax^2 + bx + c$ where $a = 1$. Not all quadratic expressions in this form will factorise into real binomial products.

Consider $x^2 + 3x - 5x + 5$:

You can take out a common factor from each pair to give $x(x + 3) - 5(x - 1)$ but it will not factorise any further. Why do some expressions with four terms factorise fully into binomial products whereas others do not?

Consider these expressions:

$x^2 - 5x + 2x - 14$	$x^2 - 4x - 3x - 11$	$x^2 + 3x - 7x - 20$
$= x(x - 5) + 2(x - 7)$	$= x(x - 4) - (3x + 11)$	$= x(x + 3) - (7x + 20)$

None of these expressions factorise into binomial products.

Consider these expressions:

$x^2 - 7x + 2x - 14$	$x^2 + x + 6x + 6$	$x^2 - 3x - 2x + 6$
$= x(x - 7) + 2(x - 7)$	$= x(x + 1) + 6(x + 1)$	$= x(x - 3) - 2(x - 3)$
$= (x - 7)(x + 2)$	$= (x + 1)(x + 6)$	$= (x - 3)(x - 2)$

All of these expressions factorise into binomial products.

In these expressions, the coefficients of the two middle terms multiply to give the constant term:

$-7 \times 2 = -14$	$1 \times 6 = 6$	$-3 \times -2 = 6$

This means that you can use the method of grouping in pairs by splitting the middle term of a monic trinomial into two parts as long as the factors of the constant term are the coefficients of the two middle terms.

Consider:

$x^2 - 5x - 14$	$x^2 + 7x + 6$	$x^2 - 5x + 6$
$= x^2 - 7x + 2x - 14$	$= x^2 + x + 6x + 6$	$= x^2 - 3x - 2x + 6$
$= x(x - 7) + 2(x - 7)$	$= x(x + 1) + 6(x + 1)$	$= x(x - 3) - 2(x - 3)$
$= (x - 7)(x + 2)$	$= (x + 1)(x + 6)$	$= (x - 3)(x - 2)$

Sometimes a quadratic trinomial has a common factor that needs to be taken out before it can be factorised as a monic trinomial.

Worked example 15

W.E. 15

Factorise these quadratic trinomials by splitting the middle term.

(a) $p^2 - 11p + 30$ **(b)** $3q^2 + 6q - 24$

Thinking	Working
(a) 1 Find the factors of the constant term (30) that add to give the coefficient of the middle term (-11). (Here, the numbers are -5 and -6.)	(a) $p^2 - 11p + 30$ $\quad \times \rightarrow +30,\ + \rightarrow -11 \}\ -5, -6$
2 Rewrite the middle term as two terms using the two numbers found as the coefficients of the variable.	$= p^2 - 5p - 6p + 30$
3 Factorise by removing a common factor from each pair using 'two and two' grouping.	$= p(p - 5) - 6(p - 5)$
4 Take out the common factor.	$= (p - 5)(p - 6)$

(b)	**1** Factorise by taking out the HCF first. (HCF = 3)	(b) $3q^2 + 6q - 24$ $= 3(q^2 + 2q - 8)$
	2 Find the factors of the constant term inside the brackets (-8) that add to give the middle term inside the brackets (+2). (Here, the numbers are +4 and -2.) Rewrite the middle term as two terms using the number found as the coefficients of the variable.	$= 3(q^2 + 4q - 2q - 8)$ $\times \rightarrow -8$, $+ \rightarrow +2$ } -2, +4
	3 Factorise by removing a comon factor from each pair using 'two and two' grouping inside the brackets. Add another set of brackets.	$= 3[q(q + 4) - 2(q + 4)]$
	4 Factorise further by taking out the common factor	$= 3(q + 4)(q - 2)$

Using the cross method to factorise monic trinomials

This method is a quick way to factorise monic trinomials when you understand the factorisation process.

1 Draw a pair of crossed arrows and write the variable at the beginning of each arrow.

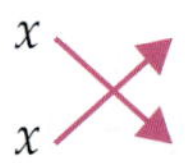

2 Identify a factor pair for the constant term.

For the expression $x^2 - 2x - 35$, the constant term -35 has factor pairs of -35, 1; -7, 5; -5, 7 and -1, 35.

3 Trial a factor pair by writing them at the end of each arrow.

For $x^2 - 2x - 35$ select a pair of factors, -7 and 5. Follow the arrows to write the two terms as a sum or difference. If this does not give the middle term, try a new pair of factors.

4 Circle the top and the bottom of the arrows as shown. Each circle contains a binomial factor.

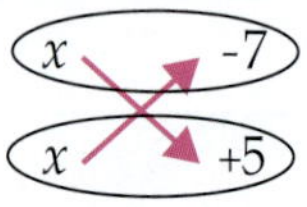

$-7x + 5x = -2x$

$-2x$ is the middle term, so this is the correct pair.

5 Write the factorisation.

$x^2 - 2x - 35 = (x - 7)(x + 5)$

Worked example 16

W.E. 16

Factorise the following quadratic expressions using the cross method.

(a) $x^2 - 10x - 24$ **(b)** $x^2 - 9x + 14$

Thinking	Working
(a) 1 Choose a pair of factors that multiply to give the constant term. If the constant term is negative, choose factors opposite in sign (try $-4 \times 6 = -24$). Use the cross method to see if they add to give the correct middle term ($6x - 4x = 2x$). This pair gives an incorrect middle term, so try another pair.	(a) $x^2 - 10x - 24$ 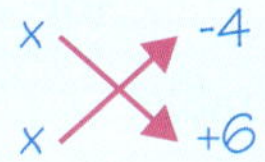 $6x - 4x = 2x$ (incorrect in magnitude and sign) 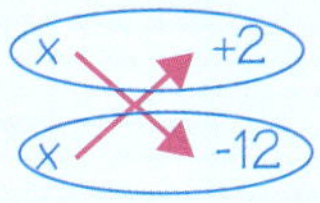 $-12x + 2x = -10x$ (correct in magnitude and sign)
2 Write the factorisation.	$(x + 2)(x - 12)$
(b) 1 Choose a pair of factors that multiply to give the constant term. If the constant term is positive, choose factors that have the same sign (try $7 \times 2 = 14$). Use the cross method to see if they add to give the correct middle term ($7x + 2x = 9x$). If this pair gives an incorrect middle term, try another pair of factors.	(b) $x^2 - 9x + 14$  $2x + 7x = 9x$ (incorrect in sign) 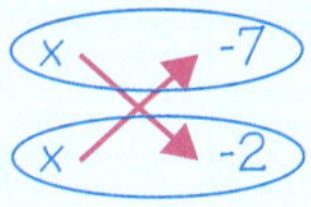 $-2x - 7x = -9x$ (correct in magnitude and sign)
2 Write the factorisation.	$(x - 2)(x - 7)$

3.5 Factorising monic quadratic expressions

Navigator

1 (columns 1–2), 2 (columns 1–2), 3, 4, 5, 6 (column 1), 7, 8, 10, 11	1 (column 2), 2 (column 2), 3, 4, 5, 6 (column 2), 7, 8, 9, 10, 11, 12	1 (column 3), 2 (column 3), 3, 4, 5, 6 (column 3), 7, 8, 9, 10, 11, 12

Answers p. 791

Fluency

W.E. 14

1 Factorise each expression by grouping in pairs.

(a) $x^2 + 2x + 4x + 8$ **(b)** $x^2 - 3x + 6x - 18$ **(c)** $x^2 - x + 7x - 7$

(d) $x^2 + 5x - 2x - 10$ **(e)** $x^2 - 4x - 3x + 12$ **(f)** $x^2 + 3x - 4x - 12$

(g) $x^2 - 2x + 4x - 8$ **(h)** $x^2 + x - 3x - 3$ **(i)** $x^2 - 2x - 8x + 16$

(j) $x^2 - 6x - 2x + 12$ **(k)** $-x^2 - 3x - 4x - 12$ **(l)** $x^2 + 5x - 6x - 30$

W.E. 15

2 Factorise these quadratic trinomials by splitting the middle term.

(a) $x^2 + 3x + 2$ (b) $y^2 + 9y + 14$ (c) $a^2 - 10a + 24$
(d) $k^2 + 11k + 24$ (e) $m^2 - 2m - 15$ (f) $x^2 - x - 30$
(g) $p^2 - 10p + 16$ (h) $d^2 - 5d + 4$ (i) $y^2 - 5y + 6$
(j) $m^2 - 4m - 12$ (k) $c^2 - 10c + 21$ (l) $g^2 + 2g - 35$
(m) $2y^2 + 14y + 24$ (n) $3m^2 + 18m + 24$ (o) $5x^2 - 5x - 100$
(p) $-4a^2 - 8a + 12$ (q) $-3k^2 + 12k + 63$ (r) $-5x^2 - 20x + 105$

3 Complete the following table.

Expanded expression	Cross method	Factorised expression
(a) $a^2 + 8a + 12$	a 2 a □	$(a + 2)(a +$ ____ $)$
(b) $x^2 - 7x + 12$	x -3 □ □	$(x - 3)($ _____ $)$
(c) $m^2 + 9m - 22$	□ 11 m □	$(m +$ ____ $)(m -$ ____ $)$
(d) $t^2 + 9t +$ ____	□ 4 t 5	$($ ____ $- 4)(t -$ ____ $)$
(e) $h^2 -$ ____ $- 30$	h -10 h 3	$($ _____ $)($ _____ $)$

4 Complete the following table.

Expanded expression	Cross method	Factorised expression
(a) $b^2 - 4b - 45$	b 5 b □	$(b + 5)(b -$ ____ $)$
(b) $p^2 - 11p + 24$	p -3 □ □	$(p - 3)($ _____ $)$
(c) $a^2 + a - 30$	□ 6 a □	$(a +$ ____ $)(a -$ ____ $)$
(d) $t^2 - 8t +$ ____	□ -4 t □	$($ ____ $- 4)(t -$ ____ $) = ($ _____ $)^2$
(e) $m^2 -$ ____ $+ 21$	m -7 m -3	$($ _____ $)($ _____ $)$

5 Find the matching pair of factors for each of the following quadratic expressions.

(a) $x^2 + 15x + 50$

A $(x + 2)(x + 25)$ B $(x - 10)(x - 5)$ C $(x + 5)(x + 10)$ D $(x - 2)(x + 25)$

(b) $t^2 - 13t + 30$

A $(t - 10)(t - 3)$ B $(t - 5)(t - 6)$ C $(t + 10)(t + 3)$ D $(t - 15)(t + 2)$

(c) $h^2 + 12h - 45$

A $(h + 9)(h - 5)$ B $(h - 3)(h + 15)$ C $(h + 3)(h - 15)$ D $(h - 9)(h - 5)$

6 Factorise these quadratic trinomials using the cross method.

W.E. 16

(a) $z^2 - 8z + 15$ (b) $k^2 + 5k - 6$ (c) $z^2 + 10z + 21$

(d) $t^2 - t - 20$ (e) $r^2 + r - 30$ (f) $t^2 - 13t + 22$

(g) $z^2 - z - 42$ (h) $m^2 - 9m + 20$ (i) $y^2 + 7y - 60$

(j) $n^2 - 3n - 28$ (k) $p^2 + 9p - 36$ (l) $a^2 - 14a + 48$

Understanding

7 If $A = 4x^2 + 6x + 29$ and $B = 1 + 9x + 5x^2$, then $A - B$ when factorised is:

A $(x - 4)(x + 7)$ B $(x + 4)(7 - x)$ C $(x + 7)(4 - x)$ D $(x - 7)(4 - x)$

8 A rectangle has a width of $(x + 3)$ cm. Find the length of the rectangle if its area is $(2x^2 + 10x + 12)$ cm^2.

9 A triangle has a base length of $(2x + 6)$ mm. Find the height of the triangle if its area is $(3x^2 - 3x - 36)$ mm^2.

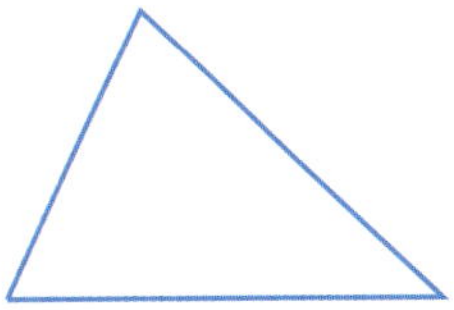

Reasoning

10 A rectangular block of land is twice as long as its width w m. A concrete path of uniform width is to be constructed around the inside border of the block and will occupy an area of $(6w - 4)$ m^2.

(a) Write an expression to represent the area that is not concrete.

(b) Factorise this expression.

(c) If the factors represent the length and width of the area that is not concrete, find the width of the path.

Hint

Rewrite your answer in (b) as the product of two algebraic factors by subtracting the same amount from the original length and width.

Open-ended

11 Write two monic quadratic trinomials that will factorise, and then factorise them.

12 If $x^2 + 5x - 24$ represents a rectangular area:

(a) write two possible sets of dimensions for the rectangle

(b) evaluate the area for each set of dimensions.

3.6 Factorising using special products

Some quadratic expressions can be factorised easily if you can recognise a special product. The special products you have seen already are the difference of two squares and the perfect squares.

The difference of two squares rule

The difference of two squares rule can be used to factorise quadratic expressions of the form $a^2 - b^2$.

$(a + b)(a - b) = a^2 - b^2$

By reversing this expansion:

$a^2 - b^2 = (a + b)(a - b)$

Worked example 17 W.E. 17

Factorise these expressions using the difference of two squares rule.

(a) $d^2 - 81$ **(b)** $180 - 5y^2$ **(c)** $4k^2 - 9m^2$

Thinking	Working
(a) 1 Look for common factors. (There are none.)	(a) $d^2 - 81$
2 Use the difference of two squares rule to factorise the expression. (Here, $a = d$, $b = 9$.)	$= d^2 - 9^2$ $= (d + 9)(d - 9)$
(b) 1 Look for a common factor and take it out. (Here, 5 is a common factor.)	(b) $180 - 5y^2$ $= 5(36 - y^2)$
2 Use the difference of two squares rule to factorise the expression. (Here, $a = 6$ and $b = y$.)	$= 5(6^2 - y^2)$ $= 5(6 + y)(6 - y)$
(c) 1 Look for common factors. (There are none.)	(c) $4k^2 - 9m^2$
2 Express both terms as squares.	$= (2k)^2 - (3m)^2$
3 Use the difference of two squares rule to factorise the expression. (Here, $a = 2k$ and $b = 3m$.)	$= (2k - 3m)(2k + 3m)$

Worked example 18

W.E. 18

Factorise $(x + 3)^2 - 9$.

Thinking	Working
1 Look for common factors. (There are none.)	$(x+3)^2 - 9$
2 Write the expression as the difference of two squares.	$= (x+3)^2 - 3^2$
3 Use the difference of two squares rule to factorise the expression. ($a = x + 3$, $b = 3$)	$= [(x+3) - 3][(x+3) + 3]$
4 Simplify by adding or subtracting like terms.	$= x(x+6)$

Surds can be used to factorise quadratics using the difference of two squares rule.

A surd is an irrational number such as $\sqrt{2}$ or $\sqrt[3]{4}$. It does not have an exact decimal equivalent. You can use surds to write an integer such as 7 as a square number $(\sqrt{7})^2 = 7$.

Worked example 19

W.E. 19

Factorise these expressions. (Your answers will include surds.)

(a) $x^2 - 3$ **(b)** $10 - 5x^2$

Thinking	Working
(a) 1 Look for common factors. (There are none.)	(a) $x^2 - 3$
2 Express both terms as squares.	$= x^2 - (\sqrt{3})^2$
3 Use the difference of two squares rule. ($a = x$, $b = \sqrt{3}$)	$= (x - \sqrt{3})(x + \sqrt{3})$
(b) 1 Look for common factors and factorise.	(b) $10 - 5x^2$ $= 5(2 - x^2)$
2 Express both terms as squares.	$= 5[(\sqrt{2})^2 - x^2]$
3 Use the difference of two squares rule. ($a = \sqrt{2}$, $b = x$)	$= 5(\sqrt{2} - x)(\sqrt{2} + x)$

Worked example 20

W.E. 20

Factorise these expressions.

(a) $(x+7)^2-(x+2)^2$ (b) $(x-5)^2-7$

Thinking	Working
(a) 1 Look for common factors. (There are none.)	(a) $(x+7)^2-(x+2)^2$
2 Use the difference of two squares rule. ($a=x+7$, $b=x+2$)	$=[(x+7)-(x+2)][(x+7)+(x+2)]$
3 Simplify. Take care with negatives.	$=(x+7-x-2)(x+7+x+2)$ $=5(2x+9)$
(b) 1 Look for common factors. (There are none.)	(b) $(x-5)^2-7$
2 Express both terms as squares.	$=(x-5)^2-(\sqrt{7})^2$
3 Use the difference of two squares rule to factorise the expression.	$=(x-5+\sqrt{7})(x-5-\sqrt{7})$

Worked example 21

W.E. 21

Factorise these expressions.

(a) $x^2-y^2+5x-5y$ (b) $25m^2-n^2-5m-n$

Thinking	Working
(a) 1 Look for common factors. (There are none.)	(a) $x^2-y^2+5x-5y$
2 Use the difference of two squares rule to factorise the first two terms (x^2-y^2) and take out a common factor of the last two terms $(5x-5y)$.	$=(x^2-y^2)+(5x-5y)$ $=(x-y)(x+y)+5(x-y)$
3 Take out the common factor $(x-y)$.	$=(x-y)(x+y+5)$
(b) 1 Look for common factors. (There are none.)	(b) $25m^2-n^2-5m-n$
2 Use the difference of two squares rule to factorise the first two terms $(25m^2-n^2)$ and take out a common factor of the last two terms $(-5m-n)$.	$=(5m+n)(5m-n)-1(5m+n)$ $=(5m+n)(5m-n-1)$

Note that there are three terms in one of the factors, so these factors are not binomial products.

The perfect square rules

$(a+b)^2=a^2+2ab+b^2$ and $(a-b)^2=a^2-2ab+b^2$

Reversing these expansions:

$a^2+2ab+b^2=(a+b)^2$ and $a^2-2ab+b^2=(a-b)^2$

You can use these rules to factorise many quadratic trinomials.

Worked example 22

W.E. 22

Factorise these expressions using a perfect square rule.

(a) $x^2 + 10x + 25$ (b) $3x^2 - 24x + 48$

Thinking	Working
(a) 1 Look for common factors. (There are none.)	(a) $x^2 + 10x + 25$
2 Factorise using a perfect square rule.	$= x^2 + 2(x)(5) + 5^2$ $= (x + 5)^2$
(b) 1 Look for common factors and factorise. (HCF = 3)	(b) $3x^2 - 24x + 48$ $= 3(x^2 - 8x + 16)$
2 Factorise further using a perfect square rule.	$= 3(x^2 - 2(x)(4) + 4^2)$ $= 3(x - 4)^2$

Special product rules

$a^2 - b^2 = (a + b)(a - b)$ (difference of two squares)

$a^2 + 2ab + b^2 = (a + b)^2$ (perfect square)

$a^2 - 2ab + b^2 = (a - b)^2$ (perfect square)

3.6 Factorising using special products

Navigator

1 (columns 1–2), 2 (columns 1–2), 3 (columns 1–2), 4 (columns 1–2), 5, 6 (column 1), 7 (a–d), 8, 9, 10 (column 1), 11 (a, b), 13, 15	1 (columns 2,–3), 2 (columns 2–3), 3 (columns 2–3), 4 (column 2), 5 (column 1), 6 (column 2), 7, 8, 9, 10 (column 2), 11 (column 1), 12, 13, 15, 16, 17	1 (column 3), 2 (column 3), 3 (column 3), 4 (column 3), 5 (column 2), 6 (column 3), 7, 9, 10 (column 3), 11, 12, 14, 15, 16, 17

Answers p. 792

Fluency

1 Factorise these expressions using the difference of two squares rule. W.E. 17

(a) $p^2 - 4$ (b) $d^2 - 100$ (c) $9 - k^2$ (d) $3p^2 - 75$

(e) $80 - 5m^2$ (f) $8k^2 - 32$ (g) $(2y)^2 - 49$ (h) $25 - (4x)^2$

(i) $(6m)^2 - (5n)^2$ (j) $9x^2 - 16$ (k) $4 - 25a^2$ (l) $81d^2 - 64$

2 Factorise these expressions. W.E. 18

(a) $(x + 2)^2 - 9$ (b) $(a - 1)^2 - 25$ (c) $16 - (k + 5)^2$

(d) $36 - (p + 1)^2$ (e) $1 - (2y - 3)^2$ (f) $1 - (3x - 2)^2$

(g) $5(x + 3)^2 - 5y^2$ (h) $7(a + 5)^2 - 7b^2$ (i) $3(x - 4)^2 - 12$

3 Factorise these expressions. (Your answers will include surds.) W.E. 19

(a) $x^2 - 2$ (b) $7 - p^2$ (c) $4k^2 - 5$

(d) $9p^2 - 2$ (e) $3a^2 - 33$ (f) $2b^2 - 6$

(g) $15x^2 - 45$ (h) $7p^2 - 49$ (i) $13k^2 - 169$

W.E. 20

4 Factorise these expressions.

(a) $(x+2)^2-(y+1)^2$ (b) $(a+2)^2-(a+3)^2$ (c) $(2k+3)^2-(m-3)^2$

(d) $(x+4)^2-(x-4)^2$ (e) $(2x+1)^2-(x+3)^2$ (f) $(5x-4)^2-(x-7)^2$

(g) $(x+2)^2-3$ (h) $(y-3)^2-6$ (i) $(x+8)^2-5$

(j) $(x-5)^2-13$ (k) $(x-1)^2-19$ (l) $(2x+1)^2-2$

W.E. 21

5 Factorise these expressions.

(a) $x^2-9+4x+12$ (b) $a^2-16+5a-20$

(c) $x^2-y^2+7x-7y$ (d) $9a^2-b^2+6a+2b$

(e) $x^2-16y^2+2x-8y$ (f) $4p^2-q^2-6p-3q$

W.E. 22

6 Factorise each expression using a perfect square rule.

(a) x^2+6x+9 (b) x^2+4x+4 (c) $x^2-14x+49$

(d) $x^2-8x+16$ (e) x^2+2x+1 (f) $x^2+12x+36$

(g) a^2-6a+9 (h) $2m^2-16m+32$ (i) $3q^2+6q+3$

(j) $4x^2+8x+4$ (k) $5x^2+30x+45$ (l) $4x^2-40x+100$

Understanding

7 When factorised, $(2a-1)^2-(a+1)^2$ becomes:

A $3(a-1)(a+1)$ B $3a(a-2)$

C $(a-1)(a+1)$ D $(2a-1)(a+1)$

8 When factorised, the expression for the blue area shown becomes:

A $4d(d+3)$ B $(4d+3)(d+1)$

C $(2d+3)(d+1)$ D $3(4d+3)$

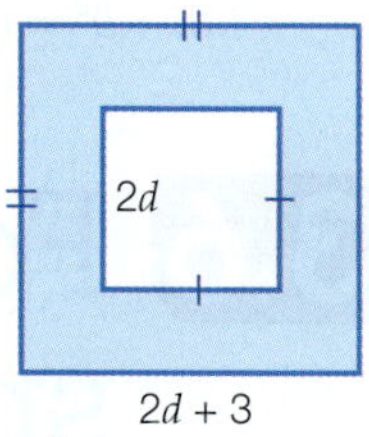

9 Factorise these expressions.

(a) $8m^2-200n^2$ (b) $6x^2-24r^2$ (c) $5a^2-45c^2$

(d) $27a^2-48b^2$ (e) $72x^2-98y^2$ (f) $175t^2-28p^2$

(g) $54k^2-24m^2$ (h) $2x^2y^2-32z^2$ (i) $3p^2q^2-12n^2$

10 Factorise these expressions.

(a) $16c^2-25d^2+10d+8c$ (b) $k^2-64+km-8m$

(c) $81-49a^2-27d+21ad$ (d) $16-9y^2-16w+12wy$

(e) $25m^2-9n^2-20m-12n$ (f) $144p^2-49r^2-12pq+7rq$

Reasoning

11 When factorised, $\left(\frac{a}{k}\right)^2-\left(\frac{b}{2k}\right)^2$ is:

A $\frac{1}{4k^2}(2a-b)(2a+b)$ B $\frac{1}{2k^2}(a-b)(a+2b)$

C $\frac{1}{4k^2}(a-3b)(a+3b)$ D $\frac{1}{k^2}(a-b)(a+b)$

12 The difference between the squares of two consecutive odd numbers x and y is 56.

(a) If x is greater than y, write the equation describing the information.

(b) Factorise the left side of the equation in part **(a)**.

(c) Explain why $x = y + 2$.

(d) Substitute $y + 2$ for x in the equation in part **(b)** and simplify.

(e) Solve the equation for y. Hence, state the value of x.

13 (a) Factorise $a^2 - b^2$.
(b) How many factors do you have?
(c) Find $\frac{a^2 - b^2}{a - b}$ for $a \neq b$.
(d) Write $a^4 - b^4$ as the difference of two squares.
(e) Factorise $a^4 - b^4$ fully.
(f) How many factors do you have?
(g) Find $\frac{a^4 - b^4}{a - b}$ for $a \neq b$.
(h) Write $a^8 - b^8$ as the difference of two squares.
(i) Factorise $a^8 - b^8$ fully using your answer from **(e)**.
(j) Find $\frac{a^8 - b^8}{a - b}$ for $a \neq b$.
(k) Use the pattern from the factorisations above to write an expression for $\frac{a^{16} - b^{16}}{a - b}$, for $a \neq b$.
(l) Use your expression from part **(k)** to show that $2^{16} - 1 = 3 \times 5 \times 17 \times 257$.

14 The shape of a house with its garage is shown.
(a) Write the expression (in m^2) for the area of the house.
(b) Write the expression (in m^2) for the area of the garage.
(c) Write the expression (in m^2) for the total area in factorised form.

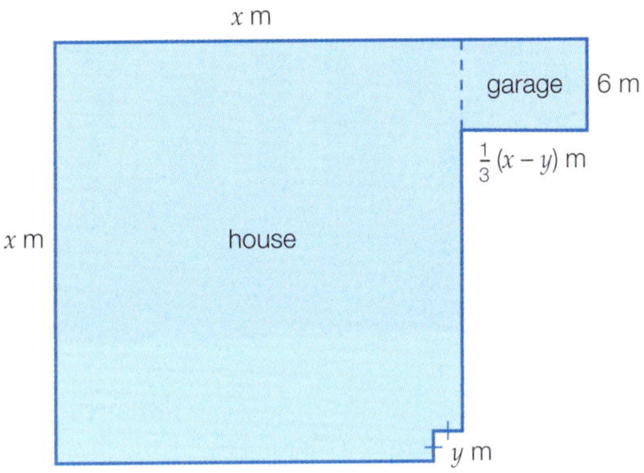

15 If $a = m - n$, $b = 2\sqrt{mn}$ and $c = m + n$, show that $c^2 = a^2 + b^2$.

16 (a) Show that if $p = kr$ then the expression $t^2 = p^2 - r^2$ becomes $t^2 = r^2(k - 1)(k + 1)$.
(b) What values of k would make $t = 0$?
(c) What values of k are not possible? Justify your answer.

Hint
Can $k = 0$?
Can k be a fraction?

Open-ended

17 Write an expression with four terms so that two terms are the difference of the squares of two variables and the other two terms are a multiple of the sum of these variables. Factorise your expression.

Problem solving

Playground swings

The length of the chains of a playground swing that takes t seconds for one swing is given by $l = \frac{5t^2}{M}$.

1 What is the difference (in m) in the chain lengths of two swings that take t_1 seconds and t_2 seconds, respectively, to make one swing? (Assume t_1 is greater than t_2.) Write your answer in factorised form.

2 If $t_1 - t_2 = \frac{3M}{10}$ and $t_1 + t_2 = \frac{M}{5}$ then find the difference in the chain lengths.

Strategy options
- Look for a pattern.
- Have I seen a similar problem?

Maths 4 Real

Staying alive

Even at a speed limit of 60 km/h, it takes longer to bring a car to a stop than you might think. The stopping distance of a car is a combination of the reaction distance and the braking distance.

Reaction distance

When driving a motor vehicle, the time you take to react to a perceived hazard (such as a pedestrian, animal, another vehicle stopping suddenly) is called the *reaction time*. This time depends on the driver's age, experience, physical fitness and whether the driver is distracted, or under the influence of alcohol or drugs. Reaction time is usually between 1 and 2 seconds.

The reaction distance d_r is the distance travelled by the vehicle during the reaction time. It depends on both the length of the reaction time and the velocity (speed) v of the vehicle.

$d_r = v \times t_r$

1 (a) Calculate the reaction distance (in metres) for the following speeds and reaction times.

(i) $v = 60\,\text{km/h}, t_r = 0.8\,\text{s}$ (ii) $v = 60\,\text{km/h}, t_r = 1.6\,\text{s}$

(iii) $v = 65\,\text{km/h}, t_r = 0.8\,\text{s}$ (iv) $v = 65\,\text{km/h}, t_r = 1.6\,\text{s}$

(v) $v = 100\,\text{km/h}, t_r = 0.8\,\text{s}$ (vi) $v = 100\,\text{km/h}, t_r = 1.6\,\text{s}$

Convert the speeds from km/h to m/s.
3.6 km/h = 1 m/s

(b) Use your answers to part (a) to answer the following.

(i) For a reaction time of 1.6 seconds, what is the difference in the reaction distance at a speed of 65 km/h, compared to 60 km/h?

(ii) What effect does doubling the reaction time have on the reaction distance at 100 km/h?

(c) If you were to graph d_r for a certain reaction time over a number of different speeds, would you produce a linear graph or a curve? Explain why. (Use some of the values calculated in (a) to draw a graph if you aren't sure).

Braking distance

The *braking distance* travelled by the vehicle is measured from when the brakes are applied to when the vehicle comes to a stop. The vehicle can be considered to be undergoing a constant rate of deceleration while braking. (Deceleration is the opposite of acceleration, i.e. the speed is decreasing instead of increasing.) The braking distance also depends on the road surface, weather conditions, and the condition of the tyres and brakes.

The formula $V_f^2 = V_0^2 - 2ad$ is used to calculate the final velocity of a vehicle (V_f) after it has undergone deceleration. V_0 is the initial velocity, a is the rate of deceleration and d is the braking distance travelled during deceleration. V_f will be zero when the car has stopped.

2 (a) Substitute $V_f = 0$ into the above equation, then rearrange it to make d the subject.

In this form, you can see that braking distance is proportional to the square of the velocity. This means a small increase in speed can have a big effect on braking distance.

(b) What would be the shape of the graph produced if braking distance (d) were graphed against velocity, (V_0)? Explain why.

3 Calculate the braking distance given a maximum deceleration of 10 m/s^2 and velocity of:

(a) 25 km/h (b) 50 km/h (c) 60 km/h

(d) 65 km/h (e) 80 km/h (f) 120 km/h

Total stopping distance

Total stopping distance = reaction time + braking distance

4 (a) Combine your formulas for reaction distance (Question 1) and braking distance (Question 2) to create a formula for total stopping distance.

(b) Using a driver reaction time of 1.6 seconds and a deceleration rate of 10 m/s^2, calculate the total stopping distance for a vehicle travelling at:

(i) 50 km/h (ii) 60 km/h

(iii) 65 km/h (iv) 100 km/h

(c) Describe the effect of doubling the speed (from 50 km/h to 100 km/h) on:

(i) reaction distance (ii) braking distance.

(d) Victoria's Transport Accident Commission (TAC) ran a safety awareness campaign with the slogan 'Wipe off 5'. Describe the effect of lowering speed from 65 km/h to 60 km/h on:

(i) reaction distance (ii) braking distance.

5 Speed limits in residential areas have been reduced from 60 km/h to 50 km/h and to 40 km/h around schools. What effect does this have on the total stopping distance when both ends of the reaction time are considered? (Assume 10 m/s^2 maximum deceleration.)

Research

6 Research the effects of alcohol or drugs on driver reaction time.

7 Find some car crash statistics for your state or local area. How many of them involved a fatality (death)? In how many crashes was speed a factor? In how many crashes was alcohol involved?

8 Many road authorities advise drivers to leave a 'safe braking distance' between their car and the car in front. Find the recommended distance. Given the braking distances you have calculated in this task, do you think the recommended distance is adequate?

Investigation

The firefighter's ladder

Equipment required: calculator

The Big Question

Can you predict how the height of the top of a ladder will change for a small change in the distance of the foot of the ladder from the wall?

Engage

A firefighter has to get into a building through a window by leaning his ladder against the wall. Given the length of his ladder and the height of the window, he will need to calculate how far from the wall he must place the bottom of his ladder.

Give exact answers for each of the following questions.

1 **(a)** For a 10 m ladder, calculate each of the following heights (h).

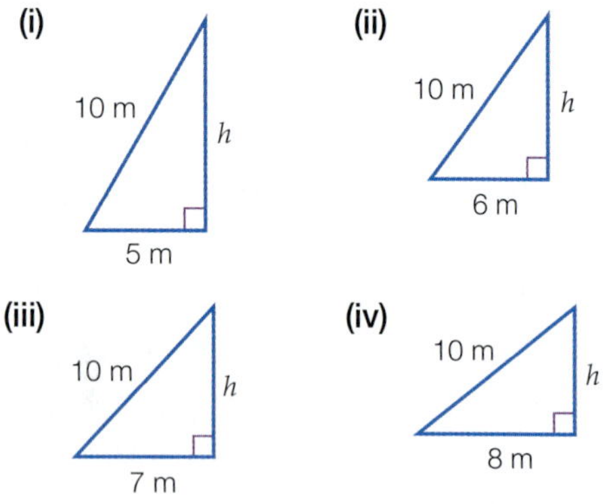

(b) If the ladder is 10 m long, find the value of h, when the distance from the base of the wall to the foot of the ladder is equal to the height to the top of the ladder on the wall.

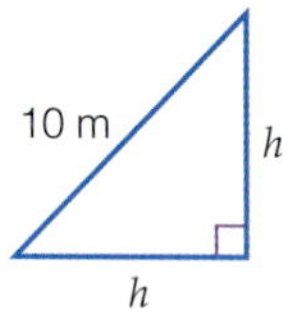

Explore

Give exact answers for each of the following questions.

2 **(a)** If the distance of the foot of the ladder to the wall and the height of the top of the ladder on the wall are both equal to 7 m, ($h = 7$ m), how long would the ladder have to be?

(b) Find the difference between the length of this ladder and 7 m.

(c) How far could this ladder move in towards the wall before it was completely upright?

(d) How far could the bottom of the ladder slip away from its position 7 m from the base of the wall before it was flat on the ground?

The base of the ladder slips p m away from the base of the wall and the top of the ladder travels a distance of n m down the wall.

(e) What is the largest possible value for:

(i) p **(ii)** n

(f) Show that $(p + 7)^2 + (7 - n)^2 = 98$.

(g) (i) Rearrange this formula and simplify to show that $(7 - n)^2 = 49 - 14p - p^2$.

(ii) Hence, show that $n = 7 - \sqrt{49 - 14p - p^2}$.

(iii) Why should you reject $n = 7 + \sqrt{49 - 14p - p^2}$?

Hint

Refer to your answer in part **(e)(ii)**.

(h) Evaluate n for all possible integer values of p.

(i) Evaluate n for $p = -16$.

Strategy options

- Make a table.
- Look for a pattern.

Explain

Use exact values in your explanations for each of the following questions.

3 (a) Does the ladder slip down the wall the same distance it moves away from the base of the wall? Explain why or why not. If not, does it slip further down or not as far down the wall as the distance away from it?

(b) Explain what happens to the distance the ladder will move up the wall when the ladder is moved closer to the wall.

(c) In **2(h)**, you could only evaluate n for integer values of $p \leq 2$. Explain why.

(d) Why can you evaluate n for $p = -16$ m when this value of p is not possible? What does $p = -16$ m mean?

Hint

Did you find n was the same for one of the p values you used in Question **2(h)**? If so, which value?

Elaborate

4 (a) If h, the distance from the base of the wall to the bottom of the ladder, is equal to the height to the top of the ladder on the wall, find an equation for n, the distance the top of the ladder moves down the wall when the foot of the ladder moves a distance of p away from the base of the wall.

(b) Use an equation to show that p can only be equal to n for $p = n = 0$.

Hint

The length of the ladder does not change.

(c) Describe what happens to the distance the ladder can reach up a wall as the bottom of the ladder slips away from the base of the wall and as the bottom of the ladder is pushed closer to the base of the wall.

Evaluate

5 (a) Did you find visualising the ladder moving up and down the wall a challenge? Did the diagrams help you?

(b) Did you find working with unknown variables more difficult than working with actual values? Did working with actual values help you to understand the situation? Explain why or why not.

(c) Why were you asked to give exact answers? Would it have been easier to understand if you had evaluated lengths and distances correct to 2 decimal places? Explain.

Extend

6 If the distance from the foot of the ladder to the base of the wall is $a - b$ and the distance up the wall to the top of the ladder is $a + b$, show that the length of the ladder can be written as $\sqrt{2(a^2 + b^2)}$.

3.7 Algebraic fractions

Multiplication and division

To simplify a fraction, look for common factors in the numerator (top) and the denominator (bottom) of the fraction. You can simplify by dividing the numerator and denominator by these common factors. These rules also apply to **algebraic fractions** (expressions involving variables in fraction form).

Worked example 23

W.E. 23

Simplify $\frac{4a}{6}$.

Thinking	Working
1 Find the factor common to the numerator and denominator (2) and simplify the fraction by dividing both by this factor.	$\frac{4a}{6}$ $= \frac{\cancel{4}^{2}a}{\cancel{6}_{3}}$
2 Write the fraction in its simplified form.	$= \frac{2a}{3}$

A common factor may be an algebraic term or expression.

Worked example 24

W.E. 24

Simplify $\frac{(x+2)(x+4)}{(x-1)(x+2)}$.

Thinking	Working
1 Write the expression and look for common factors. (Here, $(x+2)$ is common to the numerator and denominator.)	$\frac{(x+2)(x+4)}{(x-1)(x+2)}$
2 Simplify the fraction by dividing the common factors into both the numerator and the denominator.	$= \frac{\cancel{(x+2)}^{1}(x+4)}{(x-1)\cancel{(x+2)}_{1}}$ $= \frac{x+4}{x-1}$

The above example cannot be simplified further as there are no other common factors. You cannot cancel the x in both the numerator and denominator because x is *not* a factor of $x+4$ or $x-1$.

You can also simplify when multiplying or dividing two or more fractions. You may need to factorise them first to identify any common factors. As well as cancelling common factors within fractions, you can also cancel a factor that is common to the numerator of one fraction and the denominator of the other.

Worked example 25

W.E. 25

Simplify:

(a) $\frac{(x+2)(x+5)}{(x-1)(x+1)} \times \frac{(x-2)(x+1)}{(x+5)(x+3)}$ (b) $\frac{(x+2)(x+5)}{(x-1)(x+1)} \div \frac{(x+5)(x+3)}{(x-2)(x+1)}$

Thinking	Working
(a) 1 Write the expression and identify common factors of the numerators and denominators. (Here, $(x+5)$ and $(x+1)$.)	(a) $\frac{(x+2)(x+5)}{(x-1)(x+1)} \times \frac{(x-2)(x+1)}{(x+5)(x+3)}$
2 Simplify by dividing the common factors into both the numerators and the denominators.	$= \frac{(x+2)\cancel{(x+5)}^{1}}{(x-1)\cancel{(x+1)}_{1}} \times \frac{(x-2)\cancel{(x+1)}^{1}}{\cancel{(x+5)}_{1}(x+3)}$
3 Multiply the remaining numerators and denominators. Leave the fraction in factorised form.	$= \frac{(x+2)(x-2)}{(x-1)(x+3)}$
(b) 1 Write the expression.	(b) $\frac{(x+2)(x+5)}{(x-1)(x+1)} \div \frac{(x+5)(x+3)}{(x-2)(x+1)}$
2 Convert the division into a multiplication by inverting the second fraction (turning it upside-down) and simplify by cancelling the common factor.	$= \frac{(x+2)(x+5)}{(x-1)(x+1)} \times \frac{(x-2)(x+1)}{(x+5)(x+3)}$ $= \frac{(x+2)\cancel{(x+5)}^{1}}{(x-1)\cancel{(x+1)}_{1}} \times \frac{(x-2)\cancel{(x+1)}^{1}}{\cancel{(x+5)}_{1}(x+3)}$
3 Multiply the remaining numerators and denominators. Leave the fraction in factorised form.	$= \frac{(x+2)(x-2)}{(x-1)(x+3)}$

Worked example 26

W.E. 26

Simplify $\frac{4x^2-12x}{x^2-25} \times \frac{x^2+7x+10}{2x^2+4x}$.

Thinking	Working
1 Factorise expressions where possible. ($4x^2-12x$: take out a common factor of $4x$) (x^2-25: difference of two squares) ($x^2+7x+10$: factorise the quadratic) ($2x^2+4x$: take out a common factor of $2x$)	$\frac{4x^2-12x}{x^2-25} \times \frac{x^2+7x+10}{2x^2+4x}$ $= \frac{4x(x-3)}{(x-5)(x+5)} \times \frac{(x+2)(x+5)}{2x(x+2)}$

2 Cancel common factors in the numerators and the denominators. (Here, they are 2, x, $(x + 2)$ and $(x + 5)$.)	$= \dfrac{4x(x-3)}{(x-5)(x+5)} \times \dfrac{(x+2)(x+5)}{2x(x+2)}$
3 Multiply the remaining numerators and denominators.	$= \dfrac{2(x-3)}{x-5}$

Worked example 27

W.E. 27

Simplify $\dfrac{x^2 + x - 12}{3x^2 + 12x} \div \dfrac{x^2 - 9}{12x}$.

Thinking	Working
1 Change the division into a multiplication by inverting the second fraction.	$\dfrac{x^2 + x - 12}{3x^2 + 12x} \div \dfrac{x^2 - 9}{12x}$ $= \dfrac{x^2 + x - 12}{3x^2 + 12x} \times \dfrac{12x}{x^2 - 9}$
2 Factorise where possible. ($x^2 + x - 12$: factorise the quadratic) ($3x^2 + 12x$: take out a common factor of $3x$) ($4x^2 - 9$: difference of two squares)	$= \dfrac{(x-3)(x+4)}{3x(x+4)} \times \dfrac{12x}{(x-3)(x+3)}$
3 Cancel common factors in the numerators and denominators. (Here, they are $(x - 3)$, $(x + 4)$, 3 and x.)	$= \dfrac{(x-3)(x+4)}{3x(x+4)} \times \dfrac{12x}{(x-3)(x+3)}$
4 Multiply the remaining numerators. Multiply the remaining denominators. Note that it is not possible to cancel any further.	$= \dfrac{4}{x+3}$

When multiplying (or dividing) algebraic fractions:

1. Factorise if possible.
2. Change a division into a multiplication by inverting the second fraction.
3. Identify common factors in the numerators and denominators of the fractions and divide the numerators and denominators by these common factors.
4. Simplify by multiplying the remaining numerators and the remaining denominators.

Addition and subtraction

You can use the rules for the addition and subtraction of fractions to deal with algebraic fractions.

Worked example 28

W.E. 28

Simplify:

(a) $\frac{x}{4}+\frac{x}{5}$

(b) $\frac{x+7}{4}-\frac{x-3}{6}$

Thinking	Working
(a) 1 Find the lowest common denominator. (Here, the LCD of 4 and 5 is 20.)	(a) $\frac{x}{4}+\frac{x}{5}$
2 Find equivalent fractions using the lowest common denominator.	$=\frac{5x}{20}+\frac{4x}{20}$
3 Rewrite the expression as a single fraction.	$=\frac{5x+4x}{20}$
4 Simplify the numerator by adding like terms.	$=\frac{9x}{20}$
(b) 1 Find the lowest common denominator (here, the LCD of 4 and 6 is 12) and write the equivalent fractions with this (multiply the first fraction by $\frac{3}{3}$ and the second by $\frac{2}{2}$).	(b) $\frac{x+7}{4}-\frac{x-3}{6}$ $=\frac{3(x+7)}{12}-\frac{2(x-3)}{12}$
2 Rewrite the expression as a single fraction.	$=\frac{3(x+7)-2(x-3)}{12}$
3 Expand brackets in the numerator. (Notice that $-2\times -3=+6$.)	$=\frac{3x+21-2x+6}{12}$
4 Simplify the numerator by adding or subtracting like terms.	$=\frac{x+27}{12}$

When adding (or subtracting) algebraic fractions:

1 Find the lowest common denominator (LCD). If there is a factor that is common to both denominators, only use it *once* in the common denominator.

2 Rewrite each fraction as an equivalent fraction with this common denominator.

3 Add (or subtract) the new numerators.

3.7 Algebraic fractions

Navigator

Answers p.793

1 (columns 1–2), 2 (column 1), 3 (a–h), 4 (column 1), 5 (a–d), 6 (column 1), 8, 10, 12	1 (columns 2–3), 2 (column 2), 3 (column 1), 4 (column 2), 5 (column 1), 6 (column 2), 8, 10, 11, 12	1 (column 3), 2 (column 3), 3 (column 2), 4 (k–r), 5 (column 2), 6 (column 3), 7, 8, 9, 10, 11, 12

Fluency

W.E. 23

1 Simplify the following.

(a) $\frac{8x}{10}$ (b) $\frac{6k}{42}$ (c) $\frac{25p}{45}$

(d) $\frac{4(a-b)}{2}$ (e) $\frac{9(x+1)}{6}$ (f) $\frac{14(y-2)}{8}$

(g) $\frac{5y(a-4)}{10}$ (h) $\frac{8x(x+5)}{36}$ (i) $\frac{18x(x-4)}{24}$

W.E. 24

2 Simplify the following.

(a) $\frac{(x+1)(x+6)}{(x-5)(x+1)}$ (b) $\frac{(x-3)(x+2)}{(x-3)(x-2)}$ (c) $\frac{(2x-3)(x+4)}{(x+4)(2x+1)}$

(d) $\frac{(3x+1)(x-2)}{(x-7)(3x+1)}$ (e) $\frac{(a+2)(b+5)}{(b+3)(a+2)}$ (f) $\frac{(x+7)(y-2)}{(y-2)(x-1)}$

W.E. 25

3 Simplify the following.

(a) $\frac{3}{x-2}\times\frac{x-2}{9}$ (b) $\frac{15}{5-2x}\times\frac{5+2x}{5}$

(c) $\frac{x-1}{x-4}\div\frac{x-1}{x+4}$ (d) $\frac{x+4}{x+3}\div\frac{x+4}{x-3}$

(e) $\frac{x(x+7)}{2(x-3)}\div\frac{4x}{x-3}$ (f) $\frac{7x(x-4)}{10x}\div\frac{2(x-4)}{5}$

(g) $\frac{(x+1)(x-3)}{(x-4)(x+7)}\times\frac{(x+7)(x+2)}{(x-1)(x+1)}$ (h) $\frac{(x-4)(x+4)}{(x+3)(x-4)}\times\frac{(x-2)(x+3)}{(x+1)(x-2)}$

(i) $\frac{(3x-2)(x+5)}{(2x+3)(x-6)}\div\frac{(x+2)(x+5)}{(2x+3)(x-3)}$ (j) $\frac{(5x-3)(2x+3)}{(x-4)(x+7)}\div\frac{(2x+3)(5x-4)}{(x-3)(x+7)}$

W.E. 26

4 Simplify the following. Remember to factorise first.

(a) $\frac{x^2+11x+24}{x+3}$ (b) $\frac{x^2-12x+27}{x-3}$

(c) $\frac{x+2}{x^2-3x-10}$ (d) $\frac{x+5}{x^2-2x-35}$

(e) $\frac{8x^2-2}{8x+4}$ (f) $\frac{3x^2-12}{x+2}$

(g) $\frac{4x^2+12x+9}{4x+6}$ (h) $\frac{9x^2+30x+25}{6x+10}$

(i) $\dfrac{x^2+2x-3}{x^2-5x-24}$

(j) $\dfrac{x^2+3x-4}{x^2+4x-5}$

(k) $\dfrac{x^2-7x-8}{x^2-10x+16}$

(l) $\dfrac{x^2-7x+6}{x^2-2x-24}$

(m) $\dfrac{3x+6}{4x-24}\times\dfrac{8x-48}{6x+12}$

(n) $\dfrac{5x+10}{2x-8}\times\dfrac{3x-12}{9x+18}$

(o) $\dfrac{x^2-5x}{3x+3}\times\dfrac{x+1}{2x^2+6x}$

(p) $\dfrac{4x^2+8x}{x+5}\times\dfrac{2x-14}{x^2+2x}$

(q) $\dfrac{x^2-x-12}{x^2+9x+18}\times\dfrac{x^2+4x-12}{x^2-2x-8}$

(r) $\dfrac{x^2+7x+10}{x^2+x-20}\times\dfrac{x^2-7x+12}{x^2-5x-14}$

5 Simplify the following. **W.E. 27**

(a) $\dfrac{x^2-4}{x^2-9}\div\dfrac{4x+8}{12x-36}$

(b) $\dfrac{x^2-25}{x^2-49}\div\dfrac{3x-15}{5x+35}$

(c) $\dfrac{x^2-36}{x^2+6x-7}\div\dfrac{x^2-5x-6}{x^2-1}$

(d) $\dfrac{x^2-6x-16}{x^2-x-30}\div\dfrac{x^2-64}{x^2-5x-6}$

(e) $\dfrac{x^2+6x+9-y^2}{4x+6}\div\dfrac{2x^2+6x-2xy}{10x^2+15x}$

(f) $\dfrac{x^2-2x+1-4y^2}{12x^2-5x-2}\div\dfrac{3x-3+6y}{28x+7}$

6 Simplify the following. **W.E. 28**

(a) $\dfrac{x}{2}+\dfrac{x}{3}$

(b) $\dfrac{x}{5}+\dfrac{x}{2}$

(c) $\dfrac{x}{6}+\dfrac{x}{4}$

(d) $\dfrac{2x}{3}-\dfrac{x}{6}$

(e) $\dfrac{4x}{5}-\dfrac{x}{15}$

(f) $\dfrac{2x}{3}-\dfrac{5x}{4}$

(g) $\dfrac{4x}{3}-\dfrac{2y}{4}$

(h) $\dfrac{5x}{7}-\dfrac{3y}{4}$

(i) $\dfrac{11x}{8}-\dfrac{8y}{9}$

(j) $\dfrac{x+6}{2}+\dfrac{x+2}{3}$

(k) $\dfrac{x-3}{4}+\dfrac{x+4}{8}$

(l) $\dfrac{x-7}{3}+\dfrac{x+5}{9}$

(m) $\dfrac{3x+2}{6}-\dfrac{x+4}{4}$

(n) $\dfrac{3x-8}{4}-\dfrac{x-3}{10}$

(o) $\dfrac{4x-1}{9}-\dfrac{x-6}{12}$

(p) $\dfrac{3x+2y}{4}+\dfrac{5x-y}{3}$

(q) $\dfrac{7x-2y}{3}+\dfrac{x-3y}{5}$

(r) $\dfrac{8x-3y}{4}+\dfrac{5x+7y}{6}$

(s) $\dfrac{4x-3y}{5}-\dfrac{3x-2y}{6}$

(t) $\dfrac{5x-2y}{4}-\dfrac{x+y}{2}$

(u) $\dfrac{7x+3y}{2}-\dfrac{2x-y}{3}$

Understanding

7 If $4x+3y=5$, then $\dfrac{x+1}{3}+\dfrac{y+1}{4}$ simplifies to:

A $\dfrac{5}{12}$ B $\dfrac{5}{7}$ C 1 D $\dfrac{7}{12}$

8 (a) Factorise and simplify the expression

$$\frac{9b^2-9a^2}{6b^2+3bc+2ab+ac}\div\frac{3a+3b}{2b+c}.$$

(b) If $a=5$ and $b=7$, show that the expression in part **(a)** takes the value $\dfrac{3}{13}$.

9 (a) Show that $(a-b)(a^2+ab+b^2)=a^3-b^3$.

(b) Write a^6-b^6 as the difference of two cubes.

(c) Use the result from (b) to show that $\frac{a^6-b^6}{a^2-b^2}=a^4+a^2b^2+b^4$.

10 A circular disc of diameter $2w$ m is cut from a square piece of timber. The side lengths of the timber are the same as the diameter of the disc.

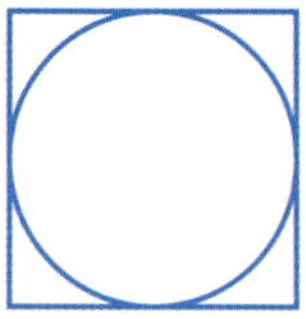

(a) Write an expression for the area of timber wasted.

(b) What fraction of the area of the uncut timber is the area of the timber wasted? Give your answer in simplified form.

Reasoning

11 The density of an object of mass m kg and volume V m^3 is given by $d=\frac{m}{V}$ kg/m^3

(a) What is the density of a sphere of mass m kg and radius r m if its volume is $\frac{4\pi r^3}{3}$ m^3?

(b) A golf ball has density d_1, radius r_1 and volume V_1. A basketball has density d_2, radius r_2 and volume V_2. Show that $\frac{d_2}{d_1}=\left(\frac{m_2}{m_1}\right)\left(\frac{r_1}{r_2}\right)^3$.

(c) If the basketball's mass is 20 times greater than the mass of the golf ball and its radius is 5 times greater than the radius of the golf ball, then what is the value of $\frac{d_2}{d_1}$?

Open-ended

12 Write the sum of two terms with different denominators that can be added to give $\frac{x+2}{3}$.

Problem solving

How to square two-digit numbers quickly—without a calculator

Start with the difference of two squares rule:

$$a^2-b^2=(a-b)(a+b)$$

Add b^2 to both sides:

$$a^2-b^2+b^2=b^2+(a-b)(a+b)$$

So: $$a^2=b^2+(a-b)(a+b)$$

This expression can be used to find the square of a two-digit number quickly, without a calculator. In the expression:

a is the number to be squared

b can be any number that makes it easy to multiply with $(a+b)$ or with $(a-b)$.

For example, to find 19^2, you have $a=19$ and can choose $b=1$.

Start with: $a^2=b^2+(a-b)(a+b)$

Substitute $a=19$ and $b=1$:

$$\begin{aligned}19^2&=1^2+(19-1)(19+1)\\&=1^2+(18)(20)\\&=1^2+360\\&=361\end{aligned}$$

Notice how the choice of 1 for b results in one of the brackets containing a multiple of 10, which makes the multiplication easier.

Use the method described above to calculate the following. Show the steps involved.

(a) 29^2 (b) 59^2 (c) 79^2 (d) 99^2

Strategy options

- Look for a pattern.
- Break problem into manageable parts.

Chapter review 3

Maths literacy

algebraic fractions	grouping in pairs	quadratic trinomial
binomial product	highest common factor (HCF)	reflection
difference of two squares	monic quadratic	transformation
dilation	perfect square	translation
factorising	quadratic expression	

Copy and complete the following using the words and phrases from the list, where appropriate. A word or phrase may be used more than once.

1 $ax^2 + bx + c$ is called a __________. When the coefficient of x^2 is 1, the expression is a __________.

2 The first thing to do to factorise $50x^2 - 2y^2$ is to take out the __________.You can then use the __________ method to factorise the expression.

3 You factorise by __________ to write the expression $2ax + 2ay + 3x + 3y$ as a product of factors $(2a + 3)(x + y)$.

4 $(3t - 2)^2$ is an example of a __________.

5 $(3a + b)(4 - 2a)$ is called a __________.

6 $x^2 - 9$ is an example of a __________.

7 The basic graph of $y = x^2$ can be transformed to produce other parabolas by using a combination of any of these __________s: __________s, __________s or __________s.

Fluency

1 Expand and simplify the following. **3.1**

(a) $5(2x + 7)$ (b) $x(x - 3) + 2(x - 3)$

(c) $(x + 6)(x + 4)$ (d) $(4x - 1)(2x + 3)$

2 Expand and simplify the following. **3.1**

(a) $(x + 6)^2$ (b) $(2x - 3)^2$

(c) $(5 - x)(5 + x)$ (d) $(3x + 7)(3x - 7)$

3 Expand and simplify the following. **3.1**

(a) $4(x - 7)(x + 2)$ (b) $-3x(x + 8)(x - 3)$

(c) $(x - 1)(x + 5)(x - 6)$ (d) $(2x + 3)(2 - x)(x + 4)$

4 Factorise these expressions. **3.2**

(a) $7x + 21$ (b) $12ab - 8a$ (c) $18x^2 + 24x$

5 Factorise these expressions. **3.2**

(a) $d(c - 4) + 6(c - 4)$ (b) $x(x + 2) - 3(x + 2)$

(c) $5a(a + 1) - 2(a + 1)$ (d) $2y(3y - 5) - 7(3y - 5)$

3.3

6 Plot the graph of $y = x^2 - 2x - 8$ for $-3 \le x \le 5$.

(a) Write the coordinates of the x- and y-intercepts.

(b) Write the coordinates of the turning point. Does the parabola have a maximum or a minimum turning point?

(c) Write the equation of the axis of symmetry.

(d) From the graph, find the value of y where $x = 3$.

(e) What are the values of x where $y = 7$?

(f) Hence, state the values of x where $x^2 - 2x - 8 = 7$.

(g) Write the values of x where $x^2 - 2x - 8 = 0$.

3.4

7 (a) The equation of the pink parabola could be:

A $y = x^2 + 2$

B $y = x^2 - 2$

C $y = (x - 2)^2$

D $y = (x + 2)^2$

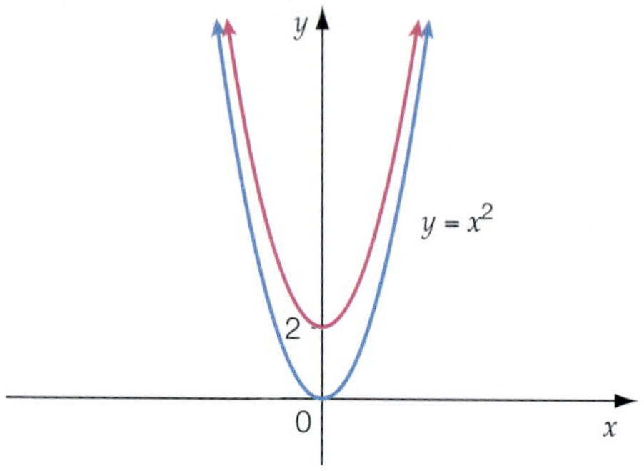

(b) The equation of the pink parabola could be:

A $y = (x - 1)^2 - 2$

B $y = (x - 1)^2 + 2$

C $y = (x + 1)^2 + 2$

D $y = (x + 1)^2 - 2$

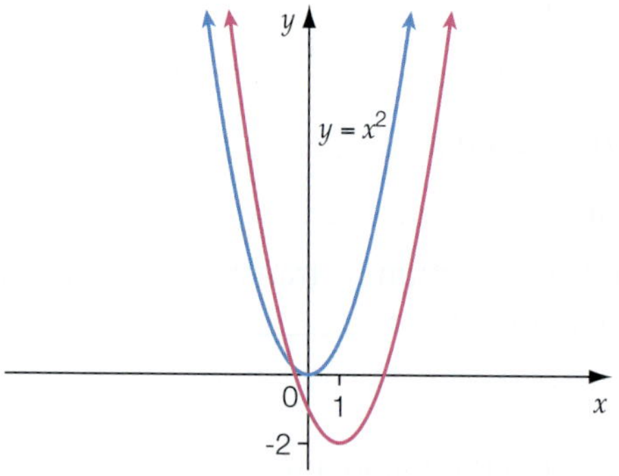

(c) The equation of the pink parabola could be:

A $y = (x - 2)^2$

B $y = (x + 2)^2$

C $y = x^2 - 2$

D $y = x^2 + 2$

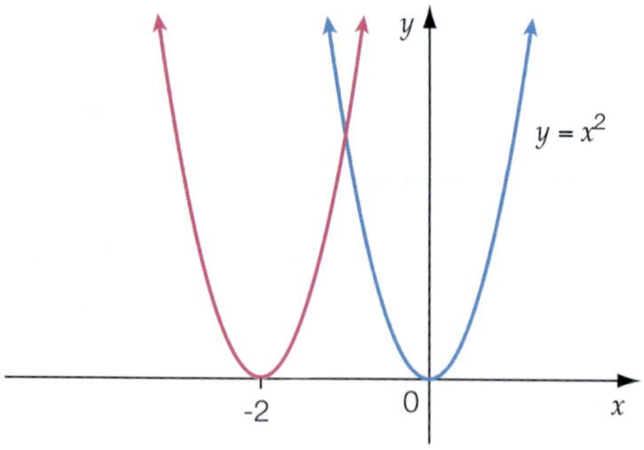

3.5

8 Factorise these expressions.

(a) $x^2 + 11x + 30$

(b) $x^2 - 10x - 24$

3.6

9 Factorise each of these quadratic trinomials.

(a) $x^2 + 10x + 25$

(b) $x^2 - 18x + 81$

(c) $3x^2 - 30x + 75$

(d) $4x^2 + 16x + 16$

3.7

10 Simplify the following.

(a) $\frac{18x}{45}$

(b) $\frac{2(a-3)}{6}$

(c) $\frac{10x(x-1)}{8x}$

(d) $\frac{12ab}{4b}$

11 Simplify the following. **3.7**

(a) $\dfrac{(x+4)(x-3)}{(x-3)(x+2)}$

(b) $\dfrac{(2x-1)(x+1)}{(x+7)(x-7)} \times \dfrac{(3x+7)(x+7)}{(x-1)(2x-1)}$

(c) $\dfrac{4x^2-4x+1}{2x-1}$

(d) $\dfrac{2x^2-8x}{x^2+3x} \div \dfrac{x^2-16}{x^2+8x+15}$

12 Simplify the following. **3.7**

(a) $\dfrac{2x}{3}+\dfrac{x}{2}$

(b) $\dfrac{x+1}{5}+\dfrac{x-2}{3}$

(c) $\dfrac{3x-2}{10}-\dfrac{x+5}{6}$

(d) $\dfrac{6x}{7}+\dfrac{3y}{2}$

(e) $\dfrac{x+5}{4}+\dfrac{y+2}{3}$

(f) $\dfrac{x-6}{5}-\dfrac{y-3}{10}$

Understanding

13 If $4x^2 - 24x + 36$ is written as $a(x-b)^2$, state the values of a and b. **3.6**

14 In January, John's weight was x kg. In April his weight was $(x+a)$ kg and in September it was $(x+b)$ kg. **3.5**

(a) If the product of John's weight in April and September is $x^2 + x - 30$, determine the values of a and b.

(b) By how many kilograms had John's weight increased by April?

(c) What was John's overall weight gain/loss by the end of September?

(d) What happened to his weight from April to September?

15 The heights in centimetres of four people are ax, $2x$, ay and $2y$. **3.2**

(a) Write the expression for their total height.

(b) Factorise the expression in part **(a)**.

(c) If $x = 65$, $y = 75$ and $a = 3$, use your factorised expression to find the average height of the four people.

Reasoning

16 (a) Write an expression (in m^2) for the area of the kitchen bench surface shown at right. Simplify the expression as far as possible. **3.5**

(b) Factorise the expression for the area.

(c) Find the area of the kitchen bench surface when the width of the bench, x, is:

(i) 1 m

(ii) 0.5 m

(iii) 0.75 m

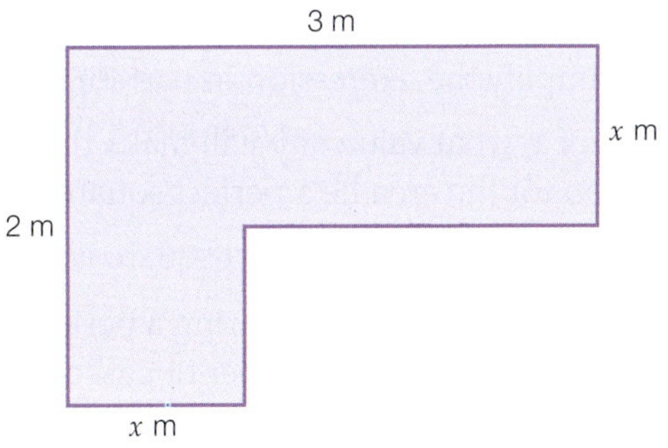

17 A biscuit box has a length 14 cm greater than its height and a width 9 cm greater than its height. **3.5**

(a) Draw a diagram showing this information. Use h to represent height.

(b) Write an expression (in factorised form) to represent the volume of the biscuit box.

(c) Expand and simplify the expression for volume.

(d) Use your answer for part **(b)** to find the volume of the biscuit box if its height is 6 cm.

(e) Use your answer for part (c) to find the volume of the biscuit box if its height is 6 cm.

(f) Which form of the volume expression (factorised or expanded) did you find easier to work with? Explain your answer.

18 A circular disc of metal expands when it is heated. When it is heated to a particular temperature, its radius increases by 2 mm. When cooled to a particular temperature below room temperature, the disc contracts and its radius decreases by 2 mm. (Do not substitute for π in any parts of this question.)

3.6

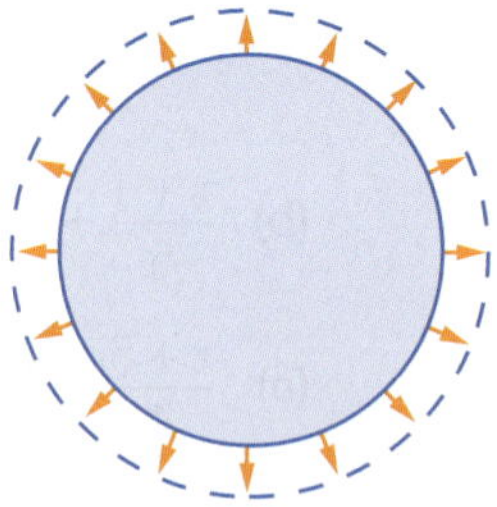

Disc expands when heated

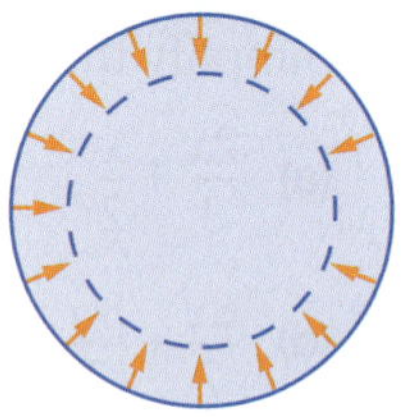

Disc contracts when cooled

(a) If the disc has a radius of x mm at room temperature (20 °C), write its area at room temperature.

(b) In terms of x, what is the radius of the heated disc?

(c) Hence, write an expression for the area of this larger disc.

(d) In terms of x, what is the radius of the cooled disc?

(e) Hence, write an expression for the area of this smaller disc.

(f) Write an expression for the difference between the area of the larger disc and that of the smaller disc.

(g) Is there a common factor that can be taken out of this expression for the difference in area? What is it?

(h) Factorise this expression for the difference in area.

(i) If the original disc (at 20 °C) had a radius of 10 mm, find the difference in area between the larger and smaller discs.

(j) If the difference in area between the larger and smaller discs is 136π mm^2, find the radius of the disc at 20 °C.

19 An exercise book is $(8x + 3)$ cm long and $(2y + 7)$ cm wide.

3.5, 3.6

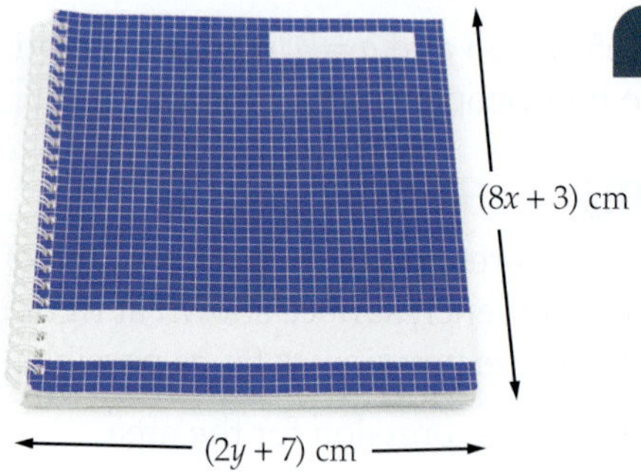

(a) Write an expression (in cm^2) for the area of the book.

(b) What is the expression for the area if $y = 4x - 5$?

(c) Expand and simplify the expression in part (b).

(d) (i) In terms of x, what value of y will make the expression for the area be a perfect square?

(ii) Expand and simplify the perfect square in (i).

20 The number of private cars observed during a period of x days is modelled by the expression $x^2 + p^2x$ where p is the number of passengers. The number of taxis observed with p passengers during y days is given by $p^2y - y^2$.

3.2, 3.6

(a) Write the expression for the total number of private cars and taxis.

(b) Factorise the expression in part (a).

(c) Explain why x must be greater than y if the number of passengers is zero.

(d) (i) How many vehicles were there in total if $x = y$?

(ii) It is found that $2px$ of these vehicles were white. Write, in factorised form, the number of non-white vehicles.

Algebra and quadratics, part 2

4

After completing this chapter you will be able to:

- factorise quadratic trinomials by completing the square
- solve quadratic equations using a variety of techniques
- solve problems that can be modelled by quadratic relationships
- sketch the graphical representations (parabolas) of quadratic relationships

10A
- factorise and solve non-monic quadratic equations using a variety of techniques.

Recall 4

Prepare for this chapter by attempting the following questions. If you have difficulty with a question, you can download a Recall Worksheet from the eBook or the Pearson Places website.

1 Expand and simplify.

(a) $3x(x + 4)$ (b) $(x - 7)(x + 3)$

(c) $(x + 8)^2$ (d) $(x - 4)(x + 4)$

2 Factorise the following.

(a) $x^2 - 25$ (b) $4x^2 - 81$

(c) $x^2 - 5$ (d) $x^2 - 8x + 16$

3 Find two numbers that:

(a) multiply to give 21 and add to give 10

(b) multiply to give -20 and add to give -8

4 Factorise each quadratic expression:

(a) $x^2 - 7x$ (b) $x^2 - 5x - 6$

(c) $3x^2 - 3x - 36$ (d) $-x^2 + 8x + 20$

Exploration Task

You can download this activity from the eBook or the Pearson Places website.

What does completing the square have to do with a square?

In this activity you will explore how 'completing the square' works and what it has to do with a square.

4.1 Solving quadratic equations

You often need to find solutions for quadratic equations when sketching graphs or when solving problems that involve quadratic relationships. The solution to a quadratic equation can be found by plotting a quadratic graph and finding the x-intercepts by hand, but this method can be slow and inaccurate if the solutions involve fractions or surds.

Algebraic techniques are used to find the exact solutions of quadratic equations. However, remember that not all quadratics cross or touch the x-axis, so not all quadratic equations will have real solutions.

To solve a quadratic equation, you can factorise the equation and use the **null factor law** to create a new linear equation, which can then be solved easily. This works wherever linear factors are multiplied together to equal zero.

If the product of two or more expressions is equal to zero, then at least one of the factors must equal zero.

Null factor law

If $ab = 0$, then $a = 0$ or $b = 0$ or $a = b = 0$.

Worked example 1

W.E. 1

Solve the following equations for x using the null factor law.

(a) $x(x - 2) = 0$ **(b)** $(x + 1)(x - 6) = 0$

Thinking	Working
(a) 1 Use the null factor law to equate each factor to zero.	(a) $x(x - 2) = 0$ $x = 0$ or $x - 2 = 0$
2 Solve each equation and state your solutions.	$x = 0$ or $x = 2$
(b) 1 Use the null factor law to equate each factor to zero.	(b) $(x + 1)(x - 6) = 0$ $x + 1 = 0$ or $x - 6 = 0$
2 Solve each equation and state your solutions.	$x = -1$ or $x = 6$

Worked example 2

W.E. 2

Solve each equation for x:

(a) $4x^2 = 0$ **(b)** $(x + 1)^2 = 0$

Thinking	Working
(a) 1 Divide both sides by the coefficient of x^2.	(a) $4x^2 = 0$ $x^2 = 0$
2 As the equation is the product of two identical linear factors, equate the linear factor to zero to find the solution.	$x = 0$
(b) 1 As the equation is the product of two identical linear factors, equate the linear factor to zero.	(b) $(x + 1)^2 = 0$ $x + 1 = 0$
2 Solve the equation and state your solution.	$x = -1$

Before you can use the null factor law, the quadratic equation should be in factorised form.

Algebraic techniques used to factorise quadratic equations can include:

- taking out all common factors, including a negative sign if appropriate
- using the difference of two squares rule (two terms in the quadratic expression)
- using the perfect square rules
- factorising monic trinomials.

After factorising the quadratic equation, apply the null factor law to find the solutions.

Some quadratics cannot be factorised using real numbers. These will have no real solutions.

If the factorised quadratic is a perfect square, then there will only be one solution. Otherwise there will be two solutions.

Worked example 3

W.E. 3

Solve each equation for x:

(a) $3x^2 - 6x = 0$ **(b)** $11x^2 = x$

Thinking	Working
(a) 1 Factorise the LHS by finding the highest common factor.	(a) $3x^2 - 6x = 0$ $3x(x - 2) = 0$
2 Apply the null factor law.	$3x = 0$ or $x - 2 = 0$
3 Solve each equation and state the solutions.	$x = 0$ or $x = 2$
(b) 1 Rearrange the equation to make one side equal to zero.	(b) $11x^2 = x$ $11x^2 - x = 0$
2 Factorise the other side of the equation by finding the HCF.	$x(11x - 1) = 0$
3 Apply the null factor law.	$x = 0$ or $11x - 1 = 0$
4 Solve each equation and state the solutions.	$x = 0$ or $11x = 1$ $x = 0$ or $x = \frac{1}{11}$

If the quadratic equation contains a two-term quadratic expression, first check for common factors, then try to factorise using the difference of two squares rule: $a^2 - b^2 = (a + b)(a - b)$.

Worked example 4

W.E. 4

Solve the following equations for x.

(a) $x^2 - 49 = 0$ **(b)** $4x^2 = 25$ **(c)** $5x^2 = 3$

Thinking	Working
(a) 1 Use the difference of two squares rule $a^2 - b^2 = (a + b)(a - b)$.	(a) $x^2 - 49 = 0$ $(x + 7)(x - 7) = 0$
2 Use the null factor law and equate each factor to zero.	$x + 7 = 0$ or $x - 7 = 0$
3 Solve each equation.	$x = -7$ or $x = 7$
4 State your solutions.	$x = \pm 7$
5 You could also solve this by rearranging the formula to make x the subject.	$x^2 - 49 = 0$ $x^2 = 49$ $x = \pm\sqrt{49}$ $= \pm 7$
(b) 1 Rearrange the equation so that the RHS = 0.	(b) $4x^2 = 25$ $4x^2 - 25 = 0$
2 Factorise using the difference of two squares rule $a^2 - b^2 = (a + b)(a - b)$.	$(2x)^2 - (5)^2 = 0$ $(2x + 5)(2x - 5) = 0$
3 Use the null factor law and equate each factor to zero.	$2x + 5 = 0$ or $2x - 5 = 0$
4 Solve each equation.	$x = -\frac{5}{2}$ or $x = \frac{5}{2}$
5 State your solutions.	$x = \pm\frac{5}{2}$
6 You could also solve this by rearranging the formula to make x the subject.	$4x^2 = 25$ $x^2 = \frac{25}{4}$ $x = \pm\sqrt{\frac{25}{4}}$ $= \pm\frac{5}{2}$
(c) 1 Rearrange the equation so that the RHS = 0.	(c) $5x^2 = 3$ $5x^2 - 3 = 0$
2 Divide each term by the coefficient of x^2.	$x^2 - \frac{3}{5} = 0$
3 Factorise using the difference of two squares rule $a^2 - b^2 = (a + b)(a - b)$.	$(x + \sqrt{\frac{3}{5}})(x - \sqrt{\frac{3}{5}}) = 0$

4 Use the null factor law and equate each factor to zero. $x + \sqrt{\frac{3}{5}} = 0$ or $x - \sqrt{\frac{3}{5}} = 0$

5 Solve each equation. $x = -\sqrt{\frac{3}{5}}$ or $x = \sqrt{\frac{3}{5}}$

6 State your solutions. $x = \pm\sqrt{\frac{3}{5}}$

7 You could also solve this by rearranging the formula to make x the subject.

$$5x^2 = 3$$
$$x^2 = \frac{3}{5}$$
$$x = \pm\sqrt{\frac{3}{5}}$$

If the quadratic equation contains a three-term quadratic expression (trinomial), first check for common factors, then try to factorise using a perfect square rule: $a^2 \pm 2ab \pm b^2 = (a \pm b)^2$. (*Note:* The symbol ± means 'plus or minus'.)

Worked example 5

W.E. 5

Solve the following equations for x.

(a) $x^2 + 10x + 25 = 0$ (b) $2x^2 - 36x = -162$

Thinking	Working
(a) 1 Recognise this quadratic as a perfect square in the form of $a^2 + 2ab + b^2$.	(a) $x^2 + 10x + 25 = 0$ $(x + 5)(x + 5) = 0$
2 Factorise using the perfect square rule $a^2 + 2ab + b^2 = (a + b)^2$.	$(x + 5)^2 = 0$
3 The result is the product of two identical factors so there is only one linear equation to solve.	$x + 5 = 0$
4 Solve the equation.	$x = -5$
(b) 1 Rearrange to make one side equal to zero.	(b) $2x^2 - 36x = -162$ $2x^2 - 36x + 162 = 0$
2 Factorise the left-hand side by: • finding the common factor • using the perfect square rule $a^2 - 2ab + b^2 = (a - b)^2$.	$2(x^2 - 18x + 81) = 0$ $2(x - 9)^2 = 0$
3 Divide both sides by the common factor.	$(x - 9)^2 = 0$
4 The result is the product of two identical factors so there is only one linear equation to solve.	$x - 9 = 0$
5 Solve the equation.	$x = 9$

If the trinomial expression is not a perfect square, then use an appropriate algebraic technique to factorise the monic trinomial remembering to take out any common factors first. Taking out a negative coefficient of x^2 often makes it simpler to factorise. You need to divide both sides of the equation by this coefficient before you factorise the trinomial.

Worked example 6

W.E. 6

Solve the following quadratic equations for x.

(a) $x^2 - 6x + 8 = 0$ (b) $2x^2 + 8x = 24$ (c) $-x^2 - 4x - 3 = 0$

Thinking	Working
(a) 1 Factorise the trinomial. (Here, find the two numbers that: × to give 8, + to give -6 } The numbers are -2 and -4.)	(a) $x^2 - 6x + 8 = 0$ $(x-2)(x-4) = 0$
2 Apply the null factor law.	$x - 2 = 0$ or $x - 4 = 0$
3 Solve each equation and state the solutions.	$x = 2$ or $x = 4$
(b) 1 Rearrange the equation to make one side equal zero.	(b) $2x^2 + 8x = 24$ $2x^2 + 8x - 24 = 0$
2 Look for a common factor. (Here, it is 2.)	$2(x^2 + 4x - 12) = 0$
3 Divide both sides by the common factor.	$x^2 + 4x - 12 = 0$
4 Factorise the trinomial. (Here, find the two numbers that: × to give -12, + to give +4 } The numbers are +6 and -2.)	$(x+6)(x-2) = 0$
5 Apply the null factor law.	$x + 6 = 0$ or $x - 2 = 0$
6 Solve each equation and state the solutions.	$x = -6$ or $x = 2$
(c) 1 Write the equation.	(c) $-x^2 - 4x - 3 = 0$
2 Look for a common factor. (Here, it is -1.)	$-1(x^2 + 4x + 3) = 0$
3 Divide both sides by the common factor.	$x^2 + 4x + 3 = 0$
4 Factorise the trinomial. (Here, find the two numbers that: × to give +3, + to give +4 } The numbers are +3 and +1.)	$(x+1)(x+3) = 0$
5 Use the null factor law.	$x + 1 = 0$ or $x + 3 = 0$
6 Solve each equation and state the solutions.	$x = -1$ or $x = -3$

These methods can be used to solve quadratic equations:

1 Equate one side to zero.
2 Factorise the expression by looking for:
   - a common factor
   - a difference of two squares $a^2 - b^2 = (a + b)(a - b)$
   - a perfect square $a^2 \pm 2ab + b^2 = (a \pm b)^2$
   - a monic trinomial where the cross method or splitting the middle term can be used to factorise the expression.

You may need to take out the coefficient of x^2 as a common factor and divide both sides by the common factor before factorising.

4.1 Solving quadratic equations

Navigator

1 (column 1), 2, 3, 4 (column 1), 5 (column 1), 6 (columns 1–2), 7 (columns 1, 2), 8, 9 (column 1), 10, 11, 12, 13, 14, 15, 16, 18	1 (column 2), 2 (b, e), 3 (b, e), 4 (column 2), 5 (column 2), 6 (column 2), 7 (column 2), 8, 9 (column 2, 3), 10, 11, 13, 14, 15, 16, 17, 18, 19, 20	1 (column 3), 2 (c, f), 3 (c–f), 4 (column 3), 5 (column 3), 6 (column 3), 7 (column 3), 9, 13, 15, 16, 17, 18, 19, 20

Answers p.795

Fluency

1 Solve the following equations for x using the null factor law. W.E. 1

(a) $x(x - 11) = 0$ (b) $x(x + 6) = 0$ (c) $x(x + 4) = 0$
(d) $2x(x - 1) = 0$ (e) $3x(x - 1) = 0$ (f) $-4x(x - 6) = 0$
(g) $x(4x + 8) = 0$ (h) $-x(15 - 5x) = 0$ (i) $(6 + 2x)(9x + 18) = 0$
(j) $(x - 3)(x + 7) = 0$ (k) $(x - 4)(x + 1) = 0$ (l) $(x - 2)(x - 6) = 0$
(m) $(4 - x)(x - 5) = 0$ (n) $(7 - x)(x + 2) = 0$ (o) $(3 - x)(x + 10) = 0$

2 Solve the following quadratic equations for x. W.E. 2

(a) $-6x^2 = 0$ (b) $3x^2 = 0$ (c) $2x^2 = 0$
(d) $(x + 7)^2 = 0$ (e) $(x + 1)^2 = 0$ (f) $(x - 9)^2 = 0$

3 Solve the following equations for x. W.E. 3

(a) $x^2 - 4x = 0$ (b) $x^2 + x = 0$ (c) $x^2 - 6x = 0$
(d) $3x^2 = 15x$ (e) $4x^2 = 2x$ (f) $4x^2 = -12x$

4 Solve the following quadratic equations for x. W.E. 4

(a) $x^2 - 25 = 0$ (b) $x^2 - 81 = 0$ (c) $x^2 - 4 = 0$
(d) $16x^2 - 25 = 0$ (e) $9x^2 - 49 = 0$ (f) $100x^2 = 81$
(g) $3x^2 - 75 = 0$ (h) $6x^2 - 5 = 0$ (i) $7x^2 - 28 = 0$
(j) $49 - x^2 = 0$ (k) $81 - x^2 = 0$ (l) $121 - x^2 = 0$
(m) $x^2 = 13$ (n) $x^2 = 24$ (o) $3x^2 - 32 = 0$

W.E. 5

5 Solve the following quadratic equations for x.

(a) $x^2 - 14x + 49 = 0$ (b) $x^2 + 8x + 16 = 0$ (c) $x^2 + 16x + 64 = 0$

(d) $x^2 - 4x + 4 = 0$ (e) $x^2 + 2x + 1 = 0$ (f) $x^2 - 6x + 9 = 0$

(g) $2x^2 - 20x + 50 = 0$ (h) $5x^2 - 10x + 5 = 0$ (i) $6x^2 + 24x + 24 = 0$

(j) $3x^2 - 18x = -27$ (k) $5x^2 - 50x = -125$ (l) $24x - 3x^2 = 48$

W.E. 6

6 Solve the following quadratic equations for x.

(a) $x^2 + 3x + 2 = 0$ (b) $x^2 + 9x + 18 = 0$ (c) $x^2 - 11x + 28 = 0$

(d) $x^2 - 10x + 9 = 0$ (e) $x^2 - x - 30 = 0$ (f) $x^2 + 2x - 8 = 0$

(g) $2x^2 - 24x + 64 = 0$ (h) $2x^2 + 4x - 30 = 0$ (i) $3x^2 - 18x - 48 = 0$

(j) $5x^2 + 20x - 60 = 0$ (k) $4x^2 - 4x - 8 = 0$ (l) $6x^2 - 12x - 18 = 0$

7 Solve the following equations. Rewrite each equation so that the right-hand side equals zero.

(a) $x^2 + 7x = -12$ (b) $x^2 + 8x = -12$ (c) $x^2 - 2x = 63$

(d) $x^2 + 5x = 24$ (e) $-x^2 - 5x = 6$ (f) $-x^2 + 2x = -15$

(g) $x^2 + 5x - 20 = 30$ (h) $x^2 - 6x - 13 = 14$ (i) $x^2 + 66 = 17x$

(j) $x^2 + 5 = 6x$ (k) $-x^2 + 16 = 6x$ (l) $-x^2 + 21 = 4x$

Understanding

8 (a) The solution/s to $x(x - 2) = 0$ are:

A $x = 2$ B $x = -2$ C $x = 0, x = -2$ D $x = 0, x = 2$

(b) The solution/s to $(x + 3)(x - 4) = 0$ are:

A $x = -3, x = 4$ B $x = 3, x = -4$ C $x = -3, x = -4$ D $x = 3, x = 4$

(c) The solution/s to $x^2 - 12x + 36 = 0$ are:

A $x = 0, x = -6$ B $x = -6$ C $x = 6, x = 0$ D $x = 6$

(d) The solution/s to $x^2 - 64 = 0$ are:

A $x = 8, x = -8$ B $x = 8$ C $x = 0, x = 8$ D $x = 0, x = -8$

9 Solve the following equations.

(a) $x^2 + 3x = 10$ (b) $x^2 - 2x = 3$ (c) $x^2 - 3x - 18 = 10$

(d) $x^2 + 8x + 12 = 5$ (e) $3x^2 - 28 = 20$ (f) $2x^2 - 15 = 3$

(g) $(x + 4)(x + 3) = 6$ (h) $(x - 5)(x + 6) = 12$ (i) $(x - 6)^2 = 36$

10 (a) The solution/s to $x^2 - 9x + 18 = 0$ are:

A $x = 3, x = 6$ B $x = -9$ C $x = 3, x = -6$ D $x = -3, x = -6$

(b) The solution/s to $x^2 + 6x + 5 = 0$ are:

A $x = 1, x = 5$ B $x = -1, x = 5$ C $x = 1, x = -5$ D $x = -1, x = -5$

(c) The solution/s to $3x^2 - 12x - 63 = 0$ are:

A $x = 3, x = 7$ B $x = -3, x = 7$ C $x = 3$ D $x = 3, x = -7$

(d) The solution/s to $-x^2 + x + 12 = 0$ are:

A $x = 3, x = 4$ B $x = -3, x = 4$ C $x = 3, x = -4$ D $x = -3, x = -4$

(e) The solution/s to $x^2 + 3x = 4$ are:

A $x = 1, x = 3$ B $x = 1, x = -4$ C $x = 1, x = 4$ D $x = -1, x = 4$

(f) The solution/s to $x^2 - 4x = -4$ are:

A $x = 2, x = -2$ B $x = 4, x = -1$ C $x = 2$ D $x = 4, x = 1$

11 The height, h centimetres, of a ball above the ground when bounced is given by $h = t(50 - t)$, where t is the time in seconds. Find the two times (in seconds) when the ball is on the ground, that is, when its height (h) is 0 cm.

12 A graph has the equation $y = (x + 1)(x - 4)$. Find the x-intercepts of this graph.

13 Determine the x-intercepts of the following quadratic relationships.

(a) $y = x^2 - 144$

(b) $y = 6x^2 + 24x$

(c) $y = x^2 - 12x + 36$

Reasoning

14 The temperature, T degrees Celsius, is measured during the day and found to follow the quadratic relationship $T = -t^2 + 10t + 5$, where t represents the time in hours after 8 am.

(a) Find the temperature when t is equal to 0.

(b) By substituting for T, find the values of t when the temperature is 21 °C.

(c) At what times is the temperature 21 °C?

15 At the zoo, a wombat enclosure is to be enlarged to accommodate recent arrivals.

(a) Write an expression for the area of the new enclosure if the same length is to be added to two sides, as in the diagram.

(b) The new enclosure is to have double the area of the existing one. Find the dimensions of the new wombat enclosure.

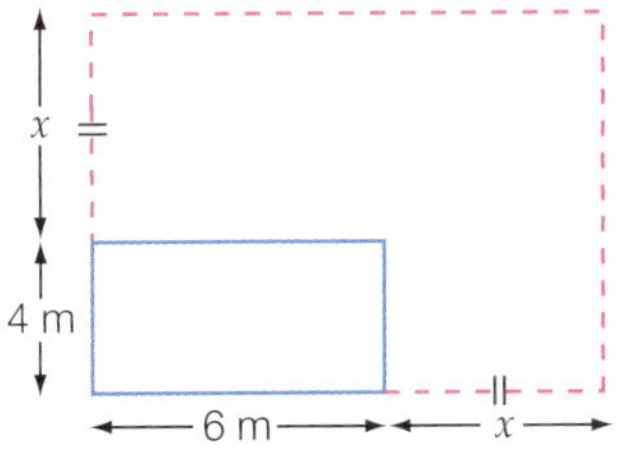

16 Gerald hit his golf ball down the fairway. The path the ball followed is parabolic in shape with the relationship:

$$h = 2d - \frac{1}{30}d^2$$

where h is the height above the ground of the golf ball when it is a horizontal distance d from him. Both h and d are in metres. How far from Gerald does the golf ball land?

17 To determine the number of diagonals (D) in a polygon containing n sides, the following quadratic equation can be used: $D = \frac{n}{2}(n - 3)$

(a) Show that a triangle contains no diagonals.

(b) Show that the given quadratic equation is true for each of these polygons—quadrilateral, pentagon, hexagon—by determining the number of diagonals algebraically and then diagrammatically.

(c) Calculate the number of diagonals in:

(i) an octagon (ii) a decagon (iii) a dodecagon.

(d) How many sides does a polygon have, if it has:

(i) 170 diagonals (ii) 299 diagonals?

(e) A similar equation can be used to determine the number of consecutive integers, beginning at 1, to add to a particular value: $S = \frac{n}{2}(n + 1)$

Use the equation to find the number of consecutive integers, beginning at 1, to add to:

(i) 105 (ii) 903

Open-ended

18 (a) Choose pairs of values for x. For each pair, write equations for which they are the solution.

(b) Choose a single value of x. Write a quadratic equation for which this x-value is the only solution.

19 Decide on two values of x that will be the x-intercepts of a quadratic graph. Write at least two possible equations for the graph.

20 Write three different quadratic equations that have two solutions between $x = -3$ and $x = 3$. At least one of the equations should produce a solution containing a surd.

Factorising by completing the square

4.2

Factorising four terms by grouping 'three and one'

Not all expressions with four terms can be factorised by grouping in pairs. An expression with three terms and one term can be factorised if the three terms form a perfect square, and the fourth term is subtracted to form a difference of two squares. Follow these steps to factorise four terms by grouping 'three' and 'one'.

- Identify a perfect square and group the three terms.
- Factorise the three terms using a perfect square rule.
- Write the expression as a difference of two squares.
- Use the difference of two squares rule to complete the factorisation.

Worked example 7

W.E. 7

Factorise $x^2 + 4x + 4 - y^2$ by grouping 'three and one'.

Thinking	Working
1 Are there any common factors? (No) Check if it can be factorised by grouping in pairs. (No)	$x^2 + 4x + 4 - y^2$
2 Identify a perfect square and group the three terms.	$= (x^2 + 4x + 4) - y^2$
3 Factorise using the perfect square rule.	$= (x + 2)^2 - y^2$
4 Use the difference of two squares rule to complete the factorisation.	$= (x + 2 - y)(x + 2 + y)$

Completing the square

If you try to factorise the quadratic trinomial $x^2 + 6x - 2$, you cannot find two numbers that multiply to give -2 and add to 6. For trinomials that cannot be factorised using binomial products, you can try the **completing the square** method of factorising. Follow these steps to factorise a quadratic expression such as $x^2 + 6x - 2$ by completing the square, as shown on the following page.

- Add a third term to the first two terms to make a perfect square.
- Subtract the amount added to keep the expression equivalent.
- Identify a 'three and one' expression.
- Factorise using a perfect square rule and the difference of two squares rule.

Consider $x^2 + 6x - 2$:

- Add a third term to the first two terms to make a perfect square. For $x^2 + 6x - 2$, add 9: $(x^2 + 6x + 9) - 2$
- Subtract the amount added to keep the expression equivalent. Subtract 9 because 9 was added: $(x^2 + 6x + 9) - 2 - 9$

A perfect square has been formed by adding 9.

Subtract 9 to keep the whole expression equivalent to $x^2 + 6x - 2$.

$x^2 + 6x - 2$ can be rewritten as $(x^2 + 6x + 9) - 9 - 2$

$$x^2 + 6x - 2 = (x^2 + 6x + 9) - 11$$

- This is now a 'three and one' expression so you can use a perfect square rule and the difference of two squares rule to factorise. Remember to use surds.

$$\begin{aligned} x^2 + 6x - 2 &= (x^2 + 6x + 9) - 11 \\ &= (x + 3)^2 - (\sqrt{11})^2 \\ &= (x + 3 - \sqrt{11})(x + 3 + \sqrt{11}) \end{aligned}$$

To complete the square of $x^2 + bx$, divide b by 2 to get $\frac{b}{2}$ and then add $\left(\frac{b}{2}\right)^2$ to make a perfect square. Subtract $\left(\frac{b}{2}\right)^2$ to keep the expression equivalent.

Worked example 8

W.E. 8

Factorise $x^2 - 8x + 9$ by completing the square.

Thinking	Working
1 Are there any common factors? (No)	$x^2 - 8x + 9$
2 By comparing the first two terms to $x^2 + bx$ identify the value of b. ($b = 8$) Add and subtract $\left(\frac{b}{2}\right)^2$ to complete the square and keep the expression equivalent. (Here, $\left(\frac{b}{2}\right)^2 = 16$, so add and subtract 16.)	$= (x^2 - 8x + 16) - 16 + 9$
3 Factorise the perfect square and simplify.	$= (x - 4)^2 - 7$
4 Use the difference of two squares rule to complete the factorisation.	$= (x - 4)^2 - (\sqrt{7})^2$ $= (x - 4 - \sqrt{7})(x - 4 + \sqrt{7})$

Always look for common factors that can be taken out of all terms. Here, you will only complete the square of a monic quadratic trinomial.

Worked example 9

W.E. 9

Factorise $3x^2 + 12x - 30$ by completing the square.

Thinking	Working
1 Take out a common factor. (In this case 3)	$3x^2 + 12x - 30$ $= 3(x^2 + 4x - 10)$
2 By comparing the first two terms to $x^2 + bx$ identify the value of b. ($b = 4$) Add and subtract $\left(\frac{b}{2}\right)^2$. Use additional brackets to separate the common factor. (Here, $\left(\frac{b}{2}\right)^2 = 4$ so add and subtract 4.)	$= 3[(x^2 + 4x + 4) - 4 - 10]$
3 Factorise the perfect square and simplify, keeping the additional brackets.	$= 3[(x+2)^2 - 14]$
4 Write as the difference of two squares using surds.	$= 3[(x+2)^2 - (\sqrt{14})^2]$
5 Factorise using the difference of two squares rule inside the additional brackets.	$= 3[(x + 2 - \sqrt{14})(x + 2 + \sqrt{14})]$
6 Remove the additional brackets as they are no longer necessary.	$= 3(x + 2 - \sqrt{14})(x + 2 + \sqrt{14})$

Using CAS to factorise

You can use your CAS to factorise algebraic expressions.

Using TI-Nspire CAS

To complete the square, select **menu** > **Algebra** > **Complete the Square** and enter the equation followed by a comma and x.

You can also factorise any expression by selecting **menu** > **Algebra** > **factor** and inserting the expression followed by a comma and x.

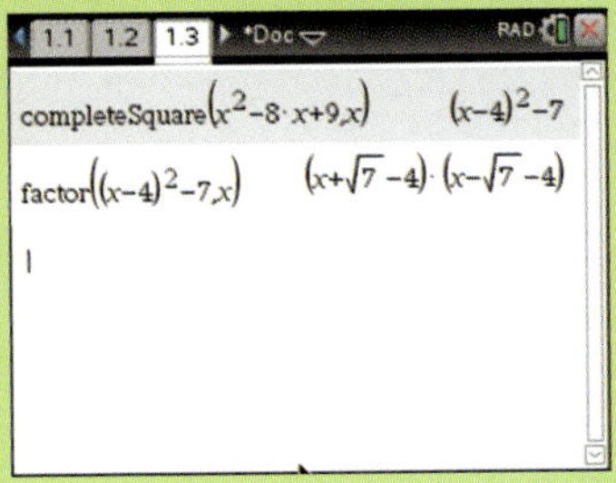

Using Casio ClassPad CAS

From the menu select **Main** and then enter the expression. Then highlight the expression and select **Interactive** > **Transformation** > **factor** > **rFactor**. This will automatically factorise expressions, including expressions that need you to complete the square.

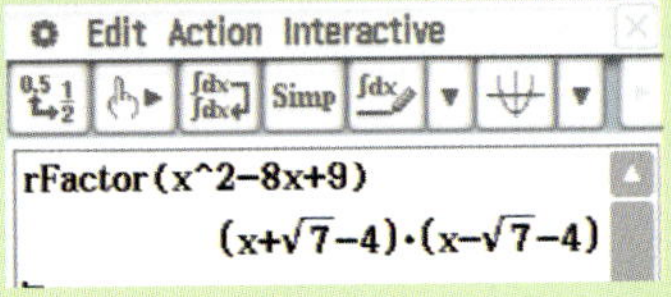

4.2 Factorising by completing the square

Navigator

Answers p.795

1 (column 1), 2, 3 (a–k), 4, 5 (a–b), 6, 8, 9	1 (column 2), 2, 3 (columns 1–2), 4, 5, 6, 7, 8, 9, 10	2 (e–j), 3 (columns 2–3), 5, 6, 7, 8, 9, 10, 11

Fluency

W.E. 7

1 Factorise these expressions by grouping 'three and one'.

(a) $x^2 + 8x + 16 - y^2$ (b) $y^2 - 4y + 4 - k^2$

(c) $y^2 + 8y + 16 - m^2$ (d) $m^2 - 10m + 25 - 4n^2$

(e) $x^2 + 2x + 1 - 9y^2$ (f) $q^2 + 6q + 9 - 16p^2$

(g) $a^2 - 6ab + 9b^2 - 16$ (h) $x^2 - 10xy + 25y^2 - 36$

(i) $4a^2 - 12a + 9 - 16b^2$ (j) $9c^2 + 30c + 25 - 4d^2$

2 Gemma is unsure why her working for this question is incorrect:

Factorise this expression by grouping 'three and one': $w^2 - 16w + 64 - 4k^2$.

$$\begin{aligned} w^2 - 16w + 64 - 4k^2 &= (w^2 - 16w + 64) - 4k^2 \\ &= (w + 8)^2 - (2k)^2 \\ &= (w + 8 - 2k)(w + 8 + 2k) \end{aligned}$$

(a) Identify the error.

(b) Write the correct solution.

W.E. 8

3 Factorise these expressions by completing the square.

(a) $x^2 + 6x + 2$ (b) $x^2 - 10x + 4$ (c) $x^2 - 8x - 3$

(d) $x^2 + 20x + 77$ (e) $x^2 - 12x + 21$ (f) $x^2 + 2x - 8$

(g) $x^2 - 14x - 2$ (h) $x^2 - 4x - 2$ (i) $x^2 + 16x + 47$

(j) $x^2 + 3x + 1$ (k) $x^2 + 7x + 9$ (l) $x^2 - x - 4$

(m) $x^2 - 5x + 5$ (n) $x^2 + 5x - 2$ (o) $x^2 + x - 5$

(p) $x^2 - 9x + 13$ (q) $x^2 - 3x - 2$ (r) $x^2 - 7x - 1$

4 Reorder these steps to show the correct sequence to factorise $2p^2 + 12p - 20$ by completing the square.

$$2p^2 + 12p - 20$$
$$= 2(p + 3 + \sqrt{19})(p + 3 - \sqrt{19})$$
$$= 2(p^2 + 6p - 10)$$
$$= 2[(p + 3)^2 - 19]$$
$$= 2[(p + 3)^2 - (\sqrt{19})^2]$$
$$= 2(p^2 + 6p + 9 - 9 - 10)$$

W.E. 9

5 Factorise each expression by completing the square.

(a) $2x^2 + 4x - 10$ (b) $4x^2 - 16x - 4$ (c) $10x^2 - 80x - 30$

(d) $3x^2 - 6x - 12$ (e) $5x^2 + 20x - 5$ (f) $7x^2 + 42x + 14$

Understanding

6 The expression $2m^2 - 12m + 4$ when factorised is:

A $2(m - 6 - \sqrt{7})(m - 6 + \sqrt{7})$ B $2(m + 7 - \sqrt{3})(m + 7 + \sqrt{3})$

C $2(m - 6 - \sqrt{2})(m - 6 + \sqrt{2})$ D $2(m - 3 - \sqrt{7})(m - 3 + \sqrt{7})$

7 Factorising $x^2 + x - \frac{5}{16}$ gives:

A $(x - \frac{1}{4})(x + \frac{5}{4})$ B $(x - \frac{5}{16})(x + 1)$ C $(x - \frac{5}{4})(x + \frac{1}{16})$ D $(x + \frac{1}{4})(x - \frac{5}{4})$

8 A seagull is diving for a fish. Its height above sea level, in metres, at time t seconds is given by the expression $t^2 - 8t + 14$.

(a) How high is the seagull at $t = 0$?

(b) Factorise the expression by completing the square.

(c) How long is the time interval between $t = 4 - \sqrt{2}$ and $t = 4 + \sqrt{2}$? Round your answer to 2 decimal places.

(d) Explain what happens when $t = 4 \pm \sqrt{2}$.

(e) Explain what the seagull is doing between $t = 4 - \sqrt{2}$ and $t = 4 + \sqrt{2}$.

9 The profit made by a magazine publisher in the first 8 months is given by $P = -5x^2 + 40x - 5$, where x is the time in months.

(a) Factorise the expression for P.

(b) Show that $P = 0$ when $x = 4 \pm \sqrt{15}$.

(c) Show what happens to the company in the first 8 months by substituting different values for x into the expression.

Reasoning

10 The number of earthworms in a compost heap after d days is given by $100 - 4d - d^2$.

(a) How many earthworms were there originally?

(b) Factorise the expression for the number of earthworms.

(c) Use your factorisation to show that when $d = 2(\sqrt{26} - 1)$, no earthworms remain.

(d) Evaluate $2(\sqrt{26} - 1)$ and round to 2 decimal places to find on which day the number of earthworms is reduced to zero.

(e) A second colony of earthworms is now in another compost heap. Their population is given by $64 - 2n - n^2$.

(i) Factorise the expression.

(ii) Use this factorisation to show that no earthworms remain when $n = \sqrt{65} - 1$.

(f) Use your answers to **(c)** and **(e) (ii)** to decide which colony stayed longer in their respective compost heaps.

Open-ended

11 (a) Write the expression $x^2 - 2ax + 2b$ in the form $(x - a)^2 + A$.

(b) Write an expression for A in terms of a and b.

(c) (i) What is the relationship between a and b if the expression is to be a perfect square?

(ii) Find a value for a and b that produces a perfect square.

Exploration CAS

Equipment required: TI-Nspire CAS or Casio ClassPad CAS

Factorisation involving surds

Factorisation by hand is often slow, especially when the factorisation involves surds. You can use a CAS calculator to factorise the following expressions (if possible) quickly and accurately.

(a) $x^2 - 13x + 12$ **(b)** $x^2 - 8x - 5$ **(c)** $x^2 - 5x - 3$ **(d)** $x^2 - 3x + 6$

Using TI-Nspire CAS

Add a **Calculator** page to your document by selecting the Calculator icon on the bottom row as shown.

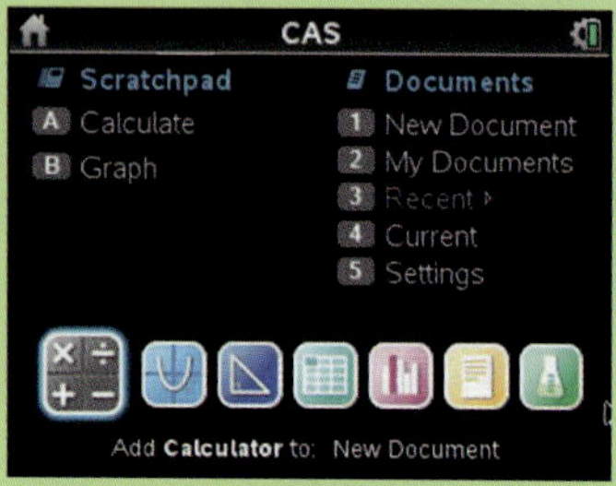

Select **menu** > **Algebra** > **Factor** and enter the expression **(a)** to be factorised inside the brackets.

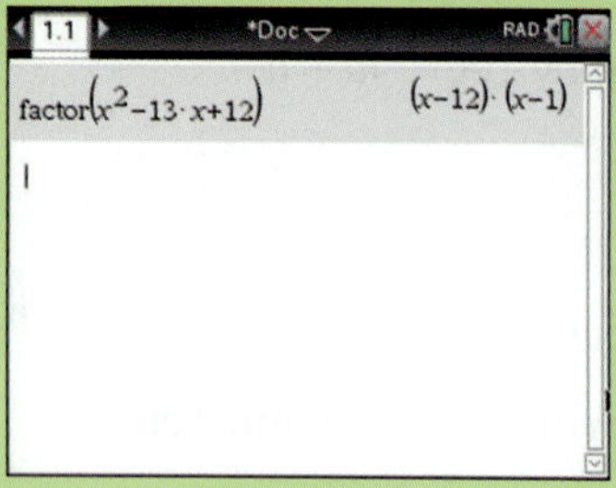

Repeat the instructions above for expression **(b)**. It does not factorise with rational factors, but you can enter the expression followed by a comma and x to give a factorisation with surds.

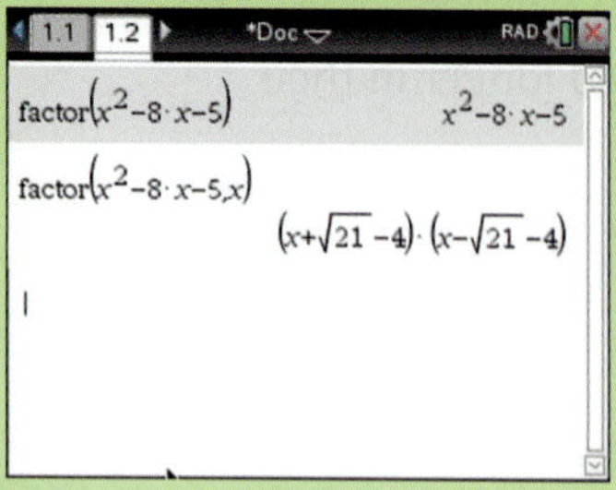

Using Casio ClassPad CAS

From the menu select **Main** and then make sure you are in **Standard** mode, with the word **Standard** shown at the bottom of the screen. If this shows **Decimal** instead, then select the word to change it to **Standard**.

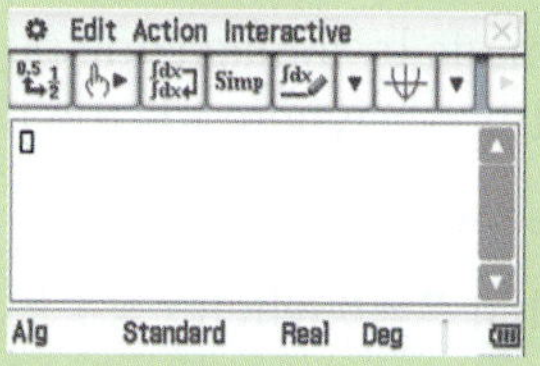

Select **Interactive** > **Transformation** > **factor** > **factor** and enter expression **(a)**.

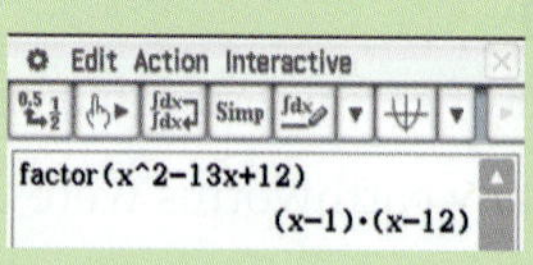

The expression will be factorised as shown.

Repeat the instructions above for expression **(b)**. It does not factorise with rational factors. However, you can use **Interactive** > **Transformation** > **factor** > **rFactor** to give a factorisation with surds.

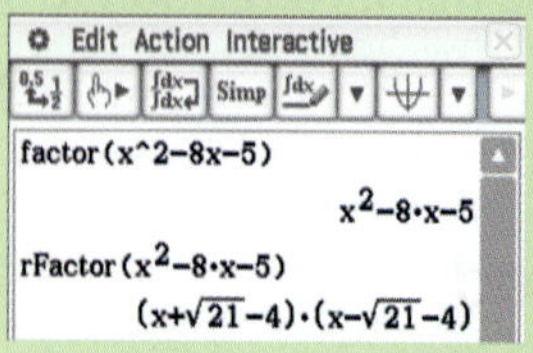

Using TI-Nspire CAS

Repeat the instructions for expression **(c)**, remembering to include the comma and x. Note that this factorisation contains fractions and surds.

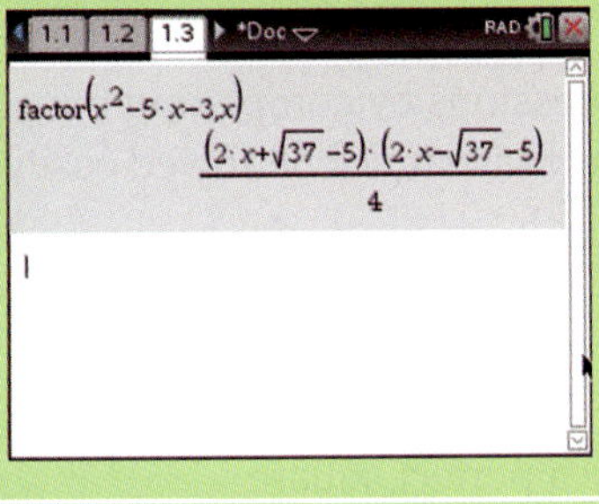

Using Casio ClassPad CAS

Repeat the instructions for expression **(c)**. It does not factorise using the **factor** command, so try using **rFactor** again. Note that this factorisation contains fractions and surds.

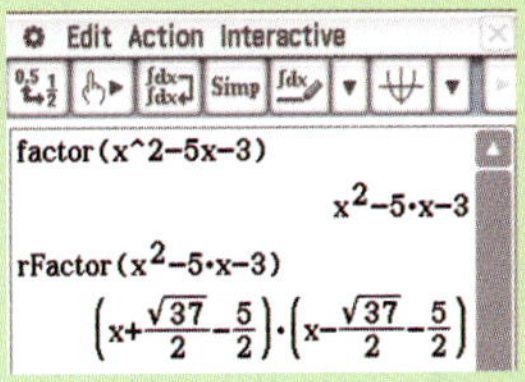

Using TI-Nspire CAS

Repeat the instructions for expression **(d)**, remembering to include the comma and x.

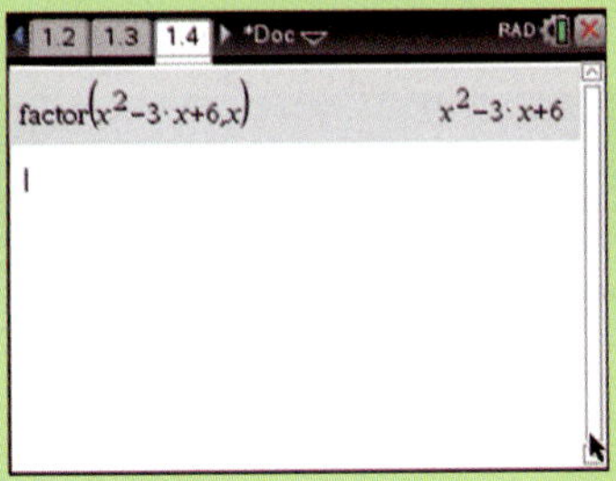

Note that this expression does not factorise, because it has no real number factors.

Using Casio ClassPad CAS

Repeat the instructions for expression **(d)**. It does not factorise using the **factor** command, so try using **rFactor** again.

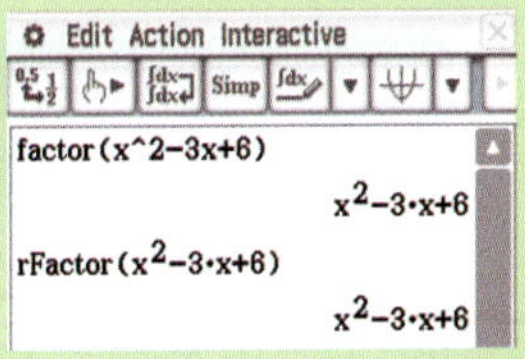

Note that this expression does not factorise, because it has no real number factors.

Use your CAS to factorise the following expressions, if possible.

1 $x^2 - 6x + 2$

2 $x^2 - 7x - 13$

3 $x^2 - 17x - 84$

4 $x^2 - 5x + 12$

Taking it further

5 Use your CAS to find three monic quadratic expressions that:

(a) have rational factors

(b) contain surds but have no fractions when factorised

(c) contain surds and fractions

(d) do not factorise.

4.3 Solving by completing the square

In the previous section you factorised quadratic expressions by completing the square. Now you will use this method to solve quadratic equations that cannot be solved easily using other methods. To do this:

- equate the expression to zero
- look for a common factor
- add and subtract terms in the expression to make a perfect square
- factorise using the difference of two squares rule
- solve using the null factor law.

Worked example 10

W.E. 10

For each of the following expressions:

(a) $x^2 - 6x + 3$ **(b)** $x^2 + 5x - 2$

(i) Complete the square and factorise.

(ii) Equate the expression to zero and use your answer for part **(i)** to solve the quadratic equation formed.

Thinking	Working
(a) (i) 1 Write the expression. Use the first two terms to create a perfect square. The term that is added and subtracted is the square of half the coefficient of the *x* term.	(a) (i) $x^2 - 6x + 3$ $= [x^2 - 6x + \left(\frac{-6}{2}\right)^2] - \left(\frac{-6}{2}\right)^2 + 3$ $= (x^2 - 6x + (-3)^2) - 9 + 3$
2 Factorise the perfect square and simplify the constant term.	$= (x - 3)^2 - 6$
3 Rewrite the expression with the constant term as a square.	$= (x - 3)^2 - (\sqrt{6})^2$
4 Factorise the expression using the difference of two squares rule.	$= (x - 3 + \sqrt{6})(x - 3 - \sqrt{6})$
(ii) 1 Equate the expression to zero.	(ii) $x^2 - 6x + 3 = 0$
2 Use the factorised expression from part **(i)** to rewrite the equation.	$(x - 3 + \sqrt{6})(x - 3 - \sqrt{6}) = 0$

3 Use the null factor law.	$(x-3+\sqrt{6})=0$ or $(x-3-\sqrt{6})=0$
4 Solve for x.	$x=3-\sqrt{6}$ or $x=3+\sqrt{6}$
5 State the solution.	$x=3\pm\sqrt{6}$
(b) (i) 1 Write the expression. Use the first two terms to create a perfect square. The term that is added and subtracted is the square of half the coefficient of the x term.	(b) (i) x^2+5x-2 $=[x^2+5x+\left(\frac{5}{2}\right)^2]-\left(\frac{5}{2}\right)^2-2$ $=(x+\frac{5}{2})^2-\frac{25}{4}-2$
2 Factorise the perfect square and simplify the constant term.	$=(x+\frac{5}{2})^2-\frac{33}{4}$
3 Rewrite the expression with the constant term as a square.	$=(x+\frac{5}{2})^2-\left(\sqrt{\frac{33}{4}}\right)^2$
4 Factorise the expression using the difference of two squares rule.	$=\left(x+\frac{5}{2}-\sqrt{\frac{33}{4}}\right)\left(x+\frac{5}{2}+\sqrt{\frac{33}{4}}\right)$
(ii) 1 Equate the expression to zero.	(ii) $x^2+5x-2=0$
2 Use the factorised expression from part **(i)** to rewrite the equation.	$\left(x+\frac{5}{2}-\sqrt{\frac{33}{4}}\right)\left(x+\frac{5}{2}+\sqrt{\frac{33}{4}}\right)=0$
3 Use the null factor law.	$\left(x+\frac{5}{2}-\sqrt{\frac{33}{4}}\right)=0$ or $\left(x+\frac{5}{2}+\sqrt{\frac{33}{4}}\right)=0$
4 Solve for x and simplify your answer.	$x=-\frac{5}{2}+\sqrt{\frac{33}{4}}$ or $x=-\frac{5}{2}-\sqrt{\frac{33}{4}}$
5 State the solution.	$x=\frac{-5}{2}\pm\sqrt{\frac{33}{4}}$ or $x=\frac{-5\pm\sqrt{33}}{2}$

Worked example 11

W.E. 11

Solve $x^2+6x-2=0$ using the method of completing the square. Leave your answer in exact surd form.

Thinking	Working
1 Write the equation.	$x^2+6x-2=0$
2 Factorise the non-zero expression by completing the square. Use the first two terms to create a perfect square. The term that is subtracted is the square of half the coefficient of the x term.	$[x^2+6x+\left(\frac{6}{2}\right)^2]-\left(\frac{6}{2}\right)^2-2=0$ $(x^2+6x+9)-9-2=0$
3 Factorise the perfect square and simplify the constant term.	$(x+3)^2-11=0$
4 Factorise the expression using the difference of two squares rule.	$(x+3)^2-(\sqrt{11})^2=0$ $(x+3-\sqrt{11})(x+3+\sqrt{11})=0$

5 Use the null factor law.	$x + 3 - \sqrt{11} = 0$	or	$x + 3 + \sqrt{11} = 0$
6 Solve the individual equations.	$x = -3 + \sqrt{11}$	or	$x = -3 - \sqrt{11}$
7 State the solution.	$x = -3 \pm \sqrt{11}$		

If there is a common factor in the quadratic equation, take it out and divide both sides by it first to form a simpler equivalent equation before you complete the square. For example:

$$2x^2 + 4x - 6 = 0$$
$$2(x^2 + 2x - 3) = 0$$
$$x^2 + 2x - 3 = 0$$

4.3 Solving by completing the square

Navigator

Answers p.796

1 (columns 1–2), 2 (columns 1–2), 3, 4 (a–c), 5, 6, 8, 10, 11	1 (columns 2–3), 2 (columns 2–3), 3, 4, 5, 6, 7, 8, 9, 10, 11	1 (columns 1–2), 2 (d–i), 3, 4, 5, 6, 7, 8, 9, 10, 11, 12

Fluency

W.E. 10

1 For each of the following expressions:

(a) $x^2 + 2x - 5$ (b) $x^2 + 4x - 7$ (c) $x^2 + 8x + 5$

(d) $x^2 + 10x + 23$ (e) $x^2 + 12x - 6$ (f) $x^2 - 14x + 39$

(g) $x^2 + 3x - 1$ (h) $x^2 + 5x + 3$ (i) $x^2 + 3x - 5$

(i) Complete the square and factorise.

(ii) Equate the expression to zero and use your answer for part **(i)** to solve the quadratic equation formed.

W.E. 11

2 Solve these quadratic equations using the method of completing the square. Leave your answers in exact surd form.

(a) $x^2 - 8x + 6 = 0$ (b) $x^2 + 6x + 4 = 0$ (c) $x^2 - 3x + 1 = 0$

(d) $x^2 - 5x - 2 = 0$ (e) $x^2 + 4x - 3 = 0$ (f) $x^2 - 20x - 12 = 0$

(g) $2x^2 - 8x - 10 = 0$ (h) $3x^2 + 12x + 3 = 0$ (i) $4x^2 + 16x - 20 = 0$

3 (a) The solution to $2x^2 + 4x - 4 = 0$ is:

A $x = 1 \pm \sqrt{3}$ B $x = -2$ or 4 C $x = -1 \pm \sqrt{3}$ D $x = -4$ or 2

(b) The solution to $x^2 - 4x - 3 = 0$ is:

A $x = 4$ or -1 B $x = 2 \pm \sqrt{7}$ D $x = -2 \pm \sqrt{7}$ E $x = -4$ or 1

(c) The solution to $x^2 - 8x + 14 = 0$ is:

A $x = -4$ or $-\sqrt{2}$ B $x = 4\sqrt{2}$ or $-4\sqrt{2}$

C $x = -4 - \sqrt{2}$ or $-4 + \sqrt{2}$ D $x = 4 + \sqrt{2}$ or $4 - \sqrt{2}$

Understanding

4 Solve the following quadratic equations.

(a) $x^2 - 4x = 2$ (b) $x^2 = 6x - 2$ (c) $3 + 4x = x^2$

(d) $x^2 = 5 + 3x$ (e) $x^2 - 5x = -2$ (f) $x^2 = 5 - 7x$

5 (a) Explain why the quadratic equation $x^2 + 4x + 7$ cannot be factorised.

(b) Explain why the quadratic equation $x^2 + 4x + 7 = 0$ has no solution.

(c) As $x^2 + 4x + 7 = 0$ has no solution, what does this tell you about the graph of $y = x^2 + 4x + 7$?

Reasoning

6 An envelope has an area of 80 cm^2. If the length is 4 cm longer than the width, find the dimensions of the envelope:

(a) exactly

(b) rounded to 1 decimal place.

7 An area of lawn as shown in the diagram is to be mown. Find the value y represents if the total area mown is 44 m^2.

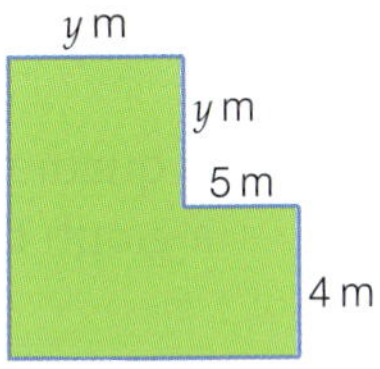

8 If the dimensions of a rectangular park are $(x + 4)$ m and $(x - 5)$ m:

(a) find the exact value of x if the area of the park is 180 m^2

(b) find the value of x rounded to 2 decimal places.

9 A photo of length l cm and width $(l - 2)$ cm is surrounded by a 2 cm rectangular frame. Altogether, the photo and the frame occupy an area of 215 cm^2.

(a) Write the dimensions of the photo and frame in terms of l.

(b) Write an equation in terms of l for the area of the photo and frame.

(c) Find the exact dimensions of the photo.

(d) Find the dimensions of the photo and frame correct to 1 decimal place.

Open-ended

10 Write a quadratic equation that gives solutions containing surds. Solve your equation.

11 Write a quadratic equation that has no solution and demonstrate how you know this.

12 The solution to a quadratic equation $x = a \pm \sqrt{3}$ is found by completing the square.

Choose a value for a between 0 and 5, and write a possible quadratic equation that could have this solution.

4.4 Sketching parabolas

The graph of a quadratic equation in the form of $ax^2 + bx + c = 0$ can be plotted using a table of values. However, a sketch graph shows us the general shape of the graph and its important features without having to complete a table of values. A sketch graph is not drawn on graph paper but needs to be neat and show the main features as described below.

In this section, you will sketch graphs of $y = ax^2 + bx + c$, where $a = \pm 1$. To make a sketch of a quadratic graph, you will need to identify the following:

1 The shape of the graph

- A quadratic equation with a positive coefficient of x^2 will produce a graph that is a positive parabola.

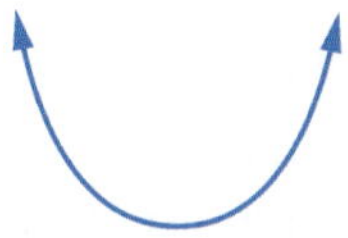

- A quadratic equation with a negative coefficient of x^2 will produce a graph that is a negative parabola.

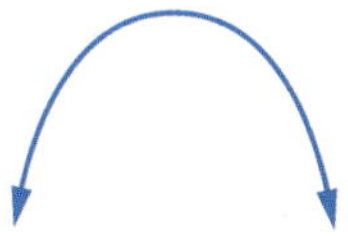

2 The y-intercept

This is where the graph crosses the y-axis and so $x = 0$. It is found by substituting $x = 0$ into the quadratic equation. There is always a y-intercept.

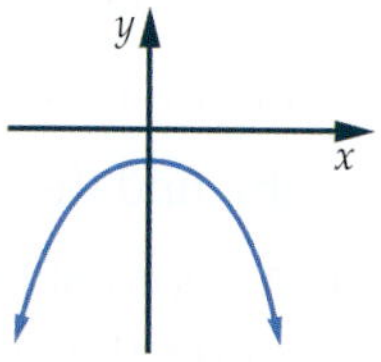

3 The x-intercept(s)

This is where the graph crosses the x-axis and so $y = 0$. There may be no x-intercepts, one x-intercept or two x-intercepts.

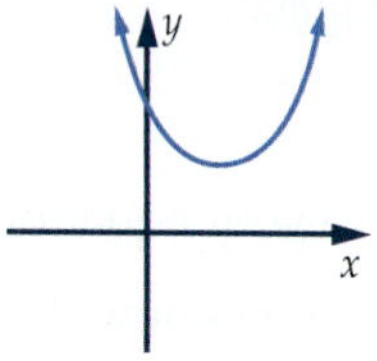

no x-intercepts

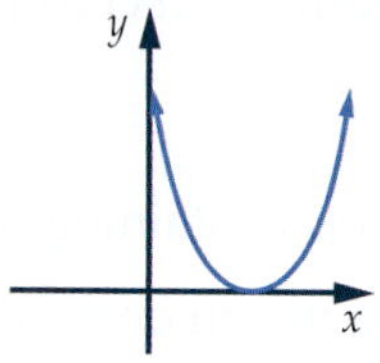

one x-intercept

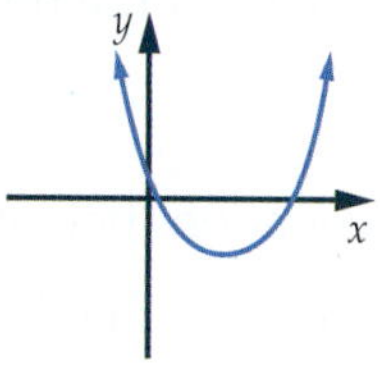

two x-intercepts

The x-intercepts are found by equating the quadratic to zero and solving.

- If there is no solution, there are no x-intercepts.
- If there is one solution, the parabola touches the x-axis. There is only one x-intercept.
- If there are two solutions, there are two distinct x-intercepts. These may involve surds.

4 The turning point

- The turning point can be found by completing the square to transform the equation into **turning point form**.
- The graph of $y = a(x - h)^2 + k$ has a turning point at (h, k).

- When a graph has two x-intercepts the x-coordinate of the turning point is halfway between the two x-intercepts. The corresponding y-coordinate can be found by substituting the x-value into the quadratic equation.

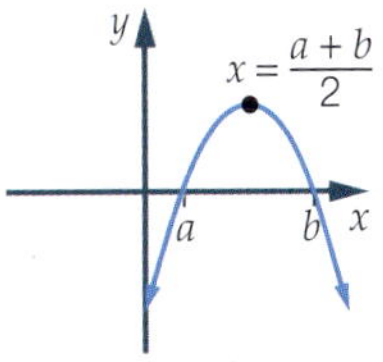

- When a graph has only one x-intercept, the turning point is the x-intercept.
- The axis of symmetry is a vertical line that passes through the turning point, dividing the graph into mirror image halves.

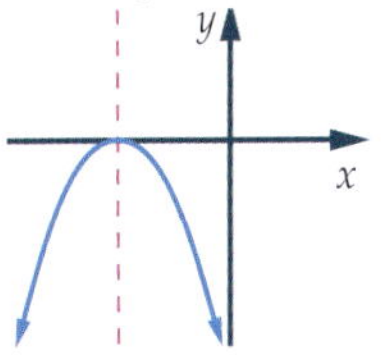

Worked example 12

W.E. 12

Sketch the graph of the quadratic relationship $y = x^2 + 2x - 8$ by finding:

(a) the shape of the graph
(b) the y-intercept
(c) the x-intercepts
(d) the turning point
(e) the axis of symmetry.

Thinking	Working
(a) Determine the shape of the graph by identifying whether the coefficient of x^2 is positive or negative.	(a) This is a  shaped parabola as the coefficient of x^2 is positive.
(b) Find the y-intercept. Substitute 0 for x in the equation and solve.	(b) $x = 0$, $y = 0^2 + 2 \times 0 - 8$ $y = 0 + 0 - 8$ $y = -8$ The y-intercept is (0, -8).
(c) Find the x-intercepts. Substitute 0 for y in the equation. Swap sides if it makes the equation easier to solve. Factorise. Use the null factor law. Solve the equations.	(c) $y = 0$, $0 = x^2 + 2x - 8$ $x^2 + 2x - 8 = 0$ $(x + 4)(x - 2) = 0$ $x + 4 = 0$ or $x - 2 = 0$ $x = -4$ or $x = 2$ The x-intercepts are (-4, 0) and (2, 0).
(d) Find the x-coordinate of the turning point by finding the midpoints of the x-intercepts.	(d) For the turning point: $x = \frac{-4 + 2}{2}$ $= -1$
Find the y-coordinate of the turning point by substituting the x-coordinate into the equation.	$x = -1$, $y = (-1)^2 + 2 \times -1 - 8$ $y = 1 - 2 - 8$ $y = -9$
Write the coordinates of the turning point.	The turning point is at (-1, -9).
Alternatively, complete the square and write in turning point form.	$y = x^2 + 2x - 8$ $y = x^2 + 2x + 1 - 8 - 1$ $y = (x + 1)^2 - 9$ The turning point is at (-1, -9).

(e) Draw and label axes.

Label the intercepts and turning point. (*Note:* No other values need to be shown on the axes.)

Draw a smooth symmetrical curve for the parabola. Arrows show that the curve continues. Label the graph. Mark in the axis of symmetry with a dotted line and label it.

(e)

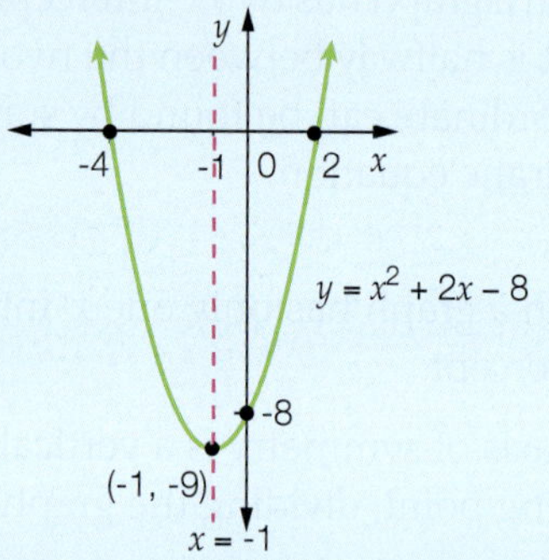

Worked example 13

W.E. 13

Sketch the parabola $y = x^2 + 2x - 15$, showing the x- and y-intercepts, turning point and the axis of symmetry. Find the coordinates of the turning point by writing the equation in turning point form.

Thinking	Working
1 State the equation.	$y = x^2 + 2x - 15$
2 Determine the shape of the graph by identifying whether the coefficient of x^2 is positive or negative.	This is a ∪ shaped parabola as the coefficient of x^2 is positive.
3 Find the y-intercept by substituting $x = 0$ into the equation.	$x = 0$, $y = 0 + 0 - 15$ $= -15$ The y-intercept is $(0, -15)$.
4 Find the x-intercepts by substituting $y = 0$ into the equation.	$y = 0$, $0 = x^2 + 2x - 15$ $x^2 + 2x - 15 = 0$
Factorise.	$(x + 5)(x - 3) = 0$
Use the null factor law.	$x + 5 = 0$ or $x - 3 = 0$
Solve the individual equations.	$x = -5$ or $x = 3$ The x-intercepts are $(-5, 0)$ and $(3, 0)$.
5 Find the turning point.	
Transform the equation into turning point form by completing the square on the first two terms.	$y = x^2 + 2x - 15$ $y = (x^2 + 2x + 1) - 1 - 15$ $y = (x + 1)^2 - 16$
Write the coordinates of the turning point.	The turning point is at $(-1, -16)$.
6 Rule and label the x- and y-axes. **7** Mark in the known points. **8** Draw a smooth curve through the points in the shape of a parabola. (Check that $y = x^2$ has been moved 1 unit left and 16 units down.) **9** Mark in the axis of symmetry with a dotted line and label it.	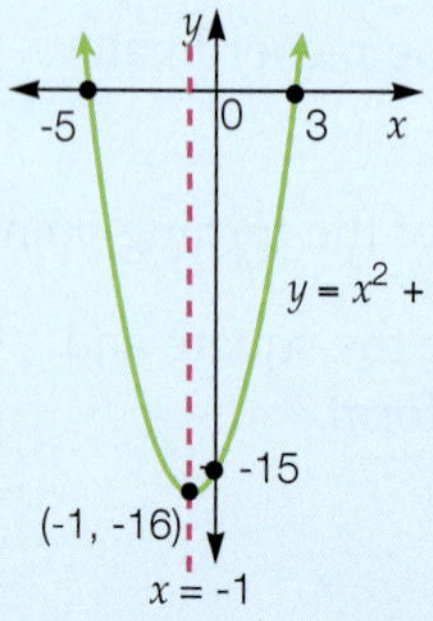

Transforming an equation into turning point form is useful when there are no x-intercepts. For example, $y = (x + 1)^2 + 2$.

This graph has no x-intercepts but has a turning point at (-1, 2).

Worked example 14

W.E. 14

Sketch the graph of the equation $y = 4 - x^2$ showing all the key features.

Thinking	Working
1 State the equation.	$y = 4 - x^2$
2 Determine the shape of the graph by identifying whether the coefficient of x^2 is positive or negative.	This is a ∩ shaped parabola as the coefficient of x^2 is negative.
3 Find the y-intercept by substituting $x = 0$ into the equation.	$x = 0,\ y = 4 - 0^2$ $y = 4$ The y-intercept is (0, 4).
4 Find the x-intercepts by substituting $y = 0$ into the equation.	$y = 0,\ 0 = 4 - x^2$ $4 - x^2 = 0$
Factorise using difference of two squares.	$(2 + x)(2 - x) = 0$
Solve using the null factor law.	$2 + x = 0$ or $2 - x = 0$ $x = \pm 2$ The x-intercepts are (-2, 0) and (2, 0).
5 Find the x-coordinate of the turning point by finding the midpoint of the x-intercepts.	For the turning point: $x = \frac{-2 + 2}{2}$ $= 0$
Find the corresponding y-value.	$x = 0,\ y = 4 - 0$ $y = 4$
Write the coordinates of the turning point.	The turning point is at (0, 4).
6 Sketch the graph. (Notice that the turning point is at the same point as the y-intercept for this graph.) **7** Mark in the axis of symmetry with a dotted line and label it.	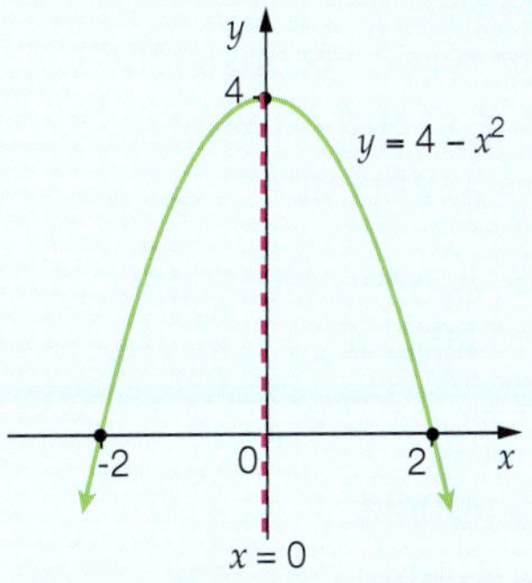

Worked example 15

W.E. 15

Sketch the graph of the equation $y = -(x-2)^2 + 5$, showing all the key features.

Thinking	Working
1 Write the equation.	$y = -(x-2)^2 + 5$
2 Compare the equation to $y = (x-h)^2 + k$ to find (h, k), the turning point. $y = -(x - 2)^2 + 5$ inverted (no dilation) · 2 → · 5 ↑	The graph of $y = x^2$ needs to be inverted and translated 2 units right and 5 units up. $\therefore h = 2$ and $k = 5$ The turning point is at (2, 5).
3 Find the y-intercept by substituting $x = 0$ into the equation.	$x = 0$, $y = -(0-2)^2 + 5$ $= -(4) + 5$ $= 1$ The y-intercept is (0, 1).
4 Find the x-intercepts by substituting $y = 0$ into the equation. Solve the equation.	$y = 0$, $0 = -(x-2)^2 + 5$ $(x-2)^2 - 5 = 0$ $(x - 2 + \sqrt{5})(x - 2 - \sqrt{5}) = 0$ $x = 2 - \sqrt{5}$ or $x = 2 + \sqrt{5}$ $x = 2 \pm \sqrt{5}$ The x-intercepts are $(2 - \sqrt{5}, 0)$ and $(2 + \sqrt{5}, 0)$.
5 Use a calculator to find approximate values for x. This may help to sketch the parabola.	$x \approx 4.2$ and -0.2
6 Rule and label the x- and y-axes. 7 Mark in the known points using exact values. 8 Draw a smooth curve through them to produce a parabola. (Check that your result is consistent with inverting $y = x^2$ and moving it 2 units right and 5 units up.) 9 Mark in the axis of symmetry with a dotted line and label it.	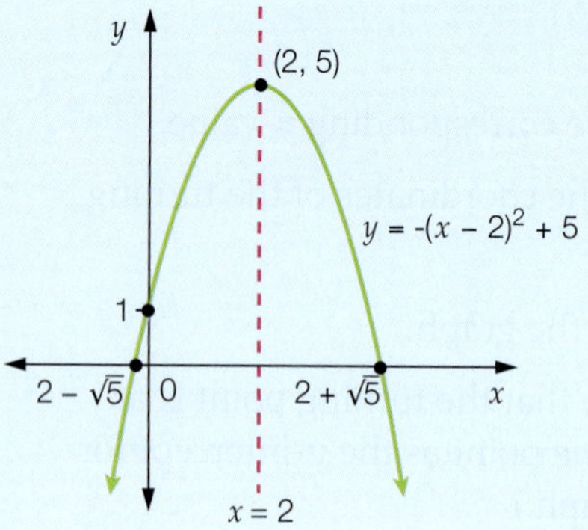

Key features of a parabola

When sketching a parabola, the key features to show are:

- the y-intercept
- the x-intercepts, if they exist
- the turning point, or vertex
- the axis of symmetry.

4.4 Sketching parabolas

Navigator

1 (columns 1–2), 2 (columns 1–2), 3 (column 1), 4 (column 1), 5 (column 1), 6, 7, 8, 9, 10, 13, 14, 17, 18	1 (column 2), 2 (column 3), 3 (column 2), 4 (column 2), 5 (column 2), 6, 7, 8, 9, 10, 11, 13, 14, 16, 17, 18	1 (column 3), 2 (column 2), 3 (column 3), 4 (column 3), 5 (column 3), 8, 9, 11, 12, 13, 15, 16, 17, 18

Answers p.797

Fluency

1 Sketch the graph of each of the following quadratics by finding **(i)** the shape of the graph, **(ii)** the y-intercept, **(iii)** the x-intercepts, **(iv)** the turning point and **(v)** the axis of symmetry. W.E. 12

(a) $y = x^2 - 8x + 15$ **(b)** $y = x^2 - 6x + 8$ **(c)** $y = x^2 - 10x + 21$

(d) $y = -x^2 + 4x + 12$ **(e)** $y = -x^2 + 2x + 3$ **(f)** $y = -x^2 + 2x + 8$

(g) $y = x^2 - 8x + 12$ **(h)** $y = x^2 + 2x - 3$ **(i)** $y = x^2 + 4x - 5$

2 Sketch the following parabolas, showing the x- and y-intercepts, the turning point and the axis of symmetry. Find the coordinates of the turning point by writing the equations in turning point form. W.E. 13

(a) $y = x^2 + 2x - 8$ **(b)** $y = x^2 + 8x + 15$ **(c)** $y = x^2 + 3x + 2$

(d) $y = x^2 - 4$ **(e)** $y = 9 - x^2$ **(f)** $y = x^2 - 7$

(g) $y = x^2 - 2x$ **(h)** $y = x^2 + 3x$ **(i)** $y = x^2 + 2x$

3 Sketch the graphs of the following equations, showing all the key features. W.E. 14

(a) $y = x^2 - 25$ **(b)** $y = x^2 - 9$ **(c)** $y = x^2 - 16$

(d) $y = 81 - x^2$ **(e)** $y = 4 - x^2$ **(f)** $y = 1 - x^2$

(g) $y = x^2 - 4x$ **(h)** $y = x^2 + 6x$ **(i)** $y = 3x^2 + 6x$

4 Sketch the graphs of the following equations, showing all the key features. W.E. 15

(a) $y = (x + 2)^2 - 9$ **(b)** $y = (x + 5)^2 - 1$ **(c)** $y = -(x + 1)^2 + 5$

(d) $y = (x - 3)^2 + 2$ **(e)** $y = (x - 2)^2 + 4$ **(f)** $y = (x - 4)^2 + 2$

5 Sketch the graphs of the following equations, showing all the key features.

(a) $y = x^2 + 10x + 16$ **(b)** $y = x^2 + 8x + 7$ **(c)** $y = -x^2 - 6x - 5$

(d) $y = -x^2 + 4x - 3$ **(e)** $y = x^2 + 6x + 9$ **(f)** $y = x^2 + 10x + 25$

(g) $y = -x^2 + 3x$ **(h)** $y = -x^2 - 5x$ **(i)** $y = -x^2 + 7x - 12$

6 **(a)** When expressed graphically, which of the following equations would have a shape like the parabola shown on the right?

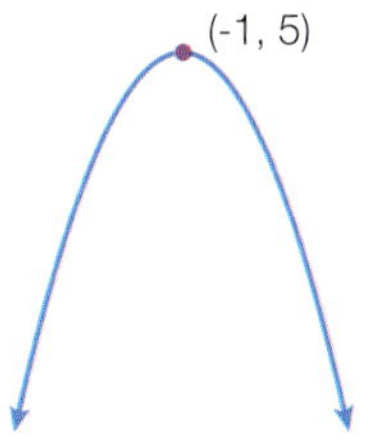

A $y = x^2 + 2x - 5$ B $y = -x^2 - 2x + 4$

C $y = 1 - x^2$ D $y = 6 - 2x - x^2$

(b) Assuming there is no dilation, the equation of the parabola shown could be:

A $y = (x - 6)^2 - 9$ B $y = (x + 6)^2 - 0$

C $y = (x - 3)^2 - 9$ D $y = (x + 3)^2 - 9$

7 (a) The coordinates of the y-intercept of $y = x^2 + 4x + 3$ are:

A (0, 8) B (-1, 0) C (3, 0) D (0, 3)

(b) The parabola with the equation $y = (x - 2)^2 - 3$ has a y-intercept of:

A (0, -2) B (0, 1) C (0, -3) D (0, 2)

(c) The coordinates of the x-intercepts of $y = x^2 - 9$ are:

A (0, -3), (0, 3) B (-3, 0), (3, 0) C (-9, 0), (9, 0) D (0, -9), (0, 9)

(d) The quadratic equation $y = x^2 - x - 12$, when factorised becomes $y = (x - 4)(x + 3)$. The coordinates of the x-intercepts are:

A (-4, 0) and (3, 0) B both (4, -3) C (0, 4) and (0, -3) D (4, 0) and (-3, 0)

8 (a) The x-values of the x-intercepts for the parabola $y = x^2 + 5x - 6$ are:

A -2 and 3 B -3 and 2 C -6 and 1 D -1 and 6

(b) If the x-intercepts of a parabola are (-1, 0) and (5, 0), then the turning point could be:

A (3, 7) B (2, 4) C (1, -6) D (-2, 2)

(c) For the equation $y = (x + 5)^2 + 3$, the turning point is at:

A (-5, 3) B (5, 3) C (3, 5) D (3, -5)

(d) A parabola has a turning point at (1, 1), y-intercept at (0, 2) and no x-intercepts. Assuming there is no dilation, the equation for this parabola could be:

A $y = x^2 + 2x + 2$ B $y = x^2 - 2x + 2$ C $y = (x - 1)^2$ D $y = (x + 1)^2 + 1$

(e) The axis of symmetry for a parabola is given as $x = 3$ and one of the x-intercepts is at (5, 0). What are the coordinates of the other x-intercept?

A (-5, 0) B (3, 5) C (0, 5) D (1, 0)

(f) The axis of symmetry for the parabola $y = -x^2 - 6x - 5$ is given as $x = -3$. What are the coordinates of the turning point?

A (4, -3) B (-3, 4) C (0, 5) D (1, 0)

(g) If the x-intercepts for the equation $y = -x^2 - 8x - 7$ are (-1, 0) and (-7, 0), then the turning point would be:

A (4, 9) B (-4, 9) C (9, 4) D (9, -4)

Understanding

9 A ball is thrown into the air and follows a path given by the relationship $h = -t^2 + 10t$, where h is the height of the ball in metres after t seconds of flight. Sketch the graph of the relationship. Use the sketch graph to find:

(a) the total time the ball was in flight

(b) the maximum height the ball reached.

10 (a) The equation of parabola A is:

A $y = (x - 2)^2 + 5$

B $y = -(x + 2)^2 + 5$

C $y = -(x - 2)^2 + 5$

D $y = -(x + 2)^2 - 5$

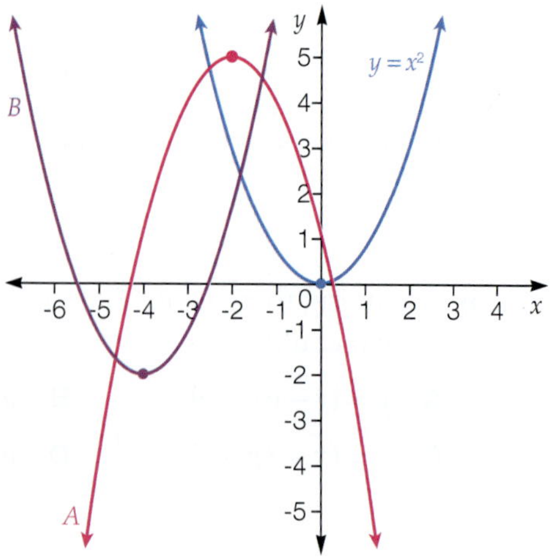

(b) The equation of parabola B is:

A $y = (x - 4)^2 - 2$

B $y = (x + 4)^2 - 2$

C $y = (x - 4)^2 + 2$

D $y = (x + 4)^2 + 2$

11 (a) In turning point form, the quadratic equation $y = x^2 - 8x + 18$ can be written as:

A $y = (x - 4)^2 + 34$ B $y = (x + 4)^2 + 2$

C $y = (x - 4)^2 - 2$ D $y = (x - 4)^2 + 2$

(b) Hence, write the coordinates of the turning point for the parabola $y = x^2 - 8x + 18$.

12 (a) Sketch the graph of the equation $y = x^2 - 6x + 5$ showing all its key features.

(b) Use the graph to find the solution to the equation $x^2 - 6x - 7 = 0$.

Hint

What can you add to -7 to equal 5?

13 Over a period of 8 hours, the temperature of a room follows the relationship $T = h^2 - 6h + 15$, where T is the temperature in degrees Celsius, h hours after commencing an experiment.

(a) Write the equation in turning point form and hence sketch its graph.

(b) What is the initial temperature?

(c) Is the temperature increasing or decreasing after **(i)** 2 hours and **(ii)** 6 hours?

(d) Find the minimum temperature and when it occurs.

(e) Would it be realistic to use the equation over a 24-hour period?

Reasoning

14 Determine the equation of a quadratic graph that has x-intercepts of -3 and 5 and a coefficient of x^2 equal to 1.

15 A rectangular area of land is to be fenced off using 44 m of continuous fencing.

(a) Write an equation for the area enclosed by the fencing if x is the side length of the rectangle in metres.

(b) Sketch a graph of this quadratic graph showing all key features.

(c) What is the **(i)** smallest and **(ii)** largest possible value for x?

(d) What dimensions give the maximum possible area enclosed by the fencing?

Open-ended

16 Explain how you could determine whether x-intercepts will exist for a quadratic graph.

17 Archie was trying to graph $y = x^2 - 2x - 8$ and drew the graph shown below.

(a) Explain why the graph is wrong and how the mistakes could have been avoided.

(b) Draw the correct graph.

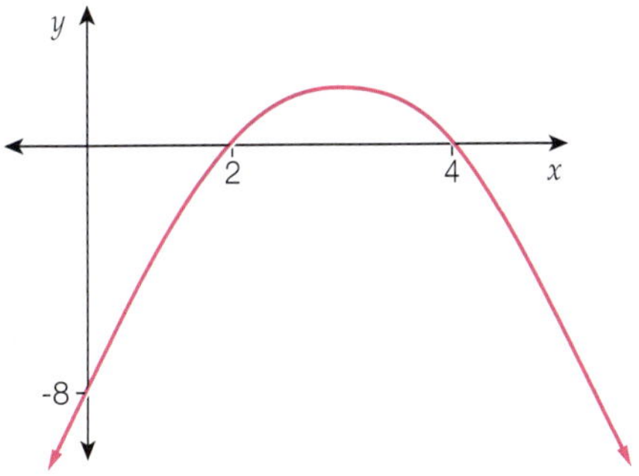

18 Write the equations of three parabolas that have $x = 5$ as an axis of symmetry. For each parabola, find the turning point.

Maths 4 Real

What goes up must come down

'What goes up must come down!' This popular saying can be used for the football team at the top of the ladder, but it more appropriately refers to the force of gravity that causes objects thrown up in the air to fall down.

The movement of objects launched (or 'projected') into the air is called 'projectile motion'. Because of gravity, the object (or 'projectile') will follow a curved path through the air. The shape of this curved path is a parabola and can be described by a quadratic equation.

Daredevil motocross riders like Evel Knievel need to have a good understanding of projectile motion when the rider (the projectile) wants to land in the right place.

Suppose Evel launches his bike upwards off a tower 58.8 m high, with a vertical velocity of 19.6 m/s. Due to the force of gravity, the height (h metres) of the bike above the ground at time (t seconds) after it has been launched can be calculated by:

$$h = -4.9t^2 + 19.6t + 58.8$$

1 At what time (t) is the bike at ground level?

> The height at ground level will be 0 m.

2 There are two theoretically possible solutions to Question **1**. Why is one of them not applicable in this situation?

3 Draw a graph of the equation:

$$h = -4.9t^2 + 19.6t + 58.8$$

Label the parts of the graph that show the ground, the height of the tower, the highest point reached by the bike, and the point where it lands.

The equation $h = -4.9t^2 + 19.6t + 58.8$ has the form of all quadratic equations: $y = ax^2 + bx + c$. Here, $a = -4.9$, which is 0.5 times the acceleration due to the force of gravity, 9.8 m/s^2. (The value is multiplied by 0.5 due to the factors of kinematics.) The acceleration is negative because it is a force downwards, which is opposite to the initial movement (upwards).

4 What factors affecting the path of the bike do b and c represent?

5 **(a)** What general equation could you use to describe this motion? (Remember that the value for the effect of gravity is the same, $a = -4.9$.)

(b) What is the maximum height and at what time after launch is it reached?

When an object is in the air, you may need to know how much time it is above a certain height.

6 An object is launched from ground level directly upward at 39.2 m/s. For how much time is the object at or above a height of 34.3 m? Form an equation and solve it for $h = 34.3$. Use the solutions to answer this question.

Comparing motion

Sometimes two objects can be launched in different ways at the same time and you want to determine when they will land.

For example, consider two identical objects thrown from a building 53 m high. The first is simply dropped (there is zero launch velocity) and the other is thrown down with a vertical velocity of 16 m/s. Which one will hit the ground first and by how much time?

The equation for the object that is dropped would be:

$$h_1 = -4.9t^2 + 53$$

The equation for the second object that is thrown down at 16 m/s would be:

$$h_2 = -4.9t^2 - 16t + 53$$

These two equations could be graphed using technology and the difference in their t-values when their height is zero could be found.

Alternatively, you could find the time when each object lands on the ground and find the difference between the two values of time.

7 Using either method, find the difference in the times taken to reach the ground for the two objects described above.

Exploration Spreadsheet

Equipment required: Microsoft® Excel or similar spreadsheet software. (For Casio ClassPad CAS or TI-Nspire CAS, you can download instructions from the eBook or the Pearson Places website.)

Factorising using 'zero-opposites'

The expression $x^2 + 3x - 10$ has factors $(x + 5)(x - 2)$.

The graph of $y = x^2 + 3x - 10$ is shown below.

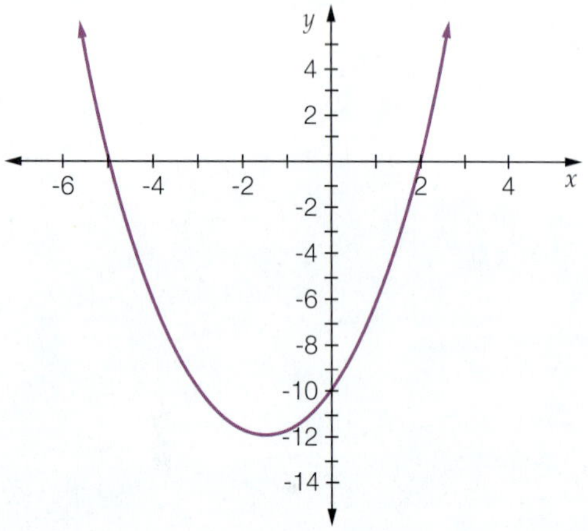

From the graph you can see that the x-intercepts are at $x = -5$ and $x = 2$, where $y = 0$. These values are the solutions to the equation $x^2 + 3x - 10 = 0$.

1 Explain the connection between the factors and the solutions.

You can use this connection to find factors for other quadratic expressions. You will use a spreadsheet to check the value of the expression for many values of x. A value of 0 indicates a solution. But be careful how you write the factor!

The spreadsheet at right shows two solutions for the original expression above.

2 **(a)** Identify the two solutions.

(b) Write the factors of $x^2 + 3x - 10$.

3 **(a)** Reproduce the spreadsheet shown using the following instructions.

In cell **A2** a starting value is entered. Cell **A3** has the formula **=A2+1**, which is copied to **Fill Down** the column.

The value of the quadratic expression is calculated in cell **B2** using the formula **=A2^2+3*A2-10**, which is also copied to **Fill Down** its column. (You should recall that **^** indicates 'to the power of'.)

(b) Change the formulas in your spreadsheet to factorise the following. You may need to include more values of x to find the intercepts.

(i) $x^2 - 32x - 144$

(ii) $x^2 - 15x + 56$

(iii) $x^2 - 17x - 84$

fx =A2^2+3*A2-10

	A	B	C
1	x-value	value of: x^2 + 3x – 10	
2	-10	60	
3	-9	44	
4	-8	30	
5	-7	18	
6	-6	8	
7	-5	0	
8	-4	-6	
9	-3	-10	
10	-2	-12	
11	-1	-12	
12	0	-10	
13	1	-6	
14	2	0	
15	3	8	
16	4	18	

4 Use a similar technique to find the factors for $x^2 - x - 20$. Use x-values between -10 and 10 for this expression.

Taking it further

How can you use this technique for expressions whose factors are not whole numbers?

Set up your spreadsheet for $8x^2 - 2x - 3$ with x-values from -3 to +3.

Notice that the values of the expression change from positive to negative and back to positive again.

The solutions for the equation $8x^2 - 2x - 3 = 0$ will be where the values of the expression change sign.

In this case, there is a solution between $x = -1$ and $x = 0$ and another between $x = 0$ and $x = 1$.

fx =8*A2^2-2*A2-3

	A	B	C
1	x-value	value of: 8x^2 – 2x – 3	
2	-3	75	
3	-2	33	
4	-1	7	
5	0	-3	
6	1	3	
7	2	25	
8	3	63	

To find this, you can change the starting value in cell **A2** to -1 and make the **A3** formula **=A2+0.5**, for example. You can then **Fill Down** to get the screen shown and find the first solution.

fx =A2+0.5

	A	B	C
1	x-value	value of: 8x^2 – 2x – 3	
2	-1	7	
3	-0.5	0	
4	0	-3	
5	0.5	-2	
6	1	3	
7	1.5	12	
8	2	25	

5 What is the first factor for $8x^2 - 2x - 3$?

So, how do you find the second factor? It occurs between $x = 0.5$ and $x = 1$, so change the increment to show values in this interval.

As the value of the expression gets closer to zero, you can make the change in x even smaller if necessary (e.g. 0.01). Continue in this way until you can find a value of zero for the expression.

6 What is the second factor for $8x^2 - 2x - 3$?

7 Use a similar technique to find the factors for each of the following.

(a) $10x^2 + 21x - 10$ **(b)** $100x^2 - 79x + 3$

Half-time 4

4.1 **1** Solve these equations using the null factor law.

(a) $(x + 2)(x - 1) = 0$ (b) $(x + 9)(x - 9) = 0$ (c) $3x(2x - 5) = 0$

4.1 **2** Solve each of the following.

(a) $x^2 + 4x = -4$ (b) $x^2 - 5x = 6$ (c) $x^2 + 8 = 9x$

(d) $x^2 - 4x = 5$ (e) $2x^2 + 6x = -4$ (f) $3x^2 + 9x = 30$

4.1 **3** Solve each of the following.

(a) $x^2 - 64 = 0$ (b) $(x + 1)^2 = 0$ (c) $(x - 2)^2 = 4$

4.1 **4** Solve these quadratic equations.

(a) $x^2 + 6x + 5 = 0$ (b) $x^2 - 9x = 0$ (c) $3x^2 + 9x = 30$

(d) $x^2 + x - 6 = 0$ (e) $x^2 - 4x + 3 = 0$ (f) $x^2 - 6x + 9 = 0$

4.4 **5** Sketch each parabola, stating the transformations made to the graph of $y = x^2$.

(a) $y = 2(x + 3)^2 - 4$ (b) $y = \frac{1}{2}(x - 1)^2 + 3$ (c) $y = (x - 2)^2 - 4$

4.2 **6** Factorise each of the following by completing the square.

(a) $x^2 + 6x + 7$ (b) $x^2 - 4x + 1$ (c) $2x^2 + 8x + 3$

4.3 **7** Solve each of the following equations by completing the square.

(a) $x^2 + 2x - 15 = 0$ (b) $x^2 - 14x + 24 = 0$ (c) $x^2 - 3x = 40$

4.4 **8** A softball thrown by an outfielder follows a path where the softball's height h metres at time t seconds is described by the relationship:

$h = -t^2 + 6t + 1$

(a) Complete the square to transform the equation into turning point form and write the coordinates of the turning point.

(b) Sketch the graph of this relationship.

(c) What is the height of the softball after 2 seconds?

(d) Use your graph to estimate when the ball is first at a height of 7 m above the ground, to 1 decimal place.

(e) During which time interval is the ball above a height of 9 m?

(f) What is the maximum height of the softball during its flight?

(g) Use your graph to estimate how many seconds the ball is in flight, if it is caught at a height of 1.6 m above the ground on its downward path.

Factorising non-monic quadratic trinomials

4.5

Before now, the only quadratic trinomials you have factorised have had a coefficient of 1 for the squared term, unless there was a common factor that could be taken out of the expression. This section shows you how to factorise **non-monic quadratic** trinomials in the form of $ax^2 + bx + c$ for any values of a.

Expand $(2x - 5)(3x + 7)$.

$$\begin{aligned} &(2x - 5)(3x + 7) \\ &= 6x^2 + 14x - 15x - 35 \\ &= 6x^2 - x - 35 \end{aligned}$$

Reversing the process:

$$\begin{aligned} &6x^2 - x - 35 \\ &= 6x^2 + 14x - 15x - 35 \\ &= 2x(3x + 7) - 5(3x + 7) \\ &= (2x - 5)(3x + 7) \end{aligned}$$

To factorise $6x^2 - x - 35$ you follow the same process as you used for factorising monic quadratics, but you now need to consider the coefficient of the squared term as well as the constant term.

10A

Method 1: The cross method

Identify factor pairs for the coefficient of the squared term:

For $6x^2 - x - 35$, 6 has factor pairs of 6, 1 and 3, 2 (do not consider the negative pairs).

Choose a factor pair that gives the squared term ($3x \times 2x = 6x^2$).

Identify factor pairs for the constant term:

For $6x^2 - x - 35$, -35 has factor pairs of 5, -7; 1, -35; 7, -5 and -1, 35.

Use the cross method to trial factor pairs:

For $6x^2 - x - 35$, the coefficient of the middle term is -1 so try factors that are close to each other in magnitude.

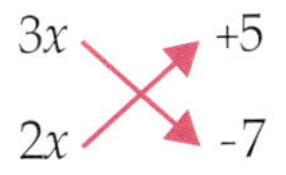

Middle term: $-21x + 10x = -11x$

(not correct because you need $-x$)

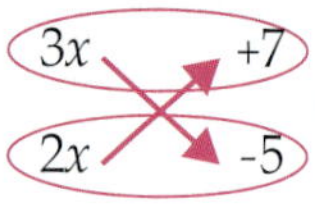

Middle term: $-15x + 14x = -x$

(correct because you need $-x$)

The circled expressions give the factors.

$6x^2 - x - 35 = (3x + 7)(2x - 5)$

Worked example 16

W.E. 16

Factorise the following quadratic expressions using the cross method.

(a) $3x^2 - 11x - 4$ (b) $5x^2 - 43x + 24$ (c) $12x^2 + 11x - 36$

Thinking | **Working**

(a) 1 Choose a factor pair that multiply to give the squared term ($3x \times x = 3x^2$). Choose a factor pair that multiply to give the constant term ($-4 \times 1 = -4$). Use the cross method to find the correct factors.

(a) $3x^2 - 11x - 4$

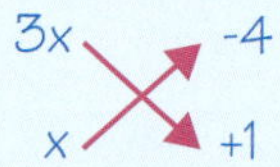

$3x - 4x = -x$

(incorrect middle term)

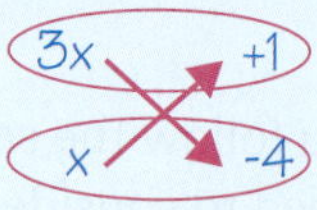

$-12x + x = -11x$

(correct middle term)

2 Write the factorisation.

$3x^2 - 11x - 4$
$= (3x + 1)(x - 4)$

(b) 1 Choose a factor pair that multiply to give the squared term ($5x \times x = 5x^2$). Choose a factor pair that multiply to give the constant term ($-8 \times -3 = 24$). If the middle term is negative use negative factor pairs, and if the middle term is quite large use factors that are not too close. Use the cross method to find the correct factors.

(b) $5x^2 - 43x + 24$

$-15x - 8x = -23x$

(incorrect middle term)

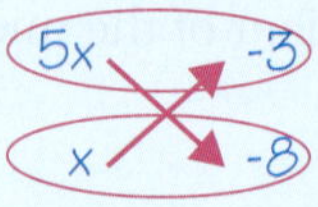

$-40x - 3x = -43x$

(correct middle term)

2 Write the factorisation.

$5x^2 - 43x + 24$
$= (5x - 3)(x - 8)$

(c) 1 Choose a factor pair that multiply to give the squared term ($4x \times 3x = 12x^2$). Choose a factor pair that multiply to give the constant term ($-9 \times 4 = -36$). Use the cross method to find the correct factors. If the pairs chosen do not give the correct magnitude, try another pair. If the magnitude is correct but the sign is incorrect, swap the signs on the factors.	(c) $12x^2 + 11x - 36$ $3x$ $+4$ $4x$ -9 $-27x + 16x = -11x$ (incorrect middle term) $3x$ -4 $4x$ $+9$ $27x - 16x = +11x$ (correct middle term)
2 Write the factorisation.	$12x^2 + 11x - 36$ $= (4x + 9)(3x - 4)$

Note that in part **(c)**, the factors $(4x - 3)(3x + 9)$ were not tried as there would have been a common factor in both brackets. You also use your knowledge of odds and evens to reject unsuitable factor pairs.

Method 2: Split the middle term

Multiply the coefficient of the squared term a by the constant c:

For $\quad ax^2 + bx + c$
$\quad = 6x^2 - 7x - 3$

$a = 6, b = -7, c = -3$

$ac = 6 \times -3$
$\quad = -18$

List factor pairs of -18:

$1 \times -18;\ 2 \times -9;\ 3 \times -6;\ -1 \times 18;\ \ldots$

This can be a lengthy process, but you can simplify it by using some simple strategies.

For a *positive* constant term:

- pairs will have the same sign as the middle term
- if the magnitude of the coefficient of the middle term is small the factors will be close to each other
- if the magnitude of the coefficient of the middle term is large the factors will be far apart.

For a *negative* constant term:

- pairs will be opposite in sign
- if the magnitude of the coefficient of the middle term is small, choose factors that are close in magnitude but opposite in sign.
- if the magnitude of the coefficient of the middle term is large, choose factors that are far apart in magnitude but opposite in sign.

As $b = -7$, you are looking for a pair of factors that have a difference of -7 such as -9 and 2.

You can now write $6x^2 - 7x - 3 = 6x^2 - 9x + 2x - 3$ and factorise by grouping in pairs as shown previously.

Worked example 17

W.E. 17

Factorise the following quadratic expressions by splitting the middle term.

(a) $3x^2 - 11x - 4$ (b) $5x^2 - 43x + 24$ (c) $12x^2 - x - 20$

Thinking		Working
(a)	1 Multiply the coefficient of x^2 and the constant term ($3 \times -4 = -12$) and choose a factor pair that adds to give the coefficient of the middle term ($-12 + 1 = -11$). Rewrite the expression using these factors as coefficients of two middle terms to replace the middle term.	(a) $3x^2 - 11x - 4$ $= 3x^2 - 12x + x - 4$ $\left.\begin{array}{l}\times \text{ to give } -12 \\ + \text{ to give } -11\end{array}\right\} -12, 1$
	2 Factorise using grouping in pairs.	$= 3x(x - 4) + 1(x - 4)$
	3 Take out the common factor.	$= (x - 4)(3x + 1)$
(b)	1 Multiply the coefficient of x^2 and the constant term ($5 \times 24 = 120$) and choose a factor pair that adds to give the coefficient of the middle term ($-40 + -3 = -43$). Rewrite the expression using these factors as coefficients of two middle terms to replace the middle term.	(b) $5x^2 - 43x + 24$ $= 5x^2 - 40x - 3x + 24$ $\left.\begin{array}{l}\times \text{ to give } 120 \\ + \text{ to give } -43\end{array}\right\} -40, -3$
	2 Factorise using grouping in pairs.	$= 5x(x - 8) - 3(x - 8)$
	3 Take out the common factor.	$= (x - 8)(5x - 3)$
(c)	1 Multiply the coefficient of x^2 and the constant term ($12 \times -20 = -240$) and choose a factor pair that adds to give the coefficient of the middle term ($-16 + 15 = -1$). Rewrite the expression using these factors as coefficients of two middle terms to replace the middle term.	(c) $12x^2 - x - 20$ $= 12x^2 - 16x + 15x - 20$ $\left.\begin{array}{l}\times \text{ to give } -240 \\ + \text{ to give } -1\end{array}\right\} -16, 15$
	2 Factorise using grouping in pairs.	$= 4x(3x - 4) + 5(3x - 4)$
	3 Take out the common factor.	$= (4x + 5)(3x - 4)$

Completing the square with non-monic quadratic expressions

As you have seen previously, not all quadratic trinomials factorise with binomial products, and some do not factorise at all. You can use 'completing the square' with non-monic quadratics to see if the expression can be factorised.

To complete the square, the coefficient of the quadratic must be 1, so you will need to take the coefficient of x out as a common factor before you can complete the square.

Worked example 18

W.E. 18

Factorise the following expressions, if possible, by completing the square

(a) $3x^2 - 6x - 5$ **(b)** $2x^2 - 5x - 2$ **(c)** $2x^2 - 4x + 6$

Thinking	Working
(a) 1 Take out the coefficient of the squared term as a factor of the expression. (Write 3 outside the brackets and divide every term inside the brackets by 3.)	(a) $3x^2 - 6x - 5$ $= 3(x^2 - 2x - \frac{5}{3})$
2 Complete the square inside the brackets and simplify (add and subtract 1).	$= 3(x^2 - 2x + 1 - 1 - \frac{5}{3})$
3 Factorise inside the brackets using the difference of two squares and surds.	$= 3[(x-1)^2 - \frac{8}{3}]$ $= 3(x - 1 - \sqrt{\frac{8}{3}})(x - 1 + \sqrt{\frac{8}{3}})$
(b) 1 Take out the coefficient of the squared term as a factor of the expression. (Write 2 outside the brackets and divide every term inside the brackets by 2.)	(b) $2x^2 - 5x - 2$ $= 2(x^2 - \frac{5}{2}x - 1)$
2 Complete the square inside the brackets and simplify (add and subtract $\frac{25}{16}$).	$= 2[(x^2 - \frac{5}{2}x + \frac{25}{16}) - \frac{25}{16} - 1]$
3 Factorise inside the brackets using the difference of two squares and surds.	$= 2[(x - \frac{5}{4})^2 - \frac{41}{16}]$ $= 2(x - \frac{5}{4} + \frac{\sqrt{41}}{4})(x - \frac{5}{4} - \frac{\sqrt{41}}{4})$
(c) 1 Take out the coefficient of the squared term as a factor of the expression. (Write 2 outside the brackets and divide every term inside the brackets by 2.)	(c) $2x^2 - 4x + 6$ $= 2(x^2 - 2x + 3)$
2 Complete the square inside the brackets and simplify (add and subtract 1).	$= 2[(x^2 - 2x + 1) - 1 + 3]$
3 As you do not have a difference of two squares you cannot factorise further.	$= 2[(x-1)^2 + 2]$ The expression cannot be factorised further.

10A

4.5 Factorising non-monic quadratic trinomials

Navigator

Answers p.801

1 (columns 1–2), 2 (columns 1–2), 3, 4 (a–e), 5 (column 1), 6, 8	1 (columns 2–3), 2 (columns 2–3), 4 (columns 1–2), 5 (column 2), 6, 7, 8	1 (columns 2–3), 2 (columns 2–3), 4 (columns 2–3), 5 (column 3), 6, 7, 8, 9

Fluency

W.E. 16

1 Factorise these expressions using the cross method.

(a) $x^2 + 7x + 6$ (b) $x^2 - 6x - 55$ (c) $x^2 + 2x - 15$

(d) $4x^2 + 8x + 3$ (e) $10y^2 + 13y + 4$ (f) $9c^2 - 19c + 2$

(g) $6w^2 + 17w + 12$ (h) $15y^2 + 7y - 30$ (i) $8x^2 - 9x - 14$

W.E. 17

2 Factorise these expressions by splitting the middle term.

(a) $x^2 + 7x + 12$ (b) $x^2 - 5x - 24$ (c) $x^2 + 4x - 45$

(d) $2x^2 + 5x + 2$ (e) $2y^2 + 9y + 7$ (f) $5b^2 + 8b + 3$

(g) $3k^2 - 11k + 6$ (h) $11a^2 - a - 12$ (i) $5x^2 - 11x - 12$

3 Find the values of a and b by completing the square.

(a) $x^2 + 2x + a = (x + b)^2$ (b) $x^2 - 6x + a = (x - b)^2$

(c) $x^2 - 4x + a = (x - b)^2$ (d) $2x^2 + 8x + a = 2(x + b)^2$

(e) $3x^2 - 12x + a = 3(x - b)^2$ (f) $-2x^2 + 4x + a = -2(x - b)^2$

10A

W.E. 18

4 Factorise the following expressions, if possible, by completing the square.

(a) $2x^2 - 6x - 1$ (b) $3x^2 - 12x - 7$ (c) $2x^2 - 8x - 3$

(d) $3x^2 - 9x + 4$ (e) $5x^2 - 10x + 3$ (f) $2x^2 - 7x + 2$

(g) $2x^2 - 4x + 5$ (h) $2x^2 - 8x + 9$ (i) $3x^2 - 6x + 5$

5 Factorise the following expresssions using an appropriate method.

(a) $2m^2 + 3m - 9$ (b) $7x^2 - 23x + 18$ (c) $11c^2 - 25c + 6$

(d) $5y^2 + 19y + 12$ (e) $3p^2 + 8p - 16$ (f) $2x^2 + 3x - 20$

(g) $3a^2 - 12a + 1$ (h) $2x^2 + 3x - 4$ (i) $3x^2 + 2x - 9$

(j) $3x^2 + 8x + 5$ (k) $7y^2 - 18y + 8$ (l) $3t^2 + t - 4$

(m) $2y^2 - 10y + 3$ (n) $4x^2 + 20x - 3$ (o) $5b^2 + 2b - 5$

(p) $9d^2 + 6d - 8$ (q) $6x^2 - x - 22$ (r) $18n^2 + 31n - 35$

(s) $24y^2 - 17y + 3$ (t) $15x^2 + 46x + 16$ (u) $12x^2 - 11x - 15$

Understanding

6 The floor (in m^2) of a rectangular room of height x m is given by the expression $6x^2 - 7x - 3$.

(a) Factorise the expression.

(b) Assuming that the two factors represent the length and the width of the room and that the length is 11.5 m, find the height of the room.

(c) What is the width of the room?

Reasoning

7 The number of bird nests in a park depends on the number of trees t (estimated to the nearest multiple of 25).

Plume Park has $4t^2 + 6t + 1$ bird nests.
Byrd Park has $12t^2 - 12t + 10$ bird nests.

(a) Write an expression in terms of t for the difference in the number of bird nests in the two parks. (Assume that the number of bird nests in Byrd Park is greater than the number of bird nests in Plume Park for all values of t.)

(b) (i) Simplify the expression in part **(a)**.

(ii) Factorise the expression in **(i)**.

(c) Use your answer from **(b) (ii)** to show that the parks will have equal numbers of bird nests when there are 75 or 150 trees.

8 Paul and Selina each have $\$x$. Later, each person has more money and the product of their money is $12x^2 - 8x - 15$.

(a) Factorise the expression.

(b) Find how much money each person had initially if Paul now has \$65 and has more money than Selina.

(c) How much money does Selina have now?

Open-ended

9 (a) Find a non-monic quadratic trinomial that will factorise to give binomial products. Factorise it.

(b) Find a non-monic quadratic trinomial that will factorise but not as binomial products. Factorise it.

(c) Find a non-monic quadratic trinomial that will not factorise. Explain why it cannot be factorised.

10A

4.6 Solving non-monic quadratics

You have already solved monic quadratic equations. Now that you can factorise non-monic expressions, you can use similar techniques to solve non-monic quadratic equations.

- Write the equation as $ax^2 + bx + c = 0$ where $a \neq 0$ or 1.
- Factorise the expression $ax^2 + bx + c$ using the cross method, by splitting the middle term or by completing the square.
- Use the null factor law to solve the equation, by equating each of the factors to zero.

Worked example 19

W.E. 19

Solve each of the following equations. Factorise first using either the cross method or by splitting the middle term.

(a) $6x^2 - 17x + 12 = 0$ **(b)** $3x^2 - 11x - 4 = 0$

Thinking	Working
(a) 1 Write the quadratic equation.	(a) $6x^2 - 17x + 12 = 0$
2 Find factor pairs for the first term. (Here, use $3x$ and $2x$). Try factor pairs for the last term. (Here, try -4 and -3. Do not try -3 and -4, as this contains a common factor.) Do not try positive factor pairs, as the middle term is negative.	$3x$ ↘↗ -4 $2x$ ↗↘ -3 $-9x - 8x = -17x$
3 Write the equation in factorised form.	$(3x - 4)(2x - 3) = 0$
4 Use the null factor law to solve the equation.	$3x - 4 = 0$ or $2x - 3 = 0$ $3x = 4$ or $2x = 3$ $x = \frac{4}{3}$ or $x = \frac{3}{2}$
(b) 1 Write the quadratic equation.	(b) $3x^2 - 11x - 4 = 0$
2 Factorise the equation. (Here, factorise by splitting the middle term.) Identify a, b and c.	$a = 3$, $b = -11$, $c = -4$
3 Find the product ac.	$ac = 3 \times (-4) = -12$
4 Find two numbers that multiply to give ac and add to give b.	× to give -12 + to give -11 } -12, 1

10A

5 Rewrite the LHS of the equation using these numbers as coefficients of two middle terms. $3x^2 - 12x + x - 4$

6 Factorise the LHS by grouping in pairs. $3x(x - 4) + 1(x - 4) = 0$
$(x - 4)(3x + 1) = 0$

7 Use the null factor law to solve the equation. $x - 4 = 0$ or $3x + 1 = 0$
$x = 4$ or $x = -\frac{1}{3}$

Worked example 20

W.E. 20

Solve $2x^2 - 7x + 2 = 0$. Factorise first by completing the square.

Thinking	Working
1 Write the quadratic equation.	$2x^2 - 7x + 2 = 0$
2 Take out the coefficient of x^2 as a factor of each term.	$2\left(x^2 - \frac{7}{2}x + 1\right) = 0$
3 Divide both sides by this factor.	$x^2 - \frac{7}{2}x + 1 = 0$
4 Complete the square. (Here, add and subtract $\left(\frac{-7}{4}\right)^2$.)	$x^2 - \frac{7}{2}x + \left(\frac{-7}{4}\right)^2 + 1 - \left(\frac{49}{16}\right) = 0$
5 Use the perfect square rule to factorise the first three terms and simplify the remaining terms.	$\left(x - \frac{7}{4}\right)^2 - \frac{33}{16} = 0$
6 Factorise further by using the difference of two squares rule and surds. (Here, $\frac{33}{16} = \left(\frac{\sqrt{33}}{4}\right)^2$.)	$\left(x - \frac{7}{4} + \frac{\sqrt{33}}{4}\right)\left(x - \frac{7}{4} - \frac{\sqrt{33}}{4}\right) = 0$
7 Use the null factor law to solve the equation.	$x - \frac{7}{4} + \frac{\sqrt{33}}{4} = 0$ or $x - \frac{7}{4} - \frac{\sqrt{33}}{4} = 0$ $x = \frac{7}{4} - \frac{\sqrt{33}}{4}$ or $x = \frac{7}{4} + \frac{\sqrt{33}}{4}$ $x = \frac{7 \pm \sqrt{33}}{4}$

10A

The quadratic formula

When solving a quadratic equation you can use the quadratic formula to factorise any quadratic expression. The quadratic formula is developed by completing the square for the general quadratic equation $ax^2 + bx + c = 0$.

Deriving the quadratic formula:

1 Write the equation.

$$ax^2 + bx + c = 0$$

2 Take out a as a common factor.

$$a\left(x^2 + \frac{b}{a}x + \frac{c}{a}\right) = 0$$

3 Divide both sides by a.

$$x^2 + \frac{b}{a}x + \frac{c}{a} = 0$$

4 Complete the square and balance the equation by adding and subtracting $\left(\frac{1}{2} \times \frac{b}{a}\right)^2$.

$$\left(x^2 + \frac{b}{a}x + \left(\frac{b}{2a}\right)^2\right) - \left(\frac{b}{2a}\right)^2 + \frac{c}{a} = 0$$

5 Identify the first three terms as a perfect square and factorise them.

$$\left(x + \frac{b}{2a}\right)^2 - \frac{b^2}{4a^2} + \frac{c}{a} = 0$$

6 Write the other two terms with a common denominator.

$$\left(x + \frac{b}{2a}\right)^2 - \frac{b^2}{4a^2} + \frac{4ac}{4a^2} = 0$$

7 Combine these terms with a common denominator. Be careful to change the sign in the numerator.

$$\left(x + \frac{b}{2a}\right)^2 - \frac{b^2 - 4ac}{4a^2} = 0$$

8 Express as the difference of two squares.

$$\left(x + \frac{b}{2a}\right)^2 - \left(\sqrt{\frac{b^2 - 4ac}{4a^2}}\right)^2 = 0$$

9 Factorise as the difference of two squares.

$$\left(x + \frac{b}{2a} + \sqrt{\frac{b^2 - 4ac}{4a^2}}\right)\left(x + \frac{b}{2a} - \sqrt{\frac{b^2 - 4ac}{4a^2}}\right) = 0$$

$$\left(x + \frac{b + \sqrt{b^2 - 4ac}}{2a}\right)\left(x + \frac{b - \sqrt{b^2 - 4ac}}{2a}\right) = 0$$

10 Use the null factor law.

$$x + \frac{b + \sqrt{b^2 - 4ac}}{2a} = 0 \text{ or } x + \frac{b - \sqrt{b^2 - 4ac}}{2a} = 0$$

11 Solve each equation.

$$x = \frac{-b - \sqrt{b^2 - 4ac}}{2a} \text{ or } x = \frac{-b + \sqrt{b^2 - 4ac}}{2a}$$

12 Write the solution in a shorthand form.

The solution to $ax^2 + bx + c = 0$ is

$$x = \frac{-b \pm \sqrt{b^2 - 4ac}}{2a}$$

10A

Worked example 21

W.E. 21

Use the quadratic formula to solve $3x^2 - 4x - 5 = 0$. Give your answers in exact form and rounded to 2 decimal places.

Thinking	Working
1 Write the quadratic expression.	$3x^2 - 4x - 5 = 0$
2 Identify a, b and c.	$a = 3$, $b = -4$, $c = -5$
3 Write the quadratic formula.	$x = \dfrac{-b \pm \sqrt{b^2 - 4ac}}{2a}$
4 Substitute the values for a, b and c.	$x = \dfrac{-(-4) \pm \sqrt{(-4)^2 - 4 \times 3 \times (-5)}}{2 \times 3}$
5 Simplify.	$x = \dfrac{4 \pm \sqrt{16 + 60}}{6}$
6 State your answer in exact form.	$= \dfrac{4 \pm \sqrt{76}}{6}$
7 Use your calculator to evaluate.	$x = 2.12$ or $x = -0.79$

When you study surds, you will learn to simplify surds so $\dfrac{4 \pm \sqrt{76}}{6}$ can be written as $\dfrac{2 \pm \sqrt{19}}{3}$.

10A

Solving non-monic quadratic equations: $ax^2 + bx + c = 0$

Factorise non-monic quadratic expressions by:

- using the cross method or by splitting the middle term
- completing the square.

Solve non-monic quadratic equations by:

- using the null factor law
- using the quadratic formula.

The solution to $ax^2 + bx + c = 0$ is: $x = \dfrac{-b \pm \sqrt{b^2 - 4ac}}{2a}$

4.6 Solving non-monic quadratics

Navigator

Answers p. 802

1 (column 1), 2 (a, b), 3, 4, 5, 6 (column 1), 7, 8, 9, 11, 12, 13, 15, 17 (a–c), 18, 20	1 (column 2), 2 (a–d), 3, 5, 6 (column 2), 7, 8, 9, 10, 11, 12, 13, 14, 15, 16, 17, 18, 20	1 (column 2), 2, 3, 5, 6 (column 2), 7, 8, 11, 12, 13, 14, 15, 16, 17, 18, 19, 20

Equipment required: calculator for Question **3**

Fluency

W.E. 19

1 Solve each quadratic equation. Factorise first using either the cross method or by splitting the middle term.

(a) $3x^2 - 4x - 4 = 0$ **(b)** $3x^2 + 8x + 4 = 0$

(c) $5x^2 - 11x + 6 = 0$ **(d)** $4x^2 + 3x - 10 = 0$

(e) $4x^2 - 25x + 6 = 0$ **(f)** $2x^2 + 5x - 12 = 0$

(g) $5x^2 - 17x + 12 = 0$ **(h)** $3x^2 + 8x + 5 = 0$

(i) $6x^2 - x - 2 = 0$ **(j)** $8x^2 + 6x - 9 = 0$

(k) $6x^2 - 11x - 10 = 0$ **(l)** $8x^2 + 2x - 15 = 0$

10A

W.E. 20

2 Solve each equation. Factorise first by completing the square.

(a) $2x^2 - 12x + 11 = 0$ **(b)** $3x^2 - 12x + 7 = 0$

(c) $2x^2 - 5x + 2 = 0$ **(d)** $3x^2 + 10x + 1 = 0$

(e) $5x^2 - 2x - 6 = 0$ **(f)** $6x^2 + 9x - 2 = 0$

W.E. 21

3 Use the quadratic formula to solve each equation. Give your answers in **(i)** exact form and **(ii)** rounded to 2 decimal places.

(a) $3x^2 - 5x - 4 = 0$ **(b)** $5x^2 + 10x + 1 = 0$

(c) $2x^2 - 11x + 6 = 0$ **(d)** $x^2 + 5x - 3 = 0$

4 For each of these quadratic equations:

(a) $x^2 + 5x + 2 = 0$ **(b)** $2x^2 - 3x + 1 = 0$

(c) $-3x^2 - x + 3 = 0$ **(d)** $5x^2 + 4x = 0$

(i) list the values that a, b and c represent for the form $ax^2 + bx + c = 0$

(ii) find the solution to each equation using the quadratic formula.

5 The solutions to the equation $10x^2 - 23x + 12 = 0$ are:

A $x = 3$ or $x = 4$ **B** $x = \frac{2}{3}$ or $x = \frac{5}{4}$

C $x = 1.5$ or $x = 0.8$ **D** $x = -1.5$ or $x = -0.8$

Understanding

6 Solve each quadratic equation using an appropriate method.

(a) $3x^2 - 5x + 2 = 0$
(b) $5x^2 + 10x + 2 = 0$
(c) $4x^2 - 9 = 0$
(d) $x^2 + 7x - 9 = 0$
(e) $6x^2 - 7x - 10 = 0$
(f) $4x^2 + 4x + 1 = 0$
(g) $7x^2 - 6 = 0$
(h) $8x^2 + 6x - 4 = 0$
(i) $36x^2 - 60x + 25 = 0$
(j) $3x^2 - 11x + 7 = 0$
(k) $2x^2 - 9x + 5 = 0$
(l) $3x^2 + x - 8 = 0$

7 Which of the following equations has the solutions $x = 1.5$ and $x = 2.5$?

A $x^2 - 16x + 15 = 0$
B $4x^2 - 16x + 15 = 0$
C $x^2 - 8x + 15 = 0$
D $4x^2 + 16x + 15 = 0$

8 The solutions to $x^2 + px + q = 0$ are:

A $x = \dfrac{-p \pm \sqrt{q}}{2}$
B $x = \dfrac{p \pm \sqrt{p^2 + 4q}}{2}$
C $x = \dfrac{-p \pm \sqrt{p^2 - 4q}}{2}$
D $x = \dfrac{-2p \pm \sqrt{p^2 - 4q}}{2}$

9 Rick's driving speed was measured over a 10-minute period and the following relationship was found to exist: $v = -4t^2 + 31t$, where v is the speed in kilometres per hour at t minutes. When was Rick travelling at a speed of 60 km/h?

10 The temperature inside a tent was measured over a period of time and the following quadratic relationship was found to exist: $T = -2h^2 + 11h + 21$, where T is the temperature in degrees Celsius at h hours.

(a) At what hour was the temperature 0 °C?
(b) At what hour was the temperature 26 °C?

11 A tourist, high above the ground enjoying the sights from a hot-air balloon, unfortunately drops a camera and watches it fall to the ground. The height of the camera above the ground, h (in metres), t seconds after it has been dropped, can be represented by the relationship: $h = -5t^2 + 192$.

(a) At what height above the ground is the camera dropped?
(b) How long does it take for the camera to fall to the ground? Round your answer to 1 decimal place.

Reasoning

12 Use the quadratic formula to show that none of the following equations has a solution. Write the part of the quadratic formula that indicates that no solutions are possible.

(a) $2x^2 - 3x + 2 = 0$
(b) $5x^2 + 5x + 2 = 0$
(c) $4x^2 - 6x + 9 = 0$
(d) $2x^2 + 7x + 9 = 0$

13 A police officer on a motorbike is following a car along a highway. After a short time, the driver of the car notices the police and starts to slow down, finally stopping on the side of the road.

The speed of the car during this time can be represented by the quadratic equation $v = -3t^2 + 17t + 70$, where v is the speed of the car in kilometres per hour, t minutes after the police started following.

(a) How long was the car followed before it stopped?

(b) During this time, did the driver break the speed limit of 100 km/h?

14 Spot's dog kennel will be 1 m high and twice as long as it is wide, and it will have an opening at one end. The outside of the kennel will be painted, including the base of the kennel, to keep it waterproof.

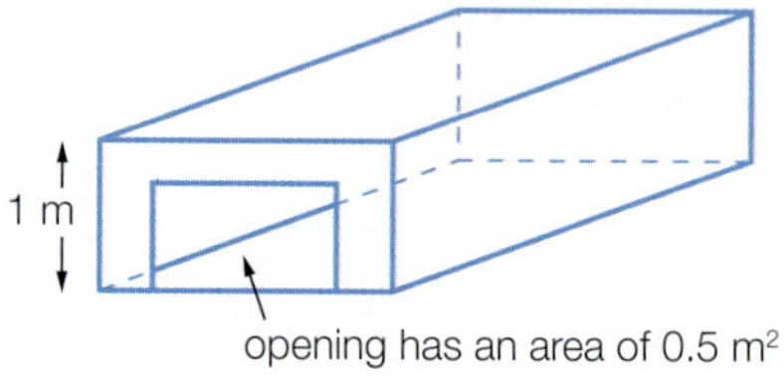

(a) Write an algebraic expression for the area of the kennel that is to be painted.

(b) A 4 L tin of paint is used, which will cover 31.4 m^2, and the kennel is given two coats of paint. Find the dimensions of the largest kennel that could be constructed. Round your answer to 1 decimal place.

15 A right-angled triangle with shorter sides x cm and $(x + 7)$ cm has an area of 60 cm^2. Find the perimeter of the triangle.

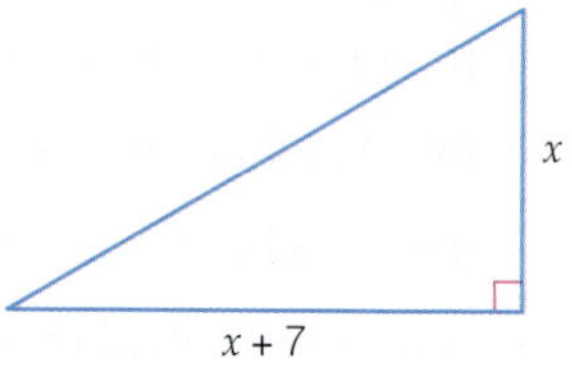

16 The sum of the first n terms of the sequence 2, 5, 8, 11 … is given by the formula $S = \frac{n}{2}[4 + 3(n - 1)]$ where n is the number of terms in the sequence. Use the formula to find the number of terms needed to make a sum of 805.

17 Solve the following quadratic equations by first multiplying by the lowest common denominator. Note that the denominator can never be zero.

(a) $2x - 1 = \frac{21}{x}$

(b) $x = \frac{9x - 14}{x}$

(c) $x = \frac{x - 9}{x - 5}$

(d) $\frac{3}{x - 1} = \frac{x}{x + 1}$

(e) $\frac{5}{x + 2} + \frac{3}{x} = 1$

(f) $\frac{x - 1}{x} + 5 = \frac{3}{x^2}$

Open-ended

18 Make up three non-monic quadratic equations that have as their solutions $x = 2.5$ and $x = -3$.

19 By using the process of completing the square, the solution $x = \dfrac{-a \pm \sqrt{5}}{2}$ was found.

By choosing a value of a between -3 and 3, make up at least three possible quadratic equations that would have this solution.

20 By choosing values for a, b and c in the general quadratic equation $ax^2 + bx + c = 0$, make up an equation that has:

(a) no solution **(b)** one solution **(c)** two solutions.

Write the solutions to your equations where they exist.

Problem solving

Projectiles

When an object is projected vertically upward with an initial velocity v metres per second, its height h metres above the ground after t seconds is approximately given by the equation $h = vt - 5t^2$.

(a) A rocket is fired vertically upward at an initial velocity of 25 m/s. At what times is the height of the rocket above ground going to be zero?

(b) The equation $h = vt - 5t^2$ can also be written as $5t^2 - vt + h = 0$.

(i) When the height of the rocket is 30 m and $v = 25$, show that the expression becomes $t^2 - 5t + 6 = 0$.

(ii) By factorising $t^2 - 5t + 6$, find at what times the rocket is at a height of 30 m.

(c) The rocket is now projected with an initial velocity of 35 m/s. When is the rocket at a height of 30 m on its way down?

(d) The rocket is again projected, but with an initial velocity of 10 m/s.

(i) Show when the left side of the equation $5t^2 - vt + h = 0$ is factorised using $h = 5$, this gives a perfect square.

(ii) For how much time is the rocket in flight?

(iii) What can you conclude about the maximum height that the rocket can reach?

Strategy options

- Draw a diagram.
- Look for a pattern.

10A

Exploration Spreadsheet

Equipment required: Microsoft® Excel or similar spreadsheet software.

Solving quadratics using iteration

Iteration is a process that involves repeating a series of steps, using the 'answer' from one step to get an improved 'answer' at the next step. This process can be very useful for solving different kinds of real-life problems. A good method for iteration is often to use a spreadsheet.

For example, if you want to find the x-intercepts for $y = x^2 - 4x + 2$, then you need to find the solutions to the quadratic equation $x^2 - 4x + 2 = 0$. By completing the square you can find the solutions: $x = 2 \pm \sqrt{2}$.

Rounded to 8 decimal places, these values are:

$x = 3.414\,213\,56$ and $x = 0.585\,786\,44$

These solutions can be used to help verify an iteration process.

1 Transform the equation so that the term containing x is on the LHS.

$$x^2 - 4x + 2 = 0$$
$$4x = x^2 + 2$$

2 Divide both sides by the coefficient of x.

$$x = \frac{x^2 + 2}{4}$$

Set up a spreadsheet as shown.

f_x =(A2^2+2)/4

	A	B	C
1	Estimate	Improved	Difference
2	1	0.75	0.25
3	0.75		

The formula in cell **B2** calculates the expression. This 'improved' value is then used for the next estimate, so the formula for cell **A3** is **=B2**.

To see how the estimate has changed from one step to the next, the difference between the two values is calculated in cell **C2** using the formula **=A2-B2**.

3 The difference column is included so that you know when you have found a solution. What value will you expect in the difference column when a solution has been found?

Now select each cell and **Fill Down** to copy the formula down its column. Do this for each column down to row 25.

4 Has this reached a solution yet? Your spreadsheet should look something like the one shown at right.

f_x =A2-B2

	A	B	C
1	Estimate	Improved	Difference
2	1	0.75	0.25
3	0.75	0.640625	0.109375
4	0.640625	0.6026001	0.0380249
5	0.6026001	0.59078172	0.01181838
6	0.5907817	0.58725576	0.00352596
7	0.5872558	0.58621733	0.00103843
8	0.5862173	0.58591269	0.00030464
9	0.5859127	0.58582342	8.927E-05
10	0.5858234	0.58579727	2.615E-05
11	0.5857973	0.58578961	7.6595E-06
12	0.5857896	0.58578737	2.2435E-06
13	0.5857874	0.58578671	6.5709E-07
14	0.5857867	0.58578652	1.9246E-07
15	0.5857865	0.58578646	5.637E-08
16	0.5857865	0.58578644	1.651E-08
17	0.5857864	0.58578644	4.8358E-09
18	0.5857864	0.58578644	1.4164E-09
19	0.5857864	0.58578644	4.1484E-10
20	0.5857864	0.58578644	1.215E-10
21	0.5857864	0.58578644	3.5588E-11
22	0.5857864	0.58578644	1.0423E-11
23	0.5857864	0.58578644	3.053E-12
24	0.5857864	0.58578644	8.9417E-13
25	0.5857864	0.58578644	2.619E-13

You should recognise the values in the **Estimate** column and the **Improved** column as being one of the solutions found by completing the square.
But what about the value in the **Difference** column? For a final solution, the difference should be zero.

5 Explain what 2.619E-13 means.

The iterations are getting closer and closer to the solution at each step. This iteration is said to *converge*.

6 Replace the initial value 1 in cell **A2** with 5. What happens to the iteration values?

For some starting values, the iteration is said to *diverge*, and does not lead to a solution.

You now need to find the other solution. However, using this iteration you cannot find it no matter what starting value is entered. This iteration formula will either converge to the solution 0.585 786… or diverge.

You can explore other ways of transposing the original equation, to try to find an iteration formula that will converge to the other solution.

Taking it further

7 Other expressions that are equivalent to x in this case are:

$$\frac{2}{4 - x}, \frac{4x - 2}{x}, \sqrt{4x - 2}.$$

(a) Show how each can be found from the original quadratic.

(b) Try these expressions in your spreadsheet. *Note:* To calculate a square root in Excel, use the formula **SQRT(**number**)**.

(c) Do any of these give the other solution?

(d) Do any of these give both solutions?

Chapter review 4

Maths literacy

completing the square	null factor law	quadratic trinomial
factorising	parabola	turning point
intercept	quadratic expression	turning point form
non-monic quadratic	quadratic formula	vertex

Copy and complete the following using the words and phrases from the list, where appropriate. A word or phrase may be used more than once.

1 $ax^2 + bx + c$ is a ________. When the coefficient of x^2 is not 1, the expression is a ________.

2 A ________ is the type of graph produced by a quadratic relationship.

3 To factorise $x^2 - 4x - 6$ as $(x - 2 - \sqrt{10})(x - 2 + \sqrt{10})$ you can use a method called ________.

4 To graph a quadratic it is important to show the x- and y-intercepts and the ________.

5 To change a quadratic that is in the form $y = x^2 + bx + c$ into ________, it is necessary to complete the square.

Fluency

1 Solve these quadratic equations using the null factor law. **4.1**

(a) $x^2 - 6x = 0$ (b) $4x^2 - 81 = 0$ (c) $x^2 - 5x - 14 = 0$

(d) $x^2 + 6x + 9 = 0$ (e) $x^2 = 8x$ (f) $x^2 + 36 = 12x$

(g) $2x(x + 4) = 0$ (h) $(x - 7)(x + 1) = 0$ (i) $x^2 - 16 = 0$

2 Solve these quadratic equations. **4.1**

(a) $x^2 + 12x + 35 = 0$ (b) $x^2 + x - 6 = 0$ (c) $-x^2 - 8x + 9 = 0$

(d) $-x^2 + 9x - 8 = 0$ (e) $3x^2 + 9x - 12 = 0$ (f) $2x^2 - 8x + 6 = 0$

3 Solve these quadratic equations. **4.1**

(a) $x^2 - 4x = 5$ (b) $x^2 - 6 = 5x$ (c) $x^2 + 12 = -7x$

(d) $3x^2 = 27$ (e) $(x - 3)(x + 7) = 11$ (f) $(x - 3)(x + 2) = 6$

4 Solve each of the following by completing the square. **4.3**

(a) $x^2 - 4x + 2 = 0$ (b) $x^2 + 8x + 9 = 0$ (c) $-x^2 - 2x + 4 = 0$

5 For each of the following: **4.4**

(i) state the changes that have been made to the graph of $y = x^2$

(ii) state the coordinates of the turning point

(iii) find the x- and y-intercepts

(iv) sketch a graph of each equation showing the turning point and intercepts.

(a) $y = x^2 - 4$ (b) $y = -2(x + 1)^2 - 4$ (c) $y = \frac{1}{3}(x - 7)^2$

6 For each of the following equations: **4.4**

(i) find the y-intercept (ii) find the x-intercepts

(iii) find the coordinates of the turning point (iv) sketch the graph.

(a) $y = x^2 - 4x - 12$ (b) $y = -(x + 1)^2 + 4$

(c) $y = 2(x - 3)^2 - 6$ (d) $y = -x^2 + 3x + 4$

7 (i) Factorise each of the following expressions. **4.1**

(ii) Equate each factorised expression to zero and then solve the equation for x using the null factor law.

(a) $x^2 - 9$ (b) $25 - x^2$ (c) $4x^2 - y^2$ (d) $49x^2 - 16$

(e) $36x^2 - 81y^2$ (f) $64 - 9x^2$ (g) $(2x - 5)^2 - 4$ (h) $(x-5)^2-(x+8)^2$

(i) $(x + 2)^2 - (x - 3)^2$ (j) $x^2 - 3$ (k) $(x + 7)^2 - 5$ (l) $(x - 4)^2 - 7$

8 Factorise these expressions by completing the square. **4.2**

(a) $x^2 + 6x - 2$ (b) $x^2 - 10x + 23$ (c) $x^2 - 4x - 7$ (d) $x^2 + 8x + 13$

10A 9 Factorise the following quadratic expressions. **4.5**

(a) $12x^2 + x - 20$ (b) $12x^2 - 4x - 5$ (c) $6x^2 + 17x - 14$

(d) $3x^2 + 5x + 1$ (e) $4x^2 - 7x - 5$ (f) $2x^2 + 6x + 3$

10A 10 Solve each quadratic equation using an appropriate method. **4.6**

(a) $2x^2 - 18 = 0$ (b) $2x^2 - 9x = 5$ (c) $(2x - 5)^2 = 3$

(d) $2x^2 + 5x + 1 = 0$ (e) $3x^2 - 6x + 2 = 0$ (f) $13x^2 + 7x - 3 = 0$

11 (a) Which of the following methods of factorising is the most appropriate for each of the expressions (i) to (vi) below? **4.2**

A factorising using common factors

B factorising using the difference of two squares rule

C factorising by grouping in pairs

D factorise by grouping (three and one)

E factorising quadratic trinomials (multiply to give …, add to give …)

F factorising by completing the square

(i) $x^2 + 5x - 24$ (ii) $25a^2 - b^2$ (iii) $x^2 + 6x + 9 - y^2$

(iv) $x^2 + 12x + 34$ (v) $16x^2 - 25x$ (vi) $4x - 4y + x^2 - y^2$

(b) Factorise each expression in part (a).

Understanding

12 A train ticket is 5 cm longer than it is wide. **4.1**

(a) Draw a diagram of the ticket, labelling the width x.

(b) Write the length of the ticket in terms of x.

(c) Write an expression in terms of x for the area of the train ticket.

(d) If each ticket requires 24 cm^2 of paper, write an equation that represents this area.

(e) Solve this equation to find x.

(f) Hence, state the dimensions of the train ticket.

(g) Did you use both of the values for x that you found using algebra to find the dimensions of the ticket? Explain your answer.

Reasoning

13 A fountain sends a jet of water on a parabolic path, which can be described by the equation $h = -2x^2 + 0.8x$, where h is the height of the water, in metres, at a horizontal distance x, in metres, from the water outlet. **4.4**

(a) Where is the jet of water again at the height that it left the outlet?

(b) If it hits the base of a pond at a horizontal distance of 0.8 m from the outlet, how far below the outlet does the water land?

(c) What is the maximum height above the base of the pond that the jet of water reaches?

(d) Sketch the graph of the path of the water.

14 A square photograph is attached to a coloured mounting and then framed. **4.1**

(a) Find the length of a side of the photo if the frame encloses a total area of 140 cm^2.

(b) Hence find the dimensions of the photo frame.

(c) Calculate the area of the photograph.

(d) Calculate the area of coloured mounting that can be seen.

(e) What area of coloured mounting would be seen if the same photograph was to have a 2 cm coloured mounting all around?

15 The sum of two numbers is 20. **4.4**

(a) The two numbers are x and y. Write an equation to represent the sum.

(b) Rearrange the equation so that y is the subject (i.e. in the form $y =$).

(c) Write an expression for the product of the two numbers using x.

(d) Sketch a graph of the equation for the product.

(e) What is the maximum product and which numbers give the maximum product?

16 The graph of $y = x^2$ is translated so that it passes through two points (3, 8) and (9, 8). **4.4**

(a) Write the x-coordinate of the turning point of the graph.

(b) Find the equation of the curve.

(c) Sketch this graph.

Mixed review A

1 Factorise the following expression fully. **3.2**

(a) $(3x - 7)(2x + 6) - 5(3x - 7)$ (b) $2x(3x - 2)(4x + 5) + 2x(3x - 2)$

2 Simplify the following. **3.7**

(a) $\frac{1}{3x} - \frac{2}{x}$ (b) $\frac{3x}{x+1} - \frac{2}{x-1}$ (c) $\frac{2x+1}{x} + \frac{3x-3}{3x}$

3 Solve each of the following equations. **1.1**

(a) $\frac{2x+3}{7} = 3$ (b) $\frac{4(3x-2)}{5} = 4$ (c) $2(4x + 5) = 26$

(d) $7(2x - 1) = 21$ (e) $6x + 2 = 7x$ (f) $8x - 9 = 7x - 4$

(g) $4(x + 5) = 2x - 4$ (h) $\frac{2(2x-4)}{3} = 4$

4 Calculate the gradient of the line that passes through each pair of coordinates. **1.2**

(a) (-3, 0) and (3, 9) (b) (-4, -1) and (-2, 5)

(c) (0, 6) and (4, -2) (d) (-3, 2) and (-1, 6)

5 Identify which of the following points are on each given line? **1.3**

(a) (-2, -1) or (2, 8) $y = 3x + 5$

(b) (1, -1), (-2, -3) or (-3, 5) $y = -2x + 1$

(c) (0, -2), (4, 1) or (-8, 0) $y = \frac{1}{4}x + 2$

6 Sketch each of the following lines using either the gradient and y-intercept method or the x- and y-intercepts method. **1.3**

(a) $y = -2x + 1$ (b) $y = 3x + 3$ (c) $y = \frac{1}{4}x - 4$ (d) $y + 6x = 9$

7 Determine whether each pair of lines is perpendicular. Write 'yes' or 'no' and justify your answer with appropriate calculations. **1.4**

(a) $2x + 4y = 8$ and $6x = 3y - 3$ (b) $2x - y = 3$ and $x + 2y = 8$

(c) $5x + 2y = 8$ and $5y = 2x + 10$ (d) $3y = 4x - 2$ and $4y = 3x - 4$

8 Determine whether each pair of lines is parallel. Write 'yes' or 'no' and justify your answer with appropriate calculations. **1.4**

(a) $6x - 3y = 9$ and $3x = 3y - 3$ (b) $x - y = 8$ and $x + y - 2 = 0$

(c) $2x + 5y = 7$ and $5y = 2x - 8$ (d) $10y = 3x + 2$ and $6x - 20y + 4 = 0$

9 Solve the following inequalities. **1.5**

(a) $2x - 1 \leq 4x + 2$ (b) $\frac{4x-1}{3} > \frac{3x+5}{2}$

10 Solve the following simultaneous equations. **1.6**

(a) $3x - 5y = 7$
$x + 5y = -11$

(b) $2x - 3y = 17$
$y = 3x - 22$

11 Find the range and the interquartile range for each of the following grouped data frequency tables. Give your answers correct to 2 decimal places. **2.1**

(a)

Class interval	Frequency
20–<25	8
25–<30	6
30–<35	3
35–<40	2
40–<45	5
45–<50	5

(b)

Class interval	Frequency
110–<115	12
115–<120	15
120–<125	17
125–<130	19
130–<135	13
135–<140	11

12 Solve each of the following equations: **4.1**

(a) $x^2 + 13x + 42 = 0$ (b) $x^2 - 100 = 0$ (c) $x^2 - 14x + 49 = 0$

13 Expand and simplify the following. **3.1**

(a) $(x - y)(3x - 2y)$ (b) $(3a - 6b)(2a - 5b)$ (c) $(5t + 4h)(3t - 9h)$

14 Fully factorise the following expressions. **3.6**

(a) $x^2 - 4x + 4$ (b) $x^2 - 9$ (c) $x^2 + 14x + 49$

(d) $100x^2 - 25$ (e) $4xy^2 + 2xy - 8x$ (f) $4x^2 + 32x + 64$

15 Simplify the following. **3.7**

(a) $\frac{x}{5} - \frac{3x}{7}$ (b) $\frac{x+3}{5} + \frac{x+5}{6}$

16 Expand and simplify. **3.1**

(a) $6(x + 11) - x(x + 11)$ (b) $4(x + 6) - x(x + 6)$

(c) $2x(x + 4) - 3(x + 4)$ (d) $7y(y - 1) + 2(y - 1)$

17 Factorise each expression. **3.5**

(a) $x^2 - 6x - 27$ (b) $x^2 - 7x + 10$ (c) $6x^2 - x - 2$

18 Expand and simplify: **3.1**

(a) $(5x - 1)^2$ (b) $(4x - 3)^2$ (c) $(8 - 3x)^2$ (d) $(6 - 5x)^2$

Understanding

19 A square piece of carpet of side length x is extended by 5 m in length and by 3 m in width. **3.1, 3.2**

(a) Write an algebraic expression for the area of the extended carpet, in factorised form.

(b) Expand your answer to **(a)** and simplify by collecting like terms.

(c) If $x = 4$, find the area of:

(i) the original (ii) the extended piece of carpet.

(d) If the original piece of carpet cost $128, suggest a comparable price for the extended piece of carpet.

Challenge A

1 Calculate $1000 - 1 + 3 - 5 + 7 \ldots$ up to 102 terms (numbers in the sequence).

2 Damo has spent \$733 on four different kinds of gifts for his 20 friends and relatives, priced at \$18, \$23, \$52 and \$69. How many of each gift did he buy?

> **Hint**
>
> Remember that integers are positive or negative whole numbers or zero.

3 A discount store has a sale on T-shirts. For every two T-shirts you bought at the regular price, you could purchase a third T-shirt for \$1. If 12 T-shirts were bought for \$120, then the regular price is:

A \$15.00 **B** \$14.50 **C** \$14.00 **D** \$10.00

4 How many integer values of x satisfy $\frac{x-1}{3} < \frac{5}{7} < \frac{x+4}{5}$?

A 1 **B** 2 **C** 3 **D** 4

5 p, q and r are three different positive integers, and $(pq)^2 = pqr$.

Which of the following is a possible value for r?

A 1 **B** 3 **C** 4 **D** 6

6 Find the area of the shaded region in the diagram.

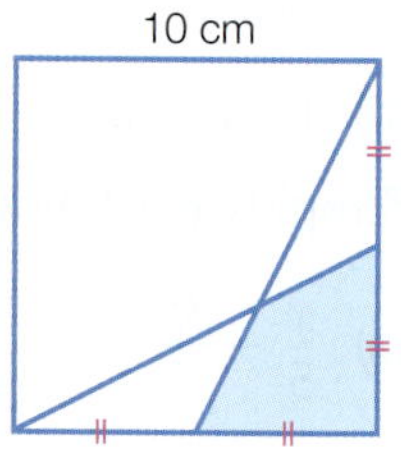

7 A square fits tightly in an equilateral triangle with its base on the triangle's base. What percentage of the area of the triangle does it occupy?

8 Find the area of the shaded trapezium shown if the regular hexagon has a side length of 40 cm.

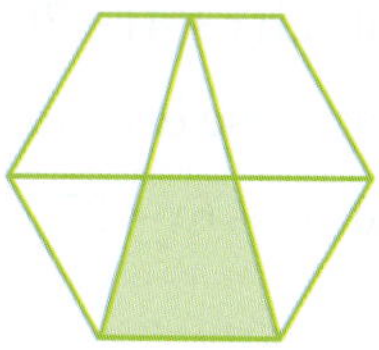

9 In the diagram, AB and CD are diameters of a circle of radius 4 cm.

If AB is perpendicular to CD, the area of the shaded region is:

A $8\pi + 4$ **B** $8\pi + 8$

C $16 + 4\pi$ **D** $16 + 8\pi$

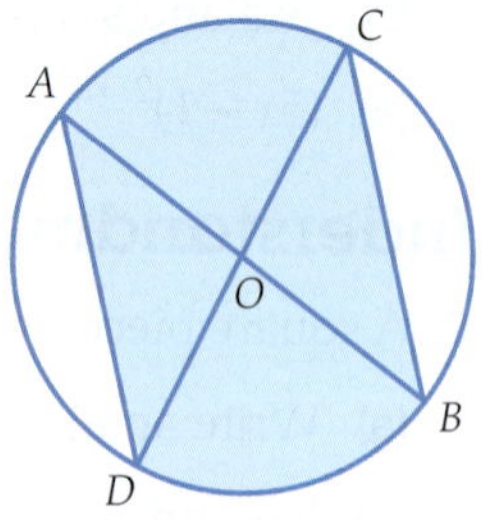

10 What is the side length of a square that just fits symmetrically within an isosceles triangle of sides 5 cm, 10 cm and 10 cm?

11 A stone pillar is 6 m high. Its cross-section is based on an equilateral triangle with side length 30 cm that has three circular sectors drawn with their centres as the vertices of the triangle. Find the pillar's volume in m^3 correct to 3 decimal places.

Exploration STEM

You can download this activity from the eBook or the Pearson Places website.

Which wheat should we eat?

Climate change impacts the world's environments. Clean water and fertilisers are both in limited supply. You live in a rural town that has been given four untested varieties of wheat to grow for food. The mayor decides to consult the town's scientists to decide which of the four wheat varieties to grow. How can this be decided? How can the scientists inform and convince the mayor?

Exploration STEM

You can download this activity from the eBook or the Pearson Places website.

Population and natural resources

Your task is to investigate the effects of future population growth on the natural resources of a country. By varying the future life expectancy and birth rates of the country, how will its future population change? Will this create problems for the country's resources, economy, industries and so on? You will need to investigate the country's size and existing resources, and the effects of population change. What strategies can you think of to help solve the potential problems?

Exploration Coding

You can access this activity from the eBook or the Pearson Places website.

The point of transformations

Develop a deeper understanding of transformations in the Cartesian plane by applying algorithmic thinking to the transformations of sets of points.

Exploration Coding

10A

You can access this activity from the eBook or the Pearson Places website.

Food aid drops

Aid workers need to carefully plan parachute drops of food into remote areas to ensure everything falls within the required location. Explore an algorithm that can decide where these food aid drops should be released.

Measurement

5

Maths on the big screen

Name: Stuart Williams
Job: Audio Visual Systems Designer
Qualifications: Bachelor of Electronic Engineering

If you designed your own home theatre system, could you connect it all together to make sure everything works? Stuart Williams can. Stuart works for Insight Systems, a company that designs and installs audio visual systems into homes, schools, universities, offices and even courtrooms. Stuart works with projectors, speakers, screens, interactive whiteboards and videoconferencing systems, which are usually linked to control panels by a complex system of cables.

'It's like a big jigsaw puzzle. I enjoy the challenge of putting it all together. I need to think logically and include every little detail, so that I can produce a useful schematic for the technicians who are installing the system.' (A schematic is an accurate scale diagram that shows how everything is connected together.)

Stuart uses a variety of mathematical skills in his daily work such as scale drawings and ratios. He applies Pythagoras' theorem to calculate the height and width of a TV or projection screen, and he uses formulas to calculate the current flowing through an amplifier.

Why learn this?

Not all shapes are rectangles, and not all solids are rectangular boxes. You need to be able to calculate the surface area and volume of pyramids, cones, spheres and complex solids made up of combinations of these. These shapes are commonly used in the design and manufacture of everyday objects, such as car components, toys, furniture and buildings. Volume and capacity are also used to describe properties of solids and liquids such as density and concentration.

After completing this chapter you will be able to:

- use area formulas of basic shapes to find the areas of composite shapes
- calculate the surface area, volume and capacity of prisms and cylinders
- rearrange formulas to make a different variable the subject

10A

- calculate the surface area, volume and capacity of pyramids, cones and spheres
- calculate the density of solids and liquids
- find the concentration of dissolved substances.

Recall 5

Prepare for this chapter by attempting the following questions. If you have difficulty with a question, you can download a Recall Worksheet from the eBook or the Pearson Places website.

1 Use your calculator to evaluate the following.

(a) 25^2 **(b)** 3.1^3 **(c)** $\sqrt{11.9025}$

2 Given the formula $A = \frac{1}{2}(a + b)h$:

(a) find A if $h = 4$, $a = 3$ and $b = 5$ **(b)** find h if $A = 30$, $a = 7$ and $b = 5$

(c) find b if $A = 48$, $h = 6$ and $a = 9$.

3 Complete the following unit conversions.

(a) 4.1 m = ______ cm **(b)** 4.25 km = ______ cm **(c)** 40 m^2 = ______ cm^2

(d) 578 cm^2 = ______ m^2 **(e)** 8.25 km^2 = ______ m^2 **(f)** 600 mL = ______ L

(g) 20 cm^3 = ______ mL **(h)** 30.8 m^3 = ______ L **(i)** 7250 cm^3 = ______ m^3

4 Find each of the following values in the sector shown, rounded to 2 decimal places (where necessary).

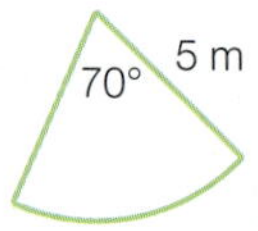

(a) the arc length **(b)** the perimeter **(c)** the area

5 **(i)** Use Pythagoras' theorem to find the perpendicular height of each of the following shapes, rounded to 2 decimal places.

(ii) Use the height to calculate the area of each shape.

(a)

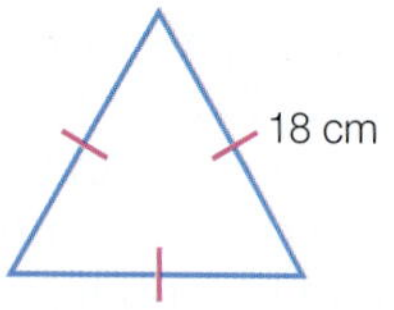

(b)

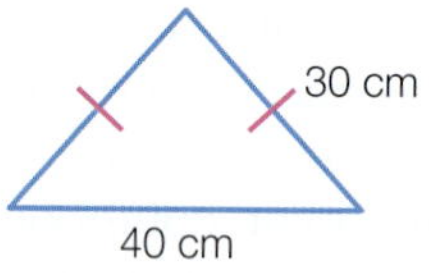

(c)

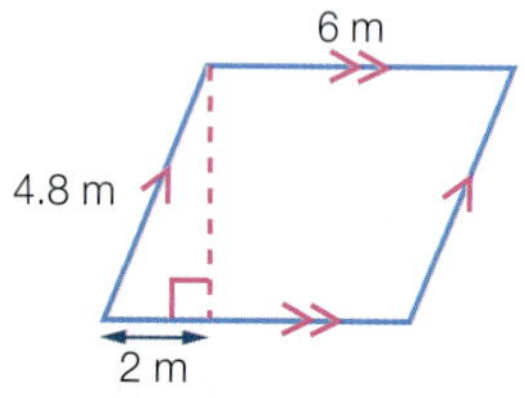

(d)

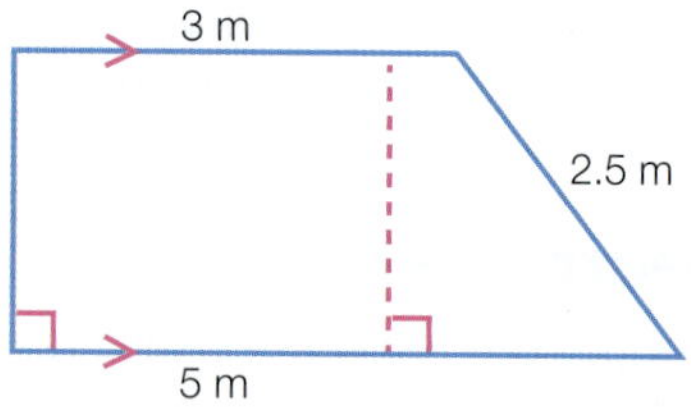

Exploration Task

You can download this activity from the eBook or the Pearson Places website.

Many paths lead to the same reward

In this activity you will explore different methods for calculating the areas of a set of composite shapes.

Area

5.1

Use these formulas to find the areas of shapes. To find the area of composite shapes, you add or subtract the areas of simpler shapes.

Rectangle	Circle	Sector	Triangle
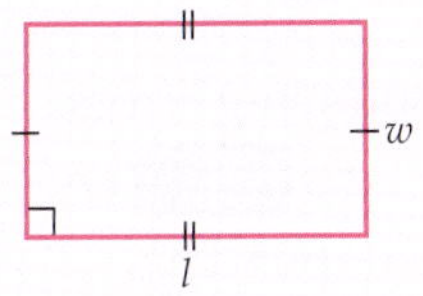	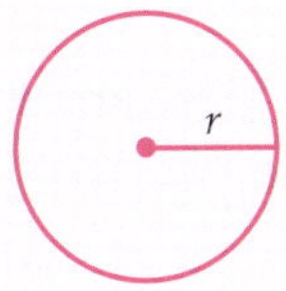	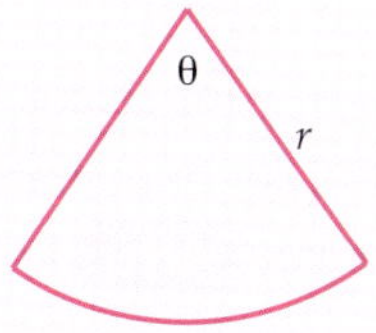	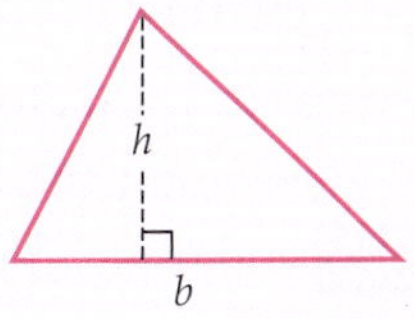
$A = lw$	$A = \pi r^2$	$A = \frac{\theta}{360} \times \pi r^2$	$A = \frac{1}{2}bh$

Trapezium	Parallelogram	Kite	Rhombus
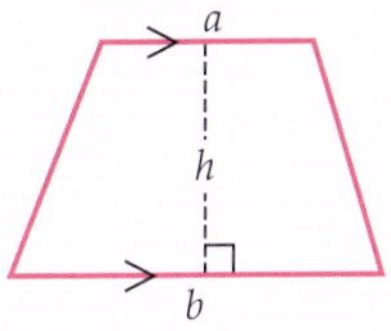	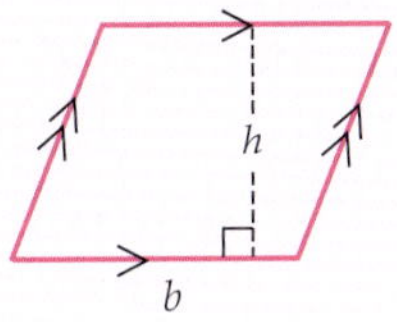	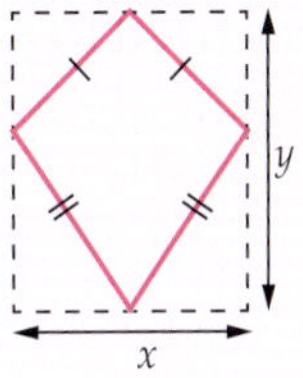	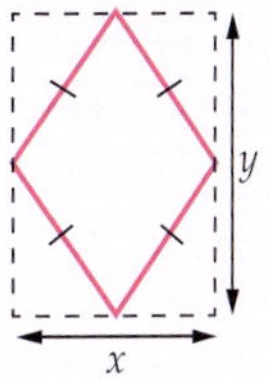
$A = \frac{1}{2}(a+b)h$	$A = bh$	$A = \frac{1}{2}xy$	$A = \frac{1}{2}xy$

Worked example 1

W.E. 1

Find the area of this shape rounded to 2 decimal places.

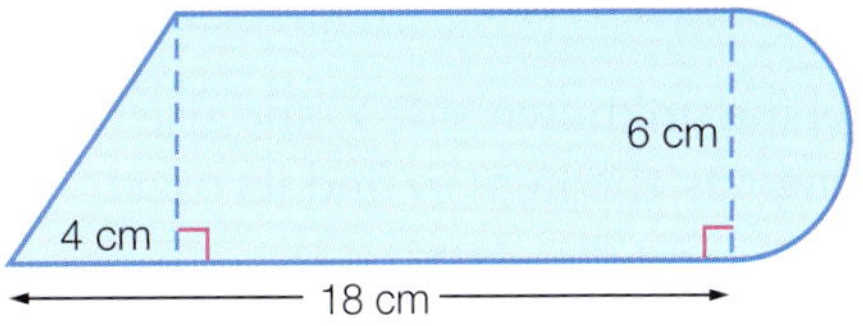

Thinking	Working
1 Divide the composite shape into smaller shapes with areas that can be calculated.	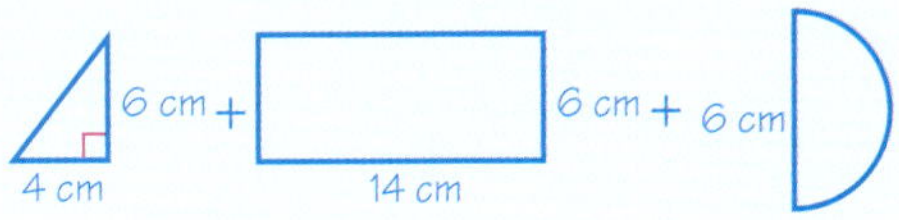
2 Write a sum with the appropriate formula for the area of each part.	$A = \frac{1}{2}bh + lw + \frac{\pi r^2}{2}$
3 Use the information given to determine any missing measurements.	$l = 18 - 4 = 14$ cm; $r = 6 \div 2 = 3$ cm
4 Substitute the measurements into each formula. Simplify the calculation where possible before using a calculator.	$A = \frac{1}{2} \times 4 \times 6 + 14 \times 6 + \frac{\pi \times 3^2}{2}$ $= 96 + \frac{9\pi}{2}$
5 Evaluate, rounding your answer to the specified number of decimal places, and include the correct unit.	$= 110.14$ cm^2 (2 d.p.)

Rounding and accuracy

The calculation in the previous example was simplified before a calculator was used. (Simplifying involves cancelling common factors and/or performing simple whole number multiplications.) The solution is only rounded once, at the very end. This gives the most accurate final answer.

If you use rounded answers as you work through a problem, then you accumulate, or build up, a small amount of error with each rounded calculation. This results in an inaccurate final answer.

To use your calculator efficiently for calculations like these and to avoid rounding errors, you must, apply the correct order of operations, making use of the bracket keys **(** and **)** and using the answer key **ANS** to recall previous answers without rounding.

Worked example 2

W.E. 2

Find the shaded area in the diagram, rounding your answer to 2 decimal places. (You may need to use Pythagoras' theorem to calculate the lengths of some unknown sides.)

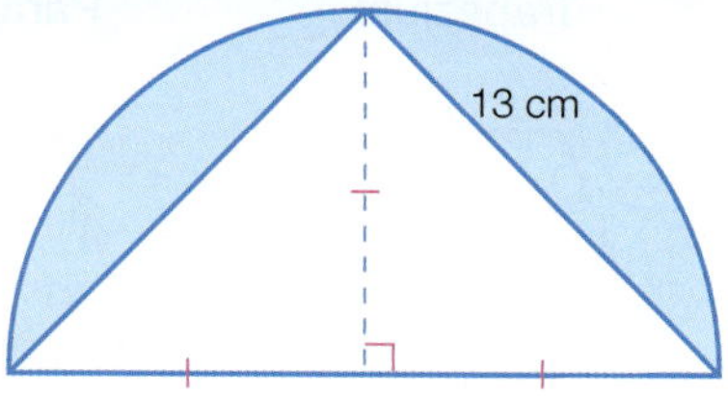

Thinking | **Working**

1 Redraw the triangle that includes the unknown side.

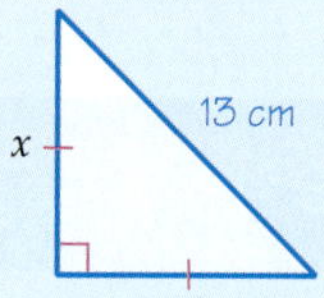

2 Label the unknown side x, and use Pythagoras' theorem to find its exact value.

$$x^2 + x^2 = 13^2$$
$$2x^2 = 13^2$$
$$x^2 = \frac{13^2}{2}$$
$$x = \sqrt{\frac{13^2}{2}}$$
$$x = \sqrt{84.5} \text{ cm}$$

3 Find the shaded area by subtracting the area of the unshaded part from the overall shape.

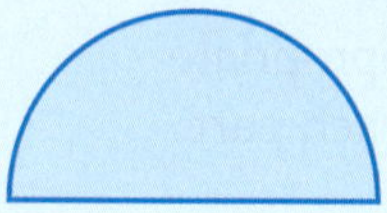 − 

4 Write the required area as the difference of two areas, using appropriate formulas for each part.

$$A = \frac{1}{2}\pi r^2 - \frac{1}{2}bh$$
$$b = 2h$$
$$A = \frac{1}{2}\pi r^2 - \frac{1}{\cancel{2}_1} \times {}^1\cancel{2}h \times h$$
$$= \frac{1}{2}\pi r^2 - h^2$$

5 Substitute the values into the formulas. (Here, r and h are equal to x. The value of x was determined in step **2**.)	$=\frac{1}{2}\times\pi\times(\sqrt{84.5})^2-(\sqrt{84.5})^2$ $=\frac{84.5\pi}{2}-84.5$ $=84.5\left(\frac{\pi}{2}-1\right)$
6 Evaluate, rounding your answer to the specified number of decimal places, and include the correct unit of area.	$=48.23\text{ cm}^2$ (2 d.p.)

Area of an annulus

An annulus is a ring shape created by two concentric circles (circles that have the same centre). To find the area, subtract the area of the smaller circle from the area of the larger circle.

Area of an annulus

$A=\pi R^2-\pi r^2$

$A=\pi(R^2-r^2)$ (factorised form)

$A=\pi(R+r)(R-r)$ (factorising using the difference of two squares rule—a useful form to work with)

You can use capital and lower-case versions of a letter to indicate the same dimension in two figures with the same shape. Here, R is the radius of the larger circle and r is the radius of the smaller circle.

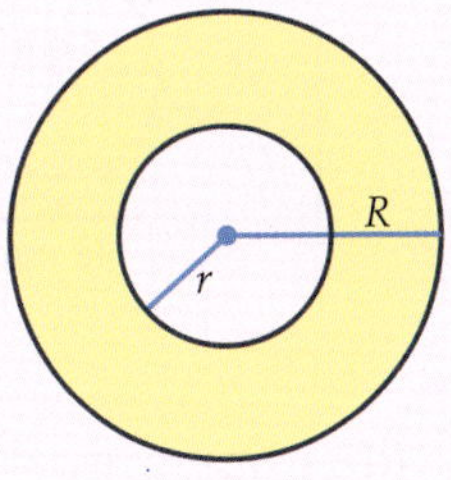

5.1 Area

Navigator

1, 2 (columns 1, 2), 3, 4, 5 (column 1), 6, 8, 9, 10, 13, 14, 15	1 (column 1), 2 (a–f), 3, 4, 5 (column 2), 6, 7, 8, 10, 11, 12, 13, 14, 15, 16	1 (column 2), 2 (columns 2–3), 3, 4, 5 (d–h), 6, 7, 8, 11, 12, 13, 14, 16

Answers p. 806

Fluency

1 Find the area of each of the following shapes. Round your answer to 2 decimal places where necessary.

W.E. 1

(a)

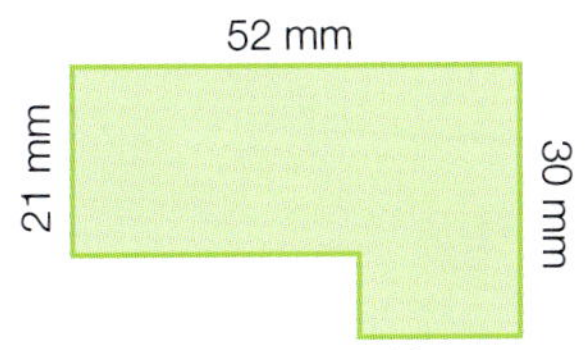

(b)

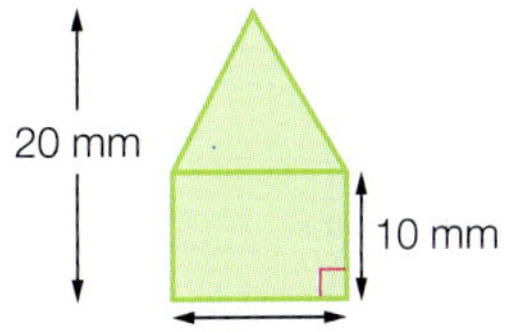

(c)

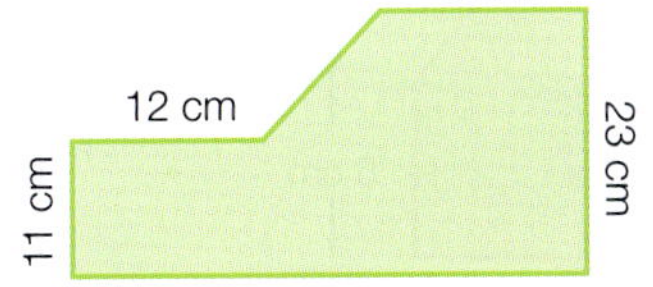

(d)

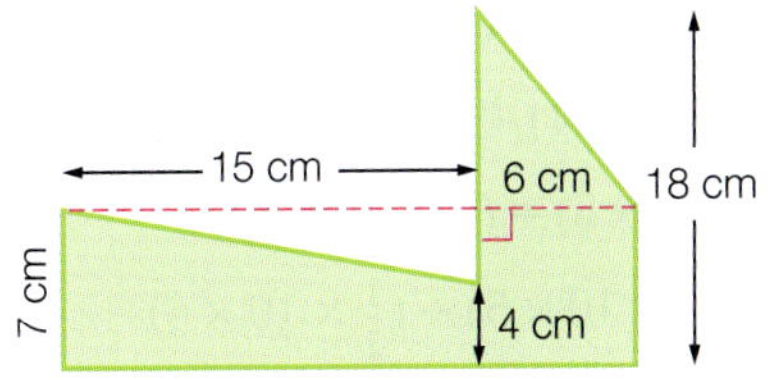

(e)

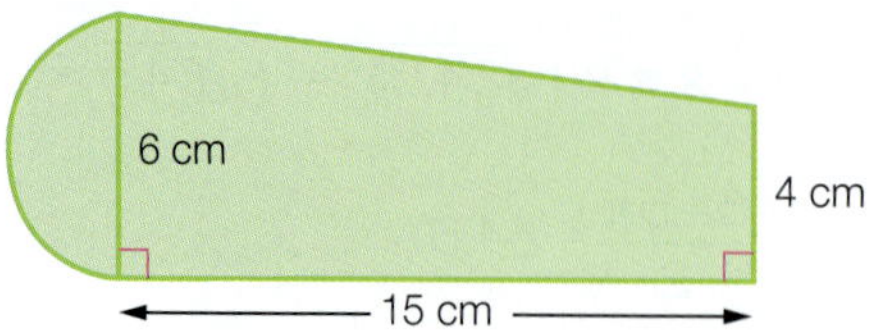

(f)

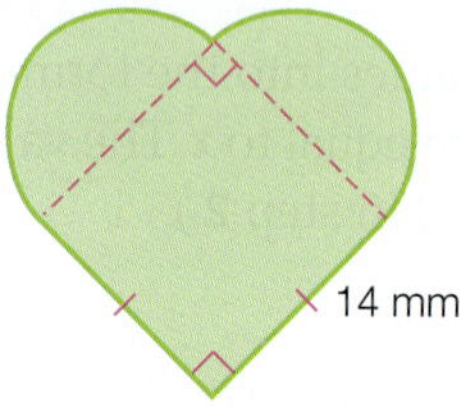

(g)

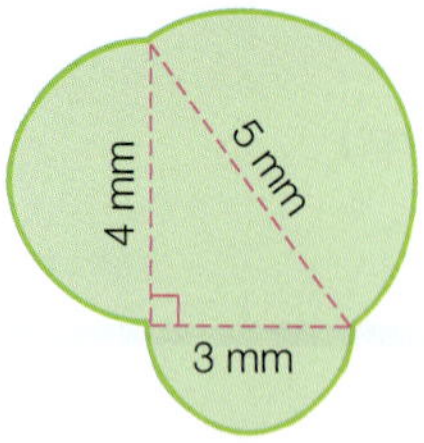

(h)

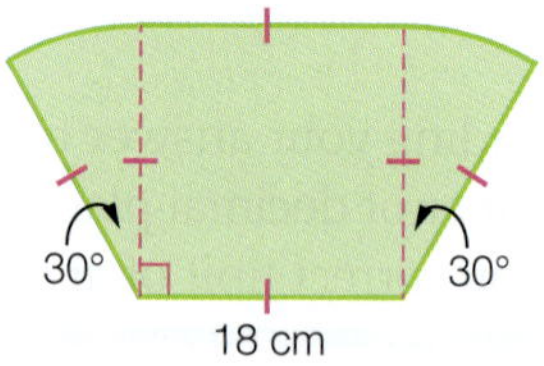

W.E. 2

2 Find the shaded area in each of the following, rounding your answer to 2 decimal places. (You may need to use Pythagoras' theorem to calculate the lengths of some unknown sides.)

(a)

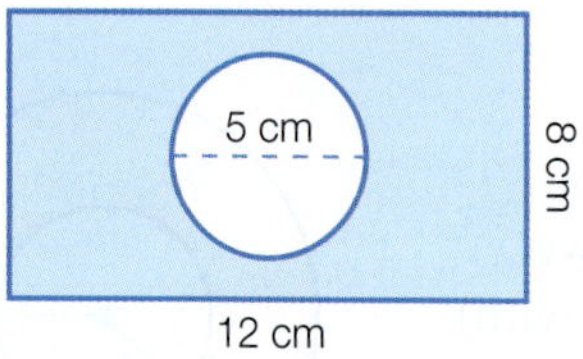

(b)

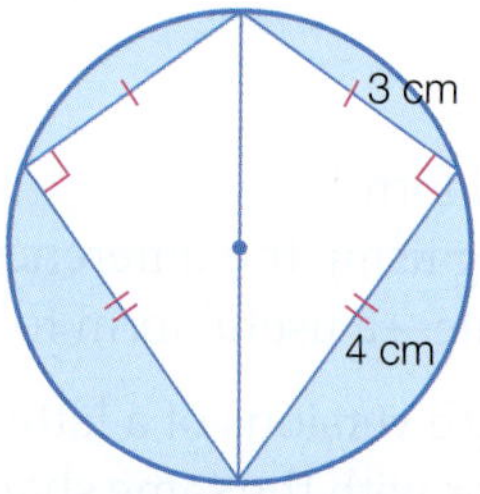

(c)

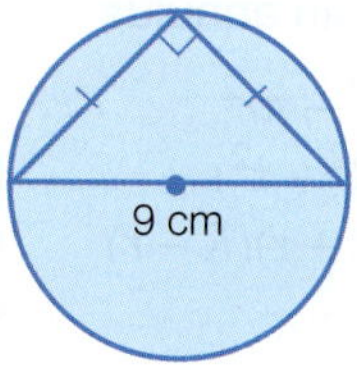

(d)

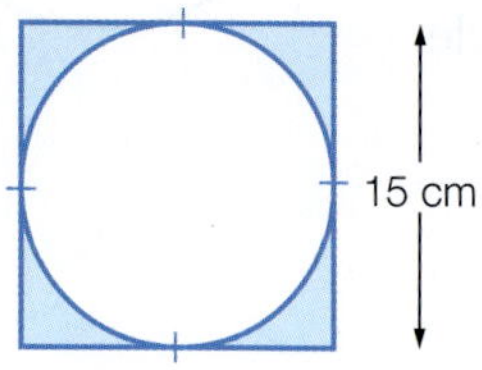

(e)

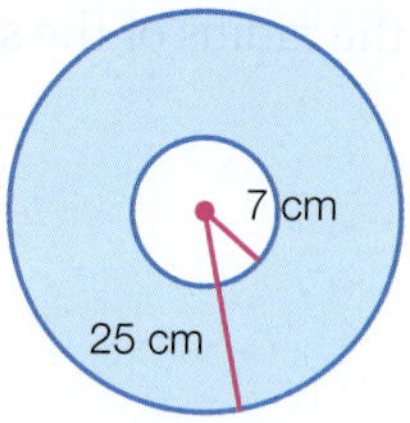

(f)

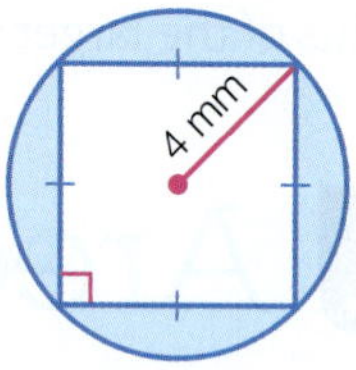

(g)

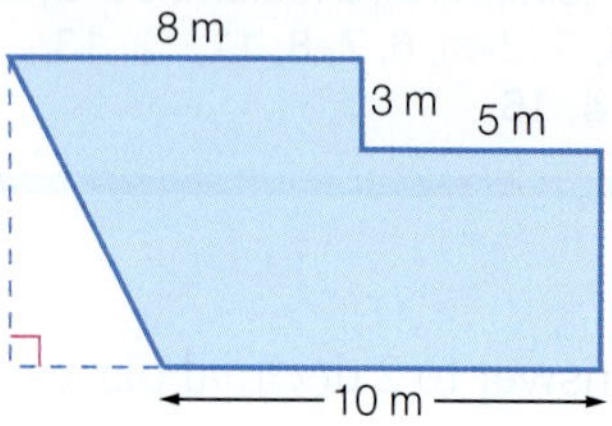

(h)

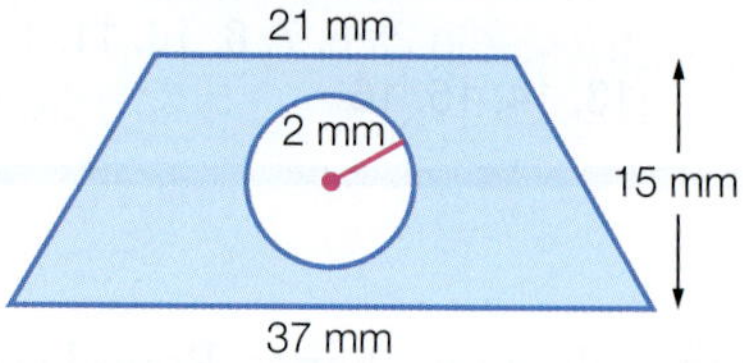

(i)

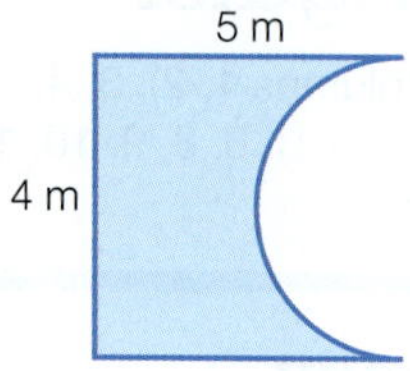

(j)

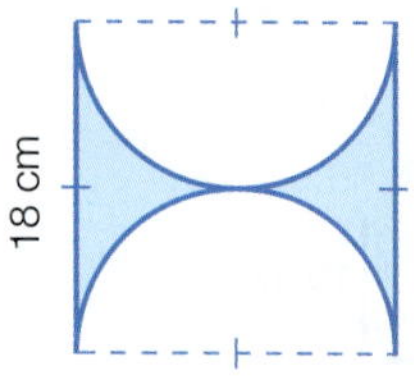

3 The area of this composite shape in cm^2 is:

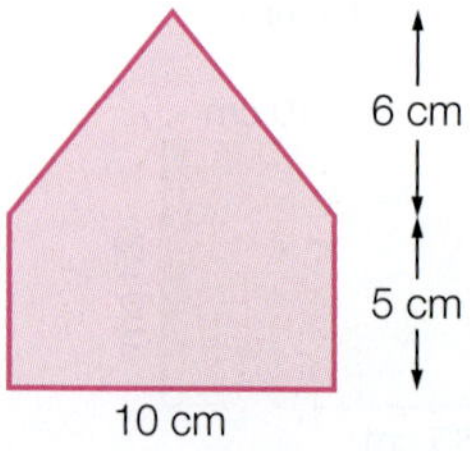

A $(10 \times 5) + (10 \times 6)$

B $(10 \times 11) + (10 \times 6)$

C $(10 \times 5) + (\frac{1}{2} \times 10 \times 6)$

D $(\frac{1}{2} \times 10 \times 5) + (\frac{1}{2} \times 10 \times 6)$

4 The area of this composite shape in m^2 is:

A $(6 \times 4) + (\frac{1}{2} \times \pi \times 6^2)$

B $(6 \times 4) - (\frac{1}{2} \times \pi \times 6^2)$

C $(6 \times 4) - (\frac{1}{2} \times \pi \times 3^2)$

D $(6 \times 4) - (\frac{1}{2} \times \pi \times 32)$

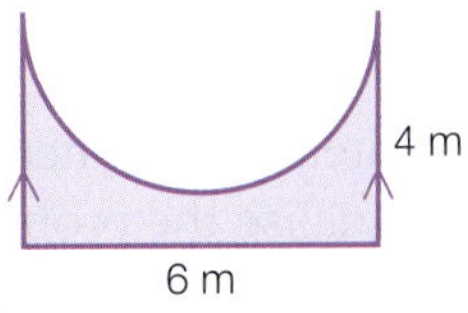

Understanding

5 Find the shaded area of each of the following, rounded to 2 decimal places.

(a)

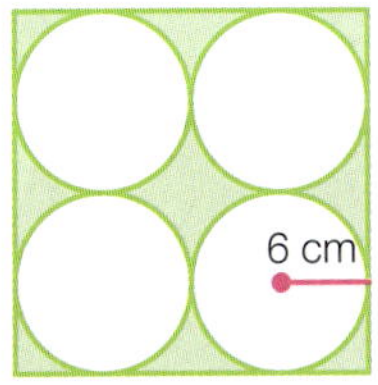

(b)

(c)

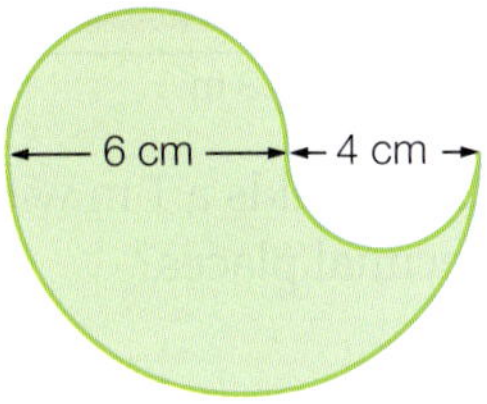

(d)

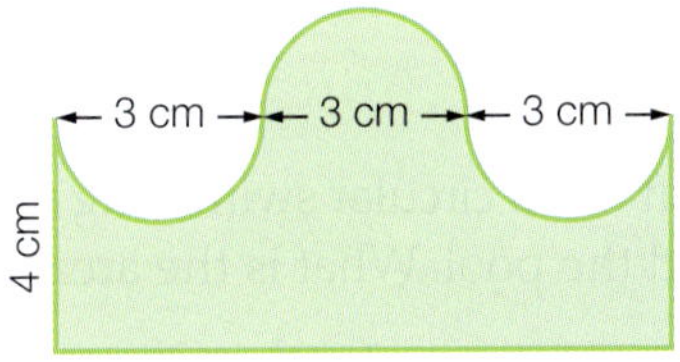

(e)

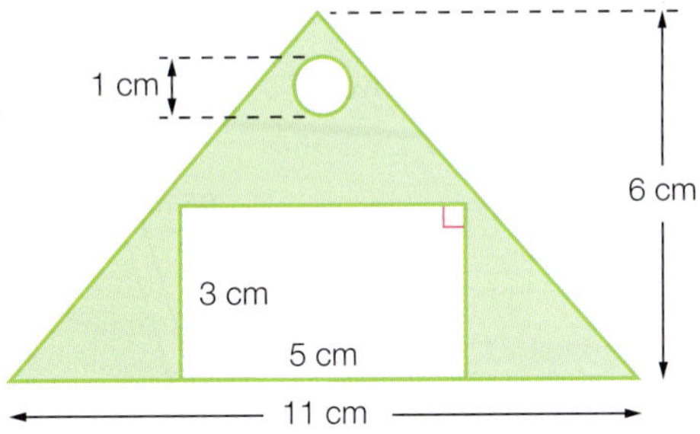

(f)

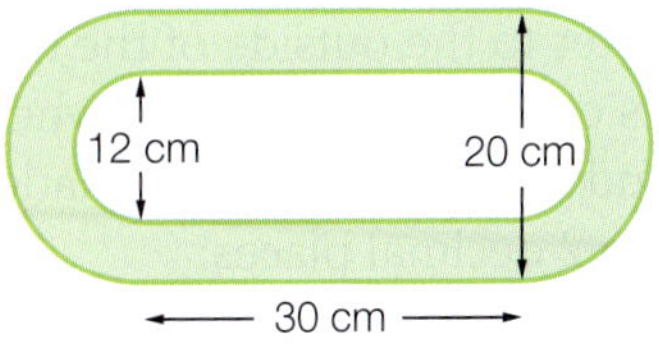

(g)

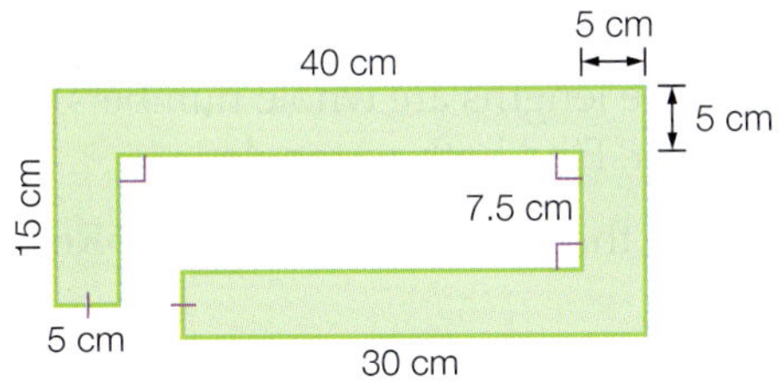

(h)

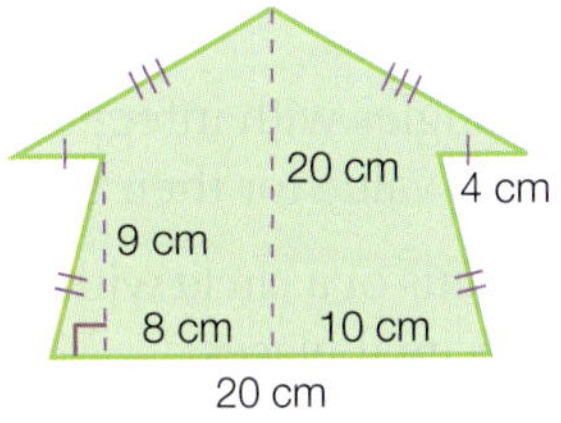

6 Find the area of both the sector and the triangle. (You will need to use Pythagoras' theorem to calculate the height of the triangle.) Write both answers correct to the nearest m^2.

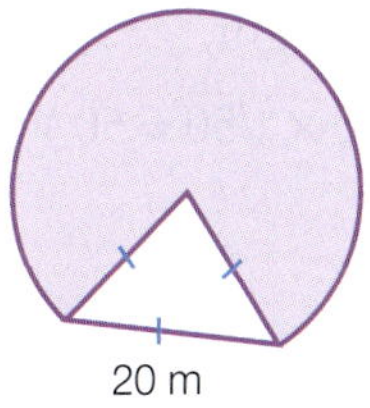

7 The Department of Roads has a machine that lays bitumen at a rate of 20 metres in length every 15 minutes. The bitumen is 6 m wide and the machine works 7 hours per day. The department is building a road 12.5 km long.

(a) How long will it take to complete the road? State your answer in both hours and days, correct to 2 decimal places.

(b) What will be the total area covered by bitumen?

Hint

Drawing a diagram can help you visualise the problem.

8 A gardener builds a 0.7 m wide concrete path around a vegetable garden. An outline of the vegetable garden bed is shown at right.

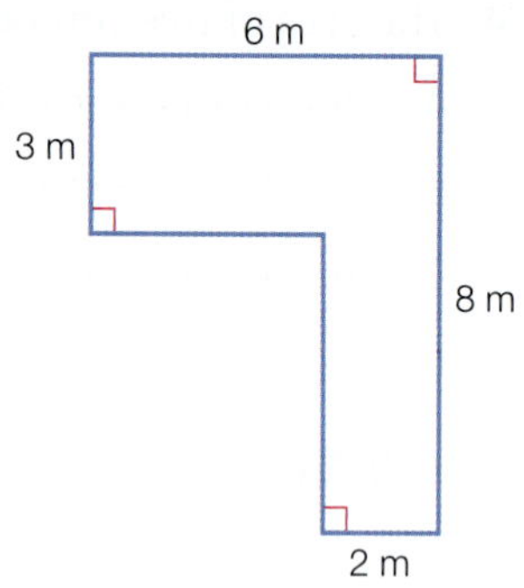

(a) Find the total area of the concrete path. (Draw a diagram to help you visualise the problem.)

(b) If one bag of concrete is required for every 2 m^2 of path, how many bags need to be purchased?

9 A table top is a trapezium, as shown at right. The surface of each table needs to be laminated. If there are 13 tables in the classroom, what area (m^2) of laminate is required? Round your answer to 2 decimal places.

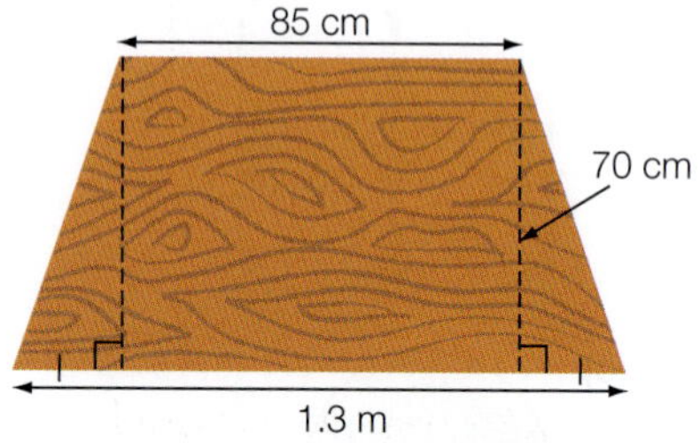

10 Elspeth has a circular swimming pool that is 7 m in diameter. She builds a 1 m wide path around the pool. What is the area of this path, rounded to 2 decimal places?

11 The throwing area for the Olympic shot put event is shown in the diagram. The shot is thrown from a circle that is 2.5 m in diameter to the outside of the rim. The rim is 6 mm wide (athletes cannot touch the rim when throwing). A legal throw must land in the zone shown shaded in green. Calculate, correct to 2 decimal places:

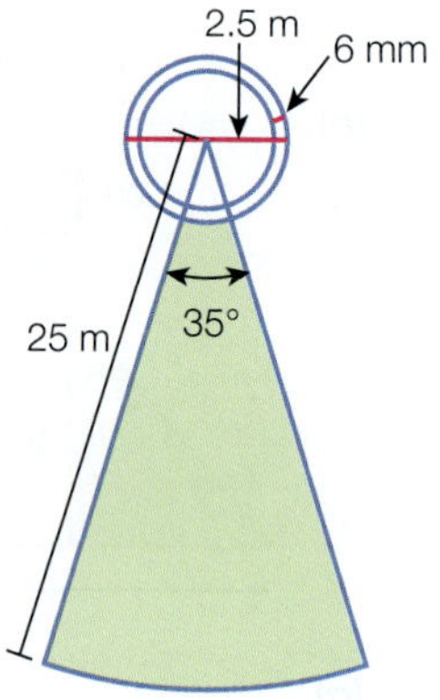

(a) the area of the rim of the circle

(b) the area of the landing zone.

Reasoning

12 Only two rectangles with integer dimensions (side lengths are whole numbers) have the same numerical value for their area and perimeter. Find both rectangles.

13 What is the radius of a circle with an area equal to that of a square 8 m wide? Round your answer to 2 decimal places.

14 Find the area of each piece of the tangram puzzle shown at right.

(Remember $\sqrt{50} \times \sqrt{50} = 50$.)

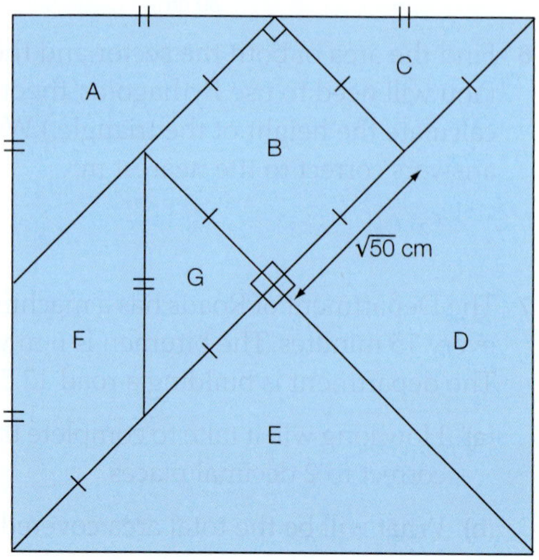

Open-ended

15 **(a)** To label his school folder, Joe cuts out the letters of his name from a 9 cm by 4 cm piece of coloured paper as shown below. What area of coloured paper do the letters cover?

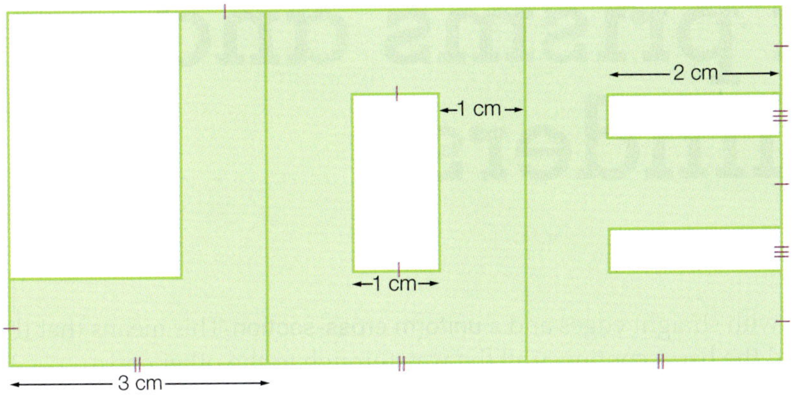

(b) Use the 9 cm by 4 cm rectangle in part **(a)** for another set of three letters. Calculate the area of paper covered by these letters. Is the area covered greater than, less than, or about the same as that covered in part **(a)**?

16 **(a)** Without doing any calculations explain why the shaded area in both shapes is equal.

(b) Draw another pair of shapes with a similar relationship to the pair shown.

(c) Draw as many such pairs of shapes as you can.

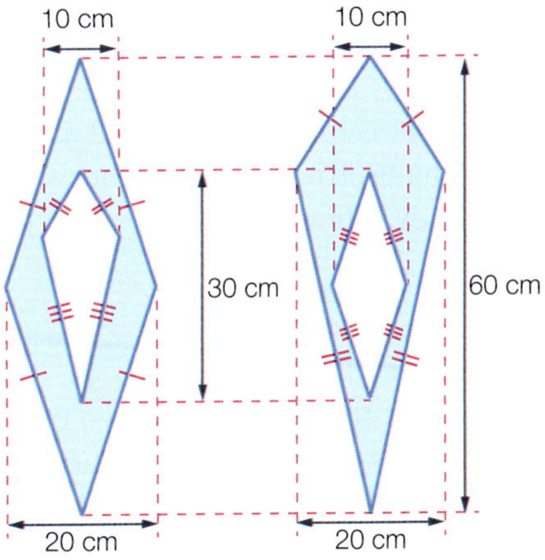

5.2 Surface areas of prisms and cylinders

Prisms

A **prism** is a solid with straight edges and a **uniform cross-section**. This means that the shape on one end (called the base) continues all the way through to the other end.

The **surface area** (SA) of a prism is the total area of all its external surfaces.

Some prisms have rectangular surfaces, such as the box shown.

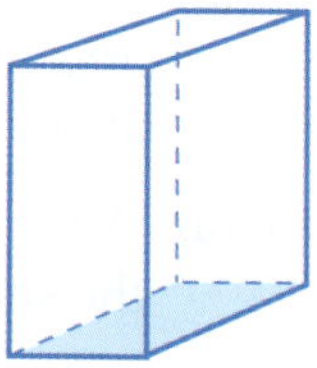

Other prisms have surfaces that are combinations of rectangles, triangles and other shapes, such as those shown here.

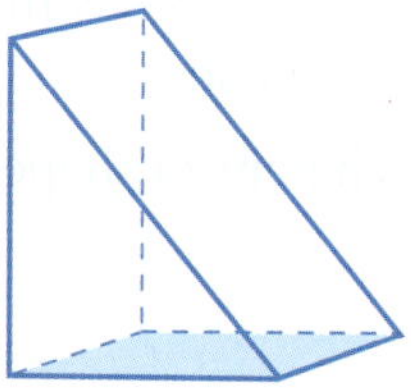

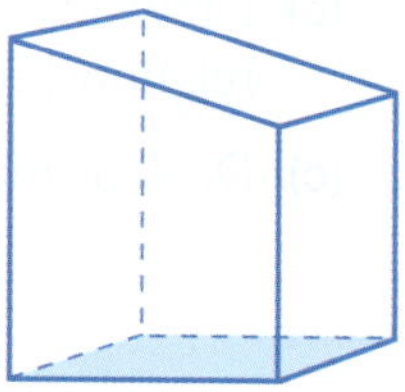

To calculate the surface area of a prism, it is often useful to draw a **net**—a 2D (flat) diagram that clearly shows each of the faces of the prism. Imagine a prism as a box, and then create the net by cutting along the edges and laying it out flat.

Worked example 3

W.E. 3

Find the surface area of the triangular prism.

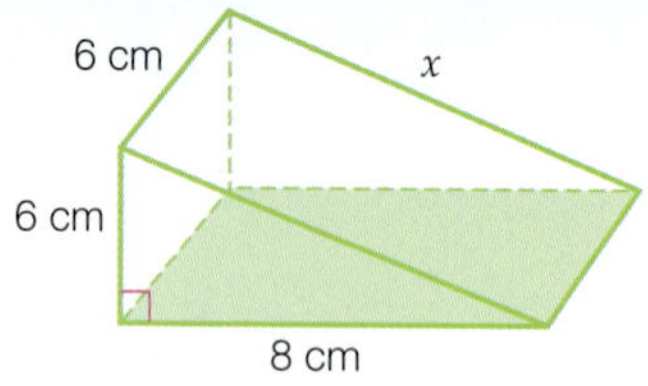

Thinking	Working
1 Draw a net of the solid that shows each of the individual faces that make up the surface. Draw each face as a flat shape. Label each of the different areas A_1, A_2, A_3 … etc. Label identical faces in the same way.	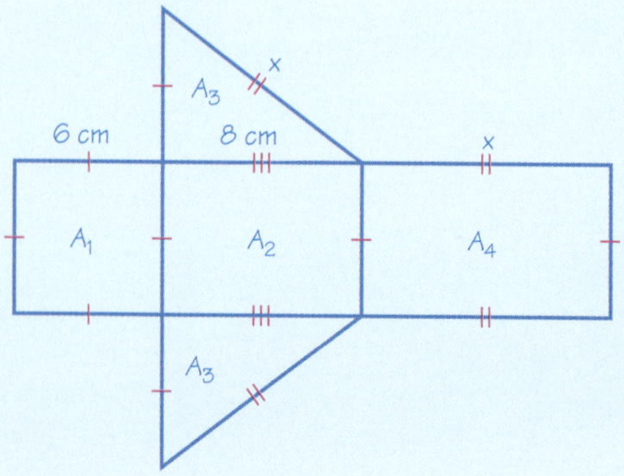

2 Use Pythagoras' theorem to find the required length.	$x^2 = 6^2 + 8^2$ $= 100$ $x = 10$ cm
3 Write the surface area as the sum of the areas of each face. Replace the labels (A_1, A_2 etc.) with the area formulas for each shape.	$SA = A_1 + A_2 + 2 \times A_3 + A_4$ $= l^2 + lw + {}^1\cancel{2} \times \frac{bh}{\cancel{2}_1} + LW$ $= l^2 + lw + bh + LW$
4 Substitute values and evaluate individual areas. Check that values are reasonable.	$= 6^2 + 8 \times 6 + 8 \times 6 + 10 \times 6$ $= 36 + 48 + 48 + 60$
5 Write your answer with the correct units of area.	$= 192$ cm^2

When you have two of the same type of shape, you can use upper and lower case letters to distinguish between their dimensions. In the previous example, the area of the smaller rectangle was written as lw, and the area of the larger rectangle as LW.

Cylinders

A **cylinder** also has a uniform cross-section, but it is not strictly a prism. The surface areas of cylinders can also be found by considering their nets.

The surface area of a cylinder is the sum of the areas of the two circular faces at each end and the area of the curved surface. The area of each circular face is $A = \pi r^2$.

When flattened, the curved surface is a rectangle with its length equal to the circumference of the circles on the ends: $C = 2\pi r$. Its width is equal to the height of the cylinder, h. The area of the curved surface is $A = 2\pi r \times h$.

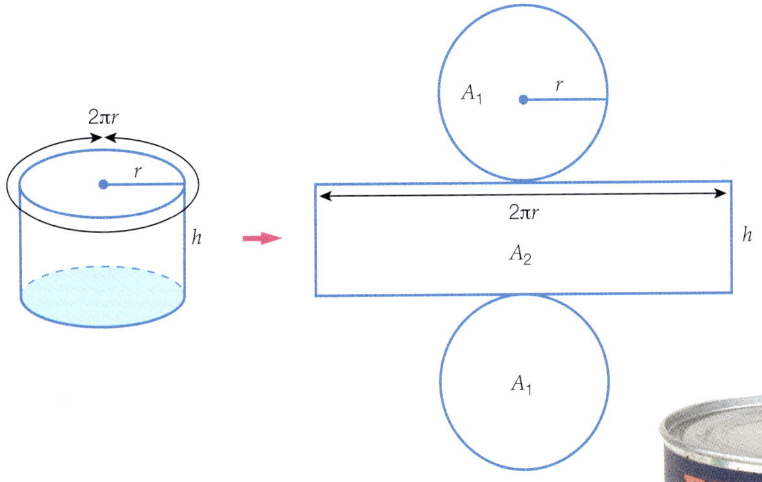

$$\begin{aligned} SA &= 2 \times A_1 + A_2 \\ &= 2 \times \pi r^2 + 2\pi r \times h \\ &= 2\pi r^2 + 2\pi r h \\ &= 2\pi r(r + h) \quad \text{(factorised form)} \end{aligned}$$

Worked example 4

W.E. 4

Find the surface area of this cylinder, rounded to 2 decimal places.

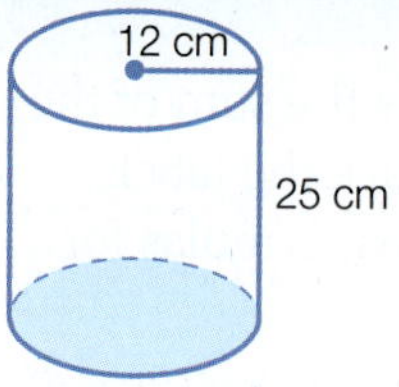

Thinking	Working
1 The surface area of a cylinder is the combined area of the two circular ends and the curved middle section. Write the formula that shows this.	$SA = 2\pi r^2 + 2\pi rh$
2 State values for r and h.	$r = 12$ cm $h = 25$ cm
3 Substitute r and h into the formula and simplify the calculation.	$SA = 2 \times \pi \times 12^2 + 2 \times \pi \times 12 \times 25$ $= 288\pi + 600\pi$ $= 888\pi$
4 Evaluate, rounding the answer to 2 decimal places.	$= 2789.73$ cm^2 (2 d.p.)

Cube

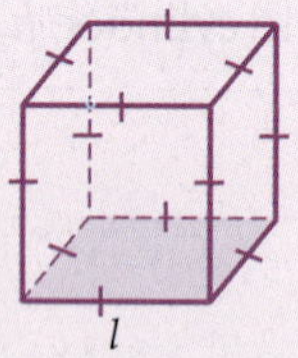

SA = area of 6 squares
$= 6l^2$

Rectangular prism

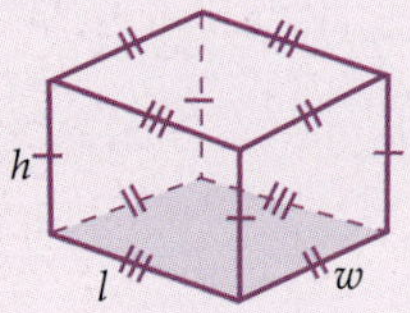

SA = area of 6 rectangles
$= 2lw + 2wh + 2lh$
$= 2(lw + wh + lh)$ (factorised form)

Cylinder

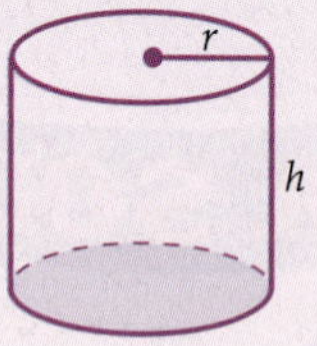

SA = area of 2 circles and area of curved surface
$= 2\pi r^2 + 2\pi rh$
$= 2\pi r(r + h)$ (factorised form)

5.2 Surface areas of prisms and cylinders

Navigator

1, 2, 3, 4, 5, 7 | 1 (column 1), 2, 3, 4, 5, 6, 7 | 1 (column 2), 2, 4, 5, 6, 7, 8

Answers p. 807

Fluency

1 Find the surface area of each solid, rounded to 2 decimal places where necessary.

W.E. 3

(a)
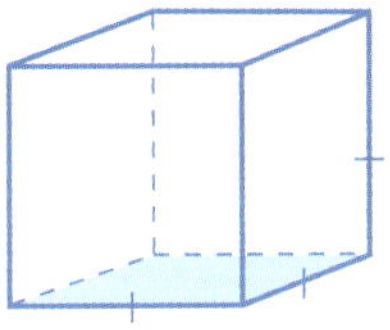

(b)
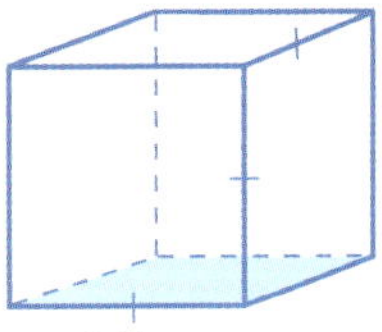

(c)
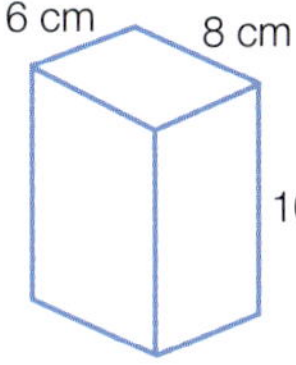

(d)
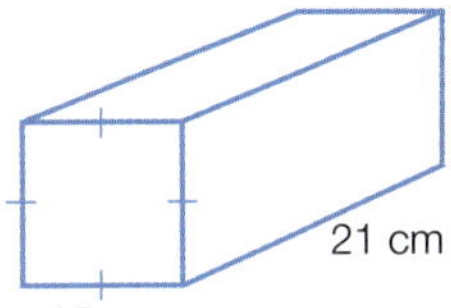

(e)
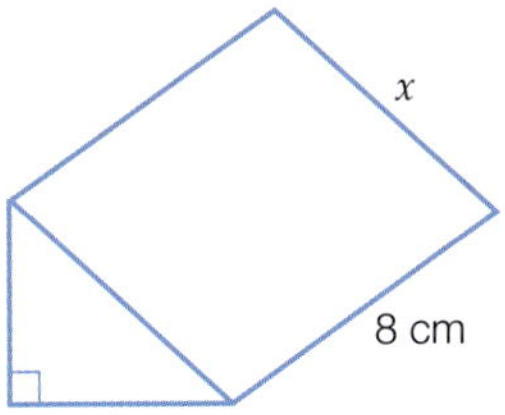

(f)
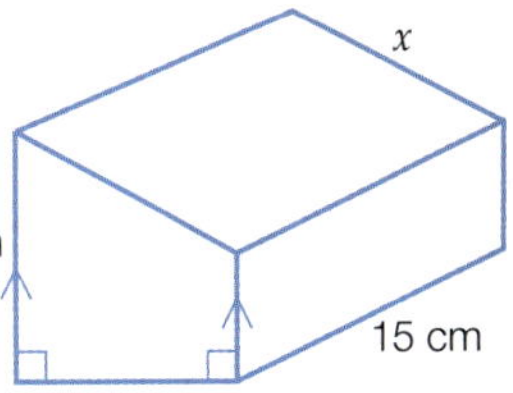

2 Find the surface area of each cylinder, rounded to 2 decimal places.

W.E. 4

(a)
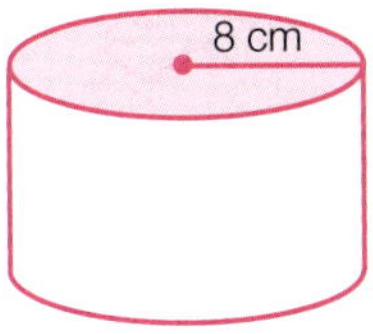

(b)
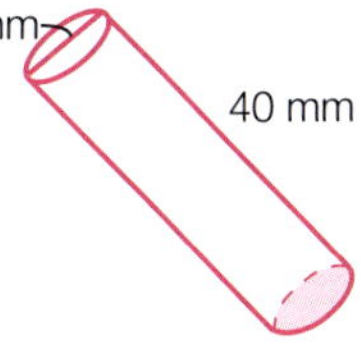

3 The calculation that gives the surface area of each of the following solids is:

(a)
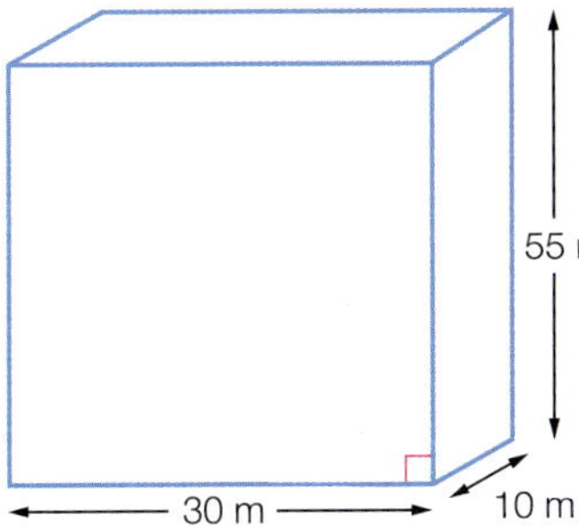

A $30 \times 10 \times 55$

B $2(30 \times 10 \times 55)$

C $2(30 + 10 + 55)$

D $2(30 \times 10) + (10 \times 55) + (30 \times 55)$

E $2(30 \times 10 + 10 \times 55 + 30 \times 55)$

(b)

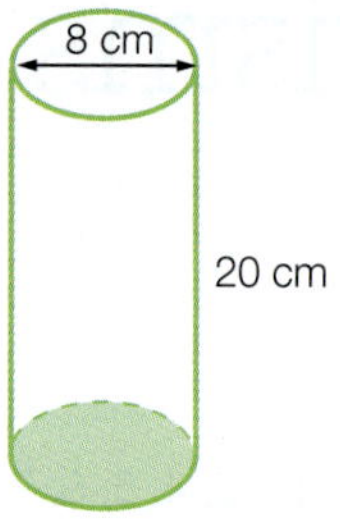

A $2 \times \pi \times 8(8 + 20)$

B $2 \times \pi \times 4(4 + 20)$

C $2 \times \pi \times 8(4 + 20)$

D $2 \times \pi \times 4(8 + 20)$

(c)

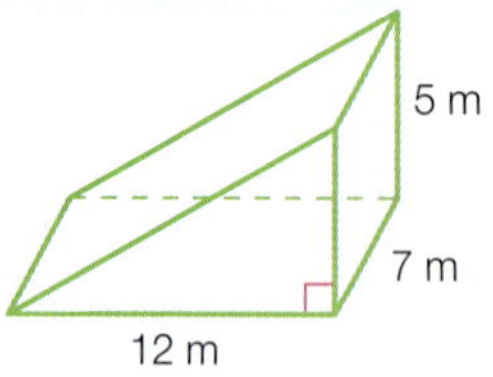

A $5 \times 12 + 7(5 + 12 + 17)$

B $\frac{5 \times 12}{2} + 7(5 + 13)$

C $\frac{5 \times 12}{2} + 7(5 + 12 + 13)$

D $5 \times 12 + 7(5 + 12 + 13)$

Understanding

4 Amanda has made a set of wooden building blocks in three sizes as shown below, as a present for her baby brother. She built ten of each block and now wants to paint them.

(a) Find the surface area of each of the three types of blocks (A, B and C).

(b) Find the total area of the 30 blocks that she has to paint.

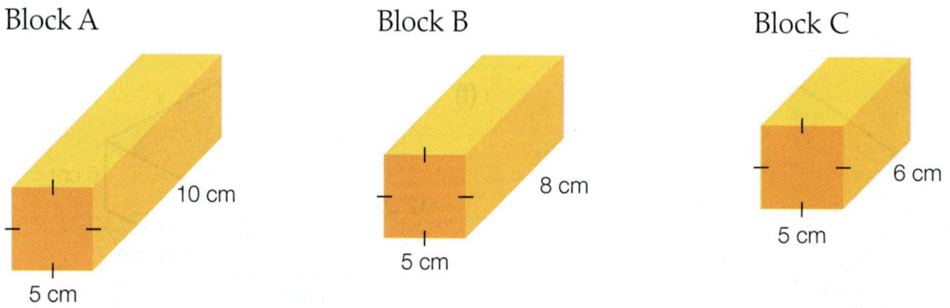

5 The diagram shows a plastic casing that has been specially made to surround a water pipe. Find the surface area of the casing, rounded to 2 decimal places.

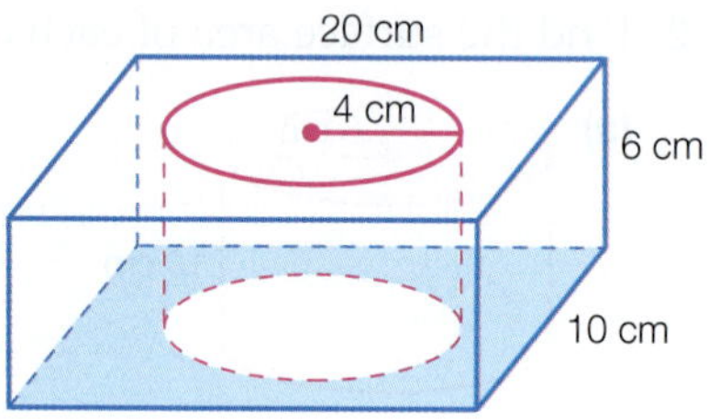

Reasoning

6 A circular podium has three steps as shown. The base of the podium has a radius of 1.5 m and the two lower steps have a width of 0.4 m. Each step is 0.25 m higher than the previous one. All visible surfaces of the podium are to be covered in carpet. Give each of the following answers correct to 2 decimal places.

(a) Calculate the area of carpet required to cover the top surface of all three steps. Hint: What is the shape of this total surface area?

(b) Calculate the area of carpet required to cover all vertical surfaces of the podium.

(c) Calculate the area of carpet required to cover all the visible surfaces of the podium.

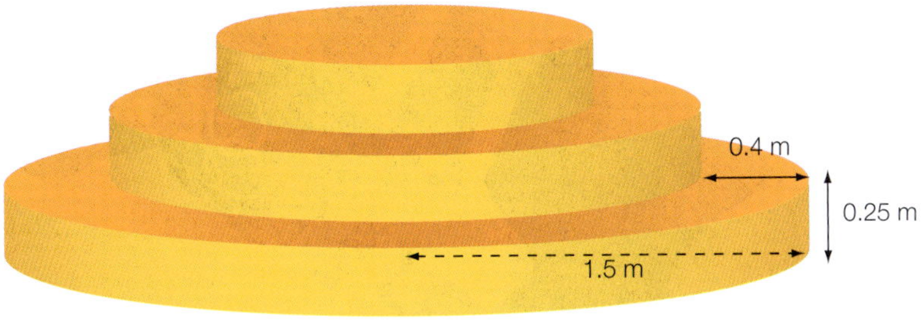

Open-ended

7 (a) James has two toy cubes to paint. The smaller cube has side lengths of 10 cm. The sides in the larger cube are twice as long as the smaller cube. If it takes 15 minutes to paint the small cube, then how much time will it take to paint the large cube?

(b) Write the formula for the surface area of a rectangular prism.

(c) By how many times does the surface area of the rectangular prism increase if all three of its dimensions are doubled? Demonstrate your answer with two suitable prisms, one of which has its length, width and height double that of the other one.

8 (a) Choose two variables to represent the radius and height of a cylinder. Find the surface area of this cylinder. Leave your answer in terms of π.

(b) Double the height of the cylinder and find the new surface area, leaving your answer in terms of π.

(c) Double the radius of the original cylinder and find the new surface area, leaving your answer in terms of π.

(d) Explain why your answers to parts **(b)** and **(c)** are not the same.

Maths 4 Real

Torah Jane Bright is a champion Australian snowboarder. Born in New South Wales, she turned pro at age 14 and in 2010 won a gold medal at the Winter Olympics in Vancouver, Canada. Torah is a master of the 'half pipe'—a half cylinder with two horizontal platforms (or 'decks').

SNOWBOA

Snowboarding terrain parks are similar to skateboard parks except that the surfaces are covered in hard-packed snow instead of concrete.

A group of professional snowboarders are designing their ideal terrain park. Here is their design for a 'half pipe' (not drawn to scale).

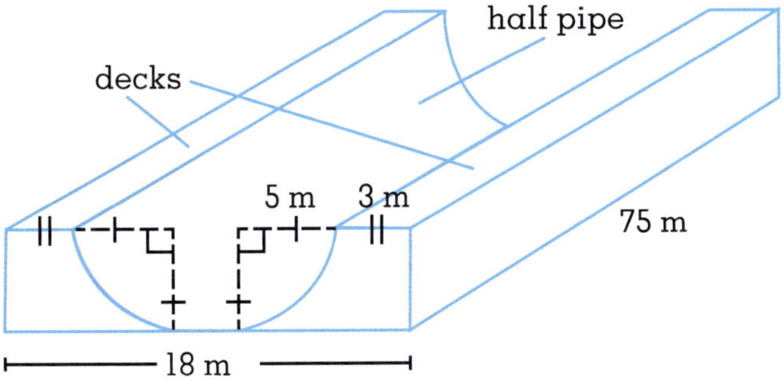

1 Find the area of the surface of the halfpipe that needs to be covered by snow (i.e. the total area of the two decks and the surface in between). Write your answer to the nearest m^2.

Each person in the group is given a 75 m by 18 m rectangle of horizontal space on which to build a feature of their own design. The cross-section of each feature is shown at right. Each feature is 75 m long.

2 Calculate the area of the surface that needs to be covered in snow for each of the three features. (You will need to calculate dimensions that are not shown, which may require techniques such as Pythagoras' theorem.) Write each answer to the nearest m^2.

Quarter pipe

Riders 'air' from it and perform tricks while airborne

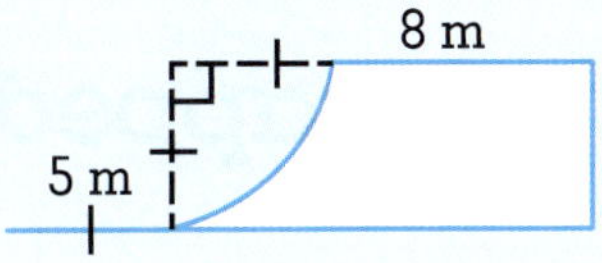

Spines

Two quarter pipes back-to-back

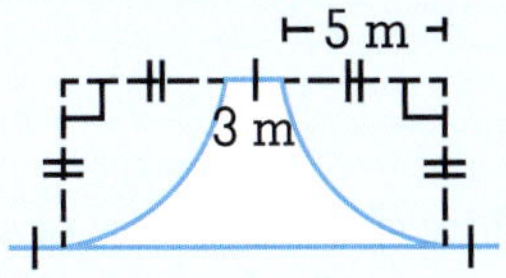

Flat bank

An angled wall to traverse

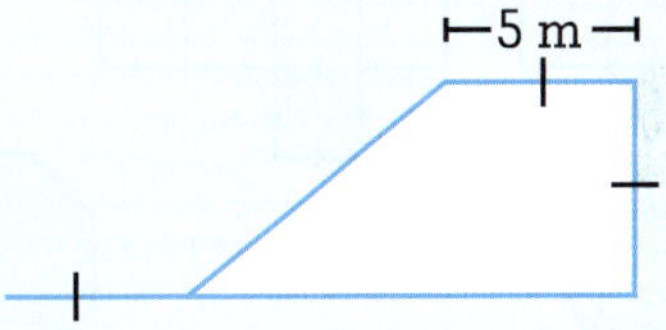

Research

3 Snowboarders who are keen to practise can purchase terrain kits and build a feature in their own backyards. Some snowboarders design and build their own features from recycled materials. Design a feature that would fit either in your backyard or in a space at your local park or school oval. Find the area of surface that would be covered in snow for your feature.

RDING TERRAIN PARK

Investigation

Prisms unwrapped

Equipment required: paper, ruler, scissors, tape (optional)

The Big Question

Can you develop a formula for the surface area of any prism?

Engage

The diagrams below show the nets of some prisms. The base of each prism is a regular polygon ('regular' means all sides and angles are equal).

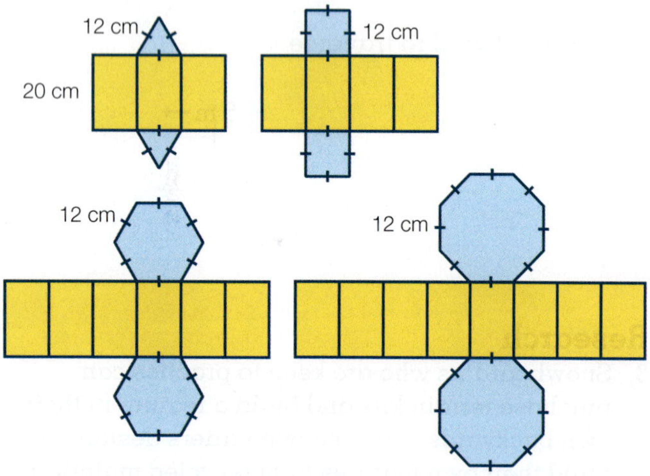

1 Sketch the shape of each prism. One way to do this is to draw two of the base shapes, one behind and slightly above the other. Then, join the matching vertices with straight lines. Label each prism (e.g. the first is an equilateral triangular prism).

2 For each net, what is the shape of the middle section (shown in yellow)?

Explore

3 If each prism has a length of 20 cm and a base side length of is 12 cm:

(a) What is the area of the middle section of each net?

(b) How can you find this area by doing only one calculation?

4 (a) To help answer the big question, draw nets of prisms with height 20 cm and bases as below:

(i) A right-angled triangle with side lengths of 6, 8 and 10 cm.

(ii) A rectangle with dimensions of 11 cm × 15 cm.

(b) (i) What is the same about the dimensions of the nets you have drawn and the nets shown on this page?

(ii) What is different about the dimensions of the nets you have drawn and the nets on this page?

Strategy options

- Draw a diagram.
- Look for a pattern.

Explain

5 Find the lengths of the 'middle section' rectangle for each net drawn. How is the length of this rectangle found from the dimensions of the base?

Elaborate

6 Copy and complete:

(a) The surface area of any prism can be found as two times the area of ___________ + the area of a rectangle whose length is the ___________ of the base and whose width is the ___________ of the prism.

(b) Write the above statement as a formula. Make sure you define each variable.

7 State your answer to the Big Question. Describe any special cases (such as when the base of the prism is a regular polygon).

8 Write a formula for the perimeter and area of the base for each of the following prisms. Hence, write a formula for the surface areas.

(a)

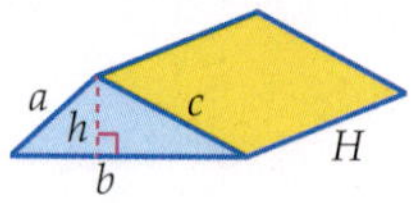

(b)

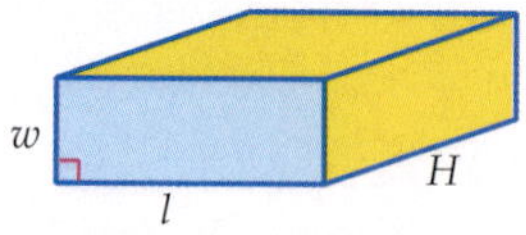

Evaluate

9 Do you think you will use what you have found in this investigation as a shortcut for surface area calculations? Why or why not?

Extend

10 A polyhedron is a solid with faces that are all polygons. Five special polyhedra, known as the 'Platonic solids' have faces that consist only of one type of regular polygon. They are the tetrahedron, cube, octahedron, dodecahedron and icosahedron. Find out what these solids look like and write a formula for their surface area.

Volumes of prisms and cylinders

5.3

Prisms

Prisms and cylinders have a uniform cross-section. Because of this, the volume of these solids is found by multiplying the area of the cross-sectional shape on the end (the base) by the height of the solid, where the height is perpendicular to the base.

Note that the base is not always the face that the prism is sitting on. The base is the face that has the uniform cross-section.

The volume of solids with constant cross-section is:

$$V = AH$$

where A is the area of the cross-section

H is the perpendicular height of the solid.

Rectangular prism

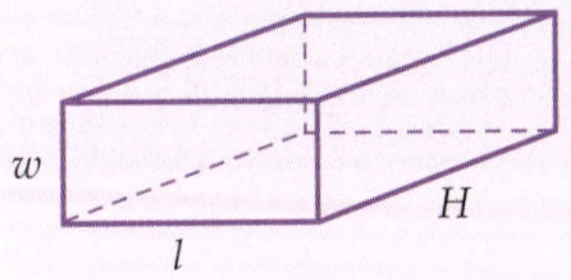

$V = AH$

$A = lw$

$V = lwH$

Triangular prism

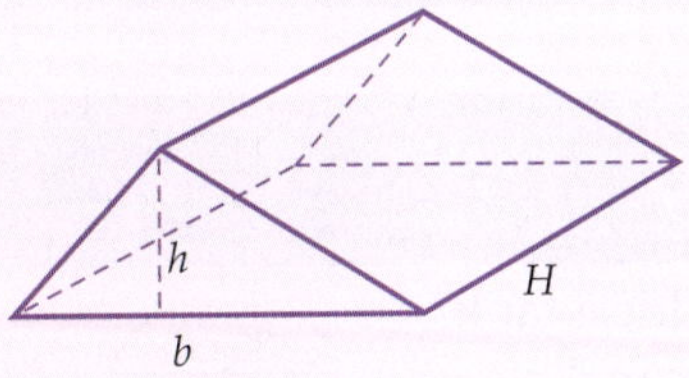

$V = AH$

$A = \frac{1}{2}bh$

$V = \frac{1}{2}bhH$

Cylinder

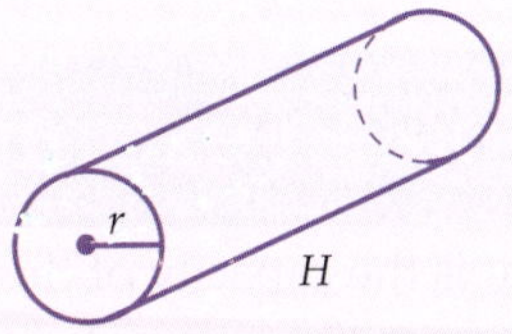

$V = AH$

$A = \pi r^2$

$V = \pi r^2 H$

In the formula for the triangular prism above, lower case h represents the height of the triangular-shaped base, and capital H represents the height of the prism.

Capacity

The **capacity** of a container refers to the volume of a substance (usually liquid or gas) that the container can hold.

The unit used to measure capacity is the litre (L).

$$1000 \text{ millilitres (mL)} = 1 \text{ litre (L)}$$

$$1000 \text{ litres} = 1 \text{ kilolitre (kL)}$$

$$1 \text{ cm}^3 = 1 \text{ mL}$$

$$1000 \text{ cm}^3 = 1 \text{ L}$$

$$1 \text{ m}^3 = 1000 \text{ L} = 1 \text{ kL}$$

Worked example 5

W.E. 5

Find the volume of the triangular prism.

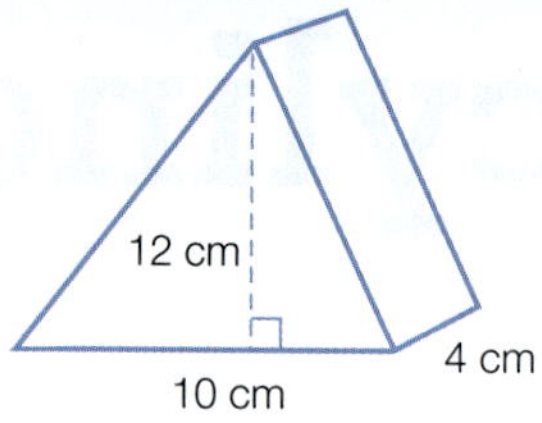

Thinking	Working
1 State the general formula.	$V = AH$
2 Substitute the formula for the area of the base. (Here, the area of the triangular base is $A = \frac{1}{2}bh$.)	$= \frac{1}{2}bhH$
3 Substitute the values.	$= \frac{1}{2} \times 10 \times 12 \times 4$
4 Write the answer with the correct unit of volume.	$= 240\ cm^3$

Worked example 6

W.E. 6

What is the capacity of this cylindrical tank?
Answer in litres, rounded to 2 decimal places.

Thinking	Working
1 Calculate the volume of the cylinder. (Here, the area of the circular base is $A = \pi r^2$.)	$V = A \times H$ $= \pi r^2 H$ $r = \frac{30}{2}$ $= 15\ cm$ $V = \pi \times 15^2 \times 50$ $= 11\,250\pi$ $= 35\,342.92\ cm^3$ (2 d.p.)
2 Convert the volume to a capacity ($1000\ cm^3 = 1\ L$) and round your answer to 2 decimal places.	Capacity = 35.34 L (2 d.p.)

5.3 Volumes of prisms and cylinders

Navigator

1, 2, 3, 4, 5, 6, 7, 8, 11 | 1, 2, 3, 4, 5, 6, 7, 8, 10, 11 | 1, 2, 3, 5, 6, 7, 8, 9, 10, 11

Answers p. 807

Fluency

1 Find the volume of each solid with constant cross-section. Round to 2 decimal places, if necessary. W.E. 5

(a)
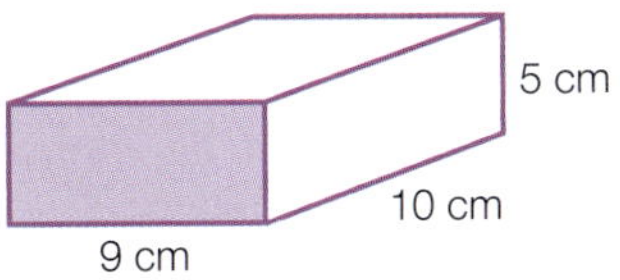

(b)
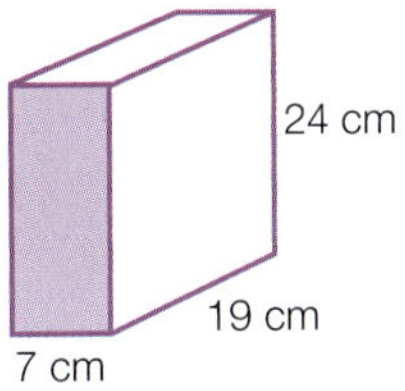

(c)
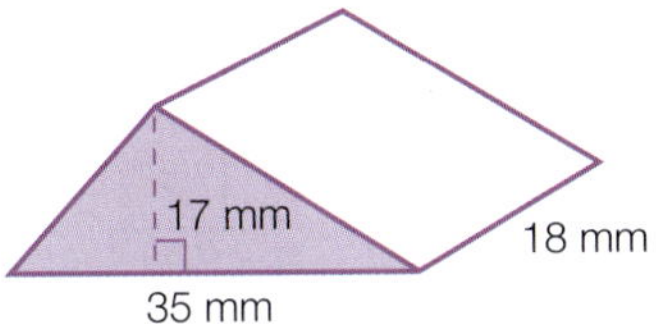

(d)
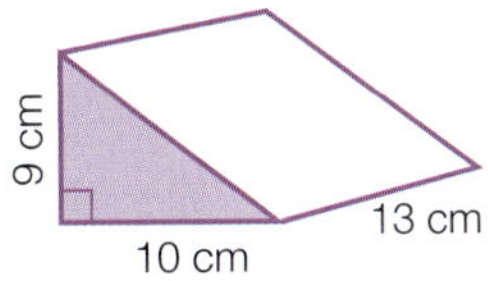

(e)
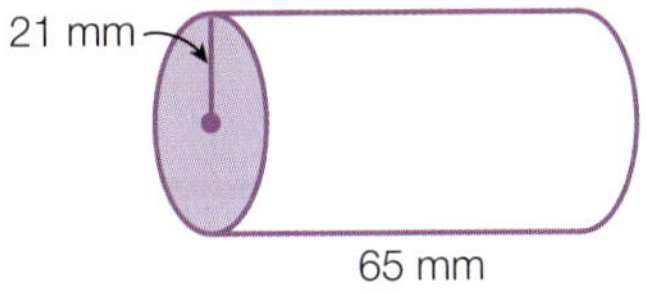

(f)
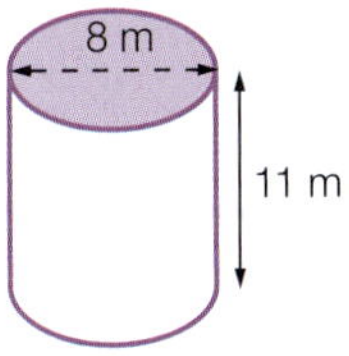

2 Find the capacity of the soft drink can in mL, rounded to 2 decimal places. Assume that the can is a regular cylinder. W.E. 6

Hint

1 cm^3 = 1 mL

3 Find the volume of each solid with constant cross-section. Round to 2 decimal places, if necessary.

(a)
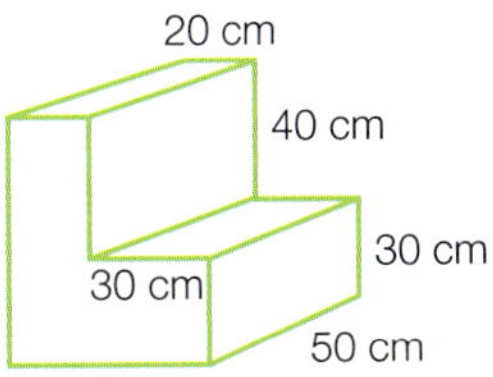

(b)
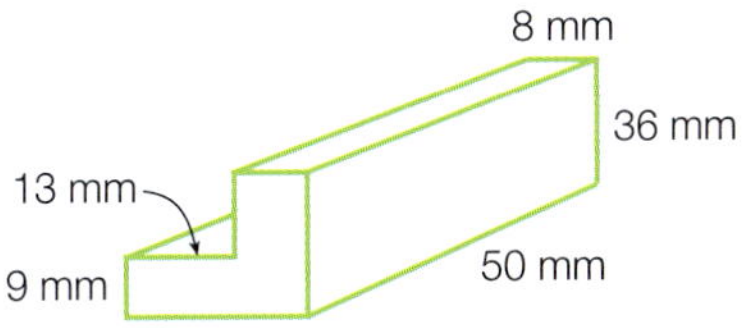

(c)

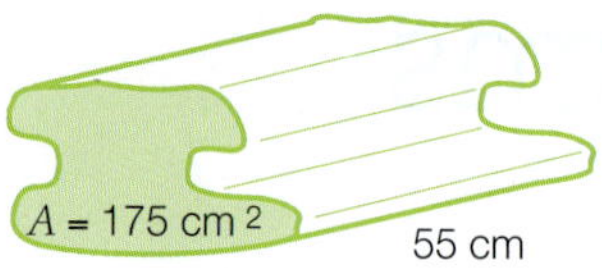

(d)

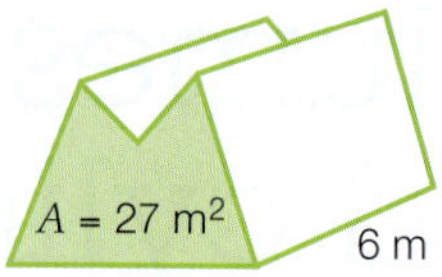

(e)

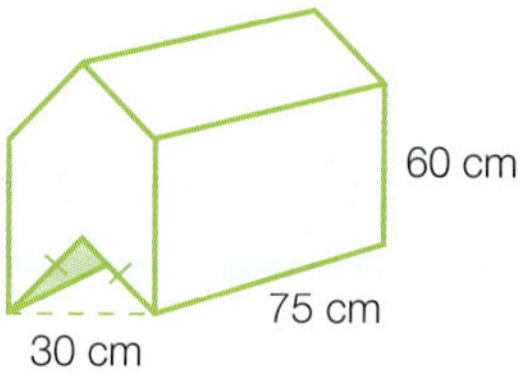

(f)

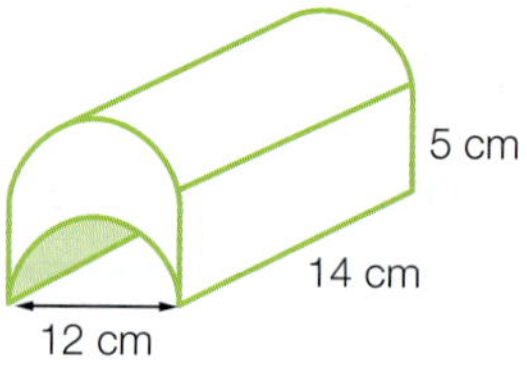

Understanding

4 **(a)** The capacity of this prism in kL can be found by:

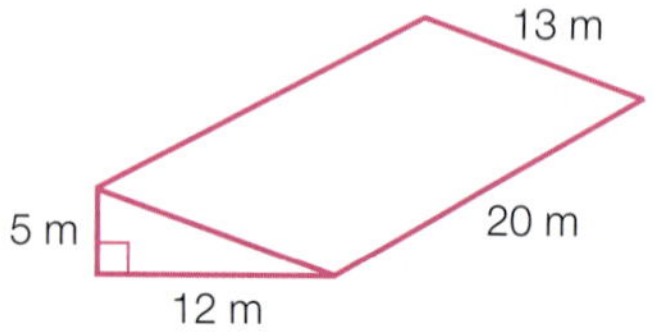

A $12 \times 5 \times 20$ **B** $\frac{1}{2} \times 5 \times 12 \times 20$

C $(\frac{1}{2} \times 5 \times 12) + (20 \times 12)$ **D** $\frac{1}{2} \times 12 \times 20 \times 5 \times 13$

(b) The capacity of this prism in mL can be found by:

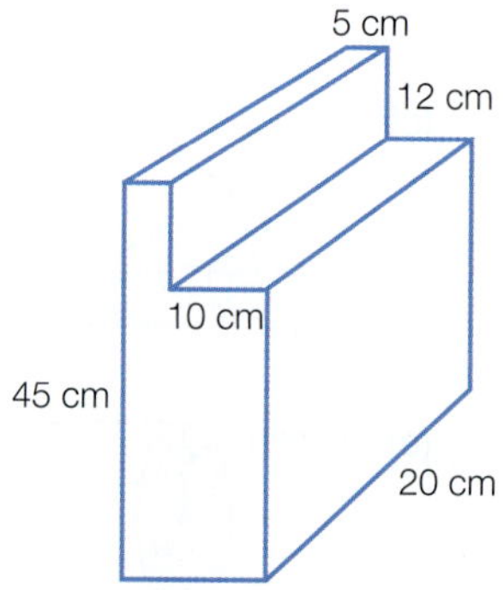

A $5 \times 12 \times 20 + 10 \times (45 - 12) \times 20$

B $5 \times 12 \times 20 + 10 \times 45 \times 20$

C $5 \times 45 \times 20 + 10 \times 12 \times 20$

D $(5 + 10) \times 45 \times 20 - 10 \times 12 \times 20$

5 Ingrid needs 25 concrete columns for her new verandah. Each column will be cylindrical in shape with a base diameter of 28 cm and a height of 85 cm.

(a) Find the volume of one of these columns of concrete, to the nearest cm^3.

(b) Find the volume of concrete required for the 25 columns in m^3, correct to 2 decimal places.

6 A 3.5 m pipe has an outer diameter of 24 cm and an inner diameter of 18 cm. Find its volume in m^3 correct to 2 decimal places.

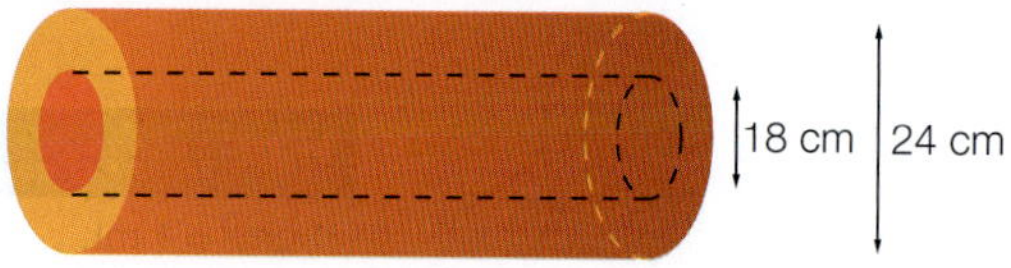

Reasoning

7 Alex's chocolate bar is a triangular-based prism with dimensions $30.5 \times 5.4 \times 5$ cm. Digby's chocolate bar is a square-based prism with dimensions $15 \times 5 \times 5$ cm.

(a) Find the volume of each bar to the nearest cm^3.

(b) Who has the bar with the greatest volume and how much bigger is the bar?

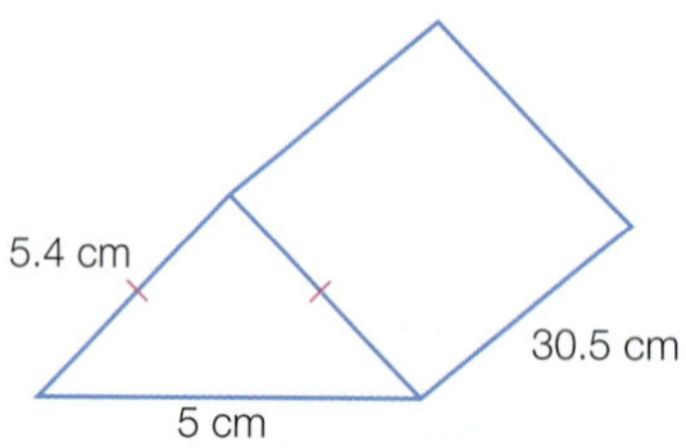

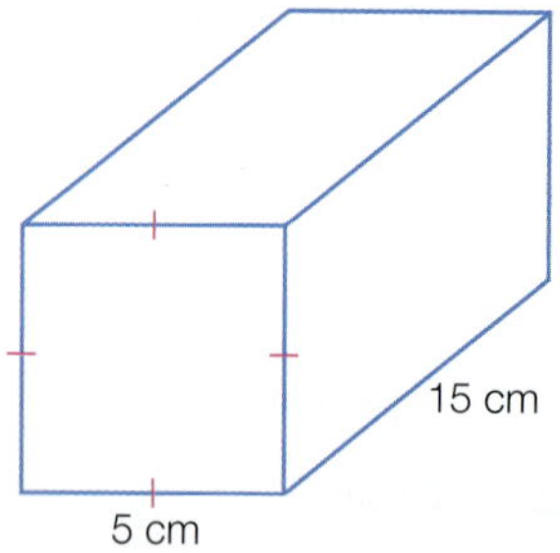

8 Gael is building an extension onto her house. She has just finished digging a series of trenches. All the trenches are 85 cm deep and 55 cm wide. The plan for the trenches is shown in the diagram. What is the volume of dirt removed? Round your answer to 2 decimal places.

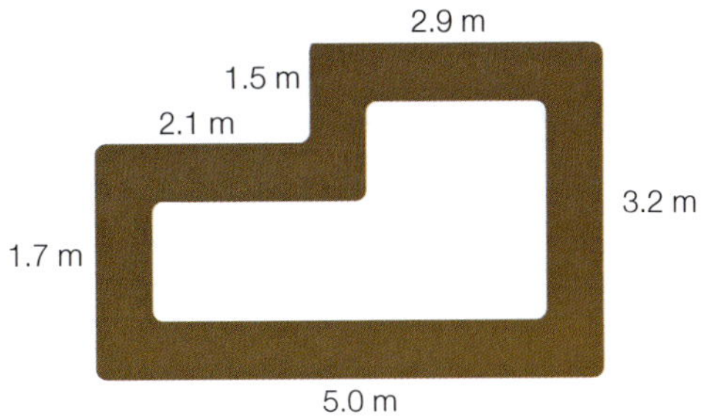

9 Tadhg's swimming pool is rectangular with two semicircular ends.

(a) Draw the two shapes that make up the area of the base of the swimming pool. Find this area correct to 2 decimal places.

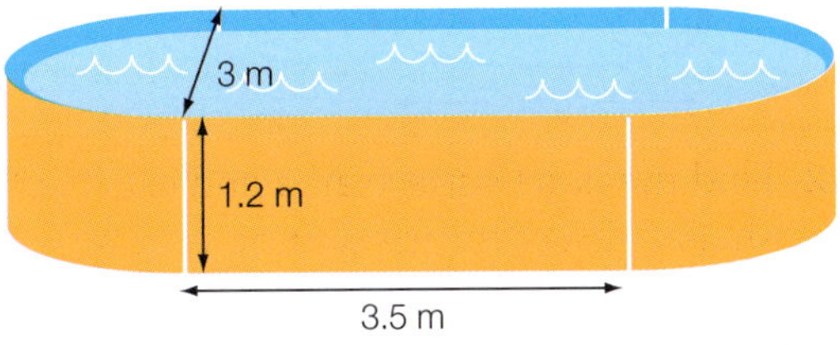

(b) Find the volume of the pool in cubic metres, and then calculate how many litres the pool will hold when full, to the nearest 1000 L. *Note:* 1 m^3 = 1000 L

Open-ended

10 A rectangular box has a volume of 200 cm^3.

(a) Find three possible whole number sets of dimensions that give the correct volume.

(b) Find three sets of dimensions with at least one non-whole number that give the correct volume.

(c) Calculate the surface area of the box for each of the three sets of dimensions found in (a).

11 Anita is attempting to find the volume of the cylinder in cm^3 correct to 2 decimal places.

(a) Identify the two mistakes in the working shown.

(b) Find the correct answer.

(c) Give Anita two points to write in her notes so she can avoid similar mistakes in the future.

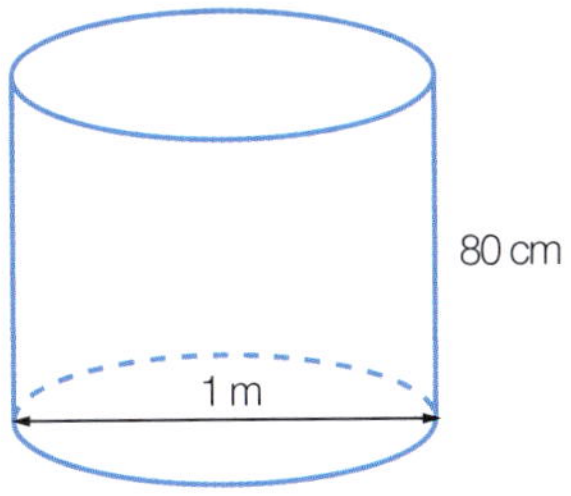

$V = AH$
$= \pi r^2 H$
$= \pi \times 1^2 \times 80$
$\approx 251.33 \text{ cm}^3$

Problem solving

Cola cost crisis

Fizzy Cola comes in cylindrical cans of diameter 5 cm and height 15 cm. The cost of one can is $2.35. The manufacturers have created a 'super-sized' can, which is the same height but twice as wide. The sales manager of Fizzy Cola is wondering at what price she should sell the promotional can. Her assistant thinks that the new can is 'twice as big' as the original can, so it must hold 'twice as much' cola and should cost 'twice the price'.

1 Suggest a price for the 'super-sized' can by comparing its volume with the original can.

2 Is the assistant's reasoning correct? Why or why not?

Strategy options

- Break problem into manageable parts.
- Draw a diagram.
- Have I seen a similar problem?

Half-time 5

5.2 1 Find the surface area of the following solids. Round your answers to the nearest whole number.

(a)

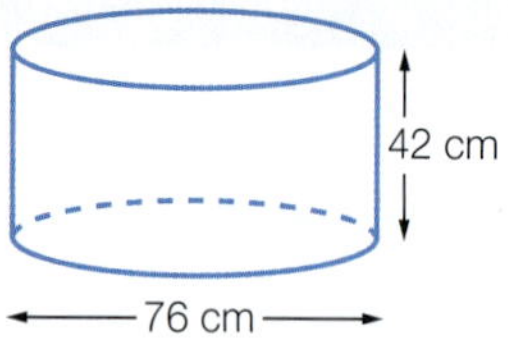

(b)

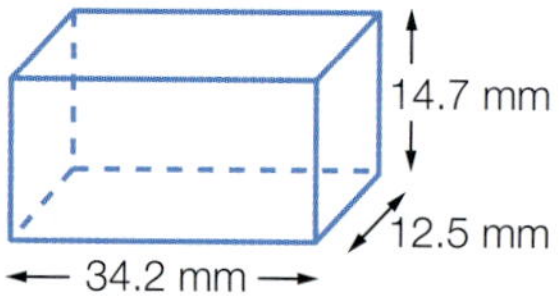

5.2 **5.3** 2 Find the exact surface area and exact volume of this composite shape. ('Exact' means writing your answer in terms of π.)

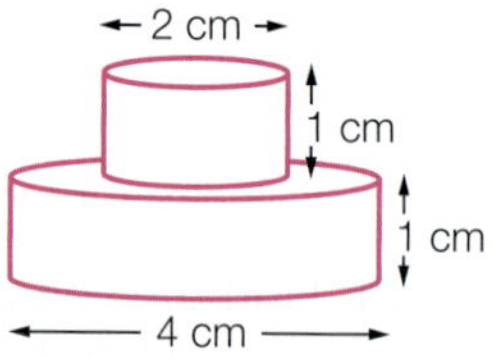

5.1 3 Find the area of the following shapes, rounded to 2 decimal places, if necessary.

(a)

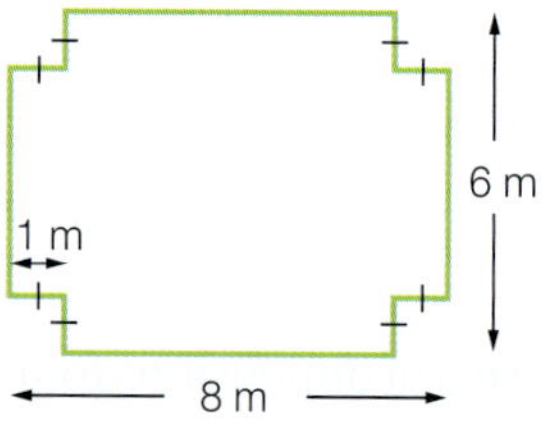

(b)

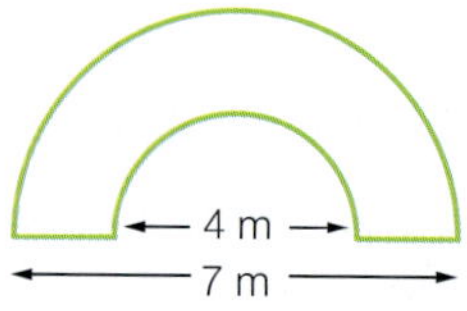

5.3 4 Which of the following solids, **A**, **B** or **C**, has the greatest volume? Justify your answer by evaluating the exact volume of each solid.

A

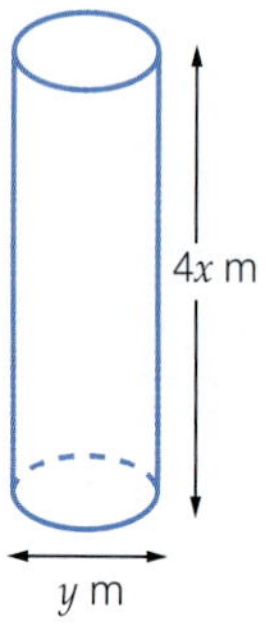

B

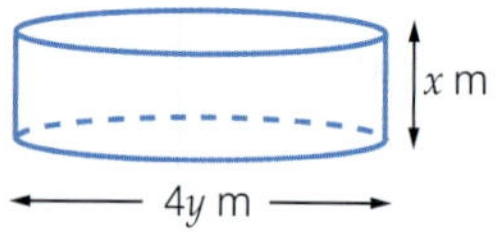

C

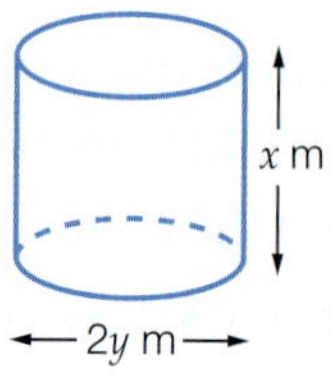

5.3 5 Find the volume of the following solids, rounded to 2 decimal places.

(a)

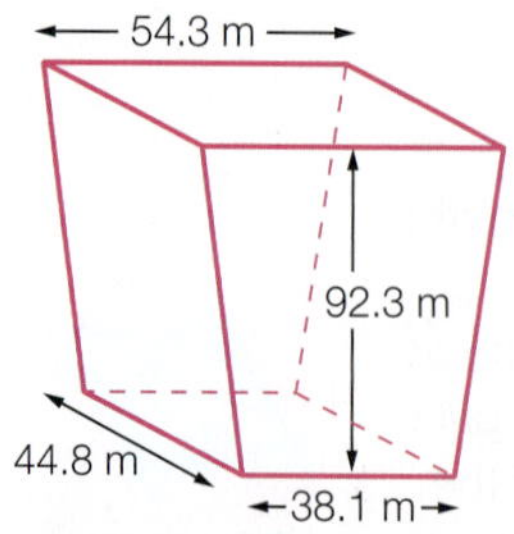

(b)

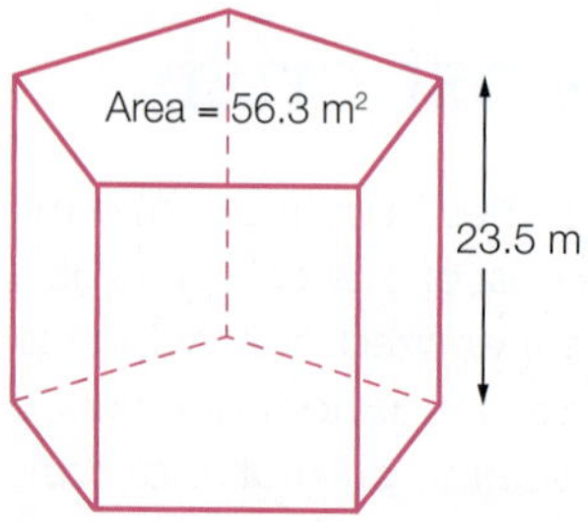

5.3 6 A cylindrical tank at a milk processing plant has a diameter of 2.6 m and a height of 3.4 m. How many 250 mL cartons could be completely filled from one tank full of milk?

Surface areas of tapered solids and spheres

5.4

10A

Tapered solids do not have a uniform cross-section. The cross-section gets smaller as you move from the base of the solid to the top. The shape of the cross-section is always a similar shape to the base—the shape is the same but the size decreases. The tapered solids shown here are **pyramids** and **cones**.

Pyramids

A pyramid is made up of a base and triangular sides that meet at a point, called the vertex. The number of triangles depends on the number of sides of the base shape.

A pyramid with a square base has four triangles, whereas a pyramid with a pentagonal base has five triangles, and so on.

The net of a square-based pyramid is shown here. The surface area is found by adding the individual areas of the base and sides. For a square based pyramid:

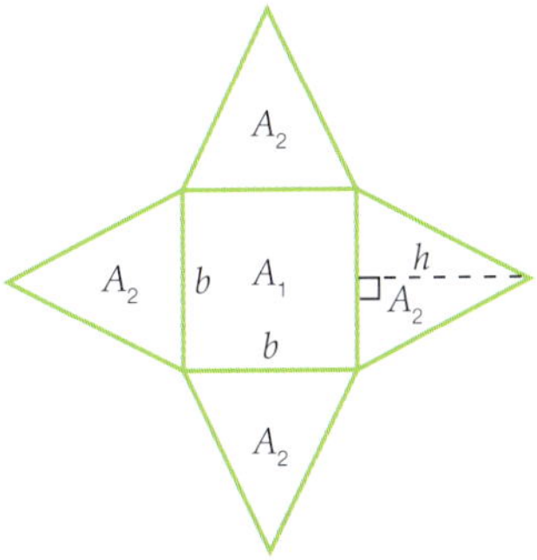

$$\begin{aligned} SA &= A_1 + 4 \times A_2 \\ &= b^2 + \cancel{4}^{2} \times \frac{bh}{\cancel{2}_{1}} \\ &= b^2 + 2bh \end{aligned}$$

Worked example 7

W.E. 7

Find the surface area of this square-based pyramid.

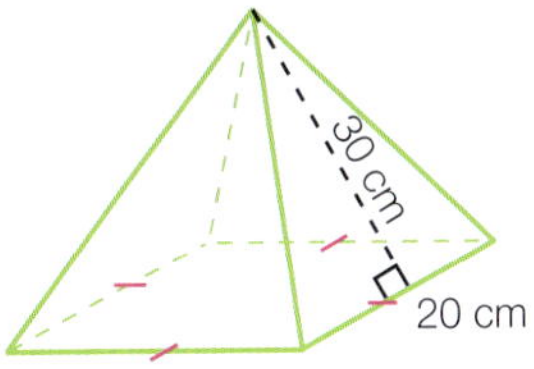

Thinking	Working
1 The surface area of a square-based pyramid is the combined area of the base and the four triangular faces.	$SA = b^2 + 2bh$
2 State the values of b and h.	$b = 20$ cm $h = 30$ cm
3 Substitute b and h into the formula and evaluate.	$SA = 20^2 + 2 \times 20 \times 30$ $= 400 + 1200$
4 Write the answer.	$= 1600$ cm^2

Cones

The surface area of a cone is the sum of the area of the circular base of radius, r, and the area of the sloping curved surface. The distance along this sloping surface from the base to the vertex is called the slant height, s.

When flattened, the curved surface is a sector of a circle that wraps around the base. This sector has a radius of s and an arc length of $2\pi r$ (i.e. the circumference of the circular base).

To find the area of the curved surface, first write the arc length in terms of r and s.

$$\text{Arc length} = \frac{\theta}{360} \times 2\pi \times s$$

$$2\pi r = \frac{\theta}{360} \times 2\pi \times s \qquad \text{(Arc length is equal to circumference)}$$

$$r = \frac{\theta}{360} \times s$$

$$\frac{r}{s} = \frac{\theta}{360}$$

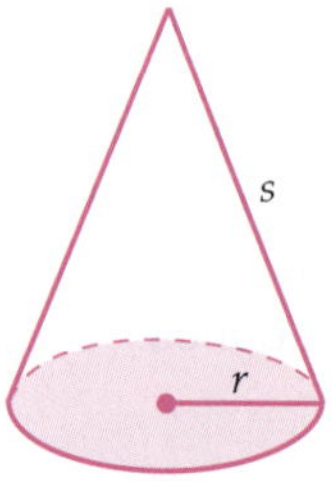

Use this expression in the formula for the area of the sector:

$$\text{Area} = \frac{\theta}{360} \times \pi s^2 \qquad (s = \text{radius of sector})$$

$$= \frac{r}{s} \times \pi \times s^2$$

$$A = \pi rs$$

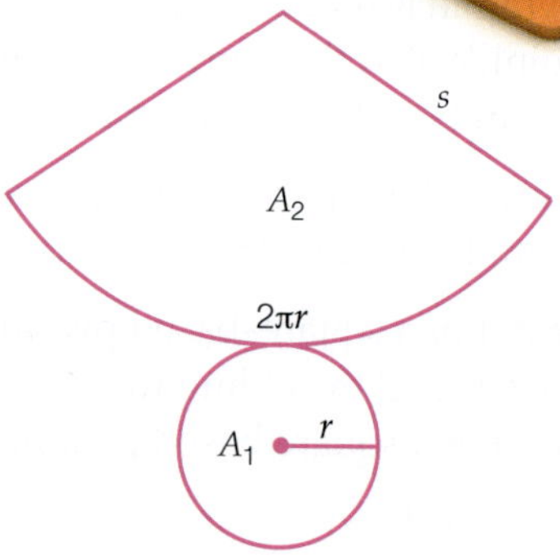

The area of the curved surface is $A = \pi rs$.

So $\quad \text{SA} = A_1 + A_2$

$$= \pi r^2 + \pi rs$$

$$= \pi r(r + s) \qquad \text{(factorised form)}$$

Worked example 8

W.E. 8

Find the surface area of this cone, rounded to 2 decimal places.

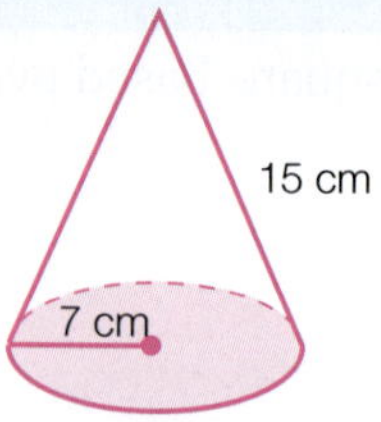

Thinking	Working
1 Write the formula for surface area.	$SA = \pi r^2 + \pi rs$
2 State the values of r and s.	$r = 7$ cm $s = 15$ cm
3 Substitute for r and s into the formula and simplify.	$SA = \pi \times 7^2 + \pi \times 7 \times 15$ $= 49\pi + 105\pi$ $= 154\pi$
4 Evaluate, rounding the answer to 2 decimal places.	$= 483.81$ cm^2 (2 d.p.)

Spheres

A **sphere** has only one single curved surface. The surface area of a sphere is equivalent to four times the area of a circle of the same radius.

$SA = 4\pi r^2$

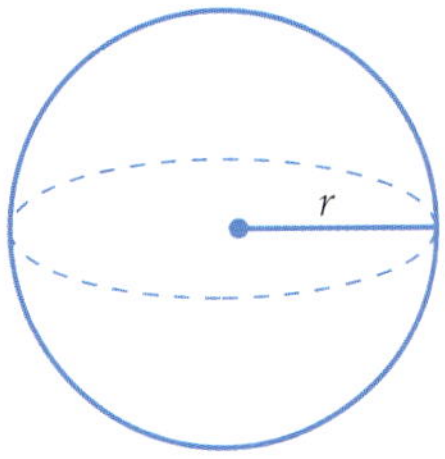

Worked example 9

W.E. 9

Find the surface area of this sphere, rounded to 2 decimal places.

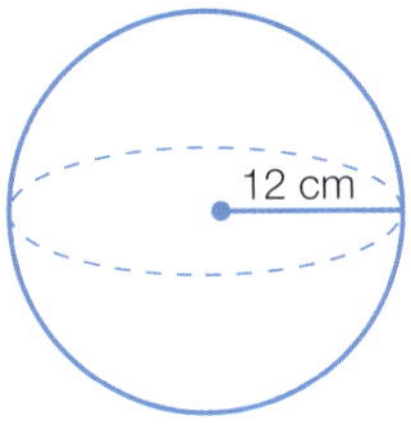

Thinking	Working
1 Write the formula for surface area.	$SA = 4\pi r^2$
2 Substitute for r and simplify.	$= 4 \times \pi \times 12^2$ $= 576\pi$
3 Evaluate, rounding the answer to 2 decimal places.	$= 1809.56 \text{ cm}^2$ (2 d.p.)

Square-based pyramid

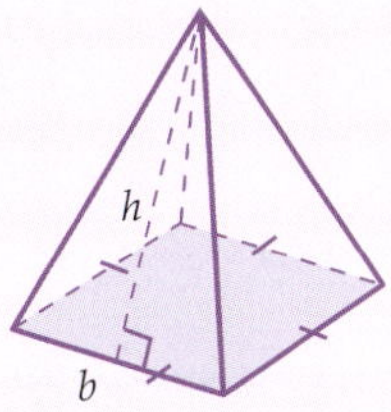

SA = area of square base and area of 4 triangles

$= b^2 + 4 \times \frac{1}{2}bh$

$= b^2 + 2bh$

Cone

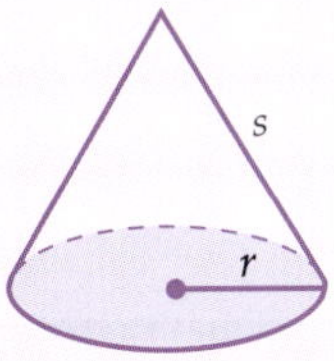

SA = area of circle and area of curved surface

$= \pi r^2 + \pi rs$

$= \pi r(r + s)$ (factorised form)

Sphere

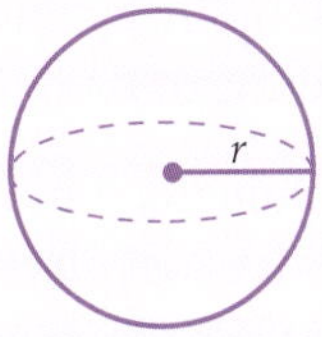

SA = area of curved surface

$= 4\pi r^2$

Composite solids

The surface area of any solid is the total area of the individual surfaces—curved or otherwise.

Worked example 10

W.E. 10

Find the surface area of each of the solids, rounded to 2 decimal places.

(a)

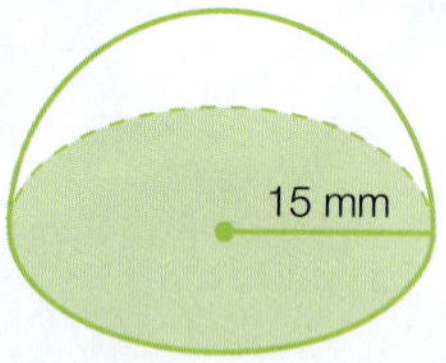

(b)

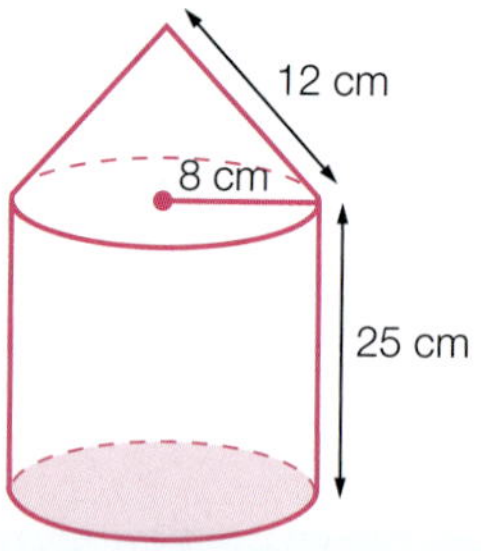

Thinking	Working
(a) 1 Identify the individual areas. (A hemisphere is half a sphere.)	(a) $SA = A_{hemisphere} + A_{circle}$
2 Write a formula for the sum of the areas. Simplify if possible.	$= \frac{1}{2} \times 4\pi r^2 + \pi r^2$ $= 2\pi r^2 + \pi r^2$ $= 3\pi r^2$
3 Substitute for r.	$= 3 \times \pi \times 15^2$ $= 675\pi$
4 Evaluate, rounding the answer to 2 decimal places.	$= 2120.58$ mm^2 (2 d.p.)
(b) 1 Identify the individual areas.	(b) $SA = A_{cone\ (curved)} + A_{cylinder\ (curved)} + A_{circle}$
2 Write a formula for the sum of the areas.	$= \pi rs + 2\pi rh + \pi r^2$
3 Substitute for r, s and h.	$= \pi \times 8 \times 12 + 2 \times \pi \times 8 \times 25 + \pi \times 8^2$
4 Simplify by finding each area in terms of π.	$= 96\pi + 400\pi + 64\pi$ $= 560\pi$
5 Evaluate, rounding the answer to 2 decimal places.	$= 1759.29$ cm^2 (2 d.p.)

5.4 Surface areas of tapered solids and spheres

Navigator

1, 2, 3, 4, 5, 6, 7, 8, 10, 15, 16	1, 2, 3, 4, 5, 6, 7, 8, 10, 11, 13, 14, 15, 16	1, 2, 3, 4, 5, 7, 8, 9, 11, 12, 13, 14, 15, 16

Answers p.807

Fluency

1 Find the surface area of each pyramid, rounded to the nearest whole number. W.E. 7

(a)

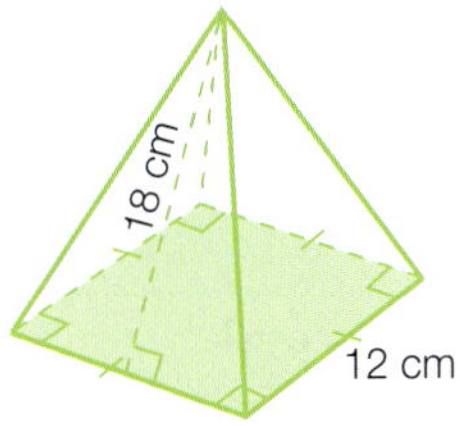

(b)

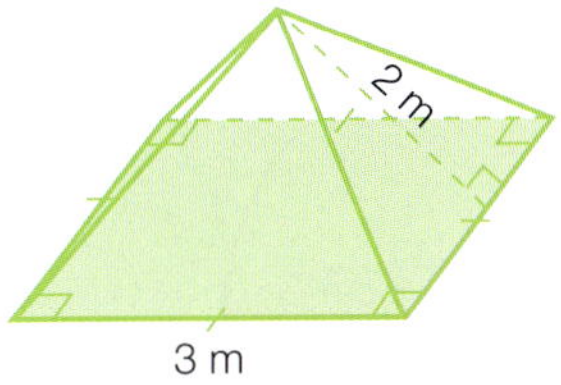

2 Find the surface area of each cone, rounded to 2 decimal places. W.E. 8

(a)

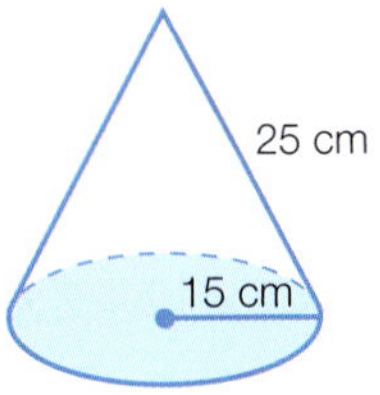

(b)

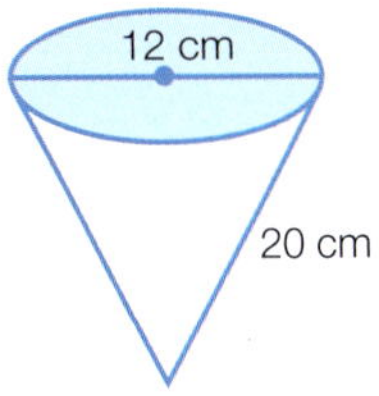

3 Find the surface area of each sphere, rounded to 2 decimal places. W.E. 9

(a)

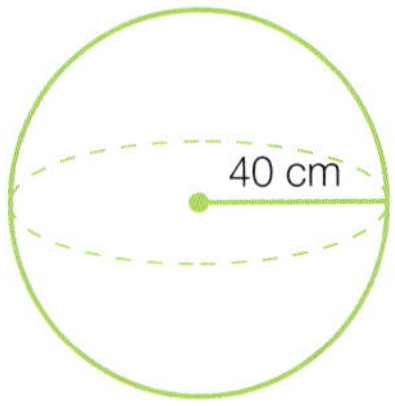

(b)

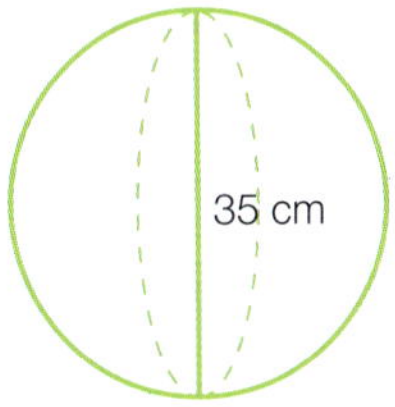

4 Find the surface area of each solid. Identify the individual areas that make up the surface area first. Round your answers to 2 decimal places. W.E. 10

(a)

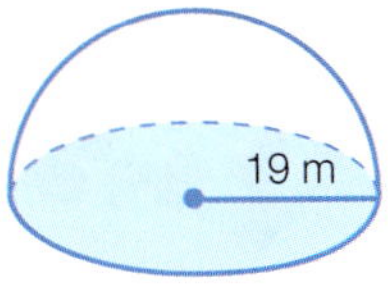

(b)

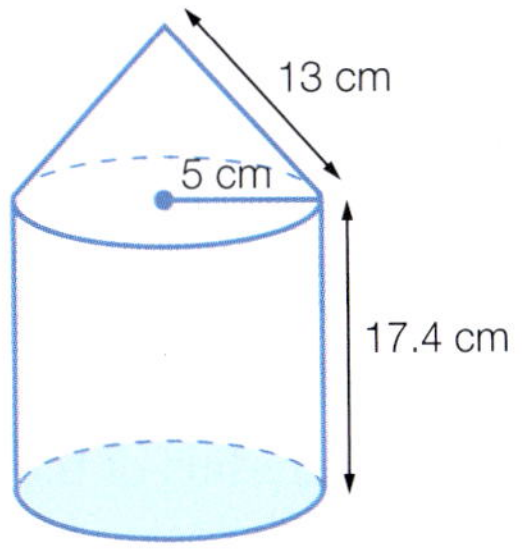

(c)

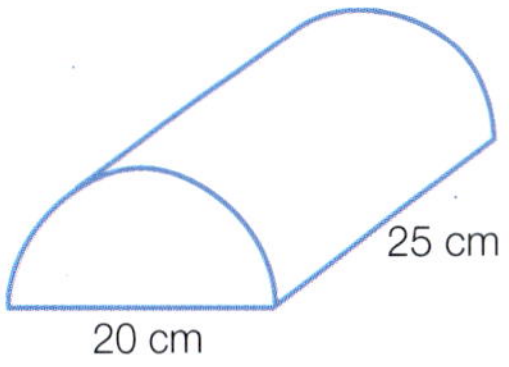

(d)

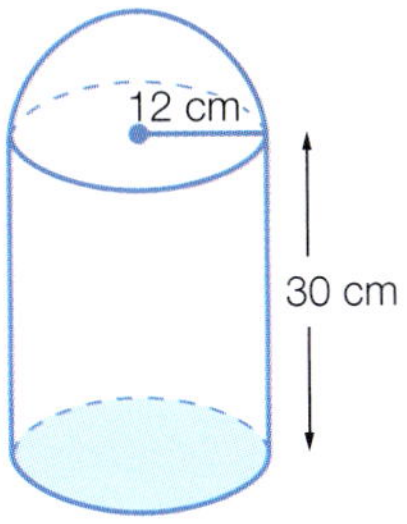

(e)

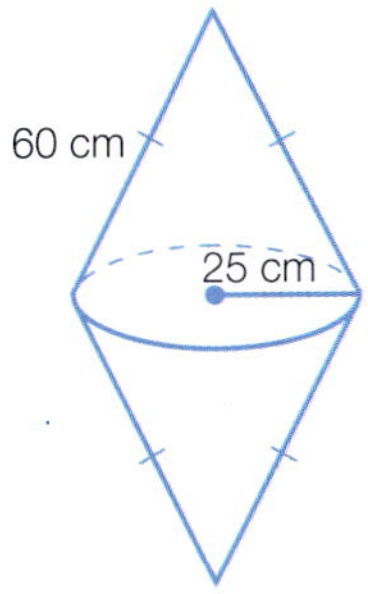

(f)

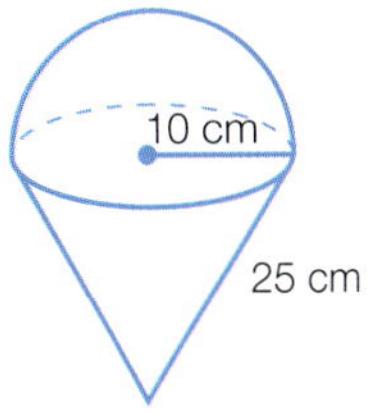

Hint

Use Pythagoras' theorem to calculate the heights of the triangles.

5 Calculate the surface area each of the following solids correct to 2 decimal places.

(a)

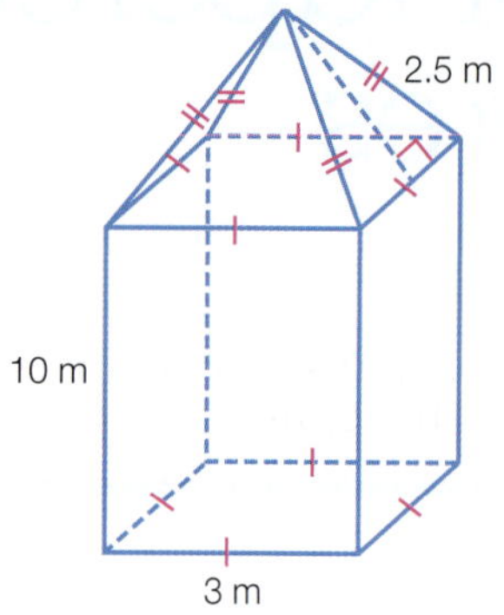

(b)

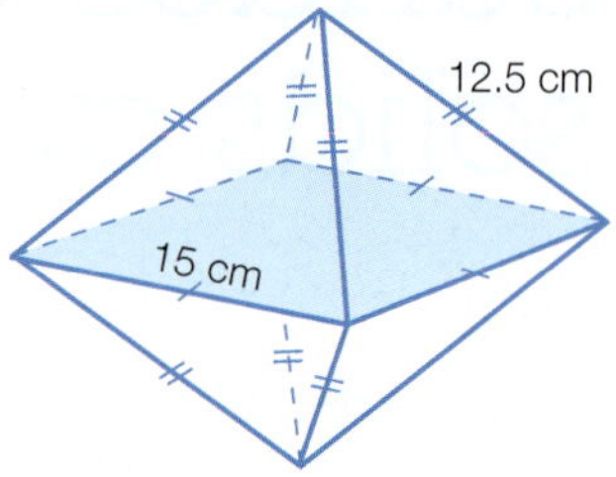

6 The calculation that gives the surface area of each of the following solids is:

(a)

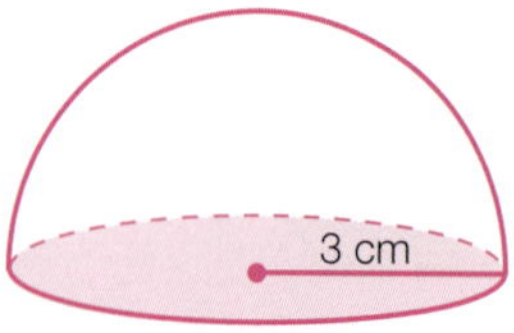

A $4 \times \pi \times 3^2$

B $\frac{1}{2} \times 4 \times \pi \times 3^2$

C $\frac{1}{2} \times 4 \times \pi \times 3^2 + \pi \times 3^2$

D $4 \times \pi \times 3^2 + \pi \times 3^2$

(b)

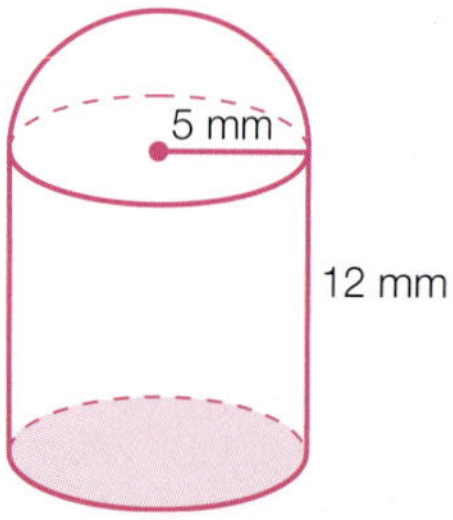

A $4 \times \pi \times 12^2 + 2 \times \pi \times 5 \times 12 + \pi \times 12^2$

B $2 \times \pi \times 5^2 + \pi \times 5 \times 12 + \pi \times 5^2$

C $4 \times \pi \times 5^2 + 2 \times \pi \times 5 \times 12 + 2 \times \pi \times 5^2$

D $\frac{1}{2} \times 4 \times \pi \times 5^2 + 2 \times \pi \times 5 \times 12 + \pi \times 5^2$

Understanding

7 Find, correct to 2 decimal places, the surface area of a capsule that has the dimensions shown below.

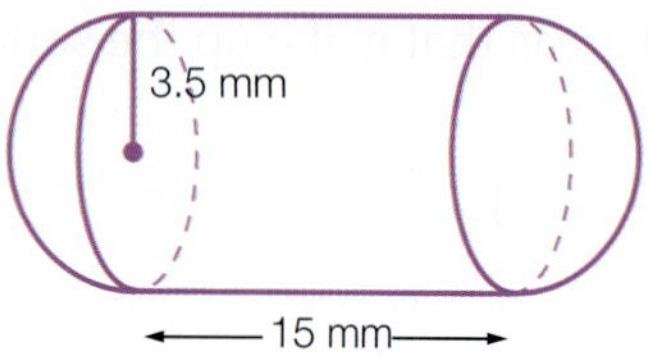

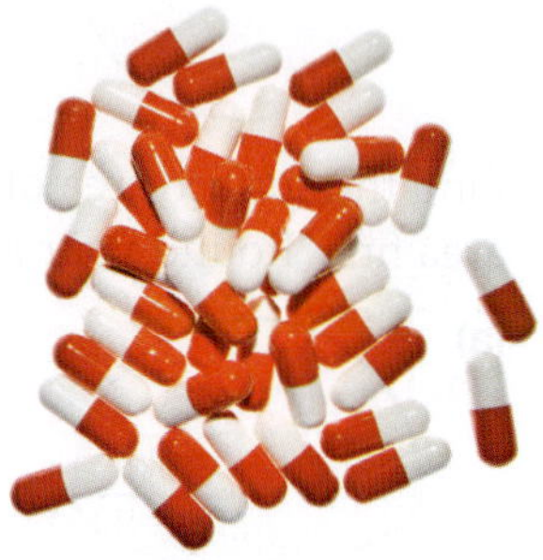

Hint

Determine the radius, then the slant height.

8 A conical party hat has a vertical height of 18 cm. The circumference of the circular base is 43 cm.

(a) Use the circumference to find the radius of the circular base, correct to 2 decimal places.

(b) Find the slant height of the cone using Pythagoras' theorem.

(c) Calculate the area of cardboard required to make the hat (assuming no overlap). Round your answer to the nearest whole number.

9 A pyramid in Egypt has a square base that covers an area of 44 100 m^2 and a height perpendicular to the base of 130 m. Find the length of the slant edge of the pyramid, then calculate the total surface area of the four triangular faces of the pyramid to the nearest square metre.

Hint

Draw a labelled diagram.

Reasoning

10 Rebecca is making a cardboard funnel for a school project. To estimate the area of cardboard required she sketches a cone and cylinder, as shown in the diagram. Note that the diagrams are not drawn to scale. The cone has a slant height of 13 cm and radius 5 cm. The cylinder has a vertical height of 3 cm and a diameter of 1 cm.

Calculate the area of cardboard that Rebecca needs to make the funnel, correct to 2 decimal places. (Assume Rebecca cuts out a whole cone.)

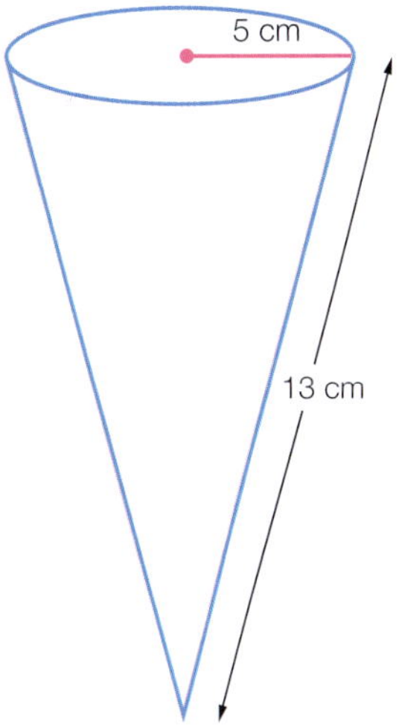

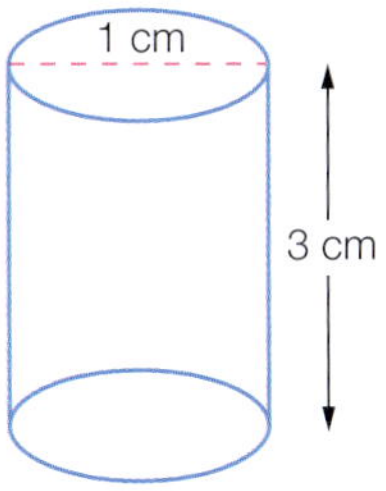

10A

11 Marcus has a new bowling ball and he wants to sew a protective bag for it. He has a rectangular piece of material 80 cm long and 40 cm wide. The bowling ball has a diameter of 30 cm.

(a) Show that the area of the material will cover the surface area of the bowling ball.

(b) If Marcus has just enough material to cover the ball, should he still consider using a longer piece of material? Explain your answer.

12 (a) Calculate the slant height of a cone of radius 25 cm and surface area 4398.23 cm^2, to the nearest centimetre.

(b) Calculate the radius of a cone with a surface area of 1809.56 cm^2 if the slant height is three times the radius.

Open-ended

13 Samarra's recipe for batter makes between 200 cm^2 and 300 cm^2 of cooked waffle. This waffle can then be shaped into a cone. Commercial waffle cones have a radius of 5 cm. Find two possible sets of dimensions for the radius and slant height of Samarra's waffle cone.

14 Explain why a cone cannot have a radius and a slant height with the same length.

15 A cone has its height doubled, but its radius stays the same.

- **(a)** Choose two values to represent the height and radius of a cone, and evaluate its surface area.
- **(b)** Evaluate its surface area after its height is doubled.
- **(c)** Choose two different values to represent another cone and evaluate its surface area.
- **(d)** Double the height of this cone and evaluate the new surface area.
- **(e)** Does doubling the height double the surface area?

16 A cone has its radius doubled, but its height stays the same.

- **(a)** Choose two values to represent the height and radius of a cone, and evaluate its surface area.
- **(b)** Evaluate its surface area after its radius is doubled.
- **(c)** Choose two different values to represent another cone and evaluate its surface area.
- **(d)** Double the radius of this cone and evaluate the new surface area.
- **(e)** Does doubling the radius double the surface area?

10A

Problem solving

Egyptian engineering

Egypt, 2020 BCE. Pharaoh Mentuhotep is constructing two massive stone obelisks on either side of the entrance to a temple. The obelisks are two tall columns that have four trapezoidal faces, narrower at the top than at the bottom. Each of these faces will be covered in hieroglyphics.

At the top of the obelisk sits a small pyramid-shaped cap, called a 'pyramidion'. The four exposed faces of the pyramidion will be covered in gold, to reflect the Sun.

Mentuhotep has given you, his chief engineer, a labelled drawing of what the obelisk should look like. Your job is to determine the surface area to be covered in:

(a) gold **(b)** hieroglyphics.

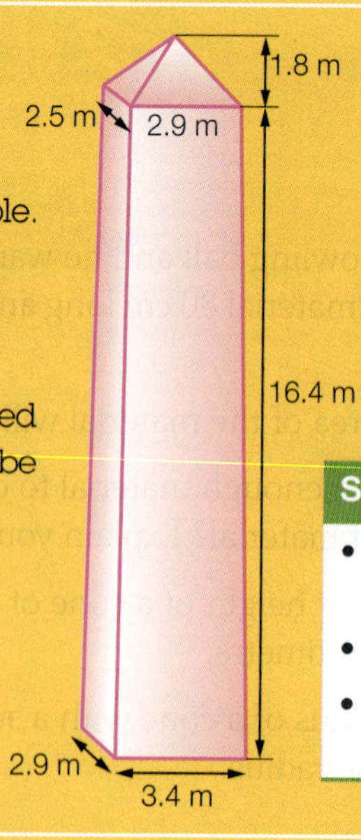

Strategy options

- Break problem into manageable parts.
- Draw a diagram.
- Have I seen a similar problem?

Volumes of tapered solids and spheres

5.5

Tapered solids

Cones and pyramids take up exactly one-third of the space of a cylinder or prism with the same base and height.

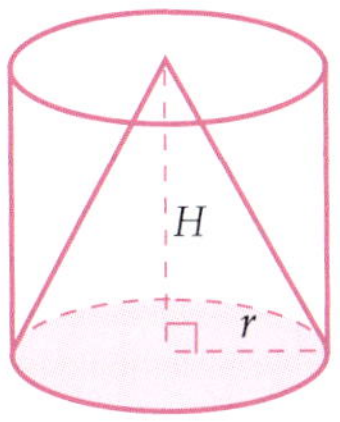

$$V_{\text{cylinder}} = AH$$

$$V_{\text{cone}} = \frac{1}{3}AH$$

The volume of a tapered solid is:

$$V = \frac{1}{3}AH$$

where A is the area of the cross-section at the base
H is the perpendicular height of the solid.

10A

Cone

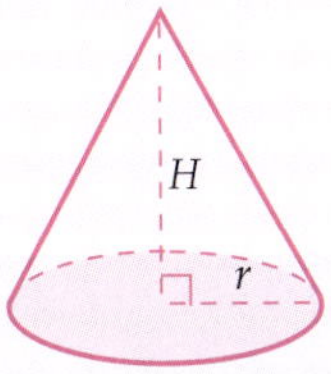

$$V = \frac{1}{3}AH$$

$$A = \pi r^2$$

$$V = \frac{1}{3}\pi r^2 H$$

Rectangular-based pyramid

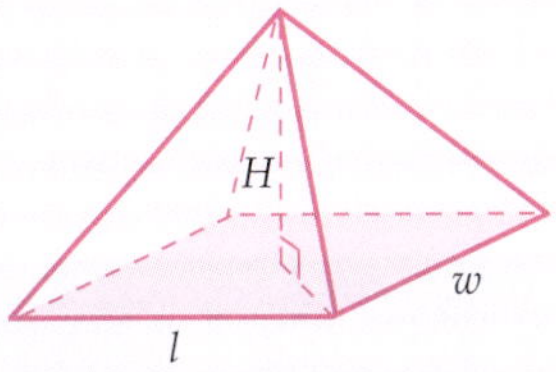

$$V = \frac{1}{3}AH$$

$$A = lw$$

$$V = \frac{1}{3}lwH$$

Worked example 11

W.E. 11

Find the volume of each of the tapered solids, rounded to 2 decimal places if necessary.

(a)

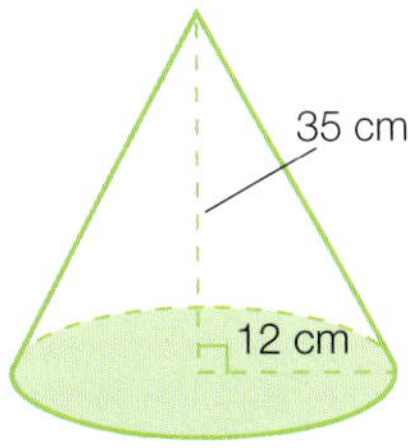

(b)

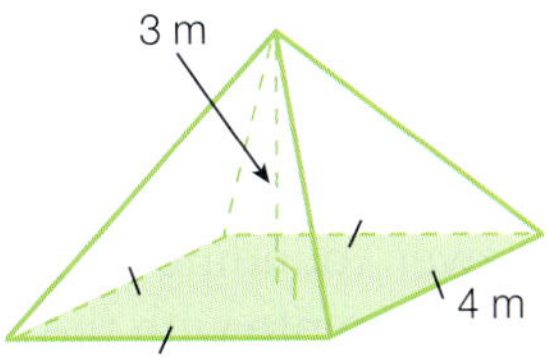

Thinking		Working
(a)	1 Write the general formula.	(a) $V = \frac{1}{3}AH$
	2 Write the specific formula for a cone.	$V = \frac{1}{3}\pi r^2 H$
	3 Substitute the values and simplify.	$= \frac{1}{3} \times \pi \times 12^2 \times 35$ $= 1680\pi$
	4 Evaluate the answer, rounding to 2 decimal places, and write with the correct unit of volume.	$= 5277.88\ \text{cm}^3$ (2 d.p.)
(b)	1 State the general formula.	(b) $V = \frac{1}{3}AH$
	2 Substitute the formula for the area of the base.	$= \frac{1}{3} l^2 H$
	3 Substitute the values.	$= \frac{1}{\cancel{3}} \times 4^2 \times \cancel{3}$
	4 Write the answer with the correct unit of volume.	$= 16\ \text{m}^3$

Spheres

Like all circular shapes, the formula for the volume of a sphere involves the value of pi (π).

Sphere

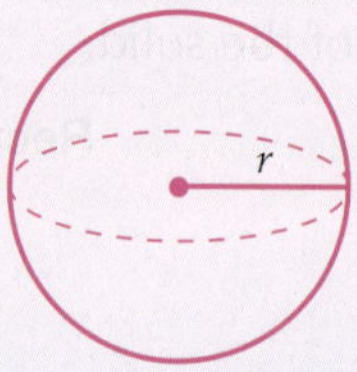

$$V = \frac{4}{3}\pi r^3$$

Worked example 12

W.E. 12

Find the volume of the sphere, rounded to 2 decimal places.

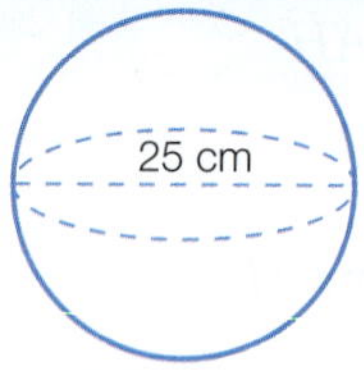

Thinking	Working
1 Write the formula.	$V = \frac{4}{3}\pi r^3$
2 Find the value of the radius.	$r = \frac{25}{2} = 12.5\ \text{cm}$
3 Substitute for r and evaluate.	$V = \frac{4}{3} \times \pi \times 12.5^3$
4 Evaluate the answer, rounding to 2 decimal places and write with the correct unit of volume.	$= 8181.23\ \text{cm}^3$ (2 d.p.)

Composite solids are made up of combinations of simpler solids. You can find their volumes by adding or subtracting the volumes of the simpler solids.

Worked example 13

W.E. 13

Find the volume of this solid, rounded to 2 decimal places.

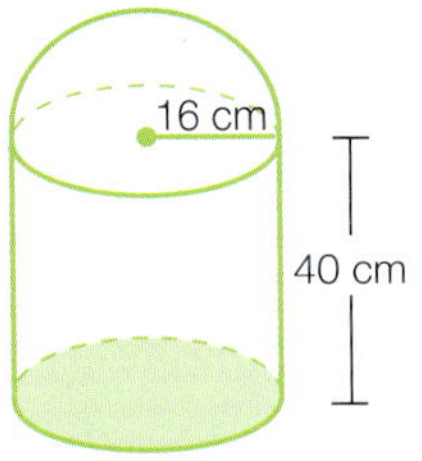

Thinking	Working
1 Identify the simple solids involved.	$V = V_{cylinder} + V_{hemisphere}$
2 Write the specific volume formula for each solid.	$= \pi r^2 H + \frac{1}{2} \times \frac{4}{3} \pi r^3$
3 Substitute the values of r and H.	$= \pi \times 16^2 \times 40 + \frac{1}{2} \times \frac{4}{3} \times \pi \times 16^3$
4 Simplify before evaluating.	$= 10\,240\pi + \frac{8192}{3}\pi$
5 Evaluate, checking that values are reasonable. Round the answer to the required number of decimal places and write with the correct unit of volume.	$= 40\,748.55$ cm^3 (2 d.p.)

10A

5.5 Volumes of tapered solids and spheres

Navigator

1, 2, 3, 4, 5, 7, 8, 9, 10, 12, 14, 15, 16	1, 2, 3, 4, 5, 6, 7, 8, 9, 10, 11, 12, 14, 15, 16	1 (column 2), 2, 3, 5, 6, 8, 9, 10, 11, 12, 13, 14, 15, 16

Answers p. 808

Fluency

1 Find the volume of each of these tapered solids, rounded to 2 decimal places.

W.E. 11

(a)

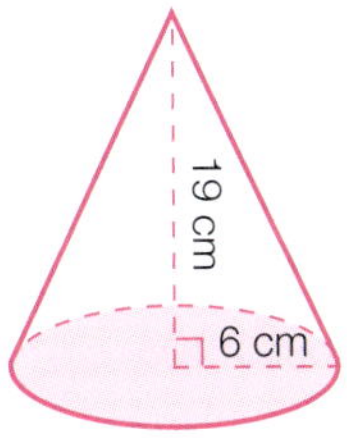

(b)

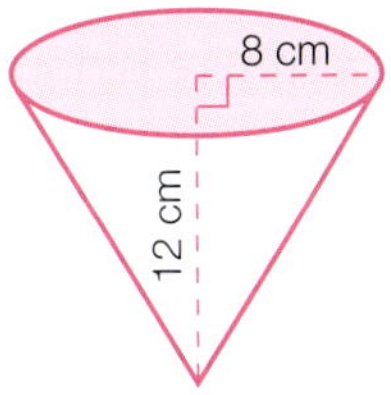

(c)

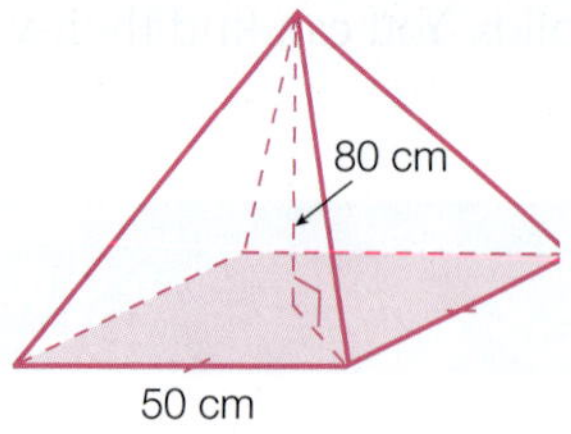

(d)

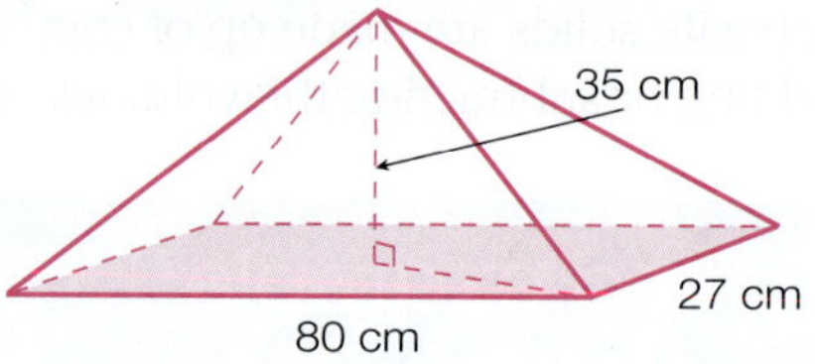

(e)

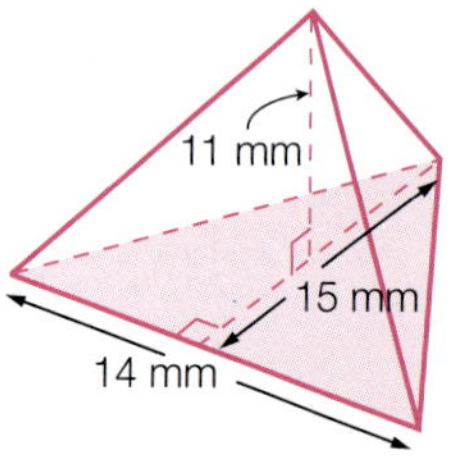

(f)

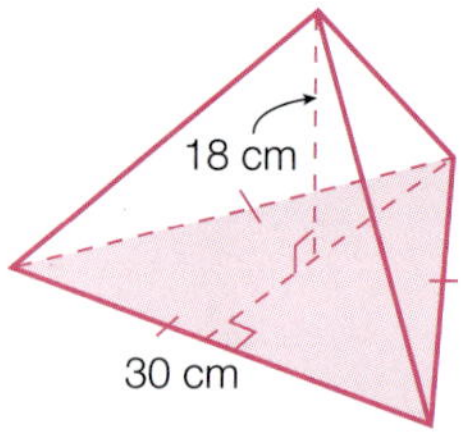

W.E. 12

2 Find the volume of each sphere, rounded to 2 decimal places.

(a)

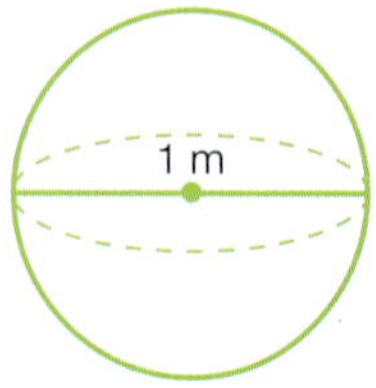

(b)

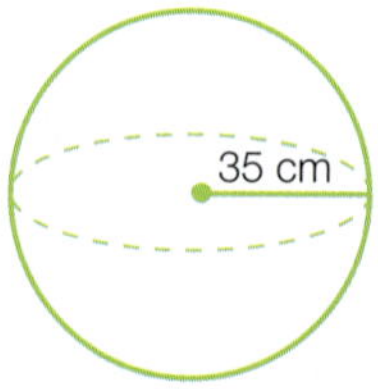

W.E. 13

3 Find the volume of each of these solids, rounded to 2 decimal places.

(a)

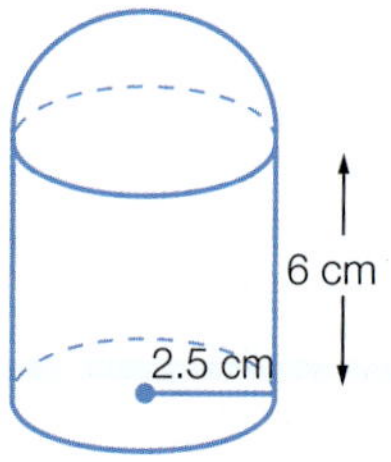

(b)

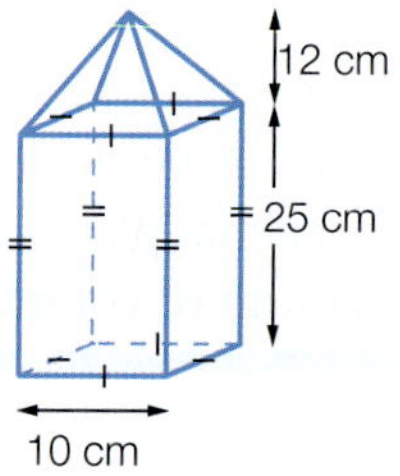

(c)

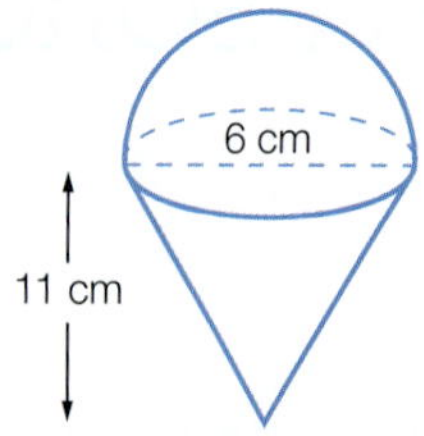

(d)

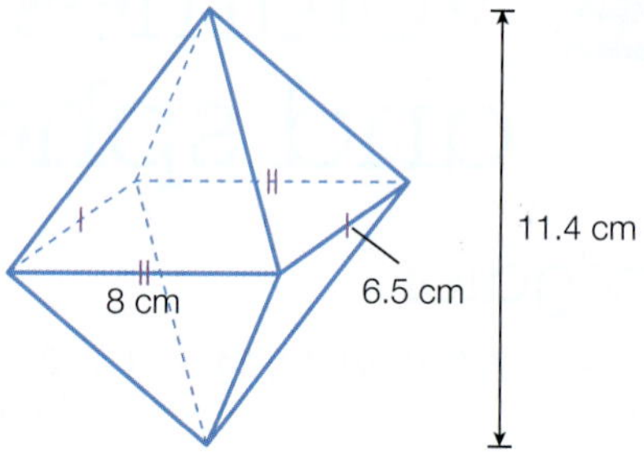

Understanding

4 Earth is assumed to be spherical with an approximate diameter of 12 800 km. What is its volume to the nearest 1000 km^3?

5 When the radius of a sphere is doubled:

A the volume is doubled

B the volume is 4 times larger

C the volume is 6 times larger

D the volume is 8 times larger

6 The radius of a cone 25 cm high with a volume of 37 969.9 cm^3 is closest to:

A 38 cm

B 121 cm

C 363 cm

D 1450 cm

7 A cone and a pyramid base have the same vertical height of 20 cm. The cone has a base diameter of 30 cm and the square base of the pyramid has a side length of 30 cm.

Find the volume of each solid to determine the difference in their volumes, to the nearest cm^3.

8 This container of fruit juice is a triangular-based pyramid with a vertical height of 9 cm. Each face is an equilateral triangle of side length 11 cm.

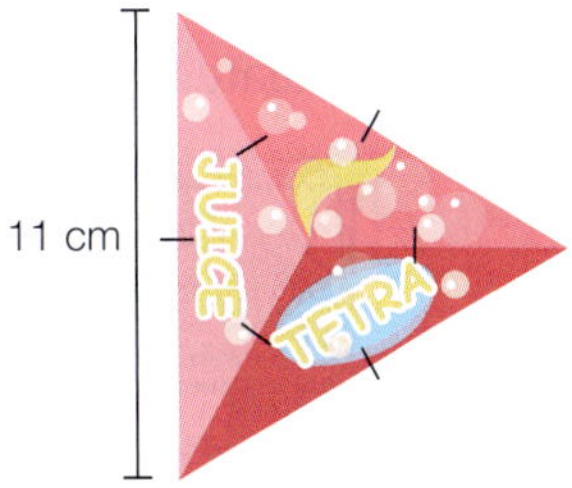

(a) Find the area of one of the triangular faces correct to 1 decimal place. Hint: Find the height of the triangle first.

(b) Find the volume of the container. State your answer as a capacity correct to the nearest mL.

9 Anica buys an ice-cream cone at her local shop. The soft-serve ice-cream fills the cone and forms a conical (cone-shaped) peak as shown in the diagram.

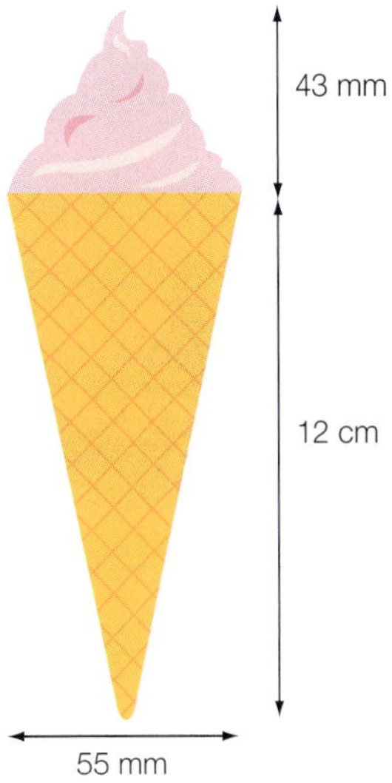

(a) Use the radius of the cone to calculate the volume of ice-cream that is inside the cone. State your answer correct to the nearest cm^3.

(b) Calculate the total amount of ice-cream in Anica's ice-cream, correct to the nearest mL.

Reasoning

10 Ball bearings are small metal spheres used to reduce friction between the moving parts of machines.

(a) Find the volume of a ball bearing with a radius of 3 mm, to the nearest mm^3.

(b) How many of these ball bearings can be made from a metal cube of side length 2 cm?

11 The table at right gives the diameters of eight planets.

(a) How many times is Jupiter's diameter bigger than Mercury's (to the nearest whole number)?

(b) Find the volumes of Mercury and Jupiter, and then calculate how many planets the same size as Mercury would fit into a planet the same size as Jupiter. Round your answer to the nearest whole number.

(c) Why are your answers to **(a)** and **(b)** so different.

(d) Calculate how many Earth-sized planets would fit into each of the larger planets.

Planet	Diameter (km)
Mercury	4 900
Venus	12 200
Earth	12 800
Mars	6 800
Jupiter	138 000
Saturn	113 000
Uranus	47 000
Neptune	50 000

10A

12 Six wheat silos, shown in the first diagram below, are full with wheat that will be transported in freight containers, as shown in the second diagram. Assume that the silos are cylinders with a cone on top and are filled to the top of the cone. The freight containers can be assumed to be trapezoidal prisms.

(a) Calculate the volumes of a wheat silo and a freight container, correct to the nearest m^3.

(b) Calculate how many freight containers will be needed to transport all the wheat.

13 (a) A cone and a cylinder have the same height and diameter. The cone is filled with water, which is then poured into the cylinder. To fill the cylinder, how many times must this be done?

(b) A second cone, the same height but twice the diameter, is filled with water, which is then poured into the cylinder. What percentage of the water in the cone will remain when the cylinder is full?

Open-ended

14 (a) Choose values for the height and base length of a square-based pyramid, and find its volume.

(b) Recalculate the volume after you double the length of the base.

(c) How many times greater is the volume of the second pyramid than the first?

(d) Repeat parts **(a)–(c)** with a different height and base. What can you conclude?

15 (a) Choose two values to represent the height and radius of a cone and evaluate its volume.

(b) Evaluate its volume after its height is doubled.

(c) Choose two different values to represent another cone and evaluate its volume.

(d) Double the height of this cone and evaluate the new volume.

(e) Can you find a relationship between the volume of a cone after the height is doubled and its original volume?

16 (a) Choose two values to represent the height and radius of a cone and evaluate its volume.

(b) Evaluate its volume after its radius is doubled.

(c) Choose two different values to represent another cone and evaluate its volume.

(d) Double the radius of this cone and evaluate the new volume.

(e) Can you find a relationship between the volume of a cone after the radius is doubled and its original volume?

10A

Rearranging formulas

5.6

A formula is an equation showing the relationship between two or more variables or unknowns. The equation $V = \pi r^2 h$ is an example of a formula used in measurement. You can rearrange this formula to make r or h the subject ($r =$ or $h =$).

To rearrange formulas you must apply the same inverse operations to both sides, using the same methods as for solving equations.

Equations involving powers

The inverse operation of squaring is taking the square root. The solution to $x^2 = 9$ is $x = 3$ or $x = -3$, because $(3)^2 = (-3)^2 = 9$. Therefore when taking the square root, you must include ± to show that the result may be + or –.

In many cases, only the positive square root is an acceptable answer. For example, variables that represent area, volume and length can only have positive values.

Worked example 14

W.E. 14

(a) Rearrange $V = \frac{1}{3}\pi r^2 h$ to make r the subject of the formula.

(b) Find r when $V = 27\pi$ and $h = 4$ are substituted into the rearranged formula.

Thinking	Working
(a) 1 Perform any inverse operations needed to make the required variable the subject of the formula. (Here, multiply by 3 and divide by πh.)	(a) $V = \frac{1}{3}\pi r^2 h$ $3V = \pi r^2 h$ $\frac{3V}{\pi} = r^2 h$ $r^2 = \frac{3V}{\pi h}$
2 Take the square root of both sides.	$r = \pm\sqrt{\frac{3V}{\pi h}}$
(b) Write the formula, substitute the values for V and h and simplify.	(b) $r = \pm\sqrt{\frac{3V}{\pi h}}$ $= \pm\sqrt{\frac{3 \times 27\cancel{\pi}}{\cancel{\pi} \times 4}}$ $= \pm\sqrt{\frac{81}{4}}$ $= \pm\frac{9}{2}$ $= \pm 4\frac{1}{2}$

Equations involving roots

To remove a square root from an equation you can square both sides.

Worked example 15 — W.E. 15

(a) Rearrange $T = 2\pi\sqrt{\frac{l}{g}}$ to make **(i)** l and **(ii)** g the subject of the formula.

(b) **(i)** Find l when $T = 0.6\pi$ and $g = 10$ are substituted into the rearranged formula.

(ii) Find g when $T = \frac{4\pi}{7}$ and $l = 0.8$ are substituted into the rearranged formula.

Thinking	Working
(a) 1 Perform any inverse operations needed to isolate the square root on one side of the equation. (Here, divide by 2π.)	(a) $T = 2\pi\sqrt{\frac{l}{g}}$ $\frac{T}{2\pi} = \sqrt{\frac{l}{g}}$
2 Square both sides of the equation.	$\left(\frac{T}{2\pi}\right)^2 = \frac{l}{g}$
3 Perform any inverse operations needed to make the required variable the subject of the formula.	$l = g \times \left(\frac{T}{2\pi}\right)^2$ (i) $l = \frac{gT^2}{4\pi^2}$ (ii) $g = \frac{4\pi^2 l}{T^2}$
(b) (i) Substitute values into the formula for l and simplify.	(b) (i) $l = \frac{gT^2}{4\pi^2}$ $= \frac{10 \times (0.6\pi)^2}{4\pi^2}$ $= \frac{10 \times \cancelto{0.09}{0.36}\cancel{\pi^2}}{\cancel{4}\cancel{\pi^2}}$ $= 0.9$
(ii) Substitute values into the formula for g and simplify.	(ii) $g = \frac{4\pi^2 l}{T^2}$ $= \frac{4 \times \pi^2 \times 0.8}{\left(\frac{4\pi}{7}\right)^2}$ $= \frac{\cancel{4}\cancel{\pi^2} \times 49 \times \cancelto{0.2}{0.8}}{\cancel{16}\cancel{\pi^2}}$ $= 9.8$

5.6 Rearranging formulas

Navigator

1 (column 1), 2, 3, 4, 5, 6, 7, 8, 10, 11	1 (column 2), 2, 3, 4, 5, 6, 7, 8, 9, 11	1 (e–i), 2, 3, 4, 5, 7, 8, 9, 11

Answers p. 808

Fluency

1 Rearrange each of the following to make the variable in brackets the subject of the formula.

(a) $A = \frac{1}{2}bh$ [h]

(b) $A = \pi r^2$ [r]

(c) $A = \frac{1}{2}r^2\theta$ [θ]

(d) $a^2 + b^2 = c^2$ [b]

(e) $y = mx + b$ [x]

(f) $x = \dfrac{b-a}{n}$ [a]

(g) $t = \sqrt{\dfrac{s^2-1}{6}}$ [s]

(h) $F = m\dfrac{d^2}{t^2}$ [t]

(i) $v = \frac{P}{4}(R^2 - 9)$ [R]

2 (i) Rearrange each of the following to make the variable in brackets the subject of the formula.

W.E. 14

(ii) Find the value of this variable when the given values are substituted into the rearranged formula.

(a) $A = s^2$ [s] $A = 27$

(b) $K = \frac{1}{2}mv^2$ [v] $K = 20$ and $m = 8$

(c) $F = \dfrac{mv^2}{r}$ [v] $F = 9$, $r = 7$ and $m = 6$

(d) $F = \dfrac{Gm_1m_2}{r^2}$ [r] $F = 16$, $G = 18$, $m_1 = 6$ and $m_2 = 5$

(e) $g_1 = \dfrac{r_2{}^2}{r_1{}^2}g_2$ [r_2] $g_1 = 9.8$, $g_2 = 10$ and $r_1 = 20$

(f) $P = \dfrac{V^2}{R}$ [V] $P = 9$ and $R = 7$

3 (i) Rearrange each of the following to make the variable in brackets the subject of the formula.

W.E. 15

(ii) Find the value of the variable in brackets when the given values are substituted into the rearranged formula.

(a) $A = \sqrt{A_x + A_y}$ [A_y] $A = 9$ and $A_x = 43$

(b) $x = \dfrac{-b \pm \sqrt{b^2 - 4ac}}{2a}$ [c] $b = 9$, $a = 4$ and $x = -1$

(c) $l = \dfrac{3r}{\sqrt{T}}$ [T] $r = 8$ and $l = 3$

(d) $x = \sqrt{y - z}$ [z] $y = 4$ and $x = 10$

(e) $x = \sqrt{\frac{y}{z}}$ [z] $y = 7$ and $x = 5$

(f) $x = \sqrt{\frac{a+b}{c}}$ [b] $c = 3, x = 2$ and $a = -4$

4 Which of the following expressions is equivalent to $S = \frac{4r^2}{\sqrt{t}}$?

A $t = \frac{16r^2}{S^2}$ B $r = \pm\sqrt{\frac{St}{4}}$ C $t = \frac{16r^4}{S^2}$ D $r = \frac{S\sqrt{t}}{4}$

Understanding

5 The formula for the area of a circle is $A = \pi r^2$.

(a) Rearrange this equation to make r the subject.

(b) Explain why there is no ± sign required.

(c) Find the radius (r) when the area (A) is 40 cm^2 correct to 2 decimal places.

6 The formula for the volume of a sphere is $V = \frac{4}{3}\pi r^3$.

(a) Rearrange this equation to make r the subject.

(b) Explain why there is no ± sign required.

(c) Find the radius (r) when the volume (V) is 125 cm^3 correct to 2 decimal places.

7 The formula for the surface area of a cone is $V = \frac{1}{3}\pi r^2 h$. Rearrange this equation to change the subject, then find correct to 2 decimal places:

(a) the height of a cone with volume 312 m^3 and radius 14 m

(b) the radius of a cone with volume 160 cm^3 and height 8 cm.

Reasoning

8 The formula $v^2 = u^2 + 2as$ is regularly used in physics to find the final velocity (v) of an object that has initial velocity (u) acceleration (a) and displacement (s). (Displacement is the distance of the object from its starting point.)

(a) Calculate the initial velocity of a car that stops in 42 m, with acceleration -20 m/s^2.

(b) What does a negative value for the acceleration mean in this situation?

9 The formula used to determine the frequency (f) of a spring, given the spring constant (k) and mass of the object (m) is $f = \frac{1}{2\pi}\sqrt{\frac{k}{m}}$. Frequency is measured in hertz (Hz), which is the number of cycles (up and down oscillations) per second. A spring with an object of mass 3.5 kg has a frequency of 3 Hz.

(a) Calculate the spring constant for this spring.

(b) An object of mass 7.2 kg is attached to the same spring. Calculate the new frequency.

10 Elizabeth tries to change the subject of the formula $g = 4h^2 - 9k$ to h by completing the following working.

$$g = 4h^2 - 9k$$

$$g + 9k = 4h^2 \quad \text{Line 1}$$

$$h^2 = \frac{g + 9k}{4} \quad \text{Line 2}$$

$$h = \sqrt{\frac{g + 9k}{4}} \quad \text{Line 3}$$

$$h = \frac{\sqrt{g} + 9k}{2} \quad \text{Line 4}$$

(a) Identify the line with the mistake.

(b) Explain why this is mathematically inaccurate.

Open-ended

11 (a) Make up a formula of your own with at least three variables. One of the variables, that is not the subject of this formula, must be squared. Rearrange the formula to make this variable the subject.

(b) Make up a formula of your own with at least three variables. Take the square root of one of the variables that is not the subject of this formula. Rearrange the formula to make this variable the subject.

Problem solving

Gravity

The formula $v = \sqrt{2gx}$ is used to calculate the velocity (v) of an object that has already fallen x metres, where g is acceleration due to gravity. Gravity is different depending on the planet that you are on, as shown in the table.

Planet	Acceleration due to gravity (m/s^2)
Mercury	3.59
Venus	8.87
Earth	9.81
Mars	3.77
Jupiter	25.95
Saturn	11.08

1 On which planet would it take the longest time for an object to fall 100 m?

Hint

The slowest velocity gives the longest time.

2 According to this formula, does it take longer for a feather or a heavy boulder to fall?

3 How much higher do you have to drop a ball when on Mars, so that the velocity when the ball hits the ground is the same as if dropped on Earth?

Strategy options

- Make a table.
- Work backwards.

Investigation

The great golf ball box dilemma

Equipment required: thin cardboard, scissors and glue (optional)

You are the packaging designer for Supreme Sports, a company that manufactures sports equipment. Your manager has asked you to design the packaging for a set of six golf balls. The packaging must be space-efficient (minimal wasted space) and as cost-effective as possible (minimal material used). The packaging should also look good and be easy to handle, pack and stack.

The Big Question

What is the best way to package six golf balls?

Engage

1 The diameter of a standard size golf ball is 4.27 cm. Calculate the volume of six golf balls.

2 You have come up with two basic designs: a long cylindrical box, packing the balls end to end, and a rectangular box holding two rows of three golf balls.

10A

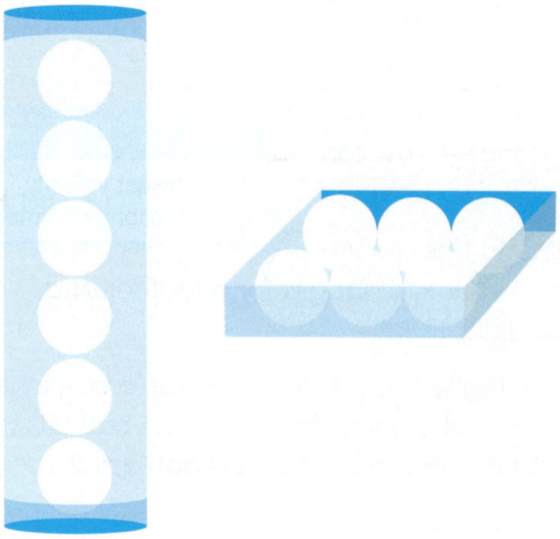

(a) Assuming that the balls are touching each other, determine the minimum length and diameter of the cylinder. Allow 1 cm at the top and 0.5 cm around the outside of the balls for clearance space.

(b) Assuming that the balls are touching each other, determine the minimum length and width of the rectangular prism. Allow 1 cm extra length, 1 cm extra width and an extra 1 cm in height, for clearance space.

Explore

3 **(a)** For the cylindrical and the rectangular boxes, calculate the volume of:

(i) the box

(ii) the empty space when the balls are inside.

(b) Find the volume of empty space as a percentage of the total volume.

4 **(a)** Draw a net for both the cylinder and the rectangular prism using the dimensions from **2**. In order for the nets to be constructed, you will need to include a minimal number of tabs (flaps) so that the sides can be stuck down. Make these tabs 1 cm wide.

(b) Determine the surface area of each net.

(c) Calculate the cost of constructing each design if the cardboard used to make them costs $\$0.005/cm^2$.

Strategy options

- Break problem into manageable parts.
- Make a table.
- Have I seen a similar problem?

Explain

5 Which design is more space-efficient? Use your results to justify your choice.

6 Which design is more cost-efficient? Use your results to justify your choice.

Elaborate

7 Considering the criteria of space and cost efficiency, which design would you recommend that the company use?

8 The packaging should be easy to handle, pack and stack. How well does your chosen design fit these criteria?

Evaluate

9 (a) Comment on the accuracy of your calculations and identify any potential sources of error.

(b) Do you think the extra room allowed for clearance space at the ends and sides was reasonable and realistic? If not, suggest some more reasonable measurements.

Extend

10 Use the nets you drew in **4** to construct a prototype of one or both of the designs. Create a company logo and use it to decorate the packaging.

11 Your manager likes your cylindrical design but thinks it would be even more efficient if the ends were rounded into the shape of a hemisphere, so that half a golf ball could fit inside.

(a) Modify your design as your manager suggested. Recalculate the required length of the packaging, allowing the same clearance space as before.

(b) Recalculate the volume and surface area of this new packaging. Is it more efficient than the other two designs?

12 Come up with a new packaging design that is an entirely different shape to those already considered. Is this design more space- or cost-efficient than the others?

10A

5.7 Applications of volume

Volume can be used to calculate other important properties of solids, liquids or gases.

Density

If two substances have the same volume, then the heavier substance has greater density. For example, a litre of water has a mass of 1 kg. A litre of cooking oil has a mass of less than 1 kg, so cooking oil has less density. This is why cooking oil floats on water.

The **density** of a substance is the amount of mass within a given unit of volume.

A hollow object, such as a ping-pong ball, is mostly made of air, so its total mass is much less than the same volume of liquid.

A ping-pong ball will float on almost any liquid.

This density-reducing 'hollowness' explains why large heavy ships float in water.

10A

Density = $\dfrac{\text{mass}}{\text{volume}}$ or $d = \dfrac{m}{V}$

where d = density in g/cm^3 (g/mL) or kg/m^3 (kg/kL)

m = mass in grams (g) or kilograms (kg)

V = volume in centimetres cubed (cm^3, or mL) or metres cubed (m^3 or kL)

Measurement units may vary, but the density units are always the mass units divided by the volume units.

Worked example 16 W.E. 16

A rock has a mass of 80 g and volume of 25 cm^3. Find its density.

Thinking	Working
1 Write the formula.	$d = \dfrac{m}{V}$
2 Substitute the values with their units.	$= \dfrac{80 \text{ g}}{25 \text{ cm}^3}$
3 Evaluate the answer and write it with the correct units (g/cm^3).	$= 3.2 \text{ g/cm}^3$

Density is like a rate that measures the change in one quantity (mass) compared to a fixed amount of another quantity (volume). A triangle diagram is helpful in rearranging the density formula to find mass and volume.

Cover the quantity you want to find. If the other two quantities are side by side, then multiply. If the two quantities are placed one over the other, then divide.

$$d = \frac{m}{V}$$

$$m = d \times V$$

$$V = \frac{m}{d}$$

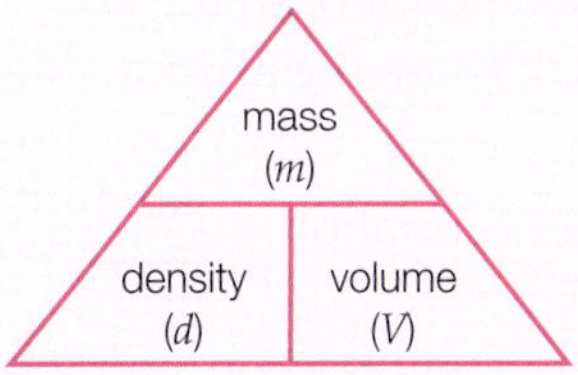

Worked example 17

W.E. 17

Find the mass of 400 cm^3 of a block of metal with a density of 1.6 g/cm^3.

Thinking	Working
1 Write the formula.	$d = \frac{m}{V}$
2 Rearrange the formula to make *m* the subject.	$m = d \times v$
3 Substitute the values and evaluate.	$m = 1.6 \times 400$
4 Write the answer giving the correct units (grams, as density was given in g/cm^3).	$= 640$ g
5 Check by recalculating the density.	Check: $\frac{640 \text{ g}}{400 \text{ cm}^3} = 1.6 \text{ g/cm}^3$

Worked example 18

W.E. 18

Find the volume of a 120 g piece of gold with a density of 19.3 g/cm^3 and round your answer to 2 decimal places.

Thinking	Working
1 Write the formula.	$d = \frac{m}{V}$
2 Rearrange the formula to make *V* the subject.	$V = \frac{m}{d}$
3 Substitute the values and evaluate.	$V = \frac{120}{19.3}$
4 Write the answer giving the correct units (cm^3, as density was given in g/cm^3).	$= 6.22 \text{ cm}^3$ (2 d.p.)
5 Check by recalculating the density.	Check: $\frac{120 \text{ g}}{6.22 \text{ cm}^3} = 19.3 \text{ g/cm}^3$ (1 d.p.)

10A

Concentration

The amount of a substance dissolved in a unit amount of liquid is the **concentration**. The dissolved substance is called the solute and the liquid it is dissolved in is called the solvent. The liquid solvent and solute together are called the solution. For example, the concentration of sugar dissolved in water is the number of grams of sugar in a litre of the solution (g/L).

The amount of solute and solution may be measured both as volume or both as mass, so mL/L or mg/kg are possible units of concentration. In medicine, g/mL is a common unit.

$$\text{Concentration} = \frac{\text{amount of solute (mass or volume)}}{\text{amount of solution (mass or volume)}}$$

Note that the concentration is the amount of solute compared to the final amount of solution, not compared to the amount of solvent added.

Worked example 19

W.E. 19

A 3 L jug of cordial solution is made with 150 mL of cordial concentrate. Find the concentration of the cordial in mL/L.

Thinking	Working
1 Write the formula.	$\text{Concentration} = \frac{\text{amount of solute}}{\text{amount of solution}}$
2 Substitute the amounts with their units.	$= \frac{150 \text{ mL}}{3 \text{ L}}$
3 Simplify the fraction so that the denominator is 1 L. (Here, divide the numerator and denominator by 3.)	$= \frac{50 \text{ mL}}{1 \text{ L}}$
4 Write the concentration as a rate.	The concentration of the cordial is 50 mL/L.

Worked example 20

W.E. 20

6 g of salt is dissolved in water to make 120 g of saline solution. Find the concentration of the solution in g/100 g.

Thinking	Working
1 Write the formula.	$\text{Concentration} = \frac{\text{amount of solute}}{\text{amount of solution}}$
2 Substitute the amounts with their units.	$= \frac{6 \text{ g}}{120 \text{ g}}$
3 Perform the division to convert to an amount per gram and then multiply by 100 to find the amount per 100 g.	$= \frac{6}{120} \times 100$
4 Write the concentration as a rate.	Concentration is 5 g/100 g.

10A

5.7 Applications of volume

Navigator

1, 2, 3, 4, 5, 6, 7, 8, 11, 12, 14	1, 2, 3, 5, 6, 7, 8, 10, 11, 12, 13, 14	1, 2, 3, 7, 8, 9, 10, 11, 12, 13, 14

Answers p. 809

Fluency

1 For each of the following, find the density: W.E. 16

(a) a metal sphere of volume 9 cm^3, and mass 63.9 g

(b) 20 cm^3 of liquid that has a mass of 128 g

(c) a solid of volume 24 cm^3 and mass 19.2 g.

2 Calculate the mass of each of the following, rounded to 2 decimal places: W.E. 17

(a) a lump of metal with a volume of 800 cm^3 and a density of 7.8 g/cm^3

(b) 450 mL of a liquid that has a density of 1.08 g/mL

(c) 2 m^3 of a gas that has a density of 1.98 kg/m^3.

3 Calculate the volume of each of the following, rounded to 2 decimal places: W.E. 18

(a) a150 g piece of lead with a density of 11.3 g/cm^3

(b) 486 g of a liquid with a density of 0.7 g/mL

(c) 13 kg of wood that has a density of 0.85 kg/m^3.

4 2.5 L of sugar solution contains 50 g of sugar. What is its concentration in g/L? W.E. 19

10A

5 500 g of weedkiller solution contains 12 g of active ingredient. Find its concentration in g/100 g. W.E. 20

6 (a) A saline solution is made from 2.5 L of water and 28 g of salt. The concentration of salt in g/L is closest to:

A 28 B 14 C 12 D 11

(b) A solution of copper sulfate is made from 1350 L of water and 1.6 kg of copper sulfate. The concentration of copper sulfate in g/L is closest to:

A 0.8 B 1 C 800 D 1000

Understanding

7 The Goulburn River carries sediment from erosion in its catchment. Barry takes a 5 L sample of water from the river. From this he extracts 38 g of sediment. What is the concentration of sediment in the water, in mg/mL?

Hint
1000 mg = 1 g

8 Marlon needs 5 L of cordial for a party. To make the cordial he adds 1 cup of cordial concentrate to 4 cups of water. He uses a standard 250 mL cup to measure.

(a) How much cordial can Marlon make from 1 cup of cordial concentrate?

(b) To make 5 L of cordial for the party, how many cups of cordial concentrate and how many cups of water are required?

(c) What is the concentration of cordial concentrate in the final solution of cordial? Write your answer in mL/L.

Hint

$1\ m^3 = 1000$ L

$1\ m^3 = 1$ kL

1 tonne = 1000 kg

9 **(a)** Crude oil is measured in units called 'barrels'. 1 barrel is equivalent to 159 litres. Calculate the mass of 1 barrel of crude oil if the density of the crude oil is $825\ kg/m^3$.

(b) Petrol has a density of $737\ kg/m^3$. A tanker truck used for transporting petrol has a capacity of 30 kL. Calculate the mass of petrol in a full tanker truck. Round your answer to the nearest tonne.

10 Jay's circular swimming pool has a diameter of 7 m and a depth of 1.6 m, but Jay only fills it to 20 cm below the rim. Calculate, correct to 2 decimal places, the chlorine concentration in g/L if he adds 2.1 kg of chlorine.

Reasoning

11 Christa pours 1200 mL of liquid fertiliser into a watering can and then fills it with water. The watering can is cylindrical with a radius of 10 cm and a height of 35 cm.

(a) Find the volume (in cm^3) of the watering can and then state its capacity to the nearest litre.

(b) Find the concentration of liquid fertiliser to the nearest mL/L in the watering can.

(c) Christa discovers she only has 500 mL of fertiliser left. It is recommended that the concentration of fertiliser is close to 120 mL/L. How many litres of water should she add to the watering can? Round your answer to the nearest litre.

12 The density of water is $1\ g/cm^3$. An object will float in water if its density is *less* than $1\ g/cm^3$.

(a) Calculate the densities of these objects and determine which will float in water.

A wooden log
Mass: 100 kg

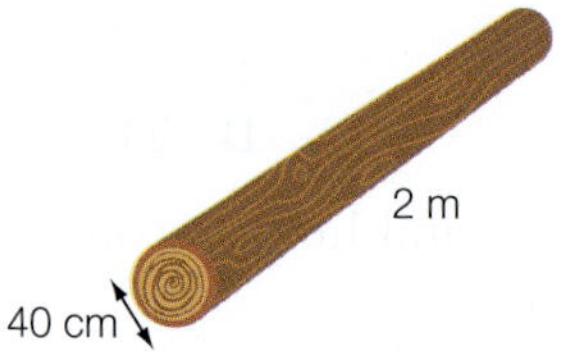

A steel rod
Mass: 0.5 kg

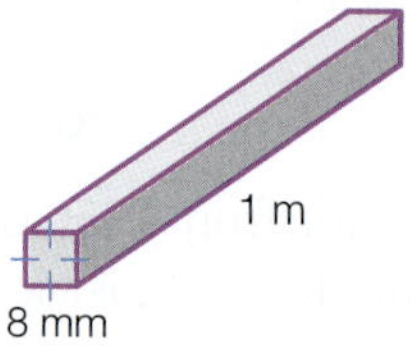

A cork
Mass: 2 g

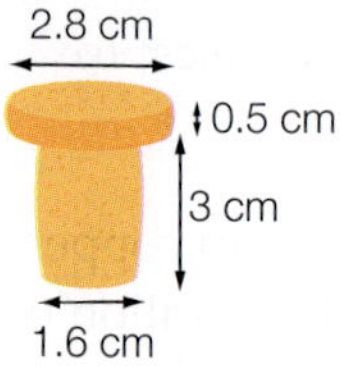

An aluminium cylinder
Mass: 200 g

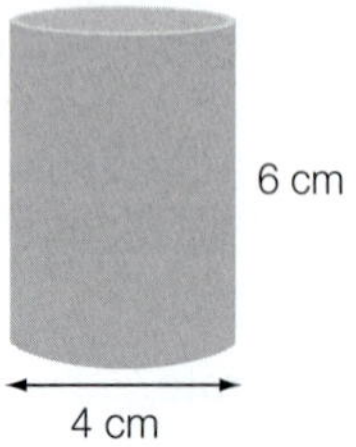

An ice block
Mass: 10 g

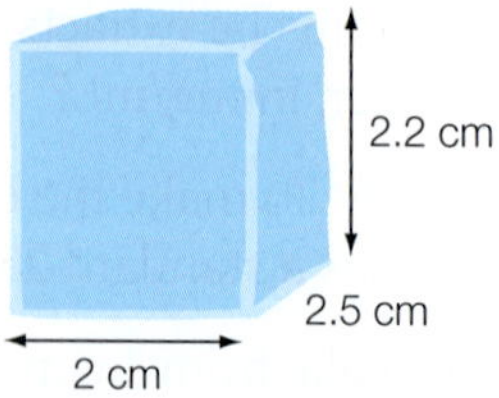

A gold bar
Mass: 16 kg

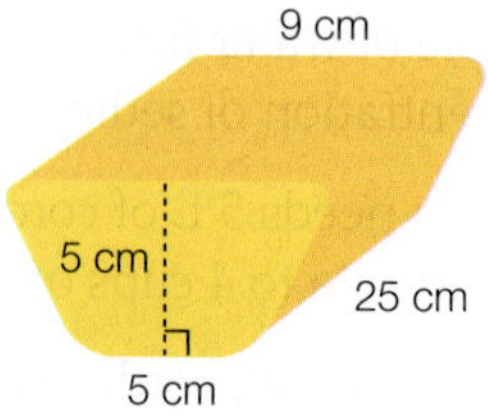

(b) Find the mass of each object if made of lead, which has a density of $11.35\ g/cm^3$. Write your answers correct to the nearest gram.

13 High phosphorus levels in water can lead to blue-green algal blooms. The town of Evanshill takes its water from Lake Farmer, which has phosphorus levels contributing to blue-green algal blooms. This means the water supply to the town must be shut down. Two main sources of phosphorus are:

- Johnstone's abattoir, which discharges 5 ML (megalitres) per day (where 1 ML = 1 000 000 L) with a phosphorus concentration of 10 mg/L
- the Arvon River, which has an average discharge of 200 ML per day at a phosphorus concentration of 0.2 mg/L.

(a) Find the mass of phosphorus (in kg) that each source contributes to Lake Farmer, to determine which source is contributing the most phosphorus.

(b) Find the actual concentration of phosphorus contributed to Lake Farmer each day in mg/L (that is, milligrams of phosphorus per litre of water). To do this, first calculate the total mass of phosphorus and the total volume of water from both sources.

Open-ended

14 The Environment Protection Authority monitors the level of the common bacteria *E. coli* at swimming beaches. If the level on any day is above 1000 organisms per 100 mL, then the beach is closed. Tess collects a 200 mL water sample from a beach. An examination of the sample shows that in 80 mL there are 700 *E. coli* organisms.

(a) Find the concentration of *E. coli* organisms per millilitre, and then calculate the number of organisms of *E. coli* in 100 mL.

(b) Is the concentration in the sample high enough to close the beach?

(c) List three sample sizes and the number of *E. coli* that would be high enough to close the beach.

10A

Problem solving

A pinch of salt

The salt concentration in the ocean is about 0.035 g/cm^3. How much salt could be extracted from a cubic kilometre of sea water?

Strategy options

- Break problem into manageable parts.
- Have I seen a similar problem?

Chapter review 5

Maths literacy

capacity	cylinder	pyramid	
composite shape	net	surface area	
cone	prism	uniform cross-section	
10A concentration	density	sphere	tapered solid

Copy and complete the following using the words and phrases from the list, where appropriate. A word or phrase may be used more than once.

1 To find the area of a __________, divide the shape into simple shapes, calculate the individual areas and then add them together.

2 Adding together the individual areas of each face of a solid will give the __________.

3 The volume of a __________ or __________ is found by multiplying the area of its uniform cross-section by its height.

4 Volume is measured in m^3 but __________ is measured in L.

10A 5 The volume and surface area of a __________, cylinder and __________ each have unique formulas involving π.

10A 6 The definition for __________ is $\frac{\text{mass}}{\text{volume}}$.

10A 7 The __________ of a dissolved substance in a solution can be expressed in many different units. These include mL/L, g/100 g, mg/L or g/L.

Fluency

1 Find the shaded area of each of these shapes, rounded to 2 decimal places. **5.1**

(a)

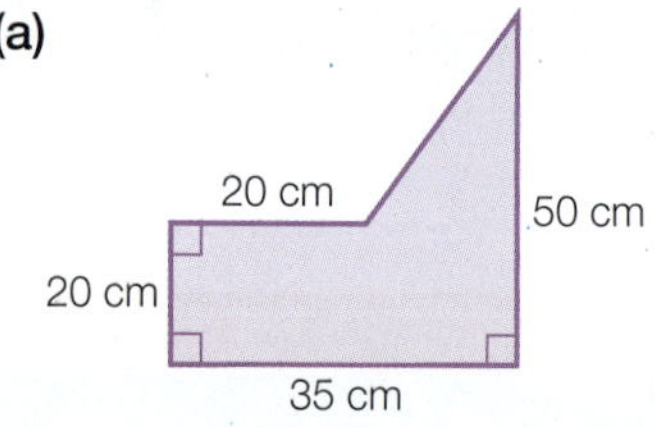

(b)

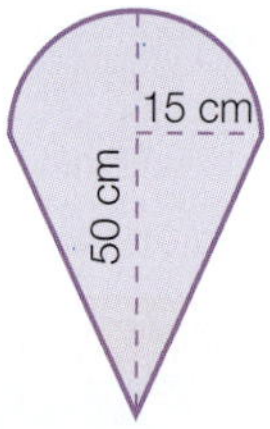

(c)

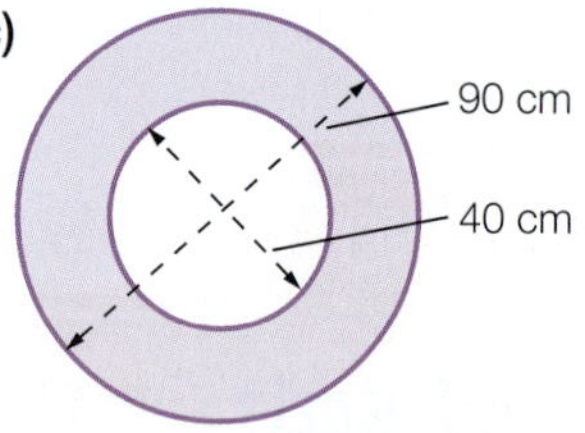

2 The area of this shape correct to 2 decimal places is: **5.1**

A 182.12 cm^2 B 945.53 cm^2

C 1143.45 cm^2 D 2145.53 cm^2

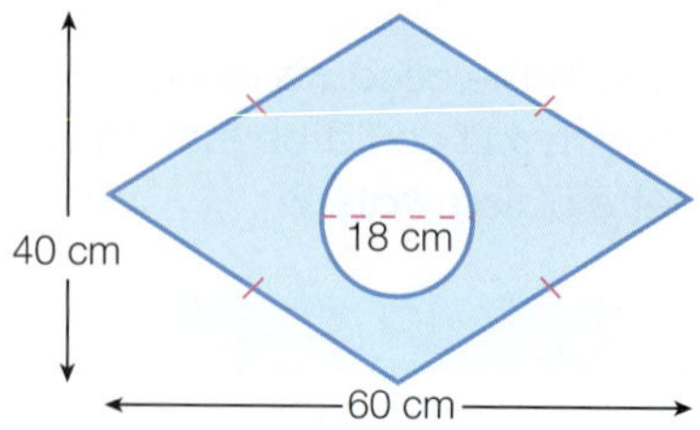

3 Find the surface area of each of these solids, rounded to 2 decimal places. **5.2**

(a)

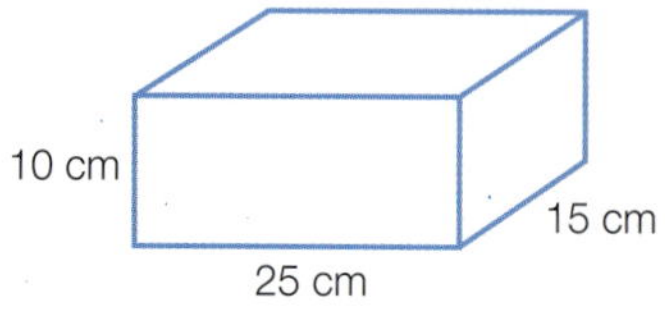

(b)

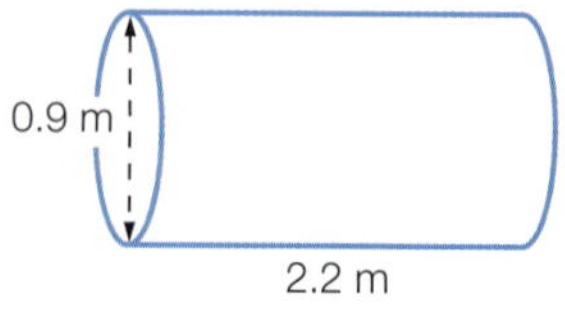

10A **4** The surface area of this solid is equal to: **5.4**

A 340 cm^2 **B** 840 cm^2

C 980 cm^2 **D** 1180 cm^2

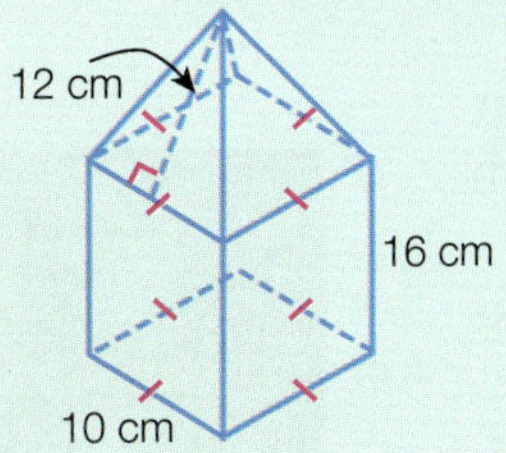

5 Find the volume of each of these solids, rounded to 2 decimal places. **5.3**

(a)

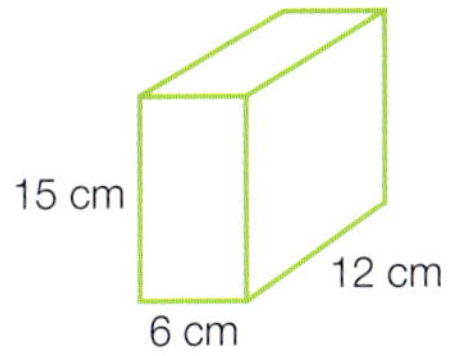

(b)

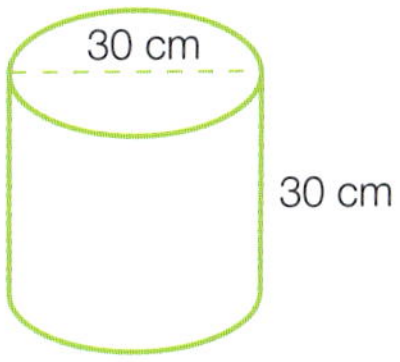

(c)

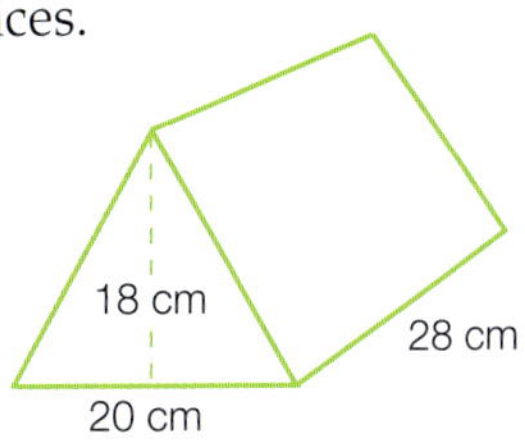

10A **6** Find the surface area of each of these solids, rounded to 2 decimal places. **5.4**

(a)

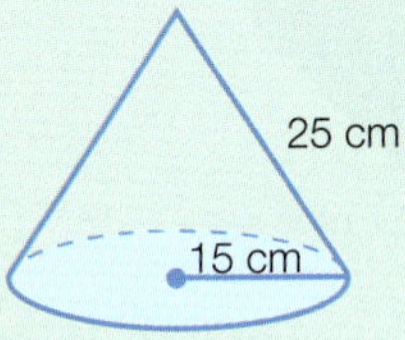

(b)

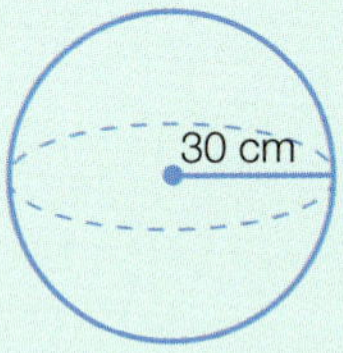

10A **7** Find the volume of each of these solids, rounded to 2 decimal places. **5.5**

(a)

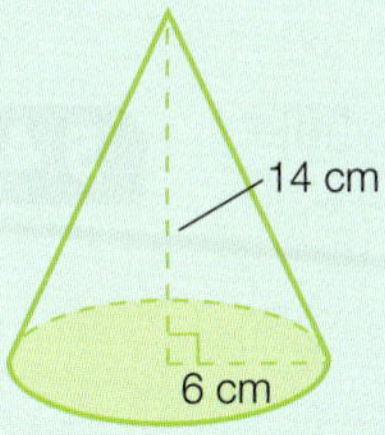

(b)

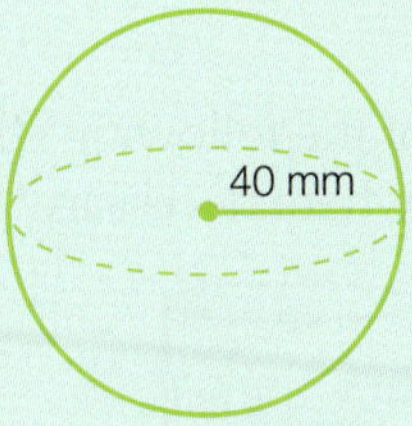

10A **8** The volume of this solid is: **5.5**

A 192 cm^3 **B** 208 cm^3

C 288 cm^3 **D** 576 cm^3

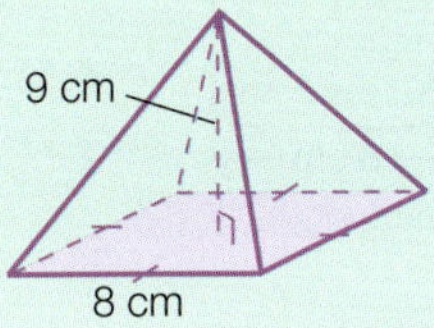

9 Rearrange each of the following to make the pronumeral in brackets the subject. **5.6**

(a) $A = lb$ $[b]$ (b) $A = 2\pi\sqrt{\frac{l}{g}}$ $[l]$

(c) $\text{SA} = 4\pi r^2$ $[r]$ (d) $a = x^2y - z$ $[x]$

10A **10** A rectangular block of wood with the dimensions of 1 m × 2 m × 2 m has a mass of 120 kg. What is its density? **5.7**

10A **11** A certain antiseptic should be mixed with warm water for best effects. If 30 mL is mixed with 120 mL of water, write this as a concentration in mL/L. **5.7**

Understanding

12 The diagram shows four garden beds, each 5.4 m long and 1.8 m wide, surrounded by paving bricks. Find the number of bricks needed to pave the area (to the nearest 100) if a brick measures 225 mm by 112 mm. **5.1**

13 **(a)** Find the capacity of a cylindrical petrol tank of height 3.2 m and diameter 2.5 m. Write your answer to the nearest litre. **5.3**

(b) The dimensions of a cylindrical rainwater tank are diameter 1760 mm and height 2100 mm. Calculate the capacity of the tank to the nearest litre.

14 Find the volume of material used to make this pipe in m^3, correct to 2 decimal places. **5.3**

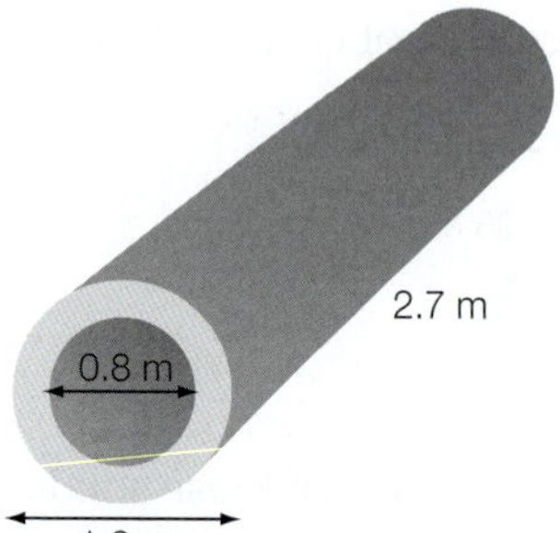

10A **15** Find the surface area and volume of the following shapes rounded to the nearest whole number. Use Pythagoras' theorem where necessary. **5.4, 5.5**

(a)

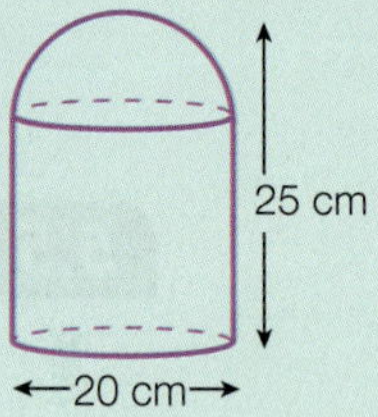

(b)

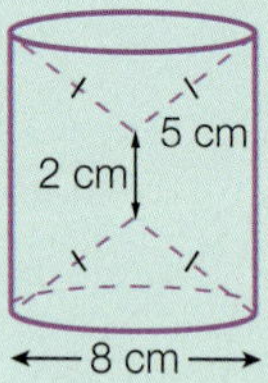

16 A snowball-throwing court has the measurements shown. **5.1**

(a) Calculate the area in which the thrower has to land the snowball, to the nearest square metre.

(b) Find the perimeter of the shaded area, to the nearest metre.

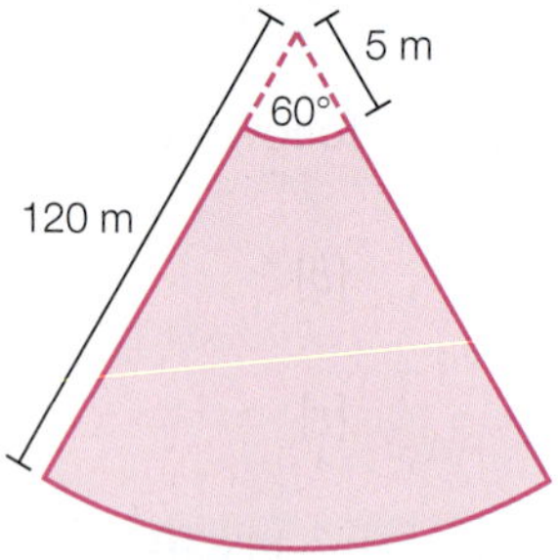

10A **17** **(a)** Find the volume of a 200 g block of Styrofoam if its density is 0.1 g/cm^3. **5.7**

(b) What mass of salt should be dissolved in 300 mL of pure water to give a concentration of 50 g/L?

10A **18** If the inside of this ice-cream cone is completely full of ice-cream, find the total volume of the ice-cream sitting in and on top of the cone, to the nearest cubic centimetre. (Assume the shape of the ice-cream on top is a hemisphere.) **5.5**

10A **19** Ayse is inflating balloons with helium gas for a birthday party. Each balloon is roughly spherical in shape and about 40 cm in diameter. If she has to inflate two dozen balloons, find the total capacity (in litres) of helium required, rounded to the nearest whole number. **5.5**

10A **20** The density of the material used to construct the pipe in Question **14** is 5.8 g/cm^3. Find the mass of the pipe. **5.7**

Reasoning

21 The local council swimming pool is shaped as shown. **5.3**

(a) Find the capacity of the pool in both litres and kilolitres.

(b) It is proposed to build a pool that is 2 m deep for its entire length. If its other dimensions remain the same, how much extra water will be required for the new pool?

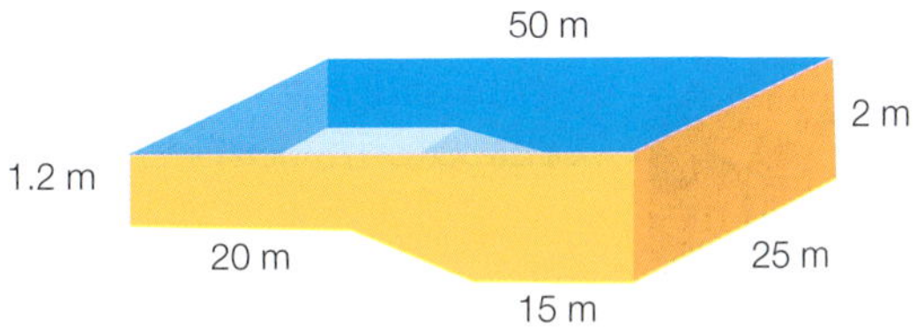

22 A cylindrical container has radius 7 cm and height 10 cm. What must be the height of a rectangular container 7 cm long and 10 cm wide if the two containers are to have the same volume? Write your answer correct to 2 decimal places. **5.3**

23 Engineers are thinking about building a tunnel through a mountain rather than having to build a road over the top. The road will cost $45 000 for every 10 m length that is laid, whether it is over the top or through the tunnel. **5.1–5.3**

(a) The tunnel would be a semicircle of radius 10 m and would be 400 m in length. Calculate the volume of earth to be removed, to the nearest cubic metre.

(b) Find the cost (in whole dollars) of tunnelling and removal of dirt if this cost is estimated to be $150/m^3.

(c) Calculate the surface area of the curved ceiling/wall, to the nearest square metre. (Assume that the entire floor will become road.)

(d) Find the cost (in whole dollars) of fortifying the ceiling wall if the cost is $100/m^2.

(e) Find the cost (in whole dollars) of laying the road through the tunnel.

(f) What is the total cost (in whole dollars) of construction of the road through the tunnel?

The road over the top of the mountain would be 1.4 km long.

(g) Find the cost of laying the road over the top of the mountain.

(h) Which is the cheaper option and by how much?

10A **24** Maris owns a large pool initially containing 1000 L of fresh water. She adds 100 L of salt water, with a concentration of 52 g/L, to the pool, causing the water to become salty. Assuming the salt spreads evenly throughout the pool, what is the concentration of salt in the pool? Write your answer correct to 2 decimal places. **5.7**

6

Trigonometry

Hands-on maths!

Name: Gavin Noble
Job: Carpenter
Qualifications: Certificate III Carpentry

Carpenters construct and maintain buildings and furniture made of wood and other products. They need good mathematical skills to ensure that they have the right amount of supplies, to build structures that are safe and to avoid delaying other tradesmen who rely on them.

As a carpenter, Gavin erects frames, hangs doors and installs kitchens. He also does home renovations, which includes pouring concrete, laying bricks and plastering walls. 'I like the variety of it and have always enjoyed working with timber and with my hands.'

Carpenters need to use mathematical skills when measuring, estimating, reading plans and scale diagrams, and producing quotes. It is crucial that carpenters are accurate with their measurements. 'I use volume to work out the amount of concrete. For plaster you can either use area to find the number of square metres required or to work out the number of sheets needed.'

Building roofs can be very complex. Carpenters can either use trigonometry or a roofing reference book containing the relevant measurements to help them with their work.

Of maths at school, Gavin says: 'I found maths hard. I find maths easier when I know why I am doing it. I would enjoy doing a maths course now, because I understand where I would use it in my job.'

Why learn this?

Trigonometry has many practical applications. Pilots and ship navigators have sophisticated equipment that uses trigonometry to plot their course and determine their position. They also need to be able to complete these calculations by hand. Surveyors use trigonometry to map areas of land, water and airspace. Medical imaging technologies such as CT scans create three-dimensional images of the body using complex mathematics including trigonometry. Engineers routinely use trigonometry to determine quantities such as the force on the supports in a building or on a bridge.

After completing this chapter you will be able to:

- use trigonometry to find unknown side lengths and angles in right-angled triangles
- use angles of elevation and depression to solve problems using trigonometry
- use compass bearings and true bearings to describe location and direction
- use bearings to solve trigonometric problems
- select and use appropriate measurements and trigonometric ratios to solve problems in two dimensions.

6 Recall

Prepare for this chapter by attempting the following questions. If you have difficulty with a question, you can download a Recall Worksheet from the eBook or the Pearson Places website.

1 Solve these equations for x. Write your answer as a decimal, correct to the number of decimal places (d.p.) given in brackets next to the equation.

(a) $3x = 24.5$ (2 d.p.)

(b) $\frac{x}{20} = 0.7825$ (1 d.p.)

(c) $\frac{12}{x} = 2.307\,02$ (2 d.p.)

(d) $0.345 = \frac{16}{x}$ (3 d.p.)

2 Determine the length of the missing side x in each pair of similar triangles.

(a)

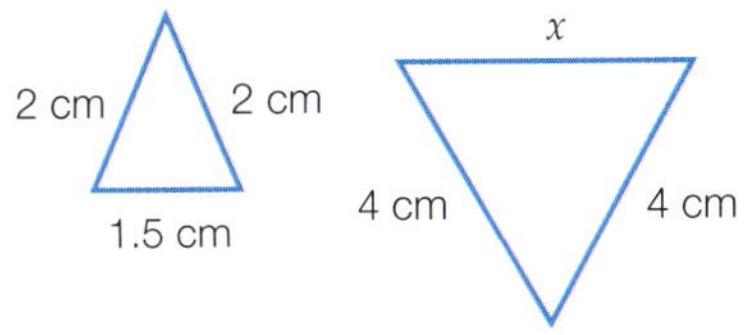

(b)

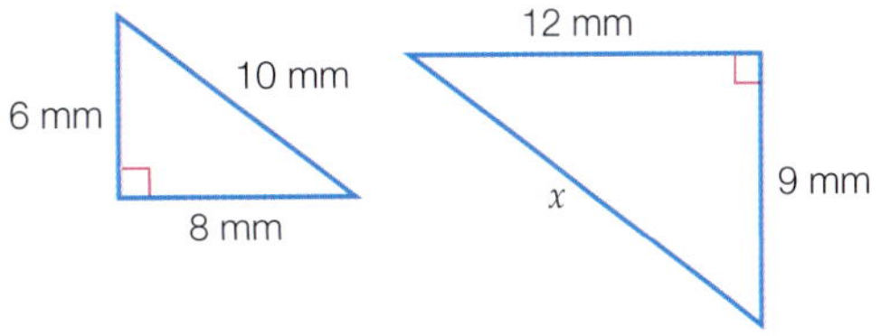

3 Use Pythagoras' theorem to calculate the value of x in each triangle. Round your answer to 2 decimal places where necessary.

(a)

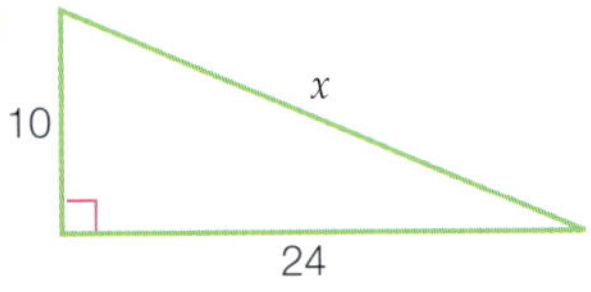

(b)

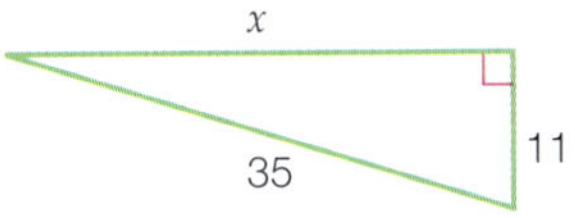

(c)

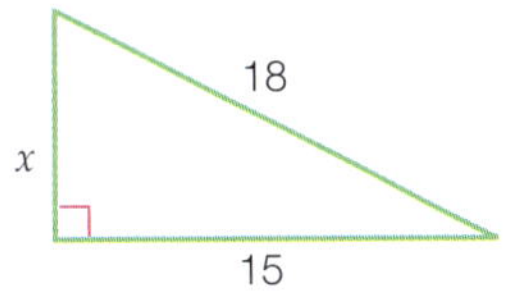

4 Copy each right-angled triangle shown below and label the hypotenuse, the opposite side and the adjacent side for the angle given.

(a)

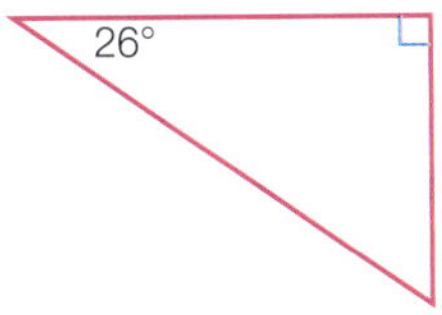

(b)

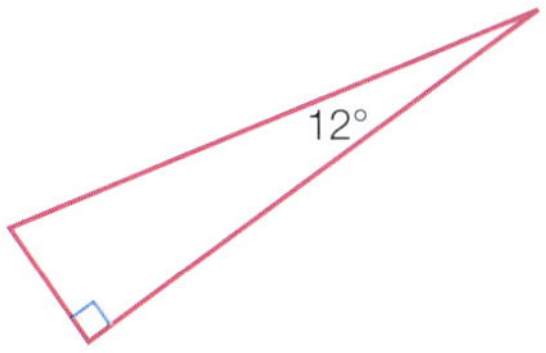

(c)

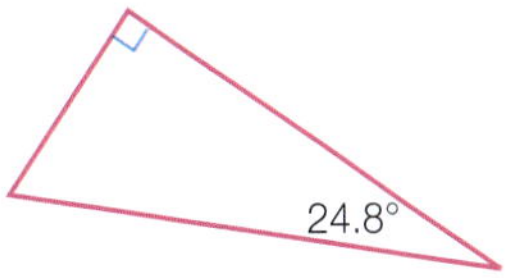

Exploration Task

You can download this activity from the eBook or the Pearson Places website.

Where do trigonometric ratios come from?

In this activity you will construct a series of right-angled triangles and explore the origins of trigonometric ratios.

6.1 The trigonometric ratios

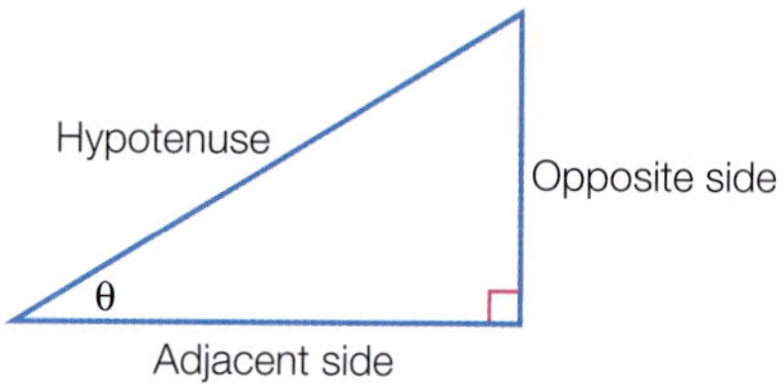

The sides of a right-angled triangle can be named using the position of a **reference angle**. This angle is often referred to as θ (the Greek letter theta) but any symbol can be used. The longest side of a right-angled triangle is the **hypotenuse**. The hypotenuse is always opposite the right angle. The side directly opposite the angle θ is the **opposite side** and the side next to θ is the **adjacent side**.

Use pairs of side lengths to calculate three **trigonometric ratios** (write the ratios in their fraction forms). The ratios compare the length of one side to the length of another side.

The ratio $\frac{\text{Opposite}}{\text{Hypotenuse}}$ is referred to as the **sine** of θ and abbreviated to sin (θ).

The ratio $\frac{\text{Adjacent}}{\text{Hypotenuse}}$ is referred to as the **cosine** of θ and abbreviated to cos (θ).

The ratio $\frac{\text{Opposite}}{\text{Adjacent}}$ is referred to as the **tangent** of θ and abbreviated to tan (θ).

In a right-angled triangle:

$$\sin(\theta) = \frac{\text{Opposite}}{\text{Hypotenuse}} \text{ or } \frac{O}{H}$$

$$\cos(\theta) = \frac{\text{Adjacent}}{\text{Hypotenuse}} \text{ or } \frac{A}{H}$$

$$\tan(\theta) = \frac{\text{Opposite}}{\text{Adjacent}} \text{ or } \frac{O}{A}$$

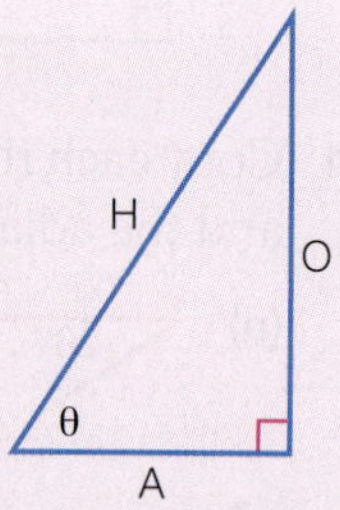

The phrase 'SOH-CAH-TOA' can help you to remember which sides belong to each ratio.

In the triangle shown at right, the sine of the given angle, θ, is $\frac{3}{5}$ or 0.6. This means that the opposite side length is three-fifths of the length of the hypotenuse.

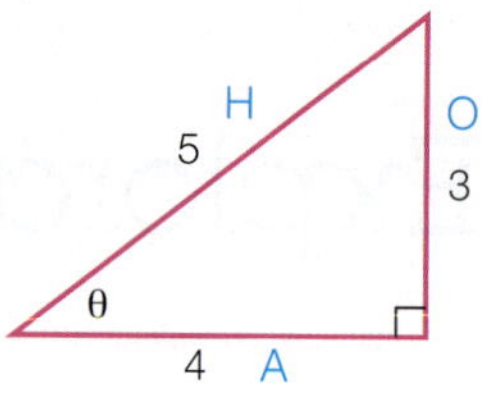

Write this as sin (θ) = 0.6.

Also in this triangle:

$$\cos(\theta) = \frac{4}{5} = 0.8$$

$$\tan(\theta) = \frac{3}{4} = 0.75$$

For a given angle θ, the values of each trigonometric ratio are the same, even if the side lengths of the triangle change. This is because two right-angled triangles with a common angle (θ) are similar triangles.

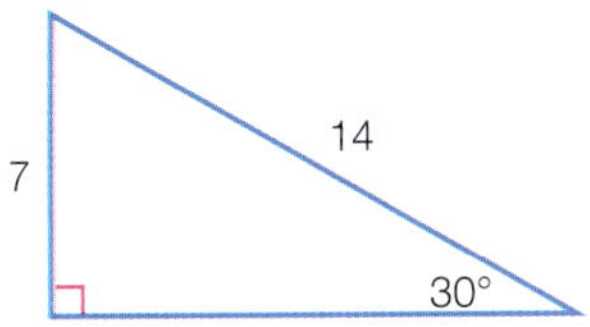

$\sin(30°) = \frac{7}{14} = \frac{1}{2}$

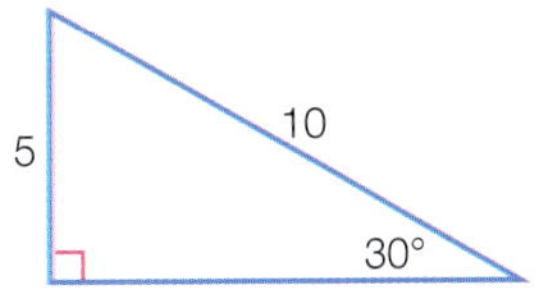

$\sin(30°) = \frac{5}{10} = \frac{1}{2}$

Worked example 1

W.E.1

For each of the following triangles, label the side indicated, relative to the angle shown.

Thinking	Working
(a) the adjacent side (A) 	The adjacent side is the side closest to the angle, but not the hypotenuse. 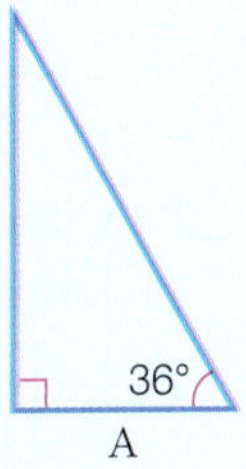
(b) the hypotenuse (H) 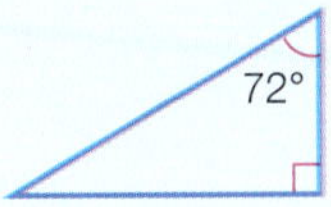	The hypotenuse is always directly opposite the right angle. 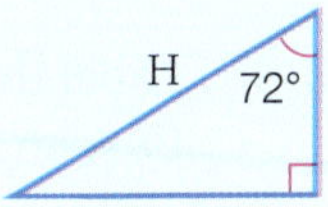
(c) the opposite side (O) 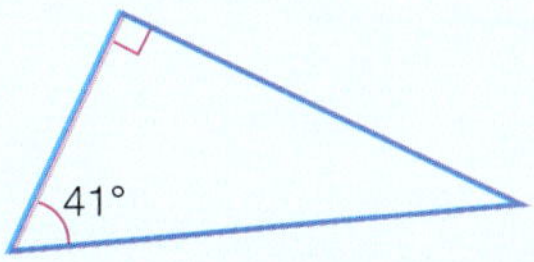	The opposite side is the side furthest from the angle. 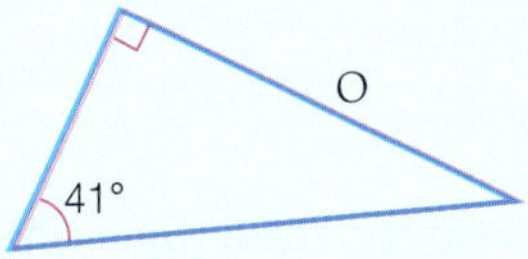

Using the ratios

The value of each of the trigonometric ratios is unique for a particular angle θ. You can use a ratio to find the length of an unknown side, if you know the value of θ and the length of one other side. Alternatively, if you know the ratio of two sides, you can determine an angle.

To identify which trigonometric ratio to use:

1. Label each side of the triangle with O, A and H, relative to the angle considered.
2. Identify which two sides are involved.
3. Decide which ratio links the two sides (SOH-CAH-TOA is helpful here).

Worked example 2

W.E. 2

Label the hypotenuse, opposite and adjacent sides on these right-angled triangles and hence decide which trigonometric ratio to use to find the unknown.

(a)

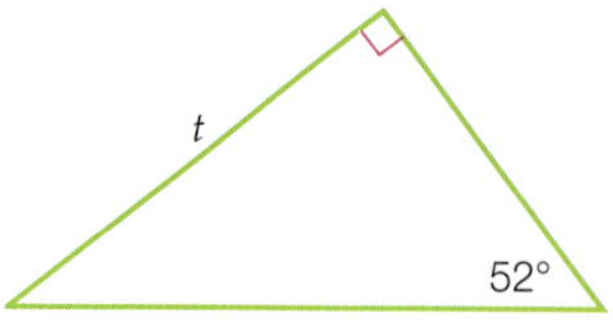

(b)

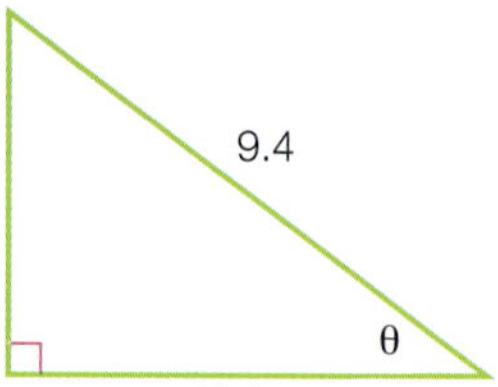

Thinking	Working
(a) 1 Label the sides of the triangle O, A and H.	(a) 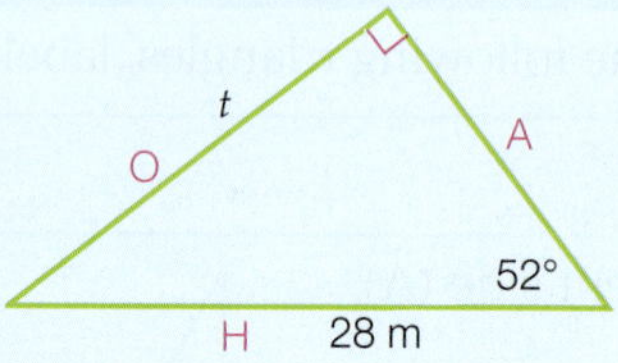
2 Choose the ratio that connects the known and unknown values. (Here, you are given the values of H and θ and need to find O, so use sin (θ) to find *t*.)	$\sin(\theta) = \frac{O}{H}$ $\sin(52°) = \frac{t}{28}$
(b) 1 Label the sides of the triangle O, A and H.	(b) 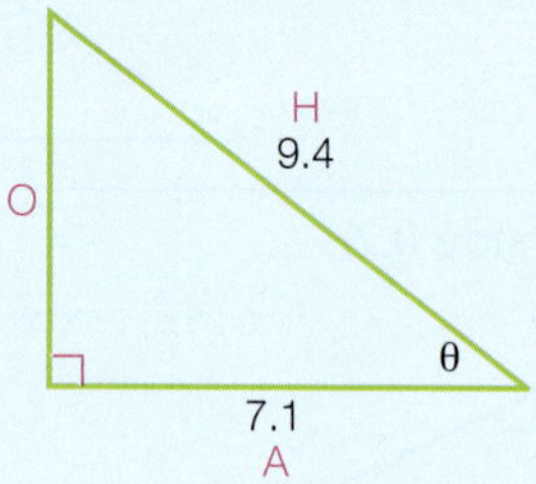
2 Choose the ratio that connects the known and unknown values. (Here, you are given the values of A and H and θ is unknown, so use cos (θ) to find θ.)	$\cos(\theta) = \frac{A}{H}$ $\cos(\theta) = \frac{7.1}{9.4}$

Using a scientific calculator or CAS technology

A scientific calculator or a CAS can also be used to find sine, cosine and tangent for a given angle.

A scientific calculator has **sin**, **cos** and **tan** keys.

The TI-Nspire CAS has a **trig** key [trig]. This displays all the trigonometric functions for use.

The Casio ClassPad CAS has trigonometric functions available via the **Trig** tab in the **Keyboard**.

Make sure that your calculator is in 'degree' mode (not 'radian' mode) for the calculations in this chapter.

Worked example 3

W.E. 3

Use a calculator to find each of the following values, rounded to 4 decimal places.

(a) sin (29°) **(b)** cos (58°) **(c)** tan (60.5°)

Thinking	Working
(a) 1 Enter the information into the calculator. Press the **sin** key, followed by the value of the angle, then **=**.	(a) sin (29°) ≈ 0.484 809 620...
2 Round the answer to 4 decimal places.	= 0.4848 (4 d.p.)
(b) 1 Enter the information into the calculator. Press the **cos** key, followed by the value of the angle, then **=**.	(b) cos (58°) ≈ 0.529 919 264...
2 Round the answer to 4 decimal places.	= 0.5299 (4 d.p.)
(c) 1 Enter the information into the calculator. Press the **tan** key, followed by the value of the angle, then **=**.	(c) tan (60.5°) ≈ 1.767 494 016...
2 Round the answer to 4 decimal places.	= 1.7675 (4 d.p.)

Angles in degrees, minutes and seconds

The units used to measure angles are degrees (°). One degree can be broken up into smaller units called **minutes** (′) and **seconds** (″). There are 60 seconds in a minute and 60 minutes in a degree.

That is: $60' = 1°$ and $60'' = 1'$

$3600'' = 1°$

An angle of 24°30′ is equal to 24.5°, because $\frac{30'}{60'} = 0.5$.

An angle of 57°45′18″ is equal to 57.755°, because $\frac{45}{60} = 0.75$ and $\frac{18}{3600} = 0.005$.

Most scientific calculators have a key marked °’ ” or D°M′S, which converts angles in degrees, minutes and seconds to decimal degrees (a degree measurement with a decimal part).

To change 24°30′ into decimal degrees using a scientific calculator, press:

2 4 °’ ” 3 0 °’ ” =

The screen will show 24.5. Pressing °’ ” switches between the two forms.

If your calculator works differently, consult its instructions.

You can also use CAS technology to convert from decimal degrees to degrees, minutes and seconds.

Using TI-Nspire CAS

First make sure you are in degrees mode. Select **doc** > **Settings & Status** > **Document Settings...** and set **Angle** to **Degree**.

To enter an angle in degrees, minutes and seconds, enter the numbers and use the π▸ key to enter the appropriate symbols for degrees, minutes and seconds. (To enter the seconds symbol " you must enter the minutes symbol ' twice.)

If you are in degrees mode, then press **ctrl** > **enter** (≈) to convert a degrees value to decimal degrees.

To convert a value to decimal degrees while in any mode, select the Catalog and scroll to select the **▶DD** command.

To convert a value to degrees-minutes-seconds, select the Catalog and scroll to select the **▶DMS** command.

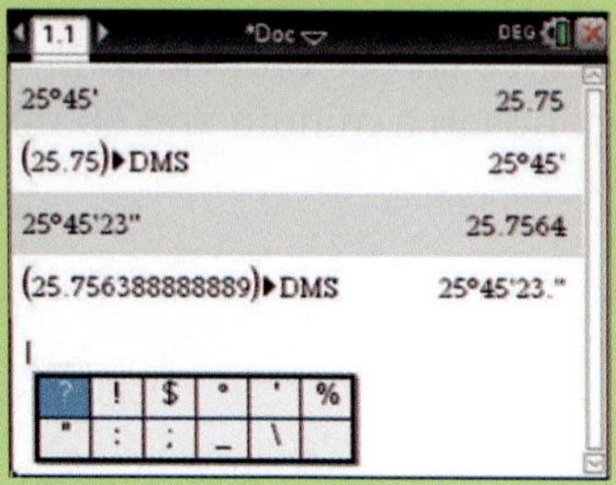

Using Casio ClassPad CAS

First make sure you are in degrees mode, with **Deg** shown at the bottom of the screen. If this shows **Rad** or **Gra** instead, then select the word **Deg** to change it to degrees.

To enter an angle in degrees, minutes and seconds, select **Interactive** > **Transformation** > **DMS** > **dms** and enter the appropriate values.

To convert a value to decimal degrees, use the **dms** command to enter the value as degrees-minutes-seconds, then press enter.

To convert a value to degrees-minutes-seconds, select **Interactive** > **Transformation** > **DMS** > **toDMS** and enter the appropriate value.

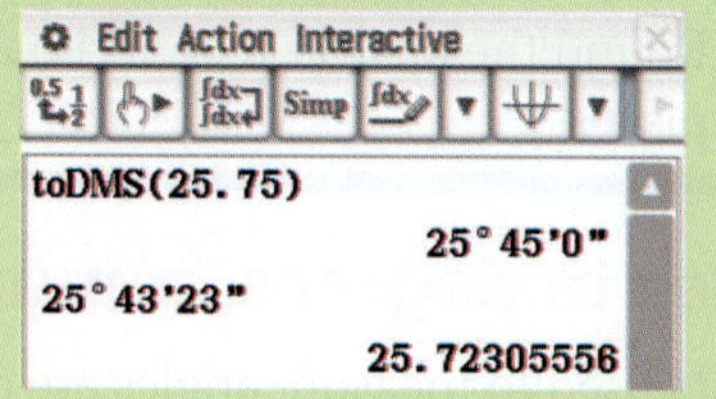

The following examples use the keys from a scientific calculator.

Worked example 4

W.E. 4

Convert the following angles to decimal degrees, rounded to 2 decimal places.

(a) $56°24'$ (b) $23°16'45''$

Thinking	Working
(a) 1 Enter the information into the calculator. Press the following keys: [5] [6] [° ′ ″] [2] [4] [° ′ ″] [=] [° ′ ″]	(a) $56°24' = 56.4°$
2 State the answer correct to 2 decimal places.	$= 56.40°$ (2 d.p.)
(b) 1 Enter the information into the calculator. Press the following keys: [2] [3] [° ′ ″] [1] [6] [° ′ ″] [4] [5] [° ′ ″] [=] [° ′ ″]	(b) $23°16'45'' \approx 23.279\,166\ldots$
2 Round the answer to 2 decimal places.	$= 23.28°$ (2 d.p.)

Worked example 5

W.E. 5

Use a calculator to find each of the following values, rounded to 4 decimal places.

(a) $\sin(46°36')$ (b) $\cos(6°12'34'')$

Thinking	Working
(a) 1 Enter the information into the calculator. Press the following keys: [sin] [4] [6] [° ′ ″] [3] [6] [° ′ ″] [=]	(a) $\sin(46°36') \approx 0.726\,574\,27\ldots$
2 Round the answer to 4 decimal places.	$= 0.7266$ (4 d.p.)
(b) 1 Enter the information into the calculator. Press the following keys: [cos] [6] [° ′ ″] [1] [2] [° ′ ″] [3] [4] [° ′ ″] [=]	(b) $\cos(6°12'34'') \approx 0.994\,133\,14\ldots$
2 Round the answer to 4 decimal places.	$= 0.9941$ (4 d.p.)

6.1 The trigonometric ratios

Navigator

Answers p. 809

1, 2, 3 (columns 1–2), 4, 5, 6, 7, 8, 10, 11, 12, 15, 16	2, 3 (columns 2–3), 4 (columns 1–2), 5 (a–c), 6, 7, 8, 9, 10, 11, 12, 14, 15, 16	3 (column 3), 4 (columns 2–3), 5 (d–f), 7, 8, 9, 10, 11, 12, 13, 14, 15, 16, 17, 18

Equipment required: ruler for Questions **7**, **16**, **17** and protractor for Question **7**

Fluency

W.E. 1

1 For each of the following triangles, label the side indicated, relative to the angle shown.

(a) the adjacent side (A)

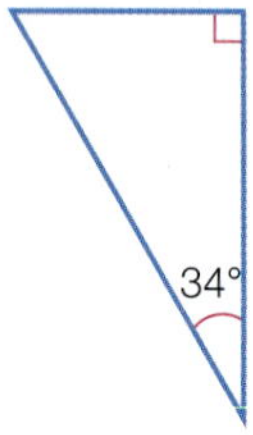

(b) the hypotenuse (H)

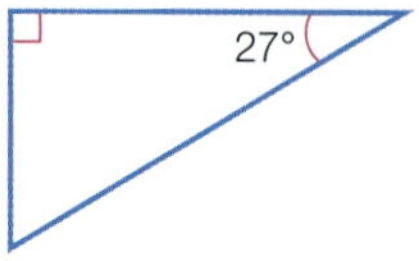

(c) the opposite side (O)

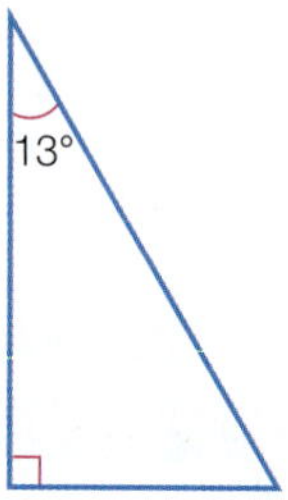

(d) the adjacent side (A)

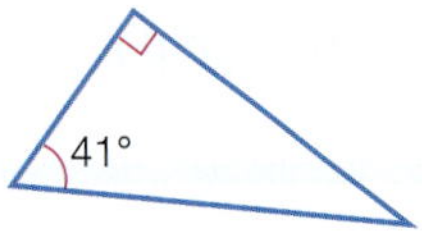

(e) the opposite side (O)

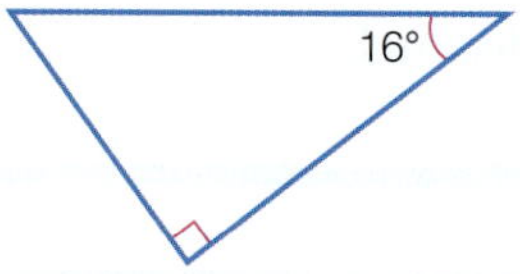

(f) the hypotenuse (H)

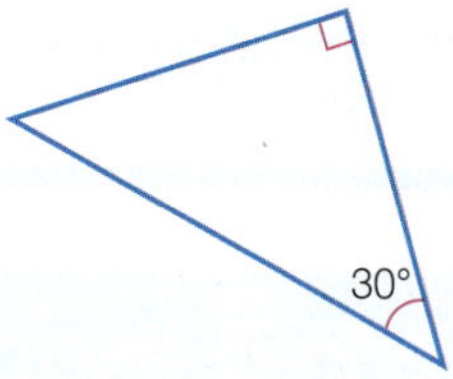

2 Label the hypotenuse, opposite and adjacent sides on these right-angled triangles and hence decide which trigonometric ratio should be used to find the unknown in each case.

(a)

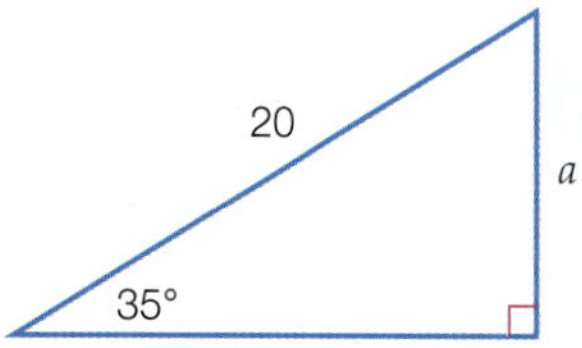

(b)

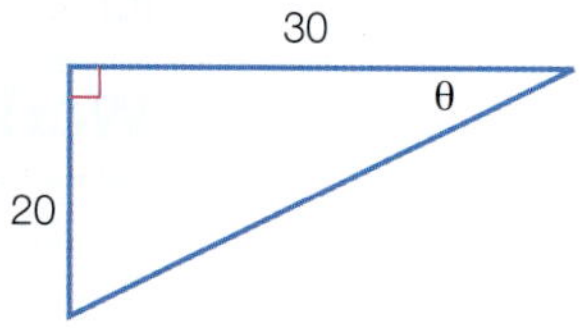

(c)

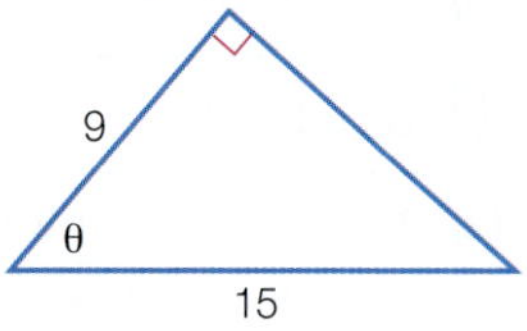

(d)

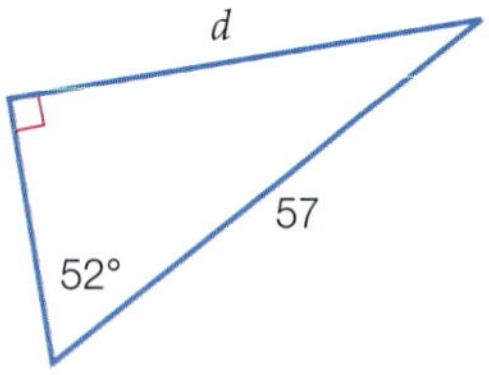

(e)

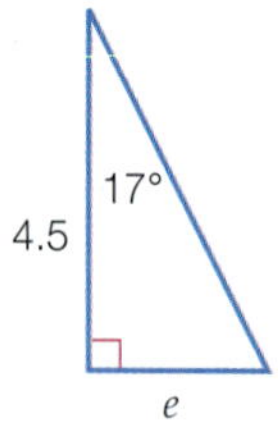

(f)

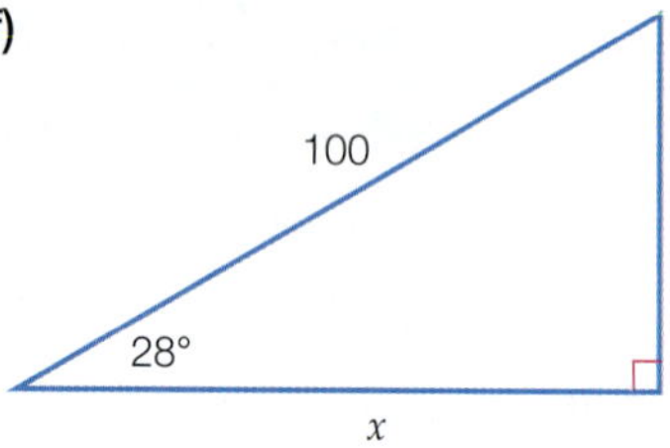

W.E. 3

3 Use your calculator to find the following values, rounded to 4 decimal places.

(a) sin (23°) **(b)** cos (17°) **(c)** tan (39°)

(d) cos (5°) **(e)** sin (41°) **(f)** cos (49°)

(g) tan (19.7°) **(h)** cos (23.5°) **(i)** tan (43.65°)

4 Convert these angles to decimal degrees. Round your answers to 2 decimal places if necessary. **W.E. 4**

(a) 15°30′ (b) 35°16′ (c) 23°10′

(d) 76°47′50″ (e) 44°59′20″ (f) 8°45′9″

5 Calculate the following, rounding values to 4 decimal places. **W.E. 5**

(a) sin (25°47′) (b) tan (88°30′) (c) cos (64°17′)

(d) tan (36°30′46″) (e) sin (20°25′39″) (f) cos (81°44′16″)

6 (a) Correct to 4 decimal places, the value of sin (37°34′) is:

A 0.6064 B 0.6065 C 0.6097 D 0.7951

(b) Correct to 4 decimal places, the value of cos (17.25°) is:

A 0.2965 B 0.3105 C 0.9542 D 0.9550

Understanding

7 Measure the side lengths of triangles *ABC* and *DEF* to the nearest millimetre.

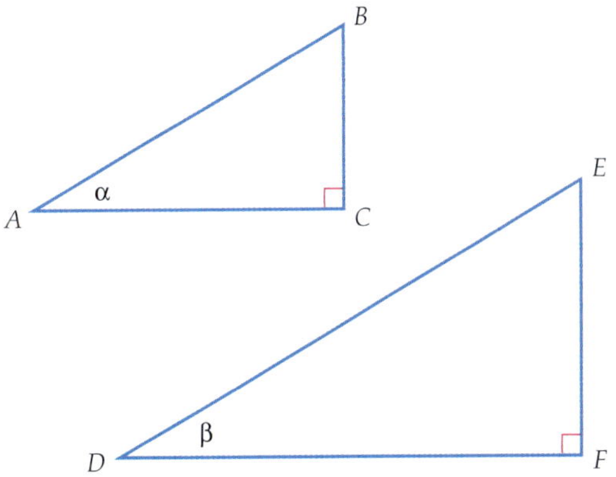

(a) Calculate the ratios of the following pairs of sides. Round your answers to 1 decimal place.

(i) $\frac{DF}{AC}$ (ii) $\frac{EF}{BC}$ (iii) $\frac{DE}{AB}$

(b) Calculate the value for sin (α).

(c) Calculate the value for sin (β).

(d) Are these sine ratios equivalent? Why?

(e) Measure the sizes of angles α and β with a protractor.

(f) Would the cosine ratios be equivalent for each of the triangles?

(g) Would the tangent ratios be equivalent for each of the triangles?

(h) Give a reason for your answers to parts **(f)** and **(g)**.

8 (a) Calculate the value of:

(i) sin (35°) (ii) cos (35°) (iii) tan (35°) (iv) $\frac{\sin(35°)}{\cos(35°)}$

(b) Comment on your results.

9 If sin (θ) = 0.755 (3 d.p.) and cos (θ) = 0.656 (3 d.p.), use your result from Question **8** to calculate the value of tan (θ), to 3 decimal places.

10 Which of the following statements is *incorrect*?

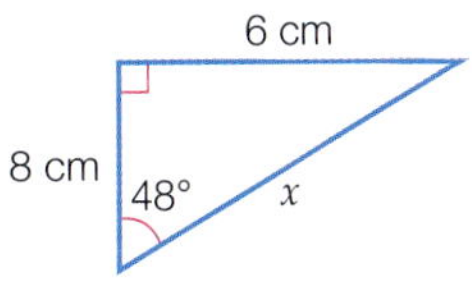

A The length of the side labelled x can be calculated using Pythagoras' theorem.

B $\sin(48°) = \frac{6}{x}$

C The unmarked angle in this triangle is 42°.

D $\cos(48°) = \frac{8}{x}$

E $\tan(48°) = \frac{x}{8}$

11 Which of the following statements is *correct*?

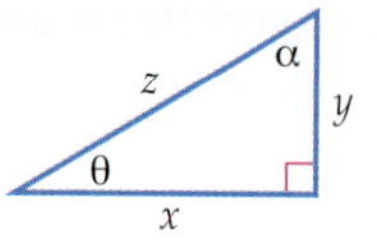

A $\sin(\theta) = \frac{x}{z}$ **B** $\cos(\alpha) = \frac{x}{z}$ **C** $\tan(\theta) = \frac{y}{x}$

D $\tan(\alpha) = \frac{z}{y}$ **E** $\sin(\alpha) = \frac{z}{y}$

Reasoning

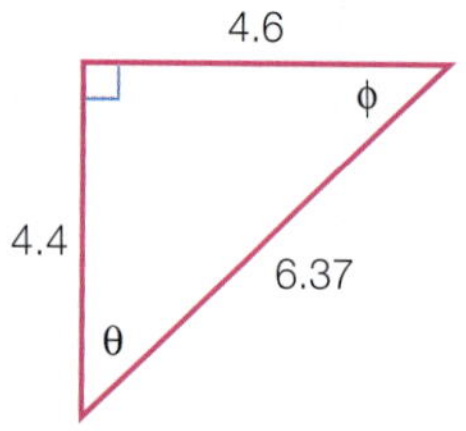

12 **(a)** Calculate the following values using the triangle shown, rounded to 3 decimal places:

(i) $\sin(\theta)$ **(ii)** $\cos(\phi)$

(b) Explain your results.

13 A right-angled triangle has an angle of 54° as one of its acute angles.

(a) Using three different pronumerals to represent the side lengths, draw a diagram of the triangle, including the measurement of the other angle.

(b) Write values of sin (54°) and cos (54°) using the pronumerals that represent the side lengths.

(c) Find the values of sin (54°) and cos (54°) using a calculator, and round to 4 decimal places.

(d) Use your answers to **(c)** to show that the value of $(\sin(54°))^2 + (\cos(54°))^2$ equals 1. (*Note*: This is normally written as $\sin^2(54°) + \cos^2(54°)$.)

(e) Use your answers to **(b)** to show that $(\sin(54°))^2 + (\cos(54°))^2 = 1$. (Hint: Use Pythagoras' theorem.)

14 The value of the tangent ratio for each of the two base angles in a right-angled isosceles triangle is 1. Explain why.

15 Only one of the trigonometric ratios can have a value greater than 1. Name it, and explain why this is not possible for the other two ratios.

Open-ended

Hint

Which values of O and A give a ratio of $\frac{7}{10}$?

16 A right-angled triangle has a tangent ratio of $\frac{7}{10}$ for one angle.

(a) Use the tangent ratio to draw the triangle. Measure the length of the hypotenuse.

(b) Calculate the other ratios, rounded to 4 decimal places.

17 Angle θ in a right-angled triangle has a sine ratio of 0.4. Draw three different triangles that could contain angle θ. Use Pythagoras' theorem to help you find any unknown side lengths.

18 **(a)** Find two angles θ and ϕ, so that $\sin(\theta) = \cos(\varphi)$.

(b) What is the relationship between θ and ϕ if $\sin(\theta) = \cos(\phi)$ for all values of θ and ϕ?

Finding lengths

6.2

For any right-angled triangle, you can find an unknown side length if you know the size of one of the acute angles and the length of one side. This is because the trigonometric ratio for a pair of side lengths (sin, cos and tan) is constant and unique for a particular angle, θ.

For example, if you know the angle θ and the hypotenuse H, you can find the length O of the opposite side using $\sin(\theta) = \frac{O}{H}$.

Worked example 6 W.E. 6

Find the value of the unknown length in each of the following. Round your answers to 2 decimal places.

(a)

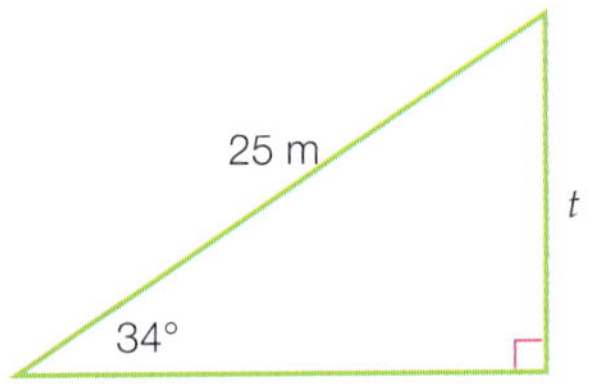

(b)

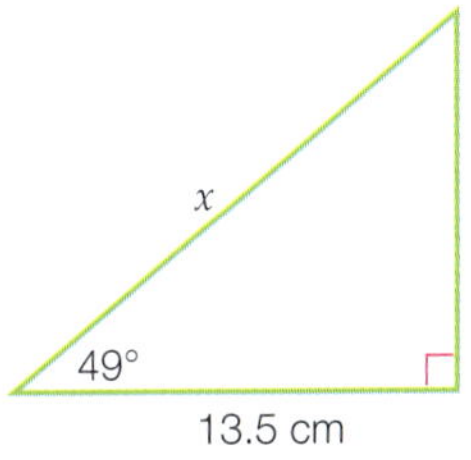

Thinking	Working
(a) 1 Copy the triangle and label the sides O, A and H.	(a) 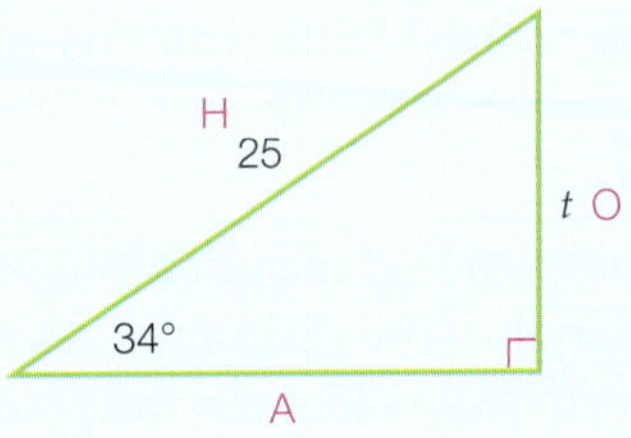
2 Write the known information.	$\theta = 34°$ $O = t$ $H = 25$
3 Identify the trigonometric ratio that connects the known information.	$\sin(\theta) = \frac{O}{H}$
4 Substitute the known values.	$\sin(34°) = \frac{t}{25}$
5 Make the unknown the subject of the equation. (Here, multiply both sides by 25.)	$t = 25 \times \sin(34°)$
6 Use your calculator to evaluate the answer correct to 2 decimal places.	$t = 13.98$ m (2 d.p.)

(b) 1 Copy the triangle and label the sides O, A and H.

(b)

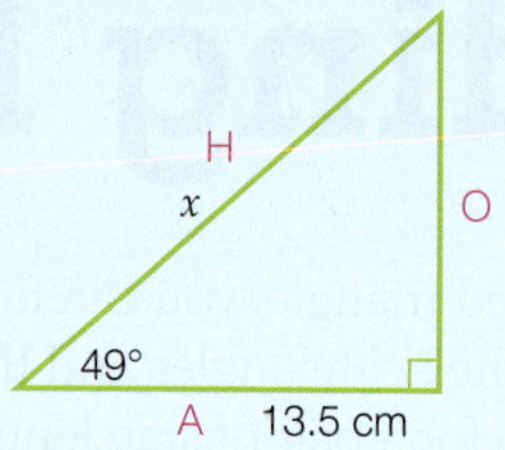

2 Write the known information.

$\theta = 49°$
$A = 13.5$
$H = x$

3 Identify the trigonometric ratio that connects the known information.

$\cos(\theta) = \frac{A}{H}$

4 Substitute the known values.

$\cos(49°) = \frac{13.5}{x}$

5 Make the unknown the subject of the equation. (Here, multiply both sides by *x*, then divide both sides by cos (49°).)

$x\cos(49°) = 13.5$

$x = \frac{13.5}{\cos(49°)}$

6 Use your calculator to evaluate the answer correct to 2 decimal places.

$x = 20.58$ cm (2 d.p.)

Worked example 7

W.E. 7

A 3.2 m ladder rests against a wall so that its angle with the wall is 65°. How high up the wall does the ladder reach? State your answer correct to the nearest centimetre.

Thinking	Working
1 Draw a diagram that includes a right-angled triangle. (The diagram does not need to be to scale.) Identify the position of the given angle, then label O, A and H. Identify and label the unknown side. (Here, label the unknown vertical distance *d*.)	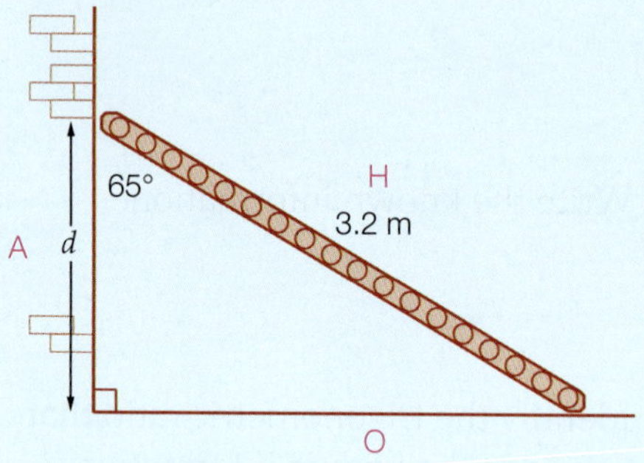
2 Write the known information.	$\theta = 65°$ $H = 3.2$ m $A = d$
3 Identify the trigonometric ratio that connects the known information.	$\cos(\theta) = \frac{A}{H}$
4 Substitute the known values.	$\cos(65°) = \frac{d}{3.2}$

Thinking	Working
5 Make the unknown the subject of the equation. (Here, multiply both sides by 3.2.)	$d = 3.2 \times \cos(65°)$
6 Use your calculator to evaluate the answer correct to 2 decimal places.	$d = 1.35$ (2 d.p.)
7 State the answer in a sentence.	The ladder rests 1.35 m up the wall.

Worked example 8

W.E. 8

A rectangular wooden frame has a diagonal piece of wood joining the opposite corners for support. The vertical height of the frame is 2.5 m. The angle between the diagonal and vertical part of the frame is 57°15′. How long is the diagonal piece of the frame?

Thinking	Working
1 Draw a diagram that includes a right-angled triangle. Identify the position of the angle and label it. Label the known sides with their values and the unknown side with a pronumeral. Label the sides O, A and H.	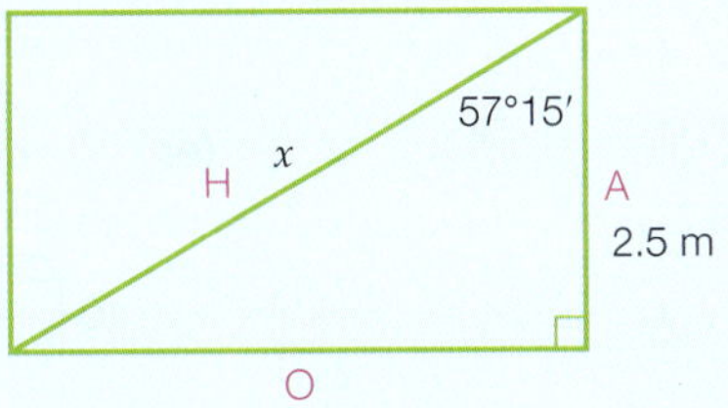
2 Write the known information.	$\theta = 57°15'$ $A = 2.5$ m $H = x$
3 Identify the trigonometric ratio that connects the known information.	$\cos(\theta) = \frac{A}{H}$
4 Create an equation by substituting the known values.	$\cos(57°15') = \frac{2.5}{x}$
5 Make x the subject of the equation. (Here, multiply both sides by x, then divide by $\cos(57°15')$.)	$x \cos(57°15') = 2.5$ $x = \frac{2.5}{\cos(57°15')}$
6 Use your calculator to evaluate the answer correct to 2 decimal places.	$x = 4.62$ (2 d.p.)
7 State the answer in a sentence.	The diagonal of the frame is 4.62 m long.

To find an unknown value in a right-angled triangle:

1 Draw the right-angled triangle, showing all known information and the unknown value.
2 Label the sides O, A and H, and write the known information.
3 Identify which trigonometric ratio to use. (SOH-CAH-TOA is helpful here.)
4 Substitute the known values, form an equation and solve for the unknown.

6.2 Finding lengths

Navigator

Answers p. 811

1 (columns 1–2), 2, 3, 4, 5, 6, 7, 8, 9, 10, 12, 14, 16	1 (column 2), 2, 3, 4, 5, 6, 7, 8, 9, 10, 11, 12, 14, 15, 16	1 (column 3), 3, 4, 5, 6, 7, 8, 9, 10, 11, 12, 13, 15, 16, 17

Fluency

W.E. 6

1 Find the value of the unknown length in each of the following. Round your answers to 2 decimal places.

(a)
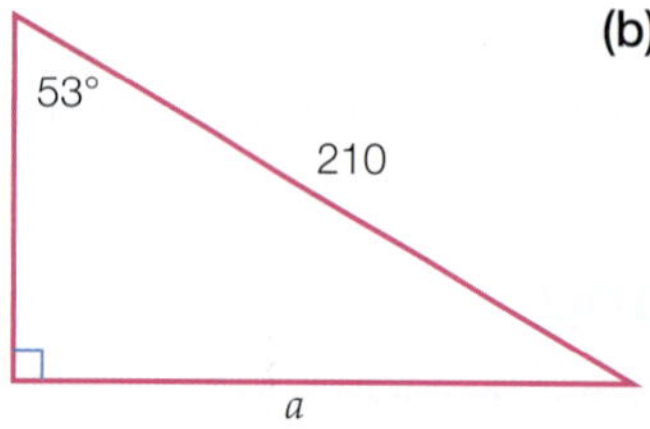

(b)
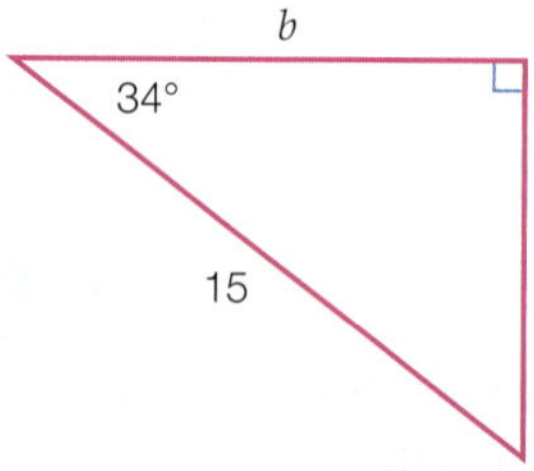

(c)
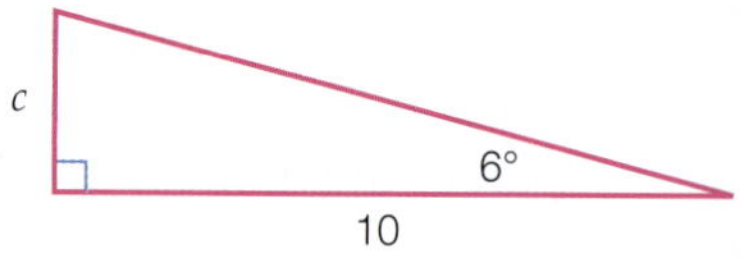

(d)
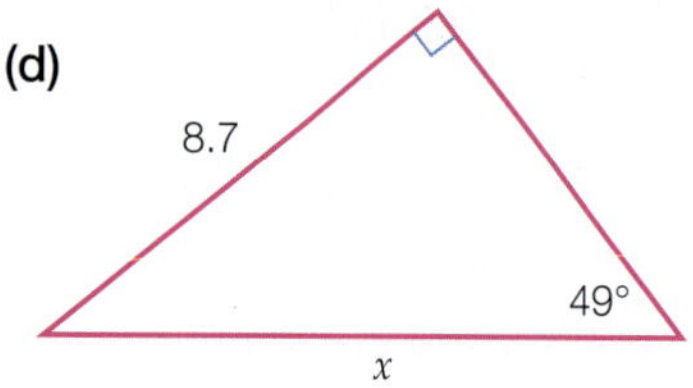

(e)
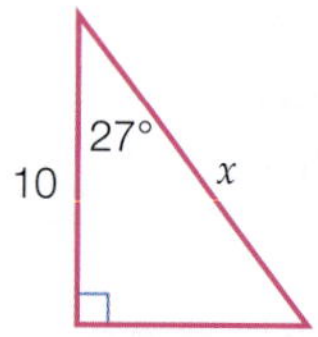

(f)
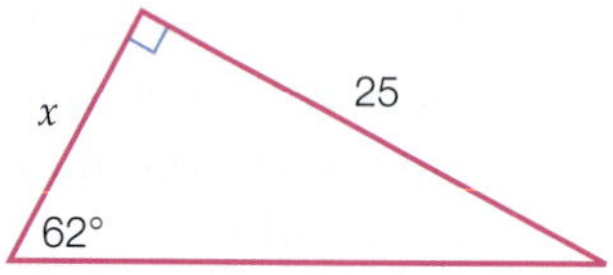

(g)
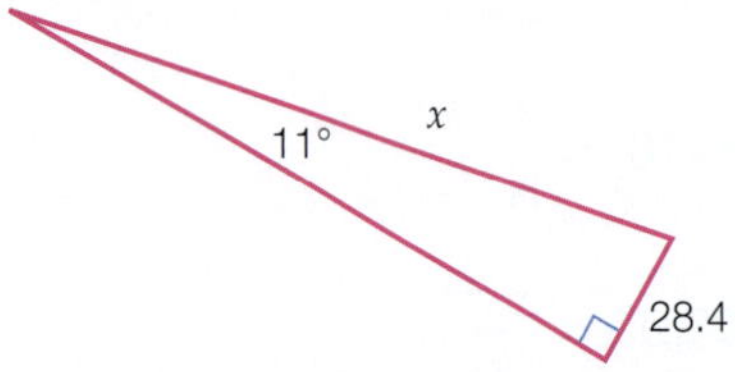

(h)
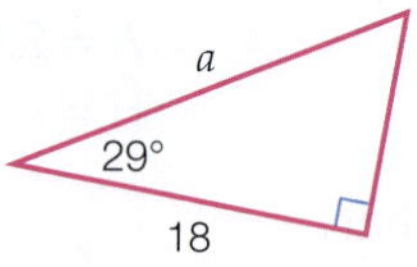

(i)
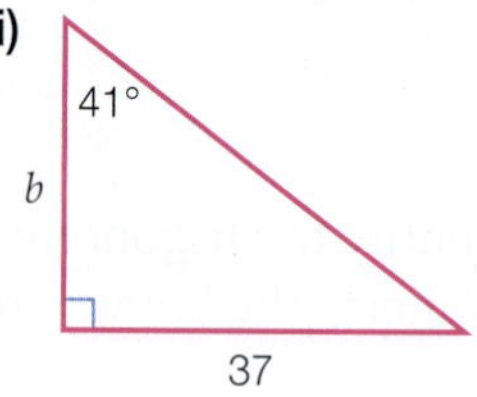

(j)
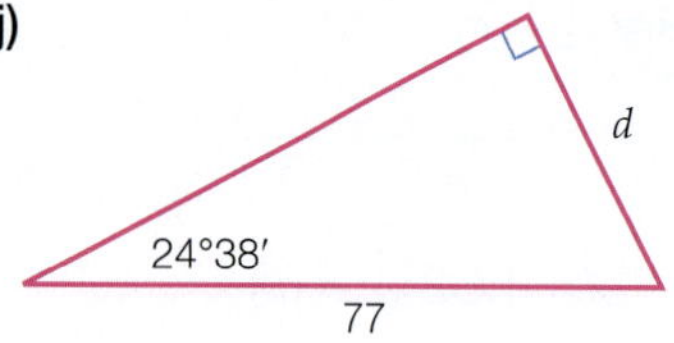

(k)
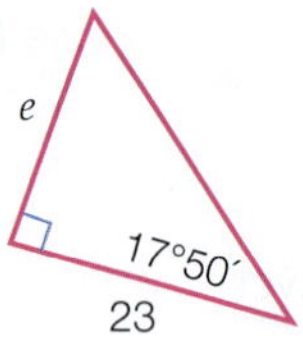

(l)
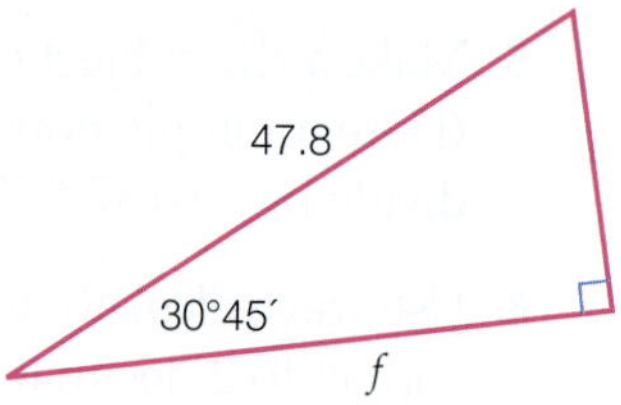

(m)
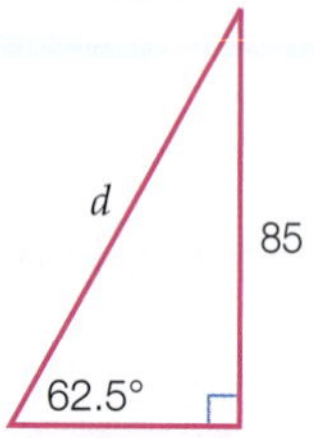

(n)
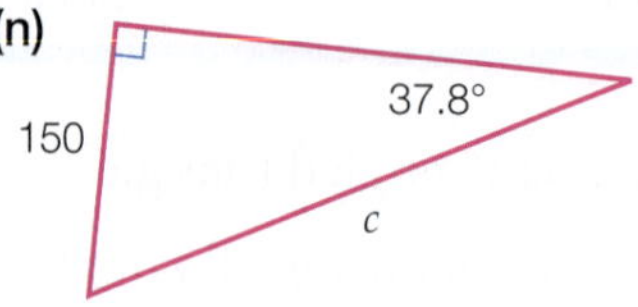

(o)
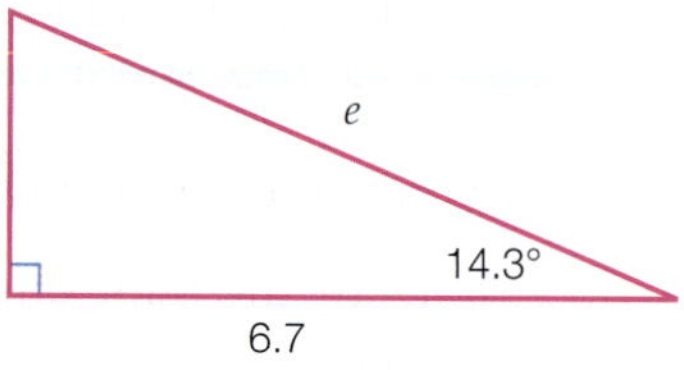

2 A 7 m ladder is placed against a wall so it makes an angle of 35° with the ground. Calculate the following values correct to 1 decimal place.

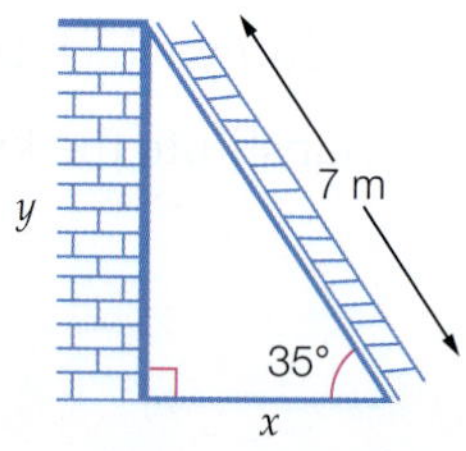

(a) Find the distance x between the foot of the ladder and the wall.

(b) Find the height y at which the top of the ladder touches the wall.

3 A kitten is looking up at a bird in a tree. The kitten's line of sight makes an angle of 61° with the bird in the tree as shown in the diagram. Assuming that the kitten's line of sight starts at ground level, calculate the following values correct to 1 decimal place.

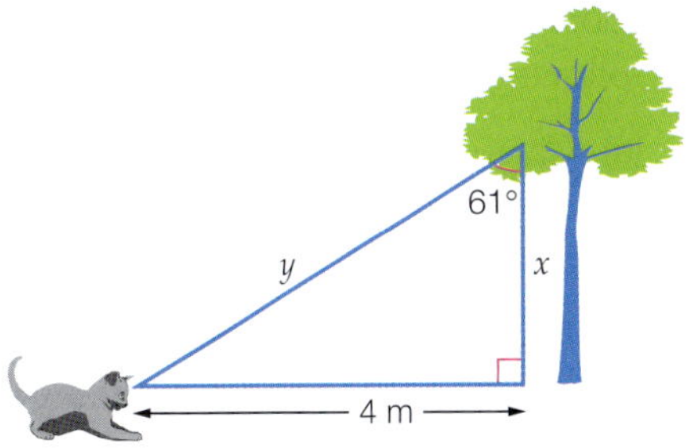

(a) Calculate the height in metres of the bird's position in the tree.

(b) Calculate the direct distance in metres between the kitten and the bird.

4 A 3 m rope is stretched from the top of a vertical tent pole and pegged into the ground so that it makes an angle of 50° with the ground. How far away from the pole is the rope pegged? Give your answer to the nearest centimetre. **W.E. 7**

5 A rectangular gate has diagonal metal supports joining the opposite corners. The supports make an angle of 25°18′ with the horizontal sections of the frame. If the gate is 1.6 m high, then how long is one of the diagonal supports, correct to 2 decimal places? **W.E. 8**

6 (a) An aircraft takes off and rises steadily at an angle of 10°48′ to the horizontal. The altitude of the aircraft after it has travelled along a flight path of 2.6 km (to the nearest metre) is:

A 472 m

B 487 m

C 2554 m

D 13 875 m

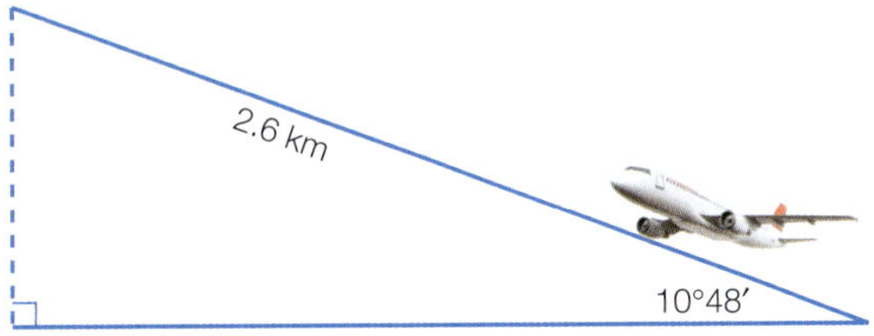

(b) The radius of the circle at right is closest to:

A 6.57 m

B 13.14 m

C 14.75 m

D 29.50 m

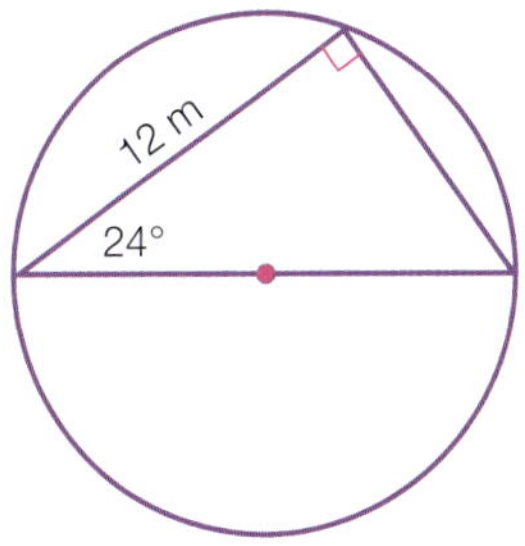

Understanding

7 Jaya wants to calculate the width of a river. She finds that lines joining her point of observation to the two posts on the opposite bank are 42° apart. Her sister Nasreen, who is on the opposite bank, measures the distance between the two posts to be 13.84 m. How wide is the river?

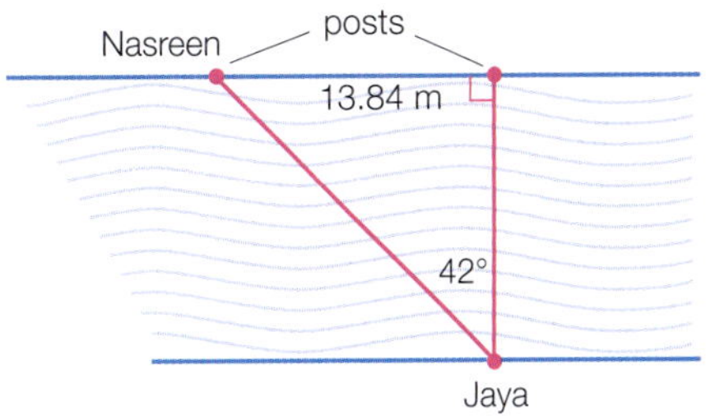

8 A hiker is resting at the top of a cliff when she spots a hot-air balloon at eye-level. Calculate the following values correct to 2 decimal places.

(a) Calculate the horizontal distance between the hiker and the hot-air balloon using the given angle.

(b) Find the unknown angle in the triangle. Use this angle and a different trigonometric ratio to calculate the horizontal distance between the hiker and the hot-air balloon.

(c) Compare your answers to parts **(a)** and **(b)**. What do you notice about the two angles?

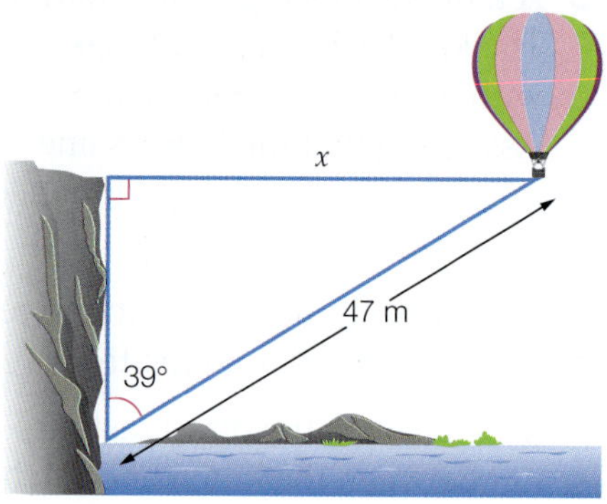

9 A flag pole is tethered to the ground using a high-tensile strength wire. Round your answers to 2 decimal places.

(a) How long is the wire?

(b) How tall is the flag pole?

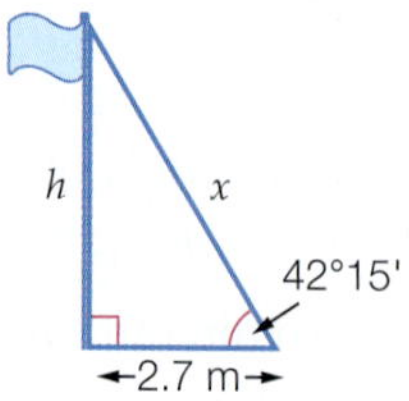

10 Jessica is flying a kite attached to a string 12 m long. She is holding the string at a height of 0.9 m from the ground. How high is the kite if it makes an angle of 48°30′24″ with the horizontal? Round your answer to 2 decimal places.

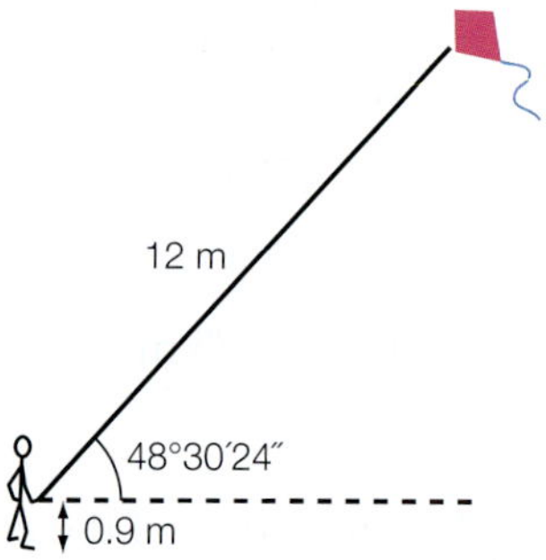

11 To measure a tall building, Charlie stands 150 m from the base of the building and holds an inclinometer (a device for measuring angles) to his eye, as shown in the diagram. He looks along the top of the inclinometer and tilts it up until he can see the top of the building. The inclinometer measures an angle of 58°. This is the angle between Charlie's line of sight to the top of the building and the horizontal. Charlie's eye-level is 1.63 m from the ground.

(a) Draw a diagram that shows all the known information.

(b) Calculate the height of the building, correct to 1 decimal place.

12 A mine shaft 350 m long is to be drilled at an angle of 21°24′ to the vertical. How far directly below the ground will the end of the shaft be, rounded to 2 decimal places?

13 A road sign is in the shape of an equilateral triangle. The line from one vertex to the midpoint of the opposite side is 34.6 cm long. How long is each side of the road sign?

Reasoning

14 James is trying to decide which room to use for a party. He has two rooms in mind but needs to use the larger one. One room is a rectangular shape with length 4 m and width 3 m. The other room is in the shape of a trapezium as shown in the diagram.

(a) Calculate the area of each of the rooms.

(b) Which room is larger?

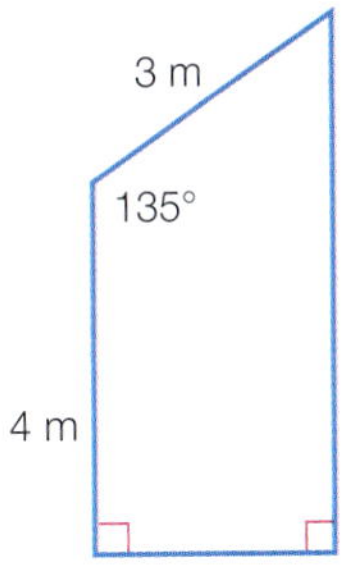

Hint

Formula for the area of a trapezium:

$A = \frac{1}{2}(a + b)h$

15 Members of a stunt team want to make a daring climb along a rope, one end of which is attached to the top of a 23 m building and the other end joins to the top of a nearby building 38 m high. The rope is inclined at an angle of 49°30′ to the horizontal. Answer the following correct to 2 decimal places.

(a) How far along the rope will each member of the stunt team need to climb to reach the other building?

(b) One of the stunt team wants to drop a ball so that a man standing midway between the two buildings can catch it. From how far along the rope does she need to drop it, and how far will it fall if the man catches it 1.8 m above the ground?

Open-ended

16 A right-angled triangle ABC has the right angle at B. If the side opposite to angle A has a length of 10 cm, write at least three possible lengths of the hypotenuse if angle A is between 35° and 40°.

17 Daredevil Dan will be shot from a cannon that sits 1.2 m off the ground, to just clear an 8 m wall.

Assume that his flight path is a straight line and the trajectory angle θ is between 30° and 50°.

Find two possible values, between 8 m and 10 m, for the horizontal distance from the cannon to the wall.

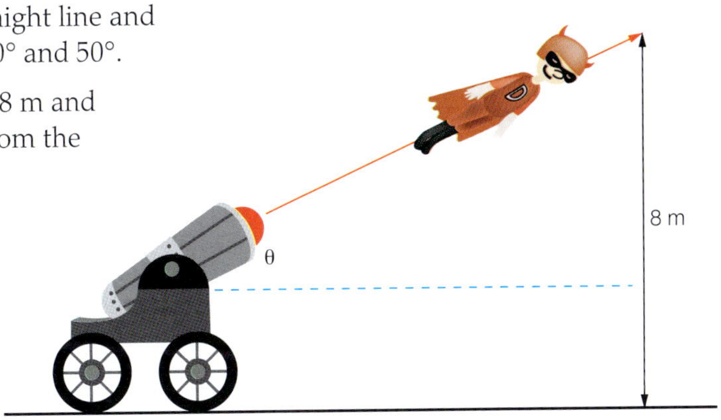

6.3 Finding angles

Sine, cosine or tangent ratios can be used to find an unknown side.

The value of a trigonometric ratio for each angle is unique so you can find a particular angle if you work backwards from a trigonometric ratio.

For example, to find the angle θ whose sine is 0.72: $\sin(\theta) = 0.72$, $\theta = ?$

To find θ you 'undo' the sine to find the angle that corresponds to the sine ratio with this value.

To 'undo' a sine ratio, you can apply the inverse, which is written as $\sin^{-1}$: $\theta = \sin^{-1}(0.72)$

You can usually do this on a calculator, by pressing a **2nd F**, **INV** or **SHIFT** key, before pressing the trigonometric ratio key **sin**.

For example, you could enter:

SHIFT **sin** **.** **7** **2** **=**

This gives the answer 46.054 480 44 (when the calculator is in degrees mode), so θ is approximately 46°.

The inverse trigonometric functions $\sin^{-1}$, $\cos^{-1}$ and $\tan^{-1}$ can all be used similarly.

On the Casio ClassPad CAS, these inverse trigonometric functions can be found on the **Trig** tab in the **Keyboard**.

On the TI-Nspire CAS, you can use the **Trig** key [trig] to enter the inverse trigonometric functions.

Worked example 9

W.E. 9

If $\cos(a) = 0.6434$, find:

(a) a in decimal degrees rounded to 2 decimal places

(b) a in degrees, minutes and seconds.

Thinking	Working
(a) 1 Make sure that your calculator is in degree mode. Enter the information into the calculator. Press the following keys: **SHIFT** **cos** **.** **6** **4** **3** **4** **=**	(a) $\cos^{-1}(0.6434) = 49.954\,181\,3$
2 Round the answer to 2 decimal places.	$a = 49.95°$ (2 d.p.)
(b) 1 Convert the answer from part **(a)** to degrees, minutes and seconds. Press **° ' "**.	(b) 49°57′15.05
2 Round the answer to the nearest second.	$a = 49°57'15''$ (nearest second)

If you know the lengths of any two sides in a right-angled triangle, you can calculate the size of any angle in it by identifying and calculating the trigonometric ratio of the two sides.

Worked example 10

W.E. 10

Find the value of θ in degrees, minutes and seconds.

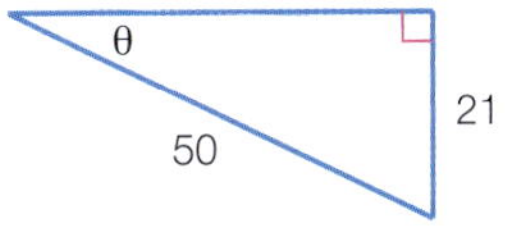

Thinking	Working
1 Label the sides of the triangle.	
2 Decide which trigonometric ratio connects the given values.	$\sin(\theta) = \frac{O}{H}$
3 Substitute the given values.	$\sin(\theta) = \frac{21}{50}$
4 Use the inverse of the ratio to find θ.	$\theta = \sin^{-1}\left(\frac{21}{50}\right)$
5 On your calculator enter SHIFT sin (2 1 ÷ 5 0) =	
6 Convert the answer to degrees, minutes and seconds. Press ° ' ".	$\theta = 24°50'4.51''$
7 Round the answer to the nearest second.	$\theta = 24°50'5''$ (nearest second)

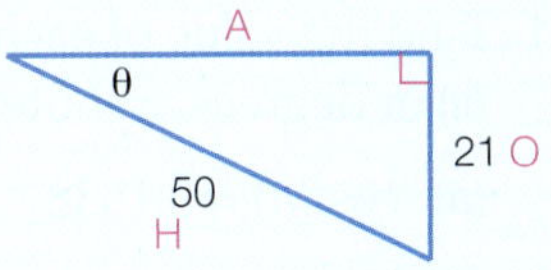

Worked example 11

W.E. 11

The top of a 1.8 m ladder that is leaning against a wall, reaches a height of 1.2 m.
Draw a diagram that shows this situation then calculate the angle that the top of the ladder makes with the wall, correct to the nearest degree.

Thinking	Working
1 Draw a diagram to represent the situation. Identify and label the given side lengths and the unknown angle.	$\theta = ?$ $A = 1.2$ m $H = 1.8$ m
2 Decide which trigonometric ratio connects the given values.	$\cos(\theta) = \frac{A}{H}$
3 Substitute the given values.	$\cos(\theta) = \frac{1.2}{1.8}$
4 Use the inverse of the ratio to find θ.	$(\theta) = \cos^{-1}\left(\frac{1.2}{1.8}\right)$
5 On your calculator enter SHIFT cos (1 . 2 ÷ 1 . 8) =	$\theta = 48.189\,685\,11$
6 Round the answer to the nearest degree.	$\theta = 48°$ (nearest degree)
7 State the answer.	The ladder makes an angle of approximately 48° with the wall.

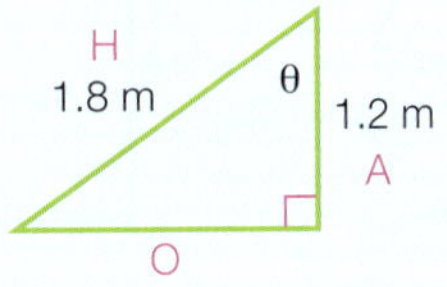

6.3 Finding angles

Navigator

Answers p. 811

1, 2, 3, 4, 5, 6, 7, 8, 9, 11, 13, 15	1 (a–c), 2 (columns 1–2), 3 (columns 2–3), 4, 5, 7, 8, 9, 10, 11, 13, 14, 15	1 (d–f), 2 (a–c), 3 (a–c), 5, 7, 8, 9, 10, 11, 12, 13, 14, 15, 16

Fluency

W.E. 9

1 Find the value of each pronumeral **(i)** in decimal degrees rounded to 2 decimal places and **(ii)** in degrees, minutes and seconds.

(a) $\cos(a) = 0.1538$ **(b)** $\sin(b) = 0.6719$ **(c)** $\tan(c) = 2.3085$

(d) $\cos(k) = 0.556$ **(e)** $\tan(m) = 4.687$ **(f)** $\sin(p) = 0.9715$

W.E. 10

2 Calculate the angle θ for each of the following triangles. Write your answer correct to the nearest degree.

(a)

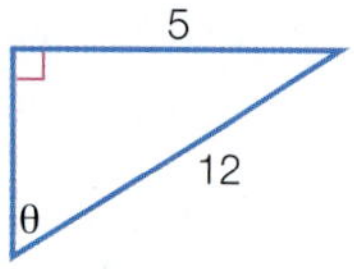

(b)

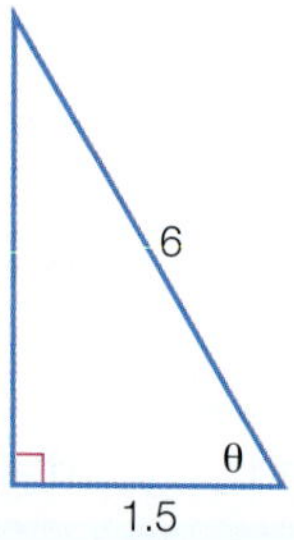

(c)

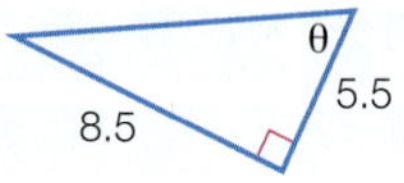

(d)

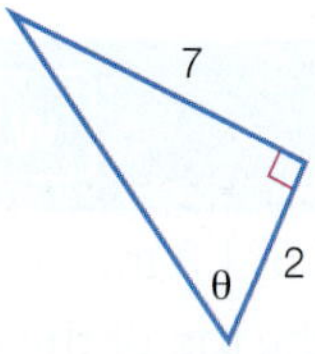

(e)

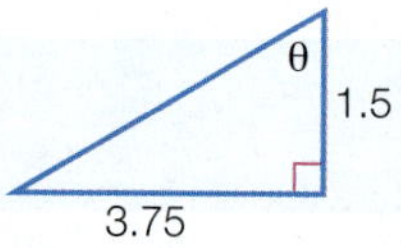

(f)

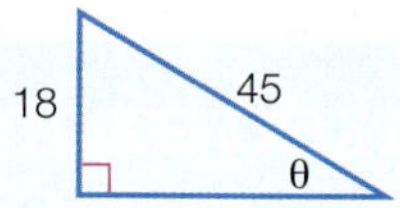

3 Find the value of θ in each triangle in degrees, minutes and seconds.

(a)

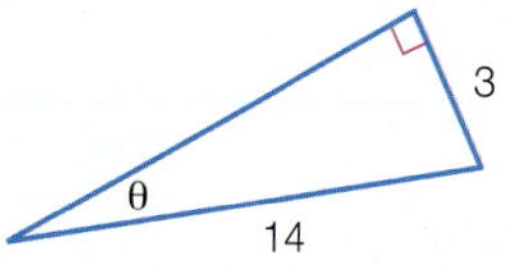

(b)

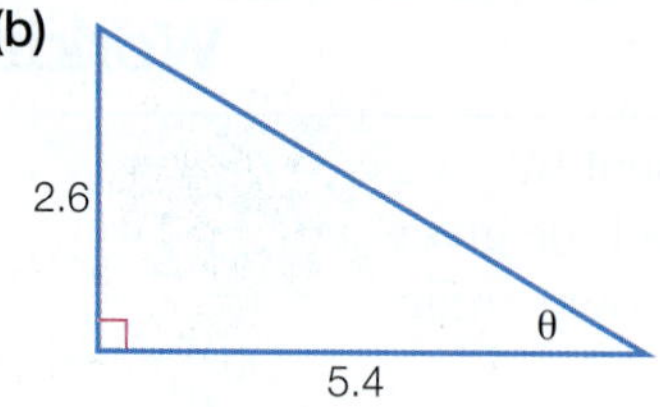

(c)

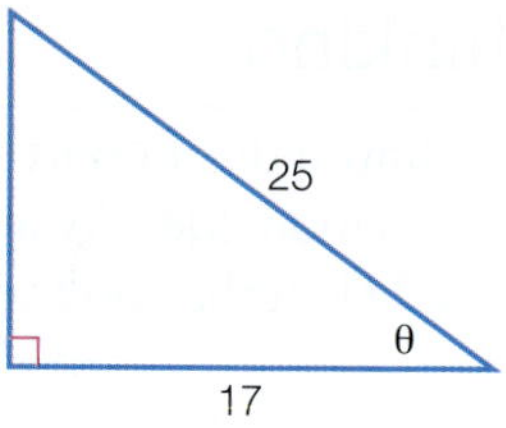

(d)

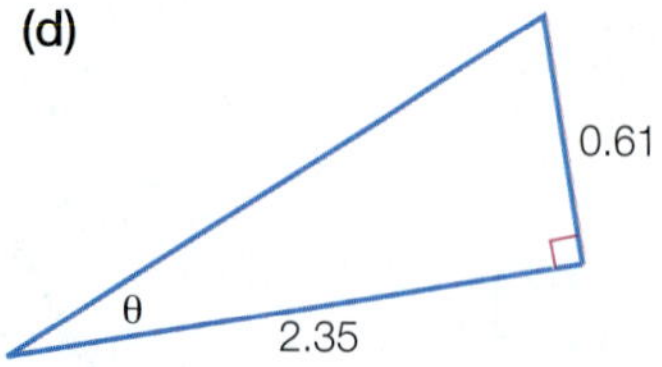

(e)

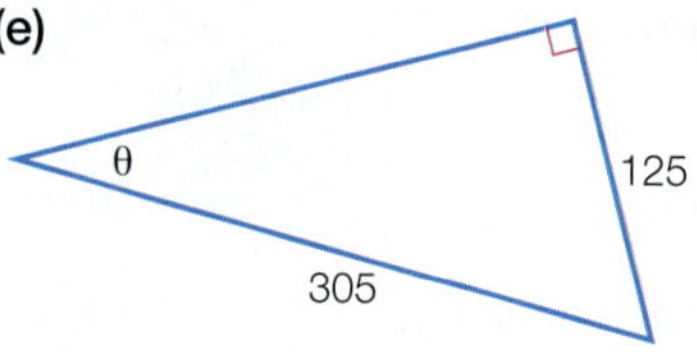

(f)

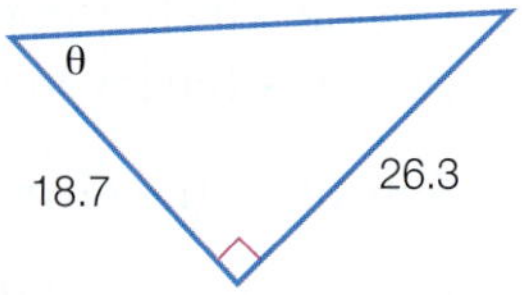

4 A rope is used to tether the top of a 2 m tall tent pole from the top of the pole to a point on the ground 7 m from the base of the pole. What angle does the rope make with the ground? Write your answer correct to 1 decimal place.

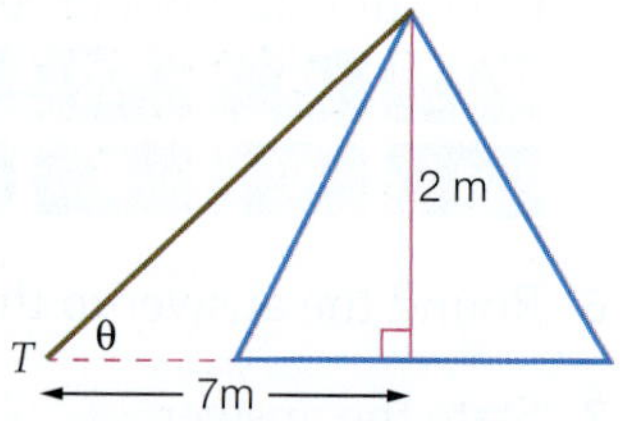

5 A sailing boat needs new sails. Only some of the measurements were recorded and given to the sailmaker, as shown in the diagram below. Calculate the values of the following angles in degrees, correct to 2 decimal places.

(a) Find angle α in sail P.

(b) Find angle θ in sail Q.

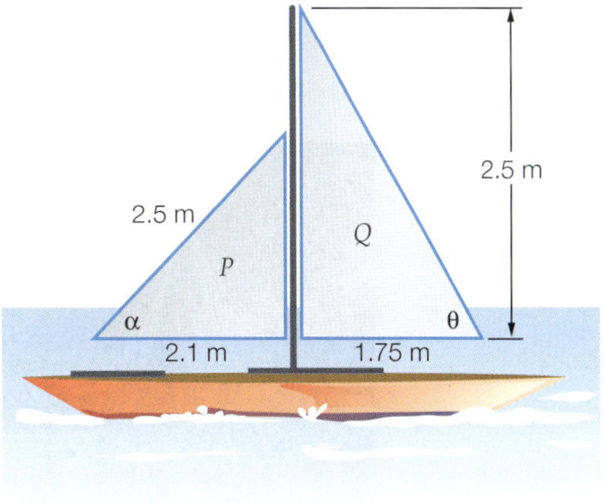

W.E. 11

6 A tree that is 8.4 m tall makes a shadow of 12 m.

(a) Draw a diagram to show this situation.

(b) To the nearest degree, calculate the angle that the Sun makes with the ground to form the shadow.

7 (a) The value of the pronumeral a in the equation $\tan(a) = 0.6395$, to the nearest degree is:

A $32°$ B $33°$ C $40°$ D $50°$

(b) The value of θ in degrees, minutes and seconds in the equation $\sin(\theta) = 0.8938$ is:

A $26°38'43''$ B $41°47'25''$ C $63°21'17''$ D $63°35'47''$

Understanding

8 Find the angle of inclination, θ, of the Sun's rays when a 9.5 m high telephone pole casts a shadow 7 m long, as shown in the photograph. Write your answer in degrees, minutes and seconds.

9 A 2.75 m long children's slide has a ladder that reaches 1.43 m vertically above the ground. To the nearest degree, the angle the slide makes with the horizontal is:

A 27° B 31° C 42° D 58°

10 On a slope, a railway rises 35 cm for every 50 m of track. Find the angle at which the track is inclined to the horizontal, correct to 2 decimal places.

11 Scott is dragging a heavy box along the floor holding the end of a 1.5 m piece of rope. The rope is attached to the top of the box and makes an angle θ with the horizontal. Scott is exactly 1.0 m in front of the box. To the nearest degree, what angle does the rope make with the horizontal?

Reasoning

12 A furniture removal company uses a ramp to wheel furniture into its vans. The back of a van is 82 cm above the ground and the ramp is 1.5 m long.

(a) To the nearest degree, what is the angle of the ramp with the ground?

(b) How much longer does the ramp need to be so that the angle is 30°?

13 A symmetrical roof frame of height 1.85 m has a span of 12.6 m as shown in the diagram at right.

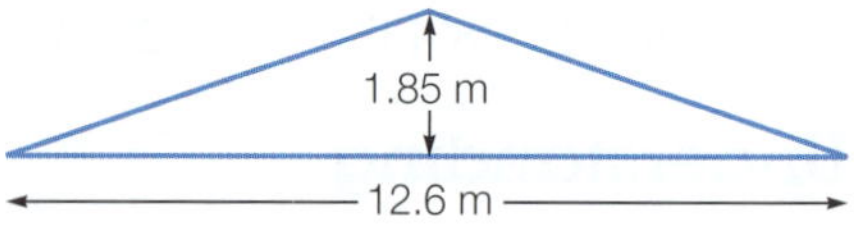

(a) Find the angle at the apex of the roof, to the nearest degree. (The apex is the highest point on the roof.)

(b) Find the total length of material needed to make the frame.

14 A 10 m ladder is to be used to reach different window sills in a building.

(a) If the heights of the window sills range from 2 m to 6 m above the ground, calculate the range of angles that the ladder makes with the ground.

(b) Is it practical to use the ladder to reach all the window sills?

Open-ended

15 A sun lounge can be adjusted from a flat, lying position through to an upright, sitting position. The adjustable back is 80 cm long. Choose three different vertical heights that the back could be raised to. For each height (x) calculate the angle the back makes with the horizontal (θ).

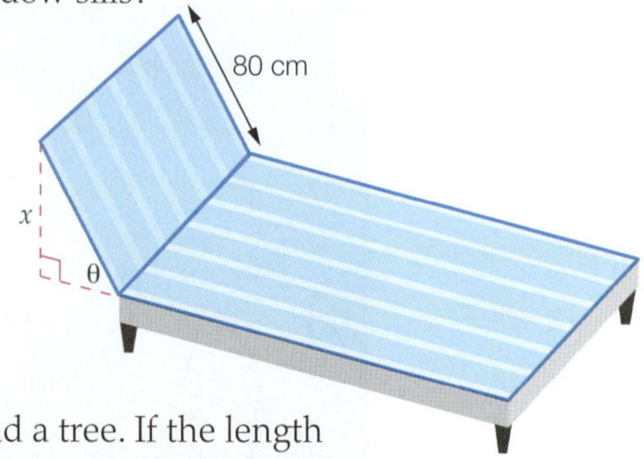

16 Bilal is 1.85 m tall and is standing near a lamp post and a tree. If the length of his shadow is 94 cm, use your own reasonable values for the heights of the lamp post and the tree to determine the length of their shadows at the same time.

Angles of elevation and depression

6.4

The **angle of elevation** from A to B is the angle that the line AB makes with the horizontal.

If you were positioned at A, you would need to *raise* your eyes from a horizontal line of sight to view an object at B.

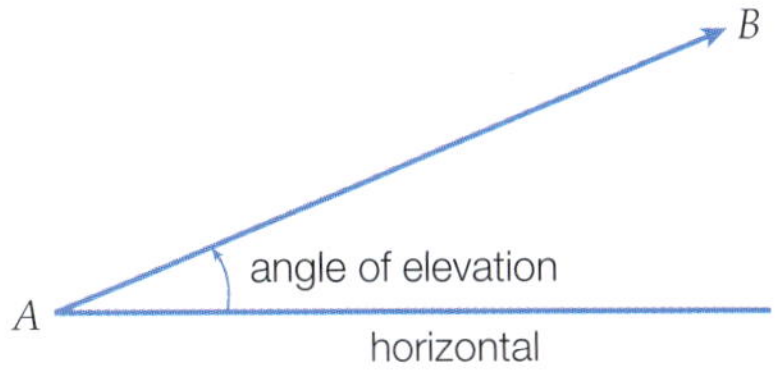

The **angle of depression** from C to D is the angle the line CD makes with the horizontal.

If you were positioned at C, you would need to *lower* your eyes from a horizontal line of sight to view an object at D.

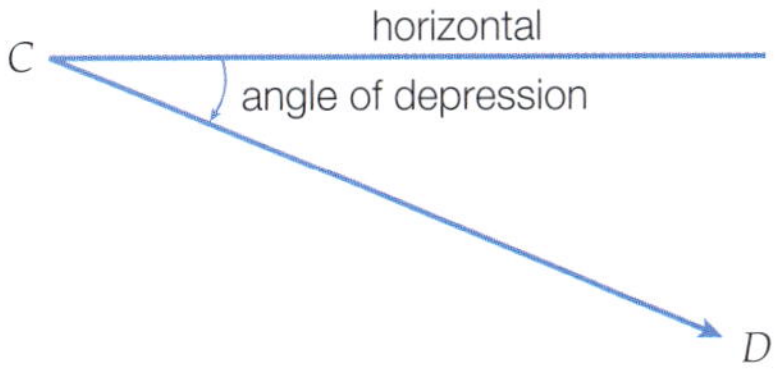

- Elevation: look up.
- Depression: look down.

The angle of depression may not always be inside the triangle you draw. However, you can use alternate angles to find an angle within the triangle.

x and y are alternate angles on parallel lines
$\therefore x = y$

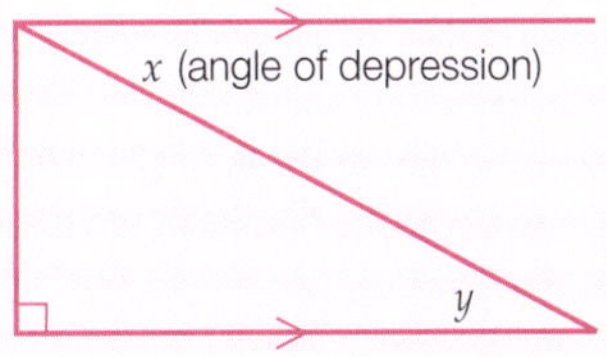

Worked example 12

W.E. 12

A marker buoy used to guide ships is 1.3 km away from the base of a 120 m cliff. To the nearest degree, what is the angle of elevation from the buoy to the top of the cliff? Answer correct to the nearest degree.

Thinking	Working
1 Draw a diagram that includes a right-angled triangle. Label the sides O, A and H.	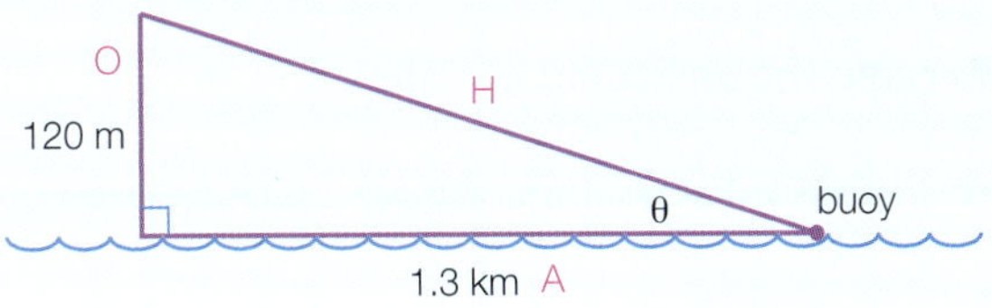
2 Decide which trigonometric ratio connects the given values.	$\theta = ?$ $O = 120$ m $A = 1.3$ km $\tan(\theta) = \frac{O}{A}$

3 Substitute the given values and form an equation. Remember to have distances in the same units.	$\tan(\theta) = \frac{120}{1300}$
4 Use the inverse of the ratio to find θ.	$\theta = \tan^{-1}\left(\frac{120}{1300}\right)$
5 Use the calculator to evaluate.	$\theta = 5.273\,89\ldots$
6 Round the answer to the nearest degree.	$\theta = 5°$ (nearest degree)
7 State the answer.	The angle of elevation from the buoy to the top of the cliff is approximately 5°.

Worked example 13

W.E. 13

A ranger at the top of a lookout tower spots a small fire at an angle of depression of 12.5°. If the tower is 76 m high, calculate the distance from the base of the tower to the fire. Round your answer to the nearest metre.

Thinking	Working
1 Draw a diagram that includes a right-angled triangle. Label the sides O, A and H. **2** Use alternate angles to find and label the angles of depression and elevation.	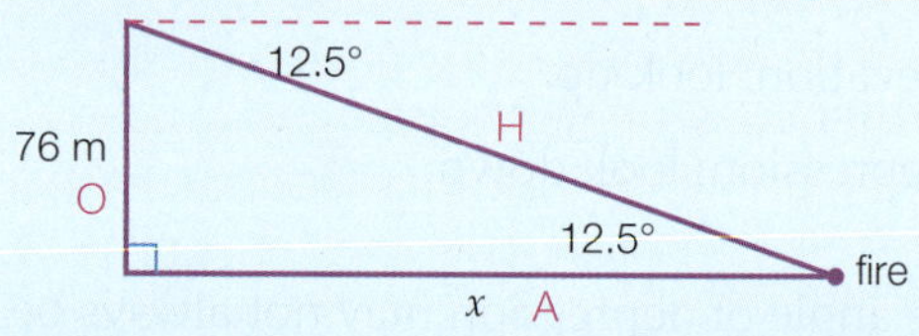
3 Decide which trigonometric ratio connects the given values.	$\theta = 12.5°$ $O = 76$ m $A = x$ $\tan(\theta) = \frac{O}{A}$
4 Substitute the given values and form an equation.	$\tan(12.5°) = \frac{76}{x}$
5 Make x the subject.	$x\tan(12.5°) = 76$ $x = \frac{76}{\tan(12.5°)}$
6 Use a calculator to evaluate and round the answer to the nearest metre.	$x = 343$ m (nearest metre)
7 State the answer in a sentence.	The fire is approximately 343 m from the base of the tower.

6.4 Angles of elevation and depression

Navigator

1, 2, 3, 4, 5, 6, 7, 8, 9, 11, 12, 13, 14, 15, 17	1, 3, 4, 5, 6, 7, 8, 9, 10, 11, 12, 13, 14, 15, 17	1, 3, 5, 6, 7, 9, 10, 11, 12, 13, 14, 15, 16, 17, 18

Answers p. 812

Fluency

1 A a swimmer struggling in the surf waves to a lifesaver in a tower. The angle of elevation from the swimmer to the lifesaver is 8° and the direct distance is 19 m, as shown in the diagram. What is the distance from the base of the lifesaver's tower to the swimmer, correct to the nearest metre?

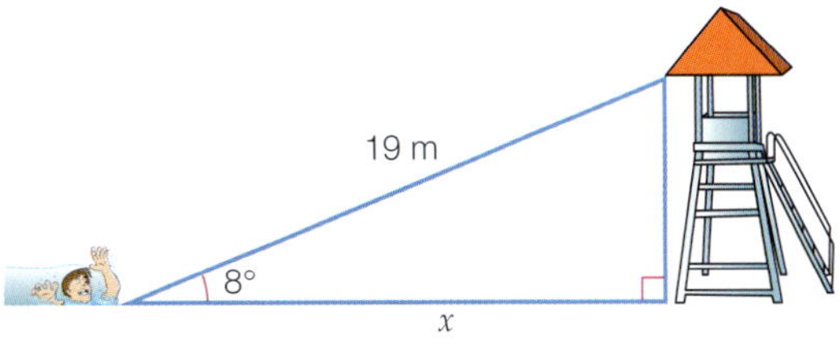

2 Layla is flying a kite on a windy morning. The length of the string from Layla to the kite is 130 m and the horizontal distance from Layla to the kite is 85 m. What is the angle of elevation of the string? Round your answer to 1 decimal place.

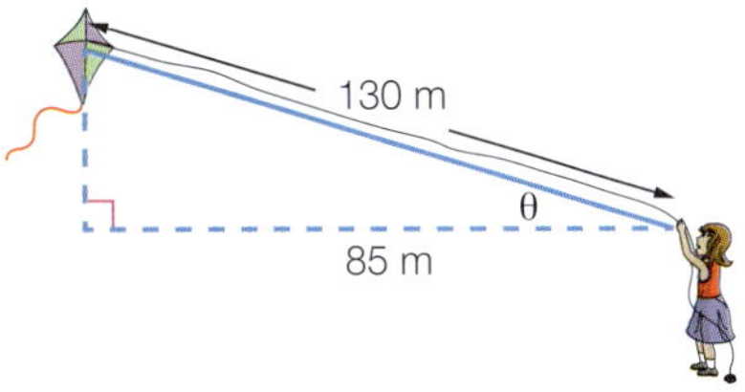

3 Jude has left his building but forgotten his keys, so he calls to his housemate to throw him his keys from the window. The apartment window is 12 m above the ground and Jude is 6 m from the building. Calculate the angle of depression from the window to the point on the ground where Jude is standing, correct to 1 decimal place.

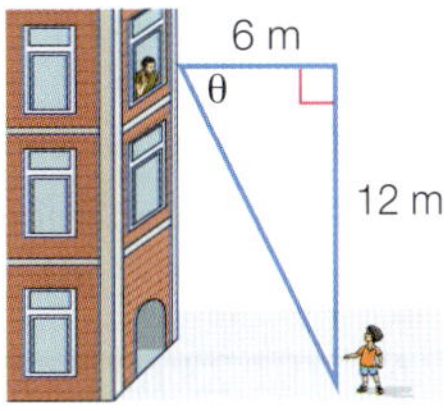

4 Driving to school a teacher sees a hot-air balloon in the sky. She estimates that the angle of elevation from the front of her car to the top of the hot-air balloon is 48° at a distance of 38 km. Calculate the horizontal distance, correct to the nearest kilometre, between the car and the centre of the hot-air balloon.

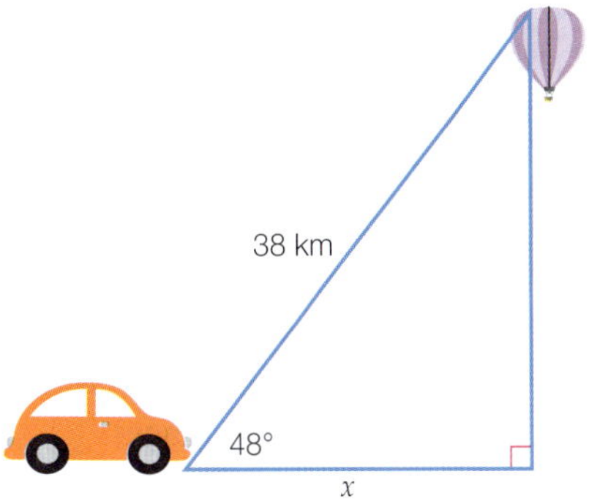

5 Hayden is standing 1.2 km away from the base of a 305 m tall building. To the nearest degree, what is the angle of depression from the top of the building to the point where Hayden is standing? W.E. 12

6 A person in a hot-air balloon 150 m above the ground spots a parked car at an angle of depression of 16.5°. Calculate the horizontal distance from the car to the point that is directly underneath the balloon, to the nearest metre. W.E. 13

7 From a lookout 3 m above the ground, a park ranger spots a potential fire 2.3 m from the base of the lookout. Find the angle of depression from the ranger to the fire. Answer to the nearest degree.

8 The angle of depression from the top of a pine tree to a point on the ground 50 m away is 38°. Find the height of the tree to the nearest metre.

9 A traffic signal 3.95 m high casts a shadow across the road. If the angle of elevation of the Sun is 24.8°, find the length of the shadow correct to 2 decimal places.

10 A helicopter is flying at a height of 160 m above a river. The pilot can see another helicopter on a pontoon along the river located at an angle of depression of 50°. Find, to the nearest metre:

(a) the direct distance between the helicopters

(b) how far the first helicopter must fly at the same altitude to be directly over the second.

11 (a) A railway track is sloped so that it rises 400 cm for every 2 km of track. From the lower end of the track, the angle of elevation (to 2 decimal places) is:

A 0.11° B 11.54° C 25° D 78.69°

(b) From a hot-air balloon, the angle of depression to a marked landing site is 6°50′30″. If the observation deck in the balloon is 190 m above the ground, the horizontal distance the balloon needs to travel to be above the landing site is closest to:

A 191.23 m B 1.58 km C 1.67 km D 1.68 km

Understanding

12 Rachel uses an inclinometer (a device for measuring angles) to measure the angle of elevation from her eye-level to the top of a billboard 25 m away and finds it to be 42°. She also measures the distance from the ground to her eye-level and finds it to be 153 cm. Find the height of the billboard to the nearest metre.

13 A television antenna is mounted on the top of a nearby hill. From a position 4.5 km away, the angle of elevation to the bottom of the antenna is 17°30′, whereas the angle of elevation to the top of the antenna is 17°50′. Find the height of the antenna in metres, correct to 2 decimal places.

Reasoning

14 A yacht is 1.8 km away from the base of a 120 m cliff. The deck of the yacht is 3 m above sea level. To the nearest degree, what is the angle of elevation from the deck of the yacht to the top of a 4.2 m flagpole on the edge of the cliff?

15 Peter views a stationary beetle on the ground, 2.6 m away, at an angle of depression of 32°. Peter's eyes are 15 cm below the top of his head. He is standing upright.

(a) Find Peter's height to the nearest centimetre.

(b) The beetle moves directly ahead to a new position. If the distance from Peter to the beetle along his line of sight is 4.7 m, find the new angle of depression, correct to 1 decimal place.

16 Sonya is standing 20 m from a hot-air balloon that is rising vertically at a speed of 1.5 m/s. When the balloon's position is at eye-level, she follows its path for 7 seconds.

(a) Calculate the angle of elevation of the balloon from Sonya's line of sight after 7 seconds.

(b) If Sonya is 175 cm tall, what is the height of the balloon above the ground after 7 seconds?

Open-ended

17 A flagpole 24 m high is to be supported by a cable that is 18 m long when pulled taut. It must be fastened at least halfway up the pole at an angle of elevation between 45° and 60°.

(a) Find three possible angles of elevation for the cable.

(b) For each angle of elevation, find the distance from the flagpole that the cable must be fastened.

18 From the top of a 20 m lookout, Ranger Sam spots a hiker 200 m due east of his position, heading due east away from him at a constant speed of 50 m/min. Ranger Sam notes the angle of depression of the hiker from his position every minute for a period of one hour. Find two values for the angles of depression that Ranger Sam recorded.

Problem solving

Outback mail run

In outback Australia, remote houses and communities receive their mail by an aircraft.

The pilot is cruising at an altitude of 4500 m when she looks out of the left window and sees one house at an angle of depression of 17°. Out of the right window, she spots another house at an angle of depression of 13°.

How far apart are the two houses, to the nearest kilometre?

Strategy options

- Draw a diagram.
- Have I seen a similar problem?
- Break problem into manageable parts.

Investigation

Viewing artwork—where should you stand?

Equipment required: scientific calculator or CAS

The Big Question

How far should you stand from artwork to have the maximum viewing angle?

Imagine you are at an art gallery to look at a famous work of art. The painting is 1 m tall, and is mounted 1 m above eye-level on a wall as shown.

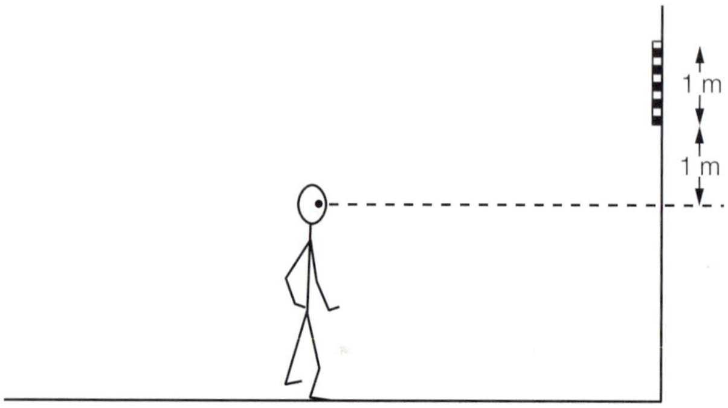

If you stand too close to the painting, the viewing angle V is too small to permit a good view (and your neck quickly gets sore).

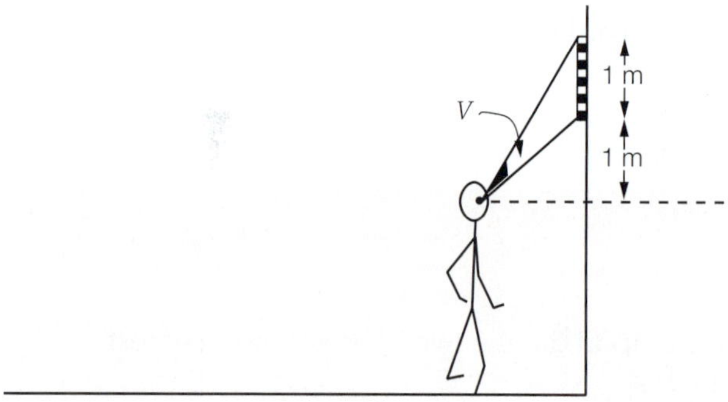

If you stand too far away the viewing angle is also too small, and you don't see any detail in the painting.

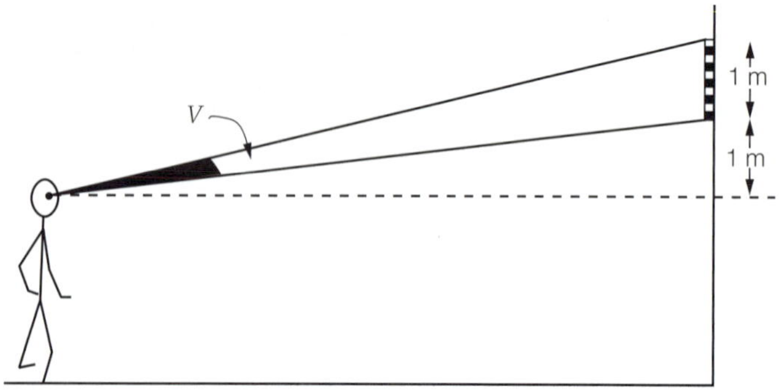

Engage

The viewing angle V is the difference between angles A and B on the diagram below.

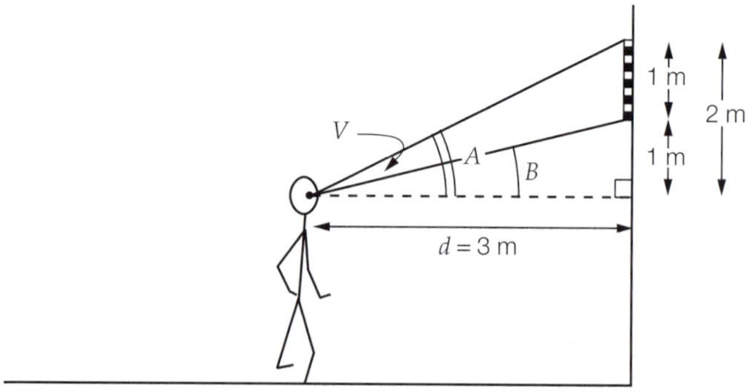

$V = A - B$.

1 Use the two triangles and the measurements in the diagram to calculate A, B and then V if the person viewing the artwork is standing a distance d of 3 m from the wall.

Explore

2 By changing the value of d, find the largest possible value for V. Try at least five values for d.

Strategy options

- Draw a table.
- Look for a pattern.
- Test all possible combinations.

Explain

3 Describe any pattern or trend you can see in the values of angles A, B, and V as you move closer to or further away from the wall.

Elaborate

4 Use different values of d to find the best viewing position for:

(a) a 0.75 m tall painting mounted 1.2 m above eye-level

(b) a 1.5 m tall painting mounted 0.8 m above eye-level

(c) a 63 cm tall painting mounted 50 cm above eye-level.

5 Summarise your findings and answer the Big Question.

Evaluate

6 (a) How confident are you that you have found the largest value for V?

(b) What could you do to improve your answer?

7 As you calculated different values of V, did you find any rules or shortcuts to use with your calculator that made the task more efficient?

Extend

8 Investigate how a person's height affects where they should stand to gain the maximum viewing angle. (Where should a short person stand compared to a tall person?)

Half-time 6

6.3 1 A ladder of length 3 m reaches 1.85 m up a vertical wall. The ladder is moved closer to the wall and now reaches 2.10 m up the wall. What is the change in the angle the ladder makes with the ground?

6.3 2 An ant travels 76 cm in a straight line, turns 90°, follows a straight path for 80 cm and then travels directly to its starting point. The smallest interior angle of the triangle formed is:

A 40.2° B 43.5° C 44.6° D 45.5°

6.1 3 Which calculation can be used to find the length of side b?

A $b = \dfrac{4.3}{\cos(35°)}$ B $b = 4.3 \times \sin(35°)$

C $b = \dfrac{\tan(35°)}{4.3}$ D $b = 4.3 \times \cos(35°)$

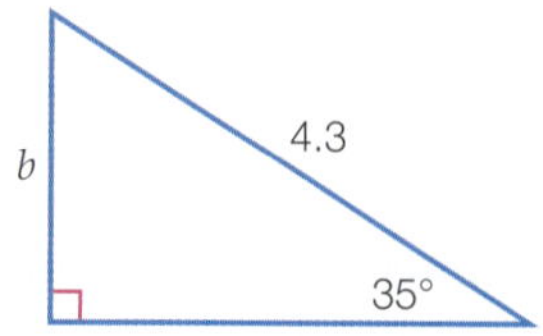

6.3 4 Find the angle θ in the small triangle and use the result to calculate the height of the building to the nearest centimetre.

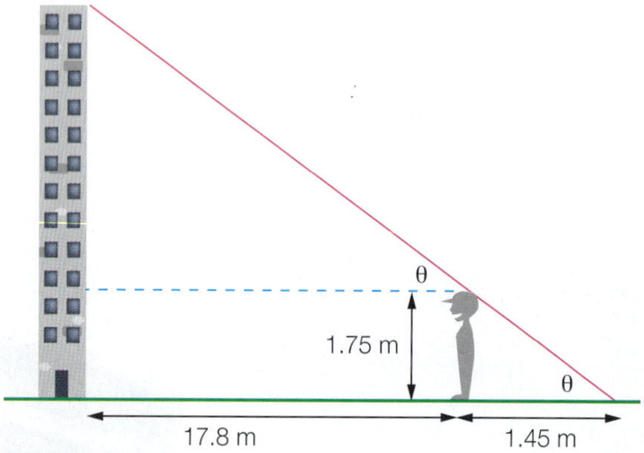

6.2 5 The value of a in the triangle on the right, rounded to 2 decimal places, is:

A 2.41 B 3.39

C 3.68 D 7.40

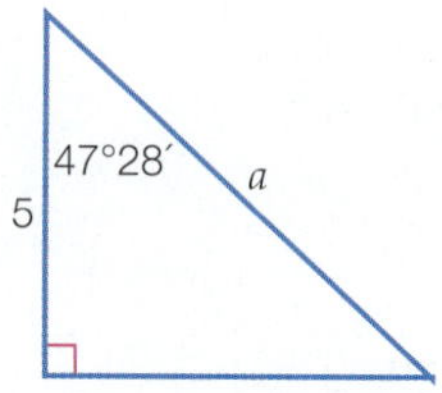

6.4 6 From the top of a lookout, a ranger spots a small scrub fire at an angle of depression of 6°14′. How far is the fire from the base of the lookout if the eye-level of the ranger is 35 m above the ground? Give your answer correct to the nearest metre.

Bearings

6.5

The direction of point A from point B can be expressed in two ways:

True bearings give the angle in degrees in a clockwise direction from north. The true bearing of A from B is 036°T. Use the letter T to denote a true bearing. True bearings are always written with three digits, so the bearing of 36° is written as 036°T.

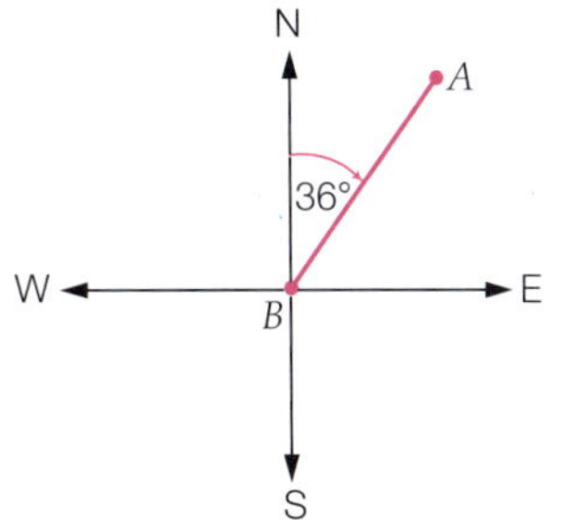

Compass bearings are directions given in terms of turning a number of degrees east or west of north or south (whichever direction is closer to the position). The compass bearing of A from B is N36°E.

The direction of southeast (SE) is shown here.

As a compass bearing, it is written as S45°E.

As a true bearing, it is written as 135°T.

Compass bearings are sometimes also called **conventional bearings**.

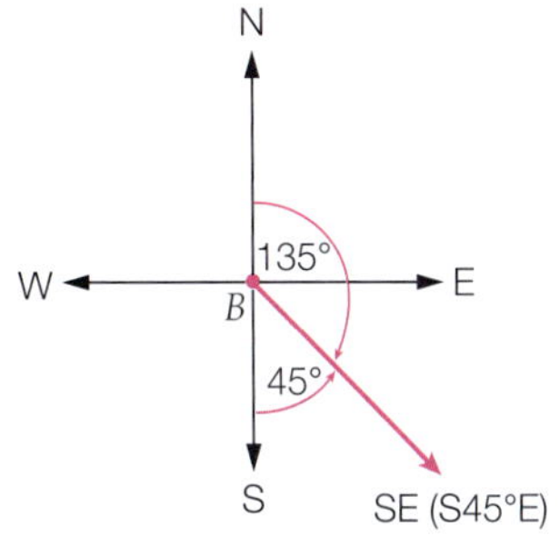

Worked example 14 W.E. 14

Find **(i)** the compass bearing and **(ii)** the true bearing of point B from point A, for each of the following.

(a)

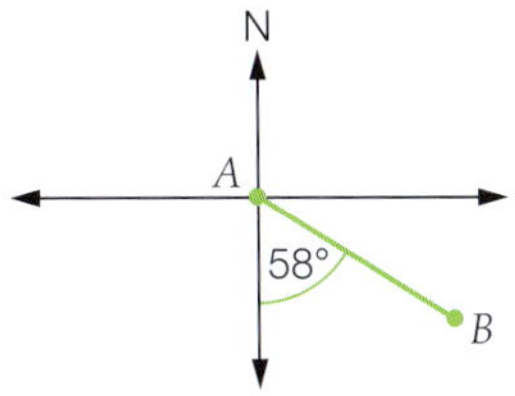

(b)

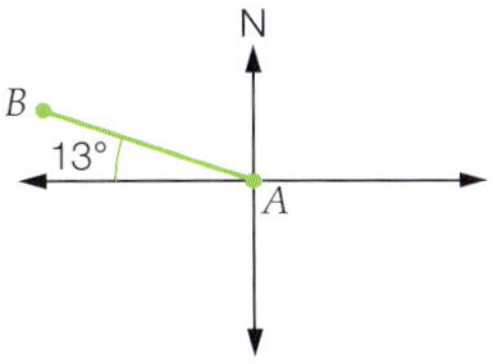

Thinking	Working
(a) (i) 1 Decide whether the point is closer to north or to south. Write either 'N' or 'S'. (Here, it is closer to south so write 'S'.)	(a) (i) S
2 Decide whether the point is closer to east or to west. Leave a space, then write either 'E' or 'W'. (Here, it is closer to east so write 'E'.)	S_E
3 How many degrees to the east or west do you need to turn? Write this number between the two letters. (Here, turn 58° east of south.)	S58°E

Thinking	Working
(ii) Find the angle of turn from north in a clockwise direction to the line joining the point to the centre of the compass. (Here, find the difference between the compass bearing and 180°.)	(ii) 180 − 58 = 122°T
(b) (i) 1 Decide whether the point is closer to north or to south. Write either 'N' or 'S'. (Here, it is closer to north so write 'N'.)	(b) (i) N
2 Decide whether the point is closer to east or to west. Leave a space, then write either 'E' or 'W'. (Here, it is closer to west so write 'W'.)	N_W
3 How many degrees to the east or west do you need to turn? Write this number between the two letters. (Here, turn 77° west of north.)	N77°W
(ii) Find the angle of turn from north in a clockwise direction to the line joining the point to the centre of the compass. (Here, add 13° to 270°.)	(ii) 270 + 13 = 283°T

Worked example 15

W.E. 15

A yacht sails 64 km on a true bearing of 215°T. How far south is the yacht from its original position? Round the answer to 2 decimal places.

Thinking	Working
1 Draw a diagram showing the known information. Let x be the distance required. Use the given bearing to show a triangle containing the angle required. (215° − 180° = 35°)	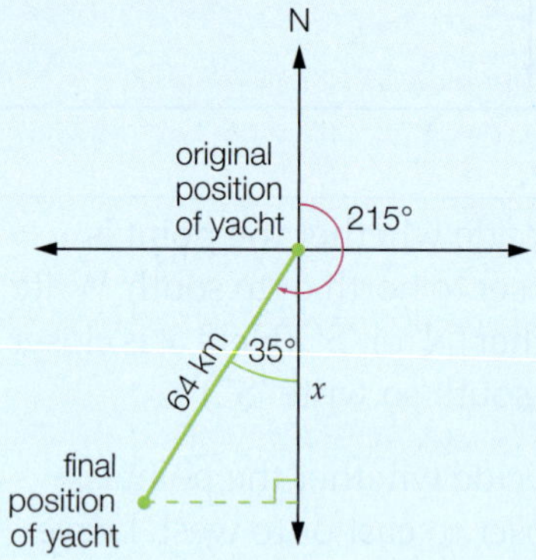
2 Write the known and unknown values and identify the trigonometric ratio that connects them.	$\theta = 35°$ H = 64 km A = x $\cos(\theta) = \frac{A}{H}$

3 Form an equation by substituting the known and unknown values into the ratio. | $\cos(35°) = \frac{x}{64}$

4 Make x the subject. | $x = 64 \times \cos(35°)$

5 Evaluate using a calculator and round the answer to 2 decimal places. | $x = 52.43$ km (2 d.p.)

6 State the answer. | The yacht is 52.43 km south from its original position.

Worked example 16 W.E.16

On a hiking trip, Sarah walks 8.3 km west and then 5.4 km north. Calculate her true bearing and compass bearing from her original position, to the nearest degree.

Thinking | **Working**

1 Draw a diagram showing the known information. Let θ be the unknown angle.

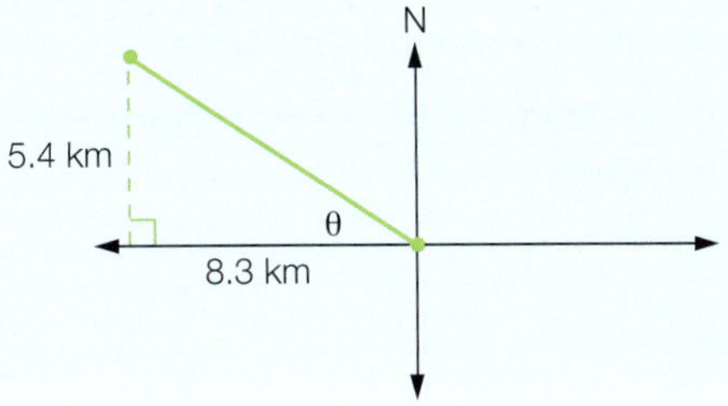

2 Write the known and unknown values and identify the trigonometric ratio that connects them.

$O = 5.4$ km
$A = 8.3$ km
$\theta = ?$

$\tan(\theta) = \frac{O}{A}$

3 Form an equation by substituting the known values into the ratio.

$\tan(\theta) = \frac{5.4}{8.3}$

4 Find θ using the inverse of the ratio and round the answer to the nearest degree.

$\theta = \tan^{-1}\left(\frac{5.4}{8.3}\right)$

$\theta = 33°$ (nearest degree)

5 Write the value as a true bearing by adding all the angle values clockwise from north.

To find the compass bearing, start at north or south (whichever is closer) and find the angle required to turn east or west. (Here, start at north and turn (90 – 33)° west.)

True bearing: $270° + 33° = 303°$T
Compass bearing: $90° - 33° = 57°$

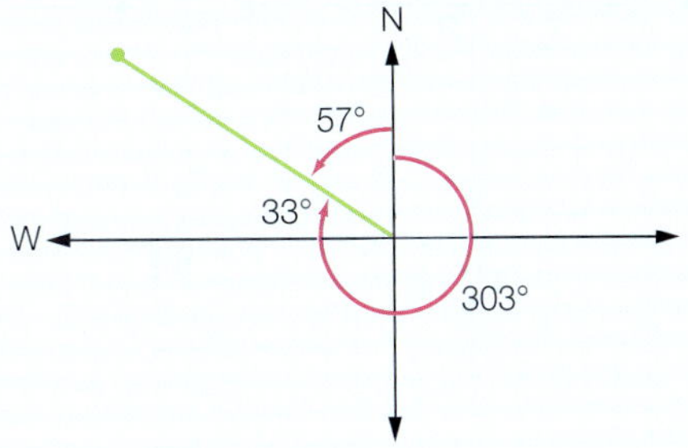

6 State the answer.

True bearing = 303°T
Compass bearing = N 57° W

The directions exactly in between the four compass points of north, east, south and west have the following bearings.

- north-east (NE) is 045°T or N45°E
- south-east (SE) is 135°T or S45°E
- south-west (SW) is 225°T or S45°W
- north-west (NW) is 315°T or N45°W

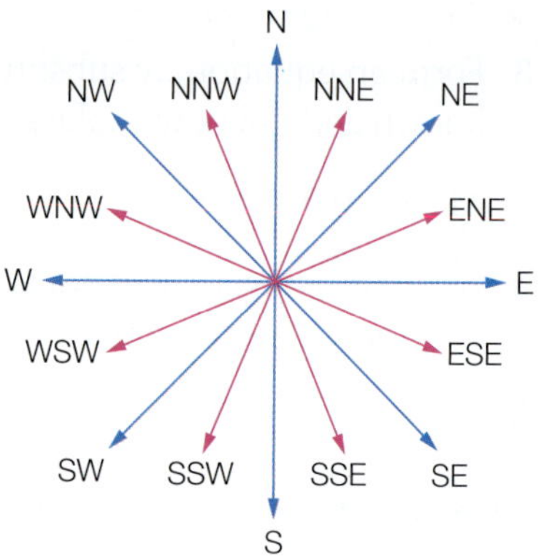

Interactive

Compass bearings

Compare how the compass bearing values and true bearing values change around the circle.

Go to the eBook or the Pearson Places website to access this interactive.

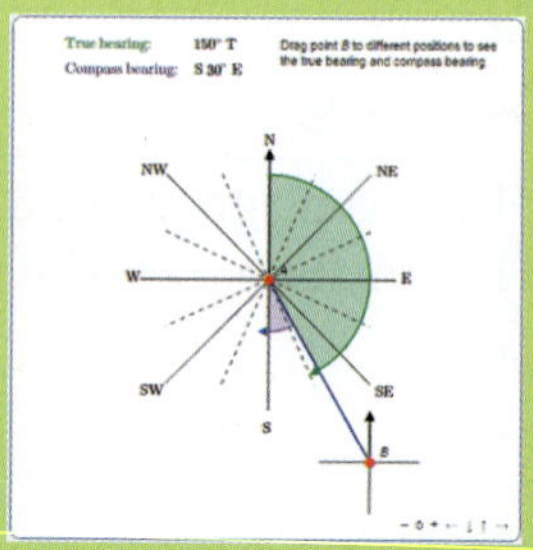

6.5 Bearings

Navigator

Answers p. 812

1, 2, 3, 4, 5, 6, 7, 8, 9, 10, 11, 12, 14, 16, 17, 18, 20, 21	1 (c–f), 2, 3, 5, 6, 8, 9, 10, 11, 12, 13, 14, 15, 16, 17, 18, 20, 21	1 (c, f), 2, 5, 6, 8, 9, 10, 12, 13, 14, 15, 16, 17, 18, 19, 20, 21

Fluency

W.E. 14

1 Find **(i)** the compass bearing and **(ii)** the true bearing for each point from O.

(a)

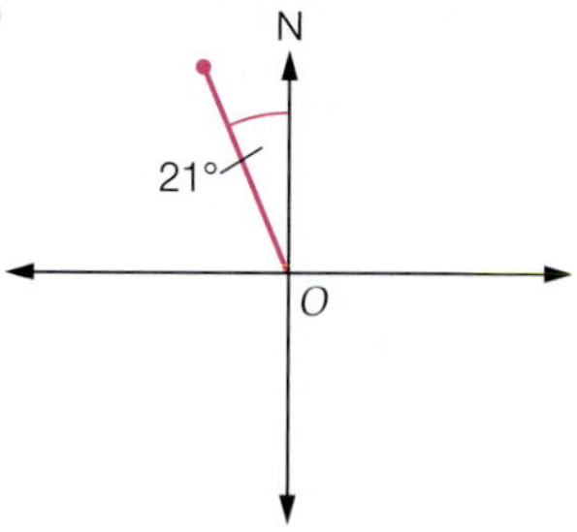

(b)

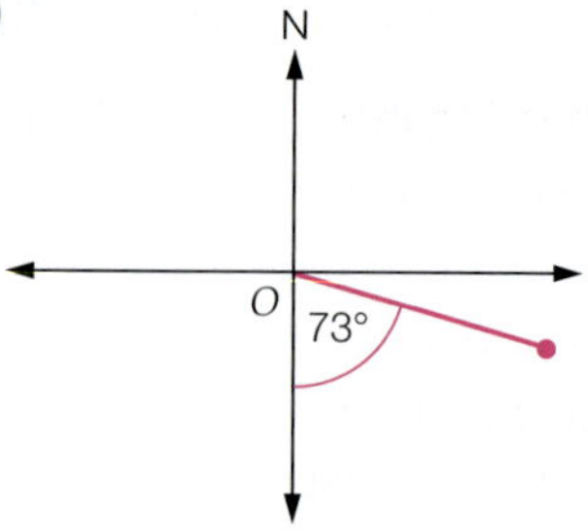

(c)

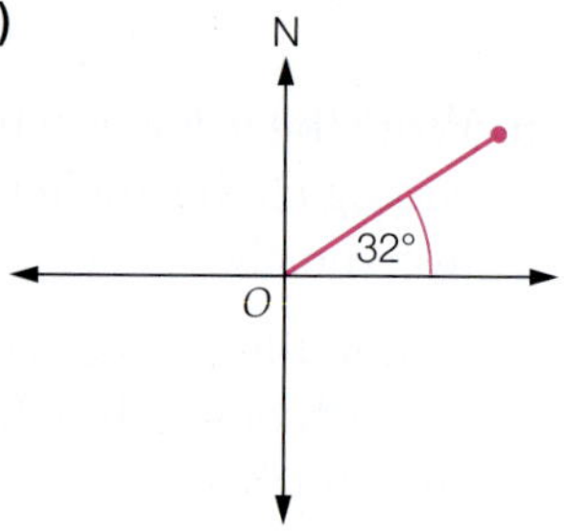

(d)

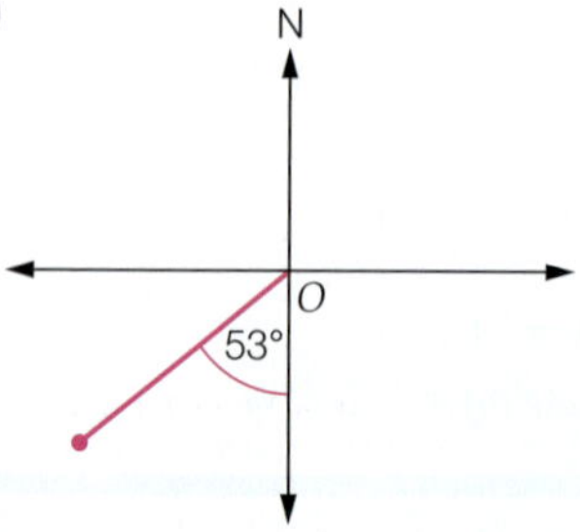

(e)

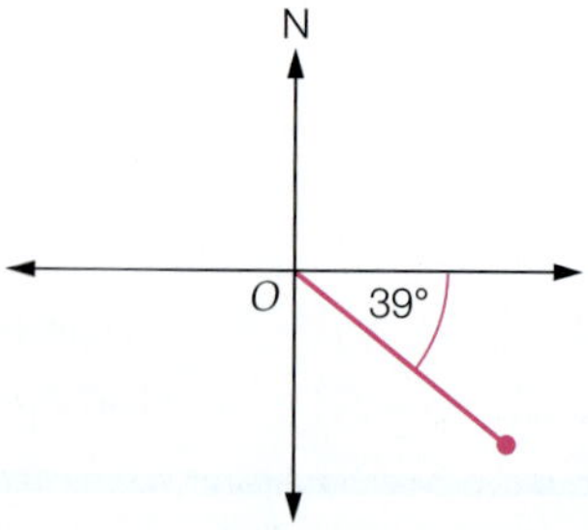

(f)

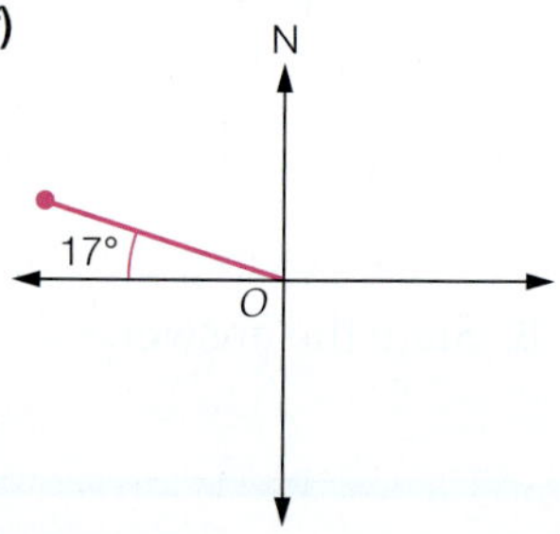

2 The compass bearing of point B from point A, as shown in the diagram, is:

A N35°E B S55°E C N35°W D S35°E

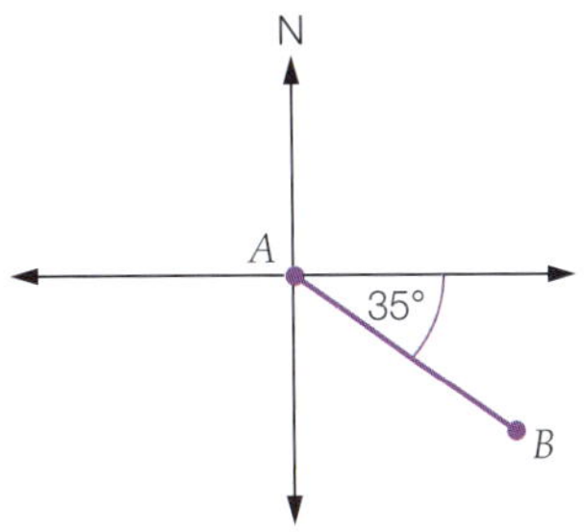

3 The true bearing of point D from point C, as shown in the diagram, is:

A 076°T B 194°T C 256°T D 284°T

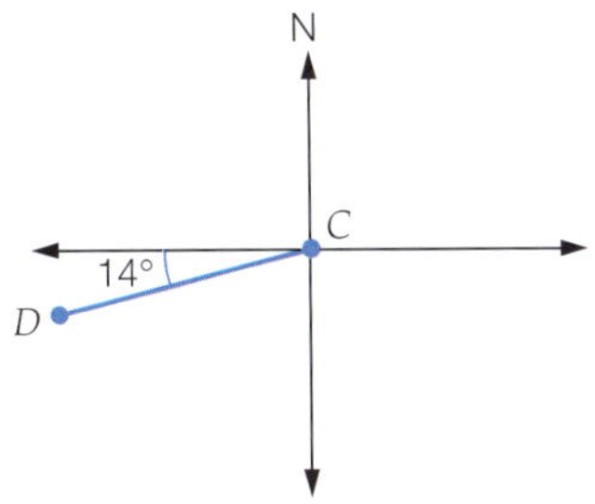

4 Angelique is training for a marathon run. Her training course takes her 5 km north and 2 km east from her home. Calculate the true bearing and the compass bearing of her finishing point from her home. Round your answer to the nearest degree.

W.E. 15

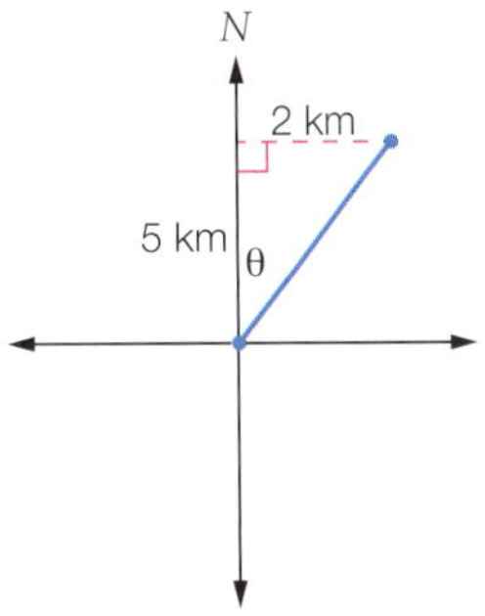

5 Devon is training his border collies for an event. He sets up a training course so that the distance from the starting point A to B is 35 m on a bearing of N29°W. Another point C is marked due south of B. The course then goes back to the starting point, A, where it ends.

W.E. 16

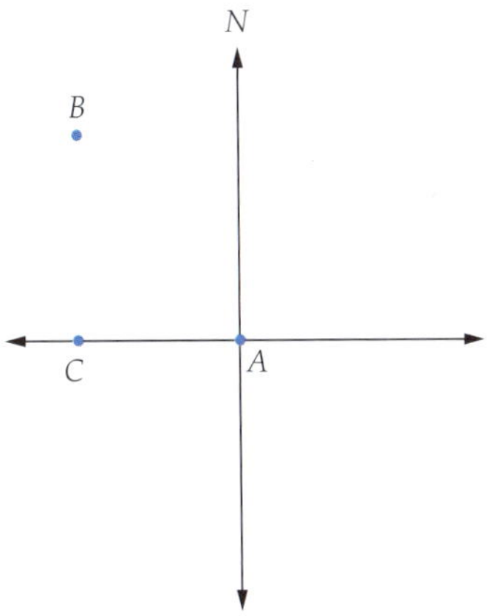

(a) Complete the diagram representing the training course by sketching and labelling with the given information.

(b) Calculate the distance between B and C. Round your answer to the nearest metre.

(c) Calculate the distance between C and the finishing point. Round your answer to the nearest metre.

(d) What is the total length of the training course? Round your answer to the nearest metre.

6 Gerry's boat is moored on the Great Barrier Reef. From her boat Gerry swims 200 m to a shipwreck on a bearing of S15°E. She then swims due west until she is due south of her boat. From that point she swims due north back to her boat.

(a) Sketch a diagram representing her swim. Label the boat's position B, the shipwreck S, and the third point P, where she changes direction.

(b) How far did Gerry swim from the shipwreck before she changed direction? Round your answer to 1 decimal place.

(c) How far did Gerry swim from P to reach the boat? Round your answer to 1 decimal place.

7 A hiker walks due north for 17 km and then due east for 12 km.

(a) Calculate the true bearing of the hiker's current position from the starting point. Round your answer to the nearest degree.

(b) Convert the true bearing to a compass bearing.

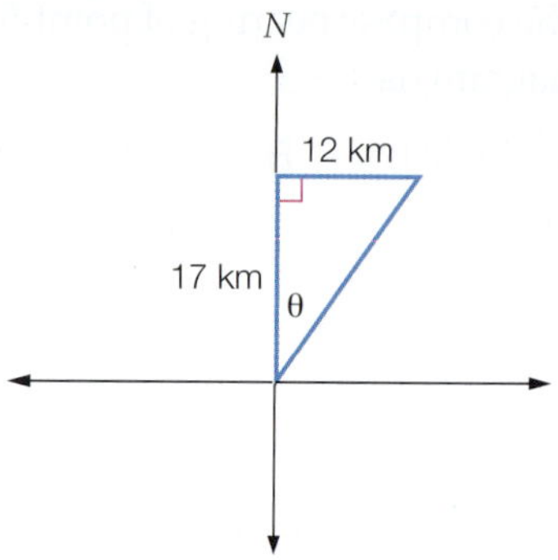

8 A fisherman rows due west for 33 km and then due north for 21 km. Calculate the bearing, rounded to the nearest degree, of the fisherman's final position from his starting point as a:

(a) compass bearing

(b) true bearing.

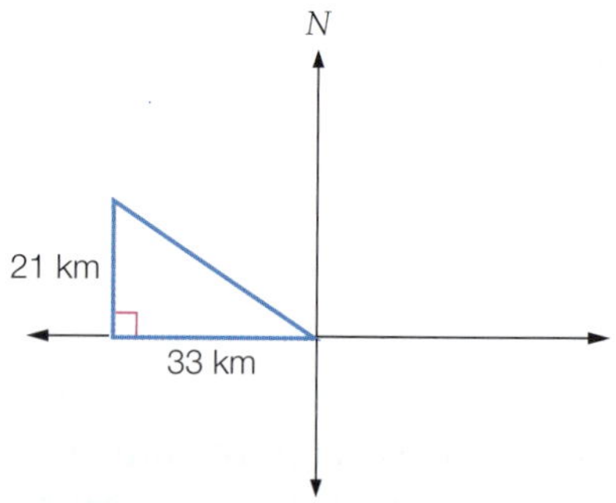

9 A ship sails for 150 km on a bearing of N40°E. How far north is it from its starting point? Round your answer to the nearest kilometre.

10 (a) Pedro sails his yacht 2.5 km east and then 4.7 km south. To the nearest degree, state Pedro's true bearing and compass bearing from his original position.

(b) A taxi driver drives 630 m south and then 410 m east. To the nearest degree, what is the bearing of the taxi from its starting position? Give your answer as a compass bearing and a true bearing.

Understanding

11 A helicopter travelling on a true bearing of 098°T is now 75 km east of its original position. How far has the helicopter travelled, to 2 decimal places?

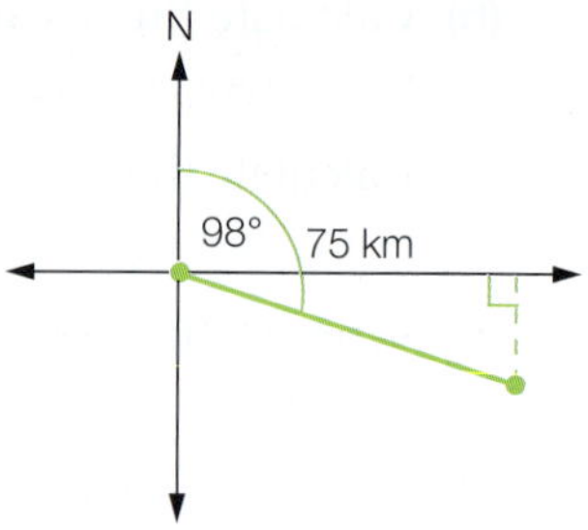

12 An airctraft flies on a true bearing of 280°T at 600 km/h after taking off at 1300 hours. To the nearest kilometre, how far north of its starting position is it at 1530 hours? Hint: First, calculate the distance that the aircraft flies in $2\frac{1}{2}$ hours.

13 Jack drives north at 80 km/h for 2 hours. He then turns and drives east at the same speed for 3 hours.

(a) To the nearest degree, what is Jack's true bearing from his original position?

(b) If Jack could drive from his current position straight back to his starting position:

(i) how far would he travel, to the nearest kilometre

(ii) what true bearing should he follow?

14 A hiker walks 5 km on a true bearing of 025°T from camp *A* to camp *B*. Round your answers to 2 decimal places.

(a) How far east has she walked from *A* to *B*?

She then continues for another 6 km on a true bearing of 078°T from camp *B* to camp *C*.

(b) (i) How far east of camp *B* is she now?

(ii) How far east of camp *A* is she now?

(c) (i) How far north of camp B is she now?

(ii) How far north of camp A is she now?

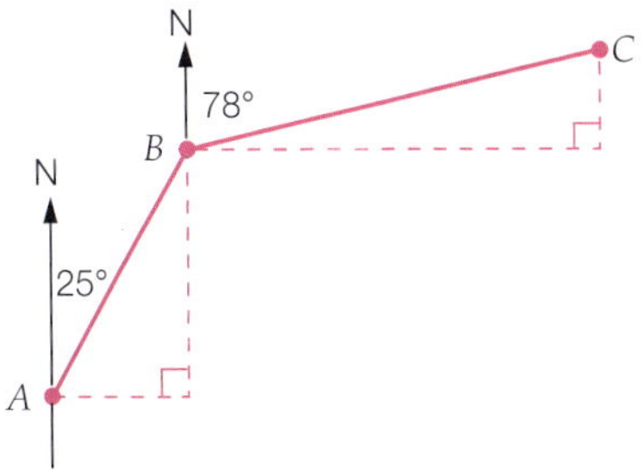

Warning

Round your final answers, but do not use rounded answers in further calculations. Use the original unrounded answers in further calculations.

15 Sasha is running an orienteering course. He runs 750 m to the first checkpoint on a bearing of S53°W. At the checkpoint he is instructed to run another 600 m on a bearing of N68°W to the next checkpoint.

(a) To the nearest metre, how far west is Sasha from his starting position?

(b) Is Sasha's current position north or south of his starting position? How far north or south?

16 A toy robot walks 16 m on a bearing of S62°W, and then turns and walks another 23 m NW. How far west is the robot from its starting point? Round your answer to 2 decimal places.

Reasoning

17 A crew row their boat from point *A* to point *B*, a distance of 200 m on a bearing of S55°E. They then row due north a further 440 m to point *C*, as shown in the diagram.

(a) Find, to the nearest metre:

(i) the distance from *B* to *Z*

(ii) the distance from *A* to *Z*.

(b) Find the true bearing of *C* from *A*, to the nearest degree.

Finally, the crew row back from *C* to *A*.

(c) What is the total distance rowed by the crew, to the nearest metre?

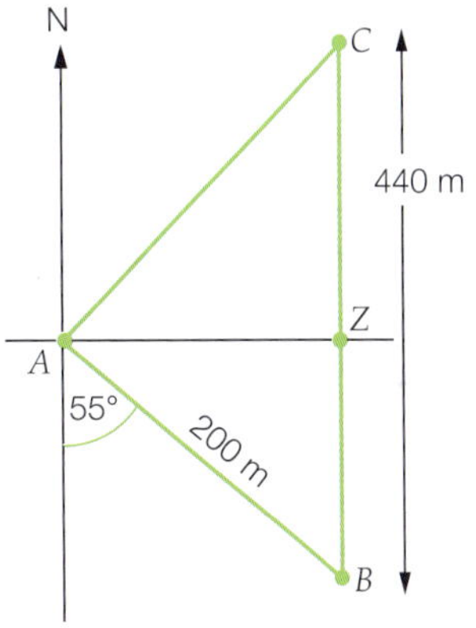

Hint

Pythagoras' theorem is helpful for finding missing sides in a right-angled triangle.

18 (a) From his home, Minh jogs 2.7 km NE and then 7.4 km SE. To the nearest degree, find his true bearing from his starting position.

(b) If Minh then jogs 6.9 km west, how far south is he from his home? Round your answer to 2 decimal places.

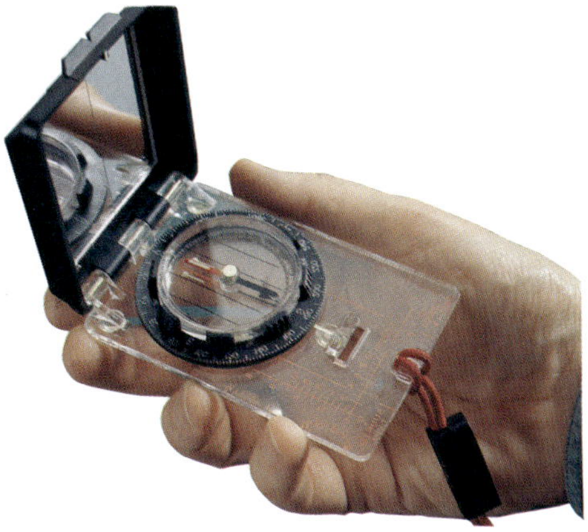

19 An air force jet takes off from an airfield and flys on a bearing of N37°E. The pilot is asked by ground staff to alter her course, 600 km into the trip, by turning 120° in a clockwise direction.

(a) State the new compass bearing of the jet after it has travelled 600 km.

(b) The jet travels for a further 800 km on its new bearing. How far east has it travelled from its starting position, to the nearest kilometre?

(c) How far north or south of its position at take-off, is the current position of the jet, to the nearest kilometre?

(d) What compass bearing should the jet now follow to return to the airstrip?

(e) How far will it travel on its return, if it follows the bearing you determined in **(d)**? Answer to the nearest kilometre.

Open-ended

20 Choose a true bearing θ less than 90° and state its true bearing and compass bearing after it is:

(a) reflected about the east-west line (the x-axis of the Cartesian plane)

(b) reflected about the north-south line (the y-axis of the Cartesian plane)

(c) rotated 180°.

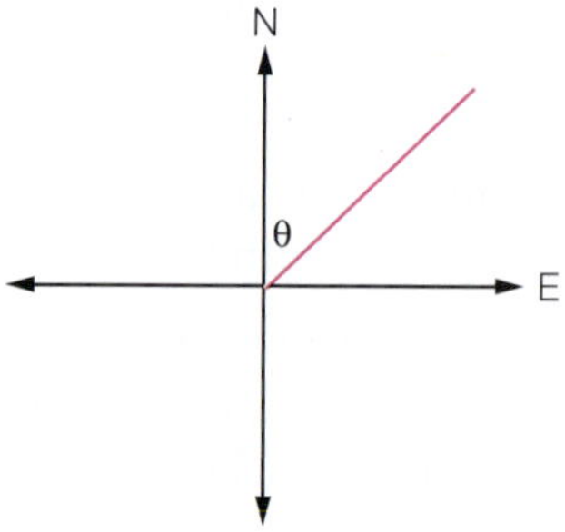

Problem solving

Seeing the light

1 A ship leaves its dock, which is 3 km due west of a lighthouse, and travels NE at a constant speed of 15 km/h.

(a) To the nearest minute, when is the ship due north of the lighthouse?

(b) At this time, how far is the ship from the lighthouse?

(c) How far has the ship travelled?

2 Rework these answers if the ship's bearing is N20°E.

Strategy options

- Draw a diagram.
- Break problem into manageable parts.

Mixed two-dimensional problems

6.6

Situations arise when more than one trigonometric ratio is required to find an unknown side or angle. In such cases, it is important to determine which ratio is used first.

It is important that you use unrounded answers in further calculations. This avoids the accumulation of errors due to rounding. Use the calculator's memory key to store values, if necessary, and then round your answer once only.

Worked example 17

W.E. 17

Find the values of x and y, rounded to 2 decimal places.

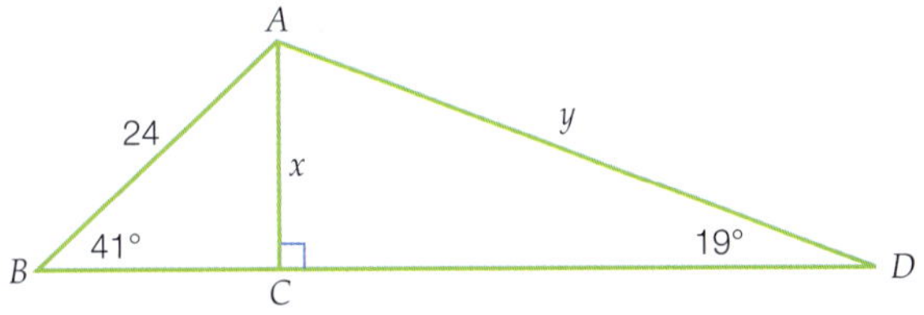

Thinking	Working
1 Identify the triangle with two known values. Use this one first to find the value of an unknown. (Here, use ΔABC to find x.) Draw the triangle and label it with known and unknown values. Label the sides O, A and H.	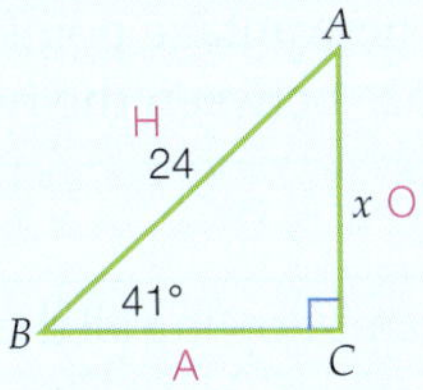
2 Write the known and unknown values, then identify the trigonometric ratio that connects them.	$\theta = 41°$ $H = 24$ $O = x$ $\sin(\theta) = \frac{O}{H}$
3 Form an equation by substituting values into the ratio.	$\sin(41°) = \frac{x}{24}$
4 Make x the subject.	$x = 24 \times \sin(41°)$
5 Perform the calculations. Do not round your answer.	$x = 15.745\,4167$

6 Draw the second triangle and show all the known and unknown values.

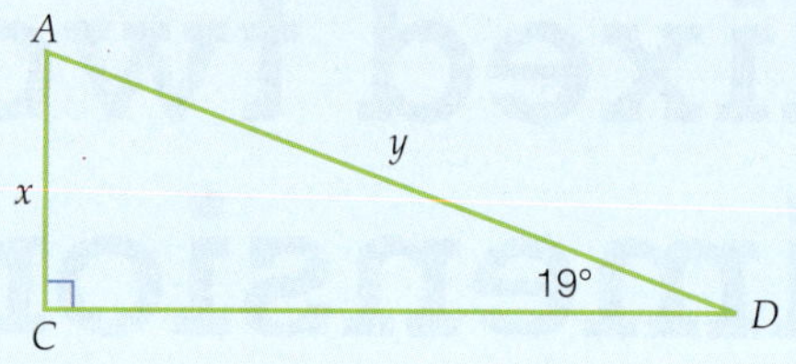

7 Write the known and unknown values, then identify the trigonometric ratio that connects them.

$\theta = 19°$
$O = 15.745\,4167$
$H = y$

$$\sin(\theta) = \frac{O}{H}$$

8 Form an equation by substituting values into the ratio.

$$\sin(19°) = \frac{x}{y}$$

9 Make y the subject of the equation. Solve using the previously found, unrounded value of x.

$$y \times \sin(19°) = x$$

$$y = \frac{x}{\sin(19°)}$$

$$y = \frac{15.745\,4167}{\sin(19°)}$$

10 Evaluate y using a calculator.

$y = 48.362\,889\,55$

Round answers to 2 decimal places.

$x = 15.75$ units $\quad y = 48.36$ units (2 d.p.)

Worked example 18

W.E. 18

Thea sights two swimmers (both on the same bearing from her) at angles of depression of 34°35′ and 55°10′ from her vantage point on the cliff. If her eye-level is 40 m above the water level, find the distance between the two swimmers, rounded to 2 decimal places.

Thinking	Working

1 Draw and label a diagram showing all the known and unknown values. (Here, Thea is at point A, the base of the cliff is point B, and the two swimmers are at points C and D.) Label the required unknown values with pronumerals. Use alternate angles to find the angles within the triangles.

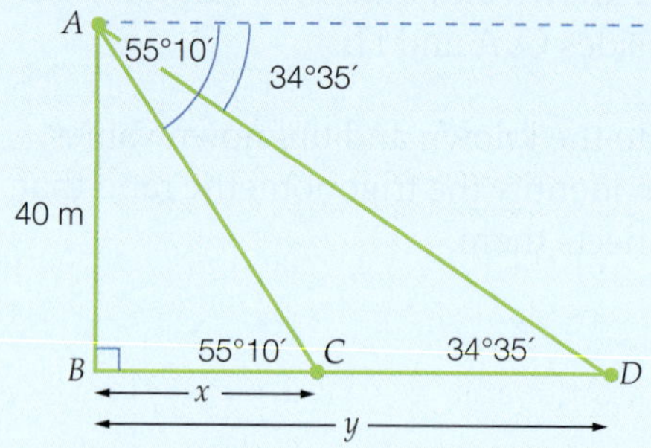

2 Draw one of the two triangles out separately. Identify the sides O, A and H.

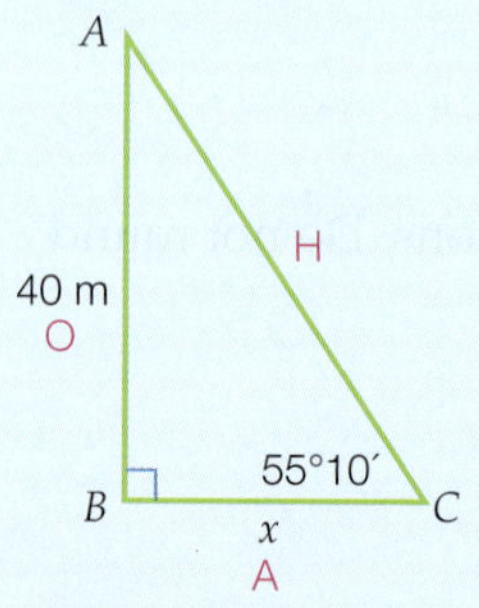

3 Write the known and unknown values, then identify the trigonometric ratio that connects them.	$\theta = 55°10'$ $O = 40$ m $A = x$ $\tan(\theta) = \frac{O}{A}$
4 Form an equation by substituting values into the ratio.	$\tan(55°10') = \frac{40}{x}$
5 Make x the subject of the equation. (Here, multiply both sides by x, then divide by tan (55°10′).)	$x = \frac{40}{\tan(55°10')}$
6 Evaluate using a calculator. Do not round the answer.	$x = 27.835\,250\,33$
7 Draw the second triangle separately. Identify the sides O, A and H.	
8 Write the known and unknown values, then identify the trigonometric ratio that connects them.	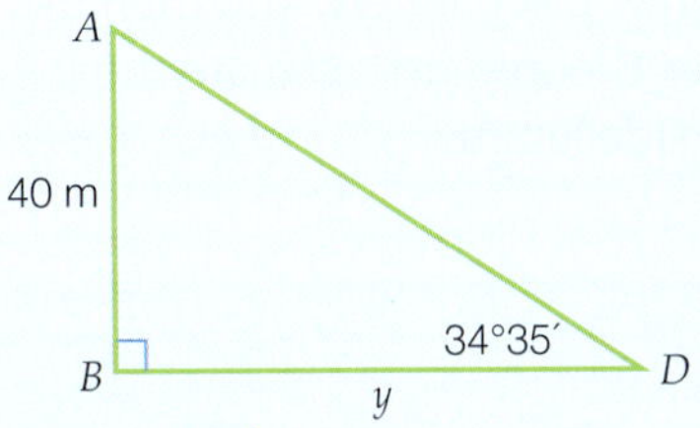 $\theta = 34°35'$ $O = 40$ m $A = y$ $\tan(\theta) = \frac{O}{A}$
9 Form an equation by substituting values into the ratio.	$\tan(34°35') = \frac{40}{y}$
10 Make y the subject of the equation. (Here, multiply both sides by y, then divide by tan (34°35′).)	$y = \frac{40}{\tan(34°35')}$
11 Evaluate using a calculator. Leave this value on the screen.	$y = 58.019\,401\,28$
12 Find the difference between the two calculated side lengths. (Here, it is $y - x$.)	$y - x = 30.184\,150\,95$
13 Round the answer to 2 decimal places.	$CD = 30.18$ m (2 d.p.)
14 State the answer.	The distance between the swimmers is approximately 30.18 m.

Using Pythagoras' theorem

In some cases it is easier to use Pythagoras' theorem as part of the calculation. The theorem states that for a right-angled triangle, $c^2 = a^2 + b^2$, where c is the length of the hypotenuse and a and b are the lengths of the two shorter sides.

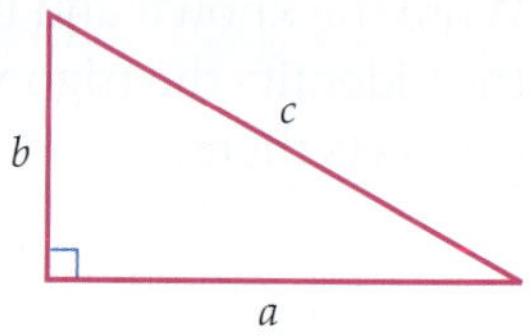

Worked example 19

W.E. 19

Calculate the value of x, rounded to 2 decimal places.

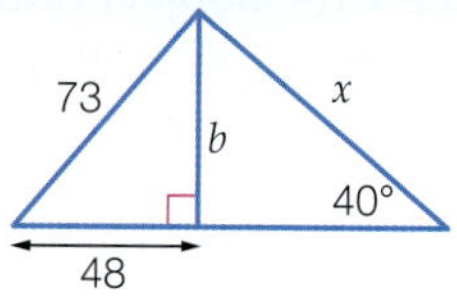

Thinking	Working
1 Identify the triangle with two known side lengths. Draw it separately, and then use Pythagoras' theorem to calculate the unknown side length.	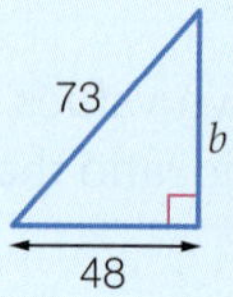 $c^2 = a^2 + b^2$, where $c = 73$, $a = 48$ $73^2 = 48^2 + b^2$ $b^2 = 73^2 - 48^2$ $b^2 = 3025$ $b = 55$
2 Draw out the second triangle. Identify the sides O, A and H. Write the known and unknown values, including those you just calculated. Use the appropriate trigonometric ratio to solve for the unknown length. Round the answer to 2 decimal places.	$\theta = 40°$ $H = x$ $O = 55$ 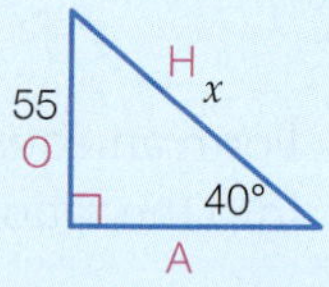 $\sin(\theta) = \frac{O}{H}$ $\sin(40°) = \frac{55}{x}$ $x = \frac{55}{\sin(40°)}$ $x = 85.56$ units

Note that in the previous example you could have used trigonometry to calculate the value of b in the first triangle, by using the two given sides to calculate the angle θ, then using the angle and a side to calculate b. However, this involves two steps, which is a less efficient method than using Pythagoras' theorem.

6.6 Mixed two-dimensional problems

Navigator

1, 2, 3, 4, 5, 6, 7, 8, 9, 10, 11, 12, 14, 17, 18, 19, 20, 23	1 (column 1), 2, 3, 4, 5, 6, 7, 8, 10, 11, 12, 14, 15, 16, 17, 18, 19, 20, 21, 23	1 (column 2), 3, 4, 5 (column 2), 6 (column 1), 7, 8, 10, 11, 12, 13, 14, 15, 16, 17, 18, 19, 20, 21, 22, 23

Answers p. 812

Fluency

1 Find the values of x and y, rounded to 2 decimal places.

W.E. 17

(a)

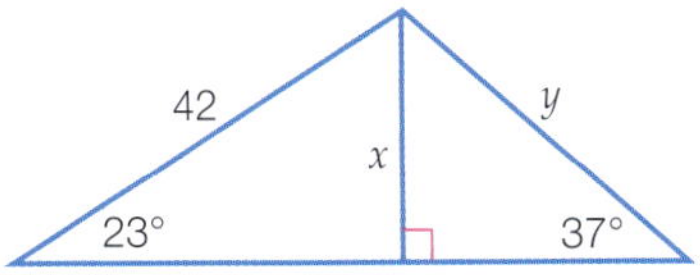

(b)

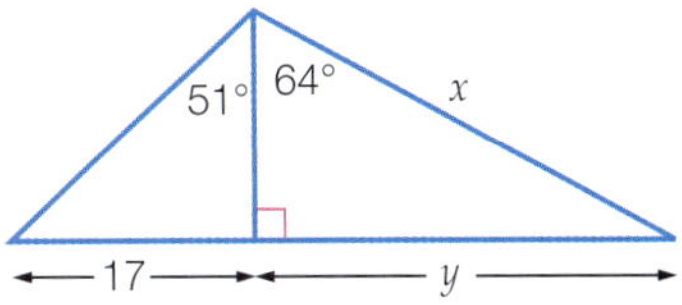

(c)

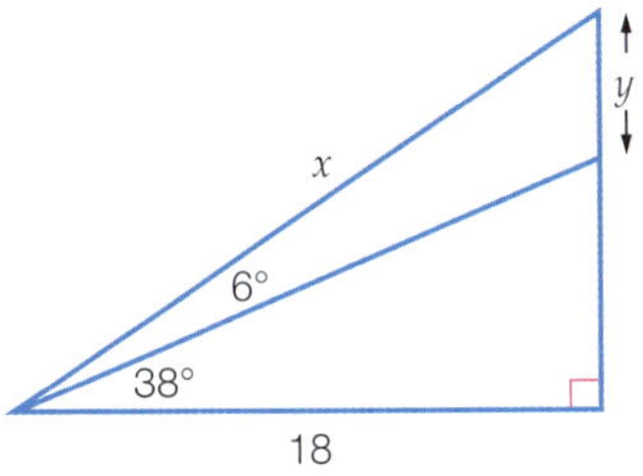

(d)

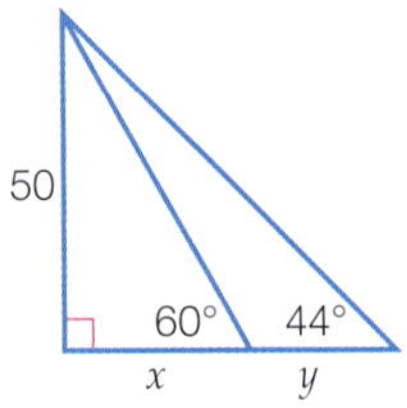

(e)

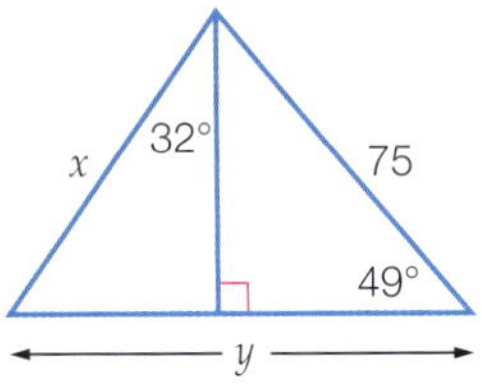

(f)

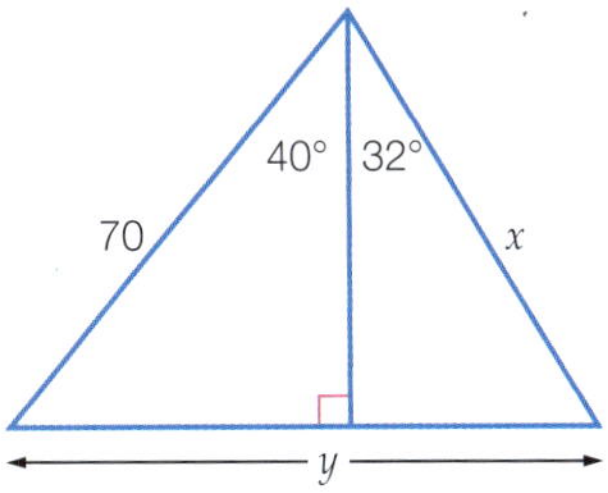

2 A ship travels 275 km on a bearing of S56°W. How far west of the starting point is the ship? Answer correct to the nearest kilometre.

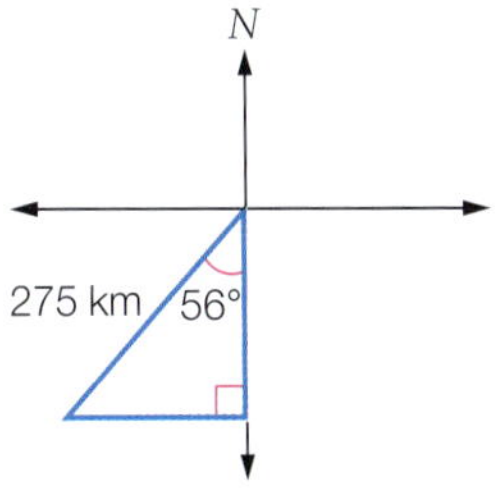

3 A farmer requires a new gate with one diagonal as shown in the diagram. The length is 1.75 m and width is 0.95 m.

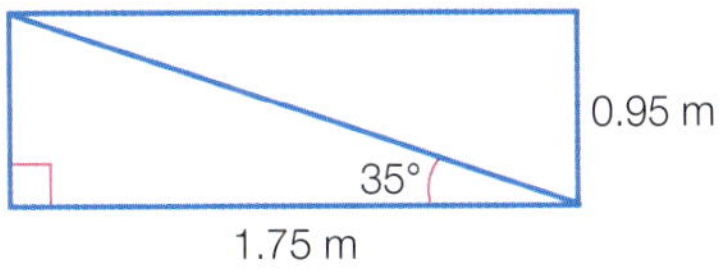

(a) Calculate the length of the diagonal. Round your answer to 2 decimal places.

(b) The wood is sold in lengths of whole metres, at $5.95 per metre. How much would the required wood cost if the farmer decides to include two diagonals in the gate?

W.E. 18

4 From the top of a lighthouse 75 m above sea level, two whales on the same bearing can be seen in the ocean at angles of depression of 28°40′ and 46°25′. Find the distance between the two whales, rounded to 2 decimal places.

W.E. 19

5 Calculate the value of x in the following, rounded to 2 decimal places.

(a)

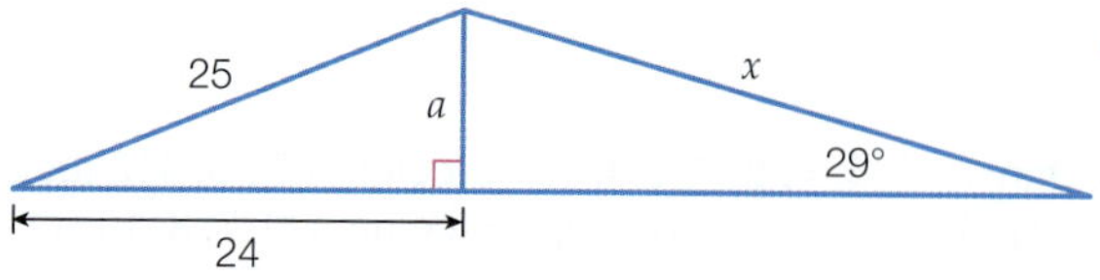

(b)

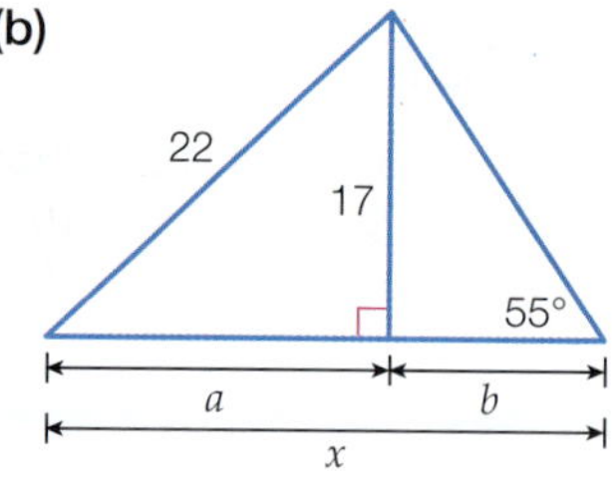

(c)

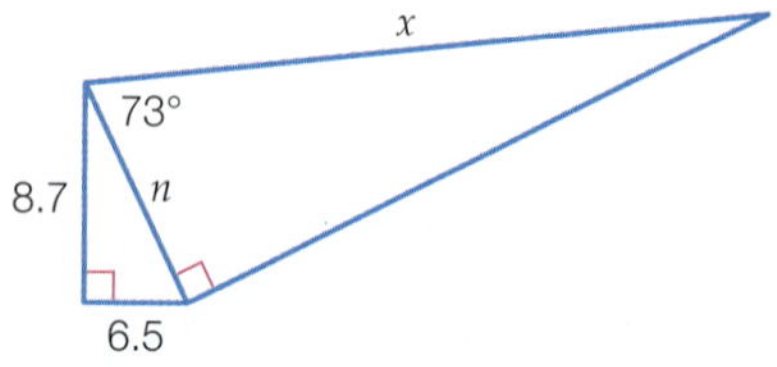

(d)

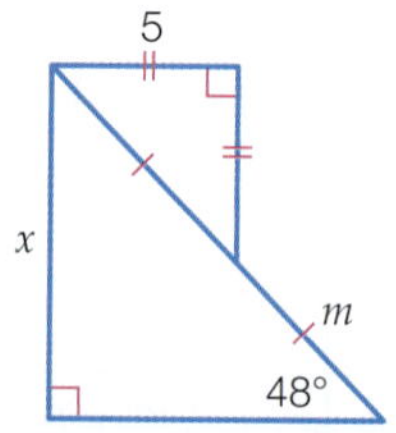

6 Find the values of the pronumerals. Round your answers to 1 decimal place.

(a)

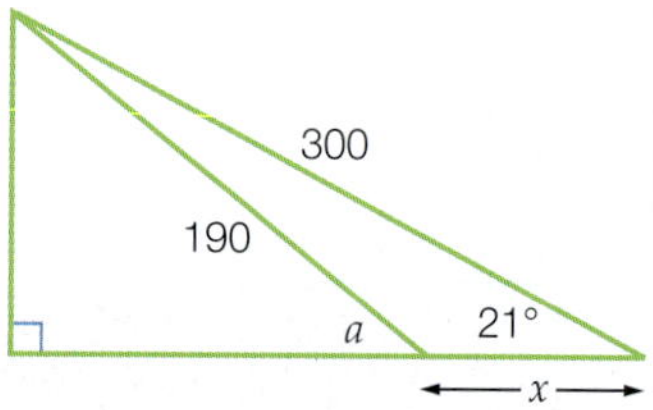

(b)

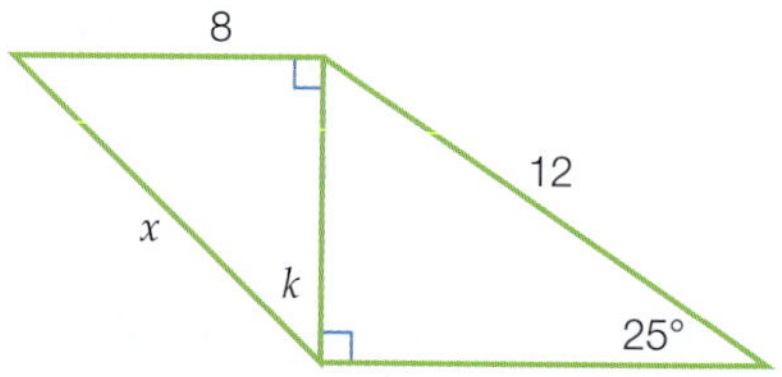

(c)

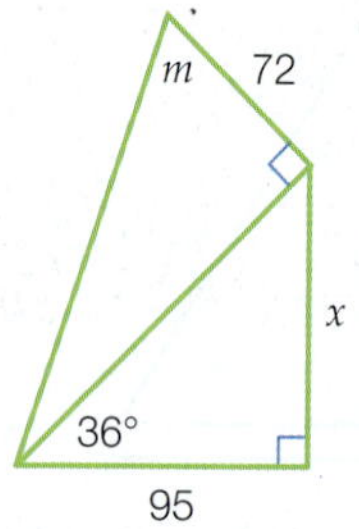

(d)

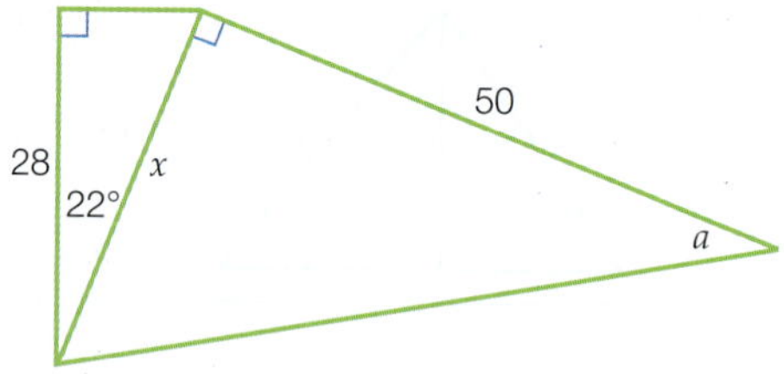

7 The length of x in the diagram is closest to:

A 15.56

B 17.96

C 31.11

D 42.04

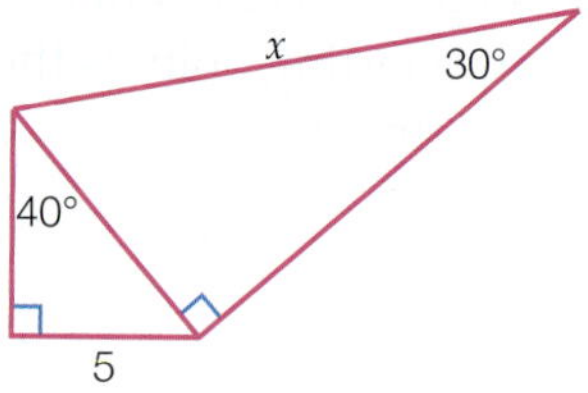

8 Andrew is building a table in the shape of a trapezium. He needs to know the angle shown, so he can adjust the angular saw machine to cut the wood at the correct angle. What is the value of θ? Round your answer to the nearest degree.

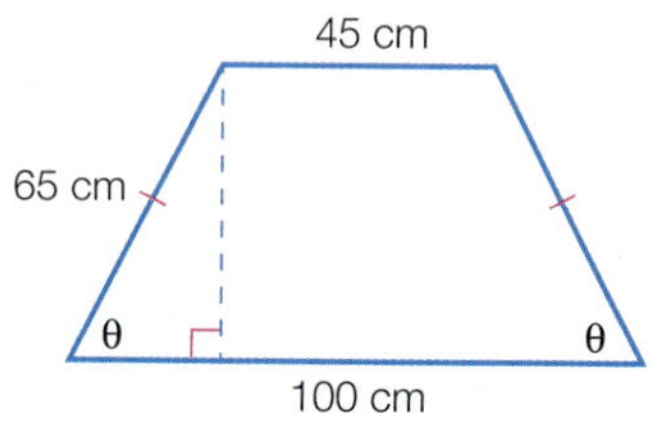

9 A wheelchair ramp is required to have an angle of elevation of only 15°. Does the following plan for a ramp comply with this requirement?

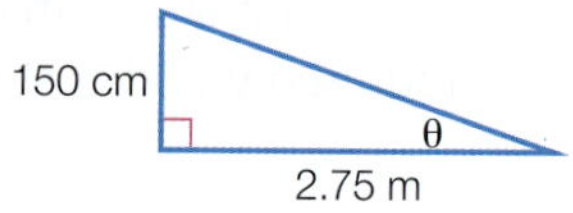

10 A slide in a playground has a vertical support and a ladder as shown in the diagram. The slide is 2.85 m long and makes an angle of 57°30′ with the vertical support. If the base of the ladder is 80 cm away from the support, the length of the ladder, to 2 decimal places, is:

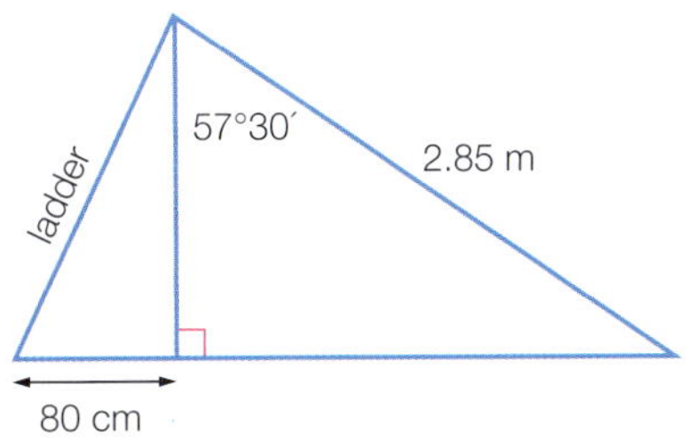

A 0.37 m B 0.42 m

C 1.53 m D 1.73 m

Understanding

11 A farm whose perimeter shape is shown requires wire fencing costing $5 per metre. Calculate the total cost of fencing, to the nearest dollar.

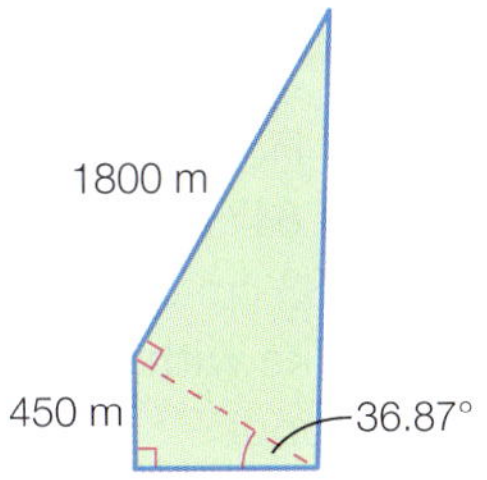

12 A red tile and a green tile are to be joined as part of a mosaic design. What angle θ is required? Write your answer in degrees and minutes.

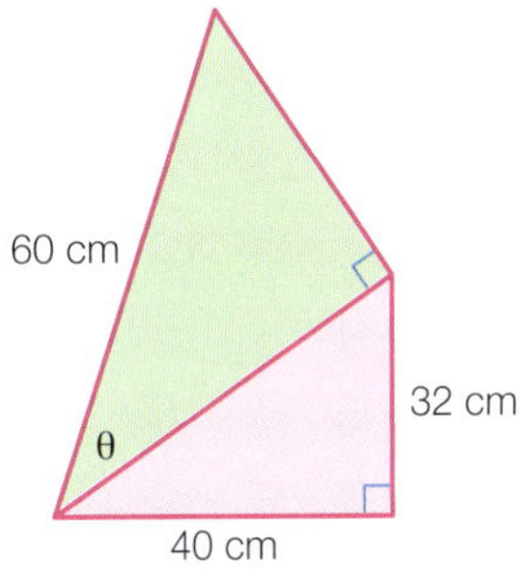

13 A surveyor is hired to peg out the garden shown. Shape *ABC* is an isosceles triangle and the fountain is positioned midway between *A* and *C*. To the nearest metre, what is the length of the direct path from the fountain to the greenhouse?

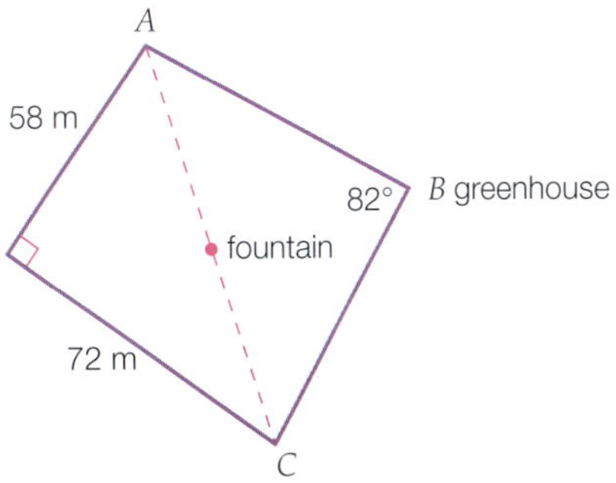

14 Miguel left Base camp and trekked to Camp 1, then to Camp 2 and finally back to Base camp. To the nearest kilometre, how far did he walk in total?

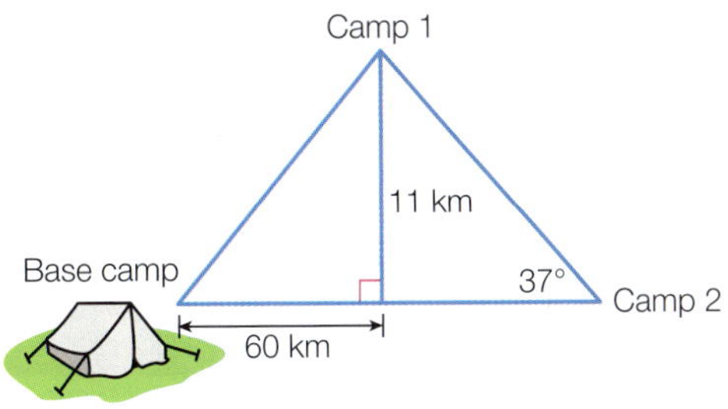

15 When Liz is 70 m away from a building she notices a stuntman at an angle of elevation of 58° from her horizontal line of sight. He is climbing up the side of the building. What will be the angle of elevation after the stuntman has climbed a further 20 m? Round your answer to the nearest degree.

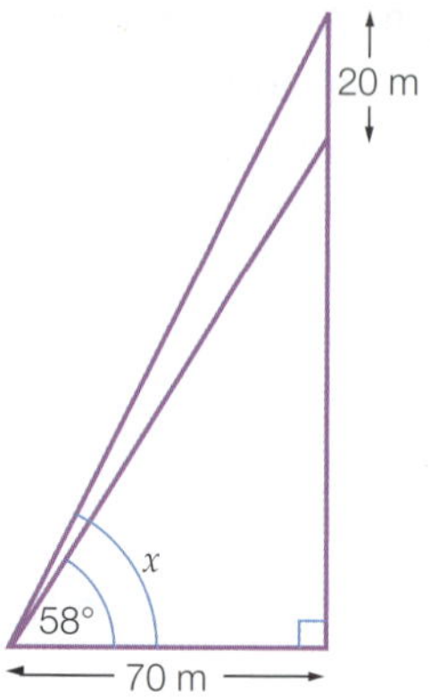

16 A 4.5 m ladder leaning against a wall makes an angle of 64°5′ with the ground. If the ladder is moved 83 cm down the wall, what angle does it now make with the ground in degrees, minutes and seconds?

17 An aircraft is coming in to land. The pilot views two lights positioned in a line on the runway at angles of depression of 18° and 22°. If the aircraft is 300 m above the runway, find the distance between the two lights. Round your answer to 2 decimal places.

Reasoning

18 An envelope has a length of 12 cm and a width of 8 cm. The back flap is an isosceles triangle with an edge that makes an angle of 35° with the top of the envelope. Find the distance between the bottom of the flap and the bottom of the envelope. Round your answer to 2 decimal places.

Hint

Draw a vertical line from the bottom of the flap to the top of the envelope.

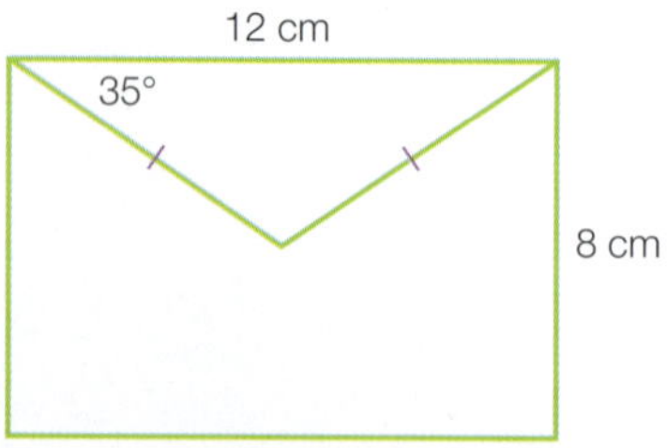

19 A chord 15 mm long forms an angle of 126° at the centre of a circle, as shown. Find the length of the diameter of the circle, correct to 2 decimal places.

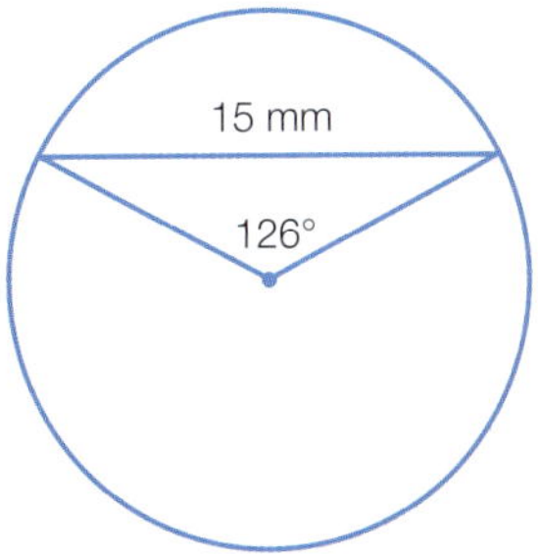

20 A cable car is supported by two wires attached to a horizontal cable. One of the wires is 5.4 m long and makes an angle of 43°20′ with the horizontal cable. The other wire makes an angle of 31°45′ with the cable. If the cable car is 2 m long, how far apart along the cable are the wires, correct to 2 decimal places?

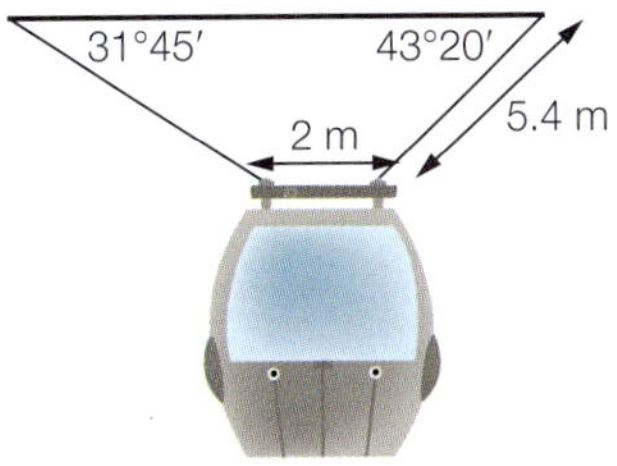

21 A jeweller has designed a medal in the shape of a circle mounted on a regular hexagon. The circular inset is plated with gold. The medallion will hang by a chain passing through the triangular shape shown at the top of the medal.

(a) Use the triangle marked on the medal to find the radius of the circular inset.

(b) Use your answer to part **(a)** to find the area of gold required to make the circular inset. Round your answer to the nearest square centimetre.

(c) What length of metal is required for the two sides of the triangular hanging top? Round your answer to 2 decimal places.

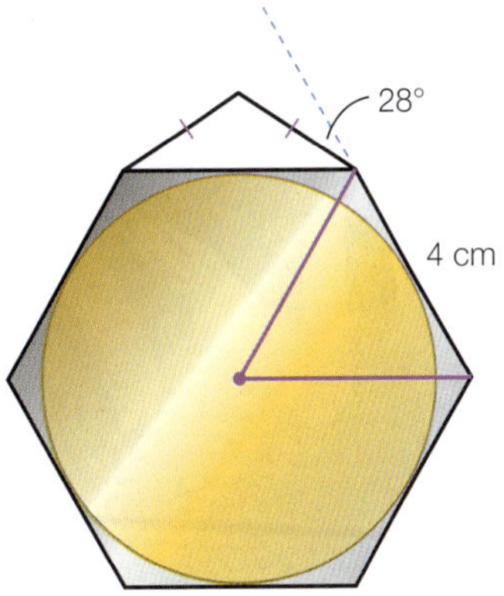

Hint

The sum of the interior angles of an n-sided polygon is $(n - 2) \times 180°$.

Open-ended

22 Determine values for a, b, c, d, θ and ϕ so that $a > b$ and $c > d$.

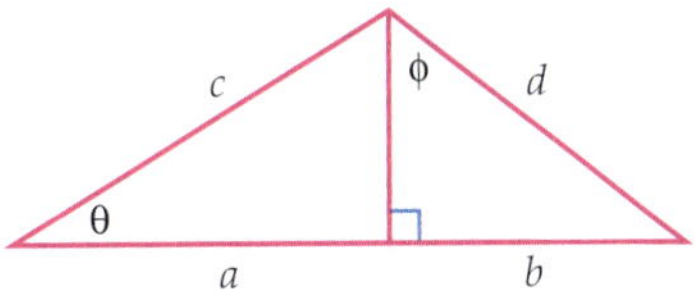

23 The cross-section of a roof (not necessarily symmetrical in shape) of height 5 m is shown. Determine values for x, y and θ so that the base has a length between 20 m and 25 m, and θ is greater than 100°.

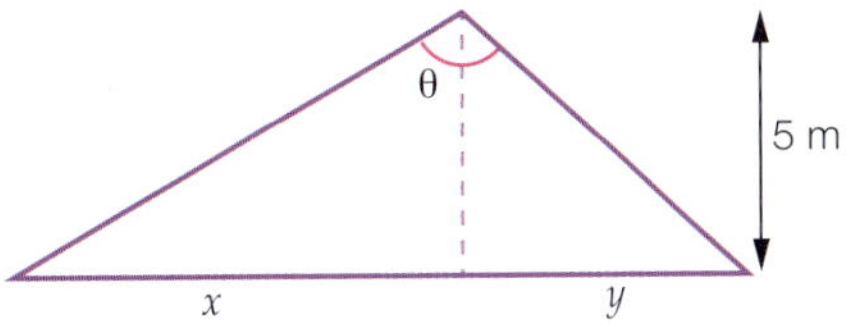

Maths 4 Real

NAVIGATION

THE DEGREE SYSTEM VS THE MILS SYSTEM

The compass that most of us are familiar with has the cardinal directions of north, south, east, west and angles given in degrees. However, this is not the only type of compass. Military forces throughout the world commonly use a compass that measures in 'mils' rather than degrees. The *mils system* gives greater accuracy in directions and makes it easier to estimate distances.

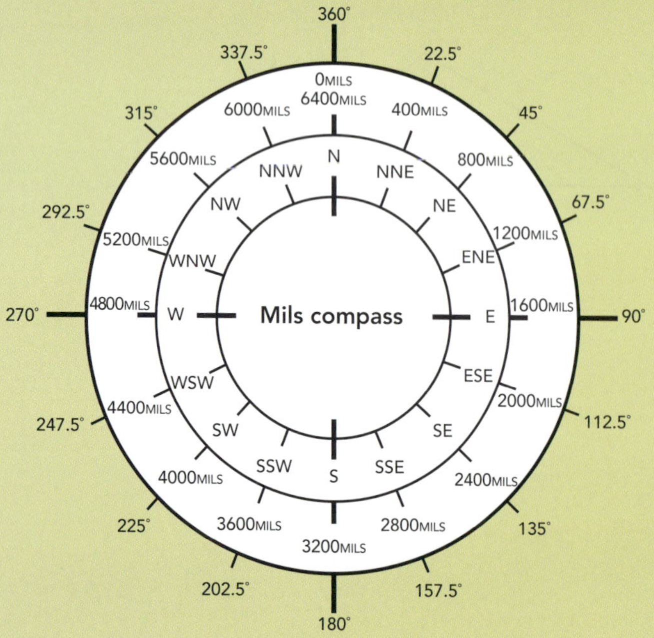

There are 6400 mils in a full 360° degree rotation.

> To convert from mils to degrees: multiply by $\frac{360}{6400}$
>
> To convert from degrees to mils: multiply by $\frac{6400}{360}$

1 Convert the following degrees to mils (round to 1 decimal place where necessary).

(a) 180° **(b)** 225° **(c)** 165°

2 Convert the following mils to degrees (round to 1 decimal place where necessary).

(a) 5400 mils **(b)** 900 mils **(c)** 40 mils

3 **(a)** How many mils are there in an anti-clockwise turn between south-east and north-east?

(b) How many mils are there in a clockwise turn between south and north-west?

(c) What is the direction in mils of the north arrow?

GRID BEARING VS MAGNETIC BEARING

The position of the magnetic north pole is not fixed but moves slowly, due to changes in Earth's molten iron core. It is important that navigators take this into account when using a compass and a map, as the compass may point in a different direction (**magnetic north**) than the map (**grid north**).

There are three different 'norths':

Magnetic north (MN)—the direction the compass points; towards the Earth's magnetic North Pole

Grid north (GN)—the direction indicated by the vertical grid lines on a map

True north (TN)—the direction to the Earth's geographic North Pole from the location

Other important words when describing direction are:

Grid magnetic angle—the angle between grid north and magnetic north

Grid bearing (GB)—a bearing measured on a map

Magnetic bearing (MB)—a bearing measured by a compass

Magnetic north can have an easterly or westerly variation from true north depending on where you are located. This variation is called the grid magnetic angle. It ranges from about 5° west of true north in parts of Western Australia to about 15° east of true north in eastern Australia. The line through the map below indicates where the magnetic variation changes from a westerly variation to an easterly variation.

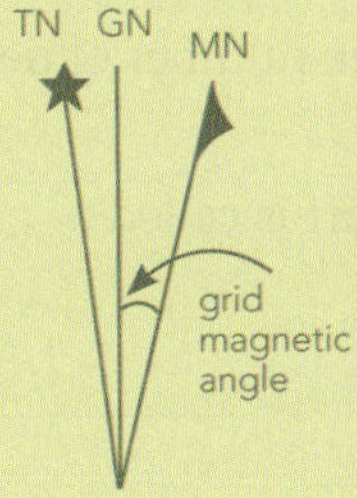

4 Use the following sample map information to answer the questions below.

Sample topographical map information for a location with easterly variation

Grid magnetic angle = 129 mils (8.4°) in 2010.

Magnetic north is correct for 2010 and moves easterly by 2 mils every 6 years.

(a) By how many mils will the grid magnetic angle change each year?

(b) What is the grid magnetic angle for 2013?

(c) What will be the grid magnetic angle for 2023?

It is important that navigators are able to convert from magnetic bearings to grid bearings in order to synchronise their use of maps and compasses.

For a location with an *easterly* magnetic variation:

To convert from *grid bearing to magnetic bearing* subtract the grid magnetic angle

GB → MB = GM − grid magnetic angle

To convert from *magnetic bearing* to *grid bearing* add the grid magnetic angle

MB → GB = MB + grid magnetic angle

For Questions **5** and **6** use the map information in Question **4** and assume the year is 2013.

5 Convert the following grid bearings to magnetic bearings.

(a) 5460 mils **(b)** 4360 mils

(c) 0200 mils **(d)** 0020 mils

6 Convert the following magnetic bearings to grid bearings.

(a) 0300 mils **(b)** 3245 mils

(c) 1834 mils **(d)** 6300 mils

Chapter review 6

Maths literacy

adjacent side	compass bearing	minute	sine	true bearing
angle of depression	cosine	opposite side	tangent	
angle of elevation	hypotenuse	second	trigonometric ratio	

Copy and complete the following using words and phrases from the list where, appropriate. A word or phrase may be used more than once.

1 The ___________ ratio is the adjacent side divided by the hypotenuse.

2 The ___________ ratio is the opposite side divided by the hypotenuse.

3 The ___________ ratio is the opposite side divided by the adjacent side.

4 When a direction is given with a measurement east or west from north or south, this is a ___________.

5 A ___________ always has three digits and shows the number of degrees from north in a clockwise direction.

6 One degree is divided into sixty equal parts. Each part is called a ___________.

7 There are 360 ___________s in one degree.

8 The angle from the horizontal as you look up is called an ___________. The angle from the horizontal as you look down is called an ___________.

Fluency

1 Use a calculator to find the following values, rounded to 4 decimal places. **6.1**

(a) sin (35°) (b) tan (51.3°) (c) cos (6.045°)

(d) tan (29°45′) (e) sin (16°24′) (f) cos (73°6′40″)

2 Find the value of k in each of these triangles, rounded to 2 decimal places. **6.2**

(a)

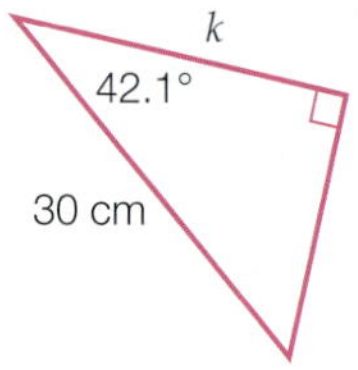

(b)

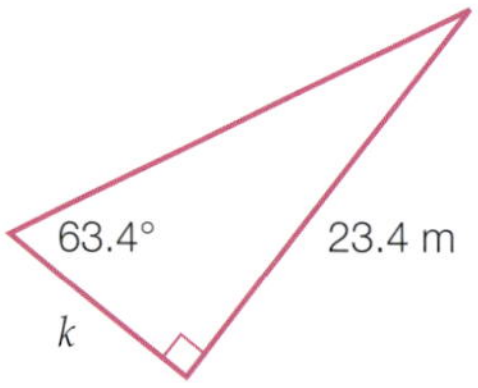

3 For each of the following trigonometric ratios, find the value of θ in: **6.3**

(i) decimal degrees, rounded to 2 decimal places

(ii) degrees, minutes and seconds.

(a) $\cos(\theta) = 0.74$ (b) $\tan(\theta) = 2.645$ (c) $\sin(\theta) = 0.4609$

4 An antenna of height 5 m is on top of the edge of a building. The base of the building is 100 m away from a drain on the road. The angle of elevation of the top of the antenna from the drain is 34°. Calculate the height of the building, rounded to 2 decimal places. **6.4**

5 An aircraft flies due north for 70 km, then due west for a further 145 km. What is the compass bearing of the aircraft from its starting point? Round your answer to the nearest degree. **6.5**

6 Find the values of the pronumerals in this diagram. Round your answers to 1 decimal place. **6.6**

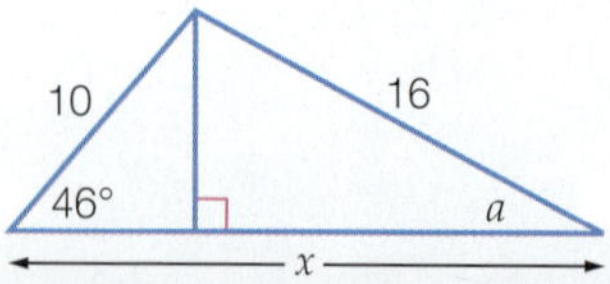

7 The sculpture, 'A study in triangles', is supported by a rigid steel rod. What is the length of the steel support? Round your answer to the nearest centimetre. **6.6**

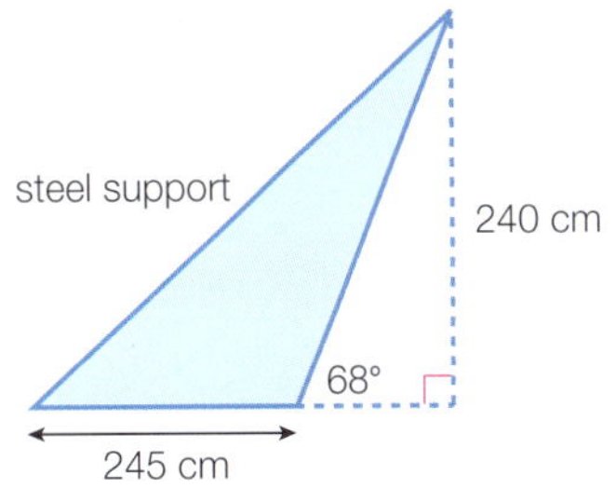

8 A conveyor belt in an underground mine is 650 m long and moves to the surface at an angle of 14°48′ from the horizontal. The depth of the mine is: **6.6**

A 166.04 m B 672.30 m C 628.44 m D 2544.57 m

9 The length of the diagonal in this rectangle is: **6.6**

A 13.25 m B 29.48 m

C 40.01 m D 47.18 m

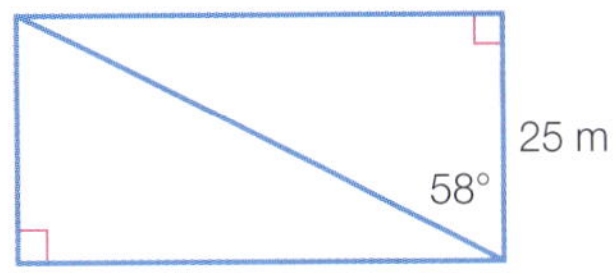

Understanding

10 A balloon on the end of a string attached to the ground is being blown by the wind. If the balloon is 12.75 m above the ground and the string makes an angle of 48° to the horizontal, find the length of the string, rounded to 2 decimal places. **6.2**

11 A ramp is 9.4 m long and joins two levels, one 3.1 m higher than the other. What angle does the ramp make with the lower level? Round your answer to 1 decimal place. **6.3**

12 A lifeguard on watch at a beach sees a shark at an angle of depression of 9°. If the lifeguard sees the shark from a height (to eye-level) of 16 m above sea-level, find the horizontal distance between the lifeguard and the shark to the nearest metre. **6.4**

13 A speedboat travels 26 km on a true bearing of 065°T. Find how far the boat is to the east of its starting point. Round your answer to 2 decimal places. **6.5**

14 An aircraft leaves an airport and flies 35 km in the direction N70°E. It then flies due east for a further 38 km. What is its distance and compass bearing from the airport? Round the distance to the nearest km and give the angle in degrees and minutes. **6.2, 6.5**

15 Decide whether the wedge shown will fit exactly into the cavity in the block of wood. Give reasons for your answer. **6.2, 6.3**

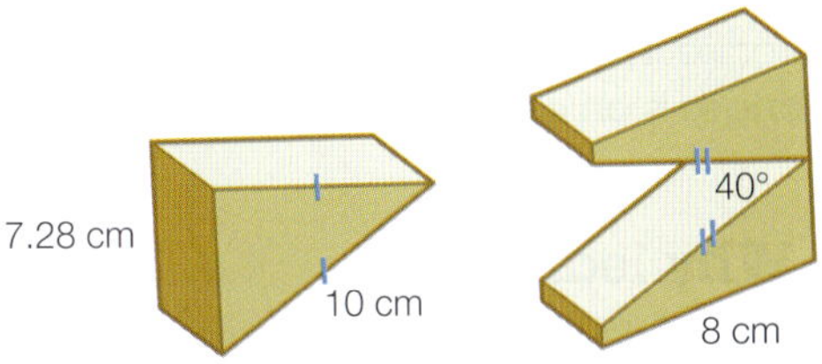

Reasoning

16 Amos leaves home and cycles for a distance of 35 km on a bearing of N30°E. He changes direction and cycles in a direction of S60°E before heading home on a bearing of S49°W. To the nearest kilometre, find the total distance Amos travelled. **6.5**

17 For an equilateral triangle, explain two different ways the height can be found using trigonometric ratios. **6.2, 6.3**

18 A case is to be constructed to hold a pendulum. The shaft of the pendulum is connected to the centre of the bob and is 67.5 cm in length. The diameter of the bob is 5 cm. If the pendulum swings through an angle of 25°, what must the minimum width of the case be? Round your answer to 2 decimal places. **6.2, 6.3**

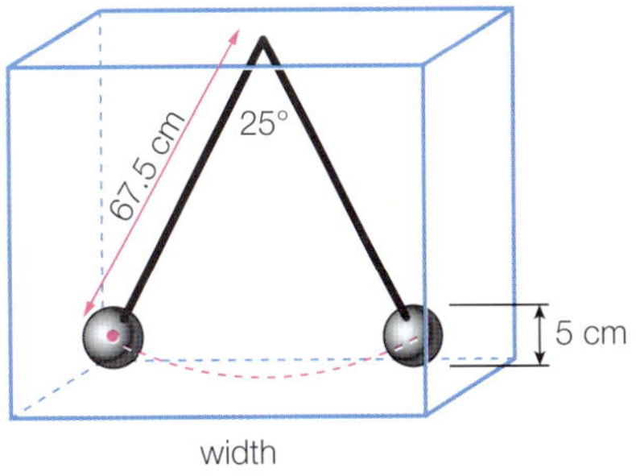

7

Advanced trigonometry

10A

In the real world not everything is parallel

Name: Michele Russo
Job: Surveyor
Qualifications: Bachelor of Geometric Engineering (Hons)

Surveyors are the 'police' of the geospatial world. Armed with technology such as GPS and a theodolite, and skills in bearings, trigonometry and triangulation, they separate land into different properties. Taking precise measurements within certain set error ranges is an essential skill. People like Michele Russo, a graduate surveyor with the Department of Environment and Sustainability, help to resolve disputes between neighbours about the ownership of land.

'As a surveyor, my job is to re-establish title boundaries on the ground. Whenever a development goes through, such as a big subdivision, I go out and locate existing boundaries using 3D measurements. In the real world not everything is parallel. We need to check how well a parcel of land fits in with the surrounding land and the accuracy of that.'

Why learn this?

The world is not flat. We live in a three-dimensional world and need to be able to calculate measurements that cannot be directly measured. Trigonometry gives you the skills to determine angles, side lengths and areas of all types of triangles, and is used extensively by engineers, surveyors and scientists. Tides and daily temperatures follow cyclic or repeating patterns that can be modelled with trigonometric graphs. The solutions to trigonometric equations can give vital information about situations with repeating patterns, from engine performance to quantum mechanics.

After completing this chapter you will be able to:

- use trigonometry to solve three-dimensional problems
- use the unit circle to define trigonometric functions and equivalent (coterminal) angles
- understand the periodic nature of trigonometric functions
- graph and solve trigonometric equations, with and without technology
- use the sine and cosine rules to find unknown values in triangles
- apply trigonometry to finding the area of a triangle.

7 Recall

Prepare for this chapter by attempting the following questions. If you have difficulty with a question, you can download a Recall Worksheet from the eBook or the Pearson Places website.

1 Calculate the value of x and round to 2 decimal places.

(a)

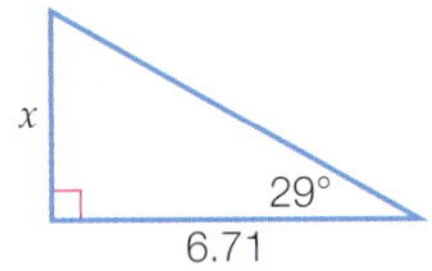

(b)

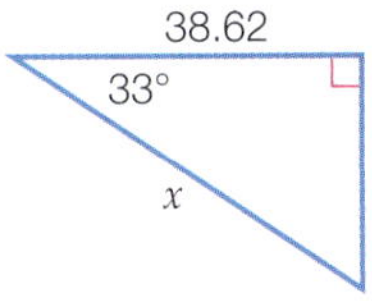

(c)

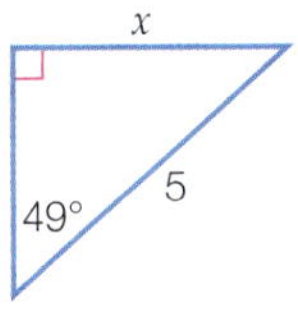

2 Determine the size of the angle θ, in degrees and minutes.

(a)

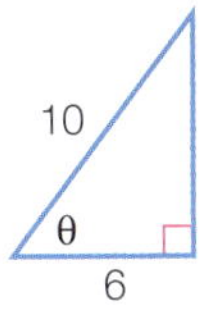

(b)

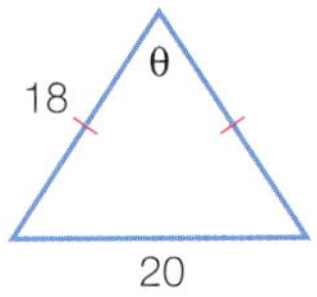

(c)

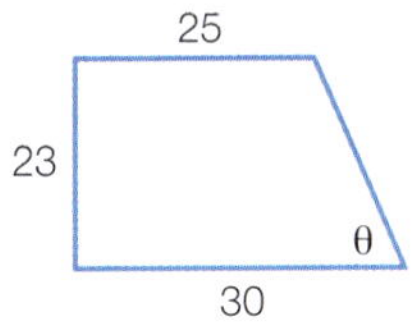

3 Find the length of x. Round your answer to 2 decimal places.

(a)

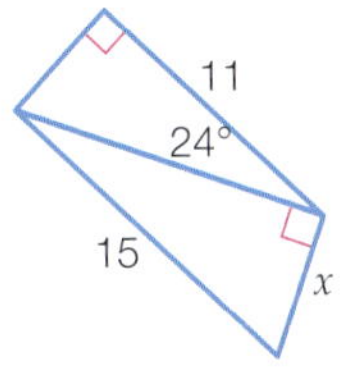

(b)

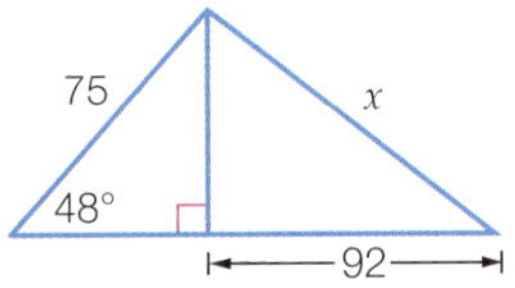

4 Jake is flying a kite on the end of a 20 m rope. If the rope has an angle of elevation from Jake's line of sight of 69°, how high is the kite above the ground, given the distance from the ground to Jake's eye-level is 1.65 m? Round your answer to 2 decimal places.

5 A hiker leaves her campsite and walks 10 km on a bearing of 143°T, then walks 6 km on a bearing of 039°T. Calculate the following and round your answers to 2 decimal places.

(a) How far is she from her campsite?

(b) What bearing must she take to walk directly back to the camp?

Exploration Task

You can download this activity from the eBook or the Pearson Places website.

Patterns and limits in trigonometry

In this activity you will explore the patterns and limitations of different values of sine, cosine and tangent ratios for angles of all sizes, including negative angles and angles greater than 360°.

7.1 Solving three-dimensional problems

A three-dimensional problem involving right-angled triangles may require the use of any combination of the sine, cosine and tangent ratios, and Pythagoras' theorem. A diagram drawn to appear three-dimensional (3D) has right-angled triangles drawn as two-dimensional (2D) shapes. Note that right angles may not appear as exactly 90° on your page in a 3D representation.

Consider this cube:

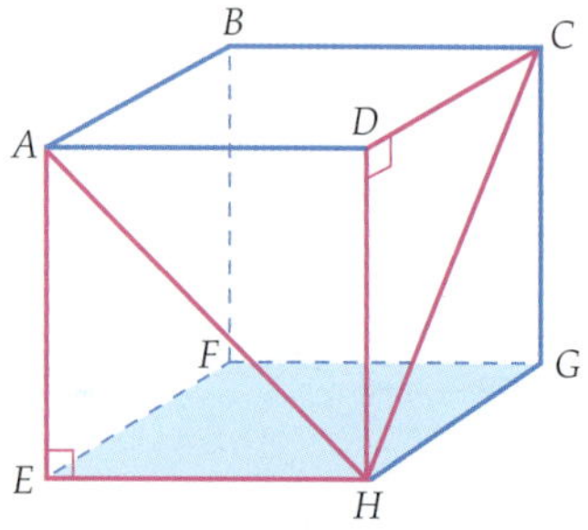

You can find right-angled triangles on the faces of the cube such as ΔAEH and ΔCDH quite easily.

10A

Within the cube you can locate right-angled triangles such as ΔAEG and ΔDBF, where the hypotenuse of the triangle is the diagonal connecting opposite vertices of the cube. (Can you find others?)

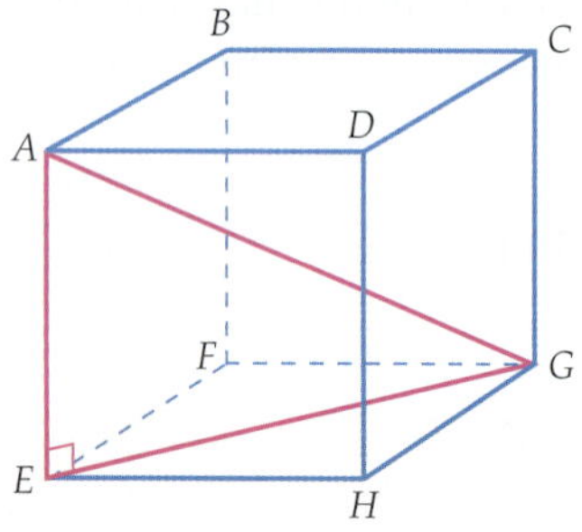

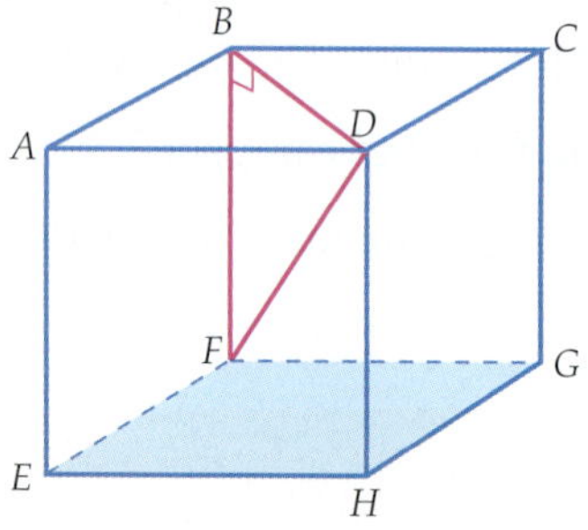

Always redraw the right-angled triangle you need to solve as a separate two-dimensional diagram.

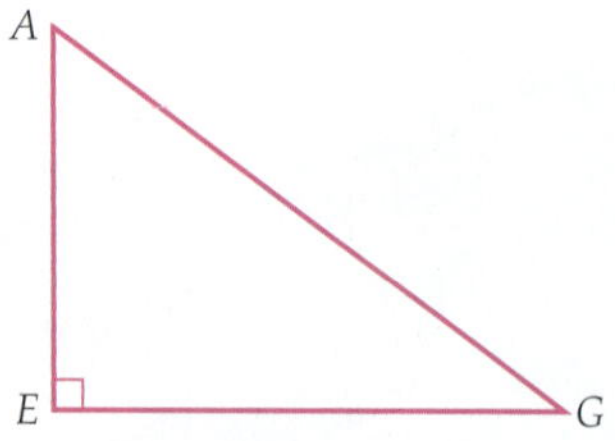

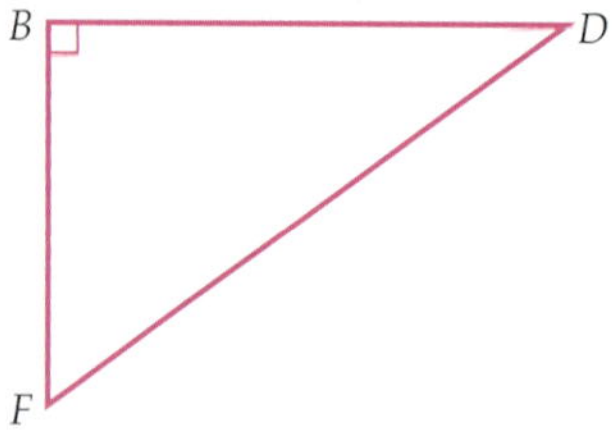

You may need to solve more than one right-angled triangle in the three-dimensional diagram to find the unknown you require.

Worked example 1

W.E. 1

A right pyramid *ABCDE* stands on a square base of edge length 10 cm. The sloping edges are each of length 15 cm. Note that in a right pyramid the vertex is directly above the middle of the base, so *AF* is the perpendicular height.

(a) Find the angle that a sloping edge makes with the base of the pyramid, rounded to the nearest degree.

(b) Hence, find the height of the pyramid, rounded to 2 decimal places.

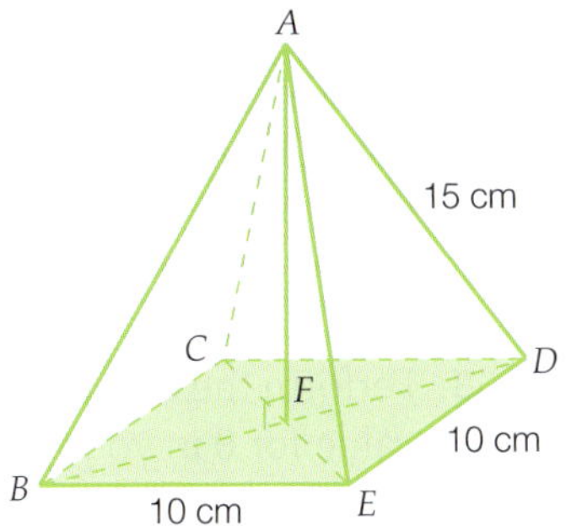

Thinking	Working
(a) **1** Let θ be the angle required. Choose a triangle to work with. Use any of the four possible triangles. (Here, use ΔADF to find θ.) Redraw this triangle, labelling known and unknown values.	(a) 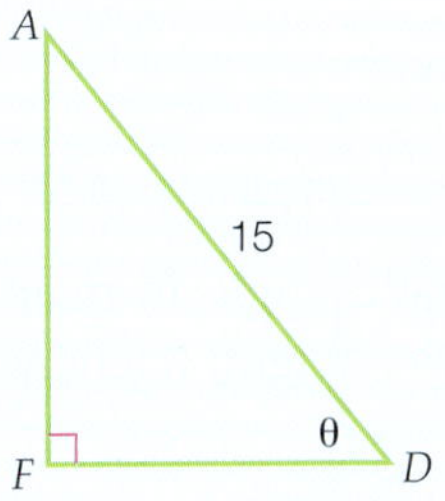
2 Identify which values need to be found first. (Here, you need to find *DF* before θ.) Redraw another triangle that includes the unknown value. (Here, use ΔBDE to find *BD* as *DF* is half of *BD*.)	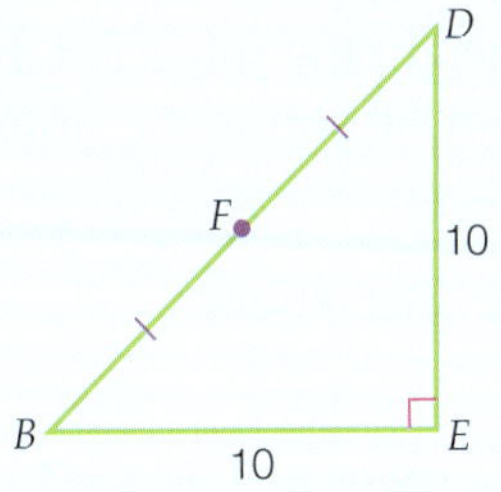
3 Use Pythagoras' theorem to find the unknown side.	$(BD)^2 = 10^2 + 10^2$ $(BD)^2 = 200$ $BD = \sqrt{200}$
4 Use the side length found in the second triangle to find the unknown side length in the first. (Here, *DF* is half of *BD*.) Store the unrounded answer in your calculator.	$DF = \dfrac{\sqrt{200}}{2}$
5 Using the first triangle (ΔADF), write the known and unknown values then identify which trigonometric ratio connects them. (Here, H = *AD* = 15 cm and A = $DF = \dfrac{\sqrt{200}}{2}$ cm.)	In ΔADF: H = 15 $A = \dfrac{\sqrt{200}}{2}$ $\theta = ?$ $\cos(\theta) = \dfrac{A}{H}$

10A

6 Form an equation by substituting values into the ratio.	$\cos(\theta) = \frac{DF}{15}$
7 Use the inverse of the ratio to find θ and substitute the stored unrounded value of the unknown length into the formula. (Here, substitute for *DF*.)	$\theta = \cos^{-1}\left(\frac{DF}{15}\right)$ $\theta = \cos^{-1}\left(\frac{\sqrt{200}}{2 \times 15}\right)$
8 Perform the calculation and store the value for θ in your calculator's memory.	$\theta \approx 61.874\,494\,3$
9 Round the answer to the nearest degree.	$\theta = 62°$
10 State the answer.	A sloping edge makes an angle of about 62° with the base of the pyramid.
(b) 1 Using your first triangle, form an equation for the height. (Here, you need *AF*.)	(b) $\sin(\theta) = \frac{O}{H}$ $\sin(\theta) = \frac{AF}{15}$
2 Make the height the subject of the equation and substitute the stored value for θ.	$AF = 15\sin(\theta)$
3 Perform the calculation.	$AF = 15\sin(61.874\,494\,3°)$ $AF = 13.228\,756\,56$
4 Round the answer to 2 decimal places.	$AF = 13.23$ (2 d.p.)
5 State the answer.	The height of the pyramid is about 13.23 cm.

To solve a three-dimensional problem involving right-angled triangles:

1. Draw a diagram that represents the problem.
2. Identify right-angled triangles that contain the unknown you want or that will help you find it.
3. Redraw these right-angled triangles.
4. Solve these triangles using trigonometric ratios and/or Pythagoras' theorem.

10A

7.1 Solving three-dimensional problems

Navigator

1, 2, 3, 4, 5, 6, 8, 10, 11, 13, 15, 17, 18, 19, 21

1, 2, 3, 4, 5, 6, 7, 8, 10, 11, 12, 13, 14, 15, 17, 18, 19, 20, 21

1, 3, 5, 6, 7, 9, 10, 11, 12, 14, 15, 16, 17, 18, 19, 20, 21, 22, 23

Answers p. 813

Fluency

1 A rectangular prism is shown at right.

(a) Calculate the length from A to B. Write your answer correct to the nearest whole number.

(b) Using your answer from part **(a)** calculate the length from A to C. Write your answer correct to the nearest whole number.

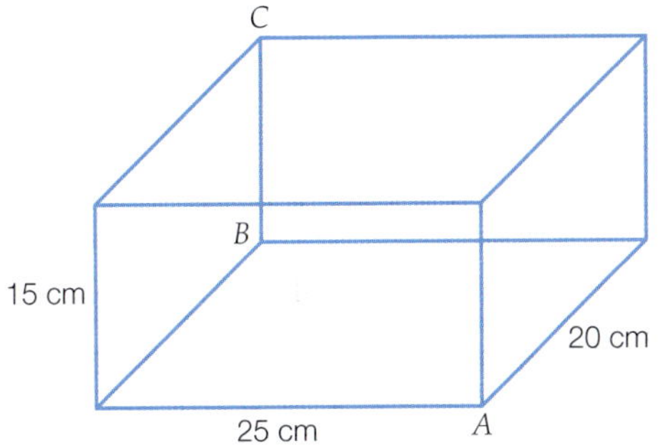

2 A cube with a side length of 7 cm is shown at right.

(a) (i) Calculate the exact length from X to Y.

(ii) Use your answer from **(i)** to calculate the length from X to Z. Write your answer correct to the nearest whole number.

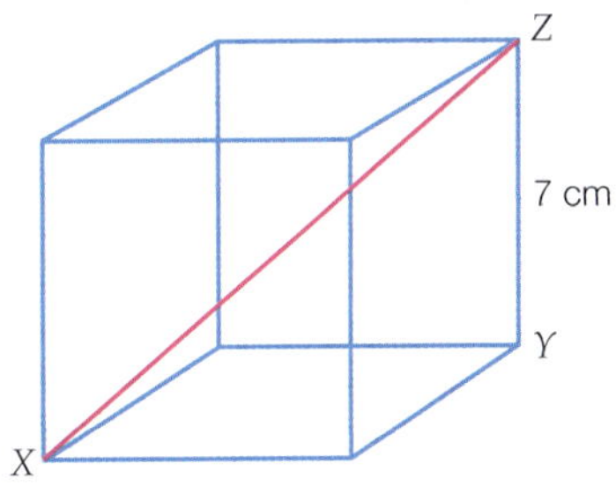

(b) Use your answers from **(a)** to calculate the angle the line XZ makes with the line XY. Write your answer to the nearest whole degree.

3 A right pyramid $ABCDE$ stands on a square base of edge length 8 cm. The sloping edges are each 14 cm long. AF is the perpendicular height.

W.E. 1

(a) Find the angle that a sloping edge makes with the base of the pyramid, rounded to the nearest degree.

(b) Hence, find the height of the pyramid to 2 decimal places.

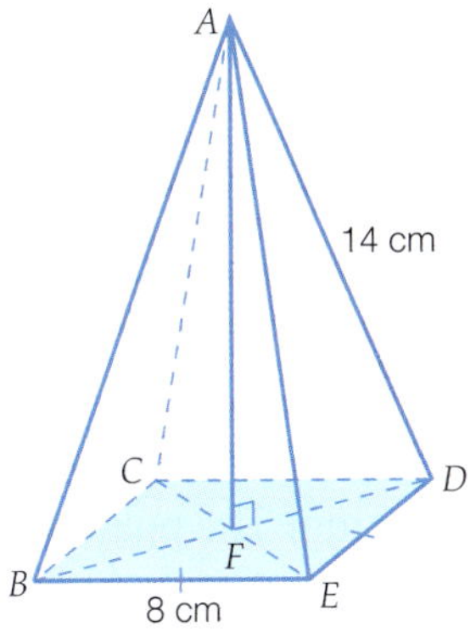

10A

4 Buffy has a pencil tin as shown. Will a 45 cm ruler fit diagonally from bottom to top?

5 Consider the wedge shown at right.

(a) Calculate the distance between *N* and *P*.

(b) Calculate the distance between *N* and *R*, correct to 1 decimal place.

(c) Calculate the height, *PR*, of the wedge. Write your answer correct to 1 decimal place.

(d) Calculate angle *QNM* correct to the nearest degree.

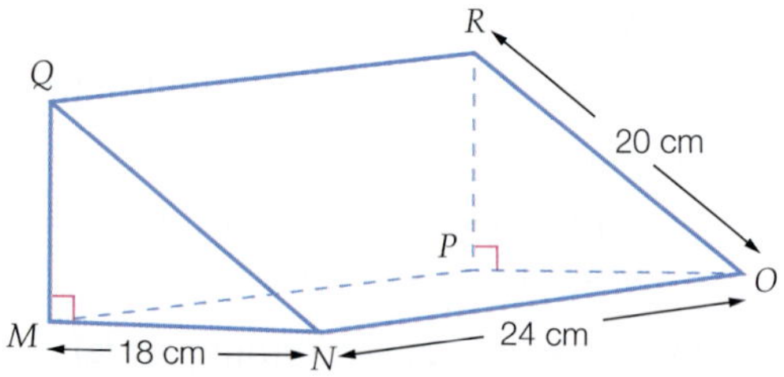

6 A tent has the dimensions shown in the diagram. Find the area of the groundsheet needed to fit inside the floor of the tent. Round your answer to 2 decimal places.

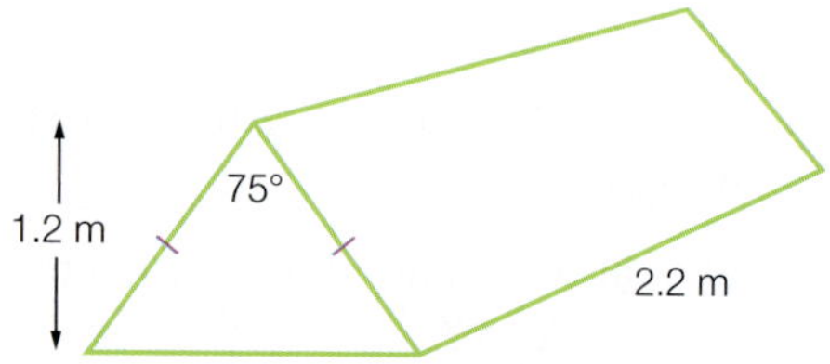

Understanding

7 (a) To find the angle that plane *ABFE* makes with plane *ABCD*, you could calculate:

A ∠*FCB* B ∠*BFC*

C ∠*DAB* D ∠*DAE*

(b) To find the length *AF*, you could use the triangle:

A *ABC* B *AEF* C *BCF* D *ABD*

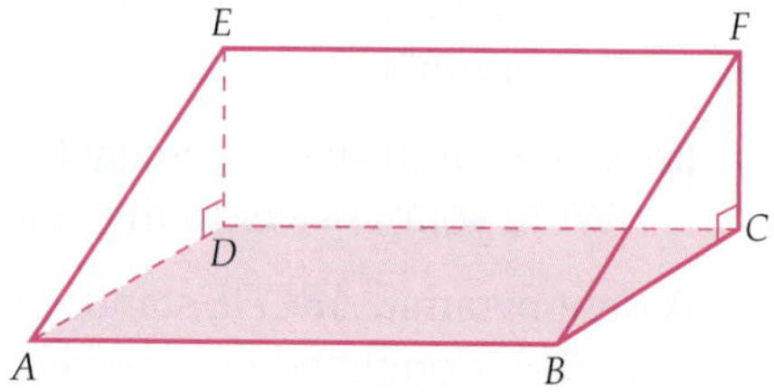

For Questions **8** to **15**, unless stated otherwise, round angles to 1 decimal place and round other answers to 2 decimal places.

8 *PQRST* is a square-based pyramid with a base of side length 14 mm and a height of 6 mm. The point *X* on the base is directly below *T* at the apex, and point *Y* is midway along *PQ*.

(a) Find the angle *TXY*.

(b) What angle does the sloping face *PQT* make with the base?

(c) Find the length of *TY*.

(d) Find the length of a sloping edge of the pyramid.

(e) Find the angle that a sloping edge makes with the base of the pyramid.

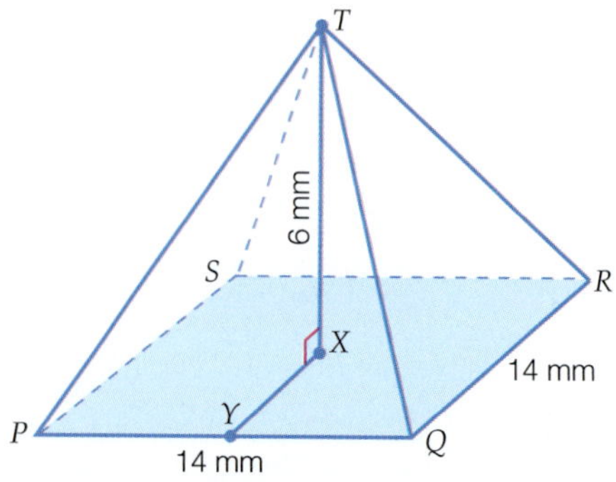

9 A new road sign that warns of merging traffic consists of a circle inside an equilateral triangle, as shown. What is the radius of the circle?

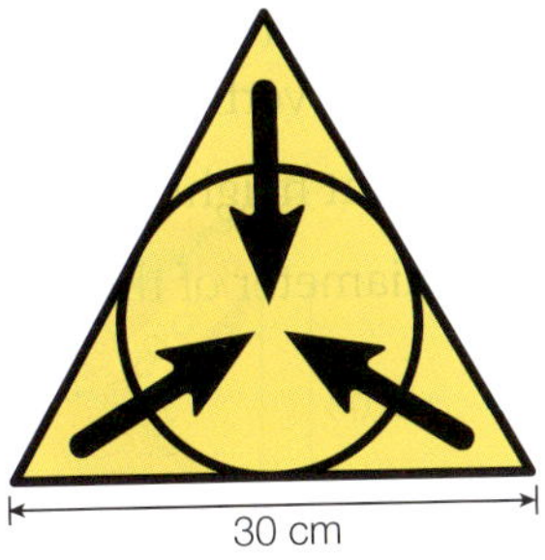

10 A hill $ABCD$ rises at an angle of 24° to the horizontal. The hill is 60 m high.

(a) If you walked straight up the hill along AB, what distance would you travel?

(b) What horizontal distance would you cover walking along AB?

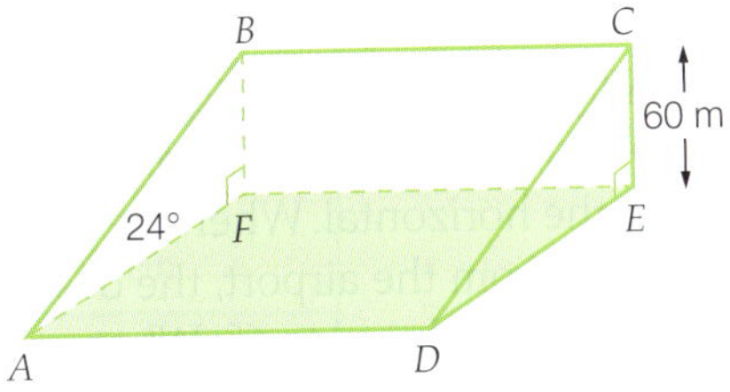

(c) Your friend walks along a track from A to C, taking 30 minutes to reach the top. Horizontally, the distance she covers is 510 m. How fast does she walk in m/min?

(d) Find the angle of the track AC to the horizontal.

(e) Gradient represents the steepness of a slope and can be calculated by using the formula $\frac{\text{rise}}{\text{run}}$ or $\frac{\text{vertical distance}}{\text{horizontal distance}}$. Calculate the gradients of AB and AC, and hence determine which track is steeper.

11 A plumber inserts a metal rod into a cylindrical pipe whose inner radius is 9 cm and discovers that there is a blockage in the centre of the inner circle of the pipe at the point shown.

(a) What is the maximum length of the rod that can be inserted inside the pipe to reach the blockage?

(b) What is the maximum angle θ possible? Give your answer in degrees, minutes and seconds.

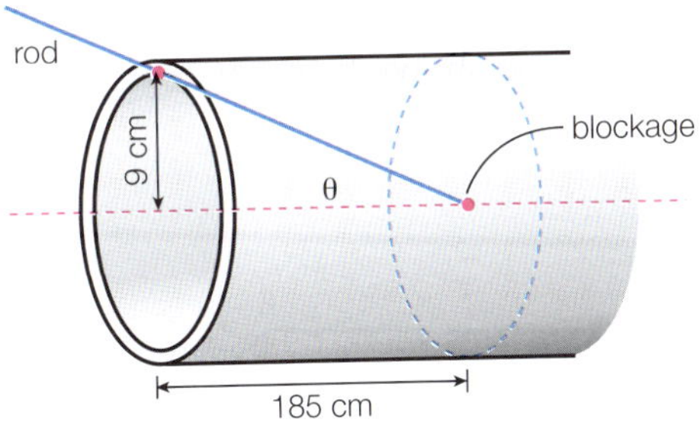

12 Val is 50 m due north of a communications tower. The top of the tower is at an angle of elevation of 41° from the ground at Val's observation point. From Con's observation point, due east of the tower, the angle of elevation of the top of the tower is 29°.

Find:

(a) the height of the tower

(b) the distance from Con to the foot of the tower

(c) the distance between Val and Con

(d) Con's true bearing from Val's observation point.

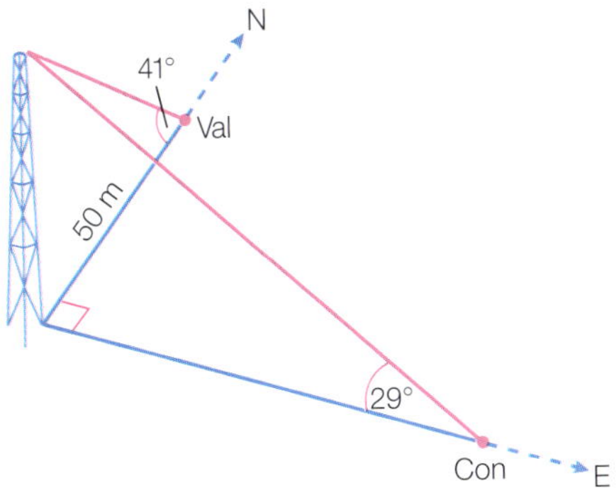

23 The cross-section of a roof is shown. a and b are the lengths of the roofing, and θ and ϕ are the angles each roofline makes with the horizontal.

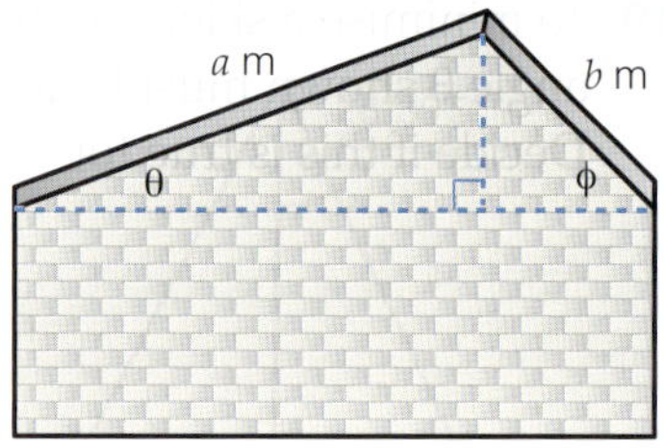

(a) Show that $\frac{a}{b} = \frac{\sin(\phi)}{\sin(\theta)}$.

(b) If $a = 2b$ and angles θ and ϕ must be greater than 20° and less than 60°, state one pair of possible values for the angles θ and ϕ. State your answer to the nearest degree.

Problem solving

The maths challenge

You are in one of two 'tribes' that are competing on a desert island. Both tribes are led to the edge of a deep, fast flowing river, where the weekly challenge is issued. Each tribe is given a large protractor, a 10 m long measuring tape, and some instructions, written on a piece of bark. You crowd around the bark with your fellow tribe members. It reads:

Calculate the height of the tree directly across the river from where you are standing. Use only the equipment you have been given, plus any stick, stone, leaf or vine from the jungle around you. The first team to determine the height of the tree accurately wins double food rations for the week.

P.S. Two triangles could be useful.

'That's impossible!' moan half the tribe members. 'We don't even know the width of the river!'

Describe the method and calculations your tribe could use to calculate the height of the tree and win the challenge.

Strategy options

- Make a model
- Act it out.
- Break problem into manageable parts.

Angles in quadrant 4

In the fourth quadrant, a right-angled triangle with an angle of θ measured clockwise from the positive direction of the x-axis will produce a reflex angle of $(360° - \theta)$ measured anticlockwise from the positive direction of the x-axis. This triangle is a reflection in the x-axis of the original triangle in the first quadrant. The two angles θ and $(360° - \theta)$ are equivalent angles that cut the circumference of the unit circle at (x, y) and $(x, -y)$ respectively.

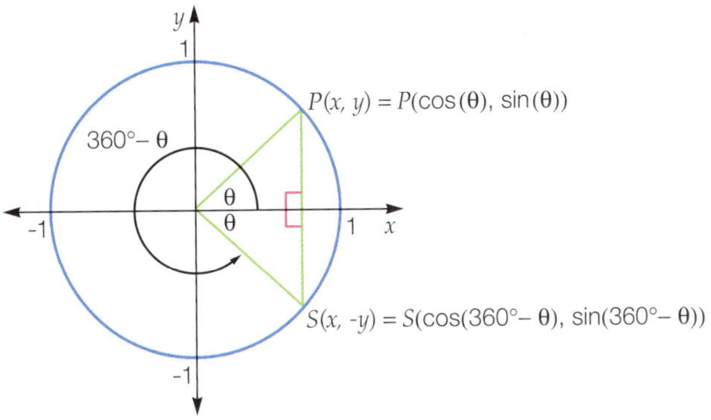

$$\sin(360° - \theta) = \frac{O}{H} = \frac{-y}{1} = -y = -\sin(\theta)$$

$$\cos(360° - \theta) = \frac{A}{H} = \frac{x}{1} = x = \cos(\theta)$$

$$\tan(360° - \theta) = \frac{O}{A} = \frac{-y}{x} = -\frac{y}{x} = -\tan(\theta)$$

In quadrant 4, only the cos ratio is positive.

$\cos(360° - \theta) = \cos(\theta)$ $\quad\sin(360° - \theta) = -\sin(\theta)$ $\quad\tan(360° - \theta) = -\tan(\theta)$

In the fourth quadrant, the angle θ varies from 270° to 360°.

As the angle changes, so do the values of $\sin(\theta)$, $\cos(\theta)$ and $\tan(\theta)$.

When θ is 360°, the point P will be back to its original position.

So $\sin(360°) = 0$, $\cos(360°) = 1$ and $\tan(360°) = 0$.

In quadrant 4, as θ increases from 270° to 360°:

- $\sin(\theta)$ increases from -1 to 0
- $\cos(\theta)$ increases from 0 to 1
- $\tan(\theta)$ is undefined at $\theta = 270°$, but then increases from a negative ratio to 0.

10A

Trigonometric ratios summary for angles in other quadrants

	Angle			
	0°	90°	180°	270°
sin (θ)	0	1	0	-1
cos (θ)	1	0	-1	0
tan (θ)	0	undefined	0	undefined

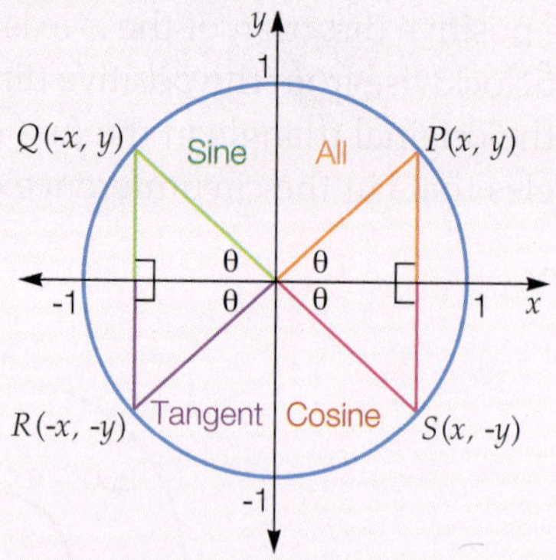

- In quadrant 1, all the ratios are positive as both x and y are positive. (All, A)
- In quadrant 2, only the sine ratio is positive. (Sine, S)
- In quadrant 3, only the tangent ratio is positive as both x and y are negative. (Tangent, T)
- In quadrant 4, only the cosine ratio is positive. (Cosine, C)

This can be remembered by using the diagram shown often called a CAST diagram. You can make up your own mnemonic (a word or phrase that you can use to help you remember).

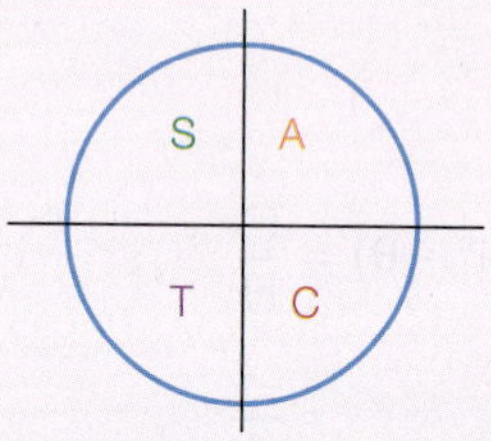

10A

Interactive

Angles and coordinates in the unit circle

Explore the trigonometric coordinates and angles of a point as it moves around the unit circle.

Go to the eBook or the Pearson Places website to access this interactive.

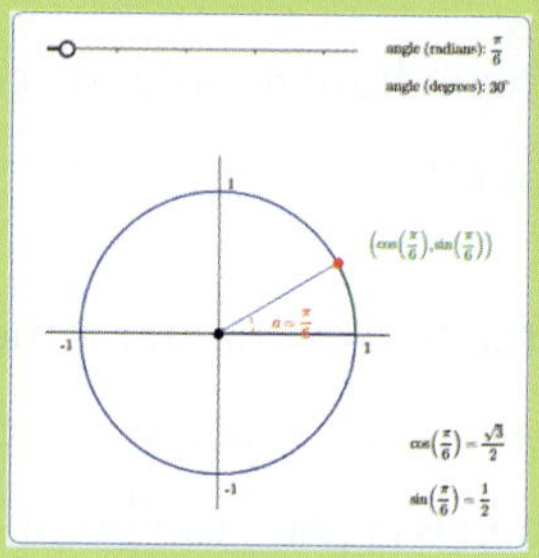

Worked example 2

W.E. 2

State the quadrant in which each of the following angles is found.

(a) 35° (b) 265° (c) 135° (d) 342°

Thinking	Working
(a) 35° is between 0° and 90°.	(a) quadrant 1
(b) 265° is between 180° and 270°.	(b) quadrant 3
(c) 135° is between 90° and 180°.	(c) quadrant 2
(d) 342° is between 270° and 360°.	(d) quadrant 4

Worked example 3

W.E. 3

For each of the following relationships, find the equivalent trigonometric ratio in quadrant 1, taking care to include the appropriate sign.

(a) tan (353°) **(b)** sin (124°) **(c)** cos (234°)

Thinking	Working
(a) 1 Draw a CAST diagram and locate the given angle. (Here, the angle is in quadrant 4.)	(a) 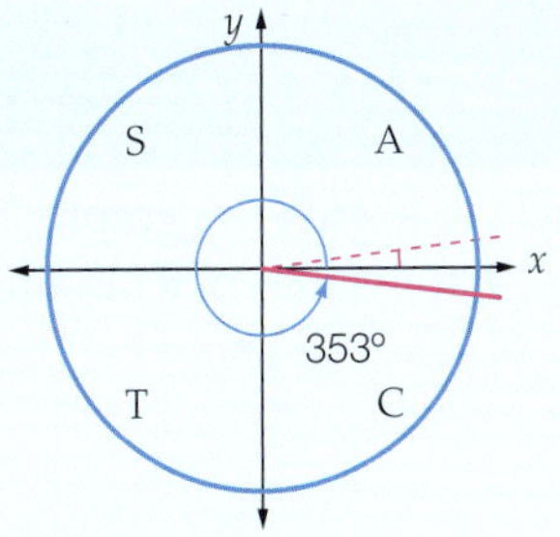
2 Find the equivalent angle in quadrant 1.	$(360 - 353)° = 7°$
3 Write the answer, taking care to use the appropriate sign for the quadrant. (tan is negative in quadrant 4)	$\tan(353°) = -\tan(7°)$
(b) 1 Draw a CAST diagram and locate the given angle. (Here, the angle is in quadrant 2.)	(b) 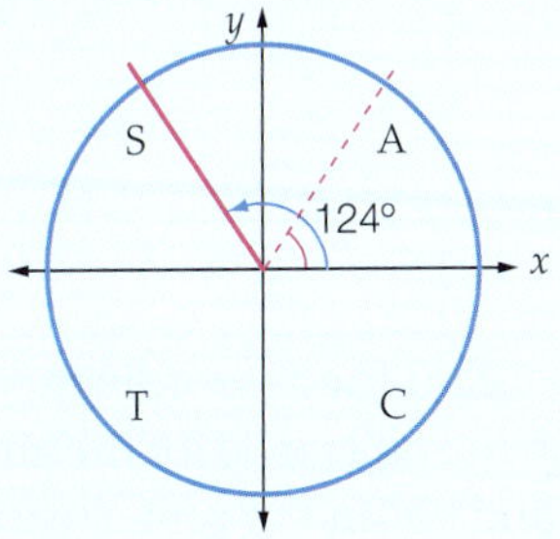
2 Find the equivalent angle in quadrant 1.	$(180 - 124)° = 56°$
3 Write the answer, taking care to use the appropriate sign for the quadrant. (sin is positive in quadrant 2)	$\sin(124°) = \sin(56°)$
(c) 1 Draw a CAST diagram and locate the given angle. (Here, the angle is in quadrant 3.)	(c) 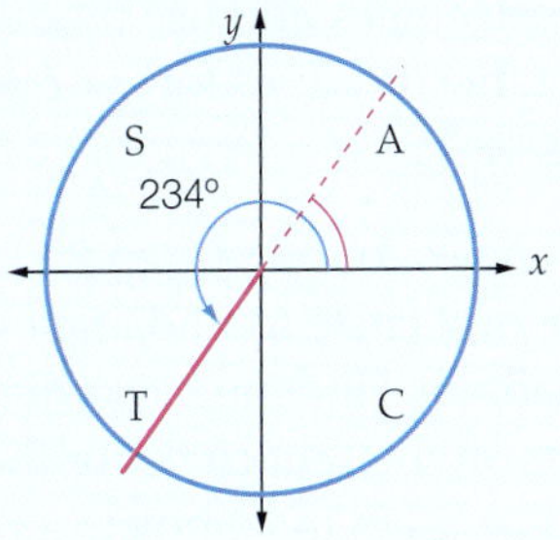
2 Find the equivalent angle in quadrant 1.	$(234 - 180)° = 54°$
3 Write the answer, taking care to use the appropriate sign for the quadrant. (cos is negative in quadrant 3)	$\cos(234°) = -\cos(54°)$

Trigonometric graphs

You can use the unit circle to sketch trigonometric graphs. This involves using symmetry properties and coordinate values of the unit circle.

In a unit circle, the radius is 1 unit in length and equal to the hypotenuse of a right-angled triangle of horizontal length x and vertical height y. Therefore, $\sin(\theta)$ is the vertical height of a point on the circumference of the circle given by the y-coordinate, and $\cos(\theta)$ is the horizontal length of the point given by the x-coordinate.

Worked example 4

W.E. 4

Use the unit circle to sketch the graph of the equation $y = \cos(\theta)$ for values of θ ranging from 0° to 360°.

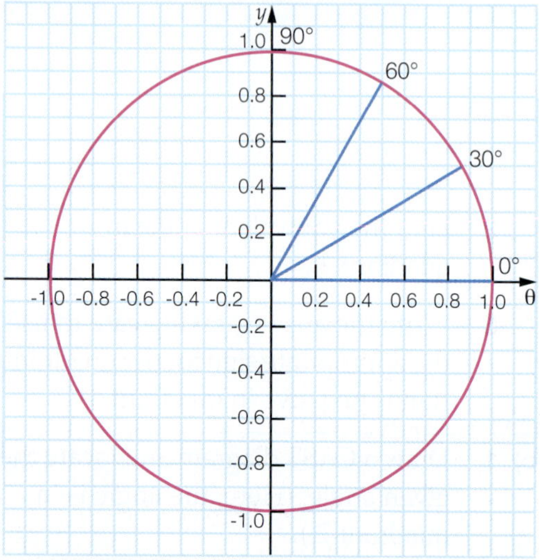

Thinking	Working

1 As $\cos(\theta) = x$, find the x-coordinates of the points on the unit circle for the angles 0°, 30°, 60° and 90° in the unit circle and use these values to complete a table of values for $y = \cos(\theta)$ for $0° \le \theta \le 90°$.

θ	0°	30°	60°	90°
cos (θ)	1	0.87	0.5	0

2 Reflect the angles 0°, 30°, 60° and 90° in the y-axis of the unit circle to find angles in the second quadrant. Find the x-coordinates of the points on the unit circle for these angles and use them to complete a table of values for $y = \cos(\theta)$ for $90° \le x \le 180°$.

θ	120°	150°	180°
cos (θ)	-0.5	-0.87	-1

3 Reflect the angles found in step **2** in the x-axis of the unit circle to find angles in the third quadrant. Find the x-coordinates of the points on the unit circle for these angles and use them to complete a table of values for $y = \cos(\theta)$ for $180° \le x \le 270°$.

θ	210°	240°	270°
cos (θ)	-0.87	-0.5	0

10A

4 Reflect the angles found in step **3** in the y-axis of the unit circle to find angles in the fourth quadrant. Find the x-coordinates of the points on the unit circle for these angles and use them to complete a table of values for $y = \cos(\theta)$ for $270^\circ < \theta \le 360^\circ$.

θ	300°	330°	360°
$\cos(\theta)$	0.5	0.87	1

5 Plot the points from the table of values found in steps **1**, **2**, **3** and **4**, and join the points with a smooth curve.

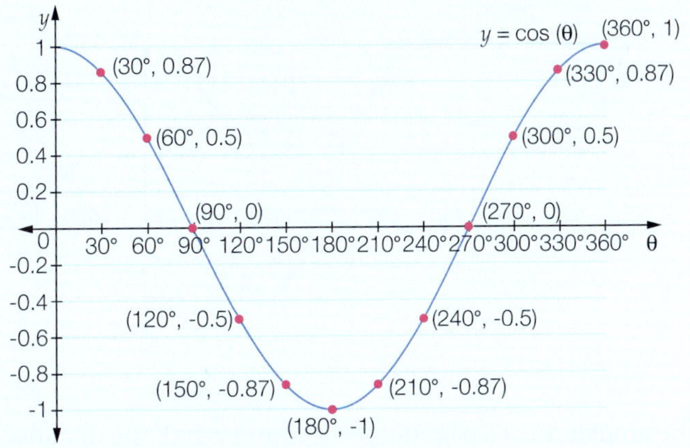

Angles of any magnitude

Up to now, only angles from 0° to 360° have been used. Notice that $y = 1$ when $\theta = 0^\circ$ and $\theta = 360^\circ$ for $y = \cos(\theta)$. This is because both angles represent the same position on the unit circle. In fact, if θ is any given angle, every multiple of 360° added to θ or subtracted from θ represents the same point on the unit circle.

For example, θ values of 25°, 25° + 360°, 25° + 720°, 25° − 360° and 25° − 720°, all occupy the same position on the unit circle and have the same trigonometric ratio values.

$\sin(385^\circ) = \sin(25^\circ)$

$\cos(745^\circ) = \cos(25^\circ)$

$\tan(-335^\circ) = \tan(25^\circ)$

Note that, by convention, positive angles are measured anticlockwise from the positive direction of the x-axis, while negative angles are measured clockwise from the positive direction of the x-axis. (So, -70° is 70° measured clockwise from the positive direction of the x-axis.)

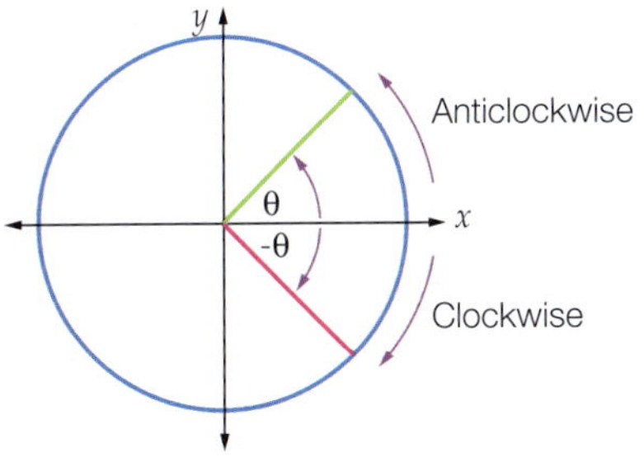

The shape of trigonometric graphs keeps repeating in cycles infinitely in both directions, as the angle values go around and around the unit circle, so they are called cyclic graphs.

Period

The **period** of a trigonometric graph is the horizontal distance it takes to complete one full cycle of the graph. For example, when the graph of $y = \cos(\theta)$ is extended by using values of θ from -720° to 720°, its graph looks like the one shown below. The period of the graph is the horizontal distance between point A and point B or between point C and point D or between point E and point F. For $y = \cos(\theta)$, the period is 360°.

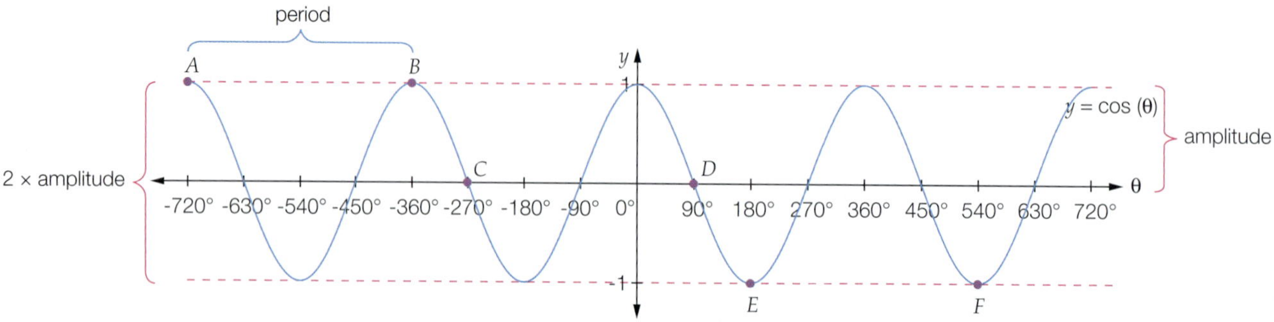

Amplitude

The **amplitude** of a trigonometric graph is half the distance between the minimum y-value and the maximum y-value. From the graph of $y = \cos(\theta)$ it can be seen that the minimum value is -1 and the maximum value is 1. The distance between them is 2 units, so the amplitude is half of this distance.

Therefore, the amplitude $= \frac{1}{2} \times 2$

$= 1$

The graphs of $\sin(\theta)$ and $\cos(\theta)$ always cycle between a minimum value of -1 and a maximum value of 1.

You cannot find the amplitude of the tangent graph because it has no bounds on its minimum or maximum value.

Using technology to sketch trigonometric graphs

Drawing trigonometric graphs by hand and reading values off a graph is inaccurate and time consuming. Instead you can use CAS technology to draw the trigonometric graph $y = \cos(\theta)$ and find values from the graph. A CAS will use x to represent θ.

Using TI-Nspire CAS

Add **Graphs** to your document. Then select **menu** > **Settings** and set the **Graphing Angle** and **Geometry Angle** settings to **Degree**.

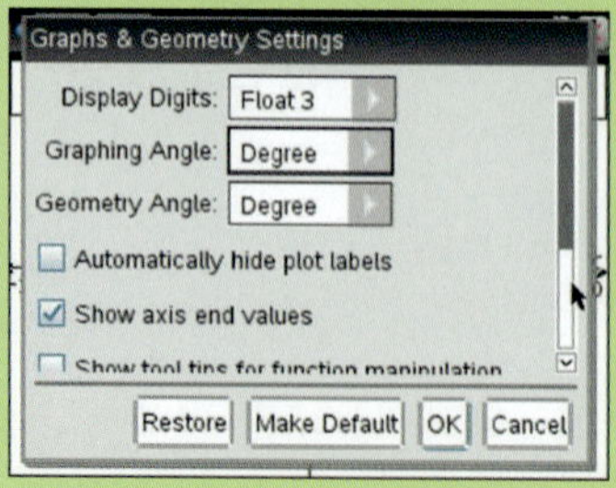

Using Casio ClassPad CAS

From the menu select **Graph & Table**.

Make sure you are in degrees mode, with **Deg** shown at the bottom of the screen. If this shows **Rad** or **Gra** instead, then select the word to change it to **Deg**.

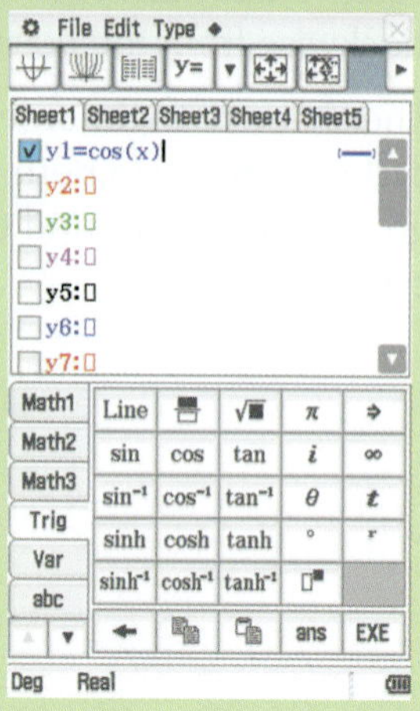

<table>
<tr><th>Using TI-Nspire CAS</th><th>Using Casio ClassPad CAS</th></tr>
<tr><td>

In the **Graphs** screen, enter the equation ***f1*(x)=cos(x)** to be graphed. Use the **trig** key [trig] to enter the cos function.

You can adjust the graph window settings by selecting **menu** > **Window/Zoom** > **Window Settings...** and then changing the values as shown.

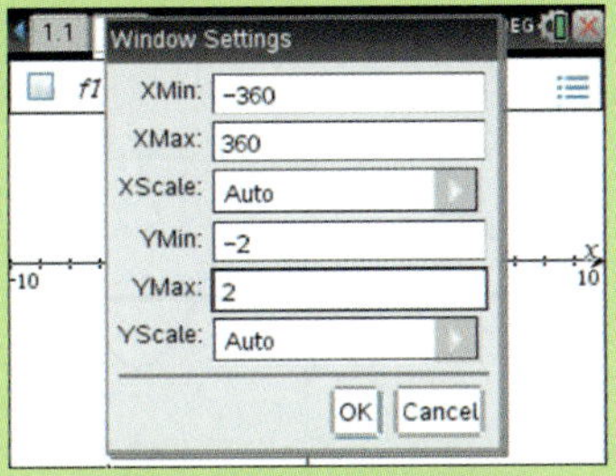

</td><td>

Enter the equation **y1=cos(x)** and tick the box next to it.

You can find the cos function on the **Trig** tab in the **Keyboard**.

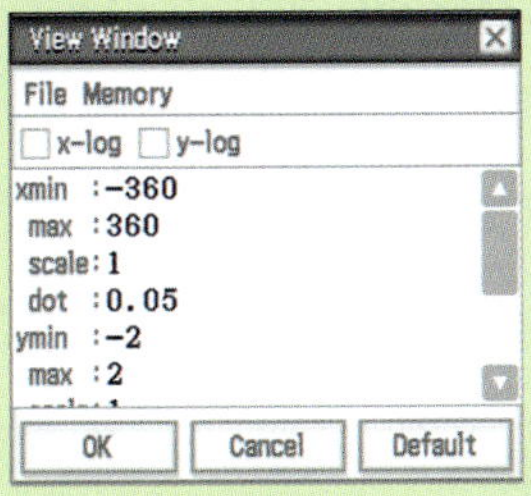

You should adjust the graph window settings by selecting the view icon and then changing the values as shown. (The **dot** value does not need to match.)

</td></tr>
<tr><td>

Press enter to display the graph.

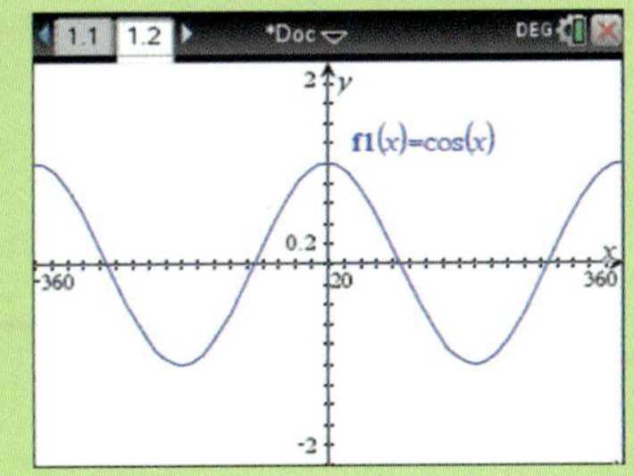

</td><td>

Select the graph icon to view the graph.

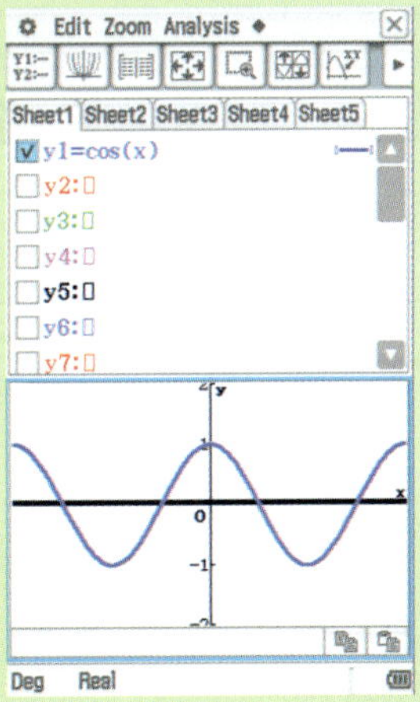

</td></tr>
<tr><td>

Select **menu** > **Trace** > **Graph Trace** and a cursor will appear on the graph that you can move, using the arrow keys, to read the values of the graph.

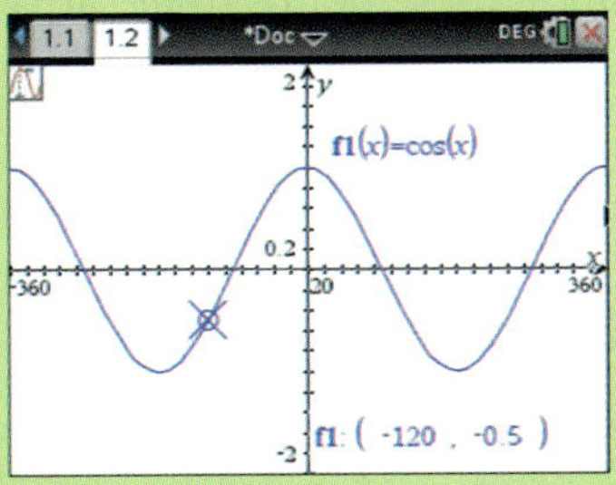

</td><td>

Select **Analysis** > **Trace** and a cursor will appear on the graph that you can move, using the arrow keys, to read the values of the graph.

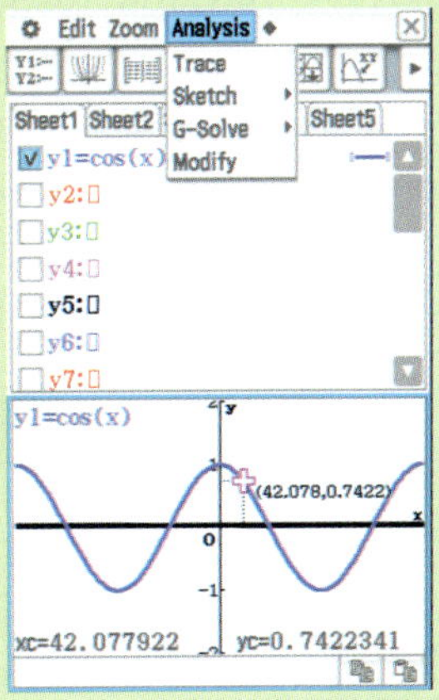

</td></tr>
</table>

Now that you have graphed $y = \cos(\theta)$ as $f1(x) = \cos(x)$, you can graph $y = \sin(\theta)$ as **f2(x) = sin(x)** in the same way. You should notice that $y = \sin(\theta)$ is exactly the same shape as $y = \cos(\theta)$ but translated 90° to the right.

Using TI-Nspire CAS	Using Casio ClassPad CAS
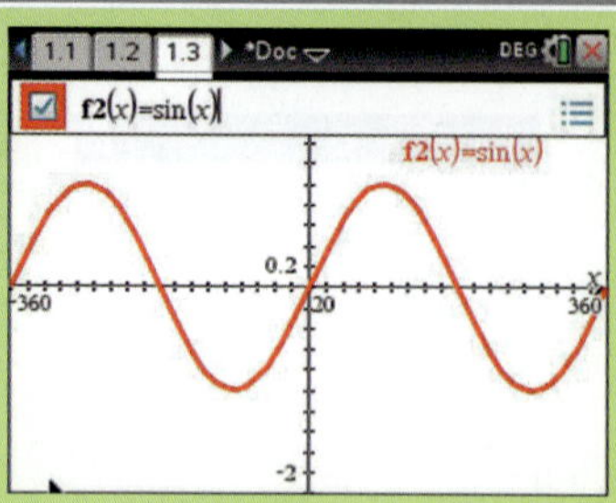	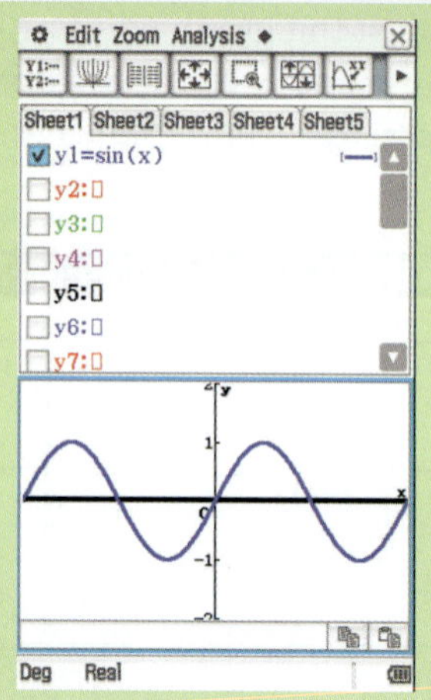

Now graph $y = \tan(\theta)$ in the same way.

Using TI-Nspire CAS	Using Casio ClassPad CAS
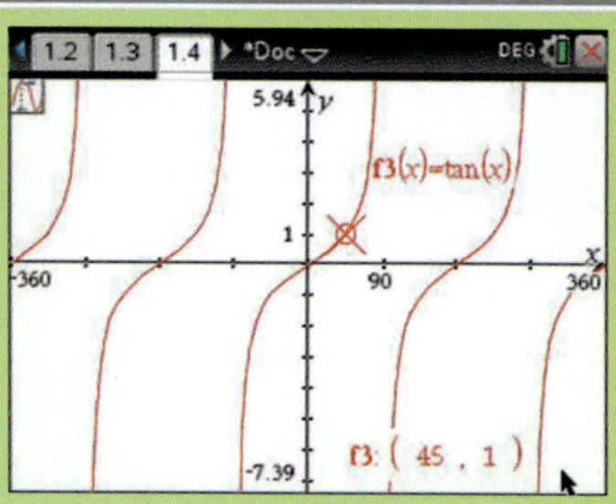	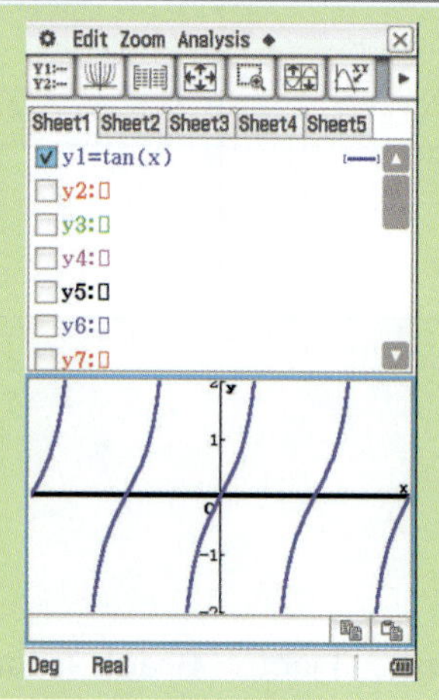

The graph of $\tan(\theta)$ is quite different to the graphs of $\sin(\theta)$ and $\cos(\theta)$. This is because it is not bounded by a maximum value of 1 and a minimum value of -1 and it has points where it is undefined. This means that the graph is not continuous at these points (there are breaks in the graph).

A tangent graph approaches **asymptotes**, where the graph is undefined. Asymptotes are lines that the graph approaches but never touches or crosses. The equations of the asymptotes are $\theta = -270°$, $\theta = -90°$, $\theta = 90°$ and $\theta = 270°$. Asymptotes occur every 180° and the graph repeats between these asymptotes, so the period of $y = \tan(\theta)$ is 180°.

	$y = \cos(\theta)$	$y = \sin(\theta)$	$y = \tan(\theta)$
period	360°	360°	180°
amplitude	1	1	undefined
asymptotes	none	none	$(90 + 180n)°$, where $n = 0, 1, 2, \ldots$

7.2 The unit circle

Navigator

1 (a–h), 2, 3, 4, 5, 6 (a–h), 7, 8, 9, 11, 12, 13, 15	1 (columns 1–2), 2, 3, 4, 5, 6 (columns 1–2), 7, 8, 9, 10, 11, 12, 13, 14, 15	1 (columns 3–4), 2, 3, 4, 5, 6 (columns 3–4), 7, 10, 11, 12, 13, 14, 15, 16

Answers p. 813

Equipment required: CAS would be useful for Questions **12**, **13**

Fluency

W.E. 2

1 State the quadrant in which each of the following angles is found.

(a) 355° (b) 222° (c) 111° (d) 102°

(e) 191° (f) 23° (g) 263° (h) 97°

(i) 286° (j) 342° (k) 156° (l) 260°

2 Which of the following statements is *incorrect*?

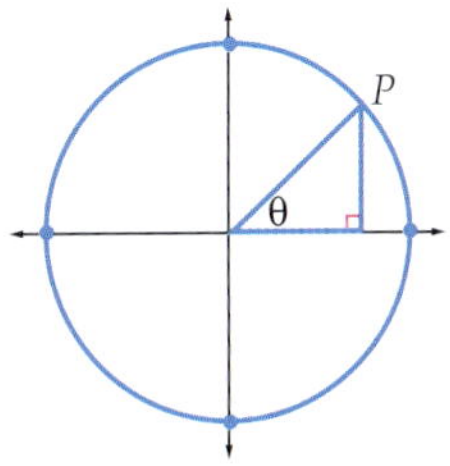

A In quadrant 1 all of the trigonometric ratios are positive.

B The angle represents the position of point *P* on the circumference of the unit circle.

C In quadrant 1 as θ increases from 0° to 90°, sin (θ) increases from 0 to 1.

D In quadrant 1 as θ increases from 0° to 90°, cos (θ) increases from 0 to 1.

3 Which of the following statements is *incorrect*?

A In quadrant 2, θ increases from 90° to 180°.

B In quadrant 2, as the angle increases, sin (θ) increases from 0 to 1.

C In quadrant 2, as the angle increases, cos (θ) decreases from 0 to 1.

D In quadrant 2, as the angle increases, tan (θ) decreases from an undefined ratio to 0.

4 Which of the following statements is *correct*?

A In quadrant 3, θ increases from 270° to 360°.

B In quadrant 3, as the angle increases, tan (θ) decreases from 1 to an undefined negative ratio.

C In quadrant 4, only the cosine ratio is positive.

D In quadrant 4, θ increases from 180° to 270°.

5 Which of the following statements is *incorrect*?

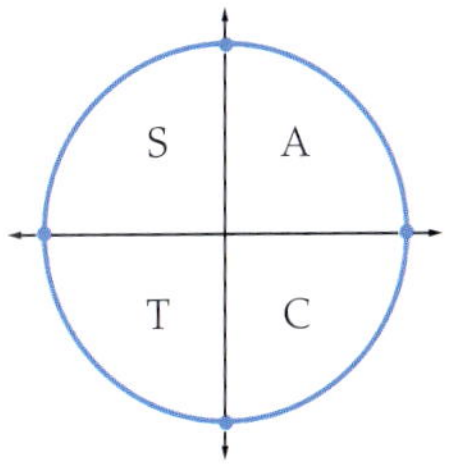

A This is an accurate CAST diagram.

B sin (90°) = 1

C cos (270°) = 0

D tan (180°) is undefined.

W.E. 3

6 For each of the following, find the equivalent trigonometric ratio in quadrant 1, taking care to include the appropriate sign.

(a) sin (135°) (b) cos (224°) (c) tan (111°) (d) sin (264°)

(e) cos (321°) (f) tan (123°) (g) sin (263°) (h) cos (97°)

(i) tan (186°) (j) sin (342°) (k) cos (166°) (l) tan (207°)

W.E. 4

7 Use the unit circle shown to sketch the graph of the equation $y = \sin(\theta)$ for values of θ ranging from 0° to 360°.

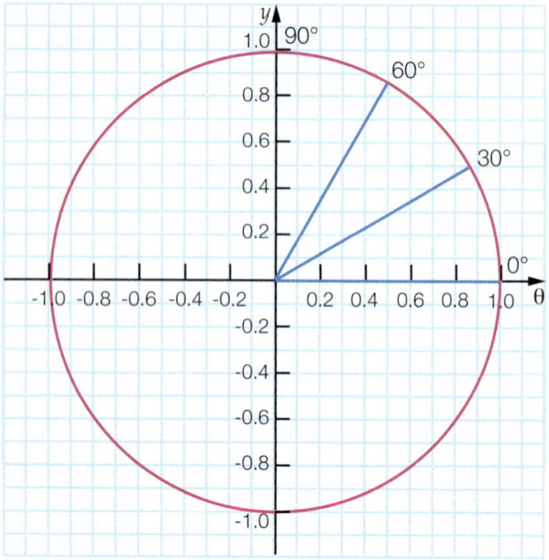

8 (a) Which of the following groups of angles are all found in quadrant 3?

A 213°, 182° and 156°

B 187°, 245° and 194°

C 98°, 124° and 156°

D 342°, 276° and 199°

(b) Find the equivalent trigonometric ratio to sin (25)°.

A sin (205°) B sin (345°)

C -sin (155°) D sin (155°)

9 The period and the amplitude of the trigonometric relationship shown is:

A 90°, 1 B 90°, 2

C 180°, 1 D 180°, 2

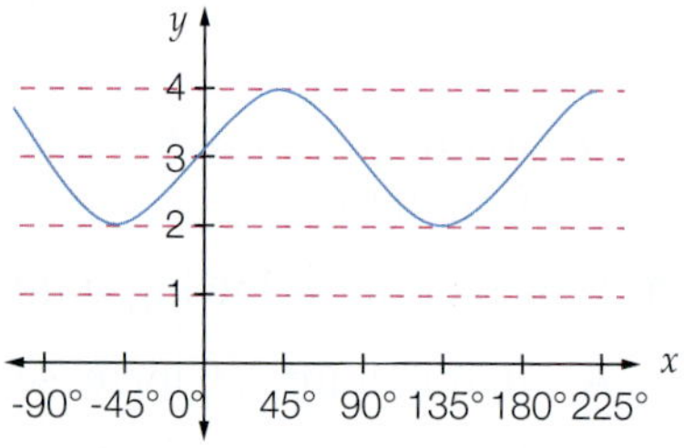

Understanding

10 (a) Use the unit circle with the tangent line shown in the diagram on the right to sketch the graph of the equation $y = \tan(x)$ for values of x from 0° to 70° in intervals of 10°.

(b) What do you think is happening to the value of $\tan(x)$ as the value of x gets closer to 90°?

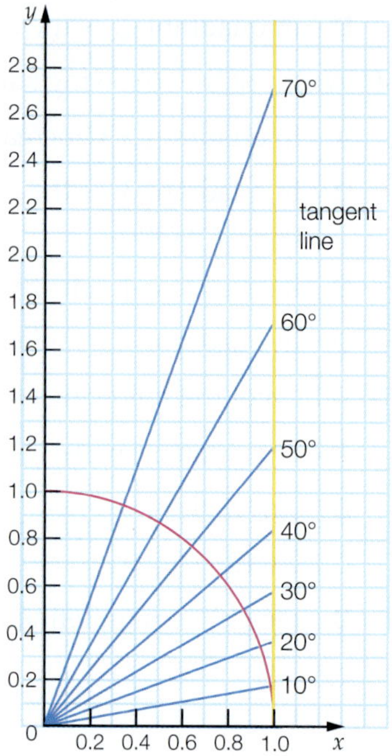

11 Jay knows the values of sine, cosine and tangent when the angles are 30°, 45° and 60°. For which other angles less than 360° can he find trigonometric ratios using what he already knows?

12 Consider the three cosine graphs shown below.

(i)

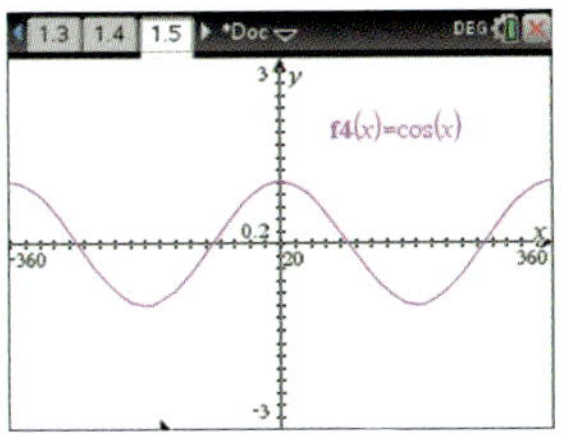

(ii)

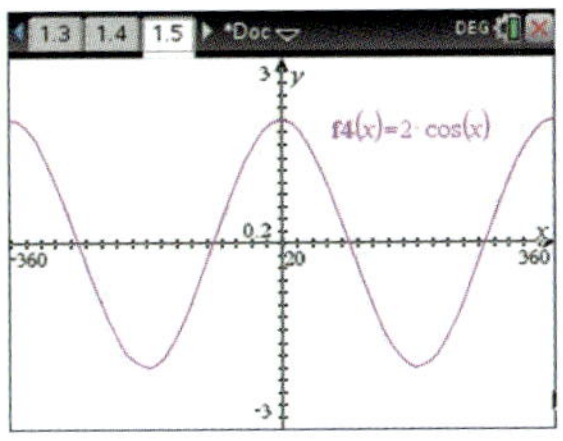

(iii)

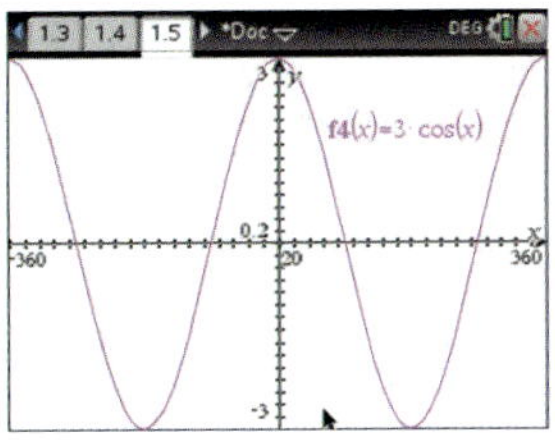

(a) State the period for each graph.

(b) State the amplitude for each graph.

(c) What conclusion can you draw about the effect of the a in $y = a \times \cos(x)$?

(d) Sketch what you think $y = 4\cos(x)$ would look like. Check your answer using a CAS if you can.

13 Consider the three sine graphs shown below, which all show values from 0° to 360°.

(i)

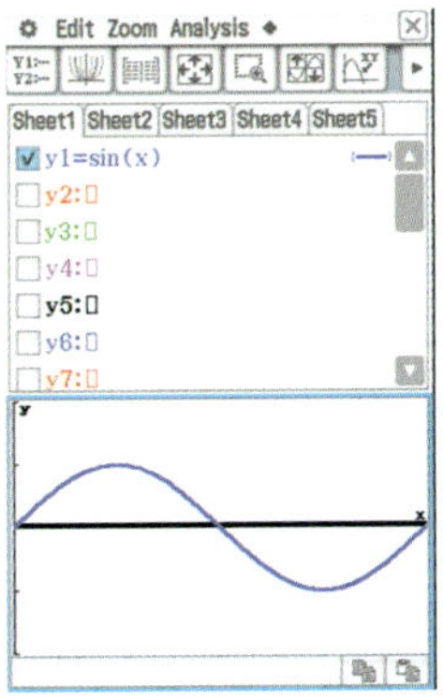

(ii)

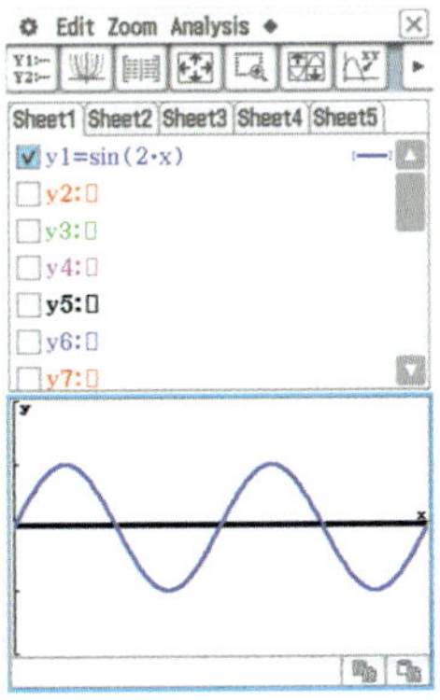

(iii)

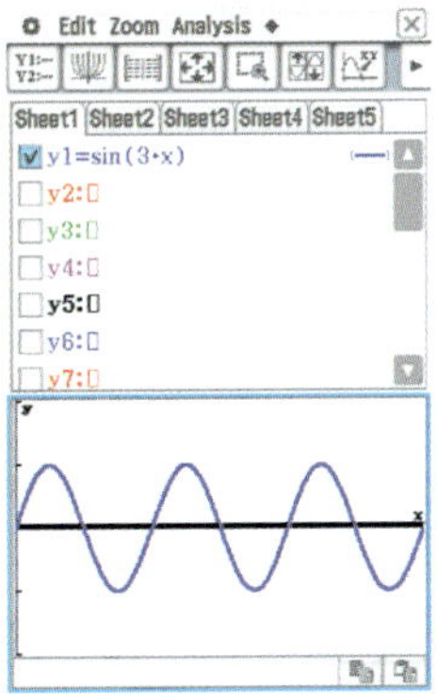

(a) State the period for each graph.

(b) State the amplitude for each graph.

(c) What conclusion can you draw about the effect of the n in $y = \sin(nx)$?

(d) Sketch what you think $y = \sin(6x)$ would look like. Check your answer using a CAS if you can.

Reasoning

14 At a harbour, a siren sounds every time the tide is at 5 m. Since tides are based on sine curves, and the siren first sounds when the angle for the sine ratio is 24°, calculate the angle when the siren will sound next.

Open-ended

15 Write two angles that are 180° apart. For each of your chosen angles, write the sign for the ratios: sin, cos and tan. Use a calculator to check that you are correct.

16 (a) Explain why $\cos(-\theta) = \cos(\theta)$ for any angle θ.

(b) Explain why $\sin(-\theta) = -\sin(\theta)$ for any angle θ.

(c) Using your answers to parts **(a)** and **(b)** and the relationship $\tan(\theta) = \dfrac{\sin(\theta)}{\cos(\theta)}$, explain why $\tan(-\theta) = -\tan(\theta)$.

7.3 Solving trigonometric equations

Just as you can solve equations such as $3x - 5 = 2$ to get $x = \frac{7}{3}$, you can use inverse operations to solve equations that involve trigonometric ratios.

Using inverse trigonometric ratios to solve equations

You can find a solution to the trigonometric equation $\sin(x) = 0.5$ by using the inverse sin function on a calculator.

$x = \sin^{-1}(0.5) = 30$, so $x = 30°$

This angle is in quadrant 1. However, there are equivalent angles in other quadrants of the unit circle.

$\sin(180° - \theta) = \sin(\theta)$, so $x = 150°$

This angle in quadrant 2 is also a solution.

In fact, there are an infinite number of solutions, because adding multiples of 360° to each solution produces more solutions.

10A

So $x = 30° + 360° = 390°$ and $x = 150° + 360° = 510°$ are another two solutions.

Worked example 5

W.E. 5

Find all solutions for x that satisfy each of the following equations.

(a) $2\cos(x) - 1 = 0$ for $0° \leq x \leq 360°$ **(b)** $3 - \tan(x) = 2$ for $0° \leq x \leq 180°$

Thinking	Working
(a) 1 Rewrite the equation to make the trigonometric ratio the subject of the equation. (Here, $\cos(x)$ is the subject.)	(a) $2\cos(x) - 1 = 0$ $2\cos(x) = 1$ $\cos(x) = 0.5$
2 Use the appropriate inverse trigonometric function on your calculator to find the solution in quadrant 1. (Here, you use $\cos^{-1}$.)	$x = \cos^{-1}(0.5)$ $x = 60°$
3 Use a CAST diagram to identify other quadrants where the ratio is positive and find an equivalent angle. (Here $\cos(x)$ is positive in quadrant 4. Subtract the first quadrant angle from 360° to find the equivalent angle.)	$\cos(x)$ is positive in quadrants 1 and 4. The equivalent angle of 60° in quadrant 4 is $360° - 60° = 300°$.
4 State the solutions.	$x = 60°, 300°$ for $0° \leq x \leq 360°$

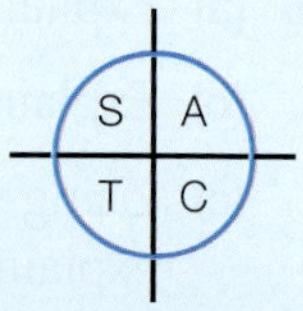

Thinking	Working
(b) 1 Rewrite the equation to make the trigonometric ratio the subject of the equation. (Here, $\tan(x)$ is the subject.)	(b) $3 - \tan(x) = 2$ $-\tan(x) = -1$ $\tan(x) = 1$
2 Use the appropriate inverse trigonometric function on your calculator to find the solution in quadrant 1. (Here, you use $\tan^{-1}$.)	$x = \tan^{-1}(1)$ $x = 45°$
3 Use a CAST diagram to identify other quadrants where the ratio is positive and find an equivalent angle. (Here, $\tan x$ is positive in quadrant 3. Add the first quadrant angle to 180° to find the equivalent angle.)	$\tan(x)$ is positive in quadrant 1 and 3. The equivalent angle of 45° in quadrant 3 is $180° + 45° = 225°$. 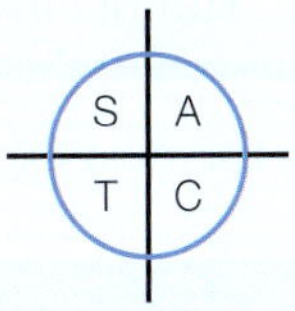
4 Check that the solutions are within the given values for x. Reject any solutions outside these values.	225° is not in $0° \le x \le 180°$ so it is not a solution.
5 State the solutions.	$3 - \tan(x) = 2$ has the solution $x = 45°$ for $0° \le x \le 180°$.

Writing coterminal angles

The trigonometric equation $\cos(x) = 0.5$ has solution $x = 60°$ in quadrant 1 and $x = 300°$ in quadrant 4. However, there can be an infinite number of solutions, as adding or subtracting multiples of 360° to an angle produces **coterminal angles** that give the same position on the unit circle.

Moving in an anticlockwise direction to add multiples of 360° to 60° in quadrant 1, produces an infinite number of solutions:

$x = 60°, x = 60° + 360°, x = 60° + 2 \times 360°, x = 60° + 3 \times 360°, \ldots$

Moving in a clockwise direction, to subtract multiples of 360° from 60° in quadrant 1, produces even more solutions.

$x = 60°, x = 60° - 360°, x = 60° - 2 \times 360°, x = 60° - 3 \times 360°, \ldots$

All the solutions, using angles coterminal with 60°, can be written in a concise way as $x = (60 \pm 360n)°, n = 0, 1, 2, 3, \ldots$

In the same way, using angles coterminal with $x = 300°$ in quadrant 4, produces solutions that can be written concisely as $x = (300 \pm 360n)°, n = 0, 1, 2, 3, \ldots$

Worked example 6 **W.E. 6**

Determine all solutions to the equation $\sin(x) = 0.5$.

Thinking	Working
1 Draw a CAST diagram and identify the quadrants where the trigonometric ratio has the correct sign. (Here, it is positive.)	$\sin(x)$ is positive in quadrants 1 and 2. 

2 Calculate the solutions for x-values from 0° to 360°.	$\sin(x) = 0.5$ $(x) = \sin^{-1}(0.5)$ $(x) = 30°$ In quadrant 2, the equivalent angle to 30° is $180° - 30° = 150°$. The solutions for $0° \leq x \leq 360°$ are $x = 30°$, $x = 150°$.
3 Find all solutions by adding and subtracting multiples of 360°.	$x = (30 \pm 360n)°$, $n = 0, 1, 2, 3, \ldots$ and $x = (150 \pm 360n)°$, $n = 0, 1, 2, 3, \ldots$

In the next example, a calculator is used to find an approximate solution.

Worked example 7

W.E. 7

Determine all solutions to the equation $2\tan(x) + 1 = 0$. Write your answers correct to 2 decimal places.

Thinking	Working
1 Rewrite the equation to make the trigonometric ratio the subject of the equation.	$2\tan(x) + 1 = 0$ $2\tan(x) = -1$ $\tan(x) = -\frac{1}{2}$
2 Draw a CAST diagram and identify the quadrants where the trigonometric ratio has the correct sign. (Here, it is negative.) Write the quadrants in which the solutions will be found.	$\tan(x)$ is negative in quadrants 2 and 4. 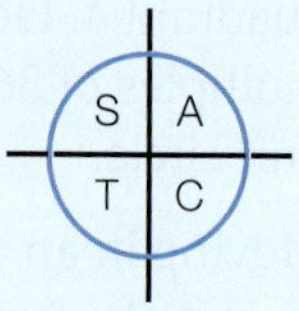
3 Find the solution for the x-value for the positive ratio in quadrant 1.	In quadrant 1, $\tan(\theta)$ is positive $x = \tan^{-1}\left(\frac{1}{2}\right)$ $x = 26.565°$
4 Calculate the solutions for x-values from 0° to 360° taking into account the correct quadrants for the sign of the given ratio.	In quadrant 2, the equivalent angle is $(180 - 26.565)° = 153.435°$ $= 153.44°$ (2 d.p.) In quadrant 4, the equivalent angle is $(360 - 26.565)° = 333.435°$ $= 333.44°$ (2 d.p.) The solutions for $0° \leq x \leq 360°$ are $x = 153.44°$ and $x = 333.44°$
5 Find all solutions by adding and subtracting multiples of 360°.	$x = (153.44 \pm 360n)°$, $n = 0, 1, 2, 3, \ldots$ and $x = (333.44 \pm 360n)°$, $n = 0, 1, 2, 3, \ldots$
6 As $y = \tan(\theta)$ has a period of 180°, write the solution more simply.	$x = (153.44 + 180n)°$, where $n = 0, 1, 2, 3, \ldots$

Using technology to solve trigonometric equations

There are several different methods for finding solutions to trigonometric equations on your CAS. Here, you will look at a graphical solution for $2\tan(x) + 1 = 0$. You can transpose this to become $2\tan(x) = -1$, then graph the two sides of the equation separately to find the point of intersection, which must be the solution.

Using TI-Nspire CAS	Using Casio ClassPad CAS
Add **Graphs** to your document. Then select **menu** > **Settings** and set the **Graphs & Geometry Settings** as shown. 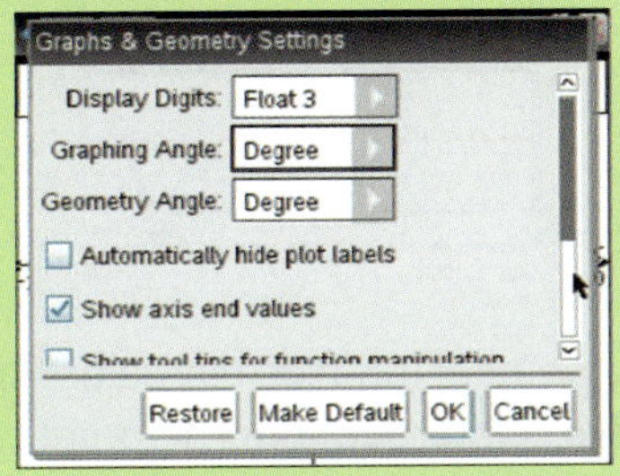 In the **Graphs** screen, select **menu** > **Graph Entry/Edit** > **Function** to enter **f1(x)=2tan(x)** and then again to enter **f2(x)=-1**.	From the menu select **Graph & Table**. Make sure you are in degrees mode, with the **Deg** shown at the bottom of the screen. If this shows **Rad** or **Gra** instead, then select the word to change it to **Deg**. If there are equations already showing, then select **Edit** > **Clear All** to clear them. Then enter **y1=2tan(x)** and enter **y2=-1**. Tick the boxes next to both. 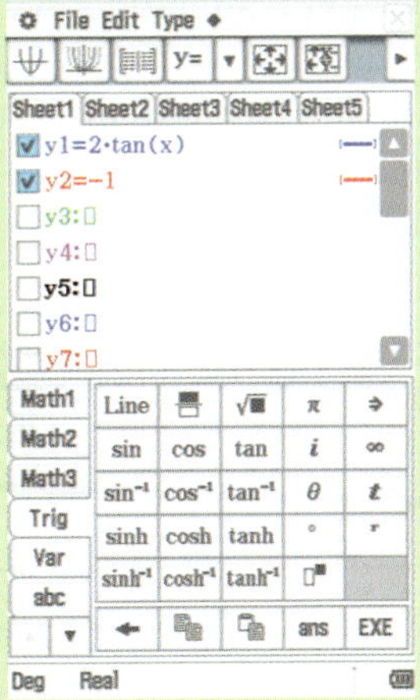
You should adjust the graph window settings by selecting **menu** > **Window/Zoom** > **Window Settings...** and then changing the values as shown. 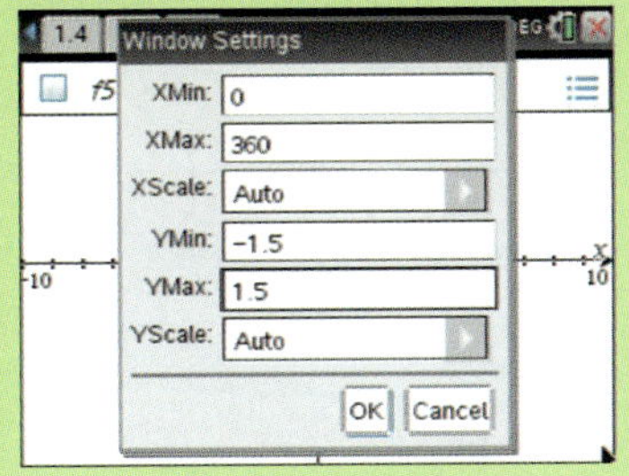	You should adjust the graph window settings by selecting the view icon and then changing the values as shown. (The **dot** value does not need to match.) 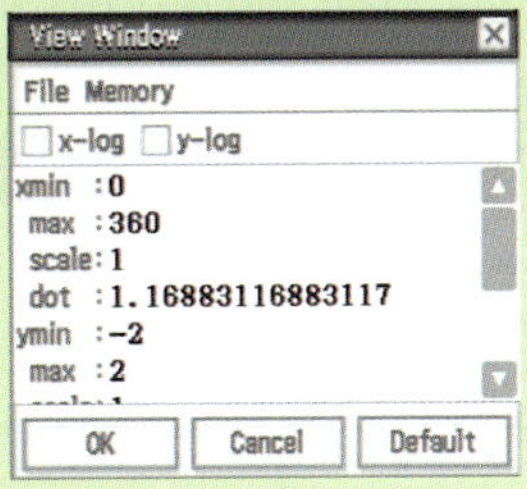

Using TI-Nspire CAS	Using Casio ClassPad CAS

Using TI-Nspire CAS

Press enter to display the graph.

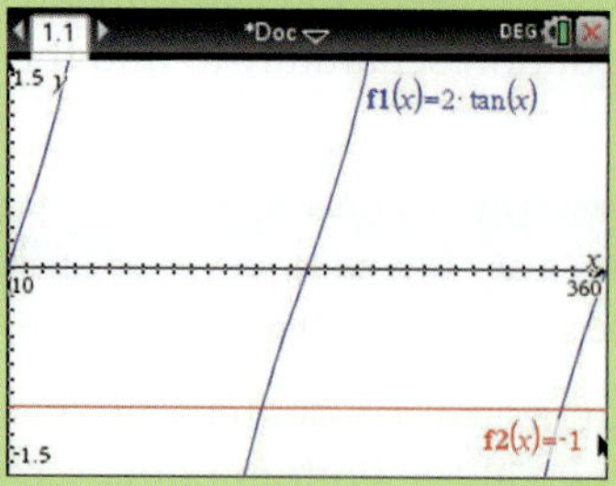

Using Casio ClassPad CAS

Select the graph icon to view the graph.

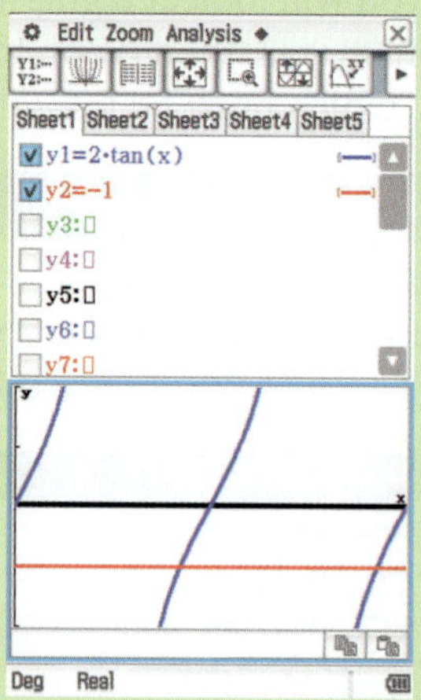

Using TI-Nspire CAS

To find a point of intersection, select **menu** > **Analyze Graph** > **Intersection**. Use the cursor to set the lower bound (a point to the left of the intersection) and then the upper bound (a point to the right of the intersection). The point of intersection will appear.

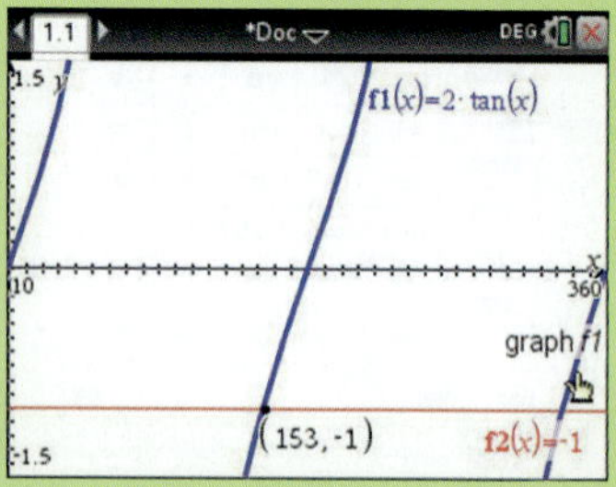

You can confirm that this is the same as the second quadrant value found earlier.

Using Casio ClassPad CAS

To find a point of intersection, select **Analysis** > **G-Solve** > **Intersection**. The point of intersection will appear.

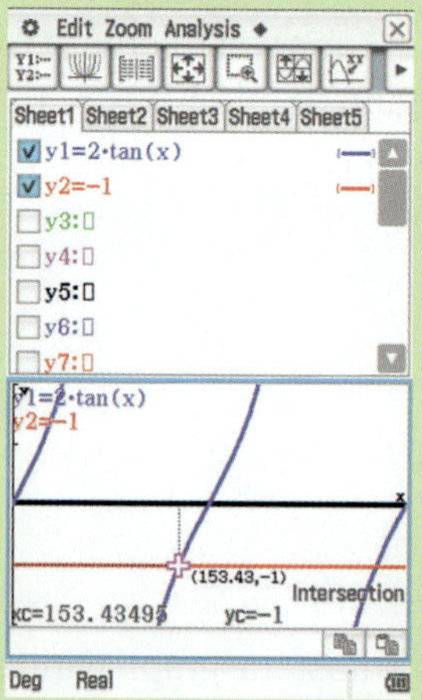

You can confirm that this is the same as the second quadrant value found earlier.

Using TI-Nspire CAS

You can then repeat the process to find the fourth quadrant solution.

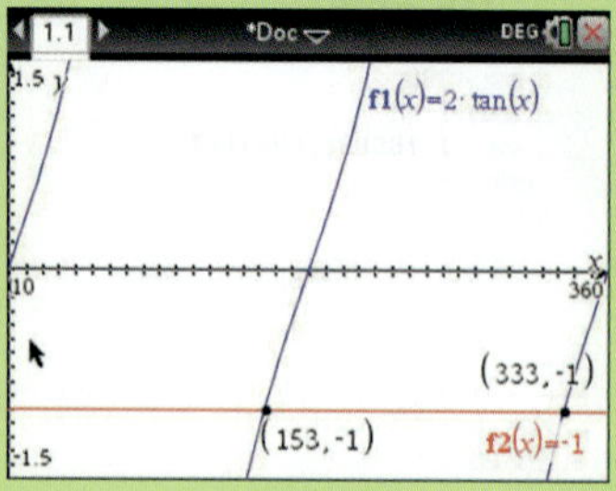

Using Casio ClassPad CAS

To find the next intersection, use the arrow keys to move across.

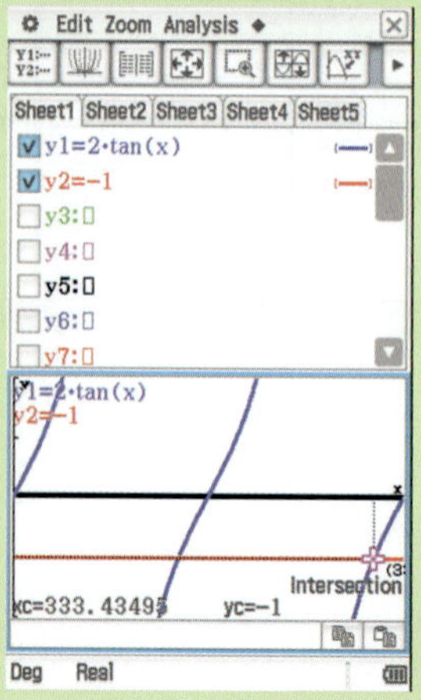

7.3 Solving trigonometric equations

Navigator

1, 2, 3, 4, 5, 6, 7, 8, 10, 11, 14, 15 | 1, 2, 3, 4, 5, 6, 7, 8, 9, 10, 11, 13, 14, 15 | 1, 2, 3, 4, 5, 8, 9, 10, 12, 13, 14, 15

Answers p. 814

Fluency

1 Find all solutions for x that satisfy each of the equations.

(a) $\cos(x) = \left(\frac{\sqrt{3}}{2}\right)$ for $0° \leq x \leq 360°$

(b) $\sin(x) = -\frac{1}{2}$ for $0° \leq x \leq 360°$

2 Find all solutions for x that satisfy $\cos(x) = \frac{\sqrt{2}}{2}$ for $0° \leq x \leq 180°$.

3 Find all solutions for x that satisfy each of the equations: W.E. 5

(a) $2\cos(x) + 1 = 0$ for $0° \leq x \leq 360°$

(b) $4 - 2\sin(x) = 3$ for $0° \leq x \leq 180°$

4 Determine all solutions to the equation $2\cos(x) + 1 = 0$. Hint: Calculate the solutions for x-values from $0°$ to $360°$, then add and subtract multiples of $360°$. W.E. 6

5 Find all solutions for x that satisfy each of the following equations.

(a) $2\cos(x) - \sqrt{3} = 0$ for $0° \leq x \leq 540°$

(b) $\sqrt{3}\tan(x) - \sqrt{3} = 0$ for $0° \leq x \leq 360°$

6 (a) Determine all solutions to the equation $\tan(x) + \sqrt{3} = 0$. W.E. 7

(b) Determine all solutions to the equation $4\sin(x) - 1 = 0$. Write all answers correct to 2 decimal places.

7 The solutions to $4 - 5\sin(x) = 0$ for x for $0° \leq x \leq 180°$ are closest to:

A $127°, 127°$ B $53°, 127°$ C $37°, 143°$ D $53°, 307°$

8 Use a calculator to find the solutions for $0° \leq x \leq 360°$ for each of the following equations. Where necessary, round your answers to 2 decimal places.

(a) $4\sin(x) - 3 = 0$ (b) $5\tan(x) + 2 = 0$ (c) $3 - 5\cos(x) = 0$ (d) $\frac{3\sin(x)}{2} - 1 = 0$

Understanding

9 The depth D metres below sea level of a submarine after T minutes is given by the trigonometric equation $D = 4\cos(T) - 4$ where $T \geq 0$. When will be the next time the submarine surfaces ($D = 0$) if it was at the surface at 3.10 pm initially?

10A

10 The graph shown at right could be used to solve which of the following trigonometric equations?

A $3\sin(x) - 1.5 = 0$

B $1.5 - 3\sin(x) = 0$

C $3\sin(x) + 1.5 = 0$

D $3\sin(x) = 1.5$

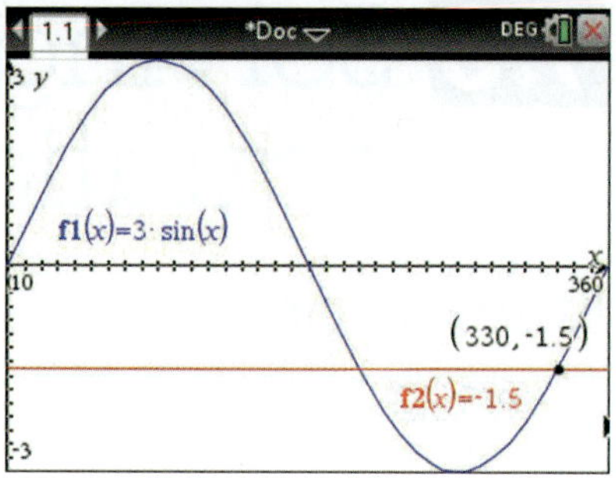

11 Considering the graph shown in Question **10** the solutions shown are in:

A quadrants 1 and 2

B quadrants 2 and 3

C quadrants 3 and 4

D quadrants 1 and 4.

Reasoning

12 A function is defined as $y = \frac{1}{2\sin(x) + 1}$. Its graph is shown at right.

The graph appears to be undefined for two values of x in the domain $[0°, 360°]$.

(a) What would cause this graph to be undefined?

(b) Find the values of x for which the graph is undefined in the domain $[0°, 360°]$.

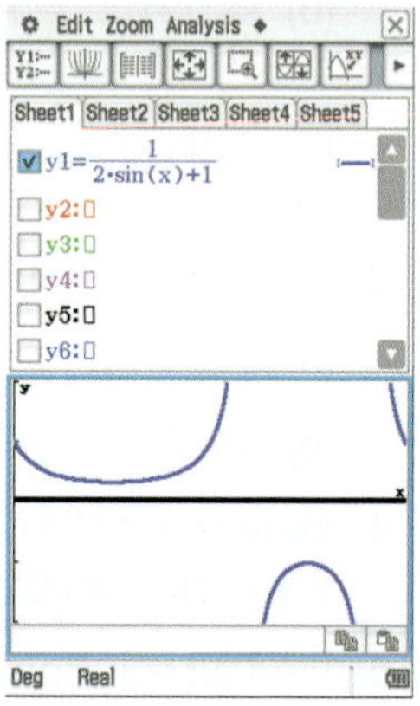

10A

Hint

Simplify the equation using $\tan(x) = \frac{\sin(x)}{\cos(x)}$ to find an equivalent equation and then factorise it using $\sin(x)$ as a common factor.

13 Find all solutions to the equation $\tan(x) = \sin(x)$ for $0° \le x \le 360°$.

Open-ended

14 For each of the sine, cosine and tangent ratios, write at least one equation similar to those in Question **8** and solve it for $0° \le x \le 360°$ using a calculator.

15 **(a)** Write any trigonometric equation that has solutions in:

(i) quadrants 1 and 3

(ii) quadrants 3 and 4.

(b) How do you know which trigonometric ratio to use in each of parts **(i)** and **(ii)**?

Half-time 7

1 A rectangular prism is shown. **7.1**

(a) Calculate $\angle ILP$, by first finding the length of LJ.

(b) Calculate the length of IL.

Round both answers to 2 decimal places.

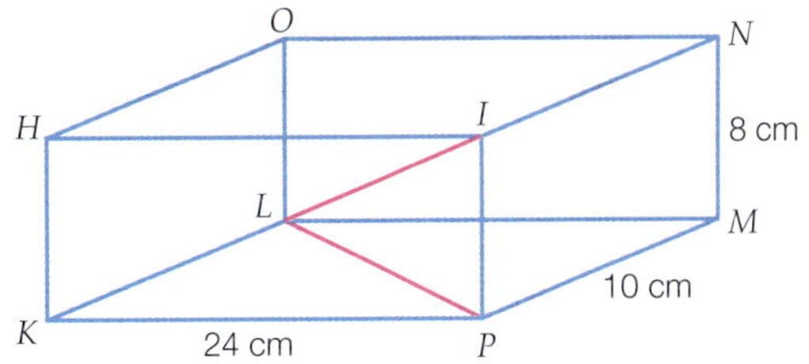

2 Find all values for x, correct to 1 decimal place, that satisfy the following equations and conditions: **7.3**

(a) $3\sin(x) - 2 = 0$, $0° \le x \le 180°$

(b) $5\tan(x) + 1 = 3$, $0° \le x \le 360°$

3 Find the equivalent first quadrant ratio for each of the following, taking care to include the appropriate sign. **7.2**

(a) $\sin(135°)$

(b) $\cos(98°)$

(c) $\tan(200°)$

(d) $\sin(247°)$

(e) $\cos(300°)$

(f) $\tan(339°)$

4 A right pyramid $ABCDE$ has a square base of side length 9 cm. The sloping edges are 15 cm in length. **7.1**

(a) Find the angle that the sloping edge makes with the base of the pyramid, to the nearest degree.

(b) Using your answer to part **(a)**, or any other method, find the height of the pyramid, h. State your answer correct to 1 decimal place.

(c) The side length of the square base of the pyramid is increased by 3 cm, but the length of the sloping edges remains the same.

(i) Will the angle you found in part **(a)** increase or decrease?

(ii) By how many degrees will it increase or decrease?

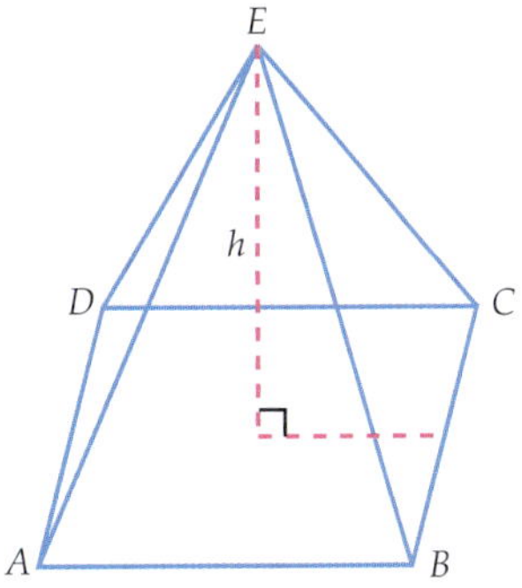

10A

5 The period and amplitude of the trigonometric graph shown is: **7.2**

A 360°, 10

B 180°, 5

C 360°, 5

D 180°, 10

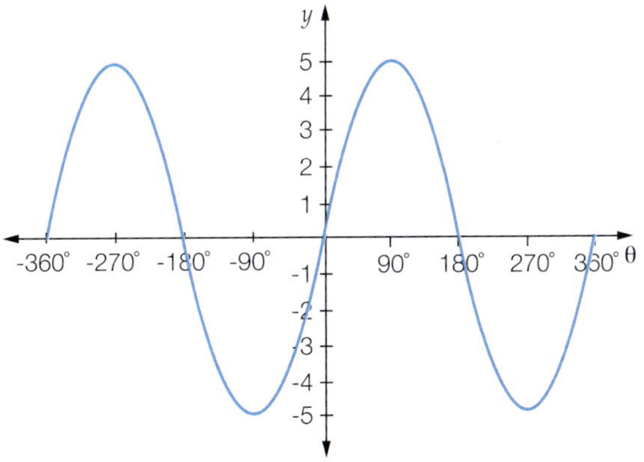

6 Determine all solutions to the following equations. State your solution correct to 2 decimal places, where necessary. **7.3**

(a) $\cos(x) = 0.5$

(b) $3\sin(x) - 1 = 0$

7.4 The sine and cosine rules

The trigonometric ratios $\sin(\theta) = \frac{O}{H}$, $\cos(\theta) = \frac{A}{H}$ and $\tan(\theta) = \frac{O}{A}$ apply to right-angled triangles. For situations that involve a triangle that is not right angled, the **sine rule** or **cosine rule** may be used.

10A

Sine rule

Consider a triangle with vertices at A, B and C and side lengths a, b, c, where a is opposite angle A, b is opposite angle B and c is opposite angle C. It consists of triangle CAD and triangle DBC, each of height h.

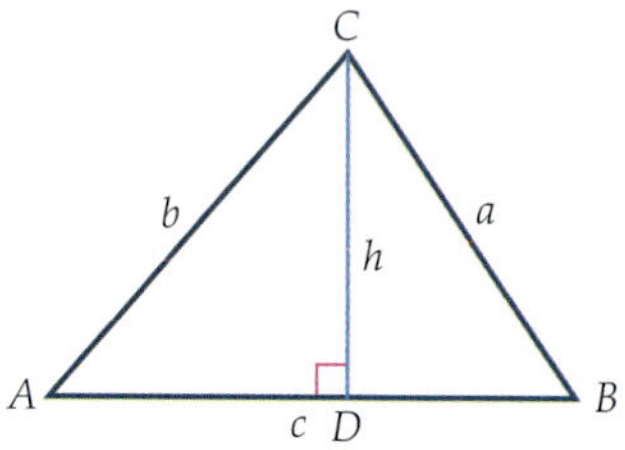

$\sin(A) = \frac{h}{b}$ (from ΔCAD)

$\therefore h = b\sin(A)$ [1]

$\sin(B) = \frac{h}{a}$ (from ΔDBC)

$\therefore h = a\sin(B)$ [2]

$b\sin(A) = a\sin(B)$ equating [1] and [2]

$\frac{\sin(A)}{a} = \frac{\sin(B)}{b}$ [3] rearranging this expression

Triangle ABC can also be divided as shown.

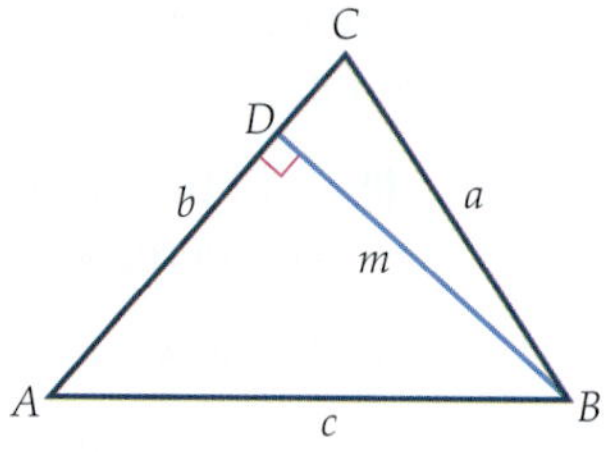

$\sin(A) = \frac{m}{c}$ (from ΔBAD)

$\therefore m = c\sin(A)$ [4]

$\sin(C) = \frac{m}{a}$ (from ΔDCB)

$\therefore m = a\sin(C)$ [5]

$c\sin(A) = a\sin(C)$ equating [4] and [5]

$\frac{\sin(A)}{a} = \frac{\sin(C)}{c}$ [6] rearranging this expression

Together, [3] and [6] give the sine rule:

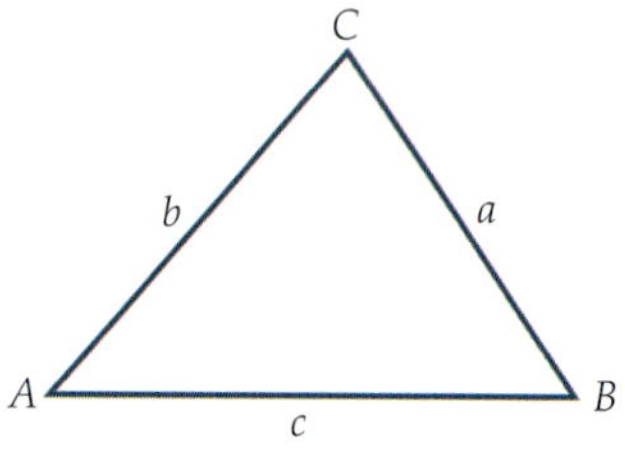

$$\frac{\sin(A)}{a} = \frac{\sin(B)}{b} = \frac{\sin(C)}{c}$$

The sine rule (also known as the law of sines or the sine formula) is sometimes given in reciprocal form:

$$\frac{a}{\sin(A)} = \frac{b}{\sin(B)} = \frac{c}{\sin(C)}$$

The sine rule

$$\frac{\sin(A)}{a} = \frac{\sin(B)}{b} = \frac{\sin(C)}{c}$$

Use to find an unknown angle when two sides and one angle are known.

$$\frac{a}{\sin(A)} = \frac{b}{\sin(B)} = \frac{c}{\sin(C)}$$

Use to find an unknown side when two angles and one side are known.

Only two equated expressions can be used at any one time.

Worked example 8

W.E. 8

Calculate the value of x correct to 2 decimal places.

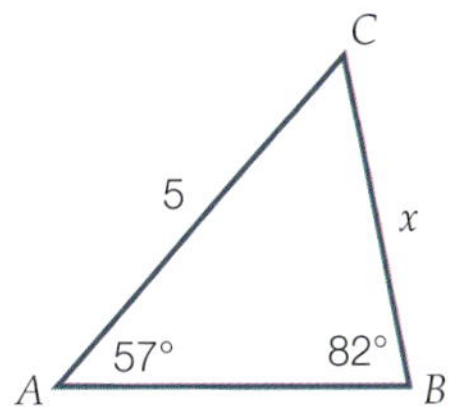

Thinking	Working
1 Draw and label the diagram with the information provided.	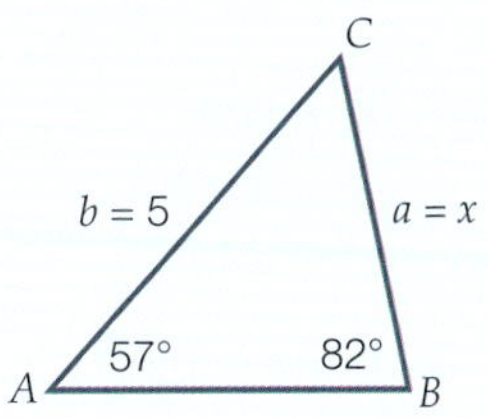
2 Determine the rule to be used. (Here, use the sine rule.)	$\frac{a}{\sin(A)} = \frac{b}{\sin(B)}$
3 Substitute the information given.	$\frac{x}{\sin(57°)} = \frac{5}{\sin(82°)}$
4 Rearrange to make the unknown value the subject.	$x = \frac{5 \times \sin(57°)}{\sin(82°)}$
5 Use a calculator to evaluate the unknown to the required number of decimal places.	$x = 4.23$ units (2 d.p.)

10A

Worked example 9

W.E. 9

Calculate the value of $\angle CAB$ in degrees, rounded to 2 decimal places.

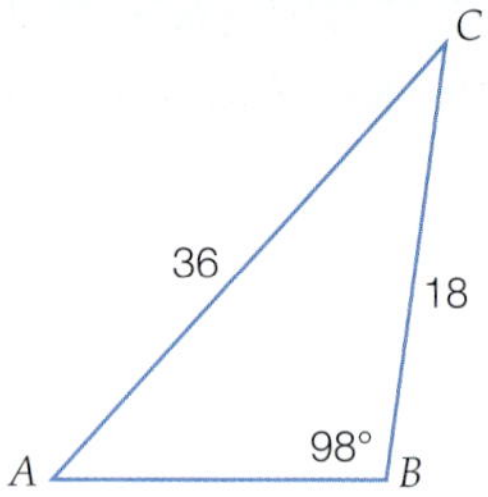

Thinking	Working
1 Draw and label the diagram with the information provided.	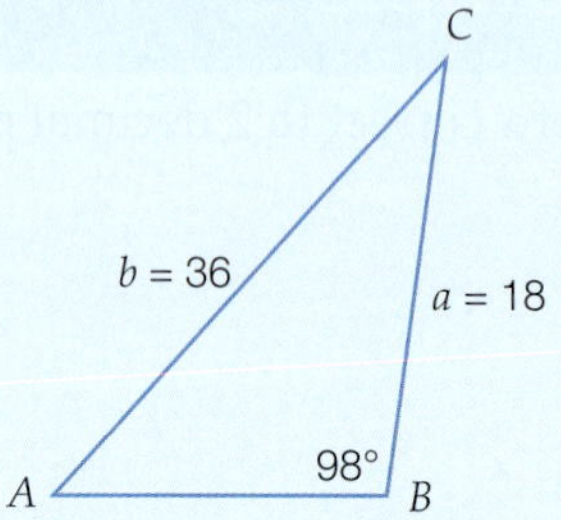
2 Determine the rule to be used. (Here, use the sine rule.)	$\frac{\sin(A)}{a} = \frac{\sin(B)}{b}$
3 Substitute the information given.	$\frac{\sin(A)}{18} = \frac{\sin(98°)}{36}$
4 Make the unknown value the subject.	$\sin(A) = \frac{18 \times \sin(98°)}{36}$ $A = \sin^{-1}\left(\frac{18 \times \sin(98°)}{36}\right)$
5 Use a calculator to evaluate the unknown to the required number of decimal places.	$\angle CAB = 29.68°$ (2 d.p.)

The ambiguous case of the sine rule

Consider ΔCAB, in which you know two sides a, b and a non-included angle θ.

Reflecting side BC about the vertical line OC produces a triangle CAB' that also satisfies the information and $\angle AB'C = 180° - \angle ABC$.

This means that there are two different possible triangles that can share this same angle and two side lengths. The sine rule must be able to produce two possible sets of solutions in this case.

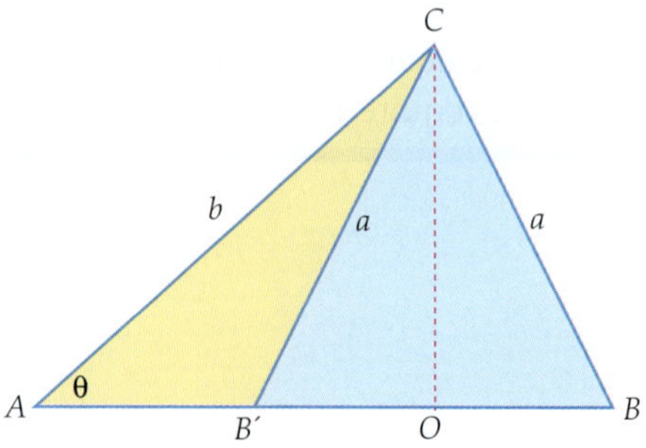

Worked example 10

W.E. 10

Calculate the values of $\angle ABC$ and $\angle AB'C$, correct to 2 decimal places.

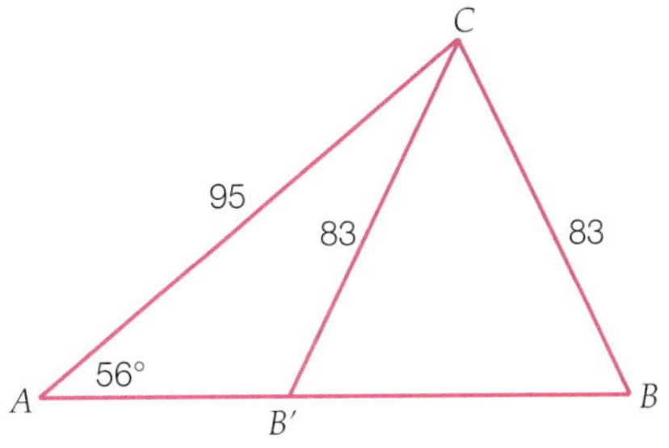

Thinking	Working
1 Draw a labelled diagram for the acute-angled triangle and choose the rule to be used.	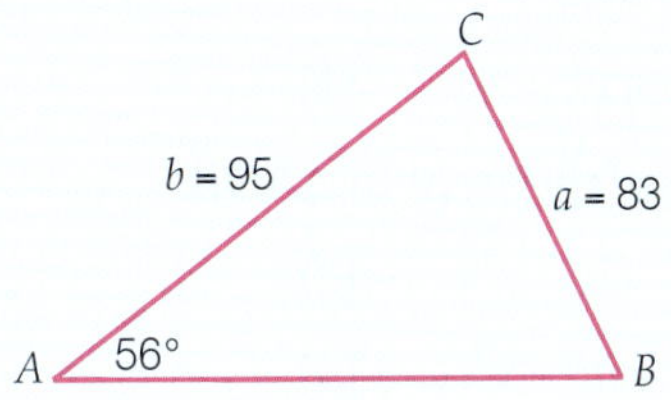 $\frac{\sin(A)}{a} = \frac{\sin(B)}{b}$
2 Substitute the information given.	$\frac{\sin(56°)}{83} = \frac{\sin(B)}{95}$
3 Rearrange the equation to make the unknown value the subject.	$B = \sin^{-1}\left(\frac{95 \times \sin(56)°}{83}\right)$
4 Use a calculator to evaluate the value of $\angle ABC$ to the required number of decimal places.	$\angle ABC = 71.60°$
5 Draw a separate labelled diagram for the obtuse-angled triangle.	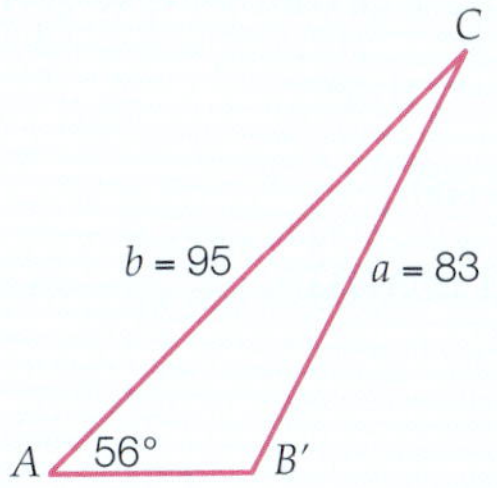
6 Find $\angle AB'C$ by subtracting the acute angle $\angle ABC$ found above from 180°.	$\angle AB'C = 180° - 71.60°$ $= 108.40°$ (2 d.p.)

10A

Cosine rule

To find the length c in a triangle using the sine rule you need to know $\angle C$, and one of the other angles $\angle A$ or $\angle B$ and its corresponding side length a or b.

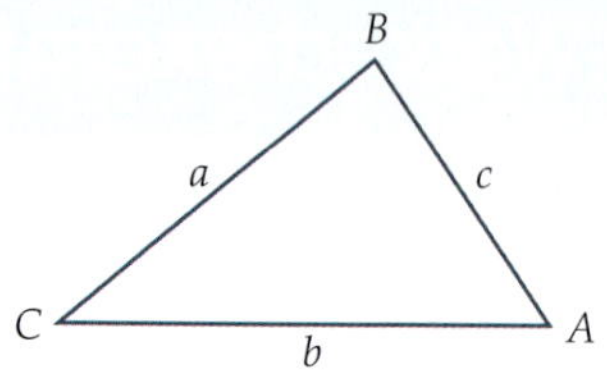

If only $\angle C$ is known and side lengths a and b, the cosine rule can be used.

To derive the formula, find expressions for c from the triangle on the right. Draw the perpendicular height, h, from B to O on side b.

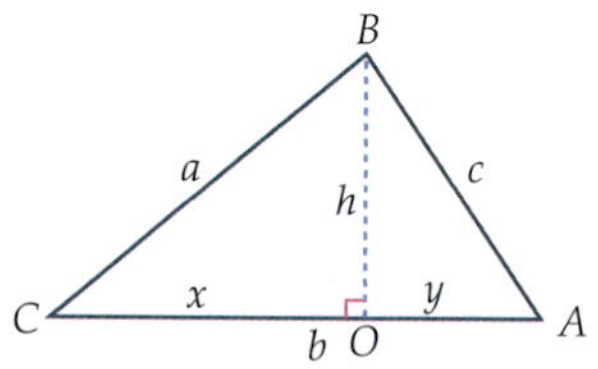

$b = x + y$

$y = b - x$

Squaring both sides:

$y^2 = b^2 - 2bx + x^2$ [1]

Using Pythagoras' theorem in ΔAOB:

$c^2 = h^2 + y^2$ [2]

In ΔCOB: $h^2 = a^2 - x^2$ [3]

Substituting for h^2 using equation [3] in equation [2]:

$c^2 = a^2 - x^2 + y^2$ [4]

Substituting for y^2 using equation [1] in equation [4]:

$c^2 = a^2 - x^2 + b^2 - 2bx + x^2$

$c^2 = a^2 + b^2 - 2bx$ [5]

In ΔCOB, $x = a\cos(C)$ [6]

Substituting for x using equation [6] in equation [5]:

$c^2 = a^2 + b^2 - 2ab\cos(C)$

This simplifies to Pythagoras' theorem when $C = 90°$.

Similarly, it can be shown that:

$b^2 = a^2 + c^2 - 2ac\cos(B)$

and $a^2 = b^2 + c^2 - 2bc\cos(A)$

The cosine rule

Depending on the angle to be used, the cosine rule can be written in three ways.

$c^2 = a^2 + b^2 - 2ab\cos(C)$

$b^2 = a^2 + c^2 - 2ac\cos(B)$

$a^2 = b^2 + c^2 - 2bc\cos(A)$

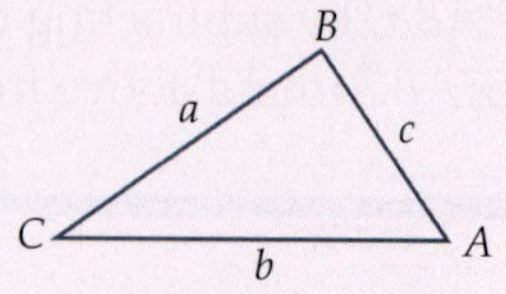

10A

Worked example 11

W.E. 11

Calculate the length of AB correct to 2 decimal places.

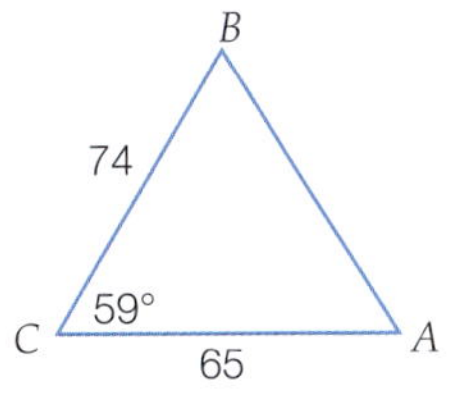

Thinking	Working
1 Draw and label the triangle with the information provided.	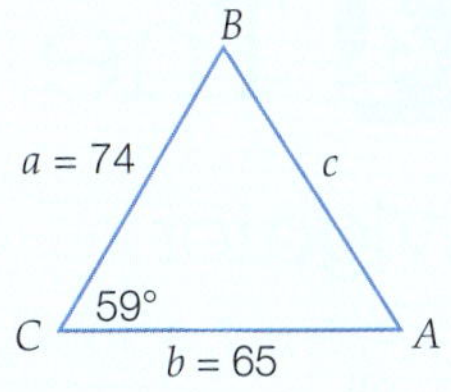
2 Determine the rule to be used. (Here, use the cosine rule.)	$c^2 = a^2 + b^2 - 2ab\cos(C)$
3 Substitute the information given into the rule. Do not evaluate c^2.	$c^2 = 74^2 + 65^2 - 2 \times 74 \times 65 \times \cos(59°)$ $c^2 = 9701 - 9620\cos(59°)$
4 Calculate the value to the required number of decimal places.	$c = 68.89$ units (2 d.p.)

Worked example 12

W.E. 12

Calculate the value of $\angle ACB$ in degrees correct to 2 decimal places.

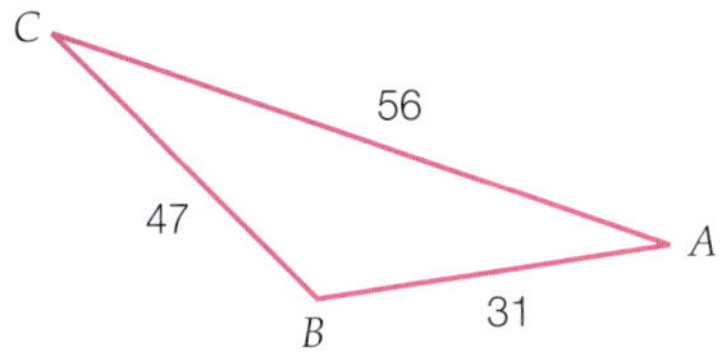

Thinking	Working
1 Draw and label the triangle with the information provided.	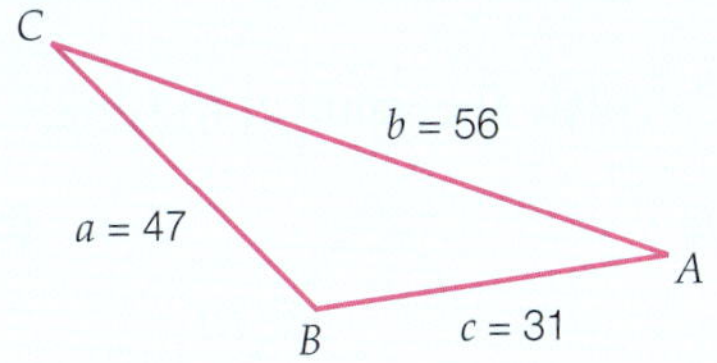
2 Determine the rule to be used. (Here, use the cosine rule.)	$c^2 = a^2 + b^2 - 2ab\cos(C)$
3 Substitute the information given into the rule.	$31^2 = 47^2 + 56^2 - 2 \times 47 \times 56 \times \cos(C)$

4 Simplify and make $\cos(C)$ the subject. Do not calculate the $\cos(C)$ value.

$$961 = 5345 - 5264 \times \cos(C)$$
$$-4384 = -5264 \times \cos(C)$$
$$\cos(C) = \frac{-4384}{-5264}$$

5 Solve to find the angle in the required format using the inverse of $\cos(C)$.

$$\angle ACB = \cos^{-1}\left(\frac{-4384}{-5264}\right)$$
$$\angle ACB = 33.61°$$

7.4 The sine and cosine rules

Navigator

Answers p. 815

1, 2, 3, 4, 5, 6, 7, 8, 10, 11, 12, 16, 18, 20	1, 2, 3, 4, 5, 6, 7, 8, 9, 10, 11, 12, 13, 16, 18, 19, 20	1 (a–b), 2, 3 (a–b), 4, 5, 6 (a–b), 7 (a–b), 8, 9, 11, 12, 13, 14, 15, 16, 17, 18, 19, 20, 21

Fluency

W.E. 8

1 Calculate the value of x in each of the following.

(a)

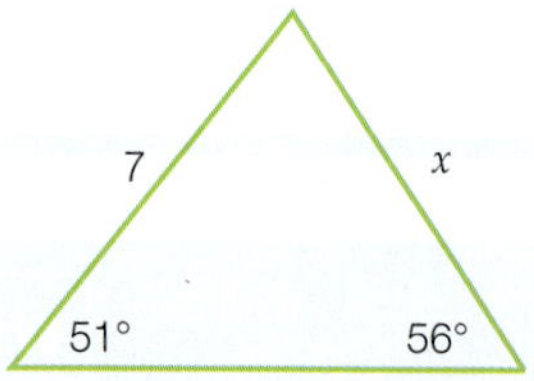

(b)

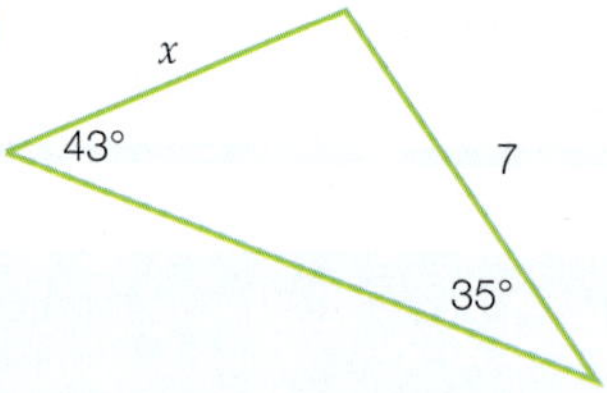

(c)

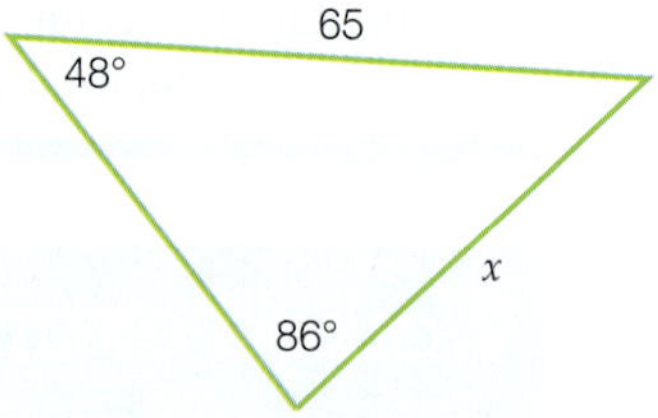

2 For each of the following triangles, find the missing angle θ and use this to calculate the length of side x. Give your answers correct to 2 decimal places.

(a)

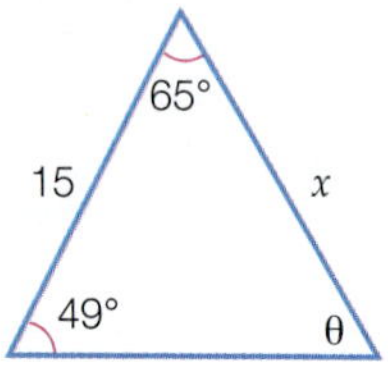

(b)

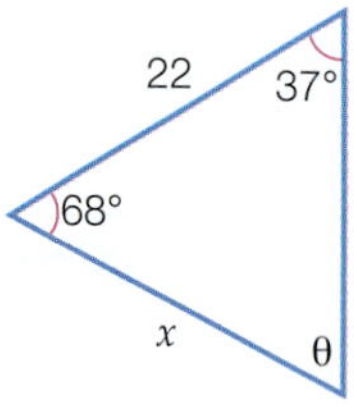

(c)

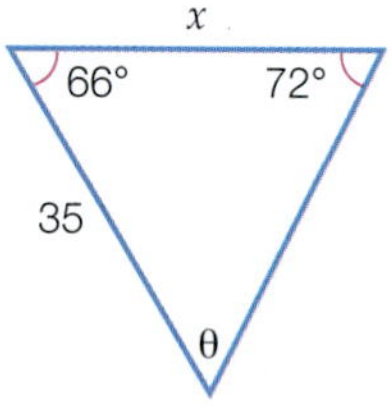

W.E. 9

3 Calculate the value of the angle in each of the following.

(a)

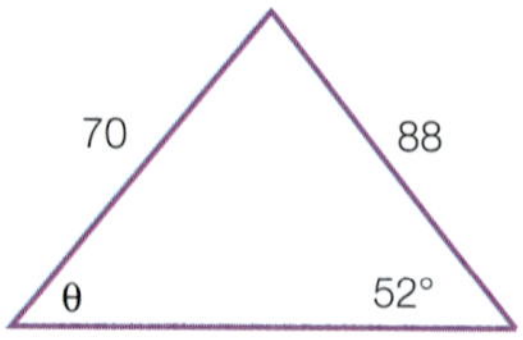

(b)

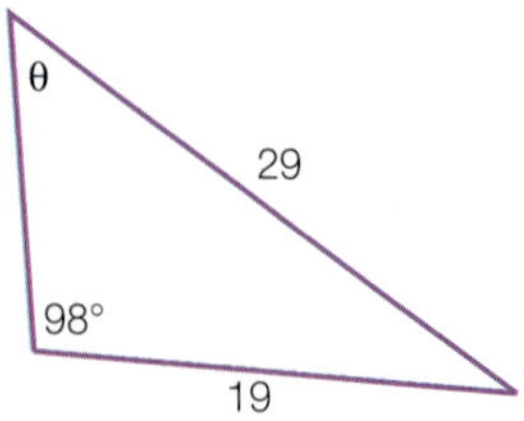

(c)

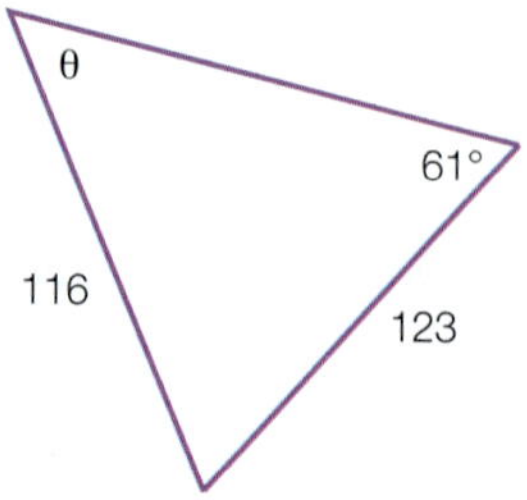

10A

4 Calculate the value of θ. Give your answers correct to the nearest degree.
Hint: You may need to calculate some of the other missing values first.

(a)
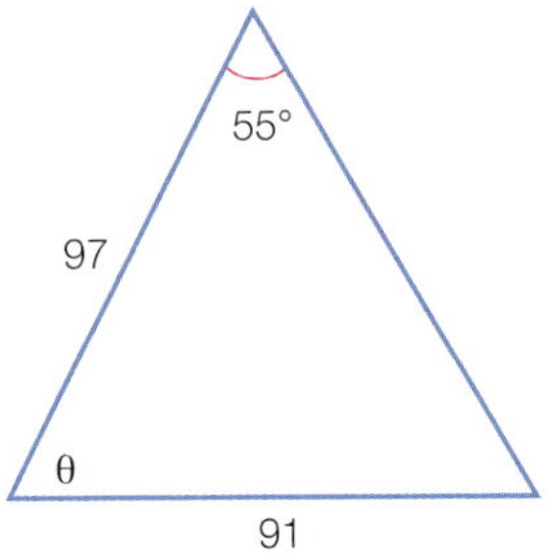

(b)
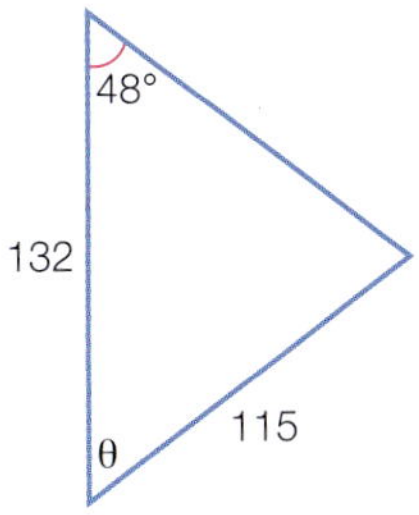

(c)
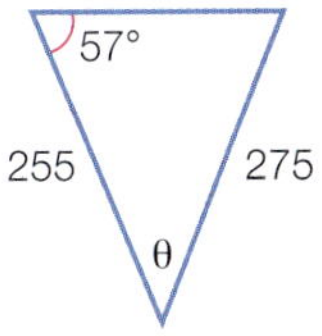

5 Calculate the values of $\angle ABC$ and $\angle AB'C$ correct to 2 decimal places.

W.E. 10

(a)
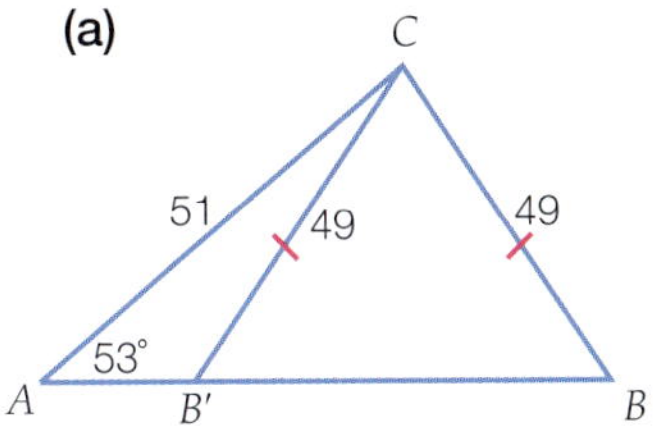

(b)
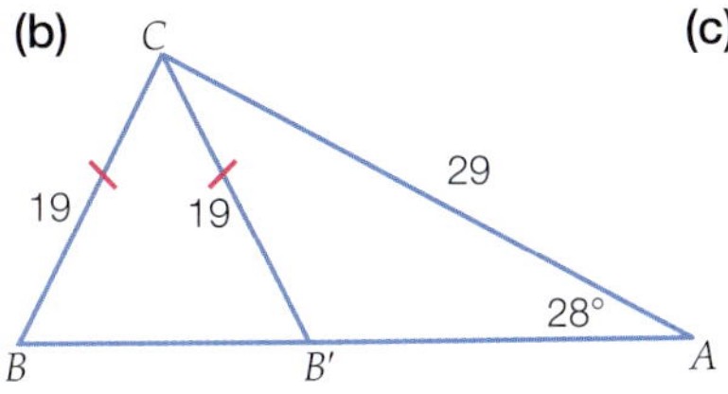

(c)
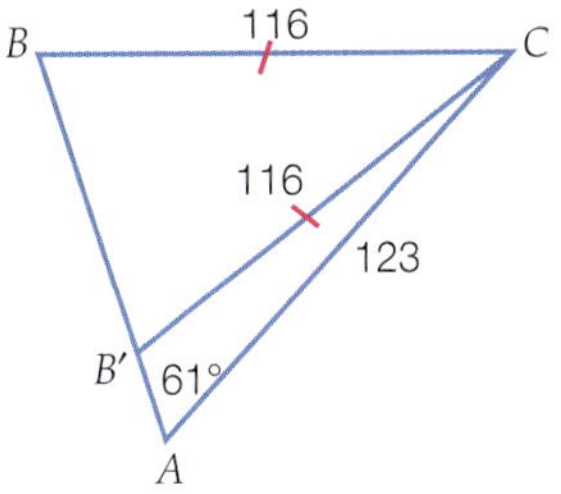

6 Calculate the length AB correct to 2 decimal places.

W.E. 11

(a)
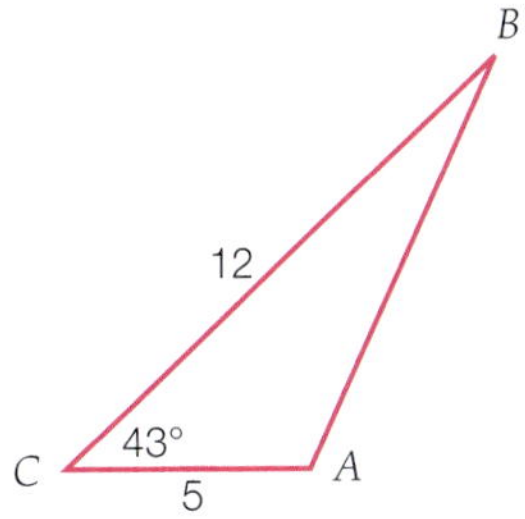

(b)
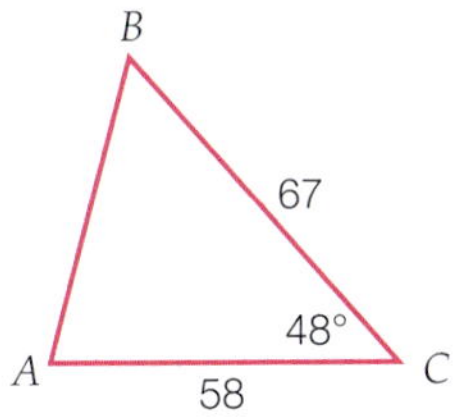

(c)
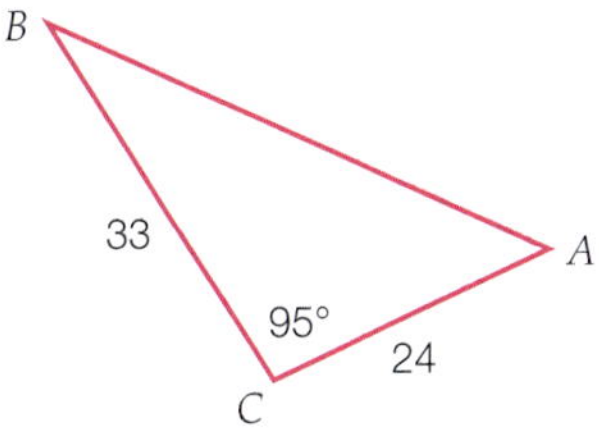

7 Calculate the value of angle $\angle ACB$ correct to 2 decimal places.

W.E. 12

(a)
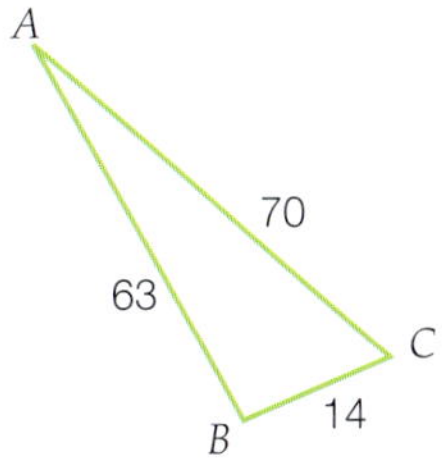

(b)
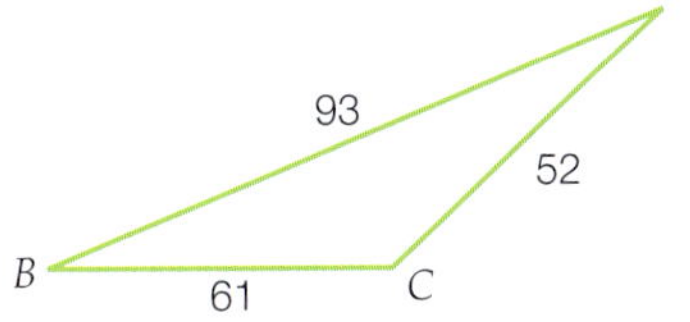

(c)
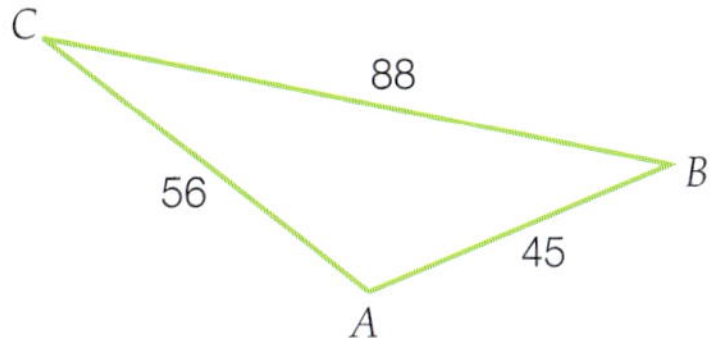

8 Draw a diagram to represent each of the following triangles, and then find all of the missing lengths and angles. Write all lengths correct to 1 decimal place and all angles correct to the nearest angle.

(a) ΔABC has angle $A = 52°$, angle $C = 55°$ and side $AB = 25$.

(b) ΔABC has angle $A = 62°$, side $AB = 102$ and side $AC = 115$.

Understanding

9 Use the cosine rule to show that $d = 2c$.

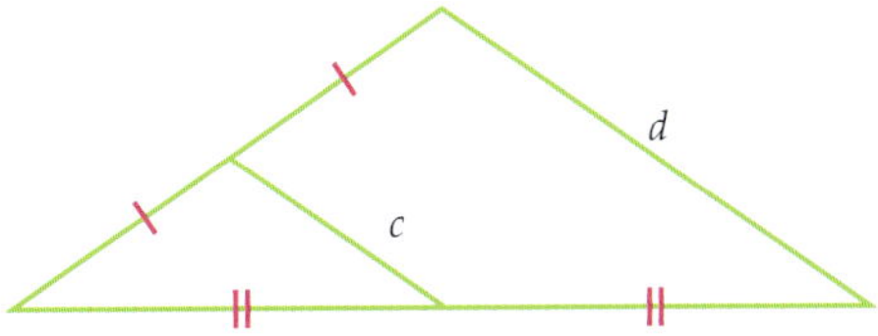

10 Calculate the value of θ, giving your answer as a decimal correct to 2 decimal places.

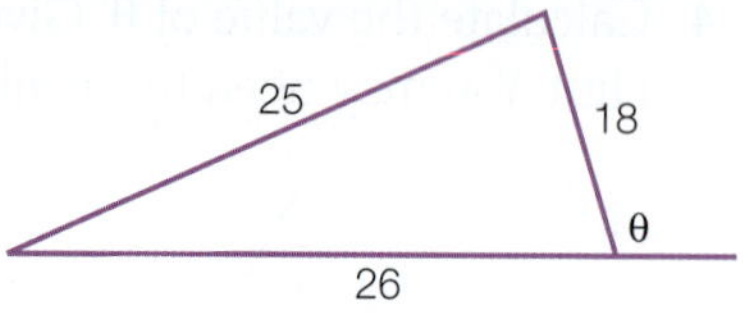

11 Calculate the size of angle θ to the nearest degree.

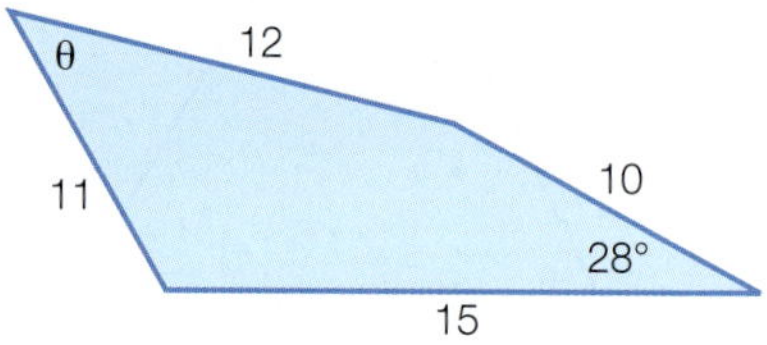

12 Calculate the value of x and round to 2 decimal places.

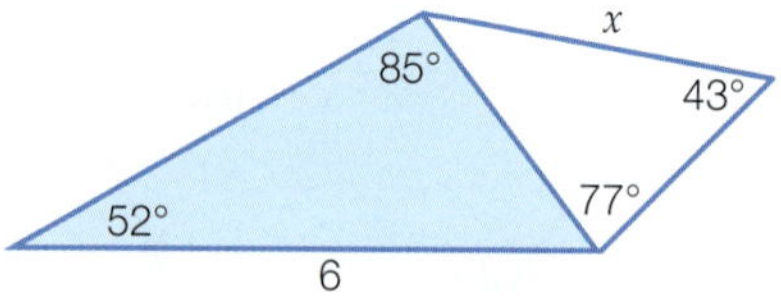

13 Show that $x = \dfrac{y\sin(\theta - \phi)}{\sin(\phi)}$.

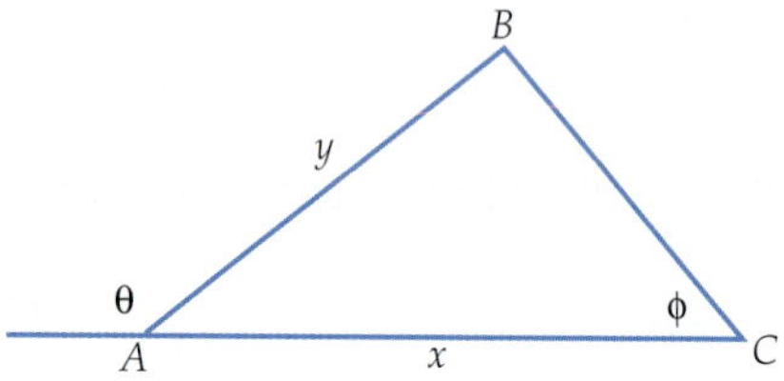

14 For the triangle shown, which is the correct statement?

A $a^2 = b^2(2 - \cos(\theta))$

B $b^2 = 2a^2(1 + \cos(\theta))$

C $a^2 = b^2(2 + \cos(\theta))$

D $b^2 = 2a^2(1 - \cos(\theta))$

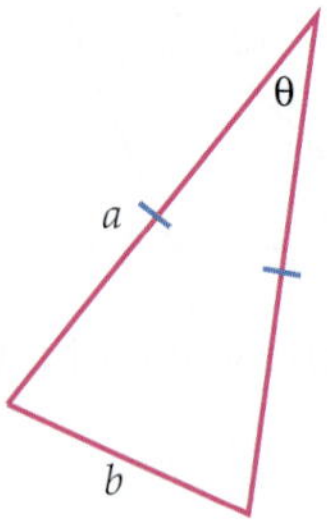

15 Calculate the value of x rounded to the nearest whole number.

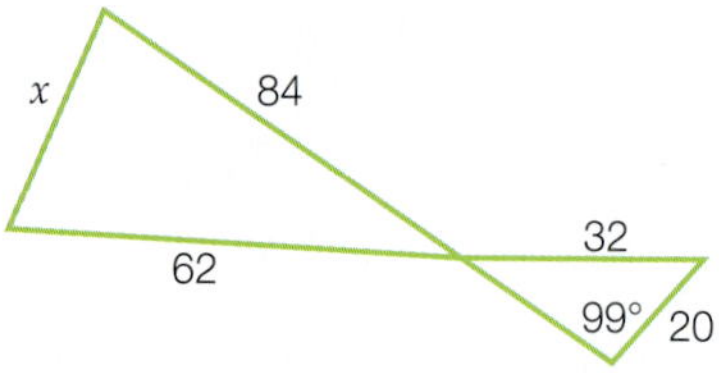

16 The value of y is:

A $7x^2$ **B** $\sqrt{7x}$ **C** $\sqrt{7x^2}$ **D** $49x$

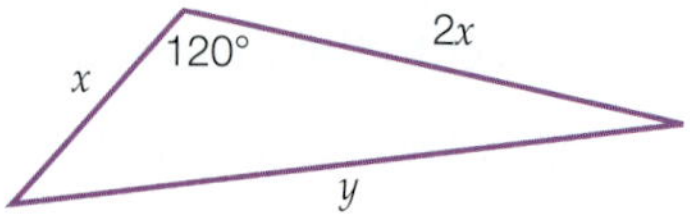

17 Show that $\sin(D) = \dfrac{EC}{AC}$.

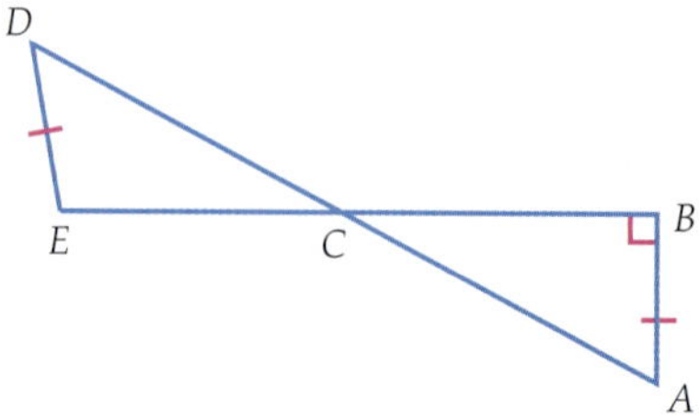

10A

Reasoning

18 Show that the sine rule reduces to $\sin(\theta) = \frac{\text{opposite side}}{\text{hypotenuse}}$ when applied to a right-angled triangle.

19 Apply the cosine rule to an equilateral triangle to show that each interior angle is 60°.

Open-ended

20 **(a)** Using the cosine formula $c^2 = a^2 + b^2 - 2ab\cos(C)$, show that if $a = b$ and $c = ka$, then $\cos(C) = \frac{2 - k^2}{2}$.

(b) State two values for k and use them to evaluate $\cos(\theta)$, rounded to 2 decimal places.

(c) What are the minimum and maximum possible values for k?

21 Determine the angle θ in the following triangle if sides b and c are positive integer multiples of a.

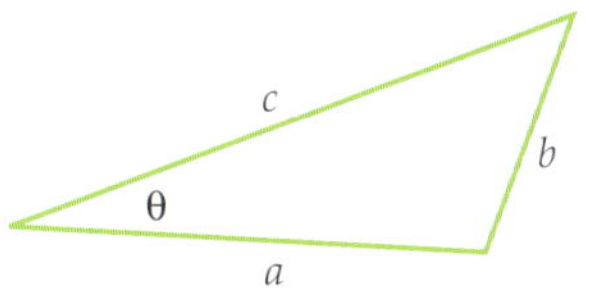

Problem solving

Trig chimps

Equipment needed: calculator

For all calculations round your answers to 2 decimal places.

A certain group of monkeys are renowned for their longevity.

The ages of three chimps, Syn, Coss and Ten satisfy the following conditions.

- The sum of their ages is 158.
- Coss's age is a multiple of 13 and is an odd number.
- Expressed in degrees, Ten's age is in quadrant 2.
- Expressed in degrees, Syn's age and Coss's age are in quadrant 1.
- The tangent of Ten's age is 15.08 less than the cosine of Coss's age.
- The two digits after the decimal point in the cosine of Coss's age are consecutive digits.

What are the ages of the three chimps?

Strategy options

- Guess and check.
- Test all possible combinations.

10A

Exploration CAS

Equipment required: TI-Nspire CAS or Casio ClassPad CAS

Demonstrating the sine and cosine rules

The sine rule states $\dfrac{a}{\sin(A)} = \dfrac{b}{\sin(B)} = \dfrac{c}{\sin(C)}$ for any triangle. In this notation, the angle A is opposite the side labelled a, and so on. You can use your CAS to test this rule.

Using TI-Nspire CAS

Add **Geometry** to your document. Then select **menu** > **Settings...** and set the **Graphs & Geometry Settings** as shown.

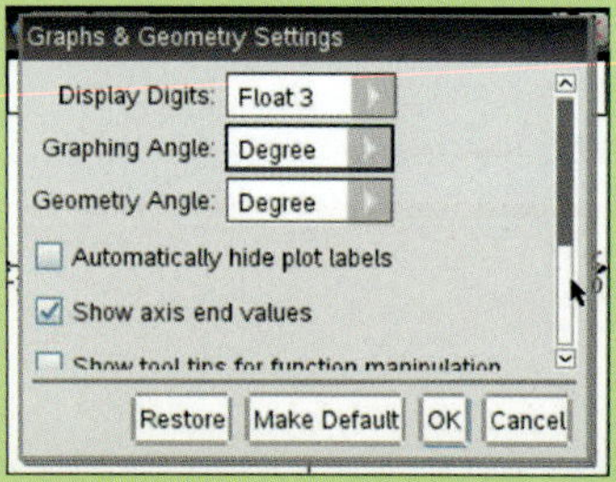

Select **menu** > **Shapes** > **Triangle**, then move the cursor to select the points where you want each vertex of the triangle to appear. Don't make your triangle too small.

Press **esc** to stop drawing.

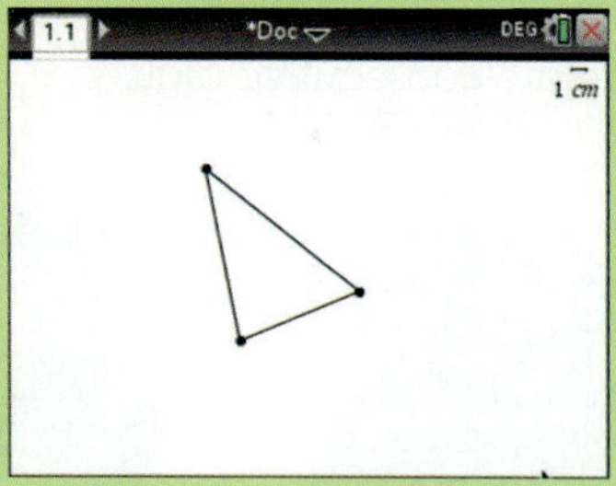

Using Casio ClassPad CAS

From the menu, select **Geometry**. Select ⚙ > **Geometry Format** and change the settings as shown.

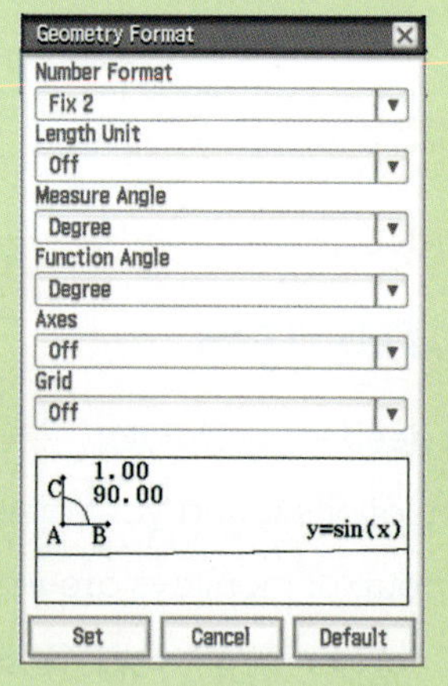

Select the scalene triangle icon and move the cursor to choose where you want the shape to appear. Don't make your triangle too small.

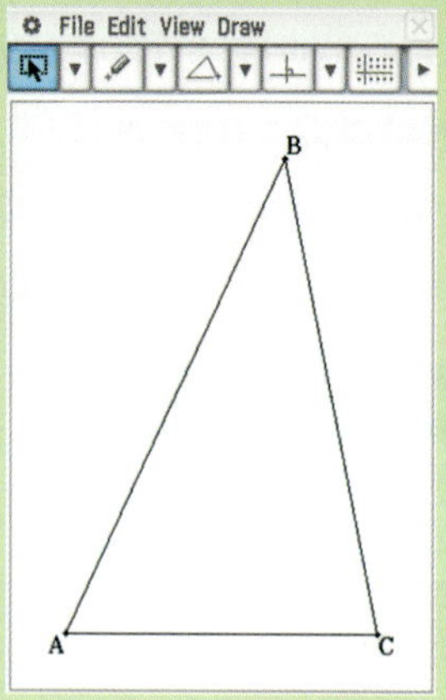

To test the sine rule, you need to measure any two angles and the corresponding two sides.

Using TI-Nspire CAS	Casio ClassPad CAS
To measure an angle select **menu** > **Measurement** > **Angle** and then select the three vertices, in order, that form one of the angles. Repeat this process to measure the second angle, then press **esc** to stop measuring angles. The measurements that appear are text objects that you can select and move around the screen. You can also edit the text by selecting **menu** > **Actions** > **Text** and selecting the text. Edit the text to label the angles α and β as shown. (You might have different angles to those shown here.) 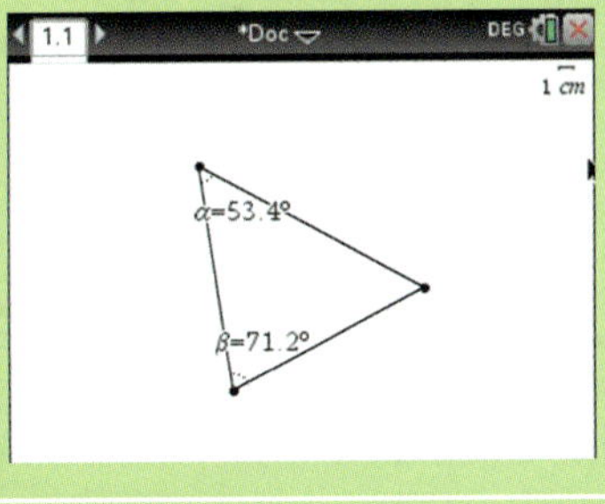 	To measure an angle, select the two sides adjacent to the angle and then select **Draw** > **Attached Angle**. The angle value will appear. (You might have different angles to those shown here.) 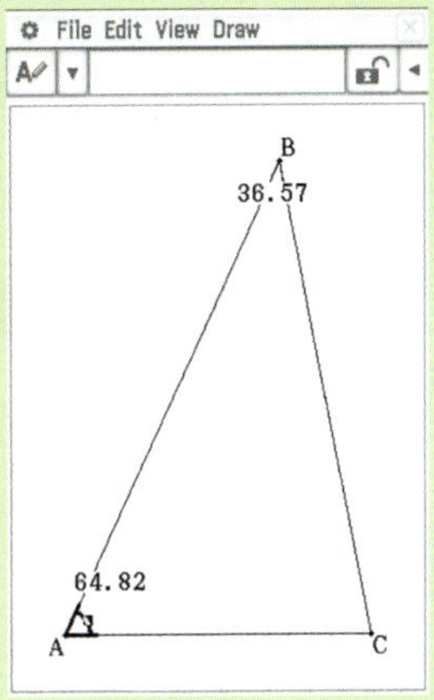
To measure a side opposite an angle, select **menu** > **Measurement** > **Length** and then select the two vertices that form the side. Edit the text as before to label the sides a (opposite the angle α) and b (opposite the angle β). 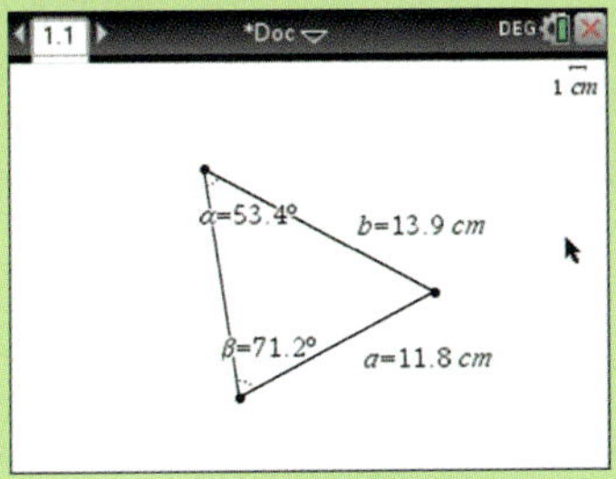 	To measure a side, select the side and then select **Draw** > **Measurement** > **Length**. The length value will appear. You may need to drag the label down so that it is closer to its side. Measure the lengths of the sides opposite the measured angles. 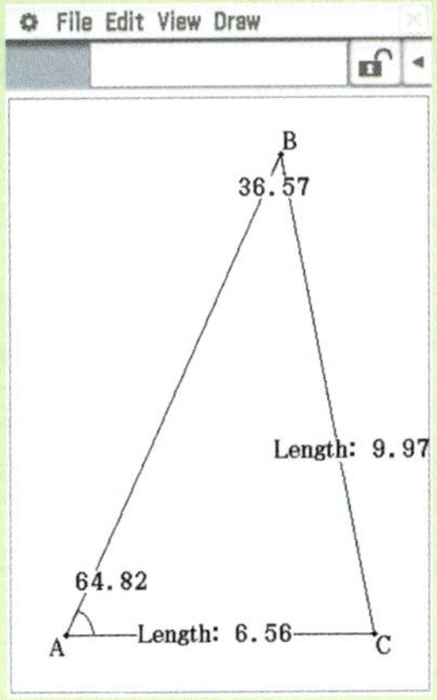

10A

Using TI-Nspire CAS

To calculate the ratio values for the sine rule, first select **menu** > **Actions** > **Text** and enter the expression $\frac{a}{\sin(\alpha)}$. Then, select **menu** > **Actions** > **Calculate** and select this expression. You will be prompted to choose the values for ***a*** and then for α, which you can do by selecting their labels on the screen. After you do this, the calculated ratio will appear.

Repeat this process to find $\frac{b}{\sin(\beta)}$.

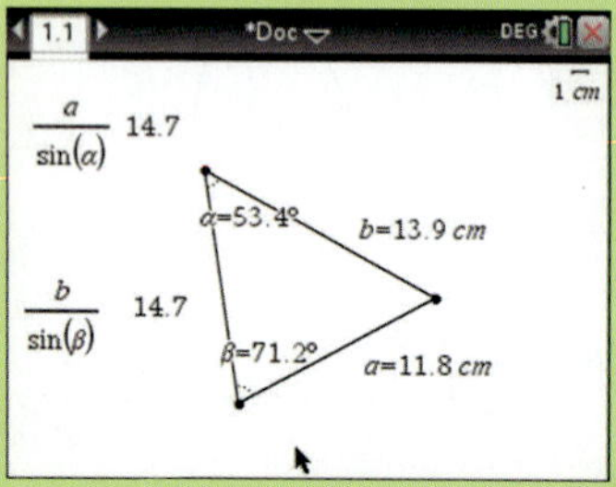

Casio ClassPad CAS

To calculate the ratio values for the sine rule, select **Draw** > **Expression**, then select the right arrow in the top right corner of the screen. This will display the box where you can enter a calculation.

Each of your side and angle measurements should now have a number label attached. To enter a measurement into the sine rule calculation **a/sin(*α*)**, select the measurement's attached number label. This will appear in the calculation as an **@** symbol with the number label, as shown.

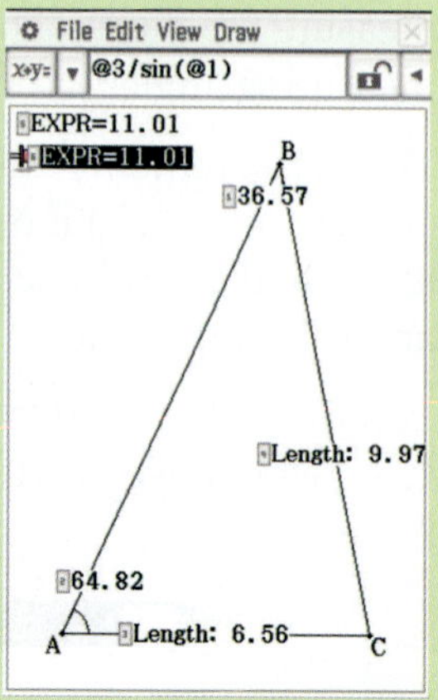

Repeat this calculation for the different side lengths and angles.

When you select **menu** > **Actions** > **Calculate**, you can choose different values for ***a*** and α or for ***b*** and β. As long as the chosen side is opposite the chosen angle, the calculated value should be the same, showing that the sine rule is true for this triangle.

If you select a vertex of the triangle and move it around, then the calculated ratios should stay equal to each other even though the triangle values change. This demonstrates that the sine rule is still true.

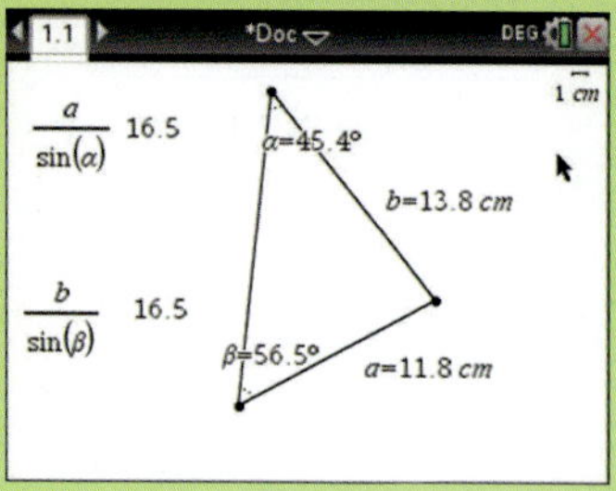

As long as the chosen side is opposite the chosen angle, then the calculated value of the sine rule ratio should be the same. This shows that the sine rule is true for this triangle.

If you select a vertex of the triangle and move it around, then the calculated ratios should stay equal to each other even though the triangle values change. This demonstrates that the sine rule is still true.

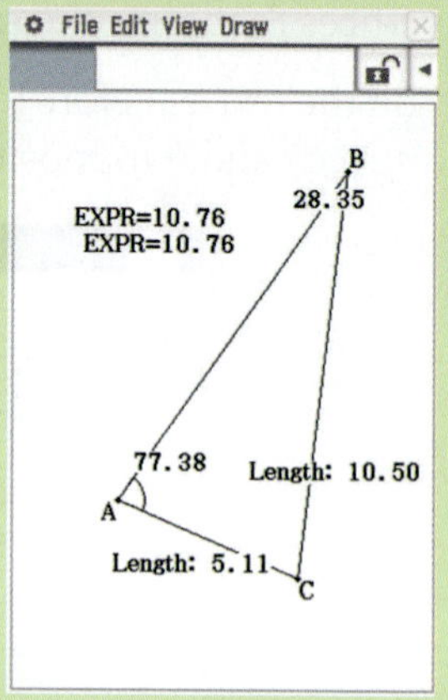

Now that you have demonstrated the sine rule, use your CAS to demonstrate the cosine rule:

$c^2 = a^2 + b^2 - 2ab\cos(C)$

Applications of the sine and cosine rules

7.5

The formulas from the previous section can be applied to a range of practical situations. In surveying, for example, a total station is an electronic instrument used to measure angles of rotation and distances. This sophisticated equipment has limitations. It needs a clear line of sight to the object, which must have reflective properties for the reflected signal to be recorded.

Worked example 13

W.E. 13

A tree growing at an angle of 104° to the horizontal is in danger of falling. It has been supported by a rope tied to the tree 3.7 m along the tree trunk and pegged to the ground at an angle of 47° as shown in the diagram. Find the length of rope used, correct to 1 decimal place.

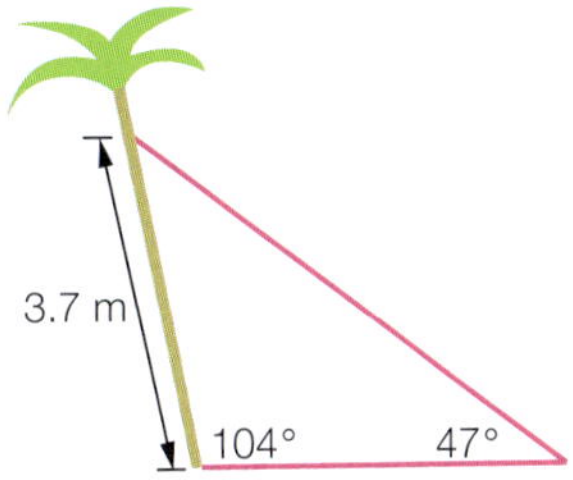

10A

Thinking	Working
1 Draw a triangle and show the information on the diagram.	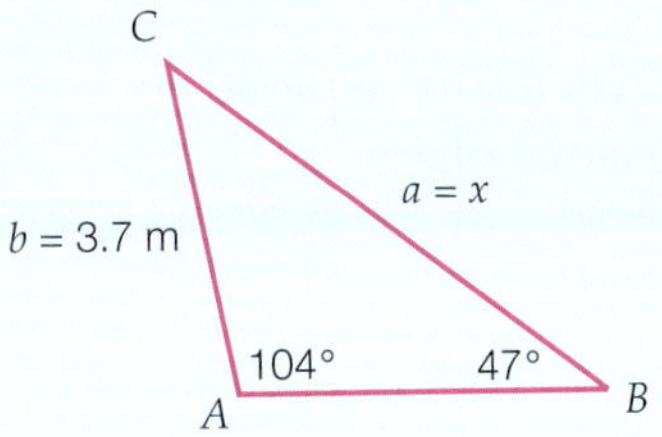
2 Choose the expression to be used and the values to be used.	$\frac{a}{\sin(A)} = \frac{b}{\sin(B)}$ $a = x, \angle A = 104,$ $b = 3.7, \angle B = 47$
3 Substitute the information given.	$\frac{x}{\sin(104°)} = \frac{3.7}{\sin(47°)}$
4 Rearrange to make the unknown value the subject.	$x = \frac{3.7 \times \sin(104°)}{\sin(47°)}$
5 Calculate the value to the required number of decimal places.	$x = 4.9$
6 State the answer.	The length of rope is 4.9 m.

Worked example 14

W.E. 14

A walkway is to be constructed across a pond from A to C. There is an existing walkway of length 16 m from A to B. A rope measuring 23 m is attached to point B and point C. The angle between the rope and the existing walkway is 42°. Calculate the length of the new walkway correct to 2 decimal places.

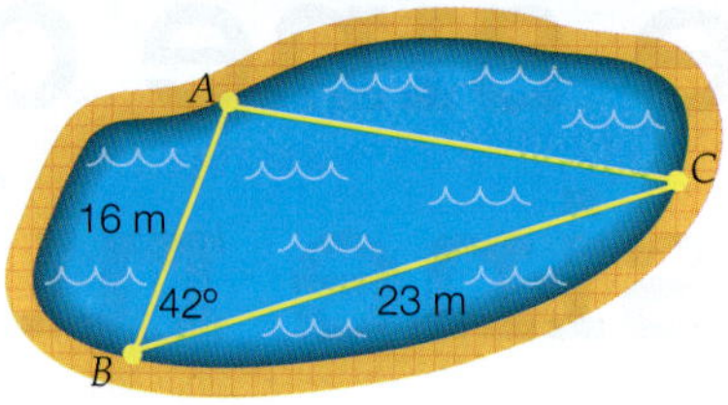

Thinking	Working
1 Draw a diagram showing the information and identify the unknown.	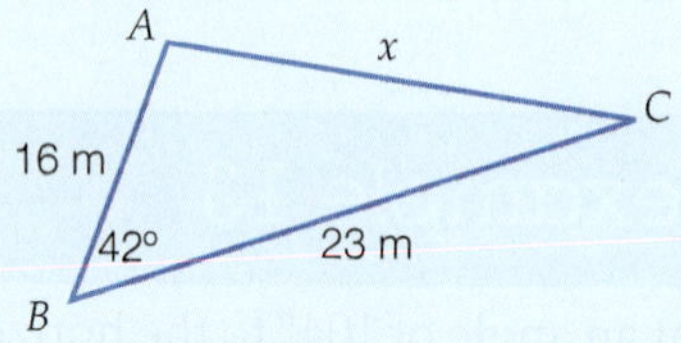
2 Identify the given information and any unknowns.	$b = x$, $a = 23$, $c = 16$, $B = 42°$
3 Decide which formula to use. (Here, use the cosine rule.)	$b^2 = a^2 + c^2 - 2ac\cos(B)$
4 Substitute the information given.	$x^2 = 23^2 + 16^2 - 2 \times 23 \times 16 \times \cos(42°)$
5 Use a calculator to evaluate the unknown value. (Do not round.)	$x^2 = 238.045\,408\,4$ $x = 15.428\,720\,25$
6 State the answer to the required number of decimal places.	The required length for the walkway is 15.43 m.

10A

Worked example 15

W.E. 15

A helicopter leaves base and flies for 12 km on a bearing of 332° T. It then heads due south for a certain distance before flying back to base on a bearing of 45° T. What is the total flight distance? Round your answer to 2 decimal places.

Thinking	Working
1 Draw a diagram showing the information and identify the unknown.	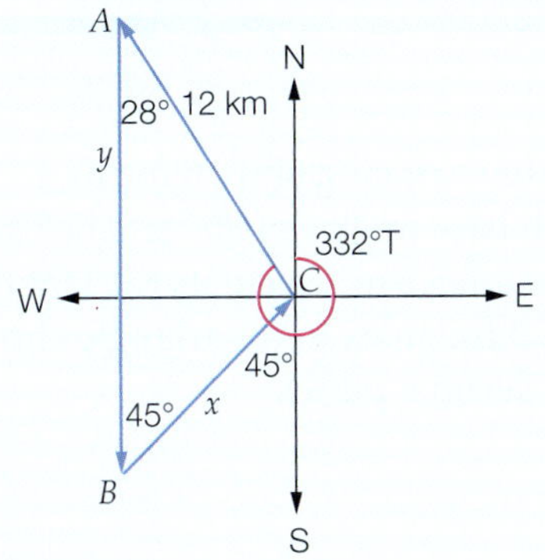
2 Use the true bearing and geometry to find triangle angles. (Here, ∠A is found using alternate angles and ∠C by 180° – (∠A + ∠B).)	$360 - 332 = 28°$ $\angle A = 28°$ $\angle B = 45°$ $\angle C = 180 - (28 + 45) = 107°$
3 Identify given information and any unknowns.	$a = x$ $b = 12$ $c = y$
4 Decide which formula to use. (Here, use the sine rule.)	$\frac{a}{\sin(A)} = \frac{b}{\sin(B)} = \frac{c}{\sin(C)}$
5 Substitute the information given to find the first unknown.	$\frac{x}{\sin(28°)} = \frac{12}{\sin(45°)}$
6 Rearrange to make the unknown value the subject.	$x = \frac{12 \times \sin(28°)}{\sin(45°)}$
7 Use a calculator to evaluate the unknown value. (Do not round.)	$x = 7.967\,196\,615$
8 Substitute the information given to find the second unknown.	$\frac{y}{\sin(107°)} = \frac{12}{\sin(45°)}$
9 Rearrange to make the unknown the subject.	$y = \frac{12 \times \sin(107°)}{\sin(45°)}$
10 Use a calculator to evaluate the unknown value. (Do not round.)	$y = 16.229\,029\,87$
11 Calculate the perimeter using the unrounded values.	Perimeter $= 12 + 16.229\,029\,87 + 7.967\,196\,615$ $= 36.196\,226\,49$
12 State the answer to the required number of decimal places.	The total flight is 36.20 km.

10A

7.5 Applications of the sine and cosine rules

Navigator

Answers p. 815

1, 2, 3, 4, 5, 6, 7, 8, 10, 12, 13, 16, 18	1, 2, 3, 4, 5, 6, 7, 8, 10, 11, 12, 13, 16, 17, 18	1, 4, 5, 6, 7, 8, 9, 11, 12, 13, 14, 15, 16, 17, 18

Fluency

W.E. 13

1 Part of the ship's rigging shown forms a triangle with base angles of 80° and 66° and rigging of length 5.3 m. Calculate the length of the section of the mast in the triangle shown correct to 2 decimal places.

W.E. 14

2 A walkway is to be constructed across a pond from *A* to *C*. There is an existing walkway of length 21 m from *A* to *B*. A rope measuring 13 m is attached to point *B* and point *C*. The angle between the rope and the existing walkway is 148°. Calculate the length of the new walkway correct to 2 decimal places.

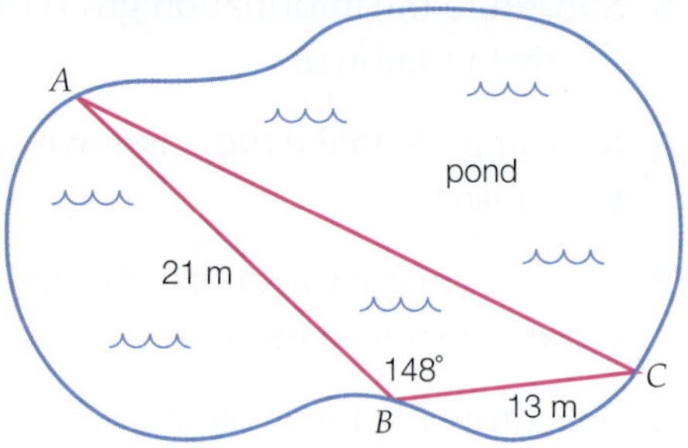

3 A pipe is laid across a section of a water dam from *A* to *B*. A 100 m rope is attached from point *A* to point *C* and the angles between the rope and *AB*, and the rope and *CB* are measured. If ∠*A* is 10° and ∠*C* is 20°, what length of pipe is required? State the answer to the nearest metre.

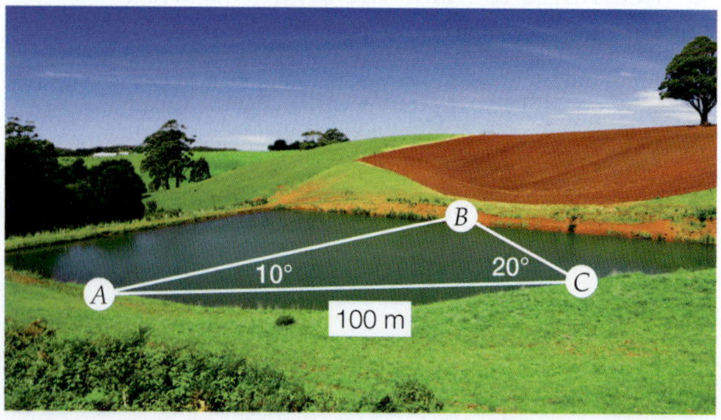

4 The fins of this classic car are triangular with side lengths of 26 cm, 39 cm and 52 cm as shown. For aerodynamic stability, it is recommended that the angle θ be less than 105°. Is this true for this car?

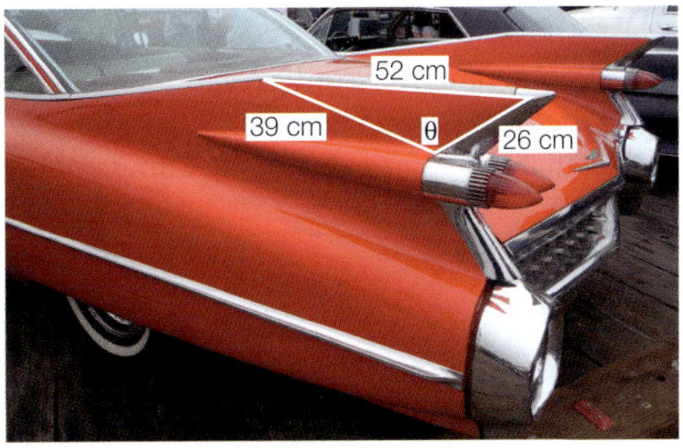

5 A children's slide is of length 3.24 m and makes an angle of 35° with the ground. If the ladder is 2.56 m, what angle θ does it make with the ground? Round your answer to 2 decimal places.

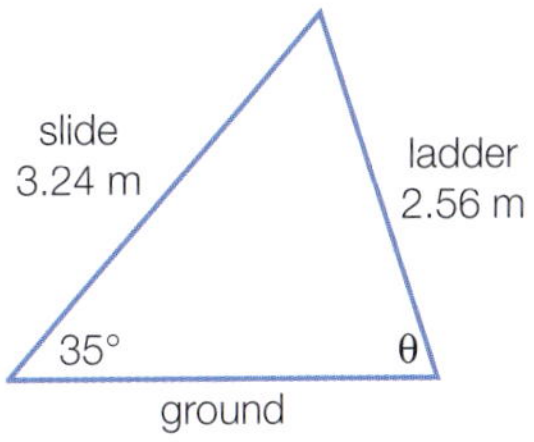

6 A ski lift is to be constructed at point *B* on a mountain. A surveyor recorded the distance from *A* to the point *C*, the top of the mountain, as 385 m and the distance from *A* to *B* as 316 m. If the angle between *AB* and *AC* is 41°, find the distance between *B* and the top of the mountain to the nearest metre.

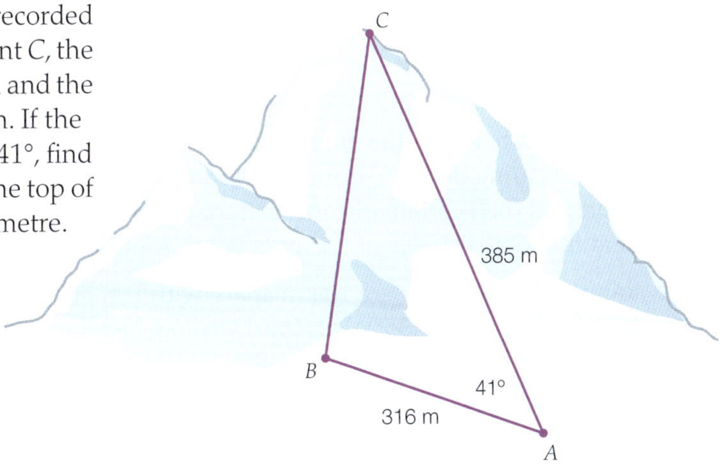

7 To the nearest cm, the length of the chain stay in the bicycle frame is:

A 44 cm

B 47 cm

C 51 cm

D 62 cm

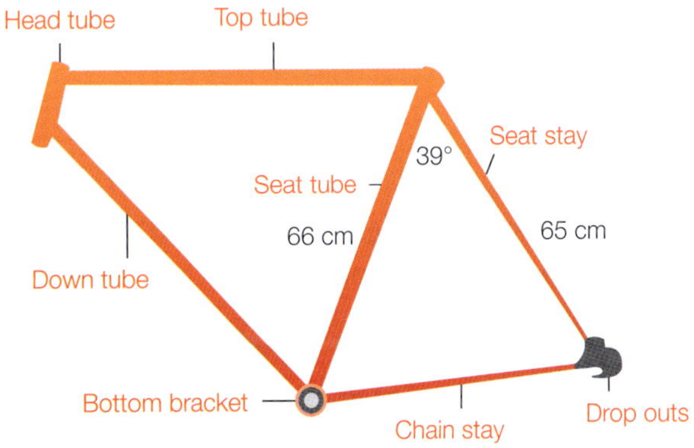

Understanding

8 A boomerang is mounted on a triangular frame as shown. If the two shorter sides of the frame are 15 cm and 17 cm respectively and the included angle is 134°, calculate the perimeter of the frame, rounded to 2 decimal places.

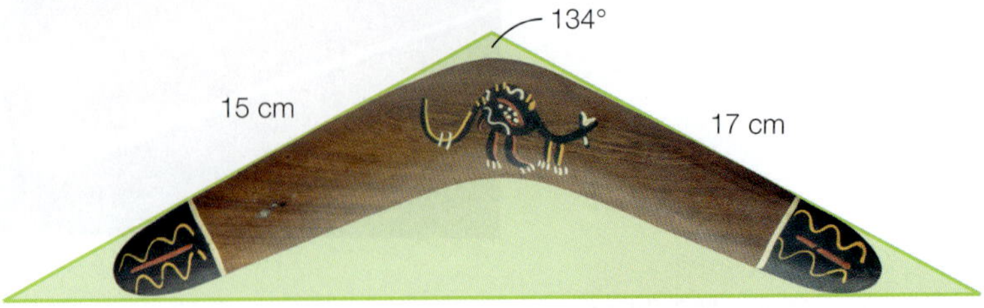

9 Two hot-air balloons are stationary at the same height above the ground. They are secured by ropes as shown.

(a) Calculate the horizontal distance between the base of each balloon, correct to 2 decimal places.

(b) What is the total length of rope required to secure the balloons?

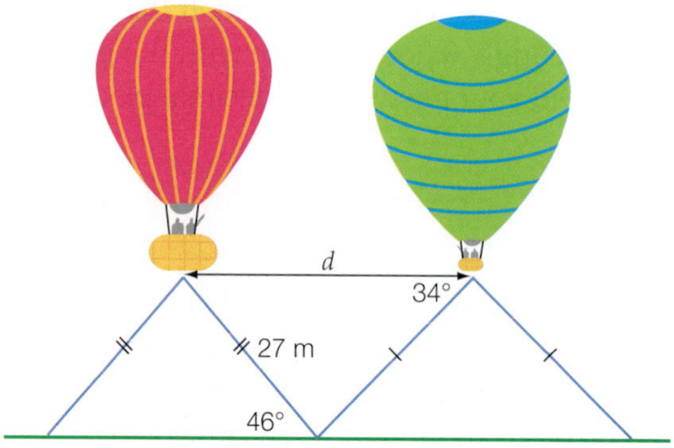

W.E. 15

10A

10 Postman Pete leaves the post office and makes deliveries for 2.5 km on Aiden Rd, which is on a bearing of 072°T from his starting point. Pete then heads due north on Sari Boulevard for a certain distance before returning to the post office on a bearing of 210°T. Find the total distance of Postman Pete's route, rounded to 2 decimal places.

11 Rangers Tim and Zoe are 10 km apart, with Tim due west of Zoe. They notice a fire that is N50°E of Tim and N35°W of Zoe. Decide which ranger is closer to the fire by finding the distance each ranger is from the fire.

Hint

Be sure to find all possible solutions.

12 The dial on a sundial casts a shadow as shown. The dial is 24 cm long, the angle between the dial and the base is 38° and there is 15 cm between the top of the dial and the end of the shadow. Find the length of the shadow correct to 2 decimal places.

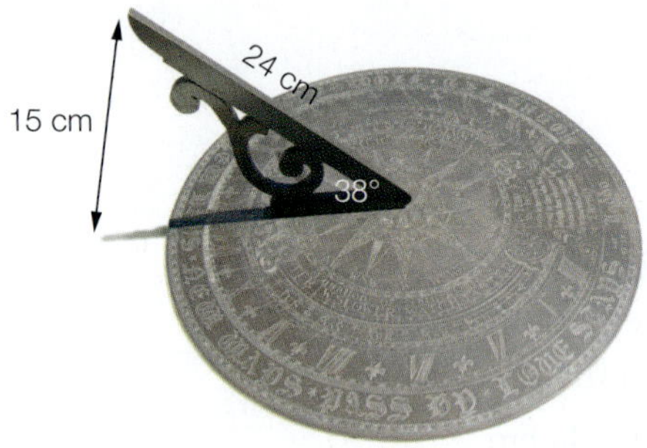

13 A new highway is being built across a paddock, from point A to point D as shown on the diagram. This highway currently runs in a straight line from A to C through B, 3050 m from A. The angle between AB and BD is 115°. The distance from B to C is a further 2160 m and the angle between BC and CD is 68°. Find the length of the new highway AD to the nearest metre.

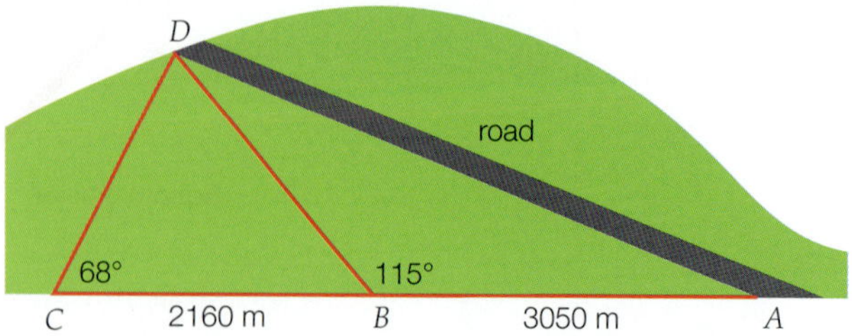

Reasoning

14 Ken and Sofia set out in different directions from their campsite. After Ken walks 2150 m on a bearing of 316° T and Sofia walks for 2000 m, the distance between them is 2696 m. What are the two possible bearings that Sofia could have used? Give your answer correct to the nearest degree.

15 The longest diagonal of a 50 cent coin is 31.65 mm. Calculate the length AB correct to 2 decimal places.

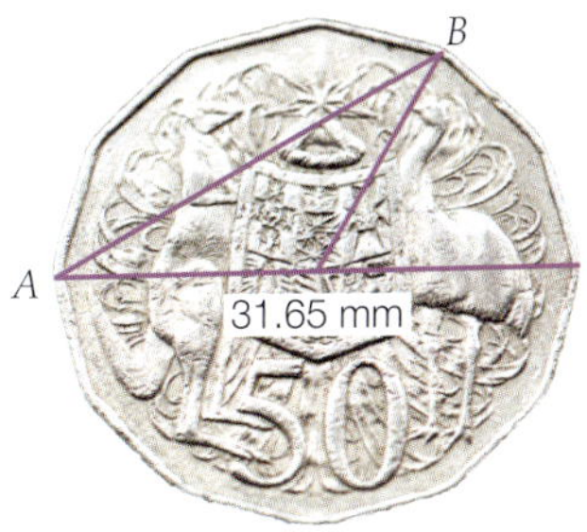

Hint

A 50c coin has the shape of a regular dodecagon (a 12-sided polygon).

16 A pendant is in the shape of a scalene triangle and its two smaller interior angles are 25° and 28°. If the longest side is of length 4 cm, the lengths of the two other sides (in cm) correct to 2 decimal places are:

A 2.26, 2.35 **B** 2.12, 2.35 **C** 2.12, 2.18 **D** 2.11, 2.35

Open-ended

17 For maximum strength and stability, the angle θ between the support wires in the hang glider must be no more than 40° and the distance x between the wheels must be between 80 cm and 1.2 m. Determine two sets of allowable values for θ and x.

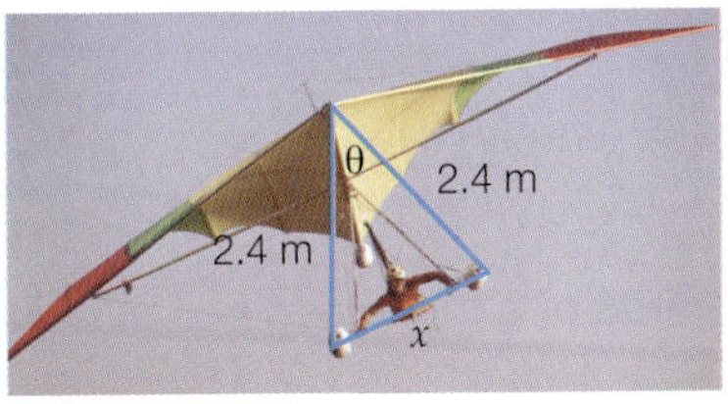

18 The length y of one of the bridge's support cables is less than 40 m and the angle θ is in the interval $15° < \theta < 75°$. Choose two reasonable values for θ, then find possible values for the lengths x and y as shown in the following diagram. Make sure that your values are within the specified ranges.

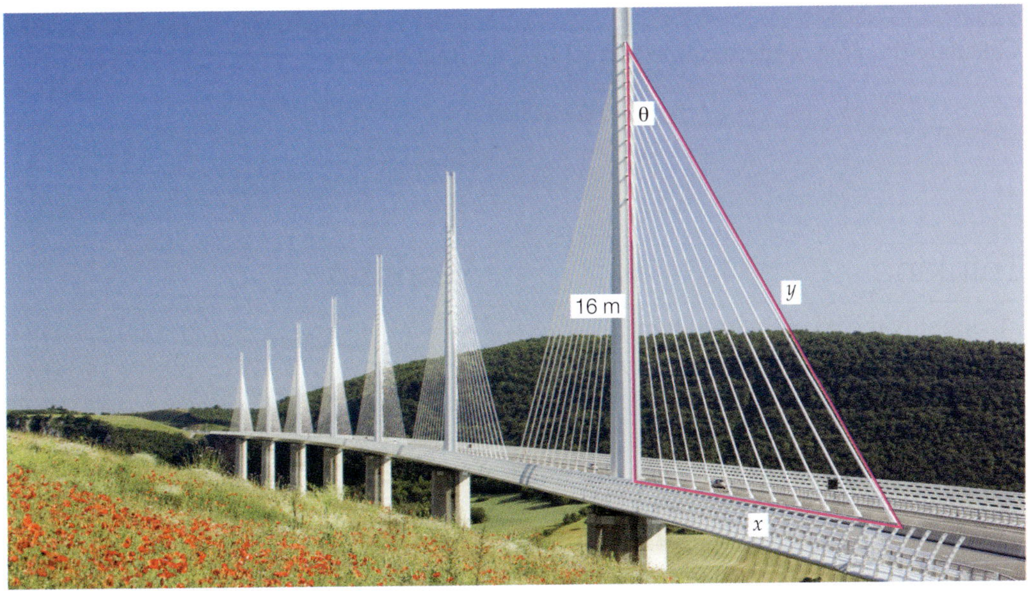

10A

7.6 Areas of triangles using trigonometry

The area of any triangle can be found using the formula $A = \frac{1}{2}bh$ where b is the base and h is the perpendicular height.

If h is not known but two sides and the angle included between them are known, a new formula can be used.

In ΔBOC, $h = a\sin(\theta)$ so you can substitute $a\sin(\theta)$ for h in the area formula.

$$A = \frac{1}{2} \times b \times h$$

$$= \frac{1}{2} \times b \times a\sin(\theta)$$

$$= \frac{1}{2}ab\sin(\theta)$$

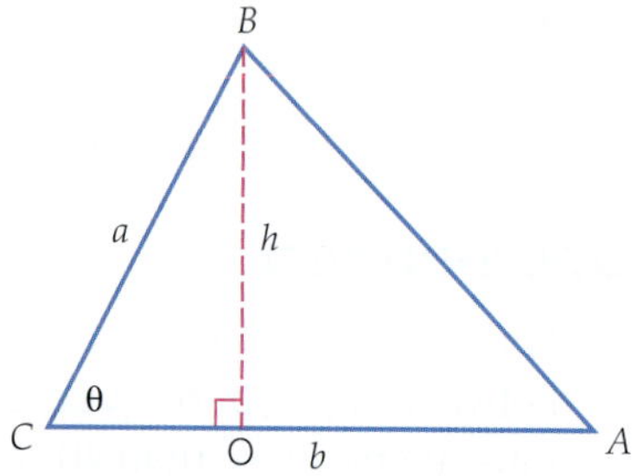

10A

Area of a triangle

$A = \frac{1}{2}ab\sin(\theta)$, where θ is the angle included between sides a and b.

Worked example 16

W.E. 16

Calculate the area of the triangle correct to 2 decimal places.

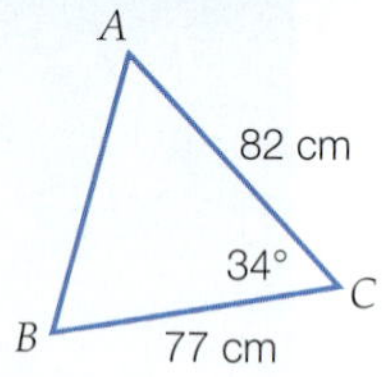

Thinking	Working
1 Identify the given information. (Here, you have a, b and θ.)	$a = 77$ cm, $b = 82$ cm, $\theta = 34°$
2 Write the formula for the area of the triangle.	Area $= \frac{1}{2}ab\sin(\theta)$
3 Substitute the given values into the formula.	$= \frac{1}{2} \times 77 \times 82 \times \sin(34°)$
4 Use a calculator to evaluate the answer. (Do not round.)	$= 1765.371\,996$
5 State the answer to the required number of decimal places.	The area is 1765.37 cm^2

Worked example 17

W.E. 17

Find the size of the angle required if the area of the triangle shown is 195 m^2. Round the answer to the nearest degree.

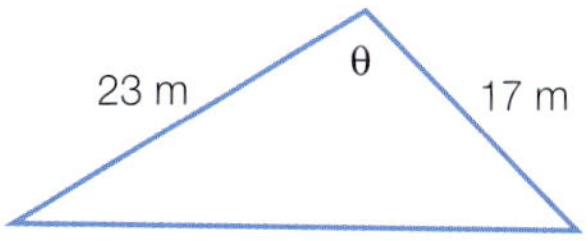

Thinking	Working
1 Identify the given information. (Here, you have *a*, *b* and Area.)	Area = 195 m^2, $a = 23$ m, $b = 17$ m
2 Write the formula for the area of the triangle.	$\text{Area} = \frac{1}{2}ab\sin(\theta)$
3 Substitute the given values into the formula.	$195 = \frac{1}{2} \times 23 \times 17 \times \sin(\theta)$
4 Use a calculator to evaluate the answer. (Do not round.)	$\sin(\theta) = \frac{195 \times 2}{23 \times 17}$ $= 0.9974$ $\theta = \sin^{-1}(0.9974)$ $= 85.901\,340\,36$
5 State the answer to the nearest degree.	The angle is 86°

7.6 Areas of triangles using trigonometry

10A

Navigator

1, 2, 3, 4, 5, 6, 7, 8, 9, 10, 11, 12, 15, 16

1, 2, 3, 4, 5, 6, 7, 8, 9, 10, 11, 12, 13, 15, 16, 17

1 (a, b), 2 (a, b), 3, 4 (a, b), 5, 6, 7, 9, 10, 12, 13, 14, 15, 16, 17, 18

Answers p. 815

Fluency

1 Calculate the area of each of the following triangles correct to 2 decimal places. Remember to include units with your answer.

W.E. 16

(a)

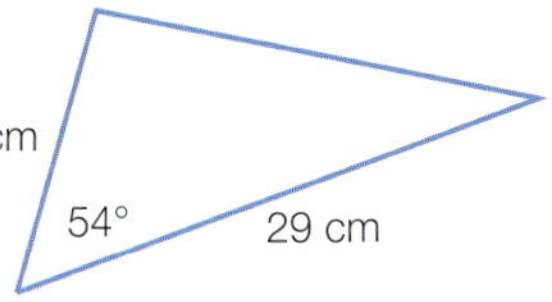

(b)

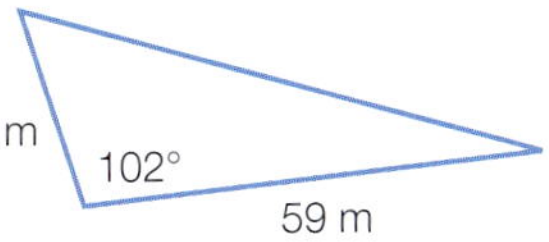

(c)

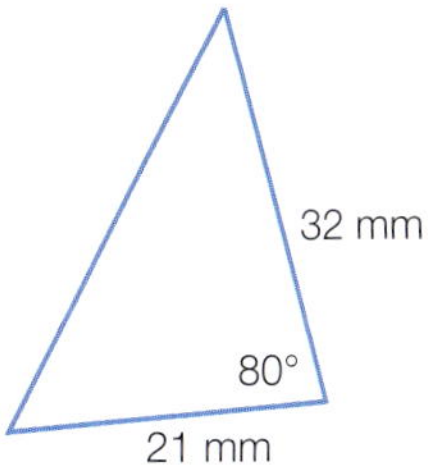

2 Calculate the area of each of the following triangles, correct to 2 decimal places.

(a)
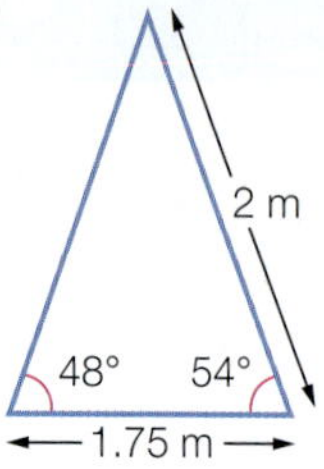

(b)
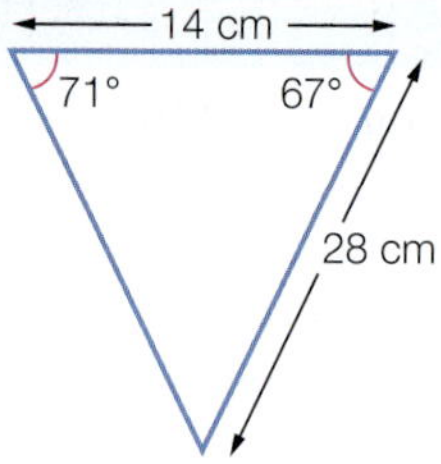

(c)
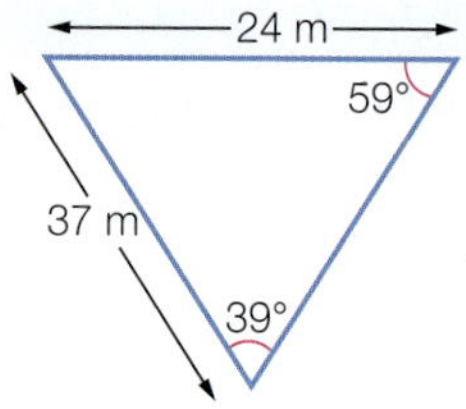

W.E. 17

3 Find the size of the angle required if the area of the triangle shown is 529 m^2. Write the answer correct to the nearest degree. Assume θ is an acute angle.

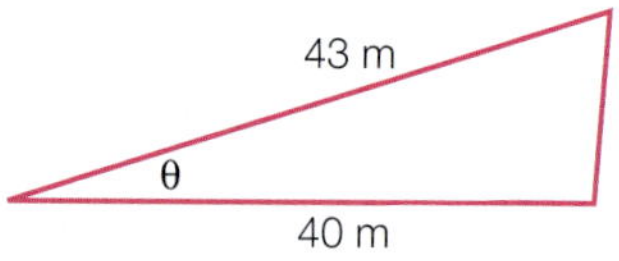

4 Calculate the area of the following triangles correct to 2 decimal places.

(a)
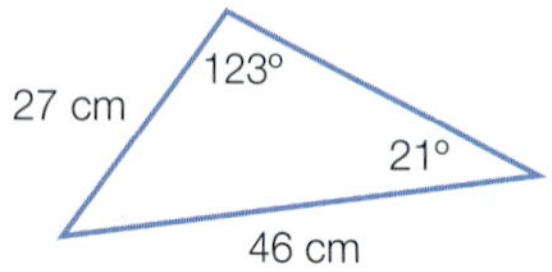

(b)
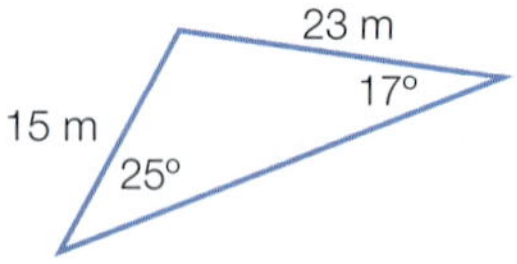

(c)
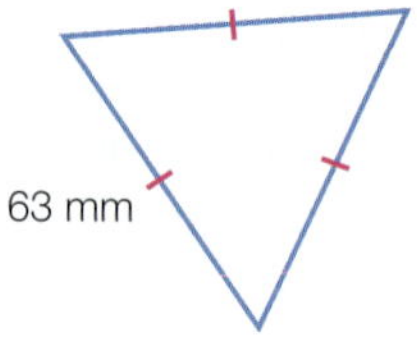

5 The area of the triangle in square metres is closest to:

A 11.47 **B** 13.12

C 19.41 **D** 20.00

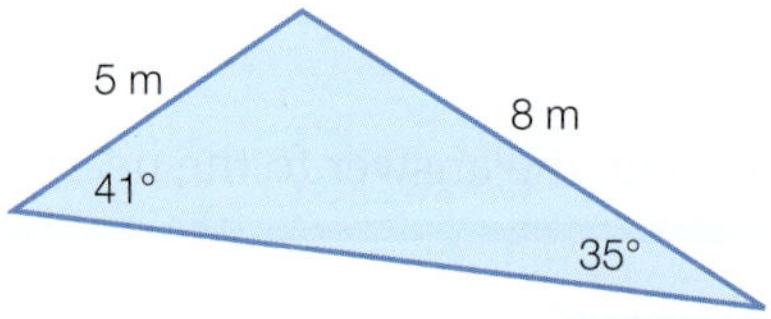

Understanding

6 Calculate the area of the shape correct to the nearest square kilometre.

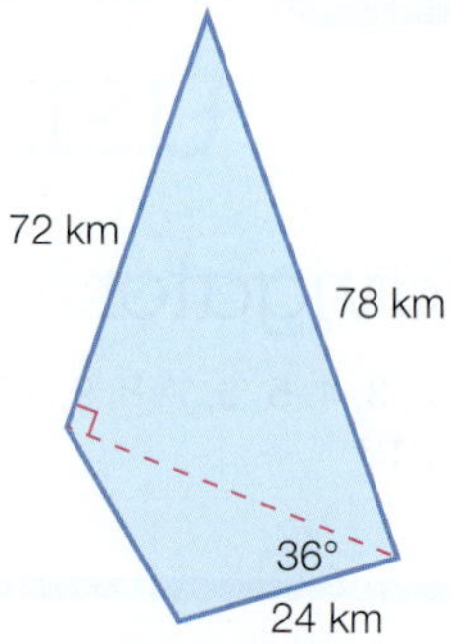

Hint

How many equilateral triangles fit together to make a hexagon?

7 The cells of a bee's honeycomb are shaped like regular hexagons. If each edge length is 1.5 mm, what is the area of the end of one cell? Round your answer to 2 decimal places.

8 A factory produces sheet metal, which is cut into various sizes and shapes for different art products. If the metal is cut into the shape shown, what is the area of metal used?

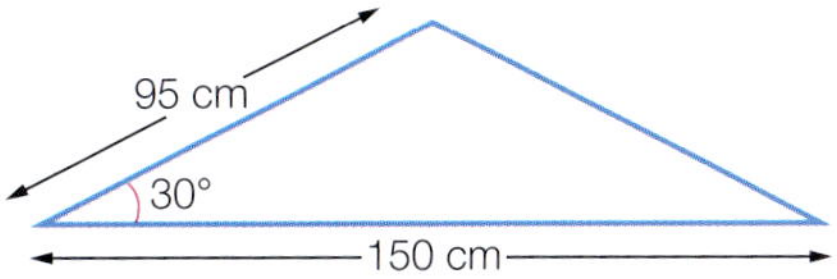

9 A cycling competition is held on the a triangular course as shown. Calculate the area covered by the course correct to 1 decimal place.

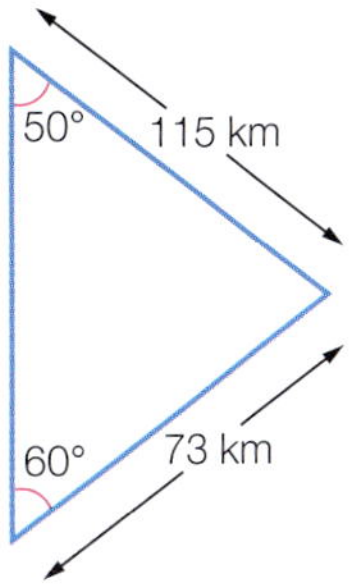

10 ΔABC has the following dimensions. Angle $B = 57°$, $AB = 42$ cm and $BC = 57$ cm. Calculate the area of this triangle. Write your answer correct to the nearest square centimetre.

11 Calculate the area of the white material that forms the back of this frame correct to the nearest square centimetre.

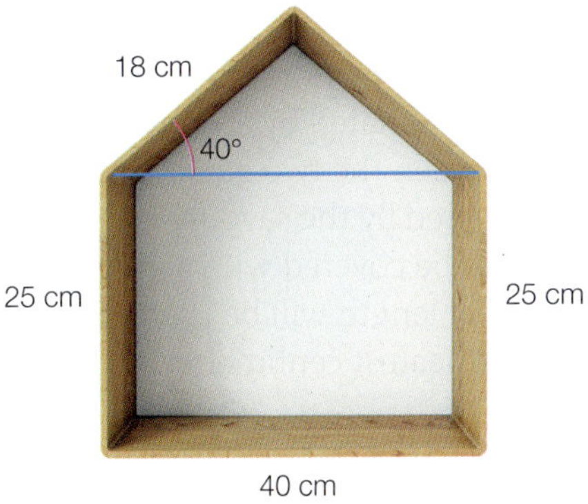

Reasoning

12 Calculate the area of the following triangles correct to 2 decimal places. Do not assume that the diagrams have been drawn accurately or are to scale.

Hint

Are there any ambiguous cases?

(a)

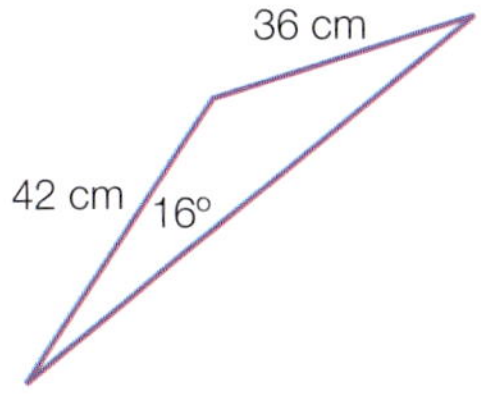

(b)

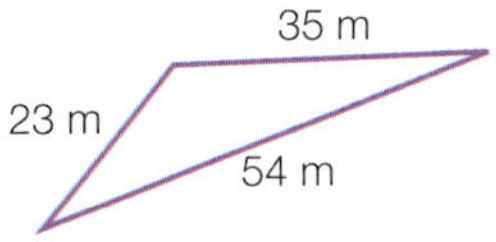

(c)

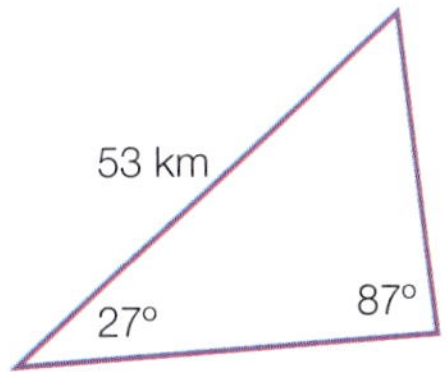

13 A plot of land is in the shape of an isosceles triangle.

(a) Show that the area of land is $L^2 \sin(\theta)\cos(\theta)$ km^2.

(b) What is the area when $L = 5.6$ km and $\theta = 45°$.

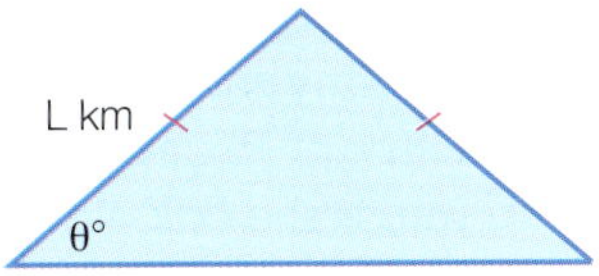

Hint

Divide this triangle into half and use trigonometry to find the base lengths and heights of the small triangles.

14 (a) Show that the area of the red stained glass, shown in the diagram on the right, is $\frac{1}{2}ac\sin(\theta)$ mm^2.

(b) For what value of θ will the area be a maximum? What is this area?

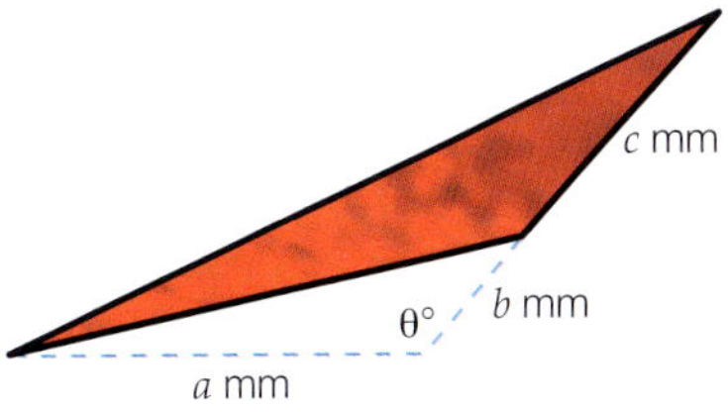

Hint

Use the sine or cosine rule to find the unknown sides or angles.

15 The sail shown, consisting of two layers of cotton fabric, is to be replaced at a cost of \$135 per square metre. If the length of the sail along the mast is 3.8 m and the two angles at the base of the sail are 85° and 57° as shown in the diagram, calculate the total replacement cost of the sail. Answer to the nearest dollar.

16 (a) In the formula for the area of a triangle, $\frac{1}{2}ab\sin(\theta)$, what is the interval of values possible for the angle θ?

(b) Explain why $\sin(\theta)$ cannot take on values less than zero.

17 The shape of the cheese savoury can be approximated by an isosceles triangle. The angle between the two equal sides is 88° and the area covered by the savoury is 21 cm^2. If the perimeter is to be covered with a strip of cheese flavouring, what length will be used? Round your answer to the nearest centimetre.

10A

Open-ended

18 Find values for the lengths of two equal sides of an isosceles triangle and the size of the angle between them, so that the area is a square number. The triangle must not be right angled.

Problem solving

Triangles to the max

Servane has a square piece of cardboard, which is 30 cm × 30 cm. She wants to cut out as many small triangles as possible that have one side of 12 cm, another of 15 cm and the angle between them is 30°. Using area formulas she calculates that she will get 20 triangles. Her friend, Sophie, disagrees and says that 16 triangles are the most she can cut.

Are either of them correct? Explain your answer. Draw a diagram to show the maximum number of triangles you could cut out from a piece of cardboard with the same dimensions.

Strategy options

- Make a model.
- Act it out.

Chapter review 7

Maths literacy

amplitude	coterminal angles	quadrant
asymptote	equivalent angles	sine rule
cosine rule	period	unit circle

Copy and complete the following using words and phrases from the list, where appropriate. A word or phrase may be used more than once.

1 One half the difference between the minimum value and the maximum value of the graph of a trigonometric ratio is the __________.

2 The __________ relates the angle between two sides of a triangle and the side length opposite the angle.

3 Angles in quadrant 1 that have the same values for trigonometric ratios in the other quadrants are __________.

4 The trigonometric ratios are derived from the __________.

5 The smallest interval after which the graph of a trigonometric relationship repeats is the __________.

6 The __________ can be used when two angles in a triangle and a side length opposite one of the angles is known.

7 Each of the four sections of the unit circle is a __________.

8 Each __________ on the graph $y = \tan(\theta)$ is a vertical line for which $\tan(\theta)$ is not defined.

9 __________ are in the same position on the unit circle so their trigonometric ratios are exactly the same.

10A

Fluency

1 Calculate the length of x in each shape. Round your answers to 2 decimal places. 7.1

(a)

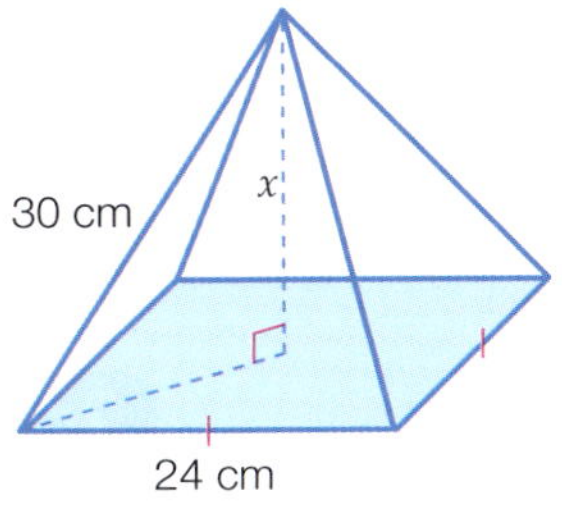

(b)

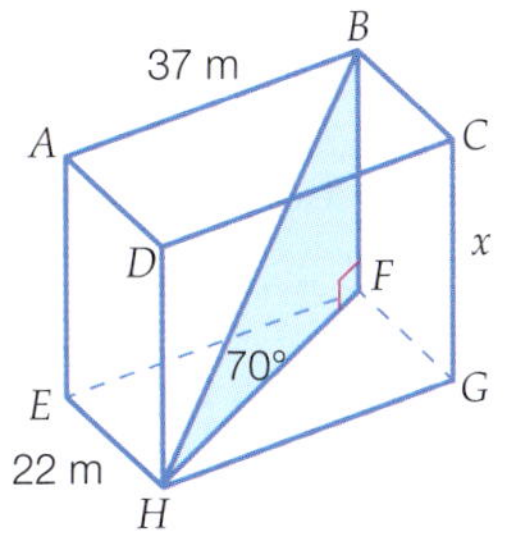

2 To the nearest centimetre, what is the length of the longest possible timber pole that can fit in a shed in the shape of a rectangular-based prism of length 24 m, width 10 m and height 4 m? 7.1

3 In which quadrant are each the following angles? 7.2

(a) 37° (b) 274° (c) 93° (d) 182°

4 Find the equivalent ratio to each of the following in quadrant 1 using the correct sign. 7.2

(a) $\cos(173°)$ (b) $\sin(254°)$ (c) $\tan(312°)$

5 The equivalent trigonometric ratio in quadrant 1 of $\cos(253°)$ is: 7.2

A $\cos(17°)$ B $-\cos(73°)$ C $-\cos(107°)$ D $\cos(37°)$

6 Find all solutions to $\tan(x) + 1 = 0$ for $0° \le x \le 360°$. 7.3

7 Determine all solutions to the equation $4\cos(x) - 1 = 0$. Round your answers to 2 decimal places. 7.3

8 In degrees, the angle θ shown is closest to: 7.4

A 70.42 B 72.10 C 73.56 D 75.78

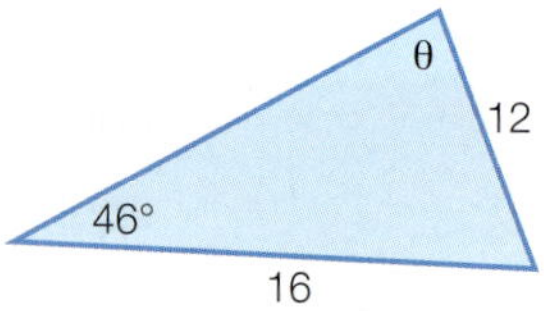

9 Calculate the value of c rounded to 2 decimal places. 7.4

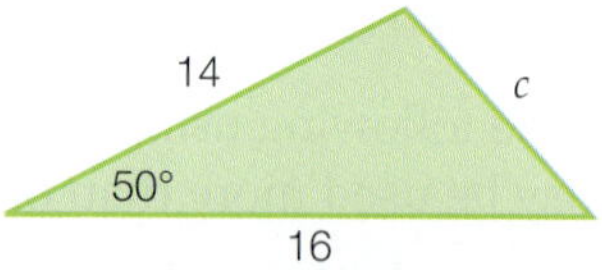

10 A roof consists of rooflines of 5.3 m as shown in the diagram. The rooflines determine the pitch, which is the angle formed where the rooflines meet. Calculate the pitch θ of the roof correct to the nearest degree. 7.5

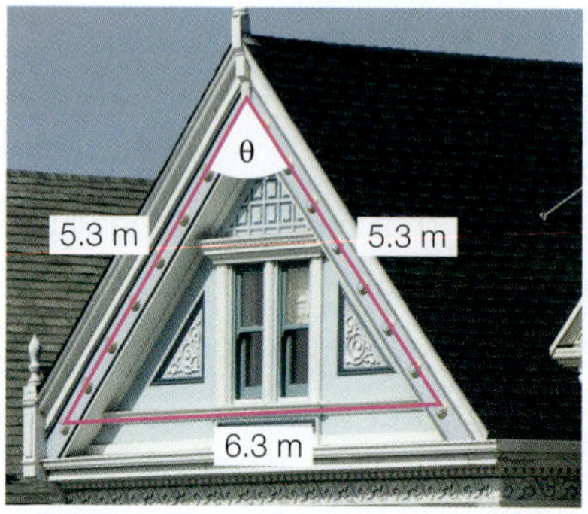

11 Calculate the area of the shape rounded to 2 decimal places. 7.6

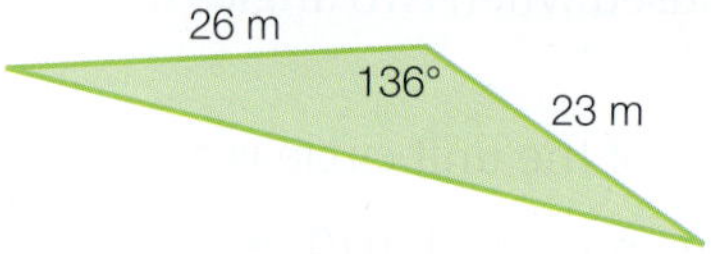

10A Understanding

12 P is the midpoint of AB. The angle θ is closest to: 7.1

A 22.6° B 33.7°

C 42.2° D 50.9°

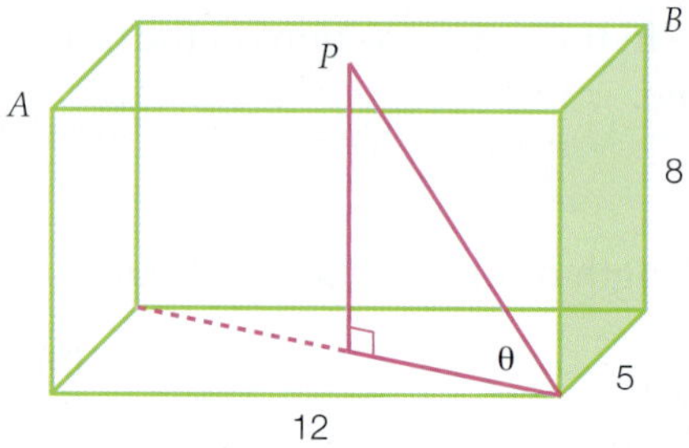

13 State the period and the amplitude, where appropriate, of each relationship. 7.2

(a) $y = 7\sin(2x)$ (b) $y = -6\cos(5x)$ (c) $y = 3\tan(4x)$

14 The period of $\tan(nx)$ is 15°. The value of n must be: 7.2

A 10 B 12 C 15 D 24

15 Sketch the graph of $y = 2\cos(x)$ for $-180° \le x \le 180°$ using the graph of the basic function $y = \cos(x)$ as a guide. 7.2

16 (a) In which quadrant is 400°? 7.2

(b) What is the first angle measured clockwise from the positive x-axis that will be coterminal with 400°?

17 The lengths of the hour and minute hand of a clock face are 5 cm and 6 cm respectively. To the nearest degree, what is the size of the obtuse angle formed between the hands when the tips of the hands are 9.63 cm apart. 7.4

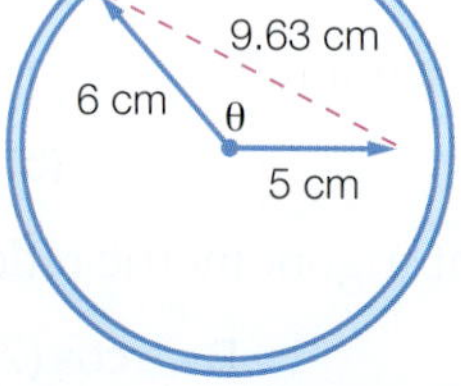

18 Calculate the area of an isosceles triangle where the angle between the two equal sides each of length 10 cm is 67°. State your answer correct to 2 decimal places. **7.6**

19 Calculate the value of x rounded to 2 decimal places. **7.4**

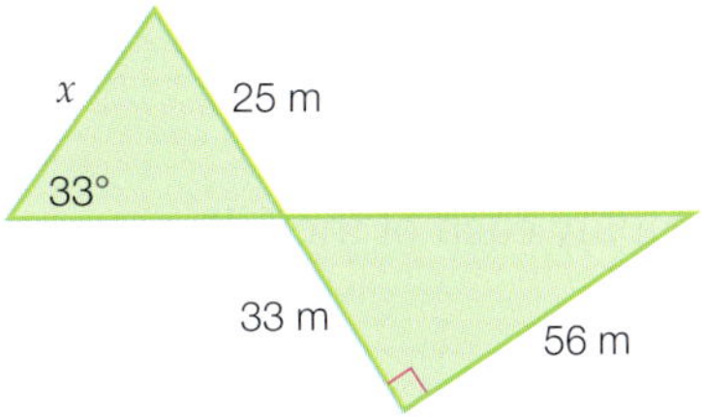

Reasoning

20 The longest drinking straw that can fit into a cylindrical soft-drink can is 15.7 cm. If the can has a radius of 36 mm, find: **7.1**

(a) the angle, correct to 1 decimal place, that the straw makes with the base of the can

(b) the height of the can rounded to 2 decimal places.

21 From a point K on top of a 200 m high building, the angle of depression to a point X on the ground due south of K is 31° and the angle of depression to a point Y on the ground due east of K is 22°. **7.1**

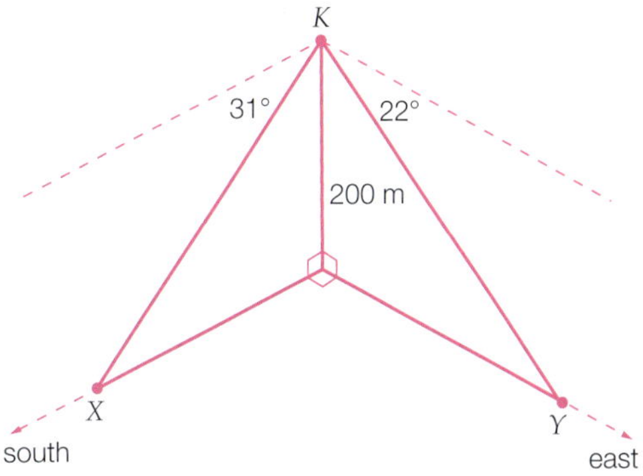

(a) Find, correct to 2 decimal places, the distance between:

(i) the building and point X

(ii) the building and point Y.

(b) Hence, find the distance between X and Y rounded to 2 decimal places. (Use Pythagoras' theorem.)

(c) What is the true bearing of Y from X, to the nearest degree?

22 Which angles in the unit circle do not have equivalent angles? Use angles from 0° to 360°. **7.2**

23 (a) (i) For given values for a, θ and ϕ explain how the sine rule can be used to find the value of x in the triangle at right. **7.5**

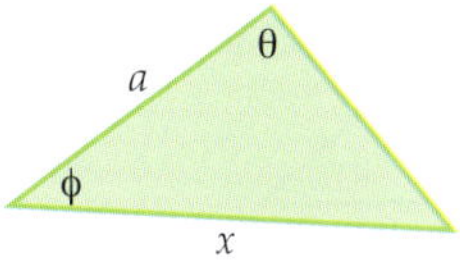

(ii) Find the value of x when $\phi = 41°$, $\theta = 93°$ and $a = 10$. Write your answer correct to 2 decimal places.

(b) (i) For given values for a and θ, explain how the sine rule can be used to find the value of x in the triangle at right.

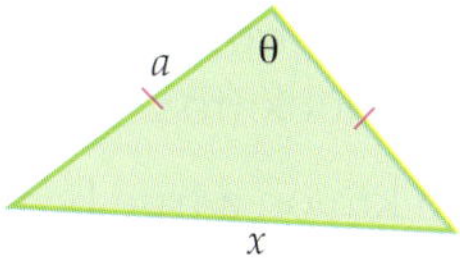

(ii) Find the value of x when $\theta = 93°$ and $a = 10$. Write your answer correct to 2 decimal places.

24 A fishing boat leaves port on a bearing of 136°T. Following this bearing, it drops x m of netting. The vessel then travels north-east, dropping y m of netting along the way. Finally, it heads directly back to its starting point, dropping netting as it moves. The expression that gives the area in square metres enclosed by the netting is: **7.6**

A $\frac{1}{2}xy\sin(44°)$ B $\frac{1}{2}xy\sin(45°)$ C $\frac{1}{2}xy\sin(46°)$ D $\frac{1}{2}xy\sin(89°)$

25 The shadow of a triangular sculpture is shown. If $y = 1.5x$, show that $2\tan(\theta) = 3\tan(\varphi)$. **7.2**

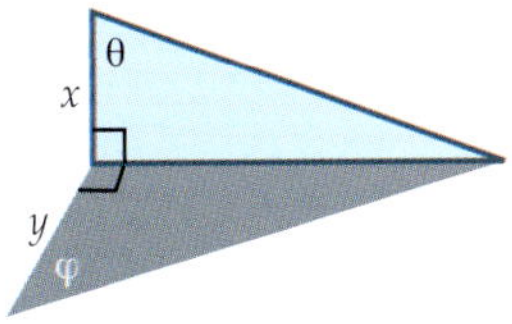

10A

Mixed review B

Fluency

1 Expand and simplify each of the following. **3.1**

(a) $(3x - 4)(x + 2)$ (b) $(2x + 5)(3x - 1)$ (c) $(4x - 3)(2x - 5)$

2 Factorise the following quadratic trinomials. **3.5**

(a) $x^2 - 3x - 4$ (b) $x^2 - 14x + 45$ (c) $x^2 - 7x + 12$

3 Find the shaded areas of the following shapes correct to 2 decimal places. **5.1**

(a)

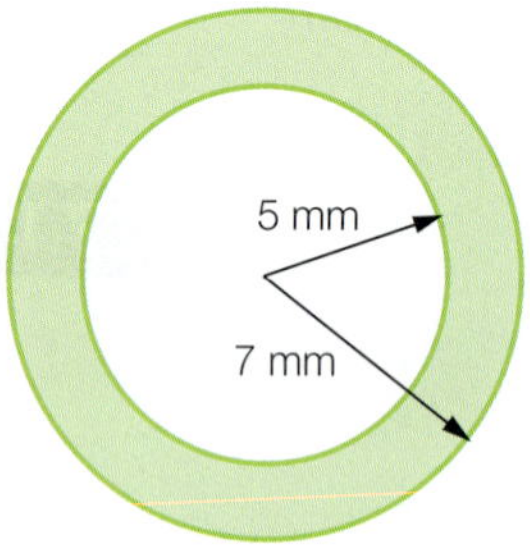

(b)

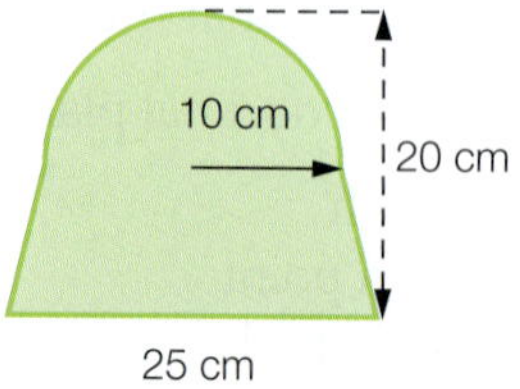

10A 4 Find the total surface area, correct to 2 decimal places, of a shape consisting of a cube of side length 7 cm with a pyramid of perpendicular height 4 cm sitting on top of it. **5.4**

10A 5 Find the volume of a square pyramid with base side length 12 cm and perpendicular height 15 cm. **5.5**

10A 6 A metal sphere has a volume of 12 cm^3. If its mass is 76.8 g, what is its density? **5.7**

7 Find the values of the pronumerals correct to 2 decimal places. **6.2**

(a)

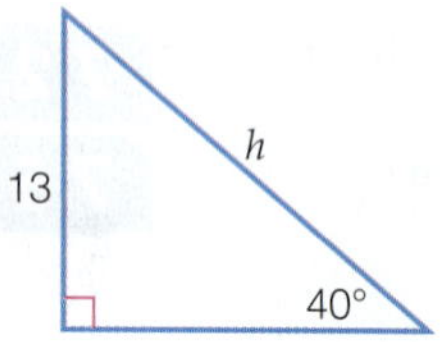

(b)

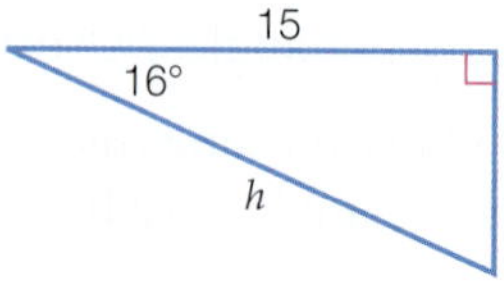

(c)

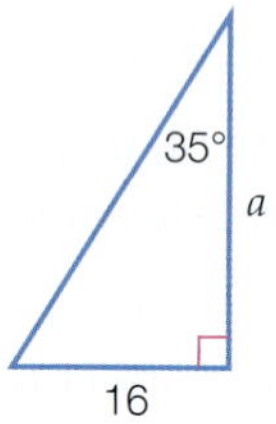

8 A man standing 8 m from a flagpole looks up to the top of it. Through what angle of elevation does he look (to 2 decimal places) if the top is 16 m higher than his eye-level? **6.4**

9 A scout walks 3 km on a true bearing of 050°T and then 5 km on a true bearing of 077°T. What is the scout's distance from the starting position (to 2 decimal places)? **6.5**

10 A proposed section of highway that was to connect the Apian Freeway to the town of Deswin will now connect the Merton Bypass to Deswin as shown. What will be the length, x km, of the new section of highway? Write your answer to the nearest kilometre. **6.6**

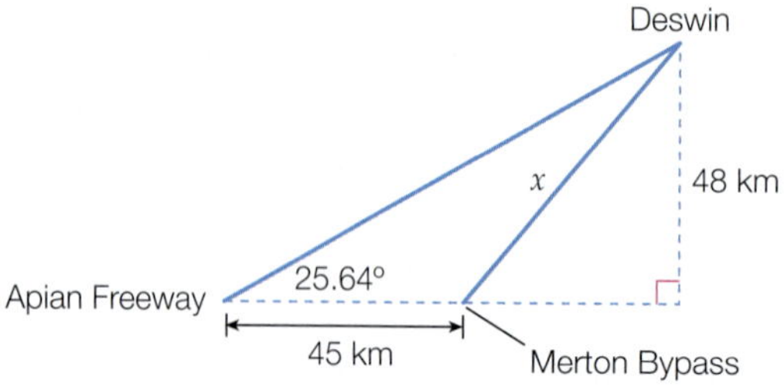

10A **11** A right pyramid *ABCDE* stands on a square base of edge length 10 cm. The sloping edges are each 15 cm long. *AF* is the perpendicular height.

(a) Find, correct to 1 decimal place, the angle that a sloping edge makes with the base of the pyramid.

(b) Hence, find the height of the pyramid, correct to 1 decimal place.

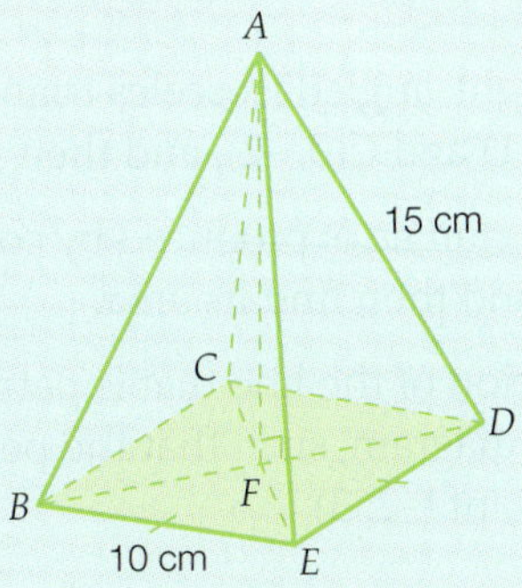

7.1

12 Find the value of x in the following triangles correct to 2 decimal places. Do not assume that the diagrams have been drawn accurately, so check whether there are any ambiguous cases. 10A

7.5

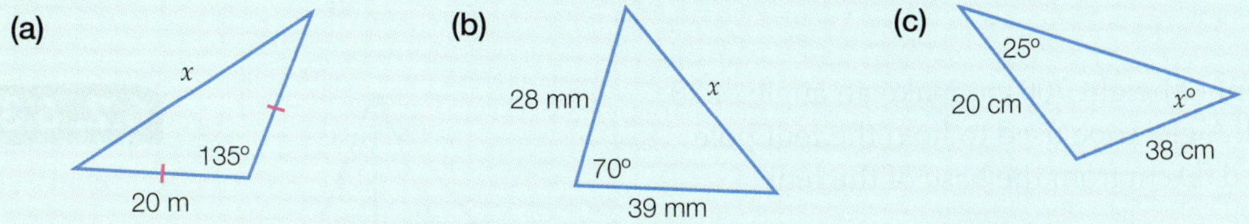

10A **13** Find the area of each of the triangles in Question **12** correct to 2 decimal places.

7.6

Understanding

14 The height of a particular young pine tree is found to increase in a linear manner each month in the first few years after planting. What was the initial height of the tree, if after 2 months the height is 34 cm, and after 5 months the height is 52 cm?

1.2

15 The sides of a rectangle differ by 4 m. If all sides are increased by 2 m, the area increases by 52 m^2. What are the dimensions of the original rectangle?

3.1

16 Determine whether the following points are collinear: $A(1, -1)$, $B(3, 3)$, $C(6, 9)$.

1.2, 1.3

17 A traffic cone, used on roads to divert traffic, rises from the outside edge at an angle of 20° for a horizontal distance of 8 cm then rises to its top at an angle of 70° for a further horizontal run of 17 cm. Find h, the height of the traffic cone, correct to 2 decimal places.

6.2

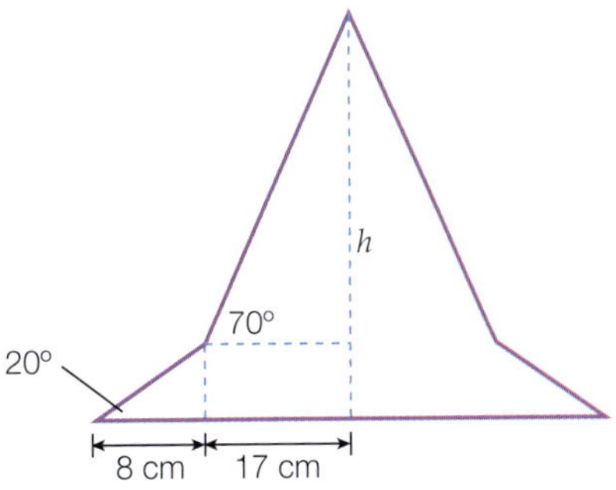

10A **18** Which of the following equations cannot be used to find x in the diagram given?

7.5

A $\dfrac{x}{\sin(41°)} = \dfrac{49}{\sin(65°)}$

B $\dfrac{x}{\sin(41°)} = \dfrac{52}{\sin(74°)}$

C $x^2 = 52^2 + 49^2 - 2 \times 52 \times 49 \cos(41°)$

D $\dfrac{x}{\sin(41°)} = \dfrac{52}{\sin(65°)}$

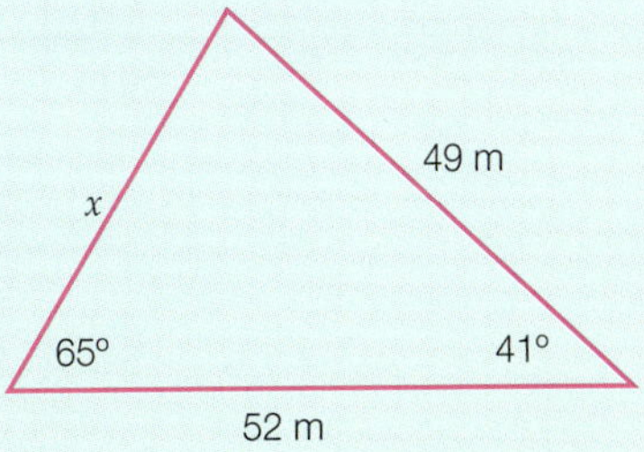

Reasoning

19 Prove that a right-angled isosceles triangle is produced by connecting the points $A(-1, 3)$, $B(5, -1)$ and $C(0, -2)$, and then find its area. **5.1, 6.6**

20 Prove that a rectangle is produced by connecting the points $A(-6, 0)$, $B(3, -6)$, $C(5, -3)$ and $D(-4, 3)$, and then find its area. **5.1, 6.6**

21 Find the equation of the line that is perpendicular to the line shown on the diagram, such that the perpendicular line passes through the point (-2, 3). **1.4**

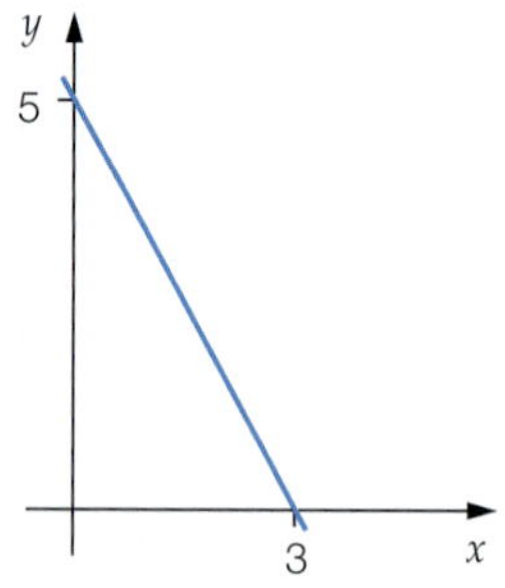

22 The sides of a tent of height 700 cm make an angle of 70° with the ground. A guy rope used to hold the tent pole upright is secured 30 cm from the base of the tent. **6.2, 6.3**

(a) Find the angle it makes with the ground correct to 2 decimal places.

(b) Find the length of the rope to the nearest centimetre.

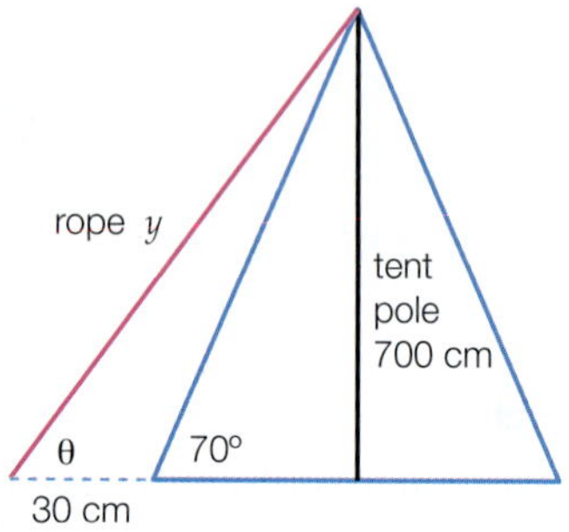

23 A point A on one side of a river is at a bearing of 037°T from point B on the other side of the river. From point C, 50 m due east of point B, the bearing of point A is 336°T. How wide is the river, in metres, correct to 2 decimal places? **6.5**

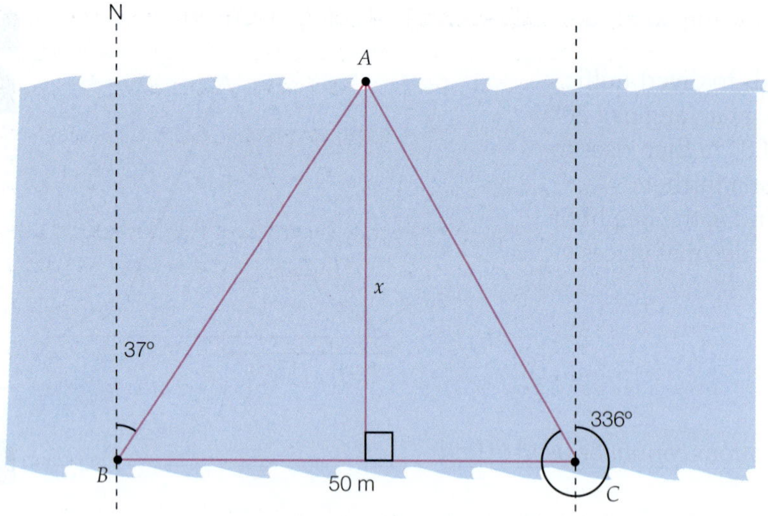

Challenge B

1 The line L crosses the x-axis at (-8, 0). The area of the shaded region is 16. What is the slope of the line L?

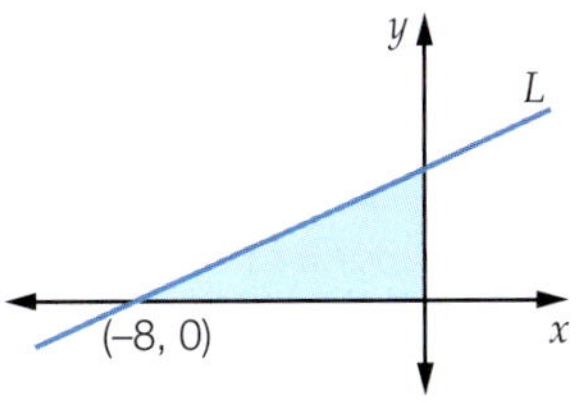

A $-\frac{1}{2}$ B $\frac{1}{2}$

C 2 D 4

2 In the rectangle shown, what is the value of $a - b$?

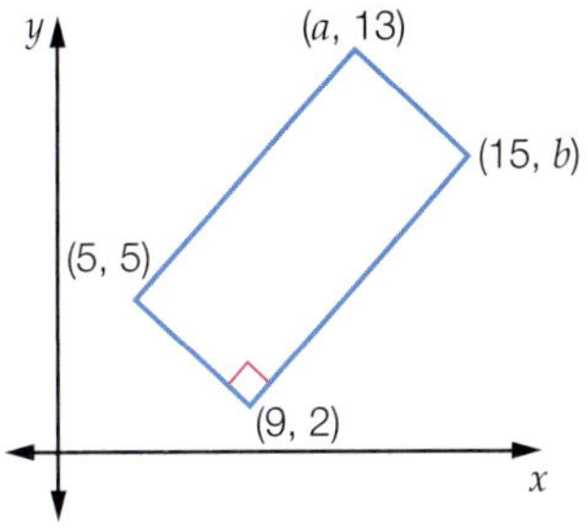

3 The lines $y = -2x + 8$ and $y = \frac{1}{2}x - 2$ meet at (4, 0), as shown. The area of the triangle formed by these two lines and the line $x = -2$ is:

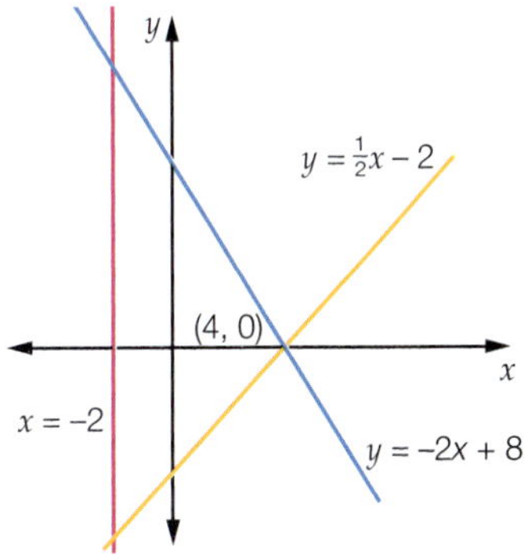

A 15 B 27

C 36 D 45

4 ΔABC has vertices $A(0, 0)$, $B(9, 0)$ and $C(0, 6)$. The points P and Q lie on the side AB such that $AP = PQ = QB$. Similarly, the points R and S lie on the side AC so that $AR = RS = SC$. The vertex C is joined to each of the points P and Q. The vertex B is joined to R and S.

(a) Find the equation of the line through the points R and B.

(b) Find the equation of PC.

(c) The line segments PC and RB intersect at X, and the line segments QC and SB intersect at Y. Show, by calculation, that the points A, X and Y lie on the same straight line.

5 A line with slope -3 intersects the positive x-axis at A and the positive y-axis at B. A second line intersects the x-axis at $C(7, 0)$ and the y-axis at D. The lines intersect at $E(3, 4)$.

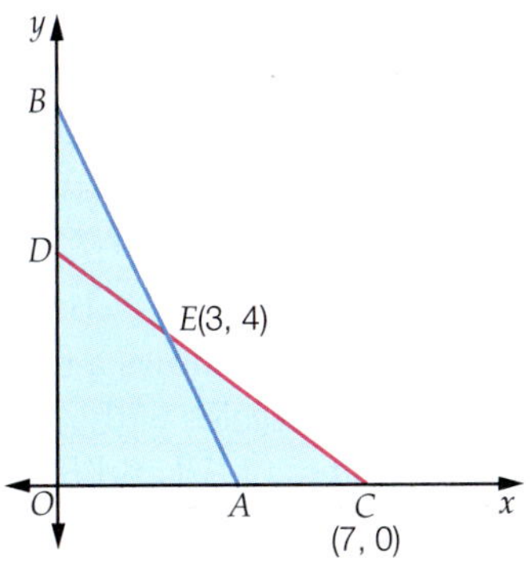

(a) Find the slope of the line through C and E.

(b) Find the equation of the line through C and E, and the coordinates of the point D.

(c) Find the equation of the line through A and B, and the coordinates of the point B.

(d) Determine the area of the shaded region.

6 The point Y is on the line $y = 5x + 3$. The coordinates of the point Z are (3, -2). If X is the midpoint of YZ, then X must be on the line:

A $y = \frac{5x}{2} - \frac{7}{2}$ B $y = 5x + 1$ C $y = \frac{5x}{2} + \frac{1}{2}$ D $y = 5x - 7$

7 In the regular hexagon $ABCDEF$, the lines BD and CF intersect at G. The ratio of the area of quadrilateral $DEFG$ to ΔBCG is:

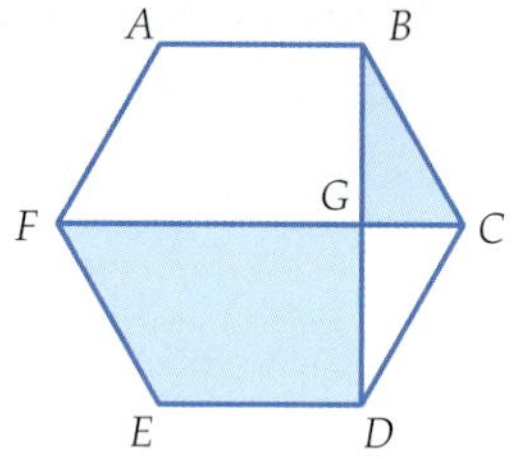

A $3\sqrt{3}:1$ **B** $5:1$

C $4:1$ **D** $2\sqrt{3}:1$

8 A flagpole AB is 20 m tall. From a point C on flat ground, the angle of elevation to the top of the flagpole is 40°. If $CD = DA$, calculate the size of $\angle ADB$, rounded to the nearest degree.

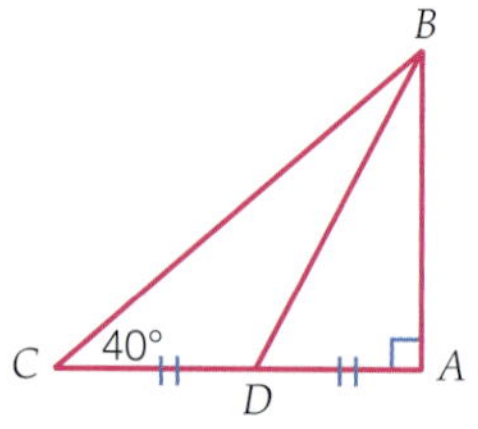

9 Find the area of this triangle if n is a whole number and the area of the triangle is a whole number of square centimetres. All the sides are between 10 cm and 20 cm. You could use Heron's formula for the area of a triangle $A = \sqrt{s(s-a)(s-b)(s-c)}$, where $s = \frac{1}{2}(a+b+c)$ and a, b and c are the sides of the triangle.

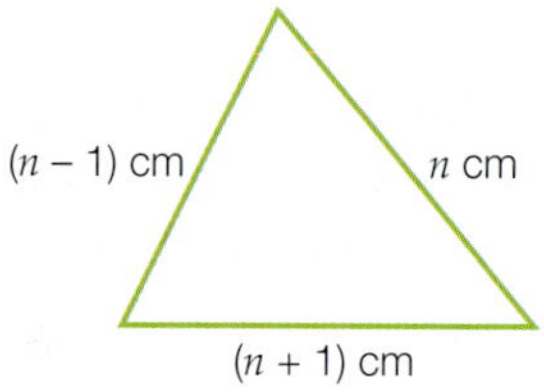

10 Triangle PQR is right-angled at Q and has side lengths $PQ = 14$ and $QR = 48$. M is the midpoint of PR. Draw the interval MS, where S is on PQ, such that $MS \parallel RQ$. Find the value of the cosine of $\angle MQP$ correct to 2 decimal places.

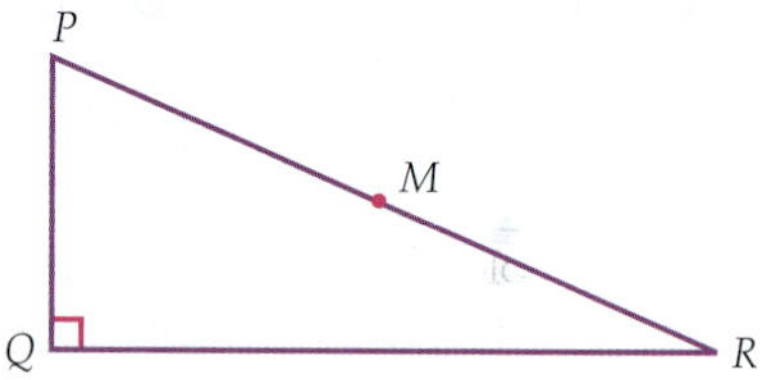

11 ΔABC is an equilateral triangle. The radius of the circle just touching the sides of ABC is 1 unit. A larger circle is drawn through the vertices of the rectangle $ABDE$. Calculate the diameter of the larger circle.

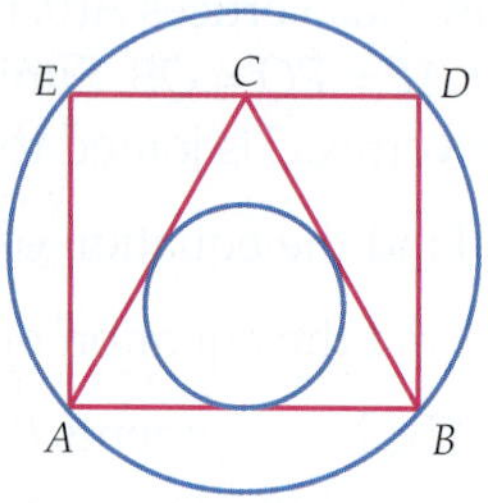

12 In the diagram, E has coordinates (0, 2), and B is on the positive x-axis so that $BE = 7$. The point C lies on the positive x-axis so that $BC = OB$. If D is in the first quadrant such that $\angle CBD = 30°$ and $\angle BCD = 90°$, what is the length of ED, correct to 2 decimal places?

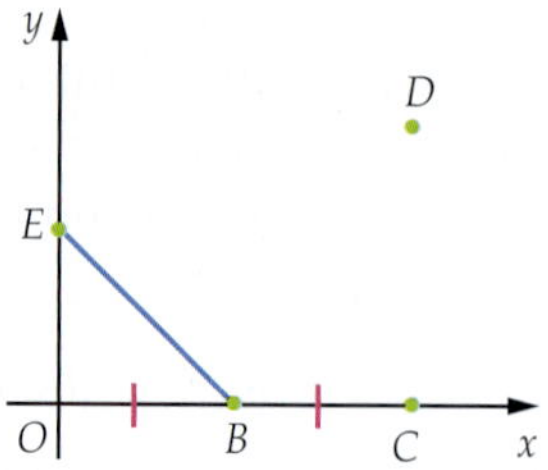

13 The diagram shows a roof truss. Triangle ABC is isosceles with $AB = BC$. The midpoints of AB and BC are Y and Z respectively. $AC = 6.0$ m and $BX = 1.2$ m.

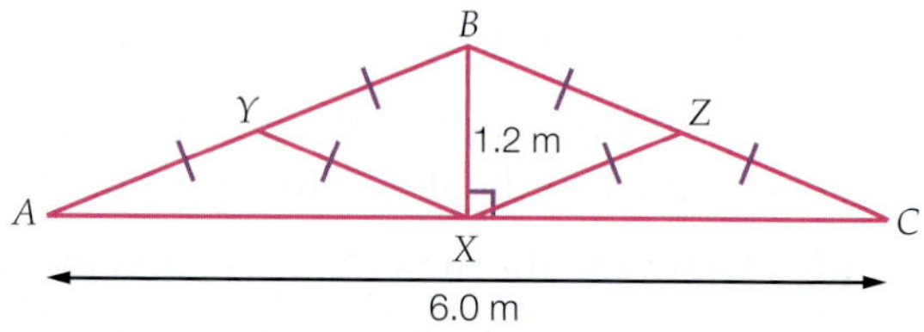

(a) Calculate the length of AB correct to 1 decimal place.

(b) Find the total length of timber required to make the truss.

(c) Calculate the size of $\angle BAX$ and hence, find the size of $\angle BYX$. Round both angles to the nearest degree.

Polynomials

8

10A

The mathematics of DNA

Name: Dr Tony Papenfuss
Job: Laboratory Head, Bioinformatics
Qualifications: PhD in Applied Mathematics (astrophysics)

Bioinformatics is the analysis of biological data using mathematics and computer science. It is a field that has grown as our ability to isolate and sequence DNA has improved, meaning that genetic data can now be generated quickly.

Dr Tony Papenfuss is a Laboratory Head in Bioinformatics at the Walter & Eliza Hall Institute of Medical Research. He works with a team of researchers to analyse sequences of DNA, to try and discover more about cancer, malaria and other diseases.

'A strong basis in mathematics is crucial in this role, as well as a keen interest in biology and an interest in working on real problems. I enjoy the collaborative aspect of the job, working closely with biologists, clinicians and with my own team to find solutions to these problems.

The data analysis is usually done on large computers. The mathematics can be simple but powerful at the same time. I would advise anyone looking to move into this field to continue with maths—it's important in lots of different areas of scientific research.'

Tony has worked on the Tasmanian devil facial tumour disease, and his work has contributed to some important discoveries about what type of cancer it is. Mathematics and medicine are not as far apart as they might seem.

Why learn this?

Graphs can communicate a lot of information in a clear and precise way. Graphs of polynomials can be used to model many types of relationships between variables with recognisable patterns, including many real-life relationships between different quantities. Because of this engineers, statisticians, surveyors and research scientists use them extensively to understand and interpret the world.

After completing this chapter you will be able to:

- recognise a polynomial
- recognise graphs in the form of $y = ax^n$
- recognise and create graphs in the form of $y = a(x - h)^n + k$ as transformations of graphs in the form of $y = ax^n$
- use the remainder theorem to find remainders after polynomial division
- use the factor theorem to factorise polynomials
- sketch graphs of polynomials.

Recall 8

Prepare for this chapter by attempting the following questions. If you have difficulty with a question, you can download a Recall Worksheet from the eBook or the Pearson Places website.

1 Plot the points in each of the following tables of values and join them with a straight line or a smooth curve to draw the graph of each relationship.

(a)

x	-3	-2	-1	0	1	2	3
y	-6	-4	-2	0	2	4	6

(b)

x	-3	-2	-1	0	1	2	3
y	11	6	3	2	3	6	11

2 State whether the relationship is linear, quadratic or neither.

(a) $y = x^3 + 2x^2 + 3$ **(b)** $b = 2c + 3$ **(c)** $z = 4 - 3w^2$

3 **(a)** Complete this table of values for $y = x^2 + 2x - 8$.

x	-5	-4	-3	-2	-1	0	1	2	3
y									

(b) On graph paper, plot the graph of the parabola $y = x^2 + 2x - 8$.

(c) From the graph, write the:

(i) y-intercept

(ii) x-intercepts

(iii) coordinates of the turning point.

(d) Decide whether the graph has a maximum or a minimum turning point.

(e) What is the equation of the axis of symmetry?

10A

Exploration Task

You can download this activity from the eBook or the Pearson Places website.

Next move? Polynomials!

In this activity you will investigate linear and non-linear graphs using technology, then explore how to apply what you have learnt to polynomials.

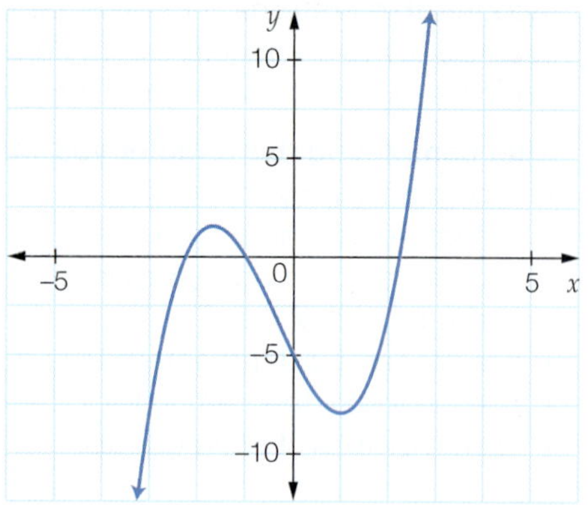

Equations of the form $y = ax^n$

8.1

The straight line with linear equation $y = 2x$, the parabola with the quadratic equation $y = x^2$, the **cubic** function given by $y = x^3$ and the **quartic** function given by $y = 5x^4$ all belong to a class of equations described by $y = ax^n$, where a is a constant and n is an integer. You will investigate the properties of graphs of $y = ax^n$ for different non-zero values of a and for $n = 1$, 2, 3 and 4.

Linear equations: $y = ax$

For $n = 1$, $y = ax^n$ gives a linear equation $y = ax$.

Its graph is a straight line that passes through the origin (0, 0) and has gradient a.

- If $a > 0$, the gradient is positive.
- If $a < 0$, the gradient is negative.

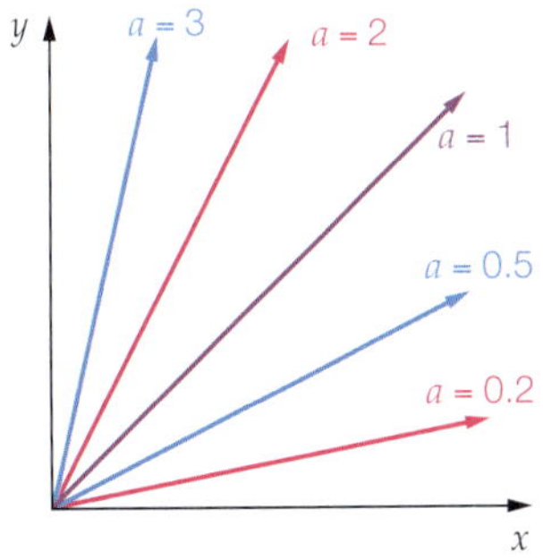

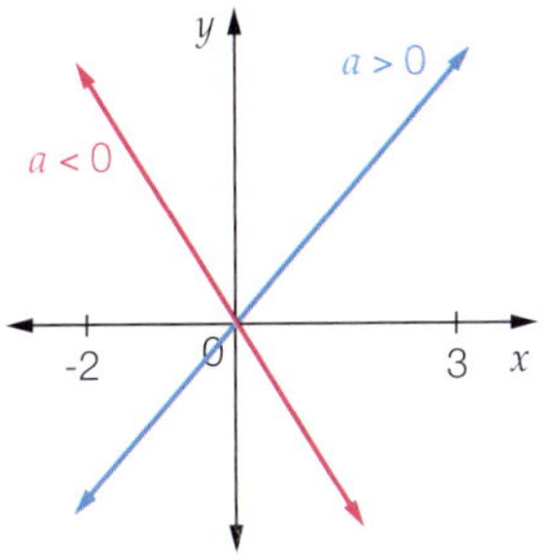

The coefficient of x is the **dilation factor**. The larger the size of a (ignoring the sign), the steeper the line. This means that every y-value of the graph of $y = x$ is multiplied by a to find the y-values of $y = ax$.

Quadratic equations: $y = ax^2$

For $n = 2$, $y = ax^n$ gives a quadratic equation $y = ax^2$.

Its graph is a parabola that passes through the origin and is symmetrical about the y-axis. The axis of symmetry is the line $x = 0$.

The coefficient of x^2, a, is the dilation factor.

- If $a > 0$, the graph has a minimum turning point at (0, 0).
- If $a > 1$, then the graph becomes narrower than the graph of $y = x^2$.
- If $0 < a < 1$, then the graph becomes wider than the graph of $y = x^2$.
- If $a < 0$, the graph is inverted. It is a reflection in the x-axis of the graph for $a > 0$ and it now has a maximum turning point at (0, 0).

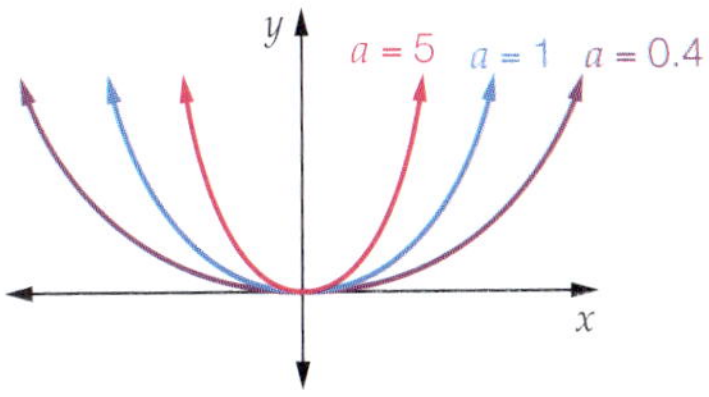

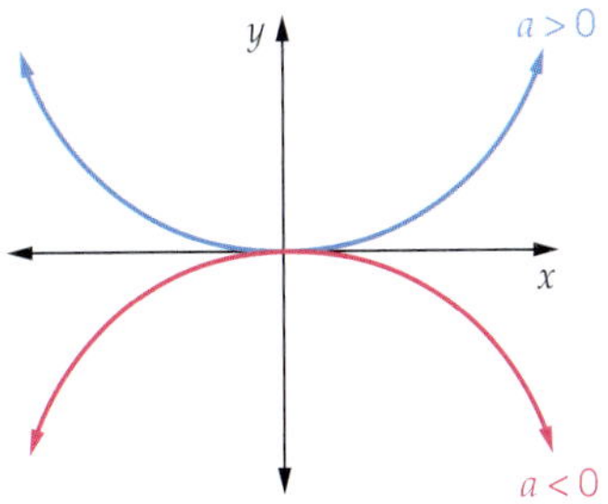

10A

Cubic equations: $y = ax^3$

For $n = 3$, $y = ax^n$ gives a cubic equation $y = ax^3$.

Its graph is a cubic that passes through the origin. It does not have a turning point (maximum or minimum) where the graph changes direction but it does have a special point at the origin called a **point of inflection**.

A cubic graph is not symmetrical about the x-axis or the y-axis but the graph on one side of the origin is reflected in both the x- and y-axes to produce the other half of the graph.

- If $a > 1$, the graph becomes steeper (narrower) than the graph of $y = x^3$.
- If $0 < a < 1$, the graph becomes flatter (wider) than the graph of $y = x^3$.

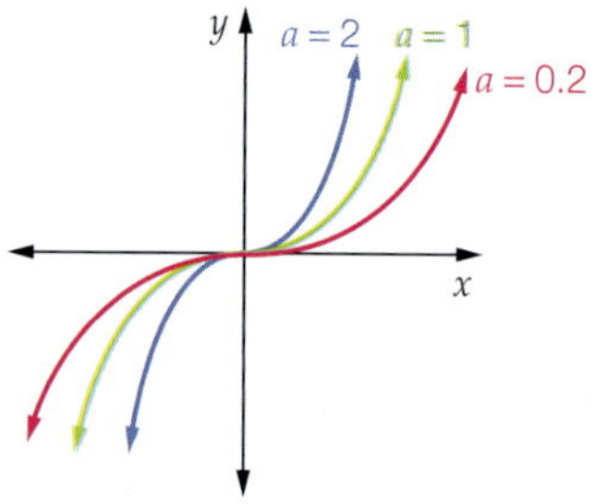

- If $a < 0$, the graph is a reflection in the x-axis of the graph for $a > 0$.

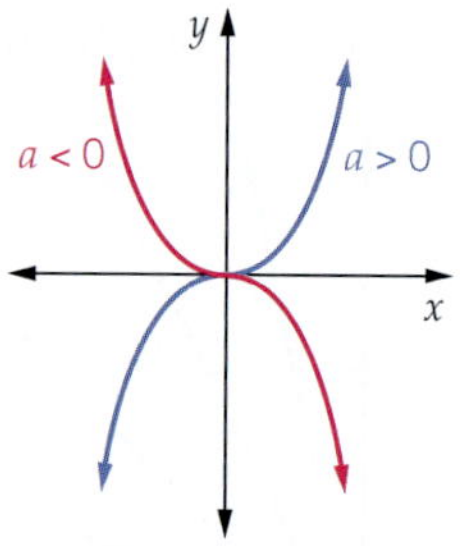

Quartic equations: $y = ax^4$

For $n = 4$, $y = ax^n$ gives a quartic equation $y = ax^4$.

Its graph is a quartic that passes through the origin and is symmetrical about the y-axis.

A quartic graph is steeper than a quadratic graph with the same dilation factor for $x > 1$ and $x < -1$. It is squarer than a quadratic graph with the same dilation factor for $-1 < x < 1$.

- If $a > 0$, the graph has a minimum turning point at (0, 0).
- If $a > 1$, the graph becomes narrower than the graph of $y = x^4$.
- If $0 < a < 1$, the graph becomes wider than the graph of $y = x^4$.

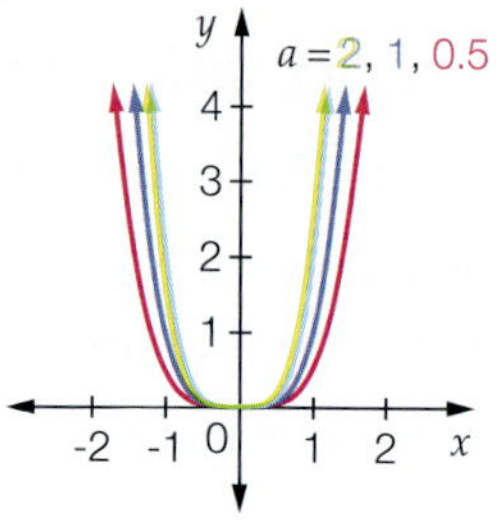

- If $a < 0$, the graph is a reflection in the x-axis of the graph for $a > 0$ and it now has a maximum turning point at (0, 0).

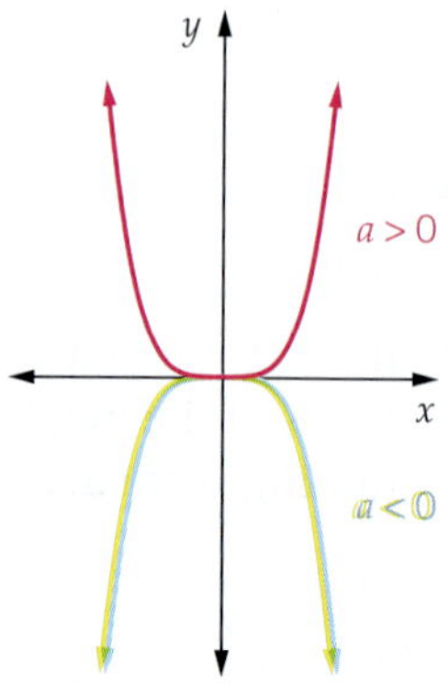

Worked example 1

W.E. 1

Complete the table of values for each of the following equations.

(a) $y = 4x^2$

x	-3	-2	-1	0	1	2	3
y		16			4		

(b) $y = \frac{1}{2}x^3$

x	-3	-2	-1	0	1	2	3
y	-13.5					4	

(c) $y = -x^4$

x	-3	-2	-1	0	1	2	3
y			-1				-81

Thinking	Working
(a) 1 Substitute each x-value into the equation to find the corresponding y-value. A selection is shown here.	(a) Substitute $x = -3$. $y = 4x^2$ $= 4 \times (-3)^2$ $= 36$ Substitute $x = -1$. $y = 4x^2$ $= 4 \times (-1)^2$ $= 4$ Substitute $x = 0$. $y = 4x^2$ $= 4 \times 0^2$ $= 0$ Substitute $x = 2$. $y = 4x^2$ $= 4 \times 2^2$ $= 16$ Substitute $x = 3$. $y = 4x^2$ $= 4 \times 3^2$ $= 36$
2 Fill in the table.	see table below

x	-3	-2	-1	0	1	2	3
y	36	16	4	0	4	16	36

10A

(b) 1 Substitute each x-value into the equation to find the corresponding y-value. A selection is shown here.

(b) Substitute $x = -2$.

$y = \frac{1}{2}x^3$

$= \frac{1}{2} \times (-2)^3$

$= -4$

Substitute $x = 3$.

$y = \frac{1}{2}x^3$

$= \frac{1}{2} \times 3^3$

$= 13.5$

2 Fill in the table.

x	-3	-2	-1	0	1	2	3
y	-13.5	-4	-0.5	0	0.5	4	13.5

(c) 1 Substitute each x-value into the equation to find the corresponding y-value. A selection is shown here.

(c) Substitute $x = -3$

$y = -x^4$

$= -(-3)^4$

$= -81$

Substitute $x = 0$.

$y = -x^4$

$= -0^4$

$= 0$

2 Fill in the table.

x	-3	-2	-1	0	1	2	3
y	-81	-16	-1	0	-1	-16	-81

10A

Worked example 2

W.E. 2

Construct a table of values for $x = -3$ to 3 and then plot the graph of the following equations.

(a) $y = x^2$ (b) $y = x^3$ (c) $y = x^4$

Thinking	Working

(a) 1 Construct the table of values.

(a)

x	-3	-2	-1	0	1	2	3
y	9	4	1	0	1	4	9

2 Draw the axes using a suitable scale. Plot the points and join them with a smooth curve.

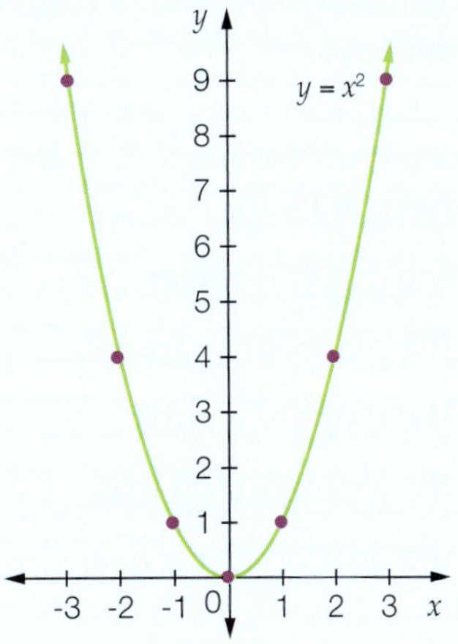

(b) 1 Construct the table of values.

(b)

x	-3	-2	-1	0	1	2	3
y	-27	-8	-1	0	2	8	27

2 Draw the axes using a suitable scale. Plot the points and join them with a smooth curve.

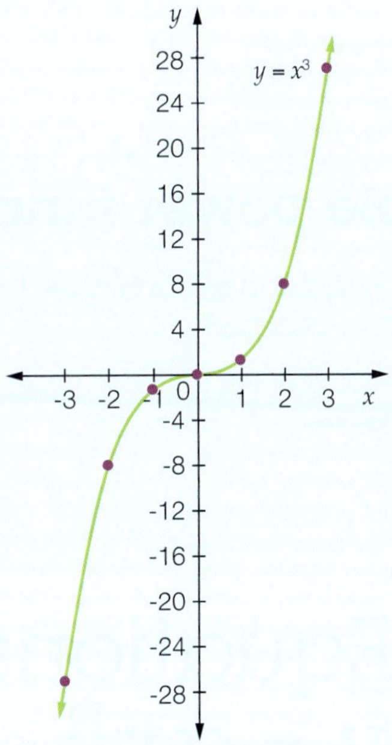

(c) 1 Construct the table of values.

(c)

x	-3	-2	-1	0	1	2	3
y	81	16	1	0	1	16	81

2 Draw the axes using a suitable scale. Plot the points and join them with a smooth curve.

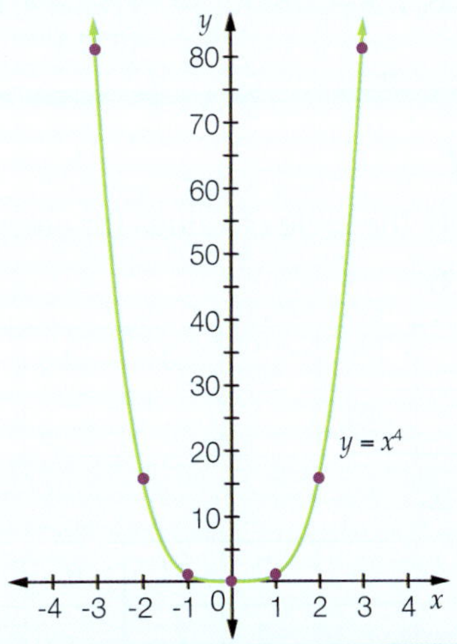

These graphs could also be drawn using a computer drawing program such as the graphing function on your CAS.

The graphs of $y = ax^n$ for $n = 1, 2, 3$ and 4 pass through the origin and have an x-intercept and a y-intercept at (0, 0).

- For $n = 1$, the graph is linear.
- For $n = 2$, the graph is a parabola.
- For $n = 3$, the graph is a cubic.
- For $n = 4$, the graph is a quartic.
- For $a > 0$, the dilation factor a narrows the graph as a increases and widens the graph as a decreases.
- For $a < 0$, the dilation factor a widens the graph as a increases (as a approaches 0) and narrows the graph as a decreases (as a becomes more negative).
- The graph of $y = -ax^n$ is the reflection of $y = ax^n$ in the x-axis, $a > 0$.

Interactive

Simple power functions

Explore the shape of a graph of the form $y = x^n$ when the variables are changed.

Go to the eBook or the Pearson Places website to access this interactive.

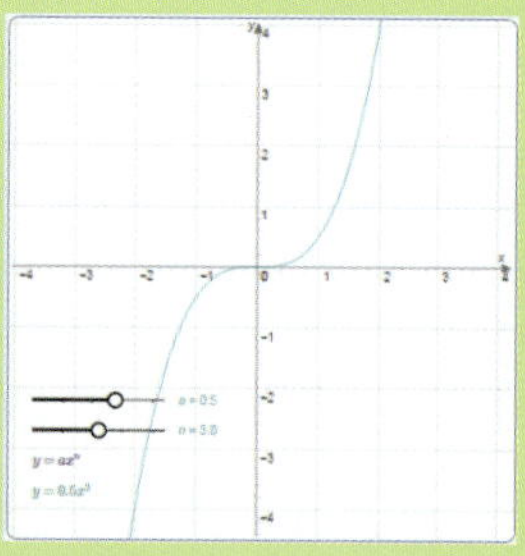

10A

8.1 Equations of the form $y = ax^n$

Navigator

Answers p.817

1, 2, 3, 4, 5, 6, 7, 8, 9, 10, 13, 14	1, 2, 3, 4, 5, 6, 7, 8, 9, 10, 12, 13, 14	2, 3, 4, 5, 6, 7, 8, 9, 10, 11, 12, 13, 14

Fluency

W.E. 1

1 Complete the table of values for each of the following equations.

(a) $y = 3x^3$

x	-3	-2	-1	0	1	2	3
y	-81					24	

(b) $y = -x^2$

x	-3	-2	-1	0	1	2	3
y		-4					-9

(c) $y = \frac{1}{2}x^4$

x	-3	-2	-1	0	1	2	3
y	40.5				0.5		

2 Construct a table of values for $x = -3$ to 3 and then plot the following equations.

W.E. 2

(a) $y = 2x^2$ (b) $y = 2x^3$ (c) $y = 2x^4$

3 A possible equation to describe the graph shown is:

A $y = -8x^2$

B $y = x^3$

C $y = -x^2$

D $y = -8x^3$

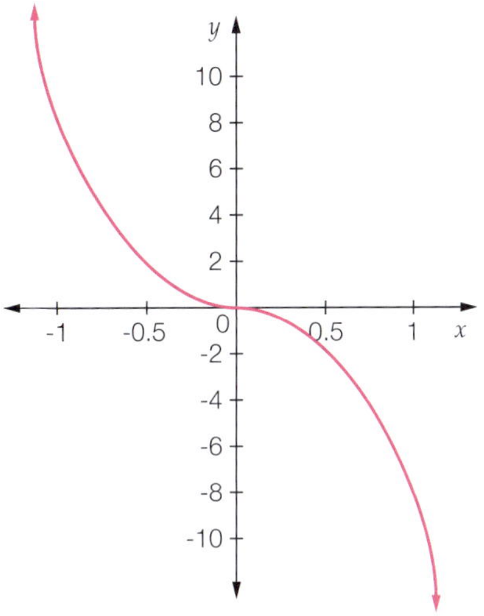

4 Match each equation to its graph.

(a) $y = -\frac{2}{3}x$

(b) $y = -6x^2$

(c) $y = \frac{x}{2}$

(d) $y = -\frac{1}{10}x^2$

(e) $y = 0.01x^2$

(f) $y = \left(-\frac{1}{2}x\right)^3$

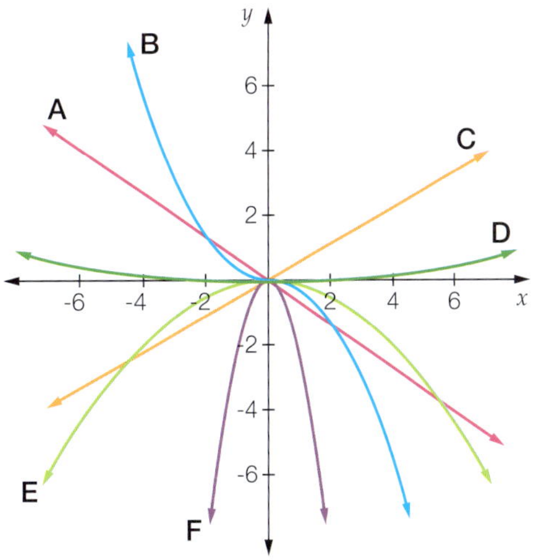

5 (a) Plot the graphs of each line on the same set of axes. Use x-values from -3 to 3.

(i) $y = x$ (ii) $y = 2x$ (iii) $y = 4x$

(b) What can you conclude about the steepness of the line in relation to the coefficient of x?

(c) Does your answer to part **(b)** apply to the graphs of $y = -x$, $y = -2x$ and $y = -4x$?

6 (a) Plot the graphs of each parabola on the same set of axes. Use x-values from -3 to 3.

(i) $y = 2x^2$ (ii) $y = 3x^2$ (iii) $y = 4x^2$

(b) Describe what happens to the shape of the parabolas as the coefficient of x^2 increases.

(c) Does your answer to part **(b)** apply to the graphs of $y = -2x^2$, $y = -3x^2$ and $y = -4x^2$?

Warning

What does the minus sign do to the graph?

10A

7 **(a)** Plot the graphs of each cubic equation on the same set of axes. Use x-values from -3 to 3.

(i) $y = x^3$ **(ii)** $y = 2x^3$ **(iii)** $y = 3x^3$

(b) Describe what happens to the shape of the cubic graphs as the coefficient of x^3 increases.

(c) Does your answer to part **(b)** apply to the graphs of $y = -x^3$, $y = -2x^3$ and $y = -3x^3$?

8 **(a)** Plot the graph of each quartic equation on the same set of axes. Use x-values from -3 to 3.

(i) $y = 0.5x^4$ **(ii)** $y = x^4$ **(iii)** $y = 3x^4$

(b) Describe what happens to the shape of the quartic graphs as the coefficient of x^4 increases.

(c) Does your answer to part **(b)** apply to the graphs of $y = -x^4$, $y = -3x^4$, $y = -0.5x^4$?

Understanding

9 The straight line shown is reflected in the x-axis and then reflected in the y-axis. The equation describing the new graph is:

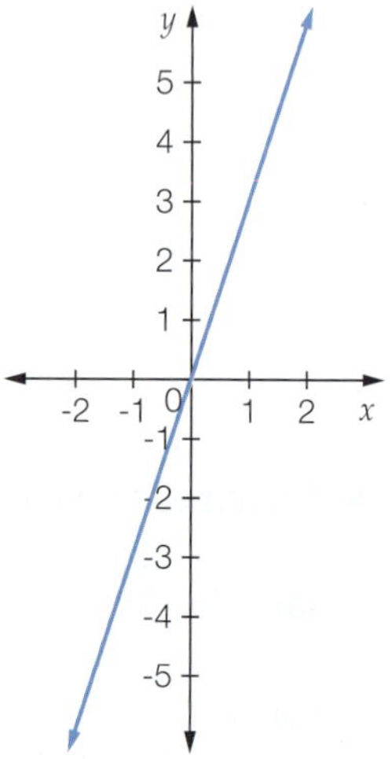

A $y = \frac{1}{3}x$

B $y = 3x$

C $y = -\frac{1}{3}x$

D $y = -3x$

Reasoning

10 The shape of a skateboard ramp is described by the equation $y = \frac{1}{25}x^4$, for $0 \le x \le 3$, where x and y are the horizontal and vertical distances respectively, in metres, from the bottom left corner of the ramp.

(a) Construct a table of values for $y = \frac{1}{25}x^4$ for $x = 0$ to 3.

(b) Use the table of values to plot the graph of $y = \frac{1}{25}x^4$, which models the skateboard ramp.

(c) A new skateboard ramp can be modelled by the equation $y = \frac{1}{50}x^4$, for $0 \le x \le 3$. Is this new ramp steeper or flatter than the original ramp? How do you know?

(d) Construct a table of values for the new skateboard ramp, for $x = 0$ to 3, and then plot the graph on the same set of axes as the graph of the original ramp.

11 The shape of a flat weathervane is defined by the two equations $y = 4x$ and $y = 2x^2$ for x-values from 0 to 2.

(a) Sketch the two graphs on the same set of axes using x-values from 0 to 2.

(b) If the area under the curve $y = 2x^2$ from $x = 0$ to $x = 2$ and bounded by the x-axis is $5\frac{1}{3}$ square metres, what is the area covered by one face of the weathervane?

(c) The wind changes direction so that the weathervane swings around 180° about the origin. What are the equations describing the shape of the weathervane now? State the values of x that apply.

12 (a) On the same set of axes draw the curves $y = x^3$ and $y = \frac{1}{2}x^4$ for $x = -2$ to 2.

(b) Shade in the region that is bounded by these two curves and the lines $x = 0$ and $x = 2$.

(c) Reflect this region in the y-axis and draw it on your graph.

(d) What is the equation of the top boundary of this new region?

Open-ended

13 (a) Find two consecutive whole number values for c so that the gradient of the straight line described by the equation $y = -\left(\frac{c-9}{1-c}\right)x$ is positive.

(b) What can be concluded about the gradient when $c = 1$? What does this tell you about the line?

14 Consider the graphs of $y = ax^2$ and $y = 2a - ax^2$. Draw the graphs on the same set of axes for $a = 1, 2$. What can you discover about the points of intersection of the two graphs? Repeat for negative values of a and comment on your answer.

10A

Problem solving

Different paths

Mark wants to build a path from his front door to the gate. He draws a grid and determines that his front door is at grid reference (0, 0) and the gate is at (5, 12). As he has been learning about graphs of the type $y = ax^n$, he decides to see how many different graphs he can find that fit these two points.

1 Find the equations of the curves that he would need for $n = 1, 2, 3$ and 4.

2 For each value of n, determine at least one other grid reference that he should use when making the path.

Strategy options

- Draw a diagram.
- Test all possible combinations.
- Look for a pattern.

8.2 Transformations of $y = ax^n$

Graphs of $y = a(x - h)^n$

The graph of $y = a(x - h)^n$ is a translation of the graph of $y = ax^n$ by h units horizontally.

- If $h > 0$ the translation is to the right.
- If $h < 0$ the translation is to the left.

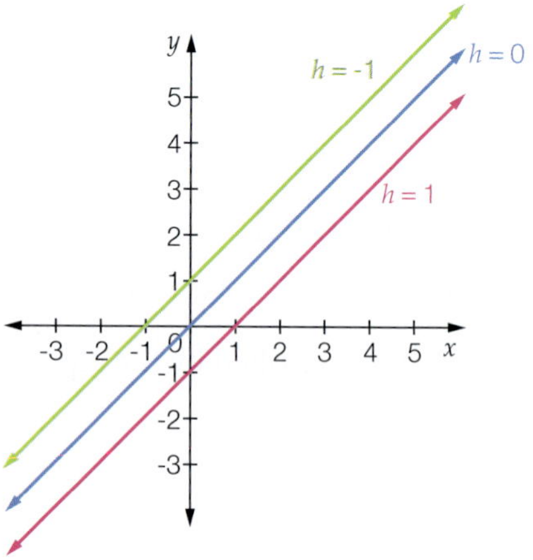

For $a = 1$, $n = 1$ and $h = -1, 0, 1$, $y = a(x - h)^n$ gives the equations $y = x - 1$, $y = x$ and $y = x + 1$. The graphs of these equations are straight lines.

The x-values of the x-intercepts are 1, 0 and -1.

The y-values of the y-intercepts are -1, 0 and 1 respectively.

The value of h represents the x-coordinate of the x-intercept and it translates the graph of $y = ax$ a distance of h units in the positive direction of the x-axis.

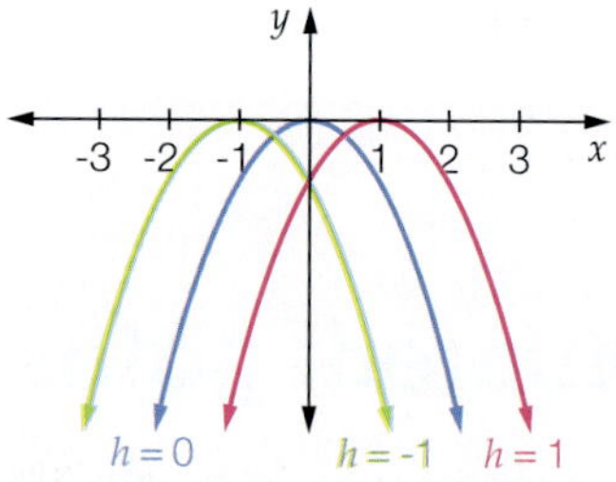

For $a = -3$, $n = 2$ and $h = -1, 0, 1$, $y = a(x - h)^n$ gives the equations $y = -3(x + 1)^2$, $y = -3x^2$, $y = -3(x - 1)^2$. The graphs of the equations are parabolas, symmetrical about the line $x = h$.

The y-value of the y-intercept is $a(-h)^2$, found by substituting $x = 0$ in $y = a(x - h)^2$.

The value of h represents the x-coordinate of the x-intercept and it translates the graph of $y = ax^2$ a distance of h units in the direction of the x-axis.

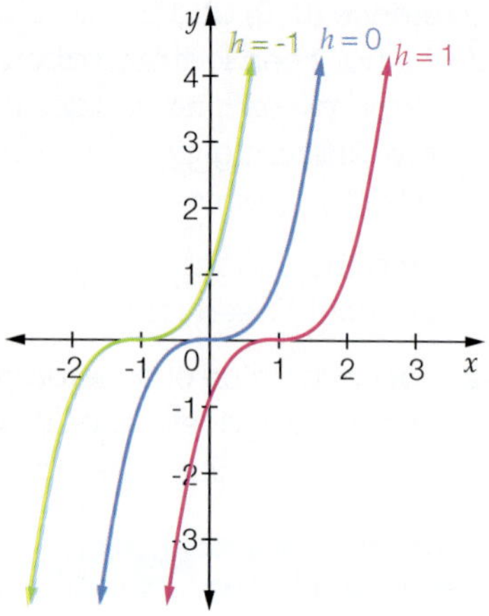

For $a = 1$, $n = 3$ and $h = -1, 0, 1$, $y = a(x - h)^n$, gives the cubic equations $y = (x + 1)^3$, $y = x^3$ and $y = (x - 1)^3$.

The y-value of the y-intercept is $a(-h)^3$, found by substituting $x = 0$ in $y = a(x - h)^3$. This type of cubic has only one x-value of the x-intercept.

The value of h represents the x-intercept and it translates the graph of $y = ax^3$ a distance of h units in the direction of the x-axis.

The usual order for graphing transformations is:

1. dilations
2. reflections
3. translations

10A

Worked example 3

W.E. 3

Sketch the graph of $y = -4(x - 1)^3$ by transforming the basic graph of $y = x^3$.

Thinking | **Working**

1 Sketch the graph of $y = x^3$.

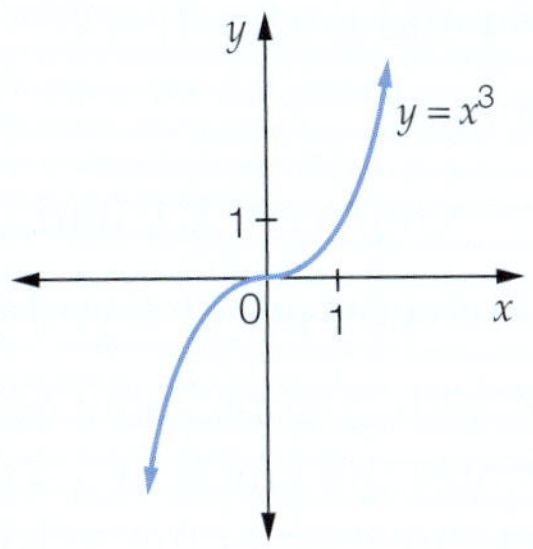

2 Dilate the graph. Find a reference point to help shape the graph. (Here, the dilation factor is 4 so the graph of $y = 4x^3$ will be narrower. For the reference point, substitute $x = 1$ to obtain the point (1, 4).)

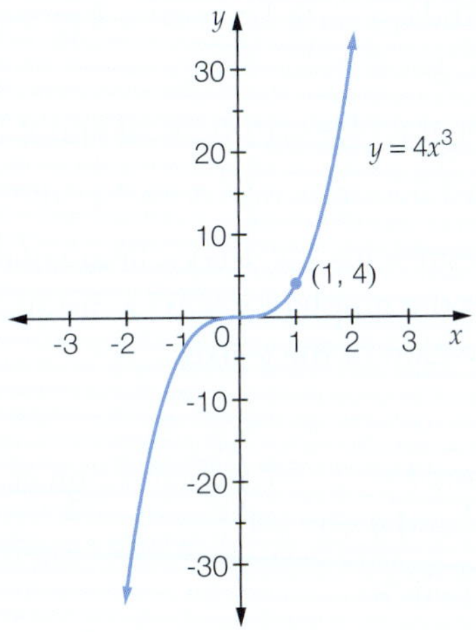

3 Reflect the graph. (Here, reflect the graph in the x-axis to obtain the graph $y = -4x^3$. When reflected in the x-axis the reference point (1, 4) becomes (1, -4).)

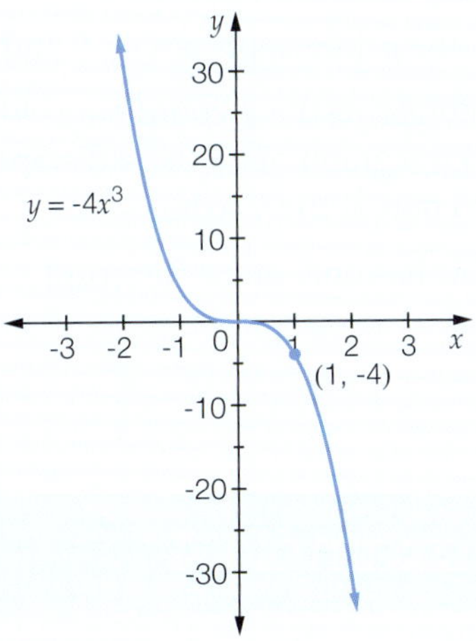

4 Translate the graph horizontally. Locate the y-intercept. (Here, translate 1 unit to the right to obtain the graph $y = -4(x - 1)^3$ and substitute $x = 0$ to find the y-intercept at the point (0, 4). The x-intercept (0, 0) becomes (1, 0).)

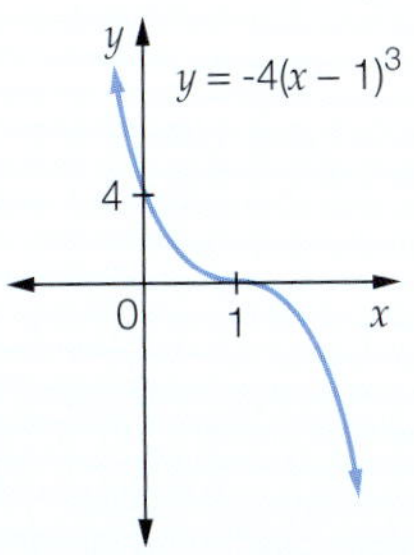

10A

Graphs of $y = ax^n + k$

The graph of $y = ax^n + k$ is a translation of the graph of $y = ax^n$ by k units vertically.

- If $k > 0$ the translation is up.
- If $k < 0$ the translation is down.

For $a = 1$, $n = 1$ and $k = -1, 0, 1$, $y = ax^n + k$ gives the equations $y = x - 1$, $y = x$ and $y = x + 1$.

The graphs are straight lines.

The x-values of the x-intercepts are 1, 0 and -1.

The y-values of the y-intercepts are -1, 0 and 1 respectively.

The value of k represents the y-coordinate of the y-intercept and it translates the graph of $y = ax$ a distance of k units in the positive direction of the y-axis.

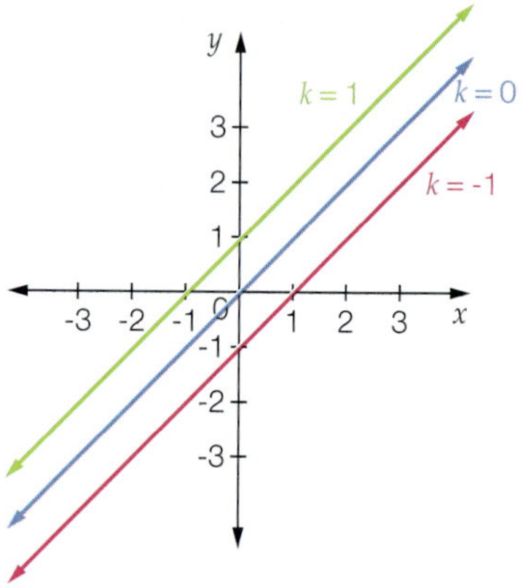

For $a = 1$, $n = 2$ and $k = -1, 0, 1$, $y = ax^n + k$ gives the equations $y = x^2 - 1$, $y = x^2$, $y = x^2 + 1$.

The graphs are parabolas symmetrical about the line $x = h$.

The graphs may have zero, one or two x-intercepts.

The value of k represents the y-coordinate of the y-intercept and it translates the graph of $y = ax^2$ a distance of k units in the positive direction of the y-axis.

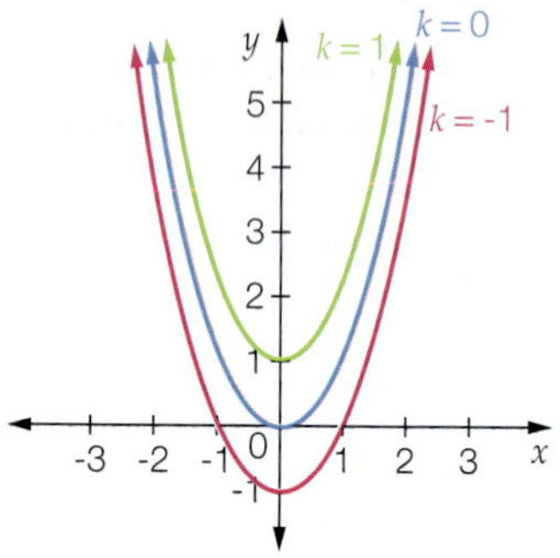

For $a = 1$, $n = 3$ and $k = -1, 0, 1$, $y = ax^n + k$ gives the equations $y = x^3 - 1$, $y = x^3$ and $y = x^3 + 1$.

The graphs are cubics.

The x-values of the x-intercepts are 1, 0 and -1.

The y-values of the y-intercepts are -1, 0 and 1 respectively.

The value of k represents the y-coordinate of the y-intercept and it translates the graph of $y = ax^3$ a distance of k units in the positive direction of the y-axis.

This type of cubic has only one x-intercept.

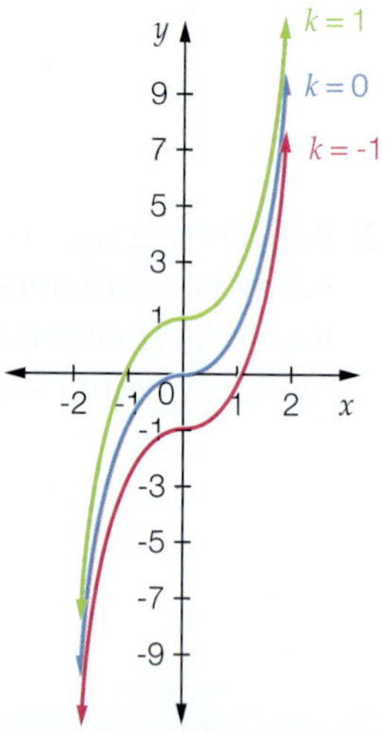

10A

Worked example 4

W.E. 4

Sketch the graph of $y = -3x^4 + 2$ by transforming the basic graph of $y = x^4$.

Thinking	Working
1 Sketch the graph of $y = x^4$.	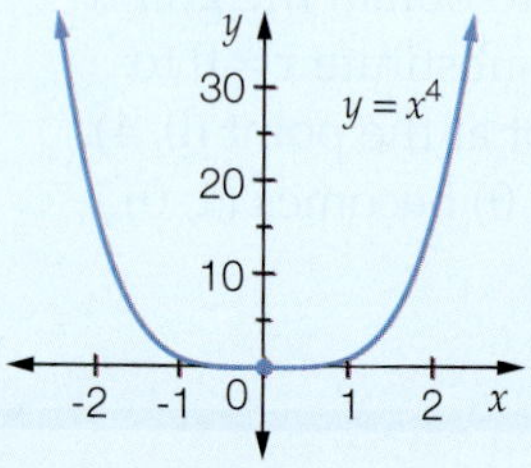

2 Dilate the graph. Find a reference point to help shape the graph. (Here, the dilation factor is 3 so the graph of $y = 3x^4$ will be narrower. For the reference point, substitute $x = 1$ to obtain the point (1, 3).)

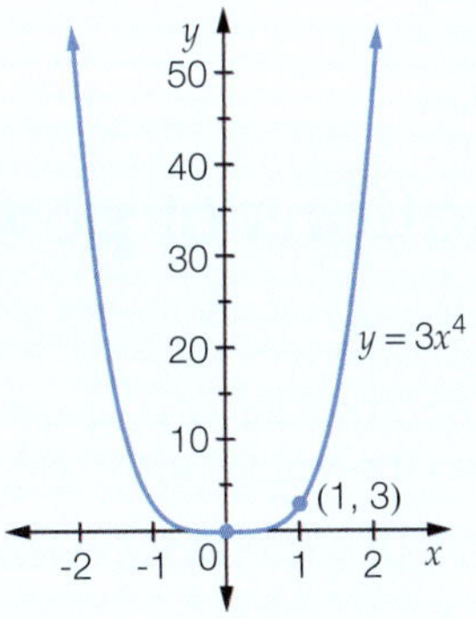

3 Reflect the graph. (Here, reflect the graph in the x-axis to obtain the graph $y = -3x^4$. When reflected in the x-axis, the reference point becomes (1, -3).)

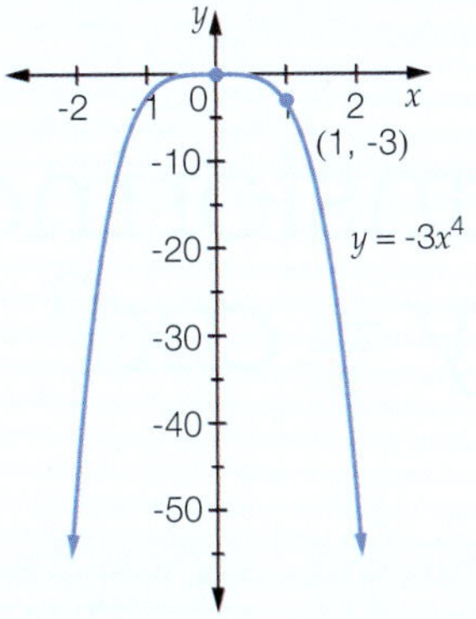

4 Translate the graph vertically. Locate the y-intercept. (Here, translate 2 units up to obtain the graph $y = -3x^4 + 2$. Substitute $x = 0$ to find the y-intercept (0, 2).)
The reference point becomes (1, -1).

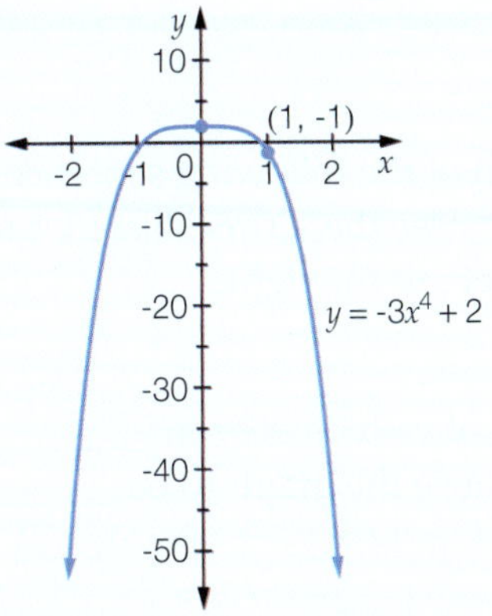

Graphs of $y = a(x - h)^n + k$

Transformations are applied in the order of dilations, reflections and translations.

The graph of $y = a(x - h)^n + k$ is a translation of the graph of $y = ax^n$ by h units horizontally and k units vertically.

- If $h > 0$, the translation is to the right and if $h < 0$, the translation is to the left.
- If $k > 0$, the translation is up and if $k < 0$, the translation is down.

10A

Interactive

Transformations of power functions

Explore the shape and position of a graph of the form $y = (x - h)^n + k$ when the values are changed.

Go to the eBook or the Pearson Places website to access this interactive.

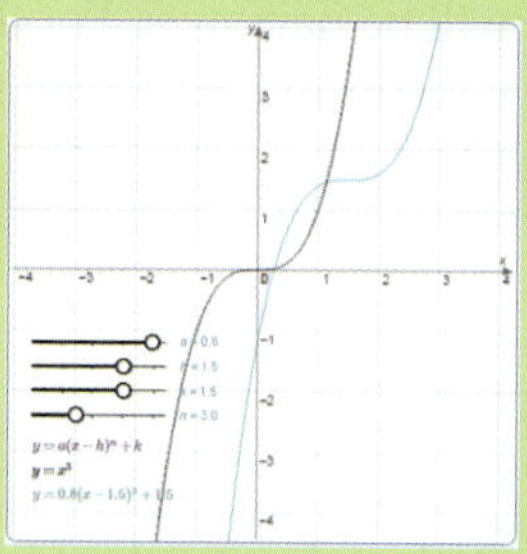

8.2 Transformations of $y = ax^n$

Navigator

Answers p. 819

1, 2, 3, 4, 5, 6, 7, 8, 9, 10, 11, 12, 13, 14	1 (a, c, e), 2, 3, 4, 5, 6, 7, 9, 10, 11, 12, 13, 14	1 (e, f), 2, 3, 4, 5, 6, 7, 9, 10, 11, 12, 13, 14

10A

Fluency

1 Complete each of the following sentences to describe the transformation(s) from the graph of $y = x^2$. Use the terms: dilated, translated, reflected, wider, narrower, x, y, up, down, to the left, to the right.

(a) $y = 3x^2$

The graph of $y = x^2$ has been ________________ by a factor of 3 parallel to the y-axis. This will cause the graph to be ________________ than that of $y = x^2$.

(b) $y = -x^2$

The graph of $y = x^2$ has been ________________ in the __-axis.

(c) $y = x^2 + 9$

The graph of $y = x^2$ has been ________________ 9 units ________________.

(d) $y = (x + 6)^2$

The graph of $y = x^2$ has been ________________ 6 units ________________.

(e) $y = 2(x - 4)^2 + 7$

The graph of $y = x^2$ has been ________________ by a factor of 2 parallel to the y-axis, ________________ 4 units ________________ and 7 units ________________.

(f) $y = -\frac{1}{3}(x + 1)^2 - 4$

The graph of $y = x^2$ has been ________________ in the __-axis, ________________ by a factor of $\frac{1}{3}$ parallel to the y-axis, ________________ 1 unit ________________ and 4 units ________________.

2 Sketch the graph of $y = 4x - 2$ by transforming the graph of $y = x$.

3 Sketch each of the following graphs by transforming the graph of $y = x^2$.

(a) $y = x^2 - 1$ (b) $y = (x - 2)^2$ (c) $y = 4x^2 + 1$

(d) $y = -x^2 + 1$ (e) $y = \frac{1}{2}(x - 3)^2$ (f) $y = -2(x - 1)^2 - 4$

4 Sketch each of the following graphs by transforming the graph of $y = x^3$.

(a) $y = -x^3 - 2$ (b) $y = -3(x + 2)^3$

5 Sketch each of the following graphs by transforming the graph of $y = x^4$.

(a) $y = 3(x - 1)^4$ (b) $y = -\frac{1}{10}(x + 2)^4$

Understanding

6 The graph of the equation $y = x^2$ is transformed to the graph of the equation $y = -2(x + 2)^2 - 1$. The graph of the transformed equation is:

A
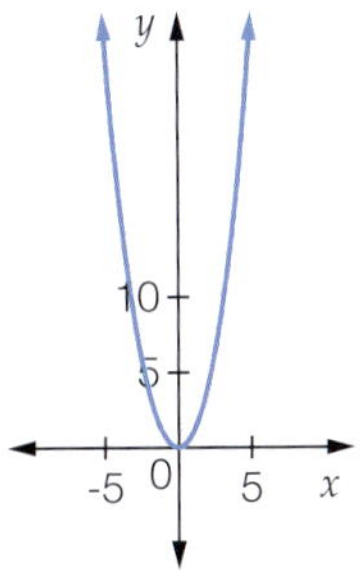

B
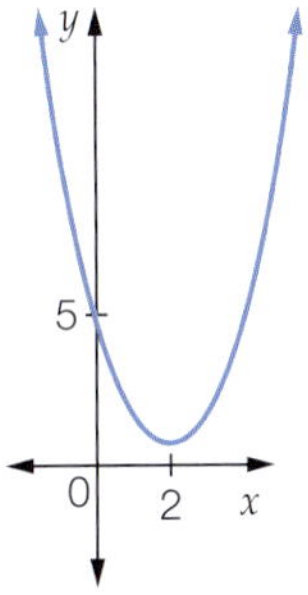

C
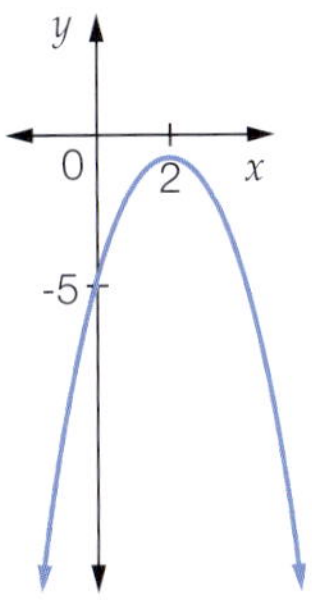

D
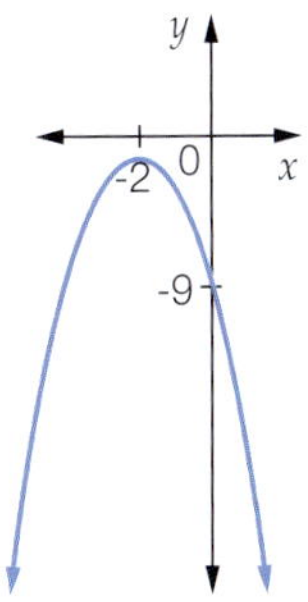

7 To obtain the graph of $y = -3(x + 2)^3 + 1$ from $y = x^3$:

A Dilate by a factor of 3 in the y-direction, reflect in the x-axis, translate 2 units right and translate one unit up.

B Dilate by a factor of 3 in the y-direction, reflect in the x-axis and translate 2 units left.

C Dilate by a factor of 3 in the y-direction, reflect in the x-axis, translate 2 units left and translate one unit up.

D Dilate by a factor of 3 in the y-direction, translate 1 unit left and translate 2 units down.

8 The y-intercept of the graph of the equation $y = (16x - 3)^3$ is:

A (0, 54) B (0, -24) C (0, -27) D (0, 48)

Reasoning

9 The graph of $y = x^3$ has been dilated and translated in both the x- and y-directions.

(a) State the values of h and k, and then substitute into the general form of the equation $y = a(x - h)^3 + k$.

(b) Find the value of a by substituting the coordinates of the point (0, -20) into the equation from your answer to part **(a)**. Hence, state the equation of the graph.

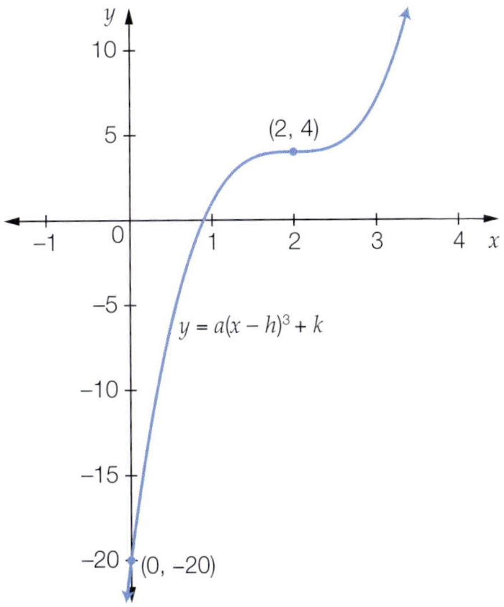

10A

10 A quadratic in the form $y = a(x - h)^2 + k$ has a turning point at (1, 2) and passes through the point (2, -1).

(a) State the values of h and k.

(b) Will the value of a be positive or negative? Give a reason for your answer.

(c) Find the value of a.

11 The shape and position of a snake's body at time t seconds is described by the equation $y = \frac{(-1)^t x^3}{10} + 100t$ where x and y are in centimetres and $-10 \le x \le 10$.

(a) What is the equation describing the snake's initial shape? (Initial means $t = 0$.)

(b) Sketch the graph of the equation in part **(a)** using the basic graph $y = x^3$.

(c) On the same axes sketch the snake's shape and position:

(i) 1 second later

(ii) 2 seconds later.

(d) Compare the position of the snake after 1 second with its initial position. State the y-value about which the snake's new position is a reflection of its initial position.

12 The cross-section of the entrance to a tunnel has the shape of the parabola $y = 5 - 0.2(x - 5)^2$, where x and y are given in metres and $0 \le x \le 10$. A second tunnel with the same basic shape is 5 m higher and has its centre 15 m to the right of the centre of the first tunnel.

(a) What is the equation describing the second tunnel's entrance?

(b) Sketch the graphs of the two equations on the same number plane.

(c) Use Pythagoras' theorem to calculate the distance, to 1 decimal place, between the tops of the entrances to the tunnels.

13 The shape of the front leg of an antique chair is described by the equation $y = (x - 2)^3 + 8$, where x and y are given in centimetres. The second front leg is a reflection of the first leg in the line $x = 12$.

(a) Sketch the graph of the equation describing the shape of the first leg for $0 \le x \le 6$.

(b) Sketch the graph of the second leg on the same set of axes as your answer to part **(a)**.

(c) State the equation describing the shape of the second leg.

(d) What is the minimum distance between the two legs?

Open-ended

14 Starting with the equation $y = x^3$, write two cubic equations so that each equation has a reflection in the x-axis, a dilation factor between 1 and 10, a horizontal translation to the left and a vertical translation downwards.

Polynomials

8.3

A **polynomial** (in one variable x) is an expression for which each term in the expression contains a variable raised to a different positive integer power. Each term has a numerical coefficient. The coefficient of any term may be zero, except for the coefficient a_n of the term with the highest power x^n. The terms are written in descending powers of the variable.

If the coefficient of a term is zero, then its value is zero, so the term is not written. For example, $6x^5 + x$ is a polynomial where all the coefficients of the powers of x are zero except for the coefficient of x^5, which is 6, and the coefficient of x, which is 1.

Consider the expression $5x^6 - 8x^5 + 7x^4 + 2x^2 - 3x + 9$.

- This is a polynomial in x, as all the powers are positive integers.
- $P(x)$ can be used to represent a polynomial in x.

  e.g. $P(x) = 5x^6 - 8x^5 + 7x^4 + 2x^2 - 3x + 9$

You can also use other capital letters such as $Q(x)$ or $R(x)$ to represent polynomials.

In a general equation of a polynomial, the highest power of the variable is n. The first term, called the **leading term**, has a power of n. The coefficient of the leading term can be labelled a_n and is called the **leading coefficient**.

In the polynomial $5x^6 - 8x^5 + 7x^4 + 2x^2 - 3x + 9$:

- There is no x^3 term. This means the coefficient of x^3 is 0.
- The power of the last term is zero. As $x^0 = 1$, this makes the last term a constant.

$5x^6$	$-8x^5$	$+7x^4$	$+0x^3$	$+2x^2$	$-3x$	$+9$
$a_n = a_6 = 5$	$a_{(n-1)} = a_5 = -8$	$a_{(n-2)} = a_4 = 7$	$a_{(n-3)} = a_3 = 0$	$a_{(n-4)} = a_2 = 2$	$a_{(n-5)} = a_1 = -3$	$a_{(n-6)} = a_0 = 9$

In a polynomial where x^n is the leading term, there are $(n + 1)$ terms. However, as many of the coefficients may be zero, fewer than $(n + 1)$ terms may appear in the expression. The **degree** of the polynomial is the highest power of x that appears in the expression. In a polynomial of $(n + 1)$ terms, the polynomial is of degree n.

- A polynomial can be written as

  $P(x) = a_n x^n + a_{(n-1)} x^{(n-1)} + a_{(n-2)} x^{(n-2)} + \ldots + a_2 x^2 + a_1 x^1 + a_0$

  where $a^n \neq 0$, n is a positive integer and $a_0, a_1, a_2, \ldots a_n$ are coefficients of the powers of x.
- The polynomial $P(x)$ is of degree n (the highest power of x).
- For $P(x)$, the leading coefficient is a_n and the constant term is a_0 (the term without an x).
- The coefficient of a monic polynomial is equal to 1, $a_n = 1$.

10A

Worked example 5

W.E. 5

State the degree, leading coefficient and constant term of each polynomial.

(a) $5x^4 + 3x^3 - x^2 + x - 2$ (b) $12 + 6x - x^3$

Thinking	Working
(a) 1 Write the polynomial in decreasing order of powers.	(a) $5x^4 + 3x^3 - x^2 + x - 2$
2 Identify the leading term, the first term in your ordered polynomial.	The leading term is $5x^4$.
3 Write the degree, the power of the leading term.	The degree is 4.
4 Write the leading coefficient, the coefficient of the leading term.	The leading coefficient is 5.
5 Write the constant term.	The constant term is -2.
(b) 1 Write the polynomial in decreasing order of powers.	(b) $-x^3 + 6x + 12$
2 Identify the leading term, the first term in your ordered polynomial.	The leading term is $-x^3$.
3 Write the degree, the power of the leading term.	The degree is 3.
4 Write the leading coefficient, the coefficient of the leading term.	The leading coefficient -1.
5 Write the constant term.	The constant term is 12.

Substitution into a polynomial

To substitute $x = 2$ into $P(x) = 5x^6 - 8x^5 + 7x^4 + 2x^2 - 3x + 9$, you write:

$$\begin{aligned} P(2) &= 5(2)^6 - 8(2)^5 + 7(2)^4 + 2(2)^2 - 3(2) + 9 \\ &= 187 \end{aligned}$$

In general, if you substitute $x = a$ into $P(x)$ you write this as $P(a)$.

$P(a)$ is the value of $P(x)$ when $x = a$.

Addition and subtraction of polynomials

Polynomials can be added and subtracted by combining like terms (terms of the same degree).

Worked example 6

W.E. 6

$P(x) = 2x^3 - x^2 + 4x - 7$ and $Q(x) = x^3 + 3x^2 - x + 1$

(a) Evaluate $P(2)$ and $Q(-2)$. (b) Write the expression for $P(x) + 2Q(x)$.

10A

Thinking		Working
(a)	1 Write the polynomial for $P(x)$.	(a) $P(x) = 2x^3 - x^2 + 4x - 7$
	2 Substitute $x = 2$ and simplify.	$P(2) = 2 \times 2^3 - 2^2 + 4 \times 2 - 7$ $= 16 - 4 + 8 - 7$ $= 13$
	3 Write the polynomial for $Q(x)$.	$Q(x) = x^3 + 3x^2 - x + 1$
	4 Substitute $x = -2$ and simplify.	$Q(-2) = (-2)^3 + 3 \times (-2)^2 - (-2) + 1$ $= -8 + 12 + 2 + 1$ $= 7$
(b)	1 Write the rule.	(b) $P(x) + 2Q(x)$
	2 Expand and simplify by collecting like terms.	$= 2x^3 - x^2 + 4x - 7 + 2(x^3 + 3x^2 - x + 1)$ $= 2x^3 - x^2 + 4x - 7 + 2x^3 + 6x^2 - 2x + 2$ $= 4x^3 + 5x^2 + 2x - 5$

Multiplication by a linear expression

A linear expression is a polynomial of degree 1. It has the form $a_1x + a_0$.

Its product with a polynomial of degree n is a polynomial of degree $(n + 1)$.

Worked example 7

W.E. 7

If $P(x) = 3x - 2$ and $Q(x) = 4x^2 + x - 5$, write $P(x).Q(x)$ as a polynomial.

Thinking	Working
1 Write the polynomial product in brackets.	$P(x).Q(x) = (3x - 2)(4x^2 + x - 5)$
2 Expand the brackets by multiplying each term in the second bracket by each term in the first bracket.	$= 3x(4x^2 + x - 5) - 2(4x^2 + x - 5)$ $= 12x^3 + 3x^2 - 15x - 8x^2 - 2x + 10$
3 Simplify by adding like terms.	$= 12x^3 - 5x^2 - 17x + 10$

Multiplication by a non-linear expression

The product will be a polynomial whose degree is the sum of the degrees of the two expressions being multiplied.

Worked example 8

W.E. 8

If $P(x) = x^2 - 2x - 3$ and $Q(x) = 4x^3 + x^2 - 6$, write $P(x).Q(x)$ as a polynomial.

Thinking	Working
1 Write the polynomial product in brackets.	$P(x).Q(x) = (x^2 - 2x - 3)(4x^3 + x^2 - 6)$
2 Expand the brackets by multiplying each term in the second bracket by each term in the first bracket.	$= x^2(4x^3 + x^2 - 6) - 2x(4x^3 + x^2 - 6)$ $- 3(4x^3 + x^2 - 6)$ $= 4x^5 + x^4 - 6x^2 - 8x^4 - 2x^3 + 12x - 12x^3$ $- 3x^2 + 18$
3 Simplify by adding like terms.	$= 4x^5 - 7x^4 - 14x^3 - 9x^2 + 12x + 18$

10A

Numerical long division

When the improper fraction $\frac{20}{7}$ is written as $2\frac{6}{7}$ it means that 7 divides into 20 twice and leaves a remainder of 6.

Before performing division of polynomials you need to be familiar with numerical long division without using a calculator. Now $25 \div 4 = 6\frac{1}{4}$, which means $25 = 6 \times 4 + 1$. Here, 25 is the **dividend**, 4 is the **divisor**, 6 is the **quotient** and 1 is the **remainder**.

Now consider $625 \div 4$. This can be evaluated using short division. Division by integers up to 12 can be done with short division and our tables. Always try to cancel factors and reduce the divisor to an integer from 1 to 12.

If the divisor is greater than 12 and has no common factors with the dividend, you need a method called long division. The process is the same as short division but you write all the steps instead of doing them mentally.

Worked example 9

W.E. 9

Evaluate each expression using long division. Do not use a calculator.

(a) $3689 \div 17$ **(b)** $16\,319 \div 23$

Thinking	Working
(a) 1 Write the divisor divided into the dividend in new notation. (Here, 17 is divided into 3689.)	(a)
2 Divide the divisor into the first number formed by the leading digits of the dividend that will give a value greater than zero but less than or equal to 9. Write this answer above the last digit used in the dividend. (Here, 36 divided by 17 is 2, with a remainder. Write a 2 above the 6.)	$\begin{array}{r} 217 \\ 17\overline{)3689} \\ -34 \\ \hline 28 \\ -17 \\ \hline 119 \\ -119 \\ \hline 0 \end{array}$
3 Multiply this answer by the divisor and write this product under the digits used in step **1**. (Here, multiply 2 by 17 to get 34 and write 34 under 36.)	
4 Subtract these numbers to find a remainder. (Here, subtract 34 from 36 to get 2.)	
5 Bring down the next digit and write it after the remainder found in step **4**. (Here, bring down the 8 to make 28.)	
6 Divide the divisor into the number you have formed at step **5** and write the answer after the first digit of your quotient. (Here, 28 divided by 17 is 1, with a remainder.)	
7 Multiply this answer by the divisor and write this product under the digits you divided into in step **6**. (Here, $1 \times 17 = 17$ and write the 17 under 28.)	
8 Subtract the numbers to find a remainder ($28 - 17 = 11$).	
9 Bring down the next digit and write it after the remainder in step **8**. (Here, bring down 9 to give 119.)	

10A

10 Divide the divisor into the number you have formed in step **9** and write the answer after the second digit in your quotient. (Here, 119 divided by 17 is 7.)

11 Multiply the divisor by this answer and write this product under the digits you divided into in step **10**. $(17 \times 7 = 119)$

12 Subtract the numbers to find a remainder. $(119 - 119 = 0)$ There are no other digits to bring down, so there is no remainder in this division. If there is no remainder, the divisor is a factor of the dividend. (17 is a factor of 3689.)

13 Write your answer.

$3689 \div 17 = 217$

(b) 1 Write the divisor divided into the dividend in new notation. (Here, 23 is divided into 16 319.)

(b)

$$\begin{array}{r} 709 \\ 23\overline{)16319} \\ -\underline{161} \\ 219 \\ -\underline{207} \\ 12 \end{array}$$

2 Divide the divisor into the first number formed by the leading digits of the dividend that will give an answer and write this answer above the last digit used in the dividend. (Here, 163 divided by 23 is 7, with a remainder. Write 7 above the 3.)

3 Multiply this answer by the divisor and write this product under the digits used in step **1**. (Here, multiply 7 by 23 to get 161. Write 161 under 163.)

4 We subtract these numbers to find a remainder. (Here, subtract 161 from the 163 to get 2.)

5 Bring down the next digit and write it after the remainder found in step **4**. (Here, bring down the 1 to make 21.)

6 Divide the divisor into the number from step **5** and write the answer after the first digit of your quotient. (Here, 21 divided by 23 is 0, with a remainder.) Write a 0 after the first digit in the quotient, bring down the next digit and write it after the remainder found in step **4**. (Here, bring down 9 to make 219.)

7 Divide the divisor into the number you have formed in step **6** and write the answer after the third digit in your quotient. (Here, 219 divided by 23 is 9, with a remainder.)

8 Multiply the divisor by this answer and write this product under the digits you divided into in step **7**. $(23 \times 9 = 207)$

9 Subtract the numbers to find a remainder. $(219 - 207 = 12)$ The remainder is 12.

10 Write your answer.

$16\,319 \div 23 = 709\frac{12}{23}$

10A

Polynomial division

When a polynomial of degree n, $P(x)$, is divided by a linear expression $D(x)$, the result is a polynomial $Q(x)$ of degree $n - 1$ with a remainder $R(x)$. $\frac{P(x)}{D(x)} = Q(x) + \frac{R(x)}{D(x)}$ or $P(x) = D(x) \times Q(x) + R(x)$, where $P(x)$ is the dividend, $D(x)$ is the divisor, $Q(x)$ is the quotient and $R(x)$ is the remainder. If $D(x)$ is linear, $R(x)$ is an integer. To find $Q(x)$ and $R(x)$ you use a method similar to long division of whole numbers.

Worked example 10

W.E. 10

Use long division to find the quotient and remainder for $\frac{x^2 + 6x + 10}{x + 2}$.

Thinking	Working
1 Write the expression in the required form. (Here, the dividend is $x^2 + 6x + 10$ and the divisor is $x + 2$.)	$x+2\overline{)\,x^2+6x+10}$
2 Divide the first term of the dividend by the leading term of the divisor and place the result above the second term in the dividend aligning the powers of x. (Here, $x^2 \div x = x$. Write x above $6x$.)	$\begin{array}{r} x \\ x+2\overline{)\,x^2+6x+10} \end{array}$
3 Multiply each term of the divisor by the first term in the quotient. Write the result underneath. (Here, $x(x + 2) = x^2 + 2x$.)	$\begin{array}{r} x \\ x+2\overline{)\,x^2+6x+10} \\ x^2+2x \end{array}$
4 Insert brackets and subtract the like terms.	$\begin{array}{r} x \\ x+2\overline{)\,x^2+6x+10} \\ -(x^2+2x) \\ \hline 0+4x \end{array}$
5 Bring down the next term from the dividend to the result of the subtraction.	$\begin{array}{r} x \\ x+2\overline{)\,x^2+6x+10} \\ -(x^2+2x) \\ \hline 0+4x+10 \end{array}$
6 Divide the first term of the new dividend by the first term of the divisor and place the result above the original dividend. (Here, $4x \div x = 4$. Write 4 above 10.)	$\begin{array}{r} x+4 \\ x+2\overline{)\,x^2+6x+10} \\ -(x^2+2x) \\ \hline 4x+10 \end{array}$ original dividend ($x^2+6x+10$); new dividend ($4x+10$)
7 Multiply each term of the divisor by the second term in the quotient. Write the result underneath. (Here, $4(x + 2) = 4x + 8$.)	$\begin{array}{r} x+4 \\ x+2\overline{)\,x^2+6x+10} \\ -(x^2+2x) \\ \hline 4x+10 \\ 4x+8 \end{array}$

8 Insert brackets and subtract the like terms.

$$\begin{array}{r} x + 4 \quad \text{quotient} \\ x+2\overline{)\,x^2 + 6x + 10} \\ -(x^2 + 2x) \\ \hline 4x + 10 \\ -(4x + 8) \\ \hline 0 + 2 \quad \text{remainder} \end{array}$$

9 Stop when the degree of the remainder is less than the degree of the divisor. State the result. You are dividing by a linear expression so the remainder will be a constant.

$$\frac{x^2 + 6x + 10}{x + 2} = x + 4 + \frac{2}{x + 2}$$

You can write this result as $x^2 + 6x + 10 = (x + 2)(x + 4) + 2$.

You could also consider the long division process in the previous worked example in the following form:

$$\begin{aligned} & x^2 + 6x + 10 \\ &= x^2 + 2x + 4x + 10 \\ &= x(x + 2) + 4x + 8 + 2 \\ &= x(x + 2) + 4(x + 2) + 2 \\ &= (x + 2)(x + 4) + 2 \end{aligned}$$

All the powers of x in the dividend (up to the degree of the polynomial) should be shown in the calculations. This may require the insertion of coefficients that are zero.

The remainder must be zero or have a degree less than the degree of the divisor.

Worked example 11 **W.E. 11**

Use long division to find the quotient and remainder for $\dfrac{8x^3 + 4x - 1}{2x + 3}$.

Thinking	Working
1 Write the expression in the required form inserting a zero coefficient for any powers of x that are missing.	$2x+3\overline{)\,8x^3 + 0x^2 + 4x - 1}$
2 Divide the first term of the dividend by the leading term of the divisor. Multiply each term of the divisor by this term in the quotient. Insert brackets and subtract the like terms.	$\begin{array}{r} 4x^2 \\ 2x+3\overline{)\,8x^3 + 0x^2 + 4x - 1} \\ -(8x^3 + 12x^2) \\ \hline 0 - 12x^2 \end{array}$
3 Bring down the next term from the dividend. Repeat step **2**.	$\begin{array}{r} 4x^2 - 6x \\ 2x+3\overline{)\,8x^3 + 0x^2 + 4x - 1} \\ -(8x^3 + 12x^2) \\ \hline 0 - 12x^2 + 4x \\ -(-12x^2 - 18x) \\ \hline 0 + 22x \end{array}$

4 Bring down the next term from the dividend. Repeat step **2**. Stop when the degree of the remainder is less than the degree of the divisor.

$$
\begin{array}{r}
4x^2 - 6x + 11 \\
2x+3\,\overline{)\,8x^3 + 0x^2 + 4x - 1} \\
-(8x^3 - 12x^2) \\
\hline
0 - 12x^2 + 4x \\
-(-12x^2 - 18x) \\
\hline
0 + 22x - 1 \\
-(22x + 33) \\
\hline
0 - 34
\end{array}
$$

5 State the result.

$$\frac{8x^3 + 4x - 1}{2x + 3} = 4x^2 - 6x + 11 - \frac{34}{2x + 3}$$

The result can also be written as $8x^3 + 4x - 1 = (2x + 3)(4x^2 - 6x + 11) - 34$.

The following diagram may help you to remember the terminology for polynomial long division:

Quotient
Divisor) Dividend
Remainder

This can also be written as:

Dividend = Divisor × Quotient + Remainder

- When adding and subtracting polynomials, only combine like terms.
- When expanding the product of binomial expressions, work through the expression in order, being very careful with negative signs.
- When dividing polynomials, make sure that all the powers less than the degree of the dividend appear in your expression. This may mean the introduction of zero coefficients.

8.3 Polynomials

Navigator

Answers p. 821

1, 2, 3, 4, 5, 6, 7, 8, 9, 11, 12, 13, 14, 15, 16, 17, 20, 21	1, 2 (a–c), 3, 4, 5 (columns 1–2), 6, 7, 8, 9, 10, 11, 12, 13, 14, 15, 16, 17, 18, 21, 22	1 (a–b), 2 (b–c), 3, 4, 5 (columns 2–3), 6, 7, 8, 10, 11, 12, 13, 14, 15, 16, 17, 18, 19, 21, 22

Fluency

W.E. 5

1 State the degree, leading coefficient and constant term of each polynomial.

(a) $2x^3 + 5x^2 - x + 2$ (b) $5 - 4x^4 - 3x^3 + 7x^2$ (c) x (d) $-6 - x - x^2$

W.E. 6

2 If $P(x) = 3x^3 - 2x^2 - x + 2$ and $Q(x) = -x^3 - 4x^2 - 5x + 6$, evaluate:

(a) $P(3)$ (b) $Q(4)$ (c) $P(-2)$ (d) $Q(-1)$

W.E. 7

3 Write each expression as a polynomial if $P(x) = 5x + 1$, $Q(x) = x^2 + 2x - 3$ and $R(x) = 2x^3 - x^2 - x + 6$.

(a) $P(x).\,Q(x)$ (b) $P(x).\,R(x)$

4 If $P(x) = x^2 - 3x + 4$ and $Q(x) = 2x^3 + 3x^2 - 5$, write $P(x).\ Q(x)$ as a polynomial. **W.E. 8**

5 Evaluate each expression by performing long division. Do not use a calculator. **W.E. 9**

(a) $3715 \div 25$ (b) $4061 \div 12$ (c) $5924 \div 17$

(d) $2018 \div 19$ (e) $14\,797 \div 29$ (f) $10\,946 \div 123$

6 For $3x^3 + 31x^2 + 21x - 16 = (3x + 4)(x^2 + 9x - 5) + 4$, state the divisor, dividend, quotient, and remainder.

7 Use long division to find the quotient and remainder for: **W.E. 10**

(a) $\dfrac{x^2 + 6x + 10}{x + 3}$ (b) $\dfrac{x^2 + 3x - 1}{x - 2}$

(c) $\dfrac{5x^2 - 2x - 9}{x - 3}$ (d) $\dfrac{6x^2 - 10x - 4}{3x + 1}$

8 Use long division to find the quotient and remainder for: **W.E. 11**

(a) $\dfrac{x^3 + 6x + 5}{x + 2}$ (b) $\dfrac{x^4 - 3x^2 + 1}{x - 3}$

(c) $\dfrac{3x^3 - x^2 + 8x - 9}{x - 2}$ (d) $\dfrac{x^4 - 7x^3 + 5x^2 - x + 3}{x - 1}$

9 Write $A(x)B(x) + 2C(x)$ as a polynomial if $A(x) = x + 1$, $B(x) = x^2 - 3x$ and $C(x) = 5x^2 + 4$.

Understanding

10A

10 Write $-3A(x)\ B(x) + C(x)$ as a polynomial if $A(x) = -x + 4$, $B(x) = 4x^2 - 2x + 1$ and $C(x) = -x^2 + 3x - 8$.

11 Find the quotient and remainder for $\dfrac{P(x)}{Q(x)}$ if:

(a) $P(x) = x^2 + 7x + 13$, $Q(x) = x + 3$ (b) $P(x) = 4x^2 + 4$, $Q(x) = 2x - 1$

(c) $P(x) = 2x^2 - 3x + 1$, $Q(x) = x + 1$ (d) $P(x) = 3x^3 + 2x - 4$, $Q(x) = x - 2$

12 If $P(x) = -5x^3 - 8x^2 - 3x + 2$ and $Q(x) = x^3 + 4x^2 - x + 1$ and $R(x) = 2P(x) - 2Q(x)$, the leading coefficient of $R(x)$ is:

A -12 B -5 C 1 D 8

13 Show by long division that the remainder is zero when $4x^2 + x - 3$ is divided by $4x - 3$.

14 The remainder when $x^3 + x^2 + x + 1$ is divided by $x - 1$ is:

A -5 B -3 C 2 D 4

15 Calculate $\dfrac{A(x)}{B(x)}$ when $A(x) = x^4 - 6x^3 + 5$ and $B(x) = x - 5$.

16 $P(x) = 3x^3 - 2x^2 - x + 2$ and $Q(x) = -x^3 - 4x^2 - 5x + 6$.

(a) If $R(x) = 5P(x) - Q(x)$, write the polynomial $R(x)$ and evaluate $R(2)$.

(b) If $R(x) = 2P(x) + 3Q(x)$, write the polynomial $R(x)$ and evaluate $R(-3)$.

(c) If $6P(x) = 2R(x) - 10Q(x)$, write the polynomial $R(x)$ and evaluate $R(-1)$.

Reasoning

17 (a) Calculate $\frac{A(x)}{B(x)}$ when $A(x) = 2x^5 - 3x^4 - 6x^2 - 3x + 18$ and $B(x) = 2x - 3$.

(b) What is the value of $\frac{A(x)}{B(x)}$ for $x = 2$?

(c) Evaluate $A(2)$ and $B(2)$ and find $\frac{A(2)}{B(2)}$. Compare this to your answer in part **(b)**.

18 Alana has $P(x)$ dollars, Bruno has $Q(x)$ dollars and Cien has $\frac{P(x)}{Q(x)}$ dollars, where $P(x) = 12x^4 + 16x^3 + 3x^2 + 28x + 32$ and $Q(x) = 3x + 4$.

(a) Write the polynomial that describes the amount of money Cien has.

(b) (i) Dylan has x dollars. What is the ratio of Cien's amount of money to Dylan's?

(ii) Use your answer to part **(i)** to explain why x cannot be zero.

19 A sheet of pastry is $(2x^2 - 9x)$ cm long and $(4x + 3)$ cm wide.

(a) Write an expression for the perimeter of the sheet of pastry.

(b) Write an expression for the area of the sheet of pastry.

(c) (i) The sheet of pastry is divided into small rectangles of side lengths x cm and $(x - 1)$ cm. Write an expression for the largest number of small rectangles that could be cut out of the pastry if any unused pastry is rerolled to the same thickness to maximise the number of squares that could be formed.

(ii) If all the pastry is used (there is no remainder), what is the smallest possible value of x and how many rectangles can be formed?

10A

Open-ended

20 Write a non-monic polynomial of degree 5 that contains four terms and has a negative constant term.

21 Find a polynomial $P(x)$ of degree 3 so that $\frac{P(x)}{x-3}$ leaves no remainder.

22 $P_n(x)$ is a polynomial of degree n. It has 1 for all its coefficients so that, for example, $P_2(x) = x^2 + x + 1$ and $P_3(x) = x^3 + x^2 + x + 1$.

(a) Write an expression for $\frac{P_n(x)}{x+1}$ for $n = 1, 2, 3, 4, 5$.

(b) By considering your results in part **(a)**, show that if n is odd, $\frac{P_n(x)}{x+1}$ is a polynomial of degree $n - 1$ consisting of even powers of x.

(c) By considering your results in part **(a)**, show that if n is even, $\frac{P_n(x)}{x+1}$ is a polynomial of degree $n - 1$ consisting of odd powers of x plus a remainder of $\frac{1}{x+1}$.

Half-time 8

1 Graph each equation by constructing a table of values for $x = -3$ to 3. **8.1**

(a) $y = 4x^2$ (b) $y = \frac{1}{4}x^3$

2 Find the quotient and remainder for $\frac{3x^3 + 4x^2 - 20x - 5}{3x + 1}$. **8.3**

3 If $P(x) = 3x - 4$ and $Q(x) = 3x^2 - 2x + 1$, write an expression for: **8.3**

(a) (i) $P(4)$ (ii) $Q(-2)$ (iii) $P(-1).\, Q(-1)$

(b) $2P(x) + 3Q(x)$

(c) $P(x).Q(x)$

(d) $Q(x) - xP(x)$

4 (a) Sketch the graph of $y = 2x^2 + 3$ by starting with $y = x^2$. **8.2**

(b) Sketch the graph of $y = -3(x - 2)^3 + 1$ by starting with $y = x^3$.

(c) Sketch the graph of $y = 2(x + 1)^4$ by starting with the graph of $y = x^4$.

5 Use polynomial division to find the remainder produced by each division. **8.3**

(a) $\frac{9x^3 + 6x^2 - 7x - 3}{x + 1}$

(b) $\frac{x^4 - x^3 + x^2 + x - 1}{3x - 1}$

10A

6 The vertical cross-section of a bottle is described by the equations $y = (x - 4)^3 + 1$ and $y = -(x - 14)^3 + 1$. **8.2**

(a) Graph both equations on the same axes using x-values from 1 to 17.

(b) State the equation of the axis of symmetry of the bottle.

7 Write $x^3 + x - 5 = (x - 1)(x^2 + x + 2) - 3$ as a division of one polynomial divided by another. **8.3**

8 By performing a division, show that: **8.3**

(a) $x - 1$ is a factor of $x^3 - 6x^2 - 9x + 14$

(b) $x - 2$ is a factor of $x^4 - 2x^3 - 3x^2 + 8x - 4$

8.4 Remainder and factor theorems

Consider the polynomial $P(x) = x^2 + x + 2$. It can also be written in the form $P(x) = x^2 + x - 6 + 8$.

It is possible to factorise part of the expression if the constant is split into two terms:

$$\begin{aligned} P(x) &= x^2 + x + 2 \\ &= x^2 + x - 6 + 8 \\ &= (x+3)(x-2) + 8 \end{aligned}$$

So $\dfrac{P(x)}{x-2} = (x+3) + \dfrac{8}{x-2}$ and $\dfrac{P(x)}{x+3} = (x-2) + \dfrac{8}{x+3}$

When $P(x)$ is divided by $(x - 2)$ or $(x + 3)$ the remainder is 8.

If you substitute $x = 2$ into $P(x)$, then $P(x) = 8$.

$$\begin{aligned} P(2) &= 4 + 2 + 2 \\ &= 8 \end{aligned}$$

If $P(x)$ is divided by $(x - 2)$, then $P(2) = 8$ is the remainder.

If you substitute $x = -3$ into $P(x)$, then $P(x) = 8$.

10A

$$\begin{aligned} P(-3) &= 9 - 3 + 2 \\ &= 8 \end{aligned}$$

If $P(x)$ is divided by $(x + 3)$, then $P(-3) = 8$ is the remainder.

Worked example 12

W.E. 12

If $P(x) = x^3 + 3x^2 - 4x + 5$ and $Q(x) = x - 2$:

(a) find $\dfrac{P(x)}{Q(x)}$ **(b)** evaluate $P(2)$ **(c)** comment on the result.

Thinking	Working
(a) Set up and perform the long division.	(a) $x - 2 \overline{) x^3 + 3x^2 - 4x + 5}$ with quotient $x^2 + 5x + 6$: $-(x^3 - 2x^2)$ $5x^2 - 4x$ $-(5x^2 - 10x)$ $6x + 5$ $-(6x - 12)$ 17
(b) Evaluate $P(2)$.	(b) $P(2) = (2)^3 + 3 \times (2)^2 - 4 \times (2) + 5$ $= 8 + 12 - 8 + 5 = 17$
(c) Compare the answers.	(c) The remainder when $P(x)$ is divided by $x - 2$ is the same as the value of $P(2)$.

Remainder theorem

The **remainder theorem** can be used to find the remainder when a polynomial $P(x)$ is divided by a linear factor $(ax - b)$.

- If the polynomial $P(x)$ is divided by $(x - a)$, then the remainder is $P(a)$.
- If the polynomial $P(x)$ is divided by $(ax - b)$, then the remainder is $P\left(\frac{b}{a}\right)$.

Worked example 13

W.E. 13

Use the remainder theorem to find the remainder for:

(a) $\dfrac{2x^2 - x + 1}{x - 2}$ (b) $\dfrac{-4x^3 + x^2 - x + 5}{2x + 3}$

Thinking	Working
(a) 1 Write $P(x)$.	(a) $P(x) = 2x^2 - x + 1$
2 Write the value of a.	$a = 2$
3 Evaluate $P(a)$. (Here, $a = 2$).	$P(2) = 8 - 2 + 1 = 7$
4 Write the remainder.	The remainder is 7.
(b) 1 Write $P(x)$.	(b) $P(x) = -4x^3 + x^2 - x + 5$
2 Write the value of $\frac{b}{a}$.	$\frac{b}{a} = \frac{-3}{2}$
3 Evaluate $P\left(\frac{b}{a}\right)$.	$P\left(\frac{-3}{2}\right) = \frac{27}{2} + \frac{9}{4} + \frac{3}{2} + 5$
(Here, you evaluate $P\left(-\frac{3}{2}\right)$.)	$= 22\frac{1}{4}$
4 Write the remainder.	The remainder is $22\frac{1}{4}$.

Factor theorem

The **factor theorem** can be used to find factors of a polynomial $P(x)$ in the form of $(ax - b)$.

If the remainder is 0, then the expression that you are dividing by is a factor of the polynomial. For the polynomial $P(x)$:

- If $P(a) = 0$, then $(x - a)$ is a factor of $P(x)$.
- If $P\left(\frac{b}{a}\right) = 0$, for the polynomial $P(x)$, then $(ax - b)$ is a factor of $P(x)$.

10A

Worked example 14

W.E. 14

Use the factor theorem to fully factorise $x^3 + 6x^2 + 11x + 6$.

Thinking	Working
1 Find a value of x that makes $P(x) = 0$. Try the factors of the constant term of $P(x)$.	$P(x) = x^3 + 6x^2 + 11x + 6$ Try $x = 1$ $P(1) = 1 + 6 + 11 + 6 = 24$ $\therefore (x - 1)$ is not a factor
As all the terms have a positive coefficient, a negative value is needed.	Try $x = -1$ $P(-1) = -1 + 6 - 11 + 6 = 0$ $(x + 1)$ is a factor.
2 Divide $P(x)$ by the first factor found.	$$\begin{array}{r} x^2 + 5x + 6 \\ x+1 \overline{) x^3 + 6x^2 + 11x + 6} \\ -(x^3 + x^2) \\ \hline 5x^2 + 11x \\ -(5x^2 + 5x) \\ \hline 6x + 6 \\ -(6x + 6) \\ \hline 0 \end{array}$$
3 Write $P(x)$ as the product of the factors.	$P(x) = (x + 1)(x^2 + 5x + 6)$
4 Factorise the quadratic factor.	$x^2 + 5x + 6 = (x + 2)(x + 3)$
5 Write $P(x)$ as the product of all its factors.	$P(x) = (x + 1)(x + 2)(x + 3)$

Worked example 15

W.E. 15

Use the factor theorem to fully factorise $x^4 + x^3 - 7x^2 - x + 6$.

Thinking	Working
1 Find a value of x that makes $P(x) = 0$. Try the factors of the constant term of $P(x)$.	$P(x) = x^4 + x^3 - 7x^2 - x + 6$ Try $x = 1$ $P(1) = 1 + 1 - 7 - 1 + 6 = 0$ $(x - 1)$ is a factor.
2 Divide $P(x)$ by the first factor found.	$$\begin{array}{r} x^3 + 2x^2 - 5x - 6 \\ x-1 \overline{) x^4 + x^3 - 7x^2 - x + 6} \\ -(x^4 - x^3) \\ \hline 2x^3 - 7x^2 \\ -(2x^3 - 2x^2) \\ \hline -5x^2 - x \\ -(-5x^2 + 5x) \\ \hline -6x + 6 \\ -(-6x + 6) \\ \hline 0 \end{array}$$

3 Write $P(x)$ as the product of the factors.	$P(x) = (x - 1)(x^3 + 2x^2 - 5x - 6)$
4 Define the non-linear factor $M(x)$.	$P(x) = (x - 1)M(x)$
5 Find a value of x so that $M(x) = 0$. Try the factors of the constant term of $M(x)$.	Try $x = -1$ $M(x) = x^3 + 2x^2 - 5x - 6$ $M(-1) = -1 + 2 + 5 - 6 = 0$ $(x + 1)$ is a factor.
6 Now divide $M(x)$ by $(x + 1)$ to find the quadratic factor and then factorise.	$M(x) = (x + 1)(x^2 + x - 6)$ $= (x + 1)(x - 2)(x + 3)$
7 Write $P(x)$ as the product of its factors.	$P(x) = (x - 1)(x + 1)(x - 2)(x + 3)$

Other methods can be used to find the factors. You could try all integer factors of 6: $\pm 1, \pm 2, \pm 3$. Then find $P(1) = 0$, $P(-1) = 0$, $P(2) = 0$, $P(-3) = 0$ so the factors of $P(x)$ are $(x - 1)$, $(x + 1)$, $(x - 2)$ and $(x + 3)$.

8.4 Remainder and factor theorems

Navigator

1, 2, 3, 4, 5, 6, 7, 8, 9, 10, 11, 13 (column 1), 16	1, 2 (column 1), 3, 4, 5, 6 (column 1), 7, 8, 9, 10, 11, 12, 13 (column 2), 14, 15, 16	1, 2 (column 2), 3, 4, 5, 6 (column 2), 7, 8, 9, 10, 11, 12, 13, 14, 15, 16

Answers p.823

10A

Fluency

1 If $P(x) = x^3 + 4x^2 - 5x + 3$ and $Q(x) = x - 2$, find $\frac{P(x)}{Q(x)}$, evaluate $P(2)$ and comment on the result. W.E. 12

2 Use the remainder theorem to find the remainder for: W.E. 13

(a) $\frac{x^2 - 2x + 1}{x - 2}$ (b) $\frac{-3x^3 + 2x^2 - x + 2}{2x + 6}$

(c) $\frac{x^3 + 4x^2 + 7x - 9}{2x - 1}$ (d) $\frac{2x^2 - 5x + 7}{4x + 1}$

3 State a linear factor given that:

(a) $P(2) = 0$ (b) $P(-5) = 0$ (c) $P\left(\frac{2}{5}\right) = 0$

4 Use the factor theorem to fully factorise $x^3 - 6x^2 + 11x - 6$. W.E. 14

5 Use the factor theorem to fully factorise $x^4 - x^3 - 7x^2 + x + 6$. W.E. 15

6 Use the remainder theorem to find the remainder if:

(a) $4x^2 + 3x - 1$ is divided by $x - 3$ (b) $5x^3 - 2x^2 + 6x - 10$ is divided by $x + 1$

(c) $x^4 + x^3 - 3x^2 - x - 1$ is divided by $2x + 1$ (d) $x^5 - 1$ is divided by $2x - 1$

(e) $(6x - 8)(4x + 5)$ is divided by $x - 5$ (f) $49x^2 - 144$ is divided by $7x + 2$

7 Use the factor theorem to factorise the following polynomials.

(a) $x^3 + x^2 - 10x + 8$
(b) $x^3 - 5x^2 + 3x + 9$
(c) $x^3 - 6x^2 + 3x + 10$
(d) $x^3 + 3x^2 - 16x + 12$
(e) $x^3 + 6x^2 + 12x + 8$
(f) $x^3 - 3x^2 - 9x - 5$

Understanding

8 Calculate the remainder for $\frac{P(x)}{Q(x)}$ given:

(a) $P(x) = 4x^3 + 2x^2 - x + 1$, $Q(x) = 5x + 15$
(b) $P(x) = -x^4 - 3x^2 + 2x + 1$, $Q(x) = -x - 1$
(c) $P(x) = -x^3 - 9x^2 + 10x - 12$, $Q(x) = -2x$
(d) $P(x) = x^5 - x^4 - x^3 + 2x^2 - x - 1$, $Q(x) = 4x + 8$

9 Use the factor theorem to factorise the following polynomials.

(a) $x^4 + 6x^3 + 7x^2 - 6x - 8$
(b) $x^4 - 7x^3 + 14x^2 - 8x$
(c) $x^4 - x^3 - 7x^2 + 13x - 6$
(d) $x^4 + 2x^3 - 12x^2 + 14x - 5$
(e) $x^4 - 4x^3 + 3x^2 + 4x - 4$
(f) $x^4 + 2x^3 - 13x^2 - 14x + 24$

10 When $2x^3 - 4x^2 + x - 1$ is divided by $x + \frac{1}{2}$, the remainder is:

A $1\frac{1}{2}$
B $-2\frac{3}{4}$
C $1\frac{3}{4}$
D $-3\frac{1}{2}$

Reasoning

11 If $\frac{x^3 - 4x^2 - 4x + 16}{x - 4} = ax^2 + bx + c$ the values of a, b and c are:

A 1, -1, -4
B -3, 4, 0
C 4, 3, -1
D 1, 0, -4

12 If $x + 2$ is a factor of $6x^3 + 19x^2 + 11x + a$, find the value of a.

13 Use the factor theorem to fully factorise each polynomial. Then solve the polynomial equation by using the null factor law.

(a) $x^3 - 6x^2 + 11x - 6 = 0$
(b) $x^3 + x^2 - 4x - 4 = 0$
(c) $x^3 + 9x^2 + 26x + 24 = 0$
(d) $x^4 + x^3 - 7x^2 - x + 6 = 0$
(e) $x^4 - 5x^2 + 4 = 0$
(f) $x^4 - x^3 - 6x^2 + 6x = 0$

14 A toy top is moving along the floor as it spins. Its distance, d cm, from a fixed point at time t seconds, is given by the equation $d = t^4 - 6t^3 + 13t^2 - 12t + 4$.

(a) Factorise the polynomial.
(b) How far is the top from the fixed point initially?
(c) At what times is the top at the fixed point?
(d) By inspection of your answer to part **(a)**, or by sketching the graph, describe the motion of the top after 2 seconds.

Open-ended

15 Give two monic polynomials of degree 3 that leave a remainder of -2 when they are divided by $x - 3$.

16 A polynomial equation of degree 4 has solutions $x = 0$, $x = -1$, $x = 2$, $x = 3$. Write two possible equations.

Exploration CAS

Equipment required: TI-Nspire CAS or Casio ClassPad CAS

Sketching polynomials with CAS

Sketching a graph by hand is time-consuming and inaccurate. The exact values for the x-intercepts are often difficult to find, and turning points and points of inflection require mathematical techniques beyond this course. Using a CAS efficiently to sketch a polynomial graph and find relevant information enables you to interpret situations and solve problems using polynomials more effectively.

1 Use a CAS to sketch $y = 2x^4 - x^3 - 5x^2 + 2x + 2$. Identify the x- and y-intercepts and any turning points and/or points of inflection. Round to 2 decimal places, where necessary.

Using TI-Nspire CAS

Add **Graphs** to your document and enter the equation **f1(x)=2x⁴–x³–5x²+2x+2** to display the graph.

Adjust the window settings by selecting **menu** > **Window/Zoom** > **Window Settings...** and then change the values to show the key features of the graph.

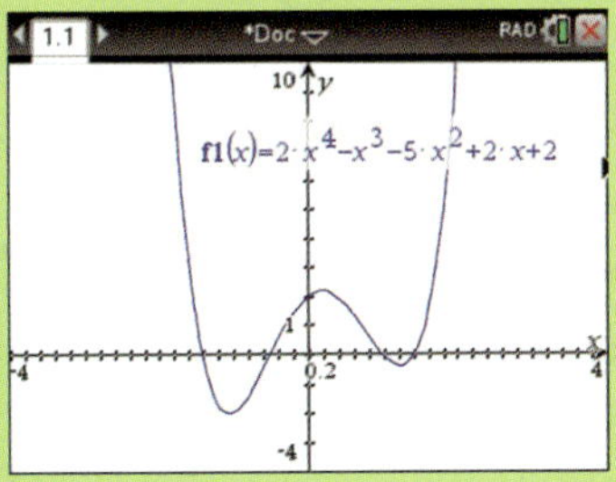

To find the y-intercept:

Select **menu** > **Trace** > **Graph Trace** and the y-intercept will be displayed.

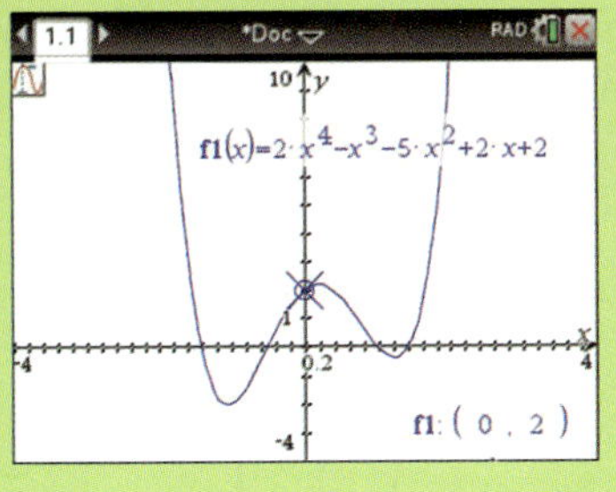

The y-intercept is at (0, 2).

Using Casio ClassPad CAS

From the menu select **Graph & Table**. Enter **y1 = 2x^4–x^3–5x^2+2x+2**, tick the box next to it, and then select the Graph icon to display the graph.

Adjust the window settings by selecting the view icon and change the values to show the key features of the graph.

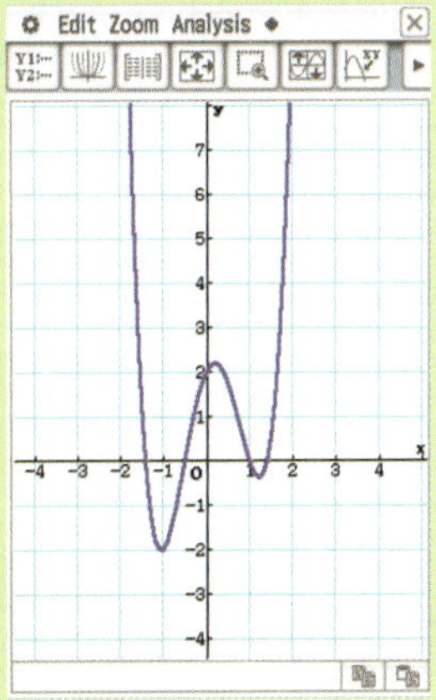

To find the y-intercept:

Select **Analysis** > **G-Solve** > **y-Intercept** and the y-intercept will be displayed.

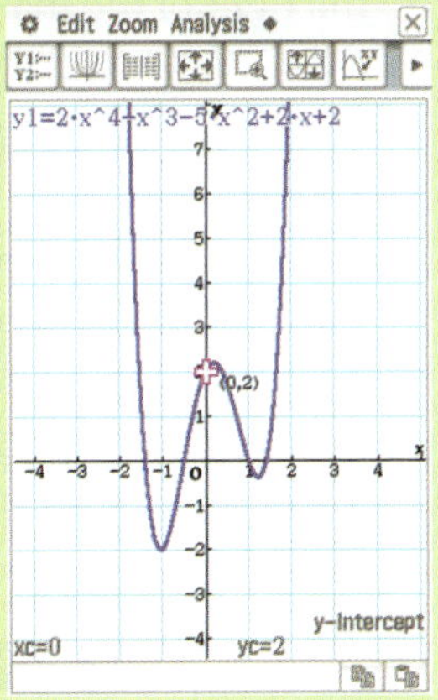

The y-intercept is at (0, 2).

Using TI-Nspire CAS	Using Casio ClassPad CAS

To find the x-intercepts:

Select **menu** > **Analyse Graph** > **Zero**. Use the cursor to set the lower bound (a point to the left of the first x-intercept) and then the upper bound (a point to the right of the first x-intercept). The coordinates of the first x-intercept will appear as (-1.41, 0).

Repeat this (with different lower and upper bounds) to find the other three x-intercepts: (-0.5, 0), (1, 0) and (1.41, 0).

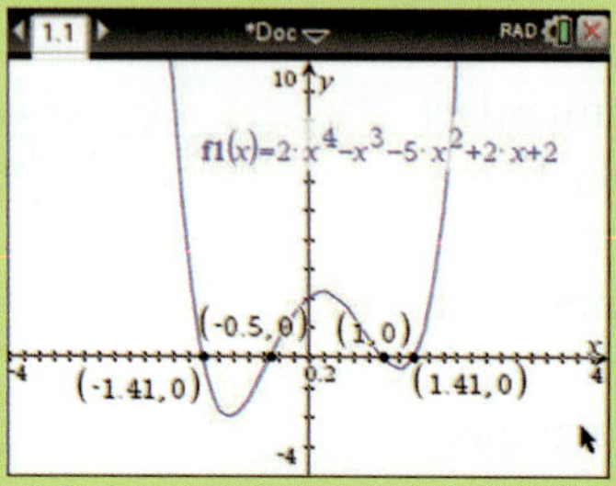

The x-intercepts are (-1.41, 0), (-0.5, 0), (1, 0) and (1.41, 0).

To find the x-intercepts:

Select > **Analysis** > **G-Solve** > **Root**. The cursor will blink on the first x-intercept at (-1.414, 0). Use the right arrow to jump to the other three x-intercepts at (-0.5, 0), (1, 0) and (1.414, 0).

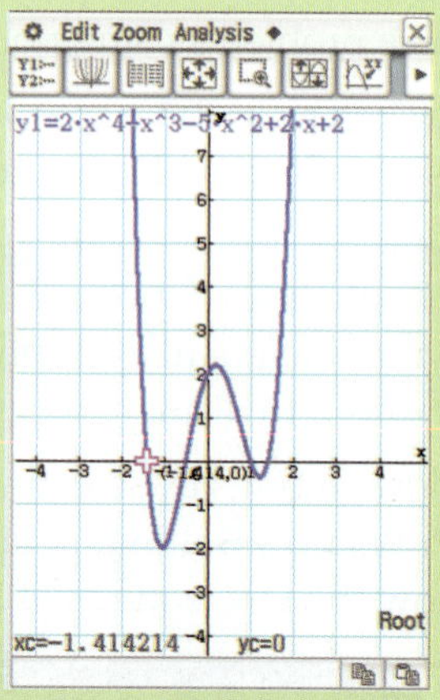

The x-intercepts are (-1.41, 0), (-0.5, 0), (1, 0) and (1.41, 0).

10A

To find the turning points:

Select **menu** > **Analyse Graph** > **Maximum**. Use the cursor to set the lower bound (to the left of the maximum turning point) and then the upper bound (to the right of the maximum turning point). The maximum turning point will appear as (0.195, 2.2).

Select **menu** > **Analyse Graph** > **Minimum**. Use the cursor to set the lower bound (to the left of the minimum turning point) and then the upper bound (to the right of the minimum turning point). The minimum turning point will appear as (-1.05, -2.02).

Repeat this (with different lower and upper bounds) to find the other minimum turning point at (1.23, -0.388).

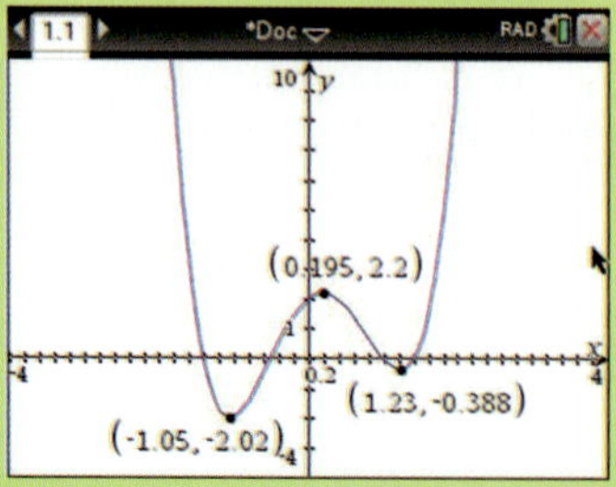

The turning points are at (0.20, 2.20), (-1.05, -2.02) and (1.23, -0.39).

To find the turning points:

Select **Analysis** > **G-Solve** > **Max**. The coordinates of the maximum turning point (0.195, 2.195) will be displayed.

Select **Analysis** > **G-Solve** > **Min**. The coordinates of the first minimum value (-1.047, -2.024) will be displayed.

Use the right arrow to jump to the next minimum turning point at (1.227, -0.388).

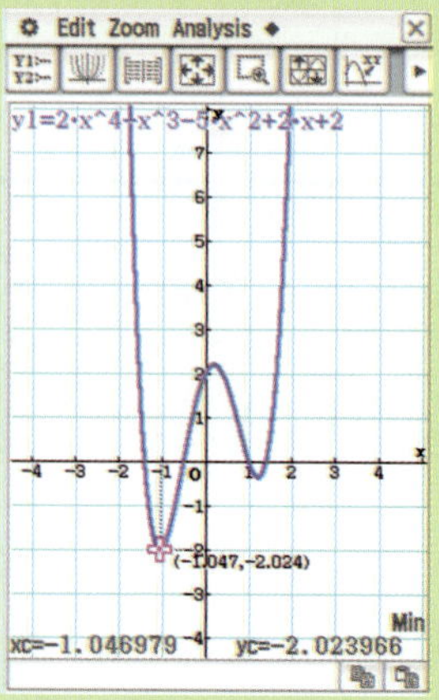

The turning points are at (0.20, 2.20), (-1.05, -2.02) and (1.23, -0.39).

You now have enough information to sketch your graph by hand, and can use your graph to find particular x-values and y-values.

2 If $y = P(x)$, find $P(2)$ and $P(-3)$.

Using TI-Nspire CAS

Evaluating polynomials:

Add **Calculator** to your document and enter **f1(2)** to evaluate the polynomial for $x = 2$. The answer 10 will be displayed.

Similarly enter **f1(-3)** to evaluate the polynomial for $x = -3$.

$P(2) = 10$

$P(-3) = 140$

Using Casio ClassPad CAS

Evaluating polynomials:

Select **Analysis** > **Trace**. Enter 2 as the **x-value** and the coordinates of the point where $x = 2$ will be displayed.

Similarly enter -3 as the **x-value** to find the coordinates of the point where $x = -3$.

$P(2) = 10$

$P(-3) = 140$

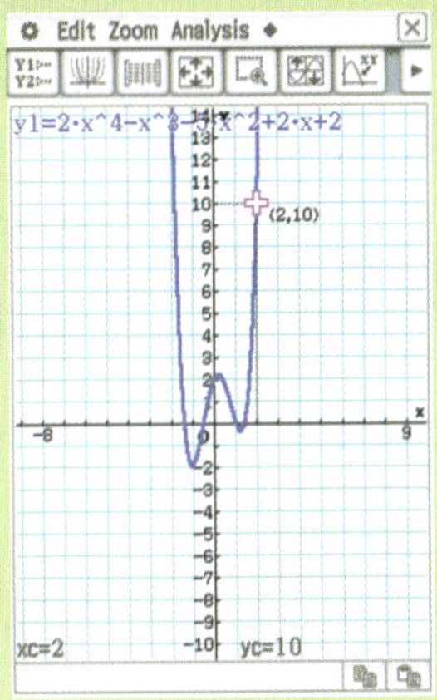

3 Solve $2x^4 - x^3 - 5x^2 + 2x + 2 = 2$. Write your answer correct to 2 decimal places.

Using TI-Nspire CAS

Solving polynomials:

Add **Calculator** to your document and select **menu** > **Algebra** > **Solve**, followed by the equation and variable to be solved as shown. The four solutions will be displayed.

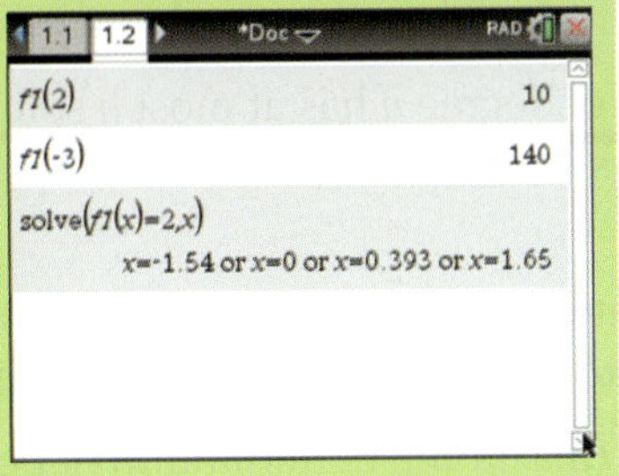

$x = -1.54, 0, 0.39, 1.65$

Using Casio ClassPad CAS

Solving polynomials:

Select **Analysis** > **G-Solve** > **x-Cal/y-Cal** > **x-Cal**. Enter 2 as the **y-value** and the coordinates of the point where the polynomial value is 2 will appear: (-1.542, 2).

Use the right arrow to jump to the next solutions (0, 2), (0.3934, 2) and (1.6485, 2).

$x = -1.54, 0, 0.39, 1.65$

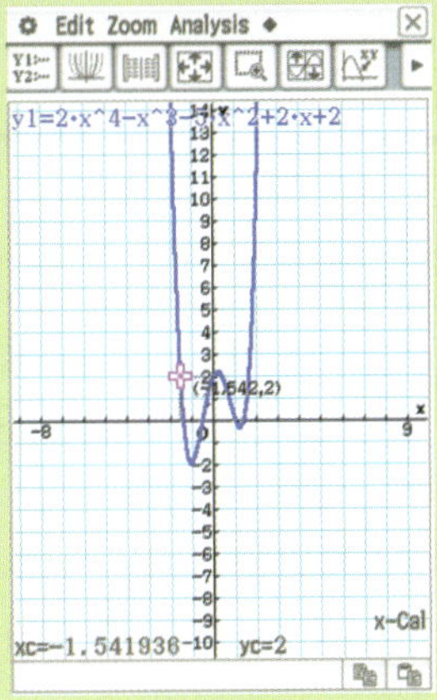

4 (a) Use a CAS to sketch $y = 6x^4 + 7x^3 - 9x^2 - 7x + 3$. Identify x- and y-intercepts and any turning points and/or points of inflection.

(b) If $y = Q(x)$, find Q(2)and $Q(-3)$.

(c) Solve $6x^4 + 7x^3 - 9x^2 - 7x + 3 = 2$. Write your answer correct to 2 decimal places.

10A

8.5 Sketching polynomials

Before you can sketch a polynomial you need to identify its key features.

Shape: Identify the shape of $y = a(x - h)^2 + k$ using transformations of $y = ax^n$.

***y*-intercept:** Find the y-intercept. All polynomial graphs have a y-intercept. Find the y-value of the y-intercept by substituting $x = 0$ into the polynomial.

***x*-intercept:** Find the x-intercepts. When a polynomial $P(x)$ is written as the product of its factors, the solution of $P(x) = 0$ can be found by equating each factor to zero and solving for x. The solutions are the x-values of the x-intercepts of the graph of $y = P(x)$.

- If the degree of $P(x)$ is 1, the linear polynomial $P(x) = mx + b$ has one solution for $P(x) = 0$. There is always exactly one x-intercept.
- If the degree of $P(x)$ is 2, the quadratic polynomial $P(x) = ax^2 + bx + c$ has at most two solutions for $P(x) = 0$. Therefore the graph of $P(x)$ may have zero, one or two x-intercepts. The x-values of the x-intercepts may not be rational.
- If the degree of $P(x)$ is 3, the cubic polynomial $P(x) = ax^3 + bx^2 + cx + d$ has at most three solutions for $P(x) = 0$. There will always be at least one x-intercept. The x-values of the x-intercepts may not be rational.
- The polynomial $P(x)$ of degree n has at most n solutions for $P(x) = 0$. If n is odd, there will always be at least one x-intercept. If n is even, there may be no x-intercepts.

10A

Note: You can solve any polynomial equation, if real solutions exist, by using the factor theorem to factorise and then the null factor law to find any x-intercepts of the graph of the polynomial.

Turning points: Find the turning points (maximum or minimum) or points of inflection where the graph changes direction. At a point of inflection (h, k), a reflection in both the lines $x = h$ and $y = k$ of one half of the graph will give the other half.

Quadratics

The general equation of a quadratic is $P(x) = ax^2 + bx + c$ where a, b and c are constants.

The shape of a quadratic is called a parabola.

Features of the parabola:

- Every parabola is symmetrical about the axis of symmetry, which is a vertical line with equation in the form $x = d$, where d is a constant.
- The y-value of the y-intercept can be found by substituting $x = 0$ into the quadratic equation.
- The x-intercepts, if they exist, can be found by solving $P(x) = 0$. You can solve quadratic equations either by factorising and using the null factor law or by substituting into the quadratic formula. These methods are preferable to the factor theorem for quadratics.
- Every parabola has a turning point where the graph changes direction.
- A parabola can have a minimum or maximum turning point. A quadratic equation written in turning point form $P(x) = a(x - h)^2 + k$ has a turning point at (h, k).

The graphs of a quadratic fit one of the following three types for $a > 0$. If $a < 0$, the graphs will be reflected in the x-axis.

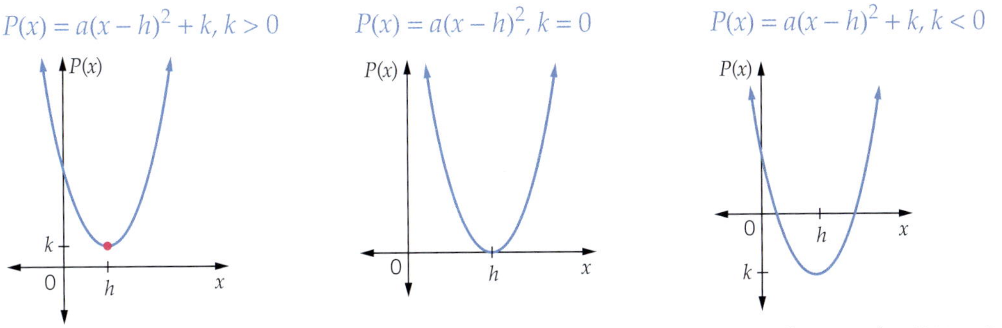

- no solution for $P(x) = 0$
- no x-intercept

- 1 solution for $P(x) = 0$
- 1 x-intercept

- 2 solutions for $P(x) = 0$
- 2 x-intercepts

Worked example 16 W.E.16

Sketch the graph of each quadratic polynomial showing the x- and y-intercepts and the coordinates of the turning point.

(a) $P(x) = 2x^2 + 2x + 1$ (b) $P(x) = 3x^2 + 6x + 3$ (c) $P(x) = -6x^2 + x + 2$

Thinking	Working
(a) 1 Write the quadratic equation.	(a) $P(x) = 2x^2 + 2x + 1$
2 Find the y-intercept by letting $x = 0$.	$x = 0, P(0) = 2(0)^2 + 2(0) + 1$ $= 1$ The y-intercept is $(0, 1)$.
3 Find the x-intercept by letting $P(x) = 0$. Solve the quadratic equation if possible. If there are no obvious factors, use the quadratic formula.	There are no obvious factors so use the quadratic formula. $a = 2, b = 2, c = 1$ $x = \frac{-b \pm \sqrt{b^2 - 4ac}}{2a}$ $= \frac{-2 \pm \sqrt{4 - 8}}{2 \times 2}$ $= \frac{-2 \pm \sqrt{-4}}{4}$ This gives no solution so the graph has no x-intercepts.
4 Identify whether the graph has a maximum or minimum turning point.	a is positive, so the graph has a minimum turning point.

10A

5 Identify the turning point using $x = -\frac{b}{2a}$.	$x = -\frac{b}{2a}$ $= -\frac{2}{4}$ $= -\frac{1}{2}$ $x = -\frac{1}{2},\ y = \frac{2}{4} - \frac{2}{2} + 1$ $= \frac{1}{2}$ The turning point is at $\left(-\frac{1}{2}, \frac{1}{2}\right)$.
6 Sketch the graph and label it showing the turning point and y-intercept.	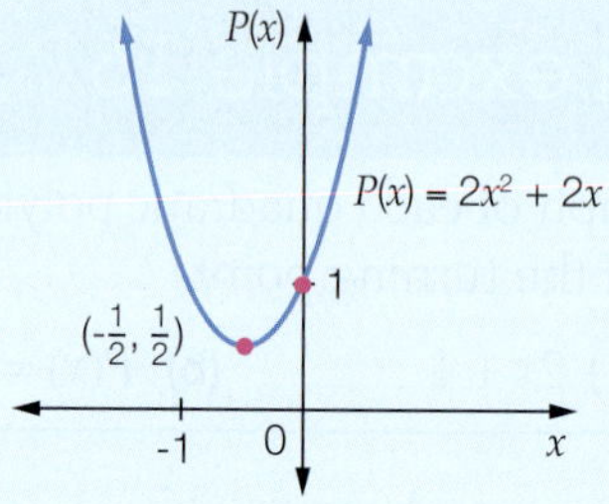
(b) 1 Write the quadratic equation.	(b) $P(x) = 3x^2 + 6x + 3 = 0$
2 Find the y-intercept by letting $x = 0$.	$x = 0,\ P(0) = 3(0)^2 + 6(0) + 3$ $= 3$ The y-intercept is $(0, 3)$.
3 Find the x-intercept by letting $P(x) = 0$. Identify the common factor and a perfect square. Solve the quadratic equation.	For $P(x) = 0,\ 3x^2 + 6x + 3 = 0$ $3(x^2 + 2x + 1) = 0$ $3(x + 1)^2 = 0$ $x = -1$ The x-intercept is $(-1, 0)$.
4 Identify whether the graph has a maximum or minimum turning point.	a is positive, so the graph has a minimum turning point.
5 Identify the turning point using symmetry as there is only one x-intercept.	The turning point is at $(-1, 0)$.
6 Sketch the graph and label it showing the turning point and y-intercept.	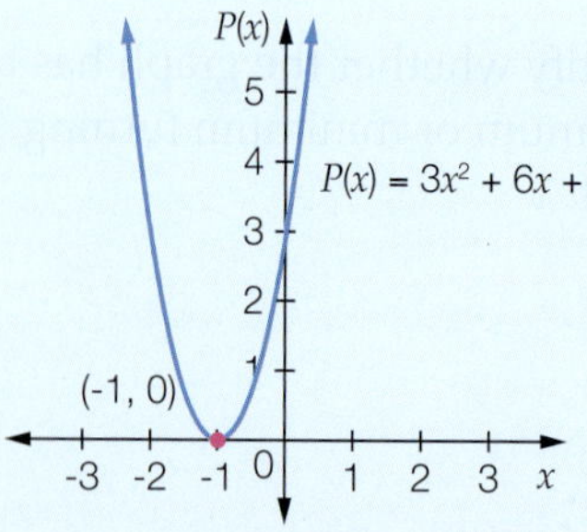

(c) 1 Write the quadratic equation.	(c) $P(x) = -6x^2 + x + 2$
2 Find the y-intercept by letting $x = 0$.	$x = 0,\ P(0) = -6(0)^2 + 0 + 2$ $= 2$ The y-intercept is (0, 2).
3 Find the x-intercept by letting $P(x) = 0$. Factorise the trinomial and use the null factor law to solve the quadratic equation.	$P(x) = 0$ $-6x^2 + x + 2 = 0$ $-(6x^2 - x - 2) = 0$ $-(3x - 2)(2x + 1) = 0$ $x = \frac{2}{3}$ or $x = -\frac{1}{2}$ The x-intercepts are $\left(\frac{2}{3}, 0\right)$ and $\left(-\frac{1}{2}, 0\right)$.
4 Identify whether the graph has a maximum or minimum turning point.	As a is negative, the graph has a maximum turning point.
5 Identify the turning point using $x = -\frac{b}{2a}$ or use symmetry as the x-coordinate is halfway between the two x-coordinates of the x-intercepts. (Here, $x = \frac{1}{2} \times \left(-\frac{1}{2} + \frac{2}{3}\right) = \frac{1}{2} \times \frac{1}{6} = \frac{1}{12}$)	$x = -\frac{b}{2a}$ $= -\frac{1}{2 \times -6}$ $= \frac{1}{12}$ $y = -6\left(\frac{1}{12}\right)^2 + \left(\frac{1}{12}\right) + 2$ $= -6 \times \frac{1}{144} + \frac{1}{12} + 2$ $= 2\frac{1}{24}$
6 Sketch the graph and label it showing the turning point, the y-intercept and the x-intercepts.	The turning point is at $\left(\frac{1}{12}, 2\frac{1}{24}\right)$.

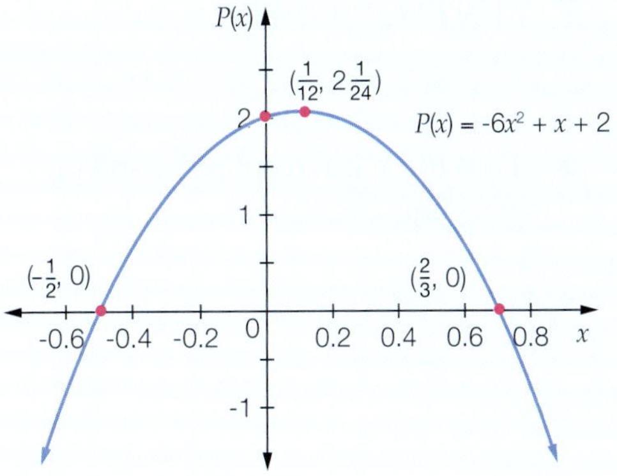

Cubics

A cubic polynomial expressed in the form $P(x) = a(x - r)(x - s)(x - t)$ is easy to sketch as the x-intercepts are given by $x = r$, $x = s$ and $x = t$. Depending on the values of r, s and t there will be one, two or three solutions to the equation $P(x) = 0$.

$P(x) = a(x - r)^3$

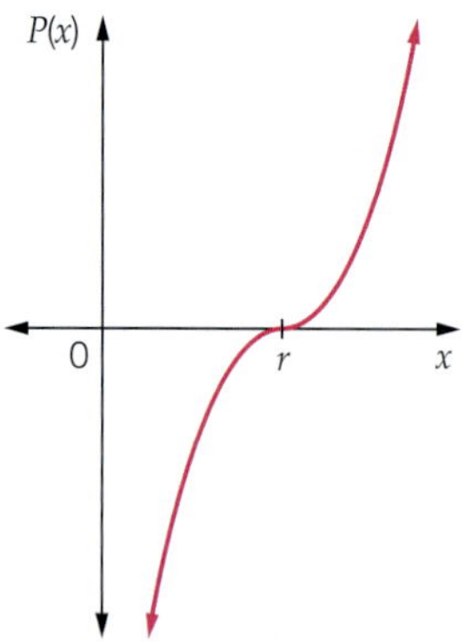

- r, s and t all have the same value
- 1 solution for $P(x) = 0$
- cuts the x-axis once (point of inflection) at $(r, 0)$

$P(x) = a(x - r)^2(x - t)$

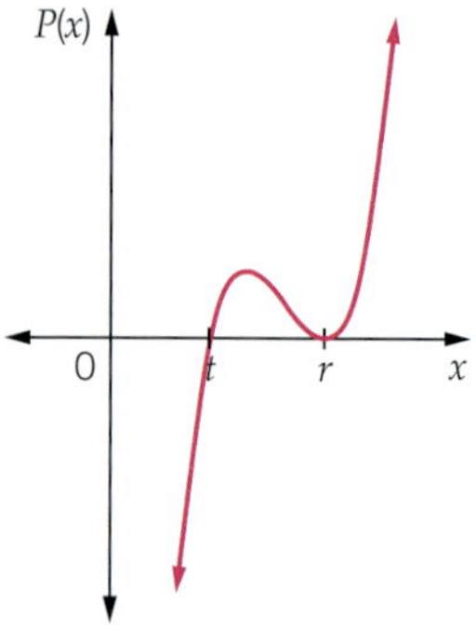

- r and s have the same value
- 2 solutions for $P(x) = 0$
- cuts the x-axis once at $(t, 0)$ and touches the x-axis once at $(r, 0)$

$P(x) = a(x - r)(x - s)(x - t)$

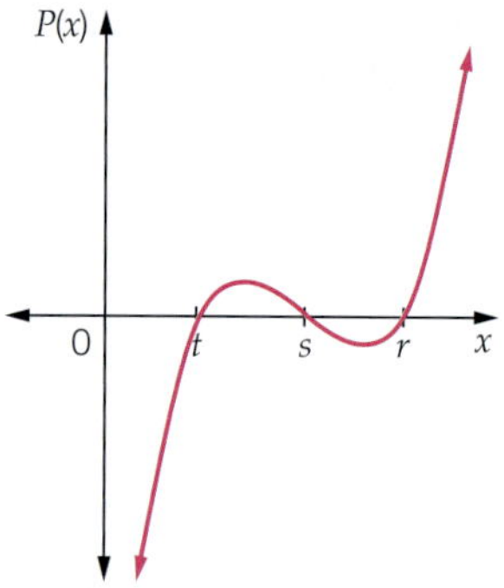

- r, s and t all have distinct values
- 3 solutions for $P(x) = 0$
- cuts the x-axis 3 times at $(t, 0)$, $(s, 0)$, $(r, 0)$

Notice that the ends of a cubic graph point in opposite directions.

10A

Worked example 17

W.E. 17

Sketch the graph of each cubic polynomial showing all x- and y-intercepts. Use a CAS to find the coordinates of the turning points or points of inflection, if necessary.

(a) $P(x) = -4(x - 1)^3$ **(b)** $P(x) = (x + 2)(x - 1)^2$ **(c)** $P(x) = x^3 - 3x^2 - x + 3$

Thinking	Working
(a) 1 Write the cubic equation.	(a) $P(x) = -4(x - 1)^3$
2 Find the y-intercept.	$x = 0$, $P(0) = -4 \times (-1)^3$ $= 4$ The y-intercept is $(0, 4)$.
3 Find the x-intercept(s) by solving the cubic equation.	$P(x) = 0$ $-4(x - 1)^3 = 0$ $x - 1 = 0$ $x = 1$ is the only solution The curve cuts the x-axis at $(1, 0)$.

4 Sketch the graph and label it showing the x- and y-intercepts.

Alternatively you could have sketched the graph of $y = x^3$, applied a dilation factor of 4, reflected it in the x-axis and translated it 1 unit to the right.

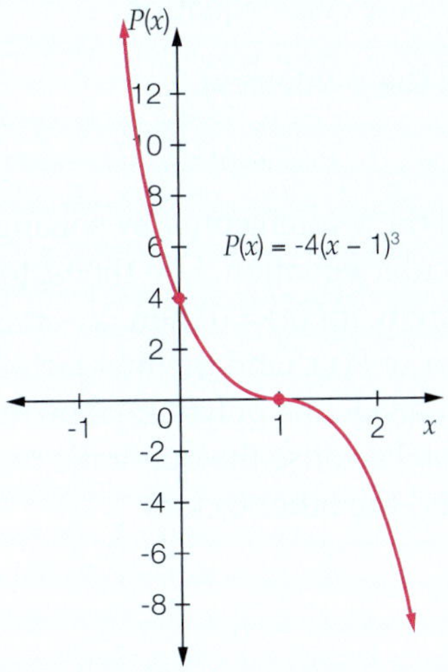

(1, 0) is a point of inflection.

(b) 1 Write the cubic equation.

(b) $P(x) = (x + 2)(x - 1)^2$

2 Find the y-intercept.

$x = 0$, $P(0) = 2 \times (-1)^2$
$= 2$

The y-intercept is (0, 2).

3 Find the x-intercept(s) by solving the cubic equation $P(x) = 0$. Any repeated factor indicates where the graph will only touch the x-axis. (Here, a repeated factor of $(x - 1)^2$ indicates that $P(x)$ touches the x-axis at $x = 1$.)

$P(x) = 0$
$(x + 2) = 0$ or $(x - 1)^2 = 0$
$x = -2$, $x = 1$

The curve cuts the x-axis at (-2, 0) and touches it at (1, 0).

4 Sketch the graph and label it showing the x- and y-intercepts. The turning points can be found with a CAS.

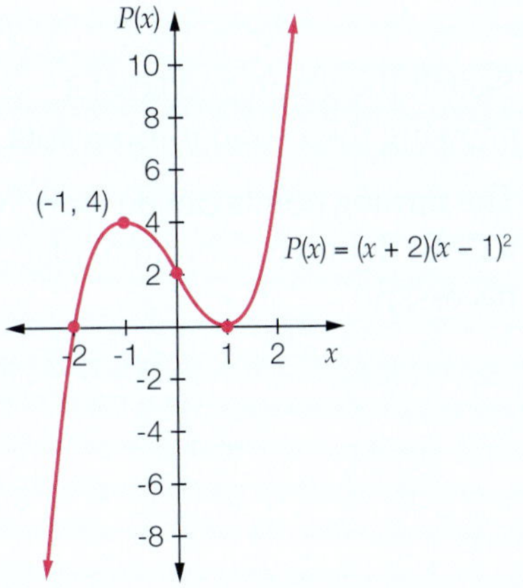

(1, 0) is a turning point.

There is another turning point at (-1, 4).

A CAS can be used to find this point.

(c) **1** Write the cubic equation.

(c) $P(x) = x^3 - 3x^2 - x + 3$

2 Find the y-intercept.

$x = 0,\ P(0) = 3$

The y-intercept is $(0, 3)$.

3 Find the x-intercept(s) by solving the cubic equation. Use the factor theorem. If $P(a) = 0$ then $(x - a)$ is a factor of $P(x)$. Find the first factor, then divide the polynomial by this factor. Factorise the quadratic result to find the other factors.

$x^3 - 3x^2 - x + 3 = 0$

$P(-1) = -1 - 3 + 1 + 3$
$= 0$

$(x + 1)$ is a factor.

$$\begin{array}{r} x^2 - 4x + 3 \\ x+1 \overline{)\, x^3 - 3x^2 - x + 3} \\ -(x^3 + x^2) \\ \hline -4x^2 - x \\ -(-4x^2 - 4x) \\ \hline 3x + 3 \\ -(3x + 3) \\ \hline 0 \end{array}$$

$x^2 - 4x + 3 = (x - 3)(x - 1)$

$x^3 - 3x^2 - x + 3 = (x + 1)(x^2 - 4x + 3)$
$= (x + 1)(x - 3)(x - 1)$

Let $P(x) = 0$

$(x + 1)(x - 1)(x - 3) = 0$

$x = -1,\ x = 1,\ x = 3$

The curve cuts the x-axis at $(-1, 0)$, $(1, 0)$ and $(3, 0)$.

4 Sketch the graph and label it showing the x- and y-intercepts.

The turning points can be found with a CAS. Round to 2 decimal places if necessary.

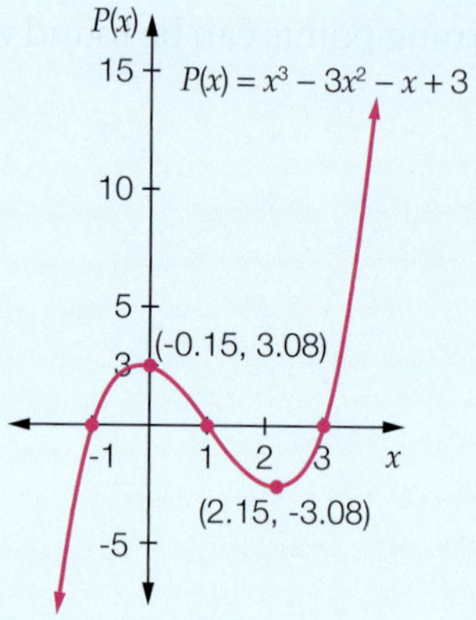

Quartics

Some quartic polynomials of the form $P(x) = ax^4 + bx^3 + cx^2 + dx + e$ can be written in the form $P(x) = a(x - r)(x - s)(x - t)(x - u)$. In this case, there can be up to four solutions for $P(x) = 0$.

Five possibilities are shown for $a > 0$. If $a < 0$ then the graphs are reflected in the x-axis and both ends would be pointing downwards.

$P(x) = a(x - r)(x - r)(x - r)(x - r)$
$= a(x - r)^4$

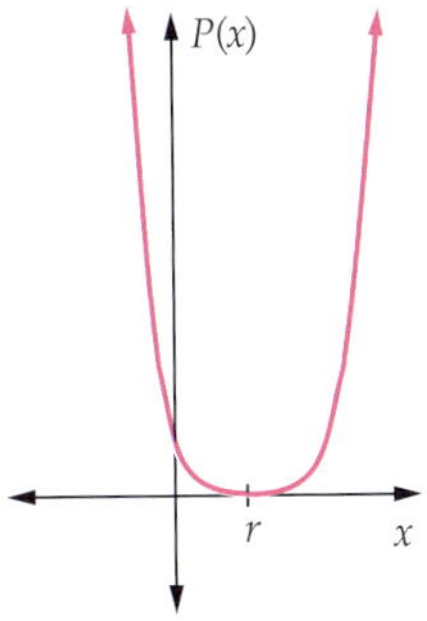

- r, s, t and u all have the same value
- 1 solution for $P(x) = 0$
- touches the x-axis at $(r, 0)$

$P(x) = a(x - r)(x - r)(x - r)(x - u)$
$= a(x - r)^3(x - u)$

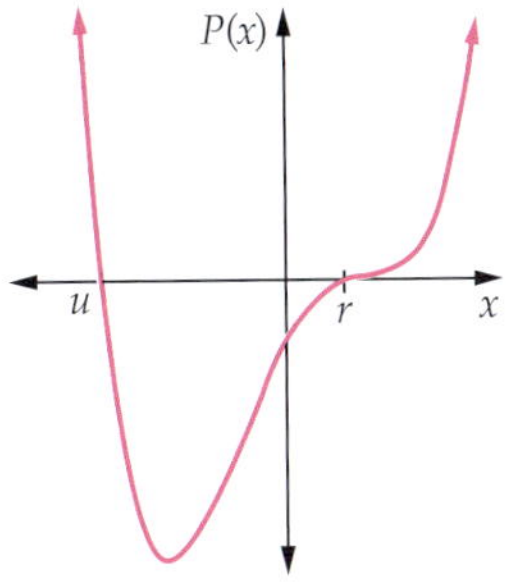

- r, s and t all have the same value
- 2 solutions for $P(x) = 0$
- cuts the x-axis at $(u, 0)$ and at $(r, 0)$ (point of inflection)

$P(x) = a(x - r)(x - r)(x - u)(x - u)$
$= a(x - r)^2(x - u)^2$

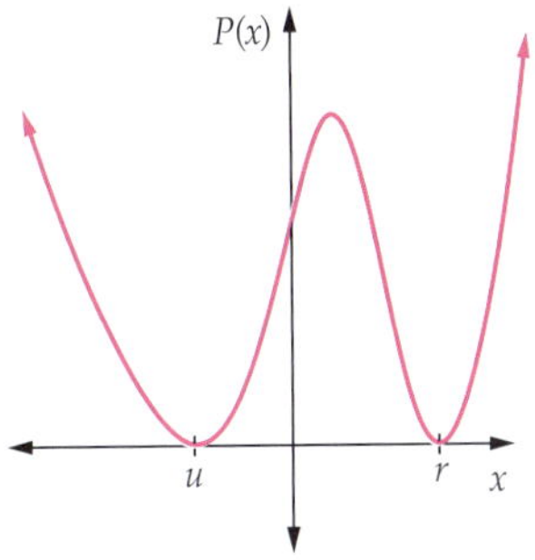

- r and s have the same value, t and u have the same value
- 2 solutions for $P(x) = 0$
- touches the x-axis at $(r, 0)$ and $(u, 0)$

$P(x) = a(x - r)(x - r)(x - t)(x - u)$
$= a(x - r)^2(x - t)(x - u)$

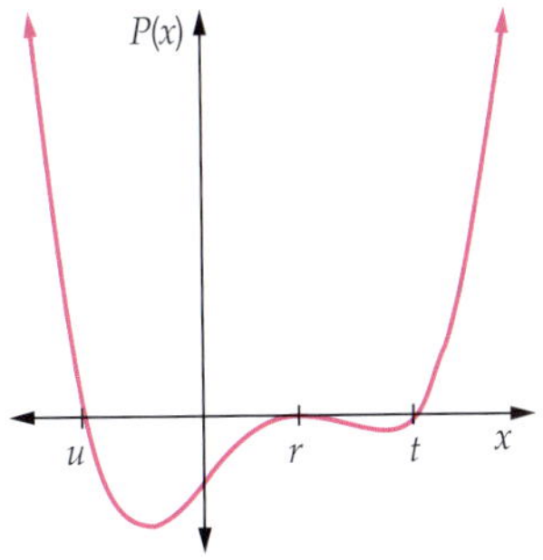

- r and s have the same value
- 3 solutions for $P(x)$
- touches x-axis at $(r, 0)$ and cuts it at $(t, 0)$ and $(u, 0)$

$y = a(x - r)(x - s)(x - t)(x - u)$

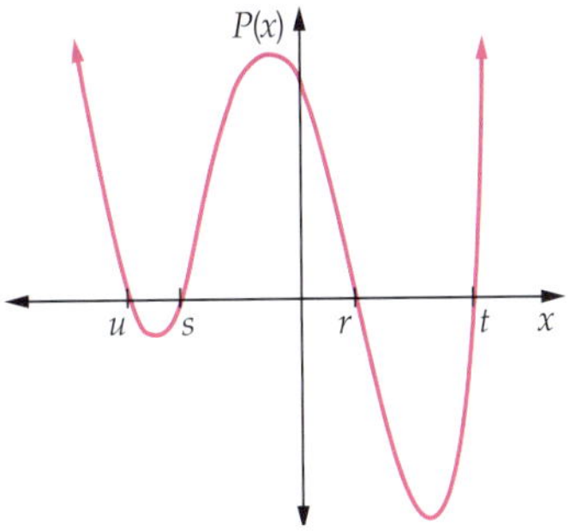

- r, s, t and u all have different values
- 4 solutions for $P(x)$
- cuts the x-axis at $(r, 0)$, $(s, 0)$, $(t, 0)$ and $(u, 0)$

- If a factor occurs only once, it corresponds to a point where the curve cuts the x-axis.
- If a factor is repeated an even number of times, it corresponds to a point where the curve touches the x-axis, creating a turning point.
- If a factor is repeated an odd number of times, it corresponds to a point of inflection, where the graph cuts the x-axis as it flattens out.

A quartic cannot have only two turning points, because the ends of the graph must point in the same direction.

Like parabolas, some quartics also have no solutions. These quartics cannot be factorised using real numbers.

10A

Worked example 18

W.E. 18

Sketch the graph of $P(x) = x^4 - 9x^2 + 4x + 12$ showing all x- and y-intercepts. Use a CAS to find the turning points or points of inflection, if necessary.

Thinking	Working
1 Write the polynomial.	$P(x) = x^4 - 9x^2 + 4x + 12$
2 Find the y-intercept.	$P(0) = (0)^4 - 9(0)^2 + 4(0) + 12$ $= 12$ The y-intercept is $(0, 12)$.
3 Find the x-intercept(s). Factorise $P(x)$ and then solve $P(x) = 0$. Use the factor theorem to find a factor.	Factorise using the factor theorem. $P(-1) = 1 - 9 - 4 + 12$ $= 0$ $(x + 1)$ is a factor.
Perform a polynomial division.	$$\begin{array}{r} x^3 - x^2 - 8x + 12 \\ x+1 \overline{) x^4 + 0x^3 - 9x^2 + 4x + 12} \\ -(x^4 + x^3) \qquad\qquad\quad \\ -x^3 - 9x^2 \qquad\quad \\ -(-x^3 - x^2) \qquad\quad \\ -8x^2 + 4x \quad \\ -(-8x^2 - 8x) \quad \\ 12x + 12 \\ -(12x + 12) \\ 0 \end{array}$$
Write as a product of factors.	$P(x) = (x + 1)(x^3 - x^2 - 8x + 12)$
Use the factor theorem to find another factor.	$M(x) = x^3 - x^2 - 8x + 12$ $M(2) = 8 - 4 - 16 + 12$ $= 0$ $(x - 2)$ is a factor.
Perform another division.	$$\begin{array}{r} x^2 + x - 6 \\ x-2 \overline{) x^3 - x^2 - 8x + 12} \\ -(x^3 - 2x^2) \qquad\qquad \\ x^2 - 8x \qquad \\ -(x^2 - 2x) \qquad \\ -6x + 12 \\ -(-6x + 12) \\ 0 \end{array}$$
Write as a product of factors.	$P(x) = (x + 1)(x - 2)(x^2 + x - 6)$
Factorise the quadratic.	$P(x) = (x + 1)(x - 2)(x - 2)(x + 3)$ $P(x) = (x + 1)(x + 3)(x - 2)^2$ For $P(x) = 0$, $x = -1$, $x = -3$, $x = 2$ (repeated solution)
Write the x-intercepts.	The curve cuts the x-axis at $(-3, 0)$ and $(-1, 0)$ and touches it at $(2, 0)$ (repeated factor).

10A

4 Sketch the graph showing the x- and y-intercepts. Use a CAS to find the turning points. Round to 2 decimal places if necessary.

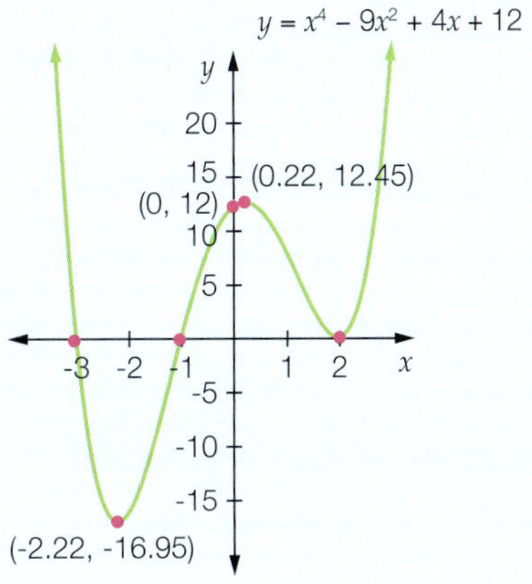

The curve cuts the x-axis twice, touches it once and has three turning points.

- If possible, factorise polynomials before sketching.
- Find the y-intercept.
- Find the x-intercept(s) by solving the equation $P(x) = 0$. Use the factor theorem, polynomial division and the quadratic formula if necessary.
- Identify whether the polynomial cuts or touches the x-axis. A repeated factor with an even power indicates that the graph touches the x-axis. A repeated factor with an odd power indicates a point of inflection on the x-axis.
- Use technology to find turning points and points of inflection.

10A

8.5 Sketching polynomials

Navigator

1, 2, 3, 4, 5 (columns 1–2), 6 (a–d), 7 (a–d), 8, 9, 10, 11, 12, 13, 15, 16, 17	1, 2, 3, 4, 5 (columns 2–3), 6 (column 1), 7 (column 1), 8, 9, 10, 11, 12, 13, 15, 16, 17	1, 2, 3, 4, 5 (column 3), 6 (column 2), 7 (column 2), 8, 9, 12, 13, 14, 15, 17

Answers p. 823

Equipment required: CAS or similar technology

Fluency

1 Find the x- and y-intercepts of each polynomial.

(a) $P(x) = 3x^2 + 7x + 2$ (b) $P(x) = 2x^3 - 3x^2 - 23x + 42$

(c) $P(x) = x^3 + 6x^2 + 11x + 6$ (d) $P(x) = x^4 + x^3 - 3x^2 - 5x - 2$

2 Sketch the graph of each quadratic polynomial showing all x- and y-intercepts. Use a CAS to find the coordinates of the turning points, if necessary. W.E. 16

(a) $P(x) = 2x^2 + 6x + 5$ (b) $P(x) = 3x^2 + 12x + 12$ (c) $P(x) = -5x^2 - 6x + 8$

3 Sketch the graph of each cubic polynomial showing all x- and y-intercepts. Use a CAS to find the coordinates of the turning points or point of inflection, if necessary. W.E. 17

(a) $P(x) = -2(x + 1)^3$ (b) $P(x) = (x - 2)(x + 1)^2$ (c) $P(x) = x^3 + 3x^2 - x - 3$

4 Sketch the graph of $P(x) = x^4 - 9x^2 - 4x + 12$ showing all x- and y-intercepts. Use a CAS to find the coordinates of the turning points or points of inflection, if necessary. W.E. 18

5 Sketch the graph of each polynomial.

(a) $P(x) = x^2 - 5x - 14$ (b) $P(x) = x^2 - 6x + 8$ (c) $P(x) = x^2 - 7x - 12$

(d) $P(x) = x^2 - 12x + 36$ (e) $P(x) = x^2 - 15x + 56$ (f) $P(x) = x^2 - x - 90$

6 Sketch the graph of each polynomial.

(a) $P(x) = x^3 + 7x^2 + 7x - 15$ (b) $P(x) = -x^3 + 9x^2 - 27x + 27$

(c) $P(x) = x^3 - 12x + 16$ (d) $P(x) = x^3 + x^2 - 12x$

(e) $P(x) = x^3 - 2x^2 - 11x + 12$ (f) $P(x) = x^3 + 5x^2 - x - 5$

7 Sketch the graph of each polynomial.

(a) $P(x) = x^4 - 4x^3 + 2x^2 + 4x - 3$ (b) $P(x) = x^4 + 2x^3 - 3x^2 - 4x + 4$

(c) $P(x) = x^4 - 2x^3 + x^2$ (d) $P(x) = x^4 + 2x^3 - 7x^2 - 8x + 12$

(e) $P(x) = x^4 - 2x^3 - 13x^2 + 14x + 24$ (f) $P(x) = x^4 + 3x^3 - 5x^2 - 3x + 4$

Understanding

8 Match the following polynomials to their graphs.

(a) $P(x) = ax^3$ (b) $P(x) = (x + b)(x - a)^2$ (c) $P(x) = (x - a)^3$

(d) $P(x) = (x + a)(x - b)(x - c)$ (e) $P(x) = -(x + a)(x - b)(x - c)$

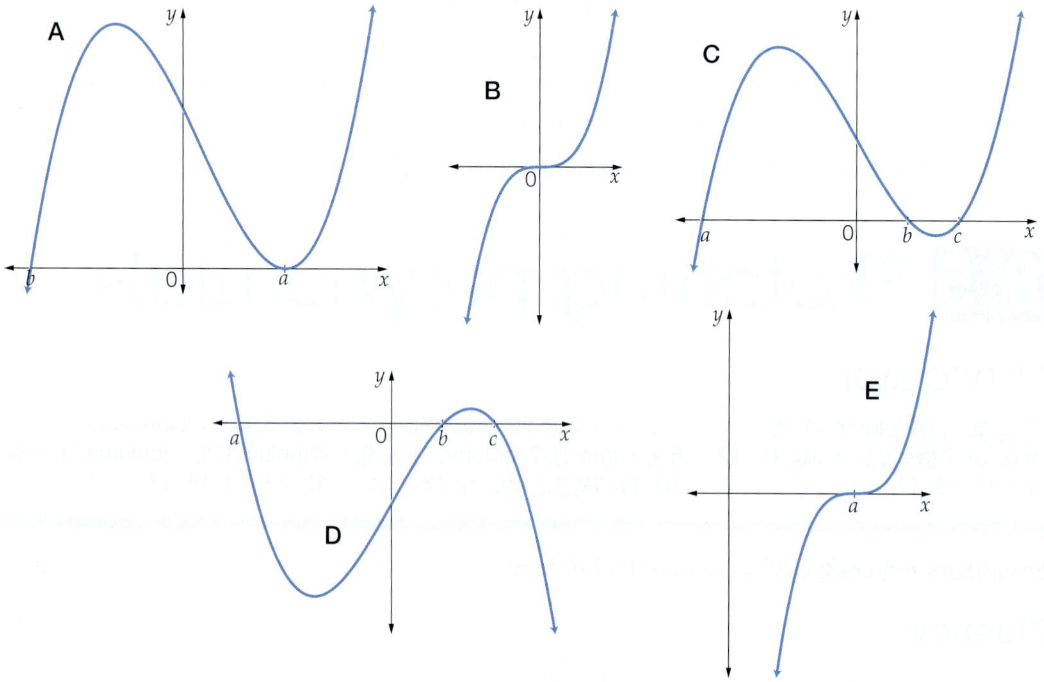

9 The equation $x^3 - 7x^2 + 11x - 5 = 0$ has:

A no solutions B only one solution at $x = 1$

C three solutions D two solutions

10 The solutions of the polynomial $P(x) = x(x - 8)(x - 2)(x - 8)$ are:

A $x = 0$ and $x = 8$ B $x = 0$ and $x = 2$

C $x = 2$ and $x = 8$ D $x = 0$, $x = 2$ and $x = 8$

11 $P(x) = 2x^3 + 5x^2 + x - 20$ and $Q(x) = 2x^3 + 4x^2 - x + 15$. Find the solutions to $P(x) = Q(x)$.

12 $P(x) = 5x^4 + 4x^3 - 3x^2 - 5x + 5$ and $Q(x) = 3x^4 + 2x^3 - x^2 - 2x + 2$. Find the solutions to $2P(x) = 3Q(x)$.

10A

Reasoning

13 A flying fish, swimming under water, suddenly leaps above the surface before re-entering the water. Its motion is given by $P(t) = -t^3 + 3t^2 - 2t$, where $P(t)$ is the vertical distance of the fish from sea level at time t seconds, with $0 \leq t \leq 2$.

(a) Factorise the polynomial and sketch the graph for values of t from 0 to 2.

(b) For how long was the fish swimming until it broke the sea surface? Approximately what was its greatest depth?

(c) For how long was the fish above the surface and approximately what was its maximum height?

14 A plumber inserts a long flexible rod in a length of water pipe to test for blockages. The pipe is described by the equation $y = x^3 - 6x^2 + 9x$, where x is the horizontal distance and y is the vertical distance from the rod's insertion point (0, 0) and the units are in metres.

(a) Factorise the expression and sketch the graph for values of x from 0 to 4.

(b) There are blockages located at the turning points of the graph. What are the coordinates of the blockages?

(c) To avoid the first blockage, the plumber decides to install a new system of pipes starting at (0, 0) and ending at the second blockage. The pipes take the shape of a parabola. Give one equation for this parabola.

(d) While installing the new pipes, the plumber is informed that there is an underground power cable located at (2.5, 1.2), which must be avoided. Can the plumber still use the parabola pipe layout?

10A

15 A dam is located at the bottom of a hill. If $y = 0$ is taken as ground level, the cross-section (west to east) of both shapes can be described by the polynomial $y = x^3 - 5x^2 - 2x + 24$ for $-2 \leq x \leq 5$, where each unit of x represents 100 m and each unit of y represents 1 m.

(a) Factorise the polynomial.

(b) Sketch the graph of the polynomial for $-2 \leq x \leq 5$.

(c) What is the horizontal distance between the western base of the hill and the eastern bank of the dam?

Open-ended

16 Find two monic quartic polynomials that have three x-intercepts, and sketch their graphs.

17 A quartic polynomial has factors of $(x + 1)$ and $(x - 3)$. Using at least these factors, write two possible quartic polynomials that:

(a) have exactly two x-intercepts

(b) have exactly three x-intercepts

(c) have four x-intercepts

(d) are non-monic with only one turning point.

Chapter review 8

Maths literacy

cubic	factor theorem	polynomial	remainder theorem
degree	leading coefficient	quartic	
dividend	leading term	quotient	
divisor	points of inflection	remainder	

Copy and complete the following using the words and phrases from the list, where appropriate. A word or phrase may be used more than once.

1 The highest power of a polynomial gives the ______________ of the polynomial.

2 The first term of a polynomial written with powers in descending order is called the ______________.

3 A linear factor is divided into a ______________ to find a ______________ and a ______________.

4 A polynomial can be factorised by using the ______________.

Fluency

10A

1 Graph each equation by constructing a table of values for $x = -3$ to 3. **8.1**

(a) $y = 3x^2$ (b) $y = \frac{1}{2}x^3$ (c) $y = -x^4$

2 (a) Sketch the graph of $y = 2x^2 - 3$ starting with $y = x^2$. **8.2**

(b) Sketch the graph of $y = -2(x + 2)^3 + 1$ starting with $y = x^3$.

(c) Sketch the graph of $y = -3(x + 2)(x - 4)(x + 1)$.

3 (a) If $P(x) = 4x - 3$ and $Q(x) = 3x^2 - x + 2$, evaluate: **8.3**

(i) $P(4)$ (ii) $Q(-2)$ (iii) $P(-1).Q(-1)$

(b) Write an expression for:

(i) $2P(x) + 3Q(x)$ (ii) $P(x) \times Q(x)$ (iii) $Q(x) - P(x)$

4 Find the quotient and remainder for $\dfrac{4x^3 + 6x^2 + 26x + 13}{4x + 2}$. **8.3**

5 Use the remainder theorem to find the remainder produced by each division. **8.4**

(a) $\dfrac{10x^3 + 7x^2 - 4x - 3}{x + 1}$ (b) $\dfrac{x^4 - x^3 + x^2 + x - 1}{2x - 1}$

6 Use the factor theorem to factorise: **8.4**

(a) $x^3 + 6x^2 - 9x - 14$ (b) $x^4 - 5x^2 + 4$

7 Sketch the graph of each polynomial. **8.5**

(a) $P(x) = x^3 - 8x^2 - x + 8$ (b) $P(x) = x^4 + 2x^3 - 4x^2 - 2x + 3$

Understanding

8 An equation $y = x^4$ is dilated by a factor of $\frac{1}{2}$ in the y-direction, reflected in the x-axis and translated 2 units left and 3 units up. **8.2**

(a) Write the new equation.

(b) Sketch the graph of the equation. Use a CAS to find the x-intercepts correct to 2 decimal places.

9 Find the remainder using long division when $P(x) = x^3 - 3x^2 - 6x + 1$ is divided by $x + 2$. Confirm your answer by finding the remainder using the remainder theorem. **8.4**

Reasoning

10 A section of a roller-coaster track is described by the equation $y = 0.001(x - r)(x - s)^2$, where x is the horizontal distance in metres from Daredevil Corner and y is the height of the roller-coaster in metres. The graph has intercepts at $x = 0$ and $x = 60$ and the y-intercept is zero. The turning points occur at $x = 60$ and $x = 20$. **8.5**

(a) Use this information to write the equation.

(b) Sketch the graph of the equation for $0 \le x \le 80$.

(c) The top of Heartstop Hill is a horizontal distance of 80 m from Daredevil Corner. The length of the track from Daredevil Corner to Heartstop Hill is six times the height of the roller-coaster at Heartstop Hill. The roller-coaster on average moves on the track at 32 km/h. How many minutes does it take to travel from Daredevil Corner to Heartstop Hill?

11 Substitute $m = x^2$ into $x^4 - x^2 - 12 = 0$ to solve the equation for x. **8.4**

12 The sum of a sequence is given by $S = \frac{n}{2}[2a + (n - 1)d]$ where n stands for the number of terms in the sequence. If $a = 5$ and $d = 4$, find the number of terms in the sequence needed to make a sum of 860. **8.1**

13 A bowl has a shape that can be described by the parabola $y = \frac{1}{10}x^2$ for $-10 \le x \le 10$. **8.1**

The position of a length of uncooked spaghetti that just touches the rim of the bowl is described by the line $y = x$. The coordinate (0, 0) is the middle of the bottom of the bowl and all units are in centimetres.

(a) Graph both equations on the same axes.

(b) What is the vertical distance, above the centre of the bowl, of the end of the spaghetti?

14 The vertical cross-section of a bottle is described by the equations $y = (x - 5)^3 + 2$ and $y = -(x - 15)^3 + 2$. **8.5**

(a) Graph both equations on the same axes using x-values from 2 to 18.

(b) State the equation of the axis of symmetry of the bottle.

15 A capsule less dense than water is released under water and moves upwards subject to ocean currents. Its height, h metres, from sea level at time, t seconds, is given by $h = t^3 - 9t^2 + 26t - 24$. **8.5**

(a) Factorise the equation and sketch the graph for $0 \le t \le 4$.

(b) Describe the motion of the capsule in this time period.

10A

Exploration STEM

You can download this activity from the eBook or the Pearson Places website.

Sheep and shapes

A sheep farmer has a very large triangular paddock. The farmer can make use of the entire paddock, but the large area is difficult to manage. Your task is to explore different ways to subdivide the paddock into smaller areas and improve this situation. You will need to find the best ways to increase the farmer's profit given the conditions of the farm.

Exploration STEM

You can download this activity from the eBook or the Pearson Places website.

The game of life and money

In life you must make many decisions that affect your future. Some important decisions relate to your finances and how you choose to invest your money. In this task you play a game to simulate making investments, working in teams (families) in competition against others. Given an initial amount, your aim is to grow this money through hard work, savings and shrewd investments, but every choice involves risk. Your task is to navigate your way through these risks and opportunities, thinking carefully about your strategy, while also negotiating and working with others.

Exploration STEM

You can download this activity from the eBook or the Pearson Places website.

Human activities and environmental impact

In the 21st century, our lifestyles consume natural resources and significantly affect our environment (including the air, water, soil, plants, animals and so on). Your task is to investigate the impact on the environment from human activities such as farming, industry, mining and population growth. You must analyse the activity and its impact in detail, using relevant mathematical and scientific skills. Will your analysis persuade, educate or inform?

Exploration Coding

10A

You can access this activity from the eBook or the Pearson Places website.

The Monty Hall problem

The Monty Hall problem has left many people guessing—should you hold or switch to increase your chance of winning the prize?

9.1 Congruent and similar triangles

Important notation in geometric diagrams

- Sides with the same number of line markings | are equal in length.
- Sides with the same number of arrow markings > are parallel.
- || means 'is parallel to'.
- ⊥ means 'is perpendicular to'.
- ≡ means 'is congruent to'.
- ∴ means 'therefore'.
- ~ means 'is similar to'.
- ΔABC means a triangle with vertices A, B, C.
- ∠ means 'angle'.
- $\angle BAC$ means the angle at vertex A between the line BA and the line AC. Angles may also be given letter names such as α, β, or ϕ.
- Intervals are named by their endpoints, so the interval between A and B is AB.

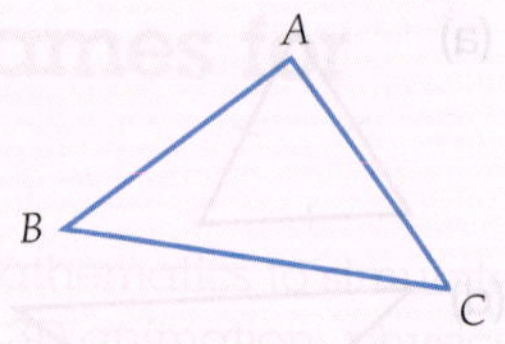

Congruency

Congruent (≡) figures are identical in size and shape, but their orientation may be different.

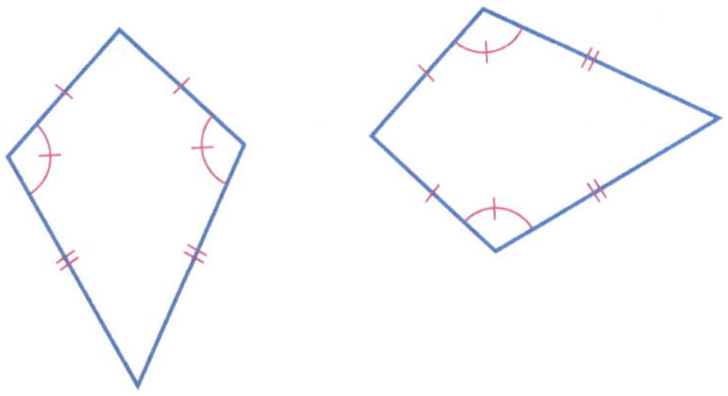

Worked example 1

W.E. 1

For the two congruent triangles shown, list the matching, or corresponding, sides for ΔABC and ΔDEF.

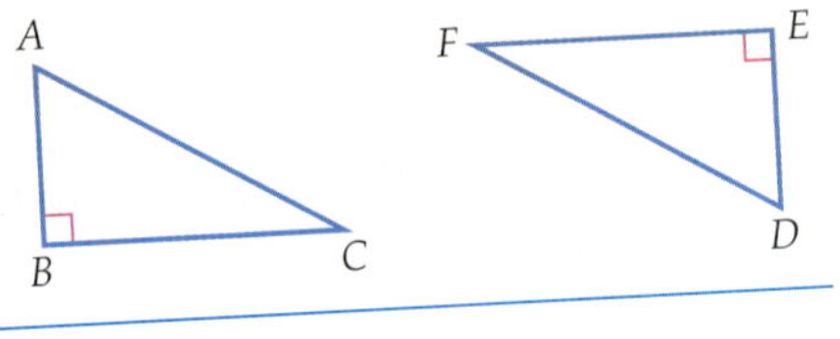

Thinking	Working
Identify the corresponding sides.	AB corresponds to DE. BC corresponds to EF. AC corresponds to DF.

It is important to recognise that shapes can be rotated, translated or flipped.

Tests for congruent triangles

Triangles are congruent (≡) if they are the same shape and size. Triangles are congruent if any one of these four sets of conditions is met:

1 **Side Side Side (SSS)**

The three pairs of corresponding sides are the same length.

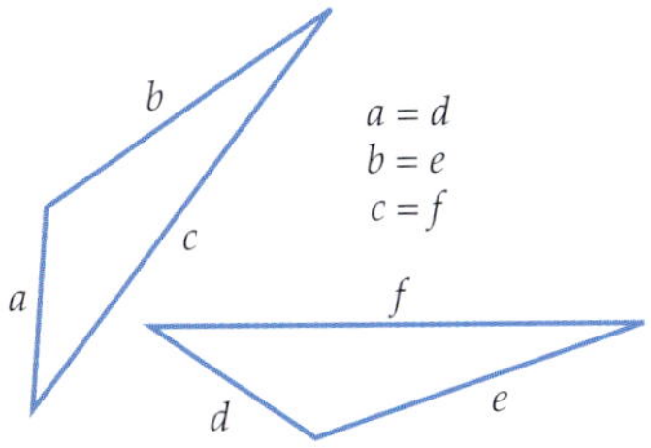

2 **Side Angle Side (SAS)**

Two pairs of corresponding sides are the same length and the **included angle** is the same.

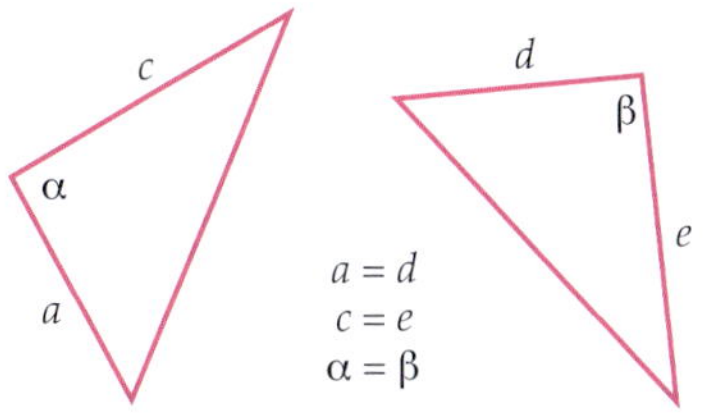

3 **Angle Side Angle (ASA)**

Two pairs of angles are the same and one pair of corresponding sides are the same length.

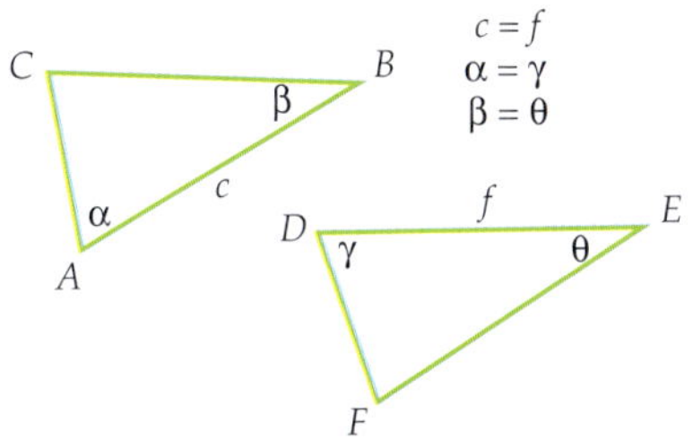

4 **Right angle Hypotenuse Side (RHS)**

Both triangles contain a right angle, the hypotenuse in both are of equal length and another pair of corresponding sides are the same length.

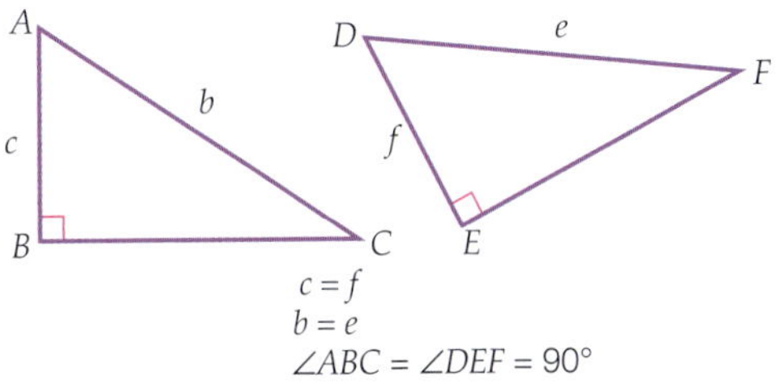

Worked example 2

W.E. 2

Show that ΔABC and ΔDFE are congruent, stating the congruency test used.

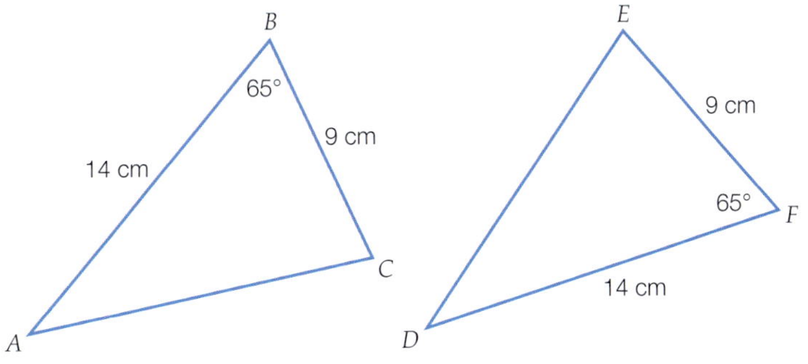

Thinking	**Working**	
1 List the corresponding sides and angles.	$AB = DF$	(both 14 cm)
	$\angle ABC = \angle DFE$	(both 65°)
	$BC = EF$	(both 9 cm)
2 Justify your answer.	The triangles are congruent due to SAS.	

Similarity

Similar (~) shapes are those where all side lengths have undergone the same dilation. Similar figures have the same shape but are different sizes. All corresponding sides of one figure have been enlarged or reduced by the same factor to form the second figure. This factor is called the 'dilation factor'.

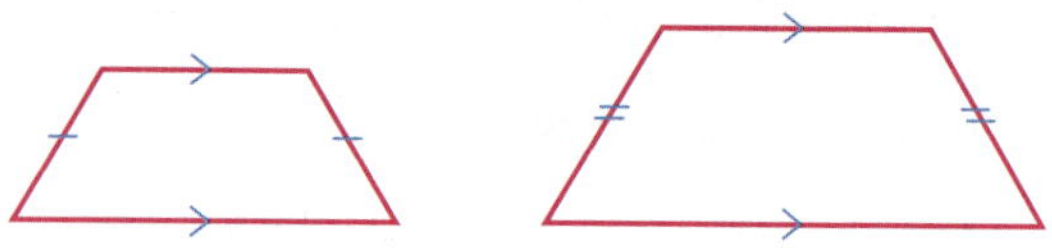

When a figure is dilated:

- corresponding angles are the same
- corresponding side lengths are in the same ratio.

Worked example 3

W.E. 3

For the similar triangles shown, identify the pairs of corresponding side lengths in ΔPQR and ΔXYZ, and show that the ratios are equivalent.

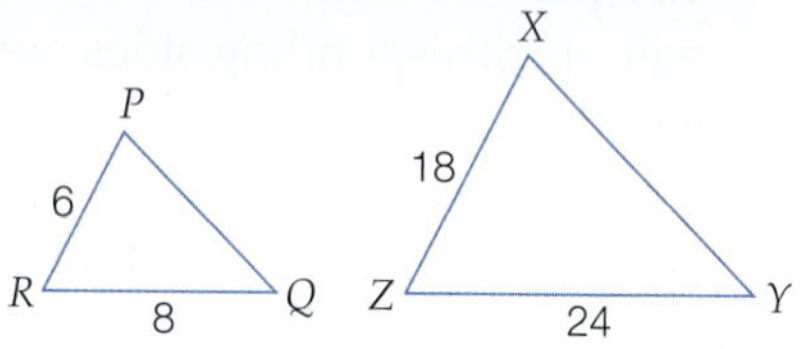

Thinking	Working
1 Identify the corresponding sides.	*PR corresponds to XZ.* *RQ corresponds to ZY.*
2 Write the lengths of the corresponding pairs of sides as ratios and simplify if possible.	$PR:XZ = 6:18$ $= 1:3$ $RQ:ZY = 8:24$ $= 1:3$
3 Confirm that both corresponding pairs of sides are in the same ratio.	*Both pairs of corresponding side lengths are in the same ratio of 1 : 3.*

Tests for similar triangles

Triangles are similar if one is proportional to the other, so that they have the same shape but are different sizes. This means that two triangles are similar if they satisfy any of the following conditions:

1 Side Side Side (SSS)

If the three sides of one triangle are proportional to the three sides of another triangle.

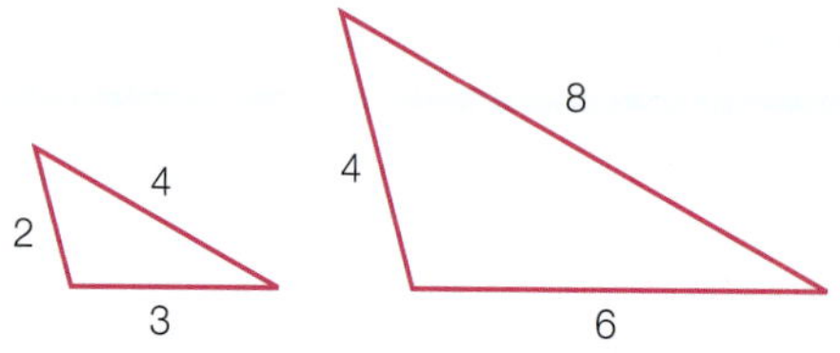

2 Side Angle Side (SAS)

If two sides of one triangle are proportional to two sides of another triangle, and the included angles are equal.

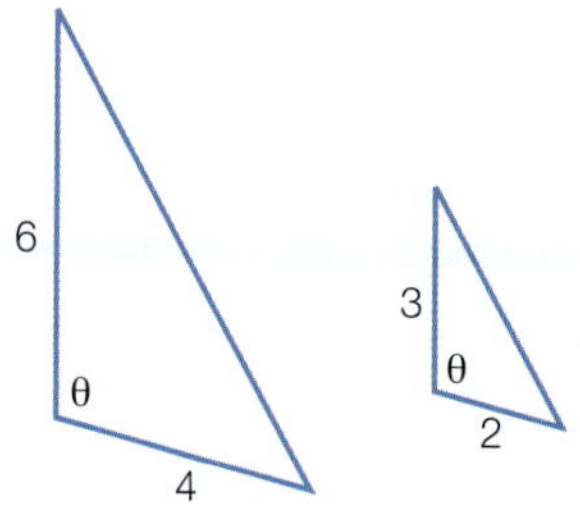

3 Angle Angle Angle (AAA)

If two angles of one triangle are equal to two angles in another triangle. All three angles are therefore equal.

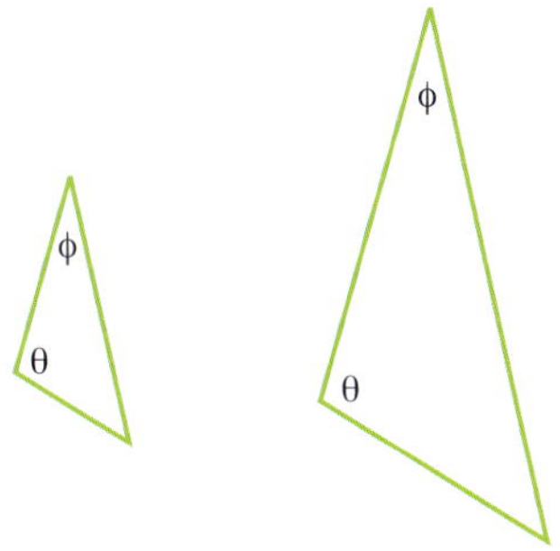

4 Right angle Hypotenuse Side (RHS)

If the hypotenuse and one side of a right-angled triangle are proportional to the hypotenuse and one side of another right-angled triangle.

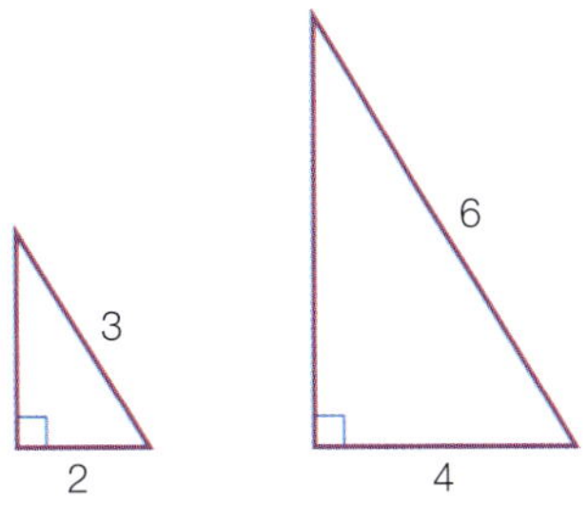

Note: These tests are similar to the four tests for congruence. The difference is that where congruence needs corresponding sides to be the same length, similarity only needs corresponding sides to be proportional.

Finding lengths using similarity

If two triangles are similar you can form an equation to solve for the unknown length using the ratio of corresponding sides. You need to identify the corresponding sides to equate the ratios.

Worked example 4 — W.E. 4

Given that these pairs of triangles are similar, find x.

(a)

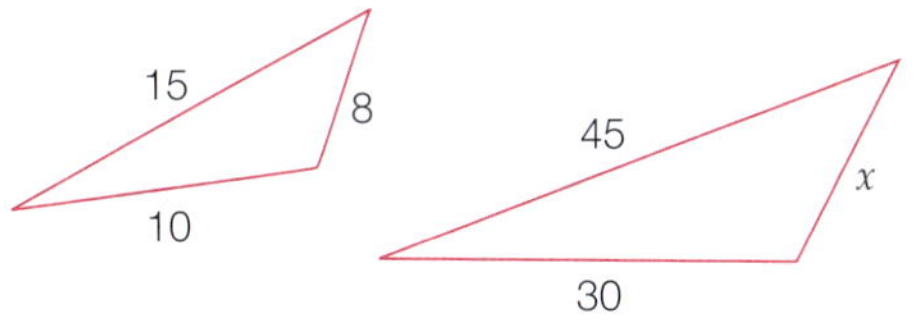

(b)

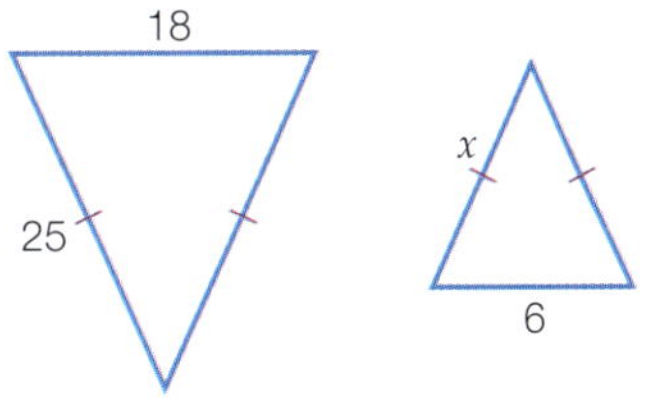

Thinking	Working
(a) 1 Identify corresponding sides and equate the two ratios.	(a) $x:8 = 45:15$
2 Write the ratios as an equation with x as a numerator and solve for x.	$\frac{x}{8} = \frac{45}{15}$ $x = \frac{45}{15} \times 8$ $x = 24$ units
(b) 1 Identify corresponding sides and equate the two ratios.	(b) $x:25 = 6:18$
2 Write the ratios as an equation with x as a numerator and solve for x. Write the length as an exact measurement.	$\frac{x}{25} = \frac{6}{18}$ $x = \frac{6}{18} \times 25$ $x = 8\frac{1}{3}$ units

To find the length of an unknown side in a similar triangle:

1. Identify two pairs of corresponding sides that include the unknown side.
2. Equate the two ratios.
3. Solve the equation to find the exact value of the unknown. Do not round decimals unless instructed to do so.

9.1 Congruent and similar triangles

Navigator

Answers p. 828

1 (a–d), 2 (a–f), 3, 4 (a–d), 5, 6, 7, 8, 9, 10, 11, 14, 15, 16	1 (column 1), 2 (column 1), 3, 4 (column 1), 5, 7, 8, 9, 10, 11, 12, 14, 15, 16, 17	1 (column 2), 2 (column 2), 3, 4 (column 2), 5, 10, 11, 12, 13, 14, 15, 17

Fluency

W.E. 1

1 List the pairs of sides of equal length in each of the following congruent triangles.

(a)

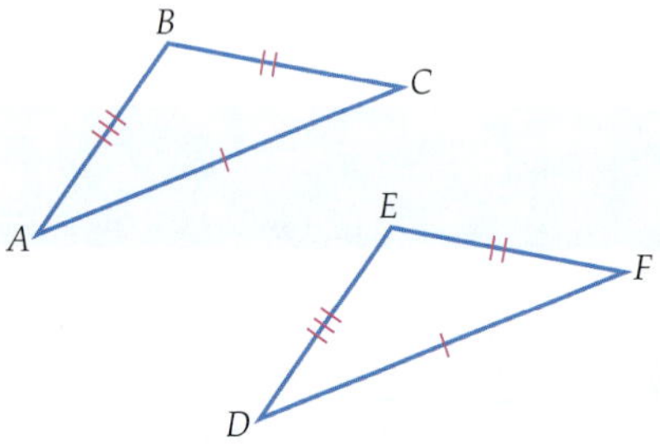

(b)

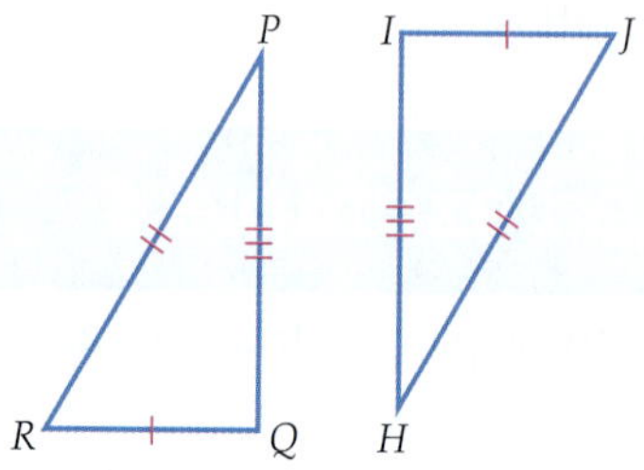

(c)

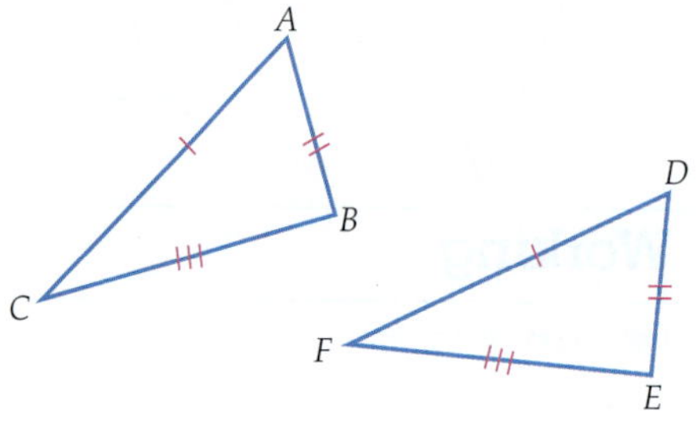

(d)

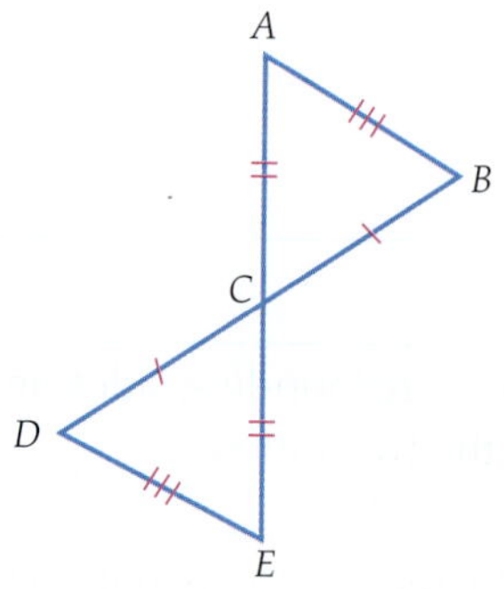

(e)

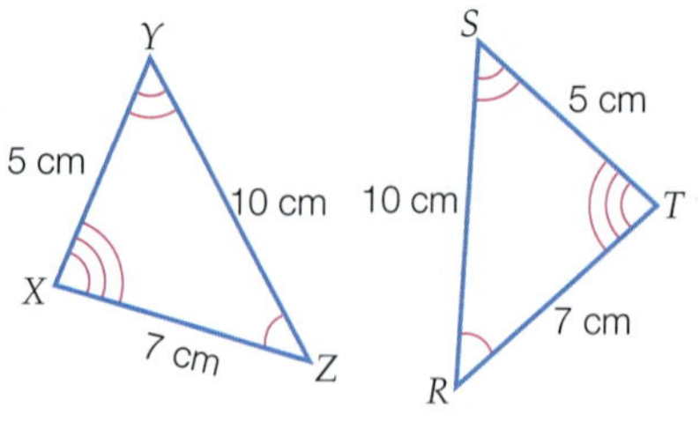

(f)

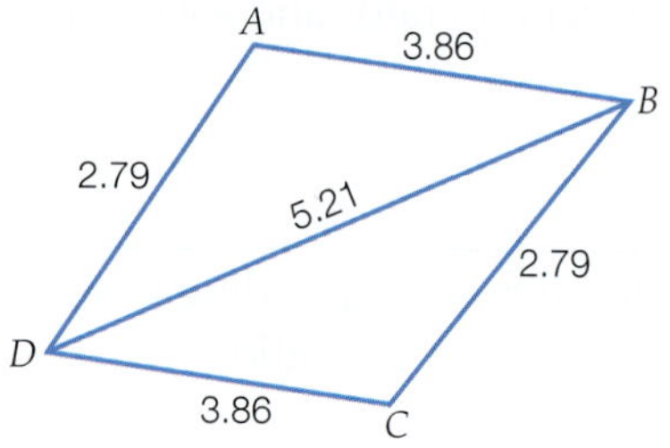

2 Show that the following pairs of triangles are congruent, stating the congruency test used. **W.E. 2**

(a)
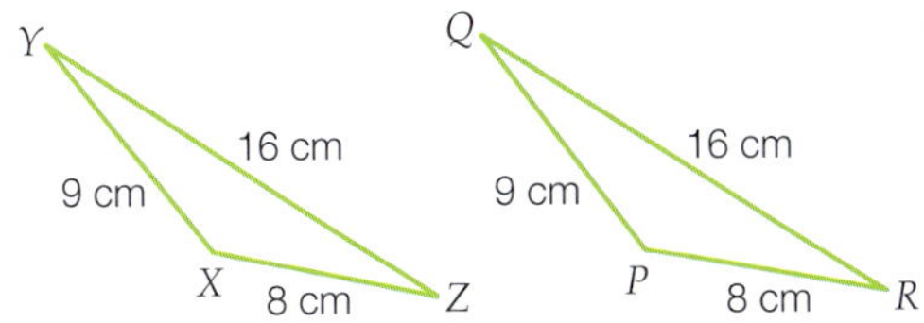

(b)
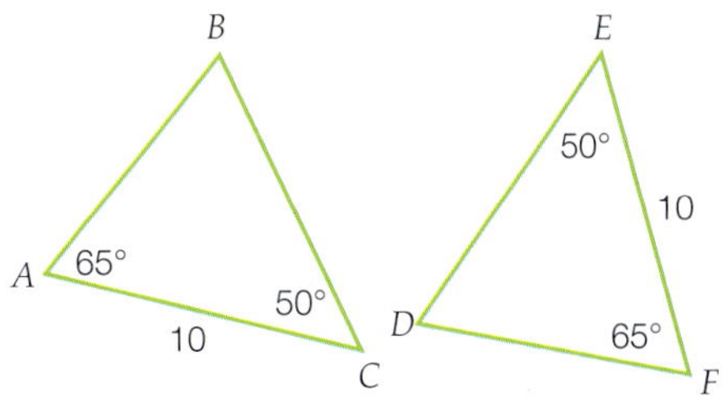

(c)
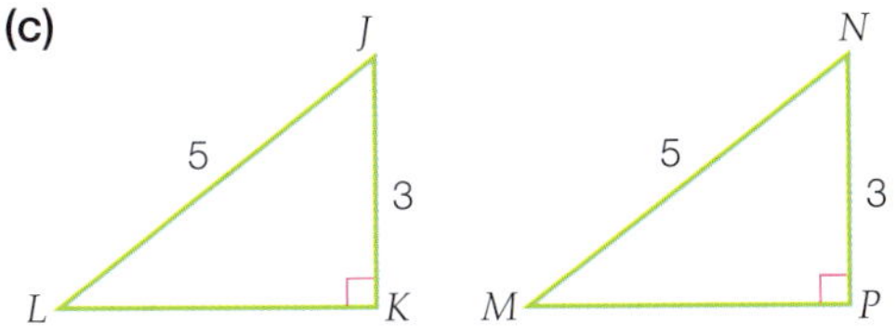

(d)
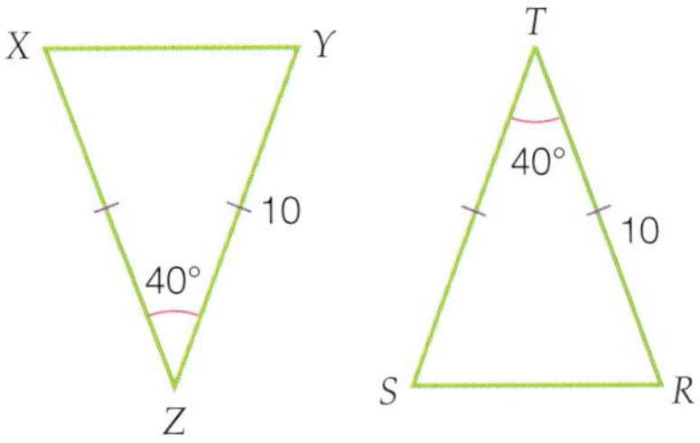

(e)
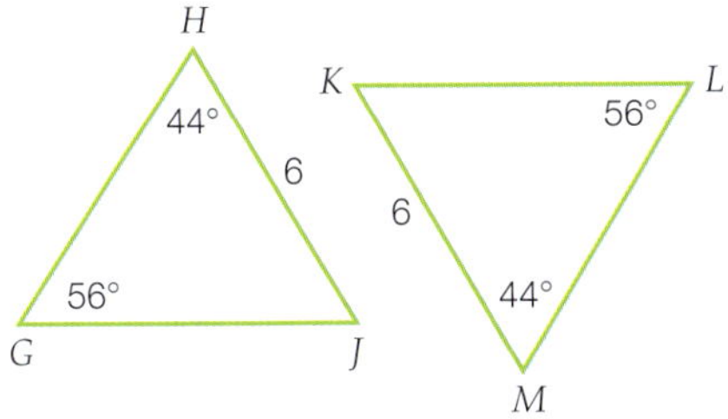

(f)
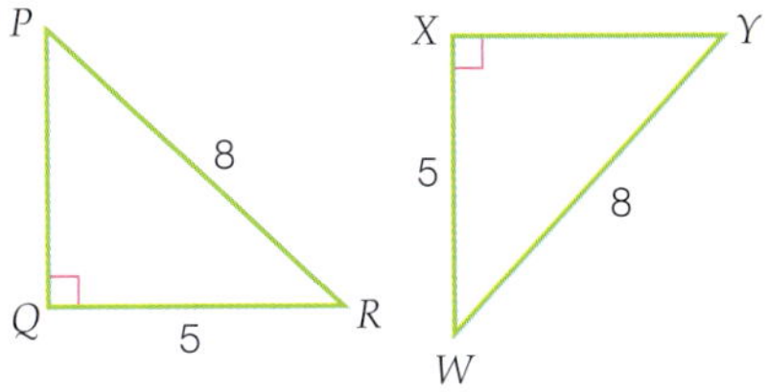

(g)
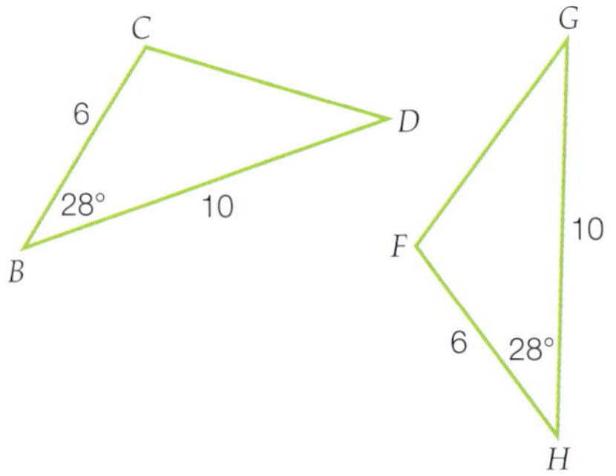

(h)
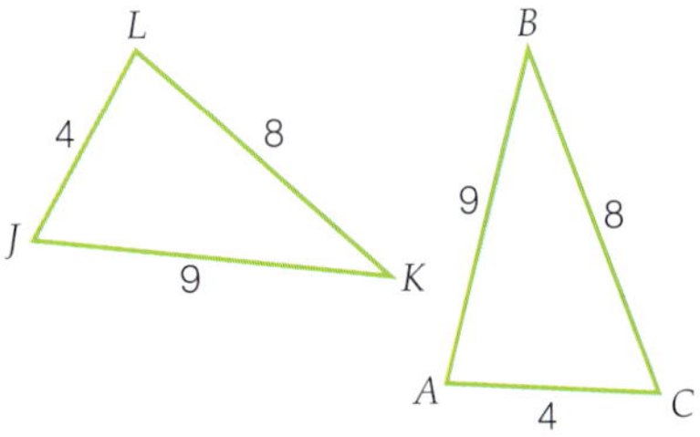

3 For each of the following pairs of similar triangles, identify the corresponding side lengths and show that the ratios are equivalent. **W.E. 3**

(a)
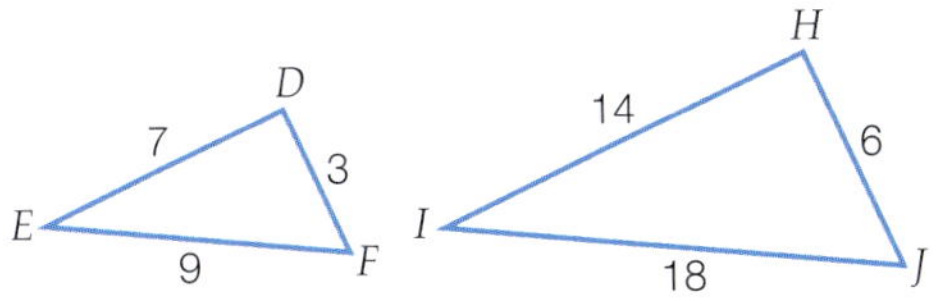

(b)
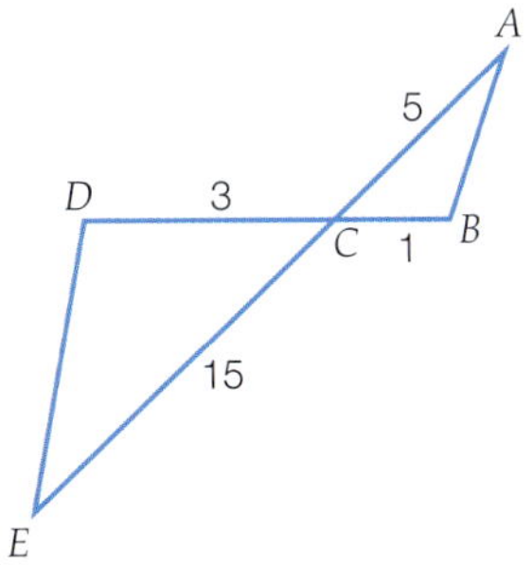

(c)
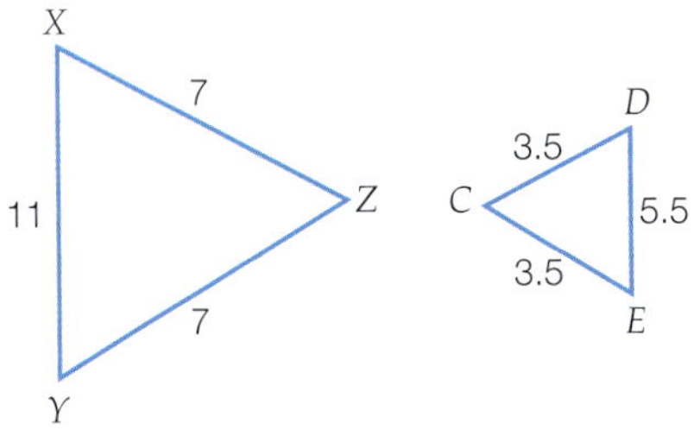

(d)
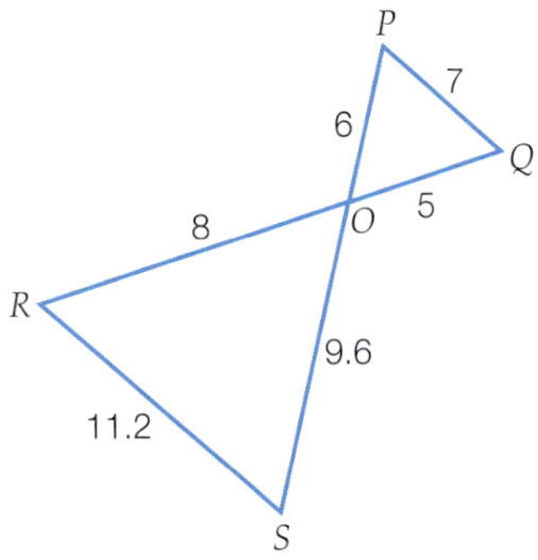

W.E. 4

4 Given that these pairs of triangles are similar, find x.

(a)

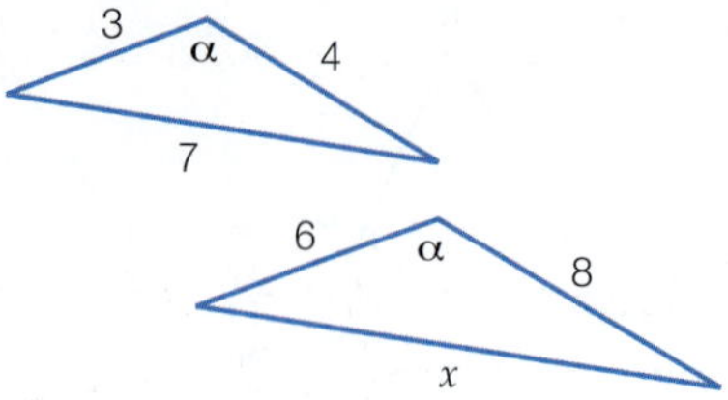

(b)

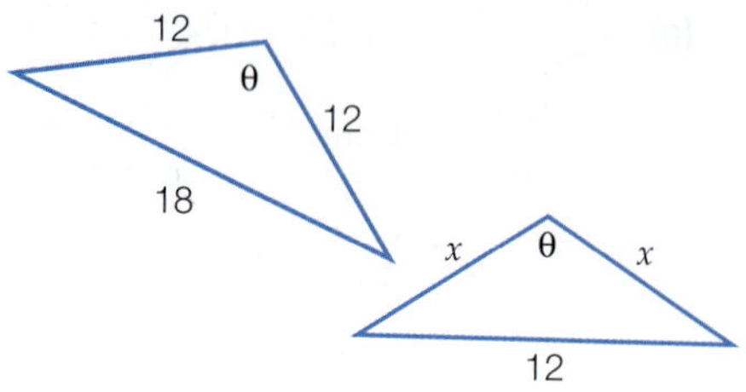

(c)

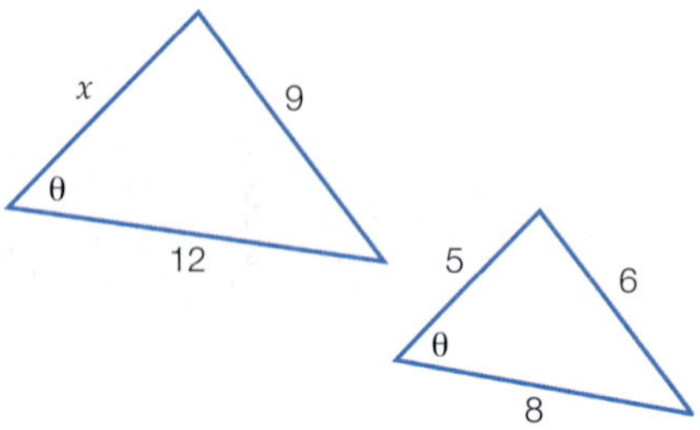

(d)

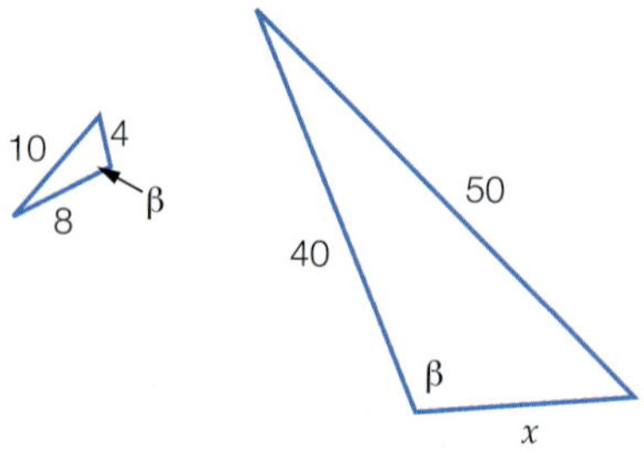

(e)

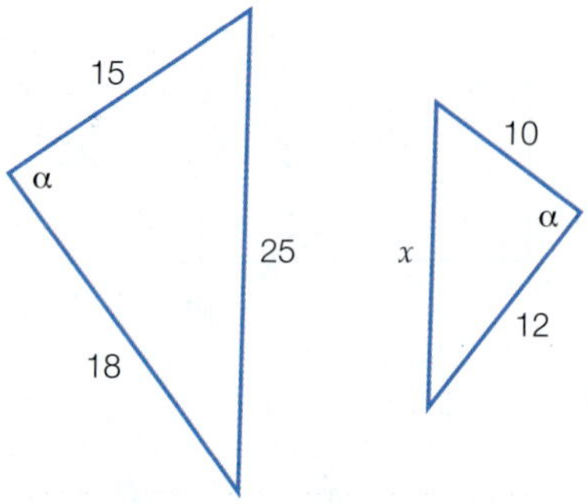

(f)

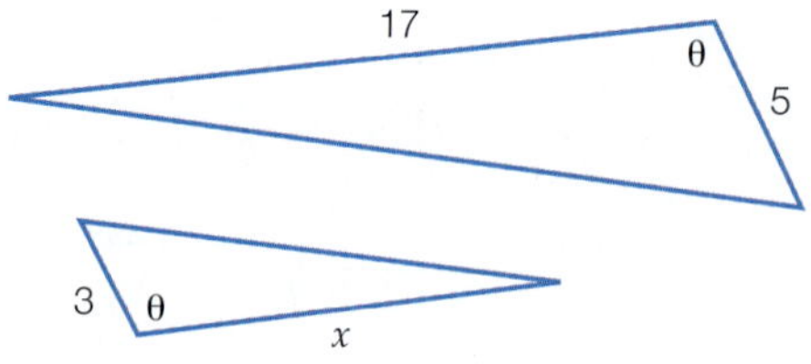

5 Match the pairs of congruent triangles given below.

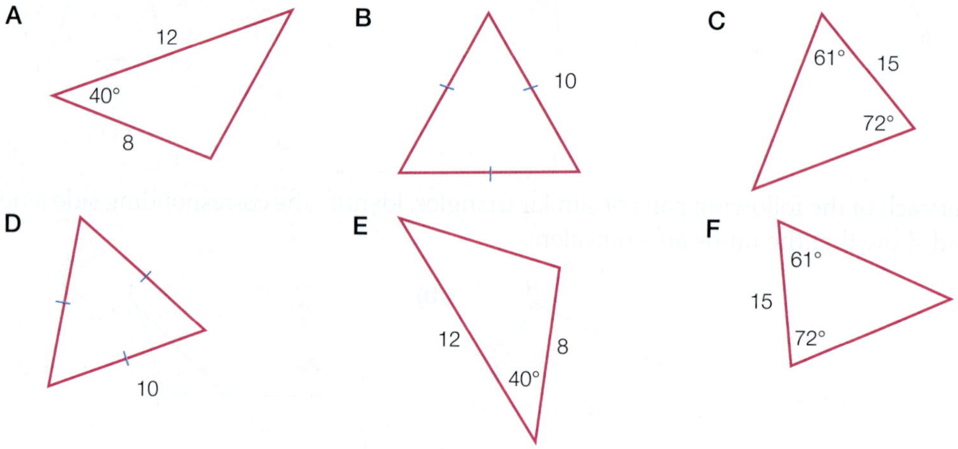

6 Which one of the following conditions does *not* ensure congruence for two triangles?

A SSS **B** SAS **C** ASA **D** AAA

7 Which condition verifies that these two triangles are congruent?

A SSS **B** SAS

C ASA **D** RHS

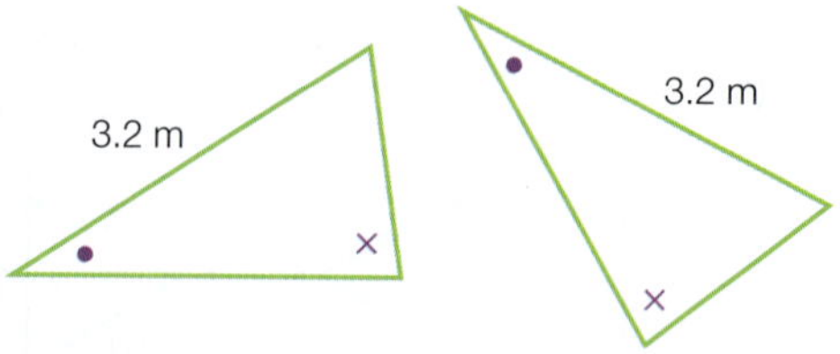

8 Given $\Delta ABC \sim \Delta EFG$:

(a) the side corresponding to AB is:

A FG B EF

C BC D AC

(b) the side corresponding to FG is:

A BC B AC

C EF D EG

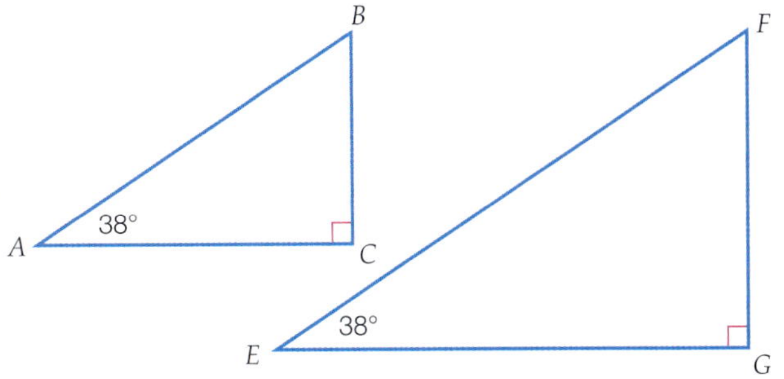

9 Given $\Delta HJK \sim \Delta LMN$:

(a) the side corresponding to LM is:

A HJ B MN

C JK D LN

(b) the side corresponding to HJ is:

A JK B LM

C HK D NL

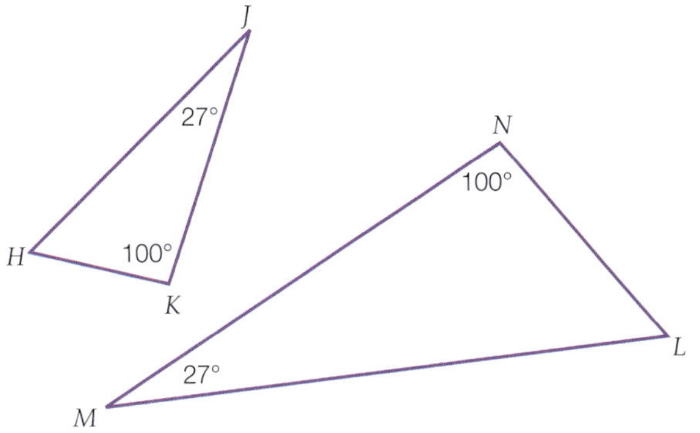

Understanding

10 In the following diagrams, first determine the similar triangles and then find the value of x.

(a)

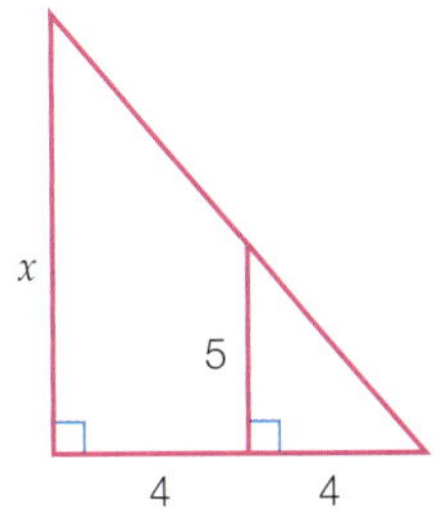

(b)

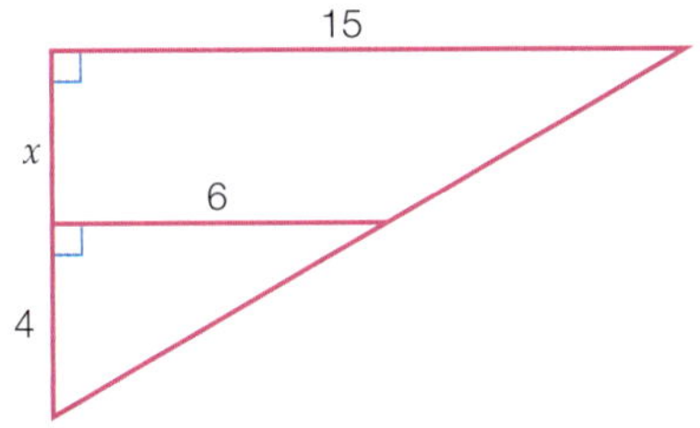

(c)

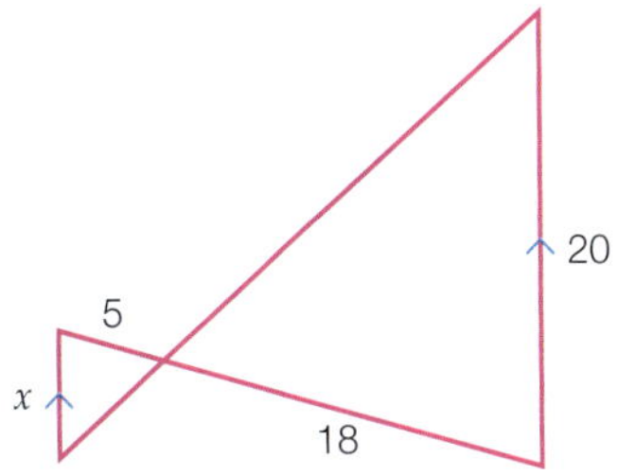

(d)

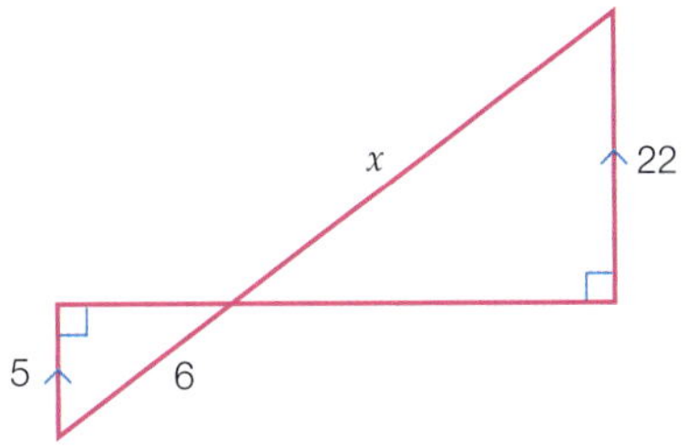

11 Dirk is climbing up a long ladder leaning against a wall. His friend Sean is holding up the ladder to prevent it wobbling. Sean is 3 m from the wall and 2 m from the base of the ladder. He holds the ladder at a point 2.1 m above the ground. How far up the wall does the ladder reach?

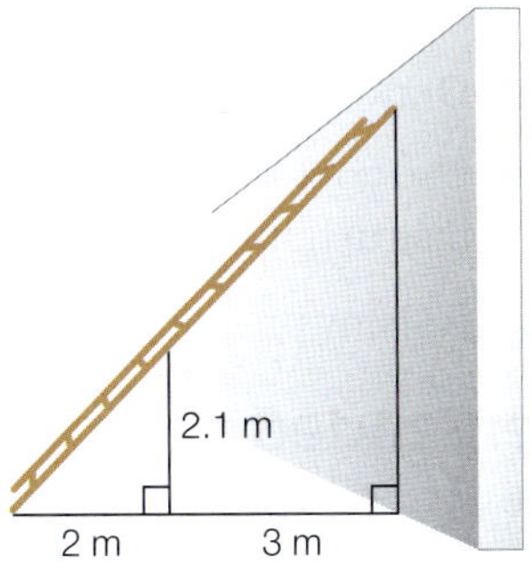

12 A light source projects the image of a photographic slide onto a screen, as shown in the diagram. The slide is 3 cm high and positioned 10 cm behind a lens. Calculate the distance between the lens and the screen if the image on the screen is 95 cm high. Give your answer correct to 2 decimal places.

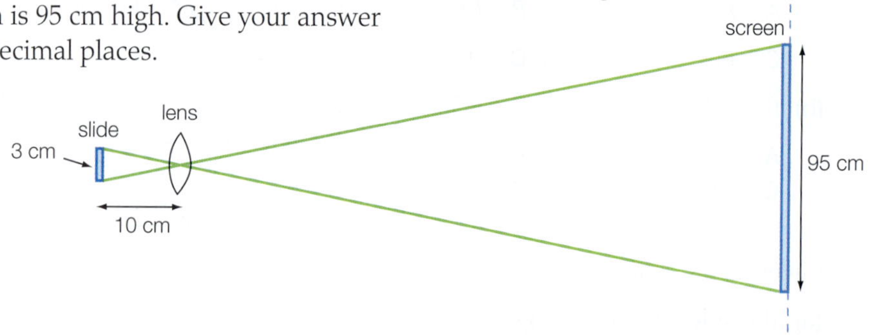

13 Ziva plays tennis at night under floodlights. When she stands 4.5 m from the base of a light, she casts a shadow 1.4 m long. If Ziva is 1.7 m tall, how high is the light?

Reasoning

14 The small triangles in the following diagram are all equilateral.

(a) Find six triangles congruent to ΔDFM.

(b) Find two pairs of congruent triangles larger than ΔDFM.

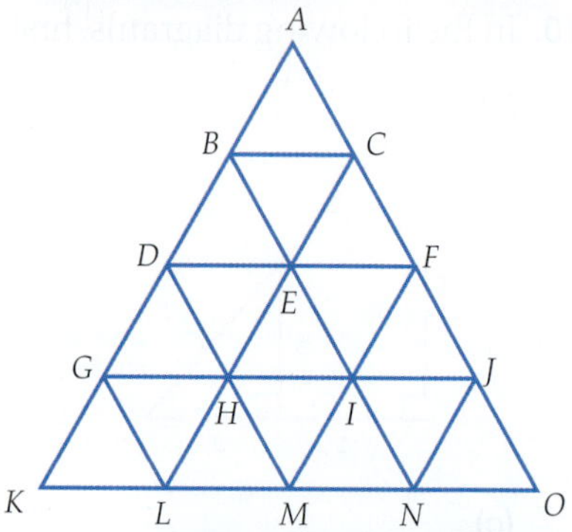

15 The following steps explain how to use a stick to estimate the height of a tree.

Step 1: Have someone hold the stick vertically on the ground.

Step 2: Lie on the ground in a position so that from your viewpoint, the top of the tree and the top of the stick are in line.

Step 3: ________________________________.

(a) Draw a diagram of the situation.

(b) Use your knowledge of similar triangles to complete the instructions for step **3**.

(c) List any practical problems you can see with this method.

Open-ended

16 Explain why two triangles with three equal angles may not necessarily be congruent. Show this by drawing diagrams.

17 $\Delta ABC \sim \Delta DEF$. The dilation factor is $\frac{3}{2}$. Draw a diagram to show both triangles and write possible values for the sides of both triangles if ΔABC is the smaller triangle and BC is equal to 4 cm.

Proving congruence and similarity

9.2

Proofs involving congruence

It is often necessary to **prove** that two triangles are congruent to complete proofs of other geometric relationships such as the special properties of parallelograms. You use one of the four tests for congruence: SSS, ASA, SAS and RHS.

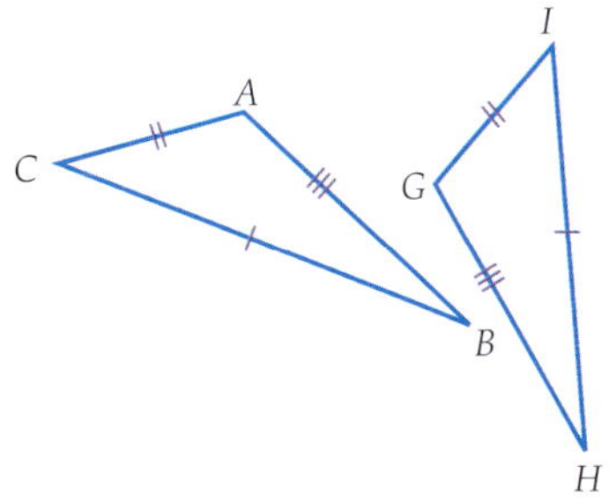

When showing that two shapes are congruent, write the matching vertices in the same order.

For example, in the congruent triangles shown here, the matching vertices are A and G, B and H, and C and I.

So, $\Delta ABC \equiv \Delta GHI$.

This allows you to easily identify corresponding sides and angles because they are in the same place in the statement.

Worked example 5

W.E. 5

Prove that ΔABC and ΔEDC are congruent.

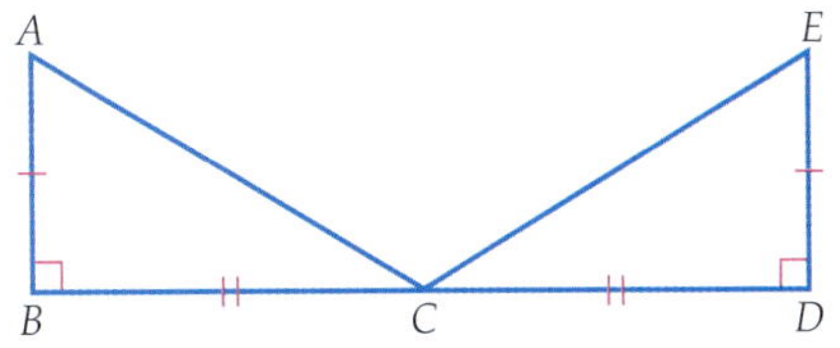

Thinking	Working
1 Name the two triangles. Ensure that the vertices are in the same order.	ΔABC and ΔEDC.
2 Write three sets of equalities, with reasons, that apply to the two triangles.	$AB = ED$ (given) $BC = DC$ (given) $\angle ABC = \angle EDC$ (given)
3 Write down the pair of congruent triangles and the test used to determine this.	$\Delta ABC \equiv \Delta EDC$ (SAS)

Worked example 6

W.E. 6

Prove that the triangles ΔMNO and ΔPQO are congruent, and hence find the value of the given variable.

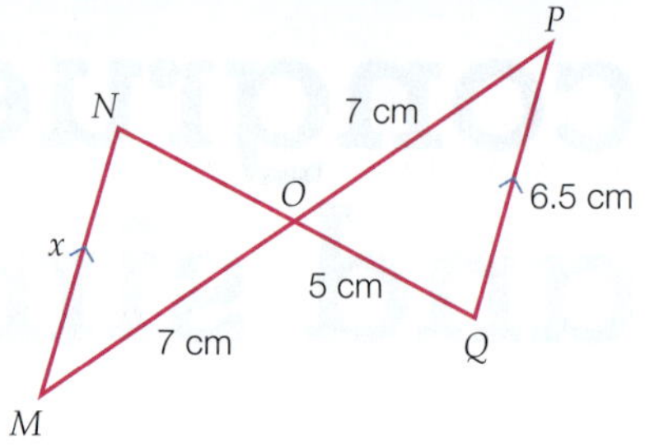

Thinking	Working
1 Name the two triangles.	$\Delta MNO \equiv \Delta PQO$
2 Write three sets of equalities, with reasons, that apply to the two triangles.	$\angle MNO = \angle PQO$ (alternate angles, $MN \parallel PQ$) $\angle MON = \angle POQ$ (vertically opposite angles) $MO = PO$ (given)
3 Write the pair of congruent triangles and the test used to determine this.	$\Delta MNO \equiv \Delta PQO$ (ASA)
4 Identify the appropriate pair of corresponding sides.	$MN = PQ$ (corresponding sides in congruent triangles)
5 State the value of x.	$x = 6.5$ cm

In more difficult examples, breaking up the composite shape into simpler shapes can be helpful for looking at congruency or similarity. Remember to look for any rotations, reflections or translations when matching sides or angles.

Proving similarity

You can use the tests for similarity to prove that triangles are similar.

Worked example 7

W.E. 7

Prove that the triangles ΔABD and ΔBCD are similar.

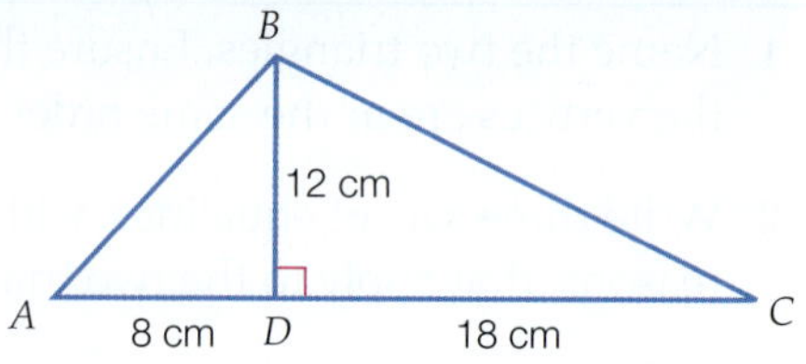

Thinking	Working
1 Redraw the two triangles with the same orientation and mark any given information on the diagrams.	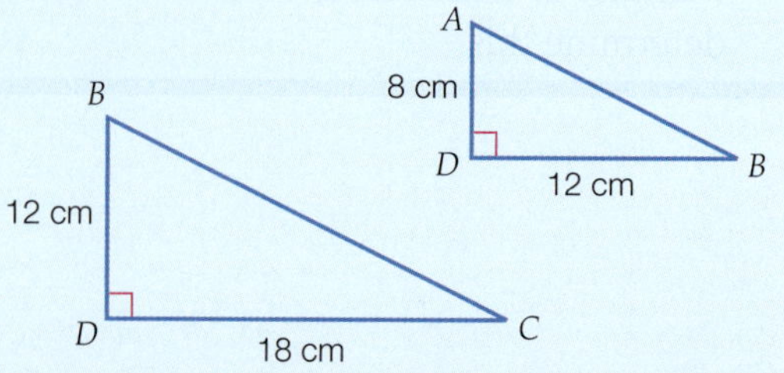

2 Check whether any of the corresponding sides given are proportional.	$\frac{BD}{CD} = \frac{12}{18} = \frac{2}{3}$ $\frac{AD}{BD} = \frac{8}{12} = \frac{2}{3}$
3 Write three statements or equations that display similarity.	*BD* and *CD* are in the ratio 2 : 3. *AD* and *BD* are in the ratio 2 : 3. $\angle ADB = \angle BDC = 90°$
4 State whether similarity exists and, if so, by which test.	$\Delta ABD \sim \Delta BCD$ because two sides of one triangle are in the same ratio as two corresponding sides of the other triangle and the included angles are equal (SAS).

Note: It can also be proved that $\angle ABC$ is a right angle, so ΔABC is similar to both ΔABD and ΔBCD (AAA). Similarity and congruency can also be proven without using numerical values.

Worked example 8

W.E. 8

Prove that $\Delta PRS \sim \Delta PQT$.

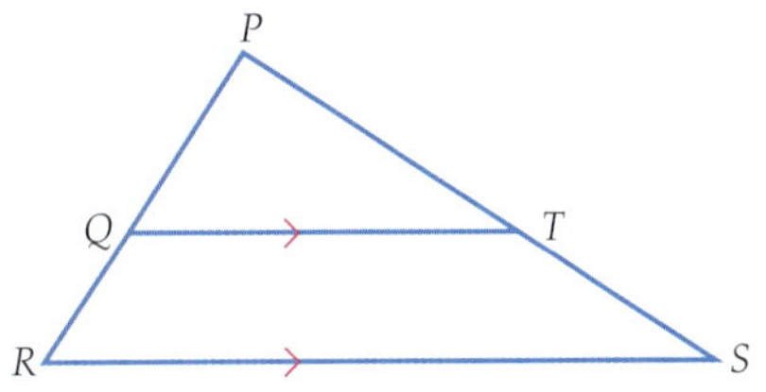

Thinking	Working
1 Name the two triangles. Note that there is a common angle.	ΔPRS and ΔPQT
2 Write three statements or equations that display similarity.	$\angle P$ is common. $\angle PRS = \angle PQT$ (corresponding angles are equal as $QT \parallel RS$) $\angle PSR = \angle PTQ$ (corresponding angles are equal as $QT \parallel RS$)
3 State a conclusion based on the statements.	All angles are equal in ΔPRS and ΔPQT.
4 Establish the side ratio.	$RS \parallel QT$ so $\frac{PS}{PQ} = \frac{PT}{PR}$
5 State similarity and the test used.	$\Delta PRS \sim \Delta PQT$ (AAA)

9.2 Proving congruence and similarity

Navigator

Answers p. 829

1, 2, 3, 4, 5, 6, 7a, 9	1, 2, 3, 4, 5, 6, 7, 8, 9	1, 2, 3, 4, 6, 7, 8, 9

Fluency

W.E. 5

1 Prove that each of the following pairs of triangles are congruent.

(a)

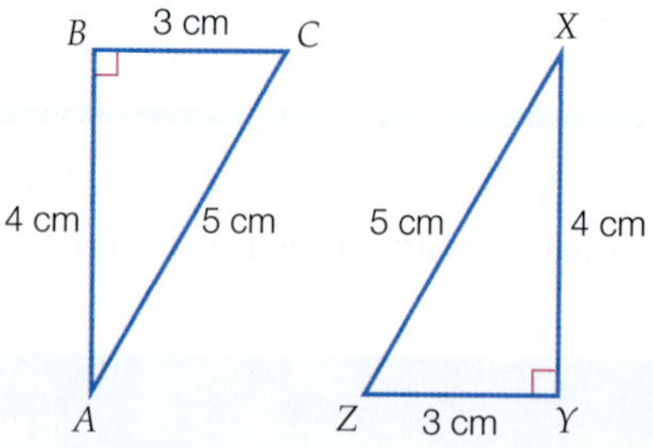

(b)

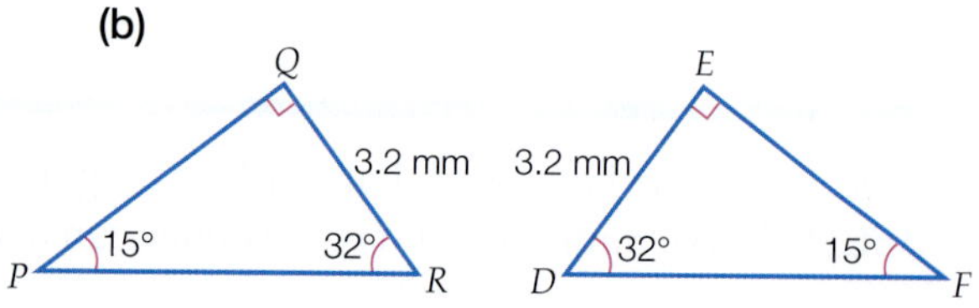

(c)

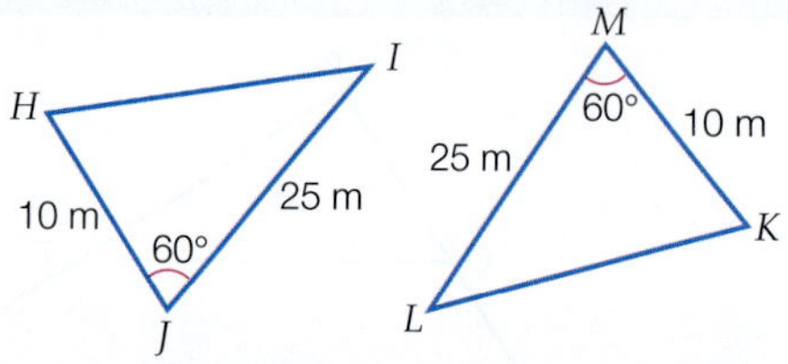

(d)

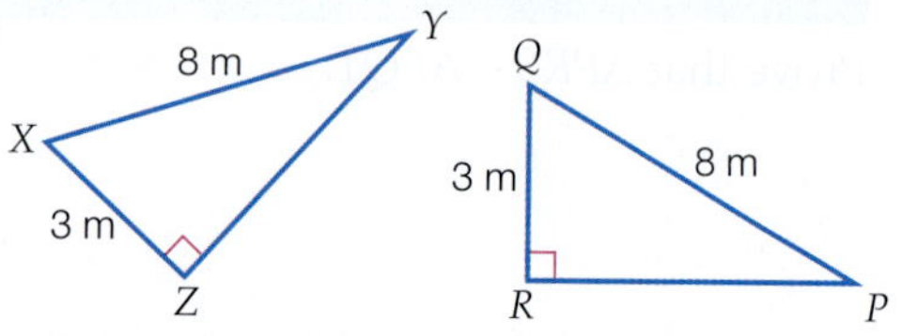

W.E. 6

2 Prove that the following pairs of triangles are congruent and hence find the value of the unknown.

(a)

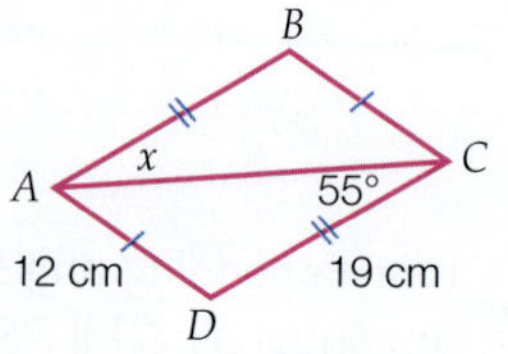

(b)

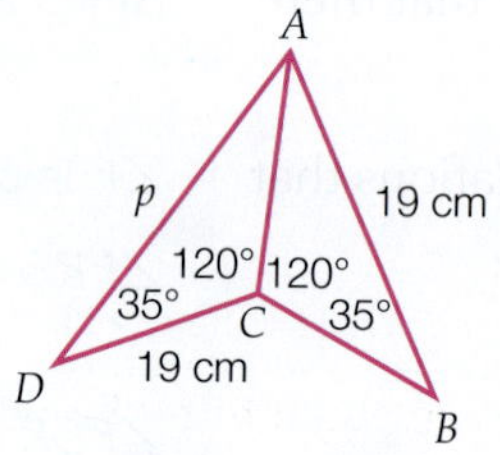

(c)

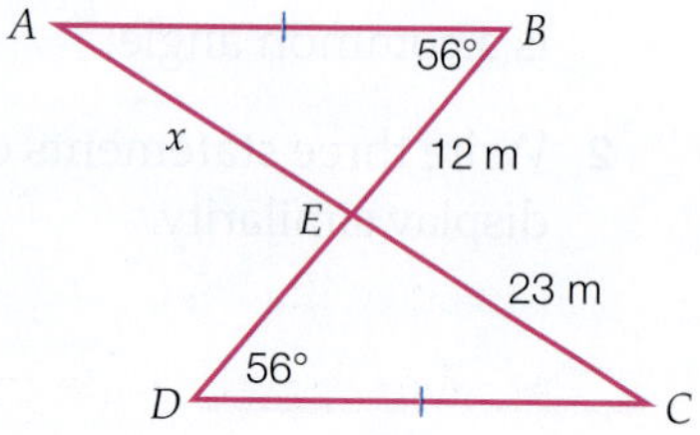

(d)

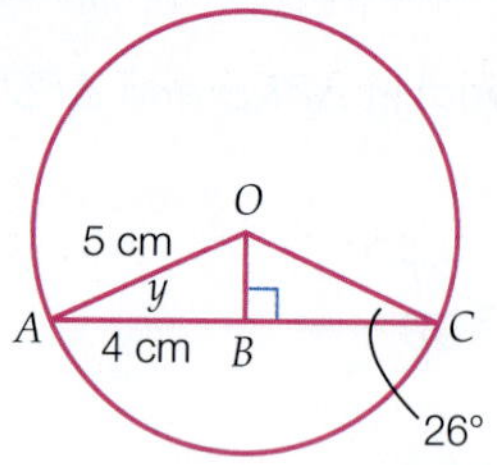

(e)

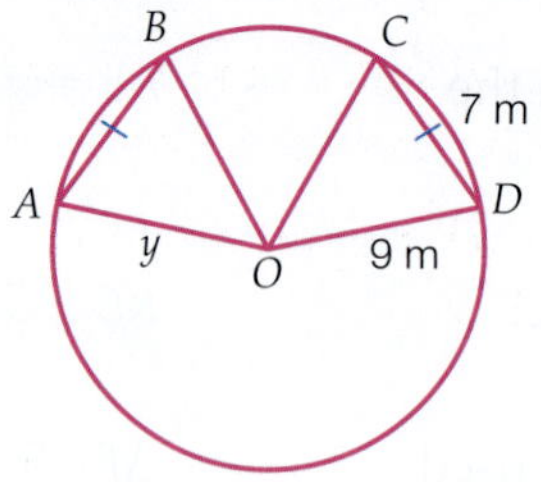

(f)

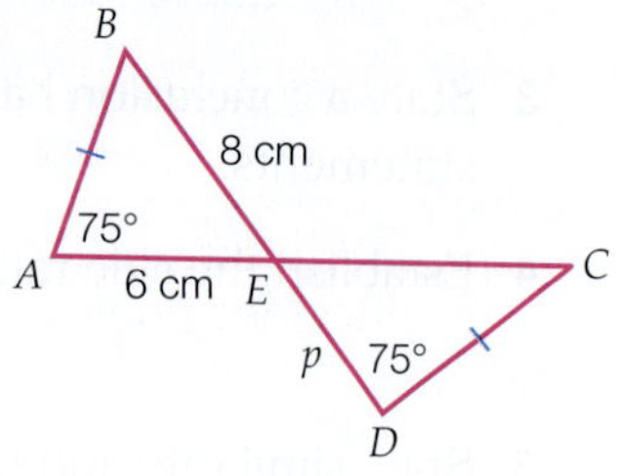

W.E. 7

3 Prove that the following pairs of triangles are similar.

(a)

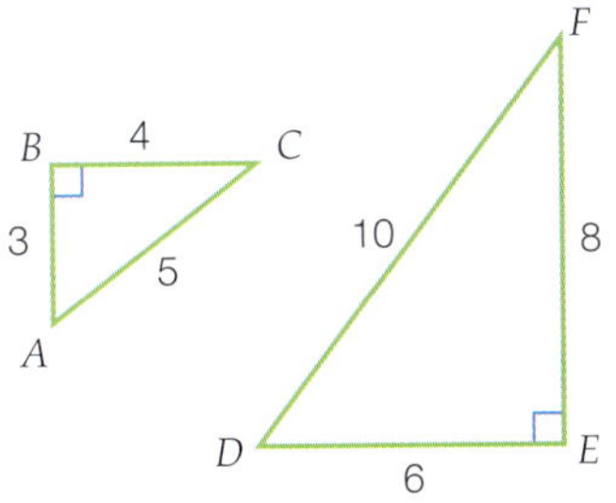

(b)

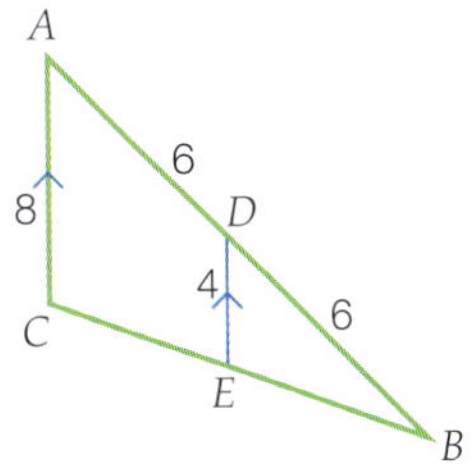

(c)

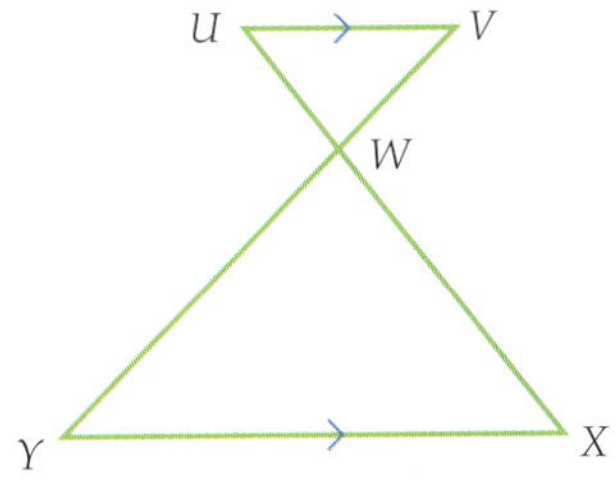

(d)

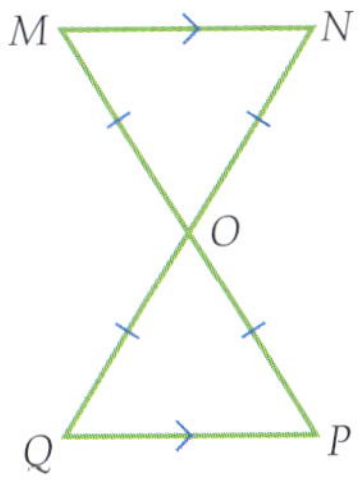

(e)

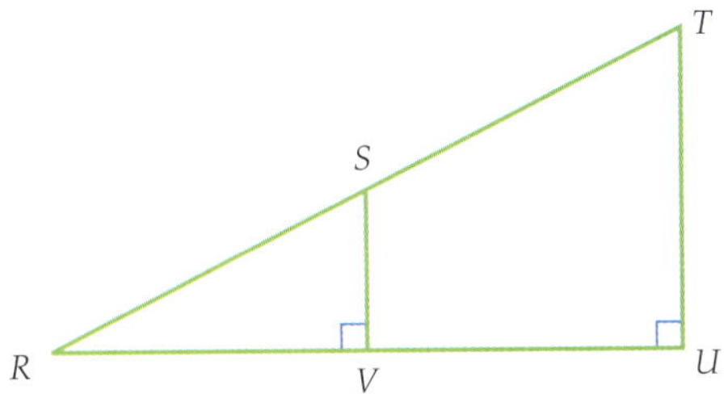

(f)

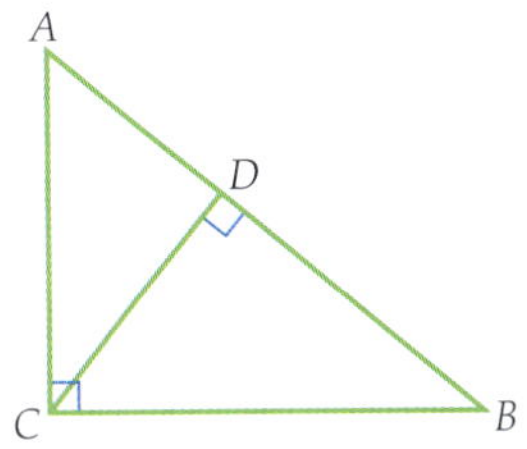

4 *DB* is the perpendicular bisector of *AC* and bisects $\angle ADC$. Which test for congruence can be used to prove that $\Delta ABD \equiv \Delta CBD$?

A SSS **B** SAS

C ASA **D** RHS

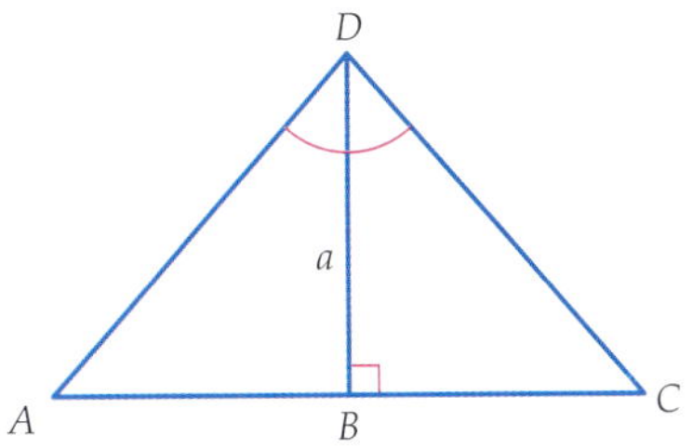

5 Which of the following is not enough to determine similarity?

A Two angles of one triangle are equal to two angles of the other triangle.

B Two sides of one triangle are proportional to two sides of another triangle.

C The hypotenuse and second side of a right-angled triangle are proportional to the hypotenuse and second side of another right-angled triangle.

D Two sides of one triangle are proportional to two sides of another triangle and the included angles are equal.

Understanding

W.E. 8

6 (a) Prove that $\Delta RUT \equiv \Delta RST$.

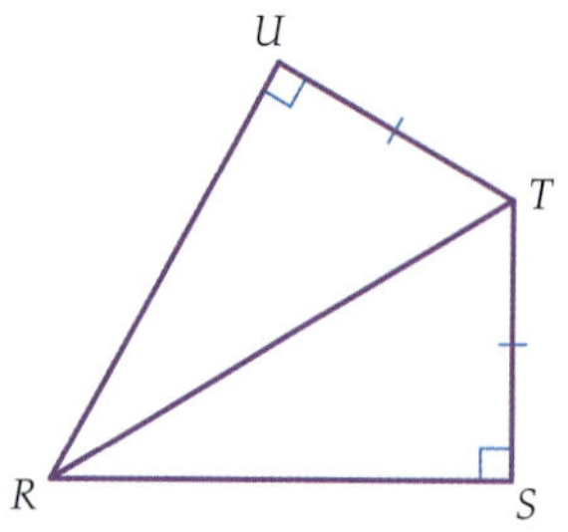

(b) $SP = RQ$ and $\angle SPQ = \angle RQP$.
Prove that $\Delta SPQ \equiv \Delta RQP$.

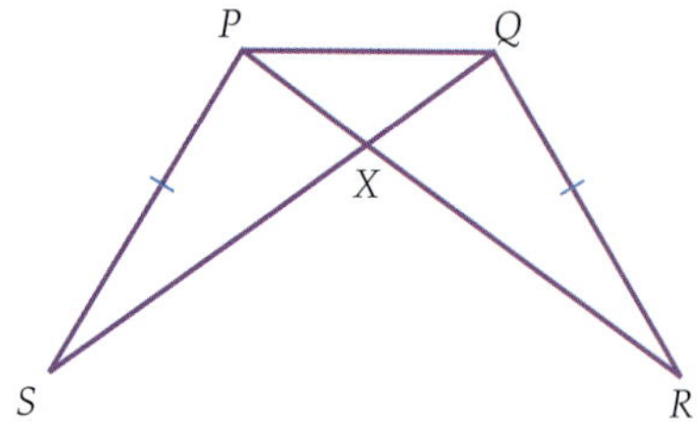

7 (a) Prove that $\Delta RQS \sim \Delta RPT$.

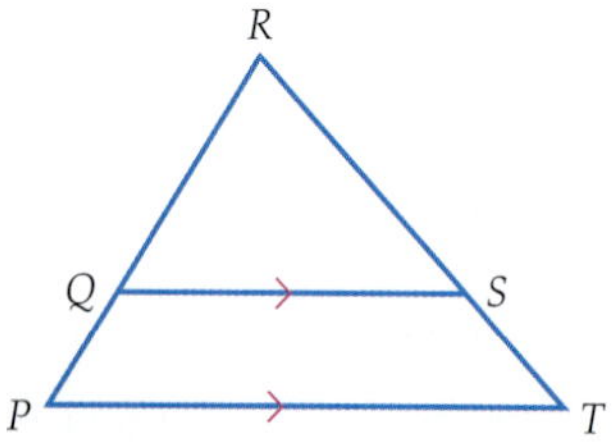

(b) Prove that $\Delta ABC \sim \Delta BDC \sim \Delta ADB$.

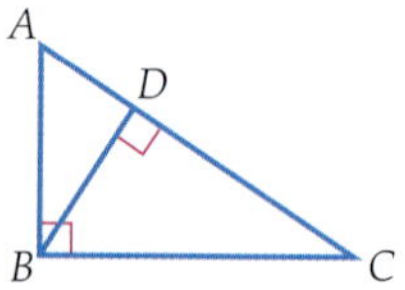

Reasoning

8 (a) Explain why or why not a pair of shapes are similar if they are any two:

(i) rectangles
(ii) squares
(iii) isosceles triangles
(iv) equilateral triangles
(v) isosceles right-angled triangles
(vi) circles.

(b) For each of the similar shapes in part **(a)**, state the conditions necessary for the pairs of shapes to be congruent.

Open-ended

9 Draw four pairs of congruent triangles and label them with enough information to show congruence. Each pair should represent a different test. Write the name of the congruence test that applies to each pair.

Problem solving

How many triangles?

How many triangles can you find in this diagram?

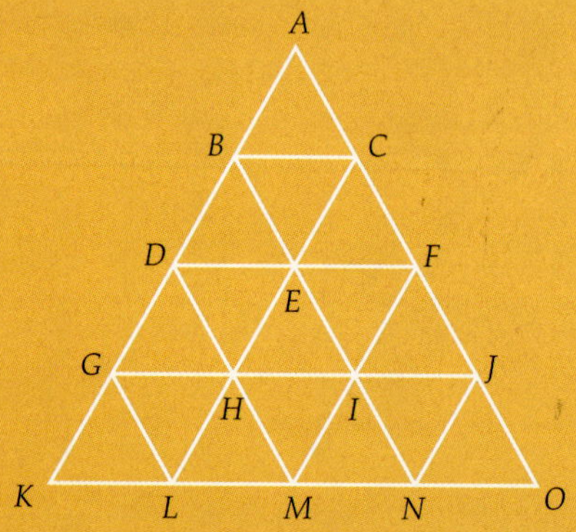

Strategy options

- Solve a simpler problem.
- Have I seen a similar problem?
- Draw a diagram.
- Look for a pattern.

Proofs using congruent triangles

9.3

A **definition** is the minimum amount of information needed to identify a particular figure.

A **property** is a fact about a particular figure that can be proved from the definition.

Definitions of triangles

1. A scalene triangle is a triangle with no two sides equal in length.
2. An isosceles triangle is a triangle with only two sides equal in length.
3. An equilateral triangle is a triangle with all sides equal in length.

Properties of triangles

Scalene triangle

A scalene triangle has no particular property as all angles are of different magnitude.

$\angle BAC \neq \angle ABC \neq \angle BCA$

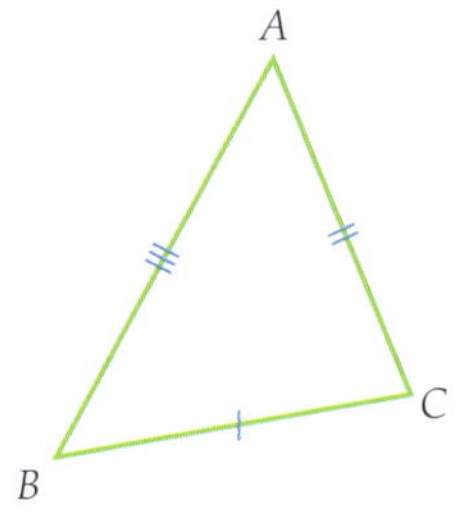

Isosceles triangle

An isosceles triangle has two angles of equal magnitude.

$\angle BAC = \angle BCA \neq \angle ABC$

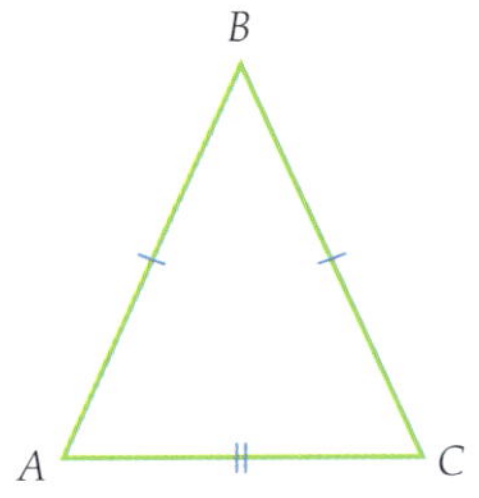

Equilateral triangle

An equilateral triangle has all angles of equal magnitude of 60°.

$\angle BAC = \angle ABC = \angle BCA = 60°$

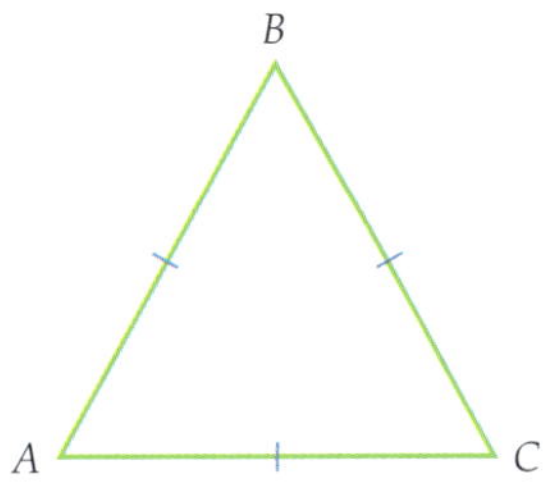

Proving properties of triangles

You can use congruent triangles to prove the properties of isosceles and equilateral triangles.

Worked example 9

W.E. 9

In ΔABC, $BA = BC$.

(a) Prove that $\angle BAC = \angle BCA$.

(b) What property of isosceles triangles have you just proved?

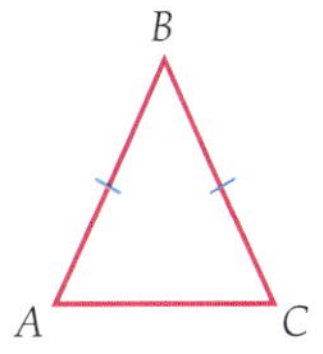

Hint

Construct a line from *B* to the midpoint of *AC*.

Thinking	Working
(a) 1 Create two triangles that you can prove are congruent by drawing a line *BD* from *B* to the midpoint of *AC*. State this construction.	(a) Line *BD* meets the midpoint of *AC*. 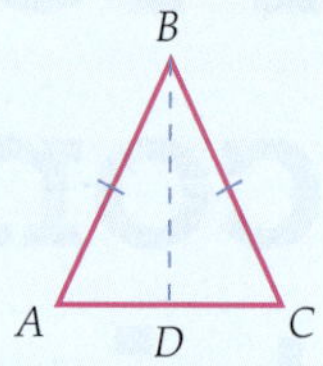
2 Name the two triangles.	ΔABD and ΔCBD
3 Write the three sets of equalities, with reasons, that apply to the two triangles.	$BA = BC$ (given) $BD = BD$ (common side) $AD = CD$ (by construction)
4 Write the pair of congruent triangles and the test used to determine this.	$\therefore \Delta ABD \equiv \Delta CBD$ (SSS)
5 Identify the appropriate pair of corresponding angles.	$\angle BAD = \angle BCD$ (corresponding angles in congruent triangles)
6 State the result you have proved.	$\therefore \angle BAC = \angle BCA$
(b) Write the answer in words.	(b) The result proved that angles opposite equal sides are equal in isosceles triangles.

9.3 Proofs using congruent triangles

Navigator

Answers p. 830

1, 2, 3, 4, 5, 6, 7, 8, 10, 12, 13, 14, 17	1, 2, 3, 4, 5, 6, 7, 8, 9, 10, 12, 13, 14, 16, 17	2, 3, 4, 6, 7, 8, 9, 10, 11, 12 (a, c), 14, 15, 16, 17

Fluency

W.E. 9

Hint

Construct a line from *B* to the midpoint of *AC* and a line from *C* to the midpoint of *AB*.

1 In ΔABC, $AB = BC = CA$

(a) Prove that $\angle ABC = \angle BCA = \angle CAB$.

(b) What property of an equilateral triangle have you just proved?

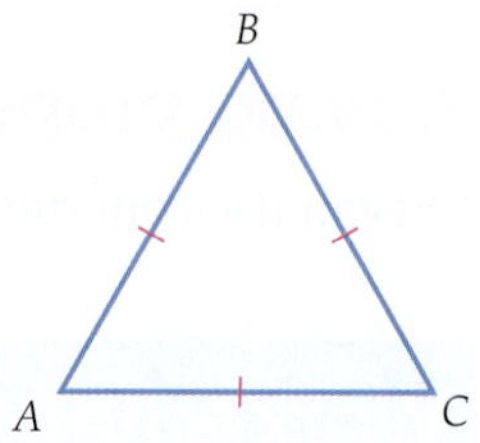

2 Use the triangles shown to prove the following properties of isosceles triangles.

(a) The bisector of the angle between the two equal sides of an isosceles triangle bisects the third side.

(b) The bisector of the angle between the two equal sides of an isosceles triangle is perpendicular to the third side.

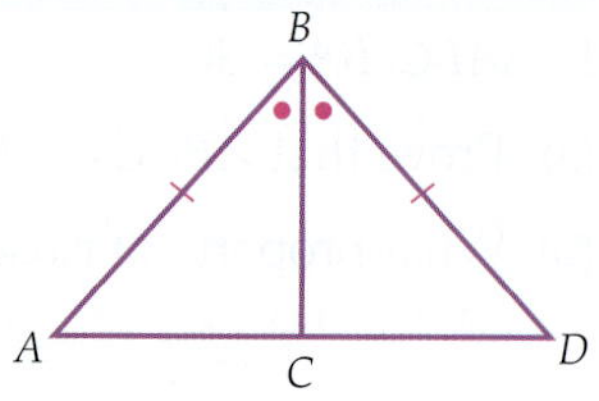

3 ΔDEF is isosceles. Prove that $\angle CDE = \angle GFE$.

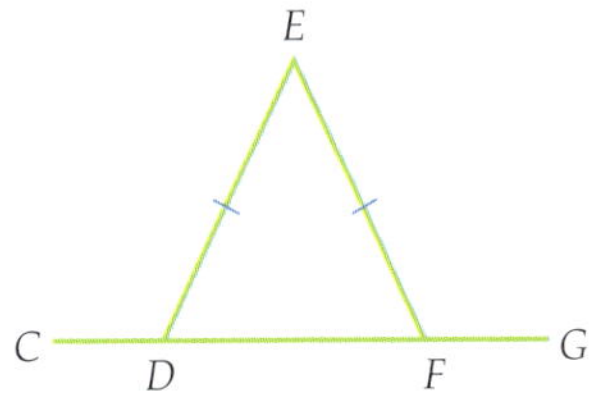

4 The diagram shows two triangles, ΔABC and ΔDCB, with $AB = DC$ and $\angle ABC = \angle DCB$.

Which of the following tests can be used to prove that $\Delta ABC \equiv \Delta DCB$?

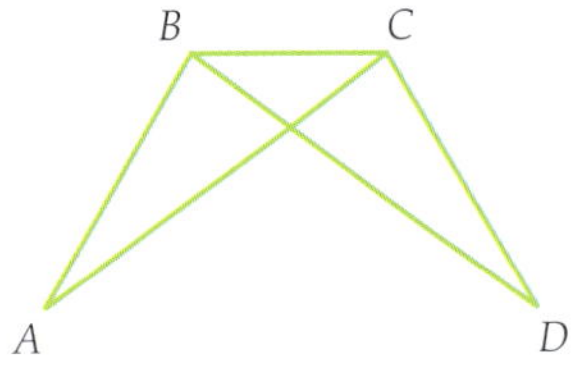

A SAS B ASA C SSS

D There is insufficient information to determine whether the triangles are congruent.

5 In the diagram AC bisects $\angle BAD$ and $\angle BCD$.

Which of the following tests can be used to prove that $\Delta ABC \equiv \Delta ADC$?

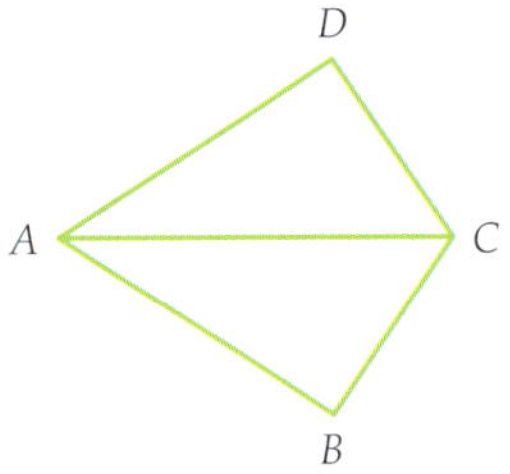

A SAS B ASA C SSS

D There is insufficient information to determine whether the triangles are congruent.

Understanding

6 For the triangle at right, show that $AC^2 = DC^2 + 2BD^2 + AD^2$.

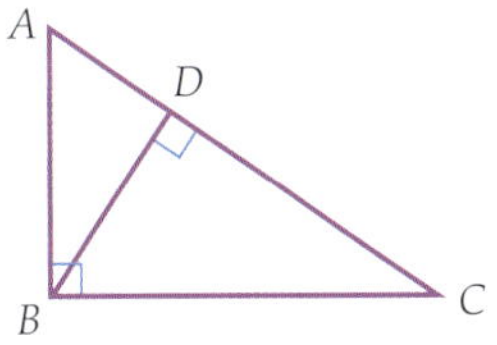

7 B is the midpoint of AC. Show that $\Delta ABF \equiv \Delta BCD$.

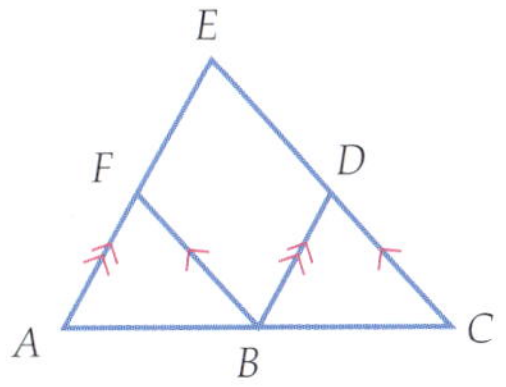

8 Prove that ΔPRT is isosceles.

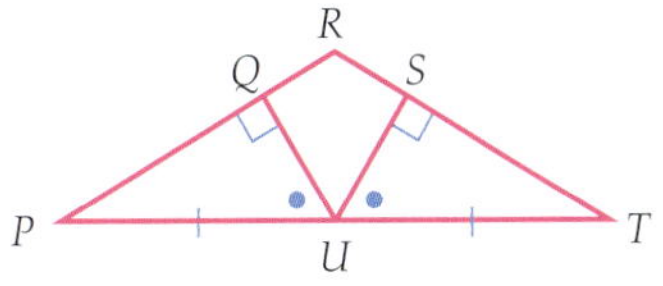

9 In ΔABC, $AB = AC$. Prove that $AM = AN$.

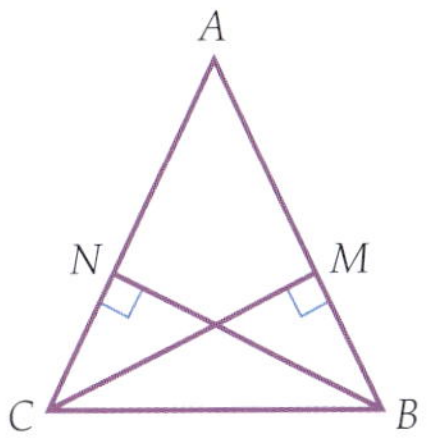

Reasoning

10 If $\angle BAC = \angle DFE$, $\angle ABC = \angle DEF$ and $BD = EC$, prove that $\Delta ABC \equiv \Delta FED$.

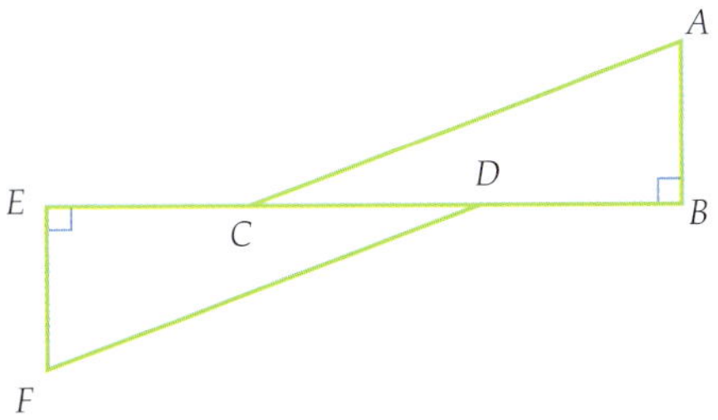

11 If $\angle ABC = \angle DEF$, $\angle ACB = \angle DFE$ and $BF = EC$, prove that $\Delta ABC \equiv \Delta DEF$.

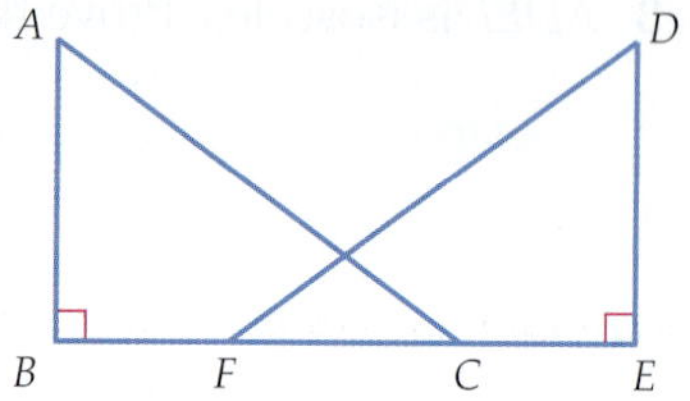

12 **(a)** In the diagram, $\angle BAD = \angle BCE$. $BA = BC$ and $AD = CE$. Prove that $\Delta BAE \equiv \Delta BCD$.

(b) Given $\angle BAD = \angle BCE$ and $\angle ABD = \angle CBE$, prove that $\Delta ABE \equiv \Delta CBD$.

(c) Given $AE = CD$ and $\angle BAD = \angle BCE$, prove that $\Delta ABD \equiv \Delta CBE$.

(d) Given $AB = CB$ and $AE = CD$, prove that $\Delta ABD \equiv \Delta CBE$.

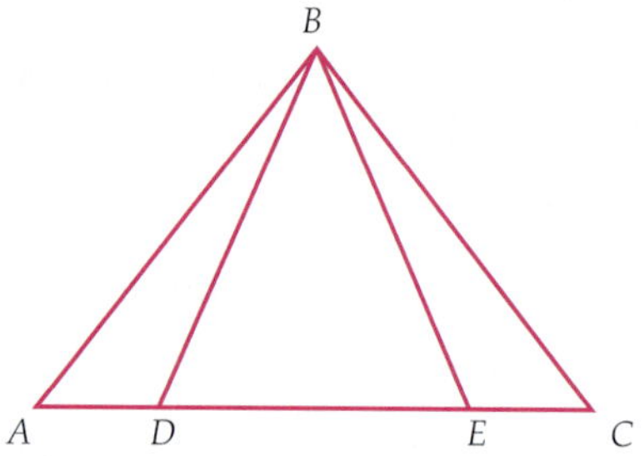

13 In the diagram, $FD = FE$, $FD \perp AB$ and $FE \perp AC$. Prove that $\Delta ADF \equiv \Delta AEF$.

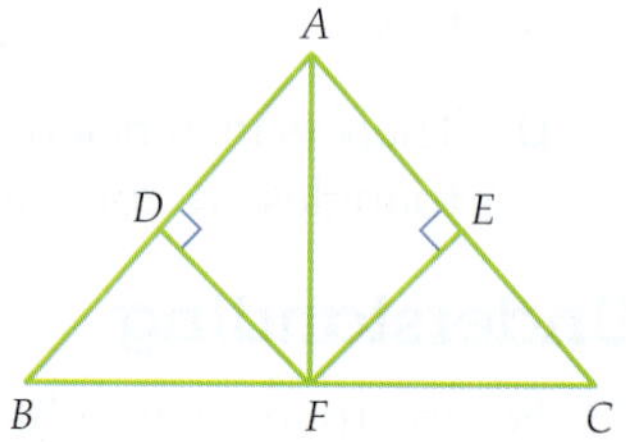

14 ΔABC and ΔCDE are right-angled triangles and $AB = CD$. If C is the midpoint of AE, prove that $\Delta ABC \equiv \Delta CDE$.

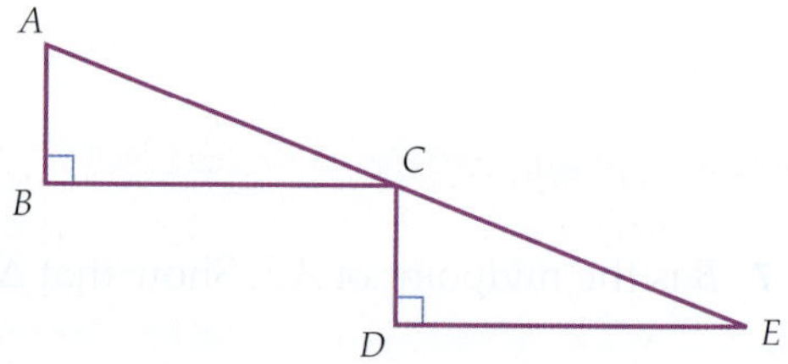

15 When you draw a line from each end of a diameter to a point on the circumference of a circle, the angle formed is called an angle in a semicircle. Prove, using isosceles triangles, that the angle in a semicircle is a right angle.

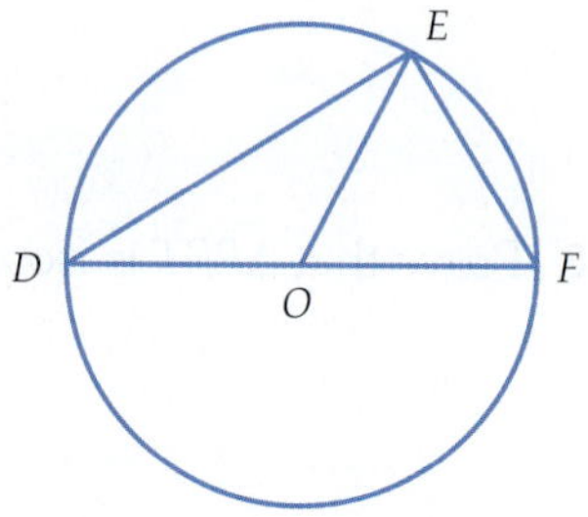

Open-ended

16 ABC is an isosceles triangle with $AB = AC$. X is a point on AB and Z is a point on AC such that XZ is parallel to BC. Y is the midpoint of XZ and $AY = YC$. Let $\angle BAY = a$.

(a) Prove $AX = AZ$.

(b) Prove $\Delta AXY \equiv \Delta AZY$.

(c) Prove $\Delta BXY \equiv \Delta CZY$.

(d) Find b, c and d in terms of a.

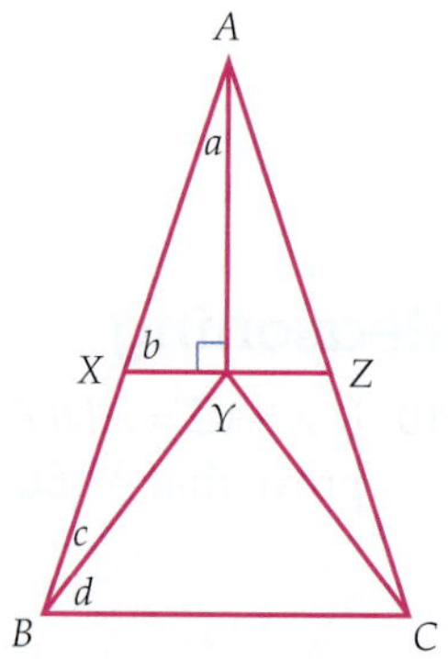

17 Draw a diagram of your own that contains two congruent triangles. Prove that they are congruent using one of the congruency tests.

Exploration CAS

Equipment required: TI-Nspire CAS or Casio ClassPad CAS

Similarity, congruence and geometrical conjecture

Congruent triangles

Using TI-Nspire CAS

Add **Geometry** to your document. Select **menu** > **Shapes** > **Triangle**, then move the cursor to select the points where you want each vertex of the triangle to appear. Press **esc** to stop drawing.

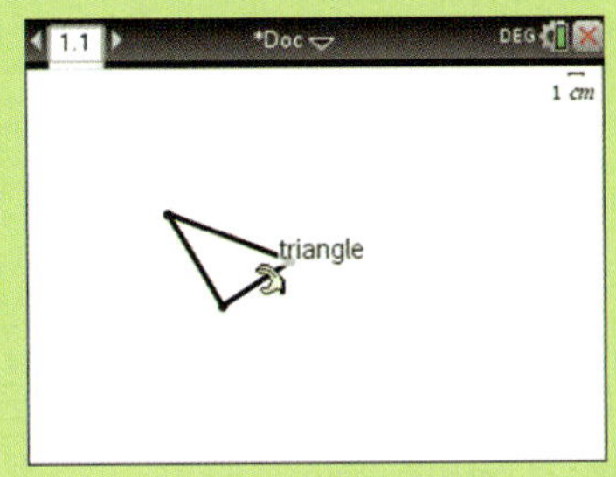

Using Casio ClassPad CAS

From the menu select **Geometry**. Select the scalene triangle icon and move the cursor to choose where you want the shape to appear. To resize the triangle, select a vertex and drag to move its position.

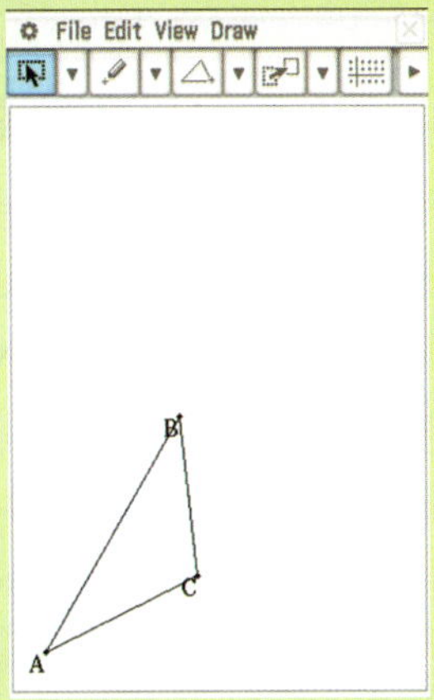

Transforming a shape using translations, rotations and reflections will produce congruent shapes.

Using TI-Nspire CAS

To translate the triangle, select **menu** > **Transformation** > **Translation**. Select the triangle, then select two points to mark the distance and direction of translation. This will create a copy of the triangle, which will be translated to the new position.

You can check that the triangle and its translation are congruent by checking that its angles and side lengths are the same.

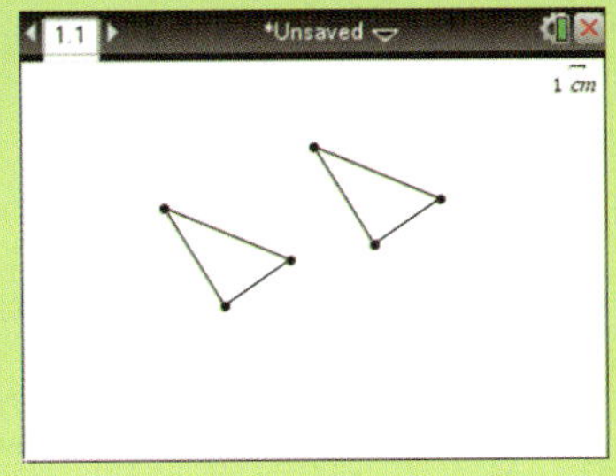

Using Casio ClassPad CAS

To translate the triangle, select all three sides and then select **Draw** > **Construct** > **Translation**. A vector will appear. Insert suitable values into the vector such as $\begin{bmatrix} 2 \\ 3 \end{bmatrix}$ to create a new copy of the shape, translated by that vector.

You can check that the triangle and its translation are congruent by checking that its angles and side lengths are the same.

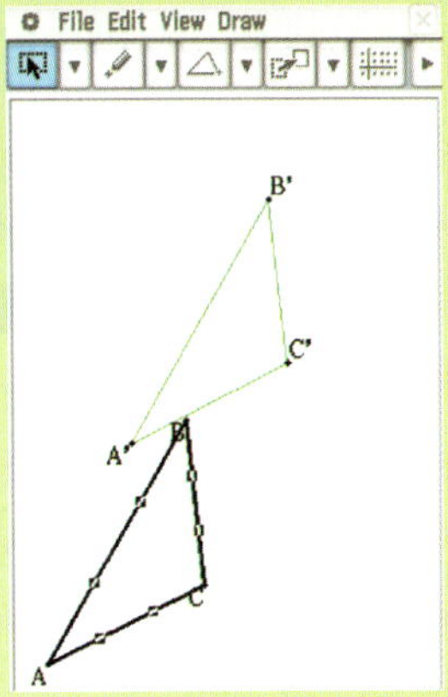

Using TI-Nspire CAS

To measure an angle, select **menu** > **Measurement** > **Angle** and then select the three vertices, in order, that form one of the angles. Press **esc** to stop measuring angles.

Repeat this process on the corresponding angle in the translated triangle to check that the two angles are the same size.

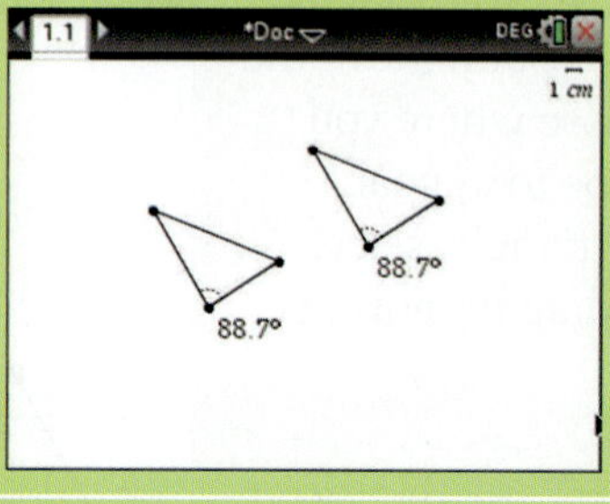

To measure a side, select **menu** > **Measurement** > **Length** and then select the two vertices that form the side. Press **esc** to stop measuring sides.

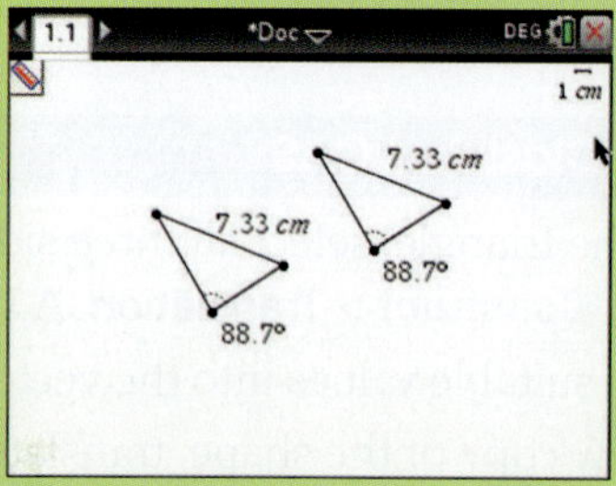

Now select one of the vertices of the original triangle and move around to change the size of your triangle. The two pairs of measurements should remain the same as each other.

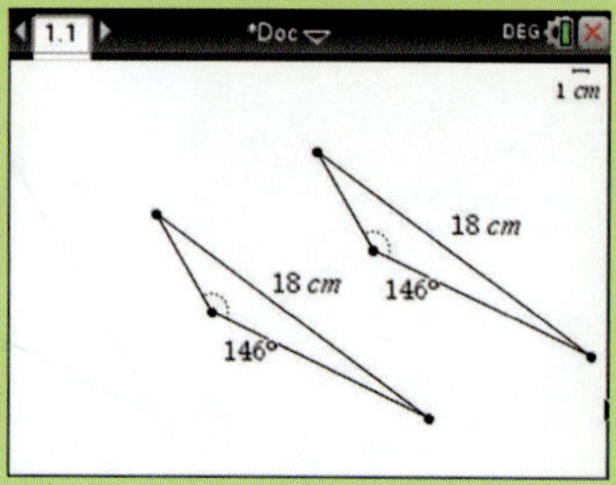

Using Casio ClassPad CAS

To measure an angle, select the two sides adjacent to the angle and then select **Draw** > **Measurement** > **Angle**. The angle value will appear. Repeat this for the corresponding angle in the translated triangle and its angle size will appear. The two angle values should be the same.

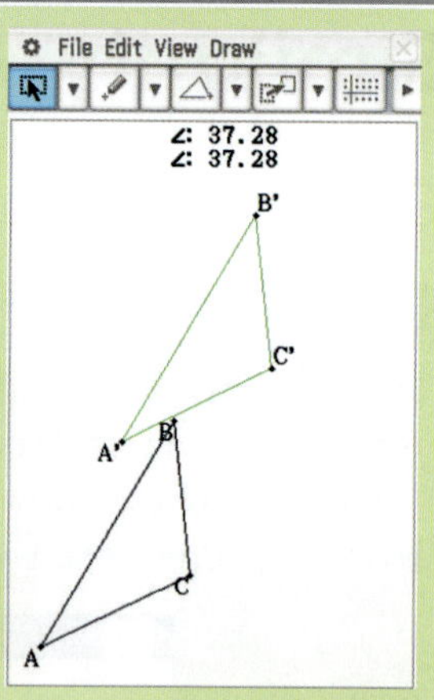

To measure a side, select the side and then select **Draw** > **Measurement** > **Length**. The length value will appear.

Repeat this for the corresponding side on the translated triangle.

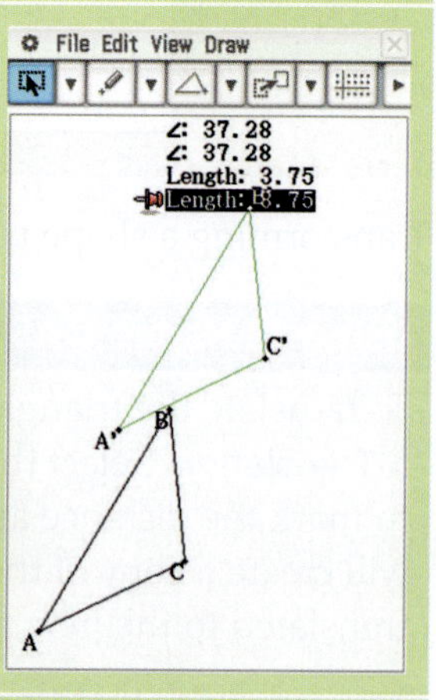

Now select one of the vertices of the original triangle and move it around. The two pairs of measurements should remain the same as each other.

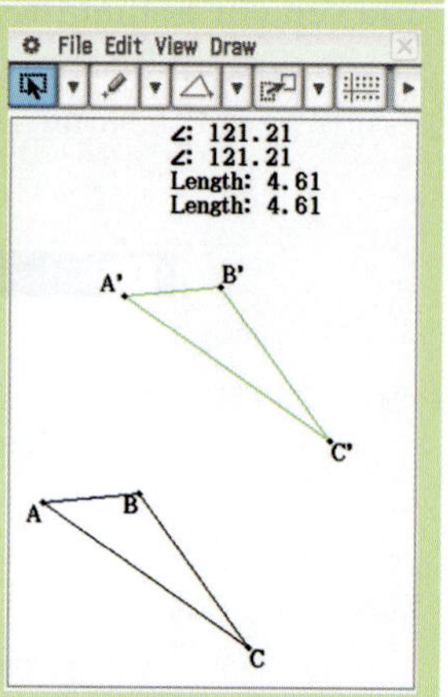

Similar triangles

Using TI-Nspire CAS

Create a new triangle, as before. Then select **menu** > **Points & Lines** > **Point** to select the centre of dilation. Now select **menu** > **Actions** > **Text** to create a text label and enter the dilation factor 1.5.

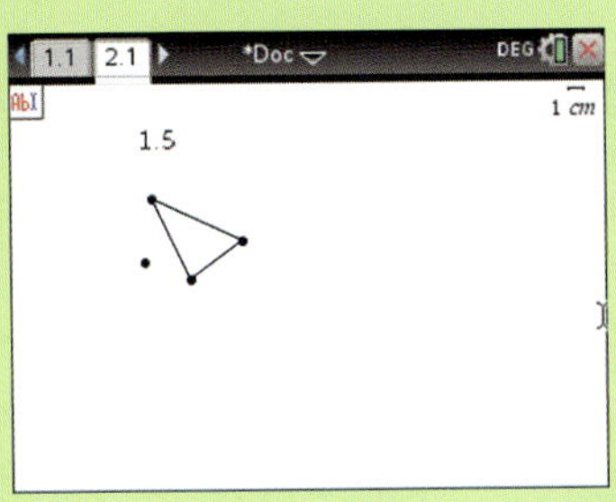

Select **menu** > **Transformation** > **Dilation**. Select the triangle, then the point and finally the dilation value. You will now have two triangles on your screen.

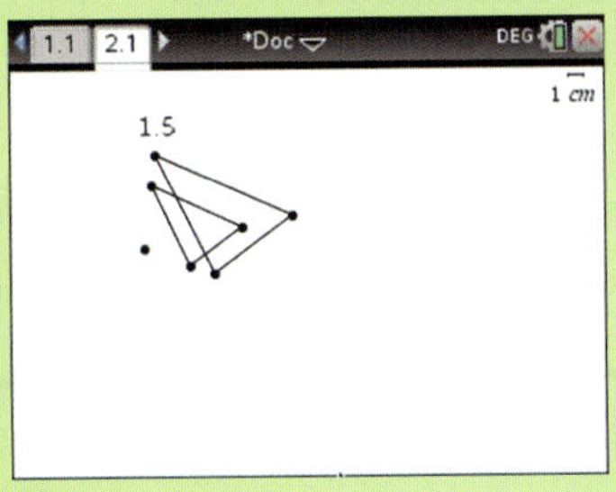

Measure the lengths of the corresponding sides of the original and the image, and compare the ratios of these sides. The ratios should all be the same.

Measure all the angles of the triangle and its image to check that corresponding angles are equal.

Now select a vertex point and change the size of your triangle. Compare side lengths and angle sizes again for the new object and image.

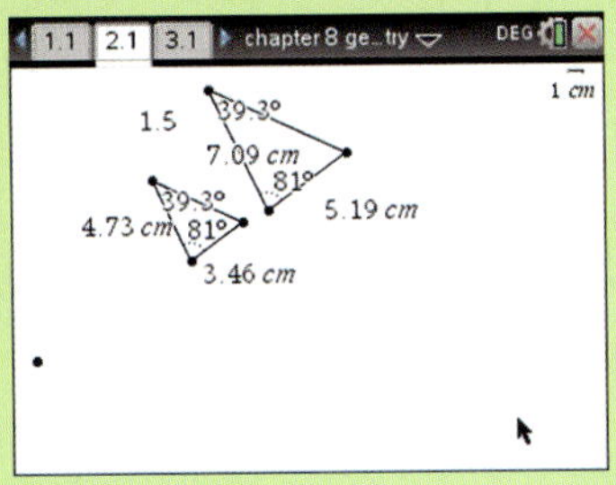

Using Casio ClassPad CAS

Create a new triangle as before and resize it if necessary. Select the three sides on the triangle and then select **Draw** > **Construct** > **Dilation**. Select a point on the screen as the centre of dilation and then enter the dilation scale value. (Unless your triangle is very small, use a value less than 1.)

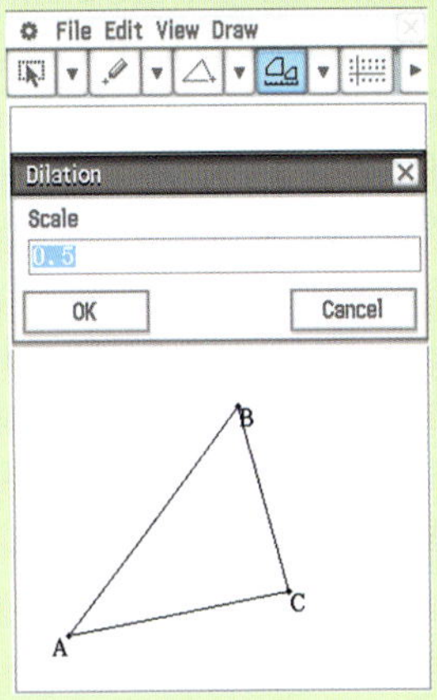

A similar triangle will appear. A point *D* will also appear, which is the centre of dilation.

Measure the lengths of the corresponding sides of the original and the image, and compare the ratios of these sides. The ratios should all be the same.

Measure all the angles of the triangle and its image to check that corresponding angles are equal.

Now select a vertex point and change the size of your triangle. Compare side lengths and angle sizes again for the new object and image.

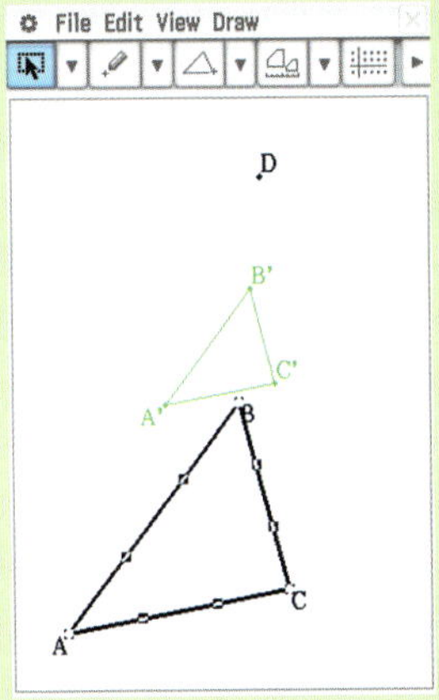

Geometrical conjecture

What is the difference between a theorem and a conjecture? A theorem has been proved to be always true, so it can then be used to prove other geometric relationships. A conjecture is not yet proved.

To establish a theorem, you need to first write a conjecture. If you can prove that the conjecture is always true, then you can use it as a theorem.

Properties of geometric shapes are examples of theorems. If you write a conjecture and use diagrams to show it is true for all possible situations, then you can attempt a formal proof. If you can draw a diagram that is not true for your conjecture, then the conjecture cannot be proved.

Consider the diagonals of a quadrilateral. Do they bisect each other? Do they intersect at right angles?
Does the answer to these questions depend on the type of quadrilateral being considered?

Below is a list of quadrilaterals with their definitions.

A **trapezium** is a quadrilateral with one pair of opposite sides parallel.	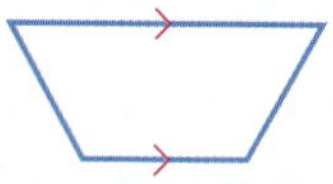
A **kite** is a quadrilateral with two pairs of adjacent sides equal in length.	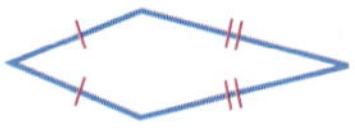
A **parallelogram** is a quadrilateral with both pairs of opposite sides parallel.	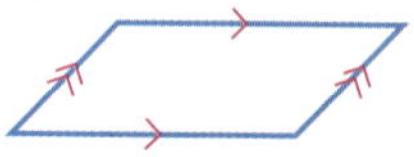
A **rhombus** is a quadrilateral with all sides equal.	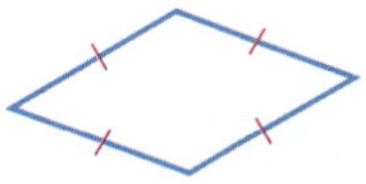
A **rectangle** is a quadrilateral in which all angles are right angles.	
A **square** is a rectangle with two adjacent sides equal.	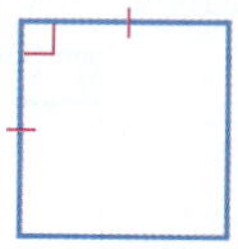

Use the list of quadrilaterals above to write as many conjectures as you can about the properties of each type of quadrilateral. Then, use your CAS Geometry features to test these conjectures.

For example, the screenshots below show how you might start to show that the diagonals of a rectangle bisect each other. Can you see what the next steps would be?

Using TI-Nspire CAS

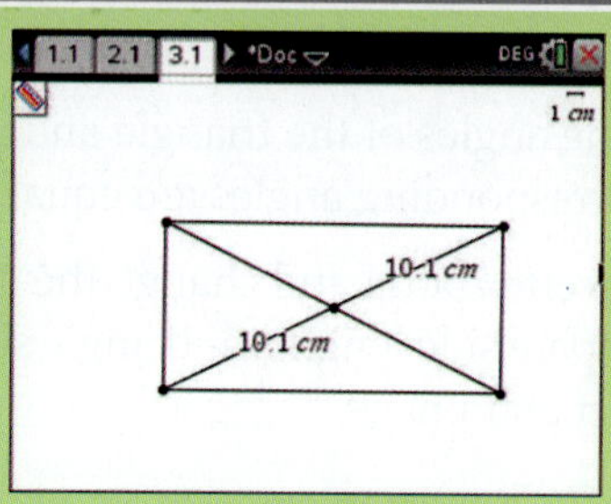

Using Casio ClassPad CAS

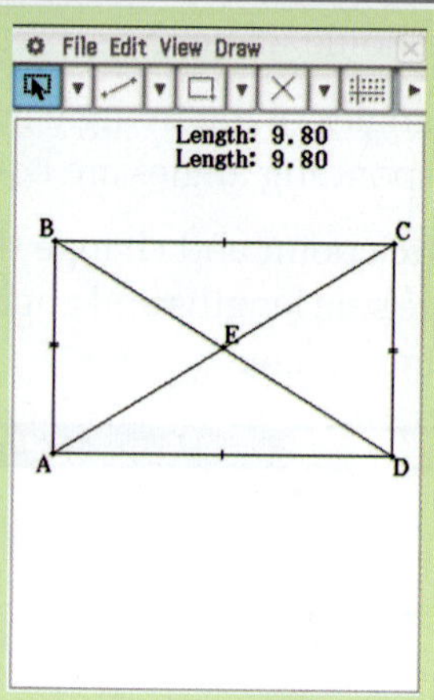

Proofs using similar triangles

9.4

The tests for similar and congruent triangles can be used to prove general results known as **theorems** about triangles and quadrilaterals. When these theorems have been proved they are known as properties of triangles and quadrilaterals, and may be used to prove other relationships in these and other geometric figures, and to solve geometric problems.

Procedure for proving theorems

Theorems are stated in general terms. This means that a triangle theorem applies to all triangles. Theorems are proved using a general example that contains some given information. To prove a theorem using similar triangles:

- draw a diagram to display the given information
- make a statement of the relationship you want to prove
- use congruence or similarity in the proof.

You can then generalise the proof as a property of the similar triangles. This generalisation becomes a theorem, which you can then use for other geometric proofs.

Midpoint theorem

The interval joining the midpoints of two sides of a triangle is parallel to the third side and is half its length.

This theorem is often used to prove relationships. Its proof is provided below.

Worked example 10

W.E. 10

In ΔABC, D is the midpoint of AB and E is the midpoint of AC.

Draw DE and prove that:

(a) DE is parallel to BC.

(b) $DE = \frac{1}{2}BC$.

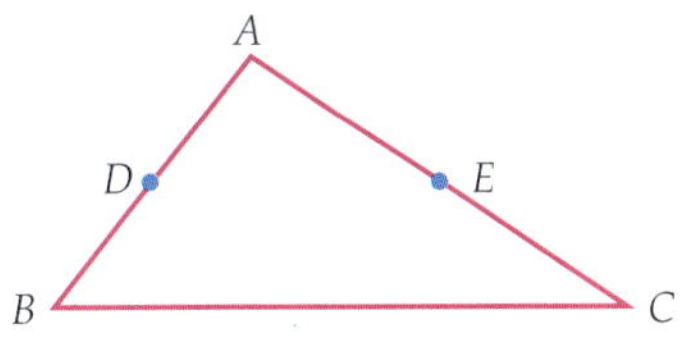

Thinking	Working
(a) 1 Copy the diagram and join D to E.	(a) 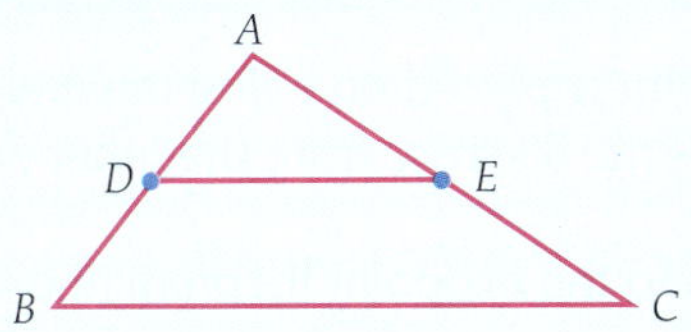
2 Name the triangles that you will be considering.	ΔADE and ΔABC

3 Write three sets of properties that will allow you to use a similarity test.	$\angle DAE = \angle BAC$ (common angles) $AD = DB$ (D is the midpoint of AB) $\therefore AB = 2AD$ $AE = EC$ (E is the midpoint of AC) $\therefore AC = 2AE$ $\therefore \frac{AD}{AB} = \frac{AE}{AC} = \frac{1}{2}$
4 Write a pair of similar triangles and the test used.	$\therefore \Delta ADE \sim \Delta ABC$ (two sides are proportional and their included angles are equal)
5 Identify pairs of equal angles.	$\angle ADE = \angle ABC$ (corresponding angles of similar triangles) $\angle AED = \angle ACB$ (corresponding angles of similar triangles)
6 Link these equal angles to parallel lines.	$\angle ADE$ and $\angle ABC$ are a pair of equal corresponding angles formed when the transversal AB cuts the lines DE and BC. $\therefore DE \parallel BC$ (Corresponding angles are equal so the lines are parallel.) $\therefore$ The line joining the midpoints of two sides of a triangle (DE) is parallel to the third side (BC).
(b) Use the property of the similar triangles from **(a)** to find the result.	(b) As the triangles are similar, the corresponding sides are proportional. $\therefore \frac{AD}{AB} = \frac{AE}{AC} = \frac{DE}{BC} = \frac{1}{2}$ $\therefore \frac{DE}{BC} = \frac{1}{2}$ $\therefore DE = \frac{1}{2}BC$ $\therefore$ The line joining the midpoints of two sides of a triangle (DE) is half the length of the third side (BC).

When a point is placed on a line interval such that the distance from one end to the point is the same as the distance from the other end to the point you say that the point **bisects** the line.

Converse to the midpoint theorem (the inverse theorem)

A line drawn through the midpoint of one side of a triangle and parallel to a second side bisects the third side.

This gives another theorem.

Intercept theorem

When a set of parallel lines make equal intercepts on one transversal, it makes equal intercepts on any other transversal.

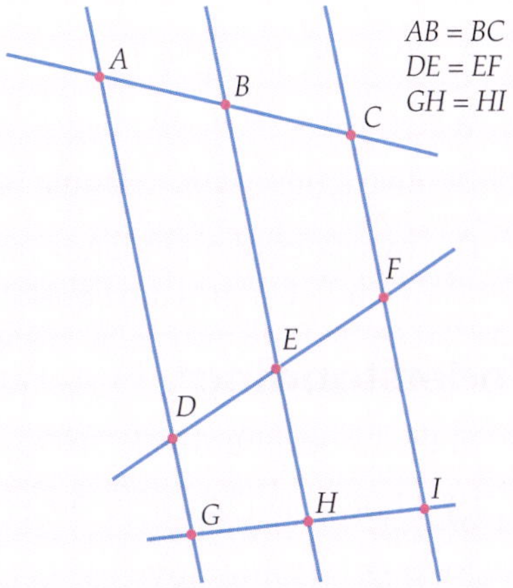

9.4 Proofs using similar triangles

Navigator

1, 2, 3, 4, 5, 6, 8, 9, 10, 12 | 1, 2, 3, 4, 5, 6, 7, 8, 9, 10, 12 | 1, 2, 3, 4, 5, 7, 8, 9, 10, 11, 12

Answers p. 832

Fluency

W.E. 10

1 The line through the midpoint of a side of a triangle parallel to another side bisects the third side. In ΔABC, D is the midpoint of AB and DE is parallel to BC, meeting AC at E. Prove that:

(a) $\Delta ABC \sim \Delta ADE$

(b) $AE = EC$

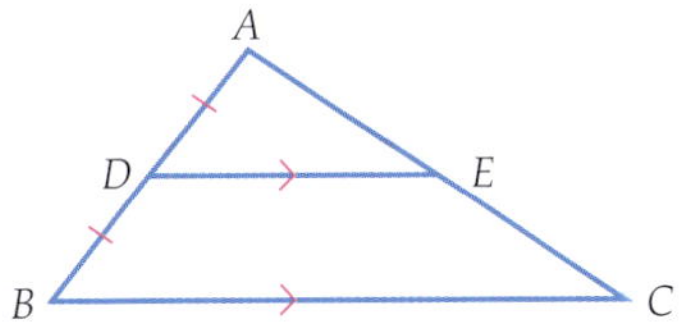

2 In a right-angled triangle, if a perpendicular is drawn from the right angle to the hypotenuse, the triangles on each side of it are similar to the whole triangle and to each other. Triangle ΔABC is right-angled at A and $AD \perp BC$. Prove that ΔADB, ΔDCA and ΔCAB are similar.

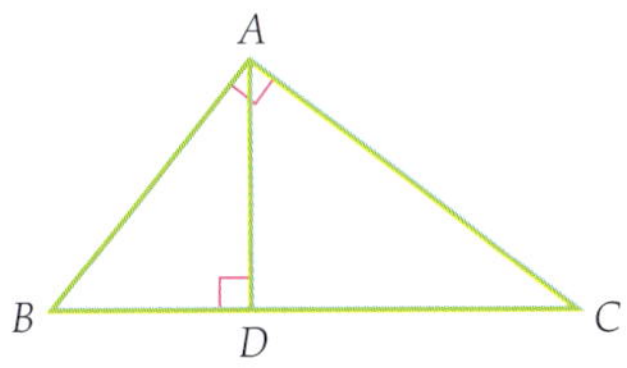

3 In a right-angled triangle, the area of the square on the perpendicular to the hypotenuse is equal to the area of the rectangle contained by the segments of the hypotenuse. Using the diagram for Question **2**, prove that $AD^2 = BD \times DC$.

4 In a right-angled triangle, the area of the square on either of the sides containing the right angle is equal to the area of the rectangle contained by the hypotenuse and the segment of the hypotenuse adjacent to that side. Using the diagram for Question **2**, prove that $BA^2 = BD \times BC$ and $CA^2 = CD \times CB$.

5 Use the result from Question **4** to prove Pythagoras' theorem: $BC^2 = BA^2 + CA^2$

6 Which of the following is not sufficient to prove that two triangles are similar?

A The two triangles have two angles the same.

B The three sides of one triangle are proportional to the three sides of the other triangle.

C Two sides of one triangle are proportional to two sides of another triangle.

D Two sides of one triangle are proportional to two sides of another triangle and the included angles are equal.

Understanding

7 In the trapezium $ABCD$, AB is parallel to DC and the diagonals intersect at O.

(a) Show that $OA:OC = OB:OD$.

(b) If $DC = 2AB$, show that O is a point of trisection of both diagonals. (That is, show that $AO:OC = BO:OD = 1:2$ and hence that $AO:AC = BO:BD = 1:3$.)

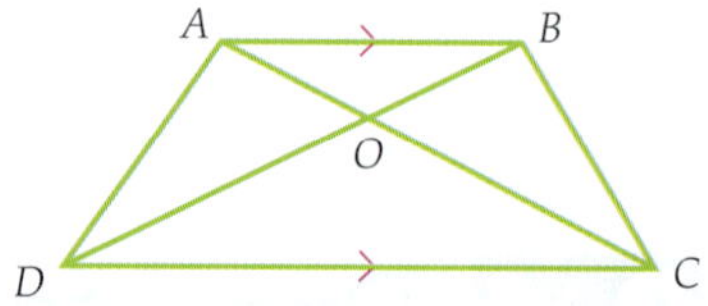

8 A straight line drawn parallel to one side of a triangle cuts the other two sides proportionally. In ΔJKL, M is a point on JK and MN is parallel to KL, meeting JL at N.

Prove that $\frac{JM}{MK} = \frac{JN}{NL}$.

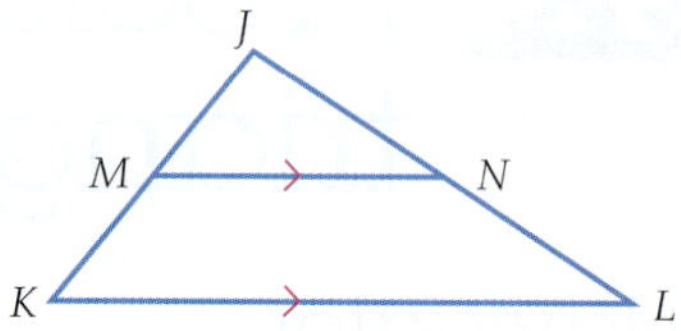

9 Conversely, if a line cuts two sides of a triangle proportionally, it is parallel to the third side.

In ΔDEF, $\frac{DG}{GE} = \frac{DH}{HF}$. Prove that $GH \parallel EF$.

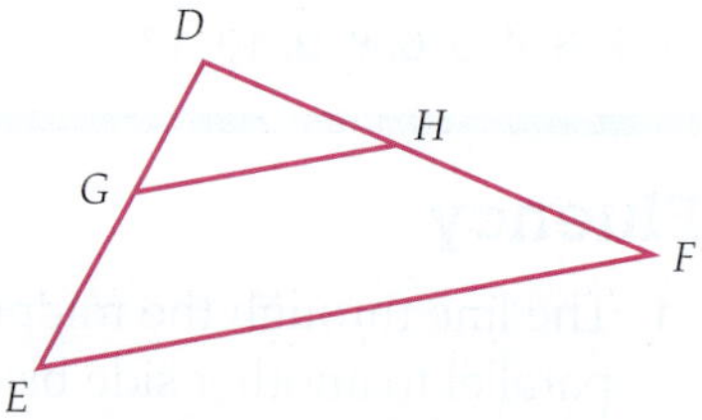

Reasoning

Hint

Extend *DA* and *CB* and let them intersect at *G*.

10 The straight line that joins the midpoints of the slanted sides of a trapezium is parallel to the parallel sides. $ABCD$ is a trapezium in which $AB \parallel DC$. E is the midpoint of AD and F is the midpoint of BC. Prove that $EF \parallel AB$.

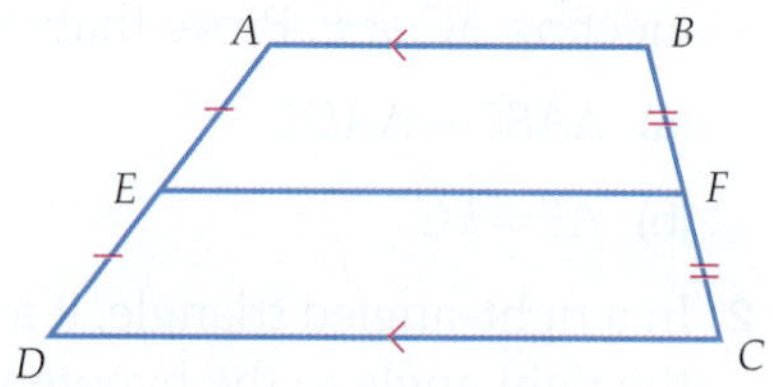

Hint

Prove $\Delta BXY \equiv \Delta CZY$.

11 ΔABC is an isosceles triangle such that $AB = AC$. X is a point on AB and Z is a point on AC such that XZ is parallel to BC. Y is the midpoint of XZ and $BX = XY$. Prove that $\Delta BXY \sim \Delta BYC$.

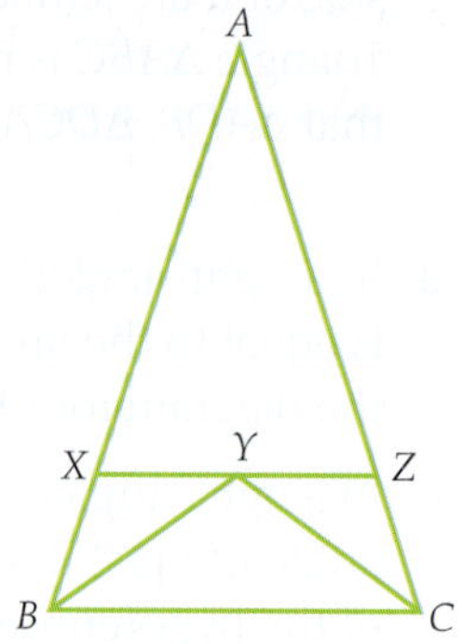

Open-ended

12 Draw a figure different from any given above that contains at least two similar triangles. Prove that your triangles are similar.

Half-time 9

1 (a) Given that $BC = DC$, prove that $\Delta ABC \equiv \Delta EDC$.

(b) Hence, prove that $AB = ED$.

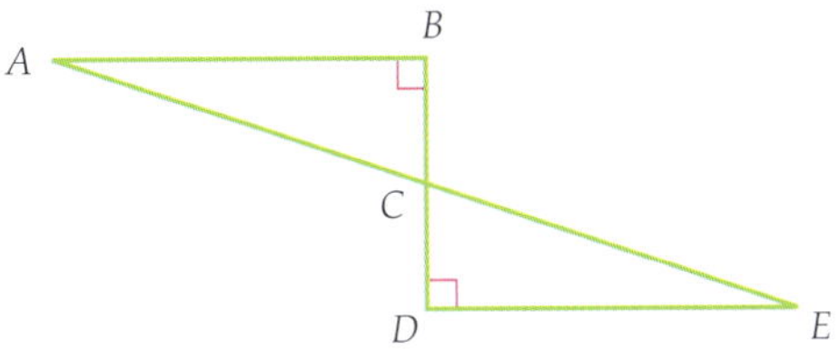

9.2

2 Given that $\angle ABC = \angle APQ$, prove that $\Delta ABC \sim \Delta APQ$.

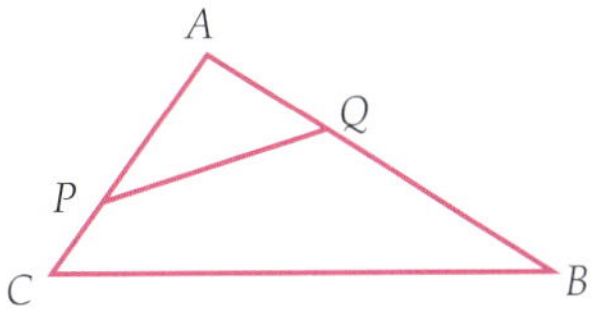

9.4

3 Give reasons why each of the following pairs of triangles is congruent.

9.1

(a)

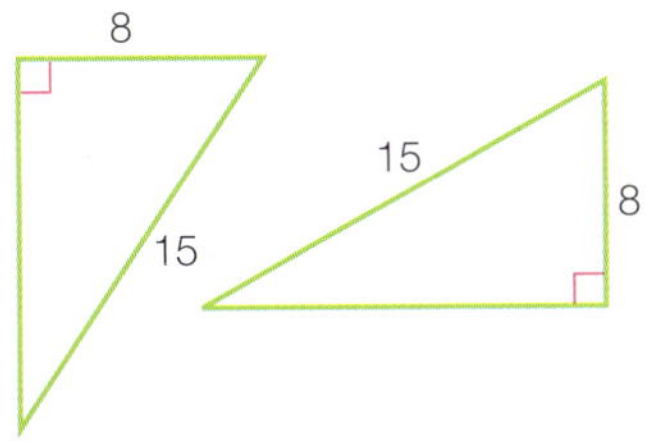

(b)

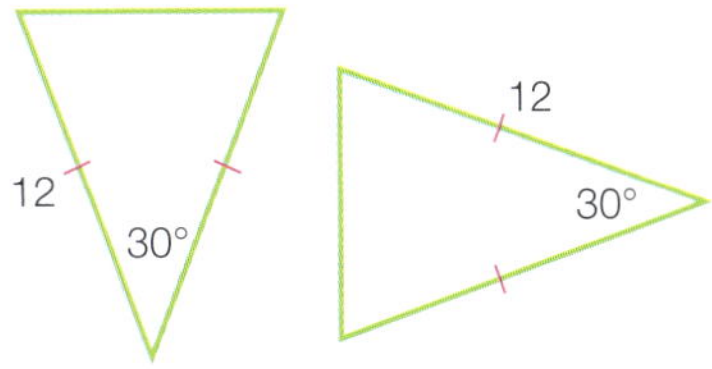

4 For each of the following, prove that the two triangles are congruent and hence find the value of the pronumeral.

9.2

(a)

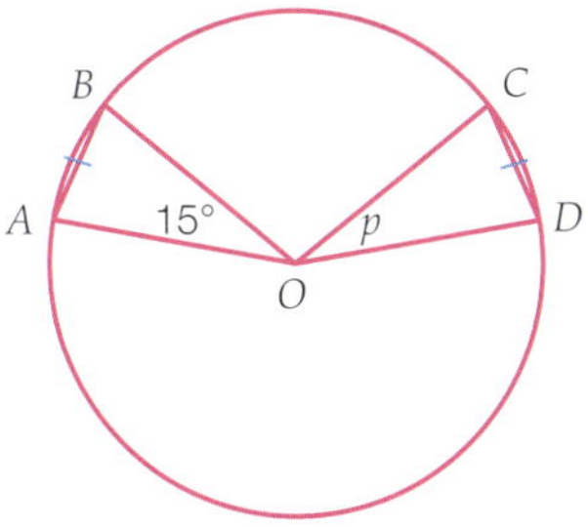

(b)

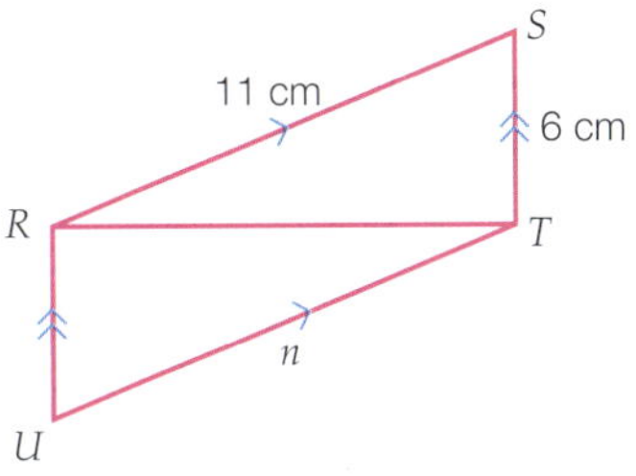

5 Given that these pairs of triangles are similar, find x.

9.2

(a)

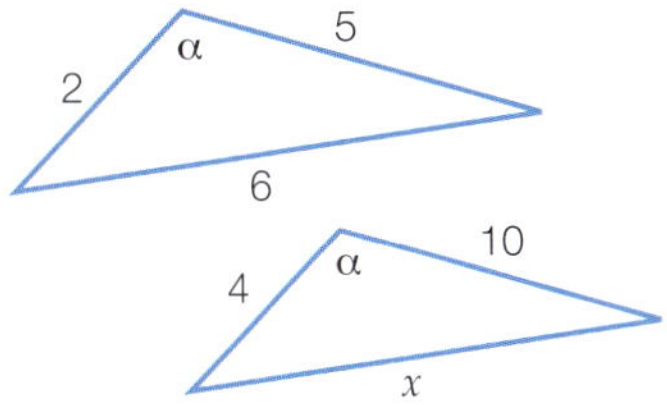

(b)

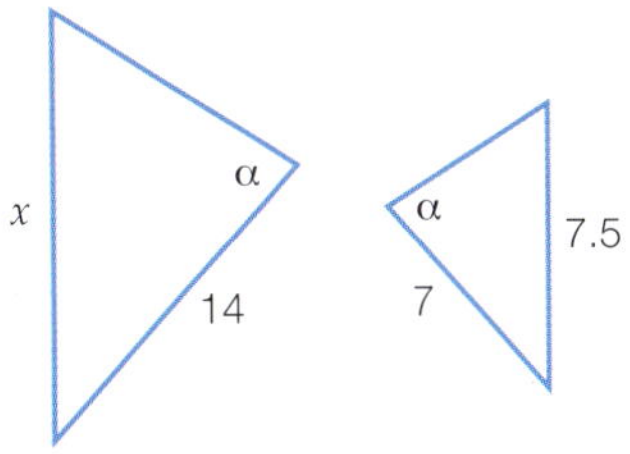

6 In ΔLMN, LX is perpendicular to MN and MY is perpendicular to LN, as shown in the diagram. The point of intersection of LX and YM is Z. Prove that $\Delta LYZ \sim \Delta MXZ$, giving reasons.

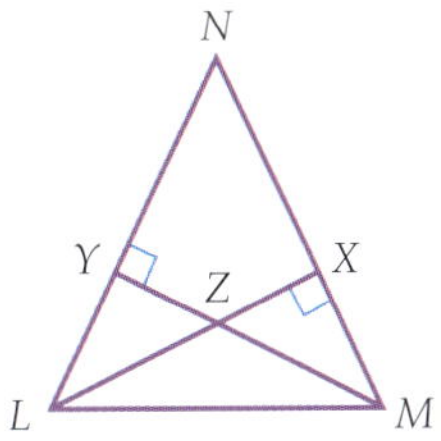

9.4

9.3

7 Given that $AC = BC$ and $AX = BX$, prove that $\Delta ACX \equiv \Delta BCX$.

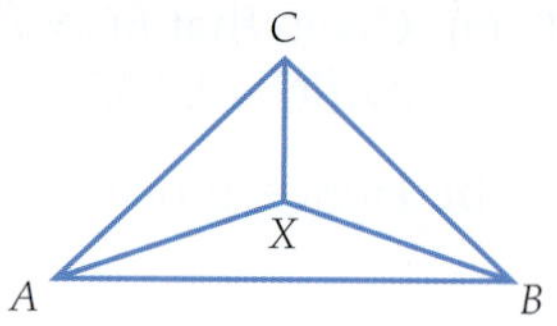

9.1

8 Find the value of x in the triangle.

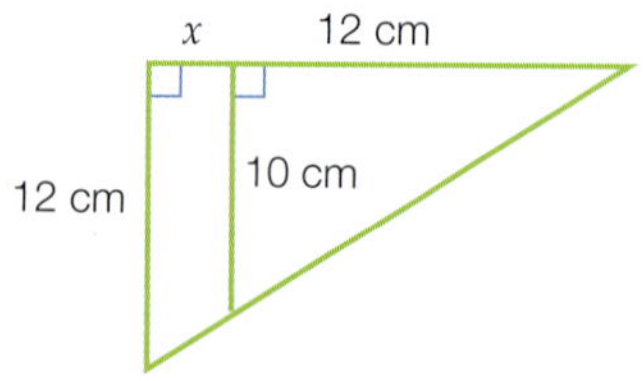

9.2

9 In the diagram, $FD = FE$. Also, $FD \perp AB$ and $FE \perp AC$. Prove that $\Delta ADF \equiv \Delta AEF$.

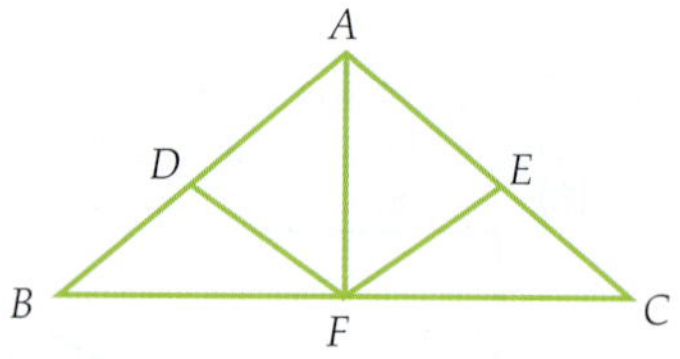

9.3

10 In the diagram, ΔABC is isosceles with $AC = BC$. X, Y and Z are the midpoints of AB, AC and BC respectively. Prove that $XY = XZ$.

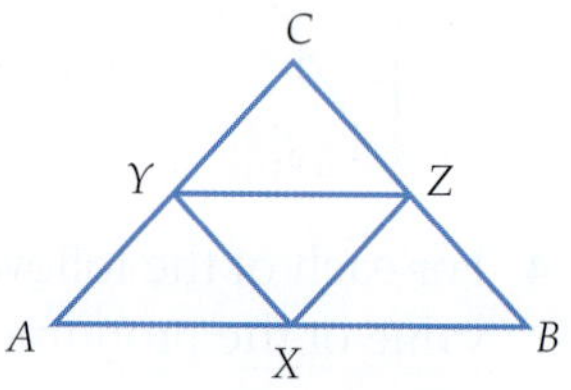

9.2

11 Given $AB \parallel DE$ and C is the midpoint of BD, prove that $\Delta ABC \equiv \Delta DEC$.

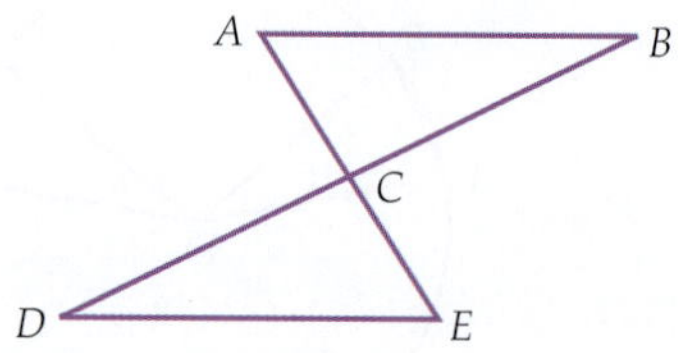

9.1

12 Draw two triangles that are not congruent using two corresponding sides and an angle that is not the included angle.

9.4

13 A student 1.5 m tall, standing 5 m in front of a 2 m wall, can just see the top of a tree growing 30 m behind the wall. How tall is the tree?

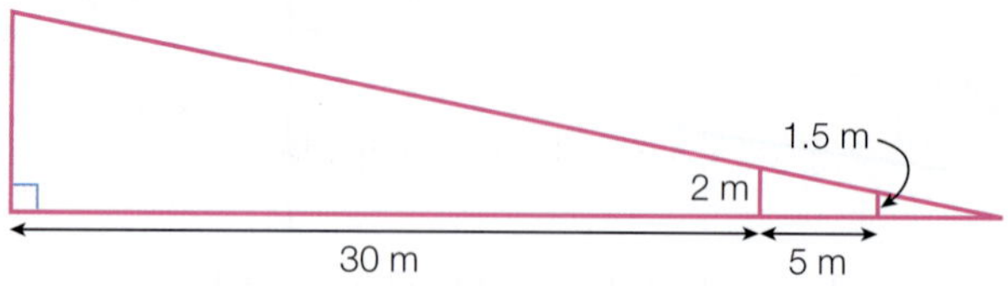

Geometric properties of special quadrilaterals

9.5

- A definition is the minimum amount of information needed to define a particular figure.
- A property is a fact that can be proven to be always true about a particular defined figure.

Definitions of the six special quadrilaterals are listed below. You can use these definitions to prove properties of these shapes.

Definitions of special quadrilaterals

- A **trapezium** is a quadrilateral with one pair of opposite sides parallel.
- A **parallelogram** is a quadrilateral with both pairs of opposite sides parallel.
- A **kite** is a quadrilateral with two pairs of adjacent sides equal in length.
- A **rectangle** is a quadrilateral in which all angles are right angles.
- A **rhombus** is a quadrilateral with all sides equal in length.
- A **square** is a rectangle with two adjacent sides equal in length, or a rhombus with all angles equal to 90°.

Properties of special quadrilaterals

Trapezium

- A trapezium, sometimes called a trapezoid, has one pair of parallel sides.

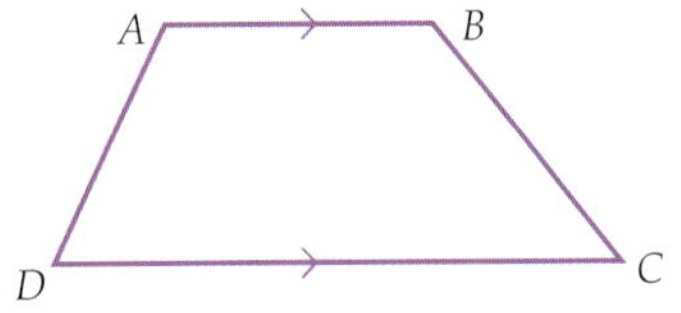

Parallelogram

- Both pairs of opposite angles are equal in magnitude: $\angle ABC = \angle ADC$ and $\angle BAD = \angle BCD$
- Both pairs of opposite sides are equal: $AB = DC$ and $AD = BC$
- Diagonals bisect each other: $AE = EC$ and $BE = ED$

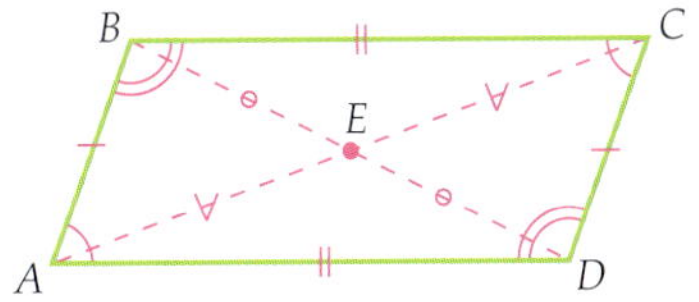

Kite

- One pair of opposite angles are equal in magnitude: $\angle ABC = \angle ADC$
- A kite has an axis of symmetry that bisects the non-equal vertex angles through which it passes.
- The axis of symmetry of the kite bisects the other diagonal at right angles.

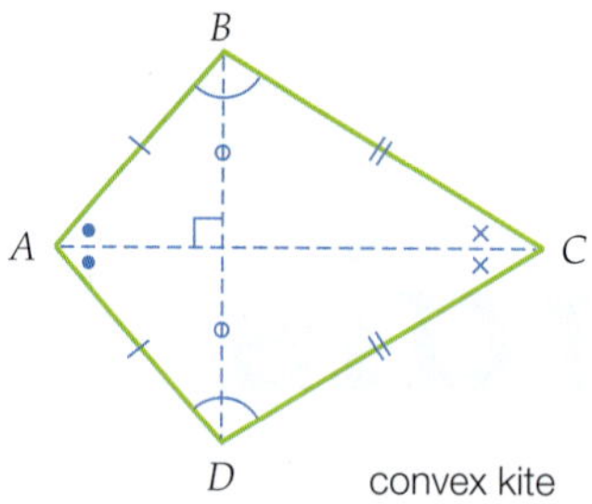

convex kite

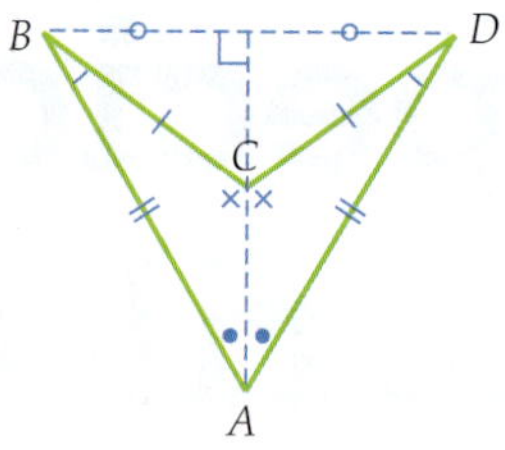

concave kite (chevron)

Rectangle

All of the properties of a parallelogram, plus:

- Diagonals are equal in length: $AC = BD$
- The lines joining the midpoints of opposite sides are lines of symmetry.

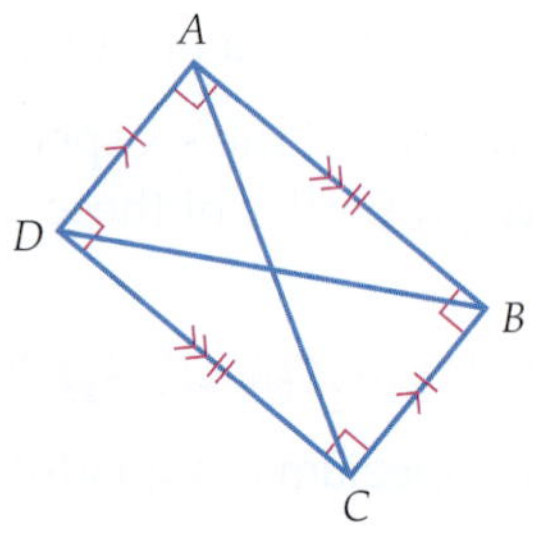

Rhombus

All of the properties of a parallelogram, plus:

- Diagonals bisect each other at right angles: $AC \perp BD$
- Diagonals bisect the vertex angles: $\angle DAC = \angle BAC = \angle DCA = \angle BCA$, $\angle ADB = \angle CDB = \angle ABD = \angle CBD$
- Both diagonals are lines of symmetry.

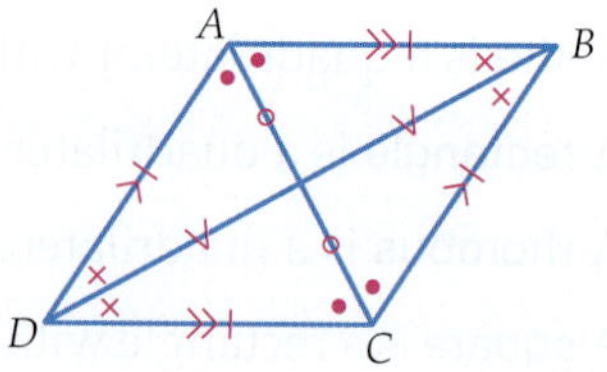

Square

All of the properties of a rectangle and a rhombus, plus:

- Diagonals make an angle of 45° with each side.
- Each diagonal has length equal to $\sqrt{2}$ times the side length.

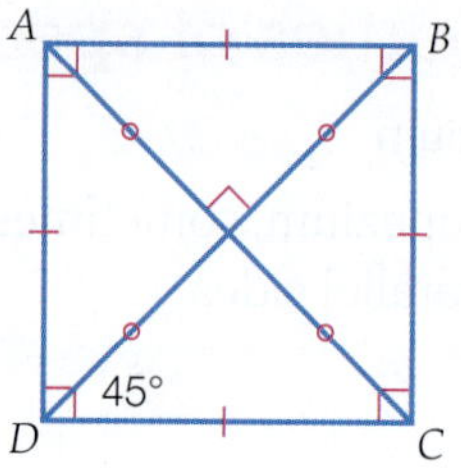

Also note:

- When two lines intersect at right angles, they are called **perpendicular**.
- When two lines intersect at right angles and one bisects the other (cuts it in half), that line is a **perpendicular bisector** of the second line.

Proving properties of special quadrilaterals

You should already be familiar with some of the properties of special quadrilaterals. You can use geometry techniques to prove some of the properties of these special quadrilaterals.

Worked example 11

W.E. 11

For each of the following shapes, write a list of the properties shown, and then use definitions to determine the type of quadrilateral.

(a) 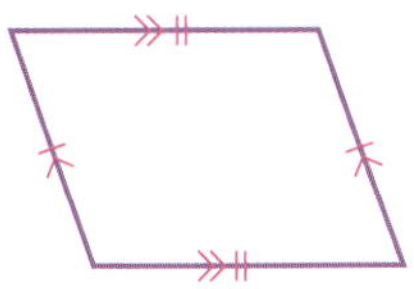

(b) 

Thinking	Working
(a) 1 Write any information displayed or given.	(a) Both pairs of opposite sides are equal and parallel.
2 Use the properties to list possible shapes, then state the type of shape and give reasons.	The shape is a parallelogram because both pairs of opposite sides are parallel.
(b) 1 Write any information displayed or given.	(b) Opposite sides are parallel. Two adjacent sides are equal in length.
2 Use the properties to list possible shapes, then state the type of shape and give reasons.	Opposite sides parallel means that the shape is a parallelogram. Two adjacent sides equal in length means that the parallelogram is a rhombus.

Worked example 12

W.E. 12

ABCD is a parallelogram.

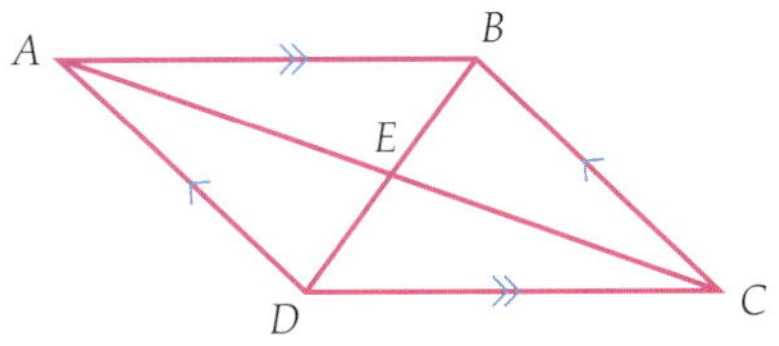

(a) By drawing the diagonal *AC*, prove that $\Delta ADC \equiv \Delta ABC$.

(b) Use these triangles to prove that $\angle ADC = \angle ABC$ and $\angle BAD = \angle BCD$.

(c) What special property of a parallelogram does this prove?

(d) By now drawing the diagonal *BD*, prove that $\Delta CDE \equiv \Delta ABE$.

(e) Use these triangles to prove that $AE = CE$ and $DE = BE$.

(f) What special property of a parallelogram does this prove?

Thinking	Working
(a) 1 Copy the diagram and complete the construction. Draw on any given side markings. (You can leave out the diagonal *BD*.)	(a) 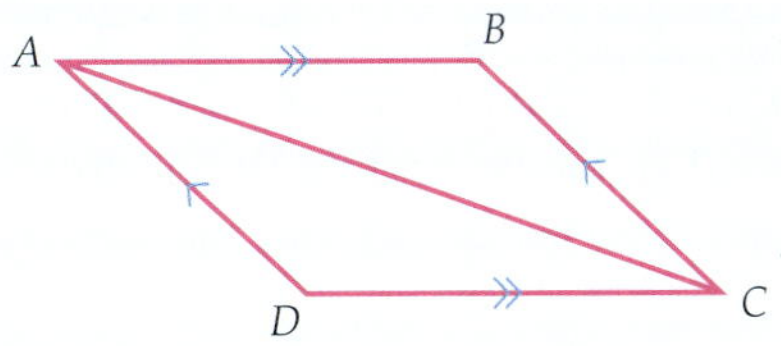
2 Name the two triangles to be considered.	ΔADC and ΔABC

3 Write three sets of equalities, with reasons, that apply to the two triangles.	$\angle DCA = \angle BAC$ (alternate angles, $AB \parallel DC$) $\angle DAC = \angle BCA$ (alternate angles, $AD \parallel BC$) AC is a common side
4 Write the pair of congruent triangles and the test used to determine this.	$\Delta ADC \equiv \Delta ABC$ (ASA)
(b) 1 Identify the appropriate pair of corresponding angles.	(b) $\angle ADC = \angle ABC$ (corresponding angles in congruent triangles)
2 Show that the other pair of angles is equal.	$\angle BAD = \angle DAC + \angle BAC$ $\angle BCD = \angle BCA + \angle DCA$ But $\angle DAC = \angle BCA$ and $\angle DCA = \angle BAC$ (from part (a) step 3) So $\angle BAD = \angle BCD$
(c) Write the answer in words.	(c) The property proved is that the opposite angles of a parallelogram are equal.
(d) 1 Name the two triangles to be considered and complete the construction on the diagram. Label point E.	(d) ΔABE and ΔCDE 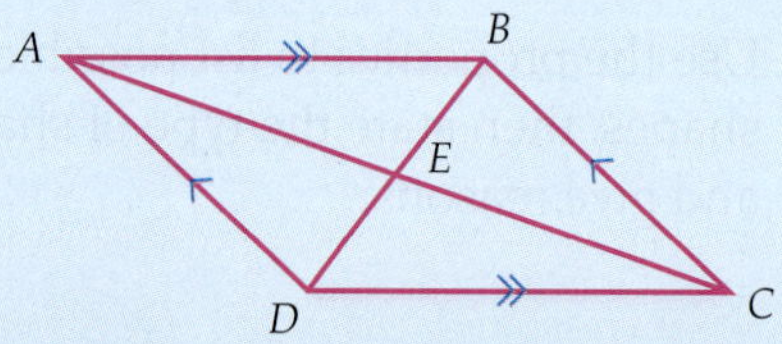
2 Write three sets of equalities, with reasons, that apply to the two triangles.	$AB = CD$ ($\Delta ADC \equiv \Delta CBA$) $\angle DCE = \angle BAE$ (alternate angles, $AB \parallel DC$) $\angle CDE = \angle ABE$ (alternate angles, $AB \parallel DC$)
3 Write the pair of congruent triangles and the test used to determine this.	$\Delta ABE \equiv \Delta CDE$ (ASA)
(e) Identify the appropriate pair of corresponding sides.	(e) $AE = CE$ and $DE = BE$ (corresponding sides in congruent triangles)
(f) Write the answer in words.	(f) The property proved is that the diagonals of a parallelogram bisect each other.

Sometimes certain properties of a shape are given. You can use this information to determine the type of shape by finding the appropriate definition.

Worked example 13

W.E. 13

The diagonals of a quadrilateral bisect each other at right angles. Prove that this quadrilateral is a rhombus.

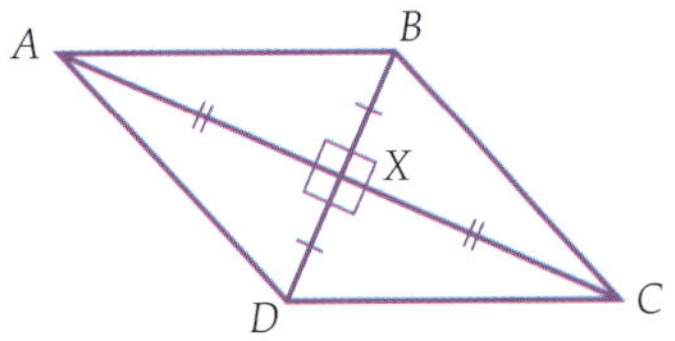

Thinking	Working
1 Use the definitions for quadrilaterals to determine the information needed to prove the shape is a rhombus.	Two adjacent sides must be proved to be equal in length and the opposite sides must be proved to be parallel.
2 Recognise that congruent triangles need to be found and identify the triangles to be considered.	Consider ΔAXD and ΔAXB
3 List three equalities, with reasons, that apply to the triangles.	$\angle AXD = \angle AXB = 90°$ (given) AX is common $DX = BX$ (given)
4 Write the congruent triangles and the test used to determine this.	$\Delta AXD \equiv \Delta AXB$ (SAS)
5 Identify the appropriate pairs of corresponding sides.	$AD = AB$ (corresponding sides in congruent triangles)
6 Consider another pair of triangles to prove congruent.	Consider ΔAXD and ΔCXB
7 List three equalities, with reasons, that apply to the triangles.	$\angle AXD = \angle CXB = 90°$ (given) $AX = CX$ (given) $DX = BX$ (given)
8 Write the congruent triangles and the test used to determine this.	$\Delta AXD \equiv \Delta CXB$ (SAS)
9 Identify equal angles and use equal alternate angles to prove the lines are parallel.	$\angle DAX = \angle BCX$ $AD \parallel BC$ (alternate angles $\angle DAX$ and $\angle BCX$ are equal)
10 Recognise that by repeating steps **6–9** with the other two triangles, you can prove the other two sides are parallel.	Similarly, $AB \parallel CD$
11 Summarise the information needed to answer the question.	$AB \parallel CD$ and $AD \parallel BC$ (opposite sides parallel) $AD = AB$ (two adjacent sides equal) So the shape is a rhombus.

9.5 Geometric properties of special quadrilaterals

Navigator

Answers p. 834

1, 2, 3, 4, 5, 6, 7, 8, 9, 10, 11, 12, 14, 15, 16	1, 2, 3, 4, 5, 6, 7, 8, 9, 10, 11, 12, 14, 15, 16, 17, 19	1, 2, 3, 4, 7, 8, 9, 10, 11, 12, 13, 16, 17, 18, 19

Fluency

W.E. 11

1 For each of the following shapes:

   (i) write a list of the properties shown,

   (ii) use the definitions for quadrilaterals to determine the type of shape.

(a)

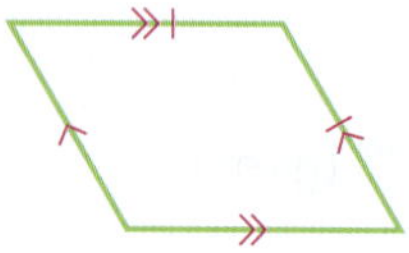

(b)

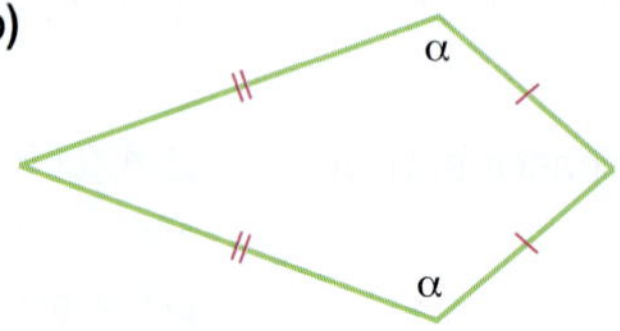

(c)

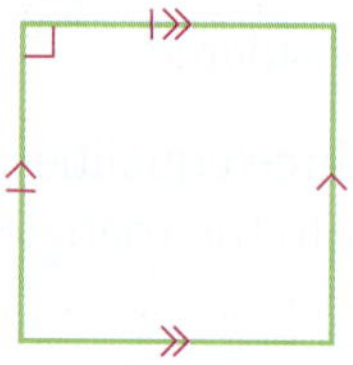

W.E. 12

2 *ABCD* is a parallelogram. Use the diagram shown to prove the following properties of parallelograms.

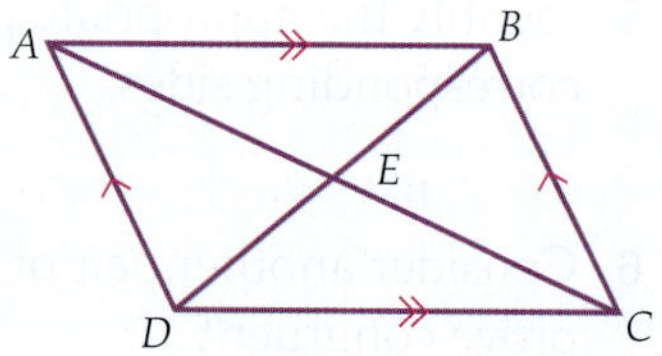

(a) By drawing the diagonal *BD*, prove that $\Delta DAB \equiv \Delta DCB$.

(b) Use these triangles to prove that $\angle DAB = \angle DCB$ and $\angle ADC = \angle ABC$.

(c) What special property of a parallelogram does this prove?

(d) By now drawing the diagonal *AC*, prove that $\Delta DAE \equiv \Delta BCE$.

(e) Use these triangles to prove that $AE = CE$ and $DE = BE$.

(f) What special property of a parallelogram does this prove?

W.E. 13

3 The diagonals of a quadrilateral are equal in length and bisect each other at right angles. Prove that this quadrilateral is a square.

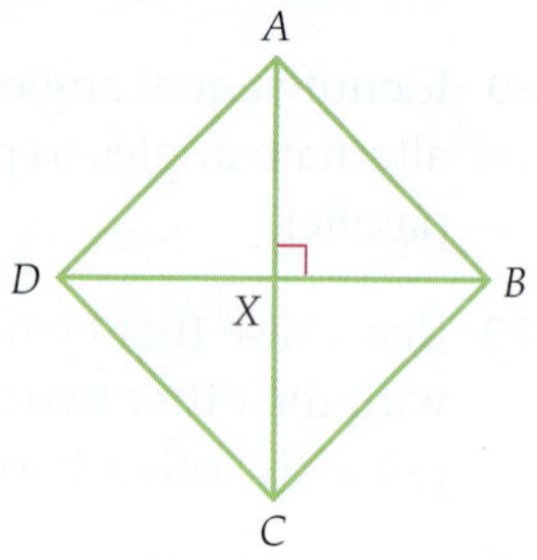

4 Use the diagram shown to prove the following properties of a kite.

(a) A kite has one pair of opposite equal angles.

(b) A kite has diagonals that cross at right angles.

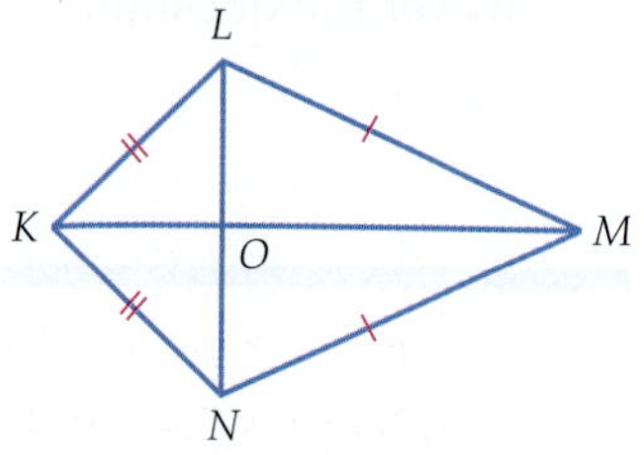

5 When proving that a quadrilateral is a trapezium but not a parallelogram, it is necessary to show:

A only one pair of parallel sides
B one pair of parallel sides and one pair of non-parallel sides
C one pair of parallel sides and one pair of equal sides
D two pairs of parallel sides.

6 A quadrilateral must be a parallelogram if one pair of opposite sides is:

A only equal in length
B only parallel
C equal in length and parallel
D parallel and the other pair of opposite sides is equal in length.

7 (a) For the diagram shown, prove $PQ = RS$.
(b) What special property of a rectangle have you just proved?

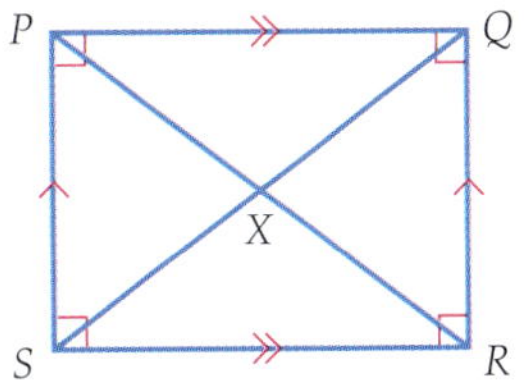

Understanding

8 Prove each of the following using the diagram shown.

(a) If a quadrilateral has all sides equal and one angle is a right angle, then it is a square.

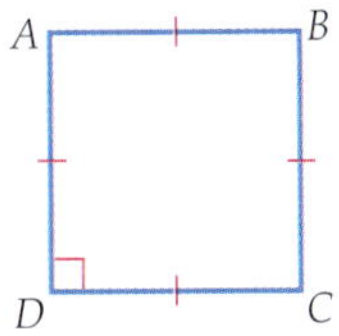

(b) If all sides of a quadrilateral are equal, then the quadrilateral is a rhombus.

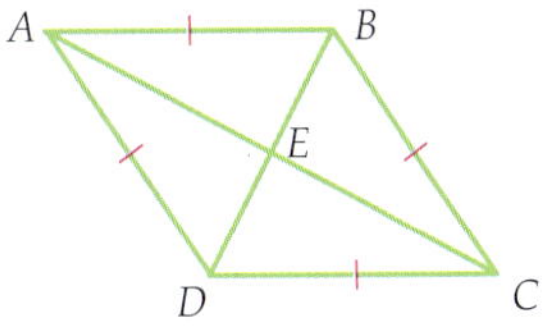

9 Prove each of the following using the diagram shown.

(a) If a pair of opposite sides in a quadrilateral is parallel and equal, then the quadrilateral is a parallelogram.

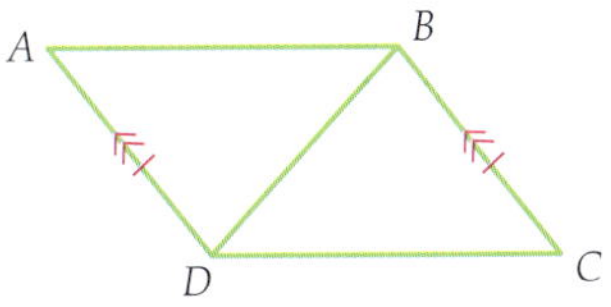

(b) If the diagonals of a quadrilateral bisect each other, then the quadrilateral is a parallelogram.

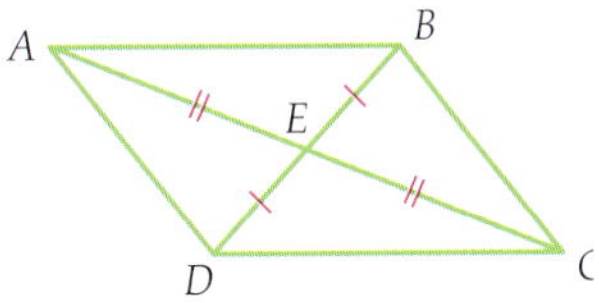

(c) If both pairs of opposite angles of a quadrilateral are equal, then the quadrilateral is a parallelogram.

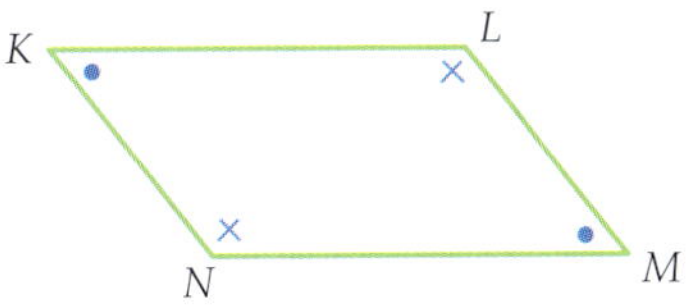

(d) If both diagonals of a quadrilateral bisect the vertex angles through which they pass, then the quadrilateral is a rhombus.

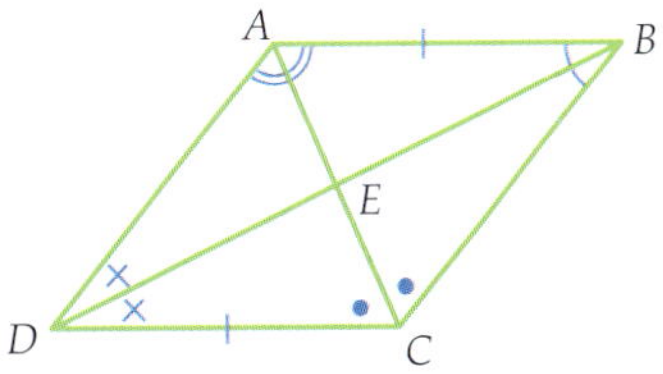

10 To prove that a parallelogram must be a rectangle, it is necessary to prove that its diagonals:

A are equal in length
B bisect each other
C bisect the angles of the parallelogram
D are perpendicular to each other.

11 *WXYZ* is a square. Show that $\Delta WMZ \equiv \Delta YNZ$.

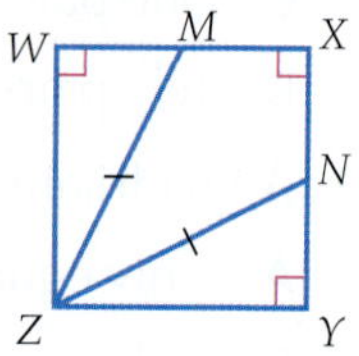

12 Copy and complete the following flowchart to show the relationships between different quadrilaterals.

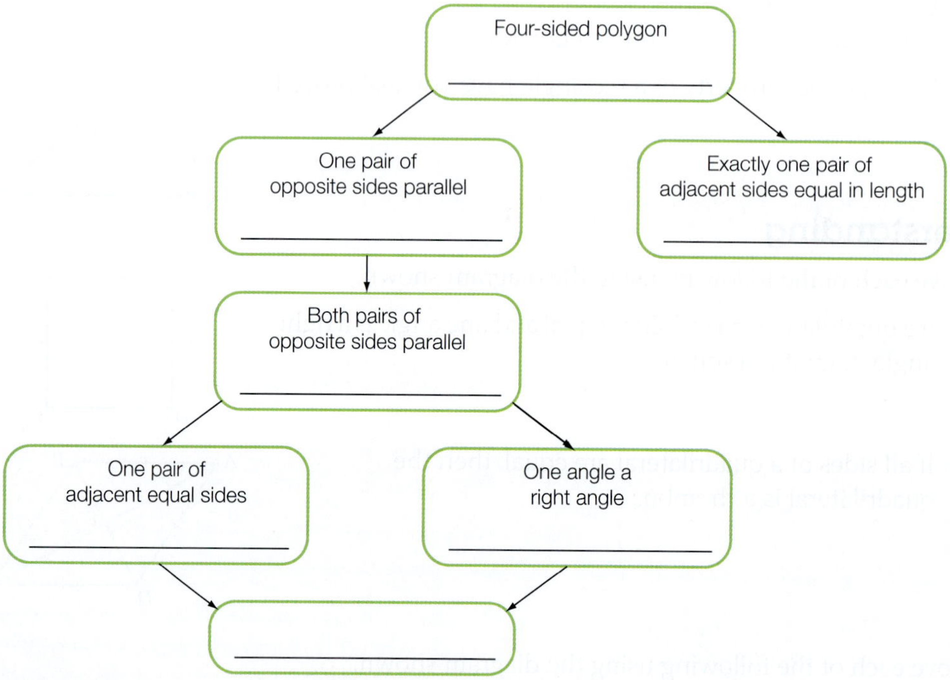

Reasoning

13 *ABCD* is a parallelogram and $AJ = CK$.
Show that $\Delta AKD \equiv \Delta CJB$.

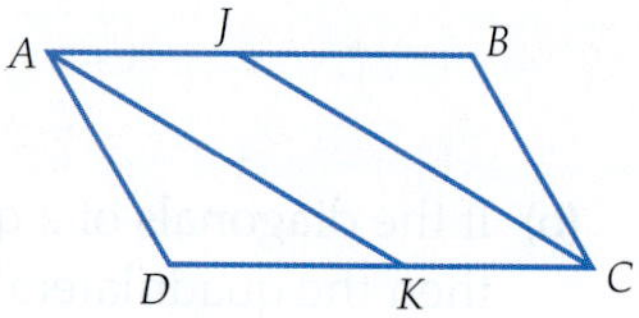

14 Explain why a square is a rhombus, but a rhombus is not necessarily a square.

15 Explain why a square is a rectangle, but a rectangle is not necessarily a square.

Open-ended

16 Adam and Alvin were discussing how many different quadrilateral shapes they could use to make biscuits. Choose three different quadrilateral shapes that they could use. The shapes need to have one of the following properties in common:

- at least one pair of parallel sides
- at least one pair of equal sides
- at least one pair of equal angles.

Draw the shapes and name them, and then identify the property you have used.

17 (a) The statement 'All squares have equal sides' is true. The converse statement is 'All shapes with four equal sides are squares'. Is this true? Explain why or why not.

(b) Write two or more true statements about special quadrilaterals and their converse statements. Examine whether or not the converse statements are true.

18 (a) Draw any three rhombuses and mark in the diagonals.

(b) Measure the angles at the centre where the diagonals intersect.

(c) What property of rhombuses does this seem to show?

(d) Is this sufficient proof of this property?

19 Gary, Matt and George are arguing about the geometric figure that could have the property that the diagonals bisect at right angles. Matt says there is only one such figure, Gary says there are two and George says there are three or more. Who is right and what shape could the figure(s) be?

Problem solving

Folding and cutting

Equipment required: some loose sheets of paper, scissors

1 You are going to fold a piece of paper twice and then cut across the dotted corner. Can you predict the shape of the resulting hole?

The shape of the resulting hole will depend on the angle at which you cut and also the angle of the second fold. See if you can produce the following shaped holes.

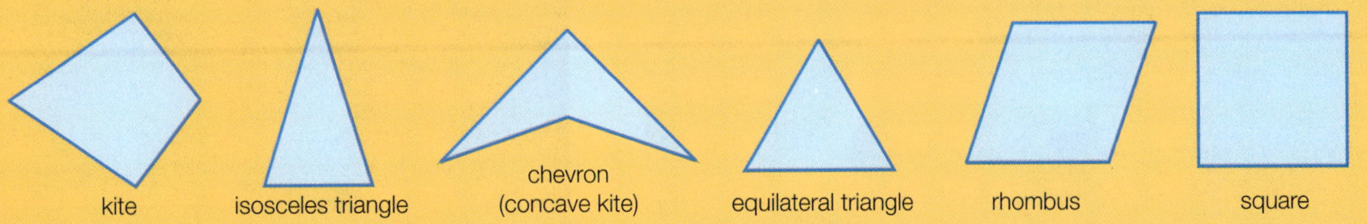

Can you find the maximum number of sides of the hole such a single cut produces for a piece of paper folded twice?

2 Now try folding the sheet of paper just once, but using two cuts to produce a hole when unfolded. What shape will your hole be?

Can you produce an equilateral triangle, an isosceles triangle and a scalene triangle?

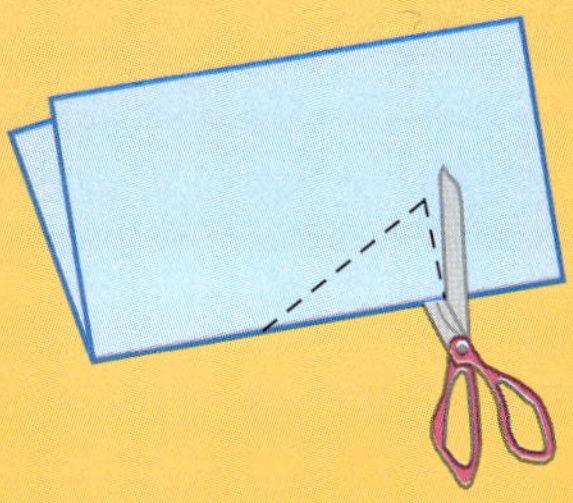

Strategy options

- Act it out
- Make a model.

Investigation

Snooker

Equipment required: centimetre graph paper

The Big Question

A player, Neil, is at a point X (6 feet from table end AB) and is aiming for point Y, which is 4 feet from A, along AB. Neil wants to determine at what distance from B the ball would next hit the cushion at point Z.

Billiards and snooker tables are designed with the 'foot' as their standard unit of measurement.

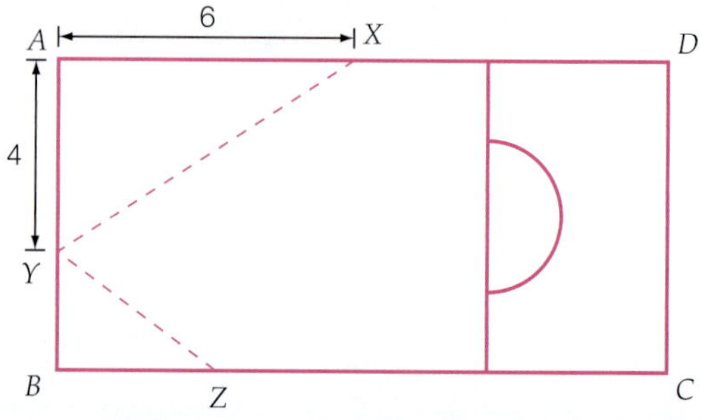

Engage

The game of billiards originated in the early 1500s. A railing was added to the original shuffleboard and then cushions were added to make the modern table. The object of the original game was to use the white ball to hit both the red ball and the spot ball with the same shot, called a cannon. The modern game was developed later and uses a 12 foot by 6 foot table (3.66 m × 1.83 m), with the usual six pockets, at each corner and in the middle of the longer sides.

Snooker was invented during the latter half of the 1800s and is now a popular sport in which world-champion professional players compete for million-dollar prizes. Snooker uses one white cue ball, 15 red balls (each worth one point), and six balls of different colours: yellow (2 points), green (3), brown (4), blue (5), pink (6) and black (7). The aim is to hit the red and coloured balls into the pockets in a specified order, to earn more points than the opposing player.

Explore

1 The angle at which a ball bounces off the cushion is equal to the angle at which it approaches the cushion.

$\angle XYA = \angle$_____

2 Draw the two triangles from **1** with the same orientation.

3 Determine the distance *BZ*. This is how far from corner *B* the ball hits the cushion *BC*.

Neil now wants to determine what happens when the ball continues its path and bounces off the cushion at point *Z*.

Strategy options

- Draw a diagram.
- Break problem into manageable parts.

Explain

4 Locate the position of the point *N*, where the ball next hits the cushion *CD*, when it bounces off point *Z*. Draw a diagram to illustrate this situation.

5 Neil is at point *X* and aiming for a point *T*, 4 feet from the corner *B* along the cushion *BC*, by bouncing off the cushion *AB* at a point *V* first. Find the position of *V*.

6 Is it possible for Neil to hit a ball from point *X* so that it bounces off all four cushions once before hitting a cushion a second time? If it is possible, describe the path of the ball.

Elaborate

7 The path the ball follows after being hit depends on the direction the ball is hit.

(a) Does the path of the ball follow any pattern?

(b) How is this pattern related to the original direction of the hit?

Evaluate

8 Did your methods change as you worked through the investigation?

9 Did you discover any shortcuts or rules that you could use to find your solutions?

Extend

10 What would happen if the table were a different shape, such as a square or an equilateral triangle? Try Neil's problems on a table with a different shape to the rectangular one above.

11 What would happen if a table had different corner angles? Is it possible to solve Neil's problems?

Maths 4 Real

THE ART OF FRACTALS

A *fractal* is a self-replicating pattern. This means that smaller parts of the whole image look like the whole image itself. Fractal-like patterns occur in nature and fractals are sometimes described as the geometry of nature.

One of the first mathematicians to work on fractals was the French mathematician Gaston Julia, early in the twentieth century. However, he was working before computers were invented, and to graph the results by hand was time-consuming, so progress was limited until decades later. The illustration at the top of this page is a graph of a 'Julia set', a class of fractals described in Julia's work.

The word *fractal* was first used by French mathematician Benoit Mandelbrot, who rediscovered the work of Julia in the 1970s. With computers able to complete the thousands of calculations needed, fractal images were now much more easy to produce. However, Mandelbrot's first images were very poor compared to those now produced by modern computers. With the use of suitable software, anybody can now produce spectacularly detailed fractals using a home computer or smartphone app.

One of the best known fractals is the *Mandelbrot set*. The shapes and colours are made by substituting the coordinates of points on the plane into a complex equation, then repeating (or 'iterating') this equation many times to determine its behaviour. Points on the plane are coloured (or not) according to how quickly the values approach infinity when iterated.

Many fractals can be produced by following simple mathematical procedures. In this task, you will first look at a particular fractal called the *Sierpinski Gasket* or *Sierpinski Triangle*.

Sierpinski Gaskets can be drawn on isometric paper:

To create a Sierpinski Gasket, draw an equilateral triangle and connect the midpoints of each side. Colour in the middle triangle. Then repeat the iteration in each of the other three triangles.

The middle diagram below is the *first iteration* of the Sierpinski Gasket.

The final diagram below is the *second iteration* of the Sierpinski Gasket.

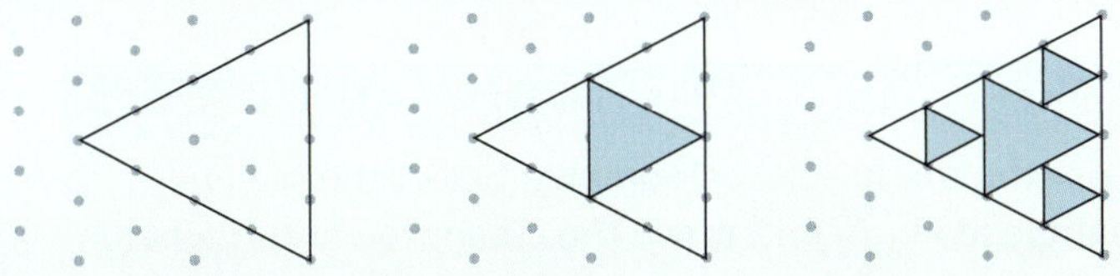

The third iteration comes from an 8-unit equilateral triangle, and so on.

1 Draw the fourth iteration of the Sierpinksi Gasket.

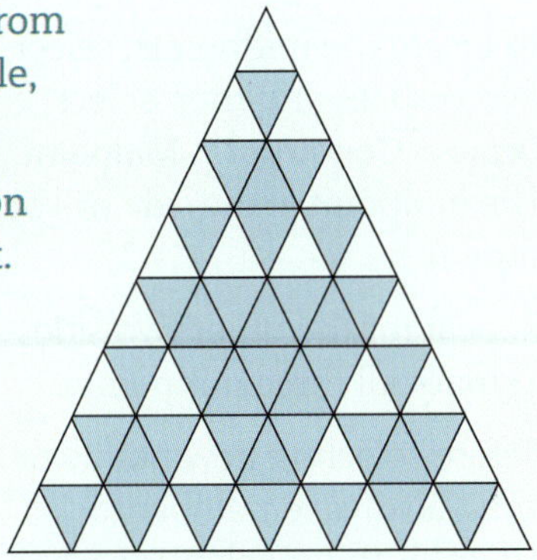

Fractals can be used to model things which occur in nature. Consider the fractal image of a fern leaf shown here.

This is an example of a type of fractal known as an *L-System*. An L-System is based on a set of rules that model growth that occurs in nature.

Consider the following example in which the rules are:

A → DB; B → C; C → D; D → E; E → A

This produces the following pattern, which models the branching growth of many simple plants.

Stage 0										A							
Stage 1						D								B			
Stage 2						E								C			
Stage 3						A								D			
Stage 4				D				B						E			
Stage 5				E				C						A			
Stage 6				A				D				D				B	
Stage 7		D			B			E				E				C	
Stage 8		E			C			A				A				D	
Stage 9		E			D		D		B		D		B			E	
Stage 10	D		B		E		E		C		E		C			A	
Stage 11	E		C		A		A		D		A		D		D		B

2 Consider the L-System that follows the rules A → B and B → AB and starts with an A.

Stage 0 is A

Stage 1 is B

Stage 2 is AB

Stage 3 is BAB

(a) Using a grid like the one shown above, write the letter outcome for each of the next four stages (i.e. Stages 4–7).

(b) Can you see a number pattern involving the number of letters in each stage?

(c) Using the pattern, or otherwise, find the number of letters in each of Stages 8, 9 and 10.

The *Koch snowflake*, which is investigated in Chapter 12 of this book, is another type of fractal. It has an infinite perimeter but a finite area. For many mathematicians this is one of the best things about fractals—they give us some contact with infinity.

Research

1 Find out what is meant by the term *fractional dimensionality* and what it has to do with fractals.

2 Produce a colourful fractal image using a computer fractals app.

Exploration CAS

Equipment required: TI-Nspire CAS or Casio ClassPad CAS

Where is the centre of a triangle?

Concurrence

Any two non-parallel lines that are in the same plane will intersect. If a third line passes through the same point of intersection, then the lines are concurrent and the point of intersection is called the point of concurrence.

The centroid of a triangle

A line drawn from a vertex of a triangle to the midpoint of the opposite side is called a median. You will construct the medians of a triangle and investigate their concurrency.

Using TI-Nspire CAS	Using Casio ClassPad CAS
Add **Geometry** to your document. Select **menu** > **Shapes** > **Triangle**, then move the cursor to select the points where you want each vertex of the triangle to appear. Press **esc** to stop drawing.	From the menu select **Geometry**. Select the scalene triangle icon and move the cursor to choose where you want the shape to appear. To resize the triangle, select a vertex and drag to move its position.
To locate the midpoints, select **menu** > **Construction** > **Midpoint**, select a vertex and then select another vertex. The midpoint of the side will be shown. Repeat for the other two sides to mark all three midpoints. 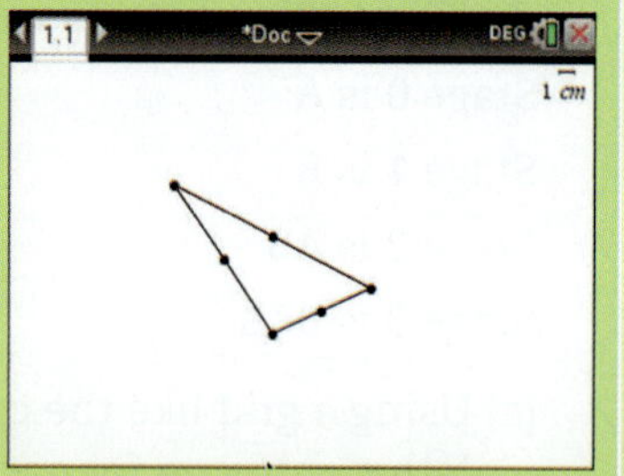	To locate the midpoints, select two vertices and then select **Draw** > **Construct** > **Midpoint**. The midpoint of the side will be shown. Repeat for the other two sides to mark all three midpoints. Before selecting new points or lines, always remember to select 'nothing' (blank space) to clear your previous selections. 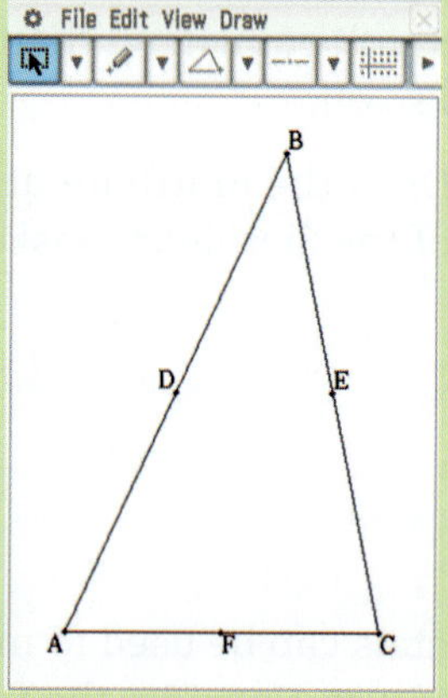
Now draw line segments to join each midpoint to the vertex opposite it. To draw a line, select **menu** > **Points & Lines** > **Segment** and select either end of each segment, as shown. 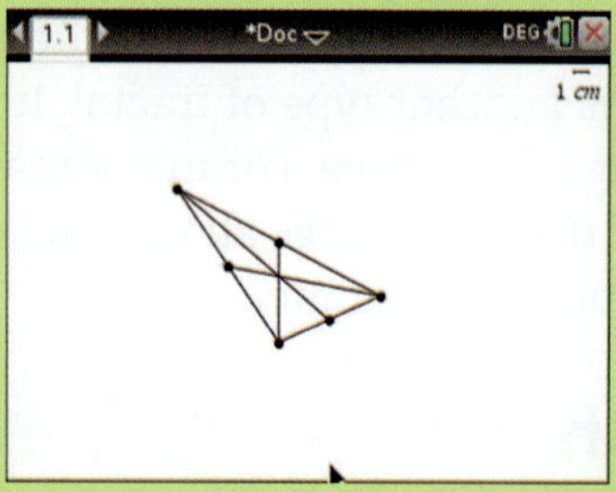 Now select a vertex of the triangle and move it around to see what happens.	Now draw line segments to join each midpoint to the vertex opposite it. To draw a line, select **Draw** > **Basic Object** > **Line Segment** and select either end of each segment, as shown. 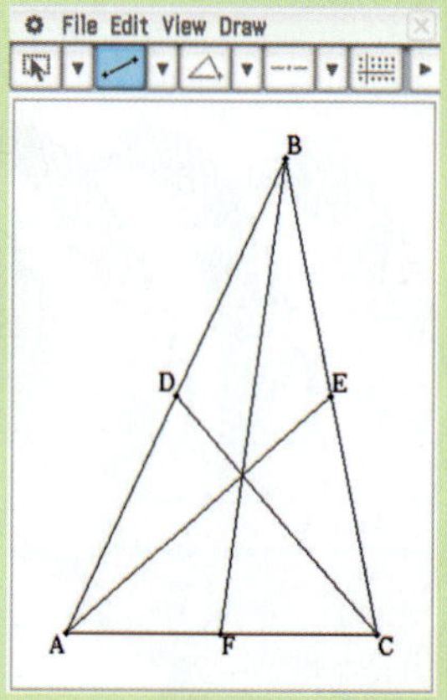 Now select the cursor tool. You can select a vertex of the triangle and move it around to see what happens.

1 Complete these sentences:

(a) Lines drawn from each vertex of a triangle to the ____________of the opposite side are concurrent.

(b) These lines are called ____________ and the point of concurrence is called the centroid of the triangle.

The orthocentre of a triangle

A line that is perpendicular to a side of a triangle and passes through the opposite vertex is called an altitude. You will construct the altitudes of a triangle and investigate their concurrency.

Using TI-Nspire CAS

Draw a new triangle on a fresh screen. Select **menu** > **Points & Lines** > **Point On** and put a point on each side of the triangle. Then select **menu** > **Construction** > **Perpendicular** and select one of the new points to construct a line perpendicular to its side.

Repeat for the other two sides to create a perpendicular line for each side.

Now select **menu** > **Construction** > **Parallel** and select one of the perpendicular lines and then its opposite vertex.

Repeat for the other two perpendicular lines and their opposite vertices.

Now select one of these new lines and drag it to make the line extend in length. Repeat for the other new lines.

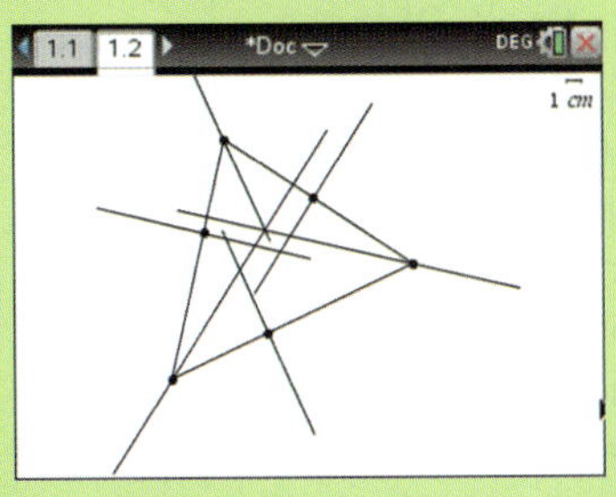

Now select one of the vertices and move it around. What happens to the point of intersection?

Using Casio ClassPad CAS

Draw a new triangle on a fresh screen. Select **Draw** > **Basic Object** > **Point** to create a new point on each side of the triangle.

Select one of the new points and its line, and select **Draw** > **Construct** > **Perpendicular**. Repeat for the other two sides to create a perpendicular line for each side.

Now select one of the perpendicular lines and its opposite vertex, and select **Draw** > **Construct** > **Parallel**.

Repeat for the other two perpendicular lines and their opposite vertices.

You can then delete the original lines through the points, so there are only perpendicular lines through each vertex. This should look something like the screen shown.

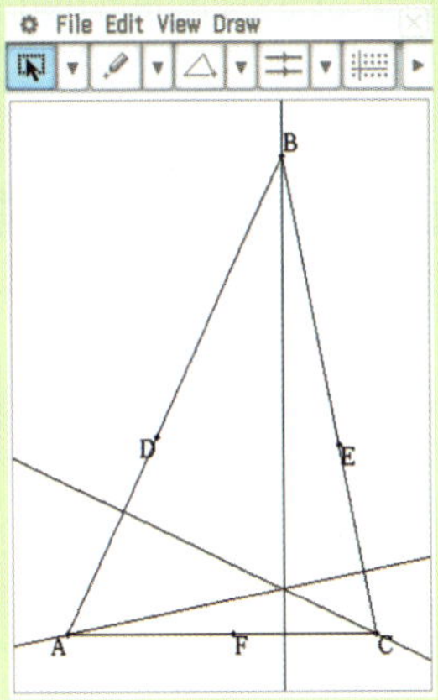

2 Complete these sentences:

(a) Lines ____________ to the side of a triangle that pass through the opposite vertex are concurrent.

(b) The lines are called ____________ and the point of concurrence is called the orthocentre of the triangle.

The incentre of a triangle

You will construct a triangle from the tangents to a circle and investigate the links to the angle bisectors of this triangle.

Using TI-Nspire CAS

Draw a new triangle on a fresh screen. Select **menu** > **Construction** > **Angle Bisector** and select each of the three vertices in turn. The angle bisector will be drawn.

Repeat this process, starting with a different vertex each time, to bisect each of the other angles.

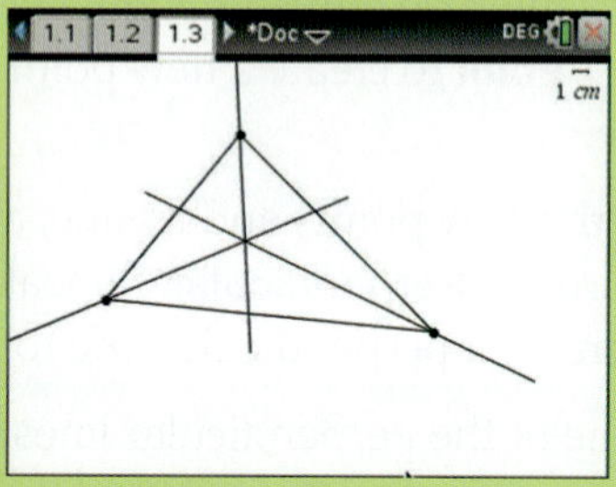

Select one of the vertices of the triangle and move it around to see what happens.

You now need to examine a circle drawn with its centre at the point of concurrence of the angle bisectors.

Select **menu** > **Shapes** > **Circle** and select the point of concurrence (where the three angle bisectors meet), to make this the centre of the circle. Then select one of the points of intersection with the sides of the triangle, to make this the edge of the circle.

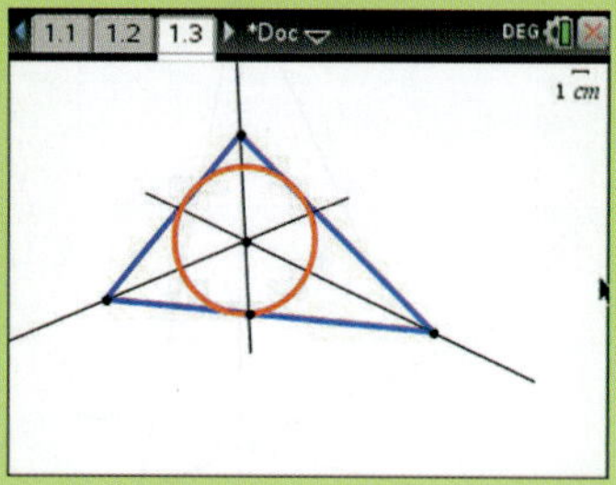

Using Casio ClassPad CAS

Draw a new triangle on a fresh screen. Select two of the sides of the triangle and then select **Draw** > **Construct** > **Angle Bisector**. The angle bisector will be drawn.

Repeat this process, using a different pair of sides each time, to bisect each of the other angles.

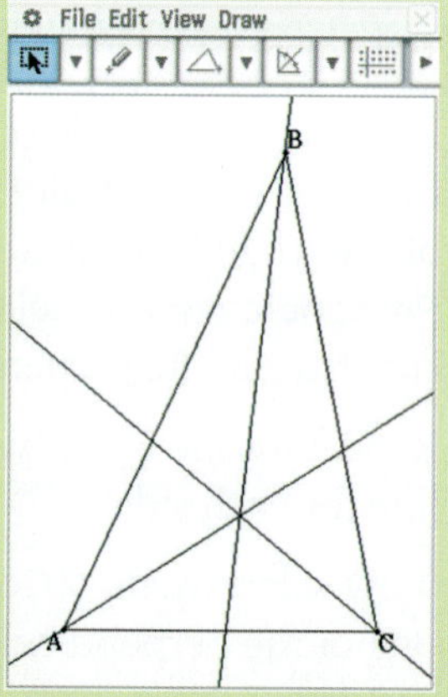

Select one of the vertices of the triangle and move it around to see what happens.

You now need to examine a circle drawn with its centre at the point of concurrence of the angle bisectors.

Select **Draw** > **Basic Object** > **Circle** and select the point of concurrence (where the three angle bisectors meet), to make this the centre of the circle. Then select one of the points of intersection with the sides of the triangle, to make this the edge of the circle.

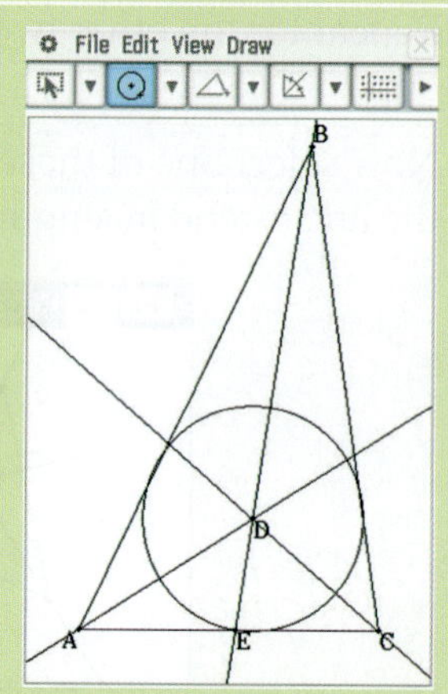

3 Complete these sentences:

(a) The lines that bisect the angles of a triangle are ____________ and the point of concurrence is called the incentre of the triangle because a circle can be drawn so that it touches all three sides.

(b) The circle is called the incircle and the sides of the triangles are ____________ to the circle.

The circumcentre of a triangle

You will construct the perpendicular bisectors of the sides of the triangle and investigate their concurrency.

Using TI-Nspire CAS

Draw a new triangle on a fresh screen. Select **menu** > **Construction** > **Perpendicular Bisector** and select one of the sides of the triangle.

Repeat this for the other two sides. If the lines are not long enough to intersect, select them in turn and drag them to make them longer.

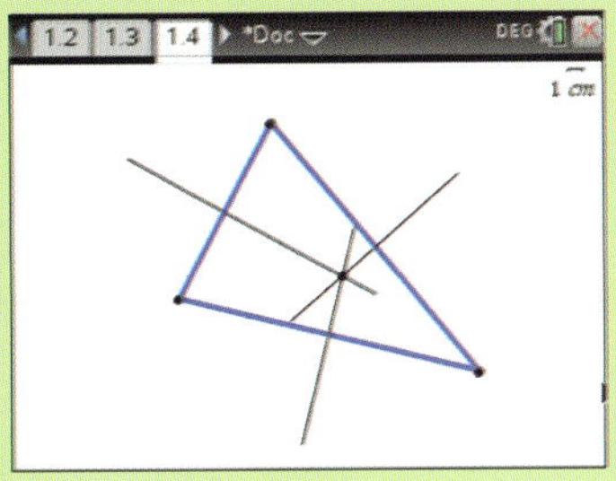

Select one of the vertices of the triangle and move it around to see what happens.

You now need to examine a circle drawn with a centre at the point of concurrence of the perpendicular bisectors.

Select **menu** > **Shapes** > **Circle** and select the point of concurrence (where the three perpendicular bisectors meet), to make this the centre of the circle.

Then select one of the vertices of the triangle, to make this the edge of the circle.

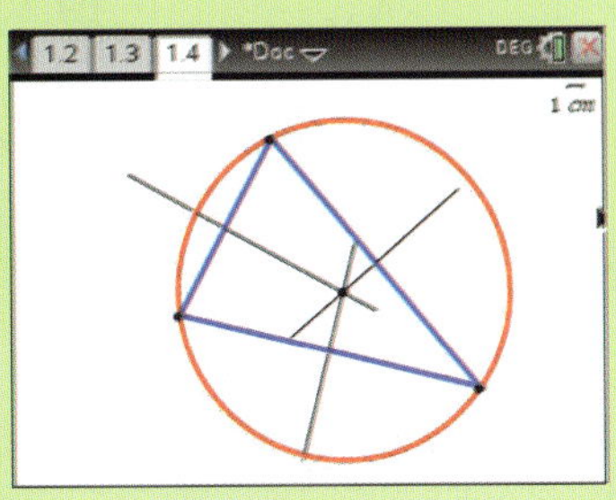

Using Casio ClassPad CAS

Draw a new triangle on a fresh screen. Select one of the sides and then select **Draw** > **Construct** > **Perp. Bisector**.

Repeat this for the other two sides.

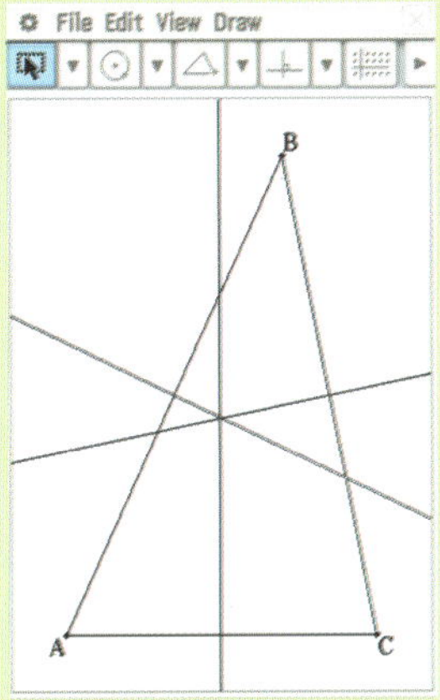

Select one of the vertices of the triangle and move it around to see what happens.

You now need to examine a circle drawn with a centre at the point of concurrence of the perpendicular bisectors.

Select **Draw** > **Basic Object** > **Circle** and select the point of concurrence (where the three perpendicular bisectors meet), to make this the centre of the circle.

Then select one of the vertices of the triangle, to make this the edge of the circle.

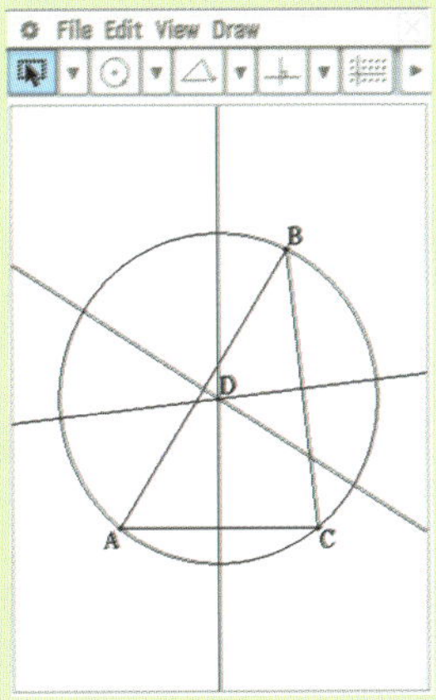

4 Copy and complete these sentences:

(a) The perpendicular bisectors of each side of a triangle are ____________.

(b) The point of concurrence is called the circumcentre of the triangle because a circle can be drawn through all three vertices using the ____________ as the centre of the circle. The circle is called the circumcircle.

9.6 Angles in circles

Terms related to circles

1 A **circle** is a plane curve consisting of all the points at a given distance (called the radius) from a fixed point in the plane (called the centre).

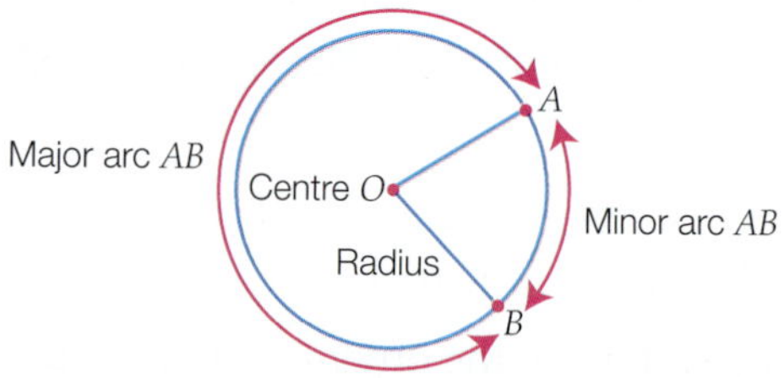

2 An **arc** of a circle is a portion of the circle bounded by two points on the circle. The shorter arc is the **minor arc** and the longer arc is the **major arc**. An arc that goes completely around the circle is called the circumference.

3 A **sector of a circle** is the plane figure bounded by two radii of the circle and one of the arcs that they form.

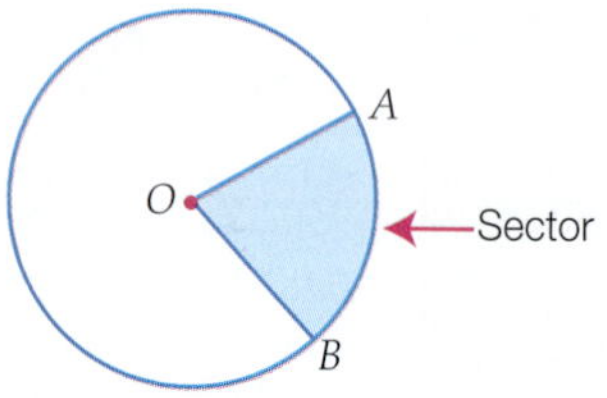

4 A **chord** of a circle is an interval joining any two points on the circle. A chord that passes through the centre of a circle is called a diameter. The length of the diameter is twice the radius. A diameter divides the circle into two semicircles.

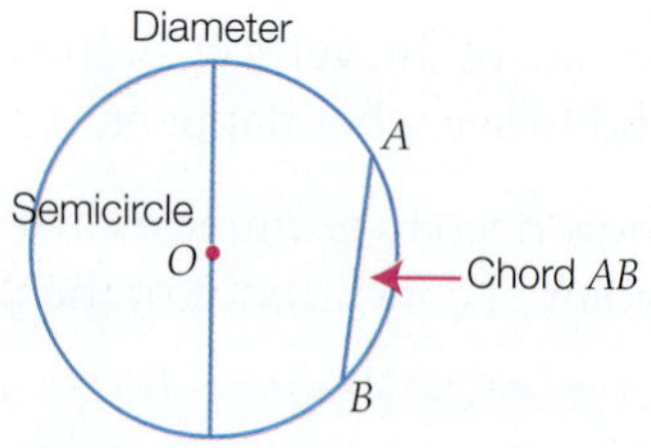

5 A **segment** of a circle is the plane figure bounded by a chord and an arc. A **minor segment** is bounded by a chord and a minor arc. A **major segment** is bounded by a chord and a major arc.

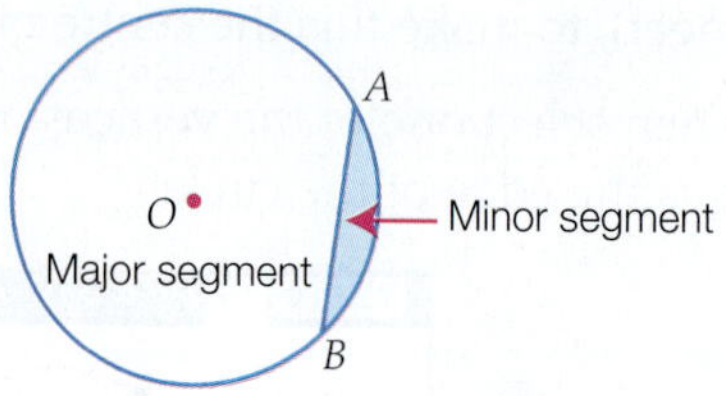

6 When an angle is **subtended**, this means that it is formed by lines that come in directly from the endpoints of an arc or chord (or other curve or line). The arc of a circle subtends an angle at the centre of the circle.

The major arc AB subtends the reflex angle AOB at the centre O of the circle.

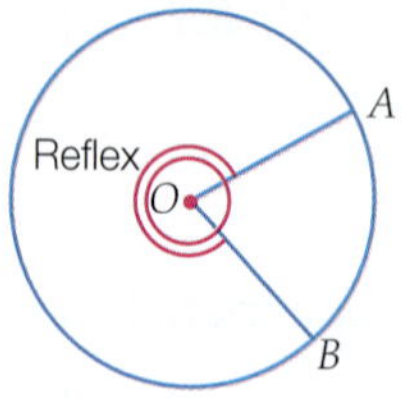

The minor arc AB subtends the acute or obtuse angle AOB at the centre O of the circle.

If you join AB, then the chord AB subtends angle AOB at the centre O.

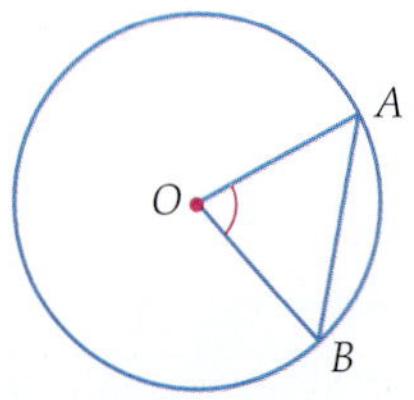

10A

7 Angles may be subtended at the centre of the circle or at the circumference. You say that they stand on the same arc or they are subtended by the same chord if the arms of the angle start from and end at the same points on the circumference of the circle.

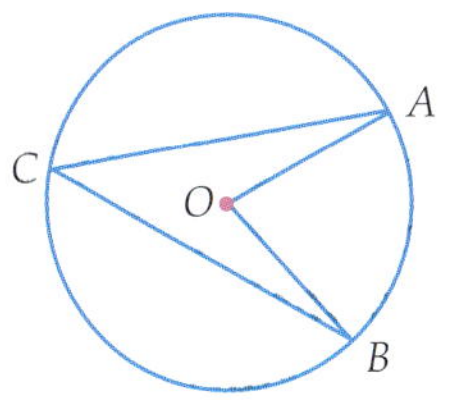

$\angle ACB$ is called the **angle at the circumference** standing on the arc AB. $\angle AOB$ is the **angle at the centre** standing on the arc AB. If the chord AB had been drawn, you would say that these angles were standing on the chord AB or they were subtended by the chord AB.

8 Angles at the circumference of a circle drawn from the same two points on the circumference are said to be standing on the same arc. If they are on the same side of the chord they are called **angles in the same segment**.

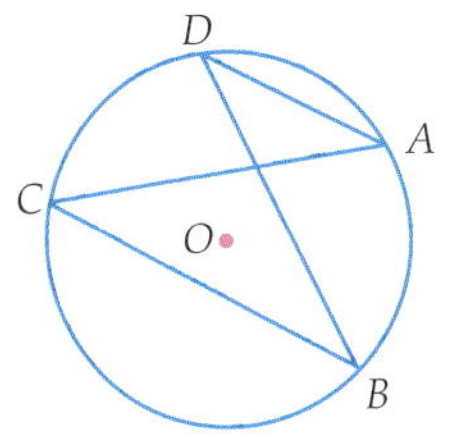

$\angle BCA$ and $\angle BDA$ are angles at the circumference standing on the minor arc AB.

They are in the same segment because they are on the same side of the chord AB.

Angles CBD and CAD are angles at the circumference standing on the minor arc CD.

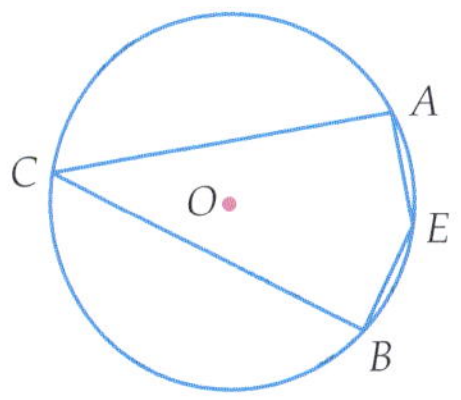

$\angle BCA$ is an angle at the circumference standing on the minor arc AB.

$\angle BEA$ is an angle at the circumference standing on the major arc AB.

They are in opposite segments as they are on opposite sides of the chord AB.

9 The **angle in a semicircle** is formed when the ends of a diameter are joined to a point on the circumference.

$\angle DEC$ is an angle in a semicircle.

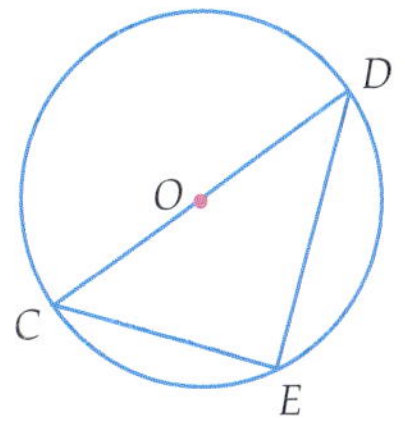

What is a theorem?

A *theorem* is a mathematical statement that can be shown to be true. A *proof* is a step-by-step argument that demonstrates the truth of a mathematical theorem.

Angles at the centre and circumference

Draw a circle with centre O with three points A, B and C on the circumference. Join AB, BC, CA, OA, OB and OC.

Measure each of the following pairs of angles: $\angle AOB$ and $\angle ACB$; $\angle BOC$ and $\angle BAC$; $\angle COA$ and $\angle CBA$. It appears that the first angle of each pair is double the second angle. Check to see for yourself.

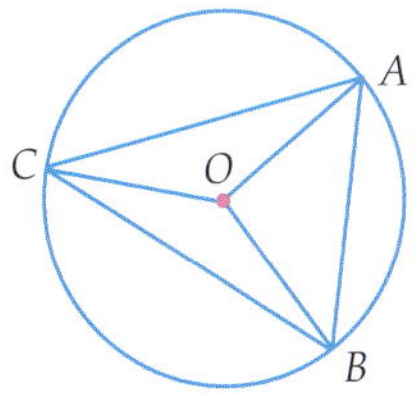

Circle theorem 1

The angle at the centre is twice the angle at the circumference standing on the same arc.

Worked example 14

W.E. 14

A, B and C are three points on the circumference of a circle with centre O.

Prove that $\angle AOB = 2\,\angle ACB$.

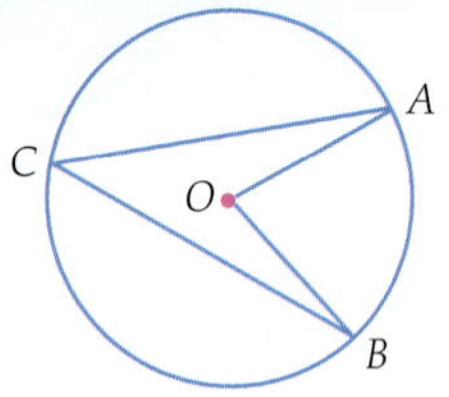

Thinking	Working	
1 Copy the diagram and join CO, extending it to D.	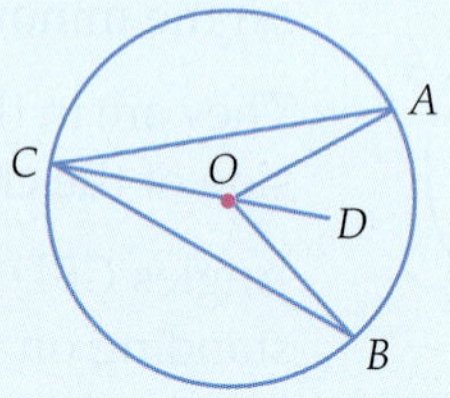	
2 Identify isosceles triangles and list pairs of equal angles.	$OC = OA$	(radii of a circle)
	$\therefore \Delta OCA$ is isosceles	
	$\therefore \angle OCA = \angle OAC$	(equal angles in isosceles triangle)
	$OC = OB$	(radii of a circle)
	$\therefore \Delta OCB$ is isosceles	
	$\therefore \angle OCB = \angle OBC$	(base angles of isosceles trianges)
3 Use this information to find $\angle AOD$ and $\angle BOD$.	$\angle AOD = \angle ACO + \angle CAO$	(exterior angle of ΔAOC equals sum of interior opposite angles)
	$\therefore \angle AOD = 2\angle ACO$	($\angle OCA = \angle OAC$)
	$\angle BOD = \angle BCO + \angle CBO$	(exterior angle of ΔBOC equals sum of interior opposite angles)
	$\therefore \angle BOD = 2\angle BCO$	($\angle OCB = \angle OBC$)
4 Find expressions for $\angle ACB$ and $\angle AOB$ and find the result to be proved.	Now $\angle ACB = \angle ACO + \angle BCO$	
	and $\angle AOB = \angle AOD + \angle BOD$	
	$\therefore \angle AOB = 2\angle ACO + 2\angle BCO$	(from above)
	$\angle AOB = 2(\angle ACO + \angle BCO)$	
	$\therefore \angle AOB = 2\angle ACB$	

10A

It is important to recognise that the point C may be placed anywhere on the circumference of the circle and the lines drawn within the circle may not necessarily form a quadrilateral. For example, they may look like the diagram shown at right.

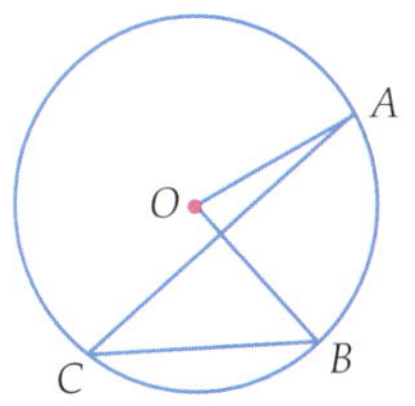

Angle in a semicircle

Draw a circle with centre O and diameter CD.

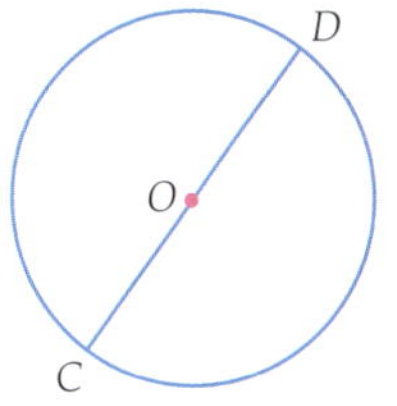

Join C and D to points K, L and M on the circumference. Measuring $\angle CKD$, $\angle CLD$ and $\angle CMD$, you find that all the angles measured are 90°.

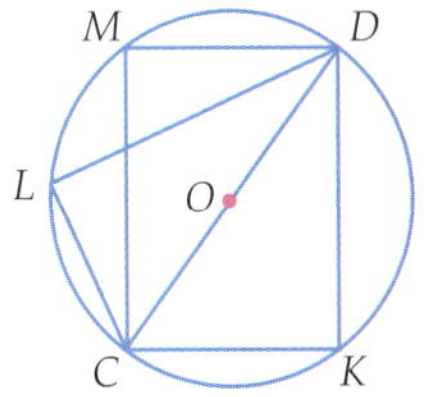

Circle theorem 2

The angle in a semicircle is a right angle.

Worked example 15

W.E. 15

CD is a diameter of a circle with centre O. E is a point on the circumference. Prove that $\angle CED$ is a right angle.

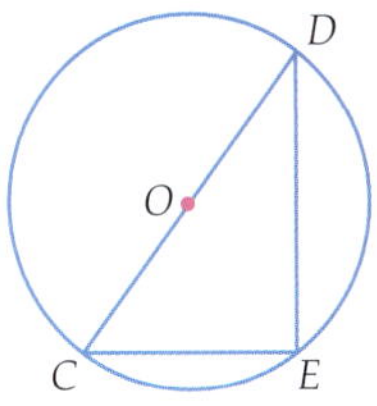

Thinking	Working
1 Copy the diagram.	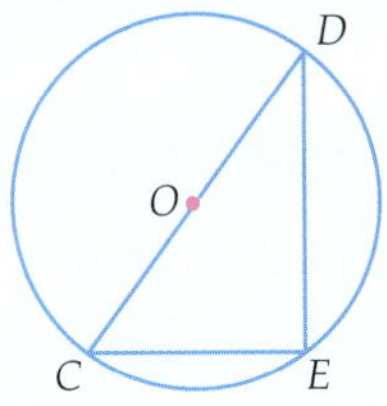
2 Look for information so that you can use the angle at the centre.	$\angle COD = 2\angle CED$ (angle at the centre is twice the angle at the circumference standing on the same arc CD) But $\angle COD = 180^\circ$ (COD is a straight line) $\therefore \angle CED = 90^\circ$
3 State the result proved.	$\angle CED$ is a right angle. $\therefore$ the angle in a semicircle is a right angle.

Angles at the circumference on the same arc

Draw a circle with the points *W*, *X*, *Y*, *Z* in order on the circumference.

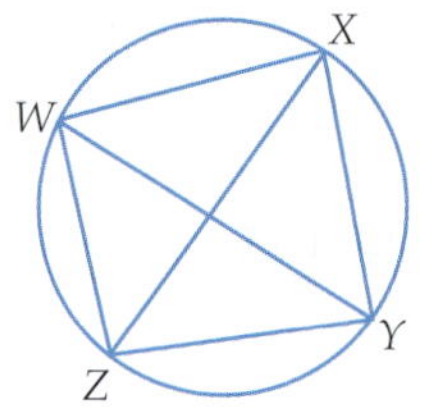

Join each point to the others, giving *WX*, *WY*, *WZ*, *XY*, *XZ*, *YZ*.

Measure $\angle WXZ$ and $\angle WYZ$; $\angle WZX$ and $\angle WYX$; $\angle XWY$ and $\angle XZY$; $\angle YXZ$ and $\angle YWZ$ etc. *What do you notice?* It appears as though the pairs of angles are equal.

Circle theorem 3

Angles in the same segment subtended by the same arc at the circumference are equal.

Worked example 16

W.E. 16

A, *B*, *C* and *D* are four points on the circumference of a circle with centre *O*.

Prove that $\angle ACB = \angle ADB$.

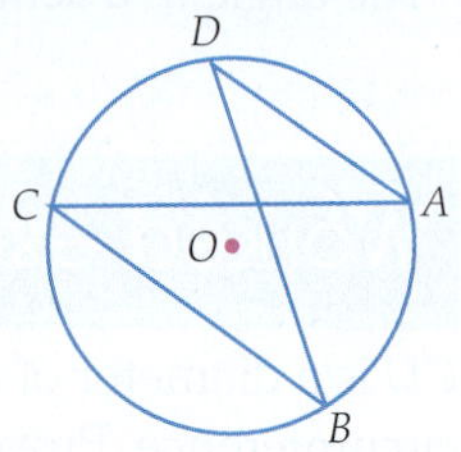

Thinking	Working
1 Copy the diagram and join *AO* and *BO*.	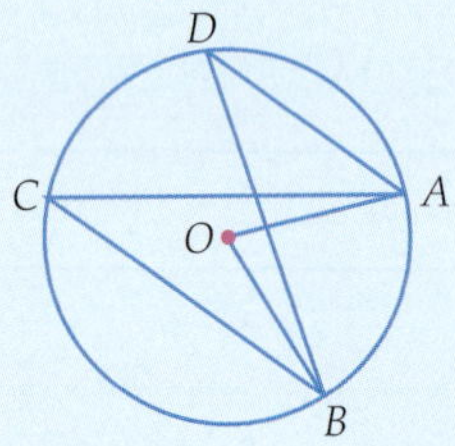
2 Look for information so that you can use the angle at the centre.	$\angle AOB = 2\angle ADB$ (angle at the centre is twice the angle at the circumference standing on the same arc) Similarly: $\angle AOB = 2\angle ACB$
3 State the result proved.	$\angle ACB = \angle ADB$ (both half of $\angle AOB$)

Worked example 17

W.E. 17

Find the size of $\angle ABC$, giving reasons for your answer.

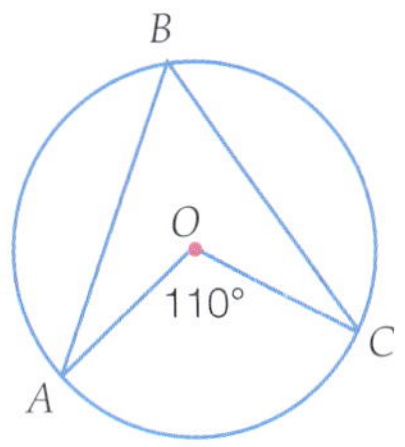

Thinking	Working
1 Write a result using angles at the centre.	$\angle AOC = 2\angle ABC$ (angle at the centre is twice the angle at the circumference standing on the same arc)
2 Consider any numerical values given in the diagram.	$\angle AOC = 110^\circ$ $\therefore 110^\circ = 2\angle ABC$
3 Calculate the size of the required angle.	$\therefore \angle ABC = 55^\circ$

Worked example 18

W.E. 18

AB is a diameter of the circle with centre O. $\angle CAB = 50^\circ$. Find the size of $\angle ABC$, giving reasons for your answer.

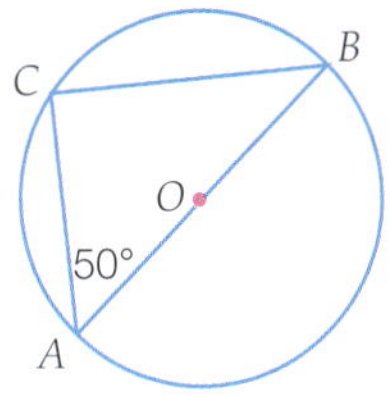

Thinking	Working
1 Write a result using diameters of a circle.	$\angle ACB = 90^\circ$ (angle in a semicircle is 90°)
2 Consider any numerical values given in the diagram.	$\angle BAC = 50^\circ$ $\angle ABC + 50^\circ + 90^\circ = 180^\circ$ (angle sum of a triangle is 180°)
3 Calculate the size of the required angle.	$\therefore \angle ABC = 40^\circ$

Worked example 19

W.E. 19

A, B, C and D are points on the circumference of the circle, AC and DB intersect at E. If $\angle CAB = 48^\circ$ and $\angle DCA = 42^\circ$, find the size of $\angle AEB$, giving reasons for your answer.

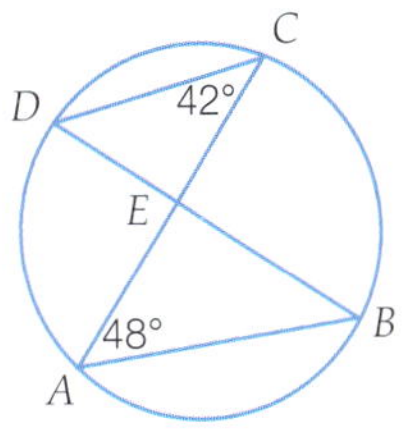

Thinking	Working
1 Write a result using the angles of a circle.	$\angle ACD = \angle ABD$ (angles at the circumference on the same arc)
2 Consider any numerical values given in the diagram.	$\therefore \angle ABD = 42°$ $\angle EAB = 48°$ (given)
3 Calculate the size of the required angle.	$\angle AEB + 42° + 48° = 180°$ (angle sum of a triangle is 180°) $\therefore \angle AEB = 90°$

Worked example 20

W.E. 20

BC is a diameter of the circle with centre *O*. *E* and *F* are points on the circumference of the circle such that $OF \parallel CE$. If $\angle OCE = 35°$, find the size of $\angle OFE$, giving reasons for your answer.

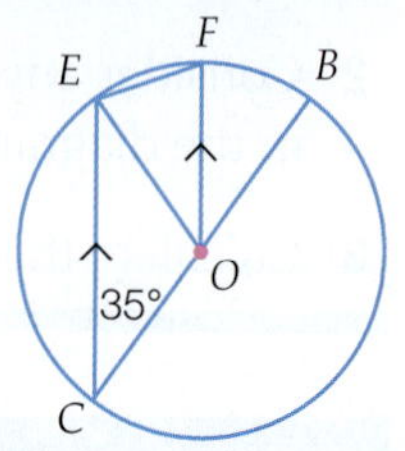

Thinking	Working
1 Write results using angle properties.	$\angle FOB = \angle ECO$ (corresponding angles, $FO \parallel EC$) $\therefore \angle FOB = 35°$ $\angle EOB = 2\angle ECB$ (angle at the centre is twice the angle at the circumference standing on the same arc) $\angle EOB = 2 \times 35° = 70°$ $\angle EOF + 35° = 70°$ $\therefore \angle EOF = 35°$
2 Use the equal radii.	ΔOEF is isosceles ($OE = OF$, radii of the circle) $\angle OEF = \angle OFE$ (base angles of an isosceles triangle) But $\angle OEF + \angle OFE + 35° = 180°$ (angle sum of a triangle) $\therefore 2\angle OFE = 145°$
3 Calculate the size of the required angle.	$\angle OFE = 72.5°$

10A

9.6 Angles in circles

Navigator

1, 2, 3, 4, 5, 6, 7, 8, 9, 10, 11, 12, 13, 17, 21, 22, 23	1, 2, 3, 4, 5, 6, 7, 8, 9, 10, 11, 12, 13, 14, 15, 17, 19, 20, 21, 22, 23, 24	1, 2, 3, 4, 5, 6, 7, 8, 9, 10 (columns 2–3), 11, 12, 13, 14, 15, 16, 17, 18, 20, 21, 22, 23, 24

Answers p. 837

Fluency

1 E, F and G are three points on the circumference of a circle with centre O. Prove that $\angle EOF = 2\angle EGF$.

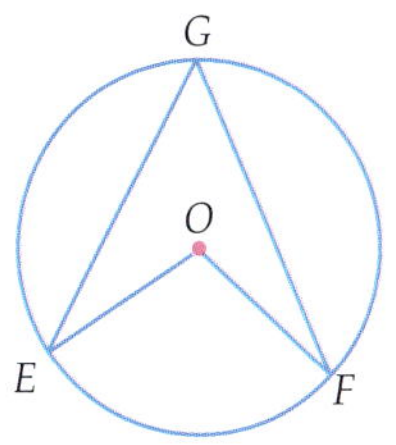

W.E. 14

2 AB is a diameter of a circle with centre O. C is a point on the circumference. Prove that $\angle ACB = 90°$.

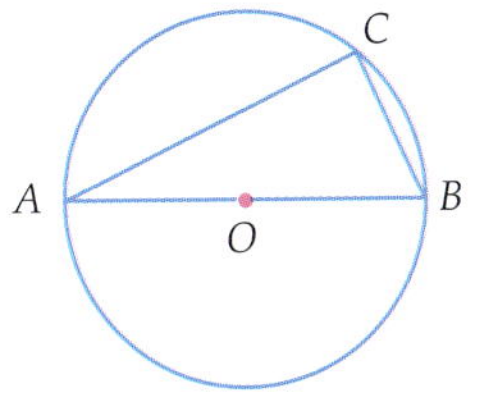

W.E. 15

3 K, L, M and N are four points on the circumference of a circle with centre O. Prove that $\angle KML = \angle KNL$.

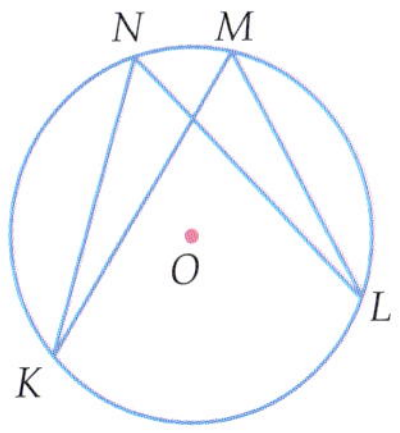

W.E. 16

4 $\angle EOF = 130°$. Find the size of $\angle EGF$, giving reasons for your answer.

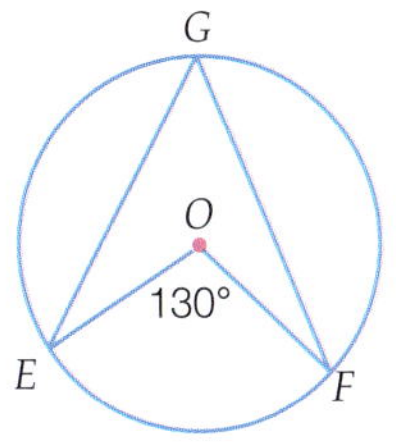

W.E. 17

5 XY is a diameter of the circle with centre O. $\angle ZXY = 35°$. Find the size of $\angle ZYX$, giving reasons for your answer.

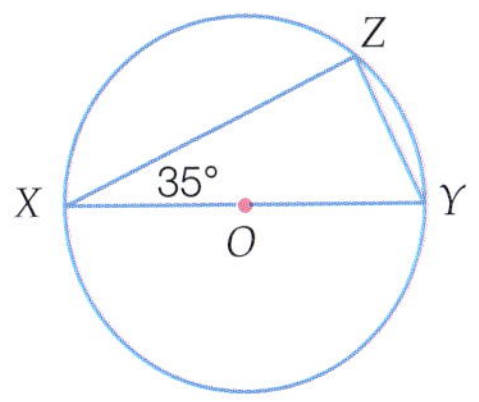

W.E. 18

6 R, S, T and U are points on the circumference of a circle. RT and SU intersect at V. If $\angle URT = 54°$ and $\angle RTS = 46°$, find the size of $\angle TVS$, giving reasons for your answer.

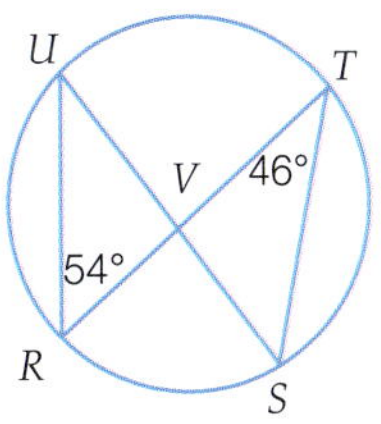

W.E. 19

10A

W.E. 20

7 *LM* is a diameter of the circle with centre *O*. *K* and *N* are points on the circumference of the circle such that $OK \parallel MN$. If $\angle OMN = 50°$, find the size of $\angle OKN$, giving reasons for your answer.

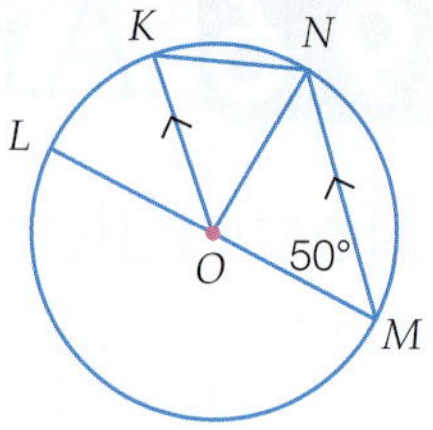

8 Find the value of the variables in each figure. Give reasons for your answer.

(a)

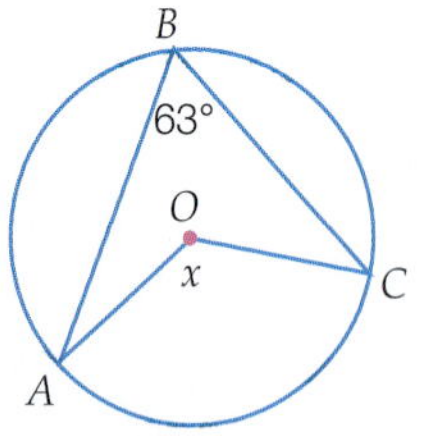

(b)

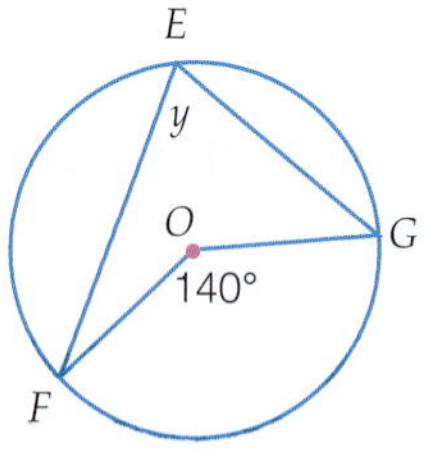

(c)

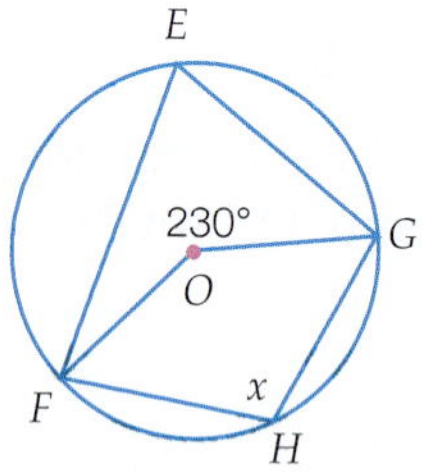

9 Find the value of the variables in each figure. Give reasons for your answer.

(a)

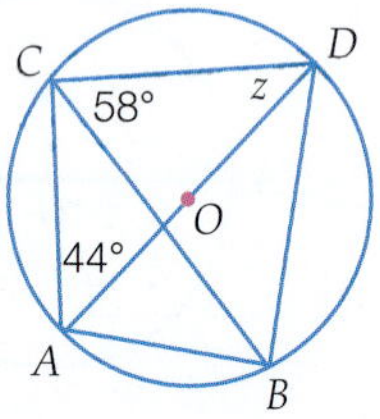

(b)

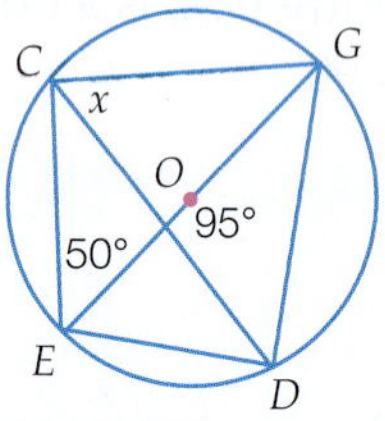

(c)

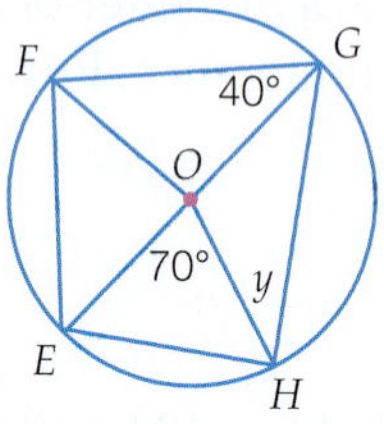

10 Find the value of the variables in each figure. Give reasons for your answer.

(a)

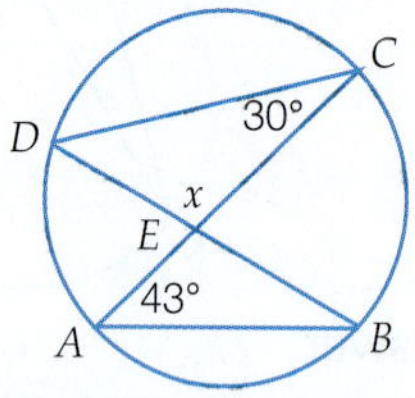

(b)

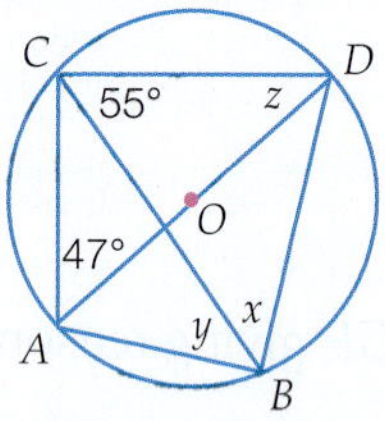

(c)

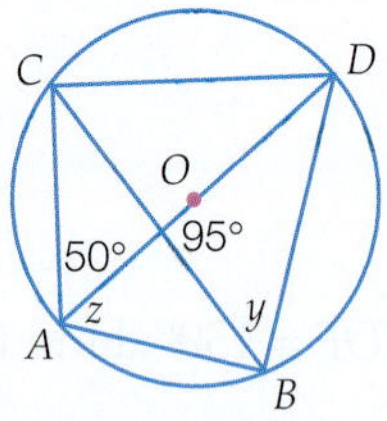

(d)

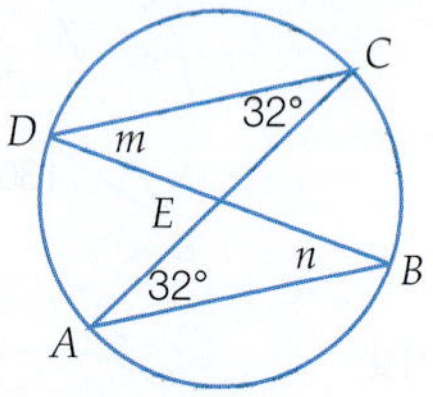

(e)

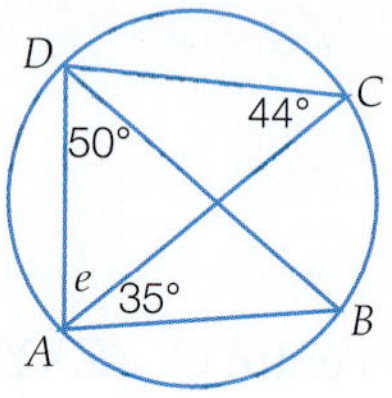

(f)

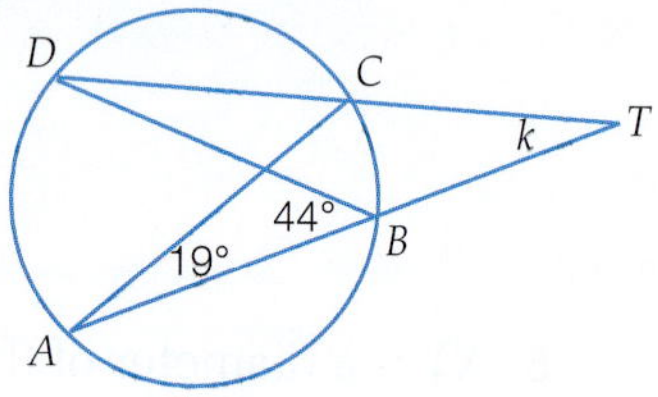

Understanding

11 Find the value of the variables in each figure. Give reasons for your answer.

(a) *EG* is the diameter of the circle

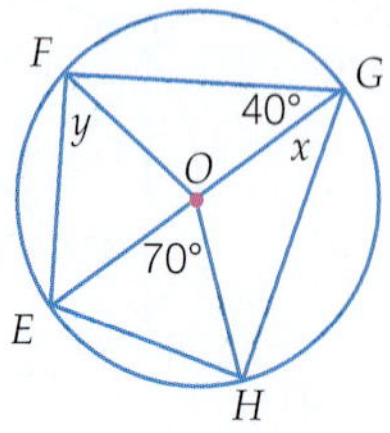

(b) $AC \parallel DB$

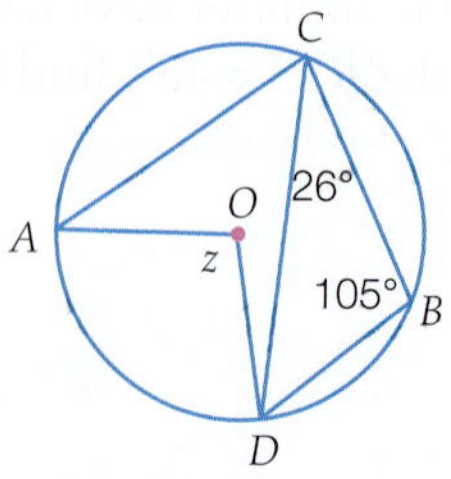

(c) $NM = MQ$, *NP* is the diameter

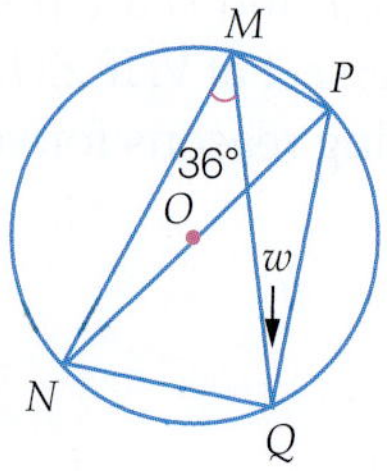

12 **(a)** O is the centre of the circle. Prove that $x + y = 90°$.

(b) If $\angle BOD = 92°$ and $\angle CDB = 62°$, calculate the size of $\angle OCD$ giving reasons for your answer.

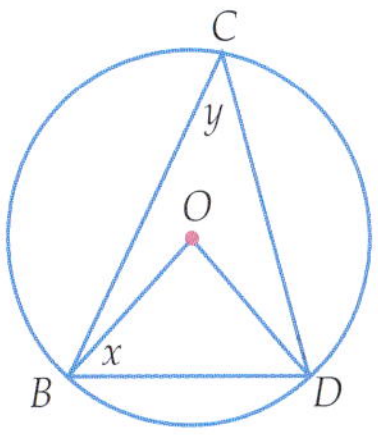

13 AC is the diameter of the circle. If $\angle BDC = 52°$ and $BD = DC$, calculate the size of $\angle ACD$ giving reasons for your answer.

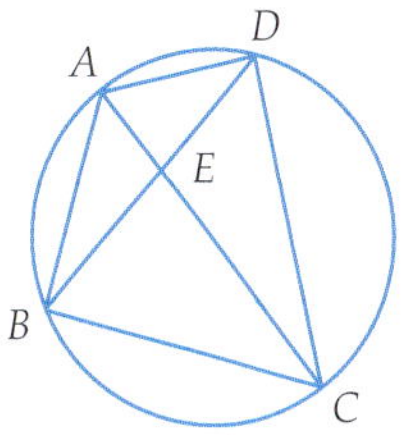

14 If $\angle FGE = 48°$, $\angle GED = 42°$ and $\angle DFE = 20°$, calculate the size of each angle in the quadrilateral $GDEF$, giving reasons for your answers. What can you say about the diagonals of this quadrilateral?

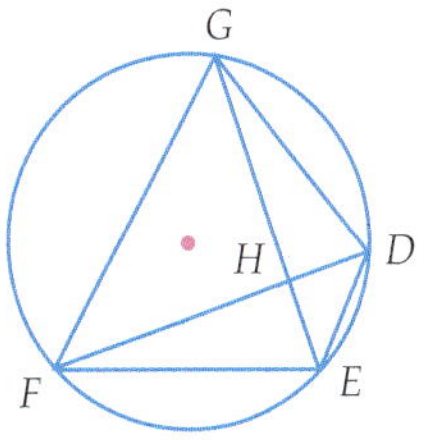

15 Find the value of the variables in each figure. Give reasons for your answer.

(a)

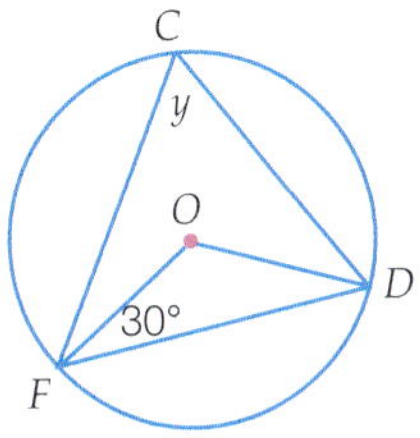

(b)

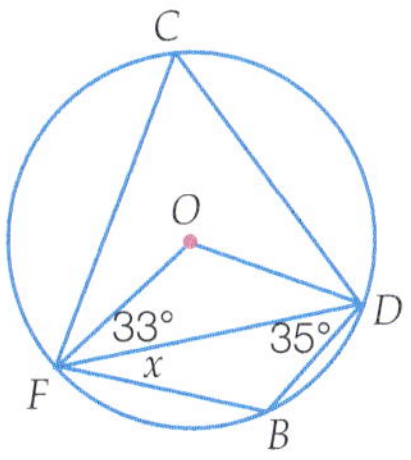

(c)

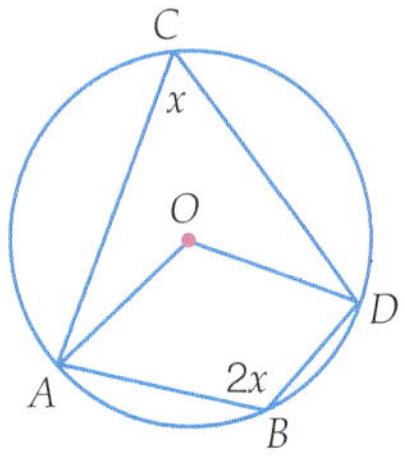

Reasoning

16 $KL = LN$ in the circle with centre O. Prove that $\angle KLM = \angle LON$.

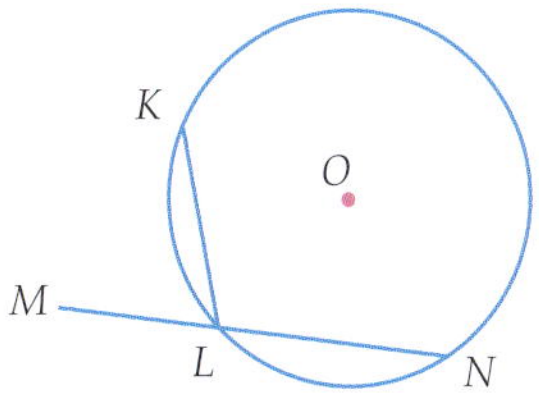

17 **(a)** Prove that $\Delta KPM \sim \Delta QNM$.

(b) If $KM = MQ$ prove that $MP = MN$.

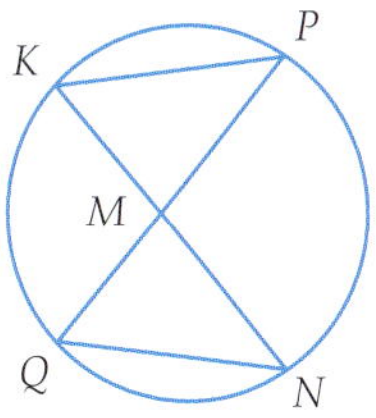

18 CF is a diameter of the circle CFA. FD is a diameter of the circle FDA. Prove that C, A and D are collinear.

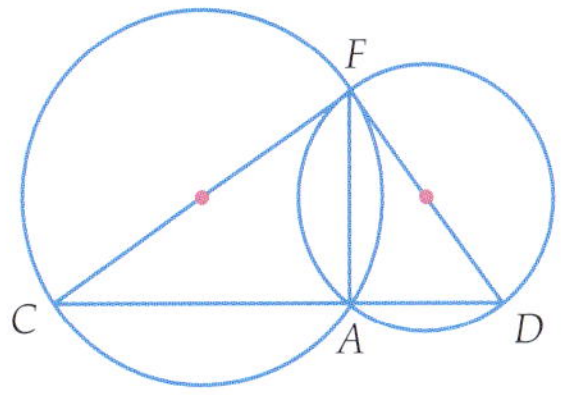

19 **(a)** VT and US intersect at W. If $UW = WT$, prove that $VS \parallel UT$.

(b) From part **(a)** it is known that $VS \parallel UT$ and $SO \perp OT$.
Calculate the size of $\angle TWU$. Give reasons for your answer.

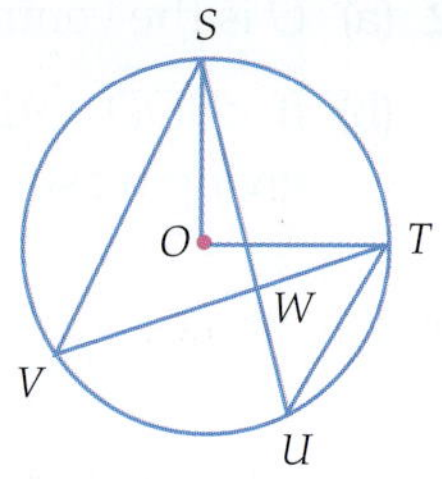

20 If $CA = AD$, prove that AF bisects $\angle EFB$.

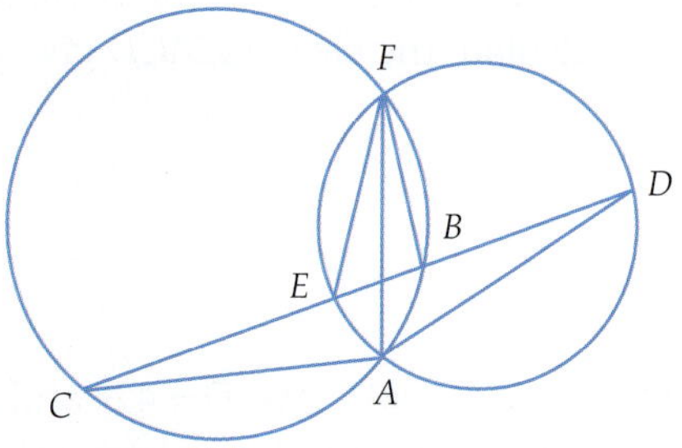

21 O is the centre of the circle KML. KL is a diameter.
OL is a diameter of the circle ONL. Prove that $KM \parallel ON$.

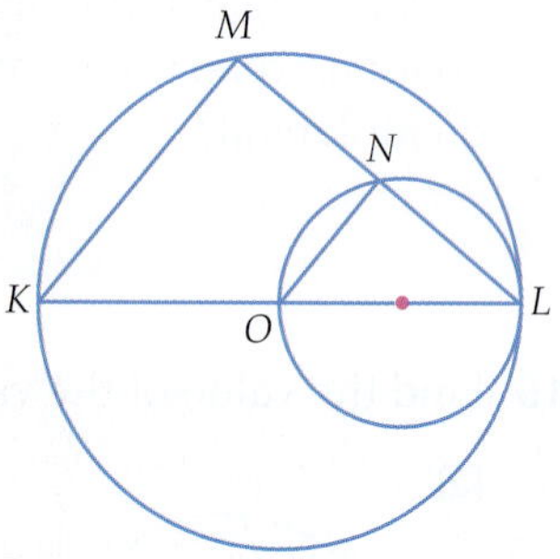

22 A cyclic quadrilateral is a quadrilateral whose vertices are on the circumference of a circle. $ABCD$ is a cyclic quadrilateral in a circle with centre O. By joining CO and OA, prove that $\angle CDA + \angle CBA = 180°$.

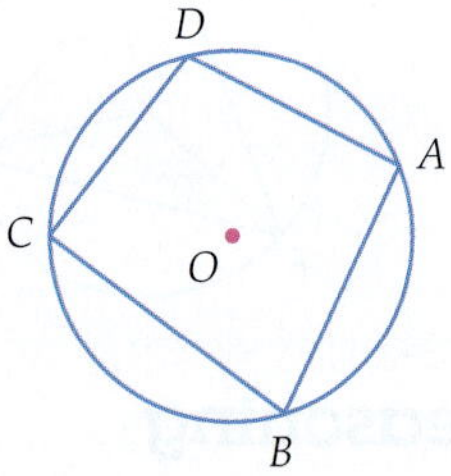

Open-ended

23 $ABCD$ is a cyclic quadrilateral whose diagonals intersect at X. $\angle DBC = 30°$ and $\angle AXD = 80°$.

(a) Find the size of:

(i) $\angle DAC$ **(ii)** $\angle ADB$ **(iii)** $\angle ACB$

(b) If $\angle BAC < 80°$ choose a size for $\angle BAC$ and then find the size of:

(i) $\angle BDC$ **(ii)** $\angle DCA$ **(iii)** $\angle DBA$

(c) If $\angle BAC = 50°$ write a true statement about:

(i) $\angle ADC$ **(ii)** $\angle ABC$

(d) If $\angle BAC = 10°$ write a true statement about:

(i) $\angle ADB$ **(ii)** $\angle BCD$

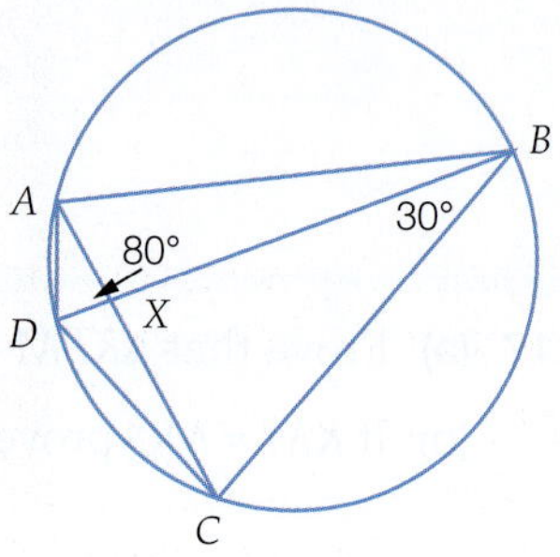

24 KM is a diameter of the circle with centre O. J and L are points on the circumference. Claire claims that $JKLM$ is a square; James claims that $JKLM$ is a kite. Who is correct, or is neither correct? Explain your decision.

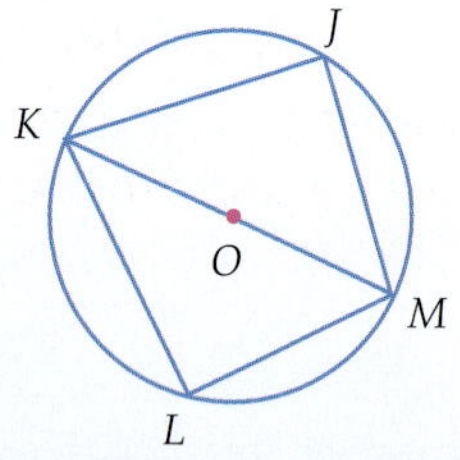

Chords of circles

9.7

The following constructions may be completed using pen and paper or using technology. Ideally, both methods should be used to help understand how the processes are different by hand and by software.

Angles at the centre subtended by chords of equal length

Draw a circle with centre *O* and mark off equal arcs *AB* and *CD*. Join *AB* and *CD* to form chords of the circle.

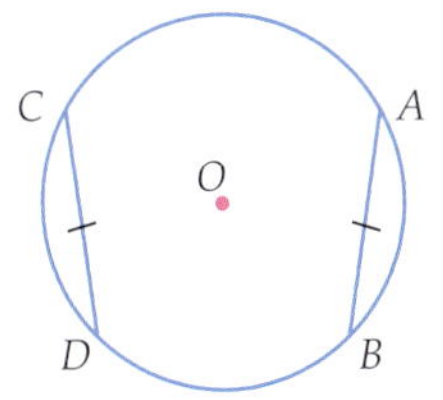

Join *AO*, *BO*, *CO* and *DO*.

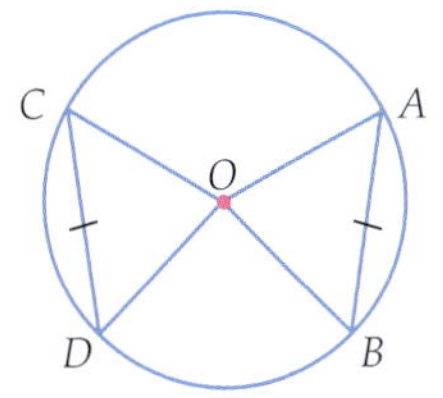

Measure ∠*AOB* and ∠*COD*. *What can you say about these two angles?*

It appears that ∠*AOB* = ∠*COD*.

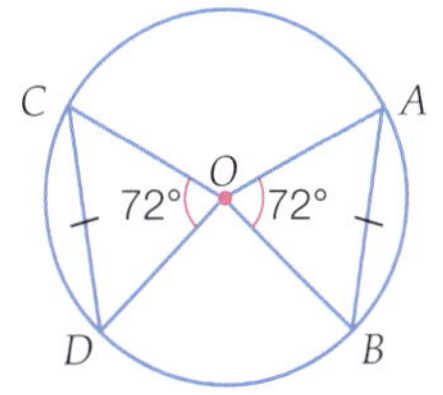

10A

Distance from the centre to chords of equal length

Draw a circle with centre *O* and mark off equal chords *AB* and *CD*.

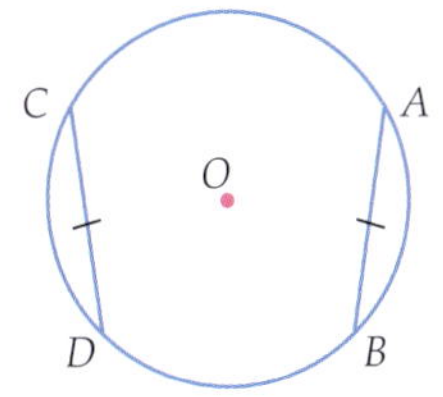

Using compasses, construct the perpendicular from *O* to each chord, cutting *AB* at *E* and *CD* at *F*.

Measure *OE* and *OF*.
It appears that *OE* = *OF*
That is, *E* and *F* are **equidistant** from *O*.

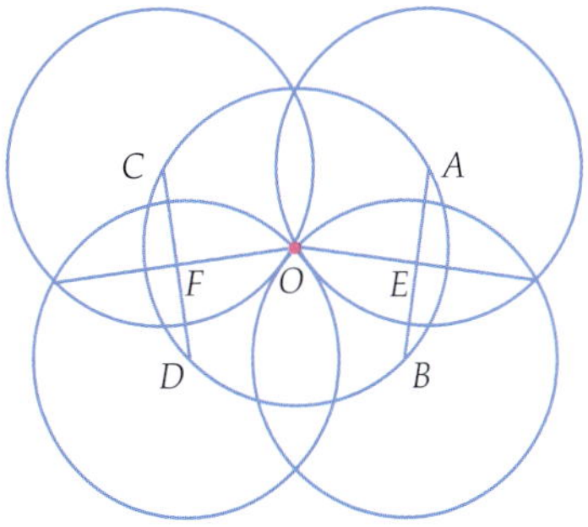

The following theorem has two parts and requires two separate proofs, one involving the angles at the centre of the circle and the other involving the perpendicular distance from the centre of the circle to each chord.

Circle theorem 4

Chords of equal length in a circle subtend equal angles at the centre and are equidistant from the centre.

Worked example 21

W.E. 21

AB and *CD* are equal chords of the circle with centre *O*.

(a) Prove that $\angle AOB = \angle COD$.

(b) *OE* and *OF* are the perpendiculars drawn from *O* to *AB* and *CD*, respectively. Prove that $OE = OF$.

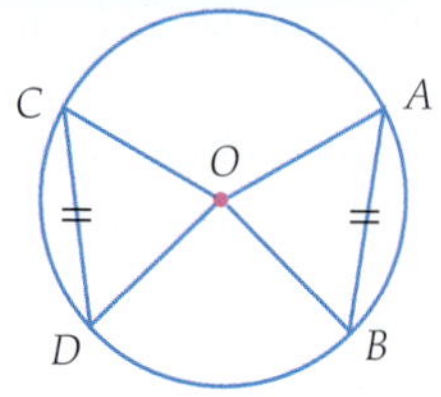

Thinking	Working	
(a) 1 Copy the diagram and mark the equal radii.	(a) 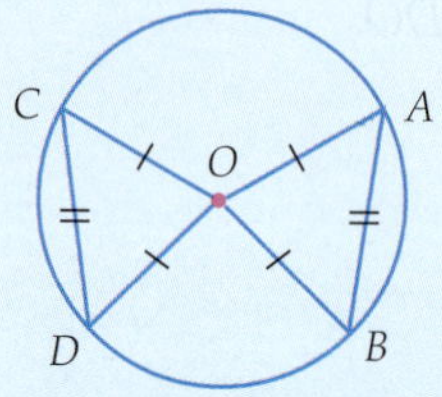	
2 Name the triangles.	ΔOAB and ΔOCD	
3 List three properties that will allow you to use a congruency test.	$OA = OC$ $AB = CD$ $BO = DO$	(radii of circle) (given, equal chords) (radii of circle)
4 Write a pair of congruent triangles and the test used.	$\Delta OAB \equiv \Delta OCD$	(SSS)
5 State your answer to the question, giving a reason.	$\angle AOB = \angle COD$	(corresponding angles of congruent triangles)
(b) 1 Copy the diagram, draw *OE* and *OF*, and mark the equal radii.	(b) 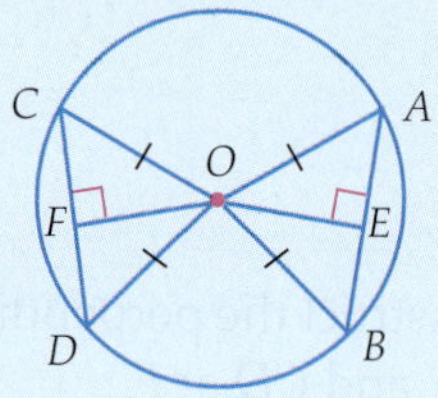	
2 Identify results from **(a)** that you need here.	$\angle BAO = \angle DCO$	(corresponding angles in congruent triangles)
3 Name the triangles.	ΔEAO and ΔFCO	
4 List three properties that will allow you to use a congruency test.	$OA = OC$ $\angle EAO = \angle FCO$ $\angle AEO = \angle CFO$	(radii of circle) (corresponding angles in congruent triangles) (both right angles, $OE \perp AB$ and $FO \perp CD$)
5 Write a pair of congruent triangles and the test used.	$\Delta EAO \equiv \Delta FCO$	(ASA)
6 State your answer to the question, giving a reason.	$OE = OF$	(corresponding sides in congruent triangles)

The perpendicular to a chord

Consider the circle with centre O and a chord AB.

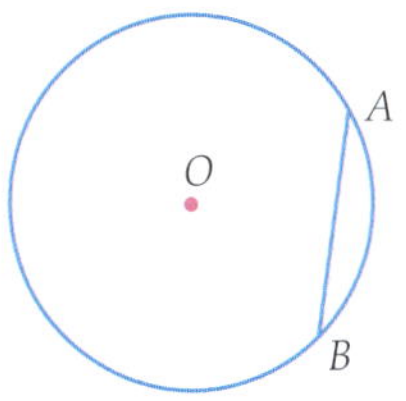

Use compasses to construct the perpendicular from O to the chord, cutting AB at E.

Measure AE and BE. *What can you say about these two intervals?*

It appears that $AE = BE$.

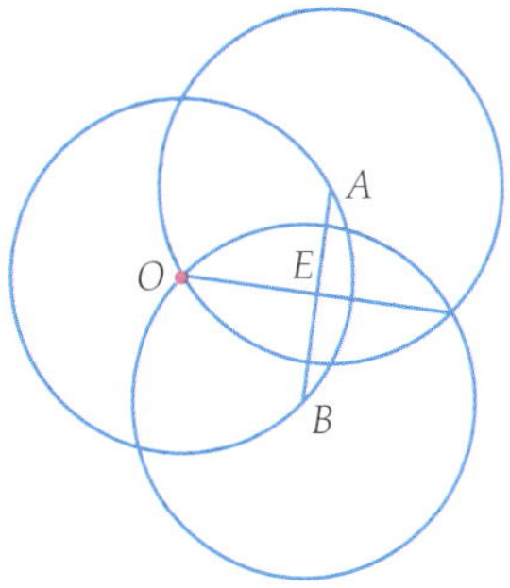

Circle theorem 5

The perpendicular from the centre of a circle to a chord bisects the chord.

Worked example 22

W.E. 22

AB is a chord of a circle with centre O. $OE \perp AB$.

Prove that $AE = EB$.

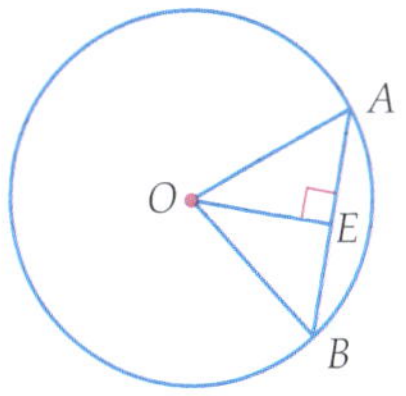

Thinking	Working
1 Copy the diagram and mark the equal radii.	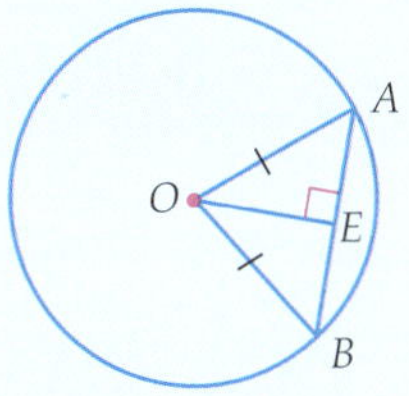
2 Name the triangles.	ΔEAO and ΔEBO
3 List three properties that will allow you to use a congruency test.	$\angle OEA = \angle OEB = 90°$ ($OE \perp AB$) $OA = OB$ (radii of circle) OE is a common side.
4 Write a pair of congruent triangles and the test used.	$\Delta EAO \equiv \Delta EBO$ (RHS)
5 State your answer to the question, giving a reason.	$AE = EB$ (corresponding sides in congruent triangles)

Midpoint of a chord to the centre

Draw a circle with centre O with a chord AB. Construct the midpoint, E, of AB.

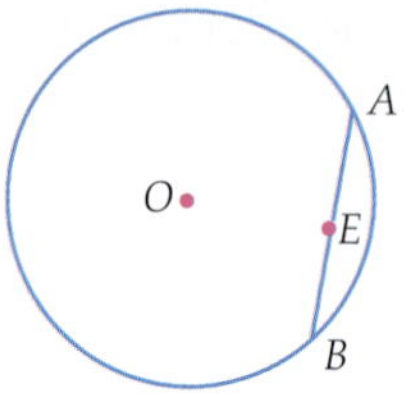

Draw OE.
Measure $\angle AEO$ and $\angle BEO$.
It appears that $\angle AEO = \angle BEO = 90°$.

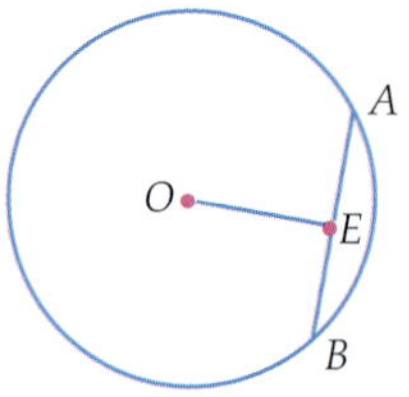

Circle theorem 6

The line from the centre of a circle to the midpoint of a chord is perpendicular to the chord.

Worked example 23

W.E. 23

AB is a chord of a circle with centre O. E is the midpoint of AB.
Prove that $OE \perp AB$.

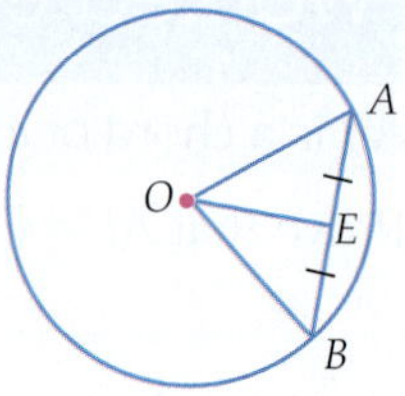

Thinking	Working
1 Copy the diagram and mark the equal radii.	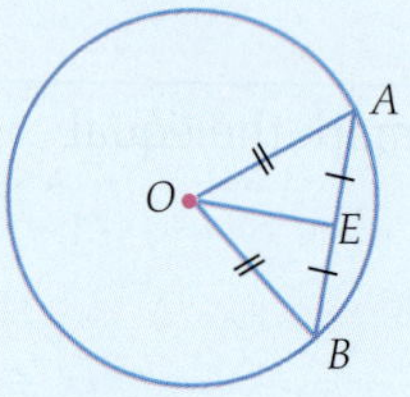
2 Name the triangles.	ΔEAO and ΔEBO
3 List three properties that will allow you to use a congruency test.	$AE = EB$ (E is midpoint of AB) $OA = OB$ (radii of circle) OE is a common side.
4 Write a pair of congruent triangles and the test used.	$\Delta EAO \equiv \Delta EBO$ (SSS)
5 State your answer to the question, giving a reason.	$\angle OEA = \angle OEB$ (corresponding angles of congruent triangles) But $\angle OEA = 180° - \angle OEB$ (straight angle) $\therefore \angle OEA = \angle OEB = 90°$ $\therefore OE \perp AB$

Perpendicular bisector of a chord

Draw a circle with centre O with several chords AB, CD etc., not of equal lengths.

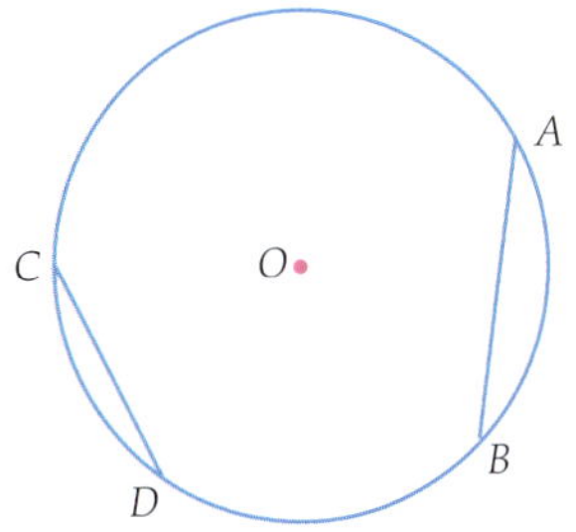

Using compasses, construct a perpendicular bisector of each chord.

What can you say about the perpendicular bisectors?

It appears that they all pass through O.

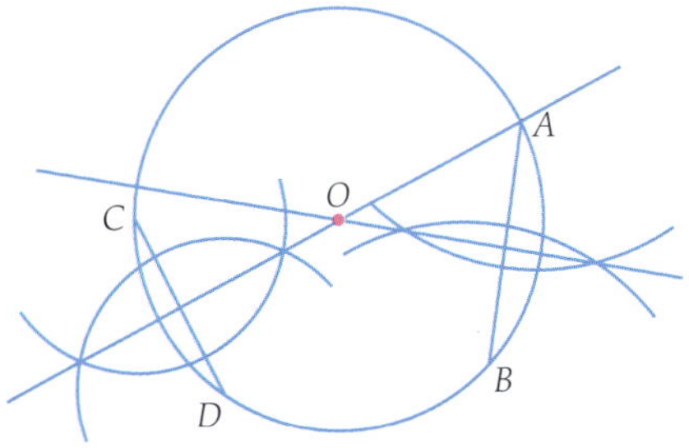

Circle theorem 7

The perpendicular bisector of a chord of a circle passes through the centre of the circle.

Worked example 24

W.E. 24

E is the midpoint of the chord AB in a circle with centre O. DE is the perpendicular bisector of AB. Prove that O is on DE.

When constructed, DE appears to pass through O. To prove this is true you need to use a different sort of proof.

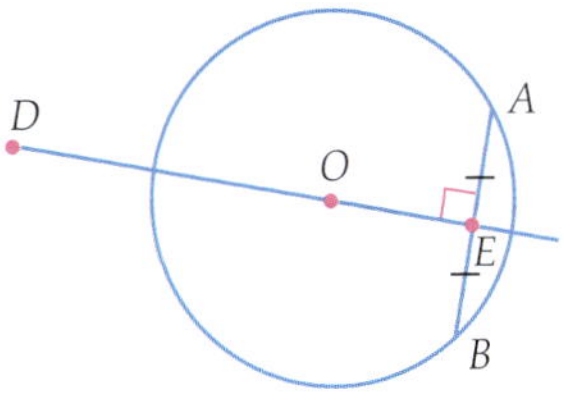

Thinking	Working	
1 Copy the diagram and join DA and DB.	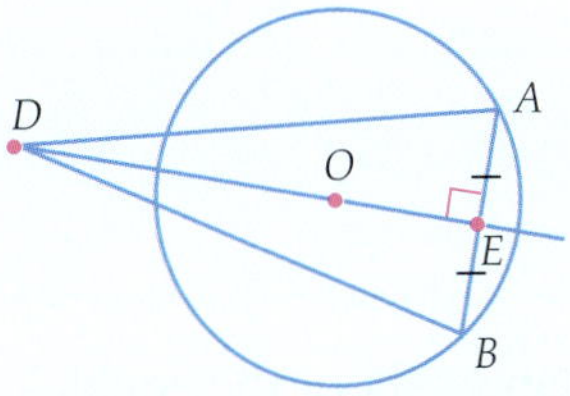	
2 Write what you know about DA and DB.	$DA = DB$	(all points on the perpendicular bisector of the interval AB are equidistant from the end points of the interval)
3 Write what you know about O.	$OA = OB$	(radii of the circle with centre O through A and B)

4 Give reasons for your answer.	O is equidistant from A and B (it must be on the perpendicular bisector of AB). $\therefore O$ is on DE. Hence, the perpendicular bisector of a chord of a circle passes through the centre of the circle.

Results like these can be verified using dynamic geometry software.

For example, you could draw a circle showing its centre. Then draw a chord of the circle. Construct the perpendicular bisector of the chord. Check to see whether the centre of the circle is on the perpendicular bisector of the chord.

Worked example 25

W.E. 25

Calculate the value of the pronumerals. Give reasons for your answers.

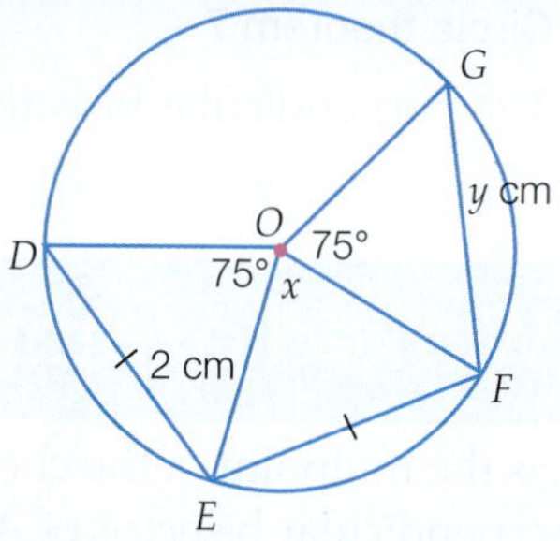

Thinking	Working
1 Copy the diagram.	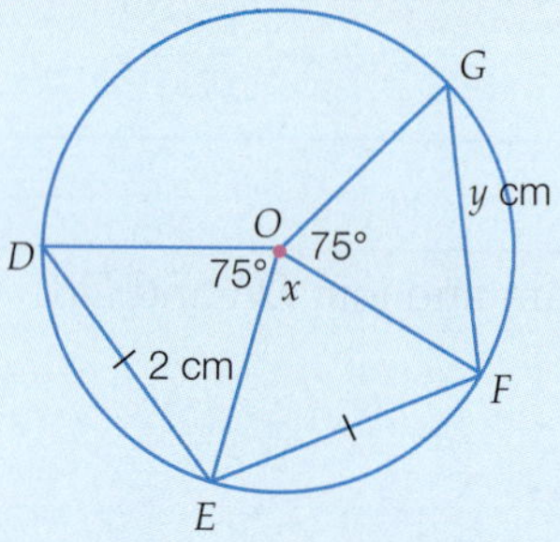
2 Write the information about the equal chords contained in the diagram.	$DE = EF$ (equal chords marked on diagram) $\therefore \angle DOE = \angle EOF$ (equal chords subtend equal angles at the centre of a circle) $\therefore x^\circ = 75^\circ$
3 Write the information about the equal chords contained in the diagram.	$\angle DOE = \angle FOG = 75^\circ$ $\therefore DE = GF$ (equal angles at the centre of a circle are subtended by equal chords) $\therefore y = 2$ cm

Worked example 26

W.E. 26

XY is a chord of a circle with centre *O*.

If $XY = 40$ cm and $OX = 29$ cm, calculate the perpendicular distance of *XY* from *O*. Give reasons for your answer.

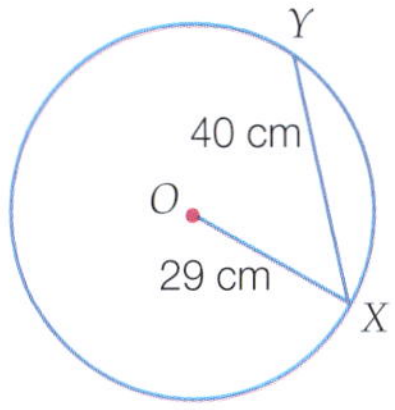

Thinking	Working
1 Copy the diagram. Let *Z* be the foot of the perpendicular from *O* to *XY*. Draw *OZ* perpendicular to *XY*.	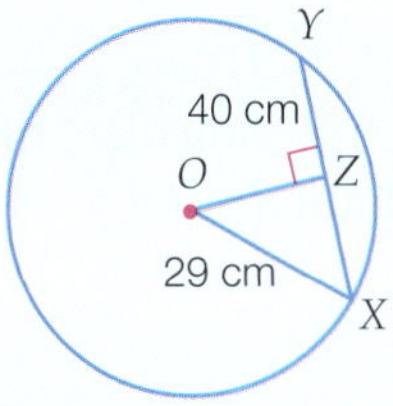
2 Identify any equal lengths.	$XZ = ZY$ (perpendicular from the centre of a circle to a chord bisects the chord) $\therefore XZ = 20$ cm (half of *XY*)
3 Calculate the length of *OZ*.	$OX^2 = XZ^2 + OZ^2$ (Pythagoras' theorem) Hence, $OZ^2 = OX^2 - XZ^2$ $\therefore OZ^2 = 29^2 - 20^2$ $= 841 - 400$ $= 441$ $\therefore OZ = 21$ cm
4 Answer the question.	*XY* is 21 cm from *O*.

10A

9.7 Chords of circles

Navigator

Answers p. 840

1, 2, 3, 4, 5, 6, 7, 8, 9, 10, 13, 14, 16, 18	1, 2, 3, 4, 5, 6, 7, 8, 9, 10, 11, 13, 14, 16, 17, 18	3, 4, 5, 6, 7 (a, b), 8 (a, b), 9 (a, b), 10 (a, b), 11, 12, 13, 14, 15, 16, 17, 18

Fluency

W.E. 21

1 WX and YZ are equal chords of a circle with centre O.

(a) Prove that $\angle WOX = \angle YOZ$.

(b) OP and OQ are the perpendiculars drawn from O to WX and YZ, respectively. Prove that $OP = OQ$.

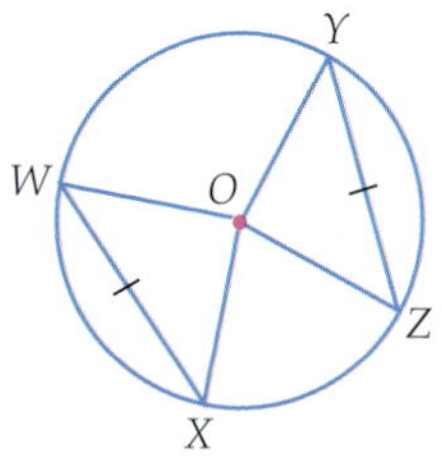

W.E. 22

2 EF is a chord of a circle with centre O. $OG \perp EF$. Prove that $EG = GF$.

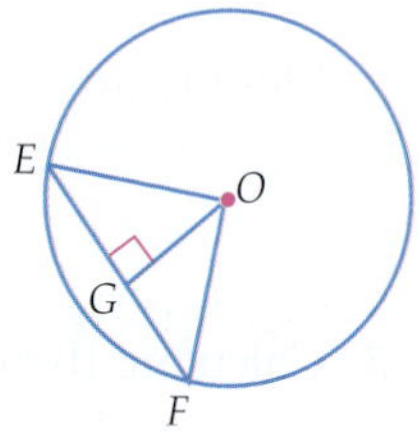

W.E. 23

3 KL is a chord of a circle with centre O. M is the midpoint of KL. Prove that $OM \perp KL$.

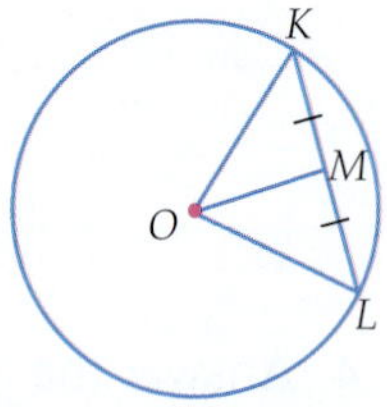

W.E. 24

4 Z is the midpoint of the chord XY in a circle with centre O. WZ is the perpendicular bisector of XY. Prove that O is on WZ.

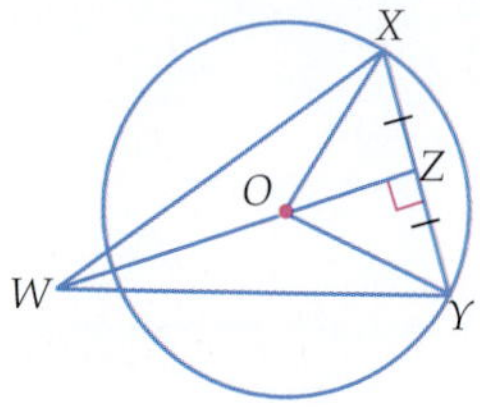

W.E. 25

5 Calculate the value of the pronumerals. Give reasons for your answers.

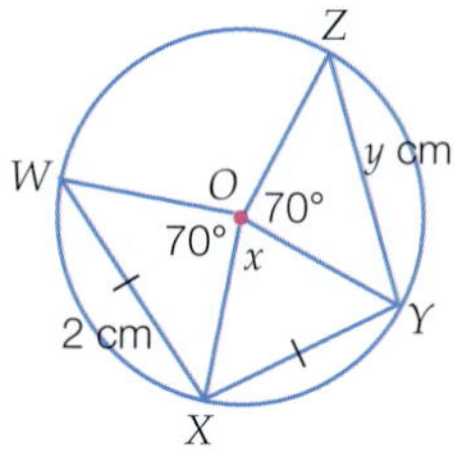

W.E. 26

6 AB is a chord of a circle with centre O. If $AB = 30$ cm and $OA = 34$ cm, calculate the perpendicular distance of AB from O. Give reasons for your answer.

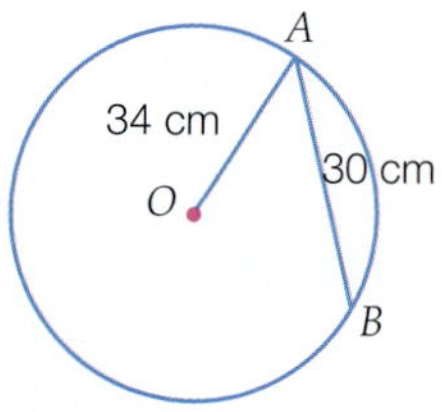

Understanding

7 Calculate the length of AB in each diagram, giving reasons for your answer.

(a) $AC = 10$ cm, $OB \perp AC$

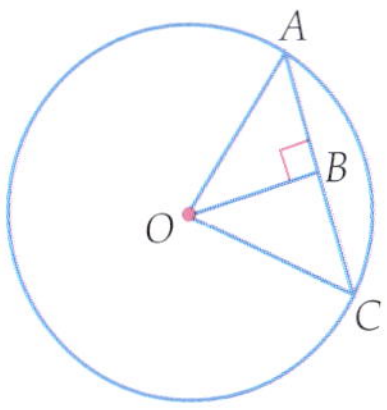

(b) $AC = 10$ cm, $OC \perp AB$

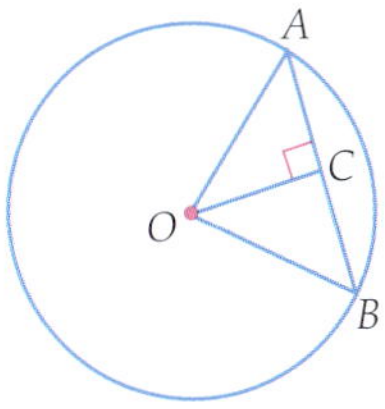

(c) $OA = 17$ cm, $OB = 8$ cm

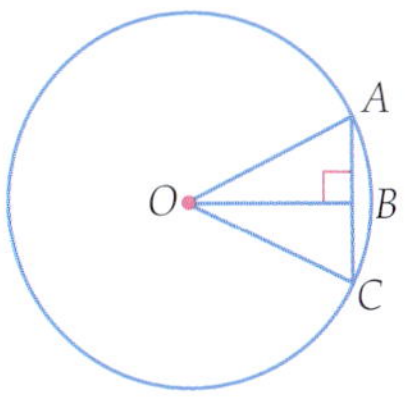

8 Calculate the size of $\angle AOB$ in each figure, giving reasons for your answer.

(a) $AB = GF$, $\angle GOF = 70°$

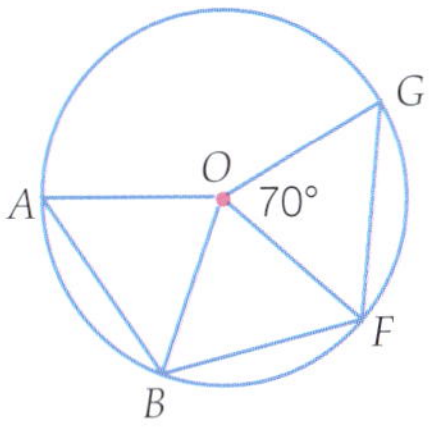

(b) $\angle OAB = 50°$

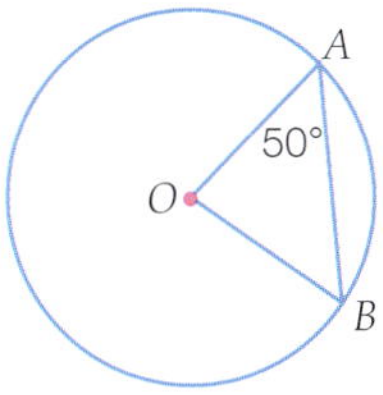

(c) $AB = GF$, $\angle OGF = 50°$

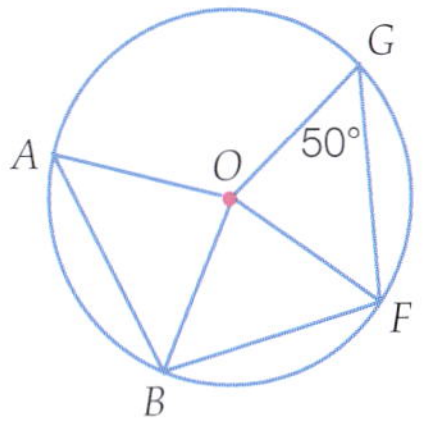

9 Calculate the length of OA in each figure, giving reasons for your answer.

(a) $AB = 48$ cm, $OE = 7$ cm, E is the midpoint of AB.

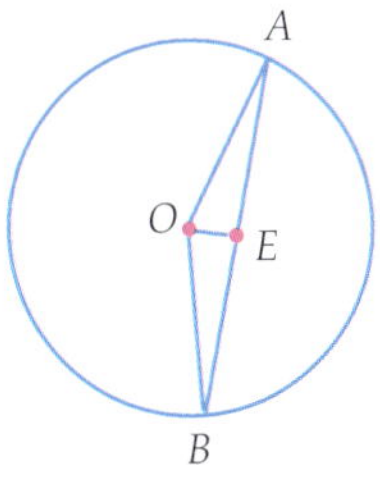

(b) $GH = HF = 21$ cm, $OH = 20$ cm, $AB = GF$

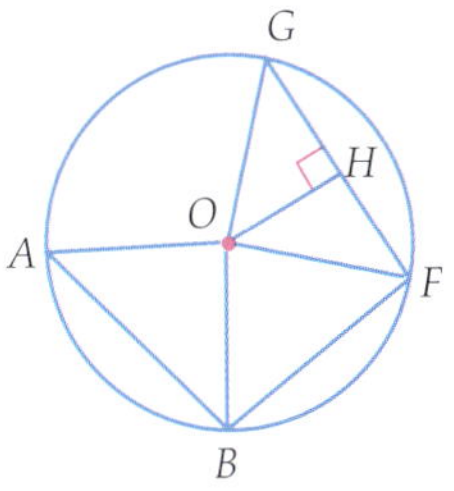

(c) $\angle OGF = 60°$, $AB = GF = 10$ cm

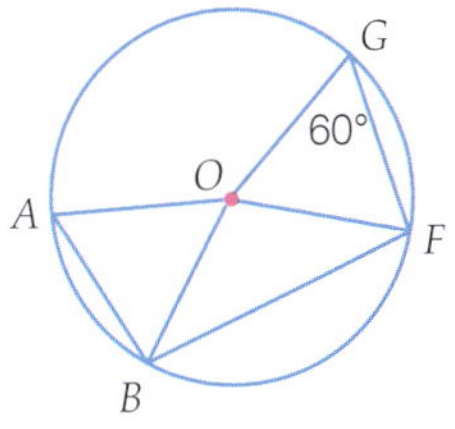

10 Calculate the length of CD in each diagram, giving reasons for your answer.

(a) $OE \perp CD$, $OC = 53$ cm, $OE = 28$ cm

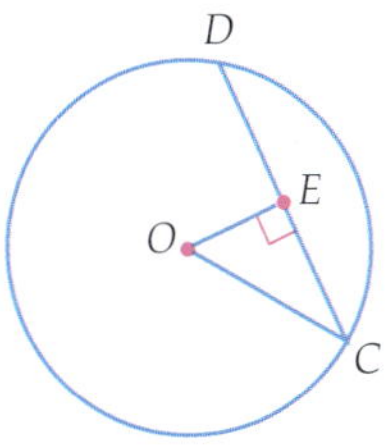

(b) $\angle FHO = 90°$, $\angle FOD = \angle DOC$, $GF = FD$, $OF = 73$ cm, $OH = 48$ cm

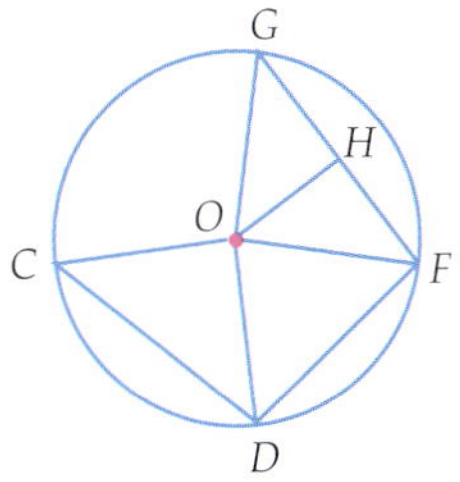

(c) $\angle GOF = \angle COD$, $GH = HF$, $OG = 85$ cm $OH = 77$ cm

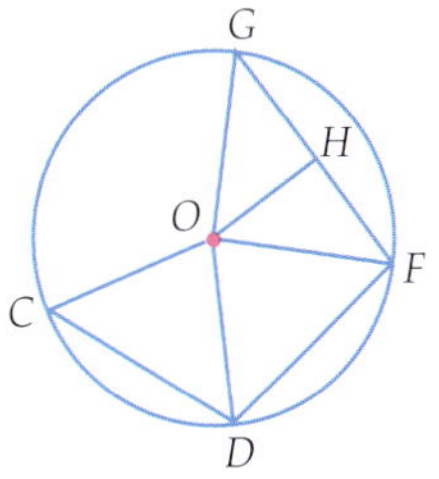

Reasoning

11 BD is the diameter of a circle with centre O. C is a point on the circumference of the circle. E is the midpoint of CD. Prove that $OE \parallel BC$, and hence that $\angle BCD = 90°$.

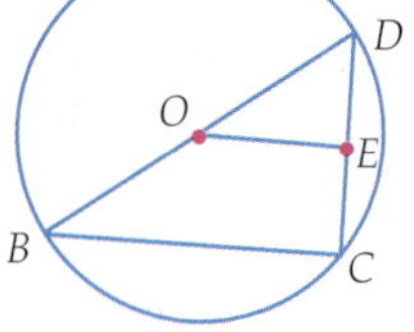

12 O is the centre of the circle ACB. $AF = FC$ and $\angle ACB = 90°$. Mark D, the midpoint of BC, and use it to prove that $BC = 2OF$.

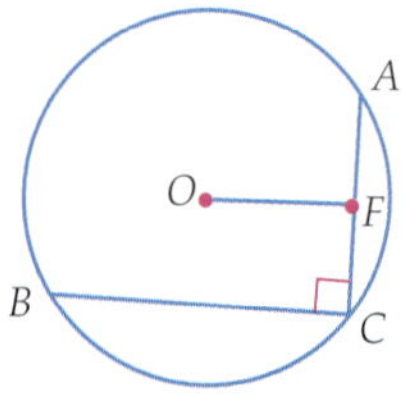

13 The circles GFH, with centre O, and FEH, with centre Q, intersect at F and H. GFE is a straight line. K is the midpoint of GF and J is the midpoint of FE. Prove that $\angle KOQ + \angle OQJ = 180°$.

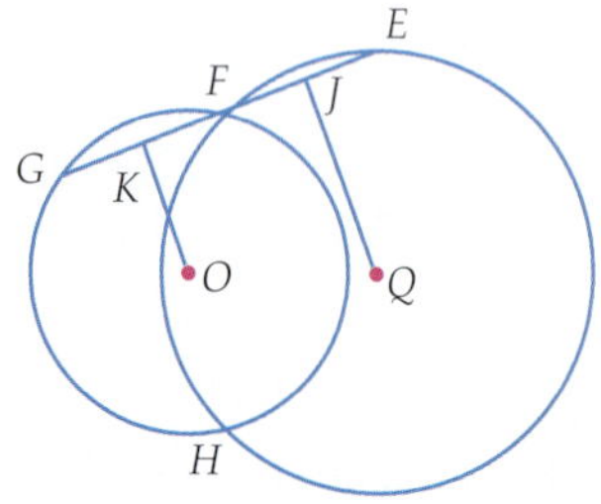

10A

14 The entrance to a shopping centre is a rectangular doorway with a circular arch above it as shown in the diagram. The doorway AB is 4 m wide and the radius of the arch is 2.5 m. Calculate the height of the arch, CH. Justify your calculations.

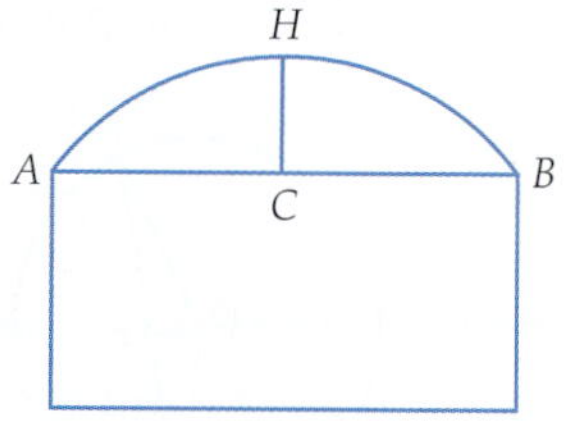

15 KLM is an isosceles triangle inscribed in a circle of radius 25 mm. If $KL = KM$ and $LM = 48$ mm, calculate the length of KL. Give reasons to justify your calculation. (A triangle inscribed in a circle has its vertices on the circumference of the circle.)

Open-ended

16 The doorways to a shopping centre are being designed. The shape is to be similar to that in the diagram for Question **14**. The rectangular doorways are to be between 4 m and 6 m wide and at least 2.5 m high. Each doorway is to have a circular arch on top of the rectangular door opening such that the total height of the door and the arch is between 3 and 4.5 m. Create two designs, giving the width and height of the rectangular doorway, the radius of the arch and the height of the arch above the centre of the doorway.

17 Construct the perpendicular bisectors of the sides KL and LM of a scalene triangle KLM. Explain why their point of intersection, X, is the centre of the circle through L, K and M.

18 Circles with centres E and F intersect at G and H. Draw this information on a diagram. Use your diagram to explain why **(i)** EF bisects GH and **(ii)** $CE \perp GH$.

Investigation

Points on polygons and circles

Equipment required: a \$1 coin and a 50c coin, an equilateral triangle, an isosceles triangle, a scalene triangle, a rectangle and a square, all cut out of cardboard or plastic

When shapes rotate, the path followed by a point on a vertex follows a path called a *locus*.

The Big Question

Can you predict the path taken by a point on a polygon or a circle as the shape rotates along a straight line?

Engage

1 Mark a dot on one vertex of the equilateral triangle. Draw a set of axes and place the equilateral triangle with its base on the x-axis and the vertex with the dot at the origin as shown.

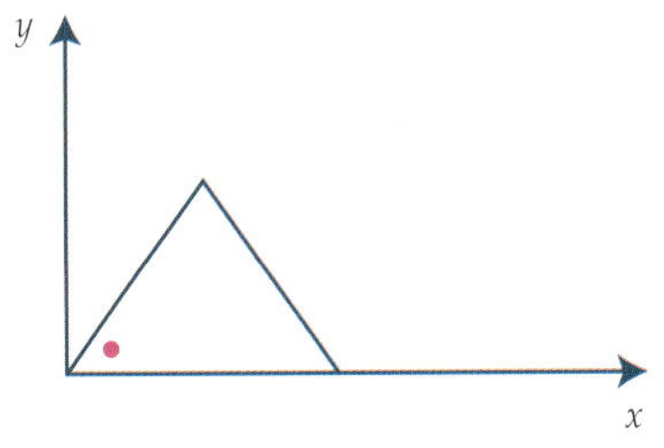

2 Rotate the triangle in a clockwise direction around the vertex on the bottom right-hand side.

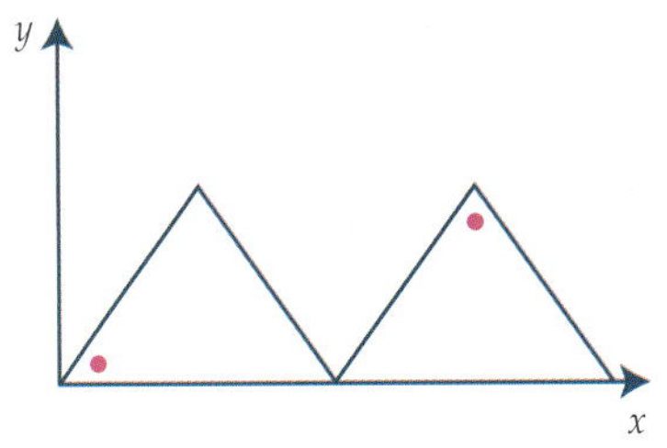

3 Rotate the triangle again in a clockwise direction around the vertex on the bottom right-hand side.

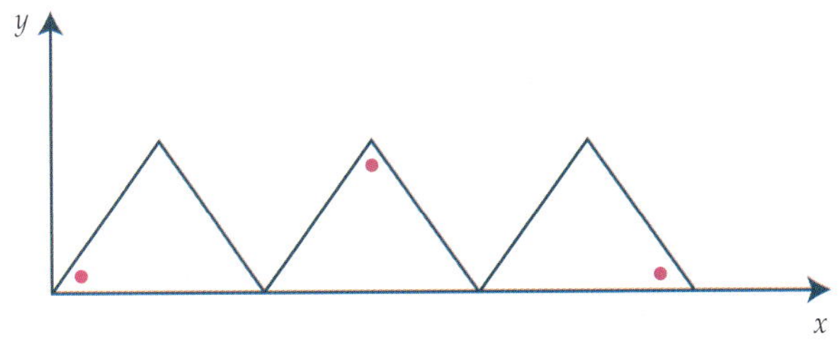

4 What path will the vertex with the dot take as the triangle rotates along the x-axis? Can you predict it? Using your triangles, draw the path that the vertex with the dot has taken as the triangle rotated.

5 Now rotate an isosceles triangle starting with the vertex with the dot at the origin and the non-equal side of the triangle along the x-axis. Again, try to predict the path first, then draw the path taken by the vertex with the dot as it is rotated along the x-axis.

6 Repeat this process for a scalene triangle. Always start with the vertex with the dot at the origin and one side of the triangle along the x-axis.

Explore

7 Now rotate a square in the same manner and draw the path taken by one corner as the shape rotates.

8 The shape of a 50c coin is a dodecahedron (12-sided polygon). Predict the path it will produce when rotated in the same manner as described above.

9 Rotate a 50c coin and draw the path.

Explain

10 Describe the path (or locus) for each of the shapes you have used so far, noting similarities and differences, and using mathematical terms such as radius and diagonal. You may find it useful to label the sides and the other angles to help you with your description. Use features of the shape to find the maximum height reached for each shape.

11 What is the locus of a point at the centre of **(i)** an equilateral triangle and **(ii)** a square?

Strategy options

- Draw a diagram.
- Make a model.
- Look for a pattern.

Elaborate

12 Rotate a $1 coin and draw the path taken by a point on the circumference. Describe this path.

13 Rotate a rectangle and draw the path taken by a point on a vertex. Describe this path.

14 What would be the locus of a point that is always equidistant from two fixed points?

Evaluate

15 Did you have any difficulty drawing the paths for any of the shapes? What could you do in future to make it easier?

16 Where do you think an understanding of paths like these would be important?

17 Did working with models make it easier for you to predict the correct path in later questions?

Extend

Choose one of the following.

18 Rotate a $1 coin around the circumference of a circle with twice the diameter of the coin and find the locus of a point on the vertex of these shapes.

19 If a goat were tethered to the midpoint of a wall d metres long with a length of rope r metres long, investigate the path the goat could take if it walked at its maximum length of rope from its tether point:

(a) if $r < \frac{d}{2}$ **(b)** if $r = \frac{d}{2}$ **(c)** if $r > \frac{d}{2}$

20 Investigate the paths formed by *curves of pursuit*.

Imagine a dog on a footpath chasing a car. As the dog heads towards the car, the car moves forward so the dog now has to change direction. As long as the car is travelling faster than the speed of the dog, the dog will not catch the car.

Exploration CAS

Equipment required: TI-Nspire CAS or Casio ClassPad CAS

Archimedes and the calculation of π

Archimedes was born around 290–280 BCE in Syracuse, Sicily, and became the best known and most influential of the ancient Greek mathematicians, as well as a great inventor.

In this investigation, you will find approximations to π using a method similar to that invented by Archimedes. Had he had access to computer technology, his investigations would have been much easier!

1 Use the formula $A = \pi r^2$ to find the area of a circle with a radius of one unit.
Note: A circle with $r = 1$ is called a unit circle.

2 To find the area of a unit circle by the method of Archimedes, start by inscribing (drawing) a square inside the unit circle. The area of the square becomes the first approximation to the area of the unit circle. A square is a regular polygon with $x = 4$ sides. If the square is changed to a regular pentagon ($x = 5$), the area of the pentagon becomes a better approximation to the area of the unit circle. As the number of sides of the polygon is increased, the area of the polygon gets closer and closer to the area of the unit circle.

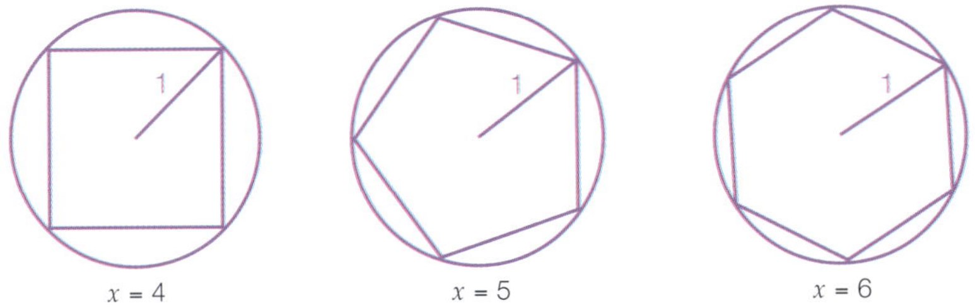

3 By finding the areas of the regular polygons, Archimedes was able to calculate an accurate approximation for π. To find the area of an inscribed polygon, divide the polygon into equal triangles with a common vertex at the centre. To find the area of one of these triangles:

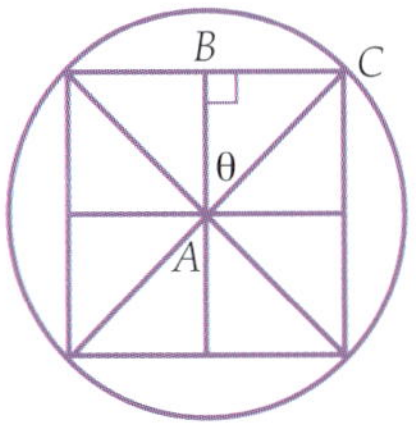

Let side BC represent the base of ΔABC and let AB represent the perpendicular height of ΔABC.

Remember, $AC = 1$ as it is a radius of the unit circle.

$$\sin(\theta) = \frac{BC}{AC} = \frac{BC}{1} = BC \quad \therefore \text{The base } BC = \sin(\theta)$$

$$\cos(\theta) = \frac{AB}{AC} = \frac{AB}{1} = AB \quad \therefore \text{The perpendicular height } AB = \cos(\theta)$$

The area of the triangle is given by:

$$\text{Area} = \tfrac{1}{2}\text{ base} \times \text{height} = \tfrac{1}{2}\sin(\theta)\cos(\theta)$$

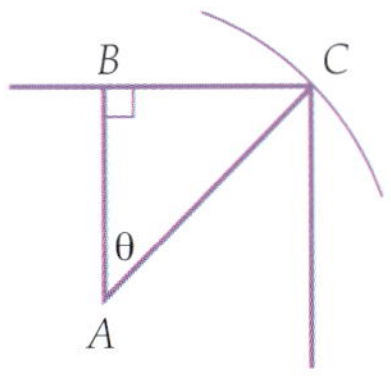

But $\theta = \dfrac{360°}{8} = \dfrac{180°}{4}$ $\quad \therefore \text{Area} = \tfrac{1}{2}\sin\left(\dfrac{180°}{4}\right)\cos\left(\dfrac{180°}{4}\right)$

As the square contains eight triangles, its area is given by:

$$\text{Area} = 8 \times \tfrac{1}{2}\sin\left(\frac{180°}{4}\right)\cos\left(\frac{180°}{4}\right) = 4\sin\left(\frac{180°}{4}\right)\cos\left(\frac{180°}{4}\right)$$

This gives Area = 2, which does not seem to be a very accurate estimate.

4 Show that the area of the inscribed pentagon ($x = 5$) is given by:

$$\text{Area} = 5 \sin\left(\frac{180°}{5}\right) \cos\left(\frac{180°}{5}\right)$$

This gives Area ≈ 2.378, a slightly better estimate than that given by the square.

5 Verify that the area of an inscribed polygon with x sides is given by:

$$\text{Area} = x \sin\left(\frac{180°}{x}\right) \cos\left(\frac{180°}{x}\right)$$

for a value of x of your own choice.

6 Use a CAS to generate a table of values that shows the number of sides of each polygon (**x**) and the area of that polygon (**f1(x)** or **y1**).

Enter the equation from **5** into your CAS. Set up your CAS to start the table with x-values at 0 and to step x up by 100 at each step.

Using TI-Nspire CAS

First make sure you are in degrees mode. Select **doc > Settings & Status > Document Settings...** and set **Angle** to **Degree**.

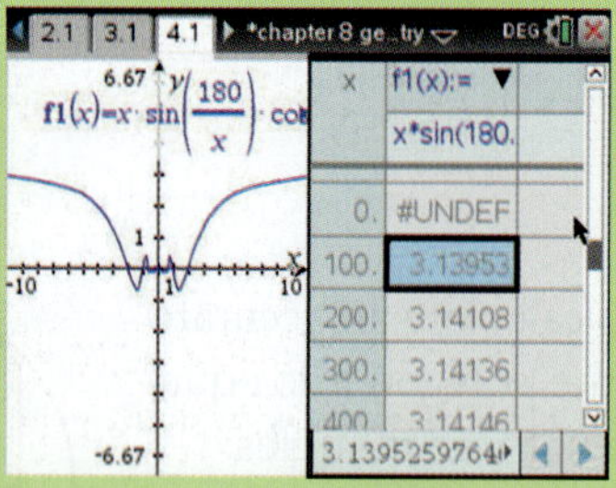

Using Casio ClassPad CAS

First make sure you are in degrees mode, with **Deg** shown at the bottom of the screen. If this shows **Rad** or **Gra** instead, then select the word to change it to **Deg**.

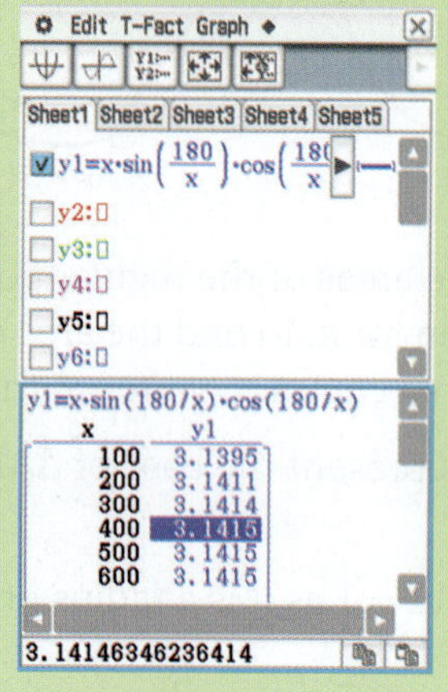

By altering the amount that the table steps each time (including using negative step values), find the lowest number of sides on a polygon required to find π correct to 4 decimal places.

Research

Archimedes established that the value of π must be between the lower and upper limits of $\frac{223}{71}$ and $\frac{22}{7}$. Do further research on Archimedes' method for determining π. Write a brief report and describe how Archimedes used polygons that circumscribed the unit circle to find an upper bound to π.

You can find the area of an x-sided circumscribed polygon using:

$$\text{Area} = x \tan\left(\frac{180°}{x}\right)$$

Derive this formula for the area. Use this formula to find an upper bound for π, correct to 2 and 4 decimal places.

Chapter review 9

Maths literacy

bisect	included angle	perpendicular bisector	similar
congruent	parallel	proof	theorem
definition	perpendicular	prove	
10A angle at the centre	arc	major segment	segment
angle at the circumference	chord	minor arc	subtend
angle in a semicircle	equidistant	minor segment	subtended
angles in the same segment	major arc	sector of a circle	

Copy and complete the following using the words and phrases from the list, where appropriate. A word or phrase may be used more than once.

1 Triangles are __________ if they are identical in all aspects; i.e. they have exactly the same shape and size.

2 When applying SAS to test for congruence it is important that the angle is the __________.

3 Triangles are __________ if they have exactly the same shape, although their sizes may be different.

4 The minimum amount of information needed to identify a particular figure is called a __________.

5 A __________ is true about a particular figure but is not enough to define that figure.

6 A line at right angles to another line that also cuts that line in half is called a __________.

7 If you divide a line or angle into two equal parts, you __________ the line or angle.

10A

8 If two radii are drawn to different points on a circle, they divide the circumference of the circle into a __________ and a __________.

9 A line joining one point on the circumference of the circle to another is called a __________.

10 An arc will __________ an angle at the circumference.

11 A chord, that is not a diameter, drawn on a circle divides the circle into a __________ and a __________.

Fluency

1 Give reasons why the following pairs of triangles are congruent. **9.1**

(a)

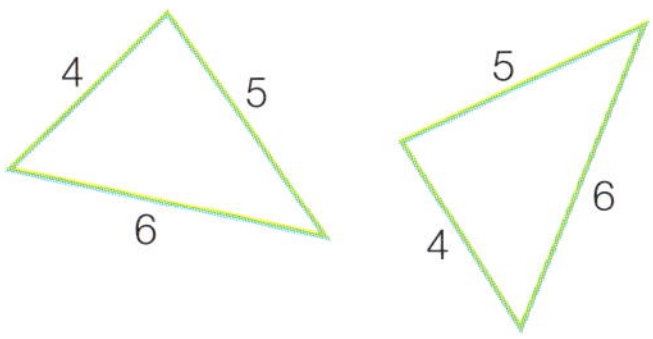

(b)

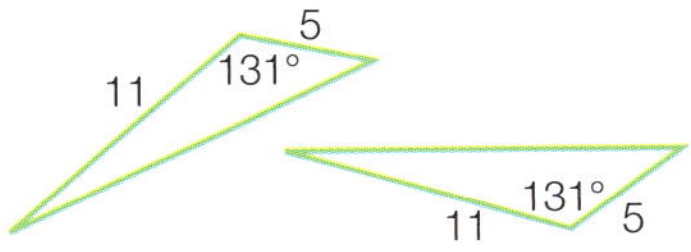

2 The triangle congruent to the triangle shown is:

9.1

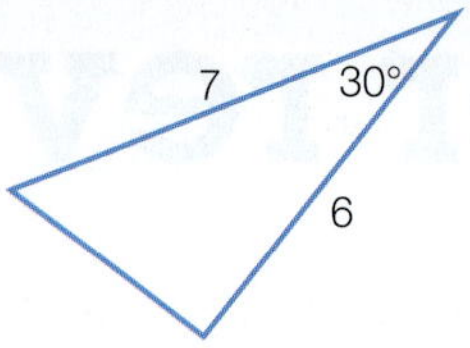

A

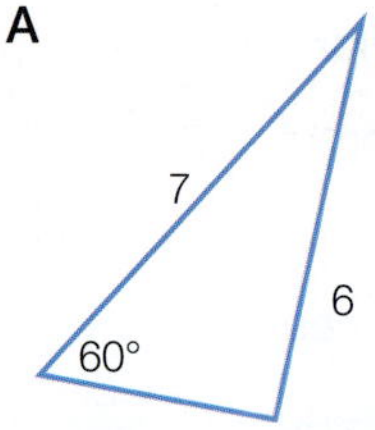

B

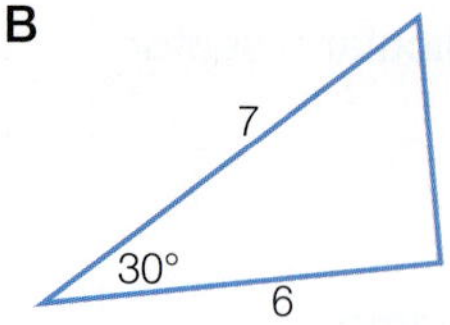

C

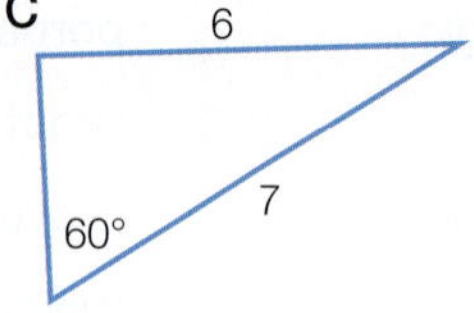

D

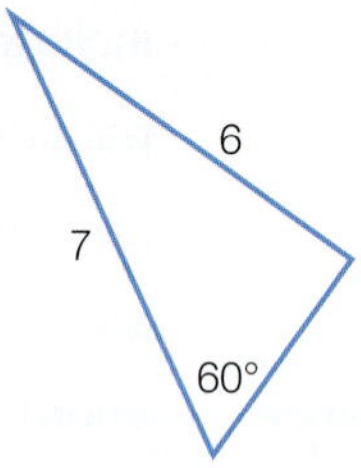

3 Find the value of x in each of the following pairs of triangles.

9.1

(a)

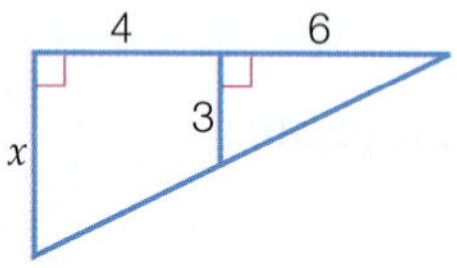

(b)

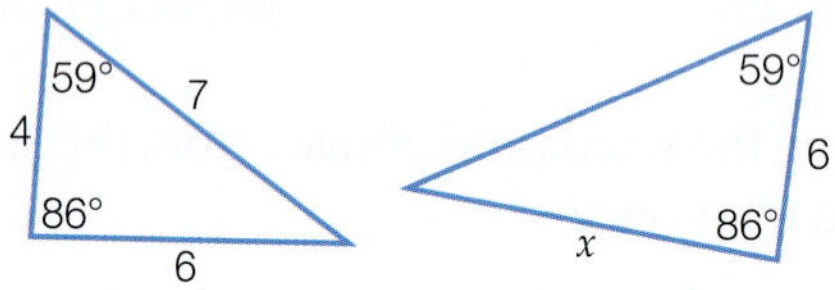

4 Prove that the following pairs of triangles are congruent and hence find the value of the pronumeral.

9.2

(a)

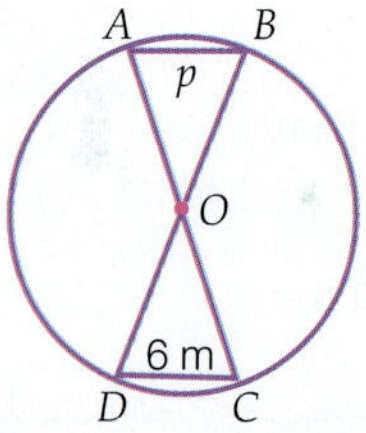

(b)

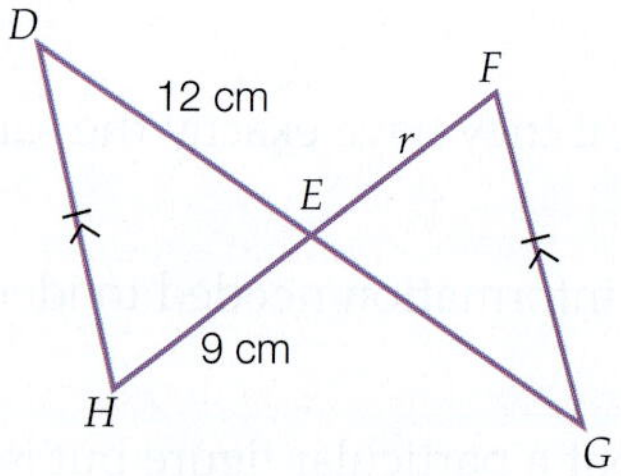

5 Prove that the following pairs of triangles, ΔABC and ΔADE, are similar.

9.2

(a)

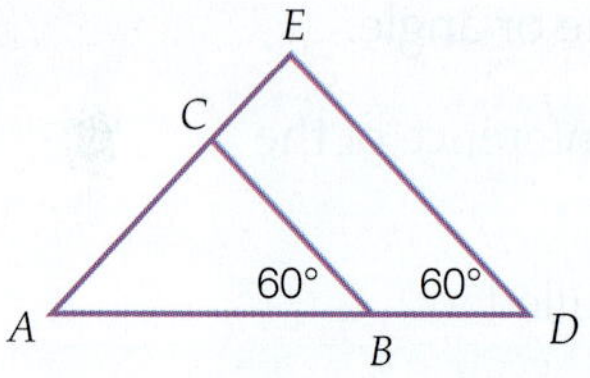

(b)

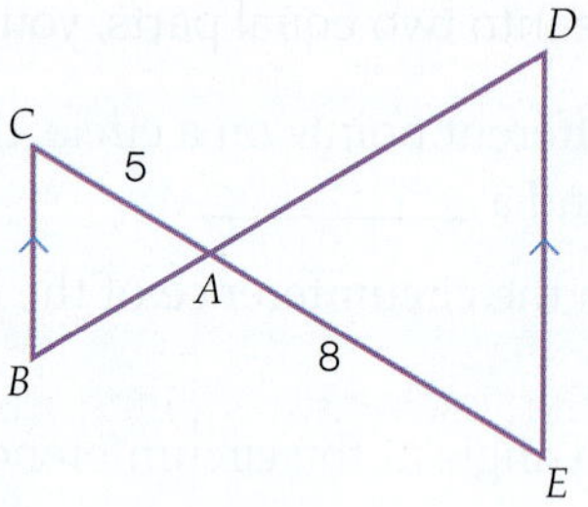

6 Given that CX bisects $\angle ACB$, and $CX \perp AB$, prove that $AX = XB$.

9.3

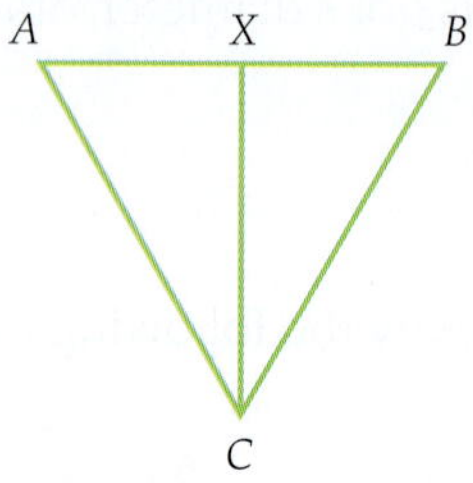

7 In ΔDEF, G is a point on DE and H is a point on DF such that $GH \parallel EF$. Prove that $\dfrac{DG}{GE} = \dfrac{DH}{HF}$.

9.4

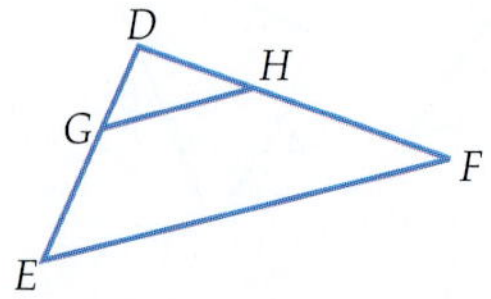

8 If both pairs of opposite sides of a quadrilateral are equal, prove that the quadrilateral is a parallelogram.

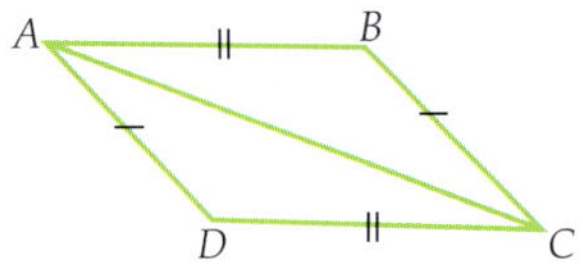

9.5

9 If the diagonals of a quadrilateral are equal and bisect each other at right angles, prove that the quadrilateral is a square.

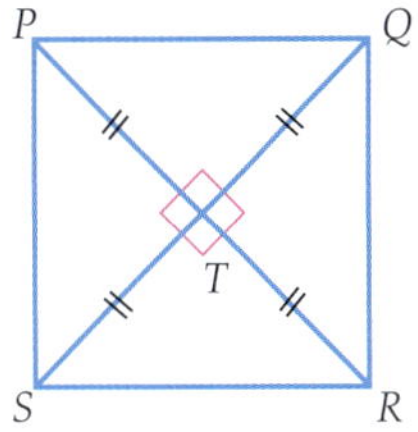

9.5

10A **10** Find the value of the pronumerals in each figure. Give reasons for your answer.

9.6

(a)

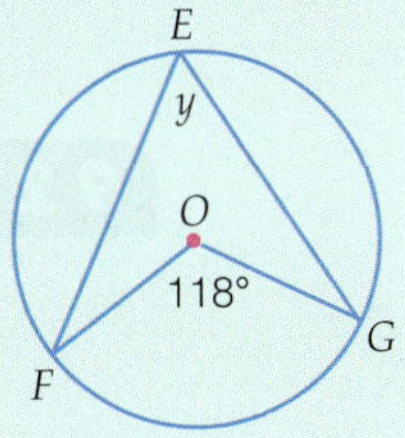

(b)

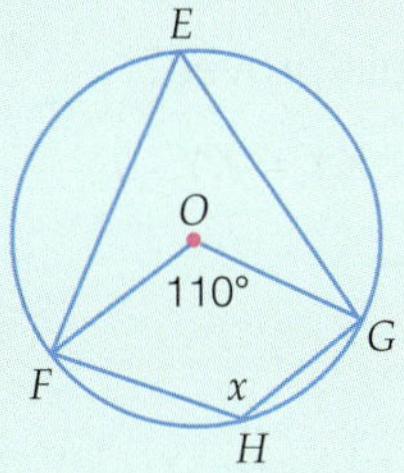

(c)

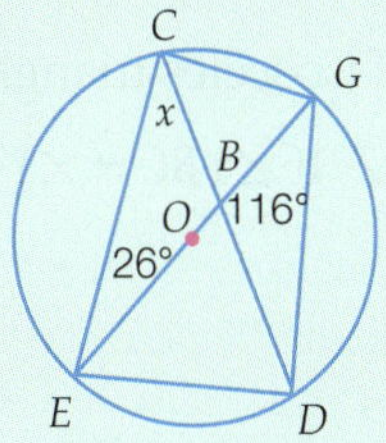

(d)

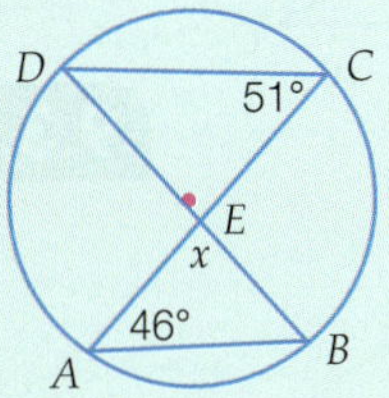

(e)

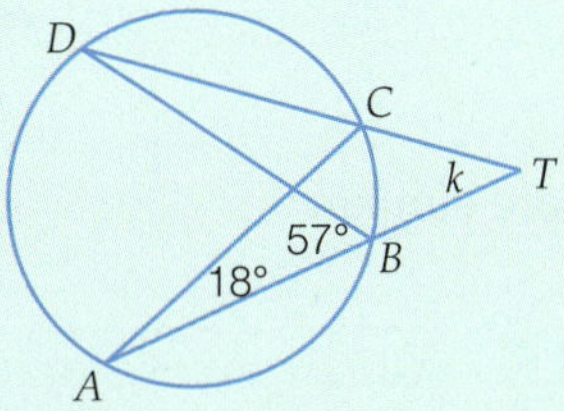

(f)

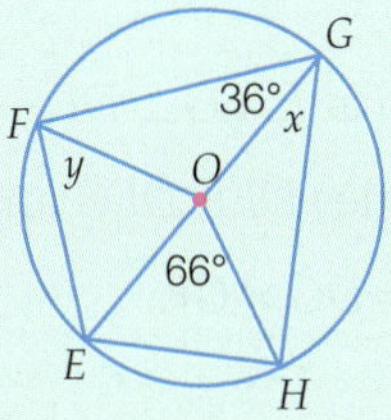

10A **11** Calculate the length of *AB* in each diagram, giving reasons for your answer.

9.7

(a)

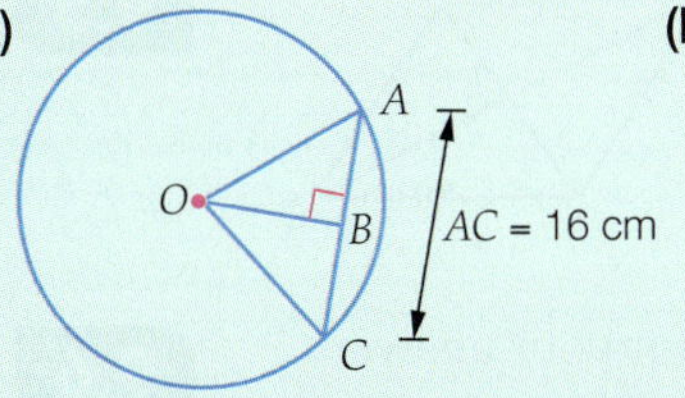

(b)

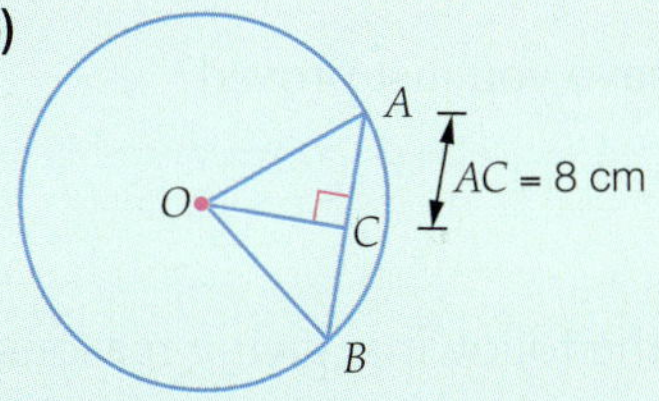

(c)

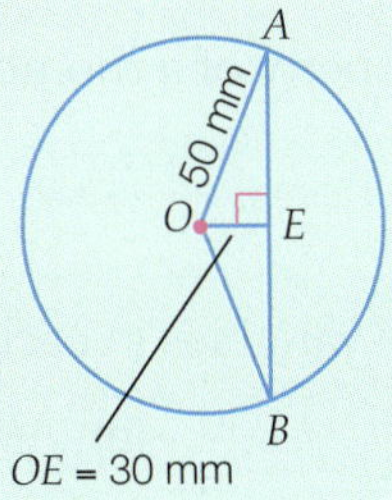

10A **12** Calculate the size of $\angle AOB$ in each figure, giving reasons for your answer.

9.7

(a)

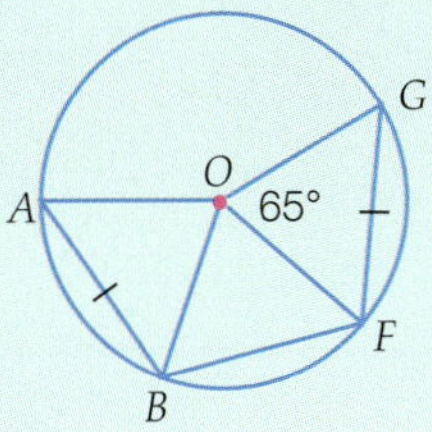

(b)

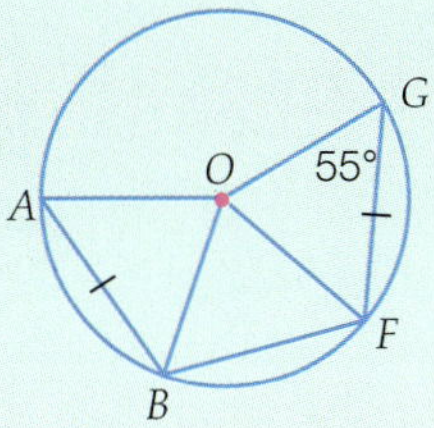

(c)

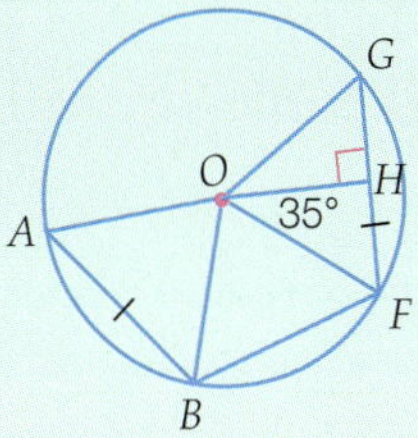

Understanding

13 Angela is on a walk through the forest when she comes to a river. She wants to know the distance across the river. She draws triangles as shown in the diagram. Find the distance across the river. **9.1**

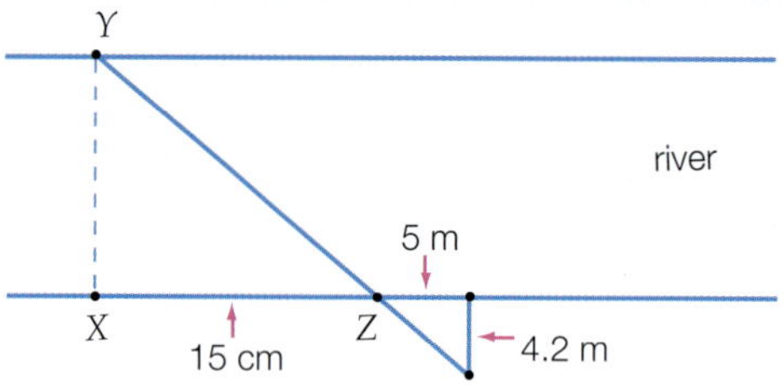

14 In ΔABC, $BA = BC$. **9.2**

(a) Prove that $\angle BAC = \angle BCA$, by drawing the line BD perpendicular to AC, meeting it at D.

(b) What property of isosceles triangles have you just proved?

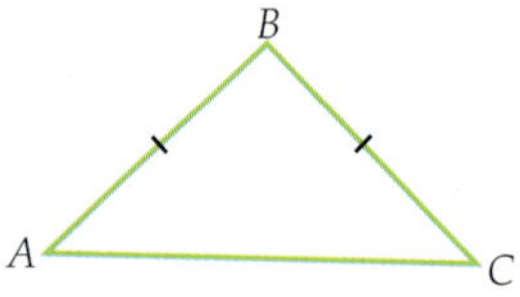

15 $AB = AC$ and $AX = AZ$. If $\angle YBC = \angle YCB$, prove that $XY = ZY$. **9.2**

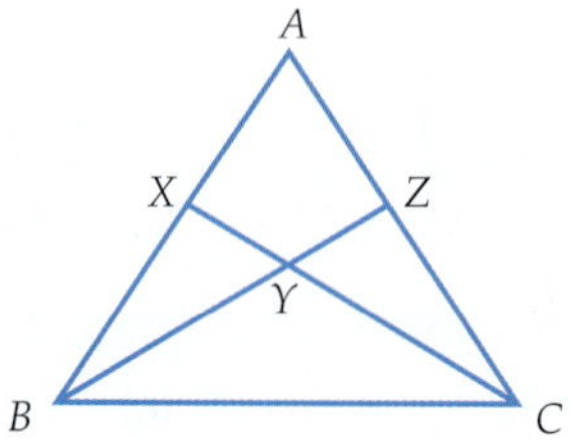

16 ΔDEF is right angled at E. $EG \perp DF$. **9.3, 9.4**

(a) Prove that triangles DEF, DEG and EGF are similar.

(b) Prove that $EG^2 = DG \times GF$.

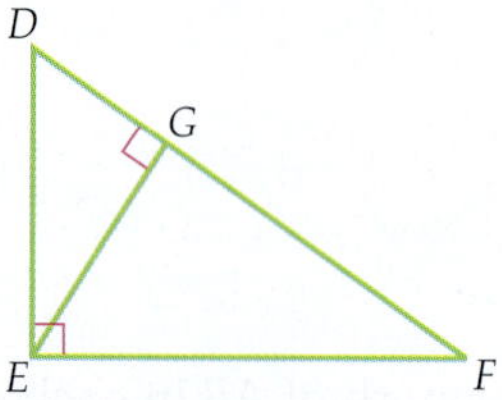

17 (a) For the diagram shown, prove $AC \perp BD$. **9.5**

(b) What special property of a rhombus have you just proved?

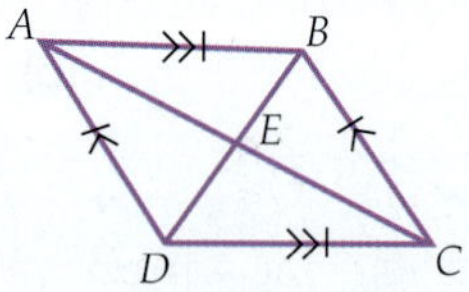

10A **18** Find the value of the variables in each of the following, giving reasons for your answers. **9.6**

(a) $GH = HJ$

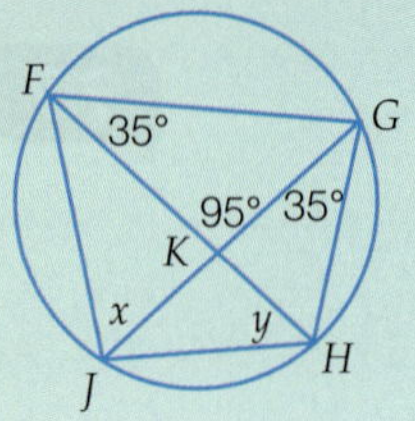

(b) $AB \parallel DC$

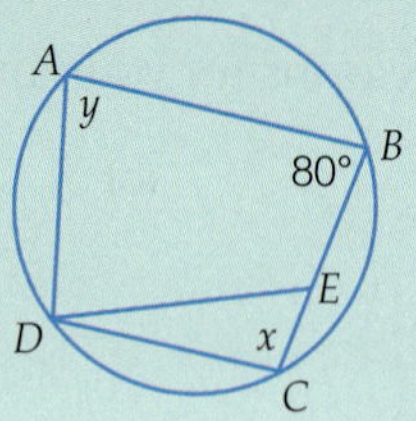

(c) $\angle YDZ = x$

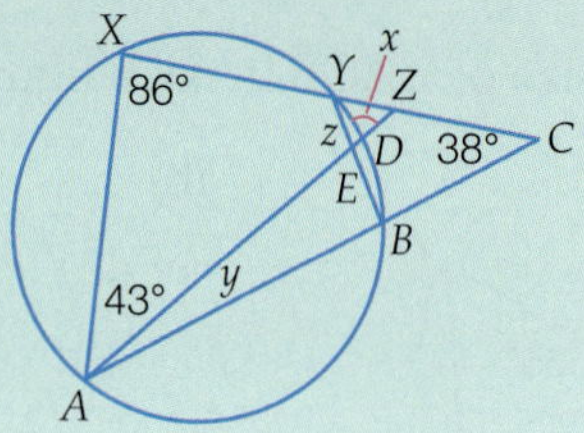

Reasoning

19 Prove that $\angle TSR + \angle TUV = 270°$. **9.1**

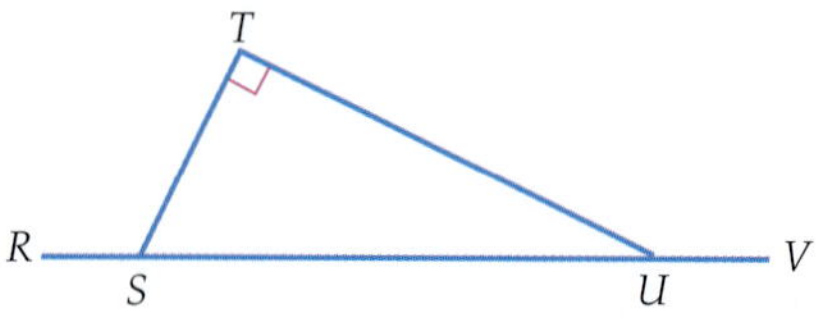

20 Use the diagram to prove that the sum of the interior angles of a trapezium is 360°.

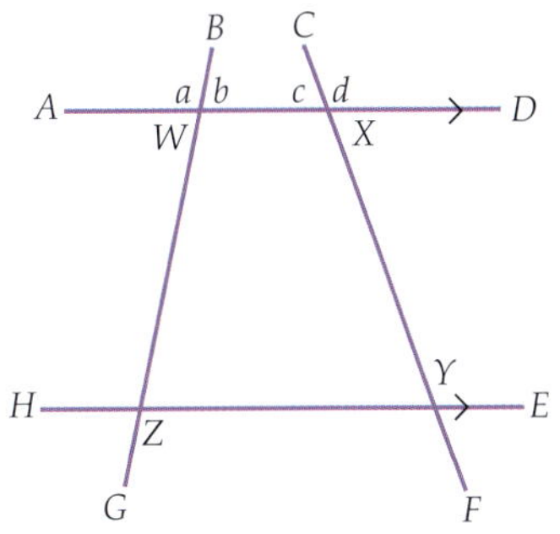

9.2

21 In triangle LMN, LX is perpendicular to NM and MY is perpendicular to LN, as shown in the diagram. The point of intersection of LX and MY is Z. Prove that $\Delta LYZ \sim \Delta MXZ$, giving reasons.

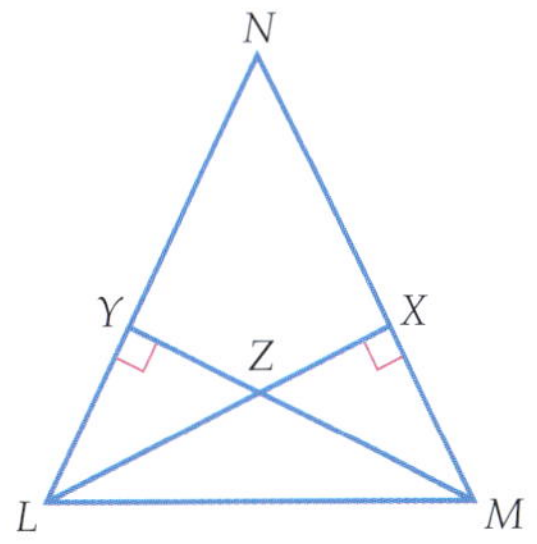

9.3

22 $UVXY$ is a rhombus and $YX = XW$. Show that $\Delta UXY \equiv \Delta VWX$.

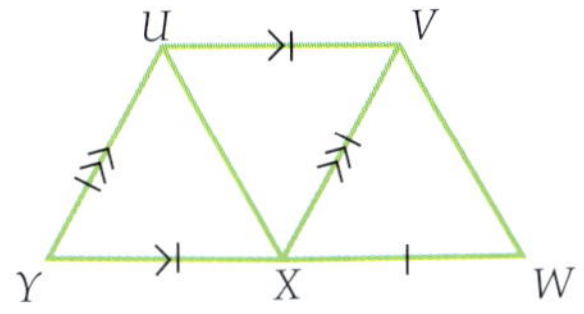

9.5

10A 23 PQR is an equilateral triangle inscribed in a circle with centre O. (P, Q, R are on the circle.) If $PQ = 6$ cm, calculate OP, the radius of the circle, giving reasons for your answer.

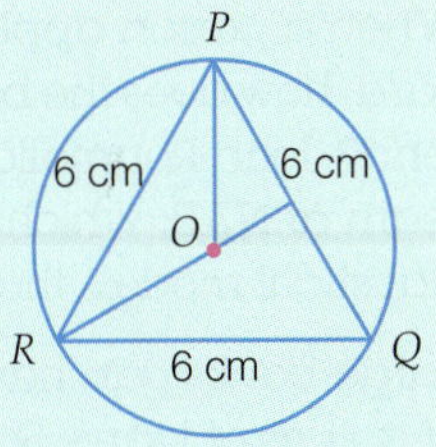

9.6

10A 24 In the circle with centre O, $SU \perp VT$ and $\angle VTU = 47°$. Find the size of $\angle SOT$, giving reasons for your answer.

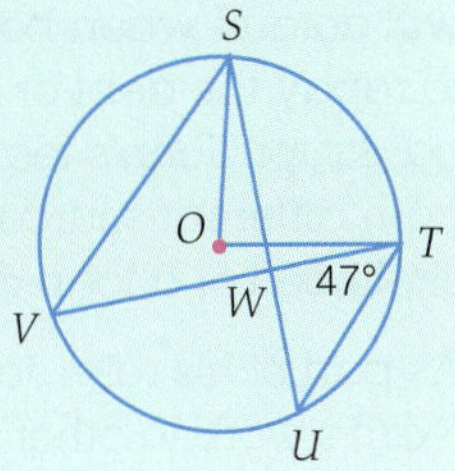

9.7

10 Probability

You want how much?

Name: John Roumeliotis
Job: Global Head of Credit Model Development
Qualifications: PhD in applied mathematics

When a person applies for a loan or credit card, how does the bank decide how much to lend? John Roumeliotis leads the credit-risk team at NAB, which develops analytical statistical models that assess credit risk.

Large banks can have hundreds of billions of dollars of loans. Some of these loans will default when borrowers are unable to repay the debt or companies become bankrupt. John's team uses data gathered from different sources to determine the likelihood of a loan defaulting.

As part of his role, John has to explain his team's work to other people. 'The trick is to take something that might be very quantitative and mathematically or statistically intensive and try to explain it in a way that makes sense and is meaningful to others. I enjoy trying to get people to understand mathematics. My team needs good communication skills to communicate their results and the methods that they used. They also need to work collaboratively, share ideas and listen to others. Some of the mathematics is high-level, but many of the skills are learnt in high school. The first thing I ask my team is whether they are comfortable with data. I expect the team to be innovative and insightful. We always need to refine our approach.'

John also appreciates the usefulness of his mathematical skills. 'The skill set I have is adaptable. Skills in mathematics give people an ability to do all sorts of things. We have mathematically trained people working in all parts of our organisation.'

Why learn this?

Probability provides a way of making judgements about future events. For example, retailers use probability when they estimate how many staff will be required to work on a particular day. Traffic engineers decide how long the traffic lights should stay green in a particular direction by predicting future traffic flows. A good knowledge of probability can also enhance our ability to play many kinds of games.

After completing this chapter you will be able to:

- estimate probabilities according to experimental results
- calculate theoretical probabilities by considering the sample space
- use Venn diagrams, tables and tree diagrams to represent the sample space and to calculate probabilities
- identify complementary events and mutually exclusive events
- describe the results of two-step and three-step chance experiments both with and without replacement
- calculate probabilities for situations involving conditional statements
- distinguish between independent events and dependent events.

10 Recall

Prepare for this chapter by attempting the following questions. If you have difficulty with a question, you can download a Recall Worksheet from the eBook or the Pearson Places website.

1 Express in simplest form.

(a) $\frac{3}{15}$ (b) $\frac{16}{52}$

2 Evaluate the following.

(a) $\frac{1}{8} + \frac{3}{4}$ (b) $\frac{2}{5} + \frac{1}{3}$ (c) $1 - \frac{3}{5}$

(d) $\frac{2}{5} \times \frac{3}{4}$ (e) 0.8×0.5 (f) 0.05×0.2

3 (a) Two students are to be chosen from their coeducational class. Complete a list of the gender pairings possible, in order first student and then second student.

(b) What are the two possible outcomes when a coin is tossed?

4 (a) A container has 25 numbered tokens: 6 are red, 10 are blue and the rest are yellow. A token is chosen at random from the container.

(i) How many different tokens can be chosen?

(ii) How many possible outcomes are there?

(iii) How many of these outcomes result in a red token being chosen?

(iv) How many of these outcomes result in a yellow token being chosen?

(v) How many of these outcomes result in a token that is not blue being chosen?

(b) A fair six-sided *die* (that is, one *dice*) is rolled.

(i) List all the possible outcomes.

(ii) How many of these outcomes are even?

Exploration Task

You can download this activity from the eBook or the Pearson Places website.

Misunderstanding probability

In this activity you will investigate the possible mistakes made on a practice test of probability questions. Can you explain what has happened and where the solutions have gone wrong?

10.1 Probability review

An understanding of **probability** helps you to make better decisions in a variety of situations. Will you need an umbrella today? Should you take a shot at goal or pass it to a team member who is in a better position? Do you need to insure your property?

Probability is the study of the chance, or likelihood, of a particular event occurring. The probability of an event is always between 0 and 1, with 0 being impossible and 1 being certain.

$0 \leq \Pr(\text{event}) \leq 1$

Estimating probability

You can estimate the probability of an outcome by examining past results. The long-term pattern of results can indicate the probability of a similar event occurring in the future. This estimation assumes that conditions will be the same for future events as in the past. The greater the number of past results (or 'trials'), the more accurate this estimate will be.

$$\text{An estimate of } \Pr(\text{success}) = \frac{\text{number of past successes}}{\text{total number of past trials}}$$

For example, Greg is playing tennis and has served 15 aces in the last 80 points. An estimate of the probability of Greg serving an ace is:

$$\Pr(\text{Greg serving an ace}) = \frac{15}{80} = \frac{3}{16}$$

If you assume this will always be the proportion of aces that Greg serves, then you can use $\frac{3}{16}$ as an estimate of the probability of the outcome over a long period of time, or a 'long run proportion' (that is, the proportion 'in the long run'). However, it is important to understand that this is only an estimate.

Probabilities are also often expressed in decimal form or as a percentage. In this case, Greg serving an ace has a probability of approximately 0.19 or 19%.

Theoretical probability

If you can list all the possible *equally likely* outcomes, then you can count them all to find the proportions and estimate the probabilities. However, it is important that the outcomes can be assumed to be equally likely. You need to examine these assumptions carefully at the beginning of each question. For example, when tossing a coin, it is reasonable to assume that heads and tails are equally likely outcomes. But if two different footballers were lined up to take a shot at goal, it probably would not be reasonable to assume that they had the same probability of scoring a goal.

Consider the situation where a \$1 coin and a \$2 coin are both tossed. There are four possible outcomes. This can be listed as what is called a **sample space**:

HH, HT, TH, TT

The proportions can be counted to find the probabilities of different outcomes, as follows:

$\Pr(\text{2 heads}) = \frac{1}{4}$, $\Pr(\text{0 heads}) = \frac{1}{4}$, $\Pr(\text{1 head}) = \frac{2}{4} = \frac{1}{2}$

Note: The total probability is $\frac{1}{4} + \frac{1}{4} + \frac{1}{2} = 1$, or 100%, because all possible events are included. It is certain that one of these events must occur.

$$\Pr(\text{success}) = \frac{\text{number of equally likely successful outcomes}}{\text{total number of equally likely possible outcomes}}$$

Complementary events

When a fair six-sided die is rolled there are six outcomes possible: 1, 2, 3, 4, 5 and 6.

It is reasonable to assume that these six events are all equally likely, so:

$\Pr(1) = \Pr(2) = \Pr(3) = \Pr(4) = \Pr(5) = \Pr(6) = \frac{1}{6}$

The sum of the probabilities of these six outcomes is 1, because this includes all possible outcomes.

All possible outcomes can also be thought of as just two events, e.g. Pr(1) and Pr(not 1).

$\Pr(1) = \frac{1}{6}$

$\Pr(\text{not } 1) = \Pr(2) + \Pr(3) + \Pr(4) + \Pr(5) + \Pr(6) = \frac{5}{6}$

Here Pr(1) and Pr(not 1) are a pair of **complementary events**. This is useful, as Pr(not 1) = 1 – Pr(1), which can simplify some of the calculations in probability problems.

You can also define rolling a 1 as 'success' and definite not rolling a 1 as 'failure', so that:

Pr(success) + Pr(failure) = 1

Pr(success) = 1 – Pr(failure)

Worked example 1

W.E. 1

Over the past 6 months Jake has been late to school on 12 occasions in 100 school days.

(a) What is the estimated probability he will be late next Wednesday?

(b) What is the estimated Pr(Jake being on time next Friday)?

Thinking	Working
(a) 1 Find the long-run proportion.	(a) Jake has been late 12 out of 100 days.
2 Express this using the formula: $\Pr(\text{success}) = \frac{\text{number of equally likely past successes}}{\text{total number of equally likely past trials}}$	$\Pr(\text{late}) = \frac{12}{100}$
3 Cancel to express the fraction in simplest form (if necessary). You can also write this as a decimal value, rounded to 2 decimal places.	$= \frac{3}{25}$ $= 0.12$
(b) As the events are complementary, use the formula: Pr(success) + Pr(failure) = 1 In this case, success = 'being late' and failure = 'being on time'.	(b) Pr(late) + Pr(on time) = 1 Pr(on time) = 1 – Pr(late) $= 1 - \frac{3}{25}$ $= \frac{22}{25}$ $= 0.88$

In the previous example, the assumption made was that 'being late' and 'being on time' covered all the possibilities. Do you think this was reasonable?

Multiple events

If you toss a coin once there are two equally likely outcomes:

- heads H
- tails T

If you toss a coin twice, there are four equally likely outcomes:

- two heads HH
- heads and tails HT
- tails and heads TH
- two tails TT

If you toss a coin three times, there are eight equally likely outcomes:

- three heads HHH
- heads, heads, tails HHT
- heads, tails, heads HTH
- heads, tails, tails HTT
- tails, tails, heads TTH
- tails, heads, tails THT
- tails, heads, heads THH
- three tails TTT

If you list all the outcomes, then you can calculate the probability of any desired outcome. For example, the probability of only one heads in three tosses can be found by counting the successes and then dividing by the total number of outcomes. This is because all of the outcomes are equally likely.

There are three successful outcomes: HTT, THT, TTH

There are eight possible outcomes: HHH, HHT, HTH, THH, HTT, THT, TTH, TTT

$\text{Pr(only 1 H)} = \frac{3}{8}$

Worked example 2

W.E. 2

Robin loves to play football with his mates after school but he also loves to play his guitar. He can't decide what to do so he tosses a coin. Heads means he will play football that day and tails means he will play his guitar. Find the probability that he:

(a) plays football 3 days in a row

(b) plays football at least twice in 3 days

(c) plays his guitar 2 days in a row

(d) alternates his activities for the 3 days

(e) doesn't play football 3 days in a row.

Thinking	Working
Make a list of all the possible outcomes and count them. Can you assume that they are equally likely?	HHH, HHT, HTH, HTT, THH, TTH, THT, TTT are all equally likely as long as the coin is fair.

(a) How many successful outcomes are listed? Use this to calculate the probability required.	(a) Pr(football 3 days in a row) $= \text{Pr(HHH)}$ $= \frac{\text{number of successful outcomes}}{\text{number of equally likely outcomes}}$ $= \frac{1}{8}$
(b) How many successful outcomes are listed? Use this to calculate the probability required.	(b) Pr(football at least twice in 3 days) $= \text{Pr(HHT or HTH or THH or HHH)}$ $= \frac{\text{number of successful outcomes}}{\text{number of equally likely outcomes}}$ $= \frac{4}{8}$ $= \frac{1}{2}$
(c) How many successful outcomes are listed? Use this to calculate the probability required.	(c) Pr(plays guitar 2 days in a row) $= \text{Pr (TTH or HTT or TTT)}$ $= \frac{\text{number of successful outcomes}}{\text{number of equally likely outcomes}}$ $= \frac{3}{8}$
(d) How many successful outcomes are listed? Use this to calculate the probability required.	(d) Pr(alternates activities for the 3 days) $= \text{Pr(HTH or THT)}$ $= \frac{\text{number of successful outcomes}}{\text{number of equally likely outcomes}}$ $= \frac{2}{8}$ $= \frac{1}{4}$
(e) Use complementary events to find the probability. Pr(failure) = 1 – Pr(success)	(e) Pr(doesn't play football 3 days in a row) $= 1 - \text{Pr(plays football 3 days in a row)}$ $= 1 - \frac{1}{8}$ $= \frac{7}{8}$

In some multiple events the number of equally likely outcomes will be different for each event. Consider tossing a coin and rolling a die.

There are two equally likely outcomes for tossing a coin: H or T

There are six equally likely outcomes for rolling a die: 1, 2, 3, 4, 5, 6

If you toss a coin and roll a die, then the equally likely outcomes are:

H1, H2, H3, H4, H5, H6, T1, T2, T3, T4, T5, T6

Worked example 3

W.E. 3

Jemma has a dilemma. She has six favourite outfits: 1 black, 1 red, 1 blue, 1 green, 1 purple and 1 multi-coloured. She has two favourite pairs of shoes: 1 has a pattern in the leather and 1 is plain. She can never decide what to wear so she rolls a die to choose her outfit and tosses a coin to choose her shoes. On the die: 1 is black, 2 is red, 3 is blue, 4 is green, 5 is purple, 6 is multi-coloured. On the coin: heads is shoes with pattern, tails is plain. Find the probability that she will wear:

(a) the black outfit with the plain shoes

(b) the green, purple or red outfit

(c) the multi-coloured outfit and the patterned shoes

(d) the red or green outfit with plain shoes

(e) the red outfit

(f) the plain pair of shoes.

Thinking	Working
Make a list of all the possible outcomes and count them. Can you assume that they are equally likely?	H1, H2, H3, H4, H5, H6, T1, T2, T3, T4, T5, T6 There are 12 equally likely outcomes.
(a) How many successful outcomes are listed? Use this to calculate the probability required.	(a) Pr(black outfit, plain shoes) = Pr(T1) $= \frac{1}{12}$
(b) How many successful outcomes are listed? Use this to calculate the probability required.	(b) Pr(green or purple or red outfit) = Pr(H2, H4, H5, T2, T4, T5) $= \frac{6}{12}$ $= \frac{1}{2}$
(c) How many successful outcomes are listed? Use this to calculate the probability required.	(c) Pr(multi-coloured outfit and shoes with pattern) = Pr(H6) $= \frac{1}{12}$
(d) How many successful outcomes are listed? Use this to calculate the probability required.	(d) Pr(red or green outfit with plain shoes) = Pr(T2, T4) $= \frac{2}{12}$ $= \frac{1}{6}$

(e) How many successful outcomes are listed? Use this to calculate the probability required.	(e) Pr(red outfit) $= \text{Pr}(T2, H2)$ $= \frac{2}{12}$ $= \frac{1}{6}$
(f) How many successful outcomes are listed? Use this to calculate the probability required.	(f) Pr(plain shoes) $= \text{Pr}(T1, T2, T3, T4, T5, T6)$ $= \frac{6}{12}$ $= \frac{1}{2}$

$$\text{Pr(successful event)} = \frac{\text{number of equally likely successful outcomes}}{\text{total number of equally likely possible outcomes}}$$

For event A:

$0 \le \text{Pr}(A) \le 1$

$\text{Pr}(A) + \text{Pr}(\text{not } A) = 1$

$\text{Pr}(A) = 1 - \text{Pr}(\text{not } A)$

Tables and tree diagrams

Duncan spins a spinner and rolls a die. What is the probability of getting a 2 with the spinner and a 5 with the die?

To find the probabilities, you need to find the total number of possible outcomes. You can show this by using a table or a tree diagram.

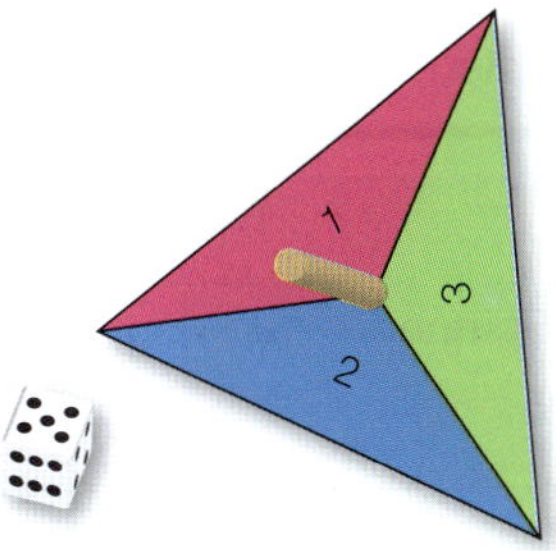

Possible outcomes

		Die					
		1	**2**	**3**	**4**	**5**	**6**
Spinner	**1**	(1, 1)	(1, 2)	(1, 3)	(1, 4)	(1, 5)	(1, 6)
	2	(2, 1)	(2, 2)	(2, 3)	(2, 4)	(2, 5)	(2, 6)
	3	(3, 1)	(3, 2)	(3, 3)	(3, 4)	(3, 5)	(3, 6)

From this table you can see that the total number of equally likely outcomes is 18. These outcomes are all equally likely, because the die has an equal chance of rolling any of its six faces and the spinner has an equal chance of showing any of its three numbers. The probability can be calculated:

$$\text{Pr(2 on the spinner, 5 on the die)} = \frac{1}{18}$$

You can also find the same result using a **tree diagram**.

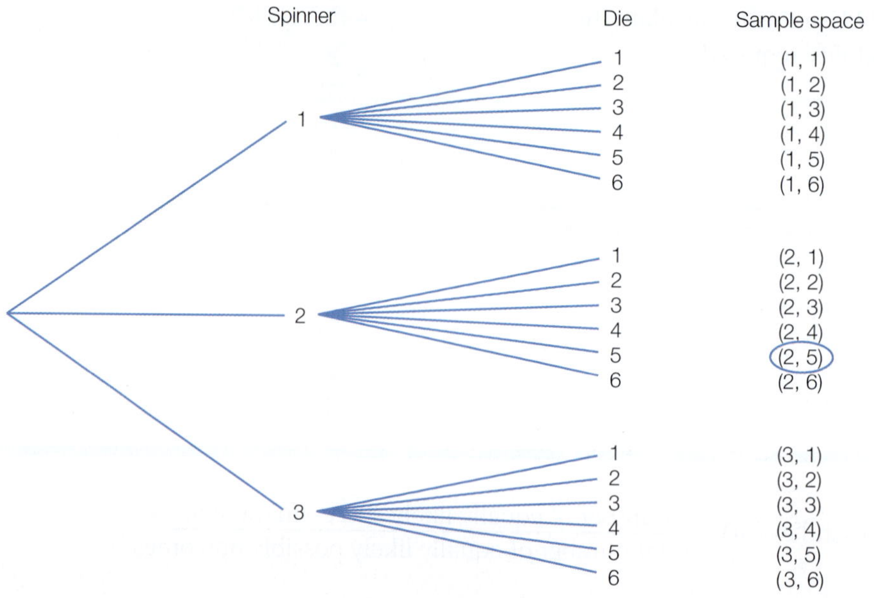

The full list of outcomes is called the sample space.

What is the probability of a total of 4 from the spinner and die? A table can be used to show all possible outcomes by adding the spinner and die numbers together, so the outcome (1, 1) is $1 + 1 = 2$. The individual outcomes are all equally likely, so each of the 18 totals in the table is equally likely to occur.

Possible totals

		Die					
		1	**2**	**3**	**4**	**5**	**6**
Spinner	**1**	2	3	4	5	6	7
	2	3	4	5	6	7	8
	3	4	5	6	7	8	9

$$\Pr(\text{total of } 4) = \frac{3}{18}$$
$$= \frac{1}{6}$$

Worked example 4

W.E. 4

Libby and Chris want to have three children. The chance of any particular child being a boy or a girl is thought to be equally likely.

(a) Represent all the possible outcomes using a tree diagram.

(b) List the sample space.

(c) Find the probability that the three children are:

- **(i)** a girl, a boy, then another girl
- **(ii)** 2 girls and a boy (in any order)
- **(iii)** 3 boys
- **(iv)** 3 children of the same gender
- **(v)** at least 2 girls.

Thinking	Working
(a) Draw the tree diagram.	(a) 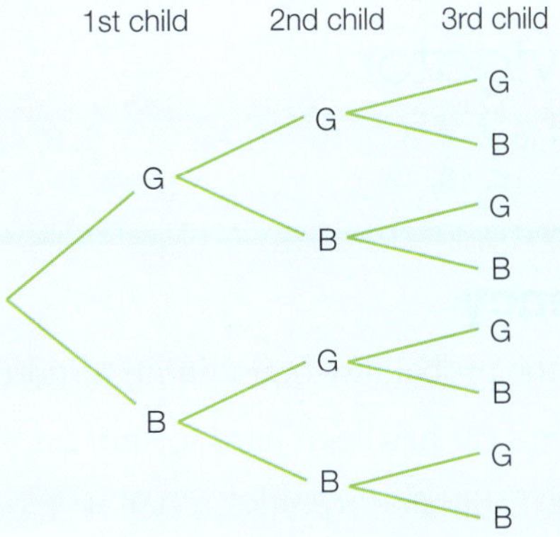
(b) Follow each path on the tree diagram to list the sample space. Comment on why these can be considered equally likely events.	(b) (G, G, G), (G, G, B), (G, B, G), (G, B, B), (B, G, G), (B, G, B), (B, B, G), (B, B, B) There are 8 equally likely outcomes. These outcomes are equally likely because each child is considered equally likely to be a boy or a girl.
(c) **(i)** List the successful outcomes. (Here, there is only 1 successful outcome out of 8 possible outcomes: (G, B, G).) Calculate the probability required.	(c) (i) Pr(G, B, G) $= \frac{1}{8}$
(ii) List the successful outcomes. (Here, there are 3 successful outcomes out of 8 possible outcomes: (G, G, B), (G, B, G), (B, G, G).) Calculate the probability required.	(ii) Pr(2 girls and 1 boy) $= \frac{3}{8}$
(iii) List the successful outcomes. (Here, there is only 1 successful outcome out of 8 possible outcomes: (B, B, B).) Calculate the probability required.	(iii) Pr(3 boys) $= \frac{1}{8}$
(iv) List the successful outcomes. (Here, there are 2 successful outcomes out of 8 possible outcomes: (G, G, G), (B, B, B).) Calculate the probability required.	(iv) Pr(3 children of the same gender) $= \frac{2}{8}$ $= \frac{1}{4}$
(v) List the successful outcomes. (Here, there are 4 successful outcomes out of 8 possible outcomes: (G, G, G), (G, G, B), (G, B, G), (B, G, G).) Calculate the probability required.	(v) Pr(at least 2 girls) $= \frac{4}{8}$ $= \frac{1}{2}$

10.1 Probability review

Navigator

Answers p.845

1, 2, 4, 5, 6, 7, 8, 9, 10, 11, 12, 13, 14, 16, 18, 19	1, 2, 3, 4, 5, 7, 8, 9, 10, 11, 12, 13, 14, 15, 16, 18, 19, 20, 21	2, 3, 4, 5, 6, 8, 9, 10, 11, 12, 13, 14, 15, 16, 17, 18, 19, 20, 21

Fluency

1 If the probability of it raining tomorrow is $\frac{2}{9}$, what is the probability of it not raining?

W.E. 1

2 Ben's car has been hard to start on 8 days over the last 4 weeks.

(a) Estimate Pr(Ben's car will be hard to start tomorrow).

(b) Estimate Pr(Ben's car will be easy to start in 3 days' time).

(c) What assumptions have you made to calculate these answers?

W.E. 2

3 Fiona is bushwalking through a national park. She finds that the path keeps dividing and she has to decide whether to take the path to the left or to the right. If she takes 3 left turns she will arrive at a beach. If she takes 3 right turns, she will reach a waterfall. If she turns left twice and right once, not necessarily in that order, she will get to a lookout. If she turns left once and right twice, not necessarily in that order, she will reach a picnic area. She can't decide where to go so she tosses a coin. Heads will mean she turns right and tails will mean that she turns left. Find the probability that she:

(a) gets to the beach

(b) arrives at the picnic area or waterfall

(c) reaches the lookout

(d) alternates the direction she takes

(e) doesn't arrive at the beach.

W.E. 3

4 Mario likes to eat at a local restaurant but he has difficulty choosing his meal as he likes 6 main courses equally and there are 2 desserts he loves. He decides to roll a die to choose his main course and toss a coin to decide his dessert. He assigns the meals this way: chicken parmigiana as 1, chilli prawns as 2, spaghetti bolognese as 3, grilled barramundi as 4, pepper steak as 5 and spicy calamari as 6. He assigns his desserts this way: sticky date pudding as H and tiramisu as T. Find the probability that he chooses:

(a) chicken parmigiana and tiramisu

(b) a seafood dish—chilli prawns, grilled barramundi or spicy calamari

(c) pepper steak and a sticky date pudding

(d) grilled barramundi or spaghetti bolognese as a main course with a tiramisu dessert

(e) chilli prawns

(f) sticky date pudding.

W.E. 4

5 A coin is tossed three times.

(a) Represent all the possible outcomes using a tree diagram.

(b) List the sample space.

(c) Find the probability of tossing:

(i) heads, then tails, then heads

(ii) heads, then heads, then tails

(iii) 3 tails

(iv) all heads or all tails

(v) heads at least twice.

6 Nikola and Brad are in a class containing 14 girls and 10 boys. To select a class representative, a person's name is chosen at random from the class roll. Find the probability that the class representative is:

(a) a girl (b) not a girl (c) Nikola

(d) Brad (e) Nikola or Brad (f) neither Nikola nor Brad.

7 Jacob and Ted are two members of a 12-person cricket team. The captain is chosen randomly each week. Find the probability that the captain is:

(a) Ted (b) not Jacob (c) Jacob or Ted (d) not Jacob or Ted.

8 The local supermarket is flooded, which destroys the labels on 12 cans of peaches, 5 cans of apples and 3 cans of beans. All the cans are put on sale for 50 cents each, but you will not know what is inside each can until it is opened.

(a) If you buy only 1 can, what is the probability of it being apples?

A $\frac{1}{5}$ B $\frac{1}{4}$ C $\frac{1}{3}$ D $\frac{5}{8}$

(b) If you buy 1 can, what is the probability of it containing fruit (not beans)?

A $\frac{3}{20}$ B $\frac{5}{20}$ C $\frac{12}{20}$ D $\frac{17}{20}$

(c) If you want to be sure of purchasing at least 1 can of peaches, what is the minimum number of cans you must purchase?

A 1 B 3 C 9 D 12

9 A spinner evenly divided and showing the numbers 1, 2, 3, 4 is spun once, and an eight-sided die showing 1 to 8 is rolled once.

(a) The probability that the total of the two numbers is 7 is:

A $\frac{3}{11}$ B $\frac{3}{32}$ C $\frac{1}{8}$ D $\frac{5}{32}$

(b) The probability that the total of the two numbers is 11 is:

A $\frac{1}{32}$ B $\frac{1}{16}$ C $\frac{1}{8}$ D $\frac{3}{32}$

(c) The probability that the total of the two numbers is a multiple of 3 is:

A $\frac{5}{16}$ B $\frac{11}{32}$ C $\frac{3}{8}$ D $\frac{15}{32}$

10 Suresh and Julian regularly play tennis against one another. Over the past 2 years, Suresh has won 16 times and Julian has won 20 times. What is the estimated probability that Suresh will win their next match?

11 Of the 125 students in Year 10, 15 ride to school every day and 30 ride sometimes. A student is chosen at random from the Year 10 students.

(a) What is the probability that the student always rides to school?

A $\frac{3}{25}$ B $\frac{9}{25}$ C $\frac{17}{25}$ D $\frac{22}{25}$

(b) What is the probability that the student never rides to school?

A $\frac{3}{25}$ B $\frac{9}{25}$ C $\frac{16}{25}$ D $\frac{22}{25}$

12 In a game of chance, 45 balls numbered from 1 to 45 are placed in a barrel. If one ball is drawn at random, find the probability that its number is:

(a) 2
(b) less than 10
(c) even
(d) odd
(e) greater than 50
(f) between 10 and 20 inclusive.

Understanding

13 Pedro has two drawers containing loose socks. There are a black, a white and a green sock in each. He is half asleep and grabs one sock, at random, from each drawer.

(a) Represent all the possible outcomes using a tree diagram.

(b) Why is it reasonable to regard these as equally likely outcomes?

(c) Find the probability that Pedro selects:

(i) a pair of black socks
(ii) one green and one white sock
(iii) a matching pair of socks
(iv) two socks that do not match.

14 Vicka and Paulo each have a bag of treats. Vicka's bag contains 3 licorice blocks, 2 jelly babies and 1 chocolate caramel. Paulo has 2 licorice blocks and 2 jelly babies in his bag. Both Vicka and Paulo offer a treat to Norma. If Norma randomly selects a treat from each bag, find the probability that she takes:

(a) 2 licorice blocks
(b) 2 jelly babies
(c) no licorice blocks
(d) no jelly babies
(e) 1 licorice block
(f) at least 1 jelly baby.

15 In a round-robin competition in bowling, Shelley plays 4 rounds. She estimates that she has an equal chance of winning or losing each round.

(a) Draw a tree diagram showing all the possible outcomes. (W = a win; L = a loss)

(b) List the sample space. How many equally likely outcomes are there?

(c) Find the probability that Shelley:

(i) wins all 4 rounds
(ii) wins only 2 of the 4 rounds
(iii) wins at least 3 rounds
(iv) wins the second round
(v) doesn't win the third round
(vi) doesn't win either the first or the fourth round.

16 In the game of Monopoly, players advance clockwise around the 40 squares of the board according to the total of the numbers shown on the roll of two dice.

(a) Use a table to represent all the possible totals from the roll of two dice.

(b) How many possible outcomes are there?

(c) Why is it reasonable to regard these as equally likely outcomes?

(d) If your token is on Whitechapel Road (on the London version of the board, as shown below), then what is the probability your next roll will result in landing on Euston Road?

(e) Suppose you land on Euston Road. Find the probability of advancing to Bow Street or further on your next roll.

(f) One way of getting out of jail in the game of Monopoly is to roll a 'double' with the two dice (that is, to roll the dice and get the same result on both). Supposing you are in jail, find the probability of getting out of jail with your next roll of the dice.

17 Two fair six-sided dice are rolled. One is numbered 1, 1, 2, 2, 3, 3 and the other is numbered 2, 4, 6, 8, 10, 12.

(a) Draw a table to show the sample space.

(b) Draw a table to show the sum of the two numbers shown.

(c) Find the following probabilities.

(i) Pr(sum is even)

(ii) Pr(sum is multiple of 5)

(iii) Pr(sum is greater than 10)

(iv) Pr(sum is not less than 11)

(d) Why is it harder to answer the questions in part **(c)** for this table than for a diagram with consecutive numbers on each side?

(e) Draw another table, this time showing the product of the two numbers.

(f) Draw a table that shows the probability for each possible product.

Reasoning

18 Lesley keeps track of lottery numbers and notices that 16 has not occurred for 12 months. Thinking the probability of 16 coming out soon must be high, she includes 16 in her lottery picks for the next two draws, but it still doesn't come out. Explain Lesley's incorrect thinking.

Open-ended

Hint

Assume that there is an equal probability of having a boy as having a girl.

19 You are given a 12-sided die, numbered 1 to 12. Write an event that would have the following probabilities.

(a) $\Pr(A) = \frac{1}{6}$ (b) $\Pr(B) = \frac{1}{3}$ (c) $\Pr(C) = \frac{5}{12}$

20 The long-term probability of Maria's team winning a game against Abbey's team at netball is $\frac{4}{5}$. List three possible sets of long-term results that would produce this figure.

21 Sharanpreet calculates the probability of a family having at least one boy in a family of three children is $\frac{3}{8}$. What is the correct answer and what mistake has Sharanpreet made?

Problem solving

A chance of freedom

A prisoner is brought before a judge who says: 'I have decided to give you a chance of freedom. Here are four red balls and four black balls. Divide the eight balls into two groups, any way you wish, and then place each group in a bag. I will hide the two bags behind my back and you will choose a bag from either my right or my left hand. You can then choose a ball at random from the bag. If it is red you can go free. If it is black your sentence will double.'

(a) How can the prisoner divide the balls to give the best chance of freedom?

(b) What division would give the worst chance?

Strategy options

- Act it out.
- Test all possible combinations.
- Solve a simpler problem.

Venn diagrams and sample space

10.2

Venn diagrams and tables are useful methods to help visualise probability questions and list the sample space.

Venn diagrams

A **Venn diagram** consists of a rectangle with one or more areas (usually circles or ovals) shown inside it. The rectangle represents the **universal set**, which is every element for the particular situation being examined. In probability, the universal set is the complete sample space. On the Venn diagram this is often labelled by ξ, X or U.

The areas on the Venn diagram represent particular outcomes or parts of the sample space, for example:

ξ = 'students in year 10'
A = 'students who travel by bus'
B = 'students with red hair'

Venn diagrams are most useful when probability problems involve the words 'and', 'or' or 'not'. For the example above:

A or B students who travel by bus or who have red hair (this includes those with both characteristics)

A and B students who travel by bus and have red hair

not A students who do not travel by bus

These can be shown on Venn diagrams as follows:

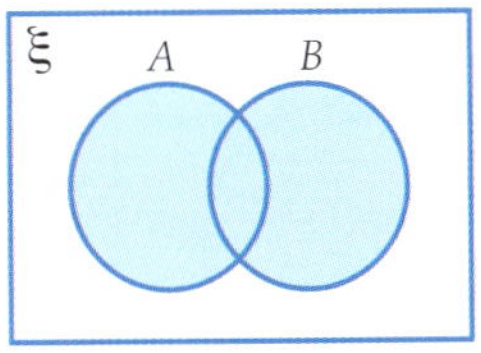

A or *B*

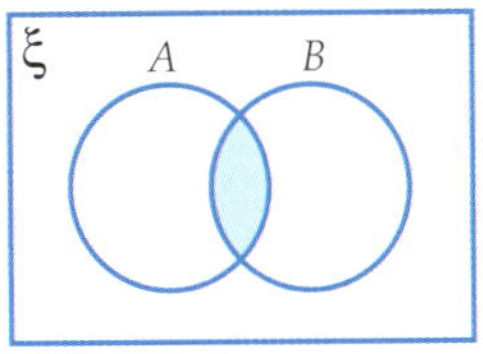

A and *B*

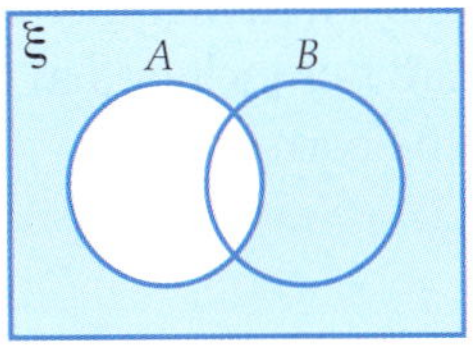

not *A*

Tables

You can use a table to show similar information. When the same set of elements can be split into categories in two different ways, this is called a **two-way table**.

	B	**not B**
A	A and B	A and not B
not A	not A and B	not A and not B

Together, the sections of the table make up the universal set (that is, the complete sample space). In addition to this, no section has any overlap with any other section.

Tables and Venn diagrams can both show the actual elements in each section, the number of elements, or the proportion of elements or data.

Worked example 5

W.E. 5

A single card is drawn from a set of 25 cards numbered from 1 to 25.

Let A = 'odd numbers' and B = 'multiples of 3'

(a) Represent this on a Venn diagram.

(b) Use the Venn diagram to find:

- **(i)** Pr(A)
- **(ii)** Pr(A and B)
- **(iii)** Pr(not B)

(c) Draw a two-way table to display the probabilities of the possible events.

Thinking / Working

(a) 1 Draw a Venn diagram that contains two overlapping circles labelled A and B.

(a)

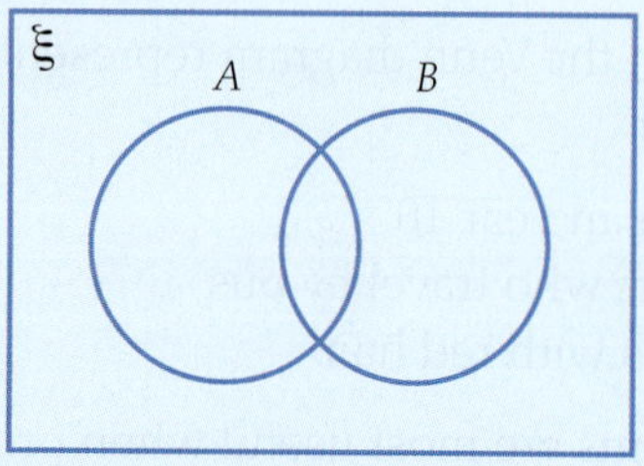

2 Place the elements that are common to both sets in the overlap of the circles.

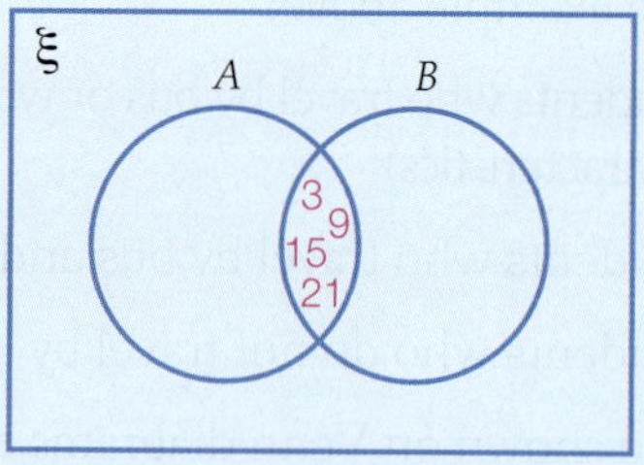

3 Put the remainder of elements that are not included in the overlap into the correct circle.

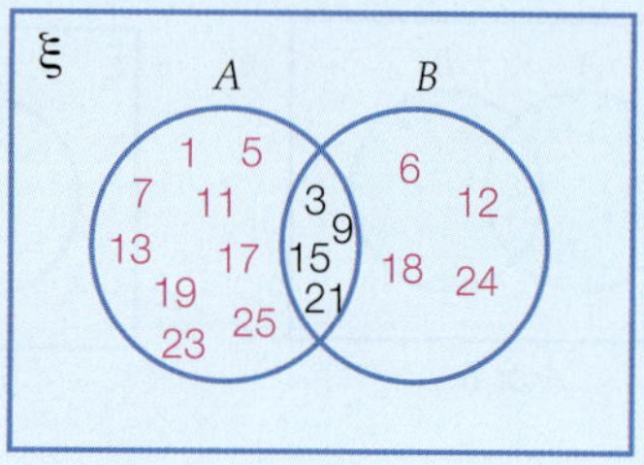

4 Place elements that do not belong inside the circles but are part of the universal set into the Venn diagram.

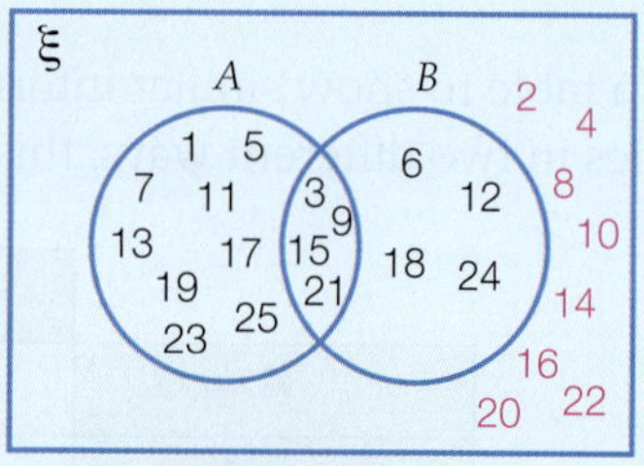

(b) **(i)** Use the formula:

$\Pr(A) = \dfrac{\text{number of elements in } A}{\text{number of elements in universal set}}$

(b) (i) $\Pr(A) = \dfrac{9+4}{25}$

$= \dfrac{13}{25}$

(ii) Use the formula:

$\Pr(A \text{ and } B) = \dfrac{\text{number of elements in } A \text{ and } B}{\text{number of elements in universal set}}$

(ii) $\Pr(A \text{ and } B) = \dfrac{4}{25}$

(iii) Use the formula:

$\Pr(\text{not } B) = 1 - \Pr(B) = 1 - \dfrac{\text{number of elements in } B}{\text{number of elements in universal set}}$

(iii) $\Pr(\text{not } B) = 1 - \Pr(B)$

$= 1 - \dfrac{8}{25}$

$= \dfrac{17}{25}$

(c) **1** Draw a table and label the possible outcomes. Enter the known probabilities from **(b)**.

(c)

	B	not B	
A	$\frac{4}{25}$		$\frac{13}{25}$
not A			
		$\frac{17}{25}$	

2 Calculate any missing probabilities. Remember that a complete set of outcomes must always add to 1. (For example, 'A' and 'not A' together must add to 1.)

	B	not B	
A	$\frac{4}{25}$	$\frac{9}{25}$	$\frac{13}{25}$
not A			
		$\frac{17}{25}$	

	B	not B	
A	$\frac{4}{25}$	$\frac{9}{25}$	$\frac{13}{25}$
not A			$\frac{12}{25}$
		$\frac{17}{25}$	

	B	not B	
A	$\frac{4}{25}$	$\frac{9}{25}$	$\frac{13}{25}$
not A		$\frac{8}{25}$	$\frac{12}{25}$
	$\frac{8}{25}$	$\frac{17}{25}$	

	B	not B	
A	$\frac{4}{25}$	$\frac{9}{25}$	$\frac{13}{25}$
not A	$\frac{4}{25}$	$\frac{8}{25}$	$\frac{12}{25}$
	$\frac{8}{25}$	$\frac{17}{25}$	

10.2 Venn diagrams and sample space

Navigator

Answers p. 847

1, 2, 3, 4, 5, 6, 7, 9, 10, 12, 13	1, 2, 3, 4, 5, 6, 7, 9, 10, 11, 12, 13	1, 3, 4, 5, 6, 7, 8, 9, 10, 11, 12, 13

Fluency

W.E. 5

1 In a non-leap year there are 28 days in February. A date in February is chosen to celebrate the 100th birthday of a city.
Let A = 'even numbered dates' and
B = 'dates that are multiples of 5'.

(a) Represent this information on a Venn diagram.

(b) Use the Venn diagram to find:

(i) Pr(A) (ii) Pr(A and B)

(iii) Pr(not B)

(c) Draw a two-way table to display the probability of possible events.

2 Use the table shown to find:

(a) Pr(A and B) (b) Pr(not A and not B)

(c) Pr(not A and B) (d) Pr(A and not B)

(e) Pr(A) (f) Pr(B)

	B	not B
A	$\frac{3}{16}$	$\frac{5}{16}$
not A	$\frac{1}{16}$	$\frac{7}{16}$

3 In a class of 20 students, 8 learn the piano, 6 take singing lessons and 10 students neither learn the piano nor have singing lessons.

(a) Represent this information in a two-way table.

(b) Use the table to find the probability of:

(i) taking singing lessons but not piano (ii) taking piano lessons but not singing

(iii) learning piano and taking singing lessons.

Understanding

4 In a group of 30 people there are 15 who like cats, 20 who like dogs and 6 who dislike both cats and dogs.

(a) Represent this information on a Venn diagram.

(b) Use the Venn diagram to find the probability that a person chosen at random is:

(i) a cat lover only (ii) a dog lover only (iii) a person who likes both.

5 Complete these tables.

(a) Pr(A) = $\frac{7}{16}$

	B	not B
A		$\frac{5}{16}$
not A	$\frac{1}{16}$	

(b) Pr(D) = 0.7

	C	not C
D	0.2	
not D		0.1

6 Use the Venn diagram at right to list the elements in each of the following.

(a) A or B

(b) A or C

(c) B or C

(d) A and B

(e) A and C

(f) B and C

(g) A and B and C

(h) not A and not B

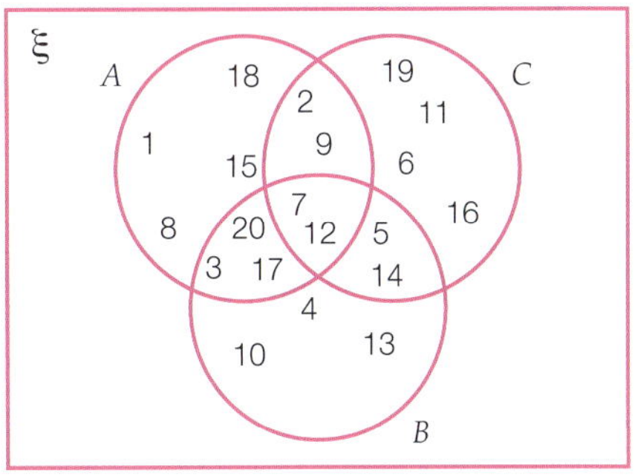

7 A poll of 2000 students revealed that 554 respondents watch TV but don't use the internet. 1200 respondents only use the internet and don't watch TV. 10 students neither watch TV nor use the internet.

(a) Draw a two-way table to represent this situation.

(b) What is the probability of a student in this survey watching TV and browsing the internet at the same time?

8 A survey of 1200 university students revealed that $\frac{3}{8}$ have a profile on Twitter, $\frac{3}{4}$ have a profile on Facebook, and $\frac{1}{4}$ have a profile on both sites.

(a) Represent the information in a two-way table.

(b) If a student is chosen at random, find the probability that the student has:

(i) no profile on either

(ii) a profile on one, but not on both.

9 Ms Lauder is creating the timetable for Year 12 students. The request forms have been entered online:

- 163 want to study mathematics, 95 physics, 124 chemistry
- 61 requested maths and physics
- 67 requested maths and chemistry
- 49 requested physics and chemistry
- 27 want to study all three subjects
- 17 want to study none of these subjects

(a) Draw a Venn diagram to illustrate the situation.

(b) How many students requested only mathematics?

(c) How many students requested only physics?

(d) How many students requested only chemistry?

(e) What was the total number of requests?

(f) What is the probability of a Year 12 student at Ms Lauder's school selecting mathematics as one of their subjects? Will the answer be the same at every school?

10 A Year 10 class has 15 boys and 12 girls. 10 of the boys and 8 of the girls say they are interested in tennis.

(a) Show this information on a Venn diagram.

(b) A student is selected at random. Use the diagram to find the probability that the student selected is:

(i) a girl who is interested in tennis

(ii) a boy who is not interested in tennis

(iii) someone who is interested in tennis

(iv) someone who is not interested in tennis.

Reasoning

11 Human blood can contain three different antigens: A, B and Rh. The following table outlines the antigens for each 'blood type'. (The actual situation is slightly more complicated for the O group.)

Blood group	A	B	Rh	% in Australia
A+	Yes	No	Yes	31
A–	Yes	No	No	7
B+	No	Yes	Yes	8
B–	No	Yes	No	2
AB+	Yes	Yes	Yes	2
AB–	Yes	Yes	No	1
O+	No	No	Yes	40
O–	No	No	No	9

The antigens present in the blood of patients at a clinic were recorded for 1 week:

- 39 patients had the A antigen
- 12 had the B antigen
- 67 had the Rh antigen
- 2 had all three antigens
- 2 had the A and B antigens
- 9 had the B and Rh antigens
- 30 had the A and Rh antigens
- 5 had no antigens.

(a) Draw a Venn diagram to show this situation.

(b) How many patients were there?

(c) What was the Pr(exactly one antigen)?

(d) What was the Pr(exactly two antigens)?

(e) Use the data to calculate the probability of each of the eight blood groups.

(f) Compare the answers from part **(e)** with those expected from long-term data as shown in the table.

12 A coin is tossed, a six-sided die is rolled and a ball is selected from a bag containing a red ball and a green ball.

M is the set of outcomes that include 'heads is tossed', N is the set of outcomes that include 'a 2 is not rolled' and P is the set of outcomes that include 'a red ball is selected'.

(a) Draw a Venn diagram to represent the situation.

(b) List the sample space.

(c) List the elements in each of the following sets.

(i) not M **(ii)** not N **(iii)** not P

Open-ended

13 Consider the following partially completed table.

	B	not B	
A	9		
not A			
			50

(a) Complete the table in two different ways.

(b) For both completed tables find:

(i) Pr(A and B) **(ii)** Pr(A or B)

Mutually exclusive events

10.3

If two events cannot occur at the same time, they are called **mutually exclusive** events.

For example, consider spinning the spinner shown at right.
Possible events include:

A = spinning an even number
= spinning a 2 or a 4
B = spinning a 3

These events are mutually exclusive because it is impossible for both of them to occur at the same time.

$\Pr(A \text{ and } B) = 0$

There is no overlap between these two events. The spinner is evenly divided into four sectors, so each number has an equal chance of occurring.

$$\Pr(A) = \frac{\text{number of successful equally likely outcomes}}{\text{total number of equally likely outcomes}}$$
$$= \frac{2}{4}$$
$$= \frac{1}{2}$$

$$\Pr(B) = \frac{\text{number of successful equally likely outcomes}}{\text{total number of equally likely outcomes}}$$
$$= \frac{1}{4}$$

$$\Pr(A \text{ or } B) = \frac{\text{number of successful equally likely outcomes}}{\text{total number of equally likely outcomes}}$$
$$= \frac{3}{4}$$

Notice that $\Pr(A) + \Pr(B) = \frac{1}{2} + \frac{1}{4} = \frac{3}{4}$

So $\Pr(A \text{ or } B) = \Pr(A) + \Pr(B)$. This is called the **addition rule for mutually exclusive events**. It can only be used when one event *or* the other event may occur, but not both, i.e. when the events are mutually exclusive.

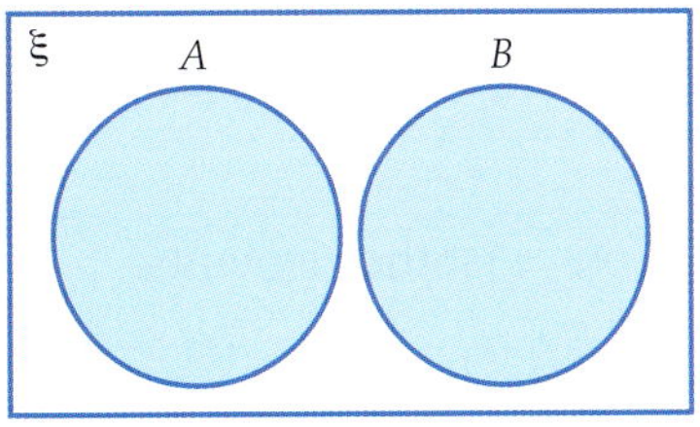

Events that are *not* mutually exclusive can happen at the same time, so on a Venn diagram they will 'overlap'. If you tried to use the addition rule for mutually exclusive events, then the overlap events would be counted twice. The **general addition rule** for any two events is:

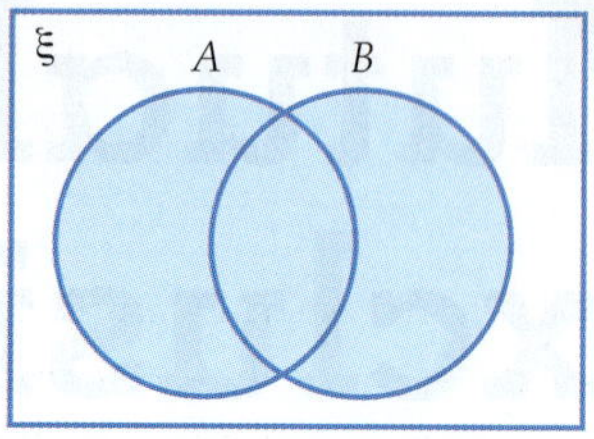

$\Pr(A \text{ or } B) = \Pr(A) + \Pr(B) - \Pr(A \text{ and } B)$

This general addition rule works for non-mutually-exclusive events. For mutually exclusive events, Pr(*A* and *B*) is zero, so this formula is also the same as the addition rule for mutually exclusive events given on the previous page.

Complementary events

Complementary events may involve one event or more than one event. This is the same for mutually exclusive events and for non-mutually-exclusive events.

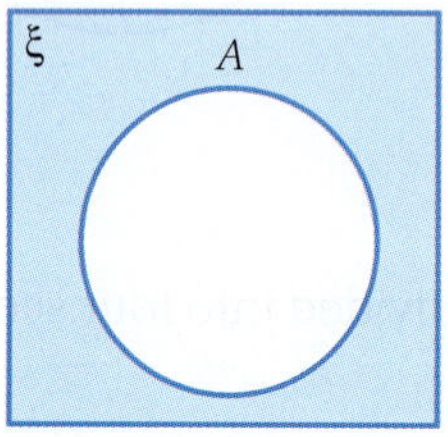

Pr (not *A*) = 1 – Pr (*A*)

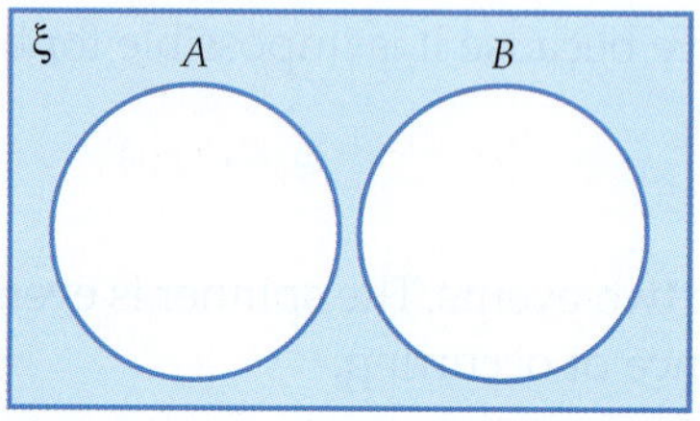

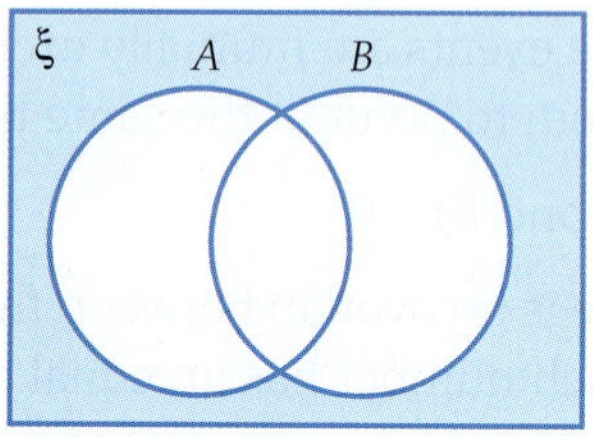

Pr (not *A* and not *B*) = 1 – Pr (*A* or *B*)

Worked example 6

W.E. 6

A standard six-sided die is rolled. Find the probability of rolling each of the following numbers.

(a) a 2 or a 5

(b) an odd number or a number greater than 3

(c) neither odd nor a 4

Thinking	Working
(a) Use the addition rule for mutually exclusive events as both events cannot occur at the same time. $\Pr(A \text{ or } B) = \Pr(A) + \Pr(B)$ Here, $\Pr(A) = \Pr(2) = \frac{1}{6}$ and $\Pr(B) = \Pr(5) = \frac{1}{6}$.	(a) Pr(2 or 5) $= \Pr(2) + \Pr(5)$ $= \frac{1}{6} + \frac{1}{6}$ $= \frac{2}{6}$ $= \frac{1}{3}$
(b) 1 List the elements.	(b) Odd numbers: 1, 3, 5 Numbers greater than 3: 4, 5, 6
2 Use the general addition rule: $\Pr(A \text{ or } B)$ $= \Pr(A) + \Pr(B) - \Pr(A \text{ and } B)$	Pr(odd number or greater than 3) $= \Pr(\text{odd}) + \Pr(>3) - \Pr(\text{odd and} >3)$ $= \frac{3}{6} + \frac{3}{6} - \frac{1}{6}$ $= \frac{5}{6}$

(c) 1	Use complementary probabilities. $\Pr(\text{not } A) = 1 - \Pr(A)$	(c) $\Pr(\text{neither odd nor a } 4)$ $= \Pr[\text{not}(\text{odd or } 4)]$ $= 1 - \Pr(\text{odd or } 4)$
2	Use the addition rule for mutually exclusive events as both events cannot occur at the same time. $\Pr(A \text{ or } B) = \Pr(A) + \Pr(B)$ Here, $\Pr(A) = \Pr(\text{odd}) = \frac{3}{6} = \frac{1}{2}$ and $\Pr(B) = \Pr(4) = \frac{1}{6}$.	$= 1 - [\Pr(\text{odd}) + \Pr(4)]$ $= 1 - \left(\frac{1}{2} + \frac{1}{6}\right)$ $= 1 - \frac{2}{3}$ $= \frac{1}{3}$

Addition rule for mutually exclusive events

If A and B are mutually exclusive events (i.e. they cannot occur at the same time), then:

$$\Pr(A \text{ or } B) = \Pr(A) + \Pr(B)$$

General addition rule for events

If A and B are events, then:

$$\Pr(A \text{ or } B) = \Pr(A) + \Pr(B) - \Pr(A \text{ and } B)$$

10.3 Mutually exclusive events

Navigator

1, 2, 3, 4, 5, 6, 7, 8, 9, 10 | 1, 3, 4, 5, 6, 7, 8, 10 | 1, 4, 5, 6, 7, 8, 9, 10

Answers p. 848

Fluency

W.E. 6

1 A fair eight-sided die, with sides numbered 1 to 8, is rolled. What is the probability that the number rolled is:

(a) a 3 or a 6

(b) an odd number or a number less than 4

(c) neither 5 nor even?

2 A spinner is coloured as shown.

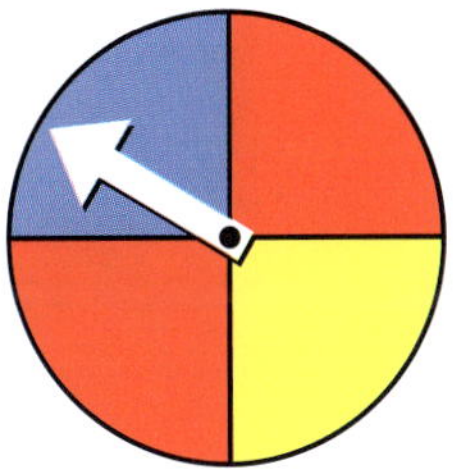

(a) What is the probability of spinning the result blue?

(b) What is the probability of spinning red?

(c) What is the probability of spinning blue or red?

(d) What is the probability of spinning blue or yellow?

3 Are the following pairs of events mutually exclusive? Answer 'yes' or 'no'.

(a) A tossed coin coming up 'heads' and a tossed coin coming up 'tails'.

(b) Choosing an odd-numbered horse in one race and a horse numbered 7 in another.

(c) Winning first prize in a raffle with a blue ticket and first prize in a raffle with a yellow ticket.

4 (a) A basket contains 3 red, 2 white and 5 black scarves. What is the probability that a scarf drawn at random is red or black?

A $\frac{4}{5}$ B $\frac{2}{3}$ C $\frac{7}{10}$ D $\frac{1}{5}$

(b) A drawer contains 6 blue socks, 4 black socks and 2 brown socks. What is the probability that a sock drawn at random is not black?

A $\frac{1}{4}$ B $\frac{1}{3}$ C $\frac{1}{2}$ D $\frac{2}{3}$

Understanding

5 The following probabilities apply to different modes of transport that students use to travel to and from a school. The modes are all mutually exclusive.

$\Pr(\text{car}) = \frac{7}{20}$ $\Pr(\text{train}) = \frac{9}{20}$

$\Pr(\text{bus}) = \frac{1}{20}$ $\Pr(\text{walking}) = \frac{3}{20}$

Find the probability that a student travels to school by the following:

(a) travels by car or train

(b) walks or catches a bus

(c) does not walk nor catch a bus

(d) does not travel by car nor by bus

(e) travels by car, train or bus.

6 A bag contains 30 balls numbered from 1 to 30. A ball is taken out of the bag. Find the probability that the number on the ball is:

(a) a multiple of 2 or a multiple of 3

(b) a factor of 12 or a factor of 30

(c) an odd number or a factor of 20

(d) not a factor of 6 and not a multiple of 6.

7 A standard pack of playing cards contains 52 cards: 26 red and 26 black. The red cards are divided into two suits: hearts and diamonds. The black cards are divided into two suits: clubs and spades. Each suit is made up of 13 cards: Ace, 2, 3, 4, 5, 6, 7, 8, 9, 10, Jack, Queen and King.

A card is drawn at random from the pack. Find the probability that the card is:

(a) red
(b) a heart
(c) an Ace
(d) the King of hearts
(e) red or the Ace of spades
(f) a Jack or a spade
(g) the 5 of clubs or the Queen of spades
(h) the Jack of diamonds, any 2, or the 7 of clubs
(i) neither an Ace nor a heart
(j) neither a black Jack nor a club
(k) neither a 6 nor a 9 of any suit
(l) neither a spade nor red.

Reasoning

8 A normal six-sided die is rolled. The following events are defined:

A = rolling an even number | B = rolling a 2 or a 6

C = rolling a 1 or a 3 | D = rolling a multiple of 3

(a) List all possible pairs of events (order does not matter).
(b) Are any of these pairs mutually exclusive?
(c) Find the probability of each event.
(d) Find the probability that either event occurs for each of the pairs listed in **(a)**.
(e) Find the probability that both of the events occur for each of the pairs listed in **(a)**.
(f) Find the sum of the probabilities of each event in all the pairs listed in **(a)**.
(g) Compare your answers to parts **(d)**, **(e)** and **(f)**. What do you find? Write an equation to summarise your findings.

9 A class conducted a traffic survey. The types of vehicles they recorded were:

sedan	station wagon	4WD	hatchback
truck	van	bus	2 door

(a) What difficulties can you see arising from the data collection?
(b) How could these difficulties have been avoided?

Open-ended

10 Consider a fair 20-sided die with sides numbered 1 to 20. The die is rolled once. For this situation:

(a) list four pairs of events that are mutually exclusive
(b) list four pairs of events that may overlap
(c) describe a pair of events, A and B, so that $\Pr(A) = \frac{1}{2}$, $\Pr(B) = \frac{3}{4}$ and $\Pr(A \text{ or } B) = \frac{17}{20}$.

Exploration CAS

Equipment required: 2 TI-Nspire CAS or Casio ClassPad CAS, two students working together

Investigating the reliability of lie detectors

A lie detector, or *polygraph*, is sometimes used to try to determine whether a person is telling the truth. Although lie detectors are used widely by authorities in the USA, their effectiveness is disputed by psychologists.

Many researchers believe that polygraphs give accurate results only 80% of the time at best.

This exploration uses CAS technology to simulate the use of a polygraph.

Assumptions:

- The lie detector is accurate 80% of the time (therefore incorrect 20% of the time).
- At any time 90% of people tell the truth (therefore 10% lie).

Working in pairs, one person uses a CAS to play the role of the detective operating the lie detector. The other person uses a CAS to play the role of the suspect being questioned.

Setting up the 'suspect' CAS

You can use the first CAS (the suspect) to generate random counting numbers from 1 to 10. If the 'suspect' CAS generates a 5, this means the suspect is lying. Any other number means that the suspect is telling the truth. To generate random counting numbers on the TI-Nspire or Casio ClassPad, see below.

Using TI-Nspire CAS

For your CAS to generate random numbers, it should be 'seeded' first. To do this, select **menu** > **Probability** > **Random** > **Seed** and enter any 4-digit number as the 'seed'. You only need to do this once.

To generate a random number, select **menu** > **Probability** > **Random** > **Integer**, then enter the lowest possible and highest possible values separated by a comma as shown (in this case, 1 and 10).

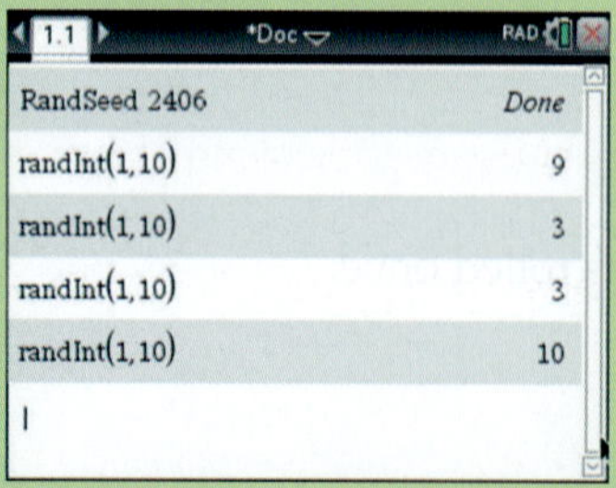

Using Casio ClassPad CAS

From the menu select **Main**. Open the **Keyboard** and go to the **Catalog** tab to find the command **rand(**.

To generate a random number, enter **rand(** followed by the lowest possible and highest possible values, separated by a comma as shown (in this case, 1 and 10).

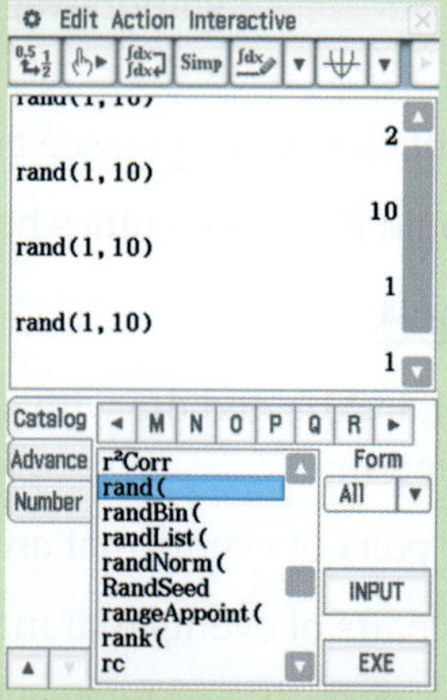

Setting up the 'lie detector' CAS

You can use the second CAS to generate random counting numbers from 11 to 20. Try to adapt the commands given for the 'suspect' (as shown above), so that the 'lie detector' CAS generates the required numbers.

If the lie detector CAS generates an 11 or 12, the result of the test is false. If the CAS generates a number from 13 to 20 inclusive, the result is true.

Recording the simulation

Copy the table below and record your results using tally marks.

	Suspect is lying CAS 1 shows 5	Suspect is telling the truth CAS 1 shows 1 to 4 or 6 to 10
Lie detector result is correct CAS 2 shows 13 to 20	*A*	*B*
Lie detector result is incorrect CAS 2 shows 11 to 12	*C*	*D*

Performing the simulation

The suspect generates a random number: 5 means the suspect has just told a lie. The detective then generates a random number to determine the truth.

Working in pairs, simulate 100 lie detector tests. When you have finished the table, combine all sets of results into a summary table. Use the combined results in the summary table to answer the following questions.

1 Calculate the probability of the following:
   - (a) a person selected at random is telling the truth
   - (b) a person selected at random is lying
   - (c) a lie detector test gives a true result
   - (d) a lie detector test gives a false result
   - (e) a randomly selected person who is telling the truth is found to be telling the truth by the lie detector test
   - (f) a randomly selected person who is telling the truth is found to be telling a lie by the lie detector test.

2 Which two outcomes in the table are undesirable? What is the probability of each?

3 Assume that 1000 people take a lie detector test. Use a tree diagram to determine the probability of a randomly selected person being in each section of your results table, and the number of people in each section.

   For example: Section *B* of table $= 0.8 \times 0.9 = 0.72$; $0.72 \times 1000 = 720$ people.

4 Use your answers to Question **3** to answer Question **1** again. Compare the theoretical and experimental probabilities.

5 Write a brief report on this activity and address the following questions. What assumptions were made? How valid are they? Do you think lie detectors should be used? Why? Can you think of other 'non-mathematical' reasons why we should or shouldn't use lie detectors?

Exploration CAS

Equipment required: TI-Nspire CAS or Casio ClassPad CAS

Investigating the 'one son' policy

China is the most populous country in the world, with more than 1.3 billion people. In an effort to reduce the population, the Chinese government for many years had a policy of one child per family. A policy like this, if strictly adhered to, would reduce the population. Can you explain why?

While this policy was in effect, some Chinese advocated a 'one son' policy. Under this policy a family would be allowed to continue to have additional children until a son is born. Possible families, excluding parents, would be: one boy only; one girl and one boy (in that order); two girls and one boy (in that order); and so on. Note that if the first born is a son, then no additional children are allowed.

Your task is to explore the implications of adopting a 'one son' policy.

1 Write what you think would be the ratio of the number of births of girls to births of boys under this policy.

2 What do you think would be the mean number of children in a family? Discuss your answers with other students and your teacher. Write your estimate.

Modelling the possible families

You can model the 'one son' policy using a random number generator, where 0 represents a girl and 1 represents a boy.

To generate random numbers (1s and 0s) on the TI-Nspire or Casio ClassPad, see below.

Using TI-Nspire CAS

For your CAS to generate random numbers, it should be 'seeded' first. To do this, select **menu** > **Probability** > **Random** > **Seed** and enter any 4-digit number as the 'seed'. You only need to do this once.

To generate a random number, select **menu** > **Probability** > **Random** > **Integer**, then enter the lowest possible and highest possible values separated by a comma as shown (in this case, 0 and 1).

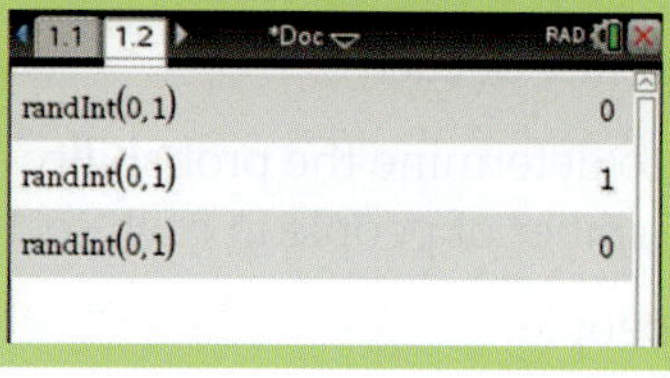

Using Casio ClassPad CAS

From the menu select **Main**. Open the **Keyboard** and go to the **Catalog** tab to find the command **rand(**.

To generate a random number, enter **rand(** followed by the lowest possible and highest possible values, separated by a comma as shown (in this case, 0 and 1).

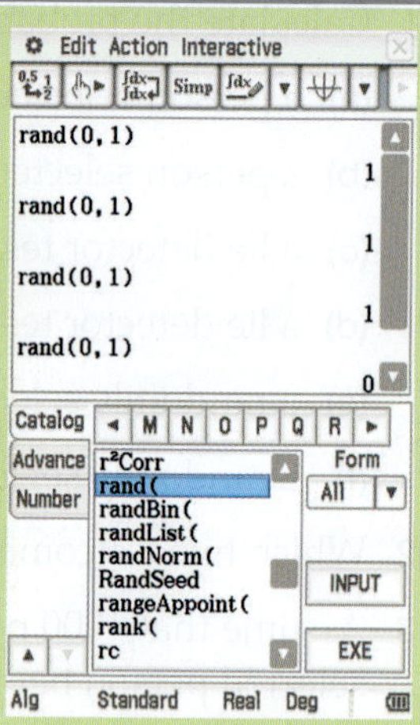

3 Work in pairs to complete the table on the following page. Each pair should aim to simulate the results for 50 families.

Generate random numbers until you get a 1, which represents the children in the first family. Continue until the next 1, which represents the children in the next family. For example, the numbers below represent 10 families:

(0 1) (0 0 1) (0 0 0 1) (1) (0 1) (0 0 1) (0 1) (1) (1) (0 1)

- 3 with 1 child only (1)
- 4 with 2 children (01)
- 2 with 3 children (001)
- 1 with 4 children (0001)

When you have finished the table, combine all the sets of results into a summary table. Use the combined results to answer the following questions.

Family type	Tally	Total number of families	Total number of boys	Total number of girls	Total number of children
1					
01					
001					
0001					
00001					
0000001					
00000001					
000000001					
0000000001					
________					
________					
	Totals:				

4 What was the ratio of births of girls to births of boys?

5 How did this ratio compare with your estimate in Question **1**?

6 Calculate the ratio of 'boy births' to 'total births'.

Finding the mean number of children per family

Suppose the following results were obtained:

Number of children	1	2	3	4	5	6	7	8	9	10
Number of families	241	130	64	32	18	7	4	3	0	1

To calculate the mean number of children per family:

$$\text{Mean number of children per family} = \frac{\text{Total number of children}}{\text{Total number of families}}$$

$$= \frac{1 \times 241 + 2 \times 130 + 3 \times 64 + 4 \times 32 + 5 \times 18 + 6 \times 7 + 7 \times 4 + 8 \times 3 + 9 \times 0 + 10 \times 1}{241 + 130 + 64 + 32 + 18 + 7 + 4 + 3 + 0 + 1}$$

$= 2.03$ children per family

7 Find the mean number of children per family using your combined results. Compare this answer to your estimate in Question **2**.

Examining assumptions

8 When you use mathematics as a model for a real-world problem, you need to make assumptions. The model above assumes that the probability of giving birth to a boy is the same as the probability of giving birth to a girl. Is this assumption valid? If not, is our model still useful?

9 If you modelled one million births, predict the mean number of families with 1, 2, 3, …, 10 children.

10 What other assumptions have been made by this model?

11 If a 'one son' policy were adopted, what would be the effect on the population?

12 If you were an adviser to the Chinese government, would you recommend the 'one son' policy? Why?

Half-time 10

10.2 **1** **(a)** Complete the following two-way table that shows the number of surveyed students who own either a smartphone or a laptop or both.

	smartphone	not smartphone	
laptop	60		75
not laptop			
	210		400

(b) Now use the completed table to state the probability of a randomly chosen student owning:

(i) a smartphone but not a laptop **(ii)** a laptop but not a smartphone.

10.1 **2** Ten balls numbered from 1 to 10 are placed in a bag and one is drawn at random. Find the probability that the number on the ball is:

(a) 1 or 2

(b) an even number

(c) a 6 or an odd number

(d) neither an odd number nor a multiple of 4.

10.1 **3** The probability of it raining on a given day of the weekend is 0.3. What is the probability that it rains on both Saturday and Sunday?

10.1 **4** In a chess tournament, each contestant must play three others. Suzanne estimates that she has an equal chance of winning or losing each game.

(a) Draw a tree diagram and list the sample space for the possible outcomes of Suzanne's three games.

(b) How many possible outcomes are there?

(c) Find the probability of Suzanne:

(i) winning all three games **(ii)** winning only one game

(iii) winning the first game **(iv)** not winning the third game

(v) winning at least two games **(vi)** not winning either of the first two games.

10.3 **5** A fair eight-sided die is numbered 1 to 8. The die is rolled once. The following events are defined:

A = 'rolling an even number'
B = 'rolling a prime number'
C = 'rolling a multiple of 5'

(a) Given that order does not matter list the pairs of events that are possible.

(b) Find the probability of each individual event.

(c) State which, if any, of the pairs of events are mutually exclusive.

Probability tree diagrams

10.4

Consider the case of tossing a fair coin three times. There are eight equally likely outcomes (where 'H' = heads and 'T' = tails): HHH, HHT, HTH, HTT, THH, THT, TTH, TTT

These outcomes can be represented using a **probability tree diagram**. Start with a dot and branch this to represent the possible outcomes from the first toss of the coin.

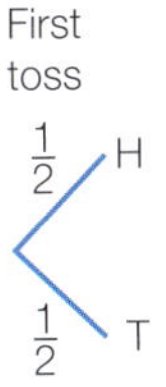

For the first toss, there are two possible outcomes (H and T) so there are two branches. The outcomes are equally likely (because the coin is fair), so each branch has a probability of $\frac{1}{2}$.

There are two equally likely outcomes.

$\Pr(H) = \Pr(T) = \frac{1}{2}$

For the second coin toss, there are again two possible outcomes (H and T) but this time there are two separate starting positions. Again $\Pr(H) = \Pr(T) = \frac{1}{2}$ so you get the probability tree diagram shown at right.

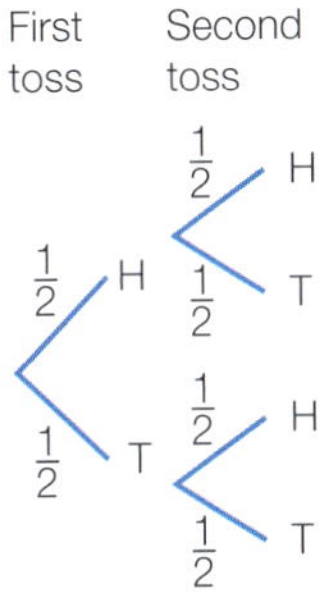

There are four equally likely outcomes.

$\Pr(HH) = \Pr(HT) = \Pr(TH) = \Pr(TT) = \frac{1}{4}$

For the third coin toss, there are again two possible outcomes (H and T) for each of the previous four outcomes.

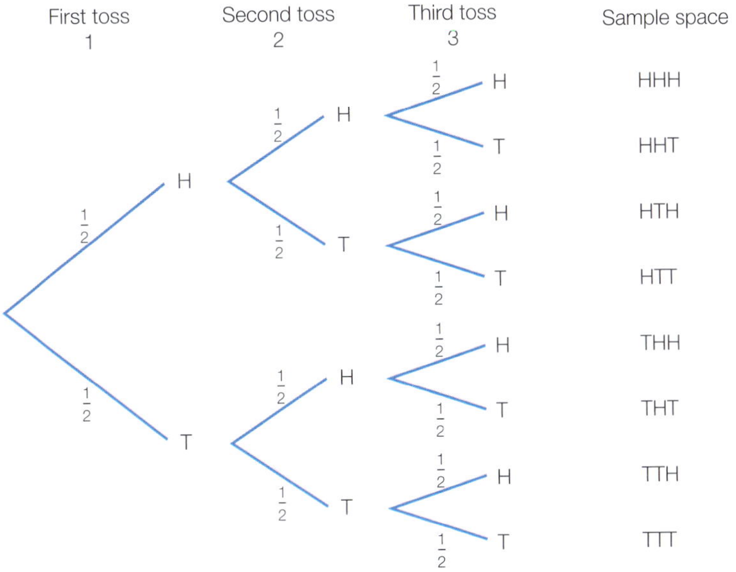

There are eight equally likely outcomes.

$\Pr(HHH) = \Pr(HHT) = \Pr(HTH) = \Pr(HTT) = \Pr(THH) = \Pr(THT) = \Pr(TTH) = \Pr(TTT) = \frac{1}{8}$

A final result with 2 heads is possible from 3 of the outcomes:

$\Pr(2H) = \Pr(HHT \text{ or } HTH \text{ or } THH)$

$= \frac{3}{8}$

Calculating probabilities using a probability tree

If you follow along the branches from the start to each possible final outcome, then the probabilities for each branch can be multiplied together to find the final probability for each outcome.

$$\Pr(HHH) = \frac{1}{2} \times \frac{1}{2} \times \frac{1}{2} = \frac{1}{8} \quad \Pr(HHT) = \frac{1}{2} \times \frac{1}{2} \times \frac{1}{2} = \frac{1}{8} \quad \Pr(HTH) = \frac{1}{2} \times \frac{1}{2} \times \frac{1}{2} = \frac{1}{8} \quad \Pr(HTT) = \frac{1}{2} \times \frac{1}{2} \times \frac{1}{2} = \frac{1}{8}$$

$$\Pr(THH) = \frac{1}{2} \times \frac{1}{2} \times \frac{1}{2} = \frac{1}{8} \quad \Pr(THT) = \frac{1}{2} \times \frac{1}{2} \times \frac{1}{2} = \frac{1}{8} \quad \Pr(TTH) = \frac{1}{2} \times \frac{1}{2} \times \frac{1}{2} = \frac{1}{8} \quad \Pr(TTT) = \frac{1}{2} \times \frac{1}{2} \times \frac{1}{2} = \frac{1}{8}$$

These probabilities correspond to the probabilities previously calculated.

The probability of 2 heads is the sum of three probabilities:

$$\begin{aligned}\Pr(HHT) + \Pr(HTH) + \Pr(THH) &= \frac{1}{8} + \frac{1}{8} + \frac{1}{8} \\ &= \frac{3}{8}\end{aligned}$$

In this example, the outcomes of all events in the sample space are equally likely. In many circumstances, however, the outcomes of events in the sample space are not equally likely.

Consider the following situation:

Peach and Mario play three games of golf. Peach's chance of winning any one game is $\frac{3}{5}$. What is the probability that Peach wins two games out of three?

Let W represent the event 'Peach wins the golf game'.

Let L represent the event 'Peach loses the golf game'.

$\Pr(W) = \frac{3}{5}$ and $\Pr(L) = \frac{2}{5}$

The probabilities of any set of branches from one point must be add to 1 (because they represent all possible outcomes at that point).

In this case $\frac{3}{5} + \frac{2}{5} = 1$.

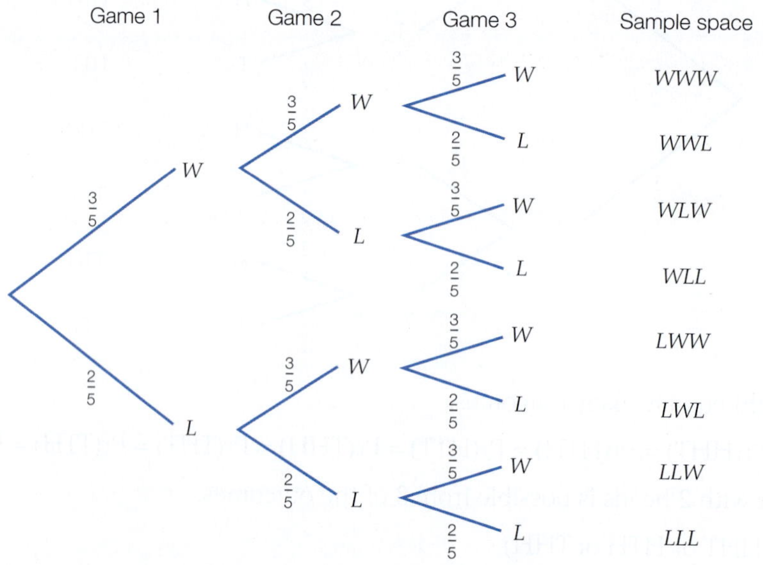

There are three possible situations where Peach wins two games: *WWL*, *WLW* and *LWW*.

$$\Pr(WWL) = \frac{3}{5} \times \frac{3}{5} \times \frac{2}{5}$$
$$= \frac{18}{125}$$

$$\Pr(WLW) = \frac{3}{5} \times \frac{2}{5} \times \frac{3}{5}$$
$$= \frac{18}{125}$$

$$\Pr(LWW) = \frac{2}{5} \times \frac{3}{5} \times \frac{3}{5}$$
$$= \frac{18}{125}$$

If you follow along the branches from the start to each possible outcome, multiplying the probabilities on each branch, this calculates the same probabilities.

These three possibilities are mutually exclusive, because they cannot occur at the same time, so you can add probabilities to find the overall probability that Peach wins two games.

$$\begin{aligned}\Pr(\text{Peach wins 2 games}) &= \Pr(WWL \text{ or } WLW \text{ or } LWW)\\ &= \Pr(WWL) + \Pr(WLW) + \Pr(LWW)\\ &= \frac{18}{125} + \frac{18}{125} + \frac{18}{125}\\ &= \frac{54}{125}\end{aligned}$$

So the probability that Peach wins two golf games out of three against Mario is $\frac{54}{125}$ or 0.432.

Worked example 7

W.E. 7

At supermarkets, cartons of eggs are inspected for broken shells before payment is made. The probability that a carton of eggs will be rejected is found to be 0.1. Two cartons of eggs are inspected.

(a) Copy and complete the following tree diagram. Evaluate the probability of each outcome.

Let R = carton of eggs is rejected and A = carton of eggs is accepted (not rejected)

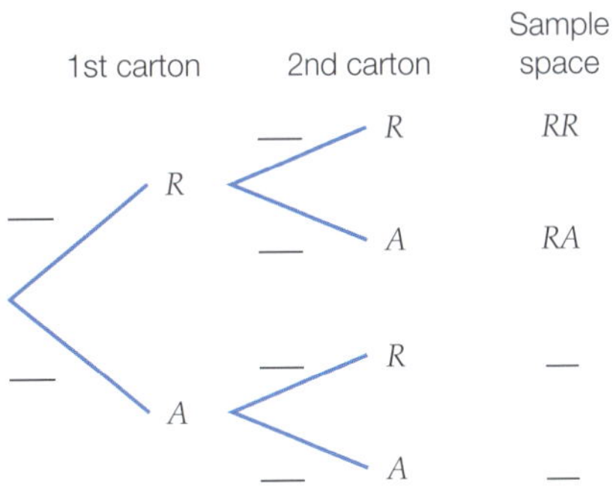

(b) Find the probability that:

- **(i)** both cartons are rejected
- **(ii)** only one carton is rejected
- **(iii)** neither carton is rejected
- **(iv)** at least one carton is rejected.

Thinking	Working
(a) **1** Use $\Pr(R) + \Pr(A) = 1$ to find $\Pr(A)$.	(a) $\Pr(R) = 0.1$, so $\Pr(A) = 0.9$
2 Copy and complete the tree diagram. Assign the probability of each branch, list the sample space and then calculate the probability of each outcome.	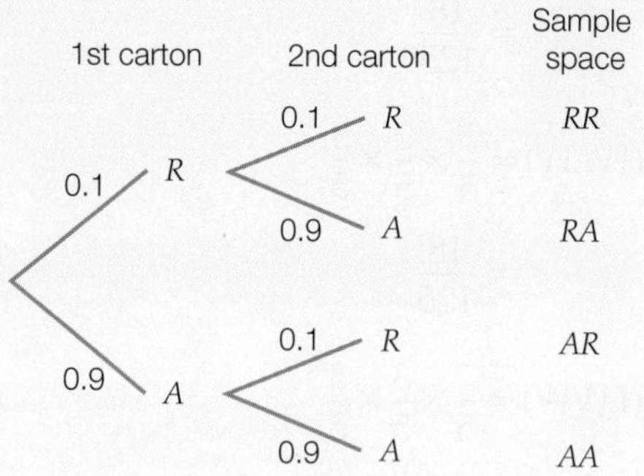 $\Pr(RR) = 0.1 \times 0.1$ $= 0.01$ $\Pr(RA) = 0.1 \times 0.9$ $= 0.09$ $\Pr(AR) = 0.9 \times 0.1$ $= 0.09$ $\Pr(AA) = 0.9 \times 0.9$ $= 0.81$
(b) **(i)** Use the sample space to find the required outcomes.	(b) (i) Pr(both cartons will be rejected) $= \Pr(RR)$ $= 0.01$
(ii) **1** Use the sample space to find the required outcomes.	(ii) Pr(only one carton will be rejected) $= \Pr(RA \text{ or } AR)$
2 These outcomes are mutually exclusive, so use the addition rule for mutually exclusive events.	$= \Pr(RA) + \Pr(AR)$ $= 0.09 + 0.09$ $= 0.18$
(iii) Use the sample space to find the required outcomes.	(iii) Pr(neither carton will be rejected) $= \Pr(AA)$ $= 0.81$
(iv) Use the equation for complementary events.	(iv) Pr(at least one carton will be rejected) = 1 − Pr(none rejected) = Pr(one carton rejected) + Pr(both cartons rejected) $= 1 - 0.81$ $= 0.19$

In the next example the probabilities change because the sampling is without replacement. This means that the set of outcomes will not be the same at each step.

Worked example 8 W.E. 8

The lucky dip at the school fete contains 16 different prizes: 10 gifts suitable for ages 4–7 and 6 gifts suitable for ages 8–11. Unfortunately, the 16 prizes have been mixed together with no labels. A 5-year-old takes two prizes. Let B represent 'a prize suitable for ages 4–7 is chosen' and N represent 'a prize suitable for ages 4–7 is not chosen'.

(a) Find Pr(both prizes are suitable for ages 4–7).

(b) Find Pr(the 5-year-old chooses at least one suitable prize).

Thinking	Working
1 Draw a tree diagram and assign a probability to each branch.	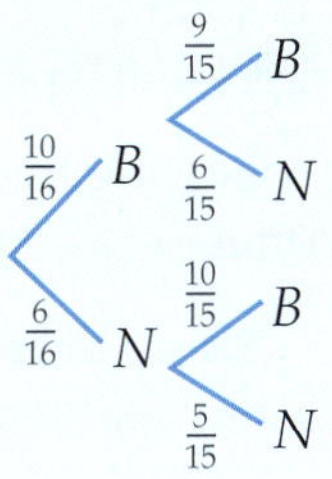
2 List the sample space.	BB, BN, NB, NN
(a) Use the sample space to find the required outcomes.	(a) Both gifts are suitable for ages 4–7: BB $\Pr(BB) = \frac{10}{16} \times \frac{9}{15}$ $= \frac{3}{8}$
(b) Use the equation for complementary events to find the probability. (Here, NN is the complementary event.)	(b) $\Pr(BB \text{ or } BN \text{ or } NB) = 1 - \Pr(NN)$ $\Pr(NN) = \frac{6}{16} \times \frac{5}{15}$ $= \frac{1}{8}$ Pr(the 5-year-old chooses at least one suitable gift) $= 1 - \Pr(\text{both gifts unsuitable})$ $= 1 - \Pr(NN)$ $= 1 - \frac{1}{8}$ $= \frac{7}{8}$

You can use a tree diagram and multiply along its branches to find the probabilities of multiple events.

10.4 Probability tree diagrams

Navigator

Answers p. 849

1, 2, 3, 4, 5, 6, 7, 8, 9, 10, 11, 12, 15	2, 3, 4, 5, 6, 7 (a, b), 8, 9, 10, 11, 12, 15	2, 3, 5, 6, 7 (b, c), 8, 9, 10, 11, 12, 13, 14, 15

Fluency

W.E. 7

1 A spinner numbered 1 to 3 is spun twice. Let T represent 'the spin result is 2' and N represent 'spin result is not 2'.

(a) Copy and complete the following tree diagram. Evaluate the probability of each outcome.

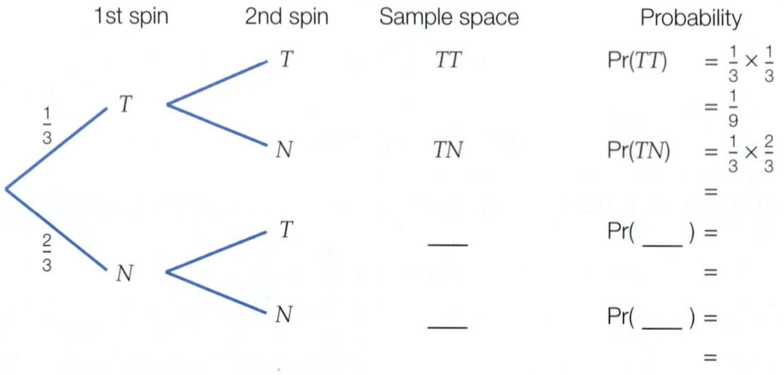

(b) Find the probability of:

(i) spinning 2 twice

(ii) spinning 2 once only

(iii) *not* spinning a 2

(iv) spinning 2 once at least.

W.E. 8

2 Two balls are taken without replacement from a bag containing 4 green marbles and 7 blue marbles. Draw a tree diagram to represent the situation and find:

(a) Pr(both marbles being blue)

(b) Pr(getting at least one blue marble).

3 The spinner described in Question **1** is now spun three times.

(a) Draw a probability tree diagram for this situation.

(b) What is the probability of spinning a 2 three times?

(c) What is the probability of *not* spinning a 2 on any of the three spins?

4 A biased coin is tossed twice. Let H represent 'result is heads' and T represent 'result is tails'. The following probability tree diagram can be drawn.

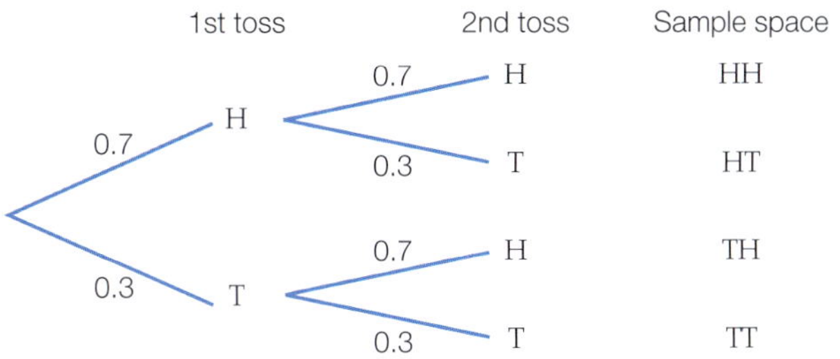

(a) The probability of getting heads twice is:

A $0.7 + 0.7$ B 0.7×0.7 C 0.25 D 0.7×0.3

(b) The probability of getting tails twice is:

A 0.6 B 0.3 C 0.21 D 0.09

(c) The probability of getting tails first, then heads:

A 0.25 B 0.21 C 0.49 D 1.0

(d) The probability of getting tails only once:

A 0.21 B 0.3 C 0.42 D 0.63

(e) The probability of getting tails at least once:

A 0.75 B 0.51 C 0.42 D 0.39

5 Two cards are drawn, without replacement, from a set of 16 cards numbered from 1 to 16.

Let M = the event that a multiple of four is drawn and N = the event that a multiple of four is not drawn.

(a) Draw a tree diagram and assign a probability to each branch.

(b) List the sample space.

(c) Find Pr(both cards are a multiple of 4).

(d) Find Pr(only one card is a multiple of 4).

(e) Find Pr(neither card is a multiple of 4).

Understanding

6 For afternoon tea, customers in a cafe can select from tea, coffee or hot chocolate, and either scones or carrot cake. The probability of a person selecting tea is 0.45, whereas the probability of a person selecting coffee is 0.35. The probability of a person selecting scones is 0.6.

(a) What is the probability of a person selecting hot chocolate?

(b) What is the probability of a person selecting carrot cake?

(c) Draw a probability tree diagram to represent this situation.

(d) Find the probability of a person:

(i) selecting tea and scones

(ii) selecting hot chocolate and carrot cake

(iii) selecting tea and scones or hot chocolate and carrot cake

(iv) *not* selecting coffee and scones.

7 For each of the following:

(i) copy and complete the probability tree diagram

(ii) write the probability of each possible outcome

(iii) find the sum of the probabilities of all outcomes.

(a) Let G = scores a goal and D = does not score a goal.

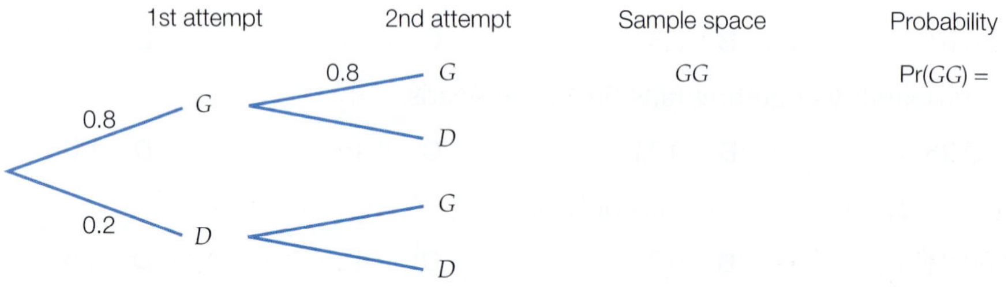

(b) P = Patrick is chosen from a group of Year 10 students.

A = Anne is chosen from a group of Year 11 students.

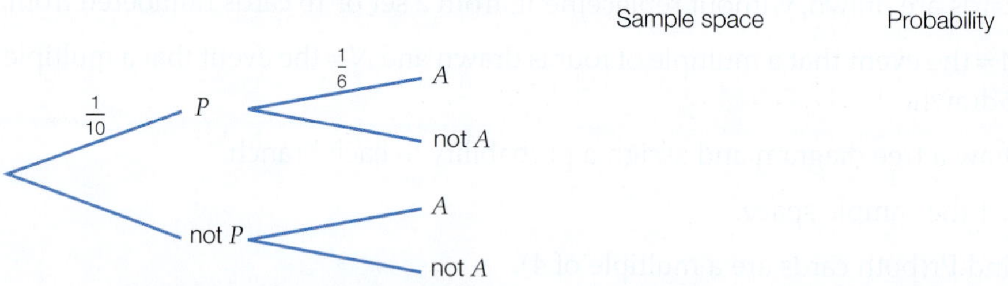

(c) Let B = student travels to school by bus and C = student does not travel by bus.

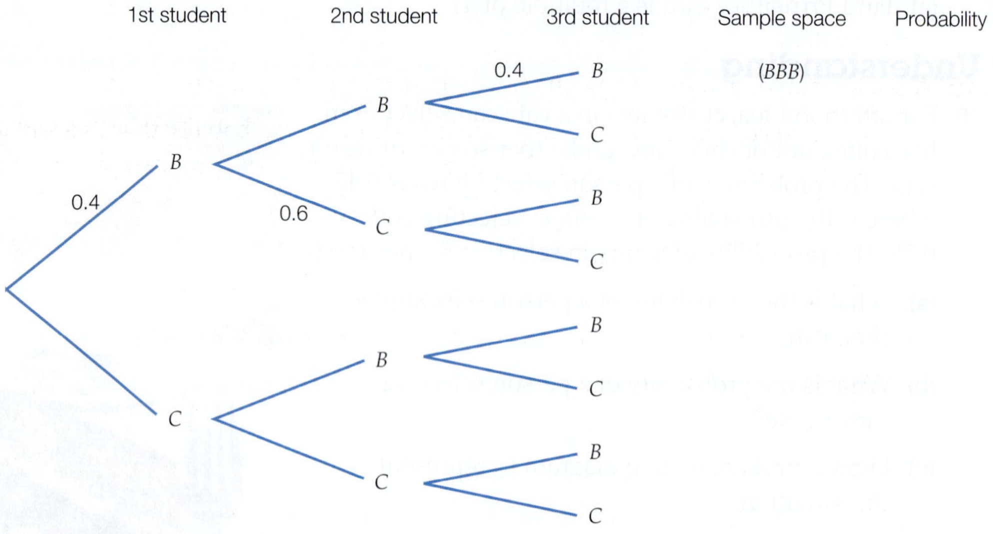

8 In a game of darts, each player has a turn to throw three darts one at a time. Seline estimates her chances of scoring a bullseye with any one throw of a dart to be 0.2. Find the probability that:

(a) Seline scores three bullseyes in her turn

(b) she scores two bullseyes only

(c) she scores at least one bullseye.

9 Two six-sided dice are rolled. Let T represent the event 'a 3 is rolled'. This is the event that a 3 is rolled on one die (not a total of 3 from the two dice added together).

(a) What does not T represent?

(b) Draw a probability tree diagram to illustrate this situation.

(c) Find the probability of:

(i) rolling two 3s

(ii) not rolling any 3s

(iii) rolling one 3 only

(iv) rolling at least one 3

(v) rolling no more than one 3.

10 A card is selected from a standard pack of 52 playing cards and then replaced. Another card is selected. Find the probability that:

(a) both cards are Aces

(b) the first card is a club and the second card is red

(c) both cards are the same colour

(d) one card is red and the other is black.

Reasoning

11 A quick quiz consists of five multiple choice questions, each with five choices for the answer. Find the probability of getting 0, 1, 2, 3, 4 or 5 correct if you randomly guess the answers.

12 A fair six-sided die is rolled again and again until the result is a 6.

(a) Copy and complete the following table to show the probability that this will take the indicated number of rolls. Round each probability to 3 decimal places.

Number of rolls	1	2	3	4	5	6
Probability						

(b) Add the probabilities you found in part (a). What sum do you get?

(c) Explain what this answer means.

13 A party game for 29 people starts by assigning everyone a number from 1 to 29. In the first round, seven numbers are drawn at random and seven people are eliminated. In the second round 6 people are eliminated, the third round eliminates 5 people, then 4, 3, 2 and finally 1, to leave the winner.

(a) Find Pr(being eliminated in round 1).

(b) Assuming you are not eliminated in the first round, find Pr(being eliminated in round 2).

(c) Assuming you are not eliminated in the previous rounds, find the probability of being eliminated in each round.

(d) Add the results for the probability of being eliminated in each round. Explain the result.

Open-ended

14 Draw a probability tree diagram that finishes with at least nine branches, where one of the final branches has a probability of $\frac{1}{9}$ and another has a probability of $\frac{1}{8}$. Explain what your tree diagram represents.

15 The estimated probability of a 25-year-old person playing cricket is 0.2 and a 25-year-old person playing golf is 0.18. Armapreet calculates that the probability of a person playing cricket or golf must be 0.38. Is she right? If not, what mistake has she made?

Problem solving

The problem of the points

The origins of modern probability theory involve a problem known as the 'problem of the points' or 'division of the stakes'.

This is about what happens in a two-person game where each person contributes the same amount to the prize pot, and the first person to reach a certain number of points will be the winner and collect the entire prize pot. But how should the prize pot be divided if the game is interrupted and must finish early?

Consider a game where the winner is the first to 10 points.

1 How would you divide the pot if the score was 7 to 5?

2 One suggestion is that the pot should be divided in proportion to the number of points. So if the score is 7 to 5, then the pot is divided $\frac{7}{12}:\frac{5}{12}$. What do you think of this solution? What would happen if the score was 1 to 0?

The mathematicians Pascal and Fermat believed that the possible ways the game might have continued were more important than what had already happened.

3 Consider a game where the score is 8 to 7. Write all the ways the game could continue and use this to decide how to divide the pot.

4 What if the score is 2 to 5? What is the problem with this sort of score?

5 What do you think would be the easiest scores to deal with if the game was interrupted?

Strategy options

- Act it out.
- Draw a diagram.

Extension

Research to find out how Pascal and Fermat solved this problem.

Conditional statements

10.5

Consider two fair six-sided dice, one green and one red. The following array shows the possible outcomes when both dice are rolled.

		Green die					
		1	2	3	4	5	6
Red die	1	(1, 1)	(2, 1)	(3, 1)	(4, 1)	(5, 1)	(6, 1)
	2	(1, 2)	(2, 2)	(3, 2)	(4, 2)	(5, 2)	(6, 2)
	3	(1, 3)	(2, 3)	(3, 3)	(4, 3)	(5, 3)	(6, 3)
	4	(1, 4)	(2, 4)	(3, 4)	(4, 4)	(5, 4)	(6, 4)
	5	(1, 5)	(2, 5)	(3, 5)	(4, 5)	(5, 5)	(6, 5)
	6	(1, 6)	(2, 6)	(3, 6)	(4, 6)	(5, 6)	(6, 6)

There are 36 equally likely outcomes, so each outcome has a probability of $\frac{1}{36}$. The number of outcomes in the sample space becomes the denominator for the probability fraction.

If you know that the number on the green die is 1, then the probability Pr(1, 1) is not $\frac{1}{36}$. The probability is $\frac{1}{6}$, because only the first column of results in the array is possible. The sample space is restricted, so only the outcomes that include a 1 on the green die are considered.

Pr[(1, 1) given green die is 1] $= \frac{1}{6}$

Similarly, you can calculate the probability that the red die is even, given that the green die is a multiple of 3. The information after the word 'given' restricts the sample space, which changes the value of the denominator for the probability. This is a **conditional statement**.

The array can be shaded to show how the sample space is restricted for Pr(red die is even, given green die is a multiple of 3).

		Green die					
		1	2	3	4	5	6
Red die	1	(1, 1)	(2, 1)	(3, 1)	(4, 1)	(5, 1)	(6, 1)
	2	(1, 2)	(2, 2)	(3, 2)	(4, 2)	(5, 2)	(6, 2)
	3	(1, 3)	(2, 3)	(3, 3)	(4, 3)	(5, 3)	(6, 3)
	4	(1, 4)	(2, 4)	(3, 4)	(4, 4)	(5, 4)	(6, 4)
	5	(1, 5)	(2, 5)	(3, 5)	(4, 5)	(5, 5)	(6, 5)
	6	(1, 6)	(2, 6)	(3, 6)	(4, 6)	(5, 6)	(6, 6)

		Green die	
		3	6
Red die	1	(3, 1)	(6, 1)
	2	(3, 2)	(6, 2)
	3	(3, 3)	(6, 3)
	4	(3, 4)	(6, 4)
	5	(3, 5)	(6, 5)
	6	(3, 6)	(6, 6)

The green shading shows the given condition: that the green die is a multiple of 3 (that is, either a 3 or a 6). There are 12 possible outcomes in the restricted sample space, shaded green, so the denominator for the probability fraction is 12.

From this restricted set, you need to count the number of outcomes with an even number on the red die. There are 6 possible outcomes, shaded red:

Pr(red die is even, given green die is a multiple of 3) $= \frac{6}{12} = \frac{1}{2}$

The order of the statements is important. Pr(red die is even, given the green die is a multiple of 3) is not the same as Pr(green die is a multiple of 3, given the red die is even).

		Green die					
		1	2	3	4	5	6
Red die	1	(1, 1)	(2, 1)	(3, 1)	(4, 1)	(5, 1)	(6, 1)
	2	(1, 2)	(2, 2)	(3, 2)	(4, 2)	(5, 2)	(6, 2)
	3	(1, 3)	(2, 3)	(3, 3)	(4, 3)	(5, 3)	(6, 3)
	4	(1, 4)	(2, 4)	(3, 4)	(4, 4)	(5, 4)	(6, 4)
	5	(1, 5)	(2, 5)	(3, 5)	(4, 5)	(5, 5)	(6, 5)
	6	(1, 6)	(2, 6)	(3, 6)	(4, 6)	(5, 6)	(6, 6)

		Green die					
		1	2	3	4	5	6
Red die	2	(1, 2)	(2, 2)	(3, 2)	(4, 2)	(5, 2)	(6, 2)
	4	(1, 4)	(2, 4)	(3, 4)	(4, 4)	(5, 4)	(6, 4)
	6	(1, 6)	(2, 6)	(3, 6)	(4, 6)	(5, 6)	(6, 6)

There are 18 possible outcomes in the restricted set, shaded red, so the denominator for the probability fraction is 18. There are then 6 possible outcomes with a 3 or 6 on the green die, which are shaded green.

Pr(green die is a multiple of 3, given the red die is even) $= \frac{6}{18} = \frac{1}{3}$

Instead of using tables, you could also use a Venn diagram to show this type of information.

Using Venn diagrams

Worked example 9

W.E. 9

A 16-sided die with faces numbered 1–16 is rolled. The following events are defined:
A = 'number rolled is less than 7', B = 'number rolled is a multiple of 4'

(a) Draw a Venn diagram to show this information.

(b) Calculate Pr(number rolled is less than 7, given it is a multiple of 4).

(c) Calculate Pr(number rolled is a multiple of 4, given it is less than 7).

Thinking	Working
(a) Draw the Venn diagram. Write the elements in the overlap (intersection) of A and B first as it makes it easier to draw the diagram.	(a) 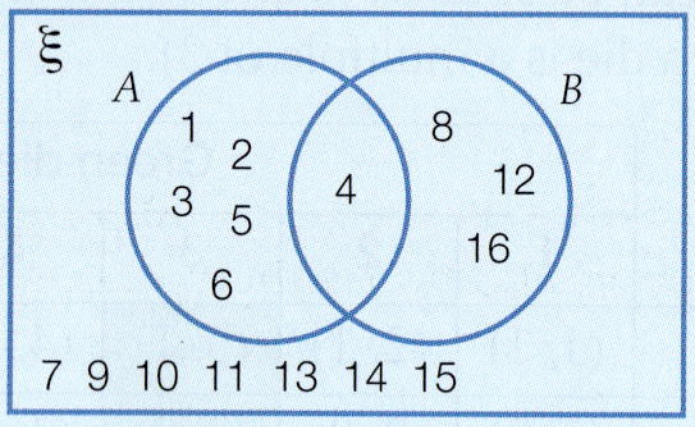
(b) 1 Identify the denominator for the probability. This is the number of elements for the set that is 'given'.	(b) The denominator is the number of elements in B (a multiple of 4): 4
2 Identify the numerator for the probability.	Find how many of the elements of B are also in A (a number less than 7): 1
3 State the probability.	Pr(a number less than 7, given it is a multiple of 4) $= \frac{n(\text{less than 7 and multiple of 4})}{n(\text{multiple of 4})}$ $= \frac{n(A \text{ and } B)}{n(B)}$ $= \frac{1}{4}$

(c) 1 Identify the denominator for the probability. This is the number of elements for the set that is 'given'.

(c) The denominator is the number of elements in A (a number less than 7): 6

2 Identify the numerator for the probability.

Find how many of the elements of A are also in B (a multiple of 4): 1

3 State the probability.

Pr(a multiple of 4, given it is a number less than 7)

$$= \frac{n(\text{multiple of 4 and less than 7})}{n(\text{multiple of 4})}$$

$$= \frac{n(B \text{ and } A)}{n(A)}$$

$$= \frac{1}{6}$$

Using tables

So far, the expression 'given that' has been used as the conditional statement. However, other expressions can also be used, such as 'knowing that', 'when' and 'if … then'. Tables can also be used to help with these probability calculations.

Worked example 10

W.E. 10

In a particular electorate, 55% of the voting population is younger than 40 years old. It is also known that 60% of the people older than 40 years old support the Progressive Conservative Party, while only 45% of the under-40s support this party.

(a) Construct a two-way table to display this information.

(b) Find the probability that a randomly selected voter from that electorate is over 40 if you know that they support the Progressive Conservative Party. Round your answer to 2 decimal places.

Thinking	Working
(a) 1 Construct the table with headings for support and non-support along the top, and for under-40 and over-40 on the left.	Total under-40s: $55\% = 0.55$ Total over-40s: $1 - 0.55 = 0.45$
2 Calculate the values from the information given.	Under-40s supporters: $0.45 \times 0.55 = 0.2475$ Over-40s supporters: $0.6 \times 0.45 = 0.27$ Total supporters: $0.2475 + 0.27 = 0.5175$ Under-40s non-supporters: $0.55 - 0.2475 = 0.3025$ Over-40s non-supporters: $0.45 - 0.27 = 0.12$ Total non-supporters: $0.3025 + 0.12 = 0.4825$

	Support PCP	Do not support PCP	
Under-40s	0.2475	0.3025	0.55
Over-40s	0.27	0.12	0.45
	0.5175	0.4825	

(b) 1 To find probability in the restricted domain: the denominator is the proportion of the 'given' group, the numerator is the proportion (of the whole population) for which the described group overlaps the given group.

Pr(over-40, given supporter of PCP)

$= \dfrac{\text{Pr(over-40 and supporter of PC)}}{\text{Pr(support PCP)}}$

2 Substitute the values from the table.

$= \dfrac{0.27}{0.5175}$

3 Calculate, rounding to 2 decimal places.

$= 0.52 \quad (2\text{ d.p.})$

Conditional probability

$$\Pr(A \text{ given } B) = \frac{n(A \text{ and } B)}{n(B)} \text{ or } \Pr(A \text{ given } B) = \frac{\Pr(A \text{ and } B)}{\Pr(B)}$$

10.5 Conditional statements

Navigator

Answers p. 851

1, 2, 3, 4, 5, 6, 7, 8, 9 (a–c), 11	1, 2, 3, 4, 5, 6, 7, 8, 9, 10, 11	1, 2, 3 (a), 4, 5, 6, 7, 8, 9, 10, 11

Fluency

W.E. 9

1 A 16-sided die with faces numbered 1–16 is rolled. The following events are defined.

(a) A = 'number rolled is less than 9' and B = 'number rolled is a multiple of 3'

(i) Draw a Venn diagram to show this information.

(ii) Calculate Pr(number rolled is less than 9, given it is a multiple of 3).

(iii) Calculate Pr(number rolled is a multiple of 3, given it is a number less than 9).

(b) A = 'number rolled is greater than 5' and B = 'number rolled is a prime number'

(i) Draw a Venn diagram to show this information.

(ii) Calculate Pr(number rolled is greater than 5, given it is a prime number).

(iii) Calculate Pr(number rolled is a prime number, given it is greater than 5).

W.E. 10

2 In a particular electorate, 65% of the voting population do not own any property. It is also known that 55% of property-owners support the Progressive Conservatives Party, but only 40% of the non-property-owners support this party.

(a) Construct a two-way table to display this information.

(b) Find the probability that a randomly selected voter from this electorate does not own property, if you know that they support the Progressive Conservative Party. Round your answer to 2 decimal places.

3 Use the information in the Venn diagram to find:

(i) Pr(A given B) (ii) Pr(B given A)

(a)

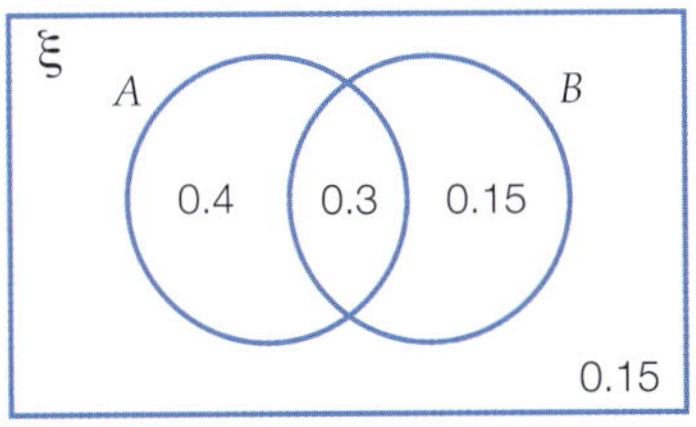

(b)

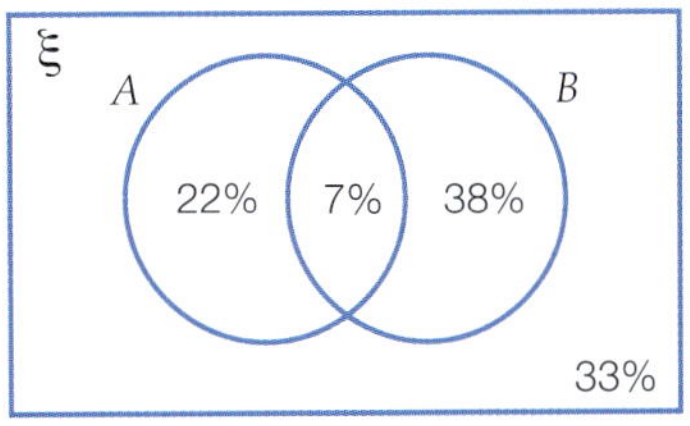

4 A coin is tossed three times.

(a) The probability that the third toss is heads, given the previous two tosses were both heads, is:

A $\frac{1}{8}$ B $\frac{1}{4}$ C $\frac{1}{2}$ D 1

(b) The probability that the third toss is heads, given at least two of the three tosses are heads, is:

A $\frac{3}{8}$ B $\frac{1}{2}$ C $\frac{2}{3}$ D $\frac{3}{4}$

5 You have a reduced pack of playing cards that includes only the 26 red cards. A card is selected at random from the pack. Find the probability that the card is:

(a) an Ace, given it is a heart

(b) a heart, given it is a picture card (King, Queen or Jack)

(c) the 5 of diamonds, given it is a diamond

(d) the 5 of diamonds, given it is 5.

Understanding

6 A new disease is known to infect 2.5% of the population. A test for the disease has been developed, but it is not completely accurate. It correctly identifies the disease in infected patients 95% of the time, but also (incorrectly) identifies the disease in 7% of the uninfected population.

(a) Construct a table to show this information.

(b) Find Pr(have disease, given test is positive) correct to 2 decimal places.

(c) Find Pr(don't have disease, given test is negative) correct to 3 decimal places.

7 A sportswear company sources its shoes from two suppliers, A and B. It is known that 5% of the shoes from supplier A are defective, whereas the figure is 9% for supplier B. The company obtains 65% of its shoes from supplier A. Where necessary, round your answers to 2 decimal places.

(a) Construct a table to show this information.

(b) Find Pr(A given the shoe is defective).

(c) Find Pr(B given the shoe is not defective).

8 A university student is studying both Economics and Literature. The probability that she will pass Economics is 0.95 and the probability that she will pass Literature is 0.8. There is a 0.75 chance she will pass both. Find:

(a) Pr(passes Economics, given that she passes Literature).

(b) Pr(passes Literature, given that she passes Economics).

(c) Pr(passes Economics, given that she passes only one of the subjects).

Reasoning

9 Ang plays a lot of table tennis. He has a 70% chance of winning any particular game. Find the probability that Ang wins a match decided by:

(a) best of three games
(b) best of three games, given that he won the first game
(c) best of three games, given that it lasts three games
(d) best of five games
(e) best of five games, given that he won the first two games.

10 A new cola is produced which the maker claims has the same taste as the market-leading cola. A test is set up whereby people are offered a drink and they say whether it is the market leader or the new cola. The new cola is correctly identified 40% of the time and the market leader is correctly identified 75% of the time. If a person has an equal chance of being given either cola and they say the cup contains the market leader, what is the probability that this is true?

Open-ended

11 A tired teacher wrote this question:

Given Pr(A) = 0.6 and Pr(B) = 0.4, find Pr(A given B).

(a) What is wrong with the question?
(b) Rewrite the question so it is correct, and then find the answer.

Dependent and independent events

10.6

Consider the situation where you have a fair coin and a spinner equally divided into three sections, labelled 1, 2 and 3.

You can find the sample space by using the following table:

		Coin	
		H	**T**
Spinner	**1**	(H, 1)	(T, 1)
	2	(H, 2)	(T, 2)
	3	(H, 3)	(T, 3)

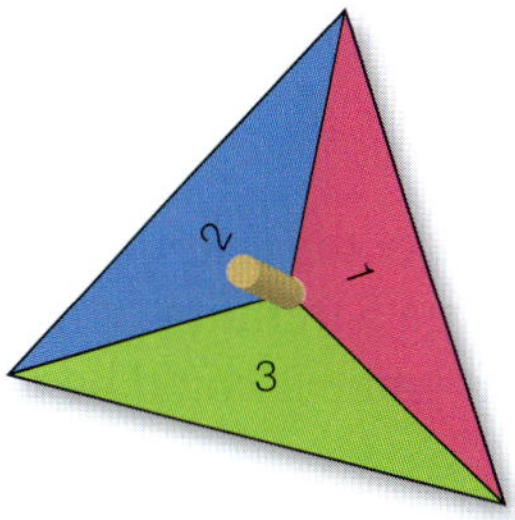

These should be equally likely outcomes, assuming that the coin is fair and the spinner is equally divided. From the table you can see that:

$$\begin{aligned}\Pr(\text{H}) &= \Pr(\text{H}, 1) + \Pr(\text{H}, 2) + \Pr(\text{H}, 3)\\ &= \frac{1}{6} + \frac{1}{6} + \frac{1}{6}\\ &= \frac{3}{6}\\ &= \frac{1}{2}\end{aligned}$$

$$\begin{aligned}\Pr(1) &= \Pr(\text{H}, 1) + \Pr(\text{T}, 1)\\ &= \frac{1}{6} + \frac{1}{6}\\ &= \frac{2}{6}\\ &= \frac{1}{3}\end{aligned}$$

What is the probability of getting heads *and* a 1? This is $\Pr(\text{H}, 1) = \frac{1}{6}$.

Note that $\Pr(\text{H}) \times \Pr(1) = \frac{1}{2} \times \frac{1}{3} = \frac{1}{6}$

So, $\Pr(\text{H and } 1) = \Pr(\text{H}) \times \Pr(1)$.

When $\Pr(A \text{ and } B) = \Pr(A) \times \Pr(B)$, this means that A and B are independent. **Independent events** are not influenced by the outcomes of other events. One event has no effect on the other event. In this case, the toss of the coin has no influence on what happens to the spinner.

Two events A and B are *independent* if the result of one event has no effect on the result of the other event.

When A and B are independent:

$$\Pr(A \text{ and } B) = \Pr(A) \times \Pr(B)$$

Dependent events are influenced by the outcome of another event. Sampling without replacement is an example of a dependent event, because each sampling probability will depend on what was sampled previously.

For example, consider two balls drawn without replacement from a box that contains 5 white balls and 3 green balls.

For the first ball drawn, Pr(white) = $\frac{5}{8}$ and Pr(green) = $\frac{3}{8}$.

However, for the second draw the probabilities depend on which colour ball was drawn first.

- If the first ball drawn was white, then Pr(white) = $\frac{4}{7}$ and Pr(green) = $\frac{3}{7}$.
- If the first ball drawn was green, then Pr(white) = $\frac{5}{7}$ and Pr(green) = $\frac{2}{7}$.

In this case the probability of the second event is dependent on the first event.

Consider the example of a sports competition in which the player who wins a game is allowed an extra 5 minutes of rest time before the next game. If a player wins game 1, then the event 'wins game 2' could be more likely because of the extra resting time.

Worked example 11

W.E. 11

Two representatives, one girl and one boy, are chosen at random from a group of 6 girls and 4 boys. If Linda and Adam are in this group, what is the probability that both Linda and Adam will be chosen?

Thinking	Working
1 Assign letters to stand for each of the basic events.	Let L = Linda chosen and A = Adam chosen
2 Are these events independent or does one result affect the other?	L and A are independent events.
3 Both must occur. Use Pr(A and B) = Pr(A) × Pr(B).	Pr(L and A) = Pr(L) × Pr(A)
4 Calculate the probability of each event happening in isolation and then calculate the probability of both.	Pr(L) = $\frac{1}{6}$ Pr(A) = $\frac{1}{4}$ Pr(L and A) = $\frac{1}{6} \times \frac{1}{4}$ = $\frac{1}{24}$
5 State the answer.	The probability that both Linda and Adam will be chosen is $\frac{1}{24}$ (i.e. 1 chance in 24).

You could have previously solved some problems like this using a tree diagram, but calculations like this makes it easier to solve more complex examples.

If you can state the individual probabilities for a chain of events, then you can multiply those probabilities to find the probability of all of the events occurring. This means that you can solve problems by multiplying probabilities instead of by drawing a tree diagram or writing out the complete sample space.

Worked example 12

W.E. 12

Two cards are taken out of a standard pack of 52 playing cards without replacement.

(a) Find Pr(drawing 2 Kings).

(b) Find Pr(both cards are the same suit).

Thinking	Working
(a) 1 Find the probability of the given event. (Here, Pr(K on first draw).)	(a) $\Pr(\text{King on 1st draw}) = \frac{4}{52} = \frac{1}{13}$
2 Find the probability of the second event knowing the first has occurred. (Here, Pr(K on second draw after K on first draw).)	$\Pr(\text{King on 2nd draw}) = \frac{3}{51} = \frac{1}{17}$
3 Find the probability of the stated event. (Here, Pr(KK), with no replacement.)	$\Pr(\text{both cards being a King}) = \frac{1}{13} \times \frac{1}{17} = \frac{1}{221}$
(b) 1 Find the probability of each successful outcome. (Here, Pr(♣♣), Pr(♠♠), Pr(♥♥), Pr(♦♦).)	(b) Pr(♣♣)) $= \frac{13}{52} \times \frac{12}{51} = \frac{1}{17}$ Pr(♠♠) $= \frac{13}{52} \times \frac{12}{51} = \frac{1}{17}$ Pr(♥♥) $= \frac{13}{52} \times \frac{12}{51} = \frac{1}{17}$ Pr(♦♦) $= \frac{13}{52} \times \frac{12}{51} = \frac{1}{17}$
2 The events are mutually exclusive so add the individual probabilities to find the probability of all successful outcomes.	Pr(both cards are from the same suit) $= \frac{1}{17} + \frac{1}{17} + \frac{1}{17} + \frac{1}{17} = \frac{4}{17}$

10.6 Dependent and independent events

Navigator

Answers p. 851

1, 2, 3, 4, 5, 6, 7, 8, 9, 10, 11, 12, 13, 14, 15, 17	1, 2, 4, 5, 6 (a–d), 7, 8, 9 (column 1), 10, 11, 12, 13, 14, 15, 16, 17, 18	1, 2, 4, 6 (e–h), 7 (c), 8, 9 (column 2), 10, 11, 12, 13, 14, 15, 16, 17, 18, 19

Fluency

W.E. 11

1 Katrina finds that on average her train is late 2 mornings out of 5, whereas Frank's bus is late 1 morning in every 4. Find the probability that on any given day both Katrina and Frank are late.

W.E. 12

2 A box contains 10 each of four different varieties of apple. Five of the 40 apples are rotten. If 2 apples are taken out without replacement, calculate:

(a) Pr(2 rotten apples)

(b) Pr(2 apples of the same variety)

3 Phillip has 2 pairs of blue jeans and 1 pair of black jeans hanging in his wardrobe. In a drawer, he has 1 red, 3 black and 5 white T-shirts. If he randomly chooses a pair of jeans and a T-shirt to wear, find the probability that he chooses:

(a) a white T-shirt and blue jeans

(b) a black T-shirt and black jeans

(c) black jeans and a T-shirt that is not black

(d) blue jeans and a T-shirt that is not red.

4 In a group of 20 people, 2 have red hair, 6 have blonde hair and the rest have dark hair. Of the group, 15 people are right-handed. If a person is chosen at random from the group, find the probability that the person:

(a) has red hair and is right-handed

(b) has blonde hair and is left-handed

(c) is right-handed and does not have red hair.

5 Label each of these pairs of events as either independent (I) or dependent (D).

(a) Tossing a coin and rolling a die.

(b) Drawing two cards from a standard pack without replacing the first before drawing the second.

(c) Drawing two cards from a standard pack replacing the first before drawing the second.

(d) Raining in two cities 100 km apart on the same day.

(e) Hours of study and test result.

(f) The number of hours of practice a learner driver does before getting their licence and the chance of having an accident in the first month of having their licence.

6 A packet of beads contains 3 red beads, 4 blue beads and 8 green beads. If two beads are taken without replacement, find:

(a) Pr(RR) (b) Pr(BB) (c) Pr(GG) (d) Pr(RB)

(e) Pr(BG) (f) Pr(GB) (g) Pr(same colour) (h) Pr(different colours)

7 (a) If a 20-cent coin is tossed in the air twice, the probability of two heads is:

A $\frac{1}{4}$ B $\frac{1}{3}$ C $\frac{1}{2}$ D $\frac{3}{4}$

(b) If a six-sided die is rolled twice, the probability of a double 6 is:

A $\frac{1}{3}$ B $\frac{1}{6}$ C $\frac{1}{12}$ D $\frac{1}{36}$

(c) If a 20-cent coin is tossed and a six-sided die is rolled, the probability of heads and a multiple of 3 is:

A $\frac{1}{6}$ B $\frac{1}{3}$ C $\frac{1}{4}$ D $\frac{1}{2}$

Understanding

8 Joshua and Tania are competing in different sections of a chess tournament. Joshua estimates that his chance of winning against his opponent is 3 in 5, whereas Tania estimates that her chance of winning against her opponent is 5 in 7. Find the probability that:

(a) both Joshua and Tania win their chess game

(b) they both lose their chess game

(c) Joshua loses his game but Tania wins hers.

9 A drawer contains 6 black socks, 8 red socks and 4 white socks. If two socks are taken without replacement, find:

(a) Pr(2 black socks) (b) Pr(2 red socks)

(c) Pr(2 white socks) (d) Pr(a pair of socks of matching colour)

(e) Pr(two odd socks) (f) Pr(one black sock)

(g) Pr(at least one black sock) (h) Pr(no white socks)

10 A spinner numbered 1 to 8 is spun twice. Find the probability of:

(a) a 5 on the first spin

(b) a 5 and a 7 in any order

(c) a 5 on the first spin and a number less than 4 on the second

(d) an even number on the first spin and an odd number on the second.

11 If the probability that person A will be alive in 20 years is 0.7 and the probability that person B will be alive in 20 years is 0.5, find:

(a) the probability that they will both be alive in 20 years

(b) the probability that only one is alive in 20 years.

12 Natalie and her team play in a weekly squash competition. The probability that her team wins in any given week is 0.7. The outcome is not affected by any previous wins or losses. Find the probability that:

(a) Natalie's team does not win in any given week

(b) her team wins 2 weeks running

(c) her team loses 3 weeks running

(d) her team loses 4 weeks running.

13 A bag contains 6 white marbles, 5 black marbles and 4 green marbles. In each draw, a marble is drawn from the bag and not replaced. In three draws, find:

(a) Pr(WBG in that order) (b) Pr(WBG in any order).

Reasoning

14 Two normal unbiased six-sided dice are rolled.

(a) Write the sample space for this experiment.

From the sample space these events are defined:

A = rolling a 6 with the first die B = rolling a 5 with the second die

C = rolling a total of 8 D = rolling a total greater than 10

(b) Find the probability for each event.

(c) List all possible pairs of events for A and B.

(d) How can you check which pairs are independent?

(e) Which, if any, of these pairs of events are independent?

15 You know that two events, A and B, are independent and that Pr(A) = 0.3 and Pr(A and B) = 0.15.

(a) Find Pr(B).

(b) Now find Pr(A or B). Explain how you calculated this value.

16 A secondary school has 900 students, of whom 475 have black hair and 870 study mathematics. There are 470 black-haired mathematics students. If a student is picked at random, show that the events 'picking a black-haired student' and 'picking a maths student' are not independent.

17 For two events, A and B, $\Pr(A) = 0.35$ and $\Pr(B) = 0.45$. Find $\Pr(A \text{ and } B)$ if:

(a) A and B are mutually exclusive

(b) A and B are independent.

Open-ended

18 Two events are defined as

A: multiples of 3 B: multiples of 4

Create a sample space so that:

(a) A and B are independent events

(b) A and B are dependent events.

19 A box contains x red balls, y yellow balls and z green balls. Three balls are drawn without replacement. Choose values for x, y and z so that:

(a) Pr(one of each colour) is between 0.1 and 0.3

(b) Pr(all three are yellow) is between 0.1 and 0.2

(c) Pr(all three are the same colour) is between 0.3 and 0.5

Problem solving

Winning the lottery

You buy a raffle ticket that has a unique six-digit number from 000001 to 999999 printed on it. The winning six-digit number is drawn at random and the ticket with that number wins first prize, \$100 000. Any ticket with the first digit wrong but the next five digits correct wins second prize, \$20 000. Any ticket with the first and second digits wrong but the last four digits correct wins third prize (\$10 000), and any ticket with the first three digits wrong but the last three digits correct wins fourth prize (\$2000).

1 What is the probability that you win the following?

(a) \$100 000

(b) \$20 000

(c) \$10 000

(d) \$2000

2 What is the total payout on the lottery, assuming every prize is claimed?

3 If every possible raffle ticket is sold, then what does the ticket price need to be for the organisers to make a profit?

Strategy options

- Break problem into manageable parts.
- Have I seen a similar problem?

Maths 4 Real

MENDEL AND THE BIRTH OF MODERN GENETICS

Your genetic code

Your genetic code is stored in the nucleus of every cell in your body, in the form of DNA molecules that contain the instructions for making your entire body. A length of DNA that contains the instructions for a particular characteristic (such as eye colour or hair colour) is called a gene. Your genes are organised into 46 packages called chromosomes, which form 23 pairs. One chromosome from each pair is from your biological mother, and the other is from your biological father. Which chromosome comes from which parent is a random and independent event.

Will it be a boy or a girl?

Although it may appear that certain families have a greater chance of having boys than girls, or girls than boys, the gender of each newborn baby is an independent event. Regardless of how many boys or girls were born previously, the probability of a baby being female is 50%, and the probability for a male is also 50%. The gender of a baby is determined by which copy of chromosome 23 it receives from its father. If it receives an 'X' chromosome, it will be a girl. A 'Y' chromosome means that the baby will be a boy. Which chromosome is passed on is a random process.

1 Determine the probability of a mother having:

(a) two boys in a row

(b) three girls in a row

(c) a girl, then a boy

(d) one of each gender, in either order.

Dominant and recessive genes

Although you have two copies of every gene (one from your mother and one from your father), usually, the physical characteristic of only one gene is actually seen, or 'expressed', in the body. This is the 'dominant' gene. The other gene is called 'recessive'.

The modern science of genetics began with an Austrian monk named Gregor Mendel. Mendel studied the effects of cross-breeding in pea plants. He was able to predict the probability of a cross-bred pea showing certain characteristics, such as seed colour.

The following diagrams are similar to tables. They show the different gene combinations that can result from each parent passing on one of their genes to their offspring.

Take two parent pea plants: one is pure-bred green GG (both genes for seed colour are green) and the other is pure-bred yellow YY (both genes for seed colour are yellow). The table below shows that all offspring are YG (one yellow gene and one green gene). Y is dominant over G, so all the offspring appear yellow.

Parent

Parent	G	G
Y	YG	YG
Y	YG	YG

Possible offspring

If, however, both parents are YG (hybrids) then the offspring will be 25% YY, 50% YG and 25% GG. The GG will be green even though its parents appeared yellow!

Parent

Parent	Y	G
Y	YY	YG
G	YG	GG

Possible offspring

2 What will be the probabilities for the seed colour of the offspring of a YY crossed with a YG, where Y (yellow) is dominant over G (green)?

Genetic diseases

Genetics has very important implications for potentially fatal human illnesses. Let us assume there is a disease, such as cystic fibrosis, where there is a dominant gene A and a recessive gene a. It is the recessive gene, a, that is responsible for the disease. You suffer the disease if you are aa. You carry it but don't suffer it if you are Aa.

If two Aa carriers have a child, there is a 1 in 4 (25%) chance of the child having the illness (aa), a 1 in 2 (50%) chance of the child being a carrier (Aa) and a 25% chance of the child not being a carrier (AA).

	A	a
A	AA	Aa
a	Aa	aa

If a carrier and a sufferer have a child, then there is a 50% chance of the child having the illness and a 50% chance of being a carrier.

	A	a
a	Aa	aa
a	Aa	aa

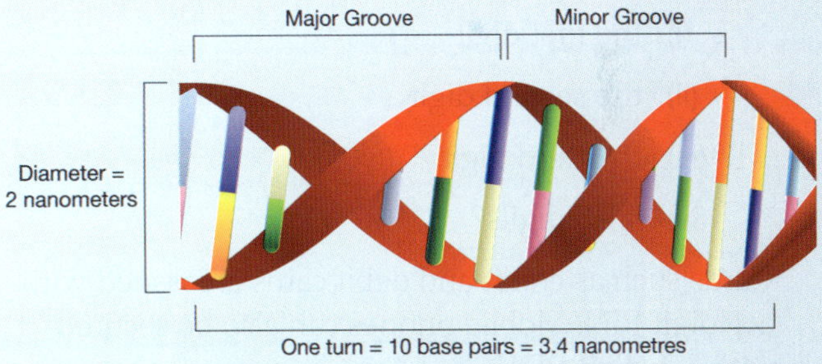

3 Blood groups are another simple example of genetics. Calculate the probabilities of each possible outcome from parents having the following characteristics. *Note*: Group A is dominant over Group O.

(a)

	O	O
A		
O		

(b)

	A	O
A		
O		

(c)

	O	O
A		
A		

Taking it further

4 Research one or more of the following topics and present your findings in a suitable manner.

(a) The life of Mendel and his work on genetics.

(b) There is a third blood group, Group B, which is co-dominant with Group A. Find out what this means.

(c) DNA fingerprinting is used to help solve crimes. Find out how it works and whether it is related to fingerprint analysis.

Investigation

Biometrics and security

Equipment required: calculator

How secure are you? It is important to protect ourselves, our money, our possessions, even our ideas from theft or danger. Keyed locks, while still useful at times, are impractical for many modern applications and too easy to break into. Security is needed at airports, when using technology, ATMs and credit cards, doors, safes and passports.

The Big Question

Do fingerprints provide better security than PIN codes or combination locks?

Engage

1 Combination locks can be a cheap and secure alternative to keyed locks. Students use them on their lockers and they are built into the fasteners of many briefcases and suitcases. The most common combination locks use a 3-digit code.

(a) Assuming a 3-digit combination lock is used, what is the probability of correctly guessing:

- **(i)** the first digit
- **(ii)** the second digit
- **(iii)** the third digit
- **(iv)** all 3 digits?

Cards such as credit and debit cards are issued with a 4-digit PIN. Mobile phones can also be secured with a 4-digit PIN.

(b) What is the probability of correctly guessing a 4-digit PIN?

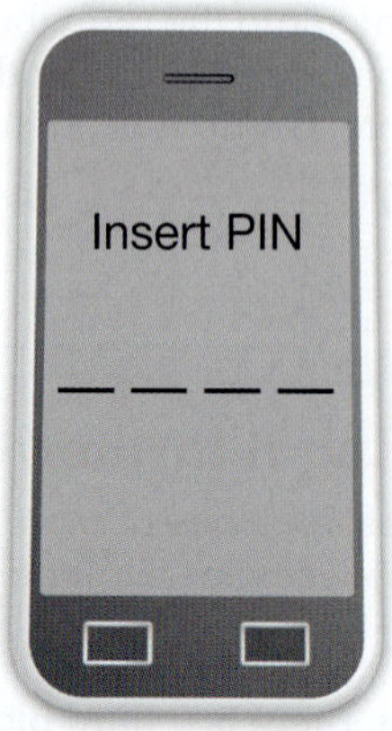

Personal computers and computer networks can be protected by using a password. A standard password must conform to certain rules. For example:

Rule 1 The password must contain at least 7 characters.

Rule 2 The password must contain three out of the following four character groups:

- Upper-case letters (A–Z). There are 26 of these.
- Lower-case letters (a–z). There are 26 of these.
- Numerics (0–9). There are 10 of these.
- Non-alphabetic (!,@,#,...). There are 32 of these on a standard keyboard.

(c) Assuming that a 7-digit password is chosen consisting of 1 upper-case letter, 4 different lower-case letters, a numeric and a non-alphabetic selected in that order, calculate the probability that this password would be guessed correctly.

Explore

People forget their passwords, and PINs can be stolen and copied. To avoid these problems, biometric security features are sometimes used.

Biometrics are measurements of human personal characteristics, which can be used to verify a person's identity. Fingerprint scanners are the most widely used and cost-effective biometric devices available. These provide an automated way of using fingerprints as a method of biometric identification and verification.

Fingerprints are always uniquely different for every person, even identical twins. To describe fingerprints, experts use 'ridge characteristics'. Some of these are shown below.

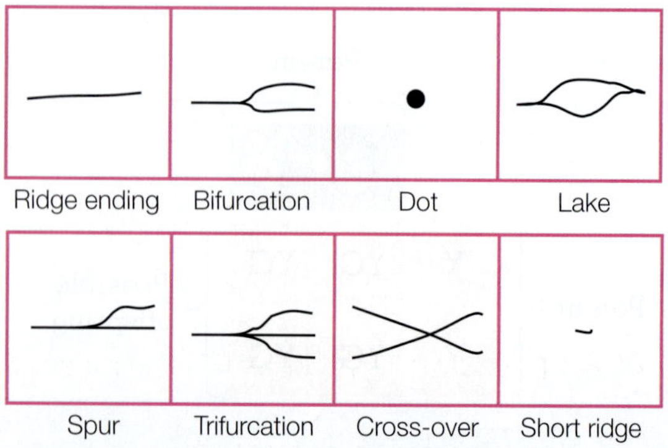

Fingerprint experts compare the pattern of ridges and furrows on the fingertips (ridge characteristics), as well as the types of minutiae points (ridge characteristics that occur when a ridge splits into two, or ends such as arch, tented arch, left loop, right loop, double loop, whorl, ...).

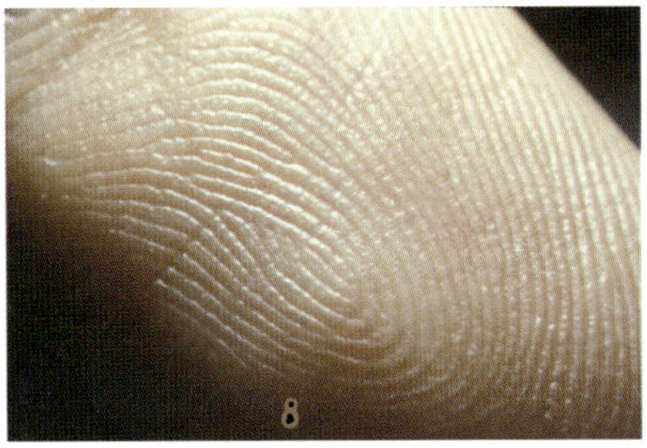

A simple method of fingerprint verification is to calculate the probability that two people will have the same ridge characteristics in the same location.

2 If the probability of having one identical ridge characteristic is assumed to be $\frac{1}{10}$, find the probability that 2 people will have:

(a) 2 identical ridge characteristics in the same location

(b) 10 identical characteristics in the same location

(c) N identical characteristics in the same location.

3 Using types of minutiae points of a fingerprint to calculate probabilities to identify individuals was first suggested by Francis Galton in 1892.

$$\text{Pr(fingerprint configuration)} = \frac{1}{16} \times \frac{1}{256} \times \left(\frac{1}{2}\right)^{24}$$

(a) Evaluate this probability writing your answer in scientific notation.

In 1930, Karl Pearson argued that this probability was more accurately found using the following calculation.

$$\text{Pr(fingerprint configuration)} = \frac{1}{16} \times \frac{1}{256} \times \left(\frac{1}{36}\right)^{24}$$

(b) Calculate this probability writing your answer in scientific notation.

A study in 1985 came up with a quite different and even more accurate formula.

Pr(fingerprint configuration) =

$\frac{N}{5} \times 0.6 \times (0.5 \times 10^{-3})^N$, where N is the number of minutiae on an average fingerprint.

(c) Evaluate this probability when $N = 12$ writing your answer in scientific notation.

(d) Evaluate this probability if $N = 36$ writing your answer in scientific notation.

Strategy options

- Look for a pattern.
- Make a table.

Explain

4 (a) Using Galton's fingerprint probability, would a 7-character password as described in **1(c)** be more secure than a fingerprint verification? Justify your answer.

(b) Would your answer be the same if Pearson's fingerprint probability was used instead? Justify your answer.

Elaborate

5 How many particular characteristics using the formula in **2 (a)** would need to be compared to be more secure than a computer password as described in **1(c)**?

Evaluate

6 (a) Why might fingerprinting be used more often for verification (e.g. access control, time and attendance at work, alarm activation/deactivation, safety deposits, club membership privileges, border control, parent entry to preschool, access to self serve facilities)?

(b) If fingerprinting is more secure than PINs or passwords, why aren't fingerscans used more for verification?

(c) Do you think that fingerprint technology will ever replace PINs and passwords? Justify your answer.

Extend

Some other methods of biometric authentication are:

- face recognition
- hand geometry biometric
- retina scan
- iris scan
- signature
- voice analysis

7 Investigate one of the above methods. What are the probabilities that no two people would have the same biometric data using these methods? What data can you find to support this?

Chapter review 10

Maths literacy

addition rule for mutually exclusive events	general addition rule for events	probability tree diagram
complementary events	independent events	sample space
conditional statement	long-run proportion	tree diagram
dependent events	mutually exclusive	two-way table
event	probability	universal set

Copy and complete the following using the words and phrases from the list, where appropriate. A word or phrase may be used more than once.

1 Every element under consideration in a particular question is called the __________.

2 If two events, A and B, are __________, then $\Pr(A) \times \Pr(B) = \Pr(A \text{ and } B)$.

3 Examples of __________ are 'having brown hair' and 'not having brown hair'.

4 A tree diagram with probabilities on each of the branches is called a __________.

5 Events that cannot occur at the same time are said to be __________.

6 A __________ or a __________ can be used to help find the sample space.

7 Spinning a spinner that has been equally divided into four pieces labelled 1, 2, 3 and 4 then calculating $\Pr(1) + \Pr(2) + \Pr(3) + \Pr(4) = 1$ is an example of the __________.

8 Words and phrases such as 'given', 'knowing that' and 'if … then' are indications you are dealing with a __________.

Fluency

1 A spinner numbered 1 to 5 is spun. Find: **10.1**

(a) Pr(spinning a 2) (b) Pr(not spinning a 2)

(c) Pr(spinning a number greater than 2) (d) Pr(spinning an even number)

(e) Pr(spinning an odd number) (f) Pr(spinning a number less than 5).

2 A box contains 15 blocks. Seven are red, five are white and the rest are blue. A block is chosen at random from the box. Find the probability that the block chosen is: **10.1**

(a) blue (b) not red (c) red or blue.

3 A standard six-sided die is rolled. What is the probability of an even number or a 5? **10.1**

4 A spinner numbered 1 to 4 is spun twice. **10.1**

(a) Represent all the possible outcomes using a table.

(b) Find:

(i) Pr(spinning a pair of 3s) (ii) Pr(spinning only one 3)

(iii) Pr(not spinning any 3s) (iv) Pr(the sum of the two numbers is 5).

5 A 50-cent coin is tossed in the air twice. Find the probability of getting two tails. **10.1**

6 This table shows the number of elements in each of three sets. **10.2**

	A and *B*	*A* and not *B*	not *A* and *B*	not *A* and not *B*	
C	1	4	3	2	10
not *C*	2	1	2	3	8
	3	5	5	5	18

Write the number of elements in:

(a) the universal set

(b) A

(c) A or B

(d) B and C

(e) not B

(f) not A and not B and not C.

7 A cutlery rack contains 6 knives, 5 forks and 2 spoons. **10.3**

(a) The probability that a piece of cutlery drawn at random will be a knife or a spoon is:

A $\frac{2}{13}$ B $\frac{2}{5}$ C $\frac{6}{13}$ D $\frac{8}{13}$

(b) The probability that a piece of cutlery drawn at random will not be a knife is:

A $\frac{6}{13}$ B $\frac{7}{13}$ C $\frac{6}{7}$ D $\frac{7}{6}$

8 Two cards are drawn, without replacement, from a set of 20 cards that are numbered 1 to 20. **10.4**

Let P = 'a prime number is drawn' and N = 'a prime number is not drawn'

(a) Draw a tree diagram to show this information.

(b) Find Pr(both cards show a prime number).

(c) Find Pr(no cards show a prime number).

(d) Find Pr(at least one card shows a prime number).

9 Use the information in the Venn diagram to find the following probabilities. Where necessary, give your answers correct to 4 decimal places. **10.5**

(i) Pr(A given B)

(ii) Pr(B given A)

(a)

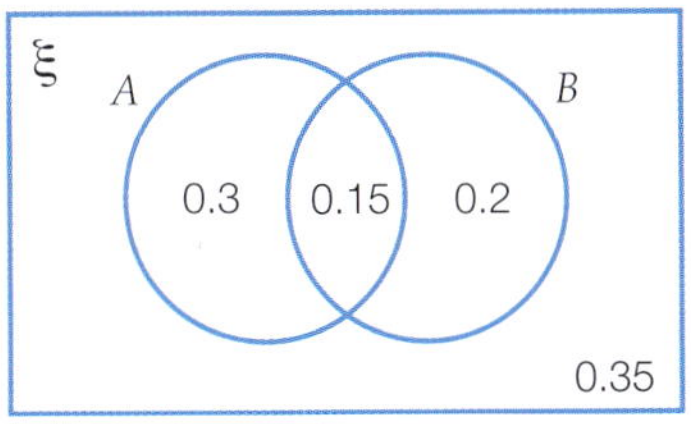

(b)

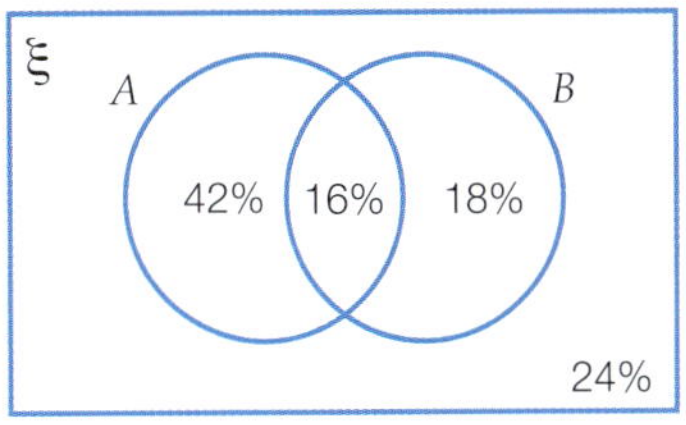

10 Rafaele and Ami are in the same class at school and both have trouble getting to school on time. Rafaele is late 2 days out of 9, whereas Ami is late 3 days out of 11. What is the probability that they will both be late on any particular day? **10.6**

Understanding

11 A fair seven-sided spinner, numbered 1 to 7, is spun. Find the probability of: **10.1**

(a) a 5
(b) an even number
(c) a number less than 4
(d) not a 6
(e) a multiple of 3
(f) not a prime number.

12 In clay pigeon shooting, a clay block in the shape of a disc is used as a target for shooting. Jackie estimates her chance of hitting the moving clay target with one shot to be 0.75. **10.4**

(a) What is the probability of her not hitting the target with one shot?

(b) Draw a probability tree diagram to represent the possible outcomes of shooting at three clay targets.

(c) Find the probability that Jackie hits:

(i) all three clay targets (ii) only one clay target (iii) at least one clay target.

13 In a survey of 60 people, it was found that 40 people liked lemonade, whereas 30 people liked cola. Two people said they disliked both. **10.2**

(a) Present this information in a two-way table.

(b) If one person is chosen at random, find the probability that this person:

(i) likes lemonade
(ii) likes lemonade only
(iii) likes cola
(iv) likes cola only
(v) likes both lemonade and cola
(vi) does not like lemonade
(vii) likes lemonade or cola
(viii) likes neither cola nor lemonade.

14 These probabilities apply to eye colour for students at a particular school. The colours are all mutually exclusive. **10.4**

$\Pr(\text{green}) = \frac{3}{16}$ $\Pr(\text{blue}) = \frac{1}{4}$ $\Pr(\text{hazel}) = \frac{3}{8}$ $\Pr(\text{brown}) = \frac{3}{16}$

A student is chosen at random. What is the probability that he or she has:

(a) blue or green eyes
(b) neither blue nor brown eyes
(c) blue, hazel or brown eyes
(d) neither hazel nor blue eyes?

15 Nerida and Ngaio play each other at badminton every week. The probability that Nerida wins any particular game is 0.65. Find, rounded to 3 decimal places where necessary, the probability that: **10.4**

(a) Ngaio wins in any particular week

(b) Nerida wins for 3 weeks in a row

(c) Ngaio wins for 4 weeks in a row

(d) over a 6-week period the winning sequence is Nerida, Ngaio, Nerida, Ngaio, Nerida, Ngaio.

16 You toss three coins: 10c, 20c and 50c. You note the results in that order. **10.1, 10.4**

(a) Find the probability of getting the following results:

(i) heads 3 times **(ii)** heads twice **(iii)** heads once **(iv)** no heads.

(b) What do the answers in part **(a)** add to? Is this an expected result? Explain your reasoning.

(c) Without doing any further calculations, explain how you can find the probability of each number of tails.

(d) Find the probability of the following (where H = heads, T = tails):

(i) HHT **(ii)** HTH **(iii)** THH

(e) Write a statement about what you have found in part **(d)**.

Reasoning

17 Justine and Kim are entered for a tennis tournament that begins with eight players. The draw profile looks like the following (because they are so good Justine and Kim are not scheduled to play each other until the final). **10.4**

Justine and Kim are very confident players and they think they have a probability of $\frac{2}{3}$ of beating any other player except for each other. If they play each other, Kim has a $\frac{3}{5}$ chance of winning. Find the following probabilities.

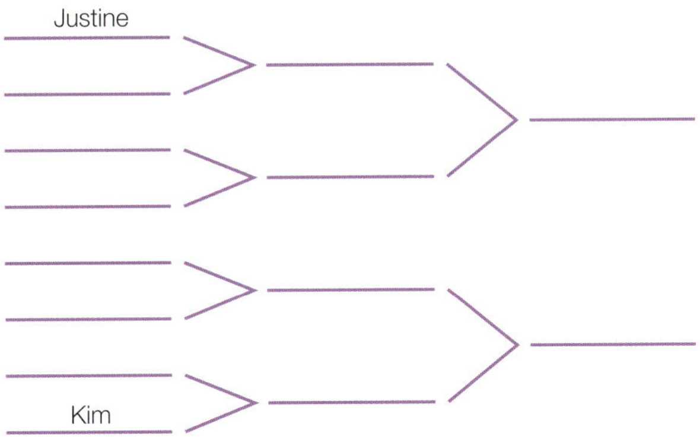

(a) The two girls play each other in the final.

(b) Kim wins the tournament (not necessarily against Justine).

(c) Justine wins the tournament (not necessarily against Kim).

18 A weather forecast predicting the chance of rain in the next 24 hours in some capital cities is recorded below. **10.6**

City	Brisbane	Sydney	Canberra	Melbourne	Hobart	Adelaide	Perth	Darwin
Chance of rain	0.15	0.4	0.3	0.1	0.2	0.4	0.4	0.1

Assuming the events to be independent, find:

(a) Pr(rain in Melbourne and Brisbane)

(b) Pr(rain in Sydney and Perth)

(c) Pr(rain in Adelaide, Hobart and Perth)

(d) Pr(rain in Sydney and Canberra).

Long-term statistics are kept and found that the probability of rain in both Sydney and Canberra on the same day is actually 0.2.

(e) Explain the difference between your answer to part **(d)** and the long-run probability.

Mixed review C

1 Factorise the following quadratic trinomials. **3.5**

(a) $a^2 - 5a - 36$ (b) $a^2 - 8a + 12$ (c) $4a^2 + 20a + 25$

2 Find the area of each of the following shapes. **5.1**

(a)

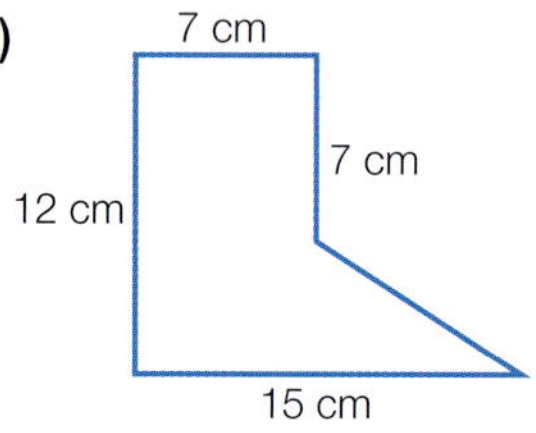

(b)

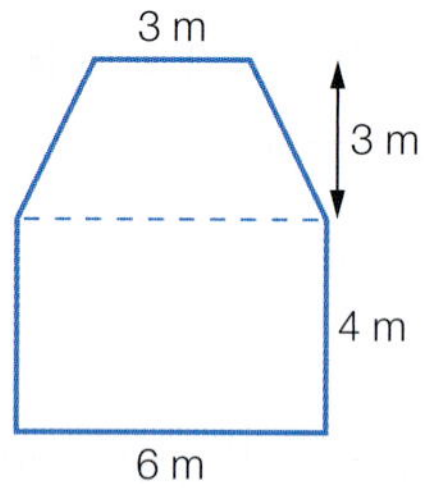

(c)

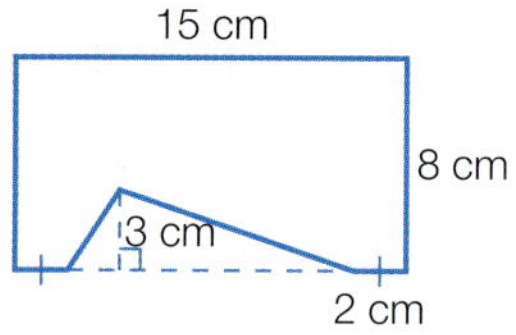

3 Solve the following equations. **1.1**

(a) $\dfrac{3a+4}{7} = -2$ (b) $\dfrac{4a+3}{2} = \dfrac{2a}{5}$ (c) $\dfrac{3a-2}{3} = \dfrac{4a-5}{2}$

4 Find the value of the pronumeral in each triangle, correct to 1 decimal place. **6.3**

(a)

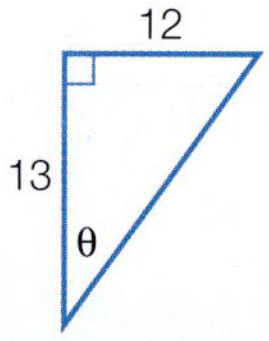

(b)

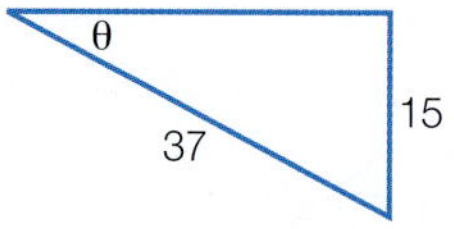

(c)

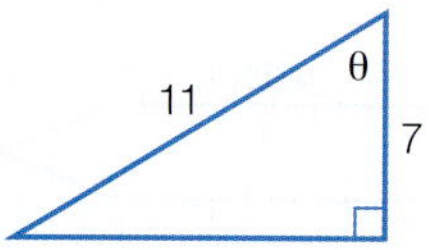

10A 5 Find the length of the third side of a scalene triangle with two sides of length 6 cm and 7 cm and an included angle of 43°. Round your answer to 2 decimal places. **7.4**

6 Prove that the diagonals of a kite intersect at right angles. **9.5**

7 Find the mean for each of the following grouped data frequency tables. Give your answers correct to 2 decimal places. **2.1**

(a)

Class interval	Frequency
20–<25	8
25–<30	6
30–<35	3
35–<40	2
40–<45	5
45–<50	5

(b)

Class interval	Frequency
111–115	12
116–120	15
121–125	17
126–130	19
131–135	13
136–140	11

8 Find the surface area of each of the following solids. Where necessary, give your answer correct to 2 decimal places. **5.2**

(a)

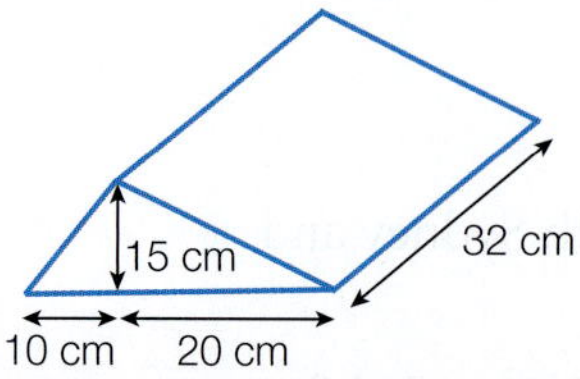

(b)

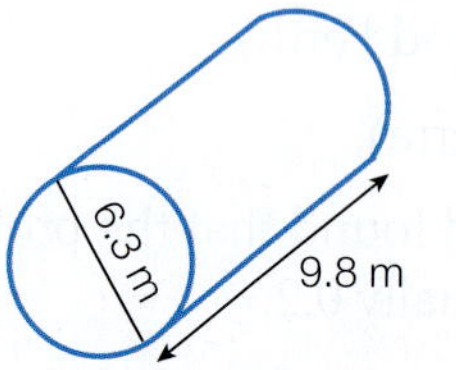

10A 9 Find the volume of each of the following solids. Round your answers to 2 decimal places. **5.5**

10 Consider the equation $y = 2x^2 + x$. **3.3**

(a) Set up a table of values with x-values from -3 to 3.

(b) Plot the points on a coordinate plane.

(c) Find the turning point of the graph.

(d) Determine the x- and y-intercepts for the graph.

11 Sketch the graph of each of the following quadratics, showing all important features. **4.2**

(a) $y = x^2 + 6x - 4$ (b) $y = x^2 - 4x + 4$ (c) $y = -2x^2 + 3x$

12 Solve each of the following for x. **4.1**

(a) $2x^2 - 50 = 0$ (b) $4x^2 + 2x = 0$ (c) $x^2 - 5x + 6 = 0$

13 Write the turning point of each of the following graphs. **3.4**

(a) $y = (x - 2)^2 + 3$ (b) $y = (x + 4)^2 - 1$ (c) $y = (x - 7)^2 - 6$

14 A smartphone screen is 2.5 cm longer than it is wide. **3.3**

(a) Write an equation for the area of the screen in terms of its width, in cm^2.

(b) Plot a graph of the area against the width.

(c) Use this graph to find the dimensions of the screen if the area of the screen is 37.5 cm^2.

15 A parabola has x-intercepts at $x = 4$ and $x = 6$. **4.4**

(a) Determine the x-coordinate of the turning point.

(b) If the x^2 coefficient is 1, determine the equation of the graph.

16 A cube has a volume of 512 cm^3. Calculate its surface area. **5.3**

17 Find the area of each of the following shapes. Where necessary, give your answer correct to 2 decimal places. **5.1**

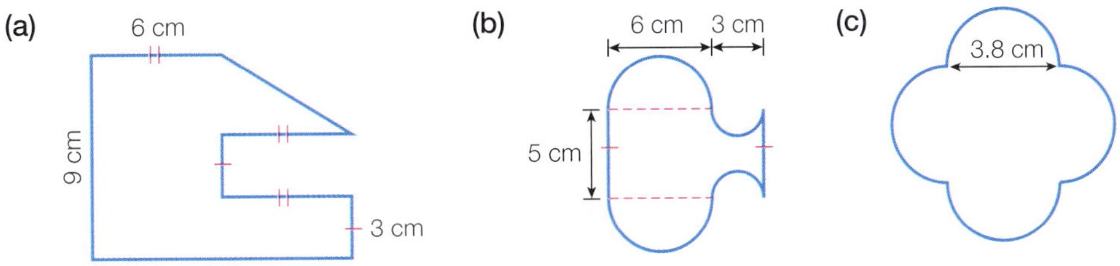

Reasoning

10A 18 Scott wants to erect a greenhouse. The support structure shown is made of metal. Using the plan shown, calculate correct to 1 decimal place: **5.4**

(a) the amount of glass needed for the walls and roof

(b) the length of metal needed for the support structure.

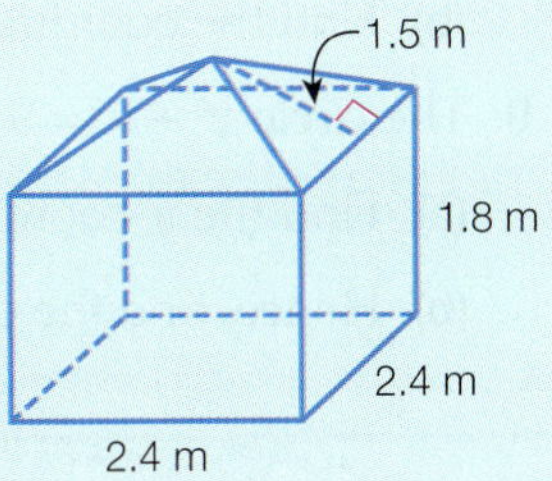

Challenge C

1 The mean of 19 consecutive integers is 99. The largest of these integers is:

A 118 **B** 117 **C** 108 **D** 107

2 Three different numbers are chosen such that when each of the numbers is added to the average of the remaining two, the numbers 65, 69 and 76 result. The average of the three original numbers is:

A 34 **B** 35 **C** 36 **D** 37

3 Find the value of x if the large triangle has been broken into two triangles of equal area. The large triangle is not marked as being right angled.

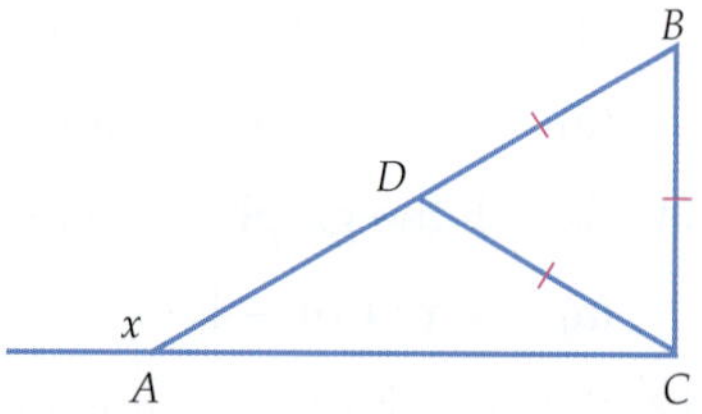

4 In the diagram, $ABC = 90°$, $CB \parallel ED$, $AB = DF$, $AD = 24$, $AE = 25$ and O is the centre of the circle. The perimeter of $CBDF$ is:

A 40 units **B** 42 units

C 43 units **D** 44 units

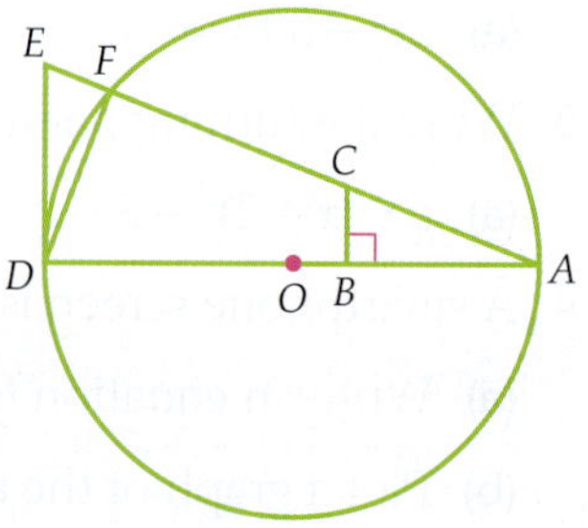

5 Find x in the following diagram.

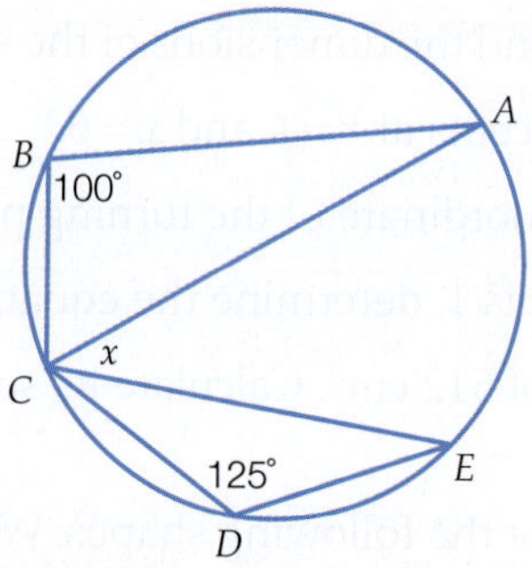

6 **(a)** Determine the values of x for which $(x + 3)(x - 6) = -14$.

(b) Determine the values of x for which $(x^2 - 3x)^2 = 4 - 3(3x - x^2)$.

7 **(a)** What is the sum of the two numbers that satisfy the equation $x^2 - 6x - 7 = 0$?

(b) What is the product of the two numbers that satisfy the equation $5x^2 - 20 = 0$?

(c) Find the average of the numbers that satisfy the equation $x^3 - 6x^2 + 5x = 0$.

8 The equation of a parabola is written in the form $y = (x - 2)(x - 8) + 7$.

(a) If $(c, 7)$ is on the parabola, find all possible values of c.

(b) Find the coordinates of the vertex of the parabola.

9 The circle $x^2 + y^2 = 9$ and the parabola $y = x^2 - 3$ intersect at three points.

(a) Find the y-coordinates of the points of intersection.

(b) Hence, find the exact values of the coordinates of the points of intersection.

Surds

11

10A

Can maths help us find alien life?

Name: Tanya Hill
Job: Astronomer
Qualifications:
Bachelor of Science,
PhD in Astronomy

Astronomers such as Dr Tanya Hill, from the Melbourne Planetarium at Scienceworks, use the power of mathematical modelling to test their theories about where planets capable of supporting life could be formed.

Mathematical modelling involves developing a formula or equation that fits a set of data. This formula can then be tested by computer simulation to see whether it produces results that are the same as those observed.

At school, Tanya 'loved the certainty of maths—knowing that you had a clear-cut answer you could work towards. I was drawn to physics and later on astronomy, because it gave you a reason for doing the maths'.

In her work at the Planetarium, Tanya gives public presentations on different areas of astronomy, including recent developments and discoveries. Part of this involves converting the units used to measure astronomical distances (light years and parsecs) into ideas that help people understand the enormous scale of the universe.

Why learn this?

Irrational numbers expressed in decimal form have an infinite number of decimal places. Some irrational numbers, however, can be written in exact form as surds. For example, the surd $\sqrt{7}$ has the decimal representation 2.645 751 311 ... For some purposes you could round this to 2.6, but mathematicians often require exact measurements. Surds allow this level of exact precision.

After completing this chapter you will be able to:

- understand different kinds of rational and irrational numbers
- add, subtract, multiply and divide surd expressions
- carry out calculations and simplifications involving surds
- know how to use the distributive law with surds
- rationalise the denominator of a surd expression.

Recall 11

Prepare for this chapter by attempting the following questions. If you have difficulty with a question, you can download a Recall Worksheet from the eBook or the Pearson Places website.

1 Write in mathematical symbols:

(a) the square root of 3

(b) the cube root of 5

(c) the number that when squared equals 7

(d) the number that when cubed equals 6

(e) the number that when raised to the power of four equals 10.

2 Evaluate each of the following. Do not use a calculator.

(a) $\sqrt{4}$ (b) $\sqrt{36}$ (c) $-\sqrt{9}$ (d) $\sqrt{1}$ (e) $\sqrt[3]{8}$

(f) $\sqrt[3]{64}$ (g) $\sqrt[3]{-27}$ (h) $\sqrt[3]{1}$ (i) $\sqrt[4]{16}$ (j) $\sqrt[5]{-1}$

3 Use a calculator to evaluate each of the following correct to 2 decimal places, where necessary.

(a) $\sqrt{11}$ (b) $-\sqrt{7}$ (c) $\sqrt[3]{15}$ (d) 34^2 (e) 3^9

4 Expand each of the following.

(a) $3(x + 4)$ (b) $-4(2y - 3)$

5 Expand and simplify each of the following.

(a) $(x - y)(x + y)$ (b) $(c - 9)(c + 9)$ (c) $(2p + 3)(2p - 3)$

10A

Exploration Task

You can download this activity from the eBook or the Pearson Places website.

Number system jigsaw

In this activity you will explore how to fix a poster for number systems that has been mixed up. How and why should the different numbers be grouped into categories?

Rational and irrational numbers

11.1

Real numbers

Real numbers are numbers that can be shown on a continuous number line. Real numbers include all rational numbers and all irrational numbers.

- A **rational number** can be written in the form $\frac{a}{b}$, where a and b are integers and $b \neq 0$. Rational numbers include whole numbers, fractions and repeating decimals.
- An **irrational number** cannot be written in the form $\frac{a}{b}$, where a and b are integers and $b \neq 0$. Irrational numbers include all surds and other special numbers such as π. When represented in decimal form, an irrational number has an infinite number of non-repeating decimal places.

Surds are numbers that can only be expressed in exact form using the 'radical' or 'root' symbol ($\sqrt{}$). However, not all numbers written with a root symbol are necessarily surds. For example, the square root of a perfect square will be a whole number, so it is not a surd. The first five perfect squares are shown below.

$\sqrt{1} = 1$ $\sqrt{4} = 2$ $\sqrt{9} = 3$ $\sqrt{16} = 4$ $\sqrt{25} = 5$

10A

A square root of a whole number that is not a perfect square will be a surd. So, $\sqrt{5}$ is a surd because 5 is not a perfect square. You cannot write $\sqrt{5}$ as a fraction. If you enter it in your calculator, then you will see 2.236 067…, a decimal with an infinite number of non-repeating decimal places.

This also applies to other roots, such as a cube root ($\sqrt[3]{}$), or a fourth root ($\sqrt[4]{}$), or a fifth root ($\sqrt[5]{}$) and so on, which may be surds. In each case, if the value is a perfect cube, or perfect fourth power, or any perfect same power of the same root then the expression is not a surd.

For example, $\sqrt[3]{8}$ is not a surd because 8 is 2^3, which is a perfect cube $\sqrt[3]{8}$. But $\sqrt[3]{9}$ is a surd, because 9 is not a perfect cube.

You may need a calculator to identify large perfect squares, perfect cubes and higher order powers.

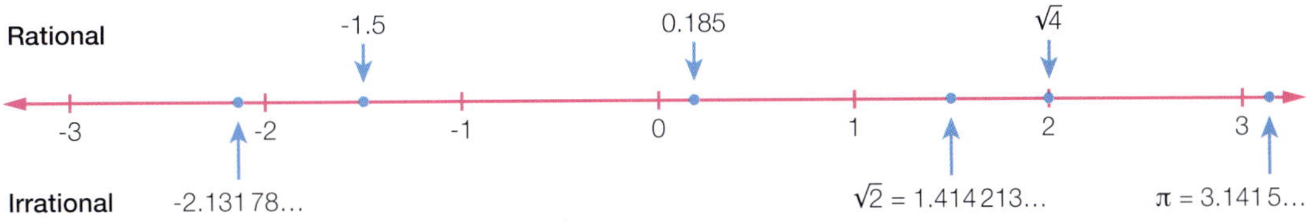

Real numbers

Rational numbers

- Integers

  $8 = \frac{8}{1}$, $\sqrt{4} = 2 = \frac{2}{1}$, $-15 = -\frac{15}{1}$

- Terminating decimals

  $0.8 = \frac{4}{5}$, $1.23 = \frac{123}{100}$, $0.0006 = \frac{6}{10\,000}$

- Recurring decimals

  $0.333\,333\ldots = \frac{1}{3}$, $0.636\,363\ldots = \frac{7}{11}$

Irrational numbers

- Surds

  $-\sqrt{3}$, $\sqrt[3]{7}$, $4\sqrt{19}$

- Other infinite non-recurring decimals

  $\pi = 3.141\,592\ldots$

Worked example 1

W.E. 1

State whether the following numbers are rational or irrational.

(a) $\sqrt{12}$ **(b)** $\sqrt{361}$

Thinking	Working
(a) Identify whether the number under the square root symbol can be written as a perfect square. If it can, the expression is rational. If not, it is irrational.	(a) 12 is not a perfect square, so $\sqrt{12}$ is irrational.
(b) Identify whether the number under the square root symbol can be written as a perfect square. If it can, the expression is rational. If not, it is irrational.	(b) $361 = 19^2$, so $\sqrt{361}$ is rational.

10A

Worked example 2

W.E. 2

Evaluate the following. Do not use a calculator.

(a) $-7\sqrt{81}$ **(b)** $3\sqrt{100} - 2\sqrt{16} + \sqrt{9}$

Thinking	Working
(a) 1 Identify and simplify the $\sqrt{\text{perfect square}}$ ($\sqrt{81} = 9$).	(a) $-7\sqrt{81}$ $= -7 \times 9$
2 Simplify.	$= -63$
(b) 1 Identify and simplify each $\sqrt{\text{perfect square}}$.	(b) $3\sqrt{100} - 2\sqrt{16} + \sqrt{9}$ $= 3 \times 10 - 2 \times 4 + 3$
2 Simplify.	$= 30 - 8 + 3$ $= 25$

In some situations, such as finding the side length of a right-angled triangle, an approximate value for a surd can be useful. You can use a calculator to find approximate values for surds.

Worked example 3

W.E. 3

Use a calculator to evaluate each of the following correct to 3 decimal places.

(a) $\sqrt{3}+\sqrt{5}$ (b) $-7\times\sqrt[3]{9}$

Thinking	Working
(a) Use your calculator.	(a) $\sqrt{3}+\sqrt{5}\approx 3.968\,118\,785$ $= 3.968$ (3 d.p.)
(b) Use your calculator.	(b) $-7\times\sqrt[3]{9}\approx -14.560\,586\,76$ $= -14.561$ (3 d.p.)

Rational numbers can be expressed as a fraction $\frac{a}{b}$, where a and b are integers and $b \neq 0$.

Irrational numbers cannot be expressed as a fraction $\frac{a}{b}$, where a and b are integers and $b \neq 0$.

11.1 Rational and irrational numbers

10A

Navigator

1 (columns 1–3), 2 (columns 1–2), 3 (columns 1–2), 4, 5, 6 (a), 7 (a), 8 (a–c), 9, 10, 11, 13	1 (columns 2–4), 2 (column 2), 3 (columns 2–3), 4, 5, 6, 7, 8, 9, 10, 11, 12, 13	1 (columns 3–4), 2 (column 3), 3 (column 3), 4, 6, 7, 8, 9, 10, 11, 12, 13, 14

Answers p. 854

Equipment required: calculator for Questions **3**, **8**

Fluency

1 For each of the following numbers, state whether it is rational or irrational. W.E. 1

(a) $\sqrt{5}$ (b) $\sqrt{36}$ (c) $-\frac{2}{7}$ (d) $\sqrt{10}$

(e) 4.6 (f) $\sqrt[3]{4}$ (g) $-\sqrt[3]{2}$ (h) $\sqrt[4]{1}$

(i) 2π (j) $\sqrt{20}$ (k) $-\sqrt{3}$ (l) 8.621

2 Evaluate the following. Do not use a calculator. W.E. 2

(a) $\sqrt{64}$ (b) $-\sqrt{25}$ (c) $-\sqrt{36}$

(d) $3\sqrt{9}$ (e) $7\sqrt{36}$ (f) $-4\sqrt{49}$

(g) $\sqrt{25+144}$ (h) $\sqrt{9}+\sqrt{16}$ (i) $7\sqrt{49}+4\sqrt{25}$

(j) $3\sqrt{16}+2\sqrt{81}$ (k) $5\sqrt{36}-7\sqrt{4}$ (l) $9\sqrt{100}-8\sqrt{16}+5\sqrt{9}$

3 Use a calculator to find the value of each of the following surds correct to 3 decimal places. W.E. 3

(a) $\sqrt{3}+\sqrt{2}$ (b) $\sqrt{8}-\sqrt{5}$ (c) $\sqrt{7}-\sqrt{13}$

(d) $\sqrt{35}$ (e) $-\sqrt{3}$ (f) $\sqrt[3]{17}$

(g) $\sqrt[3]{10}$ (h) $\sqrt{6}$ (i) $-\sqrt[3]{72}$

4 (a) Which one of the following is a surd?

A $\sqrt{9}$ B $-\sqrt{1}$ C $\sqrt{8}$ D $\sqrt[3]{8}$

(b) Which one of the following is *not* a surd?

A $\sqrt[3]{7}$ B $\sqrt{11}$ C $\sqrt[3]{16}$ D $\sqrt{25}$

(c) Which one of the following is a rational number?

A $\sqrt{3}$ B $\sqrt[3]{2}$ C $\sqrt[3]{4}$ D $\sqrt{4}$

(d) Which one of the following is *not* a rational number?

A $\frac{1}{3}$ B 1.45 C 2.717 171 ... D 3.721 452 ...

5 (a) $4\sqrt{3}$ is the same as:

A $4 + \sqrt{3}$ B $4 - \sqrt{3}$ C $4 \times \sqrt{3}$ D $\sqrt[4]{3}$

(b) $8\sqrt{16}$ is the same as:

A 8×4 B $8 \div 4$ C $8 + 4$ D 8×16

6 Do not use a calculator for this question.

(a) $\sqrt{7}$ is between $\sqrt{4}$ and $\sqrt{9}$, so $\sqrt{7}$ is between:

A 1 and 2 B 2 and 3 C 3 and 4 D 4 and 5

(b) $\sqrt{31}$ is between:

A 2 and 3 B 3 and 4 C 4 and 5 D 5 and 6

Understanding

10A

7 Do not use a calculator for this question.

(a) $2^3 = 8$ and $3^3 = 27$, so $\sqrt[3]{11}$ is between:

A 1 and 2 B 2 and 3 C 3 and 4 D 4 and 5

(b) $\sqrt[3]{100}$ is between:

A 2 and 3 B 3 and 4 C 4 and 5 D 5 and 6

8 For each of the following equations, write whether it is true (T) or false (F). Use your calculator to check your answer.

(a) $\sqrt{6+2} = \sqrt{6} + \sqrt{2}$ (b) $\sqrt[3]{-14} = -\sqrt[3]{14}$ (c) $\sqrt[3]{10^2} = (\sqrt[3]{10})^2$

(d) $6\sqrt{7} = 7\sqrt{6}$ (e) $6\sqrt{3} - 2\sqrt{5} = -(2\sqrt{5} - 6\sqrt{3})$

Reasoning

9 Find the exact side length of a square with area:

(a) 25 m^2 (b) 17 m^2

10 The area of a square vegetable garden is 60 m^2.

(a) What is the exact length of each side of the vegetable garden?

(b) A fence of chicken wire is to be constructed around the vegetable garden. What length of chicken wire must be purchased if only whole metres of wire can be purchased?

11 Find the exact side length of a cube with volume:

(a) 8 m^3 (b) 15 cm^3

12 Two identical dice have a combined volume of 14 cm^3.

(a) Find the exact side length of each die, assuming each die is a perfect cube.

(b) What is the side length of each die to the nearest mm?

Open-ended

13 (a) Write three irrational numbers greater than zero.

(b) Write three irrational numbers less than zero.

(c) Draw a number line and mark the approximate positions of these numbers.

14 Pythagoras' theorem is used to find the side lengths of right-angled triangles. Explain how to cut a piece of string to be $\sqrt{2}$ m long using a pencil, a protractor, a metre ruler and a piece of string.

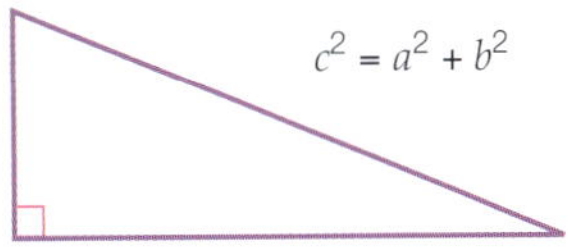

Problem solving

How much is enough?

Equipment required: graph paper

You can use straight lines to form a spiral made with ever increasing right-angled triangles. A special kind of spiral is created when you do this starting with an isosceles right-angled triangle with equal sides of length 1 cm. Give each subsequent triangle a height of 1 cm and a base that is the length of the hypotenuse of the previous triangle.

(a) What is the perimeter of the tenth right-angled triangle?

(b) Calculate the exact perimeter of the entire spiral (to ten triangles).

(c) Does the entire spiral have a rational or an irrational perimeter?

(d) Draw this spiral on graph paper. Measure the length of the perimeter of the entire spiral to check your answer to **(b)**.

(e) The first triangle has a rational area of $\frac{1}{2}\text{ cm}^2$.

(i) Which is the next triangle to have a rational area?

(ii) Find this area and compare it to the area of the first triangle.

(iii) Find the next triangle to have a rational area.

(iv) Find this area and compare it to the area of the first triangle.

Strategy options

- Look for a pattern.
- Make a table.

10A

11.2 Multiplying and dividing surds

Surds can be different kinds of roots, including cube roots ($\sqrt[3]{5}$), fourth roots ($\sqrt[4]{9}$), as well as square roots ($\sqrt{3}$). Remember that numbers such as the square roots of perfect squares (16, 36, 81, …) and the cube roots of perfect cubes (8, 27, 64, …) are rational numbers, not surds.

Multiplying surds

When multiplying surds that are perfect squares, you can see:

$$\begin{aligned}&\sqrt{4} \times \sqrt{9}\\&= 2 \times 3\\&= 6\\&= \sqrt{36}\\&= \sqrt{4 \times 9}\end{aligned}$$

The result shows that $\sqrt{4} \times \sqrt{9} = \sqrt{4 \times 9}$.

This result is a general rule that works with all surds.

10A

For example: $\sqrt{3} \times \sqrt{2} = \sqrt{3 \times 2} = \sqrt{6}$

You can check this with your calculator:

$\sqrt{3} \times \sqrt{2} = 2.449\,489\,743\ldots$ and $\sqrt{6} = 2.449\,489\,743\ldots$

> $\sqrt{a} \times \sqrt{b} = \sqrt{a \times b} = \sqrt{ab}$, where a and b are positive numbers.

Simplifying surds

Surds should always be written in their simplest form, so that the number in the root is as small as possible. This can often be done by writing the number in the root as a product of factors. If one of these factors is a perfect square, then it can be written outside the root symbol. For example:

$\sqrt{2} \times \sqrt{6} = \sqrt{12}$

The factors of 12 are 1, 2, 3, 4, 6, 12. Now 4 is a perfect square, so $\sqrt{12}$ can be simplified further:

$$\begin{aligned}\sqrt{12} &= \sqrt{4 \times 3}\\&= \sqrt{4} \times \sqrt{3}\\&= 2 \times \sqrt{3}\\&= 2\sqrt{3}\end{aligned}$$

> $2\sqrt{3}$ is simpler than $\sqrt{12}$ because the number in the root is smaller.

Worked example 4

W.E. 4

Simplify:

(a) $\sqrt{2} \times \sqrt{12}$ (b) $4\sqrt{3} \times 5\sqrt{7}$

Thinking	Working
(a) 1 Use $\sqrt{a} \times \sqrt{b} = \sqrt{a \times b} = \sqrt{ab}$ to combine the surds.	(a) $\sqrt{2} \times \sqrt{12}$ $= \sqrt{2 \times 12}$ $= \sqrt{24}$
2 Simplify if the number in the square root symbol has a factor that is a perfect square. (Factors of 24: 1, 2, 3, 4, 6, 8, 12, 24)	$= \sqrt{4 \times 6}$ $= \sqrt{4} \times \sqrt{6}$ $= 2\sqrt{6}$
(b) 1 Multiply whole numbers together and surds together.	(b) $4\sqrt{3} \times 5\sqrt{7}$ $= 4 \times 5 \times \sqrt{3} \times \sqrt{7}$
2 Simplify. (Factors of 21: 1, 3, 7, 21. 21 does not have any factors that are perfect squares, so it cannot be simplified further.)	$= 20\sqrt{21}$

Dividing surds

When dividing surds that are perfect squares you can see:

$$\sqrt{36} \div \sqrt{9} = \frac{\sqrt{36}}{\sqrt{9}} = \frac{6}{3} = 2$$

$$\sqrt{36 \div 9} = \sqrt{\frac{36}{9}} = \sqrt{4} = 2$$

The result shows that $\frac{\sqrt{36}}{\sqrt{9}} = \sqrt{\frac{36}{9}}$.

This result is a general rule that works with all surds.

For example, $\frac{\sqrt{14}}{\sqrt{2}} = \sqrt{\frac{14}{2}}$
$= \sqrt{7}$

You can check this using your calculator.

$\frac{\sqrt{14}}{\sqrt{2}} = 2.645\,751\,311\ldots$ and $\sqrt{7} = 2.645\,751\,311\ldots$

In general, $\frac{\sqrt{a}}{\sqrt{b}} = \sqrt{\frac{a}{b}}$, where a and b are positive numbers and $b \neq 0$.

Worked example 5

W.E. 5

Simplify:

(a) $\sqrt{21} \div \sqrt{7}$

(b) $\frac{12\sqrt{40}}{3\sqrt{5}}$

Thinking	Working
(a) Use $\frac{\sqrt{a}}{\sqrt{b}} = \sqrt{\frac{a}{b}}$.	(a) $\sqrt{21} \div \sqrt{7}$ $= \frac{\sqrt{21}}{\sqrt{7}}$ $= \sqrt{\frac{21}{7}}$ $= \sqrt{3}$
(b) 1 Divide the whole numbers and divide the surds.	(b) $\frac{12\sqrt{40}}{3\sqrt{5}}$ $= \frac{12}{3} \times \sqrt{\frac{40}{5}}$
2 Simplify. (Factors of 8: 1, 2, 4, 8)	$= 4\sqrt{8}$ $= 4 \times \sqrt{4 \times 2}$ $= 4 \times \sqrt{4} \times \sqrt{2}$ $= 4 \times 2 \times \sqrt{2}$ $= 8\sqrt{2}$

Sometimes it is necessary to convert a surd into an **entire surd**, which is where the entire surd is written in the root. For example:

$2\sqrt{6}$
$= 2 \times \sqrt{6}$
$= \sqrt{4} \times \sqrt{6}$
$= \sqrt{24}$

$\sqrt{24}$ is an entire surd.

10A

11.2 Multiplying and dividing surds

Navigator

1 (columns 1–2), 2 (column 1), 3 (columns 1–2), 4 (columns 1–2), 5, 6 (columns 1–2), 7 (columns 1– 2), 8, 9, 13	1 (column 2), 2 (column 2), 3 (columns 2–3), 4 (columns 2–3), 5, 6 (columns 2–3), 7 (columns 2–3), 8, 9,10, 12, 13	1 (column 3), 2 (column 3), 3 (columns 3, 4), 4 (column 4), 5, 6 (column 3), 7 (columns 3–4), 8, 9, 10, 11, 12, 13

Answers p.854

Fluency

W.E. 4

1 Simplify each of the following.

(a) $\sqrt{3}\times\sqrt{2}$ (b) $\sqrt{11}\times\sqrt{5}$ (c) $\sqrt{7}\times\sqrt{10}$

(d) $\sqrt{7}\times\sqrt{7}$ (e) $\sqrt{4}\times\sqrt{4}$ (f) $\sqrt{16}\times\sqrt{4}$

(g) $\sqrt{36}\times\sqrt{2}$ (h) $\sqrt{25}\times\sqrt{7}$ (i) $\sqrt{2}\times\sqrt{9}$

(j) $2\sqrt{3}\times 5\sqrt{2}$ (k) $\sqrt{11}\times 4\sqrt{3}$ (l) $5\times 8\sqrt{2}$

(m) $7\sqrt{5}\times\sqrt{5}$ (n) $\sqrt{6}\times 4\sqrt{6}$ (o) $5\sqrt{2}\times 4\sqrt{2}$

(p) $3\sqrt{7}\times 2\sqrt{7}$ (q) $5\sqrt{3}\times 3\sqrt{2}$ (r) $8\sqrt{6}\times 3\sqrt{12}$

W.E. 5

2 Simplify each of the following.

(a) $\sqrt{\frac{27}{3}}$ (b) $\sqrt{\frac{32}{2}}$ (c) $\frac{\sqrt{125}}{\sqrt{5}}$

(d) $\frac{\sqrt{60}}{\sqrt{15}}$ (e) $\sqrt{72}\div\sqrt{2}$ (f) $\sqrt{45}\div\sqrt{5}$

(g) $\sqrt{7}\div\sqrt{7}$ (h) $\sqrt{\frac{64}{16}}$ (i) $\sqrt{25}\div\sqrt{9}$

(j) $\sqrt{\frac{100}{49}}$ (k) $\frac{\sqrt{4}}{\sqrt{81}}$ (l) $\sqrt{\frac{36}{121}}$

(m) $\sqrt{4}\div\sqrt{64}$ (n) $\sqrt{3}\div\sqrt{27}$ (o) $\frac{2\sqrt{36}}{3}$

(p) $\frac{5\sqrt{49}}{7}$ (q) $\frac{5\sqrt{9}}{3\sqrt{25}}$ (r) $\frac{6\sqrt{9}}{9\sqrt{36}}$

(s) $\frac{18\sqrt{75}}{3\sqrt{3}}$ (t) $5\sqrt{20}\div\sqrt{5}$ (u) $18\sqrt{48}\div 3\sqrt{8}$

(v) $\frac{5\sqrt{30}}{25\sqrt{6}}$ (w) $6\sqrt{14}\div 15\sqrt{2}$ (x) $12\sqrt{30}\div 18\sqrt{10}$

3 Simplify each of the following.

(a) $\sqrt{27}$ (b) $\sqrt{20}$ (c) $\sqrt{162}$ (d) $\sqrt{96}$

(e) $\sqrt{99}$ (f) $\sqrt{500}$ (g) $\sqrt{486}$ (h) $\sqrt{405}$

(i) $5\sqrt{8}$ (j) $-4\sqrt{24}$ (k) $-5\sqrt{63}$ (l) $4\sqrt{243}$

4 Express each of the following as an entire surd.

(a) $2\sqrt{3}$ (b) $3\sqrt{7}$ (c) $6\sqrt{2}$ (d) $3\sqrt{3}$

(e) $-10\sqrt{6}$ (f) $-2\sqrt{10}$ (g) $-8\sqrt{3}$ (h) $5\sqrt{7}$

(i) $8\sqrt{2}$ (j) $7\sqrt{5}$ (k) $-6\sqrt{3}$ (l) $-5\sqrt{11}$

5 (a) $\sqrt{32}$ written in simplest form is:

A $2\sqrt{4}$ B $2\sqrt{8}$ C $4\sqrt{2}$ D $4\sqrt{8}$

(b) $\sqrt{3} \times \sqrt{7}$ simplifies to:

A $\sqrt{10}$ B $\sqrt{21}$ C $\sqrt{37}$ D $3\sqrt{7}$

(c) $\frac{\sqrt{49}}{\sqrt{4}}$ simplifies to:

A $\frac{2}{7}$ B $\frac{7}{4}$ C $\frac{7}{2}$ D 14

(d) $3\sqrt{5} \times 2\sqrt{6}$ simplifies to:

A $5\sqrt{11}$ B $6\sqrt{11}$ C $5\sqrt{30}$ D $6\sqrt{30}$

(e) $\frac{15\sqrt{6}}{3\sqrt{2}}$ simplifies to:

A $\sqrt{15}$ B $5\sqrt{3}$ C $5\sqrt{4}$ D $12\sqrt{3}$

10A

Understanding

6 Simplify each of the following.

(a) $3\sqrt{6} \times 5\sqrt{3}$ (b) $4\sqrt{2} \times 2\sqrt{14}$ (c) $-2\sqrt{6} \times 7\sqrt{10}$

(d) $-4\sqrt{8} \times 2\sqrt{10}$ (e) $-5\sqrt{3} \times \sqrt{15}$ (f) $10\sqrt{2} \times 3\sqrt{8}$

(g) $2\sqrt{12} \times 4\sqrt{3}$ (h) $2\sqrt{18} \times 3\sqrt{12}$ (i) $4\sqrt{27} \times 7\sqrt{28}$

(j) $-2\sqrt{20} \times -5\sqrt{75}$ (k) $-3\sqrt{72} \times -4\sqrt{63}$ (l) $\sqrt{54} \times \sqrt{162}$

7 Simplify each of the following.

(a) $\sqrt{\frac{24}{3}}$ (b) $\sqrt{\frac{54}{2}}$ (c) $3\sqrt{\frac{90}{2}}$ (d) $\frac{-6\sqrt{120}}{\sqrt{5}}$

(e) $\frac{\sqrt{27}}{\sqrt{4}}$ (f) $\frac{\sqrt{32}}{\sqrt{25}}$ (g) $\frac{9\sqrt{8}}{\sqrt{9}}$ (h) $\frac{15\sqrt{28}}{2\sqrt{25}}$

(i) $\sqrt{\frac{125}{36}}$ (j) $-3\sqrt{\frac{44}{81}}$ (k) $-2\sqrt{\frac{27}{100}}$ (l) $\frac{4}{5}\sqrt{\frac{147}{64}}$

8 Calculate the area of the following shapes, writing each answer as a surd in its simplest form.

(a)

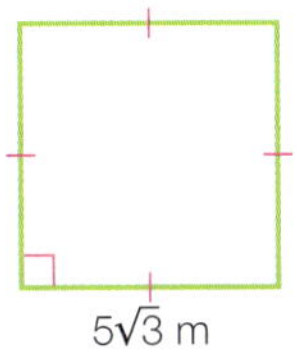

(b)

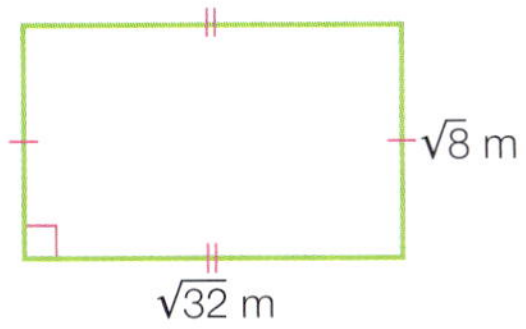

(c)

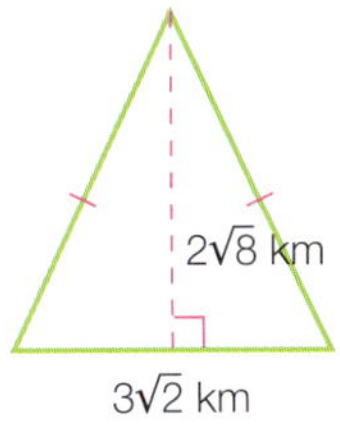

(d)

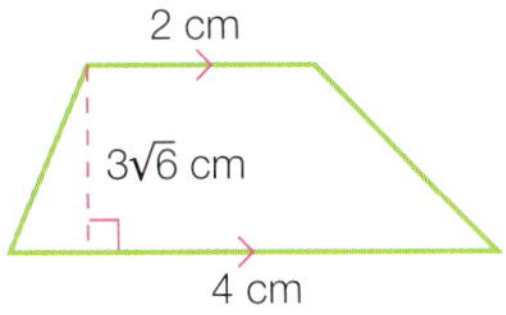

Area formulas

Square: $A = l^2$

Rectangle: $A = lw$

Triangle: $A = \dfrac{bh}{2}$

Trapezium: $A = \dfrac{(a + b)h}{2}$

Reasoning

9 (a) Chris calculated the side of a triangle and found it was $\sqrt{41}$ mm. Chris's friends' calculations gave lengths of $3\sqrt{5}$ mm and $2\sqrt{10}$ mm. If the actual length was 6 mm, which measurement was closest to this?

(b) If the area of a circular swimming pool is 45π m^2, find its *exact* circumference.

Hint

'Exact' means you should write in surd form if necessary.

10 Adrianna has $\sqrt{50}$ m of tape to go across the two diagonals of rectangular windows with side lengths 40 cm and 35 cm.

(a) What is the length of the diagonal?

(b) How many windows can be taped like this? (There might be some tape left over.)

Open-ended

11 Given $\sqrt{a} \times \sqrt{b} = 3\sqrt{c}$, what are possible values of b and c if $a = 2$?

12 Find at least three surd calculations that simplify to $2\sqrt{6}$.

13 (a) Jo multiplied two surds and got the answer 5. What could the two surds have been?

(b) Fiona divided two surds and got the answer 4. What could the two surds have been?

11.3 Adding and subtracting surds

When dealing with algebraic expressions, $3x + 9x$ can be simplified to $12x$. This is because they are *like* terms that have the same pronumeral x. The expression $8g - 3y$ cannot be simplified, because the pronumerals are different. They are not *like* terms.

The same rule is true when working with surds.

For example, consider $3\sqrt{2} + 5\sqrt{2}$.

The surd $\sqrt{2}$ is contained in each term. This means that the expression can be factorised by taking out the common factor of $\sqrt{2}$.

$$
\begin{aligned}
& 3\sqrt{2} + 5\sqrt{2} \\
&= 3 \times \sqrt{2} + 5 \times \sqrt{2} \\
&= (3 + 5)\sqrt{2} && \text{(take out common factor)} \\
&= 8 \times \sqrt{2} && \text{(simplify)} \\
&= 8\sqrt{2}
\end{aligned}
$$

So, $3\sqrt{2} + 5\sqrt{2} = 8\sqrt{2}$

> Numbers with the same surd are called *like* surds.
>
> Only terms that contain like surds can be added or subtracted.

10A

What about $4\sqrt{3} + 7\sqrt{2}$? This expression cannot be simplified, because the terms do not contain the same surd. (However, you can evaluate the expression approximately using a calculator.)

Worked example 6

W.E. 6

Simplify:

(a) $10\sqrt{3} - 4\sqrt{3}$ **(b)** $3\sqrt{5} + 2\sqrt{6} + \sqrt{5}$

Thinking	Working
(a) Add or subtract terms containing like surds.	(a) $10\sqrt{3} - 4\sqrt{3}$ $= 6\sqrt{3}$
(b) Add or subtract terms containing like surds. Different surds cannot be simplified.	(b) $3\sqrt{5} + 2\sqrt{6} + \sqrt{5}$ $= 3\sqrt{5} + 1\sqrt{5} + 2\sqrt{6}$ $= 4\sqrt{5} + 2\sqrt{6}$

To see whether there are like terms, it may be necessary to write each term in simplest form to reveal the like surds.

Worked example 7

W.E. 7

Simplify:

(a) $\sqrt{18}+\sqrt{8}$

(b) $2\sqrt{12}-\sqrt{20}-\sqrt{27}+2\sqrt{45}$

Thinking	Working
(a) 1 Simplify each term.	(a) $\sqrt{18}+\sqrt{8}$ $=\sqrt{9\times 2}+\sqrt{4\times 2}$ $=3\sqrt{2}+2\sqrt{2}$
2 Add like terms.	$=5\sqrt{2}$
(b) 1 Simplify each term.	(b) $2\sqrt{12}-\sqrt{20}-\sqrt{27}+2\sqrt{45}$ $=2\sqrt{4\times 3}-\sqrt{4\times 5}-\sqrt{9\times 3}+2\sqrt{9\times 5}$ $=2\times 2\sqrt{3}-2\sqrt{5}-3\sqrt{3}+2\times 3\sqrt{5}$ $=4\sqrt{3}-2\sqrt{5}-3\sqrt{3}+6\sqrt{5}$
2 Group like terms together.	$=4\sqrt{3}-3\sqrt{3}-2\sqrt{5}+6\sqrt{5}$
3 Add or subtract like terms. ($\sqrt{3}+4\sqrt{5}$ cannot be simplified further.)	$=\sqrt{3}+4\sqrt{5}$

11.3 Adding and subtracting surds

10A

Navigator

1 (column 1), 2 (a–j), 3, 4, 5, 6 (a), 7, 8, 10	1 (column 2), 2 (column 2), 4, 5, 6, 7, 8, 9, 10	1 (g–l), 2 (g–n), 4, 5 (c, d), 6, 7, 8, 9, 10

Answers p. 855

Fluency

1 Simplify each of the following.

W.E. 6

(a) $3\sqrt{2}+7\sqrt{2}$

(b) $8\sqrt{5}-5\sqrt{5}$

(c) $11\sqrt{3}-6\sqrt{3}$

(d) $\sqrt{7}+7\sqrt{7}$

(e) $5\sqrt{3}-9\sqrt{3}$

(f) $12\sqrt{11}-35\sqrt{11}$

(g) $5\sqrt{2}-3\sqrt{2}+9\sqrt{2}$

(h) $8\sqrt{3}-9\sqrt{3}+\sqrt{3}$

(i) $7\sqrt{5}-13\sqrt{5}+4\sqrt{5}-\sqrt{5}$

(j) $10\sqrt{6}+3\sqrt{3}-5\sqrt{6}-8\sqrt{3}$

(k) $5\sqrt{2}-7\sqrt{5}+2\sqrt{5}+2\sqrt{2}$

(l) $7\sqrt{3}+6\sqrt{11}-8\sqrt{3}-10\sqrt{11}$

W.E. 7

2 Simplify each of the following.

(a) $\sqrt{2}+\sqrt{18}$

(b) $\sqrt{28}-\sqrt{7}$

(c) $\sqrt{125}-2\sqrt{5}$

(d) $7\sqrt{18}+3\sqrt{50}$

(e) $5\sqrt{63}-7\sqrt{28}$

(f) $4\sqrt{90}-4\sqrt{160}$

(g) $\sqrt{40}+\sqrt{90}+\sqrt{160}$

(h) $2\sqrt{32}+7\sqrt{2}+4\sqrt{18}$

(i) $\sqrt{150}+5\sqrt{54}-6\sqrt{24}$

(j) $9\sqrt{48}-\sqrt{27}+2\sqrt{192}$

(k) $6\sqrt{80}-3\sqrt{180}-4\sqrt{405}$

(l) $\sqrt{8}+\sqrt{162}+\sqrt{18}+\sqrt{50}$

(m) $\sqrt{28}-\sqrt{112}+\sqrt{567}-\sqrt{175}$

(n) $2\sqrt{300}-6\sqrt{27}-3\sqrt{108}+4\sqrt{48}$

3 (a) $7\sqrt{3}-2\sqrt{3}$ simplifies to:

A 5 B $5\sqrt{3}$ C $5\sqrt{6}$ D $9\sqrt{3}$

(b) $5\sqrt{5}-3\sqrt{5}$ simplifies to:

A $2\sqrt{5}$ B 5 C $2\sqrt{7}$ D $8\sqrt{12}$

(c) $5\sqrt{3}+7\sqrt{3}+3\sqrt{2}$ simplifies to:

A $15\sqrt{3}$ B $12\sqrt{6}+3\sqrt{2}$ C $15\sqrt{8}$ D $12\sqrt{3}+3\sqrt{2}$

10A

4 (a) $\sqrt{3}+\sqrt{12}$ simplifies to:

A $\sqrt{15}$ B $2\sqrt{3}$ C $3\sqrt{3}$ D $5\sqrt{3}$

(b) $5\sqrt{2}-2\sqrt{8}$ simplifies to:

A $3\sqrt{-6}$ B $-3\sqrt{2}$ C 1 D $\sqrt{2}$

Understanding

5 Calculate the perimeter of each of the following shapes, writing your answer as a surd in its simplest form.

(a)

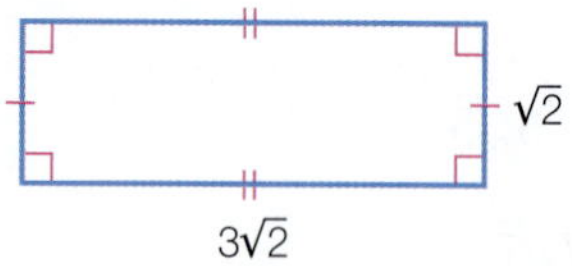

(b)

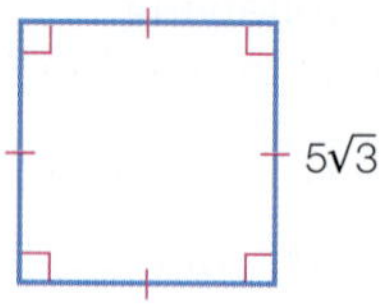

(c)

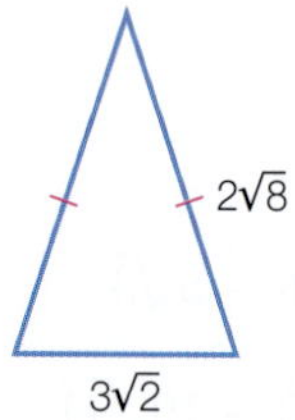

(d)

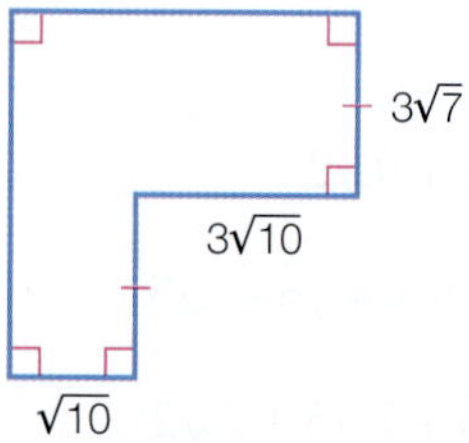

6 Find the value of the variable in each of the following shapes.

(a)

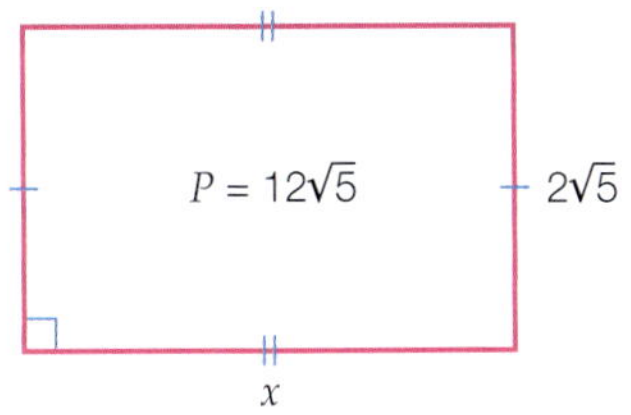

(b)

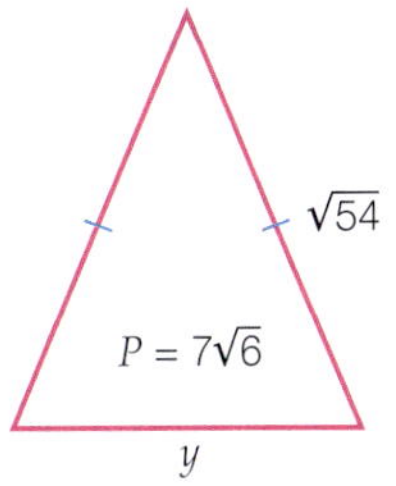

7 Find the area in exact form of one face of a circular coin with a diameter of $\sqrt{3}$ cm.

Reasoning

8 (a) Write the perimeter of the rectangle shown.

(b) What is the exact area of this rectangle?

(c) If this rectangle represents the available floor space of a bedroom, can the floor be covered by a piece of carpet with area 22 m^2?

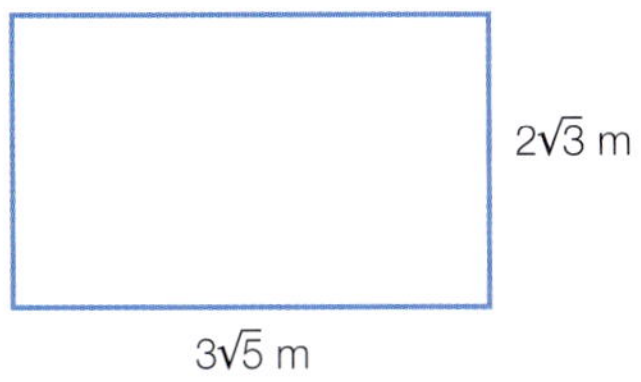

Open-ended

9 A rectangle has an area of $\sqrt{96}$ cm^2. What might its perimeter be?

10 Ruby's answers to a surds spot quiz are shown.

(i) Find the perimeter of a triangle with side lengths $2\sqrt{3}$, $2\sqrt{3}$, $3\sqrt{3}$.

$$\text{Perimeter} = 2\sqrt{3} + 2\sqrt{3} + 3\sqrt{3}$$
$$= 7\sqrt{9}$$
$$= 7 \times 3$$
$$\therefore P = 21$$

(ii) The perimeter of a rectangle is $8\sqrt{6}$. If the length of one pair of sides is $\sqrt{6}$, find the length of the other pair.

$$\text{Length of other pair of sides} = \frac{8\sqrt{6} - 2\sqrt{6}}{2}$$
$$= \frac{6}{2}$$
$$\therefore l = 3$$

(a) How would you explain to Ruby the mistake she made in each question?

(b) Write the correct answer for each question.

(c) Draw a diagram that could accompany each question.

Half-time 11

11.2 1 Simplify each of the following.

(a) $2\sqrt{3} \times \sqrt{3}$ (b) $4\sqrt{11} \times 2\sqrt{11}$ (c) $\sqrt{\frac{81}{16}}$

11.2 2 What is the area of a rectangular patio with dimensions of $5\sqrt{3}$ m and $\sqrt{6}$ m?

11.1 3 What is the exact side length of a square with an area of 45 m^2?

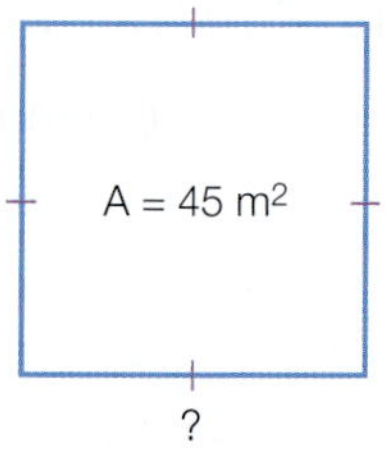

10A

11.1 4 For each of the numbers below, state whether it is rational or irrational.

(a) 3.656 565... (b) 0.924 (c) $\sqrt[3]{9}$ (d) $-\sqrt{25}$

11.3 5 Calculate the exact total area to be carpeted in the bedrooms and the living room in the house below, shaded purple.

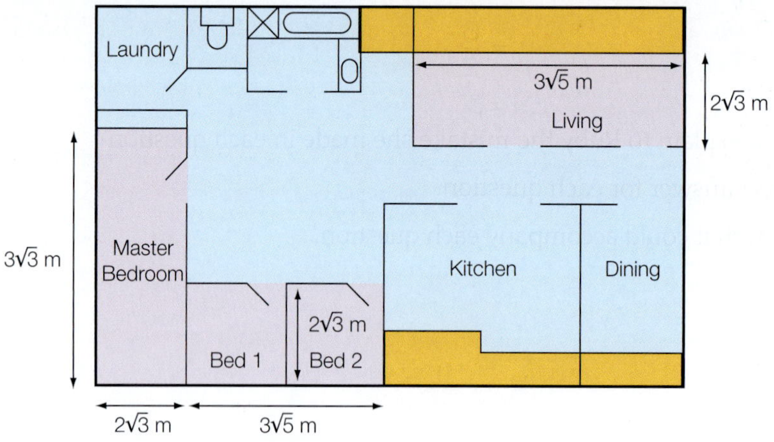

11.2 6 Simplify each of the following.

(a) $\sqrt{72}$ (b) $3\sqrt{20} \times 4\sqrt{45}$ (c) $\frac{3\sqrt{8}}{\sqrt{200}}$

11.3 7 Evaluate the following without the use of a calculator.

(a) $-5\sqrt{9}$ (b) $\sqrt{81} + \sqrt{64}$ (c) $4\sqrt{36} - \sqrt{16} + 2\sqrt{49}$

Surds and the distributive law

11.4

Consider the rectangle shown.

The side lengths *AB* and *BC* include surd terms. The two terms are not like, so they cannot be simplified to one term.

To find the area of this rectangle, the length and width must be multiplied:

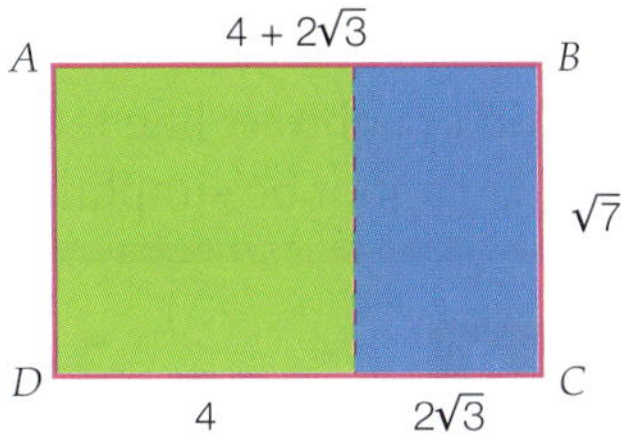

$$A = l \times w$$
$$= \sqrt{7} \times (4 + 2\sqrt{3})$$
$$= \sqrt{7}(4 + 2\sqrt{3})$$

To simplify surd expressions involving brackets, you can use the **distributive law**:

$a(b + c) = ab + ac$ or $(a + b)(c + d) = ac + ad + bc + bd$

To calculate the exact area of the rectangle, the term outside the bracket is multiplied by each term inside the bracket.

$$A = \sqrt{7}(4 + 2\sqrt{3})$$
$$= \sqrt{7} \times 4 + \sqrt{7} \times 2\sqrt{3}$$
$$= 4\sqrt{7} + 2\sqrt{21}$$

10A

Worked example 8 W.E. 8

Expand and simplify:

(a) $2\sqrt{3}(5 - \sqrt{3})$ **(b)** $(2\sqrt{2} + 1)(\sqrt{6} - 4\sqrt{3})$

Thinking	Working
(a) 1 Multiply each term inside the bracket by the term outside the bracket. $2\sqrt{3}(5 - \sqrt{3})$	(a) $2\sqrt{3}(5 - \sqrt{3})$ $= 10\sqrt{3} - 2\sqrt{9}$
2 Simplify any surds if necessary. ($\sqrt{9} = 3$)	$= 10\sqrt{3} - 2 \times 3$
3 Simplify further.	$= 10\sqrt{3} - 6$

(b) 1 Multiply each term inside the second bracket by each term inside the first bracket.

$(2\sqrt{2}+1)(\sqrt{6}-4\sqrt{3})$

2 Simplify any surds if necessary.
$(\sqrt{12} = \sqrt{4\times 3} = 2\sqrt{3})$

3 Add or subtract like terms.
($\sqrt{6}$ cannot be simplified further.)

(b) $(2\sqrt{2}+1)(\sqrt{6}-4\sqrt{3})$
$= 2\sqrt{12} - 8\sqrt{6} + \sqrt{6} - 4\sqrt{3}$
$= 2\times 2\sqrt{3} - 8\sqrt{6} + \sqrt{6} - 4\sqrt{3}$
$= 4\sqrt{3} - 8\sqrt{6} + \sqrt{6} - 4\sqrt{3}$
$= -7\sqrt{6}$

11.4 Surds and the distributive law

Navigator

Answers p. 855

1 (a–m), 2, 3, 4, 5, 7, 9, 10	1 (column 1), 2, 3, 4, 5, 6, 7, 9, 10, 11	1 (column 2), 3, 4, 5, 6, 8, 10, 11

Equipment required: calculator for Questions **5–9**

Fluency

10A

W.E. 8

1 Expand and simplify each of the following.

(a) $3(\sqrt{7}+2)$
(b) $-4(\sqrt{11}+\sqrt{3})$
(c) $9(2\sqrt{5}-\sqrt{2})$
(d) $-\sqrt{7}(5\sqrt{2}-3\sqrt{5})$
(e) $4\sqrt{2}(2\sqrt{5}+6\sqrt{3})$
(f) $3\sqrt{5}(\sqrt{2}-7\sqrt{3}+3\sqrt{7})$
(g) $\sqrt{3}(\sqrt{3}+2)$
(h) $5\sqrt{7}(3\sqrt{2}-2\sqrt{7})$
(i) $-3\sqrt{5}(2-8\sqrt{3}+\sqrt{5})$
(j) $3\sqrt{2}(\sqrt{14}-\sqrt{8})$
(k) $4\sqrt{3}(\sqrt{3}-\sqrt{6}+5)$
(l) $-4\sqrt{5}(1-2\sqrt{5}+\sqrt{15})$
(m) $(\sqrt{5}+4)(\sqrt{2}+3)$
(n) $(\sqrt{3}-\sqrt{5})(\sqrt{3}+\sqrt{7})$
(o) $(\sqrt{2}+\sqrt{3})^2$
(p) $(4\sqrt{5}-2\sqrt{2})^2$
(q) $(2\sqrt{7}+\sqrt{3})(2\sqrt{7}-\sqrt{3})$
(r) $(3\sqrt{5}-7\sqrt{2})(3\sqrt{5}+7\sqrt{2})$

2 (a) $5(2\sqrt{5}-\sqrt{2})$ is the same as:

A $10\sqrt{5}-\sqrt{10}$ **B** $10\sqrt{5}-5\sqrt{2}$ **C** $10\sqrt{5}-\sqrt{2}$ **D** $10\sqrt{3}$

(b) $4\sqrt{6}(\sqrt{2}-\sqrt{3})$ is the same as:

A $4\sqrt{8}-4\sqrt{18}$ **B** $8\sqrt{6}-12\sqrt{2}$ **C** $4\sqrt{8}-12\sqrt{2}$ **D** $8\sqrt{3}-12\sqrt{2}$

(c) $(2-\sqrt{5})(\sqrt{5}-1)$ expands to:

A $3\sqrt{5}+7$ **B** $3\sqrt{5}$ **C** $3\sqrt{5}-7$ **D** $3\sqrt{5}-3$

(d) $(\sqrt{7}+\sqrt{2})^2$ expands and simplifies to:

A $9-2\sqrt{14}$ **B** 9 **C** $9+2\sqrt{14}$ **D** $11\sqrt{14}$

Understanding

3 Find the exact area of each shape.

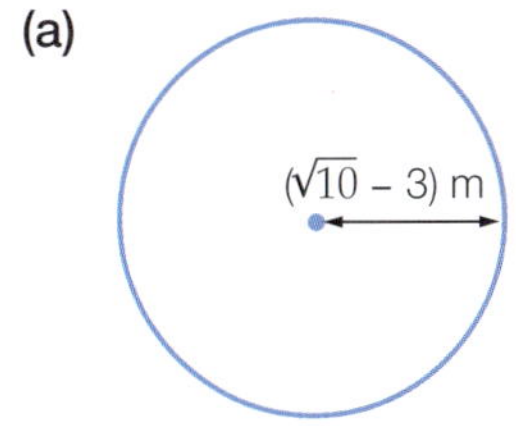

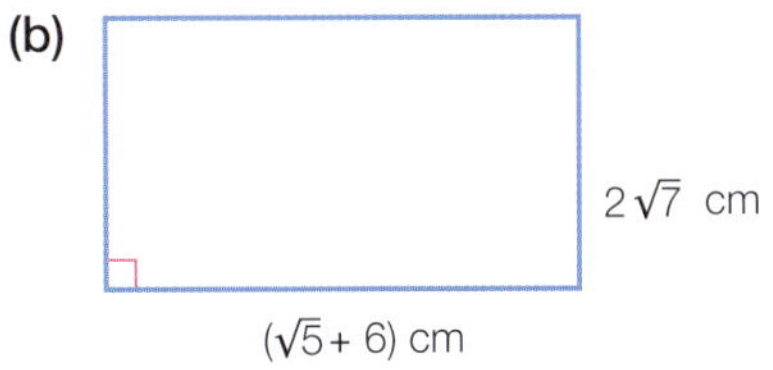

4 Find the exact area of a flat disc if it has a radius of $(4 + \sqrt{5})$ cm.

5 On a walk around the city, Luke followed the path shown in the diagram, finishing back where he started.

Hint What shape can you see?

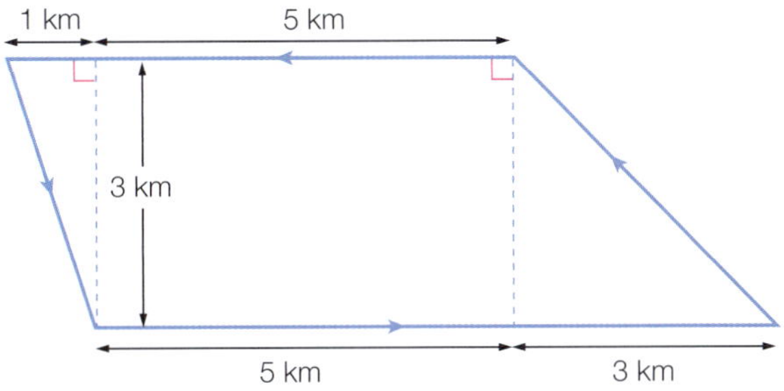

(a) Find the area enclosed by his path.

(b) Find the distance he walked. Write your answer in surd form.

(c) State the distance he walked, to the nearest metre.

6 On a scoreboard, tube lighting is set up so that any digit from 0 to 9 can be displayed, as shown in the diagram at right.

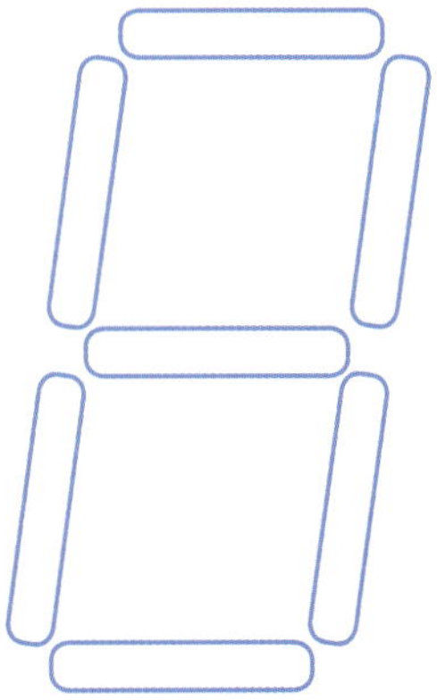

(a) If each individual section of tubing is $(3\sqrt{2} - 2)$ metres long, find the total length of tubing that would be lit to produce the following digits. Write answers in surd form.

(i) 0 (ii) 4 (iii) 5 (iv) 8

(b) Use your calculator to express each of the lengths above, correct to 2 decimal places.

10A

7 A sports medal is placed into a presentation box as shown. Give answers for parts **(a)** to **(d)** in exact form.

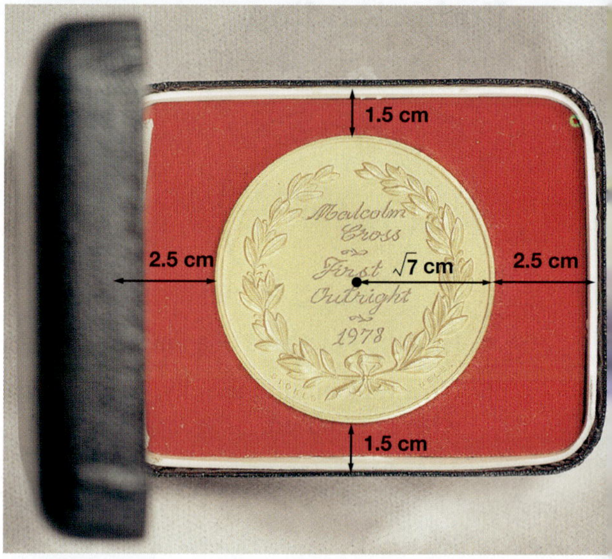

(a) What area of the base of the box does the medal take up?

(b) What are the dimensions (length and width) of the base of the box, assuming that the base is rectangular?

(c) What is the area of the base of the box?

(d) To stop the medal moving around the box, a thick layer of velvet surrounds it. What area of velvet is visible around the medal?

(e) Express your answer to part **(d)** correct to the nearest whole number.

Reasoning

8 Zev has a toy box in the shape of a cube (open at the top). He wants to cover its surface (both inside and out) with adhesive-backed plastic and needs to know how much he should buy.

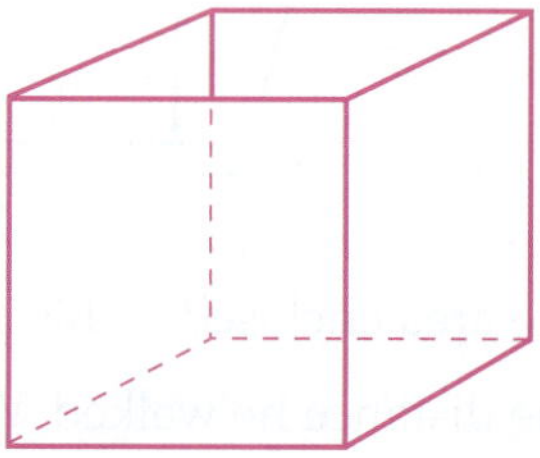

(a) What is the length of a side of the toy box, correct to 2 decimal places?

10A

(b) Use your answer to part **(a)** to find the area to be covered with plastic. Assume that the box has walls of negligible thickness. Round your answer to the nearest whole number.

(c) Find the area to be covered with plastic using the value $(8\sqrt{5} + 9\sqrt{11})$ cm and keeping your answer as a surd expression.

(d) Use a calculator to express your answer to part **(c)** correct to the nearest whole number.

(e) Compare your answers to parts **(b)** and **(d)**.

(i) Calculate the percentage difference in area caused by rounding. Round the percentage correct to the nearest whole number.

(ii) Explain your result.

9 Square biscuits are to be cut out of a length of biscuit dough. Each biscuit has a line of chocolate icing running along the 6 cm diagonal of the square.

(a) Use Pythagoras' theorem to find the side length of each biscuit in exact form.

(b) What length of biscuit dough would be needed to cut out a row of 50 biscuits? (Keep your answer in surd form.)

(c) Use a calculator to express your answer to part **(b)**, correct to 1 decimal place.

(d) Use a calculator to express the side length of each biscuit, correct to 1 decimal place.

(e) Use your answer to part **(d)** to calculate the length of biscuit dough needed to cut out a row of 50 biscuits.

(f) Compare your answers to parts **(c)** and **(e)**. What is the percentage difference in length caused by rounding early in the calculation? Round the percentage correct to the nearest whole number.

Open-ended

10 The area of a triangle is $\sqrt{48}$ cm^2. Draw at least two diagrams with different dimensions that could be the triangle.

11 A rectangle has a length of $m\sqrt{n}$ metres and a width of p metres, where m, n and p are whole numbers.

(a) Write an expression for the rectangle's perimeter in metres.

(b) If the perimeter is greater than 20 m, then what might be the values of m, n and p? Draw a diagram, showing lengths $m\sqrt{n}$ and p.

(c) Write an expression for the rectangle's area, using side lengths $m\sqrt{n}$ and p.

(d) State values of m, n and p so that the area of the rectangle is less than 10 m^2.

Problem solving

Connecting squares

The diagram below shows increasingly larger squares, connecting to each previous smaller square along one edge.

(a) If the side lengths of the smallest squares are 1 cm, then what is the area of the entire rectangle?

(b) This diagram is a spiral pattern that starts with the yellow square. Complete the spiral pattern shown on the first three squares below. If the pattern is continued, what would be the dimensions and area of the next square?

(c) Write the side lengths of each square in ascending order. Have you seen this number pattern before? Does it have a name?

(d) Explain the pattern and write the side lengths of the next five squares.

Strategy options

- Look for a pattern.
- Break problem into manageable parts.

11.5 Rationalising the denominator

Fractions are easier to work with if the denominator is a rational number. Consider the number $\frac{1}{\sqrt{3}}$. You can change the denominator of $\sqrt{3}$ to a rational number if you multiply it by itself. The numerator must also be multiplied by $\sqrt{3}$ to keep the fraction equivalent. $\left(\frac{\sqrt{3}}{\sqrt{3}} = 1\right)$

$$\begin{aligned}\frac{1}{\sqrt{3}} &= \frac{1}{\sqrt{3}} \times 1\\ &= \frac{1}{\sqrt{3}} \times \frac{\sqrt{3}}{\sqrt{3}}\\ &= \frac{1 \times \sqrt{3}}{\sqrt{3} \times \sqrt{3}}\\ &= \frac{\sqrt{3}}{3}\end{aligned}$$

So $\frac{1}{\sqrt{3}}$ can be written as $\frac{\sqrt{3}}{3}$, which has a rational denominator.

10A

This process is called **rationalising the denominator**.

Rationalising the denominator of a fraction changes the denominator to a rational number, but does not change the value of the fraction.

Worked example 9

W.E. 9

Rationalise the denominator of:

(a) $\frac{3\sqrt{5}}{\sqrt{6}}$

(b) $\frac{2\sqrt{3}}{\sqrt{20}}$

Thinking	Working
(a) 1 Identify the surd value that will rationalise the denominator ($\sqrt{6}$) and multiply the numerator and denominator by that value.	(a) $\frac{3\sqrt{5}}{\sqrt{6}}$ $= \frac{3\sqrt{5}}{\sqrt{6}} \times \frac{\sqrt{6}}{\sqrt{6}}$
2 Simplify the surd calculation.	$= \frac{3\sqrt{30}}{6}$
3 Cancel, if possible, to write your answer in the simplest form.	$= \frac{\sqrt{30}}{2}$

(b) 1 Simplify the denominator.	(b) $\frac{2\sqrt{3}}{\sqrt{20}}$
2 Identify the surd value that will rationalise the denominator ($\sqrt{5}$) and multiply the numerator and denominator by that value.	$= \frac{2\sqrt{3}}{2\sqrt{5}}$ $= \frac{2\sqrt{3}}{2\sqrt{5}} \times \frac{\sqrt{5}}{\sqrt{5}}$
3 Simplify the surd calculation.	$= \frac{2\sqrt{15}}{2 \times 5}$
4 Cancel, if possible, to write your answer in the simplest form.	$= \frac{\sqrt{15}}{5}$

Using the conjugate to rationalise denominators

If the denominator contains two different surds, then you need to use a conjugate expression to rationalise this. The distributive law can be used to show the following:

$$\begin{aligned}(\sqrt{a}+\sqrt{b})(\sqrt{a}-\sqrt{b}) &= \sqrt{a}\times\sqrt{a}-\sqrt{a}\times\sqrt{b}+\sqrt{b}\times\sqrt{a}-\sqrt{b}\times\sqrt{b}\\ &= \sqrt{a^2}-\sqrt{ab}+\sqrt{ab}-\sqrt{b^2}\\ &= a-b\end{aligned}$$

This simplifies to the difference of two squares rule:

$$\begin{aligned}(\sqrt{a}+\sqrt{b})(\sqrt{a}-\sqrt{b}) &= (\sqrt{a})^2-(\sqrt{b})^2\\ &= a-b\end{aligned}$$

The two expressions $(\sqrt{a}+\sqrt{b})$ and $(\sqrt{a}-\sqrt{b})$ are called **conjugate surds**, each being the conjugate of the other. Conjugate surds are useful because their product will not contain a surd term.

For example, the conjugate of $(3-\sqrt{2})$ is $(3+\sqrt{2})$. When these two numbers are multiplied, the result is a rational number:

$$\begin{aligned}&(3-\sqrt{2})(3+\sqrt{2})\\ &= 9+3\sqrt{2}-3\sqrt{2}-2\\ &= 9-2\\ &= 7\end{aligned}$$

or

$$\begin{aligned}&(3-\sqrt{2})(3+\sqrt{2})\\ &= 3^2-(\sqrt{2})^2\\ &= 9-2\\ &= 7\end{aligned}$$

$(\sqrt{a}+\sqrt{b})$ and $(\sqrt{a}-\sqrt{b})$ are conjugate surds.

$(\sqrt{a}+\sqrt{b})(\sqrt{a}-\sqrt{b}) = a-b$

10A

Worked example 10

W.E. 10

Express each of the following in simplest form with a rational denominator.

(a) $\dfrac{3}{3-\sqrt{2}}$

(b) $\dfrac{\sqrt{2}}{\sqrt{11}+3}$

Thinking	Working
(a) 1 Multiply both the numerator and denominator by the conjugate of the denominator $(3+\sqrt{2})$.	(a) $\dfrac{3}{3-\sqrt{2}}$ $= \dfrac{3}{(3-\sqrt{2})} \times \dfrac{(3+\sqrt{2})}{(3+\sqrt{2})}$
2 Use the difference of two squares to simplify the denominator.	$= \dfrac{3(3+\sqrt{2})}{9-2}$
3 Expand the numerator and write the final answer in the simplest way.	$= \dfrac{9+3\sqrt{2}}{7}$
(b) 1 Multiply both the numerator and denominator by the conjugate of the denominator $(\sqrt{11}-3)$.	(b) $\dfrac{\sqrt{2}}{\sqrt{11}+3}$ $= \dfrac{\sqrt{2}}{(\sqrt{11}+3)} \times \dfrac{(\sqrt{11}-3)}{(\sqrt{11}-3)}$
2 Use the difference of two squares to simplify the denominator.	$= \dfrac{\sqrt{2}(\sqrt{11}-3)}{11-9}$
3 Expand the numerator and write the final answer in the simplest way.	$= \dfrac{\sqrt{22}-3\sqrt{2}}{2}$

The denominator can be rationalised in two ways:

- Multiply the numerator and denominator by the smallest surd that will give a rational denominator (if the denominator has only one term).
- Multiply the numerator and denominator by the conjugate of the denominator (if the denominator has two terms).

10A

11.5 Rationalising the denominator

Navigator

1, 2, 3, 4, 5, 6 (a–b), 7, 9, 11, 12 | 1 (columns 2–3), 2 (columns 2–3), 3, 4, 5, 6 (a–c), 7, 9, 11, 12 | 1 (columns 3–4), 2 (columns 3–4), 3, 5, 6, 7, 8, 9, 10, 11, 12

Answers p.856

Fluency

1 Express each of the following in simplest form with a rational denominator. W.E. 9

(a) $\frac{1}{\sqrt{7}}$ (b) $\frac{4}{\sqrt{3}}$ (c) $\frac{-12}{\sqrt{6}}$ (d) $\frac{\sqrt{3}}{\sqrt{2}}$

(e) $\frac{5\sqrt{11}}{\sqrt{5}}$ (f) $3\sqrt{\frac{5}{6}}$ (g) $-\frac{3}{4\sqrt{7}}$ (h) $\frac{6\sqrt{5}}{7\sqrt{3}}$

2 Express each of the following in simplest form with a rational denominator. W.E. 10

(a) $\frac{1}{3-\sqrt{7}}$ (b) $\frac{4}{2+\sqrt{3}}$ (c) $\frac{\sqrt{3}}{\sqrt{5}-\sqrt{2}}$ (d) $\frac{2}{2\sqrt{7}+5}$

(e) $\frac{6+\sqrt{3}}{2\sqrt{3}-3}$ (f) $\frac{\sqrt{5}+1}{7-3\sqrt{5}}$ (g) $\frac{\sqrt{6}-1}{3\sqrt{2}+\sqrt{7}}$ (h) $\frac{\sqrt{2}-3}{2\sqrt{3}+\sqrt{6}}$

3 (a) $\frac{5}{\sqrt{2}}$ expressed with a rational denominator is:

A $\frac{\sqrt{10}}{2}$ B $\frac{2\sqrt{5}}{5}$ C $\frac{5\sqrt{2}}{\sqrt{2}}$ D $\frac{5\sqrt{2}}{2}$

(b) $\frac{4}{3\sqrt{3}}$ expressed with a rational denominator is:

A $\frac{4\sqrt{3}}{9}$ B $\frac{\sqrt{12}}{9}$ C $\frac{4\sqrt{3}}{3}$ D $\frac{4\sqrt{3}}{27}$

(c) $\frac{2\sqrt{2}}{3\sqrt{7}}$ expressed with a rational denominator is:

A $\frac{2\sqrt{6}}{21}$ B $\frac{2\sqrt{14}}{21}$ C $\frac{14\sqrt{2}}{3}$ D $\frac{2\sqrt{14}}{3}$

4 (a) The conjugate of $2\sqrt{3}-5$ is:

A $2\sqrt{3}-5$ B $2\sqrt{3}+5$ C $\sqrt{3}+5$ D $5\sqrt{3}+2$

(b) To rationalise the denominator of $\frac{4+\sqrt{7}}{2-3\sqrt{3}}$ it is necessary to multiply by:

A $\frac{4-\sqrt{7}}{4-\sqrt{7}}$ B $\frac{1}{2+3\sqrt{3}}$ C $2+3\sqrt{3}$ D $\frac{2+3\sqrt{3}}{2+3\sqrt{3}}$

10A

Understanding

5 Write each of the following as a single fraction with a rational denominator.

(a) $\frac{1}{3}+\frac{3}{\sqrt{2}}$ (b) $\frac{4}{\sqrt{5}}-\frac{1}{\sqrt{3}}$ (c) $\frac{2\sqrt{6}}{\sqrt{5}}-\frac{3}{2\sqrt{5}}$ (d) $\frac{3}{2\sqrt{2}}+\frac{\sqrt{10}}{4}$

6 Simplify each of the following. (Rationalise each denominator first.)

(a) $\frac{1}{\sqrt{3}+1}+\frac{1}{3-\sqrt{5}}$ (b) $\frac{1}{\sqrt{7}+1}+\frac{1}{3-\sqrt{6}}$

(c) $\frac{1}{\sqrt{6}+\sqrt{3}}-\frac{1}{4-\sqrt{11}}$ (d) $\frac{1}{\sqrt{5}+\sqrt{3}}-\frac{1}{\sqrt{7}-2}$

Reasoning

7 The floor plan of a bathroom to be tiled is shown.

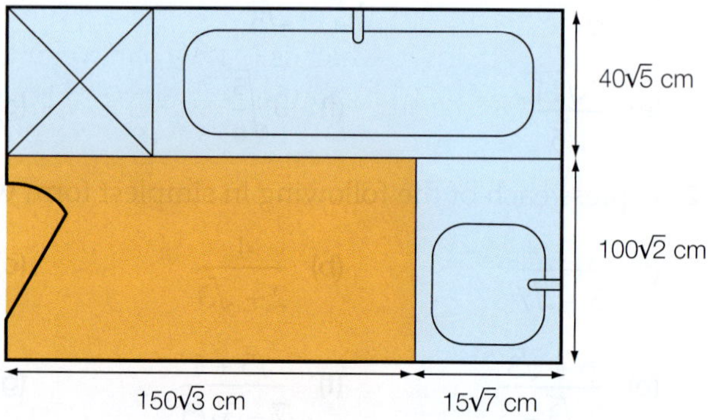

(a) Calculate the area to be tiled (shaded orange).

(b) The square tiles chosen have side lengths of $10\sqrt{3}$ cm. How many tiles are required for the bathroom floor?

(c) How many tiles would you order?

(d) Find the area of the entire bathroom in square centimetres.

8 If $\frac{3+2\sqrt{3}}{4-\sqrt{2}}a=\frac{4+\sqrt{2}}{5-6\sqrt{3}}$, find the value of a, expressing your answer with a rational denominator.

Open-ended

9 Find at least three fractions that are equivalent to $\frac{4\sqrt{2}}{\sqrt{3}}$, one of which must have a rational denominator.

10 For each of the following, find two fractions that the stated number is between, with a rational denominator formed from the original fraction.

(a) $\frac{1}{\sqrt{3}}$ (b) $\frac{3}{\sqrt{7}}$ (c) $\frac{2\sqrt{3}}{\sqrt{5}}$ (d) $\frac{2}{3\sqrt{2}}$

11 Find at least two expressions that $2+\sqrt{3}$ can be multiplied by to give a rational answer.

12 Sura is trying to rationalise the denominator of the fraction $\frac{\sqrt{2}-3}{2\sqrt{3}+\sqrt{6}}$. She is struggling to understand why the denominator can be rationalised by multiplying both the numerator and denominator by the denominator's conjugate $2\sqrt{3}-\sqrt{6}$. How could you explain this to Sura?

$$\frac{\sqrt{2}-3}{(2\sqrt{3}+\sqrt{6})}\times\frac{(2\sqrt{3}-\sqrt{6})}{(2\sqrt{3}-\sqrt{6})}$$

Investigation

Phi (Φ) and the golden rectangle

Equipment required: compass, ruler

The Big Question

What are phi, the golden ratio and the golden rectangle and how are they connected?

Engage

Square *ABCD* has side lengths of 2 units. A straight line joins point *B* with point *E*, which is halfway between *C* and *D*. The rectangle *AFGD* is formed using the radius of *BE*, as shown.

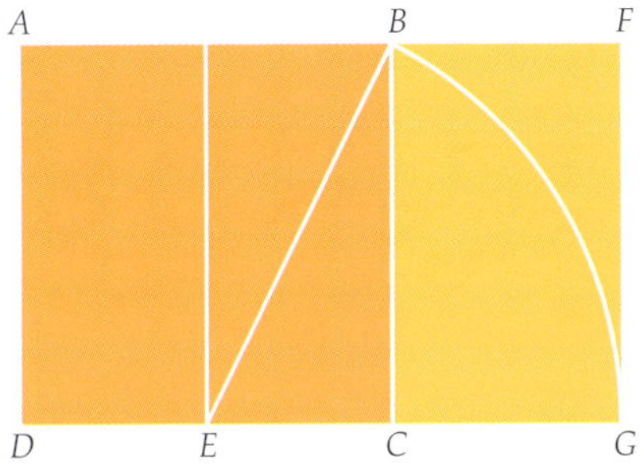

1 **(a)** What is the length *BE*, as a surd?

(b) What is the length *EG*, expressed as a surd?

(c) What is the length *DG*, as a surd expression?

(d) Find the ratio of length and width $\left(\frac{L}{W}\right)$ of rectangle *AFGD* as an expression using surds and as a decimal number rounded to 9 decimal places.

The ratio calculated in **(d)** is known as the golden ratio or phi (Φ), and is an irrational number. A rectangle with this ratio is called a golden rectangle.

2 **(a)** What is the length *CG* as a surd?

(b) Repeat **1(d)** for rectangle *BFGC*.

(c) Is your answer to **2(b)** the same as the answer to **1(d)**?

Explore

3 **(a)** Make a golden rectangle, with a ruler and pair of compasses, starting with a 10 cm square.

(b) What is the length of this rectangle? Write the length as a simplified surd. Record the length, width and ratio $\frac{L}{W}$ in a table.

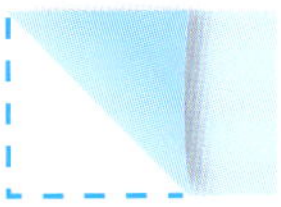

(c) Cut the rectangle into a square and a rectangle.

Prove that the smaller rectangle is also a golden rectangle by folding the rectangle into a square and a rectangle. Record this rectangle's length, width and ratio $\frac{L}{W}$ into the table created in **(b)**.

(d) Continue to divide the rectangles into the square and the rectangle, writing the results in the table. What is the smallest golden rectangle you are able to make?

Strategy options

- Make a model.
- Make a table.

Explain

4 **(a)** What did you discover about the ratio $\frac{L}{W}$ of the rectangles created?

(b) What is the exact value of phi (Φ)?

Elaborate

5 'A golden rectangle is …'. How would you finish this sentence? Use the words phi and golden ratio within your sentence.

Evaluate

6 **(a)** Describe how to create a golden rectangle using a step-by-step method.

(b) Were you surprised by the findings? Explain.

(c) Did you have any difficulties making the golden rectangle or dividing the rectangle into smaller golden rectangles? If so, what were they and how were these difficulties overcome?

Extend

7 International standard paper sizes, such as the 'A' series that includes the common sizes A4 and A3, also involve special ratios. Investigate the system of international standard paper sizes, its unique features and its benefits.

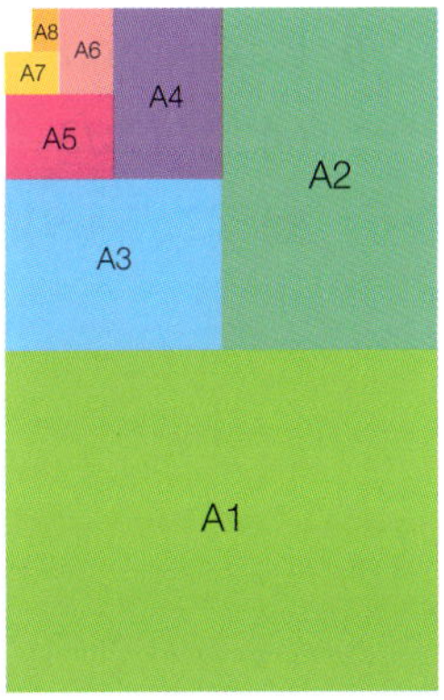

Chapter review 11

Maths literacy

conjugate surds	irrational numbers	real numbers
distributive law	rational numbers	surds
entire surd	rationalising the denominator	

Copy and complete the following using the words and phrases from the list, where appropriate. A word or phrase may be used more than once.

1 The real number system contains _________ and _________.

2 $\sqrt{2}$ and π are examples of _________.

3 _________ are numbers that contain a radical or root symbol and cannot be written as an exact number.

4 To simplify surd expressions involving brackets, you can use the _________.

5 When _________ of a fraction it is necessary to multiply both the numerator and the denominator by a surd or _________ in order to make the denominator rational.

6 The _________ $\sqrt{45}$ can be simplified to $3\sqrt{5}$.

7 $\sqrt{3} - 1$ and $\sqrt{3} + 1$ are _________.

Fluency

10A

1 Use a calculator to evaluate the following correct to 3 decimal places. **11.1**

(a) $\sqrt{17}$ (b) $\sqrt[3]{25}$ (c) $5 \times 4\sqrt{3}$ (d) $3\sqrt{7} - 4\sqrt{2}$

2 Simplify each of the following. **11.2**

(a) $4\sqrt{5} \times 3\sqrt{6}$ (b) $5\sqrt{7} \times 2\sqrt{7}$ (c) $\sqrt{\frac{64}{25}}$ (d) $\frac{18\sqrt{6}}{2\sqrt{3}}$

(e) $\sqrt{40}$ (f) $2\sqrt{125}$ (g) $3\sqrt{2} \times 5\sqrt{10}$ (h) $4\sqrt{\frac{75}{16}}$

3 Express each of the following as an entire surd. **11.2**

(a) $6\sqrt{2}$ (b) $-3\sqrt{7}$

4 Simplify each of the following. **11.3**

(a) $6\sqrt{2} - 8\sqrt{2} + 3\sqrt{2}$ (b) $5\sqrt{7} - 2\sqrt{3} - \sqrt{7} + \sqrt{3}$

(c) $4\sqrt{12} + 7\sqrt{27}$ (d) $6\sqrt{50} - 2\sqrt{24} + 4\sqrt{32} + \sqrt{54}$

5 Expand and simplify each of the following. **11.4**

(a) $\sqrt{5}(4\sqrt{3} - 2\sqrt{5} + \sqrt{2})$ (b) $-6\sqrt{2}(3\sqrt{7} - \sqrt{2} + 1)$

(c) $(2\sqrt{5} + 9)(\sqrt{5} - 2)$ (d) $(3\sqrt{7} + 4\sqrt{2})^2$

6 Express each of the following in simplest form with a rational denominator. **11.5**

(a) $\frac{5}{\sqrt{2}}$ (b) $\frac{4\sqrt{3}}{\sqrt{6}}$ (c) $\frac{7}{2\sqrt{5}}$

7 Express each of the following in simplest form with a rational denominator. **11.5**

(a) $\frac{7}{3-\sqrt{5}}$ (b) $\frac{\sqrt{2}-2}{2\sqrt{3}+\sqrt{7}}$

8 (a) $2\sqrt{6}(\sqrt{10}+3\sqrt{3})$ can be simplified to: **11.4**

A $2\sqrt{60}+5\sqrt{18}$ B $8\sqrt{15}+18\sqrt{2}$ C $4\sqrt{15}+18\sqrt{2}$ D $30\sqrt{2}+5\sqrt{18}$

(b) $8\times\sqrt{\frac{81}{64}}$ can be simplified to: **11.1**

A $\sqrt{9}$ B $24\times\sqrt{\frac{1}{8}}$ C $8\times\sqrt{\frac{9}{8}}$ D 9

Understanding

9 A round drinks coaster has a circular area of 232π cm^2. **11.2**

(a) What is its exact radius? (Leave your answer in surd form.)

(b) State the radius of the coaster correct to 2 decimal places.

10 A gift box has its four sides covered by wrapping paper. It has length $8\sqrt{10}$ cm, width 18 cm and height $(5\sqrt{6}+3\sqrt{7})$ cm. **11.1–11.4**

(a) What are the approximate length and height of the box, correct to 1 decimal place?

(b) Using your answers to part **(a)**, calculate the area to be covered by wrapping paper, correct to 1 decimal place.

(c) Using the surd expression, find the area to be covered by wrapping paper. Express your answer as a surd.

(d) Use a calculator to express your answer to part **(c)**, correct to 1 decimal place.

(e) Compare your answers to parts **(b)** and **(d)**.

10A

Reasoning

11 (a) Find the exact side length of a square with an area of 50 cm^2. **11.3**

(b) Find the exact side length of an equilateral triangle with an area of 40 cm^2.

(c) Find the exact circumference of a circular swimming pool with an area of 75π m^2.

12 If $\left(\frac{4-3\sqrt{5}}{5+2\sqrt{2}}\right)a=\frac{5-2\sqrt{2}}{4+\sqrt{5}}$, find the value of a, expressing your answer with a rational denominator. **11.5**

12

Non-linear relationships and logarithms

Maths at sea

Name: David Provis
Job: Oceanographer
Qualifications: Bachelor of Science, PhD in Mathematics

Imagine if your job paid you to study one of your favourite places. That's the case for David Provis, Senior Principal Oceanographer with engineering firm Cardno.

David always enjoyed sailing, swimming and surfing and now finds himself observing and measuring the movement of the tides, currents and sand in the marine environment.

He uses the data he collects to perform calculations to find the height and energy of waves, the volume of sand shifted by an ocean current, or how the depth of water affects the readings of his instruments. 'Maths helps me understand and describe the laws of the ocean. I have a toolbox of mathematical techniques I can use, such as geometry, trigonometry and statistics.'

The information David produces may be used by engineers when building a marine structure such as a wharf, pipeline or oil rig, or by biologists concerned about the environmental impact of such a structure. Oceanographers may also be involved in studying the effects of climate change on ocean levels, developing tsunami early warning systems, or even in designing the perfect surf break!

Why learn this?

Straight lines are not the only useful type of graph. Non-linear relationships such as exponentials and circles are useful in modelling many situations. The ripples made by a stone thrown into still water are circles. The decay of radioactive substances can be modelled by exponential equations. Using graphs to model real-life situations means that future trends can be predicted, and relationships between variables such as global temperatures and the rate of carbon emissions produced worldwide can be established.

After completing this chapter you will be able to:

- identify common types of non-linear graphs
- solve non-linear equations algebraically and graphically using technology
- identify, describe, construct and transform algebraic and graphical representations of exponential functions and circles
- solve problems that can be modelled by exponential and circle relationships

Vic
- understand algebraic and graphical representations of inverse proportion and reciprocal functions

- apply the index laws to numerical and algebraic expressions, evaluating and simplifying where necessary

10A
- describe, interpret, construct and transform rectangular hyperbolas
- understand and work with expressions that involve fractional indices, and understand their connection with surds
- understand the relationship between exponential and logarithmic expressions
- apply the logarithmic laws to numerical and algebraic expressions.

12 Recall

Prepare for this chapter by attempting the following questions. If you have difficulty with a question, you can download a Recall Worksheet from the eBook or the Pearson Places website.

1 Factorise each of the following.

(a) $x^2 + 7x + 10$ (b) $x^2 - 5x + 6$ (c) $x^2 - 13x - 14$

(d) $x^2 + 5x - 24$ (e) $x^2 + 5x - 36$ (f) $2x^2 + 14x + 12$

(g) $-x^2 - 3x - 2$ (h) $-x^2 + 7x - 10$ (i) $-2x^2 + 2x + 24$

2 Write each of the following as a number without an index.

(a) 3^3 (b) 4^{-1} (c) $5^{-2} + 3$

3 Solve the following using the null factor law.

(a) $(x - 1)(x + 3) = 0$ (b) $(x + 4)(x + 2) = 0$

(c) $(x - 1)^2 = 0$ (d) $x(x - 5) = 0$

4 Solve the following using special product rules to factorise first.

(a) $x^2 - 6x + 9 = 0$ (b) $4x^2 - 25 = 0$

(c) $x^2 + 8x + 16 = 0$ (d) $121x^2 - 36 = 0$

5 Factorise each of the following by completing the square.

(a) $x^2 + 4x + 2$ (b) $x^2 - 6x + 6$ (c) $x^2 - 3x - 5$

6 Simplify each of the following.

(a) $m^3 \times m^2$ (b) $3k^2p^4 \times 4kp^5$ (c) $\dfrac{x^3 \times x^6}{x^2}$

(d) $3x^0$ (e) $(m^3n)^4$ (f) $(x^{-3}y^2)^{-2}$

Exploration Task

You can download this activity from the eBook or the Pearson Places website.

Where will you use logarithms?

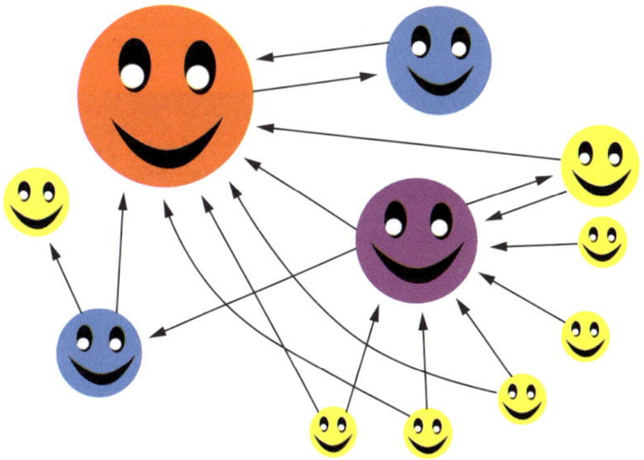

In this activity you will explore how logarithms can be useful in the world outside a classroom, investigating the example of website search rankings.

12.1 Identifying non-linear graphs and their equations

Circles

The equation of a circle with its centre at the origin (0, 0) and with radius r is given by $x^2 + y^2 = r^2$. The power of both the independent and dependent variables is 2.

Because both variables are squared, the x- and y-coordinates of a circle are often surds, so when drawing a circle graph you should find:

- the centre of the circle
- four points on the circumference (usually above , below, left and right of the centre).

For a circle with its centre at the origin and radius r these four points are $(r, 0)$, $(0, r)$, $(-r, 0)$ and $(0, -r)$.

Circle equations in the form of $(x - h)^2 + (y - k)^2 = r^2$ have the centre at (h, k) and radius r.

The four points to plot on the circumference are $(h + r, k)$, $(h, k + r)$, $(h - r, k)$ and $(h, k - r)$.

Interactive

Equation of a circle

Explore how the graph of a circle changes as the values change in the equation.

Go to the eBook or the Pearson Places website to access this interactive.

Worked example 1

W.E. 1

Plot the graphs of the following equations.

(a) $x^2 + y^2 = 25$

(b) $(x - 1)^2 + (y + 2)^2 = 36$

Thinking	Working
(a) 1 Write the equation, identify the radius (here, $r = 5$) and centre (h, k) (here, the centre is (0, 0)). Draw a table of values using $(r, 0)$, $(0, r)$, $(-r, 0)$, and $(0, -r)$. If the radius is a surd, use an approximation to 1 decimal place.	(a) $x^2 + y^2 = 25$

x	-5	0	5
y	0	±5	0

2 Draw a set of labelled Cartesian axes, plot these points and join them with a smooth curve. The scale has to be the same for both axes. Label the graph and the centre (0, 0).

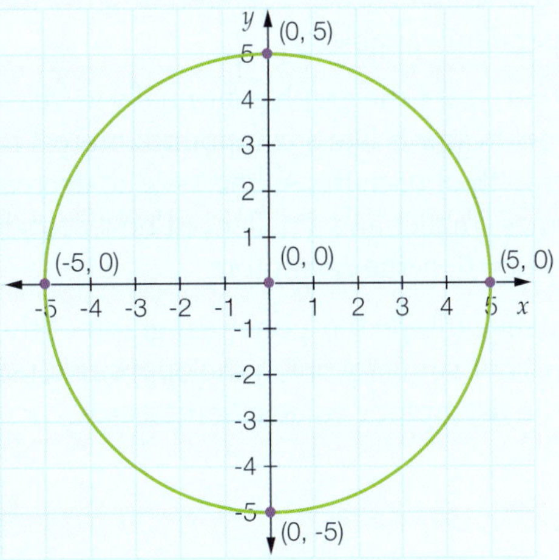

(b) 1 Write the equation and identify the radius (here, $r = 6$) and centre (h, k) (here, $h = 1$, $k = -2$). Draw a table of values using $(h + r, k)$, $(h, k + r)$, $(h - r, k)$ and $(h, k - r)$. If the radius is a surd, use an approximation to 1 decimal place.

(b) $(x - 1)^2 + (y + 2)^2 = 36$

x	-5	1	7
y	-2	-8, 4	-2

2 Draw a set of labelled Cartesian axes, plot the points and join them with a smooth curve. The scale has to be the same for both axes. Label the graph and the centre (h, k).

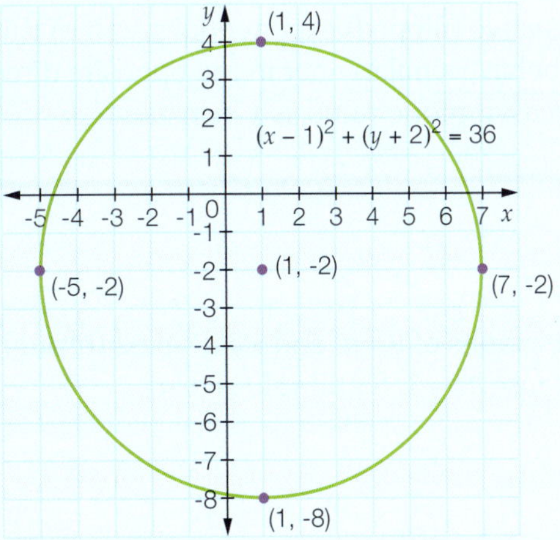

Exponentials

An **exponential** relationship is where a quantity grows or decays at a rate proportional to its current value. This relationship means that the independent variable is the index or exponent of a constant base.

All exponential graphs have a limiting value, which the graph will approach but never reach. This limiting value gives a horizontal **asymptote**. An asymptote is a line that a curve approaches but never touches.

The simplest exponential graph increases continuously. This is a model of exponential growth. Exponential growth has many examples in the real world, such as the growth of money invested with compound interest or the population growth of bacteria.

An exponential growth equation is in the form $y = a^x$ where a is the base ($a > 1$) and x is the independent variable.

For $y = a^x$, $a > 1$ the graph continually increases as x increases. This behaviour is like a compound interest investment or the growth of bacteria. As the x-value decreases, the y-value becomes smaller and smaller but never reaches zero. The x-axis ($y = 0$) is the asymptote.

A negative (flipped) exponential graph of the form $y = -a^x$ will behave similarly but in the opposite direction, decreasing from an asymptotic value as x increases.

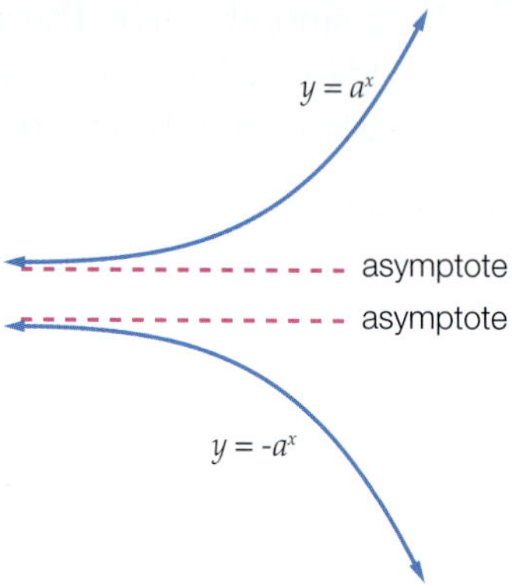

An exponential graphs of the form $y = a^{-x}$, $a > 1$, will decrease to an asymptotic value. This behaviour is like radioactive decay or an item's loss of value over time (depreciation).

As the x-value increases, the y-value becomes smaller and smaller but never reaches zero. The x-axis ($y = 0$) is the asymptote for these graphs also.

A negative (flipped) exponential graph of this form, $y = -a^{-x}$, will behave similarly but in the opposite direction, increasing to an asymptotic value as x increases.

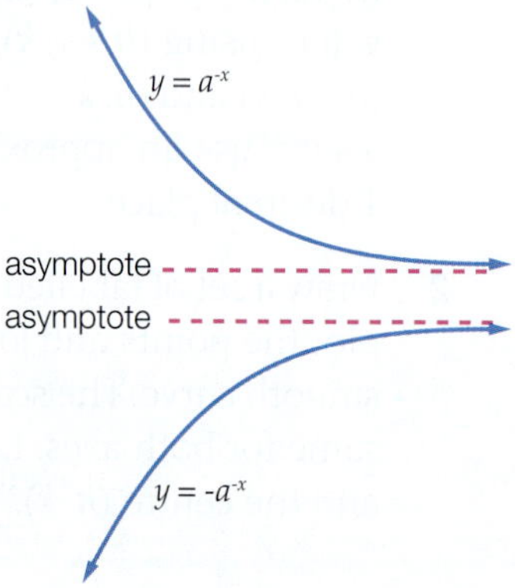

The graphs of $y = a^x$ and $y = a^{-x}$ pass through the point (0, 1) because $a^0 = 1$, for all values of a.

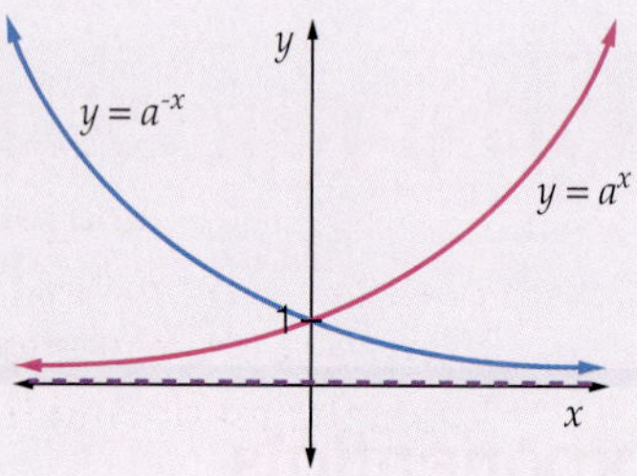

Interactive

Exponentials

Explore how the graph of an exponential changes as the values change in the equation.

Go to the eBook or the Pearson Places website to access this interactive.

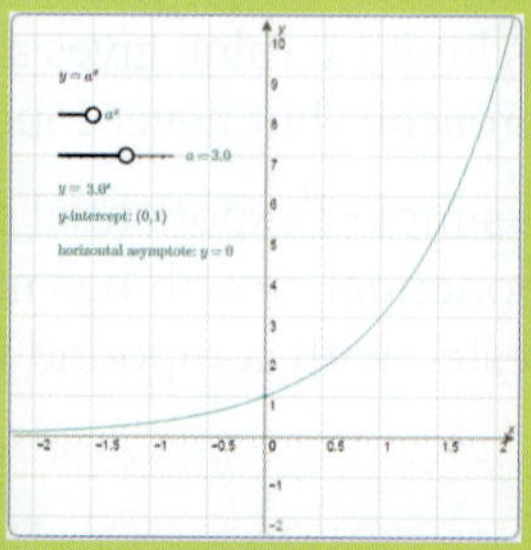

Worked example 2

W.E. 2

Plot the graphs of the following equations for $-4 \le x \le 4$.

(a) $y = 2^x$ **(b)** $y = 2^{-x} + 3$

Thinking	Working

(a) 1 Write the equation, identify the base and draw a table of values.

(a) $y = 2^x$

x	-4	-3	-2	-1	0	1	2	3	4
y	$\frac{1}{16}$	$\frac{1}{8}$	$\frac{1}{4}$	$\frac{1}{2}$	1	2	4	8	16

2 Draw a set of labelled Cartesian axes, plot the points and join them with a smooth curve. The scale does not have to be the same for both axes. Label the graph and the asymptote.

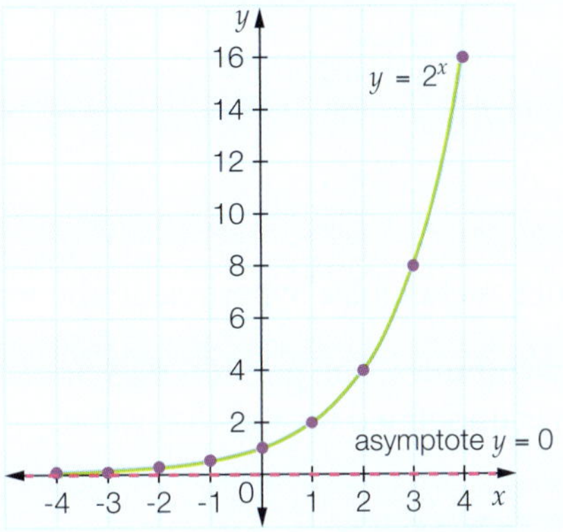

(b) 1 Write the equation, identify the base and draw a table of values.

(b) $y = 2^{-x} + 3$

x	-4	-3	-2	-1	0	1	2	3	4
y	19	11	7	5	4	$3\frac{1}{2}$	$3\frac{1}{4}$	$3\frac{1}{8}$	$3\frac{1}{16}$

2 Draw a set of labelled Cartesian axes, plot the points, and join them with a smooth curve. The scale does not have to be the same for both axes. Label the graph and the asymptote.

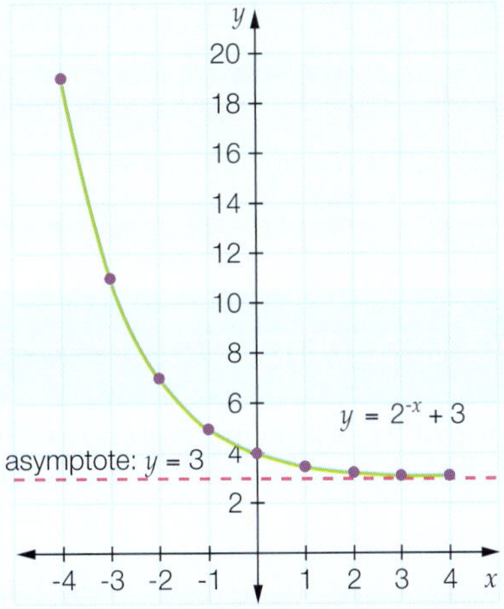

or 10A

Rectangular hyperbola

A **rectangular hyperbola** is another common non-linear relationship. In this relationship, as one variable increases, the other decreases. (For example, consider a rectangle with a constant area: as the length increases, the width decreases.)

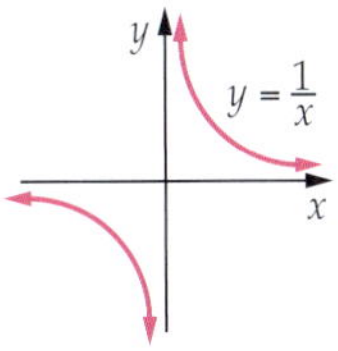

The rectangular hyperbola has two asymptotes that are at right angles to each other. The simplest form of a rectangular hyperbola equation is in the form of $y = \frac{1}{x}$, also called a reciprocal equation. The variables x and y are inversely proportional to each other.

Equations of a rectangular hyperbola in the form of $y = \frac{a}{x-h} + k$ look like this:

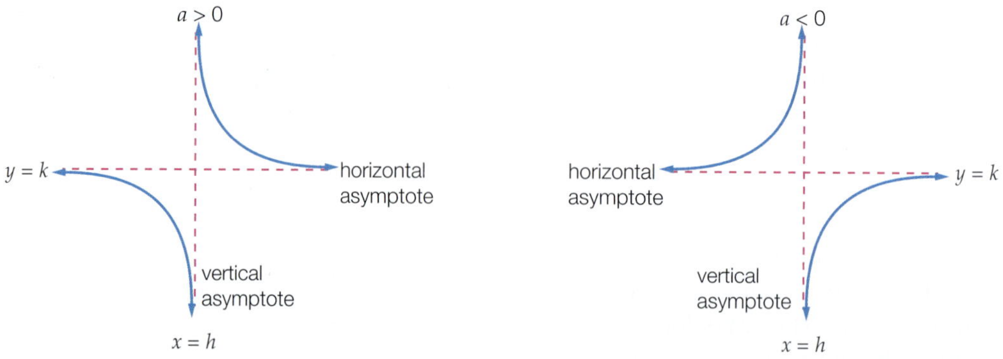

Graphs of a rectangular hyperbola in the form of $y = \frac{a}{x-h} + k$ have both horizontal and vertical asymptotes. The graph is *discontinuous* at the vertical asymptote, because the equation is undefined for x at that value.

10A

Interactive

Rectangular hyperbola

Observe in a hyperbola graph the relationship between the values and how by increasing one variable the other decreases.

Go to the eBook or the Pearson Places website to access this interactive.

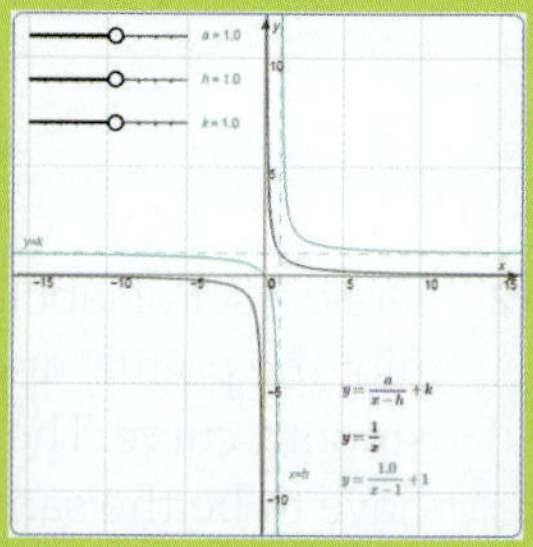

Worked example 3

W.E. 3

Plot the graphs of the following equations for $-4 \le x \le 4$.

(a) $y = \frac{4}{x}$

(b) $y = \frac{1}{x-1} + 3$

Thinking | **Working**

(a) 1 Write the equation, identify the base and draw a table of values.

(a) $y = \frac{4}{x}$

x	-4	-3	-2	-1	0	1	2	3	4
y	-1	$-\frac{4}{3}$	-2	-4	undefined	4	2	$\frac{4}{3}$	1

2 Draw a set of labelled Cartesian axes, plot the points and join them with a smooth curve. The scale does not have to be the same for both axes. Label the graph and the asymptotes.

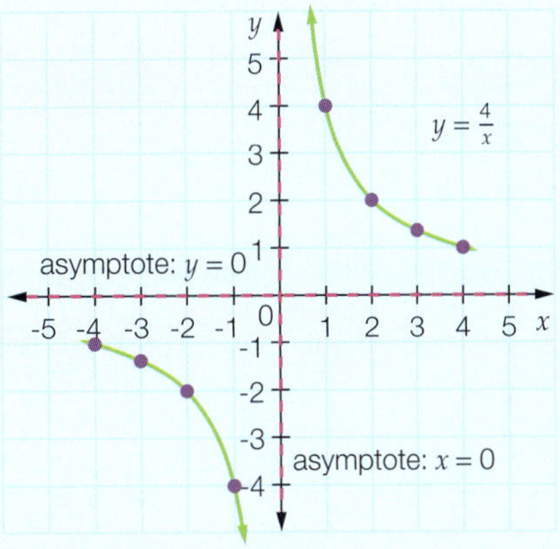

(b) 1 Write the equation, identify the base and draw a table of values.

(b) $y = \frac{1}{x-1} + 3$

x	-4	-3	-2	-1	0	1	2	3	4
y	$2\frac{4}{5}$	$2\frac{3}{4}$	$2\frac{2}{3}$	$2\frac{1}{2}$	2	undefined	4	$3\frac{1}{2}$	$3\frac{1}{3}$

2 Draw a set of labelled Cartesian axes, plot the points and join them with a smooth curve. The scale does not have to be the same for both axes. Label the graph and the asymptotes.

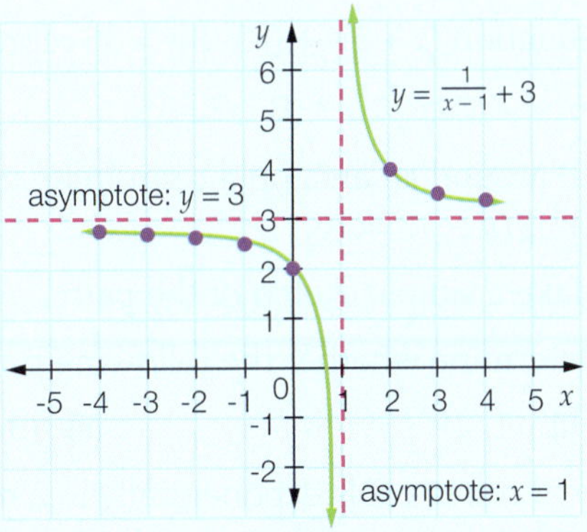

12.1 Identifying non-linear graphs and their equations

Navigator

1, 2, 4, 5, 6, 7, 8, 9, 12, 15, 16 (a), 17, 18	1 (column 1), 2 (a–d), 3, 4, 5, 6, 7, 8, 9, 10, 11, 12, 13, 14, 15, 16, 17	1 (column 2), 2 (c–f), 3, 7, 8, 9, 10, 11, 12, 13, 14, 15, 16, 17, 18

Answers p. 857

Equipment required: graph paper for Questions **1**, **2**, **3**, **8**, **12**; calculator for Questions **12**, **13**

Fluency

1 Plot the graphs of each of the following equations. **W.E. 1**

(a) $(x+3)^2 + (y+2)^2 = 16$

(b) $(x-3)^2 + (y-2)^2 = 9$

(c) $(x-1)^2 + (y+1)^2 = 25$

(d) $x^2 + (y+4)^2 = 36$

(e) $(x+4)^2 + y^2 = 8$

(f) $(x-4)^2 + (y+2)^2 = 12$

W.E. 2

2 Plot the graphs of the following equations for $-4 \le x \le 2$.

(a) $y = 2^x + 2$ (b) $y = 2^x - 1$

(c) $y = 2^x - 3$ (d) $y = 3^x - 1$

(e) $y = 3^x - 2$ (f) $y = 3^x + 4$

Vic or 10A W.E. 3

3 Plot the graphs of the following equations for $-6 \le x \le 6$.

(a) $y = \frac{2}{x}$ (b) $y = \frac{3}{x}$ (c) $y = -\frac{4}{x}$

(d) $y = \frac{1}{x-2} + 1$ (e) $y = \frac{1}{x+2} - 3$ (f) $y = \frac{1}{x+2} - 1$

Understanding

4 The graph of the circle $(x + 4)^2 + (y - 3)^2 = 49$ has a centre located at:

A (-4, 3) B (-4, -3) C (4, -3) D (4, 3)

5 The asymptote for the exponential equation $y = 3^x + 2$, exists at:

A $y = 3$ B $y = 2$ C $y = 0$ D y-axis

6 The equation $(x + 2)^2 + (y - 3)^2 = 36$ represents a circle with a radius of:

A 36 B 2 C -3 D 6

7 A path around an area on a coordinate map is described by the equation $(x + 1)^2 + (y - 3)^2 = 16$.

(a) Make a simple sketch of the path.

(b) Determine whether the following points are inside or outside the circle.

(i) (1, 2) (ii) (3, 6) (iii) (1.5, 2.5)

Hint

Will the temperature ever reach 25 °C exactly?

8 A cup of coffee cools exponentially in a room where the temperature is 25 °C. The temperature of the coffee was taken every 30 minutes and the results recorded in the table below.

t (min)	0	30	60	90	120	150
T (°C)	100	54	36.2	29.3	26.7	25.6

(a) Use graph paper to plot a graph of t against T (show t on the horizontal axis). Join the points with a smooth curve. Use your graph to answer parts **(b)** to **(e)**.

(b) What is the initial temperature of the coffee?

(c) Use the graph to find the temperature of the coffee after:

(i) 10 min (ii) 20 min (iii) 40 min (iv) 50 min

Give your answer to the nearest degree.

(d) During which 10 minute period in the first hour did the coffee cool the fastest? Justify your answer.

(e) When will the temperature be 27.5 °C?

(f) Extend the curve. Predict the temperature after 3 hours.

9 (a) Write the equation of a circle with:

(i) centre (3, 1) and radius 7 (ii) centre (-2, -4) and radius 3

(iii) centre (2, -5) and radius 2.

(b) Plot the graph for each of the equations in part **(a)**.

10A **10** The graph of the equation $y = \frac{1}{x-1} + 2$ has asymptotes at:

A $x = 2, y = 1$ **B** $x = 1, y = 2$ **C** $x = -1, y = 2$ **D** $x = 1, y = -2$

11 Which of the following points is not on the graph of $(x-3)^2 + (y+2)^2 = 25$?

A (7, 1) **B** (-1, -5) **C** (7, -5) **D** (-7, 1)

12 **(a)** For the relationship $y = 2^x$, copy and complete this table of values for x and y.

x	-2	-1	0	1	2	3	4
y							

(b) Use this table of values to plot a graph of $y = 2^x$. Join the points with a smooth curve.

(c) Use your graph to determine approximate values for:

(i) $2^{1.5}$ **(ii)** $2^{0.8}$ **(iii)** $2^{-0.5}$

(iv) $2^{-1.2}$ **(v)** $\sqrt{2}$ **(vi)** $(\sqrt{2})^3$

(d) Use a calculator to check the answers you found in part **(c)**.

(e) Use your graph to find the value of the exponent correct to 1 decimal place, when y is:

(i) 5 **(ii)** 12 **(iii)** 15

Reasoning

13 A town has an initial population of 4000, but the establishment of a new industry in the town is expected to increase the population by 6% per year for future years.

(a) What will be the predicted population 1 year after the industry is established?

(b) Complete a table of values to show the predicted population for the first 5 years after the industry is established. Round the population to the nearest whole number. Use P to represent the population after n years.

(c) Write an equation to model the predicted population over future years.

(d) Sketch a graph of this relationship using your equation.

(e) Use your graph to predict when the population will have doubled in size.

(f) Use your equation to confirm that your prediction in part **(e)** is correct.

(g) Explain why this is a first quadrant graph only.

14 The centre of a circular water fountain is 3 m east and 2 m south of the corner of a building. If the water fountain has a radius of 1 m, determine whether a tree could be planted 2.4 m east and 2.7 m south of the corner of the building. Justify your answer by using an equation to plot a graph that represents this situation.

Open-ended

15 Write an equation for a circle that does not have its centre at the origin and plot its graph.

16 Write the equation for an exponential relationship that has an asymptote at $y = -1$.

10A **17** Write an equation for a rectangular hyperbola that has asymptotes at $x = -2$ and $y = 6$.

18 Find the equations of at least two circles that pass through the point (1, 2) and have different centres.

12.2 Sketching graphs of non-linear relationships

The equations of other types of relationships can be transformed in the same way as quadratics, by translations and dilations. The graphs can then be sketched easily if the basic shape and key features are recognised.

Working backwards from this, if a graph is sketched, then its equation can be found by recognising the basic shape and writing the basic equation with the noted translations, reflections and dilations.

Quadratics

Basic equation: $y = x^2$	Basic shape is a parabola. 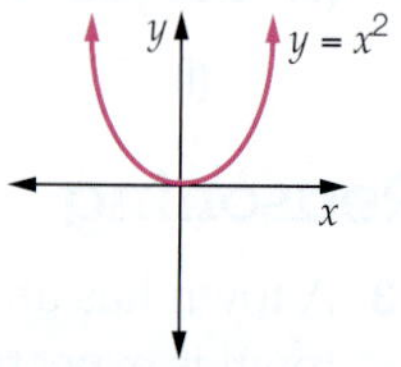
Translated equation: $y = (x - h)^2 + k$	Basic shape is moved h units to the right and k units up.
Dilated equation: $y = ax^2$	If $a < -1$ or $a > 1$, then the parabola is narrower than $y = x^2$. If $-1 < a < 1$, then the parabola is wider than $y = x^2$. If the equation of the curve has a negative coefficient, $a < 0$, then the basic shape of the parabola is reflected in the x-axis.

Circles

Basic equation: $x^2 + y^2 = r^2$	Basic shape is a circle. The centre is (0, 0) and radius is r. 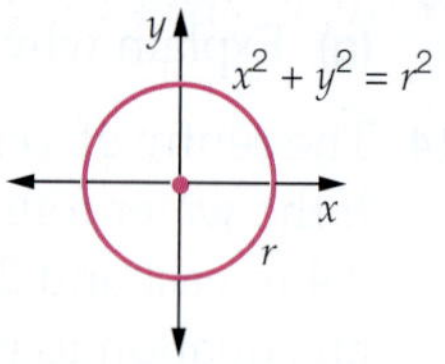
Translated equation: $(x - h)^2 + (y - k)^2 = r^2$	Basic shape is moved h units to the right and k units up. The centre is (h, k) and the radius is r.
Dilated equation: $ax^2 + ay^2 = r^2$	The radius of the circle is changed. Note that the dilation must be the same for x and y. The centre is (0, 0) and the radius is $\frac{r}{\sqrt{a}}$.

The x-intercept and y-intercept can be difficult to find by hand. Technology can help to find these values.

Note: A circle is dilated by changing its radius. It should be dilated by the same amount in both directions, or else it will not be a circle.

Worked example 4

W.E. 4

Sketch the graph of $(x-1)^2+(y+2)^2=9$ clearly showing the centre of the circle and the radius.

Thinking	Working
1 Determine the translations by comparing the equation to $(x-h)^2+(y-k)^2=r^2$ and identify the new centre of the circle.	Coordinates of the centre are (1, -2).
2 Determine the radius of the circle by comparing the equation to $(x-h)^2+(y-k)^2=r^2$. (Here, $r^2=9$.)	$r=\sqrt{9}$ $=3$ ∴ radius is 3 units
3 Sketch the graph using the centre and radius to determine four points on the circumference $(h-r, k)$, $(h, k+r)$, $(h+r, k)$ and $(h, k-r)$	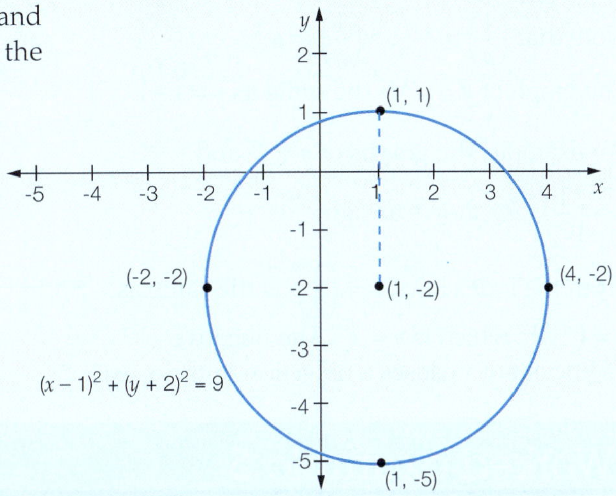

Exponential graphs

The graph of an exponential with equation $y=a^x$ has an asymptote of $y=0$ and a y-intercept of (0, 1). The value of a determines the rate of increase of the graph.

If a graph is growing exponentially by doubling, then $a=2$. If it is tripling, then $a=3$.

- As a increases, the graph becomes steeper.
- As x decreases, the graph moves closer to the line $y=0$.

Basic equation $y=a^x, a>1$	Basic shape is an exponential graph. The graph is continually increasing with an asymptote at $y=0$ and a y-intercept of (0, 1). If $0<a<1$, the graph is reflected in the y-axis. 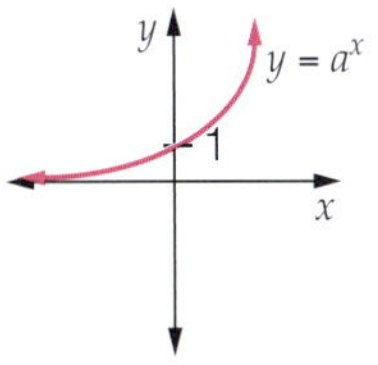
Translated equation $y=a^{(x-h)}+k$	Basic shape is moved h units to the right and k units up. The asymptote is $y=k$ and the y-intercept is $(0, a^{-h}+k)$. The point (0, 1) has moved to the point $(h, k+1)$.

The graph of $y = 2^x$ translated 3 units to the right and 4 units down is $y = 2^{(x-3)} - 4$. The translation down changes the asymptote from $y = 0$ to $y = -4$. The translation also affects the y-intercept. The original y-intercept of (0, 1) has moved 3 units right and 4 units down and so is now the point (3, -3). To calculate the new y-intercept, let $x = 0$ and evaluate:

$$\begin{aligned} y &= 2^{-3} - 4 \\ &= \left(\frac{1}{2}\right)^3 - 4 \\ &= \frac{1}{8} - 4 \\ &= -3\frac{7}{8} \end{aligned}$$

The y-intercept is at $\left(0, -3\frac{7}{8}\right)$.

For the x-intercept, let $y = 0$ and solve. This can be calculated using technology. All exponential graphs will have a y-intercept, but not all exponential graphs will have an x-intercept.

Note that $\left(\frac{1}{a}\right) = a^{-1}$ so $\left(\frac{1}{a}\right)^x = a^{-x}$.

The graph of $y = a^{-x}$ is the same as $y = \left(\frac{1}{a}\right)^x$.

For example, the graphs of $y = 2^x$ and $y = \left(\frac{1}{2}\right)^x$ are shown at right.

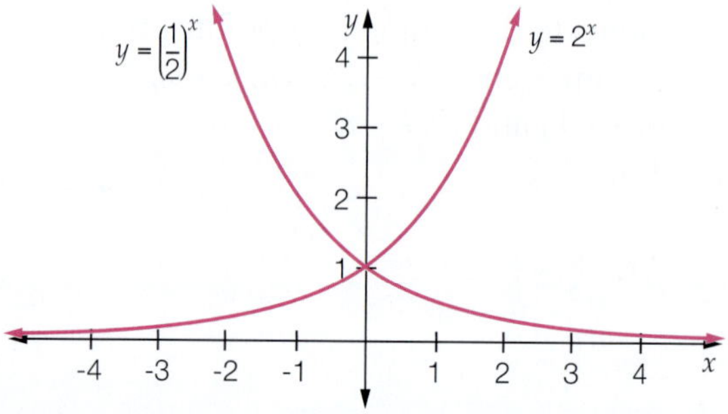

From the index laws, $y = \left(\frac{1}{2}\right)^x$ is the same as $y = (2^{-1})^x$, which is $y = 2^{-x}$. The negative coefficient of x causes a reflection in the y-axis.

Worked example 5

W.E. 5

Sketch the graph of $y = 3^{(x-1)} + 2$, clearly showing the asymptote and other key features.

Thinking	Working
1 State the basic graph and identify the translations.	$y = 3^{(x-1)} + 2$ is the graph $y = 3^x$ translated 1 unit right and 2 units up.
2 State the equation of the asymptote.	The asymptote is $y = 2$.
3 Find the y-intercept by solving with $x = 0$.	$x = 0$, $y = 3^{-1} + 2$ $= 2\frac{1}{3}$ The y-intercept is $(0, 2\frac{1}{3})$.
4 Find another point on the graph.	$x = 1$, $y = 3^{(1-1)} + 2$ $= 1 + 2$ $= 3$ Another point is (1, 3).
5 Sketch the graph, clearly showing the y-intercept and the asymptote with its equation.	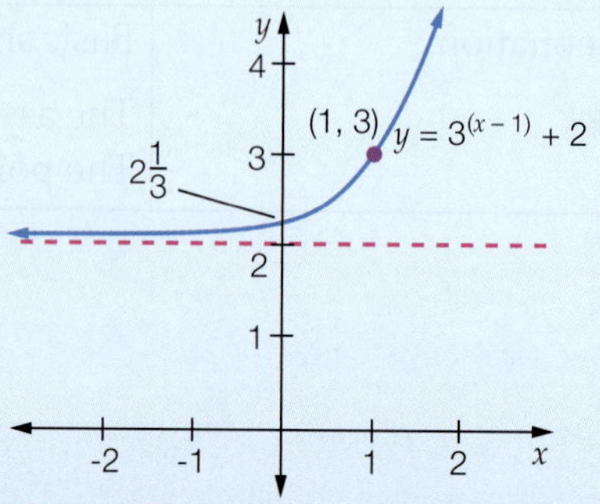

or 10A

Rectangular hyperbolas

Basic equation: $y = \frac{1}{x}$	The basic shape is a hyperbola. It has two asymptotes, the x- and y-axes. There are no x-intercepts or y-intercepts. 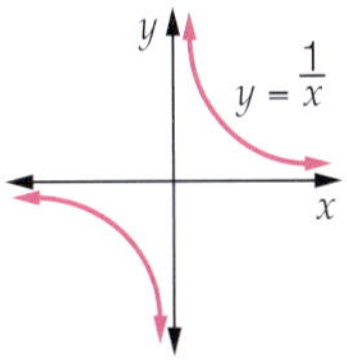
Translated equation: $y = \frac{1}{x-h} + k$	Basic shape is moved h units to the right and k units up. There are asymptotes at $x = h$ and $y = k$ and graph is discontinuous for $x = h$.

Worked example 6

W.E. 6

Sketch the graph of $y = \frac{1}{x-4} + 2$, clearly showing the key features (asymptotes, x-intercepts and y-intercepts).

Thinking	Working
1 Write the equation.	$y = \frac{1}{x-4} + 2$
2 Identify the asymptotes.	The asymptotes are $x = 4$ and $y = 2$.
3 Find the y-intercept.	$x = 0, y = -\frac{1}{4} + 2$ $= 1\frac{3}{4}$ The y-intercept is $\left(0, 1\frac{3}{4}\right)$.
4 Find the x-intercept.	$y = 0, 0 = \frac{1}{x-4} + 2$ $-2 = \frac{1}{x-4}$ $-2x + 8 = 1$ $-2x = -7$ $x = 3\frac{1}{2}$ The x-intercept is $\left(3\frac{1}{2}, 0\right)$.
5 Mark the key features (asymptotes and intercepts) on the Cartesian axes and sketch the graph with a smooth curve.	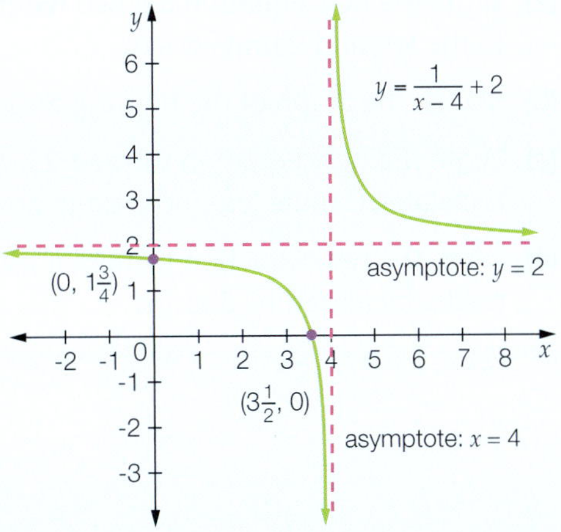

10A

12.2 Sketching graphs of non-linear relationships

Navigator

Answers p. 860

1, 2, 3, 4, 9, 12	1, 2, 3, 4, 5, 6, 8, 10, 12	1 (a–d), 2 (a–d), 3, 4, 5, 6, 7, 8, 9, 10, 11, 12

Fluency

W.E. 4

1 Sketch the graph of each of the following, clearly showing the centre of the circle and the radius.

(a) $x^2 + (y + 3)^2 = 64$

(b) $(x - 1)^2 + y^2 = 36$

(c) $(x - 2)^2 + (y + 1)^2 = 16$

(d) $(x + 3)^2 + (y - 4)^2 = 25$

(e) $(x - 2)^2 + (y - 2)^2 = 4$

(f) $(x + 4)^2 + (y + 1)^2 = 16$

W.E. 5

2 Sketch the graph of each of the following, clearly showing the asymptote and other key features.

(a) $y = 2^{(x+3)}$

(b) $y = 3^x + 2$

(c) $y = 2^{(x-2)} + 1$

(d) $y = 2^{(x+1)} + 3$

(e) $y = 2^x - 3$

(f) $y = 3^{(x-4)} - 1$

Understanding

3 What is the equation of each of the following graphs?

(a)

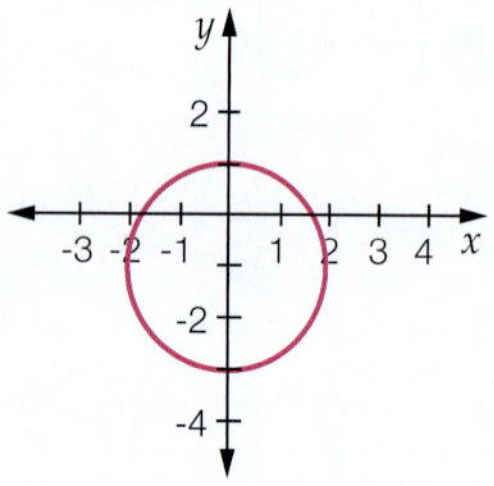

(b)

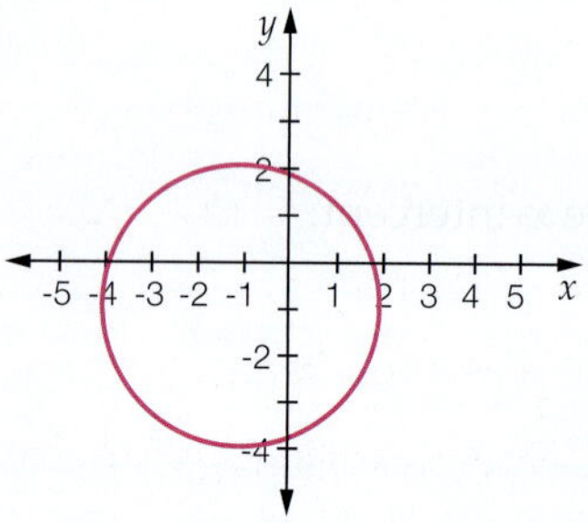

4 (a) Write the new equation formed when the circle $x^2 + y^2 = 49$ is translated 2 units to the right and 1 unit down.

(b) Sketch the graph of the new equation, showing the centre of the circle and the radius.

5 (a) Write the new equation formed when the exponential graph $y = 2^x$ is translated 1 unit to the left and 2 units down.

(b) Sketch the graph of the new equation, showing the asymptote and other key features.

6 (a) Write the new equation formed if a circle with equation $(x - 3)^2 + (y + 4)^2 = 25$ is translated 2 units left and 3 units down.

(b) Write the new equation formed if a circle with equation $(x - 3)^2 + (y + 4)^2 = 25$ has its radius increased by 3 units.

10A **7** Sketch the graph of the following, clearly showing the key features (asymptotes, x-intercepts and y-intercepts) for each graph. **W.E. 6**

(a) $y = \frac{1}{x+3} - 2$ **(b)** $y = \frac{1}{x-5} + 1$ **(c)** $y = \frac{1}{x+2} + 3$

10A **8** The equation for the graph shown at right is:

A $y = \frac{1}{x+4} + 5$ B $y = \frac{1}{x-4} - 5$

C $y = \frac{1}{x-5} - 4$ D $y = \frac{1}{x+5} - 4$

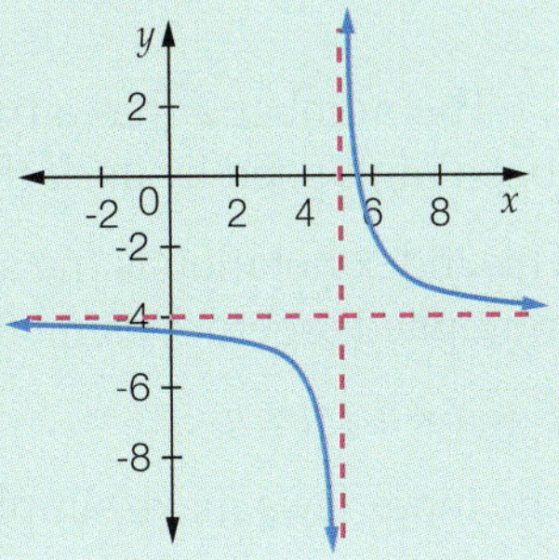

Reasoning

10A **9** **(a)** Show that $\frac{1}{x-1} + 3 = \frac{3x-2}{x-1}$.

(b) Hence, sketch the graph of $y = \frac{3x-2}{x-1}$.

10 A circular rose garden is to be planted in a rectangular lawn. The front boundary of the lawn is 25 m in length and the side boundaries are 15 m wide. A diagram of the lawn and the garden is to be drawn on a set of axes, where the origin is the front left-hand corner of the lawn, so x is the position along the front boundary and y is the position on its side boundary. The garden is designed to have a radius of 3 m with its centre in the centre of the lawn.

(a) Write the coordinates of the centre of the garden.

(b) Write the equation of the garden.

(c) Sketch the graph of the garden.

For each of the following scenarios **(i)** write the new equation and **(ii)** sketch its graph.

(d) The garden is moved 0.5 m closer to the front boundary and 6.5 m further to the left.

(e) The radius of the garden is reduced by 1 m after the move.

Open-ended

10A **11** **(a)** Write the equation of a rectangular hyperbola that is translated from the origin.

(b) Identify the asymptotes for a graph of your equation.

(c) Sketch the graph.

12 A circle whose centre is not at the origin passes through the point (1, 1). What could be the equation of this circle?

12.3 Summary of index laws

The number 2^3 has a base of 2 and a power or index of 3. This means that 2 has been multiplied by itself three times.

This index notation is the same when dealing with pronumerals. a^4 means that a has been multiplied together using four of itself:

$a^4 = a \times a \times a \times a$

The index laws make simplifying expressions with indices easier.

Index laws

$a^m \times a^n = a^{(m+n)}$	When multiplying numbers that are in index form and have the same base, keep the base and add the indices.
$\dfrac{a^m}{a^n} = a^{(m-n)}$	When dividing numbers that are in index form and have the same base, keep the base and subtract the indices.
$a^0 = 1$	Any number to the power of zero equals one, $a \neq 0$.
$(a^m)^n = a^{mn}$	When raising a number in index form to a power, keep the base and multiply the indices.
$(ab)^m = a^m b^m$	When a product in brackets is raised to a power m, each factor in the brackets is raised to the power m.
$\left(\dfrac{a}{b}\right)^m = \dfrac{a^m}{b^m}$	When a quotient in brackets is raised to a power m, both the numerator and the denominator are raised to the power m.
$a^{-m} = \dfrac{1}{a^m}$	A number raised to a negative power is the same as the reciprocal of the number raised to the positive of that power.

For more difficult expressions, you may need to use several index laws to simplify fully.

Worked example 7

W.E. 7

Simplify: $(a^4)^3b^7 \times (a^3b^{-4})^2$

Thinking	Working
1 Expand the brackets using the appropriate index laws.	$(a^4)^3b^7 \times (a^3b^{-4})^2$ $= a^{12}b^7 \times a^6b^{-8}$
2 To multiply, keep the bases and add the indices.	$= a^{18}b^{-1}$
3 Express the answer with positive indices.	$= \dfrac{a^{18}}{b}$

Worked example 8

W.E. 8

Simplify: $\left(\frac{m^2n}{m^5n^3}\right)^4 \div \frac{n^{-5}}{m^{12}}$

Thinking	Working
1 Change division to multiplication by taking the reciprocal of the second fraction.	$\left(\frac{m^2n}{m^5n^3}\right)^4 \div \frac{n^{-5}}{m^{12}}$ $= \left(\frac{m^2n^1}{m^5n^3}\right)^4 \times \frac{m^{12}}{n^{-5}}$
2 Expand brackets by multiplying indices.	$= \frac{m^8n^4}{m^{20}n^{12}} \times \frac{m^{12}}{n^{-5}}$
3 Multiply the numerators and denominators by adding the indices of like bases. ($m^8 \times m^{12} = m^{20}$, $n^{12} \times n^{-5} = n^7$)	$= \frac{m^{20}n^4}{m^{20}n^7}$
4 Divide the numerator by the denominator by subtracting the indices of like bases. ($m^{20} \div m^{20} = m^0$, $n^4 \div n^7 = n^{-3}$)	$= m^0n^{-3}$
5 Perform any further simplification. ($m^0 = 1$)	$= n^{-3}$
6 Express the answer with a positive index.	$= \frac{1}{n^3}$

When performing calculations using indices, remember to follow these steps:

1 Simplify any numerical factors (unless they are inside brackets).

2 Expand brackets by multiplying indices.

3 Simplify index numbers with the same base using the index laws. With fractions:

- multiply by keeping the base and adding indices
- divide by keeping the base and subtracting indices.

4 Apply any other index laws if necessary.

5 Express your answer with positive indices.

12.3 Summary of index laws

Navigator

1 (columns 1–2), 2 (columns 1–2), 3, 4 (a–h), 5 (columns 1–2), 6 (columns 1–2), 8, 9, 10	1 (columns 2–3), 2 (columns 2–3), 3, 4 (column 1), 5 (columns 2–3), 6 (columns 1–2), 7, 8, 9, 10	1 (j–u), 2 (d–i), 4 (column 2), 5 (e–l), 6 (columns 2–3), 7, 8, 9, 10

Answers p. 863

Fluency

W.E. 7

1 Simplify each of the following.

(a) $x^3 \times x^5$ (b) $3a^4 \times 4a$ (c) $x^2y^2 \times x^2y^5$

(d) $\frac{a^6}{a^2}$ (e) $\frac{6x^5y^3}{2xy^2}$ (f) $\frac{p^4q^6}{p^3q^9}$

(g) $(m^4)^2$ (h) $(pq^2)^5$ (i) $(a^4)^3b^5 \times (a^2b)^4$

(j) $a^3b^{-5} \times a^2b^2$ (k) $2p^3q^4 \times 3p$ (l) $\dfrac{(x^5)^{-2} \times x^8}{x^{-4}}$

(m) $\dfrac{(p^2q^3)^{-1} \times p^5q^{-2}}{p^3q^2}$ (n) $\dfrac{(c^{-3}d^{-4})^2 \times c^2d^6}{c^5d^3}$ (o) $\dfrac{4(a^2b)^3 \times 3a^{-5}b^{-2}}{6ab}$

(p) $\dfrac{p^4q^{-2}}{p^2q^{-4}} \times \dfrac{pq^5}{p^3q^6}$ (q) $\dfrac{3k^2h^4}{4k^{-3}h^2} \times \dfrac{2k^{-1}h}{5kh^4}$ (r) $\dfrac{4a^4b}{9a^{-5}b^3} \times \dfrac{3a^{-2}b^5}{2ab^{-1}}$

W.E. 8

2 Simplify each of the following.

(a) $\dfrac{k^9p^6}{k^4p^2} \div \dfrac{k^3p^2}{k^5p^4}$ (b) $\dfrac{6m^8n^6}{2m^4n^3} \div \dfrac{5m^3n^2}{2m^5n^4}$ (c) $\dfrac{7x^3y^5}{xy^{-3}} \div \dfrac{x^6}{y^2}$

(d) $\dfrac{k^{-5}p^2}{k^3p} \div \dfrac{k^5p^2}{k^6p^{-3}}$ (e) $\dfrac{2a^5b}{a^{-4}b^{-3}} \div \dfrac{3a^2b^6}{a^{-2}b}$ (f) $\dfrac{5f^4g^6}{f^{-3}g^{-1}} \div \dfrac{2f^2g^4}{f^{-4}g}$

(g) $\left(\dfrac{x^5}{y^2}\right)^2 \div \left(\dfrac{x^2}{y^2}\right)^4$ (h) $\left(\dfrac{4m^3n^4}{3mn}\right)^2 \div \dfrac{(2m^2n^3)^3}{3m^5n^3}$ (i) $\left(\dfrac{5p^2q^4}{2pq^3}\right)^2 \div \dfrac{(5p^3q^2)^3}{4p^7q^6}$

3 (a) $\dfrac{k^3 \times k^4}{k^9}$ is the same as:

A k^2 B $\dfrac{1}{k^2}$ C k^3 D $\dfrac{1}{k^3}$

(b) $(2x^3y^{-2})^3$ is the same as:

A $\dfrac{8x^9}{y^6}$ B $2x^9y^6$ C $8x^3y^6$ D $\dfrac{8}{x^9y^6}$

(c) $4(3a)^0$ is the same as:

A 0 B 1 C 4 D $12a$

4 Simplify each of the following.

(a) $\dfrac{(x^4y^{-2})^3 \times (xy^3)^2}{x^5y^{-4}}$ (b) $\dfrac{(a^3b^{-4})^5 \times (a^2b)^3}{a^9b^{-6}}$

(c) $\dfrac{(2k^{-3}p^2)^4 \times (k^2p^4)^{-2}}{(2kp)^2}$ (d) $\dfrac{(3m^4n^2)^4 \times (m^{-3}n^2)^{-2}}{(3mn)^2}$

(e) $\dfrac{(x^5y^{-3})^3}{(x^{-2}y)^2} \times \dfrac{(x^{-2}y^{-3})^2}{(x^4y^2)^{-3}}$ (f) $\left(\dfrac{m^3}{n^2}\right)^4 \times \left(\dfrac{m^{-2}}{n^4}\right)^3$

(g) $\left(\dfrac{a^{-2}b^2}{ab}\right)^3 \times \left(\dfrac{a^4b^{-3}}{a^{-1}b^2}\right)^2$ (h) $\left(\dfrac{m^{-3}n}{m^2n^2}\right)^2 \times \left(\dfrac{m^5n^{-2}}{m^{-3}n}\right)^3$

(i) $\dfrac{4x^3(y^{-2})^4}{(xy)^2} \times \left(\dfrac{x^{-4}y}{x^2y^{-3}}\right)^{-2}$ (j) $\left(\dfrac{3c^4d^3}{2cd}\right)^2 \div \dfrac{3cd}{(2c^{-2}d^3)^4}$

Understanding

5 Simplify each of these, expressing your answers with positive indices.

(a) $x^{-4} \times x^6$ (b) $a^3b^{-2} \times b^{-3}$ (c) $p^{-4}q^3 \times p^5q^{-2}$

(d) $d^7f^{-2} \times d^{-3}f^4$ (e) $x^5y^2 \times y^{-4}x^{-5}$ (f) $a^3b \times b^{-1}a^{-6}$

(g) $(m^3)^{-2}n^3 \times m^4(n^{-1})^2$ (h) $(p^4)^{-3}q^6 \times p^5(q^{-2})^2$ (i) $(a^{-2}b^3)^2 \times (a^4b^{-1})^{-3}$

(j) $(x^4y^{-3})^2 \times (x^{-5}y)^{-1}$ (k) $4(p^{-3})^5q^6 \times 3(pq)^{-2}$ (l) $(2x^2y)^{-3} \times (3x^{-4}y^5)^2$

Reasoning

6 Simplify each of these, expressing your answers with positive indices.

(a) $\dfrac{m^7n^{-3}}{m^{-2}n^{-1}}$ (b) $\dfrac{a^{-5}b^3}{a^2b^{-4}} \times \dfrac{a^3b^2}{a^{-3}b^6}$ (c) $\dfrac{x^7y^{-5}}{x^{-4}y^3} \times \dfrac{x^{-4}y^3}{x^2y^{-1}}$

(d) $\dfrac{x^{-4}y^7}{x^{-2}y^{-5}} \div \dfrac{x^3y^{-6}}{x^{-1}y}$ (e) $\dfrac{(3a^{-4}b^{-2})^2}{(a^{-2}b)^3}$ (f) $\dfrac{(2d^{-3}f^{-1})^3}{(d^{-2}f^2)^2}$

(g) $\dfrac{5(a^{-4})^3b^{-6}}{(a^2b^{-1})^4} \times \dfrac{2a^7(b^{-3})^2}{a^{-4}b^{-9}}$ (h) $\left(\dfrac{m^5n^{-2}}{m^{-3}n}\right)^{-1}$ (i) $\left(\dfrac{x^3}{y^2}\right)^{-4} \div \dfrac{x^{-10}}{(y^{-4})^2}$

7 (a) If you divide a number by its square, what is the result?

(b) You divide a room into squares that have a side length of a units. If the room has a width of $16a^2$ units and a length of $12a$ units, how many squares do you need to carpet the room?

8 Raphael and Murray were arguing over a question involving indices. Murray believes that $a^{-2} \times a^{-4}$ simplifies to $\dfrac{1}{a^8}$. Raphael disagrees.

(a) How did Murray calculate his answer? Is he correct?

(b) Why does Raphael disagree? What is Raphael's answer, assuming it is correct?

(c) How would you explain the correct answer to the students?

Open-ended

9 (a) Write two expressions involving powers and multiplication that when simplified equal $36m^4n^8$.

(b) Write two expressions involving powers and division that can be simplified to $\dfrac{4m^4}{5n^6}$.

10 For the expression a^{-b}, write three different sets of values for a and b, so that its value is less than $\frac{1}{4}$. Mark each value on a number line.

Investigation

The Koch snowflake

Equipment required: isometric paper, calculator

The Big Question

How can you describe the perimeter and area of a Koch snowflake, which is a fractal shape that repeats itself at infinitely smaller and smaller scales?

Strategy options

- Look for a pattern.
- Make a table.

Engage

Swedish mathematician Helge von Koch produced the original Koch curve in 1904. It starts with a straight line. At each step the middle third of the line is removed and replaced with two lines of the same length as the removed third.

Step 0

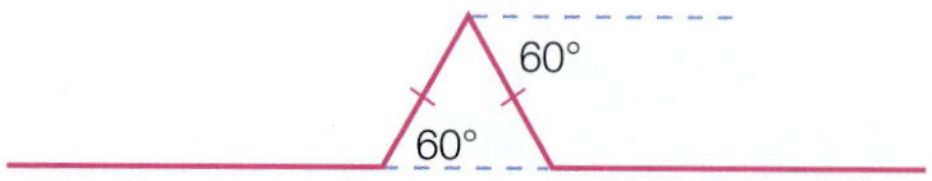

Step 1

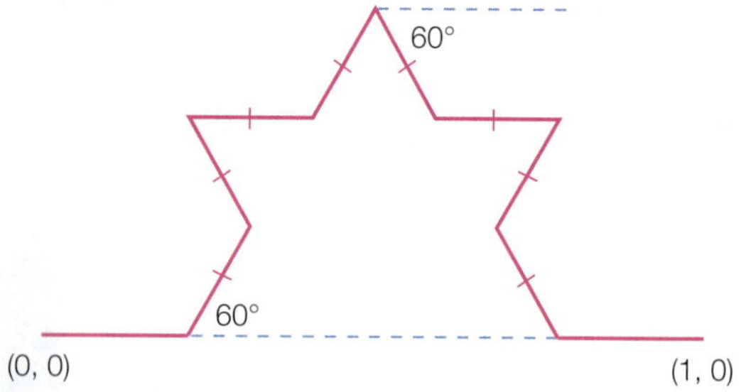

Step 2

Repeating (or 'iterating') this step over and over forever produces a special kind of geometric 'curve' with interesting and unusual properties. A shape that repeats itself like this, so that it appears to look the same even as you 'zoom in' infinitely closer, is called a fractal.

1 Follow the steps below to construct some of the early steps of what is known as the *Koch snowflake*.

(a) Begin with an equilateral triangle with 15 cm sides drawn on isometric grid paper. Call this 'step 0'.

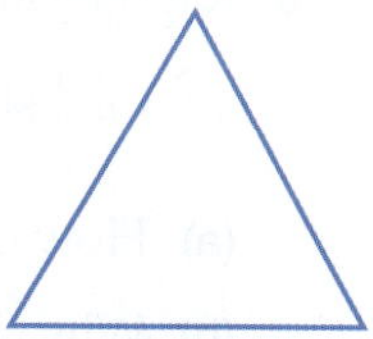

Step 0

(b) Divide each side into equal thirds and construct a new equilateral triangle on the middle third of each side. Erase the internal lines to complete the next step.

Step 1

(c) Repeat **1(b)** on the step 1 shape.

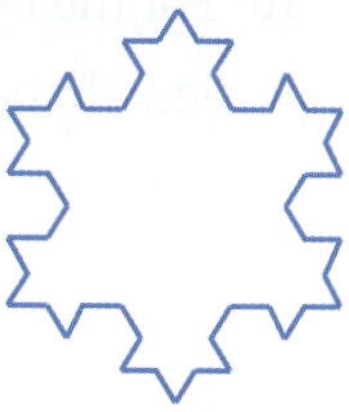

Step 2

(d) Continue this process through as many steps as possible.

(e) Describe the Koch snowflake and the process of drawing the next step.

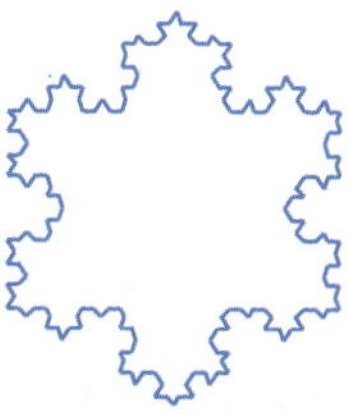

Step 3

Explore

2 (a) If the perimeter of the initial triangle is 3 units, calculate the perimeter of the Koch snowflake at steps 0, 1 and 2. Record your results in a table.

(b) If the area of the initial triangle is 1 unit^2, calculate the area of the Koch snowflake at steps 0, 1 and 2. Record your results in the table created in **2(a)**.

(c) Describe any trends in the perimeter and area at these steps. Estimate the perimeter and area for step 3.

At each step, the middle third of each side is removed and replaced with two sides of the same length or in other words, a length 1 unit long or $\frac{3}{3}$ becomes $\frac{4}{3}$ long with each step. The perimeter at any step (P_n) is equal to the initial length (P_0) multiplied by $\frac{4}{3}$ to the power of n.

(d) Write this formula.

(e) Use the formula to find the perimeter for $n = 0, 1, 2$ and 3.

Explain

3 (a) Were the perimeters you calculated in **2(a)** the same as the perimeters calculated using the formula in **2(e)**?

(b) Describe the pattern for finding the perimeter of the Koch snowflake at consecutive steps.

(c) Can a similar formula be written to describe the area at each step of the Koch snowflake? If yes, what is it? If no, why not?

Elaborate

4 (a) Does the formula for calculating the perimeter of the Koch snowflake suggest that the perimeter is finite or infinite? Give reasons for your answer.

(b) Describe the area of the Koch snowflake. Does a Koch snowflake have a finite or infinite area?

(c) Summarise your investigation of the Koch snowflake, describing what it is, and then commenting on the patterns found with the perimeter and area.

Evaluate

5 (a) Describe the method you used to calculate the perimeter and area of the Koch triangle at each step.

(b) Did you find any shortcuts to calculate the perimeter or area? If yes, what were they?

(c) Describe any difficulties you had with this investigation and the methods used to overcome them.

Extend

6 The Koch snowflake is an example of a fractal. The Sierpinski triangle, also known as the Sierpinski gasket, is another example of a fractal.

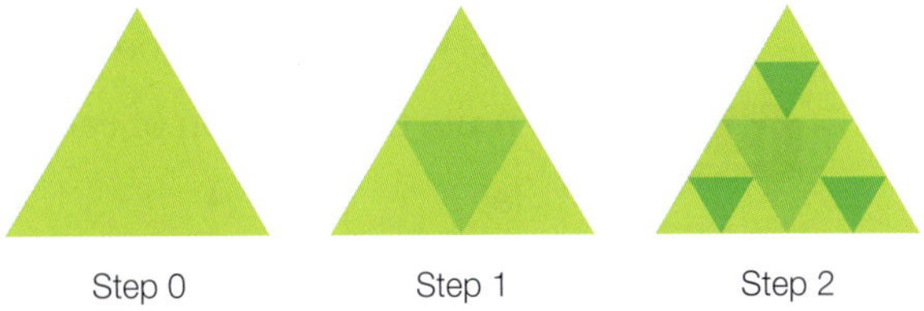

(a) Explain its pattern

(b) Draw the Sierpinski triangle at step 3.

7 Step 2 of another fractal pattern is shown. Each new set of branches are half the length of the previous set.

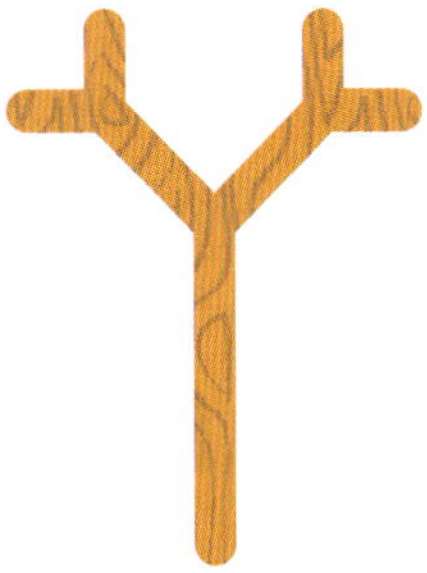

(a) Copy and complete the pattern up to step 6.

(b) Describe the appearance of this fractal pattern.

(c) Where can fractal patterns be found in nature?

Half-time 12

12.1 **1** Find the coordinates of the centre and the radius of the circle with equation $(x+1)^2+(y-2)^2=4$.

12.1 **2** Plot the graph of:

(a) $y = -2^{-x}$ for values of x from -3 to 3 inclusive

(b) $(x-2)^2+(y-3)^2=4$, showing the four points $(h+r, k)$, $(h, k+r)$, $(h-r, k)$, $(h, k-r)$ on the circumference of the circle with centre (h, k) and radius r.

10A 12.1 **3** For the non-linear equation $y = \frac{1}{x-3} + 4$:

(a) What name is given to a graph of this type?

(b) Write the equations of the x- and y-asymptotes.

12.2 **4** **(a)** Sketch the graph of $y = 2^{-x} - 1$ for values of x from -3 to 3 inclusive.

(b) What is the equation of the asymptote for the graph?

(c) What are the coordinates of the y-intercept?

12.2 **5** Write the equation of each of the following graphs:

(a)

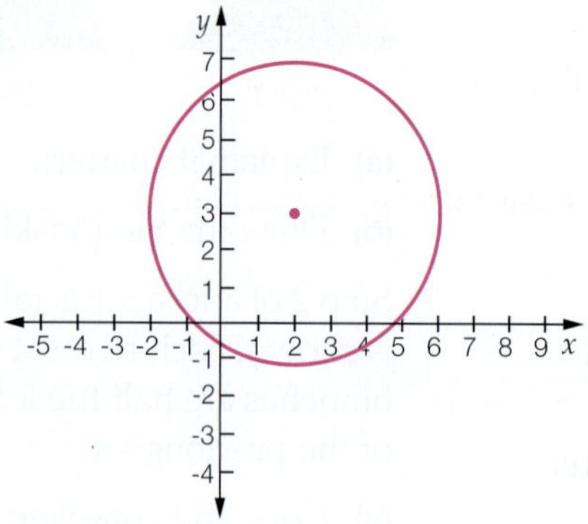

(b)

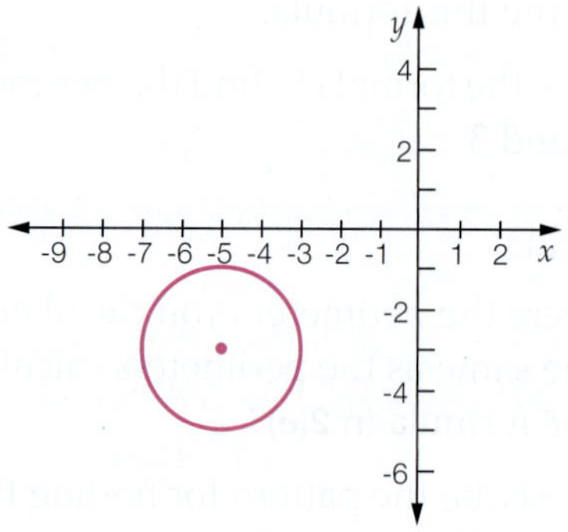

12.3 **6** Simplify each of the following:

(a) $4x^7 \times 3x^{-2}$ **(b)** $2(m^3)^2n^{-3} \times 3(mn)^4$ **(c)** $\frac{a^5b^{-2} \times a^2b^3}{a^4b^7 \times ab^{-4}}$

12.3 **7** Simplify each of the following.

(a) $\left(\frac{p^2}{q^4}\right)^3 \times \left(\frac{p^{-3}}{q^5}\right)^2$ **(b)** $\left(\frac{f^{-3}}{g^2}\right)^{-4} \div \left(\frac{f^2}{g}\right)^3$

10A 12.2 **8** **(a)** Write the equation formed if the graph of $y = \frac{1}{x}$ is translated 3 units left and 4 units up.

(b) Sketch the graph of the new equation, showing any relevant asymptotes.

Fractional indices

12.4

The index laws can also be used when the index is a fraction.

Consider $3^{\frac{1}{2}} \times 3^{\frac{1}{2}}$.

Using the index law $a^m \times a^n = a^{m+n}$, you keep the base 3 and add the powers.

$$\begin{aligned} 3^{\frac{1}{2}} \times 3^{\frac{1}{2}} &= 3^{\frac{1}{2}+\frac{1}{2}} \\ &= 3^1 \\ &= 3 \end{aligned}$$

However, what does $3^{\frac{1}{2}}$ mean? What number when multiplied by itself gives 3?

$$\sqrt{3} \times \sqrt{3} = \sqrt{9} = 3$$

So $3^{\frac{1}{2}}$ is the same as $\sqrt{3}$. In general:

$$a^{\frac{1}{2}} = \sqrt{a}, \text{ where } a \text{ is positive or zero.}$$

Similarly, consider $5^{\frac{1}{3}} \times 5^{\frac{1}{3}} \times 5^{\frac{1}{3}}$.

Using the index law $a^m \times a^n = a^{m+n}$, you keep the base 5 and add the powers

$$\begin{aligned} 5^{\frac{1}{3}} \times 5^{\frac{1}{3}} \times 5^{\frac{1}{3}} &= 5^{\frac{1}{3}+\frac{1}{3}+\frac{1}{3}} \\ &= 5 \end{aligned}$$

What number multiplied three times by itself gives 5?

$$\begin{aligned} \sqrt[3]{5} \times \sqrt[3]{5} \times \sqrt[3]{5} &= \sqrt[3]{125} \\ &= 5 \end{aligned}$$

So, $5^{\frac{1}{3}}$ is the same as $\sqrt[3]{5}$. In general:

$$a^{\frac{1}{3}} = \sqrt[3]{a}$$

A number raised to a fractional power $\frac{1}{n}$ is the nth root of the number.

$$a^{\frac{1}{2}} = \sqrt{a}, \text{ where } a \geq 0$$

$$a^{\frac{1}{3}} = \sqrt[3]{a}$$

$$a^{\frac{1}{n}} = \sqrt[n]{a}, \text{ where } a \geq 0$$

$$a^{\frac{m}{n}} = \left(a^{\frac{1}{n}}\right)^m$$

Numbers raised to a fractional power follow all the index laws.

10A

Worked example 9

W.E. 9

Simplify each of the following. Write the answers in index form.

(a) $a^{\frac{1}{2}} \times a^{\frac{1}{4}}$ (b) $\sqrt{x^6}$ (c) $\dfrac{(m^3)^{\frac{1}{3}} \times m^{-\frac{1}{3}}}{m^{\frac{1}{3}}}$

Thinking	Working
(a) Multiply by keeping the base and adding the indices. To add fractions a common denominator is needed. (LCM of 2 and 4 is 4.)	(a) $a^{\frac{1}{2}} \times a^{\frac{1}{4}}$ $= a^{\frac{1}{2}+\frac{1}{4}}$ $= a^{\frac{(2+1)}{4}}$ $= a^{\frac{3}{4}}$
(b) 1 Change to index form by using $\sqrt{a} = a^{\frac{1}{2}}$.	(b) $\sqrt{x^6}$ $= (x^6)^{\frac{1}{2}}$
2 Expand the brackets by using $(a^m)^n = a^{mn}$.	$= x^{6 \times \frac{1}{2}}$ $= x^3$
(c) 1 Expand the brackets by using $(a^m)^n = a^{mn}$.	(c) $\dfrac{(m^3)^{\frac{1}{3}} \times m^{-\frac{1}{3}}}{m^{\frac{1}{3}}}$ $= \dfrac{m^1 \times m^{-\frac{1}{3}}}{m^{\frac{1}{3}}}$
2 Simplify the numerator by keeping the base and adding the indices.	$= \dfrac{m^{1-\frac{1}{3}}}{m^{\frac{1}{3}}}$ $= \dfrac{m^{\frac{2}{3}}}{m^{\frac{1}{3}}}$
3 Divide by keeping the base and subtracting the indices.	$= m^{\frac{2}{3}-\frac{1}{3}}$ $= m^{\frac{1}{3}}$

10A

Fractional powers on a scientific calculator

Consider calculating $625^{\frac{1}{4}}$. This is the same as $\sqrt[4]{625}$.

Some calculators have a 'fractional root' key labelled [$\sqrt[x]{\ }$] or [$\sqrt[x]{y}$] or [$\sqrt[y]{x}$] or [$x^{1/y}$].

The root, which is the small number in the radical symbol ($\sqrt{\ }$), is usually entered first.

For $625^{\frac{1}{4}}$, for most calculators you would enter [4] [$\sqrt[x]{\ }$] [6] [2] [5] [=].
The result should show 5.

Worked example 10 — W.E. 10

Evaluate:

(a) $7776^{\frac{1}{5}}$ **(b)** $289^{\frac{1}{2}}$ **(c)** $1331^{\frac{1}{3}}$

Thinking	Working
(a) Use your calculator to evaluate. (Press: [5] [$\sqrt[x]{\ }$] [7] [7] [7] [6] [=])	(a) $7776^{\frac{1}{5}}$ $= \sqrt[5]{7776}$ $= 6$
(b) Use your calculator to evaluate. (Press: [$\sqrt{\ }$] [2] [8] [9] [=] or [2] [$\sqrt[x]{\ }$] [2] [8] [9] [=])	(b) $289^{\frac{1}{2}}$ $= \sqrt{289}$ $= 17$
(c) Use your calculator to evaluate. (Press: [$\sqrt[3]{\ }$] [1] [3] [3] [1] [=] or [3] [$\sqrt[x]{\ }$] [1] [3] [3] [1] [=])	(c) $1331^{\frac{1}{3}}$ $= \sqrt[3]{1331}$ $= 11$

You can also find roots by calculating them directly as fractional powers.

For example, for $7776^{\frac{1}{5}}$ you would enter: [7] [7] [7] [6] [y^x] [(] [1] [÷] [5] [)] [=]. This should produce the result 6 as in part **(a)** above.

12.4 Fractional indices

Navigator

Answers p.864

1 (columns 1–2), 2 (column 1), 3, 4, 5, 6, 7, 9, 11, 12 (a–c), 13 (a–c), 15	1 (columns 2–3), 2 (column 2), 3, 4, 5, 6, 7, 8, 9 (a–b), 10 (a–b), 11, 12, 13, 14 (a–b), 15, 16	1 (column 2), 2 (column 3), 3, 5 (column 2), 6, 7, 8, 9 (a), 10, 11, 12, 13, 14, 15, 16

Equipment required: calculator for Questions 2, 3, 4, 9, 15

Fluency

W.E. 9

1 Simplify each of the following. Write your answers in index form.

(a) $x^{\frac{1}{3}} \times x^{\frac{1}{3}}$ (b) $a^{\frac{1}{2}} \times a^{\frac{1}{3}}$ (c) $k^{\frac{1}{3}} \times k^{\frac{1}{5}}$

(d) $\dfrac{m^{\frac{1}{2}}}{m^{\frac{1}{4}}}$ (e) $\dfrac{p^{\frac{1}{2}}}{p^{\frac{1}{3}}}$ (f) $\left(y^{\frac{1}{2}}\right)^{\frac{1}{2}}$

(g) $(25p^4)^{\frac{1}{2}}$ (h) $(49x^6)^{\frac{1}{2}}$ (i) $(8ab^3)^{\frac{1}{3}}$

(j) $\sqrt{x^4y}$ (k) $\sqrt{kn^8}$ (l) $\left(\dfrac{p^4}{q^{10}}\right)^{\frac{1}{2}}$

10A

W.E. 10

2 Evaluate each of the following. Write your answer correct to 2 decimal places, where appropriate.

(a) $196^{\frac{1}{2}}$ (b) $6859^{\frac{1}{3}}$ (c) $1296^{\frac{1}{4}}$

(d) $4096^{\frac{1}{4}}$ (e) $3125^{\frac{1}{5}}$ (f) $500^{\frac{1}{3}}$

(g) $3000^{\frac{1}{2}}$ (h) $98^{\frac{1}{6}}$ (i) $145^{\frac{1}{5}}$

3 Using your calculator to help you, place these numbers in ascending order.

$2^{\frac{1}{2}}, 3^{\frac{1}{3}}, 4^{-1}, 6^{-\frac{1}{5}}$

4 (a) $20^{\frac{1}{4}}$ is closest to:

A 0.000 006 B 2.1 C 5 D 160 000

(b) $75^{\frac{1}{3}}$ is closest to:

A -25 B 0.24 C 4.22 D 421 875

5 Simplify each of the following, expressing your answers with positive indices.

(a) $\dfrac{\left(a^{\frac{1}{3}}\right)^6 \times a}{a^2}$

(b) $\dfrac{\left(d^{\frac{1}{2}}\right)^8 \times d^2}{d^3}$

(c) $(x^4y^6)^{\frac{1}{2}} \times (x^{-3}y^{-12})^{\frac{1}{3}}$

(d) $(m^6n^3)^{\frac{1}{3}} \times (m^{-8}n^{-10})^{\frac{1}{2}}$

(e) $\dfrac{(pq^3)^{\frac{1}{2}} \times p^{\frac{1}{2}}q^{\frac{1}{2}}}{pq^4}$

(f) $\left(\dfrac{x^3y^2}{y^{\frac{1}{2}}}\right)^{-\frac{1}{3}} \times \dfrac{(x^2)^{\frac{1}{3}}y}{5y^{-\frac{2}{3}}}$

6 Write each of the following powers in simplest form using a root symbol.

(a) $6^{\frac{2}{3}}$ (b) $x^{-\frac{3}{4}}$ (c) $p^{2\frac{1}{2}}$ (d) $5^{-1\frac{1}{4}}$

7 Write each of the following expressions as a single power.

(a) $\sqrt{r^3}$ (b) $\dfrac{1}{\sqrt[3]{m^2}}$ (c) $7\sqrt{7}$ (d) $\dfrac{1}{a^2\sqrt{a}}$

8 (a) Find the expression that is *not* the same as $a^{\frac{3}{2}}b^{\frac{1}{2}}$.

A $\sqrt{a^3b}$ B $(\sqrt{ab})^3$ C $(a^3b)^{\frac{1}{2}}$ D $(\sqrt{a})^3\sqrt{b}$

(b) Simplifying $\left(\dfrac{a^{\frac{1}{2}}b^3}{b^2}\right)^2 \div \dfrac{b^{\frac{3}{2}}a}{b^4}$ gives:

A $b^{\frac{9}{2}}$ B $b^{\frac{13}{2}}$ C $\dfrac{b^{\frac{7}{2}}}{a^{\frac{1}{2}}}$ D $\dfrac{b^7}{a^{\frac{1}{2}}}$

9 Evaluate each of the following correct to 3 decimal places.

(a) $8^{-\frac{1}{3}}$ (b) $37^{-\frac{1}{2}}$ (c) $54^{-\frac{1}{4}}$ (d) $100^{-\frac{1}{5}}$

Understanding

10 By simplifying first, find the value of x in each of the following equations.

(a) $\dfrac{72x^2}{\left(x \times x^{\frac{3}{2}}\right)^2} = 24$

(b) $\dfrac{\left(x^{\frac{1}{2}}\right)^5}{x^{-2}} \times \dfrac{5x^{-1}}{x^2} = 40$

(c) $\dfrac{7x^{-3}}{x^3}\left(\dfrac{x^{\frac{3}{2}}}{x}\right)^6 = 56$

11 Simplify the following.

(a) $\sqrt{x^6}$ (b) $\sqrt[3]{a^9}$ (c) $\sqrt[4]{16m^8n^4}$

10A

Reasoning

12 Find the value of x in each of the following equations.

(a) $x^{-2} = \frac{1}{4}$ (b) $3^x = 27$ (c) $3^x = \frac{1}{9}$ (d) $2 \times 5^x = 50$

13 Evaluate each of the following without using a calculator.

(a) $100^{\frac{5}{2}}$ (b) $16^{\frac{5}{4}}$ (c) $\left(\frac{1}{16}\right)^{-\frac{1}{2}}$ (d) $216^{-\frac{2}{3}}$

14 Simplify each of the following, using powers with positive indices.

(a) $3g^3 \times 4 \times \sqrt[3]{g^2}$ (b) $2a^{-2} \times 5\sqrt{a^3}$ (c) $3\sqrt{p^2q^{-3}} \times 4\sqrt{p^4q^5}$

Open-ended

15 Using your calculator, find at least three ways of writing 343 as a power. When the base number is not exact, round to 3 decimal places. For example, $343 = (18.520)^2$.

16 Write the expression $n^{\frac{3}{4}}m^{\frac{3}{5}}$ in at least three different ways.

Problem solving

Lagrange's four-square theorem

Diophantus was a Greek mathematician who lived in the 3rd century CE. He wrote a famous book called *Arithmetica*, which contained the following conjecture (a statement that has not been proven).

> Every whole number is either a square itself or the sum of two, three or four squares.

This conjecture was later proved in 1770 by Joseph-Louis Lagrange.

(a) Write a square number in each box so that each column and row adds to the total at the end.

			1^2	39
		1^2		27
	1^2			46
1^2				62
36	30	66	42	

(b) Can any of the totals in the grid be written using fewer square numbers?

(c) Starting at 1, write each whole number as a sum of four or fewer square numbers. How far can you get in 3 minutes?

Strategy options

- Make a table.
- Break problem into manageable parts.

12.5

Logarithms

Logarithm is another word for power or index. As $10^2 = 100$, you can say that the logarithm of 100 to the base 10 is 2.

logarithm

$$100 = 10^2$$

numerical value base

This can be written as $\log_{10}(100) = 2$.

In this case, 10 is the base, but any number can be used. For example, $2^3 = 8$, so the logarithm of 8 to the base 2 is 3. This can be written as $\log_2(8) = 3$.

If $y = \log_b(x)$ then $b^y = x$, where b is the base, y is the power or logarithm and x is the numerical value.

Any number can be used as the base number. The **log** key on your calculator should usually give you the base 10 logarithm for any number entered.

Worked example 11

W.E. 11

10A

(a) Write $2^6 = 64$ in its equivalent logarithmic form.

(b) Write $2^{-4} = \frac{1}{16}$ in its equivalent logarithmic form.

Thinking	Working
(a) 1 Identify the logarithm (power), the base and the numerical value.	(a) 6 is the logarithm 2 is the base 64 is the numerical value
2 Write the logarithmic form.	$\log_2(64) = 6$
(b) 1 Identify the logarithm (power), the base and the numerical value.	(b) -4 is the logarithm 2 is the base $\frac{1}{16}$ is the numerical value
2 Write the logarithmic form.	$\log_2\left(\frac{1}{16}\right) = -4$

Worked example 12

W.E. 12

Write $\log_{10}(10\,000) = 4$ in its equivalent index form.

Thinking	Working
1 Identify the logarithm (power), the base and the numerical value.	4 is the logarithm 10 is the base 10 000 is the numerical value
2 Write the index form.	$10^4 = 10\,000$

12.5 Logarithms

Navigator

Answers p. 865

1 (columns 1–2), 2 (columns 1–2), 3, 4 (a–f), 5 (a–f), 6, 7, 11	1 (columns 2–3), 2 (columns 2–3), 3, 4 (columns 1–2), 5 (columns 1–2), 6, 7, 8, 11	1 (column 3), 2 (columns 2–3), 3, 4 (columns 2–3), 5 (columns 2–3), 6, 7, 8, 9, 10, 11

Equipment required: calculator for Questions **4–5**

Fluency

10A

W.E. 11

1 Write each of the following in its equivalent logarithmic form.

(a) $2^5 = 32$ (b) $2^{12} = 4096$ (c) $3^3 = 27$

(d) $3^5 = 243$ (e) $3^8 = 6561$ (f) $10^2 = 100$

(g) $10^6 = 1\,000\,000$ (h) $2^{-2} = \frac{1}{4}$ (i) $2^{-3} = \frac{1}{8}$

(j) $2^{-4} = \frac{1}{16}$ (k) $10^{-4} = 0.0001$ (l) $3^{-2} = \frac{1}{9}$

W.E. 12

2 Write each of the following in its equivalent index form.

(a) $\log_{10}(100) = 2$ (b) $\log_{10}(100\,000) = 5$ (c) $\log_2(16) = 4$

(d) $\log_2(128) = 7$ (e) $\log_2(512) = 9$ (f) $\log_5(15\,625) = 6$

(g) $\log_{10}(0.001) = -3$ (h) $\log_7\left(\frac{1}{49}\right) = -2$ (i) $\log_{10}\left(\frac{1}{1\,000\,000}\right) = -6$

(j) $\log_2\left(\frac{1}{8}\right) = -3$ (k) $\log_2\left(\frac{1}{32}\right) = -5$ (l) $\log_2\left(\frac{1}{128}\right) = -7$

3 (a) If $\log_{10}(x) = -2$, then x is equal to:

A -200 B -0.01 C 0.01 D 200

(b) Writing $2^x = 2048$ in logarithmic form gives:

A $\log_2(x) = 2048$ B $\log_x(2) = 2048$ C $\log_x(2048) = 2$ D $\log_2(2048) = x$

Understanding

4 Solve the following for x:

(a) $3^x = 81$ (b) $3^x = 6561$ (c) $3^x = 19\,683$

(d) $5^x = \frac{1}{25}$ (e) $5^x = \frac{1}{125}$ (f) $5^x = \frac{1}{15\,625}$

(g) $6^{-x} = 216$ (h) $6^{-x} = 7776$ (i) $6^{-x} = \frac{1}{1296}$

Hint

Write the number on the RHS as an index, using the same base as the LHS.

5 Solve the following for x:

(a) $\log_2(x) = 4$ (b) $\log_2(x) = 6$ (c) $\log_2(x) = 7$

(d) $\log_3\left(\frac{1}{81}\right) = x$ (e) $\log_3\left(\frac{1}{243}\right) = x$ (f) $\log_3\left(\frac{1}{6561}\right) = x$

(g) $\log_x(625) = 4$ (h) $\log_x(343) = 3$ (i) $\log_x\left(\frac{1}{3125}\right) = -5$

Reasoning

6 (a) Solve for x in each of the following:

(i) $\log_{10}(x) = 1$ (ii) $\log_5(x) = 1$ (iii) $\log_7(x) = 1$

(b) Write a general rule based on the above, using a as your base.

7 (a) Solve for x in each of the following:

(i) $\log_{10}(x) = 0$ (ii) $\log_5(x) = 0$ (iii) $\log_7(x) = 0$

(b) Write a general rule to explain what you have just found out.

8 In the equation $\sqrt{a} = a^{\frac{1}{2}}$, why does a have to be greater than or equal to 0?

9 (a) Check that both of the following logarithmic equations are correct:

$\log_2(64) = 6$ and $\log_4(64) = 3$

(b) Now, given $\log_2(4096) = 12$, find the value of y in $\log_4(4096) = y$.

(c) Explain how you got your answer for **(b)**.

Open-ended

10 Given $\log_3(531\,441) = 12$, write two other related log statements that are true.

11 Lee wrote:

$-(6)^2 = -36$

because $-\log_6(-36) = 2$

Explain why this is incorrect and how the question should be solved.

10A

12.6 Laws of logarithms

Just as there are laws to simplify expressions with indices, there are also related laws to simplify logarithmic expressions.

Logarithm law: $\log_b(xy) = \log_b(x) + \log_b(y)$

Proof: Let $\log_b(x) = m$ and $\log_b(y) = n$

$b^m = x$ and $b^n = y$

$$xy = b^m \times b^n$$
$$xy = b^{m+n}$$

$$\log_b(xy) = m + n$$

$$\therefore \log_b(xy) = \log_b(x) + \log_b(y)$$

Worked example 13

W.E. 13

Simplify $\log_2(3) + \log_2(5)$, expressing your answer in logarithmic form.

Thinking	Working
1 Write the rule for simplifying logarithmic expressions that are added.	$\log_b(x) + \log_b(y) = \log_b(xy)$ $\log_2(3) + \log_2(5)$
2 Substitute the known values on the right-hand side.	$= \log_2(3 \times 5)$
3 Simplify the expression.	$= \log_2(15)$

10A

Logarithm law: $\log_b\left(\frac{x}{y}\right) = \log_b(x) - \log_b(y)$

Proof: Let $\log_b(x) = m$ and $\log_b(y) = n$

$b^m = x$ and $b^n = y$

$$\frac{x}{y} = \frac{b^m}{b^n}$$

$$\frac{x}{y} = b^{m-n}$$

$$\log_b\left(\frac{x}{y}\right) = m - n$$

$$\therefore \log_b\left(\frac{x}{y}\right) = \log_b(x) - \log_b(y)$$

Worked example 14

W.E. 14

Simplify $\log_2(12) - \log_2(4)$, expressing your answer in logarithmic form.

Thinking	Working
1 Write the rule for simplifying logarithmic expressions that are subtracted.	$\log_b(x) - \log_b(y) = \log_b\left(\frac{x}{y}\right)$ $\log_2(12) - \log_2(4)$
2 Substitute the known values on the right-hand side.	$= \log_2\left(\frac{12}{4}\right)$
3 Simplify the expression.	$= \log_2(3)$

Logarithm law: $\log_b(x^n) = n\log_b(x)$

Proof: Let $\log_b(x) = m$

$$\text{So } x = b^m$$
$$x^n = (b^m)^n = b^{mn}$$
$$\log_b(x^n) = mn$$
$$\log_b(x^n) = nm$$
$$\therefore \log_b(x^n) = n\log_b(x)$$

Worked example 15

W.E. 15

Simplify $\log_2(5^2)$, expressing your answer in logarithmic form without a power.

Thinking	Working
1 Write the rule for simplifying logarithmic expressions that involve a power.	$\log_b(x^n) = n\log_b(x)$
2 Simplify the expression by substituting the known values.	$\log_2(5^2) = 2\log_2(5)$

There are two additional results that help simplify logarithmic expressions.

$b^1 = b$ so $\log_b(b) = 1$
$b^0 = 1$ so $\log_b(1) = 0$

To help simplify logarithmic expressions:

$$\log_b(xy) = \log_b(x) + \log_b(y)$$
$$\log_b\left(\frac{x}{y}\right) = \log_b(x) - \log_b(y)$$
$$\log_b(x^n) = n\log_b(x)$$
$$\log_b(b) = 1$$
$$\log_b(1) = 0$$

10A

Worked example 16

W.E. 16

Simplify $\log_2(16)$, writing your answer as a basic numeral.

Thinking	Working
1 Rewrite the number inside the brackets using index notation.	$\log_2(16)$ $= \log_2(2^4)$
2 Simplify using the appropriate logarithm law.	$= 4\log_2(2)$
3 Evaluate using $\log_b(b) = 1$.	$= 4 \times 1$ $= 4$

12.6 Laws of logarithms

Navigator

Answers p. 865

1, 2, 3, 4, 5, 6, 7, 8, 9, 12	1, 2, 3, 4, 5, 6, 7, 8, 9, 10, 12	1, 2, 3, 4, 5, 6 (column 1), 7 (column 1), 8, 9, 10, 11, 12

Fluency

10A

W.E. 13

1 Simplify the following, expressing your answer in logarithmic form.

(a) $\log_2(4) + \log_2(8)$ (b) $\log_2(16) + \log_2(4)$

(c) $\log_2(8) + \log_2(32)$ (d) $\log_{10}(5) + \log_{10}(6)$

(e) $\log_{10}(7) + \log_{10}(3)$ (f) $\log_{10}(4) + \log_{10}(9)$

W.E. 14

2 Simplify the following, expressing your answer in logarithmic form.

(a) $\log_2(16) - \log_2(8)$ (b) $\log_2(20) - \log_2(4)$

(c) $\log_2(25) - \log_2(5)$ (d) $\log_{10}(70) - \log_{10}(5)$

(e) $\log_{10}(35) - \log_{10}(7)$ (f) $\log_{10}(36) - \log_{10}(6)$

W.E. 15

3 Simplify the following, expressing your answer in logarithmic form without a power.

(a) $\log_2(5^4)$ (b) $\log_2(6^3)$ (c) $\log_2(7^4)$

(d) $\log_{10}(7^3)$ (e) $\log_{10}(8^4)$ (f) $\log_{10}(11^5)$

W.E. 16

4 Simplify the following, expressing your answer as a basic numeral.

(a) $\log_2(32)$ (b) $\log_2(128)$ (c) $\log_{10}(10\,000)$ (d) $\log_3(27)$

5 (a) $\dfrac{\log_2(32)}{\log_2(4)}$ is equal to:

A $\frac{5}{2}$ B 3 C $\log_2(8)$ D $\log_2(28)$

(b) $\log_a(mn)$ can be written as:

A $\log_a(m) \times \log_a(n)$ B $\log_a(m^n)$

C $\log_a(m) + \log_a(n)$ D $\log_a(m) - \log_a(n)$

Understanding

6 Simplify, expressing each answer as a basic numeral.

(a) $\log_{10}(100) + \log_{10}(1000)$

(b) $\log_2(64) + \log_2(32)$

(c) $\log_2(4) + \log_2(8) - \log_2(16)$

(d) $\log_3(27) + \log_3(81) - \log_3(243)$

(e) $2\log_2(2) - \log_2\left(\frac{1}{8}\right) + \log_2(8)$

(f) $\log_3(3) + 2\log_3\left(\frac{1}{9}\right) - 3\log_3(9)$

7 Simplify each of the following expressions using laws of logarithms.

(a) $\log_{10}(4) + \log_{10}(250)$

(b) $\log_6(18) + \log_6(72)$

(c) $\log_{10}(8000) - \log_{10}(8)$

(d) $\log_5(875) - \log_5(35)$

(e) $\log_2(28) - \log_2(560) + \log_2(40)$

(f) $\log_{10}(30) + \log_{10}(20) - \log_{10}(6000)$

8 Simplify:

(a) $\dfrac{\log_5(x^2)}{\log_5(x)}$

(b) $\dfrac{\log_3(x^5)}{\log_3(x^6)}$

(c) $\dfrac{\log_7(x^9)}{\log_7(x^3)}$

(d) $\dfrac{\log_2(x+4)^2}{\log_2(x+4)}$

(e) $\dfrac{3\log_5(2x-1)}{\log_5(2x-1)^2}$

(f) $\dfrac{\log_7(3x+4)^4}{3\log_7(3x+4)}$

Reasoning

9 Simplify:

(a) $\dfrac{\log_x(2)}{\log_x(16)}$

(b) $\dfrac{\log_x(25)}{\log_x(5)}$

(c) $\dfrac{\log_x(49)}{\log_x(7^3)}$

10 (a) If $\log_a(y) = \log_a(4) + 3\log_a(x)$, express y in terms of x.

(b) If $\log_a(y) = \log_a(11) + 5\log_a(x)$, express y in terms of x.

11 If $5\log_2(y) = \log_2(x-1) + 2$, show that $y^5 = 4(x-1)$.

Open-ended

12 One of the questions on a test was:

Simplify $\dfrac{\log_2(x)}{\log_2(\sqrt{x})}$

Dani wrote: $\dfrac{\log_2(x)}{\log_2(\sqrt{x})} = \dfrac{\log_2(x)}{\log_2\left(x^{\frac{1}{2}}\right)}$

$= \dfrac{1}{2} \times \dfrac{\log_2(x)}{\log_2(x)}$

$= \dfrac{1}{2}$

Is Dani's answer correct? If not, what is the correct answer and how would you explain the mistake?

Maths 4 Real

MEASURING EARTHQUAKES

Earthquakes happen when forces between the tectonic plates, which make up the Earth's crust, cause energy to be released suddenly. The strength of the earthquake and the location of its epicentre (the point on the crust where the energy is released) determine the effects that the earthquake has on human life.

The strength of an earthquake is measured as a magnitude between 0 and 10 on the Richter scale. The Richter scale increases by powers of 10, which means that an increase of 1 on the Richter scale represents an earthquake 10 times stronger. An earthquake with a recorded magnitude of 5 on the Richter scale is 10 times stronger than an earthquake with a magnitude of 4. An earthquake with a magnitude of 6 on the Richter scale is $10 \times 10 = 100$ times stronger than an earthquake with a magnitude of 4.

1 (a) Complete the Richter scale chart below to show the comparative strength of earthquakes.

Richter scale	0	1	2	3	4	5	6	7	8	9
Comparative strength	1	10^1	10^2							

(b) (i) How many times stronger is an earthquake with magnitude 8 compared to an earthquake with magnitude 5?

(ii) How many times stronger is an earthquake with magnitude 6.5 compared to an earthquake with magnitude 7.5?

(c) Draw a graph showing the comparative strength of earthquakes on the Richter scale.

(d) What type of graph is this?

(e) Write an equation to describe this graph.

(f) Use the equation to find the comparative strength of an earthquake with magnitude 5.2. What does this figure mean?

2 The Earth has seven major tectonic plates that can cause earthquakes when sliding past each other. New Zealand, Indonesia, Japan, China, the west coast of the USA and South America are all directly above where two tectonic plates meet, so these places are more susceptible to frequent and severe earthquakes.

(a) Show the position of the following deadly earthquakes on the graph drawn in **1(c)**.

- Sichuan earthquake, China, 2008, magnitude of 7.9, killing 88 000 people
- The Great Kantō Earthquake, Japan, 1923, magnitude of 8.3, killing 140 000 people
- The Great Hanshin Earthquake, Kobe, Japan, 1995, magnitude of 7.2, killing 6433 people
- Java earthquake, Indonesia, May 2006, magnitude of 6.4, killing 5749 people
- Christchurch, New Zealand, February 2011, magnitude of 6.3, killing 181 people

One day after the Sichuan earthquake in China, another earthquake hit China with a magnitude of 6.1. The difference in strength between the two earthquakes can be found by calculating the ratio of the strengths of the two earthquakes using the following method:

$$\frac{\text{magnitude}_1}{\text{magnitude}_2} = \frac{10^{7.9}}{10^{6.1}} = 10^{1.8} = 63$$

The Sichuan earthquake was 63 times stronger than the earthquake the following day.

(b) Calculate the differences in strength between the following earthquakes listed in **3(a)**.

(i) The Great Kantō Earthquake and the Great Hanshin Earthquake

(ii) The 2006 Java earthquakes and the 2011 Christchurch earthquake.

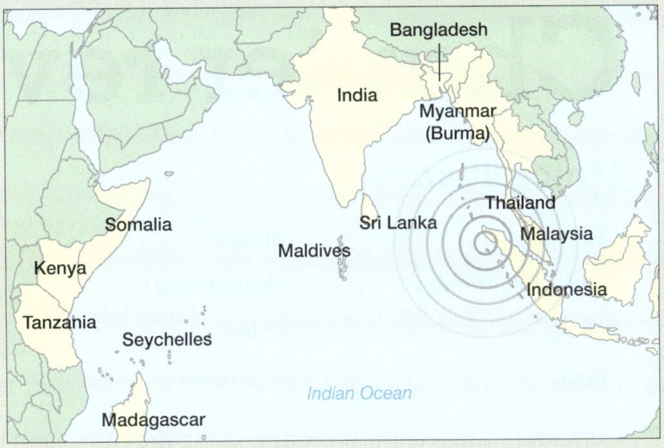

3 In 1906 an earthquake hit San Francisco, leaving 3000+ people dead and starting fires that almost flattened the whole city. This earthquake was 8 times stronger than the 1989 San Francisco earthquake that measured 6.9 on the Richter scale. The information above can be shown as:

$$\frac{\text{magnitude}_1}{\text{magnitude}_2} = \frac{10^y}{10^{6.9}} = 10^x = 8$$

$10^x = 8$ can be written as: $\log_{10}(8) = x$

Solving this on a calculator: $\log_{10}8 = 0.9$ (1 d.p.)

$$\frac{\text{magnitude}_1}{\text{magnitude}_2} = \frac{10^y}{10^{6.9}} = 10^{0.9} = 8$$

$$10^y = 10^{0.9} \times 10^{6.9}$$
$$= 10^{7.8}$$

Therefore, the magnitude of the 1906 San Francisco earthquake had a magnitude of 7.8.

Between 12 September 2007 and 13 September 2007, three earthquakes hit Indonesia, with the first registering a magnitude of 8.5. This was 4 times stronger than the second earthquake 12 hours later.

(a) What was the magnitude of the second earthquake?

(b) The first earthquake was 25 times stronger than the third earthquake. What was the magnitude of the third earthquake?

4 An earthquake in Taiwan in 2006 had a magnitude of 7.0, the same as recorded at Haiti in the 2010 earthquake. Over 230 000 people were killed in Haiti, but only 2 people were killed in the Taiwan earthquake. Give reasons for the differences in devastation caused by the two similar strength earthquakes.

Research

5 The following measures also involve exponentials:

- loudness of sound (decibels)
- apparent brightness of light
- acidity (pH readings)

Investigate how these scales are exponential and how they produce accurate and comparative readings.

Chapter review 12

Maths literacy

asymptote	dilation	indices	transformation
axis of symmetry	exponential	radius	translation
circle	index	reflection	
10A base	logarithm	rectangular hyperbola	

Copy and complete the following, using the words and phrases from the list where appropriate. A word or phrase may be used more than once.

1 An ___________ is a line on a Cartesian plane that a curve approaches but does not cut or touch for certain values of the independent variable.

2 The graph of an ___________ equation has a horizontal asymptote and may be increasing or decreasing at a rapid rate.

10A 3 When you write a statement such as '$3^4 = 81$ is equivalent to $\log_3(81) = 4$', you refer to 4 as the ___________.

Fluency

1 For each graph below, identify the type of graph and give the: **12.1**

(i) equation of the asymptote if it is an exponential

(ii) radius and coordinates of the centre if it is a circle.

(a)

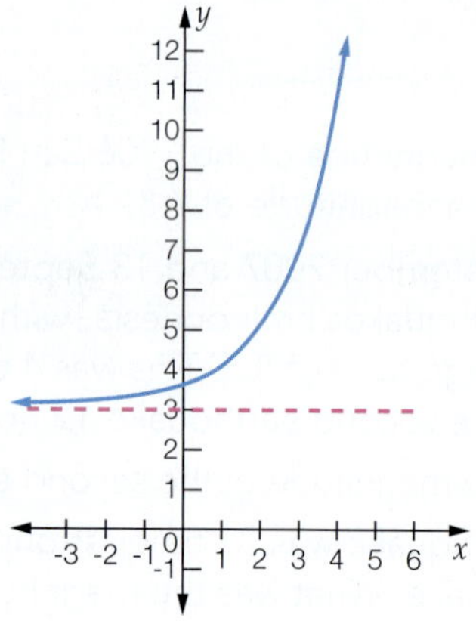

(b)

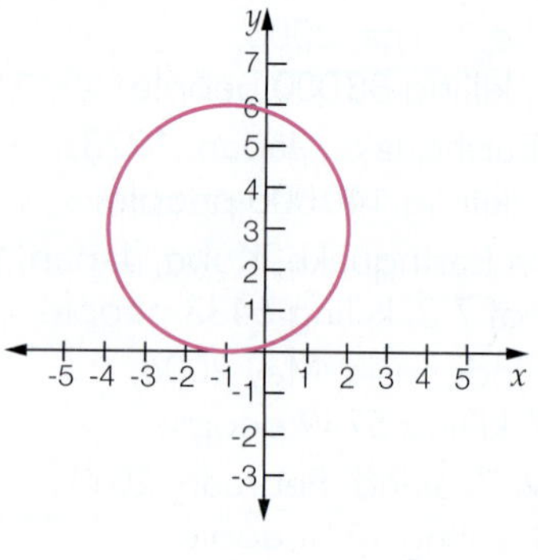

2 Sketch the graph of each of the following. **12.2**

(a) $(x+2)^2 + (y-3)^2 = 16$

(b) $x^2 + (y-3)^2 = 4$

(c) $y = 2^{(x-2)} + 1$

(d) $y = 2^{(x+2)} - 4$

10A 3 Sketch the graph of each of the following. **12.2**

(a) $y = \dfrac{1}{x-4} + 1$

(b) $y = \dfrac{1}{x+2} - 1$

4 Simplify each of the following. **12.3**

(a) $(3ab^5)^2$

(b) $\dfrac{(cd^2)^8}{(c^3d)^2}$

(c) $\left(\dfrac{t^2}{u^3}\right)^4 \times \left(\dfrac{u^2}{t}\right)^7$

(d) $\dfrac{8x^5y^6}{6x^3y^2} \times \dfrac{3x^4y^7}{2x^5y}$

(e) $\dfrac{(m^4n)^3 \times (m^2n^5)^2}{(mn)^4}$

(f) $\left(\dfrac{a^4b^2}{ab^3}\right)^2 \div \left(\dfrac{a^3b}{a^5b^2}\right)^3$

10A **5** Simplify, expressing your answers with positive indices. **12.3**

(a) $3a^{-4} \times 2a^{-2}$ (b) $(4xy^2)^{-2} \times (3x^2y)^3$ (c) $\dfrac{(m^{-3}n^2)^4}{2(mn^3)^{-2}} \times \left(\dfrac{m^2n^{-1}}{m^{-3}n^3}\right)^2$

10A **6** Simplify each of the following. **12.4**

(a) $(64m^3n)^{\frac{1}{3}}$ (b) $\sqrt{p^6q}$

(c) $\dfrac{(a^6b^3)^{\frac{1}{2}} \times a^{\frac{1}{2}}b^{\frac{1}{2}}}{a^3b^2}$ (d) $(a^2b^{-3})^{\frac{1}{2}} \times (a^{-4}b)^{\frac{5}{2}}$

(e) $\left(\dfrac{m^{\frac{1}{3}}n^{\frac{2}{3}}}{mn^{\frac{1}{2}}}\right)^2 \times \dfrac{2m^3n^5}{\left(m^{\frac{1}{2}}n\right)^4}$ (f) $\left(\dfrac{3^2m^{-2}n^2}{mn}\right)^3 \div \left(\dfrac{m^{\frac{1}{3}}n^{\frac{2}{3}}}{n^2m^{-3}}\right)^3$

10A **7** Express each of the following in index form. **12.4**

(a) $(\sqrt{k})^5$ (b) $\sqrt[4]{x^3}$ (c) $\dfrac{1}{\sqrt[3]{m^2}}$

10A **8** Write each of the following in its equivalent logarithmic form. **12.5**

(a) $4^7 = 16\,384$ (b) $2^{10} = 1024$ (c) $10^{-3} = \dfrac{1}{1000}$

10A **9** Write each of the following in its equivalent index form. **12.5**

(a) $\log_2(16\,384) = 14$ (b) $\log_3(243) = 5$ (c) $\log_{10}\left(\dfrac{1}{100\,000}\right) = -5$

10A **10** Simplify the following, expressing your answer as a basic numeral where possible. **12.6**

(a) $\log_2(8) + \log_2(64)$ (b) $\log_2(16) - \log_2(64)$

(c) $\log_3(5) + \log_3(10)$ (d) $\log_5(110) - \log_5(11)$

11 A pattern is drawn with four circles with equations: **12.2**

A $(x-3)^2 + (y-2)^2 = 16$ B $(x+1)^2 + (y-3)^2 = 4$

C $(x+5)^2 + (y+4)^2 = 9$ D $(x+2)^2 + (y+1)^2 = 16$

(a) Which two circles are the same size?

(b) By sketching the graphs find which circle intersects with the other three circles.

10A **12** Solve for x: **12.5, 12.6**

(a) $4^x = 1024$ (b) $3^x = 2187$ (c) $2^x = \dfrac{1}{64}$

(d) $\log_2(x) = -4$ (e) $\log_3(6561) = x$ (f) $\log_5\left(\dfrac{1}{625}\right) = x$

10A **13** Simplify each of the following, expressing your answer as a basic numeral where possible. **12.5**

(a) $\log_{10}(2) + \log_{10}(500)$ (b) $\log_3(81) - \log_3(27) + \log_3(9)$

(c) $\dfrac{\log_5(x^4)}{2\log_5(x)}$ (d) $\dfrac{\log_8(5x-7)^3}{4\log_8(5x-7)}$

10A **14** (a) If $\log_a(y) = \log_a(7) + 3\log_a(x)$, express y in terms of x. **12.6**

(b) If $\log_a(y) = \log_a(5) - 2\log_a(x)$, express y in terms of x.

13 Financial mathematics

Stuck for a gift idea?

Name: Naomi Simson
Job: CEO of online gifting website RedBalloon
Qualifications: Bachelor of Commerce
Awards: Winner of 2013 Pearcey Entrepreneur of the Year Award and finalist in the 2011 BRW Entrepreneur of the Year

What do learning to fly a helicopter, didgeridoo lessons or dinner in the Daintree Rainforest have in common? They are all experiences that can be purchased as gifts on the website RedBalloon.

Naomi Simson, the founder of RedBalloon, has a vision to change gifting in Australia forever. RedBalloon allows people to give the gift of an experience that will create vivid memories and stories to tell long after it is given. The business began as a side hobby for a full-time mother in 2001 and has evolved into a business providing more than 3000 experiences in every state across Australia as well as New Zealand. RedBalloon now employs 70 people and has sold more than 2.5 million experiences. Naomi has needed skills in administration, marketing, sales, customer service and strong money awareness.

'You can't run a business without financial skills because these skills enable you to understand how money moves around a business. To run a successful business you must be able to read a balance sheet and a cash flow forecast. You have to understand profit and loss and how money works.'

Why learn this?

Financial formulas are used by employers, banks, accounting firms, the tax office and other institutions. Understanding how these formulas are used can help you in many areas of your life. You can use financial formulas to check entitlements or to help choose between loans or investments. The growth formulas used in finance can also be used to help understand population growth, bacterial growth in medicine or radioactive decay in chemistry.

After completing this chapter you will be able to:

- use the simple interest formula and extend it to develop the compound interest formula
- use the compound interest formula to solve problems involving different rates and different compounding periods
- use recursive techniques to solve exponential problems such as compound interest problems
- compare effective and nominal interest rates and solve related problems
- adapt the interest formulas to solve problems such as depreciation and exponential growth and decay.

13 Recall

Prepare for this chapter by attempting the following questions. If you have difficulty with a question, you can download a Recall Worksheet from the eBook or the Pearson Places website.

1 Convert the following percentages to decimals.

(a) 17% (b) 6% (c) 19.2%

(d) 4.8% (e) 2.01% (f) $\frac{1}{4}\%$

2 Calculate:

(a) 15% of \$20 000 (b) 6% of \$10 000

(c) 9.5% of \$350 000 (d) $7\frac{1}{2}\%$ of \$45 000

3 (a) Increase 40 by:

(i) 100% (ii) 50% (iii) 25% (iv) 200%

(b) Decrease 60 by:

(i) 50% (ii) 30% (iii) 10% (iv) 75%

4 Evaluate:

(a) 7^2 (b) 2^3 (c) 10^6

(d) $(0.6)^2$ (e) $\left(\frac{1}{2}\right)^5$ (f) $(1.1)^4$

5 Convert the following to a decimal number of years. Assume a year has 52 weeks or 365 days.

(a) 6 months (b) 3 months (c) 8 years and 9 months

(d) 26 weeks (e) 13 weeks (f) 146 days

Exploration Task

You can download this activity from the eBook or the Pearson Places website.

Who knows best?

In this activity you will explore how different discounts can lead to confusing prices.

Soph: Hey, I just bought those \$200 jeans for 25% off!

Jess: SNAP, you won't believe but Mum gave me a 15% off voucher and I got an extra 10% off for using my credit card. We both got 25% off on \$200 jeans!

Soph: \$200 jeans for \$150!!

Jess: Huh? Check again. Mine cost \$153.

13.1 Interest

Borrowing and lending

There are many situations in which people or businesses borrow or lend money.

- A person might borrow money from a bank to buy a house or to start a business.
- An investor who puts money in a bank account is lending money to the bank. The bank can then lend the investor's money to other people.
- Credit cards are another way to borrow money from a bank. When you buy something using a credit card, you have to repay the money to the bank.

The original amount of money that is borrowed or loaned is called the **principal**.

The lender has risked their own money and so expects to be paid for this. The borrower eventually repays the lender the principal plus an extra amount called the **interest**.

The interest is usually shown as a percentage **per annum** (p.a.), which means 'for each year' in Latin. For example, 12% p.a. means 12% interest per year.

Calculating simple interest

Simple interest is calculated on the original principal. The amount of simple interest (I) depends on three things: the principal (P), the interest rate per annum (r), and the time in years (T). When working with financial calculations you need to know the total amount to be repaid at the end. For simple interest, the total amount repaid is the sum of the principal and the interest.

Simple interest formula

$$I = PrT$$

where I = amount of simple interest
P = principal
r = rate of interest (p.a.) as a decimal
T = time in years

Also:

$$A = P + I$$

where A = total amount repaid on the loan or earned from the investment
P = principal
I = simple interest

Simple interest is sometimes called **flat rate** interest.

Worked example 1

W.E. 1

$20 000 is borrowed for 5 years at an interest rate of 9.5% p.a. If the interest is calculated as simple interest, find:

(a) the amount of interest charged over the period of the loan

(b) the total amount to be repaid at the end of the loan.

Thinking	Working
(a) 1 Identify what needs to be found.	(a) $I = ?$
2 Write the simple interest formula.	$I = PrT$
3 Identify the values of P, r and T. Convert the value of r from a percentage to a decimal (divide by 100).	$P = 20\,000$, $r = 9.5\% = 0.095$, $T = 5$
4 Substitute these values into the formula and calculate I.	$I = 20\,000 \times 0.095 \times 5$ $= \$9500$
(b) 1 Identify what needs to be found.	(b) $A = ?$
2 Write the formula.	$A = P + I$
3 Identify the values of P and I.	$P = 20\,000$, $I = 9500$
4 Substitute these values into the formula and calculate the value of A.	$A = 20\,000 + 9500$ $= \$29\,500$

Given values for any three of these variables (A, P, r or T), you can rearrange the formula to find the value of the fourth variable.

Worked example 2

W.E. 2

Find the value of the missing variable in each of the following simple interest situations.

(a) \$200 interest is paid on an investment of \$5000 after 2 years.

(b) \$350 interest is paid on an investment at 4% p.a. after 9 months.

(c) \$1250 interest is paid on an investment of \$3000 invested at 5% p.a.

Thinking	Working
(a) 1 Write values for the known variables to identify the unknown variable. (Remember the rate is expressed as a decimal.)	(a) $I = \$200$ $P = \$5000$ $r = ?$ $T = 2$
2 Substitute the values, rearrange the formula and solve for the missing variable.	$I = PrT$ $200 = 5000 \times r \times 2$ $r = \dfrac{200}{5000 \times 2}$ $= 0.02$
3 Convert the decimal rate to a percentage (multiply by 100) and write the answer.	The rate is 2% p.a.
(b) 1 Write values for the known variables to identify the missing variable. (Remember the rate is expressed as a decimal and the time is expressed in years.)	(b) $I = \$350$ $P = ?$ $r = 0.04$ $T = \frac{9}{12} = 0.75$

	2 Substitute the values, rearrange the formula and solve for the missing variable.	$I = PrT$ $350 = P \times 0.04 \times 0.75$ $P = \frac{350}{0.04 \times 0.75}$ $= 11\,666.67$
	3 Write the answer.	The principal is \$11 666.67
(c)	1 Write values for the known variables to identify the unknown variable. (Remember the rate is expressed as a decimal and the time is expressed in years.)	(c) $I = \$1250$ $P = \$3000$ $r = 0.05$ $T = ?$
	2 Substitute the values, rearrange the formula and solve for the unknown.	$I = PrT$ $1250 = 3000 \times 0.05 \times T$ $T = \frac{1250}{3000 \times 0.05}$ $= 8.333$
	3 Write the answer.	The time is $8\frac{1}{3}$ years, which is 8 years and 4 months.

Compound interest

Compound interest is similar to simple interest, but the interest is continually added on to increase the principal, so that the amount of interest grows more every time.

Compound interest is calculated in two steps:

- Calculate the interest on the principal at the end of the time period.
- Add the interest to the principal to make the new principal for the next time period.

Consider \$1000 invested for 4 years at 6% p.a. compounded annually:

Year	Principal	Interest calculation at the end of the year $I = P \times r \times T$	Amount of investment at the end of the year $A = P + I$
1	\$1000	\$1000 × 0.06 × 1 = \$60	\$1000 + \$60 = \$1060
2	\$1060	\$1060 × 0.06 × 1 = \$63.60	\$1060 + \$63.60 = \$1123.60
3	\$1123.60	\$1123.60 × 0.06 × 1 = \$67.416	\$1123.60 + \$67.416 = \$1191.016
4	\$1191.016	\$1191.016 × 0.06 × 1 = \$71.460 96	\$1191.016 + \$71.460 96 = \$1262.476 96

The first row of the table shows the original investment of \$1000, interest calculated to be \$60, and the total amount (at the end of the year) \$1060. In the second row, this total amount is used as the new principal for the second year: interest is calculated on the new amount and again added, to make the new total for the next time period.

Compound interest continues like this, 'compounding' itself each time.

The final answer above can be rounded to \$1262.48. It is important to avoid rounding any numbers until the end of the calculation, because rounding errors can occur if you round too early.

The simple interest calculation for these values is:

$$\begin{aligned} I &= PrT \\ &= 1000 \times 0.06 \times 4 \\ &= 240 \end{aligned}$$

$$\begin{aligned} A &= P + I \\ &= 1000 + 240 \\ &= 1240 \end{aligned}$$

So, the value of this investment after 4 years of simple interest would be only \$1240. In this case, compounding the investment earned an extra \$22.48.

Considering the table of compound interest values.

1 The total value of the investment at the end of the first year is 106% of \$1000. This is the same as multiplying the initial principal by 1.06.

2 Each year, the total value is 1.06 times the value at the end of the previous year (the start of the current year).

You can rewrite the last column of the table to show this relationship more clearly.

Year	Amount owed at the end of each year (in dollars)
1	$1000 \times 1.06 = 1000 \times (1.06)^1$
2	$1000 \times (1.06)^1 \times 1.06 = 1000 \times (1.06)^2$
3	$1000 \times (1.06)^2 \times 1.06 = 1000 \times (1.06)^3$
4	$1000 \times (1.06)^3 \times 1.06 = 1000 \times (1.06)^4$

Note that the index number is the same as the year number.

If A_n is the total amount owed after n years, then for the above situation you can write:

$$A_n = 1000 \times (1 + 0.06)^n$$

(Note that these formulas are for loans or investments with an annual compounding period.)

When the interest is compounded annually:

$$A = P(1 + r)^n$$

where $A =$ amount of the loan or investment after n years

$P =$ principal

$r =$ rate of interest (per year) as a decimal

$n =$ number of years

Also:

$$I = A - P$$

where $I =$ compound interest owed or earned after a number of years

$A =$ amount of the loan or investment after the same number of years

$P =$ principal

Note that the simple interest formula calculates only the amount of interest earned during the period.

The compound interest formula calculates the total amount (principal and interest) at the end of the compounding period. This is called the **term** of the loan or investment.

Worked example 3

W.E. 3

\$12 500 is borrowed for 7 years at an interest rate of 15% p.a. If the interest is compounded annually, find:

(a) the total value of the loan after 7 years

(b) the amount of interest charged over the period of the loan.

Thinking	Working
(a) 1 Identify what needs to be found.	(a) $A = ?$
2 Write the compound interest formula.	$A = P(1 + r)^n$
3 Identify the values of P, r and n.	$P = 12\,500$, $r = 15\% = 0.15$, $n = 7$
4 Substitute these values into the formula and calculate the value of A.	$A = 12\,500(1 + 0.15)^7$ $= 12\,500 \times 1.15^7$ $= \$33\,250.25$ (rounded to the nearest cent)
(b) 1 Identify what needs to be found.	(b) $I = ?$
2 Write the formula.	$I = A - P$
3 Identify the values of A and P.	$A = 33\,250.25$, $P = 12\,500$
4 Substitute these values into the formula and solve for the value of I.	$I = 33\,250.25 - 12\,500$ $= \$20\,750.25$

Note: To evaluate numbers with powers like 1.15^7 on a calculator, you could enter $1.15 \times 1.15 \times 1.15 \times 1.15 \times 1.15 \times 1.15 \times 1.15$, but it is more efficient to use the index function on your calculator. The index key will usually look like x^y, a^x or y^x.

Appreciation

Increase in cost or value is called **appreciation**. Loans and investments are not the only things that appreciate. Land, houses, collectibles (such as antiques and memorabilia), salaries and the cost of food are also examples of things that usually appreciate over time.

The rate of increase of prices within the economy is called the **inflation rate**. This is a measure of the overall appreciation of prices.

The simple interest formula can be used to calculate a flat rate of appreciation.

The compound interest formula can be used to predict a future value if the appreciation increases with each year (compounds annually).

13.1 Interest

Navigator

1, 2, 3, 4, 5, 6, 7, 8, 9, 10, 11, 12, 13, 14, 15, 17	1, 2, 3, 4, 5, 7, 8, 9 (a), 10, 11, 12, 13, 14, 15, 16, 17, 19	1, 2, 3 (a–c), 4, 5 (d–g), 7 (a, c, e), 8, 11, 12, 13, 14, 15, 16, 17, 18, 19, 20

Answers p.866

If necessary, round answers to the nearest cent.

Fluency

1 (a) Write the formula used to calculate simple interest.

(b) Substitute the values of $P = 500$, $r = 0.016$, $T = 10$ into the simple interest formula to find the value of I.

(c) Substitute the values of $P = 2000$, $r = 0.05$, $T = 7$ into the simple interest formula to find the value of I.

2 \$1000 is invested for 5 years at 5.1% p.a.

(a) What is the principal?

(b) What is the interest rate?

(c) What is the interest rate as a decimal?

(d) What is the term?

3 (a) Substitute the values of $I = 500$, $r = 0.06$, $T = 8$ into the simple interest formula and rearrange the formula to find the value of P.

(b) Substitute the values of $I = 6750$, $r = 5\%$, $T = 35$ into the simple interest formula and rearrange the formula to find the value of P.

(c) Substitute the values of $I = 44\,070$, $P = 30\,000$, $T = 16$ into the simple interest formula and rearrange the formula to find the value of r, correct to 1 decimal place.

(d) Substitute the values of $I = 1250$, $P = 5000$, $T = 10$ into the simple interest formula and rearrange the formula to find the value of r, correct to 1 decimal place.

(e) Substitute the values of $I = 5490$, $P = 10\,000$, $r = 6.1\%$ p.a. into the simple interest formula and rearrange the formula to find the value of T.

(f) Substitute the values of $I = 4032$, $P = 7000$, $r = 3.6\%$ into the simple interest formula and rearrange the formula to find the value of T.

4 \$500 is invested for 10 years at 3.2% p.a. Answer the following:

(a) What is the value of P?

(b) What is the value of T?

(c) What is the value of r?

(d) Use the formula $I = PrT$ to find the value of I.

W.E. 1

5 For each of the following loans the interest is calculated as simple interest. Find:

(i) the amount of interest charged over the period of the loan

(ii) the total amount to be repaid at the end of the loan.

(a) \$40 000 borrowed for 1 year at 8% p.a.

(b) \$30 000 borrowed for 5 years at 7% p.a.

(c) \$40 000 borrowed for 12 years at 5% p.a.

(d) \$150 000 borrowed for 25 years at 7.5% p.a.

(e) \$300 000 borrowed for 1.1 years at 6.75% p.a.

(f) \$350 000 borrowed for 1.1 years at 7.25% p.a.

(g) \$13 000 borrowed for 5 years at $9\frac{1}{2}$% p.a.

(h) \$45 000 borrowed for $12\frac{1}{2}$ years at 5% p.a.

(i) \$10 000 borrowed for 2 years and 3 months at 15% p.a.

6 State which piece of information is missing (P, I, r or T) for each question. (You do not need to find the value of the missing variable.)

(a) \$300 for 3 years at 1.5% p.a.

(b) \$3870 paid after 5 years at 4.3% p.a.

(c) \$1783 earned from \$1100 at 5.1% p.a.

W.E. 2

7 Find the value of the missing variable in each of the following simple interest situations.

(a) \$2400 interest is paid on an investment at 20% p.a. after 3 years.

(b) \$7200 interest is paid on an investment at 15% p.a. after 6 years.

(c) \$25 200 interest is paid on an investment of \$20 000 after 9 years.

(d) \$40 788 interest is paid on an investment of \$30 900 after 12 years.

(e) \$65 000 interest is paid on an investment of \$100 000 invested at 6.5% p.a.

(f) \$135 000 interest is paid on an investment of \$150 000 invested at 7.5% p.a.

8 (a) (i) Write the formula to calculate the final amount when interest is compounded annually.

(ii) Substitute the values of $P = 10\,000$, $r = 2.5\%$, $n = 13$ into the compound interest formula and find the value of A.

(iii) Substitute the values of $P = 25\,000$, $r = 8.4\%$, $n = 25$ into the compound interest formula and find the value of A.

(b) (i) Write the formula used to calculate compound interest earned given the principal and the current value of the investment.

(ii) Substitute the values of $P = \$50\,000$ and $A = \$60\,000$ to find the value of I.

(iii) Substitute the values of $P = \$25\,780$ and $A = \$20\,000$ to find the value of I.

9 (a) Using $I = A - P$, substitute the values of $I = \$5500$ and $A = \$600\,000$. Rearrange to find the value of P.

(b) Using $I = A - P$, substitute the values of $I = \$8750$ and $P = \$50\,000$. Rearrange to find the value of A.

W.E. 3

10 For each of the loans in Question **5** the interest is now compounded annually. Find:

(i) the total amount to be repaid at the end of the loan

(ii) the amount of interest charged over the period of the loan.

11 Lily invested \$20 000 for 5 years at an interest rate of 5.25% p.a. Jarrah invested \$45 000 for 7 years at an interest rate of 9.5% p.a.

(a) Calculate the interest earned on each investment assuming:

(i) annual compound interest (ii) simple interest.

(b) Compare the compounded interest with the simple interest.

12 Calculate the total amount owing on a loan of \$7000 after 2 years, if the 16% interest p.a. is:

(a) compounded annually (b) calculated as simple interest.

13 \$4245 at 6.5% p.a. compound interest, compounded annually over 8 years, will amount to:

A $\$4245(1 + 0.065)$ B $\$4245(1 + 0.065) \times 8$

C $\$4245(1 + 0.065)^8$ D $\$4245(1 + 0.65)^8$

Understanding

14 If the simple interest on an investment over 18 years at 4% is \$93, what was the amount of the original investment?

15 For each of the following appreciation questions:

(i) state whether the problem should be modelled by the simple interest formula or by the compound interest formula

(ii) solve the problem, rounding answers to the nearest whole number or dollar.

(a) Martina currently gets \$10 pocket money a week. She has been promised a \$4 increase each year. What is the promised rate of increase? How much more pocket money will she get in 3 years than she is getting now?

(b) An antique clock appreciates in value by 11% p.a. If it was valued 10 years ago at \$350, how much is it worth now?

(c) Yang paid \$20 000 for a new motorbike. If the inflation rate is 6.5% p.a., what is the estimated cost of a similar new motorbike in 4 years' time?

16 Costs in a business are growing at an annually compounding rate of 8% p.a. If the costs are \$780 per week now, what were the costs 7 years ago?

A $\dfrac{\$780}{1.08^7}$ B $\$780 \times 1.08^7$ C $\$780 \times 1.08^7 - 780$ D $\dfrac{\$780}{1.8^7}$

Reasoning

17 Ahmed has some money to invest. He wants to at least quadruple his money.

(a) How long will it take if he invests at a simple interest rate of 3% p.a.?

(b) How long will it take if he invests at an annually compounded interest rate of 3%?

Give your answers in whole years.

Hint

Either choose a number for the amount to invest or let P be this amount.

18 An amount of money \$$P$ is deposited into a bank account with an annually compounding interest rate R. After 2 years the money has grown to \$48 951 and after 3 years it has grown to \$51 398.55. What is the interest rate?

Hint

Write two equations with P and r in them. Then use these two equations to write $\dfrac{51\,398.55}{48\,951}$ in terms of P and r.

Open-ended

19 If an investment compounds annually for 2 years, then what investment amounts and annual interest rates would give a final value within \$100 of \$15 000?

20 Calum's car is damaged during a hail storm. Jessica also has to replace her car after flood damage. Both need to borrow the same amount of money. Calum gets a flat rate loan at 10% p.a. for 8 years. Jessica has to pay the same total amount of interest as Calum, but her loan has a simple interest rate between 10% and 20% p.a. and is repaid over a shorter period of time. Find possible values for the interest rate and period of Jessica's loan.

Problem solving

The oldest cake problem

1 What percentage of the cake has been promised to Elysia by her 21st birthday?

2 How old will Elysia be when she finally gets Aunty Bee's present?

Strategy options

- Draw a diagram.
- Solve a simpler problem.

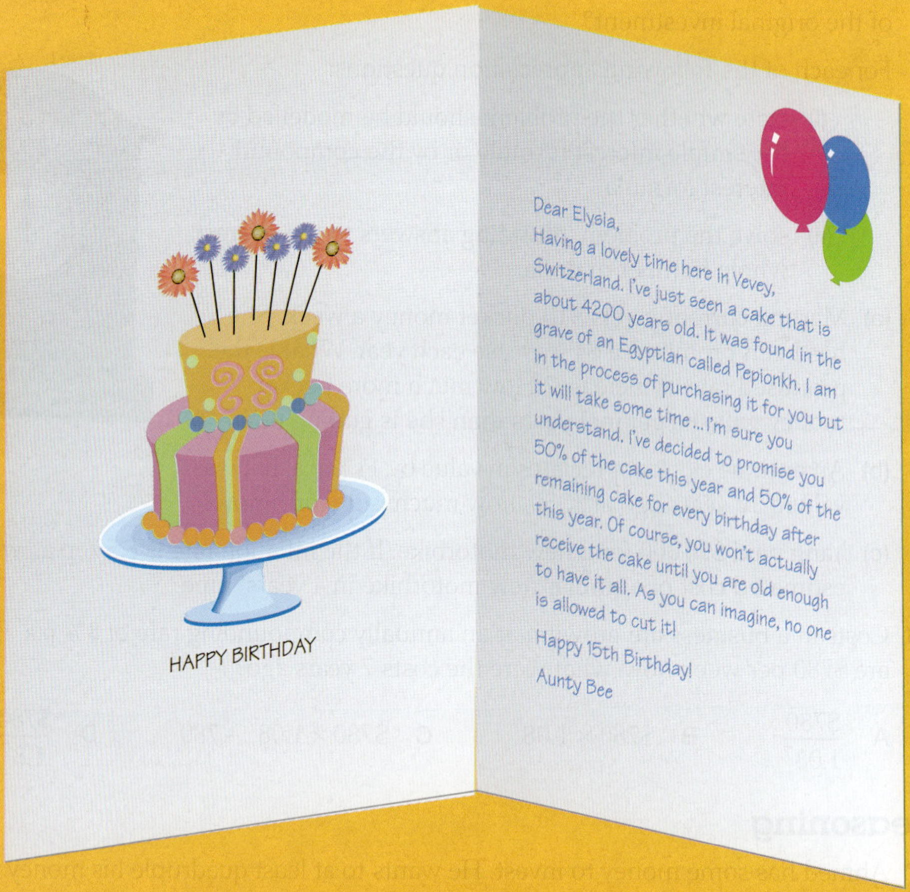

Compound interest—the general formula

13.2

In the previous section, interest was compounded annually but interest can also compound daily, monthly, quarterly or every half year.

The same formula used to calculate interest compounded annually is used for calculating interest compounded over other time periods. In the previous section, n was the number of years, but more generally n is the number of time periods (compounding periods).

In the compound interest formula, the time periods must match the given interest rate. So, if you have a monthly compounding period, you will need the monthly interest rate.

Consider a loan at 12% p.a. with different compounding periods.

- Quarterly (every 3 months): the interest rate for each period is $\frac{12}{4}\% = 3\%$ (or 0.03) per period and there will be 4 periods per year.
- Monthly: the interest rate for each period is $\frac{12}{12}\% = 1\%$ (or 0.01) per period and there will be 12 periods per year.
- Daily: the interest rate for each period is $\frac{12}{365}\% \approx 0.0329\%$ (or 0.000 329) per period and there will be 365 periods per year.

Be careful when converting a fractional rate to a decimal. If you round the decimal value, then this might cause a rounding error in your calculations, especially for daily rates. Avoid rounding the decimal values until the final answer.

Compound interest formula

$$A = P(1 + r)^n$$

where A = amount owing or earned at the end
P = principal (original amount borrowed or loaned)
r = interest rate for the period, expressed as a decimal
n = number of compounding periods

Interactive

Calculating compound interest

Explore the effects of changing values for compound interest investments.

Go to the eBook or the Pearson Places website to access this interactive.

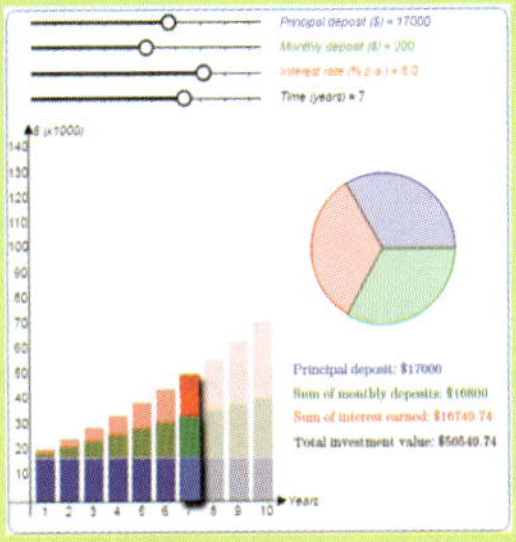

Using the compound interest formula

The general form of the compound interest formula can be used to solve problems in a variety of situations.

Worked example 4

W.E. 4

Clive borrowed $12 000 to buy a machine for his factory. He agreed to pay interest at a rate of 15% p.a. and to repay the loan in full in 3 years. Calculate the amount to be repaid if the interest is compounded:

(a) annually **(b)** half-yearly **(c)** quarterly.

Thinking	Working
(a) 1 Identify what needs to be found.	(a) $A = ?$
2 Write the compound interest formula.	$A = P(1 + r)^n$
3 Identify the values of the other variables. 'Compounded annually' means the number of time periods in a year is 1.	$P = 12\,000$ $r = \frac{0.15}{2} = 0.15$ $n = 3 \times 1 = 3$
4 Substitute these values into the formula and calculate the value of *A*.	$A = 12\,000 \times (1.15)^3$ $= \$18\,250.50$
(b) 1 Identify what needs to be found.	(b) $A = ?$
2 Write the compound interest formula.	$A = P(1 + r)^n$
3 Identify the values of the other variables. 'Compounded half-yearly' means the number of time periods in a year is 2.	$P = 12\,000$ $r = \frac{0.15}{2} = 0.075$ $n = 3 \times 2 = 6$
4 Substitute these values into the formula and calculate the value of *A*.	$A = 12\,000 \times (1.075)^6$ $= \$18\,519.62$
(c) 1 Identify what needs to be found.	(c) $A = ?$
2 Write the compound interest formula.	$A = P(1 + r)^n$
3 Identify the values of the other variables. 'Compounded quarterly' means the number of time periods in a year is 4.	$P = 12\,000$ $r = \frac{0.15}{4} = 0.0375$ $n = 3 \times 4 = 12$
4 Substitute these values into the formula and calculate the value of *A*.	$A = 12\,000 \times (1.0375)^{12}$ $= \$18\,665.45$

The more compounding periods there are, the greater the total amount to be repaid (or the greater the value of the investment).

13.2 Compound interest—the general formula

Navigator

1, 2, 3, 4, 5, 6, 7, 8, 9, 10, 12, 13, 14 | 2, 3, 4, 5, 6, 7, 8, 9, 10, 11, 12, 13, 14 | 2, 4, 5, 7, 8, 9, 10, 11, 12, 13, 14

Answers p. 867

If necessary, round answers to the nearest cent.

Fluency

1 Write the number of periods in a year if interest is compounded:

(a) quarterly (b) half-yearly (c) daily

(d) monthly (e) annually

2 Convert the annual interest rate of 9% to the corresponding interest rate for the given compounding period. Give your answers in simplest fraction form.

(a) quarterly (b) half-yearly (c) daily

(d) monthly (e) annually

3 Write the values for the pronumerals (P, r, n) in the compound interest formula $A = P(1 + r)^n$ for each of the following situations. Do not caluculate the value of A.

(a) Ryu invests \$25 000. He signs up for an interest rate of 2.5% p.a. and invests for 13 years. The loan is compounded half-yearly.

(b) Tiansha is offered a \$9000 loan at 7.2% p.a. compounded quarterly for 10 years.

(c) Bim invests \$200 000 for 35 years with an interest rate of 3.6% p.a. The loan is compounded daily.

4 (a) Write the formula used to calculate the final amount when interest is compounded annually.

(b) For each of the following loans use the formula in part **(a)** to calculate the value A.

(i) \$9500 is borrowed at a rate of 1.9% p.a. compounded monthly over 3 years.

(ii) \$62 200 is borrowed at a rate of 12.1% p.a. compounded quarterly over 5 years.

(iii) \$20 000 is borrowed at a rate of 8.5% p.a. compounded half-yearly over 15 years.

5 Calculate the amount to be repaid after 2 years on a loan of \$16 000 if the 14% interest p.a. is compounded: **W.E. 4**

(a) annually (b) half-yearly (c) quarterly.

6 Calculate how much an investor would repay on a debt of \$8000 after $1\frac{1}{2}$ years if the 12% interest p.a. is compounded:

(a) every 6 months (b) every 3 months.

7 If the interest on a loan of \$5800 is 16% p.a. compounded every half year, then over 6 years the debt would grow to:

A $\$5800(1 + 0.16)^6$ B $\$5800(1 + 0.08)^{12}$

C $\$5800(1 + 0.16)^{12}$ D $\$5800(1 + 0.8)^{12}$

8 If a \$5800 loan accrues interest at a rate of 16% p.a. compounded quarterly for 6 years, then the interest is:

A $\$5800(1 + 0.04)^6$

B $\$5800(1 + 0.16)^6 - \$5800$

C $\$5800(1 + 0.04)^{24} - \$5800$

D $\$5800(1 + 0.4)^{24} - \$5800$

Understanding

9 Calculate how much interest (to the nearest cent) is added over 10 years to an account paying 9% p.a. interest on an initial sum of \$45 800 if the interest is compounded:

(a) half-yearly (b) quarterly (c) monthly.

10 Ally invests \$32 000 over 4 years in an account offering 8.8% p.a. compounded annually.

(a) How much more interest would she earn if the interest is calculated weekly? (Assume 52 weeks in a year.)

(b) Calculate the interest earned if it is compounded daily.

11 Martina won \$5000 in the lottery and decided to invest it with a view to taking a trip to Paris some time in the future. Martina was given two options when she invested the money. She could have an interest rate of 4.5% p.a. compounded quarterly or she could have 4.4% p.a. compounded daily. Which is the better deal, and by how much, if Martina invests the money for 2 years?

Reasoning

12 (a) \$20 000 is invested for 3 years. Calculate the interest earned on each of the following to find which would give the best return.

(i) 5% p.a. compounded quarterly

(ii) 4.5% p.a. compounded every 2 months

(iii) 4% p.a. compounded monthly

(b) Do you think the size of the investment would alter which option in part **(a)** would give the best return?

(c) Prove your answer to part **(b)** algebraically by redoing part **(a)** with $\$P$ in place of \$20 000, and expressing your answer as a percentage of $\$P$.

Open-ended

13 'The more frequent the compounding, the greater the interest.' Explain what this statement means. Use a principal of \$10 000 invested at 8% p.a. for 5 years and three different compounding periods (such as weekly, quarterly or monthly) to help demonstrate your answer. Express the percentage increase correct to 1 decimal place.

14 Tyson is having trouble understanding the difference between simple and compound interest. He can't see why compound interest is higher than simple interest after the first compounding period if the principal, interest rate and overall time period are the same. How would you explain the difference to Tyson? Use an example of \$4000 invested for 3 years at 5% p.a. to help explain your answer.

Investigation

Using recursion

The Big Question

How do you program a robotic vacuum cleaner?

Engage

A recursive process requires two instructions:

1 Where to start.

2 A repeated rule for getting from one step to the next.

A repeated rule is also called a *recursive* rule.

Here are some examples of recursion:

- The set of natural numbers starts with the number 1. The recursive rule to get the next number is to add 1 to the previous number.
- The sequence: 2, 3, 5, 9, 17, 33, … starts with the number 2. The recursive rule to get the next number is to double the previous number and subtract 1.
- The sequence: 10, 14, 12, 13, 12.5, 12.75, … starts with the numbers 10 and 14. The recursive rule to get the next number is to average the last two numbers.

1 For each of the following show that the number sequence or technique is recursive.

(a) 42, 21, 10.5, 5.25, 2.625, …

(b) 1, 2, 5, 26, 677, 458 330, …

(c) Calculating the total of an investment of $5000 at 10% p.a. over several years where the interest compounds annually.

Explore

The diagram below shows the spiral cleaning path of a robotic vacuum cleaner. It can be programmed to follow this path using recursion.

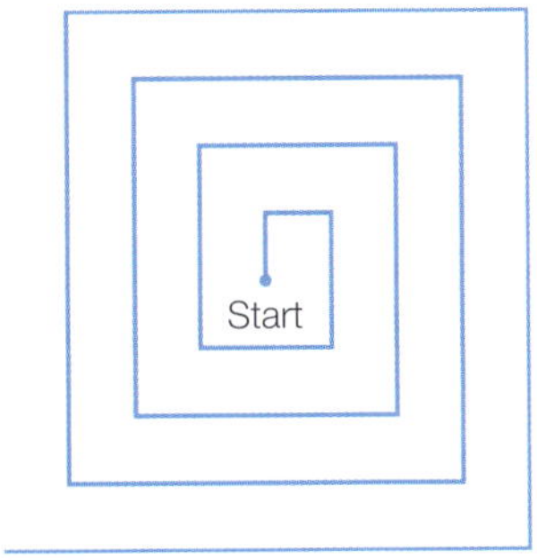

2 If you walked this path, what would you do at the end of each section?

Strategy options

- Break problem into manageable parts.
- Act it out.
- Look for a pattern.

Explain

3 Explain how a robotic vacuum cleaner can be programmed to follow the spiral cleaning path explored above (remember to specify a distance *and* direction). Organise your explanation under the headings:

- Instructions for getting started
- Instructions for the recursive step

You may like to use diagrams to make your explanation as clear as possible and to double check that your instructions work.

Elaborate

4 (a) Consider the distance travelled by the robot (at each step) when following your instructions above. Then complete this table:

Number of times the recursive step is repeated (n)	Distance travelled in this step (d_n)
0	
1	
2	
3	
4	
5	

(b) Find a formula relating n and d_n.

Evaluate

5 Is it possible to achieve a thorough cleaning of the floor with the recursion program you've developed? Explain your thinking.

Extend

The robot has completed the recursive step 16 times. It needs to cover the same area again and return to the start. Write instructions for a program that will achieve this.

Exploration Spreadsheet

Equipment required: Microsoft® Excel or similar spreadsheet software. (For Casio ClassPad CAS or TI-Nspire CAS, you can download instructions from the eBook or the Pearson Places website.)

Comparing simple and compound interest

Spreadsheets can be a useful way to calculate and display simple interest and compound interest.

Calculating simple interest

The following steps show how to calculate a simple interest investment with a starting value (principal) of $1000 and a rate of 10% p.a.

1 Open a new spreadsheet. Enter the values for the interest rate, principal and time period as shown. (The time period should be 1 because the interest rate is annual and the interest is being calculated each year.) It is important to put the values in the cells shown, because the cells must be referenced in the calculation formulas later.

2 Enter the year numbers under the heading 'Year' in Column **A**. You can do this by entering 0 in cell **A6**, then the formula **=A6+1** in cell **A7**. You can **Fill Down** this formula to copy it down the column by clicking on the lower right-hand corner of the cell and dragging down. Enter the 'Simple interest' and 'Total amount' starting values of 0 and 1000 next to 'Year 0', in cells **B6** and **C6**.

Note that you can change the column width to fit the text by double-clicking or clicking and dragging on the column headings **B** and **C** above.

fx =B$1*B$2*B$3

	A	B	C	D
1	Rate	10%		
2	Principal	1000		
3	Period	1		
4				
5	Year	Simple interest	Total amount	
6	0	0	1000	
7	1	100		
8	2			
9	3			
10	4			

3 To calculate the interest earned after one year, enter the formula **=B$1*B$2*B$3** into cell **B7**. This multiplies the values of the cells **B1**, **B2** and **B3** together (rate × principal × time period). The **$** symbol sets the value so that the cell references will not change when the formula is copied down the column.

4 **Fill Down** this interest formula to copy it down the column as described in **2**.

5 To find the new total amount for each year, you must add each year's interest to the previous year's total. For Year 1, enter the formula **=B7+C6** into cell **C7**. **Fill Down** to copy this formula down the column.

6 To show the total interest added over time, enter the heading 'Total interest' in cell **D5** and the starting value 0 in cell **D6**. To calculate the total interest for each year, enter the formula **=D6+B7** into cell **D7** and then **Fill Down** this column as before.

(a) What is the total amount of the investment after 8 years?

(b) How much of this total amount is interest?

fx =D6+B7

	A	B	C	D	E
1	Rate	10%			
2	Principal	1000			
3	Period	1			
4					
5	Year	Simple interest	Total amount	Total interest	
6	0	0	1000	0	
7	1	100	1100	100	
8	2	100	1200	200	
9	3	100	1300	300	
10	4	100	1400	400	

Compound interest

A spreadsheet for simple interest can be extended to calculate compound interest.

7 Extend the table to compare compound interest by making new headings for 'Year', 'Compound interest', and 'Total amount' as shown on the next page. In these new columns, enter the year values and the starting values for the interest and total. This can be done as previously in **2**.

8 For compound interest, you need to add the interest earned each year to the principal, then use this amount as the new principal to calculate the interest for the next year. To do this, enter a new formula as shown, to multiply the previous year's amount by the rate by the time period: **=G6*B$3*B$1**.

fx =G6*B$3*B$1

	A	B	C	D	E	F	G
1	Rate	10%					
2	Principal	1000					
3	Period	1					
4							
5	Year	Simple interest	Total amount	Total interest	Year	Compound interest	Total amount
6	0	0	1000	0	0	0	1000
7	1	100	1100	100	1	100	
8	2	100	1200	200	2		
9	3	100	1300	300	3		
10	4	100	1400	400	4		
11	5	100	1500	500	5		

9 To find the new total amount for each year, you must add each year's interest to the previous year's total. For Year 1, enter the formula **=F7+G6** into cell **G7**.

10 **Fill Down** from both cells **F7** and **G7** to copy both formulas down their columns. Highlight both cells **F7** and **G7**, put the cursor in the lower right-hand corner of **G7** and drag down to row **14**.

(a) What is the total amount of the investment after 8 years?

(b) How much of this total amount is interest?

(c) How much extra interest is calculated when it is calculated using a new amount each time, instead of the original principal?

(d) (i) Explain the difference between the simple interest formula used in **3** and the compound interest formula used in **8**.

(ii) Why was it necessary to modify the formula?

11 To show the total interest added over time for compound interest, enter the heading 'Total interest' in cell **H5** and use a suitable formula to fill the values into this column.

Spreadsheet formatting and graphs

12 This spreadsheet involves money values, so it is helpful to format the entries in your table to show this. You can select relevant cells, then select the **Currency** format option. In Excel you can find the menu for this in the **Home** tab under **Number.**

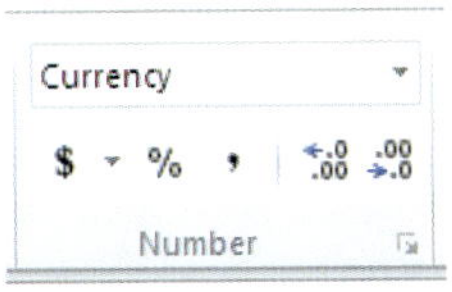

13 The difference between simple and compound interest can be seen clearly on a line graph. Highlight the data values in both 'Total interest' columns (**D** and **H**). You can hold down the **Ctrl** key to highlight more than one column. To create a graph, go to the **Insert** tab and select the line graph option as shown.

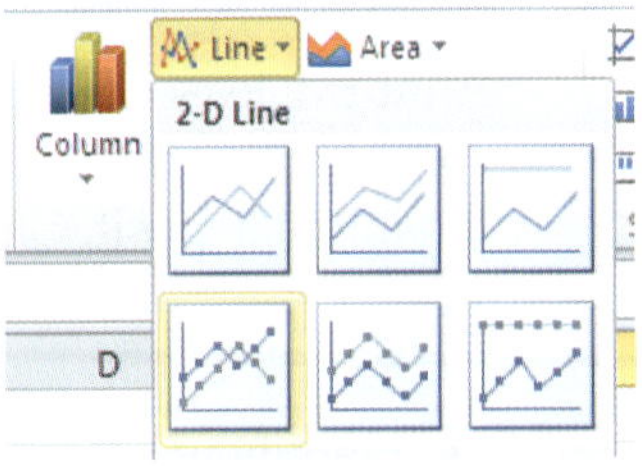

14 You can change the labels on the graph by selecting different options on the **Layout** tab.

(a) Describe and explain the shape of the simple interest graph.

(b) Describe and explain the shape of the compound interest graph.

Taking it further

15 To calculate compound interest, you calculated 10% of the principal and added this to the total. Instead of doing this, you could also have calculated 110%, to find the new amount directly.

Create a third section in your spreadsheet with the headings of 'Year' and 'Total amount'. In these new columns, enter the year values and the starting values for the interest and total, the same as previously in **2**. Then enter a formula that multiplies the principal of $1000 by 110% (or 1.1). **Fill Down** to copy this formula down the column. What do you notice about the values generated by the formula?

13.3 Compound interest—further applications

The compound interest formula $A = P(1 + r)^n$ uses four variables: A, P, r and n. If you are given values for any three of these variables, then you can rearrange the formula to find the value of the fourth variable. Remember that:

A = amount owing or earned

P = principal (original amount borrowed or loaned)

r = interest rate expressed as a decimal

n = number of compounding periods

Finding the principal

Worked example 5

W.E. 5

Aunt Eva leaves Heidi a gift of money that was invested for 10 years at 5.11% p.a. compounded daily. The value of the final payout received was \$21 669.68. Calculate the initial deposit. Give your answer rounded to the nearest dollar. (Ignore leap years.)

Thinking	Working
1 Identify what needs to be found.	$P = ?$
2 Write the compound interest formula.	$A = P(1 + r)^n$
3 Identify the values of the other variables. 'Compounded daily' means the number of time periods in a year is 365.	$A = 21\,669.68$ $r = \frac{0.0511}{365}$ $n = 10 \times 365 = 3650$
4 Substitute these values into the formula and simplify.	$21\,669.68 = P\left(1 + \frac{0.0511}{365}\right)^{3650}$
5 Transpose the equation to find P.	$P = \dfrac{21\,669.68}{\left(1 + \frac{0.0511}{365}\right)^{3650}}$
6 Evaluate P.	$P = \$13\,000.00595$ So, the principal was \$13 000.

Note: The exact value of r is used in this example. Remember that you should only round the values at the very end of your calculations, or else you may introduce rounding errors into the calculations.

Finding the rate of interest

Worked example 6

W.E. 6

In the first year of his business Jack made $228 000. This grew to $520 000 after 12 years, increasing at the same rate every year. At what annual percentage rate has the business been growing? Round your answer to 2 decimal places.

Thinking	Working
1 Identify what needs to be found.	$r = ?$
2 Write the compound interest formula.	$A = P(1 + r)^n$
3 Identify the values of *A*, *P* and *n*. (Compounding every year means the number of time periods in a year is 1.)	$A = 520\,000$, $P = 228\,000$, $n = 12 \times 1 = 12$
4 Substitute these values into the formula and simplify.	$520\,000 = 228\,000(1 + r)^{12}$
5 Rearrange the equation to find *r*. Remember: Don't round until the end. (The inverse operation to finding a power of 12 is the 12th root. See calculator note below.)	$\frac{520\,000}{228\,000} = (1 + r)^{12}$ $\sqrt[12]{\frac{520\,000}{228\,000}} = (1 + r)$ $r = \sqrt[12]{\frac{520\,000}{228\,000}} - 1$ $r = 0.071\,122\,251\,37$
6 Multiply by 100 to convert to a percentage and round the answer to the appropriate number of decimal places.	$r = 7.11\%$ p.a. The business has been growing at a rate of 7.11% p.a.

Note: To calculate roots on a calculator, the key will usually look like $\sqrt[y]{x}$ and require these steps:

1 Enter a value for x (the number under the root symbol). In the above example, this is the expression (520 000 ÷ 228 000).

2 Press the root key $\sqrt[y]{x}$.

3 Enter a value for y to find the yth root. For the above example, this is 12.

4 Press =.

Check your own calculator's instructions to find how to calculate nth roots.

Finding the time (or the number of compounding periods)

Worked example 7 W.E. 7

How many years will it take for a $452 collectible to be worth $1000, if it increases in value by 7.5% p.a.? Assume that interest is compounding annually. Give your answer to the nearest whole number of years.

Thinking	Working
1 Identify what needs to be found.	$n = ?$
2 Write the formula.	$A = P(1 + r)^n$
3 Identify the values of A, P and r.	$A = 1000$, $P = 452$, $r = 0.075$
4 Substitute these values into the formula and simplify.	$1000 = 452(1 + 0.075)^n$ $1000 = 452(1.075)^n$
5 Rearrange the equation, as far as you can, to find n.	$\frac{1000}{452} = (1.075)^n$
6 Use technology to solve for n, or else use 'guess and check' to try to find the solution.	$2.212\,389\,381 = (1.075)^n$ $n = 4,\quad (1.075)^4 = 1.335\,469$ $n = 10,\quad (1.075)^{10} = 2.061\,031$ $n = 12,\quad (1.075)^{12} = 2.381\,779$ $n = 11,\quad (1.075)^{11} = 2.215\,608$ Or, solving using technology: $n \approx 10.9799$
7 Answer the question.	It will take 11 years for the value to increase to $1000.

In this example, the compounding period is annual. If the compounding period is not annual, then to find the number of years, the value of n must be divided by the number of compounding periods in a year. So, if $n = 28$ compounding quarterly, then $\frac{28}{4} = 7$ years.

13.3 Compound interest—further applications

Navigator

Answers p.867

1, 2, 3, 4, 5, 6, 7, 8, 9 (a–b), 10 (a), 15, 16	1, 3, 4, 5, 6, 7, 8, 9, 10, 12, 14, 15, 16	1, 3, 4, 5, 6, 7, 8, 9, 10, 11, 12, 13, 14, 15, 16

Fluency

1 Substitute the following values into the compound interest formula $A = P(1 + r)^n$ and solve to find P, the principal.

(a) $A = \$1223.43$, $r = 7.7\%$ p.a. compounded monthly, for 4 years

(b) $A = \$2548.06$, $r = 2.7\%$ p.a. compounded quarterly, for 9 years

2 To answer the following questions, which variable would you need to find: P, A, r or n? (You do not need to solve the problems.)

(a) Kenji repays a total of \$320 000 on a loan over 23 years, at 2.4% p.a. compounded monthly. How much was the original loan?

(b) Arto borrows \$15 000 at 9.2% p.a. compounded annually and repays a total of \$17 530 over how many years?

(c) At what rate will Annie invest \$8700 to grow to \$10 000 over 2 years if the interest is compounded annually?

3 Write the values of P, A, r and n for the following. (You do not need to solve any equations.)

(a) Ezra borrows \$16 000 at 3.1% p.a. compounded monthly. He repays a total of \$19 266.13 over 6 years.

(b) Hoping to purchase a house, Fern signs up for a loan of \$46 000, compounded daily at 8.4% p.a. She hopes to repay the loan in 4 years and to pay a total of \$64 367.11.

(c) Zane pays back a total of \$4420.32 for a loan of \$4000, compounded monthly at 2% p.a. He pays the loan back over 5 years.

(d) Tomoe applies for a 4 year loan of \$2500 compounded quarterly at 6.3% p.a. She will repay \$4122.10 in total.

4 In each of the following, use the compound interest formula to find the principal. **W.E. 5**

(a) Calculate how much would need to be invested at 8% p.a. compounded each half year to accumulate to \$9600 in 6 years. Round your answer to the nearest dollar.

(b) How much would Li have to deposit in order to receive \$10 000 in 7 years if she places her money in an account that pays 8.8% interest, compounded quarterly? Round your answer to the nearest dollar.

(c) How much would Sylvester have to deposit to receive \$8000 in 5 years' time if he places money in an account that pays 7.8% interest, compounded half yearly? Round your answer to the nearest cent.

(d) A deposit accumulates to \$4500 in 9 months at 12% p.a. compounded quarterly. The initial deposit, in dollars, was:

A $\dfrac{4500}{1.3^3}$ B $\dfrac{4500}{1.03^3}$ C $\dfrac{4500}{1.12^{3/4}}$ D 4500×1.04^3

5 Solve the following equations using technology to find the interest rate r. Round your percentage answer to 2 decimal places.

(a) $1 + r = \sqrt[10]{\dfrac{75\,350}{62\,000}}$ (b) $1 + r = \sqrt[5]{\dfrac{635\,280}{600\,000}}$

(c) $(1 + r)^{16} = \sqrt{\dfrac{25\,940}{20\,000}}$ (d) $(1 + r)^{25} = \sqrt{\dfrac{96\,360}{50\,000}}$

6 Substitute the given values into the compound interest formula $A = P(1 + r)^n$ and solve to find the interest rate r. Round your percentage answer correct to 1 decimal place.

(a) $A = \$87\,066.08$, $P = \$80\,000$ compounded annually for 5 years

(b) $A = \$11\,653.80$, $P = \$10\,900$, compounded annually for 2 years

(c) $A = \$4939.07$, $P = \$2100$, compounded annually for 13 years

(d) $A = \$14\,913.31$, $P = \$7910$, compounded annually for 9 years

W.E. 6

7 In each of the following, use the compound interest formula to find the annual interest rate.

(a) Sales of $26 900 grow to $78 000 in 11 years. Calculate the percentage growth p.a. rounded to 1 decimal place. Assume the growth compounds annually.

(b) Costs of $6200 increase to $8000 in 8 years. Calculate the percentage increase p.a. rounded to 2 decimal places. Assume annual compounding.

(c) Calculate the rate of interest p.a. which would allow $7900 to accumulate to $9000 in 5 years if interest is compounded each half year. Round your answer to 1 decimal place.

(d) $2800 accumulates to $4500 in 3 years. The percentage rate of interest if it is compounded annually is:

A $\sqrt[3]{\frac{4500}{2800}} - 1$ B $\left(\sqrt[3]{\frac{4500}{2800}} - 1\right) \times 100$

C $\sqrt[3]{\frac{4500}{2800}} \times 100$ D $\frac{4500}{2800} \times 100$

8 Find n to the nearest whole number.

(a) $3.8182 = 1.08^n$ (b) $8.234 = 1.09^n$ (c) $\frac{9328.96}{7900} = (1 + 0.021)^n$

9 Substitute the given values into the compound interest formula $A = P(1 + r)^n$ and solve to find the number of compounding periods n. State your answer to the nearest whole number.

(a) $A = \$38\,175.34$, $P = \$22\,000$, $r = 4.7\%$ p.a. compounded annually

(b) $A = \$25\,966.38$, $P = \$20\,000$, $r = 3.8\%$ p.a. compounded annually

(c) $A = \$3473.53$, $P = \$2900$, $r = 6.2\%$ p.a. compounded annually

(d) $A = \$17\,593$, $P = \$3800$, $r = 8.4\%$ p.a. compounded annually

W.E. 7

10 In each of the following, use the compound interest formula and the method of guess and check (or a CAS) to find the number of years. In each case, assume that interest is compounding annually.

(a) How many years will it take for a $2400 sapphire ring to be worth $8000 if it increases in value by 10.6% p.a.? Round your answer to the nearest whole number of years.

(b) How many years will it take for a $40 000 block of land to be worth $90 000 if it increases in value by 8.5% p.a.? Round your answer to the nearest whole number.

(c) How many years will it take for a $3400 dinner set to be worth $5000 if it increases in value by 9.7% p.a.? Round your answer to the nearest tenth of a year.

Understanding

11 $5000 is invested for 4 years with the interest compounded quarterly. For the first 2 years the interest rate is 4% p.a. and for the next 2 years it is 4.8% p.a.

(a) How much will the investment be worth at the end of the 4 years?

(b) How much interest will be earned?

(c) How much interest would the investment earn if the rate was 4.4% p.a. for the whole 4-year period? Would the investor be better off?

(d) How much interest would be earned if the first 2 years were at 4.8% p.a. and the rest at 4% p.a.?

(e) The conditions of investment change and interest is now compounded annually. What would the interest rate need to be to earn the same amount of interest as found in part **(b)**?

(f) The interest rate is now 5% p.a. compounded annually. After how many years will the investment be at least as much as the result in part **(a)**?

12 Pene's savings are in a bank account that attracts 3.5% p.a. If her savings are to grow from $10 000 to over $20 000 in 20 years, what is the smallest number of compounding periods per year required?

Reasoning

13 A building society offered an interest rate of 7.25% p.a. on any investment over $3000. Mary deposited $3600 and received $5944 after 7 years.

(a) How often was the interest compounded?

(b) What simple interest rate would have given the same return in the same time? Round your answer to 1 decimal place.

(c) Mary was a little disappointed because she'd calculated a return of $5971. Her calculations were based on an incorrect assumption. What was that assumption?

14 If savings in an account are to be *at least* tripled over 20 years with an interest rate of 5.55% p.a., how often should the interest be compounded? Give your answer in months.

15 Jacob is planning an overseas trip that will cost $20 000. He decides to invest some money at 6.5% p.a. compounding quarterly for 1 year. Jacob made the following calculations:

How much do I need to invest?

$P = ?, A = 20\,000, n = 4, r = 0.065$

$$20\,000 = P\left(1 + \frac{0.065}{4}\right)^4$$

$$20\,000 = P(1.02)^4$$
$$20\,000 = P \times 1.08$$
$$P = \frac{20\,000}{1.08}$$
$$P = 18\,518.52$$

Checking:

$A = ?, P = 18\,518.52, n = 4, r = 0.065$

$$A = 18\,518.52\left(1 + \frac{0.065}{4}\right)^4$$

$$A = 18\,518.52(1 + 0.02)^4$$
$$A = 20\,045.04$$

When he collects his money a year later, he is almost $250 below their target. Where did he go wrong?

Open-ended

16 Find different rates and compounding periods that will result in savings being increased by at least 50% over 4 years.

Exploration CAS

Equipment required: TI-Nspire CAS or Casio ClassPad CAS

Investigating compound interest and the 'rule of 72'

The *rule of 72* can be used to estimate how much time is needed for an investment to double in value. Consider the following example.

Trinh buys a block of land for \$100 000. How many years will it take for the land to double in value, if its value increases at a rate of 8% p.a.?

Using the rule of 72, divide 72 by 8 (the interest rate) to get 9. This means that it should take approximately 9 years for the land to double in value. This can be verified using the compound interest formula:

$$
\begin{aligned}
A &= P(1+r)^n \\
&= 100\,000(1+0.08)^9 \\
&\approx 199\,900.46 \\
&\approx \$200\,000 \text{ (to the nearest thousand dollars)}
\end{aligned}
$$

If the rate of increase were 12% p.a., then it would take only 6 years for the block of land to double in value, as $72 \div 12 = 6$.

1 Verify this result using the compound interest formula.

2 Graph the rule of 72 on your CAS, as follows.

Let y represent the number of years needed for an investment to double in value. Let x represent the percentage rate of interest per year.

The relationship between y and x for the rule of 72 is $y = \frac{72}{x}$.

Set your CAS graphing window so that x ranges from 0 to 10 and y ranges from 0 to 80. Use the table of values to inspect values of x and y.

Using TI-Nspire CAS	Using Casio ClassPad CAS
Add **Graphs** to your document and enter the equation **f1(x)=72/x** to be graphed. Then select **menu** > **Table** > **Split-screen Table** to show the table of values.	From the menu select **Graph & Table**. Enter the equation **y1=72/x** and tick the box next to it, then select the Graph icon to view the graph. With the graph highlighted, select the Table icon to show the table of values.

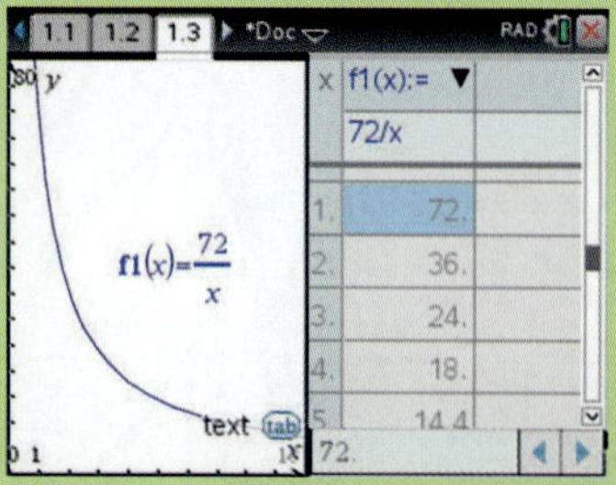

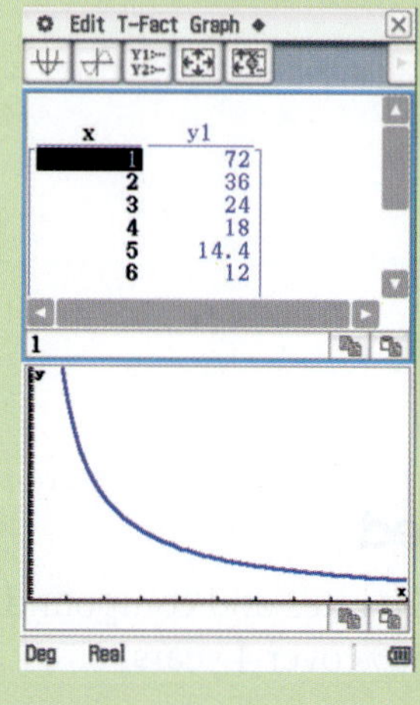

3 Use the table of values to complete the second column of the table below.

Interest rate per annum, x	Years for investment to double in value, y	Actual value of \$100 000 after y years using the compound interest formula $A = P(1 + r)^n$	Previous column subtract \$200 000 (A – \$200 000)
4			
6			
8	9	\$199 900.46	-\$99.54
9			

Between which two values of x do you think the rule of 72 gives a result of \$200 000, to the nearest dollar? Change the table step value until you have the required precision for your answer.

Taking it further

How accurate is the rule of 72?

Investigate the accuracy of the rule of 72 by creating a spreadsheet on your CAS, based on the table above. Let the x-values start at 1 and end at 50, increasing in steps of 1. Add a fifth column for the percentage error. Between which two values of x does the rule of 72 give accurate results? What level of accuracy is acceptable?

Use your CAS to find a precise answer for this.

The equation is $200\,000 = 100\,000(1.08)^n$.

This simplifies to $2 = (1.08)^n$, which can be solved on your CAS.

Using TI-Nspire CAS

From the **Calculator** screen, select **menu** > **Algebra** > **Solve** and enter the equation to be solved, followed by a comma and the variable to solve for.

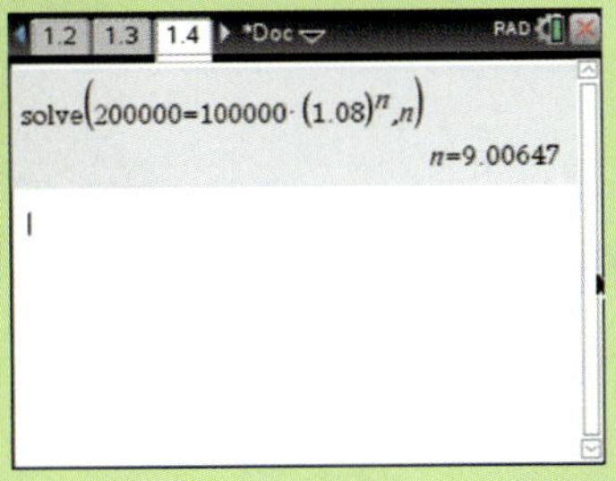

Using Casio ClassPad CAS

From the **Main** menu, select **Interactive** > **Equation/Inequality** > **solve** and enter the equation to be solved and the variable to solve for.

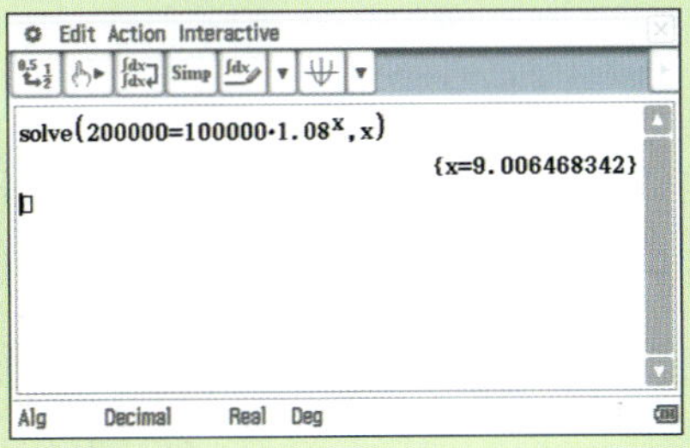

Half-time 13

Round answers to the nearest cent, unless told otherwise.

13.2 **1** If a loan of $7200 has an interest rate of 17% p.a. compounded quarterly over 5 years, the debt will grow to:

A $7200(1+0.17)^5$ **B** $7200\left(1+\frac{0.17}{4}\right)^{20}$ **C** $7200\left(1+\frac{0.17}{4}\right)^{5}$ **D** $7200(1+0.17)^{20}$

13.3 **2** Calculate how much would need to be invested at 6% p.a. compounded monthly to accumulate to $14 000 in 8 years.

13.1 **3** Calculate the total amount owing on a loan of $6000 after 3 years if the 14% interest p.a. is compounded annually.

13.3 **4** $100 000 grows to $400 000 in 1 year. The rate of growth is compounding every 6 months. Find the annual interest rate.

13.2 **5** How much interest is added over 8 years to an initial amount of $55 700, with an interest rate of 8.25% p.a., if the interest is compounded:

(a) half-yearly **(b)** monthly **(c)** daily.

13.3 **6** An item originally valued at $100 000 increased in value by a total of 55.133% over a number of years. Each year it appreciated at a rate of 5% of its value the previous year. Find the number of years, rounded to the nearest whole number.

13.1 **7** Catherine invested $20 000 for 6 years at an interest rate of 7.5% p.a. (Round your answers to the following questions to the nearest dollar.)

(a) Calculate the interest earned on the investment if it was compounded annually.

(b) Calculate the interest earned on the investment if simple interest was paid instead.

(c) Compare the compounded interest with the simple interest.

13.1 **8** A painting was purchased 11 years ago for $26 900. It has just been sold for $78 000. Calculate the flat rate of appreciation (p.a.). Round your answer to 1 decimal place.

Comparing interest rates

13.4

In the previous section, the compound interest formula was rearranged to find unknown values in compound interest situations. The formula can also be used to compare the amount of interest for different compounding time periods.

Consider the following loan: A principal of \$10 000 and annual interest rate of 6%.

If the compounding period is 1 year, then $A = 10\,000(1 + 0.06)^1 = 10\,600$. Calculating the interest earned using $I = A - P$, this gives \$600 interest.

If the compounding period is 6 months, then $A = 10\,000(1 + 0.03)^2 = 10\,609$, which represents \$609 interest. This interest can be calculated as a percentage of the principal, which gives the percentage return on the investment:

$$\frac{609}{10\,000} \times 100 = 6.09\%$$

The following table shows this information for different compounding periods. Note that simple interest over 1 year is the same as compound interest compounded annually for 1 year.

Case	Number of compounding periods	Interest in the first year	Percentage return
A	0 (simple interest)	\$600	$\frac{600}{10\,000} \times 100 = 6\%$ p.a.
B	1 (annual)	\$600	$\frac{600}{10\,000} \times 100 = 6\%$ p.a.
C	2 (6 monthly)	\$609	$\frac{609}{10\,000} \times 100 = 6.09\%$ p.a.
D	4 (quarterly)	\$613.64	$\frac{613.64}{10\,000} \times 100 \approx 6.14\%$ p.a.
E	12 (monthly)	\$616.78	$\frac{616.78}{10\,000} \times 100 \approx 6.17\%$ p.a.

In all these cases the **nominal interest rate** is 6% p.a., but the amount of interest earned is different each time.

To compare the different cases, you can calculate the **effective interest rate**. This is the interest rate that would give the same final amount, if applied as simple interest. This means it is also the same as the percentage return above. The effective interest rate is sometimes called the real interest rate.

In case C above, the effective interest rate is 6.09% p.a., in case D it is 6.14% p.a. and in case E it is 6.17% p.a.

The effective interest rate increases as the interest is compounded more frequently.

For simple interest (not compounded), the effective interest rate is just the same as the nominal rate.

Effective interest rate formula

$$r_{ef} = \frac{I_1}{P}$$

where r_{ef} = effective interest rate (per year) as a decimal
I_1 = amount of interest after 1st year
P = principal

Worked example 8

W.E. 8

Find the effective rate of interest (r_{ef}) as a percentage, if $25 000 is invested for 1 year at the following rates. Where necessary round your answers to 3 decimal places.

(a) a simple rate of 9.5% p.a.

(b) an annually compounding rate of 9.5% p.a.

(c) a monthly compounding rate of 9.5% p.a.

(d) a weekly compounding rate of 9.5% p.a.

Thinking	Working
(a) For simple interest, the effective rate is the same as the nominal rate.	(a) $r_{ef} = 9.5\%$ p.a.
(b) For annual compounding, the effective rate is the same as the nominal rate.	(b) $r_{ef} = 9.5\%$ p.a.
(c) 1 Calculate the total amount after the first year. Record your answer to 4 decimal places.	(c) $A = P(1+r)^n$ $= 25\,000\left(1+\frac{0.095}{12}\right)^{12}$ $= 27\,481.1896$
2 Calculate the amount of interest earned in the first year.	$I = A - P$ $= 27\,481.1896 - 25\,000$ $= 2481.1896$
3 Calculate the effective interest rate.	$r_{ef} = \frac{I}{P}$ $= \frac{2481.1896}{25\,000}$ $= 0.099\,247$
4 Convert to a percentage and round to 3 decimal places.	$0.099\,247 \times 100$ $= 9.925\%$ The effective interest rate is 9.925% p.a.
(d) 1 Calculate the total amount after the first year. Record your answer to 4 decimal places.	(d) $A = P(1+r)^n$ $= 25\,000\left(1+\frac{0.095}{52}\right)^{52}$ $= 27\,489.0887$
2 Calculate the amount of interest earned in the first year.	$I = A - P$ $= 27\,489.0887 - 25\,000$ $= 2489.0887$

3 Calculate the effective interest rate.	$r_{ef} = \frac{I}{P}$ $= \frac{2489.0887}{25\,000}$ $= 0.099\,563$
4 Convert to a percentage and round to 3 decimal places.	$0.099\,563 \times 100$ $= 9.956\%$ The effective interest rate is 9.956% p.a.

The effective interest rate is useful for comparing situations with different nominal rates and different compounding periods.

Converting nominal rates to effective rates

A nominal interest rate can be converted to its effective interest rate using the formula below. Multiplying by 100 converts the rate to a percentage.

Converting from nominal to effective interest rate formula

$$r_{ef} = \left(1 + \frac{r}{n}\right)^n - 1$$

where r_{ef} = effective interest rate (per year) as a percentage
r = nominal interest rate as a decimal
n = number of compounding periods (per year)

This formula is useful because you only need to know the nominal interest rate and the number of compounding periods. You do not need to know the principal or the amount of interest earned.

Worked example 9

W.E. 9

A loan attracts an interest rate of 23% compounded daily. Find:

(a) the nominal rate r and the number of compounding periods, n

(b) the effective rate r_{ef} using the values found in **(a)**.

Thinking	Working
(a) 1 The nominal rate is the stated rate.	(a) $r = 23\%$
2 The number of compounding periods will be the number of days in a year.	$n = 365$
(b) 1 Convert the nominal rate to a decimal.	(b) $r = 23\% = 0.23$
2 Write the formula.	$r_{ef} = \left(1 + \frac{r}{n}\right)^n - 1$
3 Substitute the decimal value of r and the value of n into the formula.	$r_{ef} = \left(1 + \frac{0.23}{365}\right)^{365} - 1$
4 Calculate r_{ef} as a percentage.	$= 0.2585$ $= 25.85\%$ p.a.

13.4 Comparing interest rates

Navigator

Answers p. 868

1, 2, 3, 4, 5, 6, 7, 9, 10	1, 2, 3, 4, 5, 6, 7, 8, 9, 10, 13	1 (a–b), 2, 3, 4, 6, 7, 8, 9, 10, 11, 12, 13

Fluency

1 Substitute the following values into the effective interest rate formula $r_{ef} = \frac{I_1}{P}$ to find the value of r_{ef}. State answers correct to 1 decimal place.

(a) $I_1 = \$372, P = \$4400$

(b) $I_1 = \$589, P = \$4900$

(c) $I_1 = \$623, P = \$35\,000$

(d) $I_1 = \$1935, P = \$62\,000$

W.E. 8

2 For each of the following find the effective rate of interest (r_{ef}).

(a) \$40 000 invested at a simple interest rate of 7% p.a.

(b) \$10 000 invested at an annual compounding rate of 9.4% p.a.

(c) \$100 000 invested at a monthly compounding rate of 12.75% p.a.

(d) \$35 000 borrowed at a weekly compounding rate of 19.7% p.a.

(e) \$20 000 borrowed at a fortnightly compounding rate of 16.5% p.a.

(f) \$200 000 invested at a rate of $6\frac{1}{4}$ % p.a. compounded every 5 days.

3 Substitute the following values into the effective interest rate formula $r_{ef} = \left(1 + \frac{r}{n}\right)^n - 1$ to find the value of r_{ef}, the effective interest rate. State your answers correct to 1 decimal place.

(a) $r = 0.125, n = 5$

(b) $r = 0.067, n = 15$

(c) $n = 20, r = 9.4\%$ p.a.

(d) $n = 19, r = 17.8\%$ p.a.

(e) $r = 22.4\%$ p.a. compounded quarterly

(f) $r = 9.5\%$ p.a. compounded monthly

4 The effective interest rate that corresponds to 5% p.a. compounded every 2 months is:

A $\left(1 + \frac{0.05}{2}\right)^2 - 1$

B $\left(1 + \frac{0.05}{6}\right)^6 - 1$

C $\left(1 + \frac{0.5}{6}\right)^6 - 1$

D $\left(1 + \frac{5}{2}\right)^2 - 1$

W.E. 9

5 For each of the following, find:

(i) the nominal rate r and the number of compounding periods n

(ii) the effective rate r_{ef}, using the values found in (i).

(a) 7.3% p.a. compounded 5 times a year

(b) 15.5% p.a. compounded monthly

(c) 21% p.a. compounded quarterly

(d) $16\frac{1}{4}$ % p.a. compounded every 4 months

(e) daily compounding rate of $20\frac{1}{3}$ % p.a.

(f) half-yearly compounded rate of 5.25% p.a.

Understanding

6 Two banks are currently offering the following rates on their investment savings accounts:

C Bank: 15% p.a. compounded quarterly
D Bank: 14.9% p.a. compounded daily

(a) Find the effective rates being offered by the two banks.

(b) Which bank is offering the better deal?

7 Ned needs to buy a new laptop. He is planning to take out a personal loan with an interest rate of 20.5% p.a. compounded quarterly. The alternative is to use his credit card that has an interest rate of 20.1% compounded daily.

(a) What is the effective rate of the personal loan?

(b) What is the credit card's effective rate?

(c) Which of the two options should Ned choose and why?

8 A bed can be purchased on a 'get it now, pay in 12 months' plan. The interest payment options are different depending on whether the purchase is made online or in store. The online purchase attracts a flat rate interest of 25% p.a. The interest on the in-store purchase is 23% p.a. compounded daily.

(a) What is the real or effective interest rate charged on an in-store purchase?

(b) Which is the cheaper option?

Reasoning

9 The following represent investment options. In each case, choose the better option *without calculating the effective rate* and justify your choice.

(a) 12% p.a. compounded monthly
15% p.a. compounded daily

(b) 10.5% p.a. compounded weekly
10.5% p.a. compounded monthly

(c) 11% p.a. compounded daily
12% p.a. compounded daily

(d) 15% p.a. flat rate for 2 years
15% p.a. compounded annually for 2 years

10 The options in Question **9** now represent loans. Will your choice of option change? Justify your answer.

11 Simon has saved the 10% deposit for a new car. The car company can offer him an interest-free loan on the remaining balance for 6 months and then 15% p.a. compounded daily. Alternatively, he could borrow the money elsewhere for a flat rate of 9% p.a.

(a) If he pays out the loan plus interest in a year, which option will be cheaper?

(b) If he pays out the loan plus interest in two years, which option will be cheaper?

Open-ended

12 'Nominal interest rate' and 'simple interest rate' are the same thing. Explain with an example why this statement is untrue.

13 Two investments have different nominal interest rates and different compounding periods. However, the effective interest rate for both is exactly 46.41% p.a. What might be the nominal rates and respective compounding periods for these two investments?

Maths 4 Real

Fantastic plastic?

The convenience of credit cards, coupled with the interest charged on them, has caused many people to accumulate enormous debts that they are unable to repay. In December 2015, the total amount owed by Australians on credit cards was more than $52 billion. There were approximately 16 million credit cards in circulation, and the average debt per credit card holder was $4392.

Q: What is credit?

A: 'Credit' is a way of buying goods or services without actually using cash. When you use a credit card to make a purchase, the credit card provider pays the cost, and then sends you a bill for your purchases at the end of the month.

Q: What happens if you don't pay your credit card bill by the due date?

A: If you do not pay your credit card bill in full by the due date, the bank will charge you interest: a percentage of the cost of your purchases is added to the bill.

Q: What is a debit card?

A: A debit card can be used in the same way as a credit card, except that it uses money you already have in your bank account instead of credit. Money is electronically transferred from your account to whoever is being paid.

1 List some of the advantages and disadvantages of using a debit card instead of a credit card.

Q: How is credit card interest calculated?

A: All credit cards have an annual percentage rate of interest (APR). This interest rate varies from card to card, but is usually between 15% and 20%. Periodically (usually once a month), the credit card provider issues a statement that lists all the purchases made with the card during the statement period. Interest is calculated at the end of the period and added to the amount owing.

Credit card interest is calculated as follows.

- The balance (amount owing) at the end of each day of the statement period is found. These daily balances are added together and divided by the number of days in the statement period to find the 'average daily balance', or ADB.
- A daily interest rate is calculated by dividing the annual rate by 365.
- The ADB is multiplied by the daily interest rate and by the number of days in the statement period to calculate the interest for that period.

2 Jamie applied for a credit card with an APR of 18%. The credit card statement after the first month is shown below.

Calculate the interest that would be charged on Jamie's account if the bill is not paid in full by the due date. (See if you can find a short cut to your calculations.)

Eastpac Bank

Account Number: 3478 4421 0097 1749

Statement Period: 14/5/16 to 13/6/16

Date	Purchase	Amount	Balance
	Opening balance		**$0.00**
14 May	2 movie tickets	$34.00	$34.00
19 May	1 pair jeans and 2 T-shirts	$92.90	$126.90
23 May	Camera	$181.00	$307.90
30 May	Laptop	$799.00	$1106.90
2 June	Dinner with friends	$33.00	$1139.90
7 June	Phone bill	$54.96	$1194.86
11 June	Petrol	$68.00	$1262.86
	Closing balance		**$1262.86**

3 (a) Calculate Jamie's minimum monthly repayment, if it is 5% of the closing balance.

(b) Calculate how much of the minimum repayment amount is interest.

(c) If Jamie only pays the minimum repayment amount, how much will this reduce the credit card balance?

(d) Discuss how only paying the minimum amount each month can create a problem with credit card debt.

Research

Gather some information on three or four different credit cards available. Find out the minimum requirements needed to apply for the card (such as age and income), the interest rate, any annual or monthly fees, whether there is an interest-free period, and any other special rewards or deals offered with the card.

Present your findings as a visual display. Which card would you recommend to someone who would like a credit card, but does not want a growing debt? What additional advice would you give them?

Interest-free periods

Many credit cards offer an 'interest free' period of up to 55 days. This is usually the length of time between the first day of the statement period and the due date for repayment. If the total balance is paid before the due date, then interest is not charged.

Minimum monthly repayments

Credit card holders usually do not require you to pay the full balance shown on your statement. Instead, a 'minimum monthly repayment' amount must be paid by the due date (or risk a 'late payment fee'). This amount is usually about 5% of the balance. However, not all of the minimum repayment will go towards paying off the principal of the debt (the actual amount owing). Some of the minimum repayment amount is only paying off the interest on the debt.

To calculate how much is interest:

- find the monthly interest percentage rate by dividing the APR by 12
- calculate this monthly interest percentage amount of the minimum repayment. This represents the interest in the repayment. The rest is the amount that will actually be deducted from your credit card balance (the principal).

13.5 Depreciation

Depreciation is a decrease in cost or value. Most of the objects around us are disposable items that wear out or are used up and so they depreciate in value. This includes clothes, most furniture, cars, computers, televisions, buildings, mobile phones, hardware and machines.

Depreciation is the opposite to appreciation, so its formulas are closely related to those used for appreciation, introduced earlier in section 13.1.

- The original purchase price of an item is called the **initial cost** or **prime cost**.
- As something depreciates, the amount of value lost is called the **depreciation value**.
- The value after depreciation is called the **written-down value** or **adjusted value**.

Straight-line depreciation

Straight-line depreciation or flat-rate depreciation is when items lose a constant amount of value each year. The rate of this depreciation can be written as a percentage or decimal. With straight-line depreciation, the written-down value will eventually be zero.

Consider the following situation:

A company buys a truck for $75 000. Each year the truck depreciates at 20% of its prime cost.

The depreciation each year is 20% of $75 000
= 0.2 × 75 000
= $15 000

The total depreciation after 3 years will be 20% of $75 000 multiplied by 3
= 0.2 × 75 000 × 3
= $45 000

The written-down value after 3 years will be the difference
= $75 000 – $45 000
= $30 000

After another 2 years, the written-down value will reach zero.

When dealing with straight-line depreciation:

- total depreciation = prime cost × rate of depreciation × time
- written-down value = prime cost – total depreciation

These relationships are similar to the formulas for simple interest.

Straight-line depreciation formula

$$D_T = PrT$$

where D_T = total depreciation after T years, in dollars
P = prime cost, in dollars
r = rate of depreciation per year as a decimal
T = time in years

Also:

$$V_T = P - D_T$$

where V_T = written-down value after T years, in dollars

Worked example 10

W.E. 10

The prime cost of a washing machine 2 years ago was $1950. It depreciates at 18.5% p.a. Assuming straight-line depreciation what is the:

(a) total depreciation since the time of purchase

(b) current written-down value of the machine?

Thinking	Working
(a) 1 Identify what needs to be found.	(a) $D_2 = ?$
2 Write the formula for straight line depreciation.	$D_T = PrT$
3 Identify the values of P, r and T.	$P = 1950$, $r = 18.5\% = 0.185$, $T = 2$
4 Substitute these values into the formula, and calculate D_2 (the total depreciation after 2 years).	$D_2 = 1950 \times 0.185 \times 2$ $= \$721.50$
(b) 1 Identify what needs to be found.	(b) $V_2 = ?$
2 Write the formula.	$V_T = P - D_T$
3 Identify the values of P and D_2.	$P = 1950$, $D_2 = 721.50$
4 Substitute these values into the formula, and calculate V_2 (the written-down value after 2 years).	$V_2 = 1950 - 721.50$ $= \$1228.50$

Reducing balance depreciation

Reducing balance depreciation is when items depreciate by a smaller amount each year. Each year, the amount of depreciation is calculated as a percentage of the previous year's written-down value. As the value of the item decreases, so the corresponding depreciation also decreases. With reducing balance depreciation, the written-down value will never reach zero.

Consider again the situation of the company that purchased a truck for $75 000 with a depreciation of 20% p.a. This time, depreciation is calculated on the written-down value each year rather than on the prime cost.

Initial value = $75 000

Year	Depreciation	Total depreciation	Written-down value
1	20% of $75 000 = 0.2 × 75 000 = $15 000	$15 000	75 000 − 15 000 = $60 000
2	0.2 × 60 000 = $12 000	15 000 + 12 000 = $27 000	60 000 − 12 000 = $48 000
3	0.2 × 48 000 = $9600	27 000 + 9600 = $36 600	48 000 − 9600 = $38 400
4	0.2 × 38 400 = $7680	36 600 + 7680 = $44 280	38 400 − 7680 = $30 720
5	0.2 × 30 720 = $6144	44 280 + 6144 = $50 424	30 720 − 6144 = $24 576

The table of calculations makes it easier to see that each written-down value is 80% of the previous written-down value. So, instead of calculating 0.20 × \$75 000 and subtracting that from \$75 000 to get \$60 000, you can just calculate 0.80 × \$75 000 to get \$60 000.

The written-down values can be rewritten as shown below. This makes it easier to see that this is similar to compound interest.

Year	Written-down value (\$)
0	75 000
1	$0.8 \times 75\,000$
2	$0.8 \times (0.8 \times 75\,000) = (0.8)^2 \times 75\,000$
3	$0.8 \times (0.8 \times (0.8 \times 75\,000)) = (0.8)^3 \times 75\,000$
4	$0.8 \times (0.8 \times (0.8 \times (0.8 \times 75\,000))) = (0.8)^4 \times 75\,000$
5	$0.8 \times (0.8 \times (0.8 \times (0.8 \times (0.8 \times 75\,000)))) = (0.8)^5 \times 75\,000$

After T years the written-down value is $75\,000(0.8)^T$, which can also be written as $75\,000(1 - 0.2)^T$. This is similar to the compound interest formula, except with a subtraction instead of an addition in the brackets.

Reducing balance depreciation formula

$$V_T = P(1 - r)^T$$

where V_T = written-down value after T years, in dollars
P = prime cost, in dollars
r = rate of depreciation per year as a decimal
T = time in years

This formula assumes that depreciation is annual. If it is not calculated annually, then the rate of depreciation needs to match the time period (just as for compounding interest, where the interest rate needs to match the compounding period).

Worked example 11

W.E. 11

The prime cost of a lawn mower 7 years ago was \$5600. Assuming a reducing balance depreciation of 9% p.a., calculate the:

(a) current written-down value of the mower

(b) total depreciation since the time of purchase.

Thinking	Working
(a) 1 Identify what needs to be found.	(a) $V_7 = ?$
2 Write the formula for reducing balance depreciation.	$V_T = P(1 - r)^T$
3 Identify the values of P, r and T.	$P = 5600$, $r = 9\% = 0.09$, $T = 7$
4 Substitute these values into the formula and calculate V_7.	$V_7 = 5600(1 - 0.09)^7$ $= \$2893.86$

(b)	**1** Identify what needs to be found.	(b) $D_7 = ?$
	2 Write the formula.	$D_T = P - V_T$
	3 Identify the values of P and V_7.	$P = 5600, V_7 = 2893.86$
	4 Substitute these values into the formula and calculate D_7.	$D_7 = 5600 - 2893.86$ $= \$2706.14$

Unit cost depreciation and total possible depreciation

- The **expected life** of an item is the expected amount of time, or the expected amount of work done, before it will no longer be a useful working item (e.g. the number of pages copied for a photocopier or the number of photos taken for a camera).
- The **scrap value** is the value an item may still have at the end of its expected life. This is sometimes called the **resale value**, **salvage value** or **residual value**.
- The **total possible depreciation** of an item is the difference between its original value and its scrap value.
- The **unit cost depreciation** of an item is the rate of depreciation per unit over its expected life (that is, depreciation per unit of time or depreciation per unit of work).

Total possible depreciation = prime cost − scrap value

$$\text{Unit cost depreciation} = \frac{\text{total possible depreciation}}{\text{expected life}}$$

Depreciation (after a certain amount of use)
= Unit cost depreciation × units of work done or unit of time elapsed

Worked example 12

W.E. 12

A laptop is expected to work for 6000 hours before it has a resale value or scrap value of \$100. The purchase price is \$1300. Find:

(a) the total possible depreciation

(b) the unit cost depreciation

(c) the depreciation after 1000 hours of use

(d) the written-down value after 1000 hours of use.

Thinking		Working
(a)	**1** Write the formula.	(a) Total possible depreciation = prime cost − scrap value
	2 Identify the prime cost and scrap value and substitute into the formula.	Prime cost = \$1300 Scrap value = \$100 Total possible depreciation = \$1300 − \$100
	3 Calculate the total depreciation.	= \$1200

(b) 1 Write the formula.	(b) Unit cost depreciation $= \frac{\text{total possible depreciation}}{\text{expected life}}$
2 Identify the total depreciation and expected life and substitute into the formula.	Total depreciation = \$1200 Expected life = 6000 hours Unit cost depreciation $= \frac{1200}{6000}$
3 Calculate the unit cost depreciation.	$= 0.2$ The laptop decreases in value by 20 cents for every hour of work.
(c) 1 Write the formula.	(c) Depreciation after use = unit cost depreciation × number of units of work done
2 Identify the unit cost depreciation and work done, and substitute into the formula.	Unit cost depreciation = \$0.20 per hour, number of units of work done = 1000
3 Calculate the depreciation after the given use.	Depreciation after use $= 0.2 \times 1000$ = \$200
(d) 1 Write the formula.	(d) $V = P - D$
2 Identify the values of P (original value) and D (accumulated depreciation) and substitute into the formula.	$P = \$1300$, $D = \$200$ $V = \$1300 - \$200$
3 Calculate V, the written-down value after the given use.	$V = \$1100$

13.5 Depreciation

Navigator

Answers p. 868

1, 2, 3, 4, 5, 6, 7, 8, 9, 10, 11, 12, 13, 14, 15, 16, 17, 20	1, 2, 3 (a–f), 4, 5, 6 (a–b), 7, 8, 9, 10, 11, 12, 13, 14, 16, 17, 18, 20	1 (a), 2 (a), 3 (d–h), 4, 5, 6 (a), 7, 9, 10, 11, 12, 13, 16, 17, 18, 19, 20, 21

Fluency

1 Substitute the following values into the straight line depreciation formula, $D_T = PrT$, to find the value of D_T, the total depreciation.

(a) $P = \$25\,000$, $r = 5.4\%$, $T = 5$ **(b)** $P = \$3400$, $r = 2.2\%$, $T = 15$

2 Substitute the following values into the written-down value formula $V_T = P - D_T$ to find the written-down value V_T.

(a) $P = \$25\,000$, $D_T = \$6750$ **(b)** $P = \$3400$, $D_T = \$1122$

3 For each of the following, assume straight-line depreciation. Find: **W.E. 10**

(i) the total depreciation since the time of purchase

(ii) the current written-down value.

Round all answers to the nearest cent.

(a) The initial price 5 years ago was \$30 000. $r = 7\%$ p.a.

(b) The prime cost 12 years ago was \$47 500. $r = 5\%$ p.a.

(c) The purchase price 20 years ago was \$100 000. $r = 5\%$ p.a.

(d) The initial cost 23 years ago was \$790 000. $r = 4.25\%$ p.a.

(e) The prime cost 4 years ago was \$145 000. $r = 10.5\%$ p.a.

(f) The initial price $5\frac{1}{2}$ years ago was \$300 000. $r = 6.5\%$ p.a.

(g) The purchase price 6 months ago was \$47 500. $r = 20\frac{3}{4}\%$ p.a.

(h) The prime cost 18 months ago was \$152 700. $r = 33\frac{1}{4}\%$ p.a.

4 Substitute the following values into the reducing balance depreciation formula $V_T = P(1 - r)^T$ to find the written-down value V_T.

(a) $P = \$5000, T = 2, r = 7.2\%$ p.a.

(b) $P = \$20\,000, T = 6, r = 2.6\%$ p.a.

5 Repeat Question **3** with reducing balance depreciation. Find: **W.E. 11**

(i) the current written-down value

(ii) the total depreciation since the item was purchased.

Round all answers to the nearest cent.

6 Substitute the following values into the total possible depreciation formula:
Total possible depreciation = prime cost − scrap value

(a) Prime cost = \$5200, Scrap value = \$2734

(b) Prime cost = \$37 800, Scrap value = \$24 378

(c) Prime cost = \$18 652, Scrap value = \$15 672

7 (a) A machine is expected to work for 3000 hours before it has a resale or scrap value of \$150. The purchase price is \$6150. Find: **W.E. 12**

(i) the total possible depreciation

(ii) the unit cost depreciation

(iii) the depreciation after 10 hours of use

(iv) the written-down value after 10 hours of use.

(b) A machine is expected to work for 2000 hours before it has no scrap value. The purchase price is \$1000. Find:

(i) the total possible depreciation

(ii) the unit cost depreciation

(iii) the depreciation after 1000 hours of use

(iv) the written-down value after 1000 hours of use.

Understanding

8 What would be the total depreciation after 6 years on a \$780 workstation if it is depreciated at 11.25% p.a. of its prime cost?

9 A car costs \$24 000 new. The value of the car in 3 years, if it depreciates at 25% p.a. of its prime cost each year, will be:

A $\$\dfrac{24\,000 - 24\,000 \times 0.75}{3}$

B $\$24\,000 - 24\,000 \times 0.25 \times 3$

C $\$24\,000 \times 0.75^3$

D $\$24\,000 \times 0.25 \times 3$

10 Calculate the value of a computer (initial cost \$5600) after 4 years if it is depreciated at 15% p.a of its prime cost each year.

11 Calculate the value of a computer (cost \$5600) after 4 years if it is depreciated at 15% p.a. of its written-down value at the start of each year. Compare this answer with that found for Question **10**.

12 Calculate the depreciation in each year, the total depreciation and the written-down or adjusted value for each of the first 3 years for a \$45 000 forklift that depreciates at 20% of its prime cost each year.

13 Calculate the depreciation, total depreciation and written-down value for each of the first 3 years if a \$45 000 forklift is depreciated at 20% p.a. of its written-down value at the start of each year. Compare this answer with that found in Question **12**.

14 A \$7500 machine has a scrap value of \$300, and is guaranteed for 11 000 cycles. In a year when it operates for 2630 cycles the amount of depreciation, in dollars, is:

A $\dfrac{7500 - 2630}{11\,000} \times 300$

B $\dfrac{11\,000 - 7500}{2630} \times 300$

C $\dfrac{7500 - 300}{11\,000} \times 2630$

D $\dfrac{7500}{11\,000} \times 2630$

15 **(a)** The amount of depreciation (in dollars) for a \$5000 machine with no scrap value, guaranteed for 8000 hours, in its first year when it works for 1350 hours, is:

A $\dfrac{8000}{1350} \times 5000$

B $\dfrac{8000}{5000} \times 1350$

C $\dfrac{5000}{6650} \times 1350$

D $\dfrac{5000}{8000} \times 1350$

(b) At the end of the first year the written-down value (in dollars) of the machine described in part **(a)** is:

A $5000 - \dfrac{5000}{8000} \times 1350$

B $6350 - \dfrac{8000}{5000} \times 1350$

C $8000 - \dfrac{5000}{8000} \times 1350$

D $8000 - \dfrac{1350}{8000} \times 5000$

Reasoning

16 Calculate the value of an \$11 500 screen printer with a scrap value of \$200 guaranteed for 90 000 prints when it is used 12 400 times.

17 Calculate the value of a \$3500 photocopier with a scrap value of \$400 guaranteed for 600 000 copies after it produces 210 000 copies.

18 A camera costs \$3500, has a scrap value of \$900 and is guaranteed for 8000 photos. Calculate its written-down value at the end of:

(a) the first year after taking 450 photos

(b) second year after taking 1800 photos.

19 A $30 000 car depreciates at 20% p.a. of its written-down value every year for the first 3 years. An $80 000 car depreciates at 25% p.a. of its written-down value every year for the first 3 years. After that time, both cars depreciate at 10% p.a. of their written-down value at the start of each year for every subsequent year.

(a) What is the difference in their values:

(i) at purchase (ii) after 1 year (iii) after 2 years (iv) after 3 years?

(b) When will the difference in their values be under $3000?

Open-ended

20 A boat was initially purchased for $10 000. It is now worth only $3000. Assuming that it depreciated at the same rate of its prime cost every year, what was the possible rate of decrease and corresponding elapsed time?

21 Repeat Question **20** assuming the depreciation followed a reducing-balance model.

Problem solving

Weird and wonderful numbers

When you've been up and down the stairs several times, take time out and try to answer these questions:

1 What did you find at the bottom of the stairs? What number(s) sent you back to the top?

2 Why is there a special condition attached to step one?

Strategy options

- Act it out.

3 Will you ever find something else at the bottom of the stairs?

Strategy options

- Seek an exception.
- Solve a simpler problem.

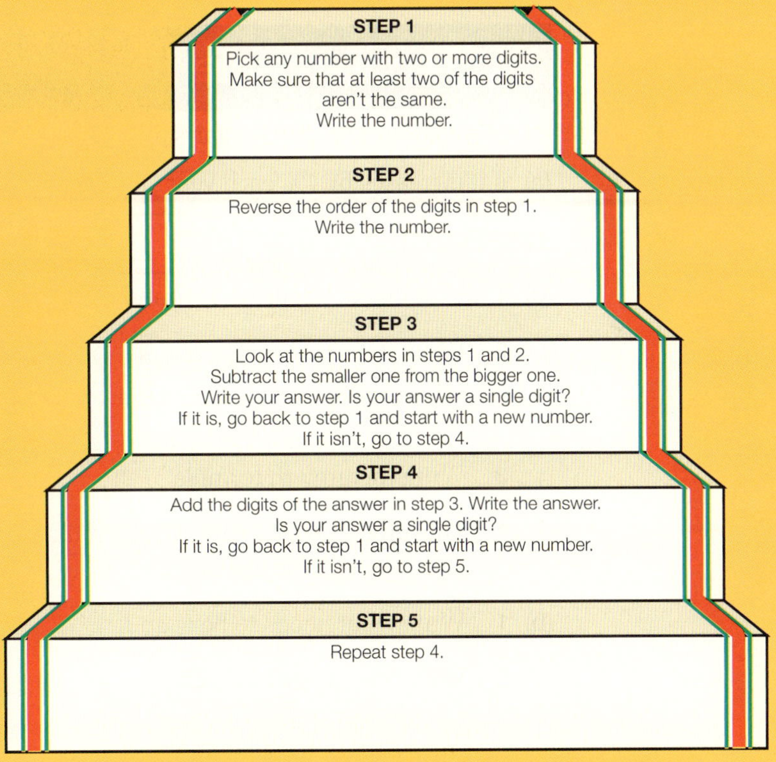

13.6 Growth and decay

Growth

In previous sections, the formula $A = P(1 + r)^n$ has been used to calculate compound interest. This formula can also be used as a model for other values that grow similarly to compound interest, such as population growth or the spread of diseases.

This common pattern of growth, where the amount of growth increases with every time period, is **exponential growth**.

> Exponential growth uses the formula $A = P(1 + r)^n$.
>
> A is the amount after a number of time periods, n.
>
> P is the initial amount. This is the value of A when $n = 0$.
>
> r is the growth rate, which is applied to the amount A from one time period to the next.

Worked example 13 — W.E. 13

The population of a district starts at 32 000. What will the population be in 7 years:

(a) if it increases by 4% each year

(b) if it has a growth rate of 1.2% per month?

Thinking	Working
(a) 1 Write what needs to be found.	(a) $A = ?$
2 Write the values of P, r and n.	$P = 32\,000$ $r = 4\%$ per year $= 0.04$ $n = 7$ years
3 Write the growth formula.	$A = P(1 + r)^n$
4 Substitute the values into the formula and calculate the answer.	$A = 32\,000(1 + 0.04)^7$ $= 42\,110$
(b) 1 Write what needs to be found.	(b) $A = ?$
2 Write the values of P, r and n. Convert the unit for time so it is consistent with the unit for rate.	$P = 32\,000$ $r = 1.2\% = 0.012$ per month $n = 7$ years $= (7 \times 12) = 84$ months
3 Write the growth formula.	$A = P(1 + r)^n$
4 Substitute the values into the formula and calculate the answer.	$A = 32\,000(1 + 0.012)^{84}$ $= 87\,159.158\,51$
5 Round down to the nearest whole number. (You cannot include less than a whole person.)	The population will be 87 159.

Decay

In previous sections the formula $V_T = P(1 - r)^n$ was used to calculate depreciation. This formula can also be used as a model for other values that decrease similarly to depreciation, such as radioactive decay, declining populations of endangered species, or the participants eliminated in 'knockout' sports tournaments.

This common pattern of decay, where the amount of decrease is smaller with every time period, is **exponential decay**.

Exponential decay uses the formula $A = P(1 - r)^n$.

A is the amount after a number of time periods, n.

P is the initial amount. This is the value of A when $n = 0$.

r is the decay rate, which is applied to the amount A from one time period to the next.

Worked example 14

W.E. 14

The initial number of subscribers to a magazine was 12 469. How many subscribers will there be after 9 months if the number of subscribers decreases by:

(a) 1% each month **(b)** 0.25% per day?

Assume 30 days in a month.

Thinking	Working
(a) 1 Identify what needs to be found.	(a) $A = ?$
2 Write the values of P, r and n.	$P = 12\,469$ $r = 1\%$ per month $= 0.01$ $n = 9$ months
3 Write the decay formula.	$A = P(1 - r)^n$
4 Substitute the values into the formula and calculate the answer.	$A = 12\,469(1 - 0.01)^9$ $= 11\,391$
(b) 1 Identify what needs to be found.	(b) $A = ?$
2 Write the values of P, r and n and change the unit for time so it is consistent with the unit for rate.	$P = 12\,469$ $r = 0.25\%$ per day $= 0.0025$ $n = 9$ months $= (9 \times 30) = 270$ days
3 Write the decay formula.	$A = P(1 - r)^n$
4 Substitute the values into the formula and calculate the answer.	$A = 12\,469(1 - 0.0025)^{270}$ $= 6343.308\,039$
5 Round up to the nearest whole number. (You cannot include less than a whole person.)	The number of subscribers will be 6344.

Combining the growth and decay formulas

The exponential growth formula has this basic structure:

Total = Initial number × (Number greater than 1)^{Time}
$A = P \times (1 + r)^n$

The exponential decay formula has this basic structure:

Total = Initial number × (Number between 0 and 1)^{Time}
$A = P \times (1 - r)^n$

These structures are very similar and can be combined into one general formula.

Exponential growth or decay formula

$$A = PR^n$$

where A = the total after n time periods
P = the initial amount (when $n = 0$)
R = the growth or decay factor
$R > 1$ for growth
$0 < R < 1$ for decay

Note that for each particular growth or decay situation:

- n and A are variables. Variables have different values and as one changes, so does the other.
- P and R are constants. These have the same fixed values, which define the conditions of the situation.

This exponential formula can be rewritten to be graphed on a Cartesian plane, by renaming the variables n and A to be instead x and y. In this case, the constants P and R are usually also renamed as k and a.

$$y = ka^x$$

Interactive

Exponential growth and decay

Observe how the exponential curve grows or decays as the values change on the graph.

Go to the eBook or the Pearson Places website to access this interactive.

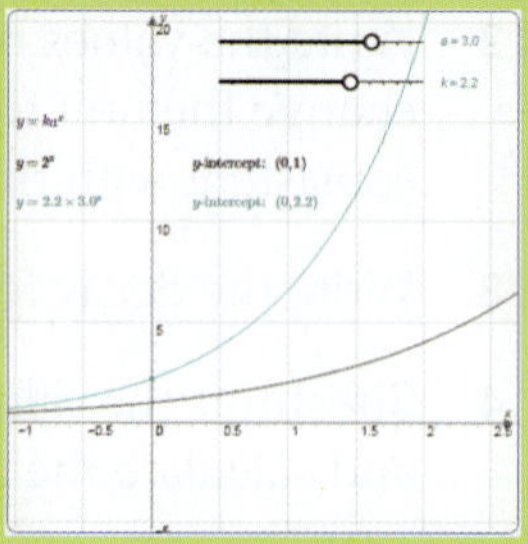

Worked example 15

W.E. 15

The number of bacterial cells (N) in a kitchen sink is increasing with time (T) hours. The bacteria count for the first 3 hours is shown below.

T (hours)	0	1	2	3	4	5	6
N	6	12	24	48			

(a) Copy and complete the table.

(b) Find the formula that models this growth.

(c) Use the formula to predict the number of bacteria in the sink after 3 hours and 20 minutes.

(d) Use the table of values to plot a graph of N against T. Join the points with a smooth curve.

(e) Use the graph to predict the approximate time when there will be 240 bacteria in the sink.

Thinking	Working
(a) Identify the doubling pattern in the values of N and complete the table.	(a) 4, 5, 6 → 96, 192, 384
(b) 1 Identify the variables.	(b) x is T = time in hours y is N = total number of bacteria
2 Identify the constants. (k is the initial value of N.) (a is the growth factor. Find the growth factor by dividing any value of N by the value of N before it.	When $T = 0$, $N = 6$ so $k = 6$ $a = \frac{12}{6} = 2$
3 Write the general form of the growth equation.	$y = ka^x$
4 Substitute the variables and constants for this question into the general form.	$N = 6 \times 2^T$
(c) 1 Convert hours and minutes into hours.	(c) $T = 3$ hours 20 minutes $= 3\frac{1}{3}$ hours
2 Substitute into the formula and find N. Round down to the nearest whole number.	$N = 6 \times 2^{\frac{4}{3}}$ $= 60.4762$ ≈ 60 bacteria

Table for (a):

4	5	6
96	192	384

(d) Plot points, using T for the horizontal axis and N for the vertical axis.

(d)

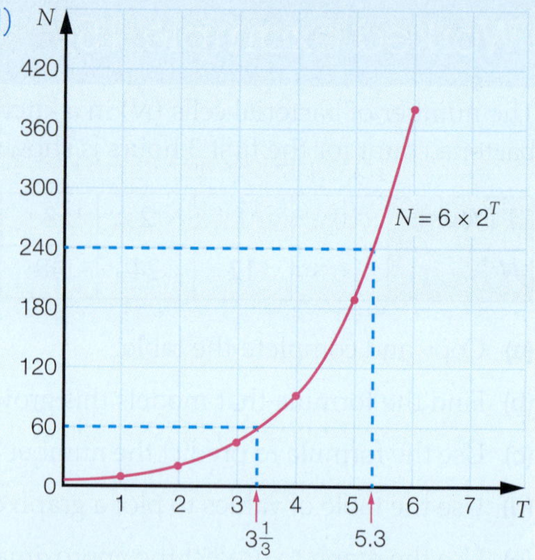

(e) Using the graph, draw a horizontal line from the given values on the vertical axis to the curve. Then draw a vertical line from this point on the curve to the horizontal axis to give the corresponding value.

(e) Approximate time = 5.3 hours

Note: The formula $N = 6 \times 2^T$ can be written in the other form of the growth formula as $N = 6(1 + 1)^T$. From this, it can be seen that $R = 1 = 100\%$. Doubling the value of N in the table, as in part **(a)**, is the same as increasing the value of N by 100%. In other words, the bacterial cells are increasing at a rate of 100% per hour.

The graph in part **(d)** is an exponential growth curve. It increases slowly at first and then very rapidly or exponentially.

An *exponential decay* curve has a similar shape, but curves downwards as its values decrease rapidly, then more slowly, as shown.

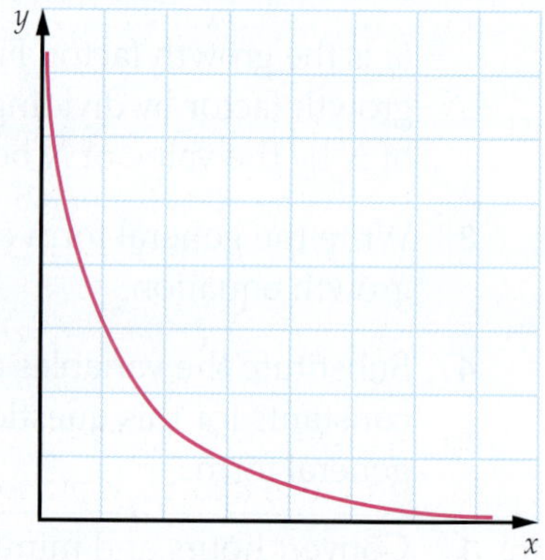

13.6 Growth and decay

Navigator

Answers p. 869

1, 2, 3, 4, 5, 6, 7, 8, 9, 10, 12, 13	2, 3, 4, 5, 6, 7, 8, 9, 10, 11, 12, 13	2, 3 (d–f), 4, 5 (d–f), 6, 7, 8, 9, 10, 11, 12, 13, 14

Fluency

1 Decide whether each of the following situations could be modelled by an exponential growth or decay formula.

(a) A population of wild rabbits doubles each month.

(b) A painting's value depreciates each year at a rate of 15%.

(c) A sample of radioactive material emits radiation as it decays.

2 Substitute the given values into the exponential growth formula for each situation below. Do not evaluate A.

(a) A population of fish in Port Phillip Bay is currently 2503 and is increasing by 12.6% each year.

(b) A vintage car is bought for \$125 000 and increases in value at a rate of 5.2% p.a.

(c) The population of a country town is 2750 in 2016 and increasing by 2% per year.

3 (a) If the population in a certain town is 10 200 and has a growth rate of 6% each year, what is the estimated population after 12 years? **W.E. 13**

(b) A colony of sea birds increases at a rate of 90% a year. If there are 8000 birds in the colony to start with, how many will there be after 1 year?

(c) The population of birds in an aviary has a growth rate of 150% p.a. If the aviary originally has 30 birds, how many birds will there be after 4 years?

(d) A house was purchased $7\frac{1}{2}$ years ago for \$190 000. Houses in this area have grown in value by 22.6% p.a. What is the estimated value of the house now, to the nearest dollar?

(e) A population increases at a rate of 1.2% per month. If the initial population is 200 000, what will be the population after 16 months?

(f) Joseph currently has 976 friends on Facebook and the number of friends is growing at a rate of 5% each year. How many friends can he expect to have after 21 months?

4 Substitute the given values into the exponential decay formula for each situation below. Do not evaluate A.

(a) A new car costs \$25 590 and its value depreciates by 17% p.a.

(b) An endangered species has a population of 3623 and is decreasing by 2.4% p.a.

(c) A new computer depreciates by 11.2% p.a. after it was originally bought for \$2400.

5 (a) The population of an endangered bird is declining at a rate of 25% a year. If there are 176 birds left now, how many will there be after 5 years? **W.E. 14**

(b) The value of a dishwasher is decreasing at a rate of 20% p.a. If it's worth \$1400 now, how much will it be worth after 2 years?

(c) At the start of a plague, the population in a medieval town is 10 200 and has a death rate of 6% each day. Estimate the population after 8 days.

(d) A population decreases at a rate of 1.2% per month. If the initial population is 200 000, what will be the population after 16 months?

(e) An eBay seller notices that the number of internet sales is declining at a rate of 7% a month. She sold 2370 items this January. If the trend continues, how many can she expect to sell in the month of July (in 6 months' time)?

(f) Cigarette sales have shown a 2.98% p.a. decay rate in recent decades. If a new company sold 5 million packets of cigarettes in their first year, how many packets can they expect to sell in their fourth year? (*Note*: The fourth year is in 3 years' time.)

6 Answer these questions for each of the following tables of values, which represent either exponential growth or exponential decay.

(i) What is the initial value (k)?

(ii) Is this an example of growth or decay?

(iii) What is the growth factor a?

(iv) Write the formula that models this growth.

(a)

T (days)	0	1	2	3	4
N	4	8	16	32	64

(b)

T (days)	0	1	2	3	4
N	1	5	25	125	625

(c)

T (days)	0	1	2	3	4
N	324	108	36	12	4

(d)

T (hours)	0	1	2	3	4
N	320	160	80	40	20

W.E. 15

7 The following table shows the number of staff *S* employed by a company after *T* years.

T	0	1	2	3	4	5	6
S	5	15	45	135			

(a) Complete the table.

(b) Find the formula that models this growth.

(c) Use the formula to predict the number of staff employed by the company after 10 years and 3 months.

(d) Use the table of values to plot a graph of *S* against *T*. Join the points with a smooth curve.

(e) Use the graph to predict approximately when the company will have 78 employees.

Understanding

8 (a) If the population of Growthland starts at 15 000 and increases by 5% each year, the population in 6 years' time will be:

A 4500 B 19 500 C 20 101 D 90 000

(b) A population decreases at the rate of 2% per month. If the intial population was 130 000, the population after 1 year is closest to:

A 98 800 B 102 013 C 127 400 D 164 871

9 Kate invests \$5000 in a bank that offers 10% interest per year.

(a) What will her investment be worth after 1 year?

(b) What percentage of her original investment will her bank balance be at the end of the first year?

(c) Express this percentage as a decimal.

(d) What will Kate's investment be worth after 2 years?

(e) What is the first year's balance multiplied by to find the second year's balance?

(f) Complete a table of values showing Kate's bank balance at the end of each year (up to 10 years).

(g) Write an equation relating the amount Kate has in the bank to the number of years of investment.

(h) Plot a graph of this exponential relationship.

(i) Use the graph to predict what Kate's investment will be worth after $3\frac{1}{2}$ years. Check this result with a calculator.

(j) Use the graph to predict when Kate's investment will exceed \$8000.

(k) How long would Kate need to have her money in the bank to double her investment?

10 The value of a car, currently estimated at \$45 000, depreciates at 5% each year.

(a) What percentage of the current value will the car be worth after 1 year?

(b) Express this percentage as a decimal.

(c) Find the equation that describes the relationship between the value of the car and the number of years of depreciation.

(d) Complete a table of values for this exponential relationship. (Consider up to 5 years.)

(e) Plot the graph of this exponential relationship on graph paper.

(f) If the car is to be sold after $2\frac{1}{2}$ years, use the graph to predict what price the owner should expect to sell it for. Use a calculator to check your result.

(g) When should the car be sold to expect a price of \$36 000?

11 Sybil was born in 2002 with an inheritance worth 1 million dollars. The money was unfortunately invested in a fund that has been decreasing by half each year.

(a) Find an equation that models the value of the inheritance (I) after T years.

(b) How much will Sybil's inheritance be worth when she is 20 years old?

(c) When will the inheritance be worth \$7812.50?

Reasoning

12 When Angie and Rashid married, some of their wedding cake was kept to celebrate the anniversary of their wedding each year for the next 6 years. On their first anniversary the family decided to eat a third of the cake. At each subsequent anniversary, it became a tradition to eat a third of the cake remaining from the previous year. There was 5 kg of cake left at the end of the wedding.

(a) Find the equation for the amount of cake (A) left after the nth anniversary.

(b) On the sixth and final anniversary, the remaining cake will be divided equally between 62 people. Find the mass of each slice. Round your answer to the nearest gram.

Open-ended

13 The population of a small town started with 10 people. After some time it reached 7290. Give some possibilities for the number of years elapsed and the annual rate of increase.

14 Ana, Ernie and Janet are looking through the newspaper for something they can use in their Year 10 Science presentation. They come across this report:

> Scientists have become aware of a radioactive compound known as Xradium in the soil of Nuketown that decays at a rate of 15% every 60 years. Nuketown will be too radioactive for human habitation for the next 850 years, until the amount of Xradium has decayed to about 12% of its current level.

Janet is convinced the reporter made a mistake. She does the following calculation:

$6\frac{2}{3} \times 15\% = 100\%$ so it should only take about $6\frac{2}{3} \times 60 = 400$ years.

Ernie says: 'You can't work it out like that, Janet! You've got to use the decay formula.'

Ana says: 'How are you supposed to use the decay formula without knowing how much of the stuff is in the soil right now?'

Is there an error in the newspaper report? Help the students to resolve their problem.

Chapter review 13

Maths literacy

adjusted value	flat rate	salvage value
appreciation	inflation rate	scrap value
compound interest	initial cost	simple interest
constants	interest	straight-line depreciation
depreciation	nominal interest rate	total depreciation
depreciation value	per annum	total possible depreciation
effective interest rate	prime cost	unit cost depreciation
expected life	principal	variables
exponential decay	reducing balance depreciation	written-down value
exponential growth	residual value	

Copy and complete the following using the words and phrases from the list, where appropriate. A word or phrase may be used more than once.

1 The original amount of money that is borrowed or loaned is called the __________.

2 Interest rates are normally expressed with p.a. after them. This stands for __________ and means for every year.

3 __________ is always higher than __________ after the first compounding period.

4 Increase in cost or value is called __________ and decrease in cost or value is called __________.

5 The stated interest rate is known as the __________. The adjusted rate that takes into account the effect of any compounding in the first year is called the __________.

6 The amount of depreciation on a certain item is \$10 each year for 7 years. This is an example of __________.

7 Depreciation that has a decreasing value from one year to the next is called __________.

8 If the depreciation is \$5 in the first year and \$4 in the second year, the __________ over the first two years is \$9.

9 The value of an item after depreciation is known as its __________ or __________.

10 The formula $V_T = P - D_T$ has three __________, V_T, P and D_T. In this formula, P stands for the __________ or __________.

11 Any value that an item still has at the end of its useful life is known as __________, __________ or __________.

12 The initial purchase price less any scrap value is the __________.

13 This value divided by the expected life gives the __________.

14 The growth of bacteria is an example of __________. The opposite process is called __________.

Fluency

1 (a) The amount an investment of \$8200 amounts to after 2 years, if 9.5% p.a. interest is compounded annually, is closest to: **13.1**

A \$1558 B \$1632 C \$7790 D \$9832

(b) The amount an investment of \$4000 amounts to after 2 years, if 3.5% p.a. interest is compounded annually, is closest to:

A \$285 B \$4192 C \$4280 D \$4285

2 Michael borrowed \$14 000 to buy a new piano. He agreed to pay interest at a rate of 16% p.a. and to repay the loan in full after 4 years. Calculate the amount to be repaid if the interest is compounded: **13.2**

(a) half-yearly (b) quarterly (c) monthly.

3 Helga deposited some money into a savings account. The account guaranteed a return of 9.25% p.a. compounded daily. Calculate the initial deposit if the savings had grown to \$47 638.38 after 5 years. Give your answer rounded to the nearest dollar. **13.3**

4 Find the effective interest rate for each of the following: **13.4**

(a) \$38 000 invested at a simple rate of 8.7% p.a.

(b) \$50 000 invested at an annually compounding rate of 9.5% p.a.

(c) the interest owed on a loan of \$700 after 1 year was \$28

(d) 17.4% p.a. compounded every 2 months

5 The initial price 5 years ago was \$60 000. The rate of depreciation is 7% p.a. **13.5**

(a) Assuming straight-line depreciation, find the current written-down value.

(b) Assuming reducing balance depreciation, find the current written-down value.

Round all answers to the nearest cent.

6 The population of a small rural community has been declining at a rate of 1% a year. If there were 461 residents in 2006, estimate the population in 2016. **13.6**

Understanding

7 Simple interest is paid on an investment of \$12 500 and yields \$19 687.50 in 5 years. What is the rate of interest being paid? **13.4**

8 Nina's salary has increased by 6% p.a. each year over the past 3 years to \$70 500 p.a. Calculate her salary 3 years ago, to the nearest dollar. **13.4**

9 Assume an inflation rate of 5.6% p.a. In 4 years, an ice-cream, currently \$1.20, will cost: **13.4**

A $\$1.20(1.014)^4$ B $\$1.20(1.056)^4$ C $\$1.20(1.014)^{16}$ D $\$1.20(1.56)^4$

10 Calculate the rate of interest p.a. (rounded to 1 decimal place) that would allow \$6700 to accumulate to \$10 000 in 5 years if interest is compounded each quarter. **13.4**

11 A dining suite was purchased $3\frac{1}{2}$ years ago for \$4600. It depreciated at a fixed percentage of its reduced value each year and has just sold for \$2800. The rate of depreciation is closest to: **13.5**

A 15.25% p.a. B 13.22% p.a. C 13% p.a. D 11.18% p.a.

12 Copy and complete the following. **13.5**

(a) If the prime cost of a photocopier is \$10 800 and it has a scrap value of \$500, then the total possible depreciation is \$__________. If it has an expected life of 206 000 pages, then the unit cost depreciation is \$__________per page. After it copies 125 000, the depreciation is \$__________ and the written-down value is \$__________.

(b) If the prime cost of a machine is \$52 500 and it has a scrap value of \$750 then the total possible depreciation is \$__________. If it has an expected life of 5000 days, then the unit cost depreciation is \$_____per day. After it does a year's work, the depreciation is \$__________ and the written-down value is \$__________.

13 The population of Metown increases by 1.1% each year. What was the population 6 years ago if its current population is 68 000? **13.5**

14 The temperature inside a room decreases by 3% every 15 minutes. Find the temperature at 7 pm, if it is 26 °C at 5 pm. Round your answer to 1 decimal place. **13.5**

Reasoning

15 How many years will it take for an investment of \$10 000 to increase by at least 50% if the interest rate of 10% p.a. is compounded half-yearly? Give your answer to the nearest tenth of a year. **13.4**

16 The following represent loan options. In each case, choose the better option *without calculating the effective rate* and justify your choice. **13.4**

(a)	11% p.a. compounded quarterly	14% p.a. compounded monthly
(b)	10% p.a. compounded daily	10% p.a. compounded monthly
(c)	17% p.a. compounded daily	17.5% p.a. compounded daily
(d)	12% p.a. flat rate for 2 years	12% p.a. compounded annually for 2 years

17 One pair of breeding rats can become a colony of 2000 in one year. Estimate the population rate of growth per month. Give your answer rounded to 1 decimal place. **13.6**

18 A new city restaurant had only 12 customers for lunch on its opening day. Business improved and the number of lunchtime customers followed an exponential growth curve. After 2 years the restaurant had reached its capacity of 120. What was the growth rate per month over these first 2 years? Round your answer to the nearest whole number. **13.6**

Mixed review D

Fluency

1 An investment of \$5000 is made at 3.5% p.a. for 6 years. Find the difference in interest earned, to the nearest cent, if interest is compounded monthly compared to annually. **13.2**

2 Find the area and perimeter, correct to 1 decimal place, of a semicircle with a diameter of 9.6 cm. **5.1**

10A 3 Write the following in simplest form. **11.2**

(a) $\sqrt{3} \times \sqrt{8}$ (b) $\sqrt{3} \times 3\sqrt{12}$ (c) $\sqrt{2} \times \sqrt{18}$

(d) $2\sqrt{5} \times 4\sqrt{10}$ (e) $2\sqrt{6} \times 3\sqrt{10}$ (f) $3\sqrt{6} \times 7\sqrt{12}$

10A 4 Simplify the following. **11.2**

(a) $\sqrt{80}$ (b) $3\sqrt{240}$ (c) $7\sqrt{525}$

5 Solve the following equations. **1.1**

(a) $\frac{3a+4}{7} = -2$ (b) $\frac{4a+3}{2} = \frac{2a}{5}$ (c) $\frac{3a-2}{3} = \frac{4a-5}{2}$

10A 6 Express each of the following in simplest form with a rational denominator. **11.5**

(a) $\frac{4\sqrt{3}}{\sqrt{5}}$ (b) $\frac{2}{5-\sqrt{3}}$ (c) $\frac{\sqrt{2}}{\sqrt{3}+\sqrt{5}}$

7 Factorise each of the following expressions. **3.2**

(a) $k(k-3)-(k-3)$ (b) $3t(t+2)-5(t+2)$

8 Simplify the expressions below: **12.3**

(a) $\frac{4a^4b}{9a^{-5}b^3} \times \frac{3a^{-2}b^5}{2ab^{-1}}$ (b) $\frac{(a^4b^2)^2}{(a^2b^3)^2} \times \frac{(a^{-3}b)^3}{(a^5b^2)^{-3}}$ (c) $\frac{k^{-8}}{n^4} \div \frac{k^9}{n}$

9 Find the value of each pronumeral, correct to 1 decimal place. **6.3**

(a) $\cos(a) = 0.1314$ (b) $\tan(b) = 2.621$ (c) $\sin(c) = 0.5231$

10 Find the gradient of the line joining: **1.2**

(a) points (1, 7) and (4, 14) (b) points (3, 6) and (7, -2).

11 Find the interest earned on an investment of \$6000 at 4.2% p.a. for 3 years, where the interest is compounded: **13.2**

(a) annually (b) half-yearly (c) quarterly.

12 A painting has increased in value, solely due to the effects of inflation, from \$2500 to \$3100 over 5 years. Find the average annual inflation rate. State your answer correct to 1 decimal place. **13.3**

10A 13 What are the asymptotes of the graph of $y = \frac{1}{x+1} - 3$? **12.1**

10A 14 Sketch the graphs of each of the following, clearly showing the turning point. **4.2**

(a) $y = (x+3)(x-2)$ (b) $y = x^2 + 2x + 5$ (c) $y = 2x^2 - 4x$

10A **15** For the rectangle shown: 11.3

(a) State its perimeter in simplest form.

(b) State its area in simplest form.

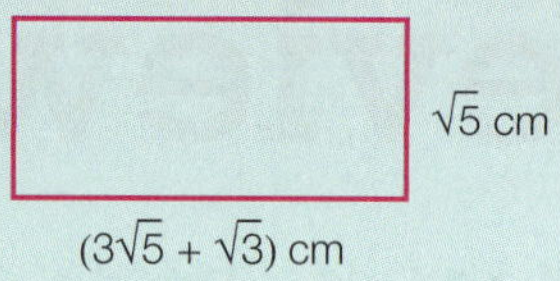

16 Write a definition for each of the following quadrilaterals. 9.5

(a) rectangle **(b)** rhombus **(c)** kite

17 Two dice, both numbered 1 to 6, are rolled at the same time. Find the probability that the sum of the two faces showing will be: 10.2

(a) 4 **(b)** 6 **(c)** greater than 9.

10A **18** Without using a calculator, determine which is the larger of the following pairs, placing <, > or = between the numbers below. 12.4

(a) 3^{-5} ____ 5^{-3} **(b)** $9^{\frac{1}{2}}$ ____ 9^{-1} **(c)** $8^{-\frac{1}{3}}$ ____ $\dfrac{2^{-3}}{2^{-2}}$

19 On a Venn diagram show the region described by not A and B and not C. 10.2

20 Describe what is needed for an equation to be a quadratic equation. 3.5

Understanding

21 Leonard wants to develop a vegetable garden in his backyard. He has 20 m of chicken wire and enough posts to erect a fence around the four sides of the rectangular garden. 4.4

(a) If the length of the rectangular garden is x m, find an expression (in terms of x) for the width of the graden.

(b) Show that the area of the garden, A m^2, is given by the quadratic expression $A = -x^2 + 10x$.

(c) Sketch the graph of the area of the vegetable garden, A m^2, against the length of the garden, x m. (Show length on the horizontal axis.)

(d) What should be the dimensions of the garden if the area is to be as large as possible?

(e) What is the area of the largest vegetable garden that Leonard can make?

Reasoning

10A **22** A square with side lengths $\sqrt{2}$ cm has been drawn using four squares with side lengths of 1 cm, as shown. What is the perimeter of a square that would be made by joining the diagonals of four squares with side lengths of 2 cm? 11.2

10A **23** **(a)** Draw an isosceles triangle with base length 10 cm and height 10 cm. 11.2

(b) Show that the length of the equal sides in the triangle in part **(a)** is an irrational number.

24 Write the equation formed when $x^2 + y^2 = 25$ is translated 4 units right and 2 units up. 12.2

25 A point that is not on the graph of $(x - 3)^2 + (y - 2)^2 = 20$ would be: 12.1

A (7, 0) **B** (5, 6) **C** (6, 5) **D** (-1, 4)

26 The graph of the equation $y = 2^{(x-1)} + 1$ has an asymptote at: 12.2

A $x = 1$ **B** $x = -1$ **C** $y = 1$ **D** $y = -1$

Challenge D

1 In a dice game, a player rolls two dice. The score is the larger of the two numbers on the dice. For example, if the player rolls 3 and 5, then the score is 5. If the player rolls 4 and 4, then the score is 4. What is the probability that the score is 3 or less?

A $\frac{5}{36}$ B $\frac{1}{6}$ C $\frac{7}{36}$ D $\frac{1}{4}$

2 A prize wheel is divided into 6 equal sectors. Each time the wheel is spun it stops on one of the sectors, resulting in a prize of the amount shown.

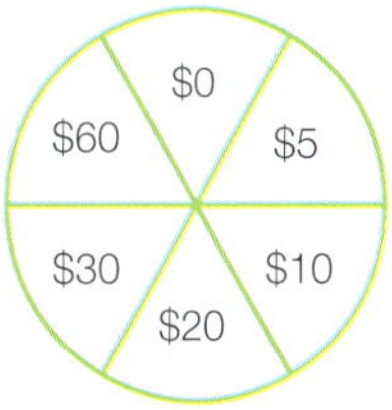

(a) On the first spin $5 is won. What is the probability of winning $5 on the second spin?

(b) The wheel is spun a total of 50 times and the results recorded in the table.

(i) Copy and complete the table by writing in the two missing pieces of information.

(ii) How much prize money was won?

Prize	Frequency	Relative frequency
$0	8	0.16
$5		0.14
$10	9	
$20	10	0.20
$30	8	0.16
$60	8	0.16
Total	**50**	**1**

(c) To play this game you have to pay $20 per spin.

(i) If each outcome was equally likely and you spun the wheel 48 times, would you expect to make a profit?

(ii) If you spun the wheel 50 times and the results were as shown in the table in part **(b)**, would you make a profit?

3 A pyramid has a square base with an area of 1440 cm^2. Each of the pyramid's triangular faces is identical and has an area of 840 cm^2. The height of the pyramid is:

A $20\sqrt{3}$ cm B 40 cm C $30\sqrt{2}$ cm D $4\sqrt{145}$ cm

4 The square *KLMN*, with side length 2 units, is inscribed in a circle as shown. Using each side of the square as a diameter, semicircular arcs are drawn. The total area of the shaded region outside the circle and inside the semicircles is:

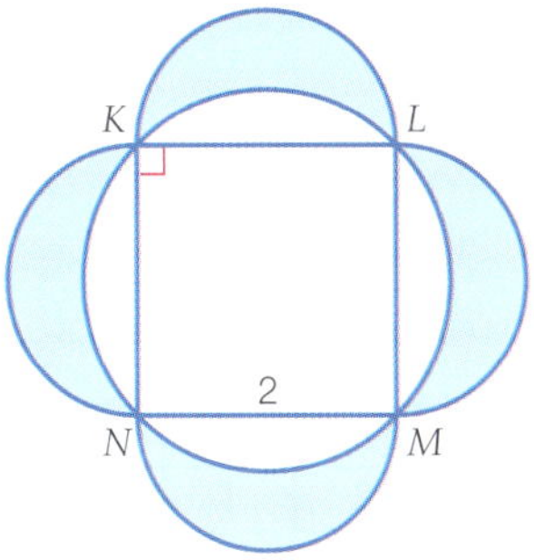

A 4 units^2 B π units^2

C 2π units^2 D $(2\pi - 4)$ units^2

5 Find the value of x in each of the following.

(a) $x^{-1} = 3^{-1} + 4^{-1}$ (b) $x + 27^{\frac{1}{3}} = 125^{\frac{1}{3}}$ (c) $2^x = 2(16^{12}) + 2(8^{16})$

(d) $2^{2x-4} = 8$ (e) $16^x = 2^{x+5} - 2^{x+4}$

Answers

Worked solutions and answers to *all* activities appear in the Teacher Companion.

Chapter 1

Recall 1

1 (a) $P = -7$ (b) $P = 3$ (c) $P = -15$ (d) $P = -3$

2 (a) $a = 7$ (b) $b = 24$ (c) $c = 10$ (d) $d = 7$

3

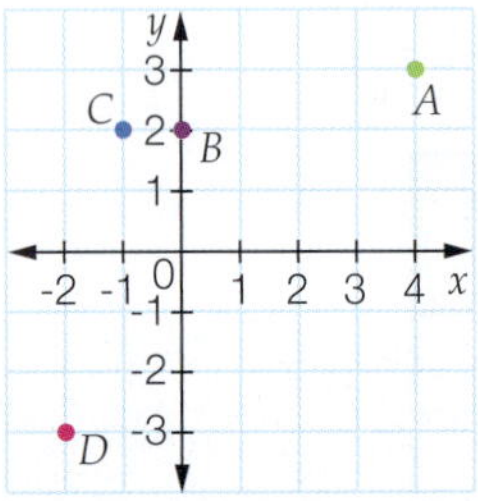

4 (a)

x	-2	-1	0	1	2
y	1	2	3	4	5

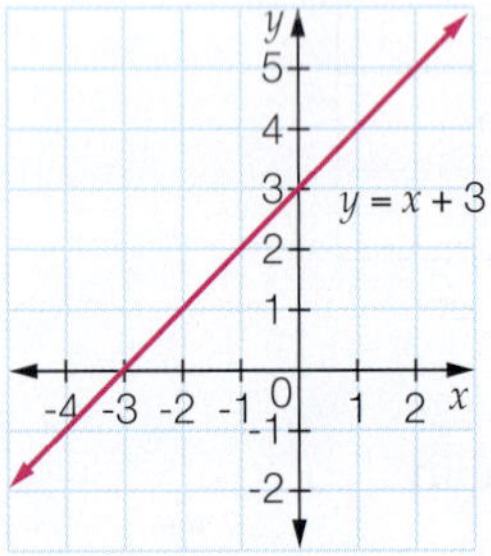

(b)

x	-2	-1	0	1	2
y	-4	-3	-2	-1	0

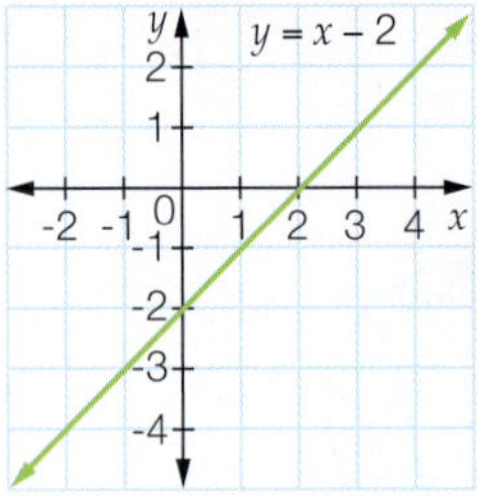

(c)

x	-2	-1	0	1	2
y	4	3	2	1	0

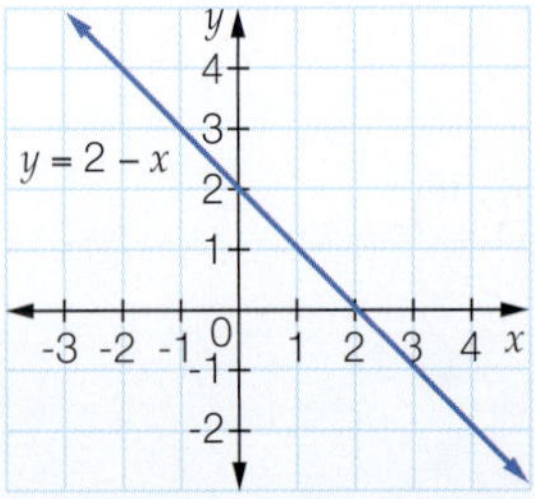

(d)

x	-2	-1	0	1	2
y	-6	-3	0	3	6

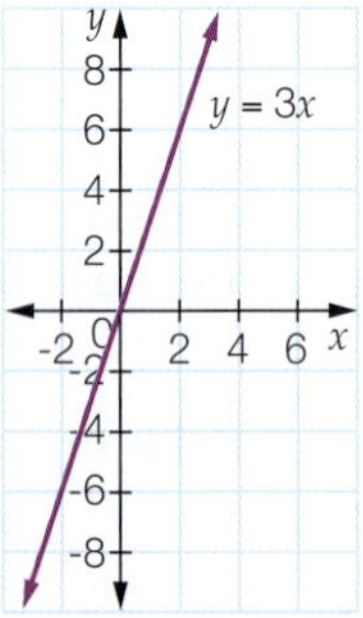

5 (a)

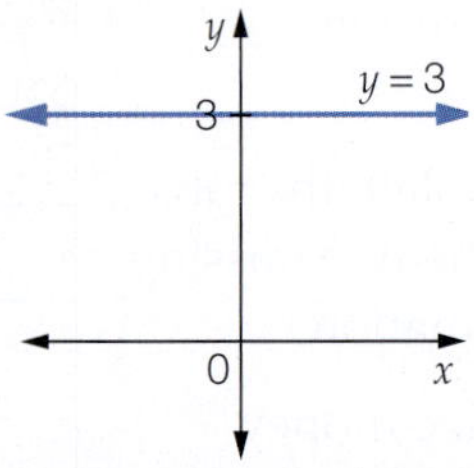

(b)

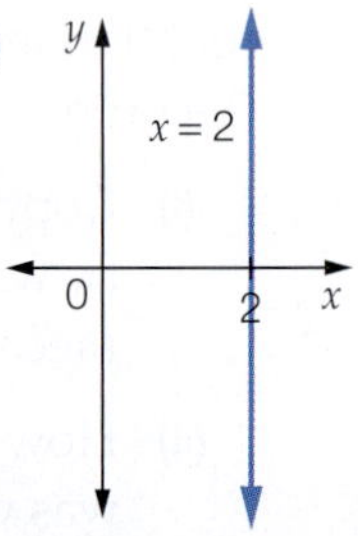

6 (a) (i) 3 (ii) $(0, 5)$

(b) (i) -2 (ii) $(0, 3)$

(c) (i) $-\frac{2}{3}$ (ii) $\left(0, \frac{4}{3}\right)$

7 (a) $-\frac{5}{3}$ (b) 1

8 (a)

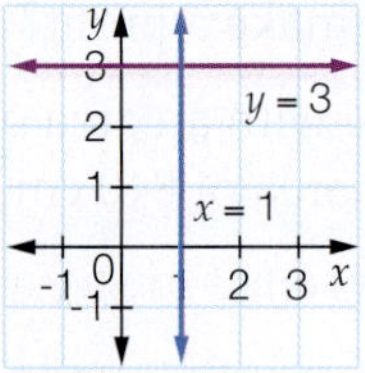

(b) $(1, 3)$

Exercise 1.1

1 (a) $x = 3$ (b) $x = 8$ (c) $c = -3$ (d) $a = 4$

(e) $x = -6$ (f) $d = -3$ (g) $m = 2$ (h) $n = 1$

(i) $x = -1$ (j) $f = 2$ (k) $e = -2$ (l) $p = \frac{1}{5}$

(m) $x = 0$ (n) $g = 21$ (o) $h = 16$ (p) $k = 3$

(q) $x = 6$ (r) $y = 0$ (s) $x = \frac{5}{4}$ (t) $x = 3$

(u) $x = 7$

2 (a) $a = 1$ (b) $d = -1$ (c) $e = \frac{5}{6}$

(d) $c = 2$ (e) $b = 4$ (f) $f = \frac{13}{6}$ or $2\frac{1}{6}$

(g) $g = 8$ (h) $h = \frac{4}{5}$ (i) $k = -1$

(j) $x = -1$ (k) $x = 1$ (l) $x = -\frac{7}{3}$ or $-2\frac{1}{3}$

(m) $x = \frac{3}{2}$ or $1\frac{1}{2}$ (n) $x = \frac{1}{4}$ (o) $x = \frac{1}{3}$

(p) $m = -\frac{11}{2}$ or $-5\frac{1}{2}$ (q) $d = \frac{17}{4}$ or $4\frac{1}{4}$ (r) $d = 7$

(s) $a = 8$ (t) $a = -\frac{16}{3}$ or $-5\frac{1}{3}$ (u) $g = \frac{8}{7}$ or $1\frac{1}{7}$

(v) $t = 0$ (w) $h = \frac{11}{2}$ or $5\frac{1}{2}$ (x) $j = -\frac{3}{11}$

3 (a) C (b) D (c) D

4 (a) 35 (b) 16 (c) 30

5 (a) $a = -22$ (b) $b = 8$ (c) $c = 9$

(d) $d = -1$ (e) $x = -2$ (f) $y = 1$

(g) $m = \frac{3}{4}$ (h) $k = \frac{3}{2}$ or $1\frac{1}{2}$ (i) $t = 2$

(j) $a = -25$ (k) $p = 15$ (l) $m = -27$

(m) $p = 5$ (n) $n = 2$ (o) $t = -18$

(p) $m = \frac{13}{22}$ (q) $k = -\frac{7}{4}$ or $-1\frac{3}{4}$ (r) $x = \frac{5}{6}$

6 (a) $x + 6 = 13, x = 7$ (b) $y - 7 = -2, y = 5$

(c) $2x = 17, x = 8\frac{1}{2}$ (d) $\frac{x}{3} = 12, x = 36$

(e) $3x + 2 = 23, x = 7$ (f) $4x - 5 = -3, x = \frac{1}{2}$

7 (a) B (b) B (c) C

8 (a) $y = 5, z = 6$ (b) $x = 11, z = 13$ (c) $y = 6, z = 8$

9 40, 42, 44

10 Nikki 7, Sam 14, Minh 21

11 length 100 m, width 50 m

12 (a) $p + 7$ (b) $2(p + 7)$ (c) 17

13 98 m

14 walk 1.5 km, train 9 km

15 15 years old

16 45

17 (a) $3x - 9 = 6$ (b) $x = 5$

(c) You would know the answer is $x = 5$ because this is just (b) multiplied by 3 on both sides.

(d) The answer is $x = 5$ because the equation is equivalent to the equations in (b) and (c).

18 24 years old

19 43 km/h and 53 km/h

Open-ended – Sample answers

20 Students to check own answers, but examples could be:

$\frac{x+8}{6} = -\frac{x}{2}$, $5x - 2 = -10 + x$, $-4x - 11 = \frac{3x}{2}$

21 (a) $2l + 10$ or $2(l + 5)$

(b) (i)

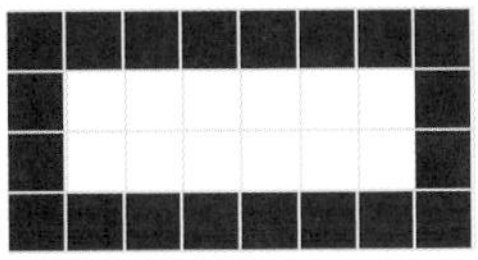

(ii) 20 black tiles needed

(iii) $2t + 8$ or $2(t + 4)$

(c) For 5×3 white tiles: $n = 20$

Exercise 1.2

1 (a) $\frac{1}{2}$ (b) -2 (c) 0 (d) $\frac{3}{0}$, undefined

2 (a) $y = 6x - 3$ (b) $y = -2x + 7$

(c) $y = 3x + 9$ (d) $y = -5x + 8$

3 (a) positive (b) negative

(c) zero (d) undefined

4 (a) $\frac{1}{3}$ (b) 2 (c) -4 (d) $-\frac{3}{2}$

(e) $-\frac{5}{4}$ (f) -2 (g) 3 (h) 2

(i) -1 (j) $-\frac{2}{3}$ (k) 0 (l) 0

(m) $-\frac{2}{5}$ (n) $\frac{1}{7}$ (o) undefined

5 (a) C (b) B (c) D

6 (a) C (b) B

(c) x can have any value, $y = -7$

(d) $x = -3$, y can have any value

7 (a) (i) (-3, 2) and (2, 2); (5, -4) and (-7, -4)

(ii) (2, 5) and (2, 7)

(iii) (4, 10) and (5, 6)

(iv) (-1, -1) and (6, 0)

(b) (0, 3) and (2, 7) parallel to (2, 1) and (-2, -7)

8 (a) 800 m (b) 600 m (c) 120

(d) 120 m/min (e) 440 m (f) 560 m

(g) $d = 200 + 120t$

(h) Alex is walking at a constant speed.

9 (a) 4800 grains (b) 1600 grains

(c) -1600 (d) 1600 grains/min

(e) decrease (f) $S = 4800 - 1600t$

(g) When $t = 2$, $s = 4800 - 3200$
$= 1600$

(h) $0 \le t \le 3$

10 (a) (i) $y = 2.5x - 1$ (ii) $y = 2.5x - 1$

(b) Yes, either point can be used as (x_1, y_1) to find the equation of the line.

11 $a = 17$

12 $-\frac{9}{35}$

13 (a) 2 (b) 2 (c) 2

(d) The line segments are all part of the same straight line, so they must have the same gradient.

14 Gradient of the line joining the points (-1, -5) and (2, 4) = 3 and the gradient of the line joining the points (2, 4) and (5, 13) = 3. As the two lines have a point in common, (2, 4) and the same gradient, the points are collinear.

15 (a) $m = \frac{5}{2}$ (b) $m = \frac{5}{2}$

(c) $y = 18$ (d) $x = -1$

Open-ended – Sample answers

16 (a) (1, 9), (2, 6), (3, 3) (b) (-2, 0), (-1, -1), (1, -3)

(c) (0, 4), (2, 3), (4, 2) (d) (2, -5), (3, -3), (4, -1)

(e) (1, 0), (2, 5), (3, 10)

17 $m = -3$, (1, 4) and (4, -5)

Exercise 1.3

1 (a) (i) (0, -3) (ii) (1, -1)

(b) (i) (0, 4) (ii) (1, 1)

(c) (i) (0, 3) (ii) (4, 6)

(d) (i) (0, -4) (ii) (5, -6)

2 (a) (i) (4, 0) (ii) (0, 3)

(iii) $-\frac{3}{4}$ (iv) $y = -\frac{3}{4}x + 3$

(b) (i) (2, 0) (ii) (0, -4) (iii) 2 (iv) $y = 2x - 4$

3 (a) $b = 2, m = 3$

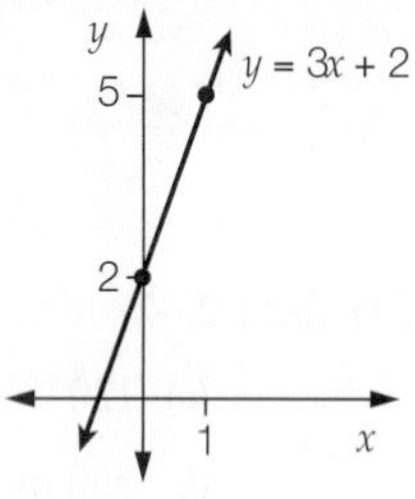

(b) $b = -3, m = 4$

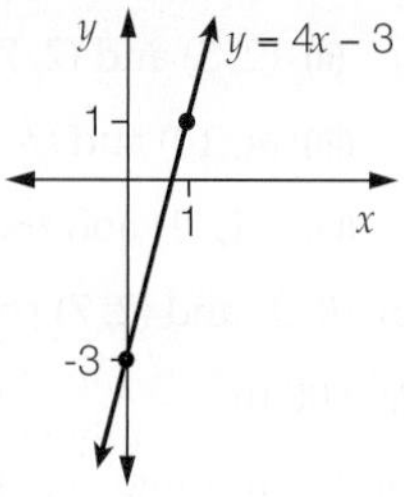

(c) $b = 1, m = -2$

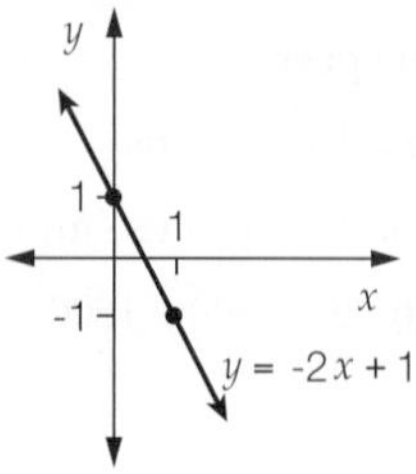

(d) $b = 2, m = -3$

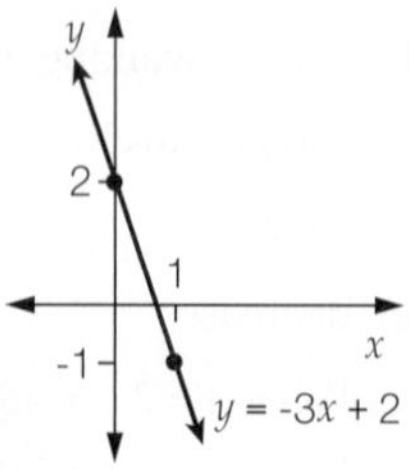

4 (a) $b = -2, m = -\frac{1}{2}$

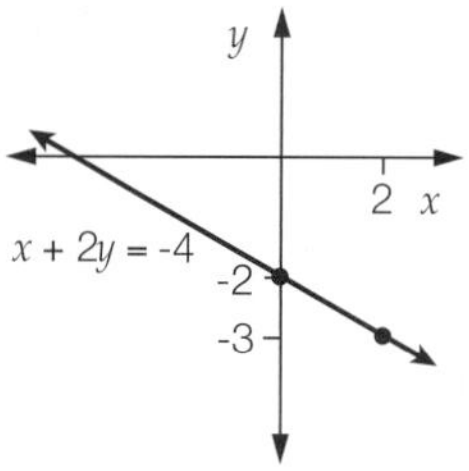

(b) $b = 1, m = -\frac{1}{3}$

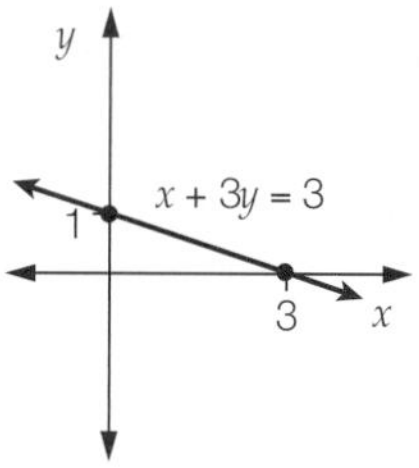

(c) $b = 2, m = -\frac{3}{4}$

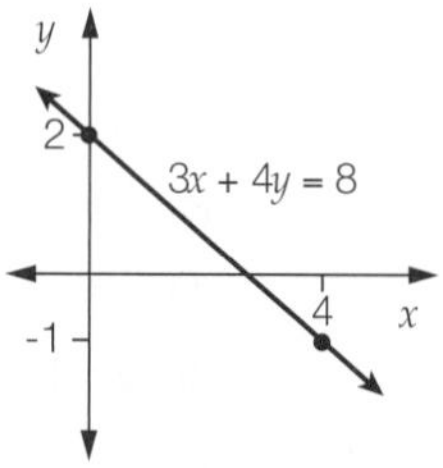

(d) $b = -1, m = -\frac{2}{5}$

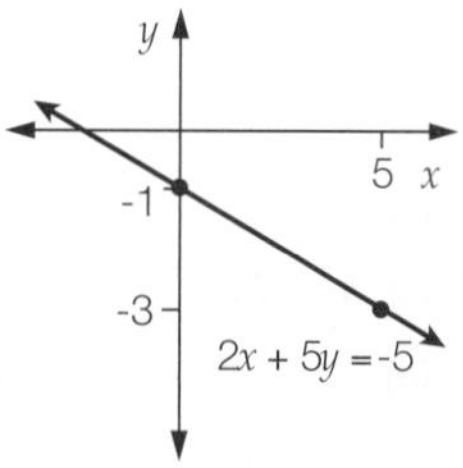

(e) $b = \frac{1}{3}, m = \frac{2}{3}$

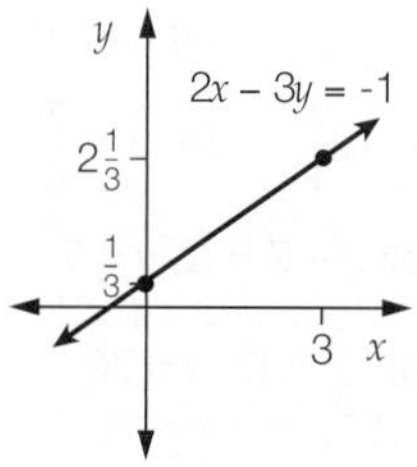

(f) $b = -1\frac{1}{2}, m = 2$

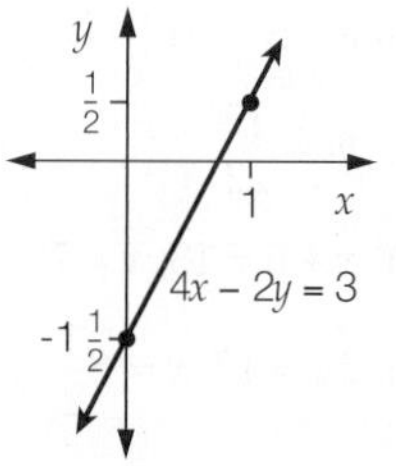

5 (a)

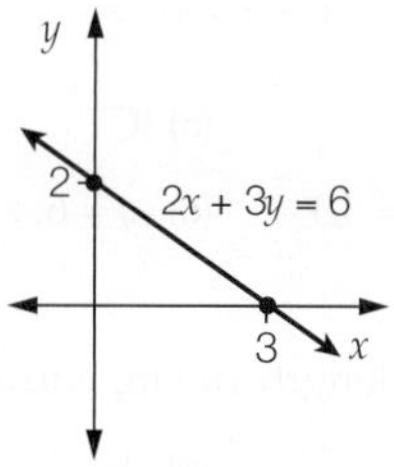

(b)

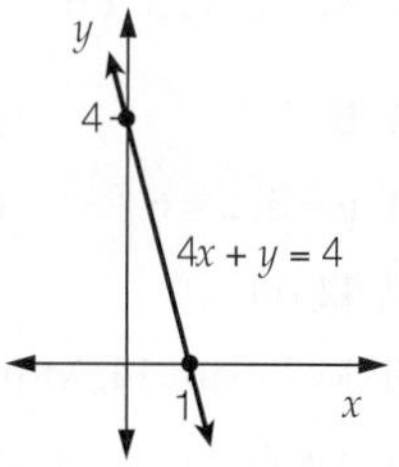

(c)

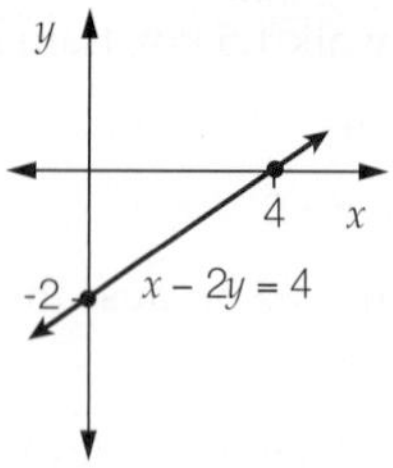

(d)

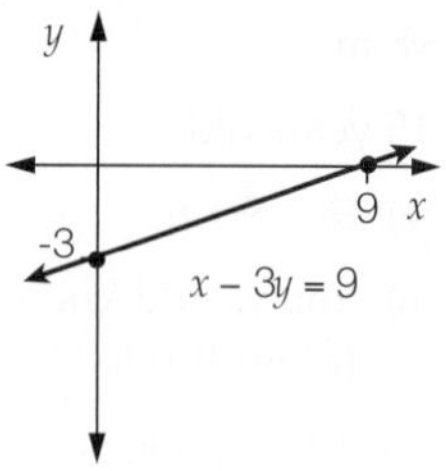

(e)

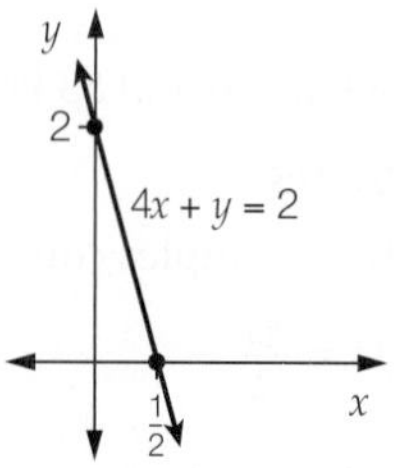

(f)

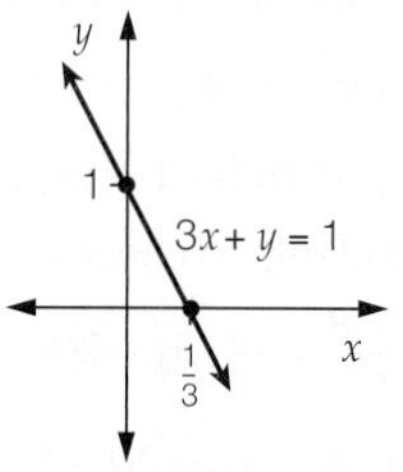

6 (a)

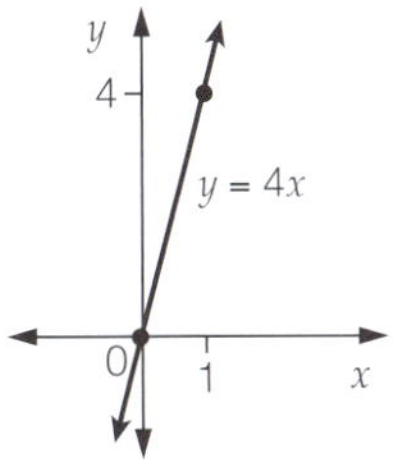

(b)

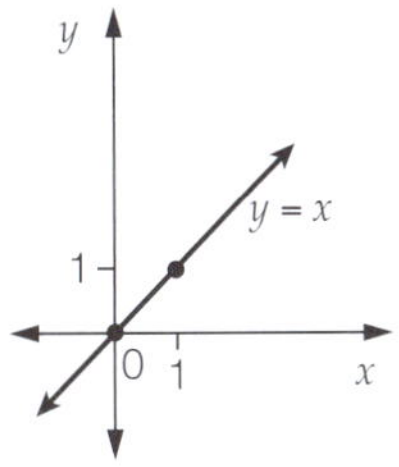

(c)

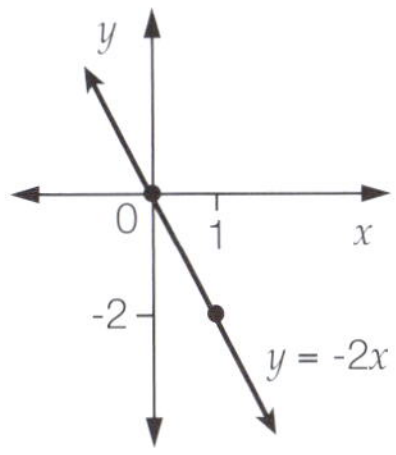

(d)

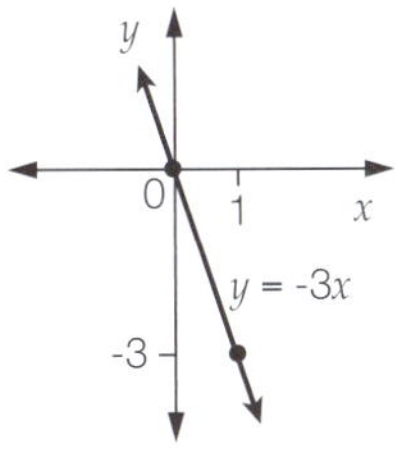

7 (a) D (b) B (c) A

8 (a) C (b) C (c) B

9 (a)

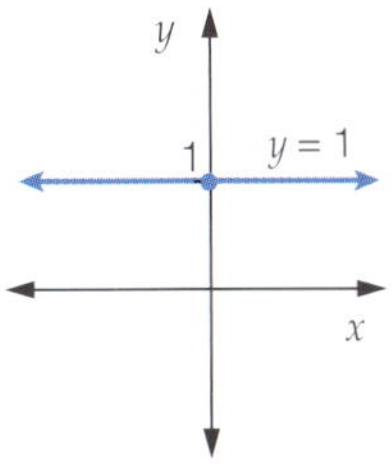

(b)

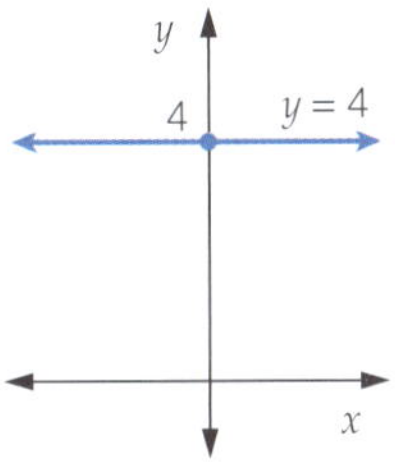

(c)

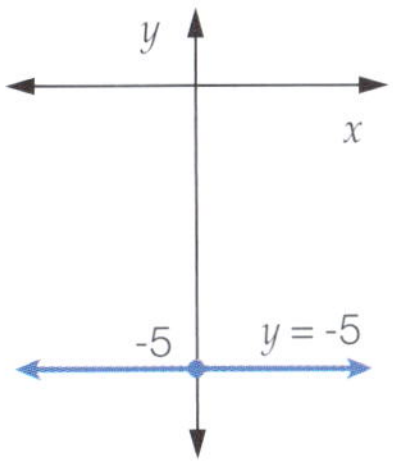

(d)

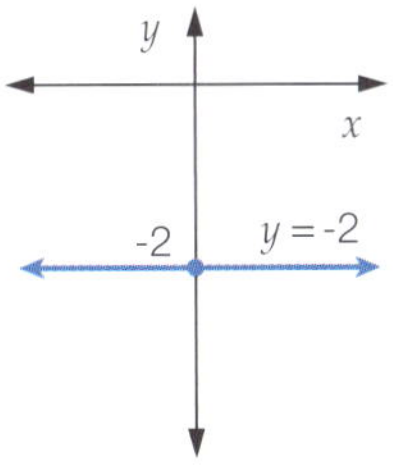

(e)

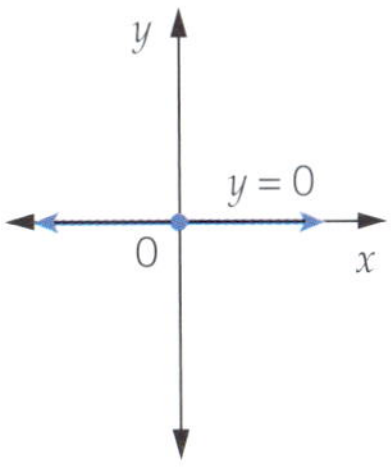

(f)

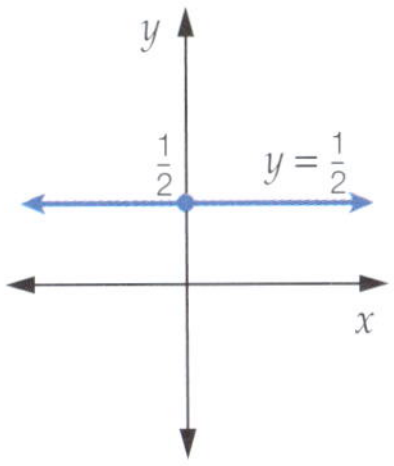

(g)

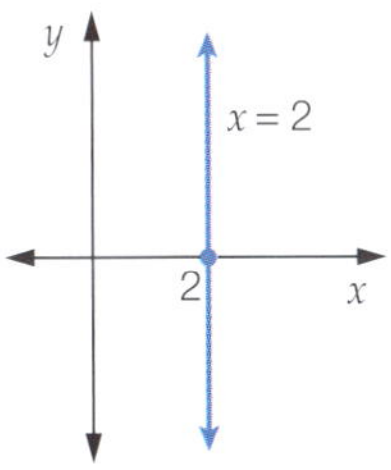

(h)

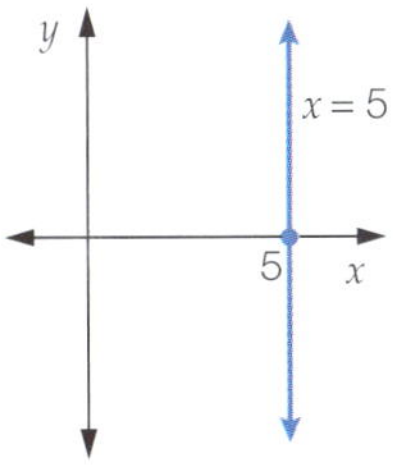

(i)

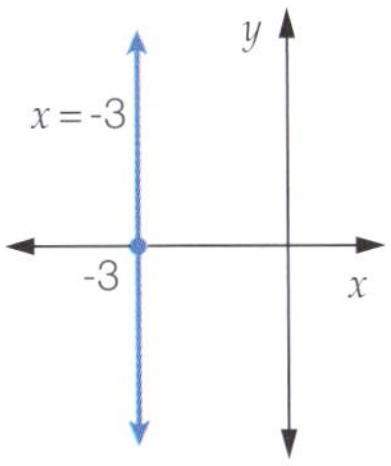

(j)

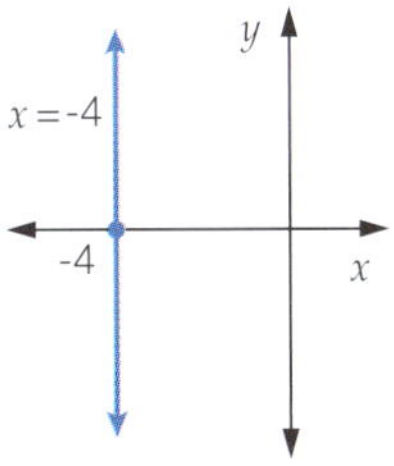

(k)

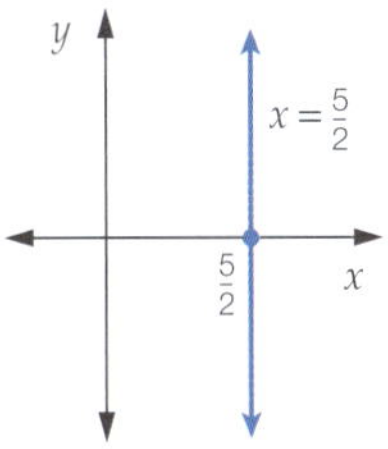

(l)

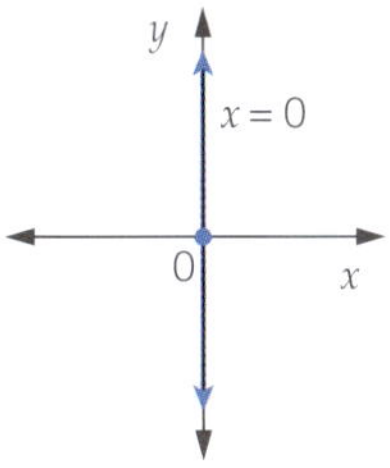

10 (a) D (b) C (c) D

11 (a) 100 km (b) 80 (c) $D = 80T + 100$

(d) 80 km/h (e) $D = 100T + 250$ (f) 93 km/h

12 (a) \$80 (b) $A = 200 - 40n$

(c) \$200 (d) 5 weeks

(e)

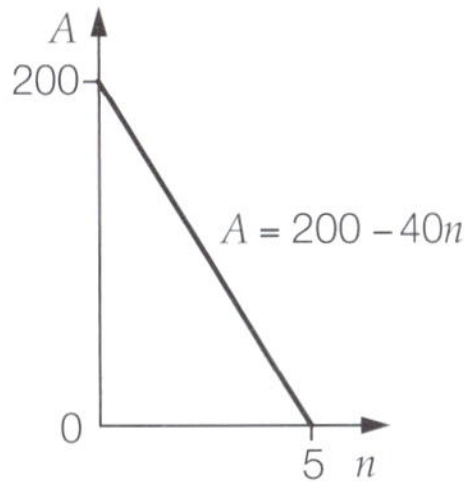

13 (a) 225π cm^2 (b) $A = 225\pi - kt$

(c) $A = 225\pi - 5k$ (d) $k = 10\pi$

(e) 22.5 min (f) The horizontal axis intercept.

14 (a)

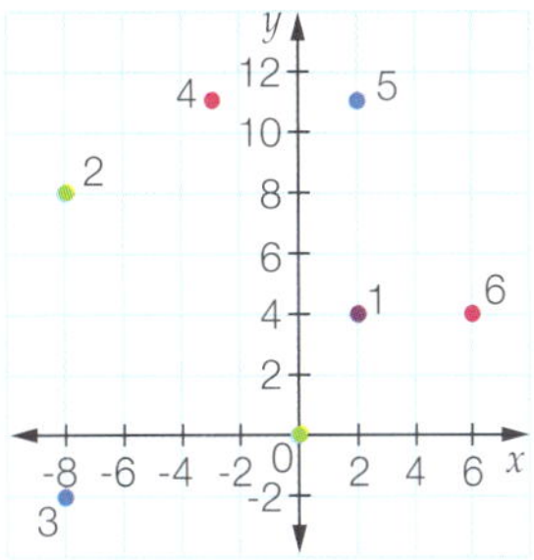

(b)

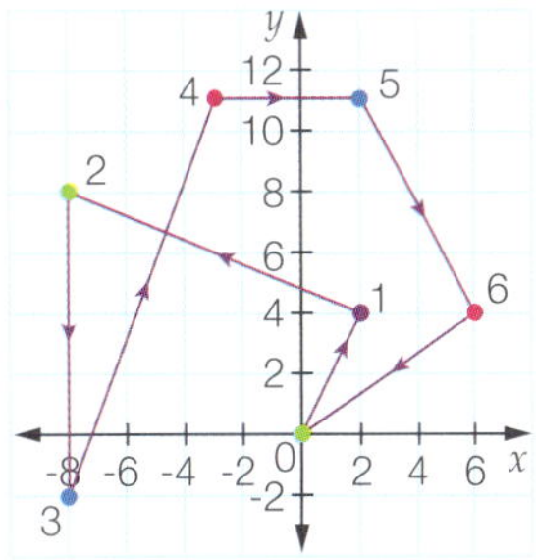

(c) (i) $y = 2x$ (ii) $y = -\frac{2}{5}x + 4\frac{4}{5}$ or $5y + 2x = 24$

(iii) $x = -8$ (iv) $y = \frac{13}{5}x + 18\frac{4}{5}$ or $5y - 13x = 94$

(v) $y = 11$ (vi) $y = -\frac{7}{4}x + 14\frac{1}{2}$ or $4y + 7x = 58$

(vii) $y = \frac{2}{3}x$

(d) Yes. It will be on the direct path between the 5th and 6th buoy.

15 (a) $V = 225 - 0.75T$ (b) 5 min

16 (a) $V = 248 - 4n$

(b) Maximum capacity = 248 million kL

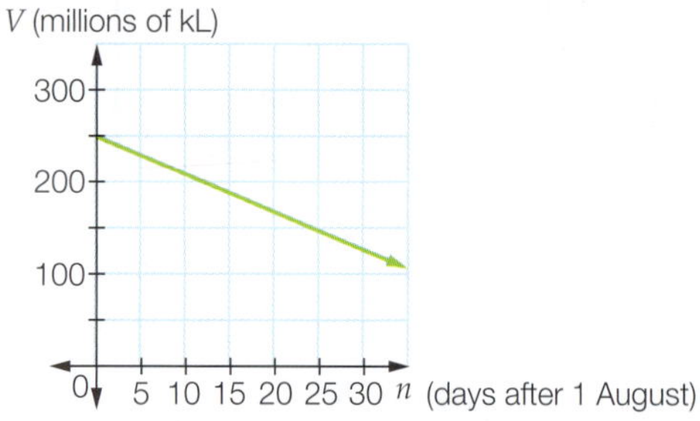

(c) It does not make sense to extend the time and volume into negative values.

17 (a) 1000 cm^3 (b) $V = 1000 - kt$

(c) 125 cm^3 (d) $k = 17.5$

(e) The volume is decreasing. (f) $V = 1000 - 17.5t$

(g) The ice block has completely melted after 57.15 minutes.

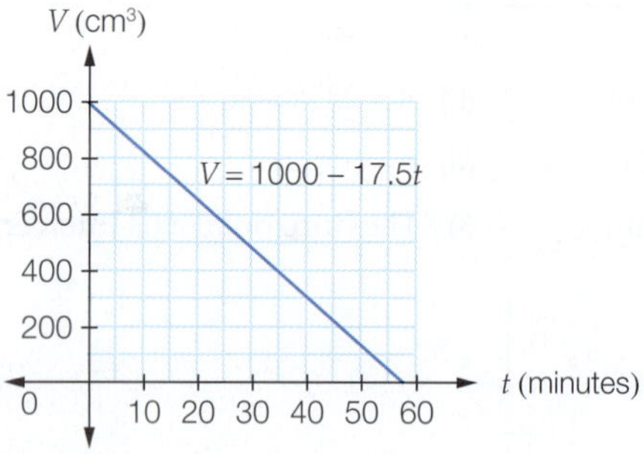

18 (a) Ringo: $R = 600 - 20t$; Sophia: $S = 100 + 30t$

(b) 15.7 weeks

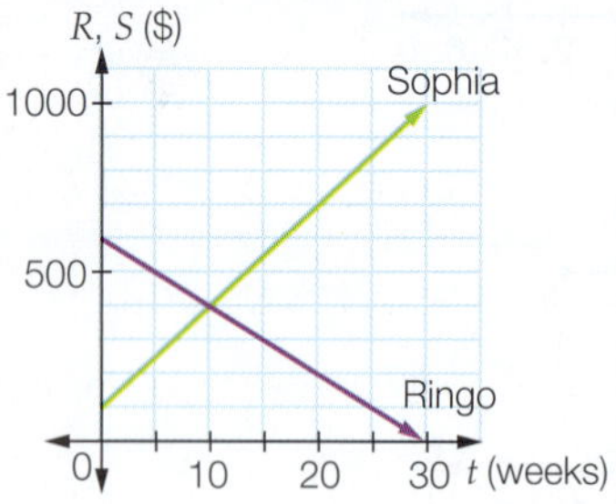

Open-ended – Sample answers

19 $y = 2x - 4$, $y = 2x - 1$, $y = 2x + 2$. The coefficient of x in all three equations is the same, indicating that they have the same gradient.

20 (a) D (b) E (c) A (d) B (e) C

21

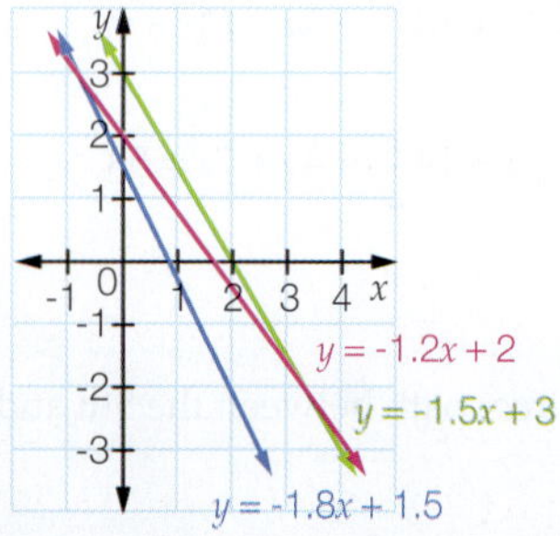

Half-time 1

1 (a)

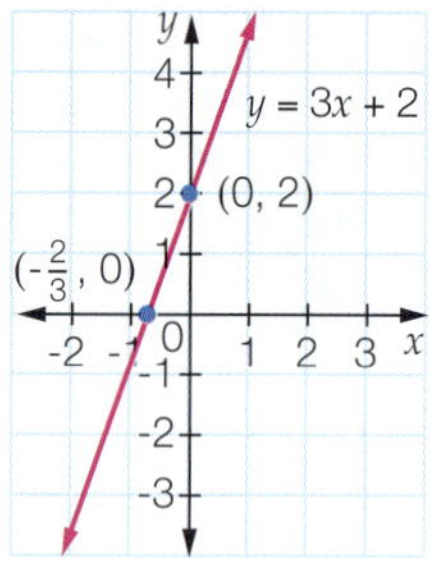

(b)

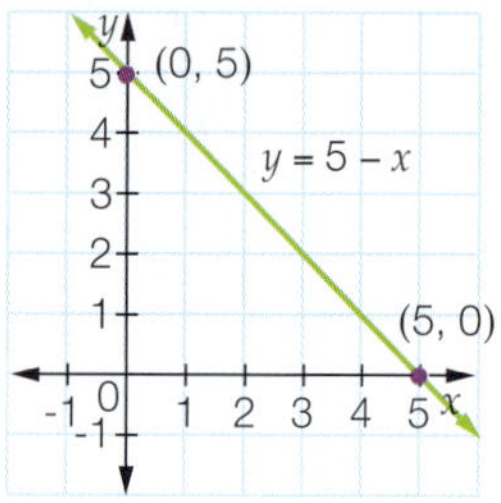

(c)

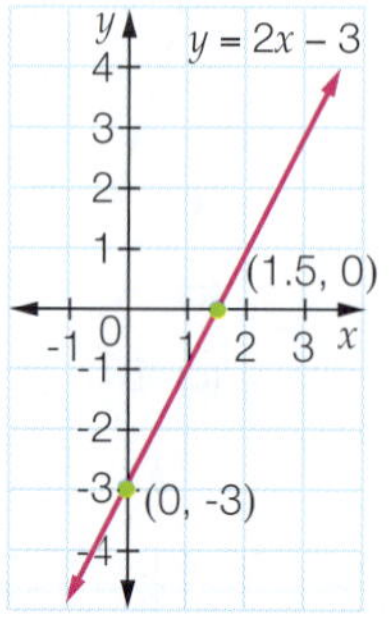

(d)

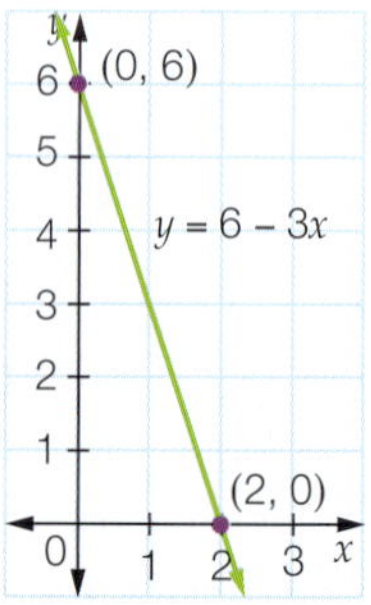

2 (a) $m = 0.5$ (b) $m = -2$ (c) $m = 0.5$ (d) $m = -8$

3 (a)

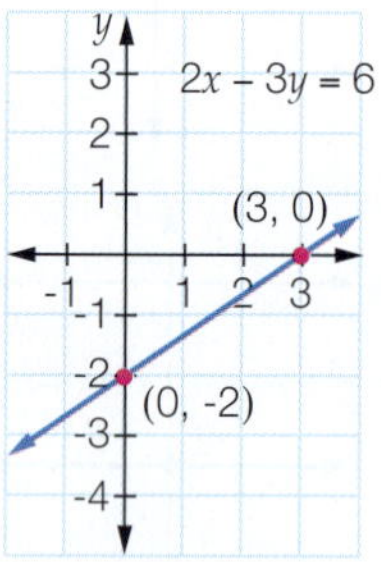

(b)

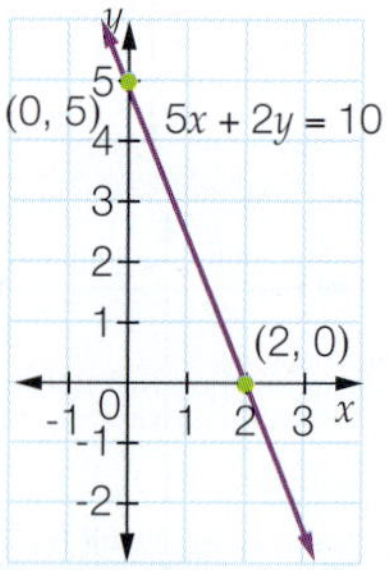

(c)

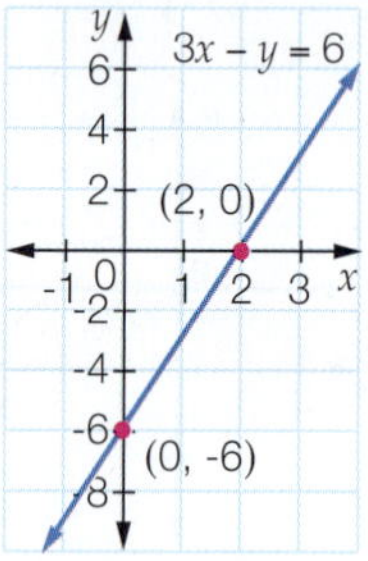

(d)

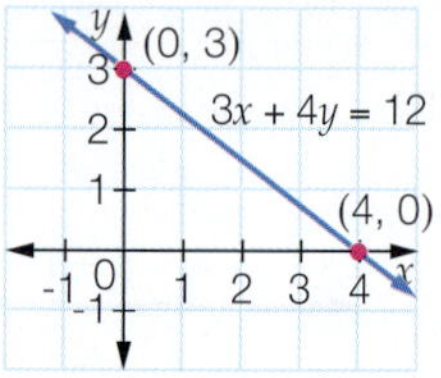

4 Using the points (32, 0) and (212, 100):

$$m = \frac{100 - 0}{212 - 32} = \frac{100}{180} = \frac{5}{9}$$

Hence, $C = \frac{5}{9}F + c$. Substituting values (32, 0):

$$0 = \frac{5}{9} \times 32 + c$$

$$c = -\frac{160}{9}$$

Therefore, $C = \frac{5}{9}F - \frac{160}{9}$

$$C = \frac{5}{9}(F - 32)$$

5 (a)

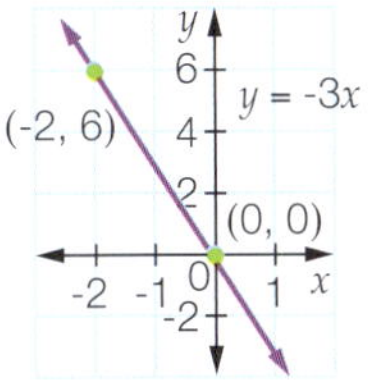

(b)

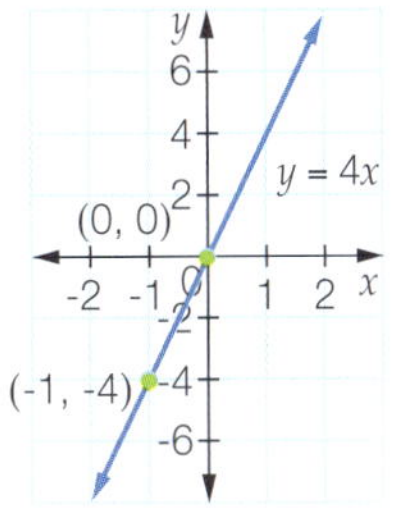

(c)

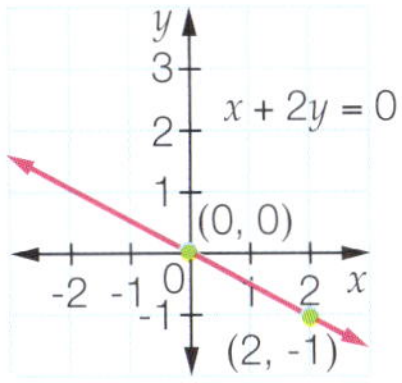

(d)

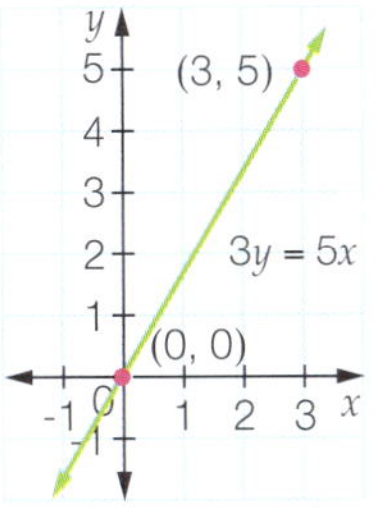

6 (a) $x = 2$ (b) $x = 2$ (c) $x = 19$

(d) $x = 3$ (e) $x = 3$ (f) $x = 4$

7 (a)

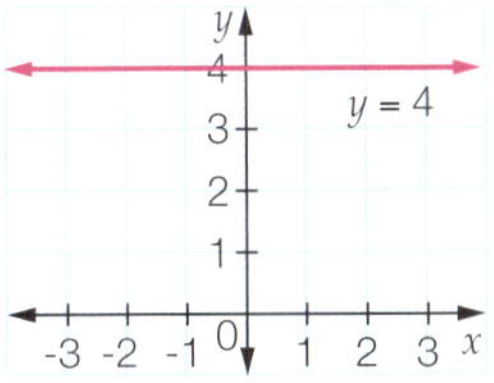

(b)

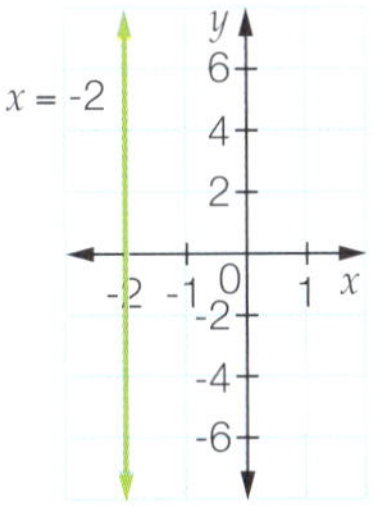

(c)

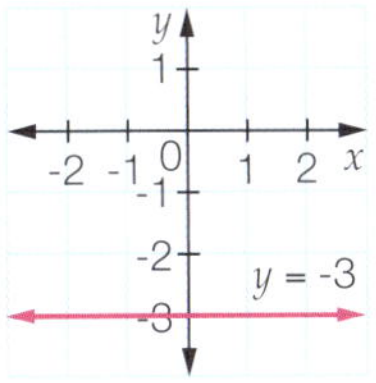

(d)

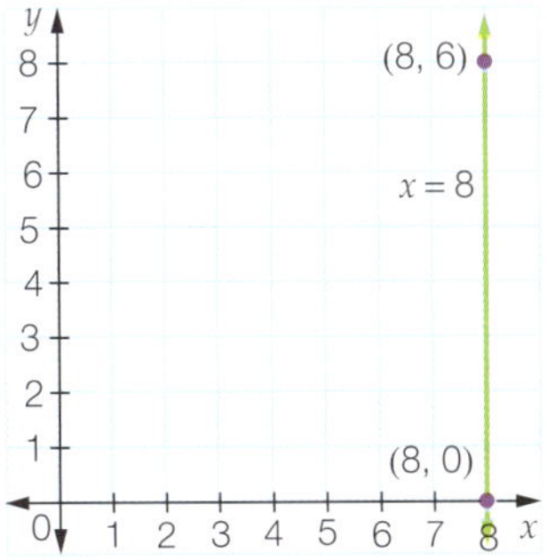

8 (a) $y = 4x + 9$ (b) $y = -x - 4$

Exercise 1.4

1 (a) (i) 3 (ii) $-\frac{1}{3}$

(b) (i) $-\frac{2}{3}$ (ii) $\frac{3}{2}$

2 (a) The gradient of both lines is 2.5, so they are parallel.

(b) The gradient of both lines is 0.5, so they are parallel.

3 (a) The gradient of $4x + 3y = 7$ is $-\frac{4}{3}$ and the gradient of $8y = 6x - 2$ is $\frac{3}{4}$. As $-\frac{4}{3} \times \frac{3}{4} = -1$, the two lines are perpendicular.

(b) The gradient of $2x - 5y - 1 = 0$ is $\frac{2}{5}$ and the gradient of $4y + 10x = 3$ is $-\frac{5}{2}$. As $\frac{2}{5} \times -\frac{5}{2} = -1$, the two lines are perpendicular.

4 (a) $m_1 = 3$, $m_2 = 3$; lines are parallel.

(b) $m_1 = 2$, $m_2 = -2$; lines are not parallel.

(c) $m_1 = 0.5$, $m_2 = 0.5$; lines are parallel.

(d) $m_1 = -4$, $m_2 = 4$; lines are not parallel.

(e) $m_1 = 5$, $m_2 = 9$; lines are not parallel.

(f) $m_1 = 2$, $m_2 = 2$; lines are parallel.

(g) $m_1 = -3$, $m_2 = -3$; lines are parallel.

(h) $m_1 = -3$, $m_2 = -4$; lines are not parallel.

(i) $m_1 = 4$, $m_2 = 4$; lines are parallel.

(j) $m_1 = -\frac{5}{3}$, $m_2 = \frac{1}{5}$ lines are not parallel.

5 (a) $m_1 = 2$, $m_2 = -\frac{1}{2}$, $2 \times -\frac{1}{2} = -1$; lines are perpendicular.

(b) $m_1 = 4$, $m_2 = -\frac{1}{2}$, $4 \times -\frac{1}{2} = -2$; lines are not perpendicular.

(c) $m_1 = \frac{1}{3}$, $m_2 = -\frac{1}{3}$, $\frac{1}{3} \times -\frac{1}{3} = -\frac{1}{9}$; lines are not perpendicular.

(d) $m_1 = \frac{1}{2}$, $m_2 = -2$, $\frac{1}{2} \times -2 = -1$; lines are perpendicular.

(e) $m_1 = \frac{3}{4}$, $m_2 = -\frac{3}{2}$, $\frac{3}{4} \times -\frac{3}{2} = -\frac{9}{8}$; lines are not perpendicular.

(f) $m_1 = \frac{4}{5}$, $m_2 = -\frac{5}{8}$, $\frac{4}{5} \times -\frac{5}{8} = -\frac{1}{2}$; lines are not perpendicular.

6 C 7 D

8 Gradient of $y = x + 10$ is 1. Gradient between the two points is $\frac{8-2}{4-(-2)} = \frac{6}{6} = 1$. Therefore, the lines are parallel.

9 Gradient of $2x + y - 10 = 0$ is -2. Gradient between the two points is $\frac{-7-5}{4-(-2)} = \frac{-12}{6} = -2$. Therefore, the lines are parallel.

10 Gradient of $y = -\frac{4}{3}x - 2$ is $-\frac{4}{3}$. Gradient between the two points is $\frac{4-1}{0-(-4)} = \frac{3}{4}$. $\frac{3}{4} \times -\frac{4}{3} = -1$. Therefore, the lines are perpendicular.

11 Gradient of $y = -\frac{5}{3}x + 10$ is $-\frac{5}{3}$. Gradient between the two points is $\frac{8-5}{9-4} = \frac{3}{5}$. $-\frac{5}{3} \times \frac{3}{5} = -1$. Therefore, the lines are perpendicular.

12 (a) A and C

(b) B and D

13 $y = 3x - 9$

14 Parallel line is $y = -2x + 5$, perpendicular line is $y = 0.5x + 5$.

15 (a) The gradient of AB is -1.

(b) The gradient of BC is 3.

(c) The gradient of AC is -3.

(d) Dakota is not correct.

16 $m_{AB} = -\frac{3}{2}$, $m_{BC} = 5$, $m_{CA} = \frac{2}{3}$. $-\frac{3}{2} \times \frac{2}{3} = -1$. Therefore, AB is perpendicular to CA and ABC must be a right-angled triangle.

17 $m_{AB} = 1$, $m_{BC} = -\frac{1}{2}$, $m_{DC} = 1$, $m_{AD} = -\frac{1}{2}$.

$AB \parallel DC$ and $BC \parallel AD$. Therefore $ABCD$ is a parallelogram.

18 $m_{AB} = -\frac{5}{2}$, $m_{BC} = \frac{2}{5}$, $m_{DC} = -\frac{5}{2}$, $m_{AD} = \frac{2}{5}$. $-\frac{5}{2} \times \frac{2}{5} = -1$, therefore $AB \perp BC$, $BC \perp CD$, $CD \perp DA$ and $DA \perp AB$. Therefore $ABCD$ is a quadrilateral with four right angles. $|AB| = |BC| = |CD| = |DA| = 29$. All four sides are of equal length, therefore $ABCD$ is a square.

Open-ended – Sample answers

19 (a) $y = 0.5x + 4$, $2y = x + 3$, $2y - x = 10$ (all have gradients of 0.5)

(b) $y = -3x - 1$, $3x + y = 10$, $3x = 5 - y$ (all have gradients of -3)

Exercise 1.5

1 (a)

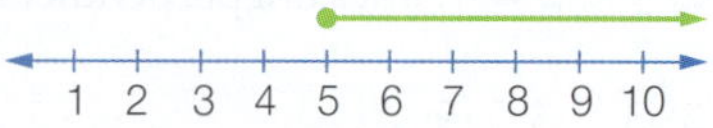

(b)

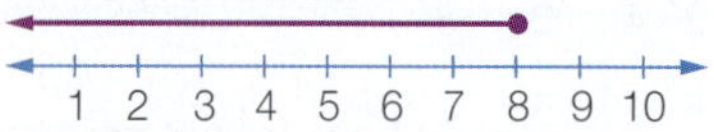

(c)

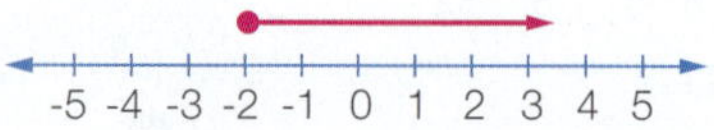

(d)

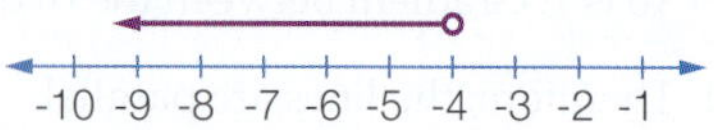

(e)

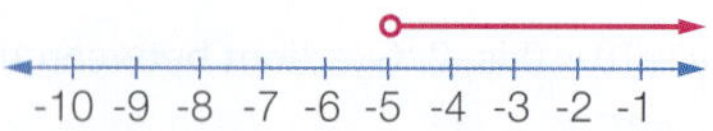

(f)

(g)

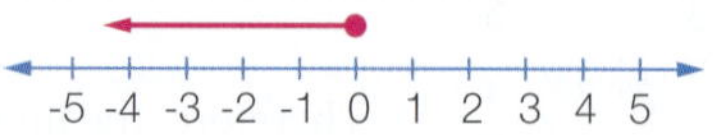

(h)

(i)

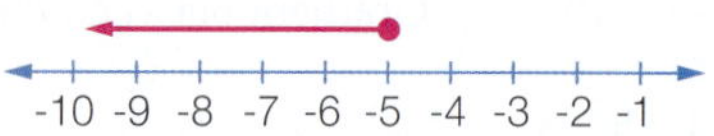

(j)

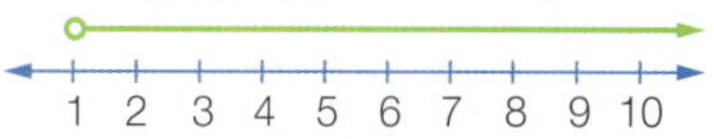

(k)

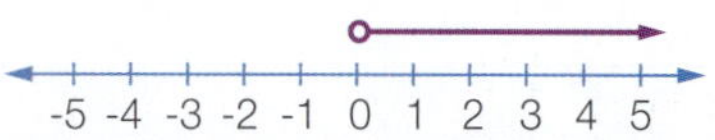

(l)

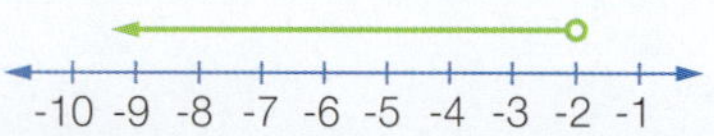

(m)

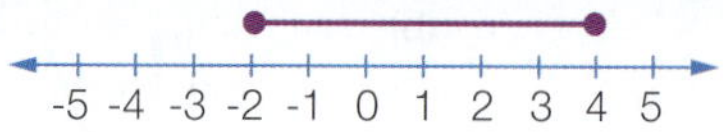

(n)

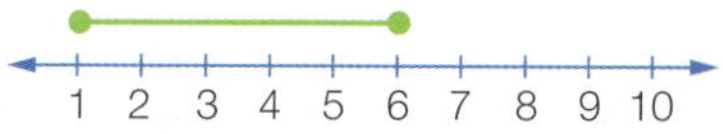

(o)

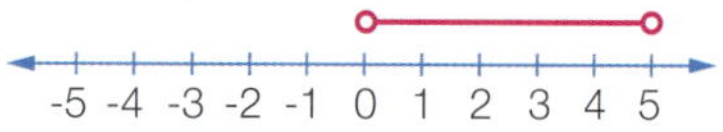

(p)

(q)

(r)

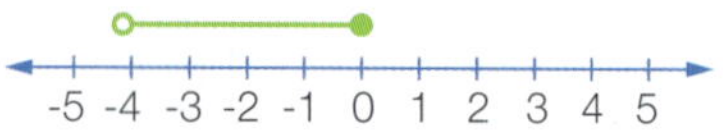

2 (a) $x > -10$ (b) $x < -5$ (c) $x \leq -10$

(d) $x \leq 35$ (e) $x \geq -9$ (f) $x > -28$

(g) $x < -\frac{16}{3}$ (h) $x > -\frac{27}{5}$ (i) $x \leq -9$

(j) $x \geq 8$ (k) $x \leq -2$ (l) $x > -8$

(m) $x > 4$ (n) $x < 11$ (o) $x \leq 4$

(p) $x \leq -8$ (q) $x > 11$ (r) $x > 10$

3 (a) $y \leq -5$ (b) $y \geq -6$ (c) $y \geq -5$

(d) $y > 12$ (e) $y < 21$ (f) $y < 12$

(g) $y \geq -8$ (h) $y \leq 2$ (i) $y \geq -4$

(j) $y \leq -\frac{1}{15}$ (k) $y \geq \frac{1}{2}$ (l) $y \leq \frac{19}{6}$

(m) $y < \frac{13}{3}$ (n) $y > \frac{23}{6}$ (o) $y < 4$

4 (a) B (b) C (c) B (d) D

5 (a) A (b) D (c) A

6 (a) A (b) B (c) D (d) D

7 (a)

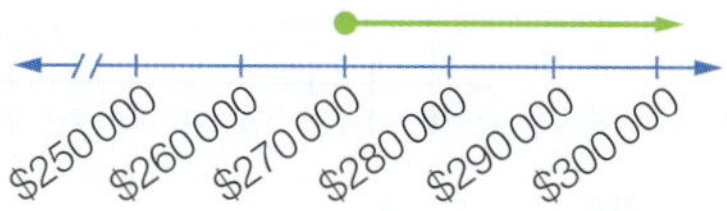

(b) $P \geq 270\,000$ (c) $270\,000 \leq P < 320\,000$

8 $x \leq 100$ 9 $n \geq 5$ 10 $j > 55$

11 (a) $x \geq \frac{2}{3}$ (b) $x \leq -\frac{3}{11}$ (c) $x < 0$

(d) $x > -2$ (e) $x \leq 4$ (f) $x \geq 12$

(g) $x > -25$ (h) $x < \frac{34}{7}$ (i) $x \geq 2$

12 (a) $a < -7$ (b) $a \geq 18$ (c) $a < \frac{22}{7}$

13 (a) $y \geq -1$ (b) $y < 3$ (c) $y \leq -\frac{7}{3}$

(d) $y > \frac{1}{3}$ (e) $y \geq 30$ (f) $y \leq 36$

(g) $y > -12$ (h) $y < \frac{6}{7}$ (i) $y < \frac{61}{13}$

14 (a) Let f be the number of cartons of full cream milk.
Let l be the number of cartons of low fat milk.
$50 \le f \le 120$

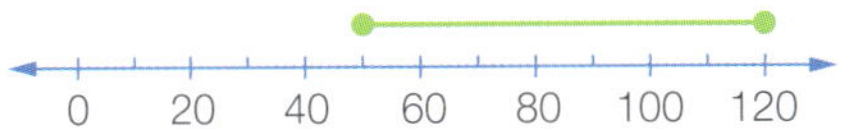

$0 \le l \le 40$

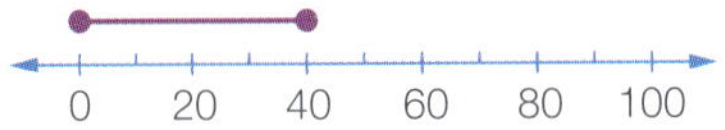

(b) $l + f \le 120$

15 (a) $20p > 428$

(b) $p > 21.4$
Tony should sell the calculators for any price higher than \$21.40 each.

16 (a) Let n be the number of liquorice straps. $0.4n \le 1.55$

(b) $n \le 3.875$
As n must be an integer, $n \le 3$.

Open-ended – Sample answers

17

18 (a) $x - 10 \ge 3x - 8$

(b) $13x - 4 \ge 9x + 8$

Exercise 1.6

1 (a) (1, 3)

(b) (-2, 2)

2 (a) (i)

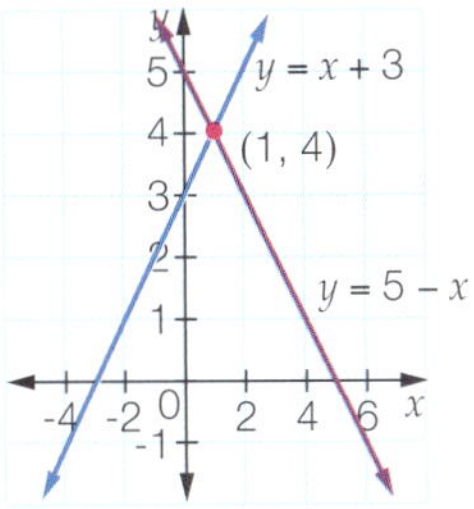

(ii) $x = 1, y = 4$

(b) (i)

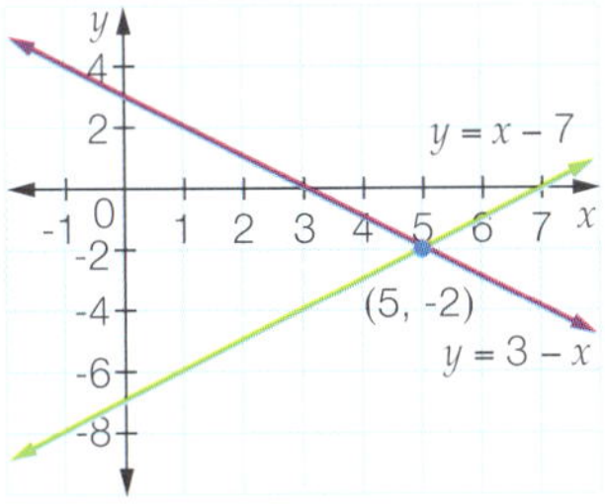

(ii) $x = 5, y = -2$

(c) (i)

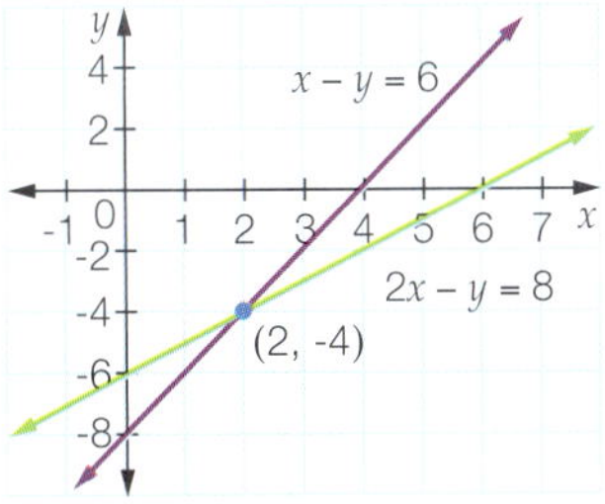

(ii) $x = 2, y = -4$

(d) (i)

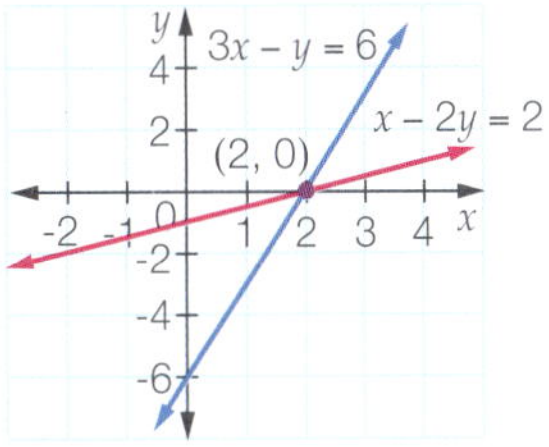

(ii) $x = 2, y = 0$

(e) (i)

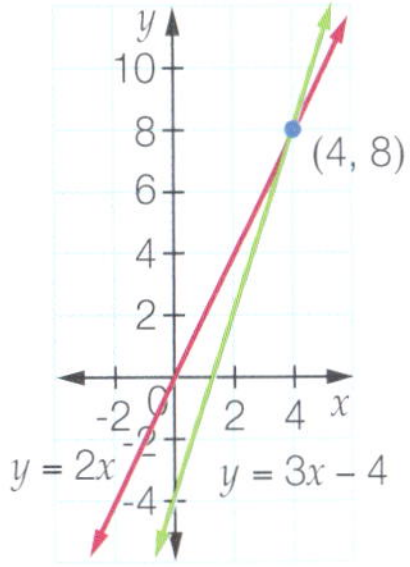

(ii) $x = 4, y = 8$

(f) (i)

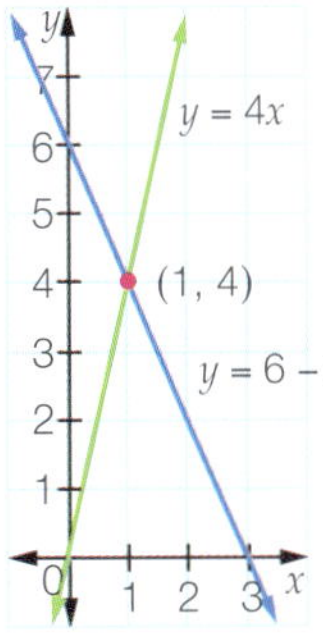

(ii) $x = 1, y = 4$

(g) (i)

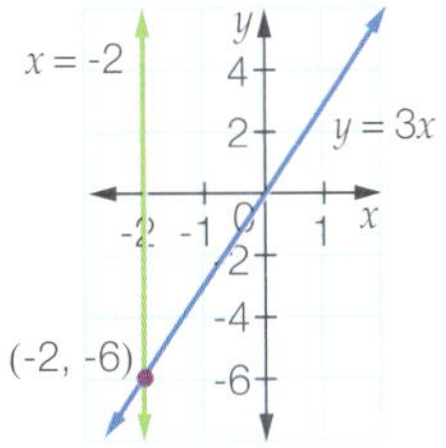

(ii) $x = -2, y = -6$

(h) (i)

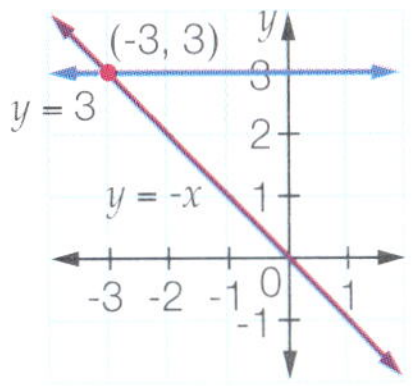

(ii) $x = -3, y = 3$

(i) (i)

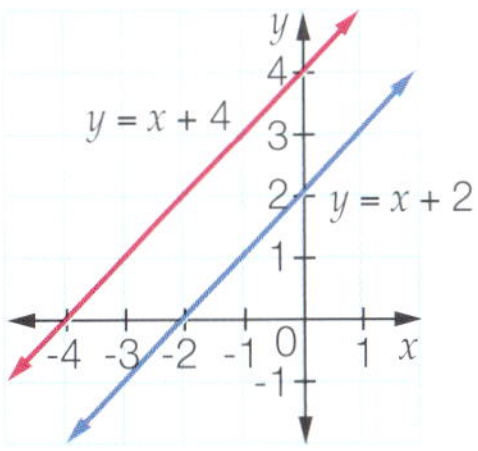

(ii) no solution

3 (a) (4, 1) (b) (5, -4) (c) (3, 1)
(d) (5, -1) (e) (1, 3) (f) (2, -6)
(g) (-2, 3) (h) (2, 2) (i) (2, 3)

4 (a) (1, -1) (b) (-3, 4) (c) (-1, 1)
(d) (2, -1) (e) (1, -1) (f) (-3, 1)
(g) (3, 1) (h) (-3, 4) (i) (1, 1)
(j) (4, -2) (k) (5, 12) (l) (9, 4)
(m) (-5, -10) (n) (-3, -8) (o) (-6, 5)

5 (a) (2, 4) (b) (-3, -9) (c) (4, -2)
(d) (1, 5) (e) (3, 2) (f) (3, -2)
(g) (-2, 5) (h) (9, -1) (i) (-1, 9)

6 (a) C (b) D

7 (a) D (b) B

8 (a) A (b) C

9 (a) D (b) D

10 (a) $5a + 3c = 65$ (b) $7a + 6c = 100$ (c) $10a + 6c = 130$
(d) Adult's ticket is $10.
Child's ticket is $5.

11 a pineapple costs $2.60, a kiwifruit costs 35 cents

12 a salad roll costs $3.60, a pie costs $2.44

13 $x = 7, y = 5$

14 7 and 31

15 128 pages and 372 pages

16 30 m by 40 m

17 three 5-cent coins, 15 10-cent coins

18 Michelle is 18 years old, Yesmin is 31 years old

19 (a) Let D = distance in km from Wes's house and t = time taken in hours. Wes: $D = 80t$, Tony: $D = 360 - 100t$.
(b) 2 hours

20 Speed of barge = 12 km/h, speed of current = 3 km/h

21 (a) $A(-3, -7)$, $B(6, 5)$, $C(4, -2)$, $D(1, 7)$
(b)

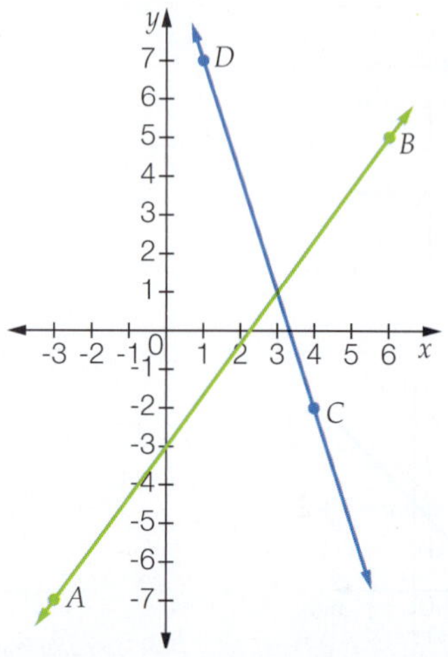

(c) (3, 1)

(d) $y_{AB} = \frac{4}{3}x - 3$ [1]
$y_{CD} = -3x + 10$ [2]
Substitute [1] into [2]:

$$\frac{4}{3}x - 3 = -3x + 10$$
$$4x - 9 = -9x + 30$$
$$13x - 9 = 30$$
$$13x = 39$$
$$x = 3$$

Substitute $x = 3$ into [2]: $y = -9 + 10 = 1$

Therefore, the treasure is at (3, 1).

Open-ended – Sample answers

22 $x + y = 7$ and $x - y = -1$, $2x - y = 2$ and $x + 2y = 11$, $5x + 2y = 23$ and $3x - y = 5$

23 3, 8 and 14. Method example: Let x, y and z be the three numbers.

$x + y = 11$ [1]
$x + z = 17$ [2]
$y + z = 22$ [3]

Using elimination, [1] – [2]:

$$\begin{array}{r} x + y = 11 \\ -\quad x + z = 17 \\ \hline y - z = -6 \end{array} \quad [4]$$

$y = z - 6$

Substitute [4] into [3]:

$$z - 6 + z = 22$$
$$z = 14$$

Substitute $z = 14$ into [2]:

$$x + 14 = 17$$
$$x = 3$$

Substitute $x = 3$ into [1]:

$$3 + y = 11$$
$$y = 8$$

Chapter review 1

1 (a) $c = 8$ (b) $x = -12.5$ (c) $v = 3$ (d) $a = 1$

2 D

3 (a) $m = 5$, y-intercept = (0, 2)
(b) $m = -2$, y-intercept = (0, 0)
(c) $m = \frac{2}{5}$, y-intercept = (0, -2)
(d) $m = -4$, y-intercept = (0, 8)

4 (a)

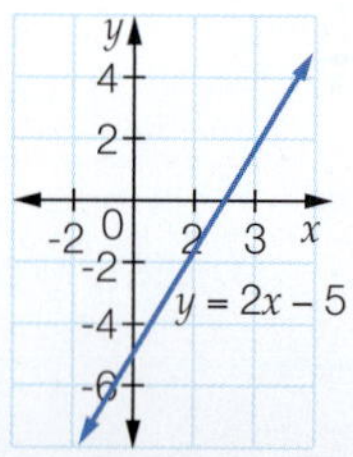

(b)

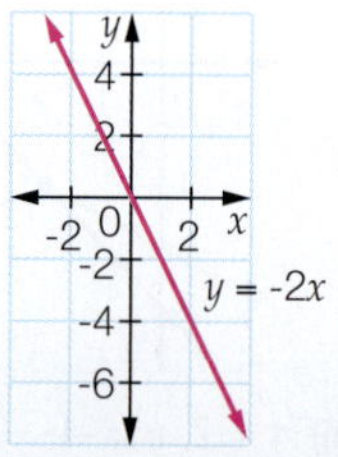

(c)

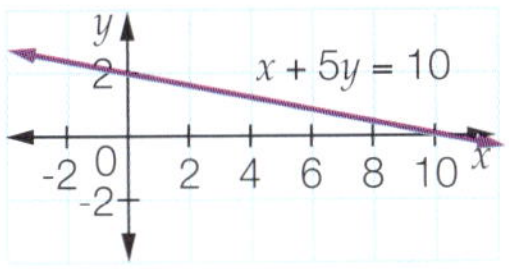

(d)

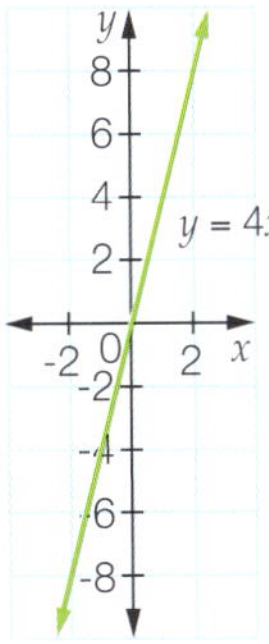

5 (a)

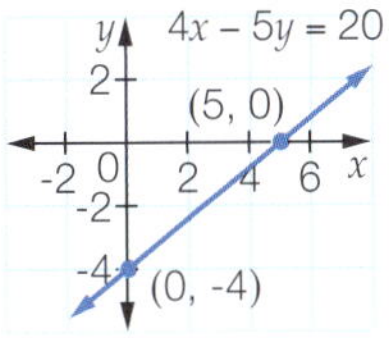

(b)

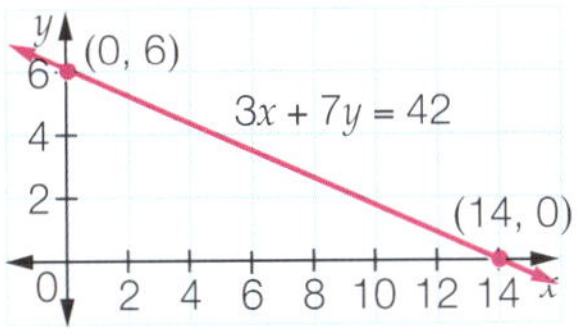

(c)

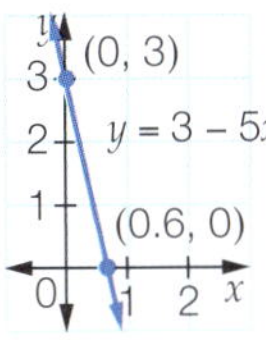

(d)

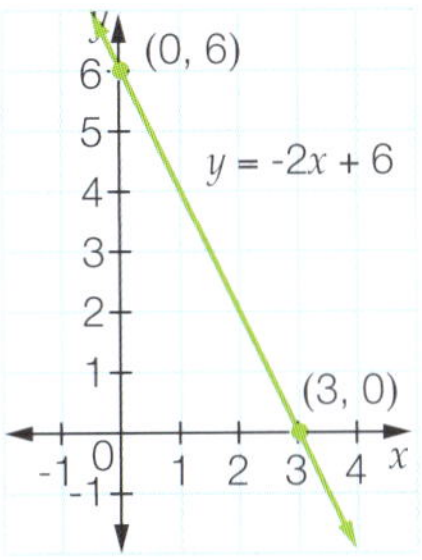

6 (a) Both have gradients of -0.5, therefore they are parallel.

(b) With different gradients of 1 and 1.5 they are not parallel.

(c) Both have gradients of -4, therefore they are parallel.

(d) Both have gradients of 4 so they are parallel.

7 (a) The lines have gradients of -1 and -2. As $-1 \times -2 = 2$, they are not perpendicular.

(b) The lines have gradients of 1 and -1. As $1 \times -1 = -1$, they are perpendicular.

(c) The lines have gradients of 0.5 and -2. As $0.5 \times -2 = -1$, they are perpendicular.

(d) The lines have gradients of $-\frac{3}{2}$ and $\frac{2}{3}$. As $-\frac{3}{2} \times \frac{2}{3} = -1$, they are perpendicular.

8 (a) $x \geq \frac{7}{3}$ (b) $x > -\frac{35}{3}$ (c) $x > \frac{5}{6}$

9 (4, 6)

10 (-1, 3)

11 18 wombats, 9 emus

12 4 hours

13 (a) 8 hours (b) 1500 L (c) 1500 (d) $V = 1500T$

14 (a) $d = 0.6t + 20$

(b)

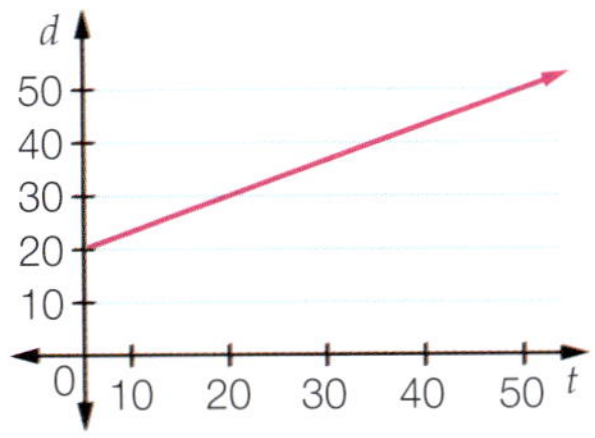

(c) Every minute of travel takes Manny 0.6 km further away from Melbourne.

(d) He began 20 km away from Melbourne.

(e) 28.4 km

(f) 46 min 40 s

15 $y = -x + 9$

16 $y = 5x - 42$

17 (a) $x < 2\frac{4}{5}$ (b) $x \leq -2\frac{2}{3}$

18 6 and -5

19 $a = 9, b = -3$

20 (a) $570\,000 \leq p \leq 620\,000$ (b) $57\,000 \leq d \leq 62\,000$

(c) $p > 670\,000$ (d) $p < 670\,000$

21 $d = 165$ km, $s = 82$ km/h

Chapter 2

Recall 2

1 (a) Mean = 3.2; Median = 4; Mode = 4; Range = 4

(b) Mean = 4.7; Median = 5.0; Mode = 5.0; Range = 2

2 (a) discrete (b) continuous (c) ordinal

3 (a)

Stem	Leaf
16	1 3
16	5 5 5 6 7 7 7 9 | 9
17	0 1 2 2 3
17	8 9
18	0 3

Key: 16 | 1 = 161

(b) Median = 169 cm

4 (a) 6.4875 m (b) 6.50–<6.75 m

Exercise 2.1

1 (a) (i) No mode (ii) Median = 7 (iii) Mean = 7

(b) (i) No mode (ii) Median = 27.5 (iii) Mean = 27.8

(c) (i) Mode = 28 (ii) Median = 28 (iii) Mean = 27.7

(d) (i) Mode = 2 (ii) Median = 2 (iii) Mean = 1.8

2 (a) (i) Modal class: 30–<40

(ii) Estimated mean: 28.4

(iii)

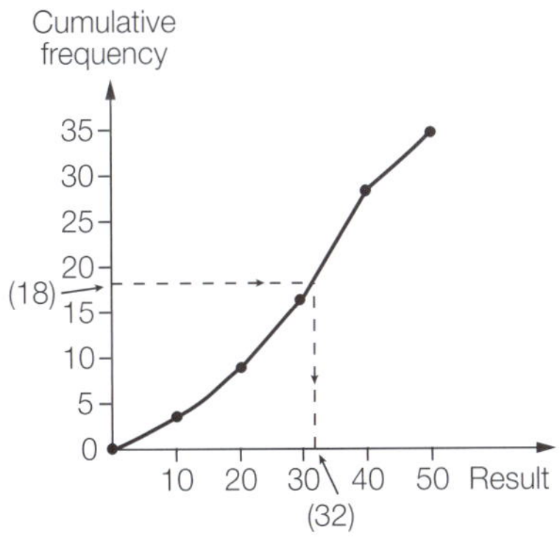

Median = 18th term ≈ 32

(b) (i) Modal class: 300–<350

(ii) Estimated mean: 346.3

(iii)

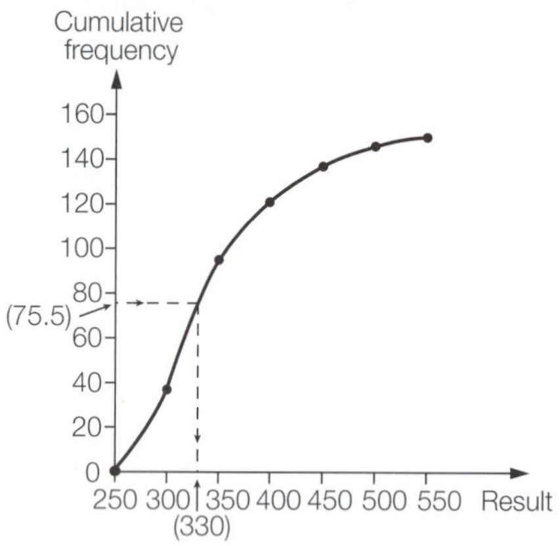

Median = 75.5th term ≈ 330

3 (a) (i) Modal class: 480–489 students

(ii) Estimated mean: 484.1 students

(iii)

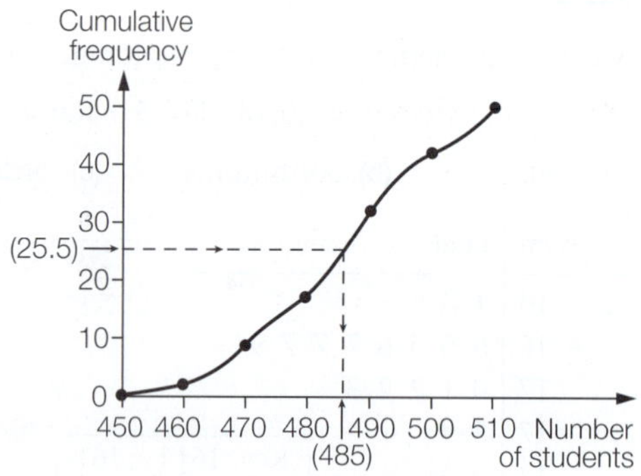

Median = 25.5th term ≈ 485 students

(b) (i) Modal class: 30–39 points

(ii) Estimated mean = 25.9 points

(iii)

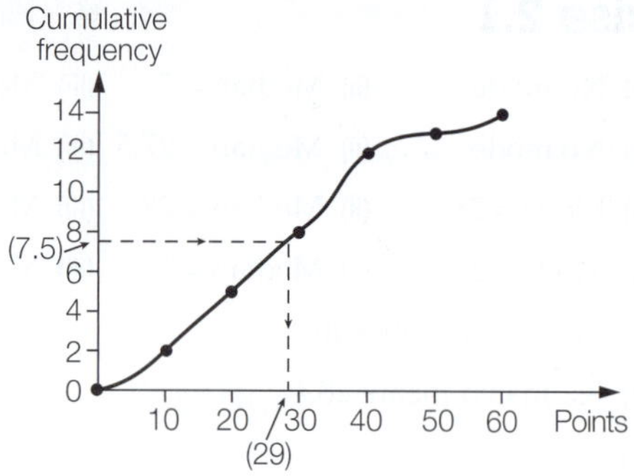

Median = 7.5th term ≈ 29 points

4

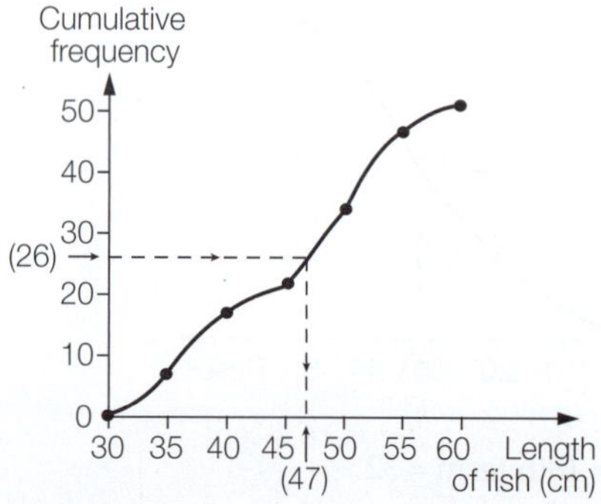

Median = 26th term ≈ 47 cm

5 (a) A (b) C (c) B

6 (a) Median = 21 people (b) 9

(c)

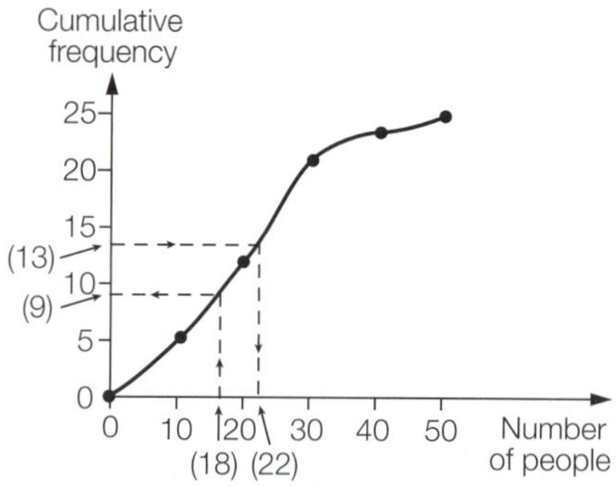

Median = 13th term ≈ 22 people

(d) Using the graph, the estimated median is 22 and there were 9 trains with fewer than 18 people.

(e) The actual median is 21 and the estimated median is 22 people, so the values are about the same.

The actual and estimated number of trains with fewer than 18 people are both 9.

7 (a) 30.9

(b) Close to the estimated median of 32.

(c) 485.7 students

(d) Close to the estimated median of 485.

Open-ended – Sample answers

8 Students' own answers.

Exercise 2.2

1 Median = 87 kg; Range = 38 kg; IQR = 11 kg

2 (a) (i) Range = 50 – 0 = 50

(ii)

Result	Cumulative frequency
<0	0
<10	4
<20	9
<30	17
<40	28
<50	35

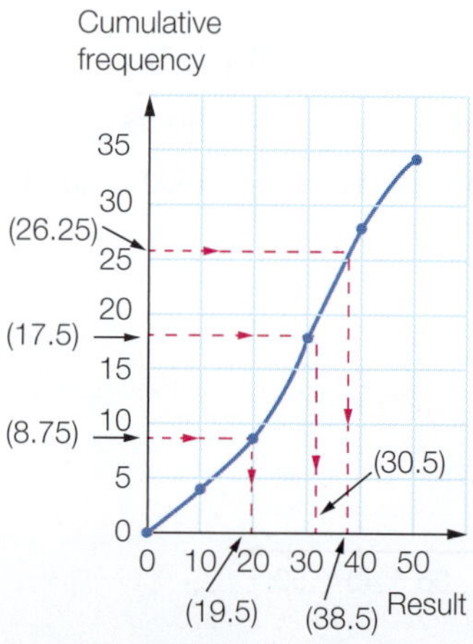

IQR = 38.5 – 19.5 = 19

(iii)

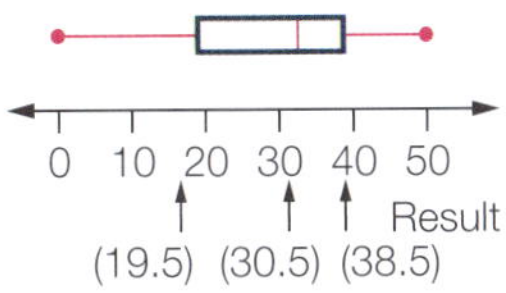

(b) (i) Range = 109 – 60 = 49

(ii)

Result	Cumulative frequency
<60	0
<70	9
<80	21
<90	40
<100	57
<110	70

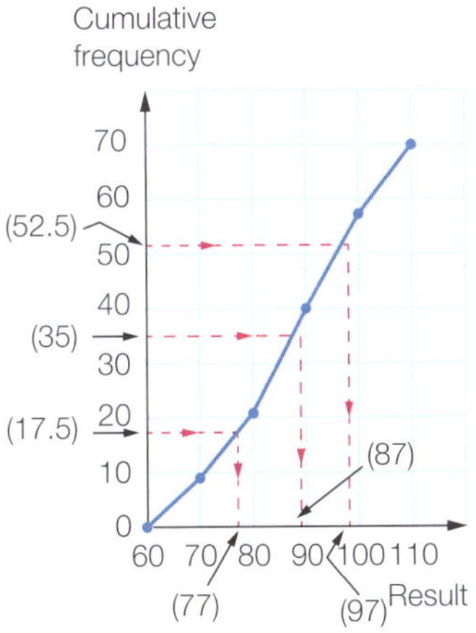

IQR = 97 – 77 = 19

(iii)

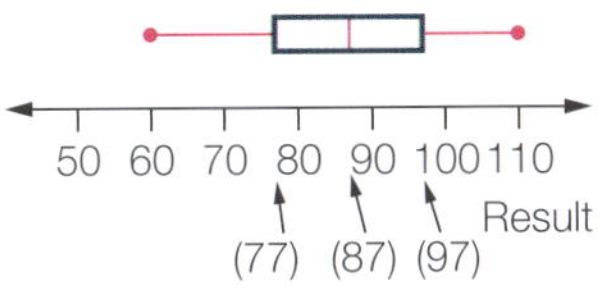

3 (a)

Result	Cumulative frequency	Percentage cumulative frequency
<15	0	0
<16	4	17
<17	9	39
<18	16	70
<19	19	83
<20	23	100

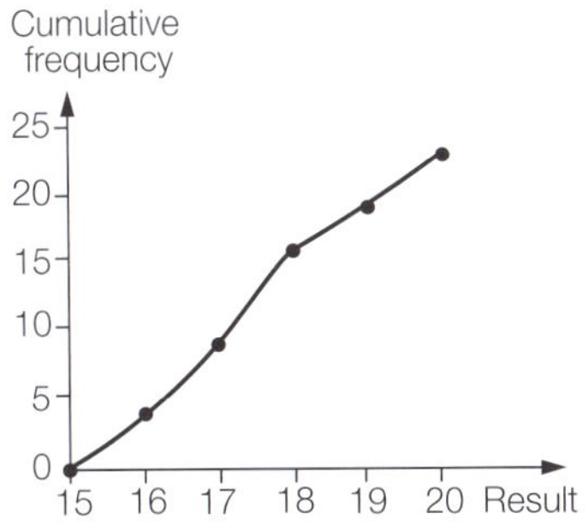

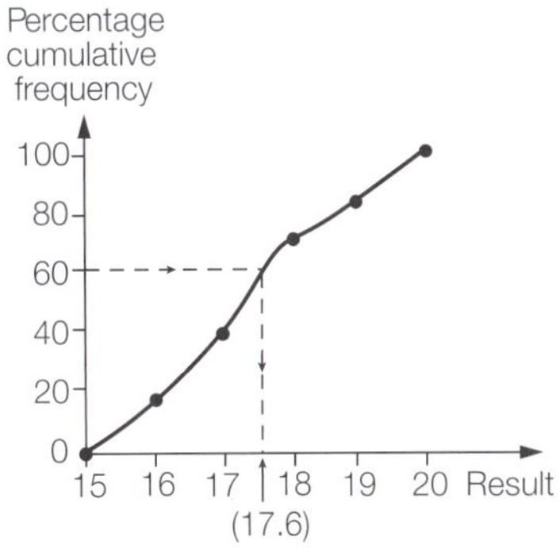

60th percentile is 17.6.

(b)

Result	Cumulative frequency	Percentage cumulative frequency
<150	0	0
<155	1	8
<160	3	23
<165	6	46
<170	10	77
<175	13	100

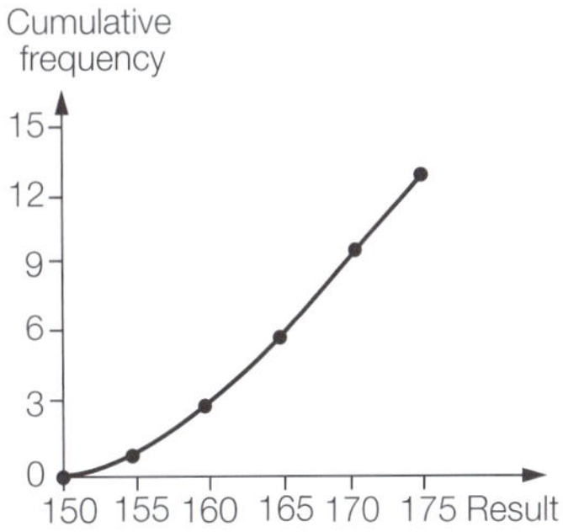

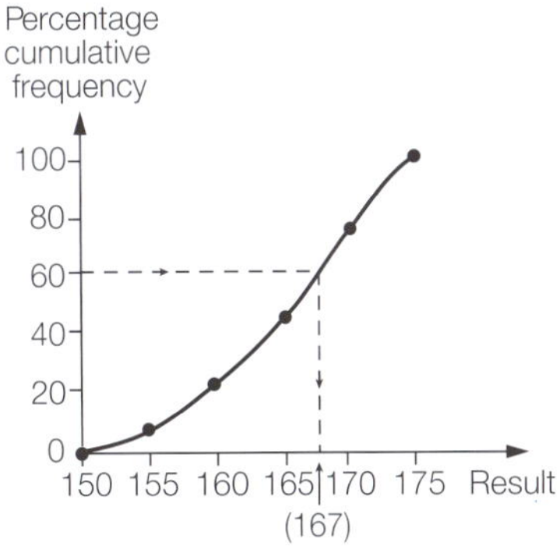

60th percentile is 167

4

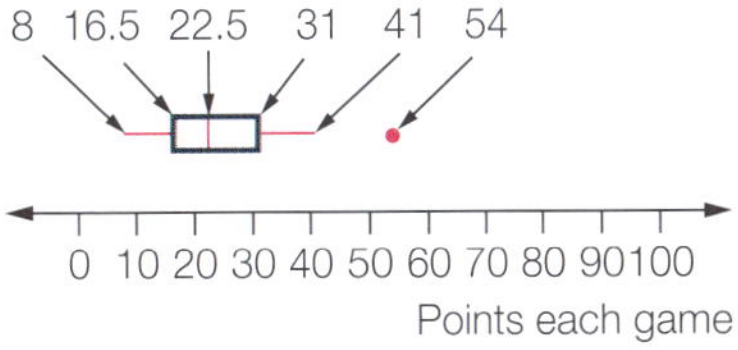

5 (a) 65

(b) $Q_L = 159.5$; $Q_U = 179.5$; IQR = 20

(c) lower than 129.5 or higher than 209.5

(d) no

6 (a) C (b) D (c) B

7 D

8 (a)

Distance run (km)	Cumulative frequency	Percentage cumulative frequency
<10	0	0
<15	4	7
<20	10	17
<25	23	39
<30	39	66
<35	50	85
<40	59	100

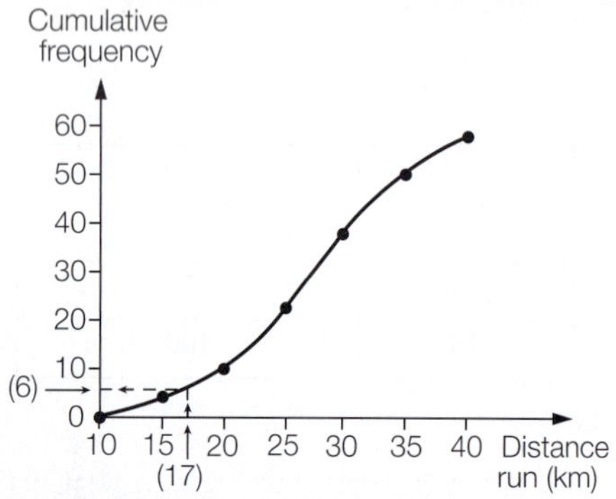

(b) Six distances run were less than 17 km.

(c)

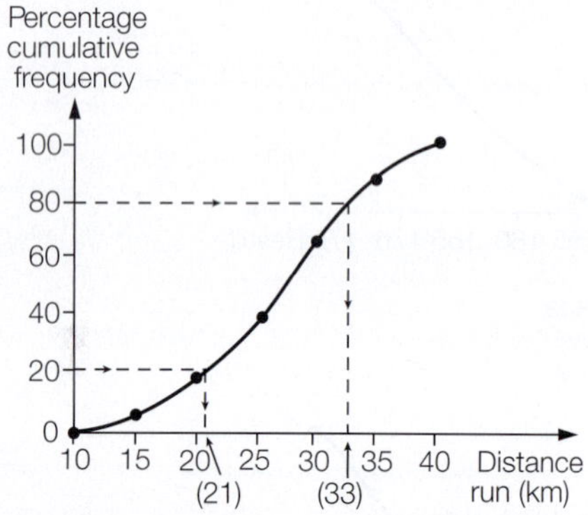

(d) 20th percentile is 21; 0.8 quantile is 33.

9 (a) 1635

(b) $Q_L = 972$; $Q_U = 1275$; IQR = 303

(c)

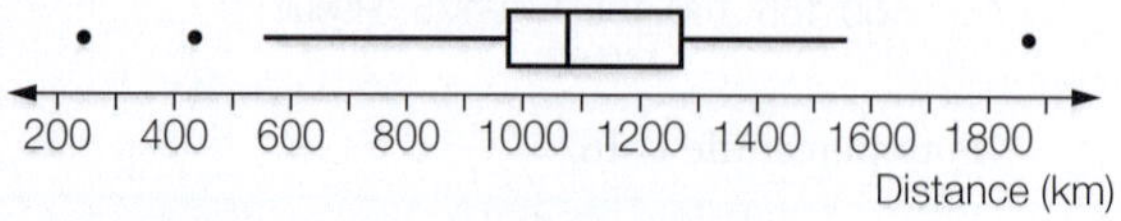

10 (a) (i) less than -5 and greater than 115

(ii) no

(b) (i) less than 32.5 and greater than 52.5

(ii) yes, at both ends

11 (a) mean = \$46 829.27, median = \$45 000

(b) range = \$50 000, IQR = \$10 104.17

(c) Yes, values above \$66 093.76, so at least the three highest values are outliers and perhaps one or both of the values in the \$65 000–< \$70 000 class interval.

(d) Removing only the three highest values gives mean = \$45 664.56, median = \$44 791.67, range = \$35 000, IQR = \$9687.50.

(e) The mean and range are more affected by the removal of outliers than the median and IQR.

Open-ended – Sample answers

12 (a) (i) 2 2 2 2 2 2 4 4 5 5
5 5 5 5 5 5 6 6 7 7

(ii) The lowest value is the same as the lower quartile so there needs to be at least six 2s in the list, assuming exactly 20 values. The median does not appear as a separate line, so it must be the same as either the lower or upper quartile. In this case the upper quartile was chosen. If the lower quartile was chosen, then there would have to be at least eleven 2s.

(b) (i) 3 3 3 3 3 3 4 5 5 6
6 6 6 6 7 7 7 7 7 7

(ii) In this case there are no whiskers, so at least six 3s and six 7s are needed. Place 6s in the middle to ensure that 6 is the median and then fill in the gaps.

13 (a) Students' own graph.

(b) Students' own graph.

14 (a) 1 2 3 4 5 6 7 8 9 10 11 12 13 14 15 16 17 18 19 20 34

IQR = 11, $Q_L = 5.5$, $Q_U = 16.5$, so any value above 16.5 + 16.5 = 33 will be an outlier. The given upper limit is 38.5.

(b) 36 50 51 52 53 54 55 56 57 58 59 60 61 62 63 64 65 66 67 68 69

IQR = 11, $Q_L = 53.5$, $Q_U = 64.5$, so any value below 53.5 – 16.5 = 37 will be an outlier. The given lower limit is 31.5.

Exercise 2.3

1 (a) Set A: (i) 30.74 (ii) 30 (iii) 18 (iv) 10
Set B: (i) 30.21 (ii) 30 (iii) 28 (iv) 10

(b) Set A: (i) 51.22 (ii) 51.5 (iii) 16 (iv) 8
Set B: (i) 51.22 (ii) 51.5 (iii) 56 (iv) 8

(c) Set A: (i) 16.82 (ii) 15.5 (iii) 15 (iv) 8
Set B: (i) 16.64 (ii) 14.5 (iii) 15 (iv) 8

(d) The mean is affected by all changes. The only time the mean did not change in these samples was when the lowest number and the highest number were adjusted by the same amount in opposite directions. The median is very stable and was only affected slightly when data in the middle of the data set was changed. The range fluctuates and is severely affected by any change in the extreme values. The IQR is also quite stable.

2 (a) 8C: 155 cm; 8D: 155 cm

(b) 8C: 24; 8D: 15

(c) 8C: 11; 8D: 8

(d)

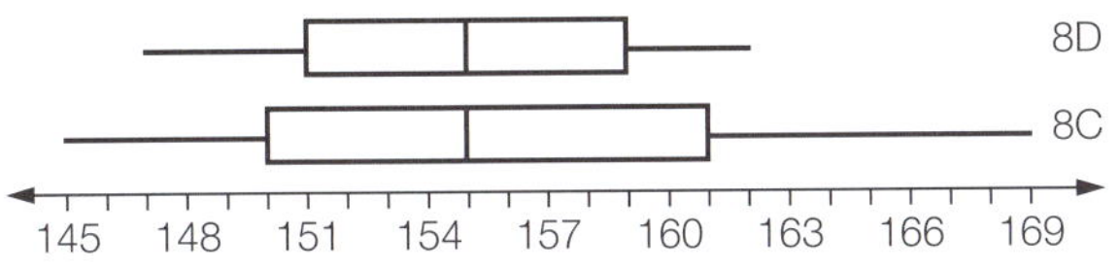

(e) 8C has a much greater range and IQR than 8D. They have the same median value. The height of students in 8D is much more consistent than for the students in 8C.

3 Brisbane Broncos

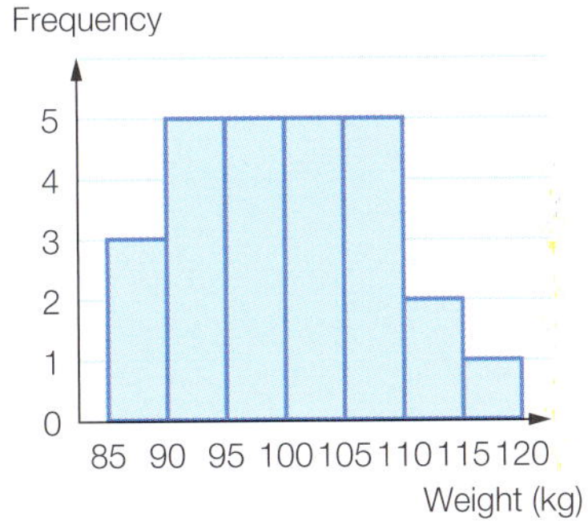

(b) min: 85 = lower quartile = 94; median = 99; upper quartile = 107; max = 115

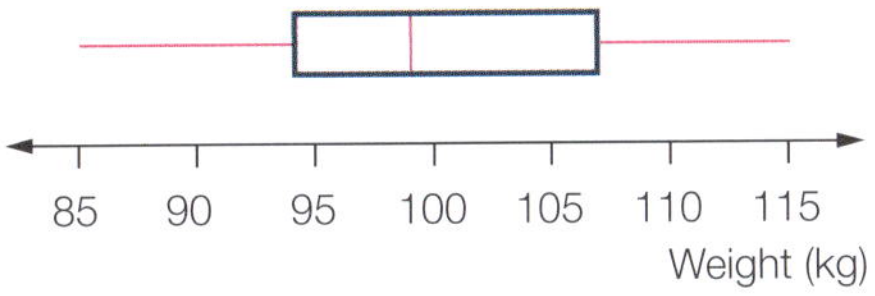

(c)

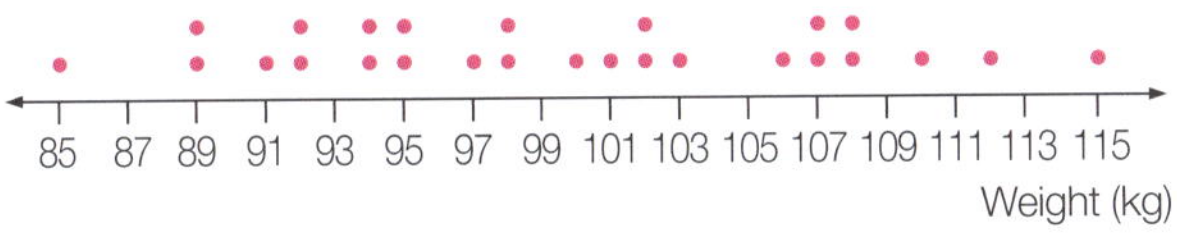

(d) Accuracy has been lost in the histogram by grouping the data, making it appear that the weights range from 85 kg to 120 kg. The box plot displays the median, range and IQR accurately. All three displays suggest the data is evenly spread with less density at the extremes. The dot plot has retained all raw data. It can be clearly seen that there is a 4 kg gap between the lightest player and the second lightest.

4 (a)

Weight (kg)	Cumulative frequency
<85	0
<90	3
<95	8
<100	13
<105	18
<110	23
<115	25
<120	26

(b) Brisbane Broncos

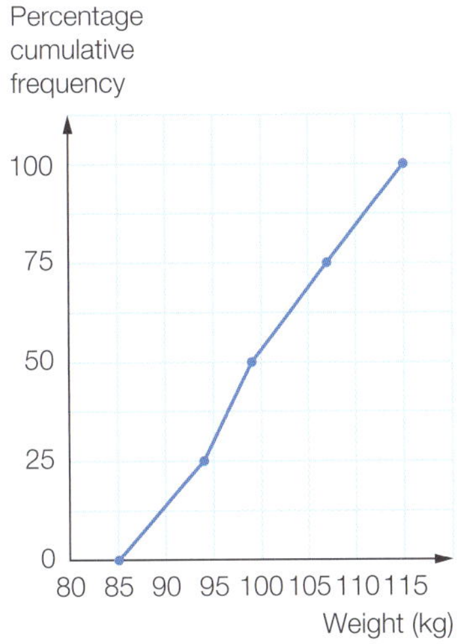

(c)

Weight (kg)	Cumulative frequency
85	1
89	3
91	4
92	6
94	8
95	10
97	11
98	13
100	14
101	15
102	17
103	18
106	19
107	21
108	23
110	24
112	25
115	26

5 (a) (i) Set A: 45; Set B: 50

(ii) Set A: 35; Set B: 35

(iii) Set A: 10; Set B: 20

(b) Although they have the same range, the two data sets are quite different. Set B has a much bigger IQR, whereas Set A has a much longer right-hand whisker. About 75% of Set A are less than the median of Set B.

6 (a) B (b) B

7 The large range and relatively large IQR for Class A means there were some students who did not perform very well on the test, as well as some who did extremely well. The fact that some students did very well increased the mean. Class B was the most consistent class. You could argue Class B did best because of its even results. However, the very highest results are unlikely to be found in Class B. The high IQR for Class C indicates that the middle results were very spread out but the other results were not far outside the middle box.

8 (a) (i) 3 (ii) 1

(b) (i) 21 (ii) 1

(c) The median and IQR give a better feel for the data set as a whole but do not give any hint that there are two extreme outliers.

9 (a) Club B seems to be taller.

(b) Mean heights: Club A is 171.75 cm and Club B is 174.1 cm

Median heights: Club A is 169 cm and Club B is 175 cm

(c) Range: Club A is 20 and Club B is 10; IQR: Club A is 8.5 and Club B is 5.

(d)

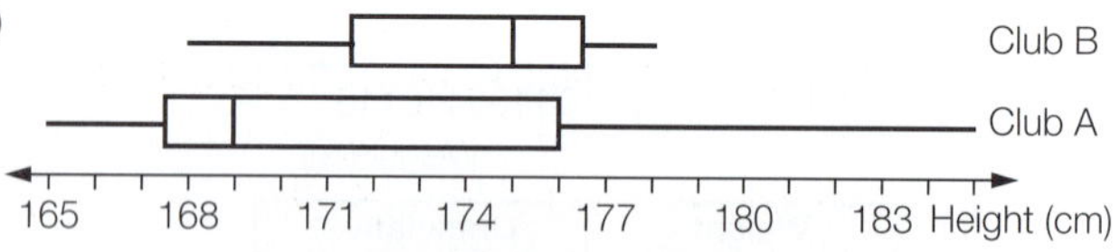

(e) Yes. Even though Club B does not have the tallest player, it has more consistently tall players, shown by the smaller range and IQR spread over higher values.

(f) The bunching of the dots for Club B shows that it will have a small IQR. The parallel dot plot clearly indicates that the range for Club A will to be much greater.

10 (a) Brisbane

Mean height = 188.8 cm

Mean weight = 86.8 kg

Essendon

Mean height = 187.3 cm

Mean weight = 84.6 kg

(b) Brisbane

Median height = 189 cm

Median weight = 85 kg

Essendon

Median height = 186.5 cm

Median weight = 83 kg

(c) Brisbane

Height: Range = 28 cm, IQR = 10.5 cm

Weight: Range = 31 kg, IQR = 9.5 kg

Essendon

Height: Range = 29 cm, IQR = 9 cm

Weight: Range = 40 kg, IQR = 14 kg

(d)

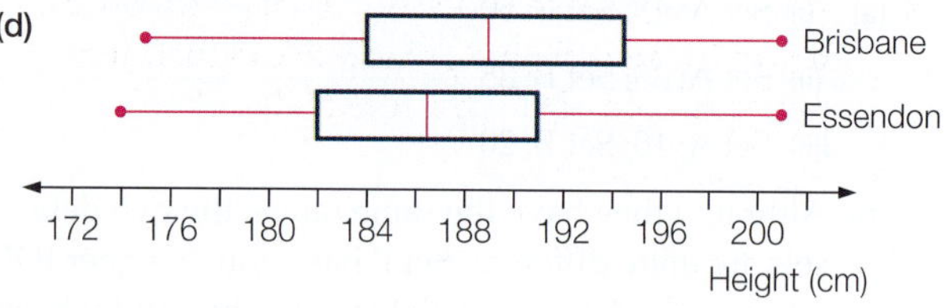

(e)

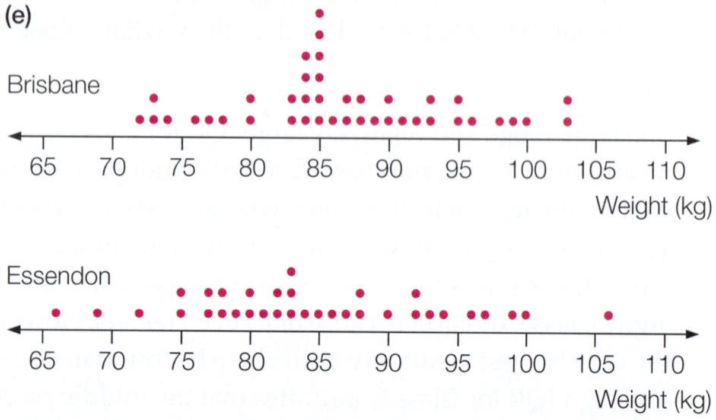

(f) Both the mean and median show that the Brisbane players are taller, on average.

(g) As you would expect with Brisbane being taller, on average, both the mean and median weights confirm that the Brisbane players are heavier, on average. The range, IQR and dot plot show a much greater spread in the weights of Essendon compared to Brisbane.

11 (a)

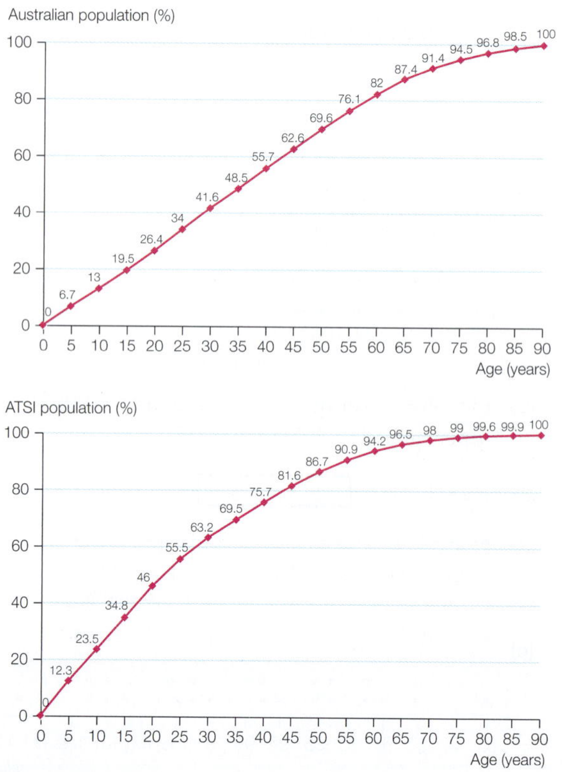

(b) Australia: min = 0; $Q_L = 18$; median = 35; $Q_U = 52$; max = 90

ATSI: min = 0; $Q_L = 10$; median = 22; $Q_U = 39$; max = 90

(c)

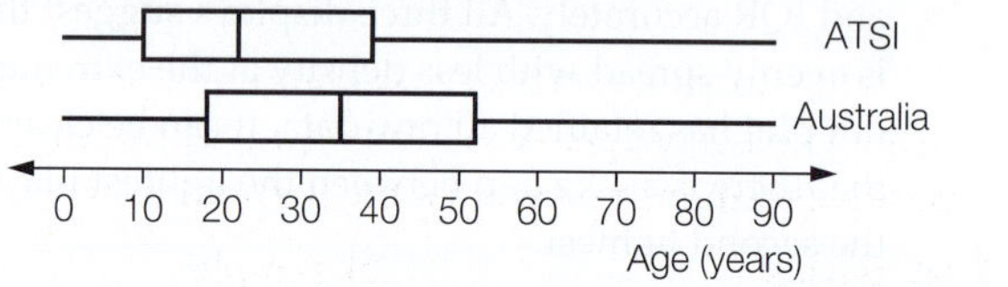

(d) The cumulative frequency curves show the difference with their very different shapes. The rapid rise at the beginning of the ATSI graph indicates that a great percentage of the population are young. This is confirmed by the box plot, which shows the median age to be 22 and the lower quartile to be 10. This means 25% of the ATSI population is less than 10 years of age, compared to an age of 18 for the Australian population as a whole. The medians are different by 13 years. The upper quartile for the ATSI population is barely bigger than the median for the Australian population.

Open-ended – Sample answers

12 (a) The distances range from 2 to 40 km with a median distance of 12 km. This means about 50% of the students travel up to 12 km to school. The IQR lies between 8 and 26 km.

(b) Students to give their own answer, but it should be greater than 12 km.

(c) Students to give their own answer, but a possible answer would be based on:

Distance (km)	Frequency
0–<4	10
4–<8	30
8–<12	40
12–<16	10
16–<20	15
20–<24	10
24–<28	15
28–<32	10
32–<36	10
36–<40	10

This data set has a mean of 16.5 km.

Exercise 2.4

1 (a) primary (b) primary

(c) secondary (d) primary

(e) secondary

2 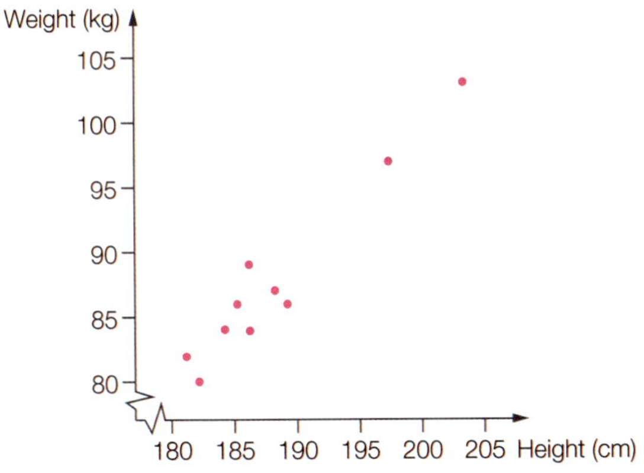

As height increases so it seems does weight. However, this relationship is not perfect so you might say there is a strong positive relationship between the two variables.

3 (a) census (b) sample

(c) census (d) census

4 (a) Inappropriate—not relevant.

(b) Probably appropriate—if not often then may not know much about sport.

(c) Inappropriate—is on the way to being appropriate but should not be a closed question.

(d) Inappropriate—not relevant.

5 (a) Biased—not a broad enough range of people.

(b) Biased—may be more likely to support environmental causes.

(c) unbiased

(d) An inappropriate method, whether biased or not is unclear.

6 (a) 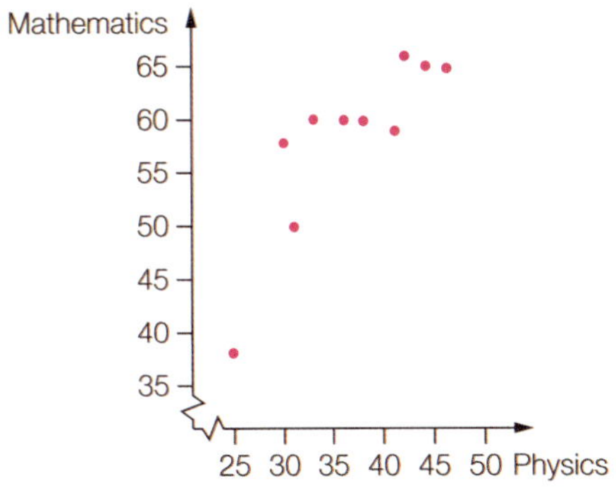

(b) For the lower physics marks it seems that as the mark increased so did the mathematics mark. However, this trend did not continue with the higher physics marks where the mathematics mark seemed to flatten out and not increase at the same rate as the physics mark. Overall, there is a moderate positive relationship.

7 (a) This is a very strong negative linear relationship.

(b) This is a strong positive linear relationship.

(c) This looks like a non-linear relationship; it has a curve associated with it. However, it does display a negative relationship.

(d) This looks like there is no relationship between the variables.

(e) This looks like a non-linear relationship. The points go up and down as you move across the horizontal axis.

(f) This looks like a weak negative linear relationship.

8 (a) It would be difficult and time consuming to get the opinions of parents compared to students and staff who are on campus.

(b) Too many to conduct a census.

(c) You need to keep the representation level the same as for students.

(d) A stratified random sample will randomly select students from each strata, in this case year levels. It is quick and relatively cheap to conduct, and representative of the structure of the school.

(e) Students' own answers.

9 (a) 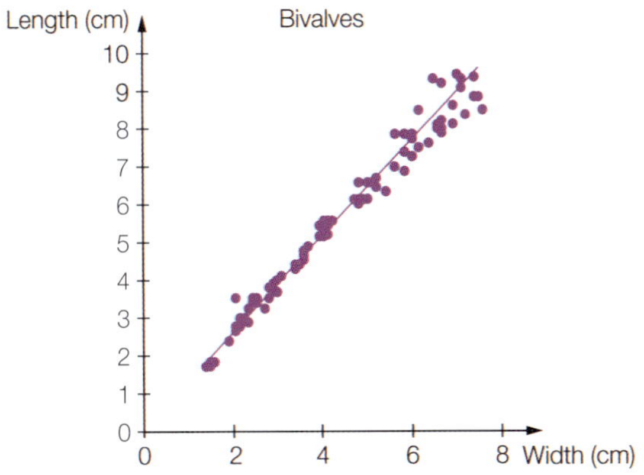

(b) As the width of the bivalve increases so does the length. These two variables seem to have a very strong positive linear relationship.

10 (a) secondary

(b) There seems to be a strong positive linear connection between the two variables. Male rates are possibly a little higher than female rates.

(c) No—a different population group entirely.

11 (a) The countries with a higher literacy rate have a higher number of radios.

(b)

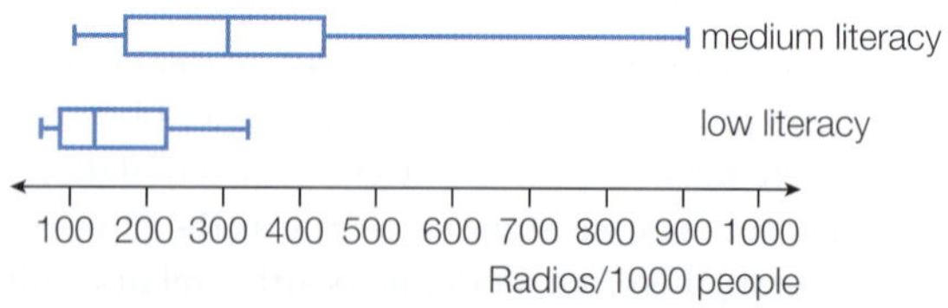

(c) The range for medium literacy countries is much higher but the lowest value for the medium countries is higher than the lower quartile for the low countries. The highest literacy value for the low countries is only just higher than the median value for the medium countries. The medium literacy countries have a very long right-hand whisker; it may be that there are one or two outliers that have made the comparisons inaccurate. The numbers of radios are clearly different between these two types of groups of countries. However, it would be incorrect to conclude that a country's literacy rate determines the number of radios.

(d) 907 radios per 1000 people in a medium literacy rate country.

Open-ended – Sample answers

12 Students' own answers.

13 Students' own answers.

Half-time 2

1 (a) Range = 30

(b)

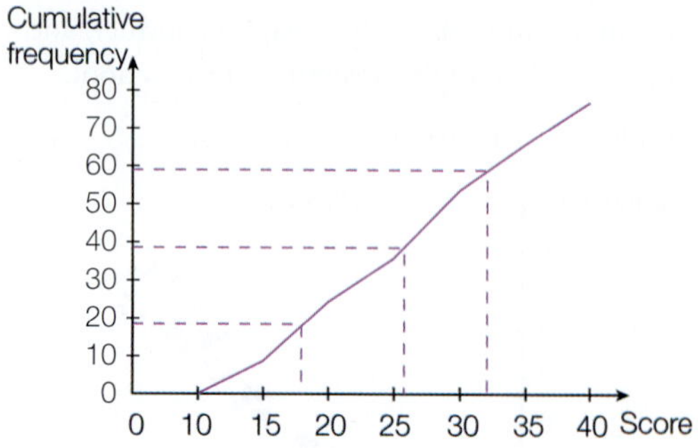

From the cumulative frequency graph $Q_L = 18$ and $Q_U = 32$, which gives an IQR of $32 - 18 = 14$.

(c)

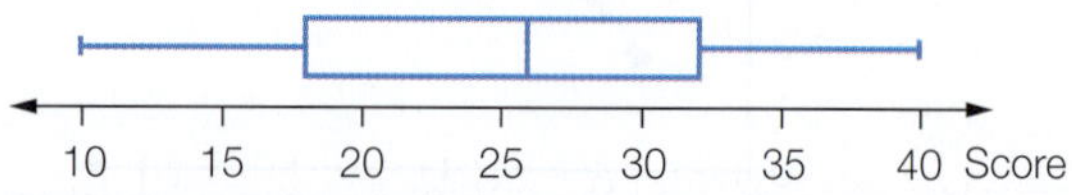

2 (a) Club B appears to be taller but this is mainly because it does not have as many players at the lower end of the height scale.

(b) Club A: mean = 178.47, median = 178, range = 18, IQR = 8

Club B: mean = 178.4, median = 178, range = 17, IQR = 6

(c)

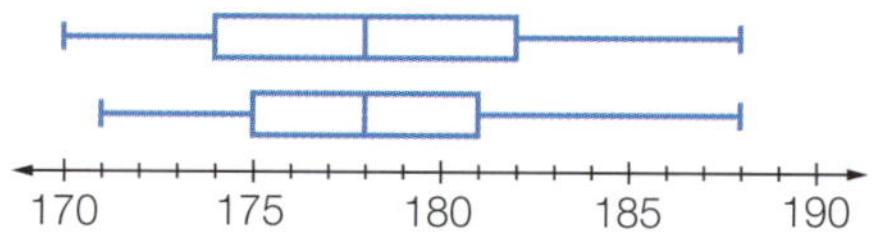

(d) Club A has a very slightly larger average height, but they are essentially the same. The IQR in Club A is also slightly larger, which shows a slightly larger spread of heights compared to Club B.

(e) The IQR, as this statistic has the greatest difference and shows how the spread of the data varies between the two clubs.

3

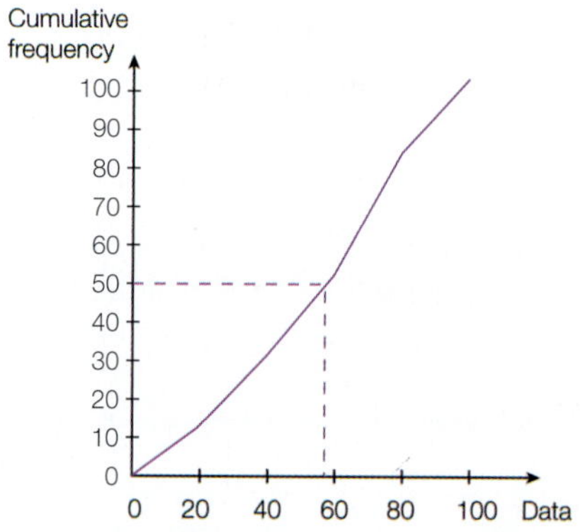

Median = 58

4 There appears to be a moderate positive linear relationship between head circumference and length in babies at birth.

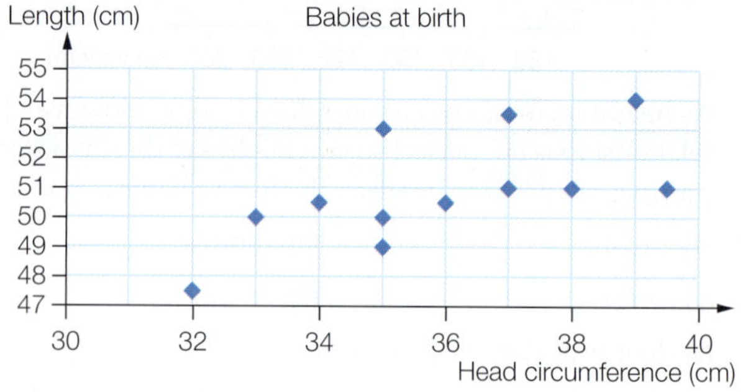

Exercise 2.5

1 (a)

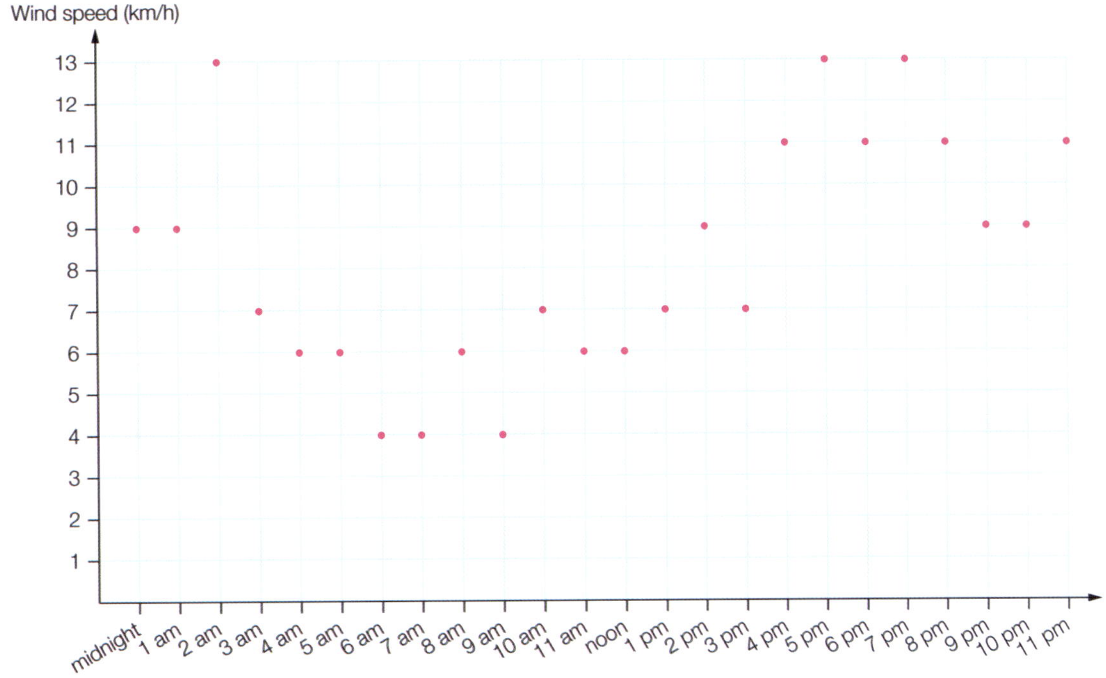

(b) It is much higher than the readings around it.

(c) The wind speed fluctuates between 11 km/h and 13 km/h.

2 (a) Hours spent studying; this is the variable under control.

(b) Age; this is the variable that is always changing—height does not necessarily change.

(c) Number of cigarettes; this is what has an impact on age at death.

(d) Number of people; change the number of people and you would expect a change in the amount of rubbish.

3 There does not appear to be a pattern in the number of bicycles. This is perhaps not an unexpected result because the weather could play a significant role in the number of cyclists on any particular day.

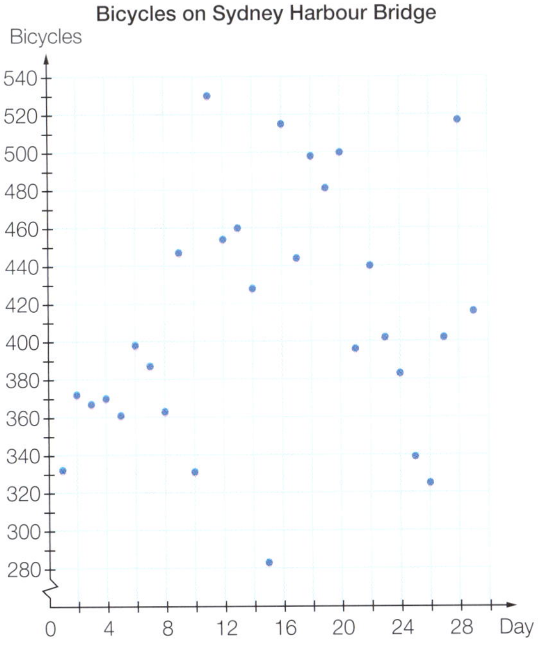

4 (a)

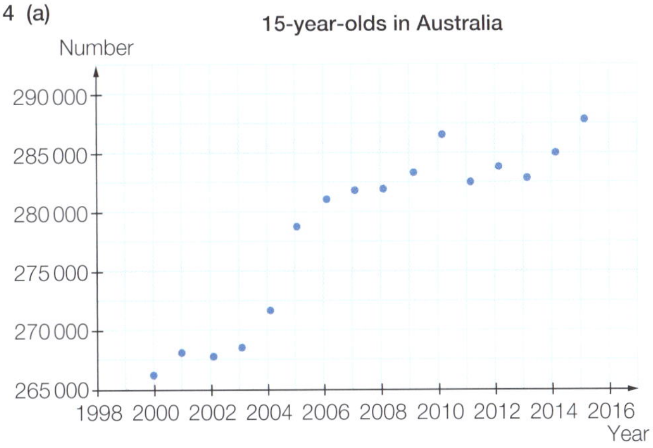

(b) The number of 15–year–olds has increased over time, even though there was a decrease in some individual years. The rate of increase accelerated during 2004–2005.

5 (a)

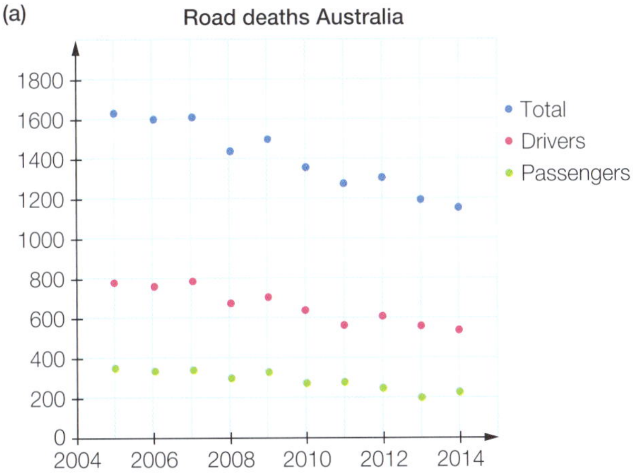

(b) The number of road deaths overall and in the two specific categories is decreasing over time (despite a 16% increase in the population in the time period given). The trend appears exaggerated for the overall figures but this is only because the actual numbers are larger.

6 (a) The population has increased consistently over this period of time. The last 40–50 years have seen the greatest rate of increase.

(b) At this scale, the inclusion would have minimal impact except perhaps in the very early years shown on the graph. The graph would have started at a higher value. Depending on how rapidly the ATSI population declined would determine whether or not the inclusion would be detectable.

(c) The population was relatively small, but the scale needs to be as it is to fit today's population into a reasonably sized graph.

(d) In the very early years by counting every individual. Records of arrivals, departures, births and deaths would also have been used.

(e) The very early years may actually be more reliable than the later ones because the later ones have some element of estimation built into them.

(f) There is a census every 5 years and population models are based on the census data collected.

7 The number of 90-year-olds is increasing at a fairly rapid rate. The average lifespan is increasing with improvements in medical care. For younger people, the same trend would probably not be seen.

Open-ended – Sample answers

8 The graph lacks a title and axes labels. There also needs to be a key to identify which colour represents which model.

9 Student's own answers.

Exercise 2.6

1 (a) Line of best fit may vary.

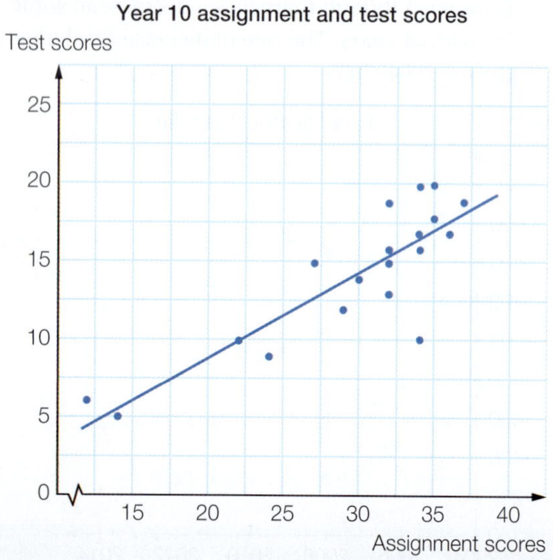

(b) Answers may vary.
$y = 0.55x - 2.25$
test result = assignment result × 0.55 – 2.25

(c) Fiona's test score: 11 or 12.
Enrico's assignment score: 20, 21 or 22.

2 (a) $y = 12.107x + 4.536$ (b) $y = 15.497x - 14.552$

(c) $y = -4.261x + 46.931$ (d) $y = -4.863xx + 39.999$

3 (a) C

(b) (i) $q = 2.69 + 1.16p$ (ii) A

4 (a)–(b)

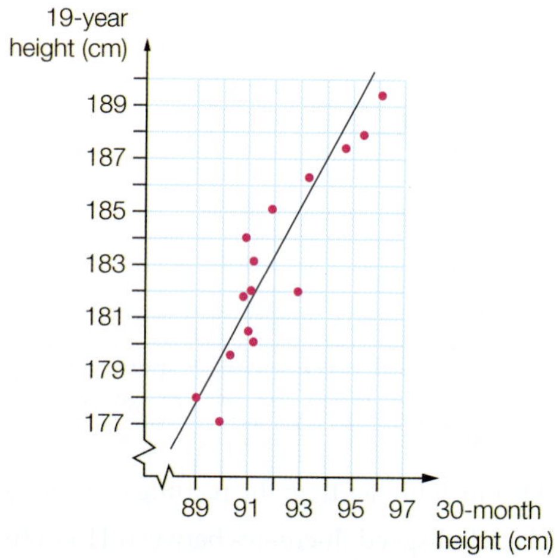

(c) The 30-month height does seem to be a reasonably good predictor of adult height.

(d) approximately 187 cm (will vary depending on the line of best fit)

5 (a) solubility = 0.05 × temperature + 35.05

(b) (i) 36.30 units (ii) 38.95 units (iii) 41.05 units

(c) temperature = 20.02 × solubility – 701.40

(d) (i) 19.32 °C (ii) 69.37 °C (iii) 199.50 °C

(e) Both of these are extrapolated values; you would need to assume that the connection between the variables remains the same.

6 (a)

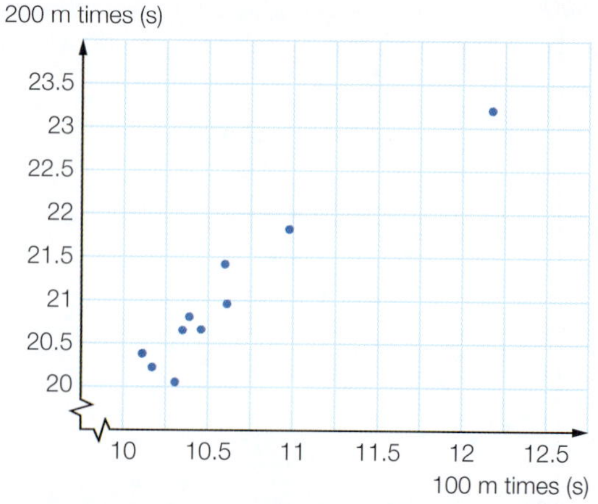

(b) There does appear to be a strong positive linear connection between these two times. In general, as one value increases, so does the other. It would seem appropriate to fit a straight line for these points.

(c)

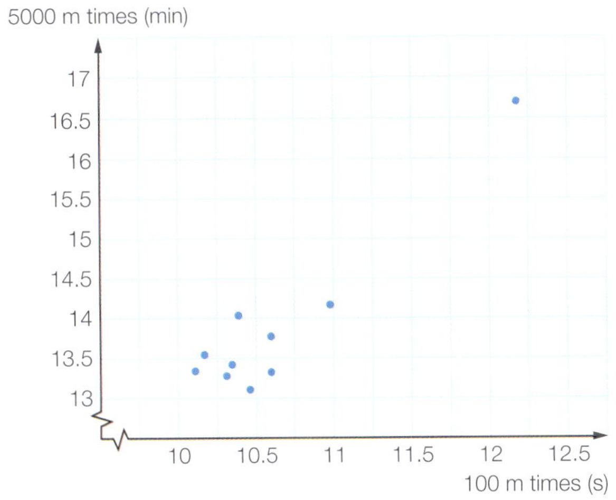

(d) There does not appear to be a strong connection between these two times. It would not seem appropriate to fit a straight line to these points.

(e) It is likely that the same person may hold the two sprint records but it will be a different person who holds the long-distance record. The two sprint times are therefore potentially linked but no such link is established for a sprint and a long-distance run.

7 (a)

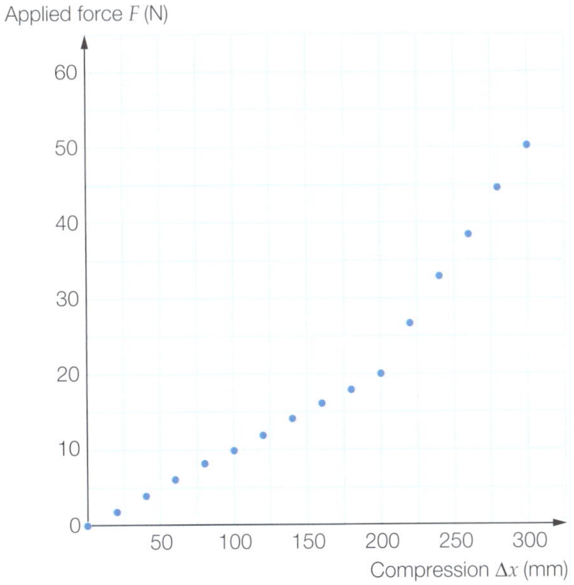

The first group ends at the point (200, 20)

(b) $F = 0.100\Delta x - 0.04,\ 0 \leq \Delta x \leq 200$ and $F = 0.294\Delta x - 37.75,\ 200 < \Delta x \leq 300$

(c) (i) 7.5 N (ii) 15.0 N (iii) 28.4 N

(d) $\Delta x = 9.965F + 0.440,\ 0 \leq F \leq 20$ and $\Delta x = 3.406F + 128.667,\ 20 < F < 50.2$

(e) (i) 50 mm (ii) 231 mm (iii) 282 mm

(f) Substituting 500 mm into the appropriate equation gives 110.4 N, which is much greater than the stated value, so assume the equation is not valid in that region. Without any other information, assuming the data is linear between (300, 50.2) and (500, 90) gives the equation $F = 0.199\Delta x - 9.5$.

Open-ended – Sample answers

8 (a) Emily has drawn the line of best fit to simply connect the first and last points. It should be drawn to represent the general trend of the points with approximately equal numbers of values either side of the line.

(b)

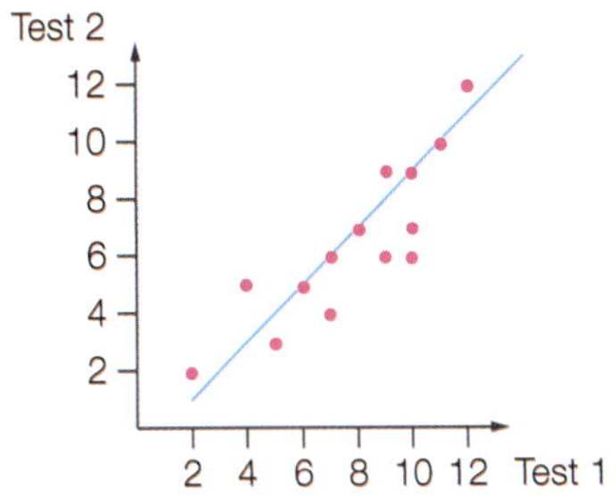

9 Students' own answers.

Exercise 2.7

1 (a) False. Although these are all people who have earned well from sport it is apparent that Hamilton has worked on at least two movies. The data given refers to 'wealth', not just earnings from sport.

(b) None dropped, but Rio Ferdinand stayed the same.

(c) Rory McIlroy: 36% increase.

2 (a) A pie graph assigns percentages of a whole but these percentages do not add to 100. The explanation is that it was possible to offer support to more than one candidate. That is fine, but a pie chart should never have been used to represent it. You would not be able to draw a correct graph from this information as you would need to know which one candidate each person supported.

(b) The lottery graph shows that fewer people pick the numbers from 0 to 100, 99. With fewer people to claim the prize pool, the winning payoff is larger.

3 (a) As the age of the children increases, the amount of coffee consumed by their parents increases.

(b) Parents of the youngest children (0–2 years) are likely to be sleep-deprived and therefore expected to be more likely to consume coffee.

(c) According to the data, older people tend to drink more coffee. As parents of older children are usually older than parents of young children, their coffee consumption would be higher.

(d) Parents drink 2.4 more cups of coffee per week.

(e) Parents of children aged 12–15 are the biggest coffee drinkers. They are also more likely to own a coffee maker.

(f) Answers may vary.

4 (a) The data comes from the Australian Census, which is only conducted every 5 years. The article compares data from 2006 and from 2011.

(b) About 9000 over 55s from Queensland moved to the coast.

About 3330 of these may have retired between 2006 and 2011.

Assuming that the 37% of over-55s who moved to the coast and then retired is independent of their origin, then Queenslanders were just as likely to retire as anyone else. This assumption may not be valid as people who haven't moved far may be more likely to keep ties with past employment.

(c) 37% of the over-55s who made the sea change were retired by 2011 and only 19% overall had done so, which means that a sea change over-55 working person is almost twice as likely to retire as the average over-55 working person.

5 (a) 90% of adolescents own, or have access to, a mobile phone.

Adolescent average of 1.5 hours per day text messaging.

Adolescent average of 7.5 hours per day on digital media.

(b) There were references to the original articles so interested readers could probably find this out. Phone records could be made available to researchers but these are likely to be from samples rather than the whole population. If the data was gathered by just a survey it is unlikely that the amounts of time would be accurate. The report was not sensational at all and there seemed no reason for the figures to be made up, so intentional bias seems unlikely.

6 (a) A 7944% increase is about 80 times as much.

A huge increase of a very small quantity may still be a small quantity.

(b) Potato exports: $3.9 million

Cabbages and cauliflower: $9000

(c) $3.45 billion

7 (a) Ms Casey is the preferred Prime Minister (48%) compared with Mr Smith (46%). Only female voters preferred Mr Smith (47%) compared with Ms Casey (41%). The heading is misleading.

(b) Telephone survey is the most convenient sampling method.

(c) The figures for female opinions (as in part (a)) give some justification for the heading 'Voters back Smith'—but this is not true for male voters.

(d) When you add the votes for either preferred Prime Minister, the total male votes equal 99% and the total female votes equal 88%. The missing 1% and 12% of votes respectively are due to the 'neither/don't know' category.

(e) (Answers may vary.) These categories attempt to identify the grounds for the voters' opinions. You do not know whether these categories are important to the voters.

(f) This poll does indicate that the preferred Prime Minister is Ms Casey (in the general population), but only slightly. The survey also covers the personal qualities of the candidates.

Open-ended – Sample answers

8 Students' answers will vary.

Exercise 2.8

1 5.72

2 English: mean = 34.3 and standard deviation = 9.22

Mathematics: mean = 34.7 and standard deviation = 7.12

Although the means are effectively the same the standard deviation for mathematics is smaller, indicating that the results in that subject are generally more close to the mean. This implies that the more extreme results will be found in the English marks.

3 mean = 3.7 and standard deviation = 2.52

4 (a) Australia: 241 m; New Zealand: 597 m; Brazil: 940 m; Sri Lanka: 628 m

(b) Australia: mean = 2105 m, SD = 76 m; New Zealand: mean = 3301 m, SD = 191 m; Brazil: mean = 2758 m, SD = 250 m; Sri Lanka: mean = 2124 m, SD = 166 m

(c) New Zealand's top 10 peaks are, on average, the highest, followed by Brazil, Sri Lanka and then Australia. As the range increases so does the standard deviation, except for Sri Lanka. Its standard deviation is less than New Zealand's despite its range being slightly higher than New Zealand's. This is possibly due to the highest peak in Sri Lanka being comparatively higher than the next value.

5 (a) (i) mean = 18.4, SD = 4.81

(ii) mean = 19.9, SD = 8.77

(iii) mean = 17.7, SD = 6.12

(iv) mean = 19.1, SD = 9.66

(b) Introducing an outlier at one end moved the mean in the same direction and increased the standard deviation. Introducing an outlier at both ends increases the standard deviation even further but the effect on the mean is partially cancelled out.

6 B

7 (a) mean = 4:22 and standard deviation = 2:12

(b) mean = 3:06 and standard deviation = 0:41

(c) mean = 4:12 and standard deviation = 2:36

(d)

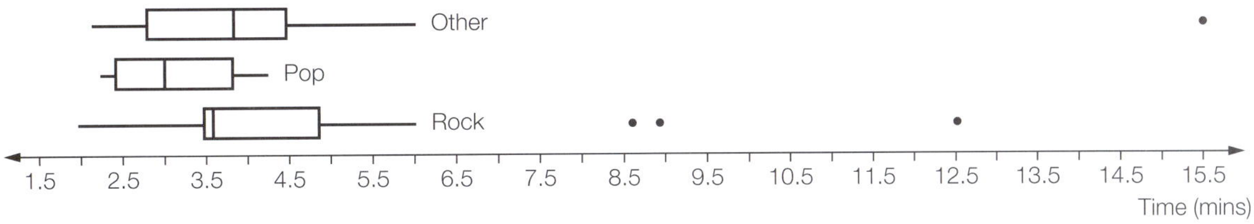

(e) From the evidence presented you could conclude that pop songs are more more consistently short than either of the other categories. The standard deviation is small, the box has no outliers and the range is small. The longest pop song is within the third quarter of each of the other box plots. This means at least 25% of each of the other categories are longer than the longest pop song. The number of pop songs was the smallest, which may have had an effect on the data. The Other category has one extreme outlier. However, there were three outliers in the Rock category.

8 (a) Mean = 5.4 g, SD = 1.23 g

(b) This is a sample answer only. The data values used are shown.

A: 6, 6, 6, 8, 6; mean = 6.4 g, SD = 0.89 g

B: 4, 5, 5, 6, 7; mean = 5.4 g, SD = 1.14 g

C: 7, 6, 3, 6, 5; mean = 5.4 g, SD = 1.52 g

D: 5, 6, 6, 6, 8; mean = 6.2 g, SD = 1.10 g

E: 5, 5, 5, 6, 5; mean = 5.2 g, SD = 0.45 g

F: 5, 5, 7, 6, 6; mean = 5.8 g, SD = 0.84 g

G: 3, 5, 5, 6, 6; mean = 5 g, SD = 1.22 g

H: 5, 7, 5, 6, 7; mean = 6 g, SD = 1 g

I: 3, 8, 7, 5, 3; mean = 5.2 g, SD = 2.28 g

J: 5, 5, 4, 5, 5; mean = 4.8 g, SD = 0.45 g

K: 5, 3, 5, 5, 6; mean = 4.8 g, SD = 1.10 g

L: 5, 7, 4, 5, 6; mean = 5.4 g, SD = 1.14 g

(c) This is a sample answer based on the samples taken in part (b).

A, B; mean = 5.9 g, SD = 1.10 g

C, D; mean = 5.8 g, SD = 1.32 g

E, F; mean = 5.5 g, SD = 0.71 g

G, H; mean = 5.5 g, SD = 1.18 g

I, J; mean = 5 g, SD = 1.56 g

K, L; mean = 5.1 g, SD = 1.10 g

(d) This is a sample answer based on the samples taken in part (b).

A, B, C, D; mean = 5.9 g, SD = 1.82 g

E, F, G, H; mean = 5.5 g, SD = 0.95 g

I, J, K, L; mean = 5.1 g, SD = 1.32 g

(e) This is a sample answer based on the samples taken in (b).

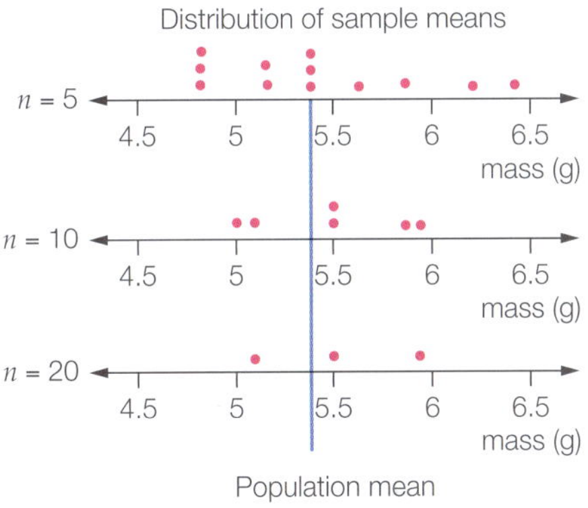

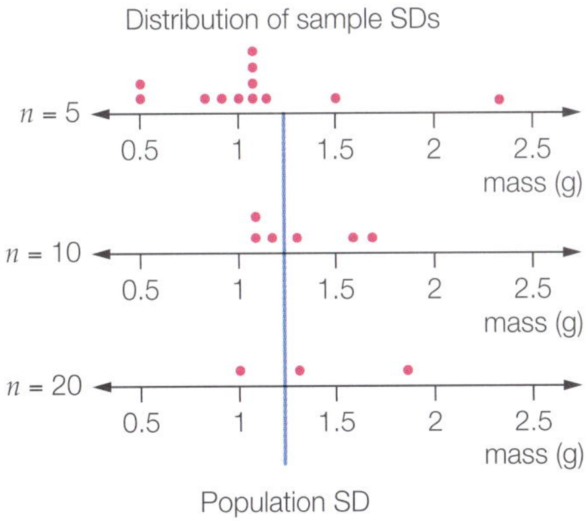

The mean and standard deviation for the samples were distributed either side of the population values. For the larger samples the sample values clustered closer to the population values. The larger the sample size, the closer the sample statistics are to the population.

Open-ended – Sample answers

9 (a) Students' own answers.

(b)–(c) The range condition means the data set cannot be mostly the same number. The first step would be to calculate what the sum of the data set would be, number of values × mean. Then, make a start on the data set so it adds to this value. Some adjustments may be required to make the standard deviation work.

10 The first step is to identify the five-number summary: min = 12, $Q_L = 13$, median = 16.5, $Q_U = 18$, max = 20. Construct the data set around these values and then find the mean and the standard deviation.

Chapter review 2

1 (a) (i) median = 20.0 (ii) IQR = 15

(iii)

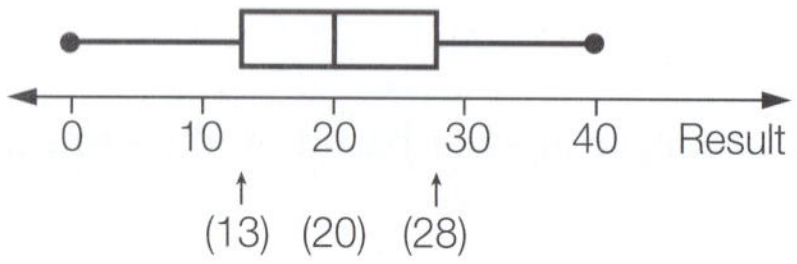

(b) (i) median = 19 (ii) IQR = 15

(iii)

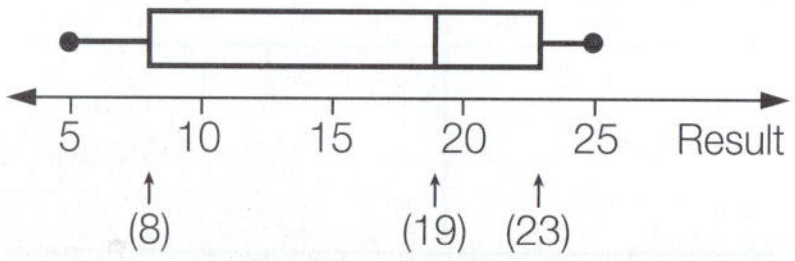

2 (a)

Result	Frequency	Percentage cumulative frequency
<120	0	0
<125	5	3
<130	17	15
<135	32	37
<140	54	74
<145	24	91
<150	13	100

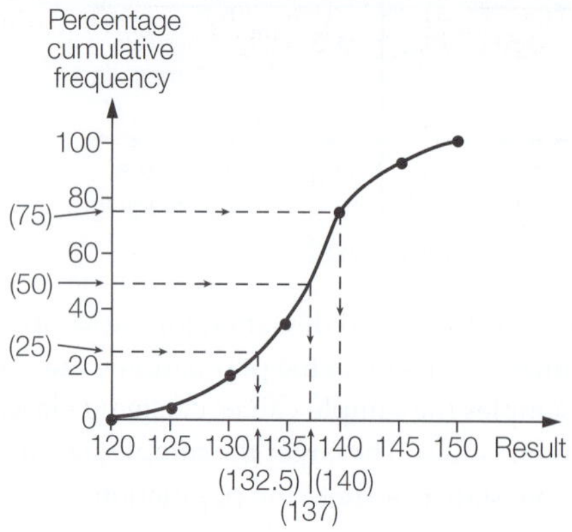

median = 137

(b) IQR = 7.5

(c)

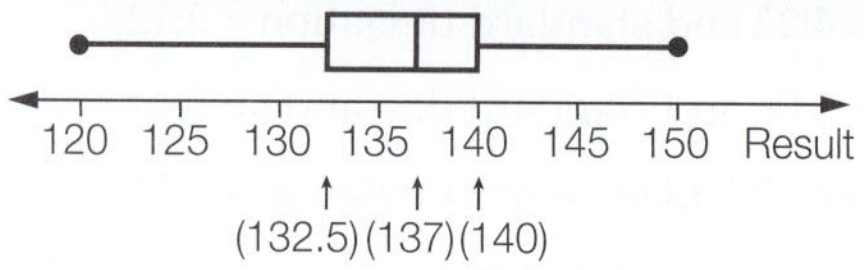

3 less than 11 cm or greater than 27 cm

4 Team A has been far more consistent than Team B, with lower values for the range and the IQR. Both teams have the same median number of goals. However, consistency in scores is not a predictor of success in team-based sports. It looks as if Team B would have lost the game when it scored 4 goals and won when it scored 24 goals but other than this it is difficult to make any firm statements about their success.

5 (a) primary (b) secondary

(c) secondary (d) primary

6 (a) 3.77 (b) 1.08

(c) 5.35

7 (a) $y = 5.536x + 2.536$ (b) $y = -2.080x + 19.428$

8 (a)

Class interval	Frequency	Cumulative frequency
20–<30	0	0
30–<40	1	1
40–<50	2	3
50–<60	7	10
60–<70	6	16
70–<80	4	20

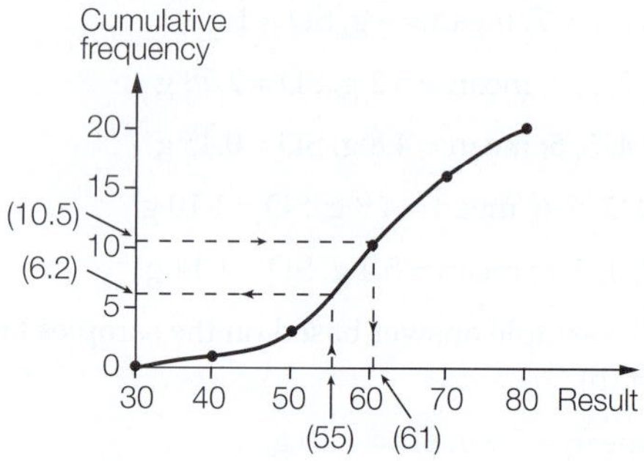

(b) min = 34.5; $Q_L = 53.55$; median = 60.85; $Q_U = 67.75$; max = 73.6

(c) From cumulative frequency curve,

Median = 10.5th term = 61, compared with

Median = $\frac{59.9 + 61.8}{2}$ = 60.85 from the raw data, so they are basically the same.

9 (a) 44

(b) min = 0; $Q_L = 3$; median = 10.5; $Q_U = 17$; max = 44

(c)

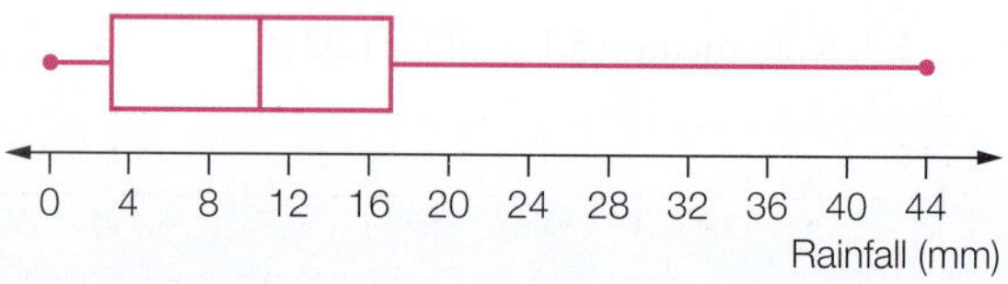

10 (a) (i) less than 8.5 and greater than 20.5

(ii) yes (at least one result greater than 20.5)

(b) (i) less than 24 and greater than 40

(ii) no (the lowest result is right on the margin)

11 (a)

Jan	Feb	Mar	Apr	May	Jun
13.0	13.3	13.0	12.1	11.9	10.9
Jul	**Aug**	**Sep**	**Oct**	**Nov**	**Dec**
10.5	10.7	10.5	11.7	12.1	12.6

(b)

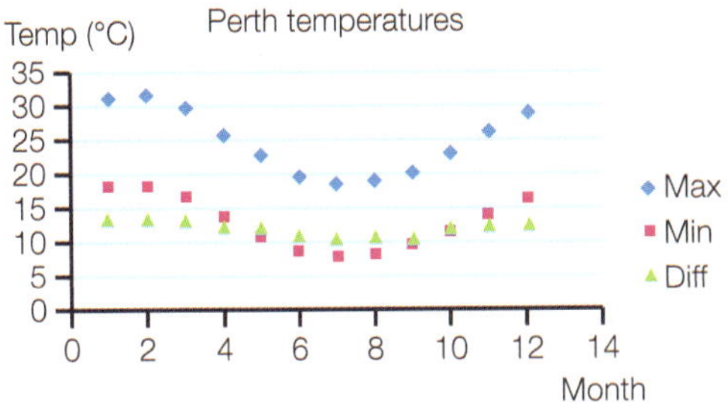

(c) Perth follows a typical temperature pattern for the southern part of Australia—warm to hot in the summer and cool to cold in the winter. As might be expected the difference graph shows the most consistency of the three, but does appear to be remarkably consistent. The smallest differences are seen in the winter months.

(d) No—temperature patterns can be quite local, and temperatures inland vary from coastal temperatures.

(e) The mean minimum for June 2010 was 7.1 °C compared to a long-term mean of 8.4 °C, whereas the mean maximum was 19.0 °C compared to the long-term mean of 19.3 °C. June 2010 was slightly more varied in temperature than normal with the minimum being lower and the maximum (slightly) higher. More data would need to be collated before any conclusions could be drawn about changing weather patterns.

12 (a) 3

(b)

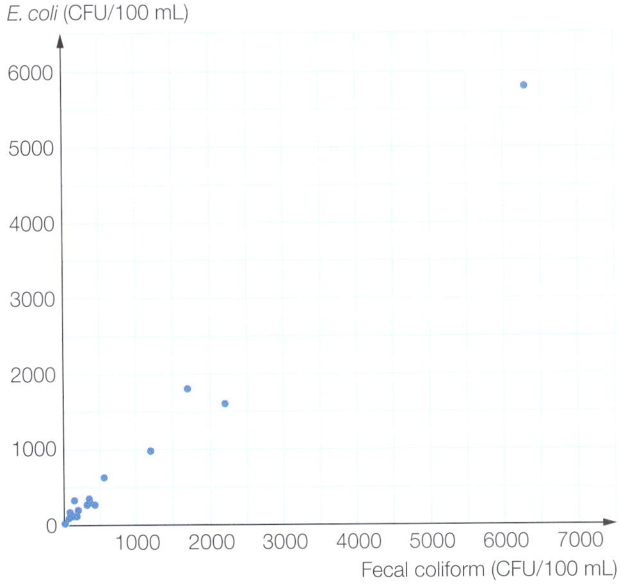

(c) $y = 0.91x + 8.02$

13 (a) The toll roads may be unpopular, but the survey does not show a clear intention for voters to vote against the Government based on this one issue. The percentage results are too close to tell.

(b) Government and Opposition voters together make up only 80% (33% + 47%) for Tullamarine, so 20% of people surveyed chose not to support the Government or Opposition party directly.

This is important for the overall judgement of the poll, because the voting preferences of this 20% could swing the overall vote in either direction.

(c) It is convenient to do the survey by telephone.

(d) Tullamarine and Essendon residents are largely affected by the new tolls, Oakleigh and Knox residents to a lesser extent. Many people avoid paying the tolls if possible whether they use the roads often or not.

14 (a)

Salary (×$10 000)	Percentage of employees (3 years ago)	Percentage of employees (currently)
1–<3	27	19
3–<5	34	21
5–<7	20	19
7–<9	11	17
9–<11	7	15
11–<13	0	10

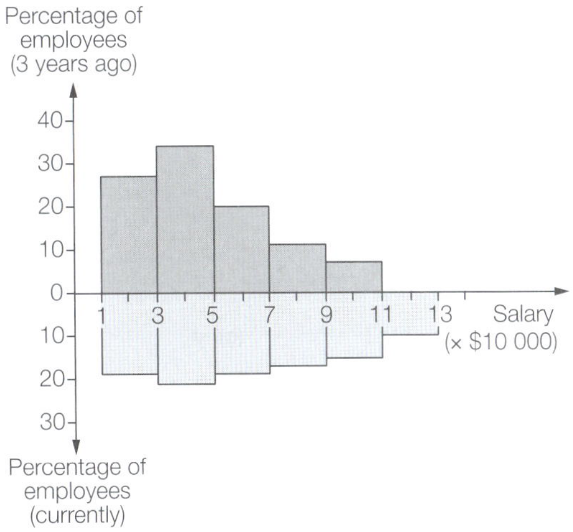

(b)

3 years ago		*Currently*	
Mean	= 4.73	Mean	= 6.38
Median	= 4.2	Median	= 6.0
IQR	= 3.3	IQR	= 5.5

(c)

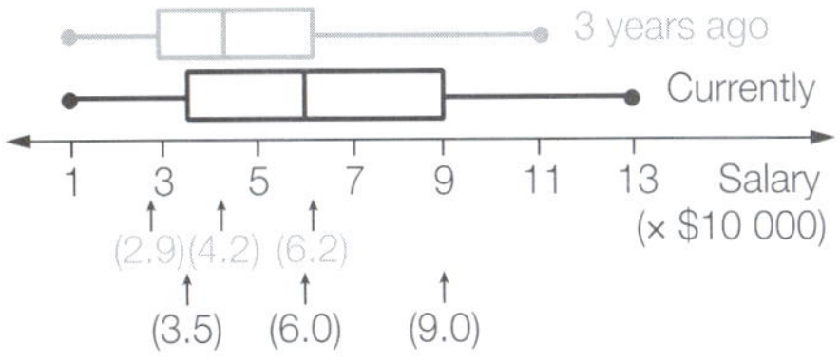

(d) Salaries have increased, on average, by \$15 000, and there is a much greater variation in the salaries paid.
mean difference = \$16 500
median difference = \$18 000

(e) total salary bill 3 years ago = \$2 080 000,
total salary bill currently = \$3 060 000;
i.e. \$980 000 increase

(f) about 47%

15 (a) There are 255 employees. The data set has most people in the range 41–50 but this is balanced by very few in 51–55. Other than this, it is relatively uniform so the mean will be near the median, say about 38.

(b) mean = 36.04 and standard deviation = 10.57

(c)

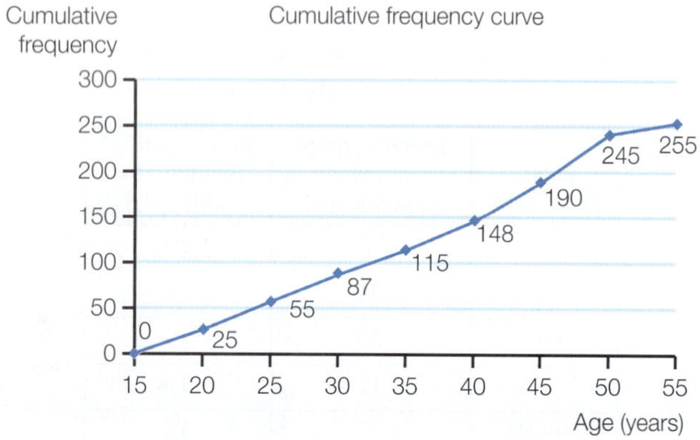

min = 16, $Q_L = 28$, median = 38, $Q_U = 48$, max = 55

(d)

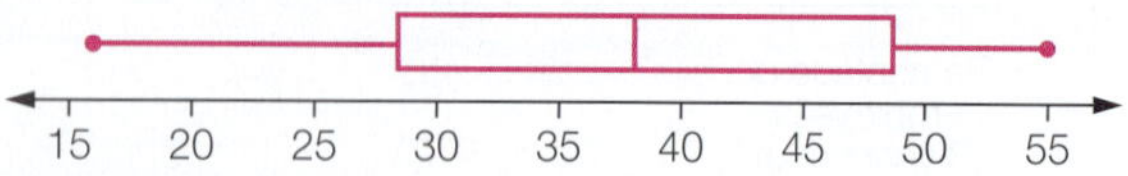

(e) (i) approx. 100 (ii) approx. 35

Chapter 3

Recall 3

1 (a) 3 (b) 2 (c) 4

2 (a) 4 (b) -6 (c) 1 (d) -12

3 (a) $3a + 5b$ (b) $24x - 8y$

4 (a) $3x + 6$ (b) $5x - 15$ (c) $5x - 8$

5 (a) $-2x - 6$ (b) $-8x + 40$ (c) $-x - 2$

6 (a) $9x^2$ (b) $4y^2$ (c) $49k^2$

7 (a) $14x^2$ (b) $-8p^2$ (c) $2y^2$

8 (a) $x = \frac{y-3}{2}$ (b) $x = 2ac + 4b$ (c) $x = \frac{3s + at^2}{5}$

9 (a) (i) 4 (ii) 5 (iii) 10

(b) (i) 10 (ii) 18 (iii) 0

10 (a) $\frac{11}{12}$ (b) $\frac{11}{24}$ (c) $\frac{5a}{3}$

11 (a) $\frac{1}{6}$ (b) $\frac{1}{2}$ (c) $\frac{3}{4}$

Exercise 3.1

1 (a) $4x + 12$ (b) $8x + 16$ (c) $12x + 132$

(d) $12x - 2$ (e) $14x - 35$ (f) $5x^2 - 4x$

(g) $-2x - 2$ (h) $-x^2 + 7x$ (i) $-3x^2 + 6x$

(j) $x^2 + 5x + 6$ (k) $2x^2 + 5x + 2$ (l) $5x^2 + 17x + 6$

(m) $2x^2 - 11x - 30$ (n) $10x^2 + 13x + 32$ (o) $12x^2 - 16x - 6$

(p) $5x^2 - 14x$ (q) $-3x^2 - 12x$ (r) $5x^3 - 8x^2 + x$

2 (a) $x^2 + 4x + 3$ (b) $x^2 + 6x + 8$ (c) $x^2 + 9x + 20$

(d) $x^2 + 5x - 24$ (e) $x^2 - 2x - 63$ (f) $x^2 + 4x - 21$

(g) $x^2 - 8x + 7$ (h) $x^2 - 9x + 20$ (i) $x^2 - 12x + 27$

(j) $2x^2 - x - 6$ (k) $4x^2 + 7x - 15$ (l) $7x^2 - 27x - 4$

(m) $6x^2 - 17x + 5$ (n) $-5x^2 + 38x - 21$ (o) $-18x^2 + 27x - 10$

3 (a) $4 - x^2$ (b) $49 - x^2$ (c) $36 - x^2$

(d) $4x^2 - 1$ (e) $9x^2 - 16$ (f) $25 - 9x^2$

(g) $x^2 + 12x + 36$ (h) $x^2 + 22x + 121$ (i) $64 + 16x + x^2$

(j) $x^2 - 6x + 9$ (k) $x^2 - 14x + 49$ (l) $x^2 - 24x + 144$

(m) $4x^2 + 20x + 25$ (n) $9x^2 + 12x + 4$ (o) $49x^2 - 28x + 4$

4 (a) A (b) B (c) A (d) C

(e) A (f) D (g) C (h) C

5 (a) $x^3 + 10x^2 + 24x$ (b) $x^3 + 4x^2 - 5x$

(c) $x^3 - 3x^2 - 4x$ (d) $-x^3 + x^2 + 6x$

(e) $-3x^3 + 9x^2 - 6x$ (f) $-4x^3 + 20x^2 + 96x$

(g) $x^3 + 11x^2 + 38x + 40$ (h) $x^3 - 4x^2 - 7x + 10$

(i) $x^3 - 11x^2 + 36x - 36$ (j) $x^3 + 5x^2 - 9x - 45$

(k) $x^3 + 7x^2 - 4x - 28$ (l) $2x^3 + x^2 - 32x - 16$

(m) $x^3 + 5x^2 - 8x - 48$ (n) $x^3 + 3x^2 - 4$

(o) $x^3 - 7x^2 - 24x + 180$

6 (a) There are two errors in line 2.

(b) $(3x - 2)(2x - 3)$

$= 3x \times 2x + 3x \times -3 + -2 \times 2x + -2 \times -3$ Line 1

$= 6x^2 - 9x - 4x + 6$ Line 2

$= 6x^2 - 13x + 6$ Line 3

7 (a) $8x^2 + 14xy + 15y^2$ (b) $35x^2 + 44xy + 12y^2$

(c) $15x^2 - 44xy + 32y^2$ (d) $4x^2 + 28xy + 49y^2$

(e) $49x^2 + 84xy + 36y^2$ (f) $25x^2 + 80xy + 64y^2$

(g) $16x^2 - 8xy + y^2$ (h) $36x^2 - 60xy + 25x^2$

(i) $81x^2 - 126xy + 49y^2$ (j) $25y^2 - x^2$

(k) $9x^2 - y^2$ (l) $4x^2 - 9y^2$

(m) $6x^3 + 7x^2 - 5x$ (n) $15x^3 - 8x^2 - 12x$

(o) $24x^3 + 78x^2 + 63x$ (p) $-15x^3 + 8x^2 + 16x$

(q) $-30x^3 - 58x^2 - 24x$ (r) $-24x^3 + 76x^2 - 60x$

8 (a) $(x-4)(3x+1) = 3x^2 - 11x - 4$

(b) $(2x+3)(2x+3) = 4x^2 + 12x + 9$

(c) $\frac{1}{2}(x+5)(x-5) = \frac{1}{2}(x^2 - 25)$ or $\frac{1}{2}x^2 - \frac{25}{2}$

(d) $\frac{x^2 - 81}{2}$ or $\frac{1}{2}x^2 - \frac{81}{2}$

(e) $7x^2 + 17x$

(f) $x^2 + 16x + 28$

9 B

10 (a) $(x-2)(x)(x+2)$

(b) $x^3 - 4x$

(c) It is easier to have x as the second number because the other two expressions, $(x-2)$ and $(x+2)$, then form a difference of two squares which is simpler to multiply.

11 (a) $(x+5)$ and $(x-2)$

(b) $A = (x+5)(x-2) = x^2 + 3x - 10$

(c) x^2 and $5x$ (d) $2x + 10$

(e) $A = x^2 + 3x - 10$

(f) The answers are the same.

12 (a) $x^3 + (a+b)x^2 + abx$

(b) (i) $a + b$ (ii) ab

(iii) $a + b = -4$ and $ab = -12$. Using the expansion from (a) this shows that $x(x-6)(x+2) = x^3 - 4x^2 - 12x$.

(c) (i) $a = 25, b = -4$ (ii) $x^3 + 21x^2 - 100x$

13 (a) $(36-x)(36+x)$ (b) 1260 cm^2

14 (a) $x^2 - 2ax$ (b) $x^2 - 2ax$

(c) They are equivalent.

15 (a) $55 - y$ m (b) $x(55-y)$ m^2

(c) $55x$ m^2 (d) $(x-30)(55-y)$ m^2

(e) $55x - xy - 1650 + 30y$ (f) $45x - 1350$ m^2

(g) $10x + 1350$

16 (a) $(x-2)^2(x+3)$ (b) 28 cm^3

(c) When $x = 5$, dimensions are 3 cm, 8 cm and 3 cm, so the volume $= 3 \times 8 \times 3 = 72$ cm^3.

(d) $x^3 - x^2 - 8x + 12$; 72 cm^3

(e) height and width 3 cm, length 8 cm

(f) $x = 3$ (g) 6 cm^3

Open-ended – Sample answers

17 $a = 2, b = 1, c = 0, d = 1; a = 1, b = 0, c = 1, d = 1$

18 (a) -4, -5 (b) 4, 5 (c) 2, -3

Exercise 3.2

1 (a) $3(3x-1)$ (b) $-7(x+2)$ (c) $x(3+x)$

(d) $b(8a-1)$ (e) $4m(4n-m)$ (f) $5xy(5x+1)$

(g) $-x^2(1+17y)$ (h) $ab(2a+9b)$ (i) $10x^2(1-10x)$

2 (a) $4(x+2)$ (b) $3(6x-5)$ (c) $3(8x-3)$

(d) $x(y+7)$ (e) $3k(4x+7)$ (f) $4b(8a+5)$

(g) $7y(x^2+y)$ (h) $3x(3x+y)$ (i) $8m(6m+1)$

(j) $-x(6+x)$ (k) $-x(4-x)$ (l) $-2a(12a-5)$

(m) $6ab(1-3a)$ (n) $8xy(5x+2y)$ (o) $2cd(5d-1)$

3 (a) $(k+1)(x+4)$ (b) $(x-4)(a+6)$ (c) $(n-7)(m+2)$

(d) $(p+3)(x-9)$ (e) $(q-2)(n-7)$ (f) $(u-4)(x-5)$

(g) $(c+5)(d+1)$ (h) $(d+2)(a-1)$ (i) $(m+7)(k-1)$

(j) $(x-3)(x+5)$

4 (a) $(y+5)(x+3)$ (b) $(b+2)(a+7)$

(c) $(n+4)(m+k)$ (d) $(y+3)(x-1)$

(e) $(3+c)(a-b)$ (f) $(m+2)(k-4)$

(g) $(5b-4)(3a-2)$ (h) $(2b-5)(3a-2)$

(i) $(2n-3)(4m-3)$ (j) $(v-a)(x-y)$

(k) $(q-n)(p-r)$ (l) $(h-d)(g-f)$

(m) $(n+t)(m^2+p)$ (n) $(ab-c)(a+b)$

(o) $(dk-r)(d-1)$ (p) $(9y-5)(3x+2)$

(q) $(4a-5)(9-c)$ (r) $(5v-3)(5w+1)$

(s) $(8a^2-3b)(4ac-5)$ (t) $(3d-5f)(7d-4h)$

(u) $(mr-5)(4p+mn)$

5 (a) true (b) false (c) false

(d) true (e) false (f) false

6 (a) $(s+10)(6+t)$ (b) $2(1-4a)$

(c) $(x-0.5)(7x+3)$ (d) $(q+6)(p-3)$

(e) $5x(5x+3)$ (f) $(4a+7)(3b-1)$

7 D 8 B 9 A

10 (a) $a(p+5) + b(p+5)$ (b) $(a+b)(p+5)$

(c) $3a(p+5)$

11 (a)

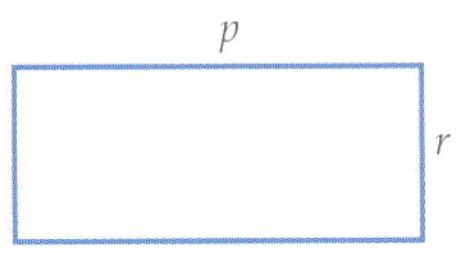

(b)

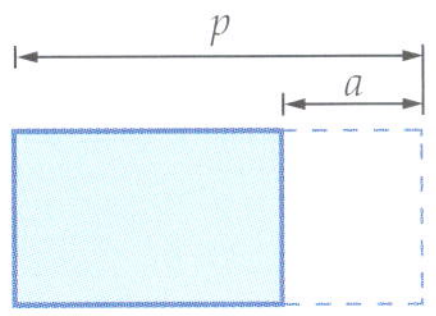

Area $= pr - ar$

(c)

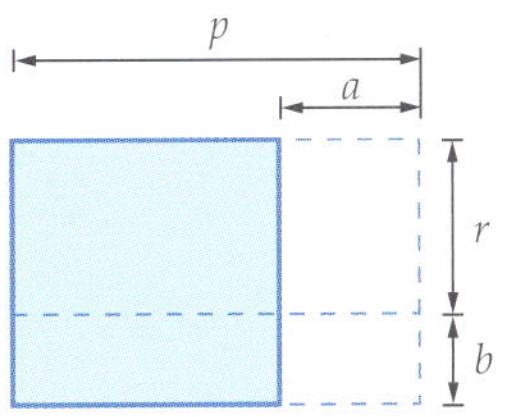

Area added $= pb - ab$

(d) Area $= pr - ar + pb - ab$

(e) Area $= (r+b)(p-a)$

(f) part (e) is the factorised form of part (d)

12 (a) $V = \dfrac{2\pi r^3}{3} + \dfrac{\pi r^2 h}{3}$

(b) $V = \dfrac{\pi r^2}{3}(2r + h)$

(c) 151 cm^3

(d) Cone volume is $\dfrac{\pi r^2(2r)}{3} = \dfrac{2\pi r^3}{3}$, which is equal to the hemispherical volume of the ice-cream.

Open-ended – Sample answers

13 (a) $2x^2 + 4xy\text{ cm}^2$ (b) $2x(x + 2y)$

(c) $x = 12, y = 3; x = 8, y = 4$

Exercise 3.3

1 (a)

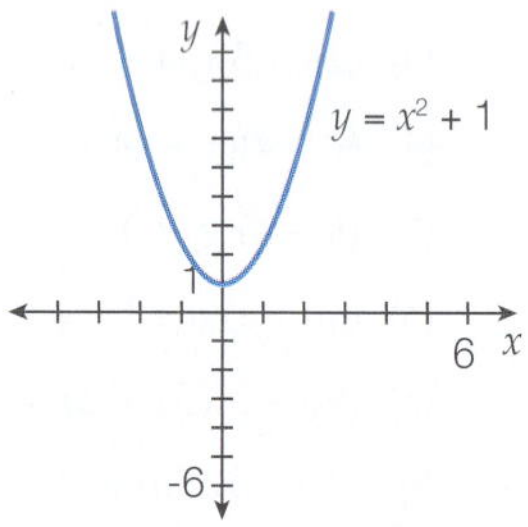

(b)

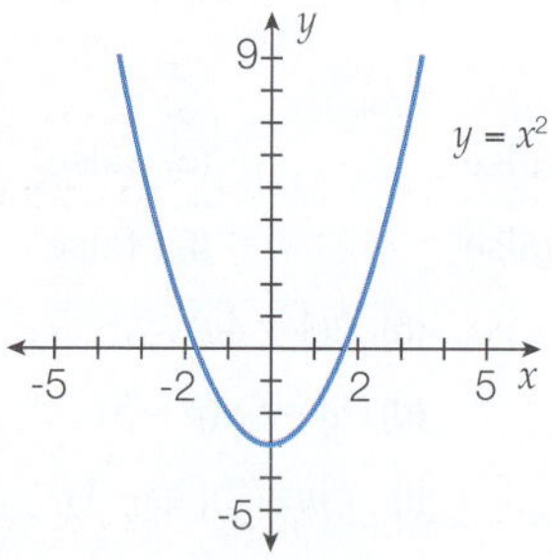

(c)

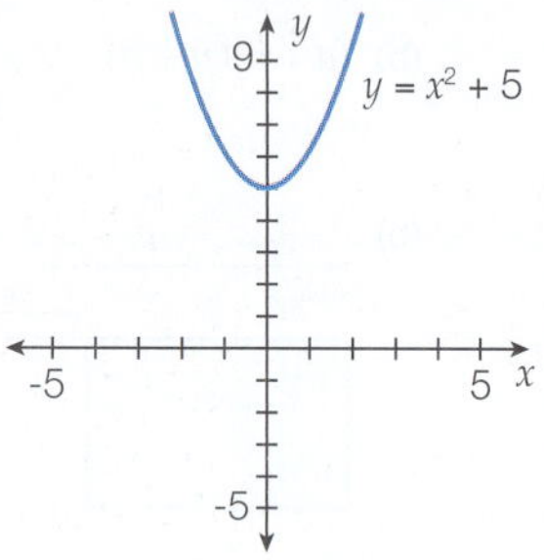

(d)

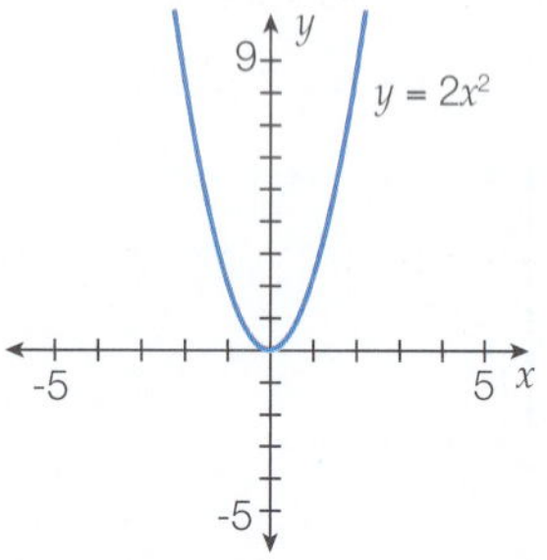

(e)

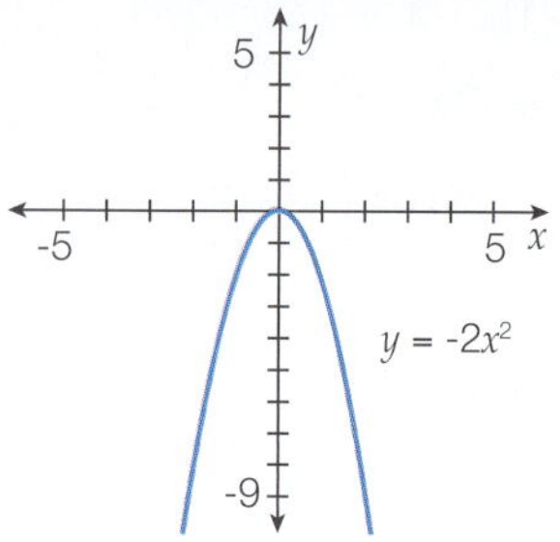

(f)

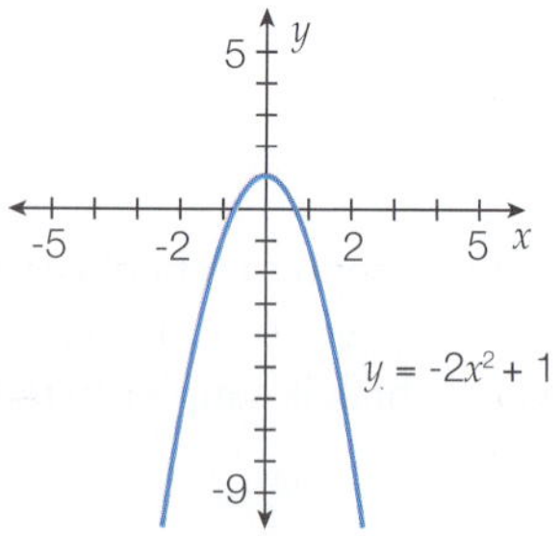

2 (a)

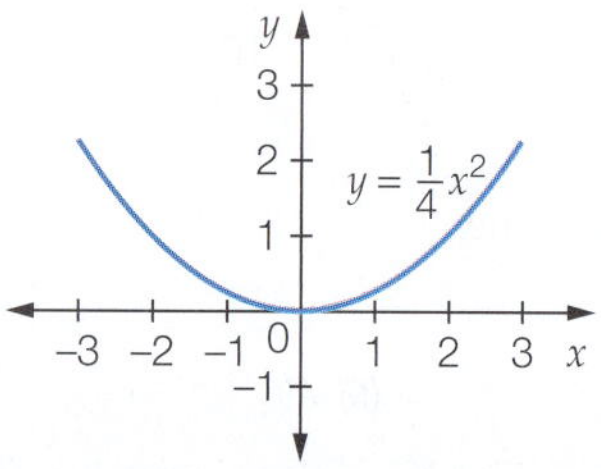

(b)

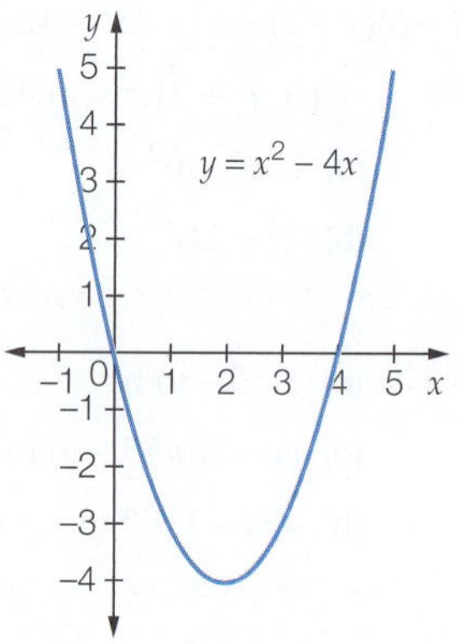

(c)

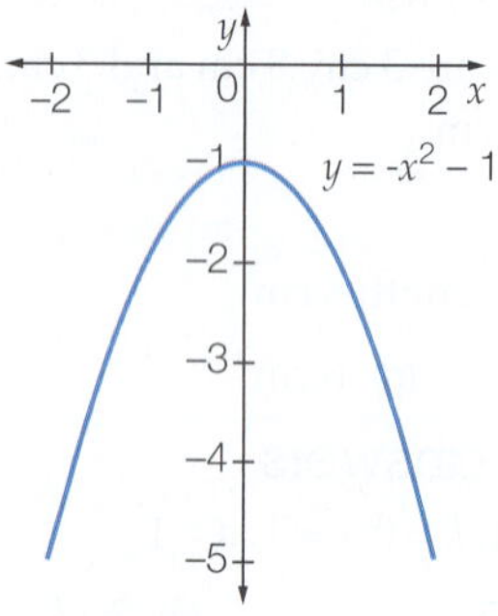

3 (a) (i) $x = 0$ (ii) $(0, 0)$

(iii) minimum turning point

(b) (i) $x = 2$ (ii) $(2, -4)$

(iii) minimum turning point

(c) (i) $x = 0$ (ii) $(0, -1)$

(iii) maximum turning point

4 (a) (i)

x	-4	-3	-2	-1	0	1	2	3	4
y	5	0	-3	-4	-3	0	5	12	21

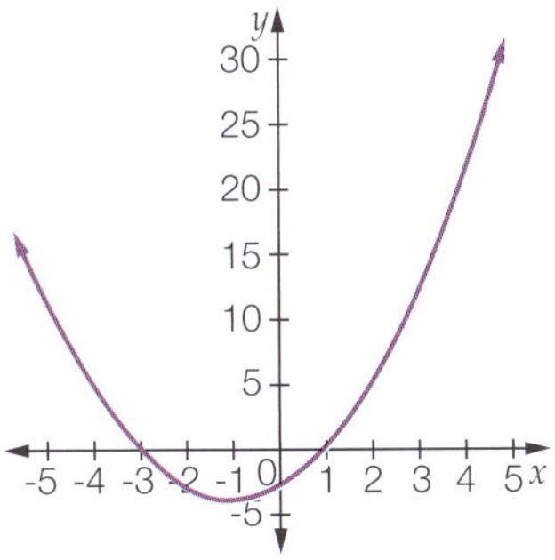

(ii)

x	-4	-3	-2	-1	0	1	2	3	4
y	4	0	-2	-2	0	4	10	18	28

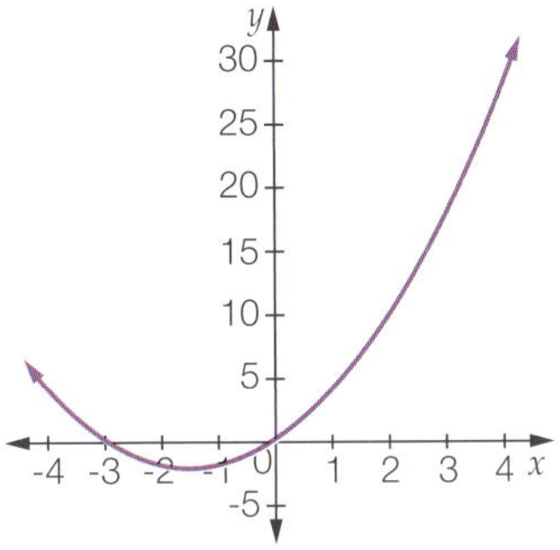

(iii)

x	-4	-3	-2	-1	0	1	2	3	4
y	0	5	8	9	8	5	0	-7	-16

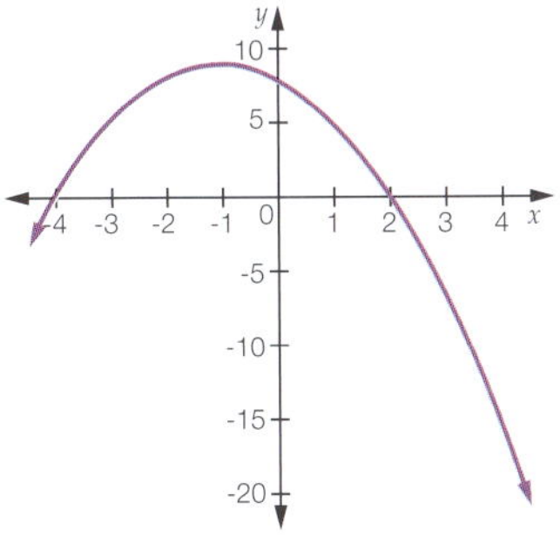

(b) (i) (-1, 4), minimum

(ii) (-1.5, -2.25), minimum

(iii) (-1, 9), maximum

5 (a) $V = \pi r^2 h$

$= \pi r^2 \left(\frac{3}{\pi}\right)$

$= 3r^2$

(b) It is a quadratic because the highest power of r is 2.

(c)

r	0	1	2	3	4	5
r^2	0	1	4	9	16	25
V	0	3	12	27	48	75

(d)

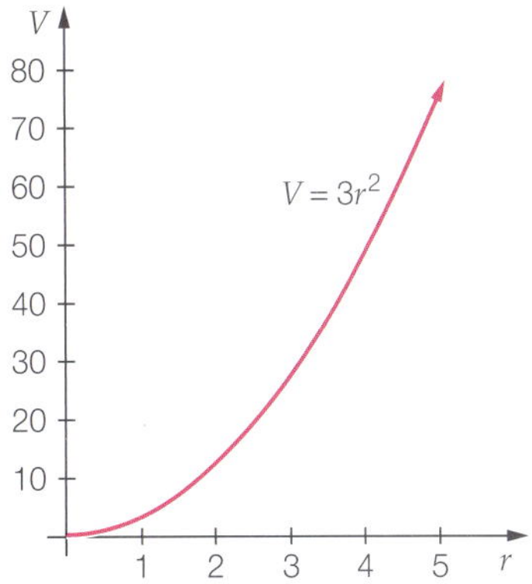

(e) (i) 18.8 m^3

(ii) 43.3 m^3

(f) (i) 3.2 m

(ii) 4.7 m

(g) 30 m^3

6 (a) $A = l(l - 0.4)$ m^2

(b)

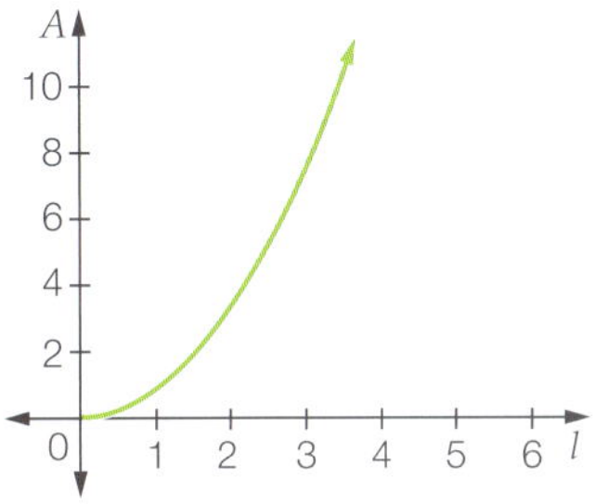

(c) 2 m × 1.6 m

7 (a)

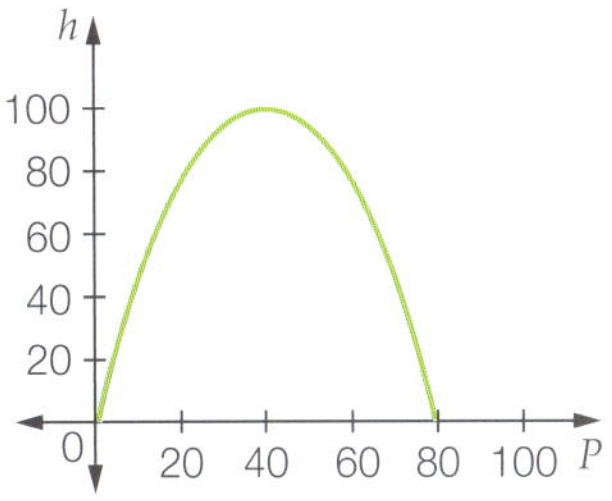

(b) 100 m

(c) 80 m

(d) 20 106.19 m^2

Open-ended – Sample answers

8 For a maximum turning point the coefficient of x^2 must be negative.

9 (a) Students to provide their own sketch.

(b) The graph should start at the origin and be symmetrical in shape with a maximum turning point half-way between the two x-intercepts.

Exercise 3.4

1 **(a)** **(i)** 2 **(ii)** narrower

(iii)

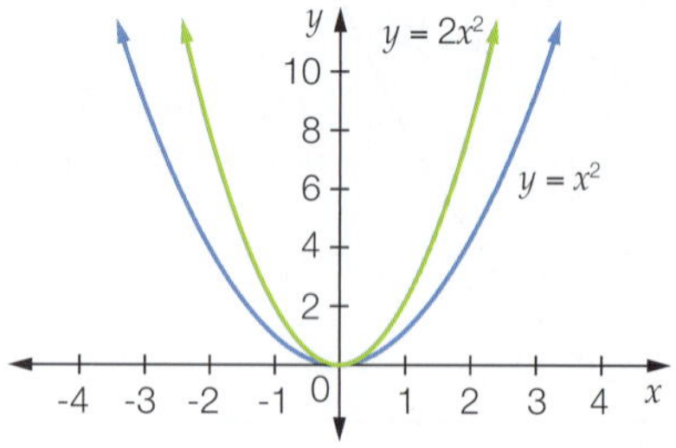

(b) **(i)** 0.5 or $\frac{1}{2}$ **(ii)** wider

(iii)

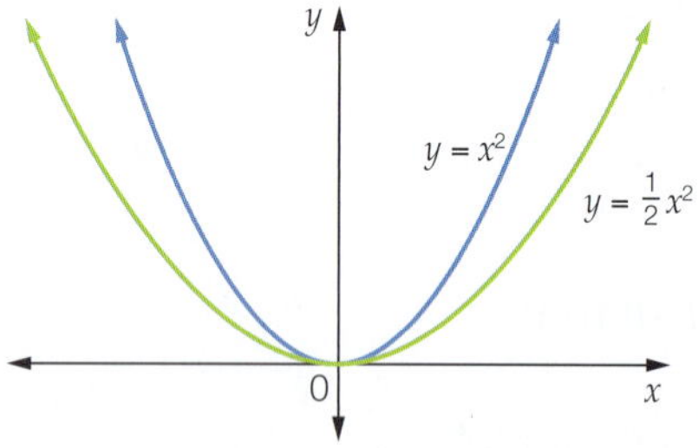

(c) **(i)** 4 **(ii)** narrower

(iii)

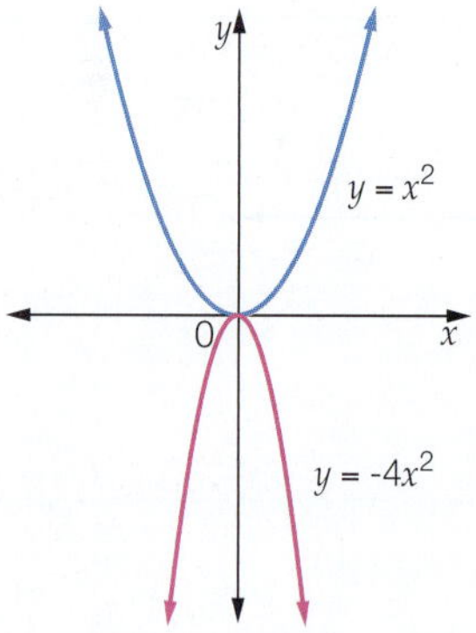

2 **(a)** **(i)**

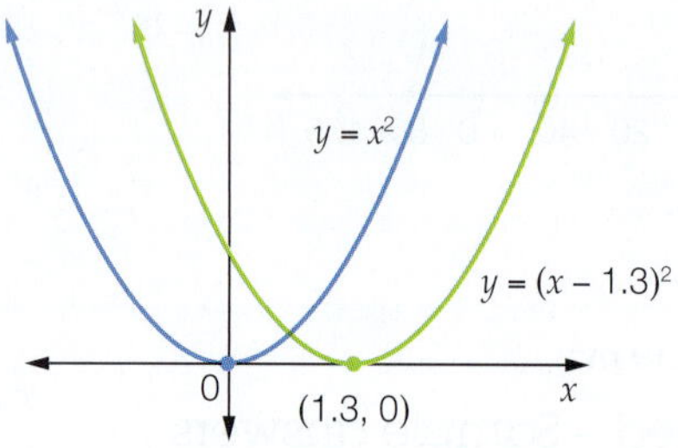

(ii) 1.3 units to the right

(b) **(i)**

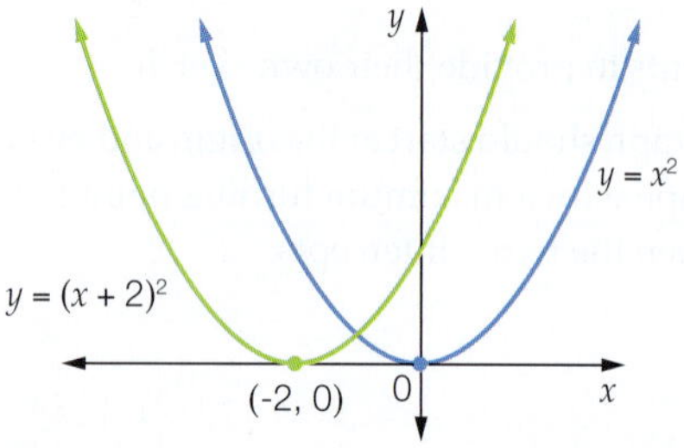

(ii) 2 units to the left

3 **(a)** **(i)**

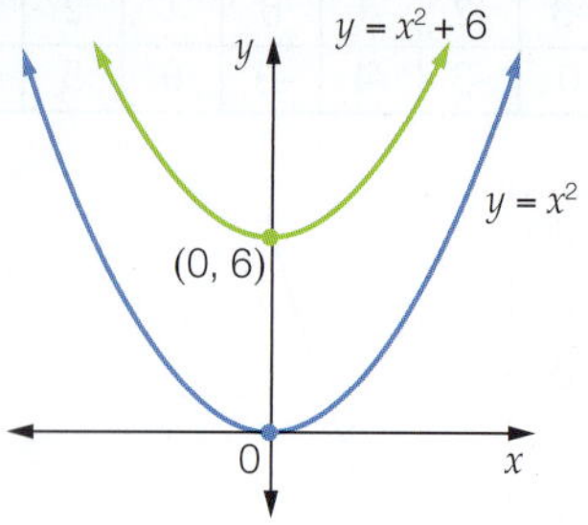

(ii) 6 units up

(b) **(i)**

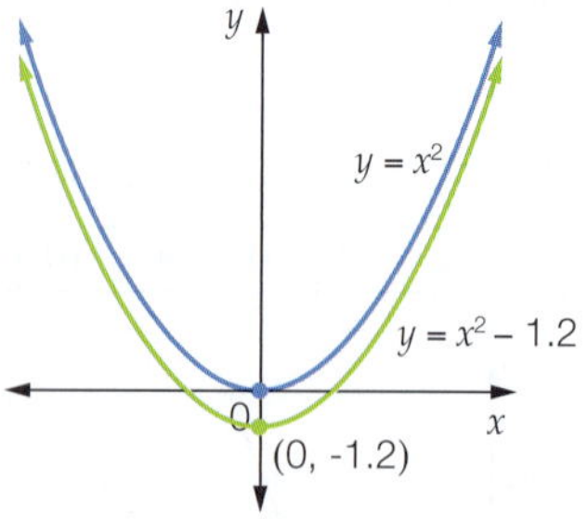

(ii) 1.2 units down

4 **(a)**

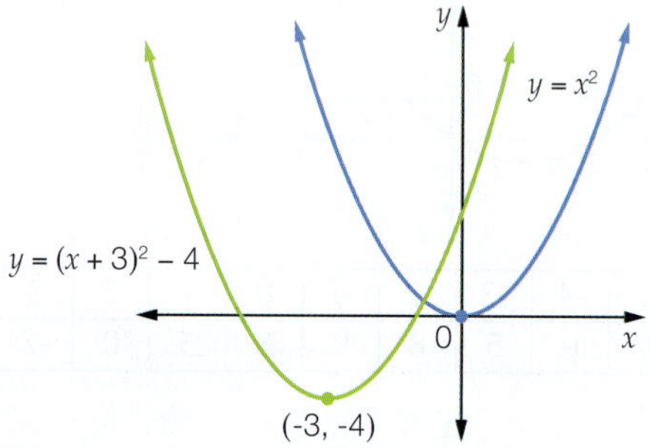

(b)

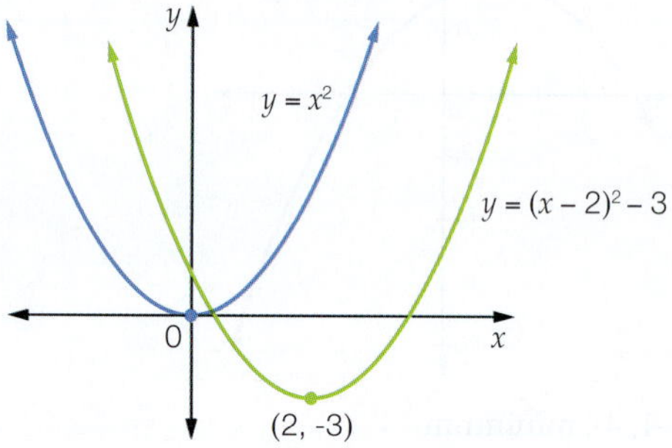

5 **(a)** $y = (x - 2)^2 - 5$ **(b)** $y = -x^2 - 1$

6 B

7 **(a)–(d)**

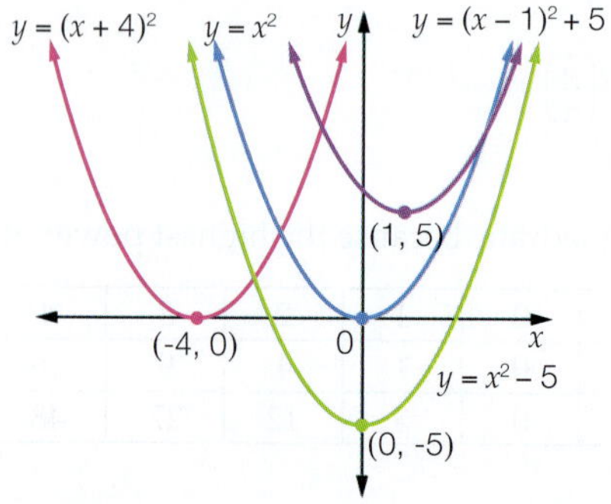

8 (a)–(d)

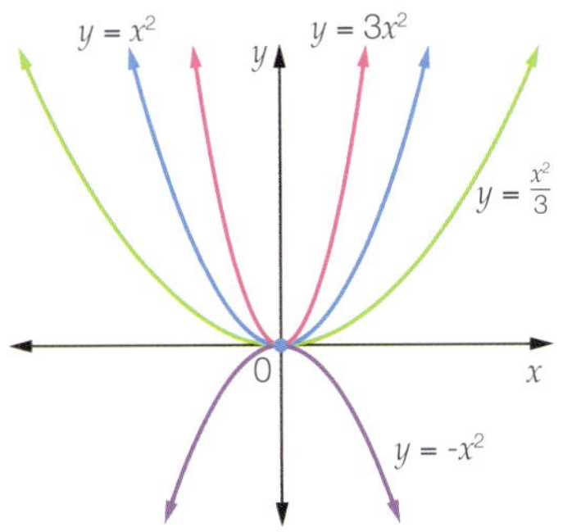

9 D

10 (a)–(b)

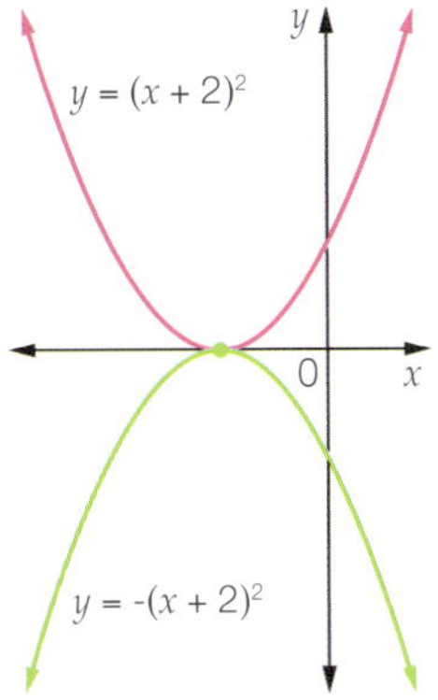

11 (a)–(b)

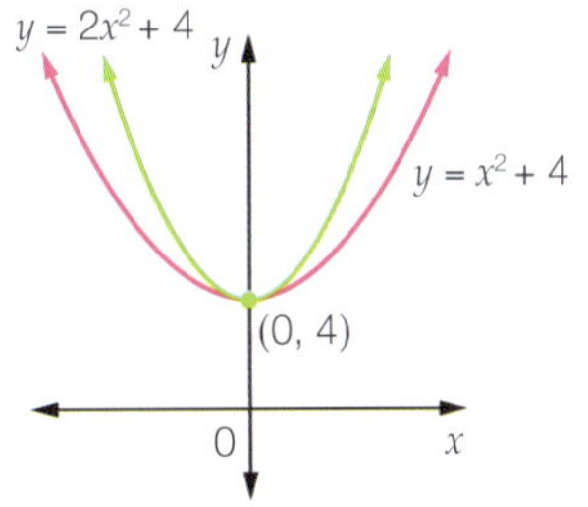

(c) $y = 2x^2 + 4$

12 (a)–(b)

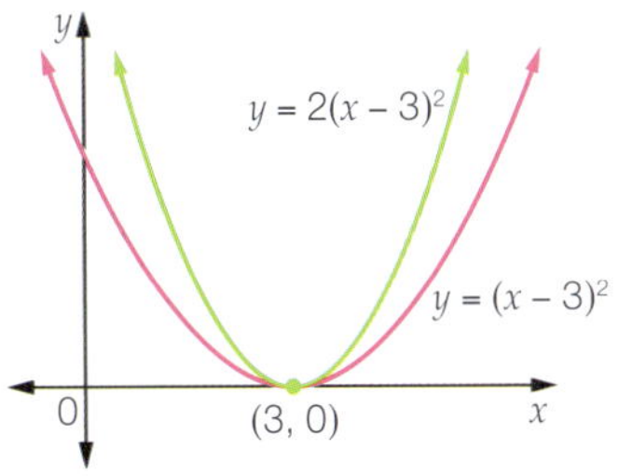

(c) $y = 2(x - 3)^2$

13 (a) $y = (x + 3)^2 + 2$

(b)

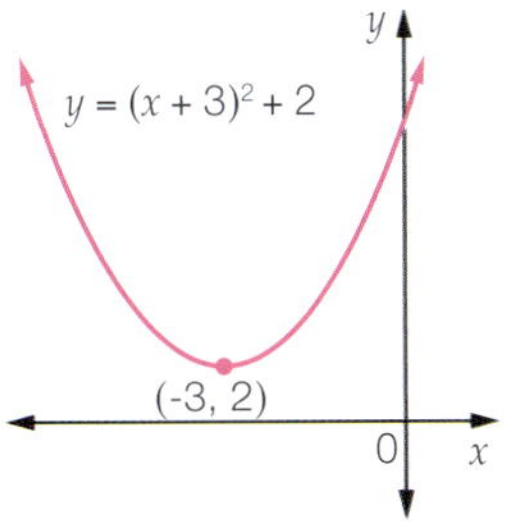

14 $y = (x + 2)^2 + 0.5$

15 (0, 0), (1, 5), (2, 8), (3, 9), (4, 8), (5, 5), (6, 0)

Open-ended – Sample answers

16 (a) The equation should be in the form $y = x^2 + k$.

(b) The equation should be in the form $y = (x - h)^2$.

17 When dealing with brackets you are looking for the value of x that will make the bracket equal zero. So, $(x + 6)$ requires x to equal -6 to make the bracket 0. This means the turning point will be in line with $x = -6$. Syd is correct in his thinking about the vertical shift.

Half-time 3

1 (a) $x^2 + 8x + 12$ (b) $x^2 + 2x - 35$

(c) $3x^2 + 13x - 10$ (d) $2x^2 + 5x - 3$

2 D

3 (a) $7(x + 2y)$ (b) $6x(3x + 4y)$

(c) $(x - 2)(3 + y)$ (d) $(x + 5)(3 + 2y)$

4 (a) $x^2 - 9$ (b) $x^2 - 64$

(c) $9x^2 - y^2$ (d) $36x^2 - 25y^2$

5 (a) $x^3 + 5x^2 + 6x$ (b) $3x^2 - 12$

(c) $63x^3 + 150x^2 - 141x - 228$ (d) $48x^3 - 156x^2 + 36x$

6 (a)

x	-3	-2	-1	0	1	2	3
y	-8	-3	0	1	0	-3	-8

(b)

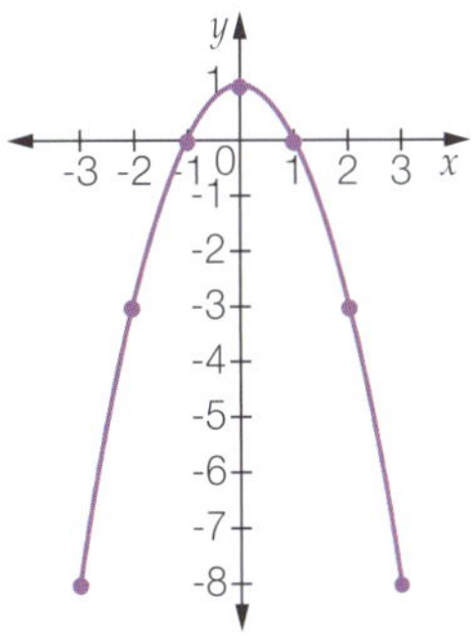

(c) (0, 1)

(d) $x = 0$

7 $y = (x - 3)^2 + 2$

8 (a) $(x + 2)(x + 8)$

(b) $18(x + 2)(x + 8)$

(c) $720

9 $16x^2 + 24xy + 9y^2$

Exercise 3.5

1 (a) $(x + 2)(x + 4)$ (b) $(x - 3)(x + 6)$

(c) $(x - 1)(x + 7)$ (d) $(x + 5)(x - 2)$

(e) $(x - 4)(x - 3)$ (f) $(x - 4)(x + 3)$

(g) $(x + 4)(x - 2)$ (h) $(x + 1)(x - 3)$

(i) $(x - 8)(x - 2)$ (j) $(x - 2)(x - 6)$

(k) $-(x + 4)(x + 3)$ (l) $(x + 5)(x - 6)$

2 (a) $(x+1)(x+2)$ (b) $(y+2)(y+7)$
(c) $(a-4)(a-6)$ (d) $(k+3)(k+8)$
(e) $(m-5)(m+3)$ (f) $(x-6)(x+5)$
(g) $(p-2)(p-8)$ (h) $(d-1)(d-4)$
(i) $(y-3)(y-2)$ (j) $(m-6)(m+2)$
(k) $(c-7)(c-3)$ (l) $(g+7)(g-5)$
(m) $2(y+3)(y+4)$ (n) $3(m+2)(m+4)$
(o) $5(x+4)(x-5)$ (p) $-4(a+3)(a-1)$
(q) $-3(k+3)(k-7)$ (r) $-5(x+7)(x-3)$

3

Expanded expression	Cross method	Factorised expression
(a) $a^2+8a+12$	a 2 a 6	$(a+2)(a+6)$
(b) $x^2-7x+12$	x -3 x -4	$(x-3)(x-4)$
(c) $m^2+9m-22$	m 11 m -2	$(m+11)(m-2)$
(d) $t^2+9t+20$	t 4 t 5	$(t+4)(t+5)$
(e) $h^2-7h-30$	h -10 h -3	$(h-10)(h+3)$

4

Expanded expression	Cross method	Factorised expression
(a) $b^2-4b-45$	b 5 b -9	$(b+5)(b-9)$
(b) $p^2-11p+24$	p -3 p -8	$(p-3)(p-8)$
(c) a^2+a-30	a 6 a -5	$(a+6)(a-5)$
(d) $t^2-8t+16$	t -4 t -4	$(t-4)(t-4)=(t-4)^2$
(e) $m^2-10m+21$	m -7 m -3	$(m-3)(m-7)$

5 (a) C (b) A (c) B
6 (a) $(z-5)(z-3)$ (b) $(k+6)(k-1)$
(c) $(z+3)(z+7)$ (d) $(t+4)(t-5)$
(e) $(r+6)(r-5)$ (f) $(t-11)(t-2)$
(g) $(z+6)(z-7)$ (h) $(m-5)(m-4)$
(i) $(y+12)(y-5)$ (j) $(n-7)(n+4)$
(k) $(p+12)(p-3)$ (l) $(a-6)(a-8)$

7 C 8 $(2x+4)$ cm
9 $(3x-12)$ mm
10 (a) $2w^2-6w+4$
(b) $(2w-2)(w-2)=2(w-1)(w-2)$
(c) 1 m

Open-ended – Sample answers

11 $x^2-x-2=(x+1)(x-2)$ and $x^2+x-42=(x+7)(x-6)$
12 (a) $x+8$ and $x-3$ are the factors so possible dimensions could be 12 units by 1 unit or 14 units by 3 units.
(b) Area $=12\times1=12$ units^2 or Area $=14\times3=42$ units^2

Exercise 3.6

1 (a) $(p-2)(p+2)$ (b) $(d-10)(d+10)$
(c) $(3-k)(3+k)$ (d) $3(p-5)(p+5)$
(e) $5(4-m)(4+m)$ (f) $8(k-2)(k+2)$
(g) $(2y-7)(2y+7)$ (h) $(5-4x)(5+4x)$
(i) $(6m-5n)(6m+5n)$ (j) $(3x-4)(3x+4)$
(k) $(2-5a)(2+5a)$ (l) $(9d-8)(9d+8)$
2 (a) $(x-1)(x+5)$ (b) $(a-6)(a+4)$
(c) $(-k-1)(k+9)$ or $-(k+1)(k+9)$
(d) $(5-p)(7+p)$ (e) $4(2-y)(y-1)$
(f) $3(1-x)(3x-1)$ (g) $5(x+3-y)(x+3+y)$
(h) $7(a+5-b)(a+5+b)$ (i) $3(x-6)(x-2)$
3 (a) $(x-\sqrt{2})(x+\sqrt{2})$ (b) $(\sqrt{7}-p)(\sqrt{7}+p)$
(c) $(2k-\sqrt{5})(2k+\sqrt{5})$ (d) $(3p-\sqrt{2})(3p+\sqrt{2})$
(e) $3(a-\sqrt{11})(a+\sqrt{11})$ (f) $2(b-\sqrt{3})(b+\sqrt{3})$
(g) $15(x+\sqrt{3})(x-\sqrt{3})$ (h) $7(p-\sqrt{7})(p+\sqrt{7})$
(i) $13(k-\sqrt{13})(k+\sqrt{13})$
4 (a) $(x-y+1)(x+y+3)$ (b) $-(2a+5)$
(c) $(2k-m+6)(2k+m)$ (d) $16x$
(e) $(x-2)(3x+4)$ (f) $(4x+3)(6x-11)$
(g) $(x+2-\sqrt{3})(x+2+\sqrt{3})$ (h) $(y-3-\sqrt{6})(y-3+\sqrt{6})$
(i) $(x+8-\sqrt{5})(x+8+\sqrt{5})$
(j) $(x-5-\sqrt{13})(x-5+\sqrt{13})$
(k) $(x-1-\sqrt{19})(x-1+\sqrt{19})$
(l) $(2x+1-\sqrt{2})(2x+1+\sqrt{2})$
5 (a) $(x+3)(x+1)$ (b) $(a-4)(a+9)$
(c) $(x-y)(x+y+7)$ (d) $(3a+b)(3a-b+2)$
(e) $(x-4y)(x+4y+2)$ (f) $(2p+q)(2p-q-3)$
6 (a) $(x+3)^2$ (b) $(x+2)^2$ (c) $(x-7)^2$
(d) $(x-4)^2$ (e) $(x+1)^2$ (f) $(x+6)^2$
(g) $(a-3)^2$ (h) $2(m-4)^2$ (i) $3(q+1)^2$
(j) $4(x+1)^2$ (k) $5(x+3)^2$ (l) $4(x-5)^2$
7 B
8 D

9 (a) $8(m + 5n)(m - 5n)$ (b) $6(x - 2r)(x + 2r)$
(c) $5(a - 3c)(a + 3c)$ (d) $3(3a - 4b)(3a + 4b)$
(e) $2(6x - 7y)(6x + 7y)$ (f) $7(5t - 2p)(5t + 2p)$
(g) $6(3k - 2m)(3k + 2m)$ (h) $2(xy - 4z)(xy + 4z)$
(i) $3(pq - 2n)(pq + 2n)$

10 (a) $(4c + 5d)(4c - 5d + 2)$ (b) $(k - 8)(k + 8 + m)$
(c) $(9 - 7a)(9 + 7a - 3d)$ (d) $(4 - 3y)(4 + 3y - 4w)$
(e) $(5m + 3n)(5m - 3n - 4)$ (f) $(12p - 7r)(12p - q + 7r)$

11 A

12 (a) $x^2 - y^2 = 56$ (b) $(x - y)(x + y) = 56$
(c) The difference between consecutive odd numbers is 2.
(d) $4(y + 1) = 56$ (e) $x = 15, y = 13$

13 (a) $(a - b)(a + b)$ (b) two
(c) $a + b$ (d) $(a^2 - b^2)(a^2 + b^2)$
(e) $(a - b)(a + b)(a^2 + b^2)$ (f) three
(g) $(a + b)(a^2 + b^2)$ (h) $(a^4 - b^4)(a^4 + b^4)$
(i) $(a - b)(a + b)(a^2 + b^2)(a^4 + b^4)$
(j) $(a + b)(a^2 + b^2)(a^4 + b^4)$
(k) $(a + b)(a^2 + b^2)(a^4 + b^4)(a^8 + b^8)$
(l) $$\begin{aligned}(2^{16} - 1) &= (2^{16} - 1^{16}) \div (2 - 1)\\ &= (2 + 1)(2^2 + 1^2)(2^4 + 1^4)(2^8 + 1^8)\\ &= 3 \times 5 \times 17 \times 257\end{aligned}$$

14 (a) $x^2 - y^2$ m^2= (b) $2(x - y)$ m^2
(c) $(x - y)(x + y + 2)$ m^2

15 $$\begin{aligned}a^2 + b^2 &= (m - n)^2 + (2\sqrt{mn})^2\\ &= m^2 - 2mn + n^2 + 4mn\\ &= m^2 + 2mn + n^2\\ &= (m + n)^2\\ &= c^2\end{aligned}$$

16 (a) $$\begin{aligned}t^2 &= (kr)^2 - r^2\\ &= (kr - r)(kr + r)\\ &= r(k - 1)r(k + 1)\\ &= r^2(k - 1)(k + 1)\end{aligned}$$
(b) $k = 1$ or -1
(c) k cannot be equal to zero as this creates the equation $a^2 = -b^2$, which is impossible for real numbers. k cannot be a fraction between -1 and 1 as this creates a similar situation where a^2 is equal to a negative fraction of b^2.

Open-ended – Sample answers

17 $a^2 - b^2 + 3a + 3b = (a + b)(a - b + 3)$

Exercise 3.7

1 (a) $\frac{4x}{5}$ (b) $\frac{k}{7}$ (c) $\frac{5p}{9}$
(d) $2(a - b)$ (e) $\frac{3(x+1)}{2}$ (f) $\frac{7(y-2)}{4}$
(g) $\frac{y(a-4)}{2}$ (h) $\frac{2x(x+5)}{9}$ (i) $\frac{3x(x-4)}{4}$

2 (a) $\frac{x+6}{x-5}$ (b) $\frac{x+2}{x-2}$ (c) $\frac{2x-3}{2x+1}$
(d) $\frac{x-2}{x-7}$ (e) $\frac{b+5}{b+3}$ (f) $\frac{x+7}{x-1}$

3 (a) $\frac{1}{3}$ (b) $\frac{3(5+2x)}{5-2x}$ (c) $\frac{x+4}{x-4}$
(d) $\frac{x-3}{x+3}$ (e) $\frac{x+7}{8}$ (f) $\frac{7}{4}$
(g) $\frac{(x-3)(x+2)}{(x-4)(x-1)}$ (h) $\frac{x+4}{x+1}$
(i) $\frac{(3x-2)(x-3)}{(x-6)(x+2)}$ (j) $\frac{(5x-3)(x-3)}{(x-4)(5x-4)}$

4 (a) $x + 8$ (b) $x - 9$ (c) $\frac{1}{x-5}$
(d) $\frac{1}{x-7}$ (e) $\frac{2x-1}{2}$ (f) $3(x - 2)$
(g) $\frac{2x+3}{2}$ (h) $\frac{3x+5}{2}$ (i) $\frac{x-1}{x-8}$
(j) $\frac{x+4}{x+5}$ (k) $\frac{x+1}{x-2}$ (l) $\frac{x-1}{x+4}$
(m) 1 (n) $\frac{5}{6}$ (o) $\frac{x-5}{6(x+3)}$
(p) $\frac{8(x-7)}{x+5}$ (q) $\frac{x-2}{x+2}$ (r) $\frac{x-3}{x-7}$

5 (a) $\frac{3(x-2)}{x+3}$ (b) $\frac{5(x+5)}{3(x-7)}$ (c) $\frac{x+6}{x+7}$
(d) $\frac{(x+2)(x+1)}{(x+5)(x+8)}$ (e) $\frac{5(x+y+3)}{4}$ (f) $\frac{7(x-1-2y)}{3(3x-2)}$

6 (a) $\frac{5x}{6}$ (b) $\frac{7x}{10}$ (c) $\frac{5x}{12}$
(d) $\frac{x}{2}$ (e) $\frac{11x}{15}$ (f) $\frac{-7x}{12}$
(g) $\frac{8x-3y}{6}$ (h) $\frac{20x-21y}{28}$ (i) $\frac{99x-64y}{72}$
(j) $\frac{5x+22}{6}$ (k) $\frac{3x-2}{8}$ (l) $\frac{4x-16}{9}$
(m) $\frac{3x-8}{12}$ (n) $\frac{13x-34}{20}$ (o) $\frac{13x+14}{36}$
(p) $\frac{29x+2y}{12}$ (q) $\frac{38x-19y}{15}$ (r) $\frac{34x+5y}{12}$
(s) $\frac{9x-8y}{30}$ (t) $\frac{3x-4y}{4}$ (u) $\frac{17x+11y}{6}$

7 C

8 (a) $\frac{3(b-a)}{3b+a}$ (b) $\frac{3(7-5)}{3 \times 7 + 5} = \frac{3}{13}$

9 (a) $$\begin{aligned}(a - b)(a^2 + ab + b^2) &= a^3 + a^2b + ab^2 - a^2b - ab^2 - b^3\\ &= a^3 - b^3\end{aligned}$$
(b) $(a^2 - b^2)(a^4 + a^2b^2 + b^4)$
(c) $$\begin{aligned}\frac{a^6 - b^6}{a^2 - b^2} &= \frac{(a^2 - b^2)(a^4 - a^2b^2 + b^4)}{a^2 - b^2}\\ &= a^4 + a^2b^2 + b^4\end{aligned}$$

10 (a) $4w^2 - \pi w^2$ (b) $\frac{4 - \pi}{4}$

11 (a) $\frac{3m}{4\pi r^3}$ kg/m^3

(b) Assuming that the golf ball and basketball are both spheres:

$$d_1 = \frac{m_1}{\left(\frac{4}{3}\pi r_1{}^3\right)}, d_2 = \frac{m_2}{\left(\frac{4}{3}\pi r_2{}^3\right)}$$

$$\frac{d_2}{d_1} = \frac{m_2}{\left(\frac{4}{3}\pi r_2{}^3\right)} \div \frac{m_1}{\left(\frac{4}{3}\pi r_1{}^3\right)}$$

$$= \frac{m_2}{\left(\frac{4}{3}\pi r_2{}^3\right)} \times \frac{\left(\frac{4}{3}\pi r_1{}^3\right)}{m_1}$$

$$= \frac{m_2}{m_1}\left(\frac{r_1}{r_2}\right)^3$$

(c) $\frac{4}{25}$

Open-ended – Sample answers

12 $\frac{x+1}{2} + \frac{1-x}{6}$

Chapter review 3

1 (a) $10x + 35$ (b) $x^2 - x - 6$

(c) $x^2 + 10x + 24$ (d) $8x^2 + 10x - 3$

2 (a) $x^2 + 12x + 36$ (b) $4x^2 - 12x + 9$

(c) $25 - x^2$ (d) $9x^2 - 49$

3 (a) $4x^2 - 20x - 56$ (b) $-3x^3 - 15x^2 + 72x$

(c) $x^3 - 2x^2 - 29x + 30$ (d) $-2x^3 - 7x^2 + 10x + 24$

4 (a) $7(x+3)$ (b) $4a(3b-2)$

(c) $6x(3x+4)$

5 (a) $(c-4)(d+6)$ (b) $(x+2)(x-3)$

(c) $(a+1)(5a-2)$ (d) $(3y-5)(2y-7)$

6

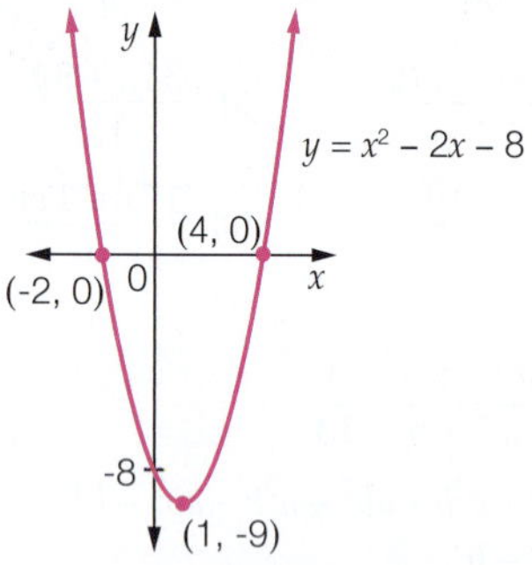

(a) x-intercepts: (-2, 0) and (4, 0); y-intercept: (0, -8)

(b) (1, -9), minimum (c) $x = 1$ (d) $y = -5$

(e) $x = 5$ and -3 (f) 5 and -3 (g) -2 and 4

7 (a) A (b) A (c) B

8 (a) $(x+5)(x+6)$ (b) $(x+2)(x-12)$

9 (a) $(x+5)^2$ (b) $(x-9)^2$ (c) $3(x-5)^2$ (d) $4(x+2)^2$

10 (a) $\frac{2x}{5}$ (b) $\frac{a-3}{3}$ (c) $\frac{5(x-1)}{4}$ (d) $3a$

11 (a) $\frac{x+4}{x+2}$ (b) $\frac{(x+1)(3x+7)}{(x-7)(x-1)}$

(c) $2x - 1$ (d) $\frac{2(x+5)}{x+4}$

12 (a) $\frac{7x}{6}$ (b) $\frac{8x-7}{15}$ (c) $\frac{4x-31}{30}$

(d) $\frac{12x+21y}{14}$ (e) $\frac{3x+4y+23}{12}$ (f) $\frac{2x-y-9}{10}$

13 $a = 4, b = 3$

14 (a) -5, 6 (b) 6 kg gain (c) 5 kg loss

(d) From April to September John lost 11 kg in weight.

15 (a) $ax + 2x + ay + 2y$ (b) $(a+2)(x+y)$

(c) 175 cm

16 (a) $5x - x^2$ (b) $x(5-x)$

(c) (i) 4 m^2 (ii) 2.25 m^2 (iii) 3.1875 m^2

17 (a)

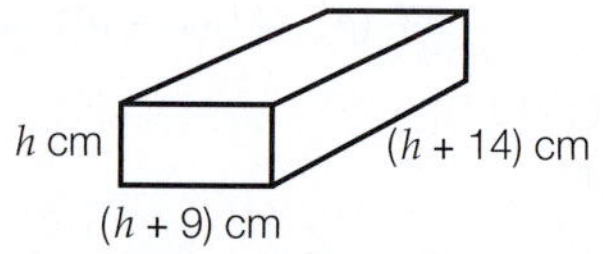

(b) $h(h+9)(h+14)$ (c) $h^3 + 23h^2 + 126h$

(d) 1800 cm^3 (e) 1800 cm^3

(f) Students' own answers.

18 (a) πx mm^2 (b) $(x+2)$ mm

(c) $\pi(x+2)$ mm^2 (d) $(x-2)$ mm

(e) $\pi(x-2)$ mm^2 (f) $\pi(x+2)^2 - \pi(x-2)$ mm^2

(g) π (h) $8\pi x$ mm^2

(i) 80π mm^2 (j) 17 mm

19 (a) $(8x+3)(2y+7)$ cm^2 (b) $(8x+3)(8x-3)$ cm^2

(c) $64x^2 - 9$ cm^2

(d) (i) $y = 4x - 2$ (ii) $64x^2 + 48x + 9$

20 (a) $x^2 + p^2x + p^2y - y^2$ (b) $(x+y)(x-y+p^2)$

(c) $(x+y)(x-y+p^2) = (x+y)(x-y)$; $x - y$ must be positive as the total result is positive, so x must be greater than y.

(d) (i) $2p^2x$ or $2p^2y$ vehicles

(ii) $2px(p-1)$ non-white vehicles

Chapter 4

Recall 4

1 (a) $3x^2 + 12x$ (b) $x^2 - 4x - 21$

(c) $x^2 + 16x + 64$ (d) $x^2 - 16$

2 (a) $(x+5)(x-5)$ (b) $(2x+9)(2x-9)$

(c) $(x+\sqrt{5})(x-\sqrt{5})$ (d) $(x-4)^2$

3 (a) 3, 7 (b) -10, 2

4 (a) $x(x-7)$ (b) $(x-6)(x+1)$

(c) $3(x-4)(x+3)$ (d) $-(x-10)(x+2)$

Exercise 4.1

1 (a) $x = 0$ or $x = 11$ (b) $x = 0$ or $x = -6$
(c) $x = 0$ or $x = -4$ (d) $x = 0$ or $x = 1$
(e) $x = 0$ or $x = 1$ (f) $x = 0$ or $x = 6$
(g) $x = 0$ or $x = -2$ (h) $x = 0$ or $x = 3$
(i) $x = -3$ or $x = -2$ (j) $x = 3$ or $x = -7$
(k) $x = 4$ or $x = -1$ (l) $x = 2$ or $x = 6$
(m) $x = 4$ or $x = 5$ (n) $x = 7$ or $x = -2$
(o) $x = 3$ or $x = -10$

2 (a) $x = 0$ (b) $x = 0$ (c) $x = 0$
(d) $x = -7$ (e) $x = -1$ (f) $x = 9$

3 (a) $x = 0$ or $x = 4$ (b) $x = 0$ or $x = -1$
(c) $x = 0$ or $x = 6$ (d) $x = 0$ or $x = 5$
(e) $x = 0$ or $x = \frac{1}{2}$ (f) $x = 0$ or $x = -3$

4 (a) $x = 5$ or $x = -5$ (b) $x = 9$ or $x = -9$
(c) $x = 2$ or $x = -2$ (d) $x = \frac{5}{4}$ or $x = -\frac{5}{4}$
(e) $x = \frac{7}{3}$ or $x = -\frac{7}{3}$ (f) $x = \frac{9}{10}$ or $x = -\frac{9}{10}$
(g) $x = 5$ or $x = -5$ (h) $x = \sqrt{\frac{5}{6}}$ or $x = -\sqrt{\frac{5}{6}}$
(i) $x = 2$ or $x = -2$ (j) $x = 7$ or $x = -7$
(k) $x = 9$ or $x = -9$ (l) $x = 11$ or $x = -11$
(m) $x = \sqrt{13}$ or $x = -\sqrt{13}$
(n) $x = \sqrt{24}$ or $x = -\sqrt{24}$ (or $x = 2\sqrt{6}$ or $x = -2\sqrt{6}$)
(o) $x = \sqrt{\frac{32}{3}}$ or $x = -\sqrt{\frac{32}{3}}$ (or $x = 4\sqrt{\frac{2}{3}}$ or $x = -4\sqrt{\frac{2}{3}}$)

5 (a) $x = 7$ (b) $x = -4$ (c) $x = -8$ (d) $x = 2$
(e) $x = -1$ (f) $x = 3$ (g) $x = 5$ (h) $x = 1$
(i) $x = -2$ (j) $x = 3$ (k) $x = 5$ (l) $x = 4$

6 (a) $x = -1$ or $x = -2$ (b) $x = -3$ or $x = -6$
(c) $x = 4$ or $x = 7$ (d) $x = 1$ or $x = 9$
(e) $x = -5$ or $x = 6$ (f) $x = -4$ or $x = 2$
(g) $x = 4$ or $x = 8$ (h) $x = -5$ or $x = 3$
(i) $x = -2$ or $x = 8$ (j) $x = -6$ or $x = 2$
(k) $x = -1$ or $x = 2$ (l) $x = -1$ or $x = 3$

7 (a) $x = -4$ or $x = -3$ (b) $x = -6$ or $x = -2$
(c) $x = -7$ or $x = 9$ (d) $x = -8$ or $x = 3$
(e) $x = -3$ or $x = -2$ (f) $x = -3$ or $x = 5$
(g) $x = -10$ or $x = 5$ (h) $x = -3$ or $x = 9$
(i) $x = 6$ or $x = 11$ (j) $x = 1$ or $x = 5$
(k) $x = -8$ or $x = 2$ (l) $x = -7$ or $x = 3$

8 (a) D (b) A (c) D (d) A

9 (a) $x = -5$ or $x = 2$ (b) $x = -1$ or $x = 3$
(c) $x = -4$ or $x = 7$ (d) $x = -7$ or $x = -1$
(e) $x = -4$ or $x = 4$ (f) $x = -3$ or $x = 3$
(g) $x = -6$ or $x = -1$ (h) $x = -7$ or $x = 6$
(i) $x = 0$ or $x = 12$

10 (a) A (b) D (c) B (d) B (e) B (f) C

11 $t = 0$ seconds and $t = 50$ seconds

12 (-1, 0) and (4, 0)

13 (a) (-12, 0) and (12, 0) (b) (-4, 0) and (0, 0)
(c) (6, 0)

14 (a) 5 °C (b) $t = 2$ and $t = 8$
(c) 10 am and 4 pm

15 (a) $(x + 4)(x + 6)$ m^2 (b) length 8 m, width 6 m

16 60 m

17 (a) $n - 3 = 0$ so $D = 0$
(b) quadrilateral: 2 diagonals

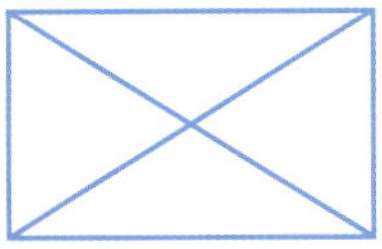

pentagon: 5 diagonals

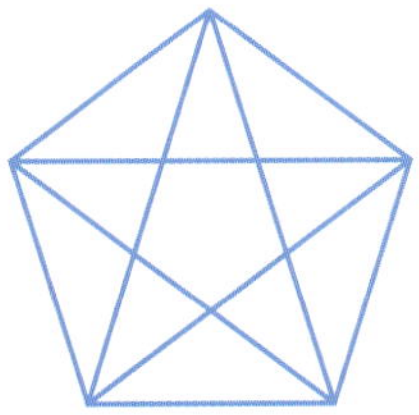

hexagon: 9 diagonals

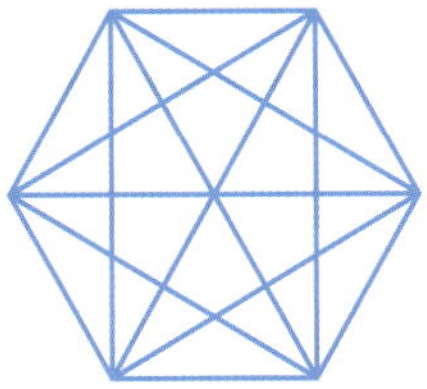

(c) (i) 20 (ii) 35 (iii) 54
(d) (i) 20 (ii) 26
(e) (i) 14 (ii) 42

Open-ended – Sample answers

18 (a) Students' own equations; they could be of the form $(x + a)(x + b) = 0$.
(b) Students' own equations; they could be of the form $(x + a)^2 = 0$.

19 Students' own equations; the easiest way to get the second equation is to make it the negative of the first.

20 $x^2 + 2x - 2 = 0$; $6x^2 - 13x - 5 = 0$; $x^2 - x - 2 = 0$

Exercise 4.2

1 (a) $(x + 4 - y)(x + 4 + y)$ (b) $(y - 2 - k)(y - 2 + k)$
(c) $(y + 4 - m)(y + 4 + m)$ (d) $(m - 5 - 2n)(m - 5 + 2n)$
(e) $(x + 1 - 3y)(x + 1 + 3y)$ (f) $(q + 3 - 4p)(q + 3 + 4p)$
(g) $(a - 3b - 4)(a - 3b + 4)$ (h) $(x - 5y - 6)(x - 5y + 6)$
(i) $(2a - 3 - 4b)(2a - 3 + 4b)$ (j) $(3c + 5 - 2d)(3c + 5 + 2d)$

2 (a) $(w^2 - 16w + 64)$ has been incorrectly factorised as $(w + 8)^2$; it should be factorised to be $(w - 8)^2$.

(b) $w^2 - 16w + 64 - 4k^2 = (w^2 - 16w + 64) - 4k^2$
$= (w - 8)^2 - (2k)^2$
$= (w - 8 - 2k)(w - 8 + 2k)$

3 (a) $(x + 3 - \sqrt{7})(x + 3 + \sqrt{7})$

(b) $(x - 5 - \sqrt{21})(x - 5 + \sqrt{21})$

(c) $(x - 4 - \sqrt{19})(x - 4 + \sqrt{19})$

(d) $(x + 10 - \sqrt{23})(x + 10 + \sqrt{23})$

(e) $(x - 6 - \sqrt{15})(x - 6 + \sqrt{15})$

(f) $(x - 2)(x + 4)$

(g) $(x - 7 - \sqrt{51})(x - 7 + \sqrt{51})$

(h) $(x - 2 - \sqrt{6})(x - 2 + \sqrt{6})$

(i) $(x + 8 - \sqrt{17})(x + 8 + \sqrt{17})$

(j) $\left(x + \frac{3}{2} - \frac{\sqrt{5}}{2}\right)\left(x + \frac{3}{2} + \frac{\sqrt{5}}{2}\right)$

(k) $\left(x + \frac{7}{2} - \frac{\sqrt{13}}{2}\right)\left(x + \frac{7}{2} + \frac{\sqrt{13}}{2}\right)$

(l) $\left(x - \frac{1}{2} - \frac{\sqrt{17}}{2}\right)\left(x - \frac{1}{2} + \frac{\sqrt{17}}{2}\right)$

(m) $\left(x - \frac{5}{2} - \frac{\sqrt{5}}{2}\right)\left(x - \frac{5}{2} + \frac{\sqrt{5}}{2}\right)$

(n) $\left(x + \frac{5}{2} - \frac{\sqrt{33}}{2}\right)\left(x + \frac{5}{2} + \frac{\sqrt{33}}{2}\right)$

(o) $\left(x + \frac{1}{2} - \frac{\sqrt{21}}{2}\right)\left(x + \frac{1}{2} + \frac{\sqrt{21}}{2}\right)$

(p) $\left(x - \frac{9}{2} - \frac{\sqrt{29}}{2}\right)\left(x - \frac{9}{2} + \frac{\sqrt{29}}{2}\right)$

(q) $\left(x - \frac{3}{2} - \frac{\sqrt{17}}{2}\right)\left(x - \frac{3}{2} + \frac{\sqrt{17}}{2}\right)$

(r) $\left(x - \frac{7}{2} - \frac{\sqrt{53}}{2}\right)\left(x - \frac{7}{2} + \frac{\sqrt{53}}{2}\right)$

4 $2p^2 + 12p - 20$
$= 2(p^2 + 6p - 10)$
$= 2(p^2 + 6p + 9 - 9 - 10)$
$= 2[(p + 3)^2 - 19]$
$= 2[(p + 3)^2 - (\sqrt{19})^2]$
$= 2(p + 3 + \sqrt{19})(p + 3 - \sqrt{19})$

5 (a) $2(x + 1 - \sqrt{6})(x + 1 + \sqrt{6})$

(b) $4(x - 2 - \sqrt{5})(x - 2 + \sqrt{5})$

(c) $10(x - 4 - \sqrt{19})(x - 4 + \sqrt{19})$

(d) $3(x - 1 - \sqrt{5})(x - 1 + \sqrt{5})$

(e) $5(x + 2 - \sqrt{5})(x + 2 + \sqrt{5})$

(f) $7(x + 3 - \sqrt{7})(x + 3 + \sqrt{7})$

6 D   7 A

8 (a) 14 m   (b) $(t - 4 - \sqrt{2})(t - 4 + \sqrt{2})$

(c) 2.83 seconds   (d) The seagull is at sea level.

(e) The seagull is below sea level, catching a fish.

9 (a) $-5(x - 4)^2 + 75 = -5(x - 4 - \sqrt{15})(x - 4 + \sqrt{15})$

(b) $P = -5(4 \pm \sqrt{15} - 4)^2 + 75$
$= -5(\pm\sqrt{15})^2 + 75$
$= -5 \times 15 + 75$
$= 0$

(c) When $x = 0$, $P = -5$, so the company initially operated at a loss. When $x = 4$, $P = 75$, so the company had made a healthy profit after 4 months. When $x = 8$, $P = -5$, so the company made a loss again after 8 months.

10 (a) 100 earthworms

(b) $-(d + 2 + 2\sqrt{26})(d + 2 - 2\sqrt{26})$

(c) When $d = 2(\sqrt{26} - 1)$, $(d + 2 - 2\sqrt{26}) = 0$, so no earthworms remain.

(d) 8.20, so the number of earthworms reduced to zero during the 9th day.

(e) (i) $-(n + 1 + \sqrt{65})(n + 1 - \sqrt{65})$

(ii) When $n = \sqrt{65} - 1$, $(n + 1 - \sqrt{65}) = 0$, so no earthworms remain.

(f) $\sqrt{65} - 1 = 7.06$, so the first colony will stay longer in their compost heap.

Open-ended – Sample answers

11 (a) $(x - a)^2 + 2b - a^2$

(b) $A = 2b - a^2$

(c) (i) $2b - a^2 = 0$ (ii) $a = 2, b = 2$

Exercise 4.3

1 (a) (i) $(x + 1 - \sqrt{6})(x + 1 + \sqrt{6})$ (ii) $x = -1 \pm \sqrt{6}$

(b) (i) $(x + 2 - \sqrt{11})(x + 2 + \sqrt{11})$ (ii) $x = -2 \pm \sqrt{11}$

(c) (i) $(x + 4 - \sqrt{11})(x + 4 + \sqrt{11})$ (ii) $x = -4 \pm \sqrt{11}$

(d) (i) $(x + 5 - \sqrt{2})(x + 5 + \sqrt{2})$ (ii) $x = -5 \pm \sqrt{2}$

(e) (i) $(x + 6 - \sqrt{42})(x + 6 + \sqrt{42})$ (ii) $x = -6 \pm \sqrt{42}$

(f) (i) $(x - 7 - \sqrt{10})(x - 7 + \sqrt{10})$ (ii) $x = 7 \pm \sqrt{10}$

(g) (i) $\left(x + \frac{3}{2} - \frac{\sqrt{13}}{2}\right)\left(x + \frac{3}{2} + \frac{\sqrt{13}}{2}\right)$ (ii) $x = \frac{-3 \pm \sqrt{13}}{2}$

(h) (i) $\left(x + \frac{5}{2} - \frac{\sqrt{13}}{2}\right)\left(x + \frac{5}{2} + \frac{\sqrt{13}}{2}\right)$ (ii) $x = \frac{-5 \pm \sqrt{13}}{2}$

(i) (i) $\left(x + \frac{3}{2} - \frac{\sqrt{29}}{2}\right)\left(x + \frac{3}{2} + \frac{\sqrt{29}}{2}\right)$ (ii) $x = \frac{-3 \pm \sqrt{29}}{2}$

2 (a) $x = 4 \pm \sqrt{10}$ (b) $x = -3 \pm \sqrt{5}$ (c) $x = \frac{3 \pm \sqrt{5}}{2}$

(d) $x = \frac{5 \pm \sqrt{33}}{2}$ (e) $x = -2 \pm \sqrt{7}$

(f) $x = 10 \pm \sqrt{112}$ (or $x = 10 \pm 4\sqrt{7}$)

(g) $x = 2 \pm \sqrt{10}$ (h) $x = -2 \pm \sqrt{3}$ (i) $x = -2 \pm \sqrt{7}$

3 (a) C (b) B (c) D

4 (a) $x = 2 \pm \sqrt{6}$ (b) $x = 3 \pm \sqrt{7}$ (c) $x = 2 \pm \sqrt{7}$

(d) $x = \dfrac{3 \pm \sqrt{29}}{2}$ (e) $x = \dfrac{5 \pm \sqrt{17}}{2}$ (f) $x = \dfrac{-7 \pm \sqrt{69}}{2}$

5 (a) $x^2 + 4x + 7$
$= x^2 + 4x + 4 + 3$
$= (x + 2)^2 + 3$

This cannot be factorised without finding the square root of -3, which is not a real number.

(b) $(x + 2)^2 + 3$ is not the difference of two squares. It is the sum of a square and a positive number, so it is always positive for all real values of x. This means it can never be equal to zero.

(c) The graph does not cut the x-axis.

6 (a) width $= -2 + 2\sqrt{21}$ and length $= 2 + 2\sqrt{21}$

(b) width = 7.2 cm and length = 11.2 cm

7 $y = -2 + \sqrt{28} \approx 3.3$ cm

8 (a) $x = \dfrac{1 + \sqrt{801}}{2}$ m (b) $x = 14.65$ m

9 (a) length $= l + 4$, width $= l + 2$

(b) $(l + 4)(l + 2) = 215$

(c) $l = 6\sqrt{6} - 3$, $w = 6\sqrt{6} - 5$

(d) 15.7×13.7 cm

Open-ended – Sample answers

10 Students' own equations.

11 Students' own equations that do not result in the difference of two squares.

12 $x = -2 \pm \sqrt{3}$, $x^2 + 4x + 1 = 0$

Exercise 4.4

1 (a) (i) parabola, min turning point
(ii) (0, 15)
(iii) (3, 0), (5, 0)
(iv) (4, -1)
(v) $x = 4$

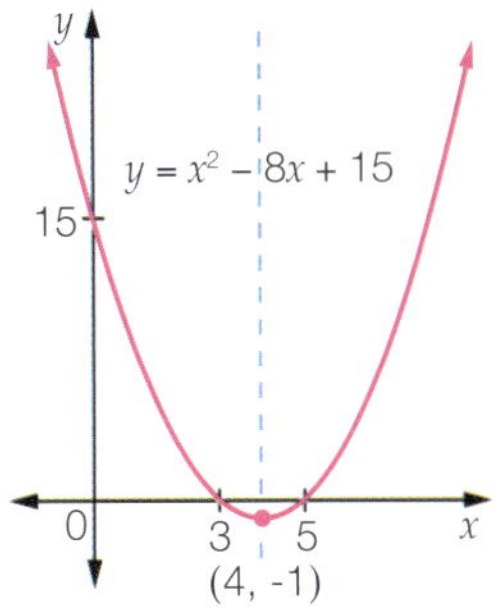

(b) (i) parabola, min turning point
(ii) (0, 8)
(iii) (2, 0), (4, 0)
(iv) (3, -1)
(v) $x = 3$

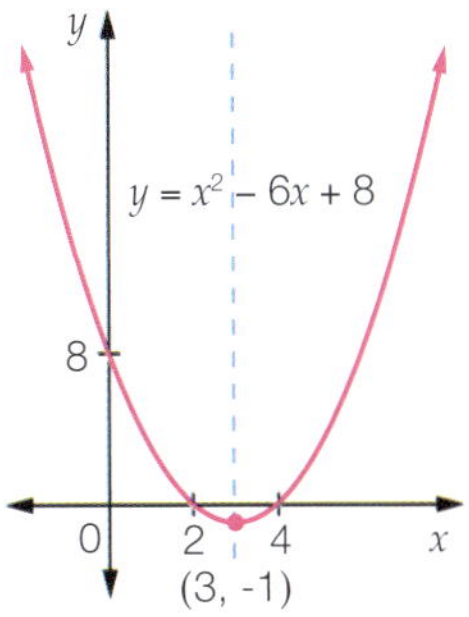

(c) (i) parabola, min turning point
(ii) (0, 21)
(iii) (3,0), (7, 0)
(iv) (5, -4)
(v) $x = 5$

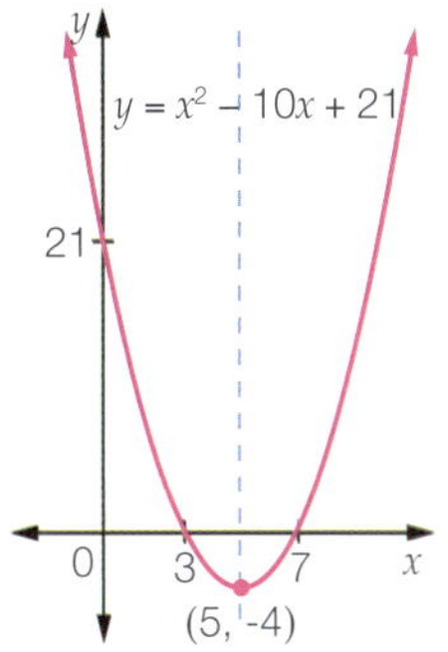

(d) (i) parabola, max turning point
(ii) (0, 12)
(iii) (-2, 0), (6, 0)
(iv) (2, 16)
(v) $x = 2$

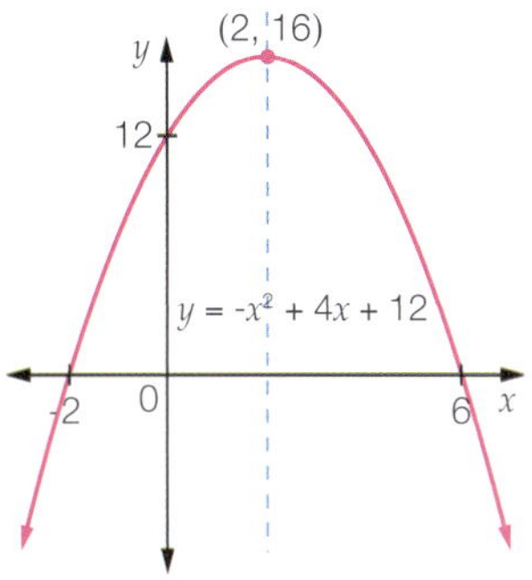

(e) (i) parabola, max turning point
(ii) (0, 3)
(iii) (-1, 0), (3, 0)
(iv) (1, 4)
(v) $x = 1$

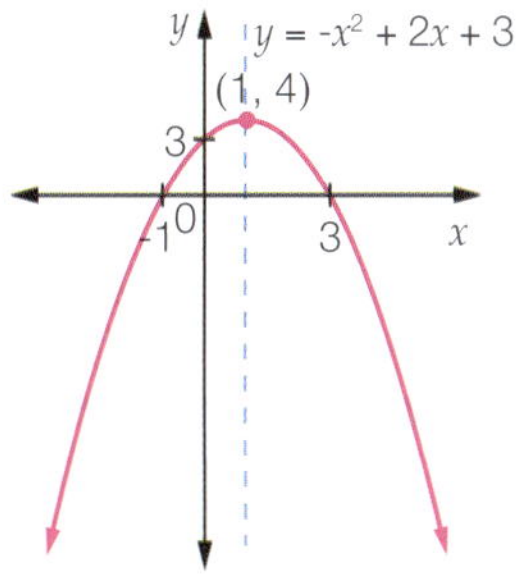

(f) (i) parabola, max turning point
(ii) (0, 8)
(iii) (-2, 0), (4, 0)
(iv) (1, 9)
(v) $x = 1$

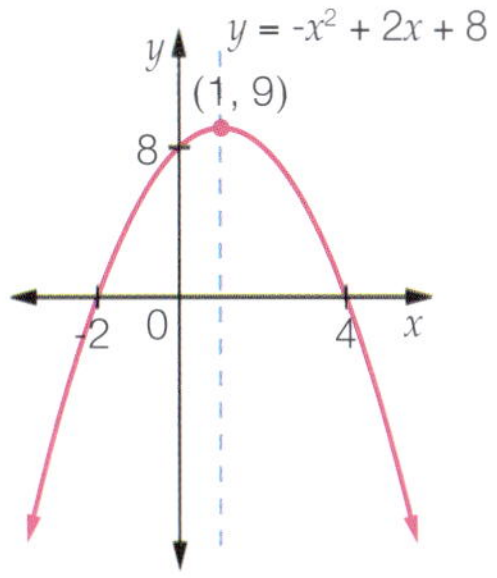

(g) (i) parabola, min turning point
(ii) (0, 12)
(iii) (2, 0), (6, 0)
(iv) (4, -4)
(v) $x = 4$

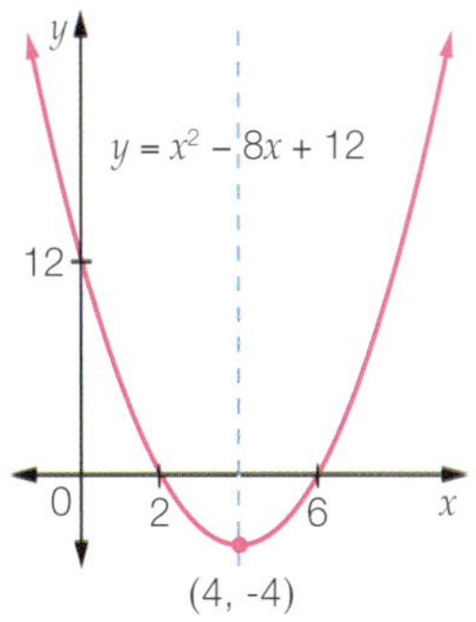

(h) (i) parabola, min turning point
(ii) (0, -3)
(iii) (-3, 0), (1, 0)
(iv) (-1, -4)
(v) $x = -1$

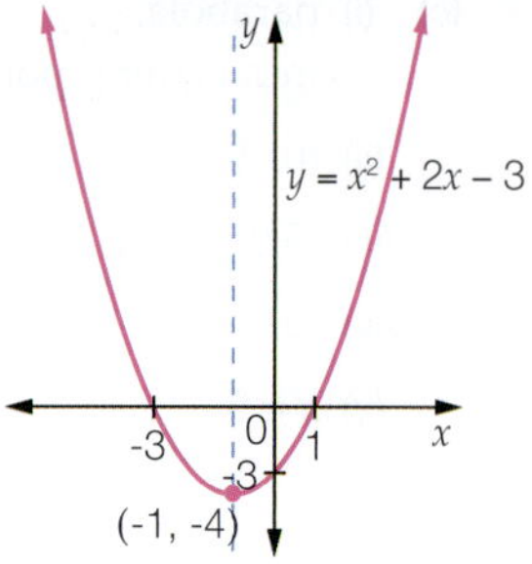

(i) (i) parabola, min turning point
(ii) (0, -5)
(iii) (-5, 0), (1, 0)
(iv) (-2, -9)
(v) $x = -2$

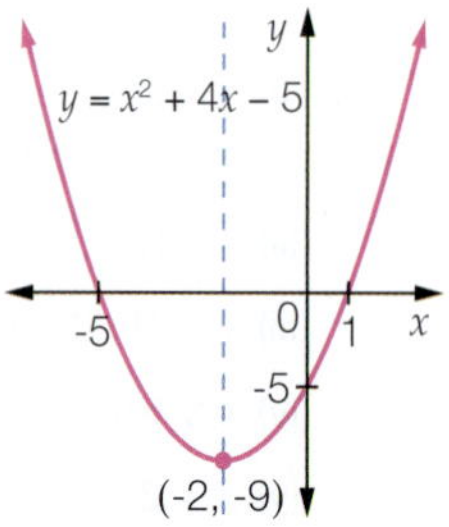

2 (a)

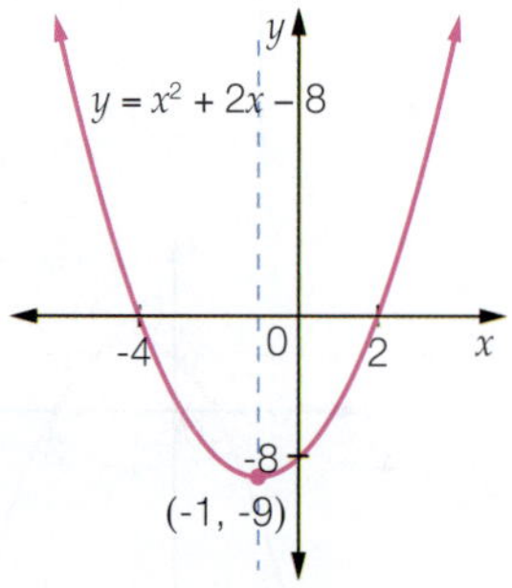

(b)

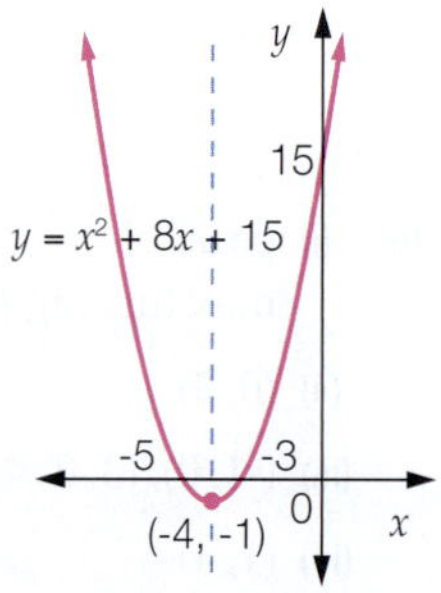

(c)

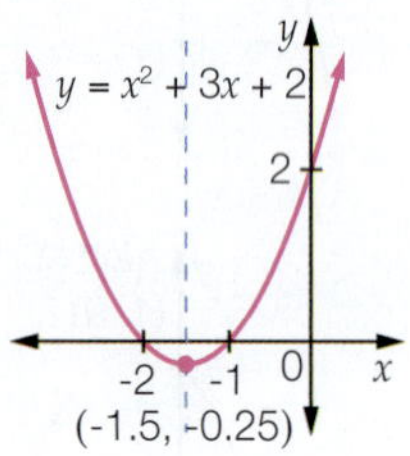

(d)

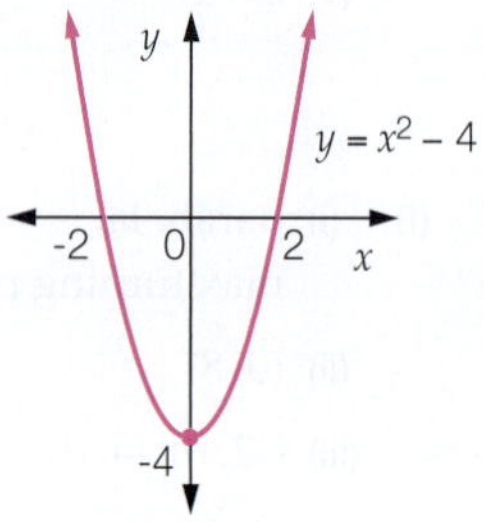

(e)

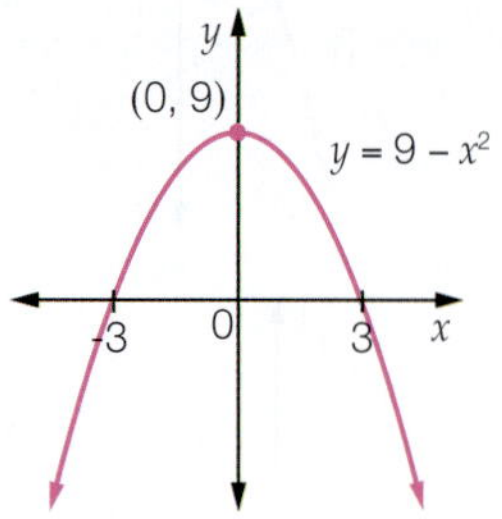

(f)

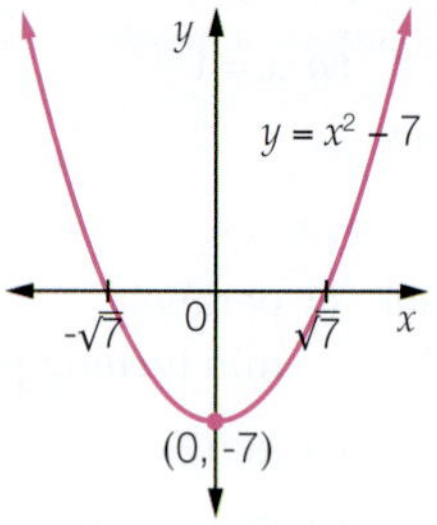

(g)

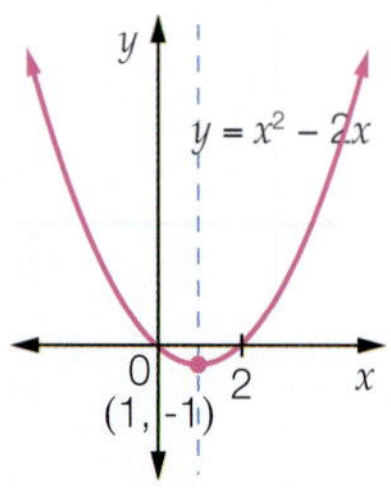

(h)

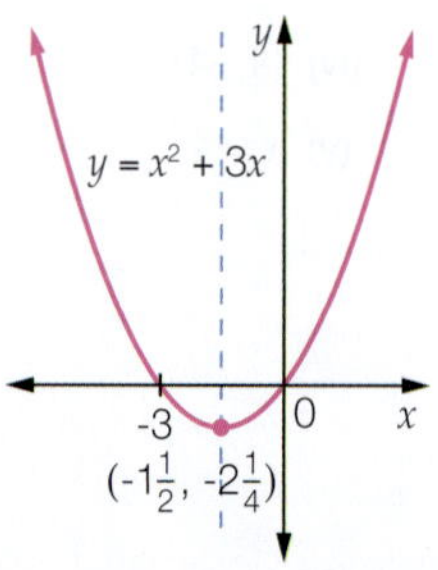

(i)

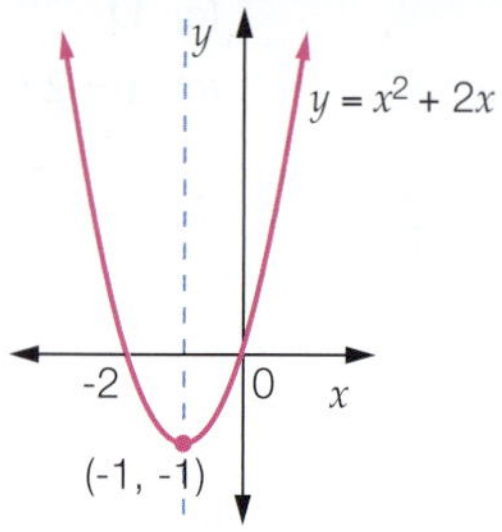

3 (a)

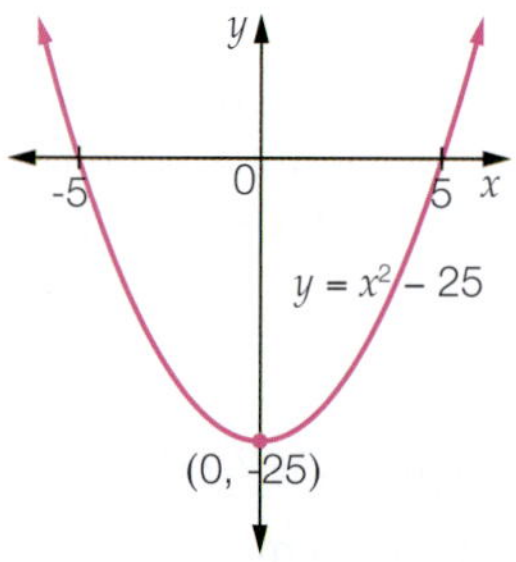

(b)

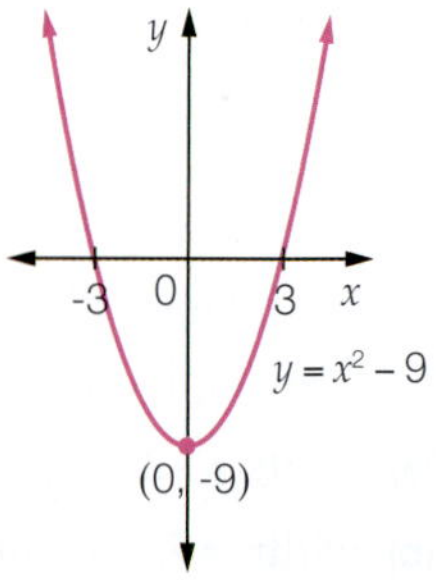

(c)

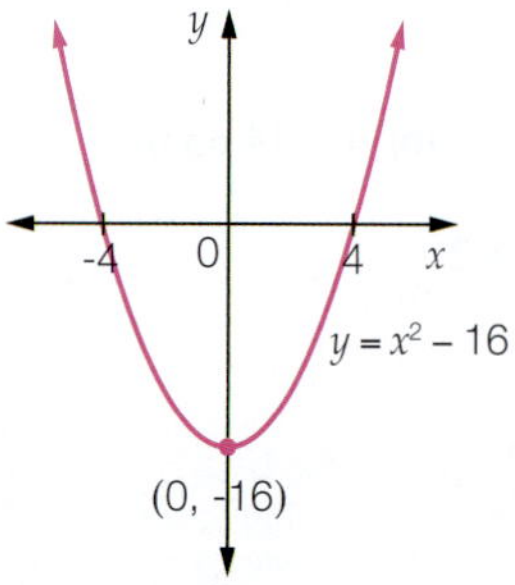

(d)

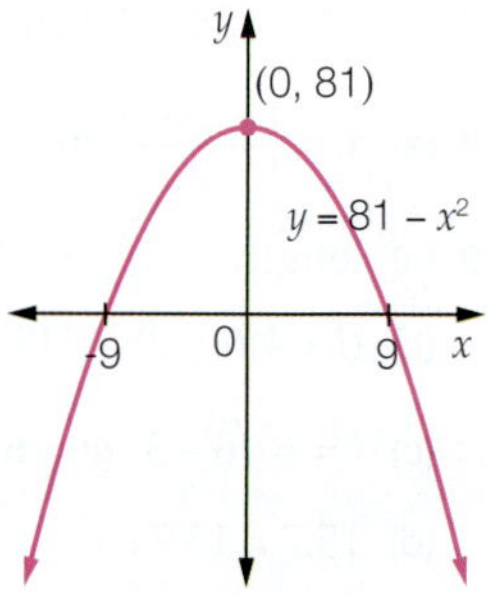

(e)

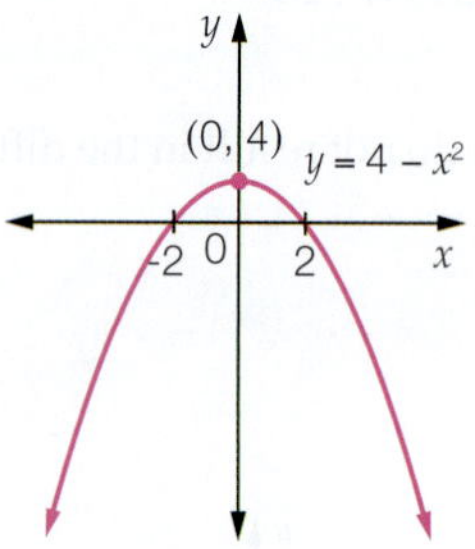

(f)

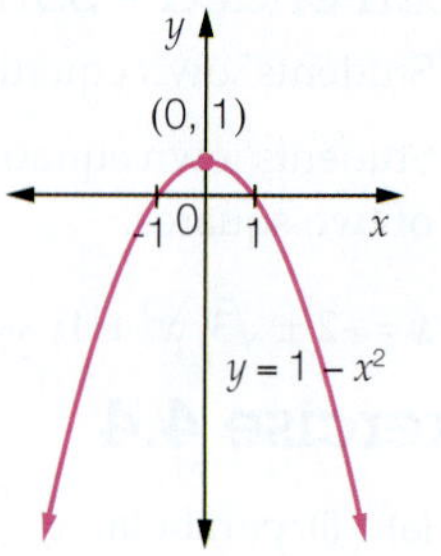

(g)

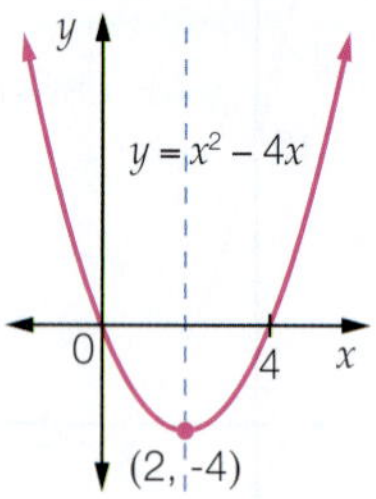

(h)

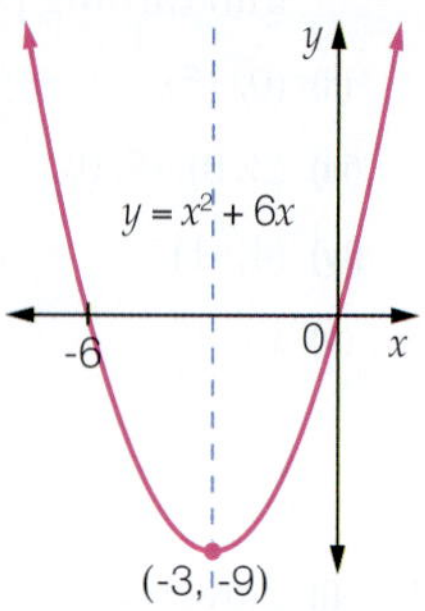

(i)

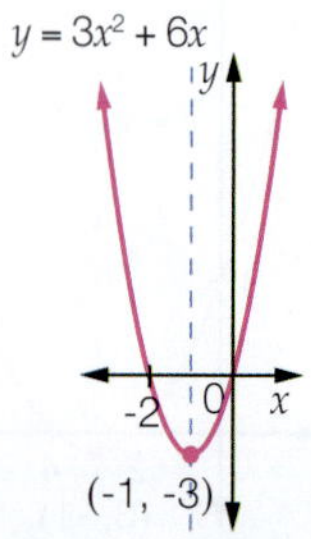

4 (a)

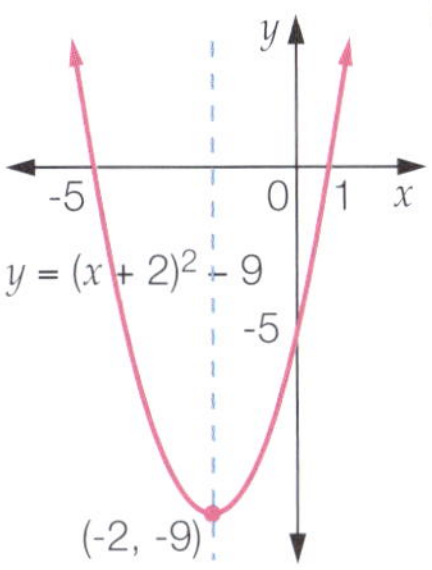

(b)

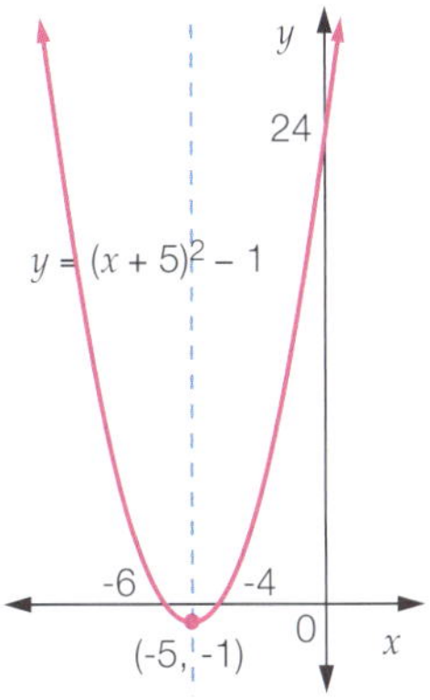

(c)

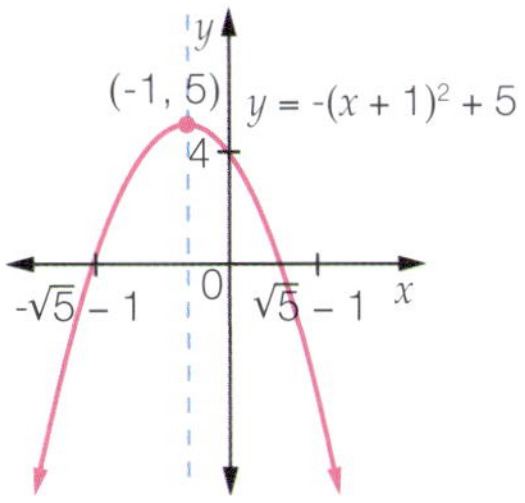

(d)

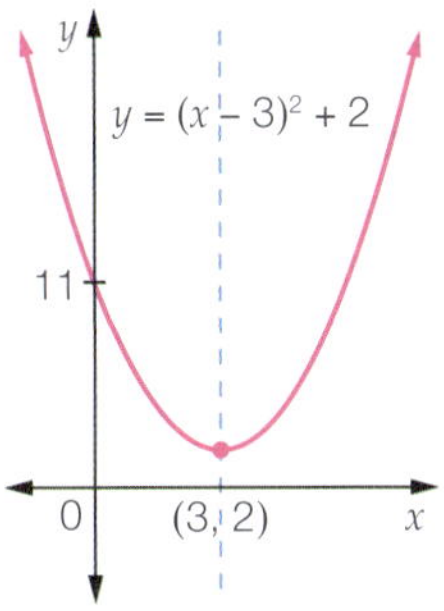

(e)

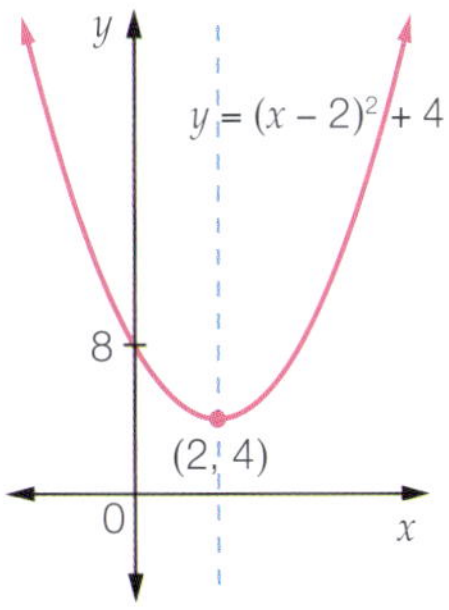

(f)

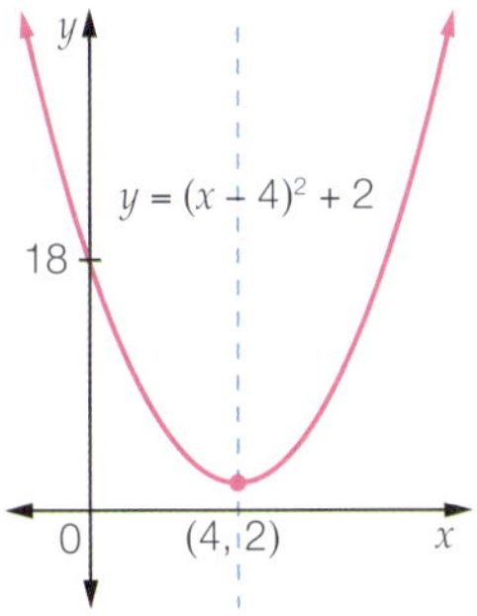

5 (a)

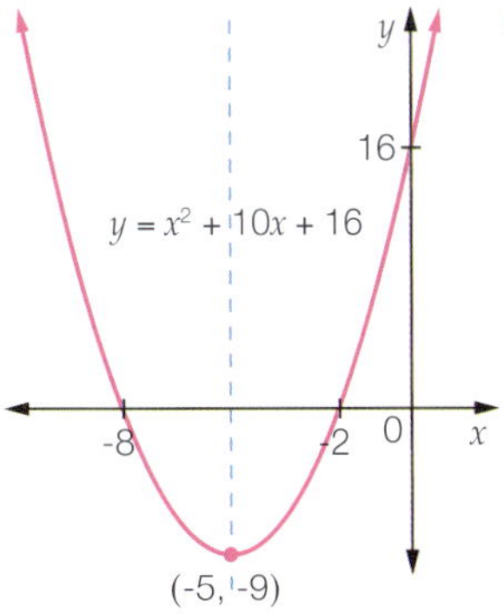

(b)

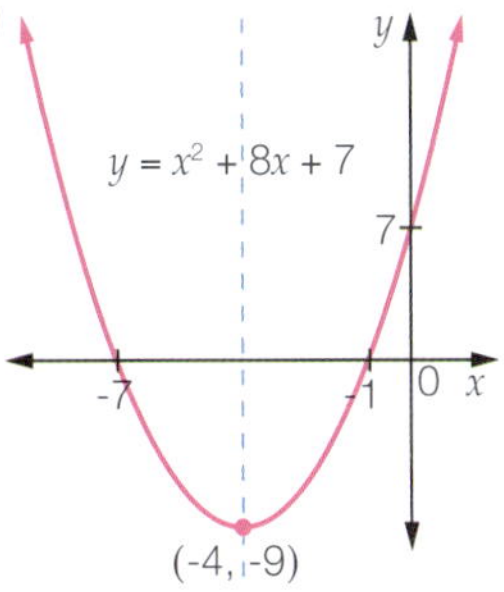

(c)

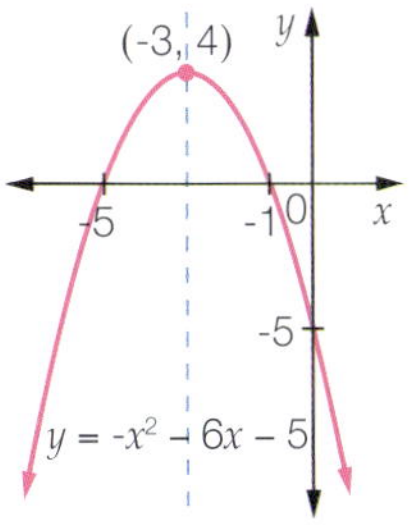

(d)

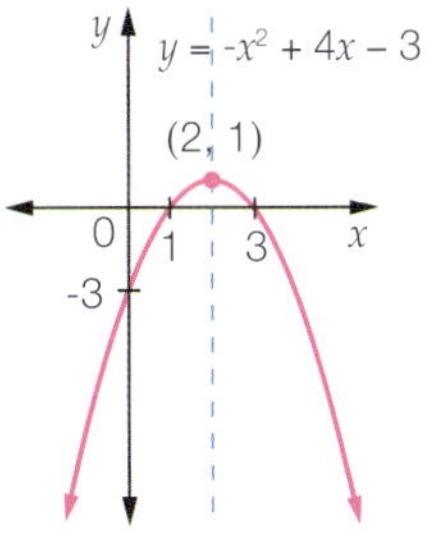

(e)

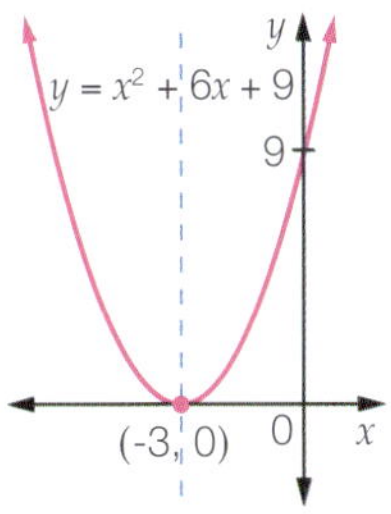

(f)

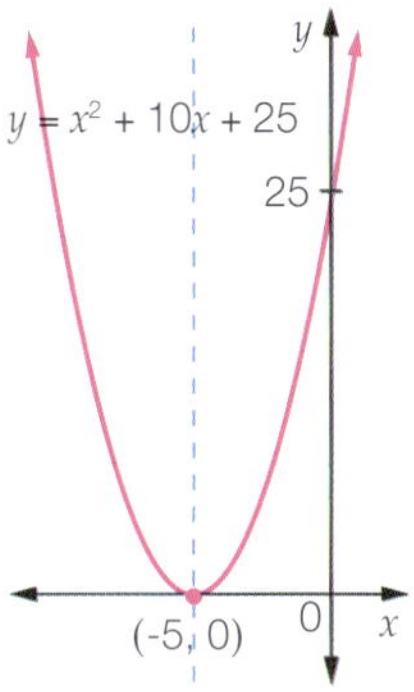

(g)

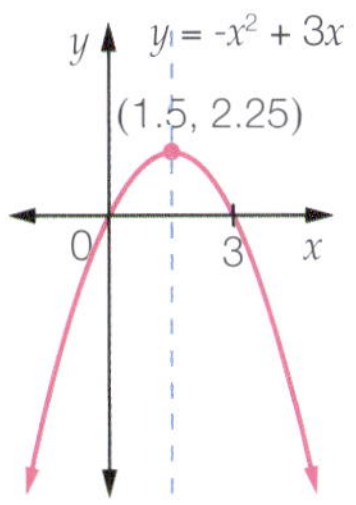

(h)

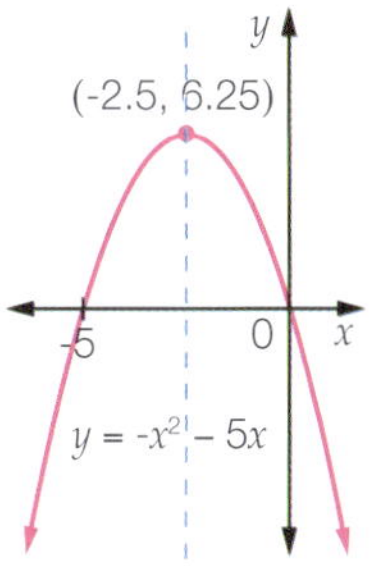

(i)

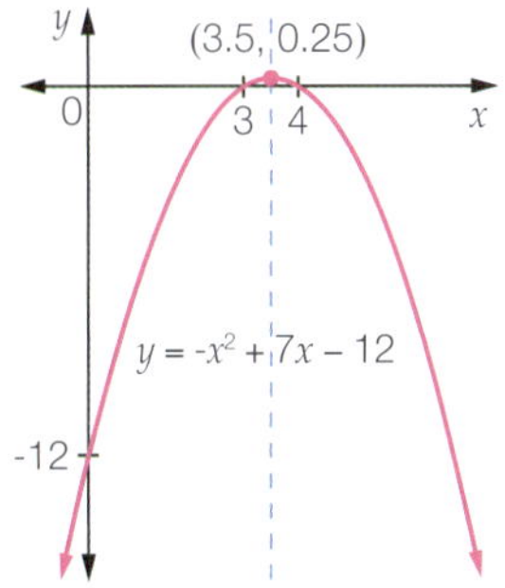

6 (a) B (b) D

7 (a) D (b) B

(c) B (d) D

8 (a) C (b) B

(c) A (d) B

(e) D (f) B

(g) B

9

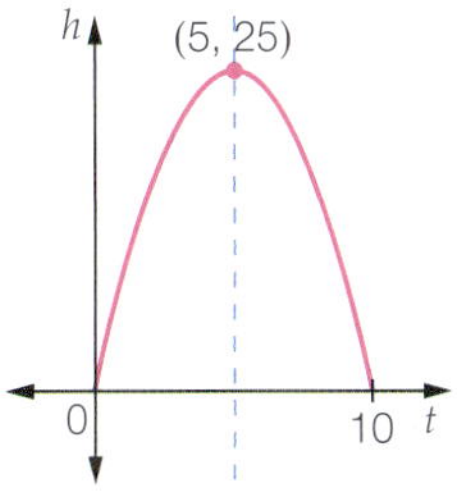

(a) 10 seconds (b) 25 m

10 (a) B (b) B

11 (a) D (b) (4, 2)

12 (a)

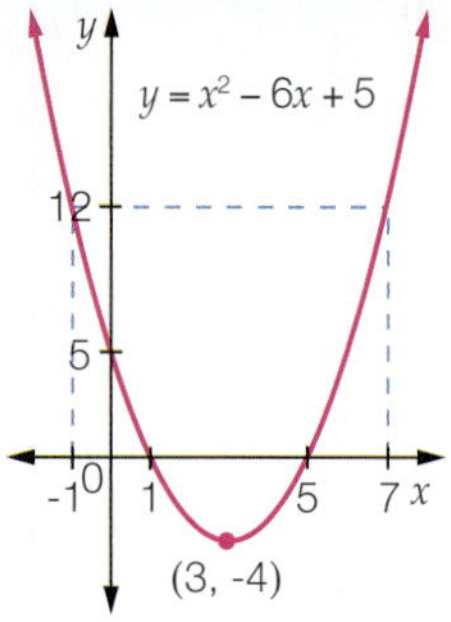

(b) $x = -1$ or $x = 7$

13 (a) $T = (h - 3)^2 + 6$

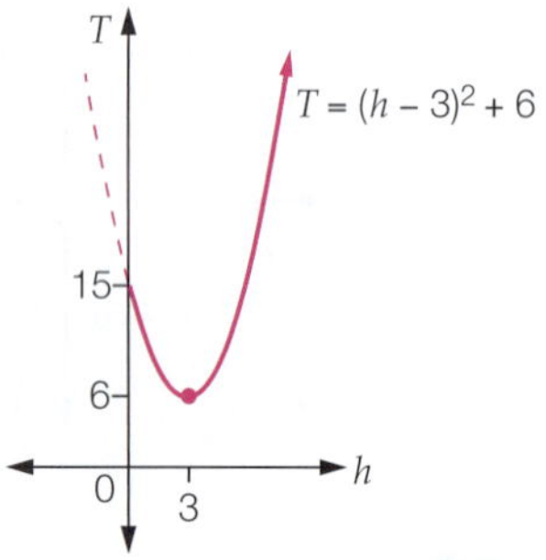

(b) 15 °C

(c) (i) decreasing (ii) increasing

(d) 6 °C after 3 hours

(e) No, the temperature of the room is unlikely to continue to increase like this. Using the equation, the temperature after 24 hours would be 447 °C.

14 $y = (x + 3)(x - 5)$, $y = x^2 - 2x - 15$ or $y = (x - 1)^2 - 16$

15 (a) $A = x(22 - x)$

(b)

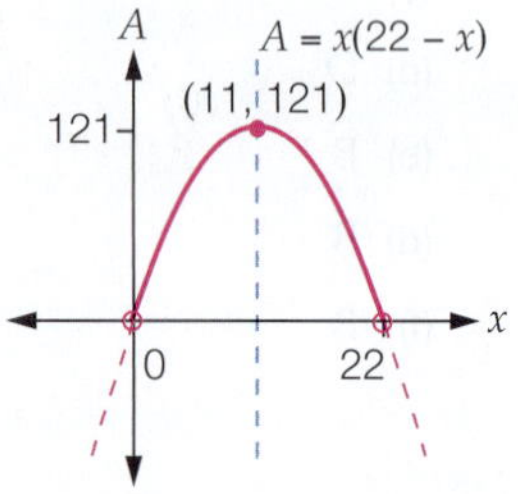

(c) (i) $x = 0$ m (ii) $x = 22$ m

(d) A square of length 11 m has a maximum area of 121 m^2.

Open-ended – Sample answers

16 If the equation is in factorised form then it will always have x-intercepts. If it is in turning point form, you can tell by inspection. If the TP is above the x-axis and it is positive quadratic then no x-intercepts; if the TP is below the x-axis and it is negative quadratic then no x-intercepts. If it is in standard form then complete the square to put it into turning point form.

17 (a) The equation factorises to $y = (x - 4)(x + 2)$. Archie may have just found factors of 8 and ignored the sign. As it is negative one of the factors will be positive and the other negative. Archie needs to be more careful with his factorising.

(b)

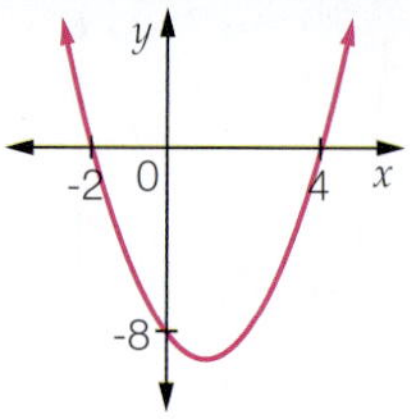

18 Students need to produce their own equations but using factorised form will make it easier. As an example, $(x - 4)(x - 6)$ has x-intercepts at 4 and 6, so the x-coordinate of the turning point is $x = 5$.

Half-time 4

1 (a) $x = -2, x = 1$ (b) $x = \pm 9$ (c) $x = 0, x = 2.5$

2 (a) $x = -2$ (b) $x = -1, x = 6$ (c) $x = 8, x = 1$

(d) $x = 5, x = -1$ (e) $x = -2, x = -1$ (f) $x = -5, x = 2$

3 (a) $x = 8, x = -8$ (b) $x = -1$ (c) $x = 4, x = 0$

4 (a) $x = -5, x = -1$ (b) $x = 0, x = 9$ (c) $x = -5, x = 2$

(d) $x = -3, x = 2$ (e) $x = 1, x = 3$ (f) $x = 3$

5 (a) dilated by a factor of 2, translated 3 units left and 4 units down

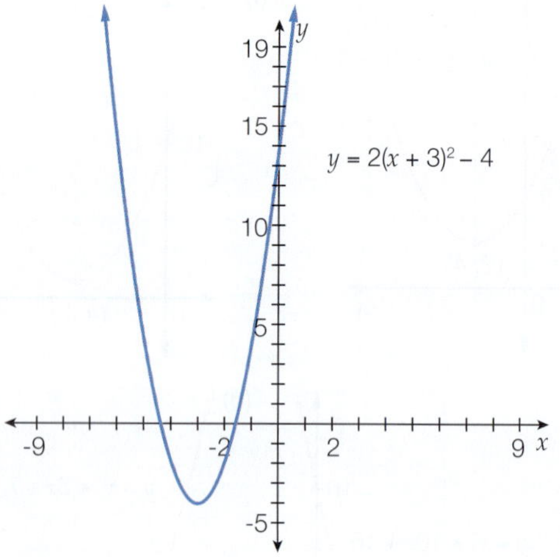

(b) dilated by a factor of a half, translated right 1 unit and up 3 units

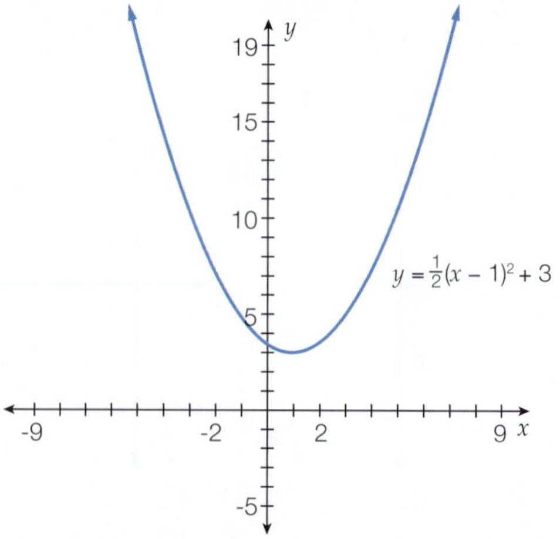

(c) translated 2 units right and 4 units down

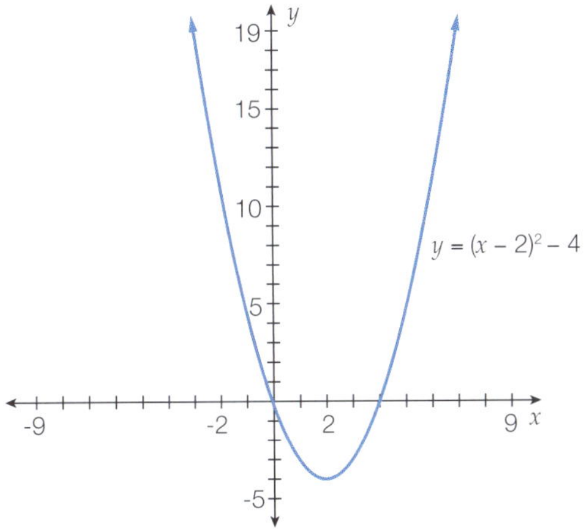

6 (a) $(x+3-\sqrt{2})(x+3+\sqrt{2})$

(b) $(x-2-\sqrt{3})(x-2+\sqrt{3})$

(c) $2(x+2+\sqrt{\frac{5}{2}})(x+2-\sqrt{\frac{5}{2}})$

7 (a) $x = \text{-}5, 3$ (b) $x = 2, 12$ (c) $x = \text{-}5, 8$

8 (a) $h = \text{-}(t-3)^2 + 10$, (3, 10)

(b)

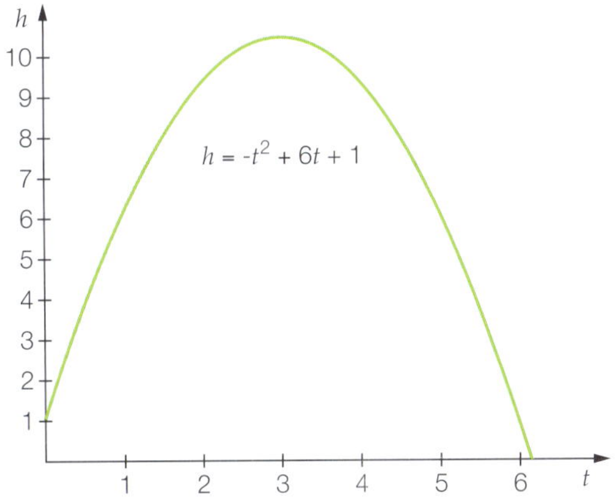

(c) 9 m (d) 1.3 seconds

(e) $2 < t < 4$ (f) 10 m

(g) 6 seconds

Exercise 4.5

1 (a) $(x+1)(x+6)$ (b) $(x+5)(x-11)$

(c) $(x+5)(x-3)$ (d) $(2x+3)(2x+1)$

(e) $(5y+4)(2y+1)$ (f) $(9c-1)(c-2)$

(g) $(3w+4)(2w+3)$ (h) $(5y-6)(3y+5)$

(i) $(8x+7)(x-2)$

2 (a) $(x+4)(x+3)$ (b) $(x+3)(x-8)$

(c) $(x-5)(x+9)$ (d) $(2x+1)(x+2)$

(e) $(2y+7)(y+1)$ (f) $(5b+3)(b+1)$

(g) $(3k-2)(k-3)$ (h) $(11a-12)(a+1)$

(i) $(5x+4)(x-3)$

3 (a) $a=1, b=1$ (b) $a=9, b=3$

(c) $a=4, b=2$ (d) $a=8, b=2$

(e) $a=12, b=2$ (f) $a=\text{-}2, b=1$

4 (a) $2\left(x-\frac{3}{2}-\frac{\sqrt{11}}{2}\right)\left(x-\frac{3}{2}+\frac{\sqrt{11}}{2}\right)$

(b) $3\left(x-2-\sqrt{\frac{19}{3}}\right)\left(x-2+\sqrt{\frac{19}{3}}\right)$

(c) $2\left(x-2-\sqrt{\frac{11}{2}}\right)\left(x-2+\sqrt{\frac{11}{2}}\right)$

(d) $3\left(x-\frac{3}{2}-\sqrt{\frac{11}{12}}\right)\left(x-\frac{3}{2}+\sqrt{\frac{11}{12}}\right)$

(e) $5\left(x-1-\sqrt{\frac{2}{5}}\right)\left(x-1+\sqrt{\frac{2}{5}}\right)$

(f) $2\left(x-\frac{7}{4}-\frac{\sqrt{33}}{4}\right)\left(x-\frac{7}{4}+\frac{\sqrt{33}}{4}\right)$

(g) not possible, no real factors

(h) not possible, no real factors

(i) not possible, no real factors

5 (a) $(2m-3)(m+3)$ (b) $(7x-9)(x-2)$

(c) $(11c-3)(c-2)$ (d) $(5y+4)(y+3)$

(e) $(3p-4)(p+4)$ (f) $(2x-5)(x+4)$

(g) $3(a-2-\sqrt{\frac{11}{3}})(a-2+\sqrt{\frac{11}{3}})$

(h) $2(x+\frac{3}{4}-\frac{\sqrt{41}}{4})(x+\frac{3}{4}+\frac{\sqrt{41}}{4})$

(i) $3(x+\frac{1}{3}-\frac{\sqrt{28}}{3})(x+\frac{1}{3}+\frac{\sqrt{28}}{3})$

(j) $(x+1)(3x+5)$ (k) $(y-2)(7y-4)$

(l) $(3t+4)(t-1)$

(m) $2(y-\frac{5}{2}-\frac{\sqrt{19}}{2})(y-\frac{5}{2}+\frac{\sqrt{19}}{2})$

(n) $4(x+\frac{5}{2}-\sqrt{7})(x+\frac{5}{2}+\sqrt{7})$

(o) $5(b+\frac{1}{5}-\frac{\sqrt{26}}{5})(b+\frac{1}{5}+\frac{\sqrt{26}}{5})$

(p) $(3d-2)(3d+4)$ (q) $(6x+11)(x-2)$

(r) $(9n-7)(2n+5)$ (s) $(8y-3)(3y-1)$

(t) $(5x+2)(3x+8)$ (u) $(4x+3)(3x-5)$

6 (a) $(3x+1)(2x-3)$ (b) 3.5 m (c) 4 m

7 (a) $12t^2-12t+10-(4t^2+6t+1)$

(b) (i) $8t^2-18t+9$ (ii) $(4t-3)(2t-3)$

(c) The difference in the number of birds must be zero. $0 = (2t-3)(4t-3)$. For this to be true either $2t-3=0$ ($t=1.5$) or $4t-3=0$ ($t=0.75$). Hence, when there are 75 or 150 trees there are equal numbers of birds.

8 (a) $(6x + 5)(2x - 3)$

(b) They each had \$10 initially.

(c) \$17

Open-ended – Sample answers

9 (a) $5x^2 + 10x = 5x(x + 2)$

(b) $60x^2 + 80x - 15 = 5(6x - 1)(2x + 3)$

(c) $2x^2 - 4x + 20$. This will not factorise completely; after factoring out the 2 and completing the square, the last term is positive, so the difference of two squares cannot be used to find real factors.

Exercise 4.6

1 (a) $x = -\frac{2}{3}$ or $x = 2$ (b) $x = -\frac{2}{3}$ or $x = -2$

(c) $x = 1$ or $x = 1.2$ (d) $x = -2$ or $x = 1.25$

(e) $x = \frac{1}{4}$ or $x = 6$ (f) $x = \frac{3}{2}$ or $x = -4$

(g) $x = 1$ or $x = 2.4$ (h) $x = -1\frac{2}{3}$ or $x = -1$

(i) $x = -0.5$ or $x = \frac{2}{3}$ (j) $x = -1.5$ or $x = 0.75$

(k) $x = \frac{5}{2}$ or $x = -\frac{2}{3}$ (l) $x = -\frac{3}{2}$ or $x = \frac{5}{4}$

2 (a) $x = 3 \pm \sqrt{\frac{7}{2}}$ (b) $x = 2 \pm \sqrt{\frac{5}{3}}$

(c) $x = 0.5$ $x = 2$ (d) $x = \frac{-5 \pm \sqrt{22}}{3}$

(e) $x = \frac{1 \pm \sqrt{31}}{5}$ (f) $x = \frac{-9 \pm \sqrt{129}}{12}$

3 (a) (i) $x = \frac{5 \pm \sqrt{73}}{6}$ (ii) $x = -0.59$ or $x = 2.26$

(b) (i) $x = \frac{-5 \pm 2\sqrt{5}}{5}$ (ii) $x = -1.89$ or $x = -0.11$

(c) (i) $x = \frac{11 \pm \sqrt{73}}{4}$ (ii) $x = 0.61$ or $x = 4.89$

(d) (i) $x = \frac{-5 \pm \sqrt{37}}{2}$ (ii) $x = -5.54$ or $x = 0.54$

4 (a) (i) $a = 1, b = 5, c = 2$ (ii) $x = \frac{-5 \pm \sqrt{17}}{2}$

(b) (i) $a = 2, b = -3, c = 1$ (ii) $x = \frac{1}{2}, x = 1$

(c) (i) $a = -3, b = -1, c = 3$ (ii) $x = \frac{-1 \pm \sqrt{37}}{6}$

(d) (i) $a = 5, b = 4, c = 0$ (ii) $x = -0.8, x = 0$

5 C

6 (a) $x = \frac{2}{3}$ or $x = 1$ (b) $x = \frac{-5 \pm \sqrt{15}}{5}$

(c) $x = -1.5$ or $x = 1.5$ (d) $x = \frac{-7 \pm \sqrt{85}}{2}$

(e) $x = -\frac{5}{6}$ or $x = 2$ (f) $x = -0.5$

(g) $x = \pm\sqrt{\frac{6}{7}}$ (h) $x = \frac{-3 \pm \sqrt{41}}{8}$

(i) $x = \frac{5}{6}$ (j) $x = \frac{11 \pm \sqrt{37}}{6}$

(k) $x = \frac{9 \pm \sqrt{41}}{4}$ (l) $x = \frac{-1 \pm \sqrt{97}}{6}$

7 B 8 C

9 $4t^2 - 31t + 60 = 0, t = 3\frac{3}{4}$ or 4. After $3\frac{3}{4}$ min and 4 min.

10 (a) after 7 hours

(b) after 0.5 hours and 5 hours

11 (a) 192 m

(b) 6.2 seconds

12 (a) $x = \frac{3 \pm \sqrt{-7}}{4}$, no real solution

(b) $x = \frac{-5 \pm \sqrt{-15}}{10}$, no real solution

(c) $x = \frac{6 \pm \sqrt{-108}}{8}$, no real solution

(d) $x = \frac{-7 \pm \sqrt{-23}}{4}$, no real solution

13 (a) $3t^2 - 17t - 70 = 0, t = \frac{17 + \sqrt{1129}}{6} = 8.4$ min

(b) Try to solve $-3t^2 + 17t + 70 = 100, 3t^2 - 17t + 30 = 0$,

$t = \frac{17 \pm \sqrt{-71}}{6}$ (no real solutions). Driver does not reach 100 km/h speed limit.

14 (a) Let kennel be x m wide, hence $2x$ m long.
Total painted area $= 4x^2 + 6x - 0.5$

(b) $4x^2 + 6x - 0.5 = 15.7, 8x^2 + 12x - 1 = 31.4$,

$20x^2 + 30x - 81 = 0, x = \frac{-30 + \sqrt{7380}}{40} \approx 1.4$.

width 1.4 m, length 2.8 m, height 1 m

15 $\frac{x}{2}(x + 7) = 60, x^2 + 7x - 120 = 0, x = 8$.

hypotenuse $= \sqrt{8^2 + 15^2} = 17$ cm
perimeter = 40 cm

16 $3n^2 + n - 1610 = 0, (3n + 70)(n - 23) = 0, n = 23$ terms

17 (a) $x \neq 0; 2x^2 - x - 21 = 0. x = -3$ or 3.5

(b) $x \neq 0; x^2 - 9x + 14 = 0, x = 2$ or 7

(c) $x \neq 5; x^2 - 6x + 9 = 0, x = 3$

(d) $x \neq \pm 1; x^2 - 4x - 3 = 0, x = 2 \pm \sqrt{7}$

(e) $x \neq -2, 0; x^2 - 6x - 6 = 0, x = 3 \pm \sqrt{15}$

(f) $x \neq 0; 6x^2 - x - 3 = 0, x = \frac{1 \pm \sqrt{73}}{12}$

Open-ended – Sample answers

18 $2x^2 + x - 15 = 0$ and multiples of this equation

19 $a = 0$; $4x^2 - 5 = 0$
$a = 1$; $x^2 + x - 1 = 0$ etc.

20 (a) $x^2 + x + 1 = 0$
(b) $4x^2 + 20x + 25 = 0$, $x = -2.5$
(c) $12x^2 + 7x - 10 = 0$, $x = \frac{2}{3}$ or -1.25

Chapter review 4

1 (a) $x = 0$ or 6 (b) $x = \pm\frac{9}{2}$ (c) $x = -2$ or 7
(d) $x = -3$ (e) $x = 0$ and 8 (f) $x = 6$
(g) $x = 0$ and -4 (h) $x = -1$ and 7 (i) $x = \pm 4$

2 (a) $x = -7$ and -5 (b) $x = -3$ and 2 (c) $x = 1$ and -9
(d) $x = 1$ and 8 (e) $x = 1$ and -4 (f) $x = 3$ and 1

3 (a) $x = -1$ and 5 (b) $x = -1$ and 6 (c) $x = -4$ and -3
(d) $x = \pm 3$ (e) $x = -8$ and 4 (f) $x = -3$ and 4

4 (a) $x = 2 \pm \sqrt{2}$ (b) $x = -4 \pm \sqrt{7}$ (c) $x = -1 \pm \sqrt{5}$

5 (a) (i) Move 4 units down. (ii) (0, -4)
(iii) (0, -4), (-2, 0), (2, 0) (iv)

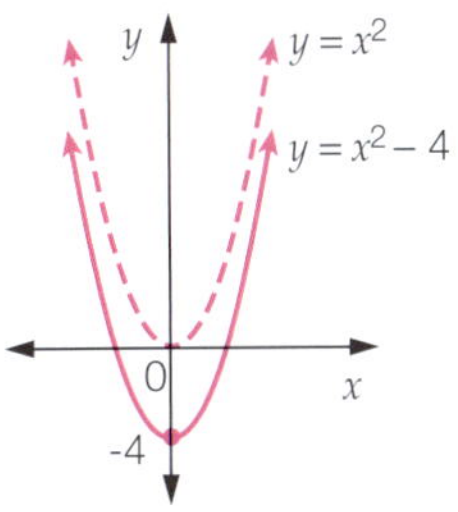

(b) (i) Reflect in x-axis, dilate by a factor of 2 (make narrower) and translate 1 unit left and 4 units down.
(ii) (-1, -4) (iii) (0, -6)
(iv)

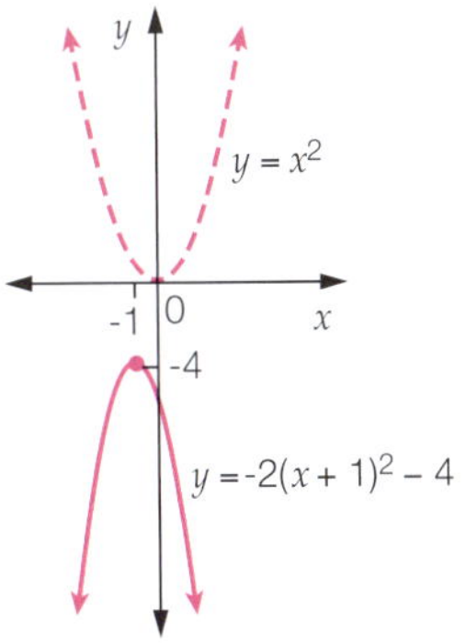

(c) (i) Dilate by a factor of $\frac{1}{3}$ (make wider) and move 7 units right.
(ii) (7, 0) (iii) $(0, \frac{49}{3})(7, 0)$
(iv)

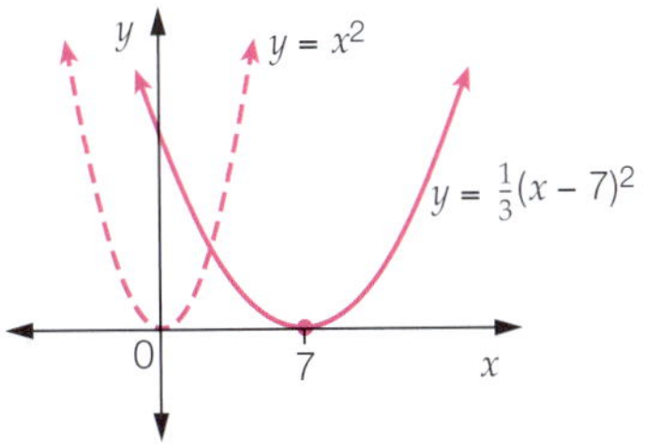

6 (a) (i) (0, -12) (ii) (-2, 0), (6, 0) (iii) (2, -16)
(iv)

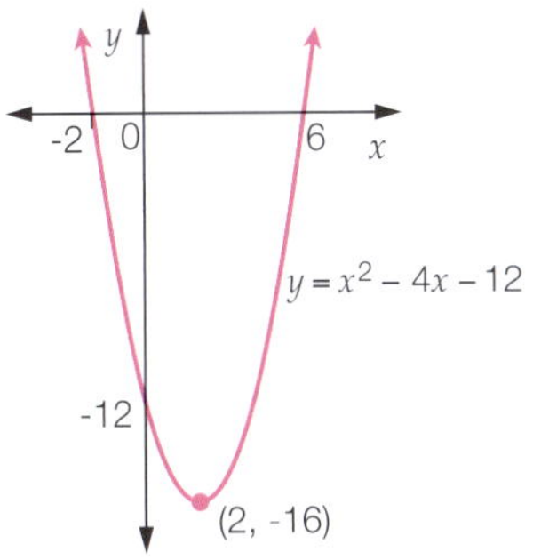

(b) (i) (0, 3) (ii) (-3, 0), (1, 0) (iii) (-1, 4)
(iv)

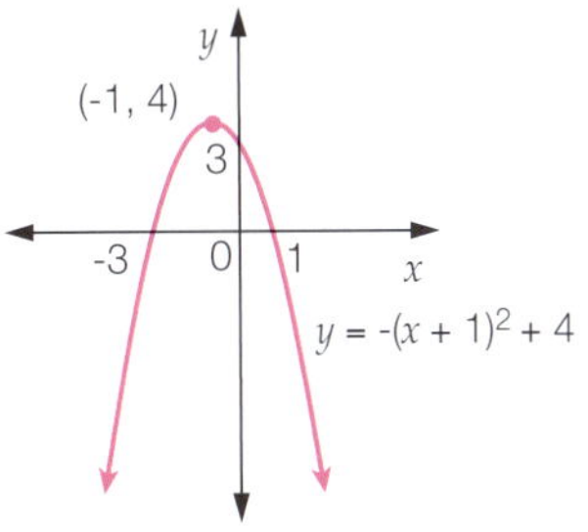

(c) (i) (0, 12) (ii) $(3 \pm \sqrt{3}, 0)$ (iii) (3, -6)
(iv)

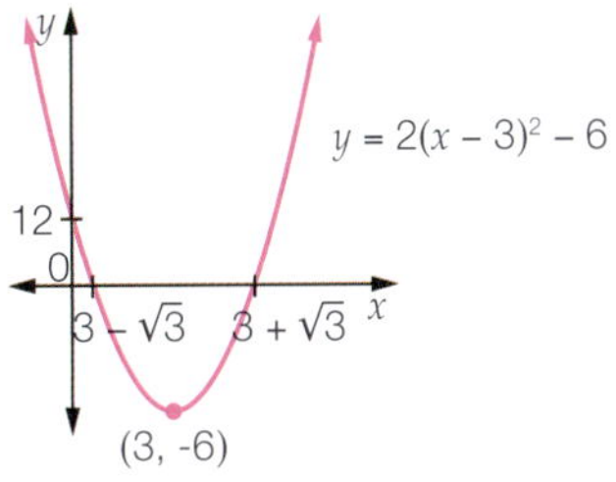

(d) (i) (0, 4) (ii) (-1, 0), (4, 0) (iii) $(\frac{3}{2}, \frac{25}{4})$
(iv)

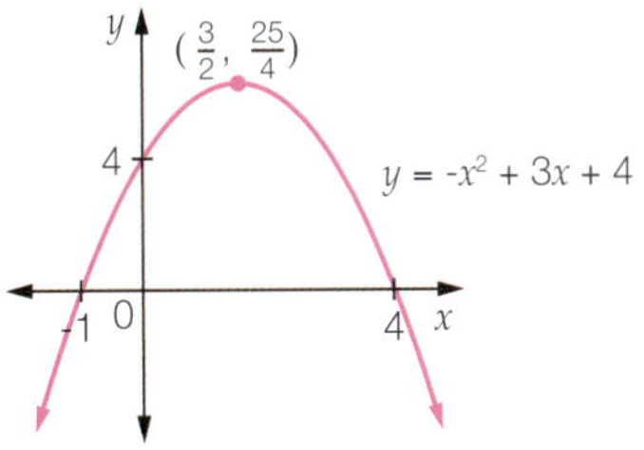

7 (a) (i) $(x - 3)(x + 3)$ (ii) $x = \pm 3$
(b) (i) $(5 - x)(5 + x)$ (ii) $x = \pm 5$
(c) (i) $(2x - y)(2x + y)$ (ii) $x = \pm\frac{y}{2}$
(d) (i) $(7x - 4)(7x + 4)$ (ii) $x = \pm\frac{4}{7}$
(e) (i) $9(2x - 3y)(2x + 3y)$ (ii) $x = \pm\frac{3y}{2}$
(f) (i) $(8 - 3x)(8 + 3x)$ (ii) $x = \pm\frac{8}{3}$
(g) (i) $(2x - 7)(2x - 3)$ (ii) $x = \frac{7}{2}, \frac{3}{2}$
(h) (i) $-13(2x + 3)$ (ii) $x = -\frac{3}{2}$

(i) (i) $5(2x-1)$ (ii) $x=\frac{1}{2}$

(j) (i) $(x-\sqrt{3})(x+\sqrt{3})$ (ii) $x=\pm\sqrt{3}$

(k) (i) $(x+7-\sqrt{5})(x+7+\sqrt{5})$ (ii) $x=-7\pm\sqrt{5}$

(l) (i) $(x-4-\sqrt{7})(x-4+\sqrt{7})$ (ii) $x=4\pm\sqrt{7}$

8 (a) $(x+3-\sqrt{11})(x+3+\sqrt{11})$

(b) $(x-5-\sqrt{2})(x-5+\sqrt{2})$

(c) $(x-2-\sqrt{11})(x-2+\sqrt{11})$

(d) $(x+4-\sqrt{3})(x+4+\sqrt{3})$

9 (a) $(3x+4)(4x-5)$

(b) $(2x+1)(6x-5)$

(c) $(2x+7)(3x-2)$

(d) $3\left(x+\frac{5}{6}-\frac{\sqrt{13}}{6}\right)\left(x+\frac{5}{6}+\frac{\sqrt{13}}{6}\right)$

(e) $4\left(x-\frac{7}{8}-\frac{\sqrt{129}}{8}\right)\left(x-\frac{7}{8}+\frac{\sqrt{129}}{8}\right)$

(f) $2\left(x+\frac{3}{2}-\frac{\sqrt{3}}{2}\right)\left(x+\frac{3}{2}+\frac{\sqrt{3}}{2}\right)$

10 (a) $x=\pm3$

(b) $(2x+1)(x-5)=0$, $x=-0.5$ or 5

(c) $x=\frac{5\pm\sqrt{3}}{2}$ (d) $x=\frac{-5\pm\sqrt{17}}{4}$

(e) $x=\frac{6\pm\sqrt{12}}{6}=\frac{3\pm\sqrt{3}}{3}$ (f) $x=\frac{-7\pm\sqrt{205}}{26}$

11 (a) (i) E (ii) B (iii) D

(iv) F (v) A (vi) C

(b) (i) $(x+8)(x-3)$ (ii) $(5a-b)(5a+b)$

(iii) $(x+3-y)(x+3+y)$ (iv) $(x+6-\sqrt{2})(x+6+\sqrt{2})$

(v) $x(16x-25)$ (vi) $(x-y)(4+x+y)$

12 (a)

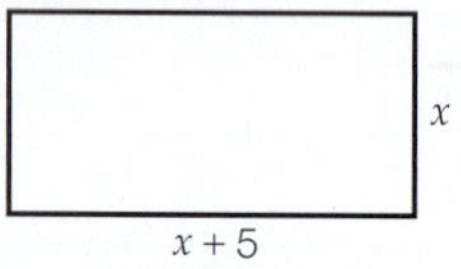

(b) $(x+5)$ cm

(c) $x(x+5)$ cm^2 (d) $x(x+5)=24$

(e) $x=-8$ or 3 (f) length 8 cm, width 3 cm

(g) $x=-8$ is not a feasible solution because a distance measurement must be positive.

13 (a) A horizontal distance of 0.4 m from the outlet.

(b) 0.64 m below the outlet

(c) 72 cm or 0.72 m

(d)

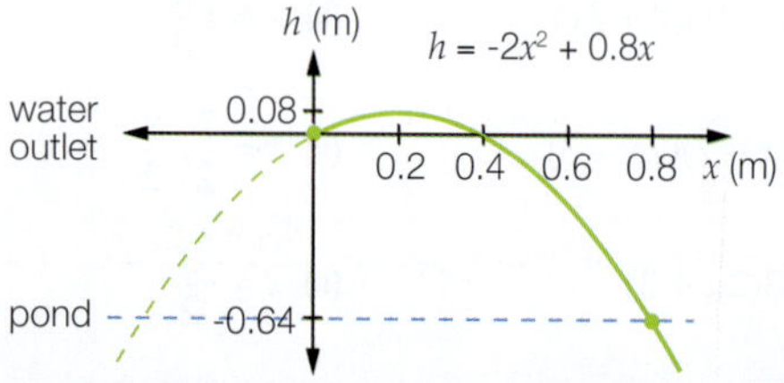

14 (a) 6 cm (b) 14 cm by 10 cm (c) 36 cm^2

(d) 104 cm^2 (e) 64 cm^2

15 (a) $x+y=20$ (b) $y=20-x$ (c) $x(20-x)=P$

(d)

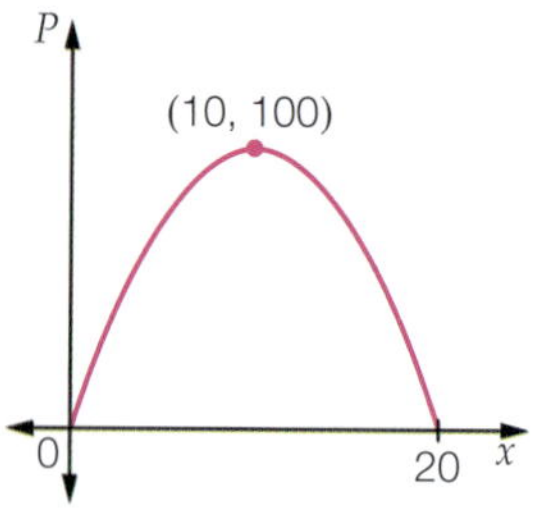

(e) The maximum product is 100 and it is obtained by using $x=y=10$.

16 (a) $(6,-1)$ (b) $y=(x-6)^2-1$

(c)

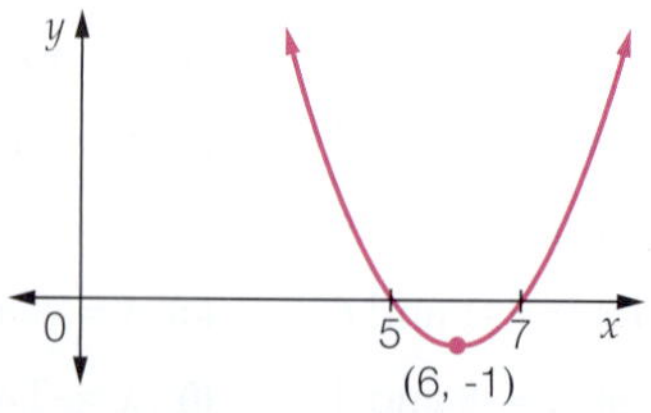

Mixed review A

1 (a) $(3x-7)(2x+1)$ (b) $4x(3x-2)(2x+3)$

2 (a) $\frac{-5}{3x}$ (b) $\frac{3x^2-5x-2}{x^2-1}$ (c) 3

3 (a) $x=9$ (b) $x=\frac{7}{3}$ (c) $x=2$ (d) $x=2$

(e) $x=2$ (f) $x=5$ (g) $x=-12$ (h) $x=5$

4 (a) 1.5 (b) 3 (c) -2 (d) 2

5 (a) $(-2,-1)$ (b) $(1,-1)$ (c) $(-8,0)$

6 (a)

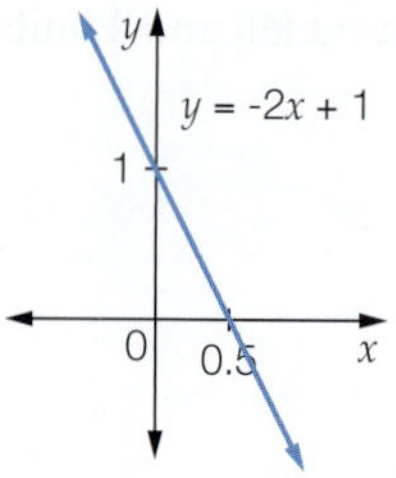

(b)

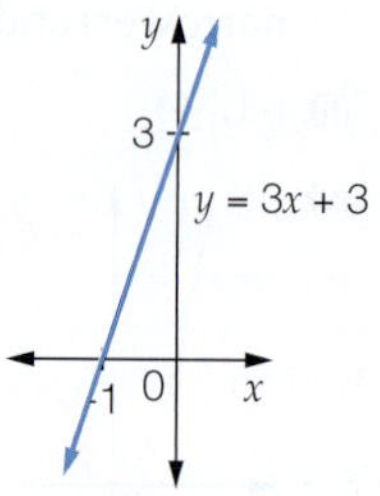

(c)

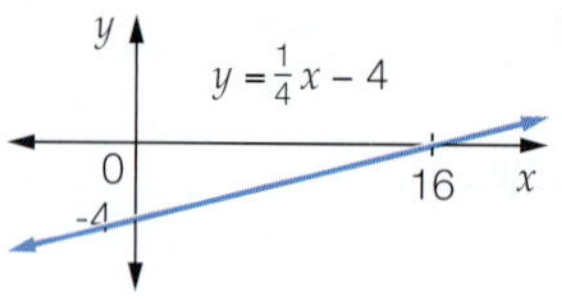

(d)

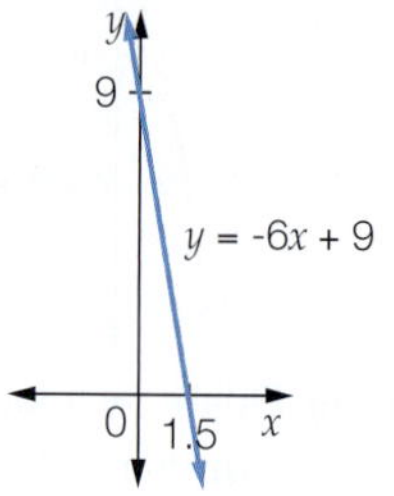

7 (a) yes
$m_1=-\frac{1}{2}$, $m_2=2$, $m_1m_2=-1$

(b) yes
$m_1=2$, $m_2=-\frac{1}{2}$, $m_1m_2=-1$

(c) yes
$m_1=-\frac{5}{2}$, $m_2=\frac{2}{5}$, $m_1m_2=-1$

(d) no
$m_1=\frac{4}{3}$, $m_2=\frac{3}{4}$, $m_1m_2=-1$

8 (a) no

$m_1 = 2, m_2 = 1, m_1 \neq m_2,$

(b) no

$m_1 = 1, m_2 = -1, m_1 \neq m_2$

(c) no

$m_1 = -\frac{2}{5}, m_2 = \frac{2}{5}, m_1 \neq m_2$

(d) yes

$m_1 = \frac{3}{10}, m_2 = \frac{3}{10}, m_1 = m_2$

9 (a) $x \geq -\frac{3}{2}$ (b) $x < -17$

10 (a) $x = -1, y = -2$ (b) $x = 7, y = -1$

11 (a) Range = 30, IQR ≈ 18 (b) Range = 30, IQR ≈ 13

12 (a) $x = -7$ and -6 (b) $x = \pm 10$

(c) $x = 7$

13 (a) $3x^2 - 5xy + 2y^2$ (b) $6a^2 - 27ab + 30b^2$

(c) $15t^2 - 33ht - 36h^2$

14 (a) $(x-2)^2$ (b) $(x+3)(x-3)$

(c) $(x+7)^2$ (d) $25(2x+1)(2x-1)$

(e) $2x(2y^2 + y - 4)$ (f) $4(x+4)^2$

15 (a) $-\frac{8x}{35}$ (b) $\frac{11x+43}{30}$

16 (a) $-x^2 - 5x + 66$ (b) $-x^2 - 2x + 24$

(c) $2x^2 + 5x - 12$ (d) $7y^2 - 5y - 2$

17 (a) $(x-9)(x+3)$ (b) $(x-5)(x-2)$

(c) $(3x-2)(2x+1)$

18 (a) $25x^2 - 10x + 1$ (b) $16x - 24x + 9$

(c) $64 - 48x + 9x^2$ (d) $36 - 60x + 25x^2$

19 (a) $(x+5)(x+3)$ (b) $x^2 + 8x + 15$

(c) (i) 16 m^2 (ii) 63 m^2

(d) $504

Challenge A

1 $1000 - 1 + 3 - 5 + 7 \ldots - 197 + 199 - 201$
$= 1000 + 50 \times 2 - 201 = 899$

2 $18x + 23y + 52z + 69w = 733$, $x + y + z + w = 20$. One of x and w must be odd, as must one of x and z. Use trial and error or a spreadsheet to find $x = 3, y = 10, z = 2, w = 5$. That is, 3 gifts at $18, 10 gifts at $23, 2 gifts at $52 and 5 gifts at $69.

3 B

4 D

$35x - 35 < 75 < 21x + 84$

$35x < 110, x < 3\frac{1}{7}$ and $-9 < 21x, x > -\frac{3}{7}$

$-\frac{3}{7} < x < 3\frac{1}{7}, x = 0, 1, 2, 3$

5 D

6 $\frac{5}{10} = \frac{x}{10-x}$, $50 - 5x = 10x$, $3x = 10$, $x = \frac{10}{3}$

Shaded area $= x^2 + x \times (5 - x) = 5x = \frac{50}{3} = 16\frac{2}{3}$ cm^2

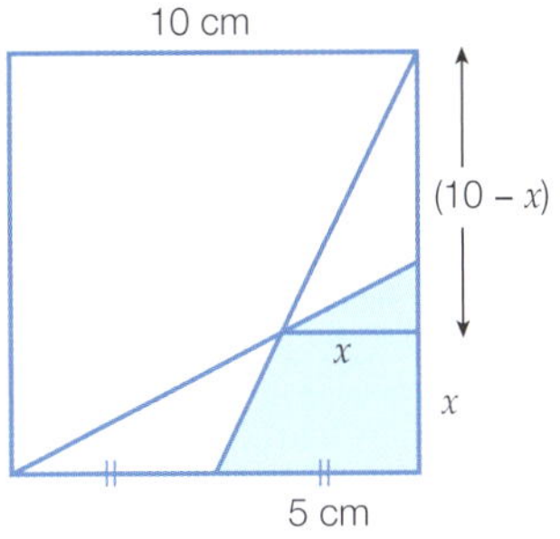

7

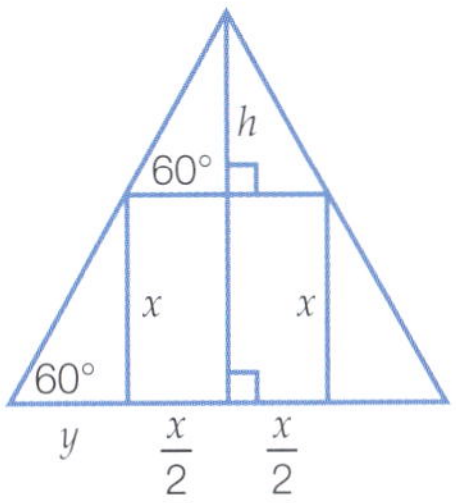

$\frac{x}{y} = \tan(60)°, \frac{x}{y} = \sqrt{3}, y = \frac{x}{\sqrt{3}}$

Base of equilateral triangle $= x + 2y = x + \frac{2x}{\sqrt{3}} = \frac{(3 + 2\sqrt{3})x}{3}$

$\frac{h}{\left(\frac{x}{2}\right)} = \tan(60)°, h = \frac{x\tan(60°)}{2} = \frac{x\sqrt{3}}{2}$

Base of equilateral triangle $= x + 2y = x + \frac{2x}{\sqrt{3}} = \frac{(2+\sqrt{3})x}{2}$

Height of equilateral triangle $= x + h = x + \frac{\sqrt{3}x}{2} = \frac{(2 + \sqrt{3}x)x}{2}$

Area of square $= x^2$

Area of triangle $= \frac{1}{2} \times \frac{(3 + 2\sqrt{3})x}{3} \times \frac{(2+\sqrt{3})x}{2}$

$= \frac{(12 + 7\sqrt{3})x^2}{12}$

Square as fraction of triangle $= x^2 \div \frac{(12 + 7\sqrt{3})x^2}{12}$

$= \frac{12}{12 + 7\sqrt{3}} \approx 49.7\%$

OR, using similar triangles, $\frac{h}{\left(\frac{x}{2}\right)} = \frac{x}{y}$ so $h = \frac{x^2}{2y}$

Base of triangle $= x + 2y$

Height of triangle $= h + x = \frac{x^2}{2y} + x = \frac{x(x+2y)}{2y}$

Area of triangle $= \frac{1}{2} \times (x + 2y) \times \frac{x(x+2y)^2}{2y} = \frac{x(x+2y)^2}{4y}$

Square as fraction of triangle $= x^2 \div \frac{x(x+2y)^2}{4y} = \frac{4xy}{(x+2y)^2}$

Now $\frac{x}{y} = \tan(60°)$ so $y = \frac{x}{\sqrt{3}}$ (or $y = 0.5774x$)

Square as fraction of triangle $= \frac{4xy}{(x+2y)^2} = \frac{4x \times \frac{x}{\sqrt{3}}}{\left(x + \frac{2x}{\sqrt{3}}\right)^2}$

$= \frac{4x^2}{\sqrt{3}x^2\left(1+\frac{2}{\sqrt{3}}\right)^2} = \frac{4\sqrt{3}}{(\sqrt{3}+2)^2} = \frac{4\sqrt{3}}{(7+4\sqrt{3})}$

$= 0.497$ or 49.7%

8

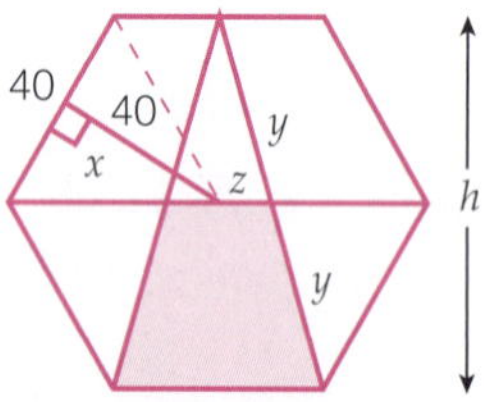

$x^2 = 40^2 - 20^2 = 1200; x = 20\sqrt{3}$

$h = 2x = 40\sqrt{3}$

$4y^2 = 20^2 + 1600 \times 3 = 5200; y^2 = 1300; y = 10\sqrt{13}$

$z^2 = y^2 - x^2 = 1300 - 1200 = 100; z = 10$

Shaded area $= \frac{20+40}{2} \times 20\sqrt{3} = 600\sqrt{3}$ cm^2

9 D

10

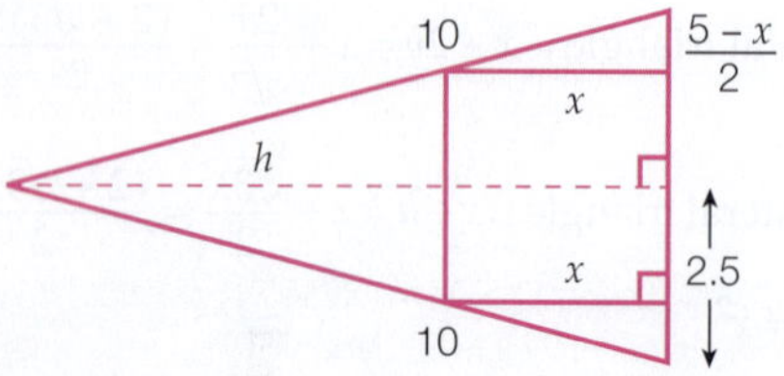

$h^2 = 10^2 - 2.5^2 = 93.75, h = \sqrt{\frac{375}{4}} = \frac{5}{2}\sqrt{15}$

From similar right-angled triangles, $\frac{h}{2.5} = \frac{x}{\left(\frac{5-x}{2}\right)}$

$x = \frac{5h}{5+h} = \frac{5 \times \frac{5\sqrt{15}}{2}}{5 + \frac{5\sqrt{15}}{2}} = \frac{5}{11}(15 - 2\sqrt{15})$ cm

11 Radius = 15 cm

Area of base $= 3 \times \frac{5}{6} \times \pi \times 15^2 + 15 \times 15 \tan 60°$

$= \frac{1125\pi}{2} + 225\sqrt{3} \approx 2157$ cm^2

Volume $= \left(\frac{1125\pi}{2} + 225\sqrt{3}\right) \div 10\,000 \times 6 \approx 1.294$ m^3

Chapter 5

Recall 5

1 (a) 625 (b) 29.791 (c) 3.45

2 (a) $A = 16$ (b) $h = 5$ (c) $b = 7$

3 (a) 410 cm (b) 425 000 cm (c) 400 000 cm^2
(d) 0.0578 m^2 (e) 8 250 000 m^2 (f) 0.6 L
(g) 20 mL (h) 30 800 L (i) 0.007 25 m^3

4 (a) 6.11 m (b) 16.11 m (c) 15.27 m^2

5 (a) (i) 15.59 cm (ii) 140.3 cm^2
(b) (i) 22.36 cm (ii) 447.21 cm^2
(c) (i) 4.36 m (ii) 26.18 m^2
(d) (i) 1.5 m (ii) 6 m^2

Exercise 5.1

1 (a) 1263 mm^2 (b) 180 mm^2 (c) 583 cm^2
(d) 157.5 cm^2 (e) 89.14 cm^2 (f) 349.94 mm^2
(g) 25.63 mm^2 (h) 493.65 cm^2

2 (a) 76.37 cm^2 (b) 7.63 cm^2 (c) 43.37 cm^2
(d) 48.29 cm^2 (e) 1809.56 cm^2 (f) 18.27 mm^2
(g) 88.5 m^2 (h) 422.43 mm^2 (i) 13.72 m^2
(j) 69.53 cm^2

3 C

4 C

5 (a) 123.61 cm^2 (b) 52.15 cm^2 (c) 47.12 cm^2
(d) 32.47 cm^2 (e) 17.21 cm^2 (f) 441.06 cm^2
(g) 437.5 cm^2 (h) 294 cm^2

6 triangle: 173 m^2; sector: 1047 m^2

7 (a) 156.25 hours = 22.32 days (b) 75 000 m^2

8 (a) 21.56 m^2 (b) 11 bags

9 9.78 m^2

10 25.13 m^2

11 (a) 0.05 m^2 (b) 190.42 m^2

12 4×4; 3×6 **13** 4.51 m

14 A, B, F: 50 cm^2; C, G: 25 cm^2; D, E: 100 cm^2

Open-ended – Sample answers

15 (a) 26 cm^2 (b) various answers

16 (a) The formulas for the overall area and the subtracted area are the same. The substituted values are the same so the calculations are identical.

(b) Possible pair of shapes:

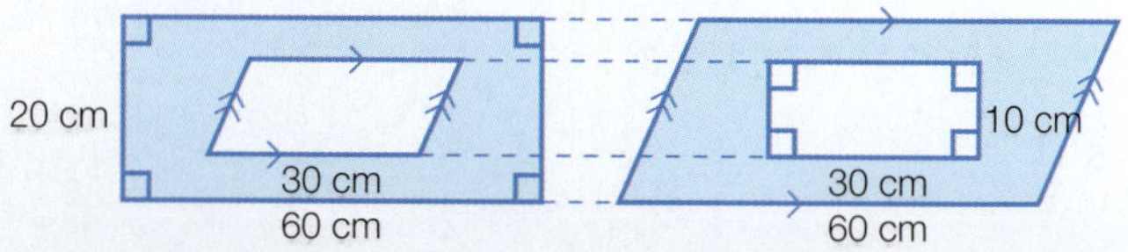

(c) Another possible pair of shapes:

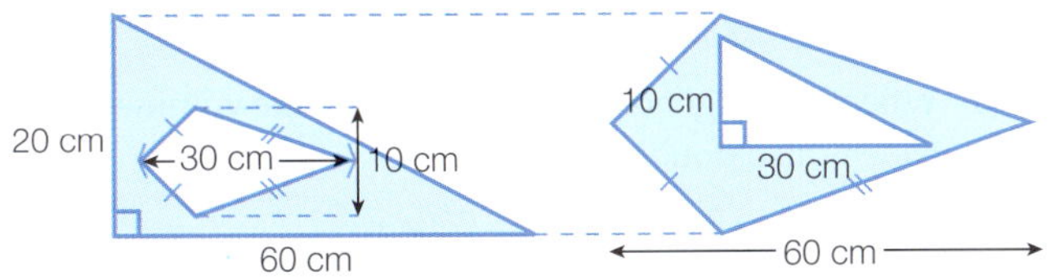

Exercise 5.2

1 (a) 766.14 m^2 (b) 82.14 m^2 (c) 376 cm^2
(d) 1710 cm^2 (e) 108 cm^2 (f) 684 cm^2

2 (a) 904.78 cm^2 (b) 1258.21 mm^2

3 (a) E (b) B (c) D

4 (a) Block A 250 cm^2
Block B 210 cm^2
Block C 170 cm^2
(b) 6300 cm^2

5 810.27 cm^2

6 (a) 7.07 m^2 (b) 5.18 m^2 (c) 12.25 m^2

Open-ended – Sample answers

7 (a) 60 minutes
(b) $2(lw + wh + lh)$
(c) Prism 1: $l = 10$ cm, $w = 5$ cm, $h = 2$ cm, $\text{SA} = 160\text{ cm}^2$
Prism 2: $l = 20$ cm, $w = 10$ cm, $h = 4$ cm, $\text{SA} = 640\text{ cm}^2$
The surface area quadruples when the dimensions are doubled.

8 (a) $r = 2, h = 5$, surface area $= 28\pi$
(b) $r = 2, h = 10$, surface area $= 48\pi$
(c) $r = 4, h = 5$, surface area $= 72\pi$
(d) Doubling the height increases the area of the curved face but not the circular ends. Doubling the radius increases the area of both the curved face and the circular ends. Also, the effect of doubling the radius is magnified, as the radius is then squared in the formula.

Exercise 5.3

1 (a) 450 cm^3 (b) 3192 cm^3 (c) 5355 mm^3
(d) 585 cm^3 (e) $90\,053.75\text{ mm}^3$ (f) 552.92 m^3

2 374.36 mL

3 (a) $115\,000\text{ cm}^3$ (b) $20\,250\text{ mm}^3$ (c) 9625 cm^3
(d) 162 m^3 (e) $135\,000\text{ cm}^3$ (f) 840 cm^3

4 (a) B (b) D

5 (a) $52\,339\text{ cm}^3$ (b) 1.31 m^3

6 0.07 m^3

7 (a) Alex: 365 cm^3, Digby: 375 cm^3
(b) Digby: 10 cm^3

8 6.64 m^3

9 (a) 17.57 m^2 (b) 21 000 L

Open-ended – Sample answers

10 (a) $4 \times 5 \times 10$ cm, $2 \times 10 \times 10$ cm, $1 \times 10 \times 20$ cm
(b) $8 \times 2.5 \times 10$ cm, $4 \times 4 \times 12.5$ cm, $0.5 \times 40 \times 10$ cm
(c) $4 \times 5 \times 10$: $\text{SA} = 220\text{ cm}^2$
$2 \times 10 \times 10$: $\text{SA} = 280\text{ cm}^2$
$1 \times 10 \times 2$: $\text{SA} = 460\text{ cm}^2$

11 (a) Anita has used the diameter value for the radius and did not convert one of the values to cm.
(b) $628\,318.53\text{ cm}^3$
(c) Point 1: Identify radius or diameter before using circle formulas; point 2: Check for matching units before substituting values into formulas.

Half-time 5

1 (a) $19\,101\text{ cm}^2$ (b) 2228 mm^2

2 Surface area $= 14\pi\text{ cm}^2$, volume $= 5\pi\text{ cm}^3$

3 (a) 44 m^2 (b) 12.96 m^2

4 Volume A $= xy^2\pi\text{ m}^3$, Volume B $= 4xy^2\pi\text{ m}^3$, Volume C $= xy^2\pi\text{ m}^3$
B has the largest volume

5 (a) $191\,038.85\text{ m}^3$ (b) 1323.05 m^3

6 72 206 cartons

Exercise 5.4

1 (a) 576 cm^2 (b) 21 m^2

2 (a) 1884.96 cm^2 (b) 490.09 cm^2

3 (a) $20\,106.19\text{ cm}^2$ (b) 3848.45 cm^2

4 (a) 3402.34 m^2 (b) 829.38 cm^2 (c) 1599.56 cm^2
(d) 3619.11 cm^2 (e) 9424.78 cm^2 (f) 1413.72 cm^2

5 (a) 141 m^2 (b) 600 cm^2

6 (a) C (b) D

7 (a) 483.81 cm^2

8 (a) 6.84 cm (b) 19.26 cm (c) 414 cm^2

9 $70\,185\text{ m}^2$ 10 213.63 cm^2

11 (a) Surface area of bowling ball $= 2827.43\text{ cm}^2$ (2 d.p.); area of material $= 3200\text{ cm}^2$. The area of material is larger than the surface area of the bowling ball.
(b) The material is not long enough to wrap around the ball, so the existing material will need to be cut and sewn to wrap it. $C = \pi d = \pi \times 30 = 94.25$ cm (2 d.p.)

12 (a) 31 cm (b) 12 cm

Open-ended – Sample answers

13 Cone A:
$r = 5$ cm
$s = 13$ cm
$\text{SA} \approx 204\text{ cm}^2$

Cone B:
$r = 6$ cm
$s = 14$ cm
$\text{SA} \approx 264\text{ cm}^2$

14 The slant height is the hypotenuse of a right-angled triangle. The radius is one of the shorter sides.

15 (a) $h = 4$ m, $r = 3$ m, total surface area = 24π m^2
(b) total surface area = $3\pi(3 + \sqrt{73})$ m^2
(c) $h = 12$ m, $r = 5$ m, total surface area = 90π m^2
(d) total surface area = $5\pi(5 + \sqrt{601})$ m^2
(e) no

16 (a) $h = 15$ m, $r = 8$ m, total surface area = 200π m^2
(b) total surface area = $16\pi(16 + \sqrt{481})$ m^2
(c) $h = 24$ m, $r = 7$ m, total surface area = 224π m^2
(d) total surface area = $14\pi(14 + \sqrt{821})$ m^2
(e) no

Exercise 5.5

1 (a) 716.28 cm^3 (b) 804.25 cm^3
(c) 66 666.67 cm^3 (d) 25 200 cm^3
(e) 385 mm^3 (f) 2338.27 cm^3

2 (a) 0.52 m^3 (b) 179 594.38 cm^3

3 (a) 150.53 cm^3 (b) 2900 cm^3
(c) 160.22 cm^3 (d) 197.60 cm^3

4 1 098 066 219 000 km^3

5 D 6 A

7 Pyramid is 1288 cm^3 larger than the cone.

8 (a) 52.4 cm^2 (b) 157 mL

9 (a) 95 cm^3 (b) 129 mL

10 (a) 113 mm^3 (b) 70 ball bearings

11 (a) 28
(b) $V_{Mercury} = 6.16 \times 10^{10}$
$V_{Jupiter} = 1.38 \times 10^{15}$
Mercury fits into Jupiter 22 338 times.
(c) Calculations for volume involve cubing the radius.
(d) 1253 Earths in Jupiter, 688 Earths in Saturn, 50 Earths in Uranus, 60 Earths in Neptune

12 (a) Silo = 848 m^3
Train truck = 15 m^3
(b) 340 trucks

13 (a) 3 times
(b) 25%

Open-ended – Sample answers

14 (a) $b = 5$ cm, $h = 9$ cm; volume = 75 cm^3
(b) Volume = 300 cm^3
(c) 4 times greater (300 ÷ 75 = 4)
(d) $b = 6$ cm, $h = 10$ cm; Volume of first pyramid = 120 cm^3
Volume of second pyramid = 480 cm^3
Conclusion: When the base of any square-based pyramid is doubled, the volume is quadrupled.

15 (a) $h = 4$ m, $r = 3$m, volume = 12π m^3
(b) volume = 24π m^3
(c) $h = 12$ m, $r = 5$ m, volume = 100π m^3
(d) volume = 200π m^3
(e) When the height of a cone is doubled, the volume of the cone is also doubled.

16 (a) $h = 15$ m, $r = 8$ m, volume = 320π m^3
(b) volume = 1280π m^3
(c) $h = 24$ m, $r = 7$ m, volume = 392π m^3
(d) volume = 1568π m^3
(e) When the radius of a cone is doubled, the volume of the cone is multiplied by 4.

Exercise 5.6

1 (a) $h = \frac{2A}{b}$ (b) $r = \pm\sqrt{\frac{A}{\pi}}$ (c) $\theta = \frac{2A}{r^2}$
(d) $b = \pm\sqrt{c^2 - a^2}$ (e) $x = \frac{y - b}{m}$ (f) $a = b - xn$
(g) $s = \pm\sqrt{6t^2 + 1}$ (h) $t = \pm d\sqrt{\frac{m}{F}}$ (i) $R = \pm\sqrt{\frac{4v}{P} + 9}$

2 (a) (i) $s = \pm\sqrt{A}$ (ii) $s = \pm 3\sqrt{3}$
(b) (i) $v = \pm\sqrt{\frac{2K}{m}}$ (ii) $v = \pm\sqrt{5}$
(c) (i) $v = \pm\sqrt{\frac{Fr}{m}}$ (ii) $v = \pm\sqrt{10.5}$
(d) (i) $r = \pm\sqrt{\frac{Gm_1m_2}{F}}$ (ii) $r = \pm\sqrt{33.75}$
(e) (i) $r_2 = \pm\sqrt{\frac{g_1r_1^2}{g_2}}$ (ii) $r_2 = \pm 14\sqrt{2}$
(f) (i) $V = \pm\sqrt{PR}$ (ii) $V = \pm 3\sqrt{7}$

3 (a) (i) $A_y = A^2 - A_x$ (ii) $A_y = 38$
(b) (i) $c = \frac{b^2 - (2ax + b)^2}{4a}$ (ii) $c = 5$
(c) (i) $T = \frac{9r^2}{l^2}$ (ii) $T = 64$
(d) (i) $z = y - x^2$ (ii) $z = -96$
(e) (i) $z = \frac{y}{x^2}$ (ii) $z = 0.28$
(f) (i) $b = cx^2 - a$ (ii) $b = 16$

4 C

5 (a) $r = \sqrt{\frac{A}{\pi}}$
(b) The radius is a distance and must be positive.
(c) 3.57 cm

6 (a) $r = \sqrt[3]{\frac{3V}{4\pi}}$
(b) Because it is a cube root and not a square root.
(c) 3.10 cm

7 (a) $h = \frac{3V}{\pi r^2} = 1.52$ m (b) $r = \sqrt{\frac{3V}{\pi h}} = 4.37$ cm

8 (a) 41 m/s (b) The car is slowing down.

9 (a) $k = 1243.57$ (b) 2.1 Hz

10 (a) Line 3 does not have ± symbol.

(b) This is mathematically inaccurate because it is possible that the square root of an expression is either negative or positive.

For example, $x^2 = 9 = (-3)^2 = 3^2$

Therefore x could be either -3 or 3.

Open-ended – Sample answers

11 (a) $a = 2b + c^2$ $c = \pm\sqrt{a - 2b}$

(b) $d = 2e + \sqrt{f}$ $f = d^2 - 4de + 4e^2$

Exercise 5.7

1 (a) 7.1 g/cm^3 (b) 6.4 g/cm^3 (c) 0.8 g/cm^3

2 (a) 6240 g (b) 486 g (c) 3.96 kg

3 (a) 13.27 cm^3 (b) 694.29 mL (c) 15.29 m^3

4 20 g/L 5 2.4 g/100 g

6 (a) D (b) B

7 7.6 mg/mL

8 (a) 1.25 L

(b) 4 cups cordial concentrate and 16 cups water

(c) 200 mL/L

9 (a) 131.175 kg (b) 22 t

10 0.04 g/L

11 (a) 11 L (b) 123 mL/L (c) 4 L

12 (a) wooden log: 0.40 g/cm^3; steel rod: 7.81 g/cm^3; cork: 0.22 g/cm^3; aluminium cylinder: 2.65 g/cm^3; ice block: 0.91 g/cm^3; gold bar: 18.29 g/cm^3

The wooden log, cork and ice block will all float on water.

(b) lead log: 2 852 562 g; lead rod: 726 g; lead cork: 103 g; lead cylinder: 856 g; lead block: 125 g; lead bar: 9931 g

13 (a) Johnstone's abattoir = 50 kg
Arvon = 40 kg
Johnstone's abattoir contributes the most

(b) 0.44 mg/L

Open-ended – Sample answers

14 (a) 875 organisms/100 mL (b) no

(c) 600 organisms/50 mL; 2000 organisms/100 mL; 800 organisms/60 mL

Chapter review 5

1 (a) 925 cm^2 (b) 878.43 cm^2 (c) 5105.09 cm^2

2 B

3 (a) 1550 cm^2 (b) 7.49 m^2

4 C

5 (a) 1080 cm^3 (b) 21 205.75 cm^3 (c) 5040 cm^3

6 (a) 1884.96 cm^2 (b) 11 309.73 cm^2

7 (a) 527.79 cm^3 (b) 268 082.57 mm^3

8 A

9 (a) $b = \frac{A}{l}$ (b) $l = \frac{A^2 g}{4\pi^2}$

(c) $r = \pm\sqrt{\frac{SA}{4\pi}}$ (d) $x = \pm\sqrt{\frac{a + z}{y}}$

10 30 kg/m^3 11 250 mL/L 12 1800 bricks

13 (a) 15 708 L (b) 5109 L

14 1.70 m^3

15 (a) 1885 cm^2, 6807 cm^3 (b) 327 cm^2, 302 cm^3

16 (a) 7527 m^2 (b) 361 m

17 (a) 2000 cm^3 (b) 15 g

18 335 cm^3 19 804 L 20 9839 kg

21 (a) 1950 kL (b) 550 kL

22 21.99 cm

23 (a) 62 832 m^3 (b) $9 424 778 (c) 12 566 m^2

(d) $1 256 637 (e) $1 800 000 (f) $12 481 415

(g) $6 300 000 (h) the road over the top by $6 181 415

24 4.73 g/L

Chapter 6

Recall 6

1 (a) $x = 8.17$ (b) $x = 15.7$

(c) $x = 5.20$ (d) $x = 46.377$

2 (a) 3 cm (b) 15 mm

3 (a) 26 (b) 33.23 (c) 9.95

4 (a)

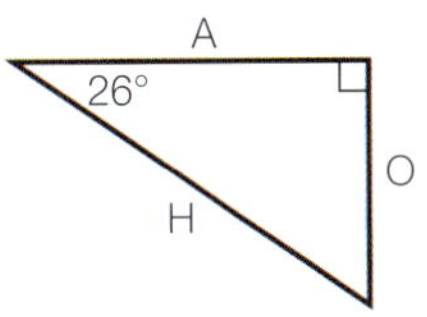

(b)

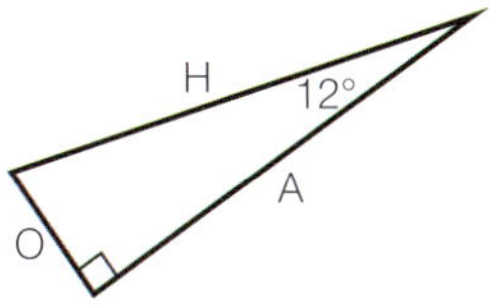

(c)

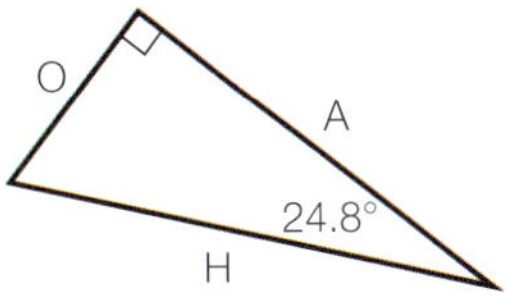

Exercise 6.1

1 (a)

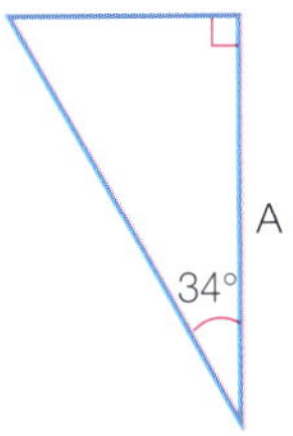

(b)

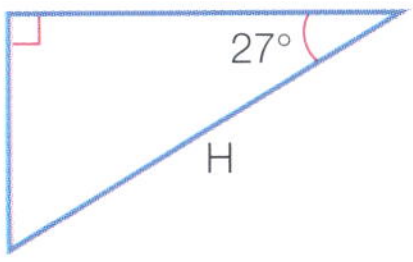

(c)

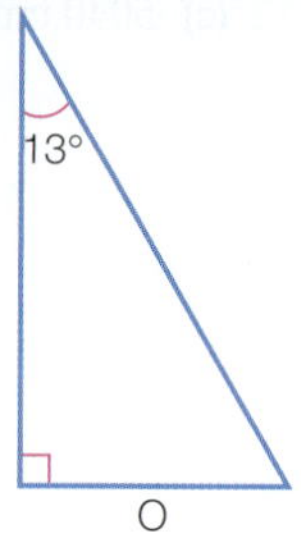

(d)

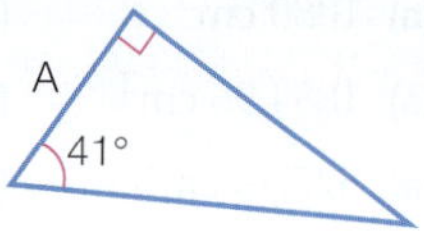

(e)

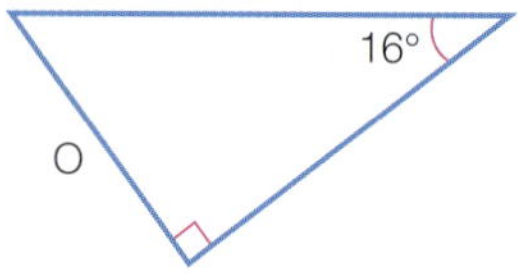

(f)

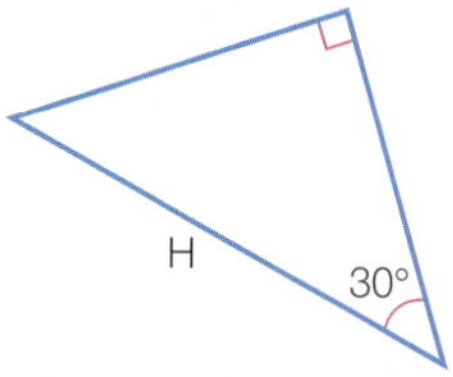

2 (a)

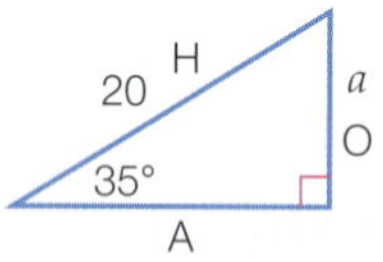

$\sin(35°) = \frac{a}{20}$

(b)

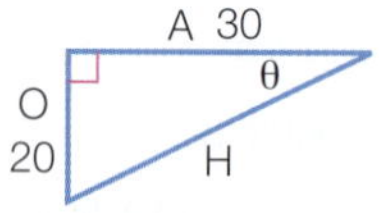

$\tan(\theta) = \frac{20}{30}$

(c)

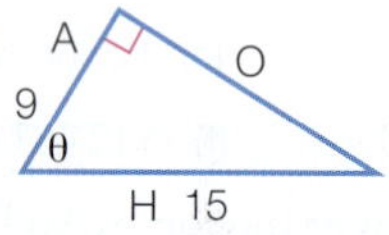

$\cos(\theta) = \frac{9}{15}$

(d)

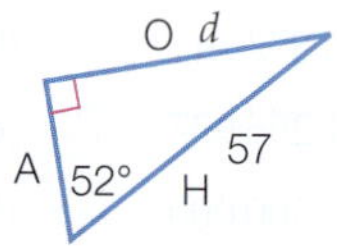

$\sin(52°) = \frac{d}{57}$

(e)

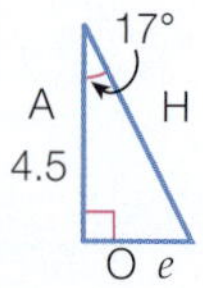

$\tan(17°) = \frac{e}{4.5}$

(f)

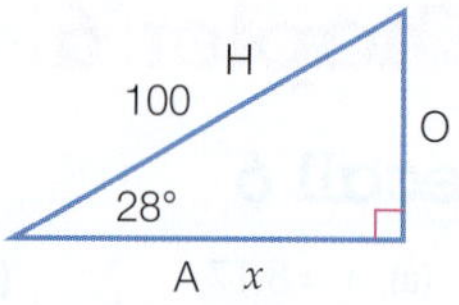

$\cos(28°) = \frac{x}{100}$

3 (a) 0.3907 (b) 0.9563 (c) 0.8098
(d) 0.9962 (e) 0.6561 (f) 0.6561
(g) 0.3581 (h) 0.9171 (i) 0.9540

4 (a) 15.50° (b) 35.27° (c) 23.17°
(d) 76.80° (e) 44.99° (f) 8.75°

5 (a) 0.4350 (b) 38.1885 (c) 0.4339
(d) 0.7403 (e) 0.3490 (f) 0.1437

6 (a) C (b) D

7 (a) (i) 1.5 (ii) 1.5 (iii) 1.5
(b) 0.5 (c) 0.5
(d) Yes, because the triangles are similar.
(e) $\alpha = \beta = 30°$ (f) yes (g) yes
(h) The triangles are similar, so the ratio between the corresponding sides of the triangles is constant. This means that the ratio of a pair of sides within one triangle is the same as the ratio of the corresponding pair of sides within the other triangle.

8 (a) (i) 0.5736 (ii) 0.8192 (iii) 0.7002 (iv) 0.7002
(b) The ratio of sin (35°) and cos (35°) is the same as tan (35°).

9 1.151 10 E 11 C

12 (a) (i) 0.722 (ii) 0.722
(b) Ratios use the same side lengths $\left(\frac{4.6}{6.37}\right)$.

13 (a)

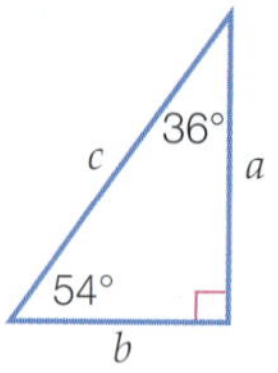

(b) $\sin(54°) = \frac{a}{c}$, $\cos(54°) = \frac{b}{c}$
(c) $\sin(54°) = 0.8090$, $\cos(54°) = 0.5878$
(d) $(\sin(54°))^2 + (\cos(54°))^2$
$= 0.8090^2 + 0.5878^2$
$= 1$
(e) $(\sin(54°))^2 + (\cos(54°))^2$
$= \left(\frac{a}{c}\right)^2 + \left(\frac{b}{c}\right)^2 = \frac{a^2 + b^2}{c^2} = \frac{c^2}{c^2} = 1$

14 An isosceles triangle has two equal sides and two equal angles. In a right-angled isosceles triangle, these two sides are the opposite and adjacent of the two base angles, which are each 45°. So:

$\tan(\theta) = \frac{O}{A}$
$= \frac{x}{x}$
$= 1$

15 The value of the tan ratio can be greater than 1. The values of the cos and sin ratios can't be greater than 1, because to calculate them you need to divide either the adjacent or opposite side by the hypotenuse, and by definition the hypotenuse will always be larger than the other two sides.

Open-ended – Sample answers

16 (a) hypotenuse = 12.2 cm

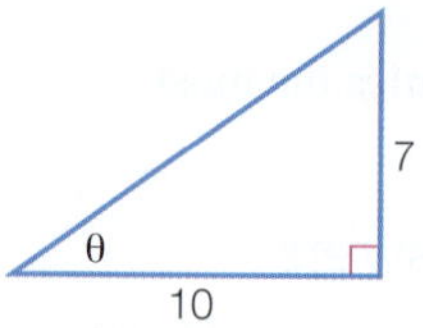

(b) $\sin(\theta) = \frac{7}{12.2}$ $\cos(\theta) = \frac{10}{12.2}$
$= 0.5738$ $= 0.8197$

17

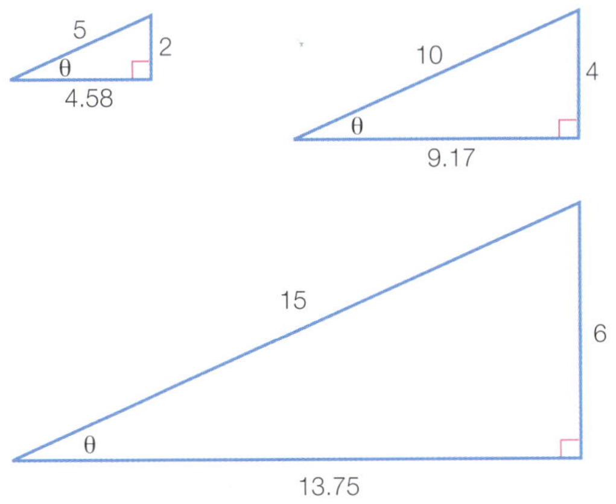

18 (a) $\theta = 20°$ and $\varphi = 70°$

(b) $\theta = 90 - \varphi$ or $\varphi = 90 - \theta$

Exercise 6.2

1 (a) $a = 167.71$ (b) $b = 12.44$ (c) $c = 1.05$

(d) $x = 11.53$ (e) $x = 11.22$ (f) $x = 13.29$

(g) $x = 148.84$ (h) $a = 20.58$ (i) $b = 42.56$

(j) $d = 32.09$ (k) $e = 7.40$ (l) $f = 41.08$

(m) $d = 95.83$ (n) $c = 244.74$ (o) $e = 6.91$

2 (a) 5.7 m (b) 4.0 m

3 (a) 2.2 m (b) 4.6 m

4 193 cm

5 3.74 m

6 (a) B (b) A

7 15.37 m

8 (a) 29.58 m (b) 51°; 29.58 m

(c) sin (39°) = cos (51°)

39° and 51° are complementary angles (add up to 90°).

9 (a) 3.65 m (b) 2.45 m

10 9.89 m

11 (a)

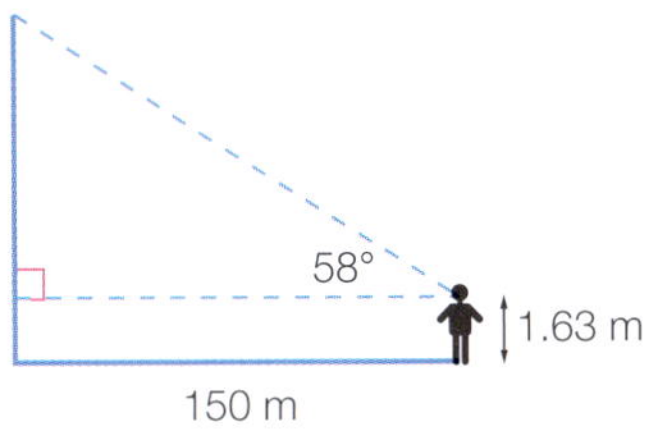

(b) 241.7 m

12 325.87 m 13 39.95 cm

14 (a) trapezium 10.74 m^2, rectangle 12 m^2

(b) rectangular room

15 (a) 19.73 m

(b) dropped from 9.86 m along rope, falls 28.7 m

Open-ended – Sample answers

16 16 cm, 16.5 cm, 17 cm

17 Taking 'just clear' to be a height of 8.1 m, two possible angles are 40° and 35°.

$$\tan(40°) = \frac{8.1 - 1.2}{x}$$

$$x = \frac{6.9}{\tan(40°)} = 8.22 \text{ m}$$

$$\tan(35°) = \frac{8.1 - 1.2}{x}$$

$$x = \frac{6.9}{\tan(35°)} = 9.9 \text{ m}$$

Exercise 6.3

1 (a) (i) $a = 81.15°$ (ii) $a = 81°9'10''$

(b) (i) $b = 42.21°$ (ii) $b = 42°12'50''$

(c) (i) $c = 66.58°$ (ii) $c = 66°34'43''$

(d) (i) $k = 56.22°$ (ii) $k = 56°13'13''$

(e) (i) $m = 77.96°$ (ii) $m = 77°57'22''$

(f) (i) $p = 76.29°$ (ii) $p = 76°17'17''$

2 (a) 25° (b) 76°

(c) 57° (d) 74°

(e) 68° (f) 24°

3 (a) 12°22′25″ (b) 25°42′36″

(c) 47°9′23″ (d) 14°33′5″

(e) 24°11′40″ (f) 54°35′10″

4 15.9°

5 (a) 32.86° (b) 55.01°

6 (a)

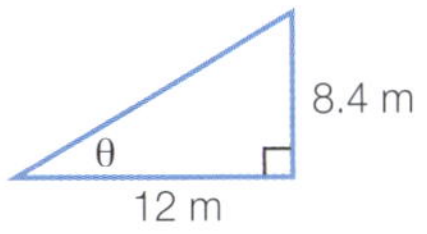

(b) 35°

7 (a) B (b) C

8 53°36′56″ 9 B 10 0.40° 11 48°

12 (a) 33° (b) 14 cm

13 (a) 147° (b) 25.7 m

14 (a) 11.54° to 36.87°

(b) The ladder is too long for the 2 m window. It would be placed 9.8 m away, at an angle of 11.5°.

Open-ended – Sample answers

15 30 cm (22°), 40 cm (30°), 50 cm (39°), 60 cm (49°), 70 cm (61°)

16 Angle of Bilal's shadow with the sun is $\theta = \tan^{-1}\left(\frac{185}{94}\right) = 63.06°$. For example, take the height of the lamp post as 5 m and the height of the tree as 12 m. Then the length of the lamp post's shadow is $\frac{5}{\tan(63.06°)} = 2.54$ m and the length of the tree's shadow is $\frac{12}{\tan(63.06°)} = 6.10$ m

Exercise 6.4

1 19 m 2 49.2° 3 63.4°
4 25 km 5 14° 6 506 m
7 53° 8 39 m 9 8.55 m
10 (a) 209 m (b) 134 m
11 (a) A (b) B
12 24 m
13 28.84 m
14 4°
15 (a) 177 cm (b) 20.2°
16 (a) 27.7° (b) 12.25 m

Open-ended – Sample answers

17 (a) 46°, 50°, 54°
(b) 12.95 m (46°), 13.79 m (50°), 14.56 m (54°)
18 0.95° (after 20 min), 0.48° (after 44 min)

Half-time 6

1 6.4° 2 B 3 B
4 $\theta = 50.36°$, height = 23.23 m
5 D 6 320 m

Exercise 6.5

1 (a) (i) N21°W (ii) 339°T
(b) (i) S73°E (ii) 107°T
(c) (i) N58°E (ii) 058°T
(d) (i) S53°W (ii) 233°T
(e) (i) S51°E (ii) 129°T
(f) (i) N73°W (ii) 287°T
2 B
3 C
4 True bearing: 022°T
Compass bearing: N22°E
5 (a)

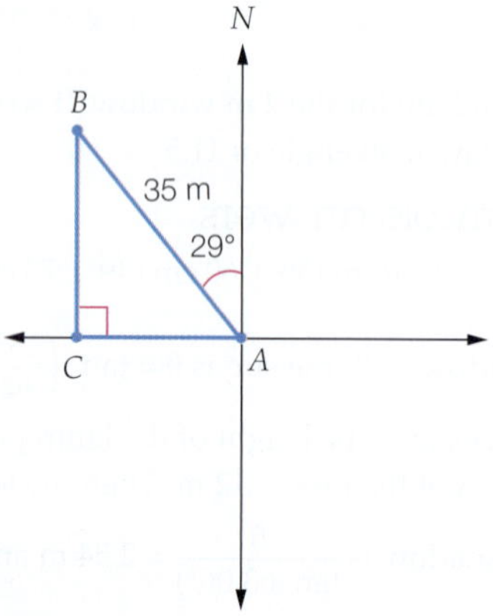

(b) 31 m
(c) 17 m
(d) 83 m
6 (a)

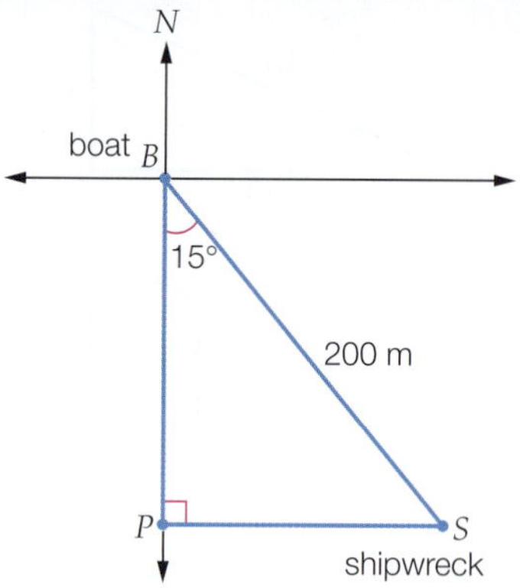

(b) 51.8 m (c) 193.2 m
7 (a) 035°T (b) N35°E
8 (a) N58°W (b) 302°T
9 115 km
10 (a) True bearing: 152°T
Compass bearing: S28°E
(b) True bearing: 147°T
Compass bearing: S33°E
11 75.74 km 12 260 km
13 (a) 056°T
(b) (i) 288 km (ii) 236°T
14 (a) 2.11 km
(b) (i) 5.87 km (ii) 7.98 km
(c) (i) 1.25 km (ii) 5.78 km
15 (a) 1155 m (b) south 227 m
16 30.39 m
17 (a) (i) 115 m (ii) 164 m
(b) 027°T (c) 1004 m
18 (a) 115°T (b) 3.32 km
19 (a) S23°E (b) 674 km (c) 257 km south
(d) N69°W (e) 721 km

Open-ended – Sample answers

20 If $\theta = 30°$, then:
(a) 150°T, S30°E (b) 330°T, N30°W
(c) 210°T, S30°W

Exercise 6.6

1 (a) $x = 16.41$, $y = 27.27$ (b) $x = 31.40$, $y = 28.23$
(c) $x = 25.02$, $y = 3.32$ (d) $x = 28.87$, $y = 22.91$
(e) $x = 66.75$, $y = 84.57$ (f) $x = 63.23$, $y = 78.50$
2 228 km
3 (a) 1.99 m (b) $59.50
4 65.80 m
5 (a) $x = 14.4$ (b) $x = 25.87$ (c) $x = 37.14$ (d) $x = 10.51$
6 (a) $a = 34.5°$, $x = 123.4$ (b) $k = 57.6°$, $x = 9.5$
(c) $m = 58.5°$, $x = 69.0$ (d) $a = 31.1°$, $x = 30.2$

7 A
8 $\theta = 65°$

9 $\theta = 29°$ (to the nearest degree)
No, this ramp doesn't comply because the angle is larger than 15°.

10 D
11 \$24 000
12 31°23′
13 53 m
14 154 km
15 62°
16 45°38′31″
17 180.78 m
18 3.8 cm
19 16.83 mm
20 11.92 m

21 (a) 3.46 cm (b) 38 cm^2 (c) 4.53 cm

Open-ended – Sample answers

22 $a = 20$, $\theta = 10°$ so $c = 20.3$; $b = 5$, $\varphi = 20°$ so $d = 14.6$

23 Let $x = 13$ and $y = 11$, then $\theta = 134.5°$.

Chapter review 6

1 (a) 0.5736 (b) 1.2482 (c) 0.9944
(d) 0.5715 (e) 0.2823 (f) 0.2905

2 (a) 22.26 cm (b) 11.72 m

3 (a) (i) 42.27° (ii) 42°16′7″
(b) (i) 69.29° (ii) 69°17′23″
(c) (i) 27.45° (ii) 27°26′43″

4 62.45 m
5 N64°W
6 $a = 26.7°$, $x = 21.2$
7 418 cm
8 A
9 D
10 17.16 m
11 19.3°
12 101 m
13 23.56 m
14 72 km, N80°25′

15 No, the length of the side of the wedge is correct, but the angle at the point is greater than 40°.

16 84 km

17 Draw a line from the top of the triangle to the midpoint of the base to create a right-angled triangle. Now either use the sine ratio with one side of the original triangle as the hypotenuse and the height as the opposite side, or use the tangent ratio with half of the base as the adjacent side and the height as the opposite side. In both cases the angle will be 60°.

18 34.22 cm

Chapter 7

Recall 7

1 (a) 3.72 (b) 46.05 (c) 3.77

2 (a) 53°8′ (b) 67°30′ (c) 77°44′

3 (a) 8.95 (b) 107.57

4 20.32 m

5 (a) 10.34 km (b) 288.74°T

Exercise 7.1

1 (a) 32 cm (b) 35 cm

2 (a) (i) $\sqrt{98}$ cm (ii) 12 cm
(b) 34°

3 (a) 66° (b) 12.81 cm

4 Yes, a 45 cm ruler will fit.

5 (a) 30 cm (b) 31.2 cm
(c) 8.7 cm (d) 26°

6 4.05 m^2

7 (a) D (b) B

8 (a) 90° (b) 40.6° (c) 9.22 mm
(d) 11.58 mm (e) 31.2°

9 8.66 cm

10 (a) 147.52 m (b) 134.76 m
(c) 17.12 m/min (d) 6.66°
(e) Gradient of *AB* is 0.45 and gradient of *AC* is 0.12, so *AB* is steeper.

11 (a) 185.22 cm (b) 2°47′7″

12 (a) 43.46 m (b) 78.41 m (c) 93.00 m (d) 122.5°T

13 (a) 5.37 cm (b) 5.13 cm

14 14 289 m

15 Yes. The distance from *A* to *B* is 49.75 cm.

16 248 m

17 (a) 67.71 m (b) 75.12 m

18 1941 cm^2
19 Yes. The angle is 75°.

20 223.07 m

Open-ended – Sample answers

21 The ratio is always 0.07.

22 Two values: 20° and 30°
(a) $BF = 4.45$ cm, 6.5 cm
(b) $DF = 12.22$ cm, 11.26 cm
$EF = 9.23$ cm, 7.92 cm
$BC = 9.15$ cm, 10.31 cm
(c) 84.50 cm^2, 81.66 cm^2
(d) As the value of θ increased, the area of *ABCD* decreased.
(e) 234.97 cm^2, 247.98 cm^2
(f) Less plastic was used when θ was smaller and *ABCD* had a larger area.

23 (a) $h = a\sin(\theta)$ and $h = b\sin(\phi)$ so $\frac{a}{b} = \frac{\sin(\phi)}{\sin(\theta)}$.
(b) Any of the following values:
$\theta = 21°, \phi = 46°$
$\theta = 22°, \phi = 49°$
$\theta = 23°, \phi = 51°$
$\theta = 24°, \phi = 54°$
$\theta = 25°, \phi = 58°$

Exercise 7.2

1 (a) 4 (b) 3 (c) 2 (d) 2 (e) 3 (f) 1
(g) 3 (h) 2 (i) 4 (j) 4 (k) 2 (l) 3

2 D
3 B
4 C
5 D

6 (a) sin (45°) (b) -cos (44°) (c) -tan (69°) (d) -sin (84°)

(e) cos (39°) (f) -tan (57°) (g) -sin (83°) (h) -cos (83°)

(i) tan (6°) (j) -sin (18°) (k) -cos (14°) (l) tan (27°)

7

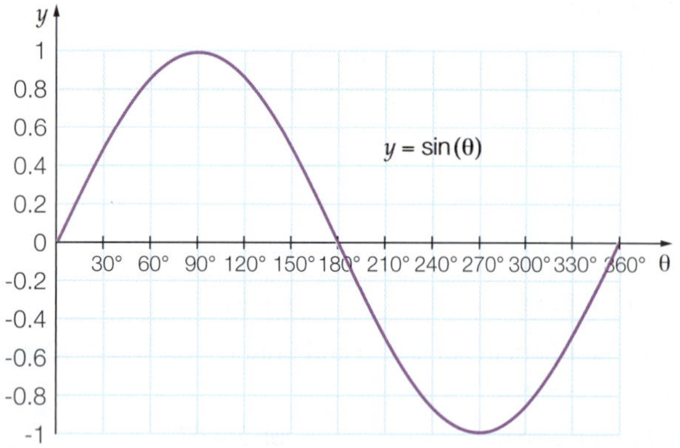

8 (a) B (b) D

9 C

10 (a)

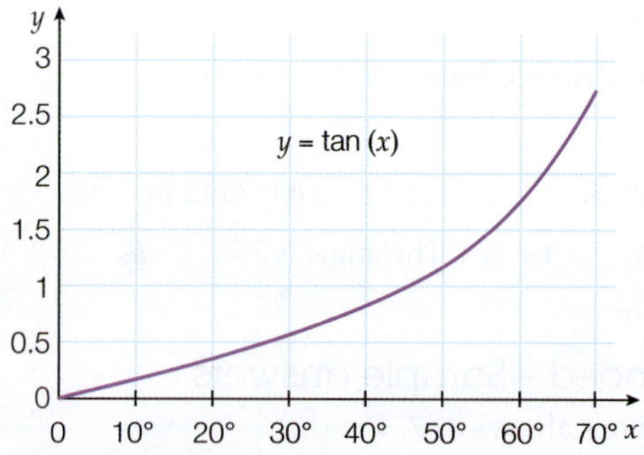

(b) The value of tan (x) becomes larger at a faster rate and approaches infinity as x approaches 90°.

11 For 30°: 150°, 210°, 330°
For 45°: 135°, 225°, 315°
For 60°: 120°, 240°, 300°

12 (a) (i) 360° (ii) 360° (iii) 360°

(b) (i) 1 (ii) 2 (iii) 3

(c) The value of a is the amplitude.

(d)

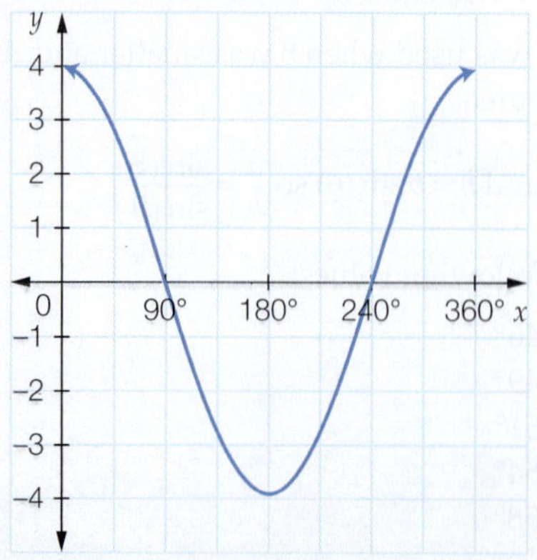

13 (a) (i) 360° (ii) 180° (iii) 120°

(b) (i) 1 (ii) 1 (iii) 1

(c) The value of n changes the period.

(In fact, the period is $\frac{360°}{n}$)

(d)

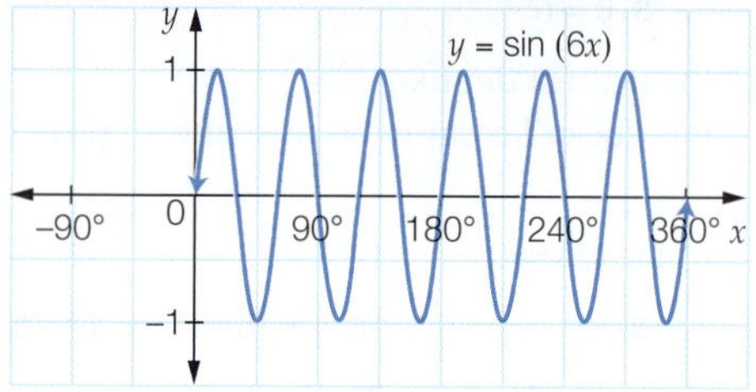

14 It will be 156°, the equivalent angle in quadrant 2.

Open-ended – Sample answers

15 For 24° and 204°

sin: + −
cos: + −
tan: + +

16 (a) If θ is in quadrant 1, -θ is in quadrant 4 (a clockwise rotation of θ). So cos (-θ) and cos (θ) are positive and of equal magnitude.

If θ is in quadrant 2, -θ is in quadrant 3 so both cos (-θ) and cos (θ) are negative and of equal magnitude.

If θ is in quadrant 3, -θ is in quadrant 2 so both cos (-θ) and cos (θ) are negative and of equal magnitude.

If θ is in quadrant 4, -θ is in quadrant 1 so both cos (-θ) and cos (θ) are positive and of equal magnitude.

(b) Using a similar method as in part (a), sin (θ) is positive in quadrants 1 and 2, and negative in quadrants 3 and 4.

(c) $\tan(-\theta) = \frac{\sin(-\theta)}{\cos(-\theta)} = \frac{-\sin(\theta)}{\cos(\theta)} = -\tan(\theta)$

Exercise 7.3

1 (a) $x = 30°$ and $x = 330°$ (b) $x = 210°$ and $x = 330°$

2 $x = 45°$

3 (a) $x = 120°$, $x = 240°$ (b) $x = 30°$, $x = 150°$

4 $x = (120 \pm 360n)°$, $n = 0, 1, 2 \ldots$ and
$x = (240 \pm 360n)°$, $n = 0, 1, 2 \ldots$

5 (a) $x = 30°$, $x = 330°$ and $x = 390°$

(b) $x = 45°$ and $x = 225°$

6 (a) $x = (120 \pm 360n)°$, $n = 0, 1, 2, 3, \ldots$
$x = (300 \pm 360n)°$, $n = 0, 1, 2, 3, \ldots$

(b) $x = (14.48 \pm 360n)°$, $n = 0, 1, 2 \ldots$ and
$x = (165.52 \pm 360n)°$, $n = 0, 1, 2 \ldots$

7 B

8 (a) 48.59° and 131.41° (b) 158.20° and 338.20°

(c) 53.13° and 306.87° (d) 41.81° and 138.19°

9 9.10 pm

10 C

11 C

12 (a) If the denominator becomes equal to 0.

(b) $x = 210°$ and 330°

13 $x = 0°$, 180° and 360°

Open-ended – Sample answers

14 $5\sin(x) + 2 = 0$; $x = 203.58°, 336.45°$
$7\tan(x) - 3 = 0$; $x = 23.20°, 203.20°$
$4\cos(x) - 3 = 0$; $x = 41.41°, 318.59°$

15 (a) Students' own answers.
(b) In (i) it had to be a positive tan equation and in (ii) it had to be a negative sin equation.

Half-time 7

1 (a) 17.10° (b) 27.20 cm

2 (a) $x = 41.8°, 138.2°$ (b) $x = 21.8°, 201.8°$

3 (a) sin (45°) (b) -cos (82°) (c) tan (20°)
(d) -sin (67°) (e) cos (60°) (f) -tan (21°)

4 (a) 65° (b) 13.6 cm
(c) (i) decrease (ii) 9°

5 C

6 (a) $x = (60 \pm 360n)°$ and $x = (300 \pm 360n)°$, $n = 1, 2, 3, \ldots$
(b) $x = (19.47 \pm 360n)°$ and $x = (160.53 \pm 360n)°$, $n = 1, 2, 3, \ldots$

Exercise 7.4

1 (a) 6.56 (b) 5.89 (c) 48.42

2 (a) $\theta = 66°$, $x = 12.49$ cm (b) $\theta = 75°$, $x = 13.71$ cm (c) $\theta = 42°$, $x = 24.62$ cm

3 (a) 82.16° (b) 40.45° (c) 68.03°

4 (a) $\theta = 64°$ (b) $\theta = 73°$ (c) $\theta = 72°$

5 (a) 56.23°, 123.77° (b) 45.77°, 134.23°
(c) 68.03°, 111.97°

6 (a) 9.01 (b) 51.50 (c) 42.46

7 (a) 54.90° (b) 110.52° (c) 26.05°

8 (a) $a = 24.1$, $b = 29.2$, $B = 73°$
(b) $a = 112.3$, $B = 65°$, $C = 53°$

9 $c^2 = a^2 + b^2 - 2ab\cos(\theta)$ and $d^2 = 4b^2 + 4a^2 - 8ab\cos(\theta)$
$d^2 = 4c^2 \Rightarrow d = 2c$

10 113.62° 11 39° 12 6.78 units

13 $\angle ABC = \theta - \phi$, so $\dfrac{x}{\sin(\theta - \phi)} = \dfrac{y}{\sin(\phi)}$

$x = \dfrac{y\sin(\theta - \phi)}{\sin(\phi)}$

14 D 15 61 units 16 C

17–19 Students' own answers.

Open-ended – Sample answers

20 (a) $c^2 = a^2 + b^2 - 2ab\cos(C)$
$k^2a^2 = a^2 + a^2 - 2a^2\cos(C)$
$k^2 = 2 - 2\cos(C)$
$2 - k^2 = 2\cos(C)$
$\cos(C) = \dfrac{2 - k^2}{2}$

(b) $k = -1$, $\cos(\theta) = 0.5$
$k = 1.5$, $\cos(\theta) = -0.125$
(c) $-2 \le k \le 2$

21 If $b = 4a$ and $c = 4a$
$\theta = 82.82°$ ($\cos^{-1}(\theta) = \frac{1}{8}$)

Exercise 7.5

1 4.92 m 2 32.76 m 3 68.40 m

4 Yes, because the angle is 104.478°.

5 46.55° 6 254 m 7 A 8 61.47 cm

9 (a) 47.55 m (b) 123.46 m

10 10.60 km

11 Tim 8.22 km, Zoe 6.45 km; so Zoe is closer.

12 16.33 cm or 21.50 cm 13 4885 m

14 037°T and 235°T 15 27.41 mm 16 B

Open-ended – Sample answers

17 Students' own answers.

18 $\theta = 45°$, $x = 16$ m, $y = 22.63$ m
$\theta = 60°$, $x = 27.71$ m, $y = 32$ m

Exercise 7.6

1 (a) 175.96 cm^2 (b) 1269.64 m^2 (c) 330.90 mm^2

2 (a) 1.42 m^2 (b) 180.42 cm^2 (c) 439.68 m^2

3 38°

4 (a) 365.01 cm^2 (b) 115.43 m^2 (c) 1718.63 mm^2

5 C 6 1292 km^2 7 5.85 mm^2

8 3562.5 cm^2 9 3944.4 m^2 10 1004 cm^2 11 1231 cm^2

12 (a) 36.38 cm^2 or 431.01 cm^2 (b) 278.60 m^2
(c) 583.30 km^2

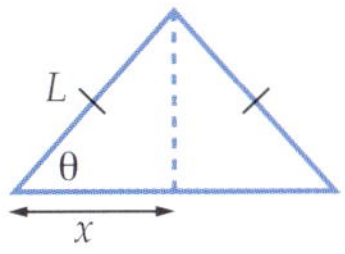

13 (a) $x = L\cos(\theta)$
Length of base $= 2L\cos(\theta)$
Area of triangle $= \frac{1}{2}L \times 2L\cos(\theta) \times \sin(\theta)$
$= L^2\cos(\theta)\sin(\theta)$

(b) 15.68 km^2

14 (a) Area $= \frac{1}{2}a(b + c)\sin(\theta) - \frac{1}{2}ab\sin(\theta)$
$= \frac{1}{2}ab\sin(\theta) + \frac{1}{2}ac\sin(\theta) - \frac{1}{2}ab\sin(\theta)$
$= \frac{1}{2}ac\sin(\theta)$

(b) 90°, $\frac{1}{2}ac$ mm^2

15 \$1426

16 (a) $0° < \theta < 180°$
(b) The area cannot be negative.

17 22 cm

Open-ended – Sample answers

18 Any triangle where the length of the equal sides is a square number and the included side is 30° (e.g. 36, 36, 30° or 16, 16, 30°).

Chapter review 7

1 (a) 24.74 cm (b) 118.27 m

2 26.31 m

3 (a) 1 (b) 4 (c) 2 (d) 3

4 (a) -cos (7°) (b) -sin (74°) (c) -tan (48°)

5 B 6 135°, 315°

7 $x = (75.52 \pm 360n)°$ and $x = (284.48 \pm 360n)°$, $n = 0, 1, 2 \ldots$

8 C 9 12.81 10 73° 11 207.70 m^2

12 D

13 (a) 180°, 7 (b) 72°, 6 (c) 45°, no amplitude

14 B

15

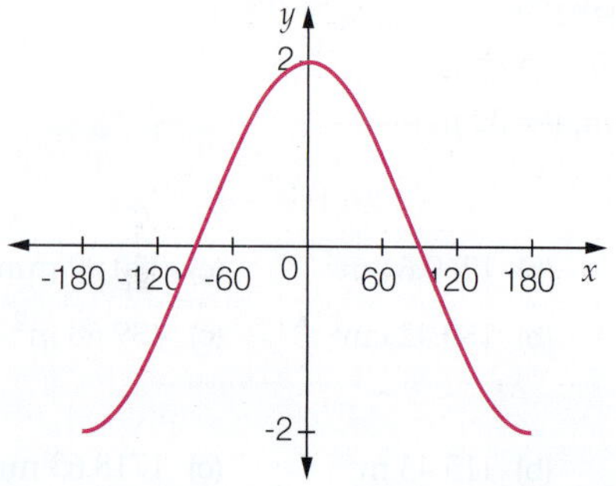

16 (a) quadrant 1 (b) 320°

17 122° 18 46.03 cm^2 19 39.55 m

20 (a) 62.7° (b) 13.95 cm

21 (a) (i) 332.86 m (ii) 495.02 m

(b) 596.52 m (c) 056°T

22 multiples of 90°

23 (a) (i) Find the third angle α and use the sine formula $\dfrac{x}{\sin(\theta)} = \dfrac{a}{\sin(\alpha)}$, where $\alpha = 180° - \theta - \phi$

(ii) 13.88

(b) (i) Each of the two other angles is $90° - \frac{1}{2}\theta$.

Use the sine formula $\dfrac{x}{\sin(\theta)} = \dfrac{a}{\sin\left(90° - \frac{1}{2}\theta\right)}$.

(ii) 14.51

24 D

25 Students' own answers.

Mixed review B

1 (a) $3x^2 + 2x - 8$ (b) $6x^2 + 13x - 5$ (c) $8x^2 - 26x + 15$

2 (a) $(x - 4)(x + 1)$ (b) $(x - 5)(x - 9)$ (c) $(x - 4)(x - 3)$

3 (a) 75.40 mm^2 (b) 382.08 cm^2

4 319.41 cm^2 5 720 cm^3 6 6.4 g/cm^3

7 (a) $h = 20.22$ (b) $h = 15.60$ (c) $a = 22.85$

8 63.43° 9 7.79 km 10 73 km

11 (a) 61.9° (b) 13.2 cm

12 (a) $x = 36.96$ m (b) $x = 39.47$ mm

(c) $x = 12.85°$ or $x = 167.15°$

13 (a) 141.42 m^2 (b) 513.07 mm^2 (c) 233.18 cm^2

14 22 cm 15 14 m × 10 m

16 Yes 17 49.62 cm 18 D

19 AB has a gradient of $\frac{-2}{3}$, BC has a gradient of $\frac{1}{5}$, CA has a gradient of -5.

BC and CA are perpendicular. Length of BC = length of $CA = \sqrt{26}$. Therefore, ABC is a right-angled isosceles triangle.

Area of ABC = 13 units^2

20 AB has a gradient of $-\frac{2}{3}$, BC has a gradient of $\frac{3}{2}$, CD has a gradient of $-\frac{2}{3}$, DA has a gradient of $\frac{3}{2}$. AB and BC are perpendicular, BC are CD are perpendicular, CD and DA are perpendicular and DA and AB are perpendicular. Therefore, $ABCD$ is a rectangle.

Area of $ABCD$ = 39 units^2

21 $5y = 3x + 21$

22 (a) 67.86° (b) 756 cm

23 41.71 m

Challenge B

1 B

2 $a = 15 - 4 = 11$, $b = 13 - 3 = 10$. $a - b = 1$

3 D

4 (a) $R(0, 2)$, BR: $2x + 9y - 18 = 0$

(b) $P(3, 0)$, PC: $2x + y - 6 = 0$

(c) X: $8y - 12 = 0$, $y = 1.5$, $x = 3.6$. $X(2.25, 1.5)$

$Q(6, 0)$, QC: $x + y - 6 = 0$

$S(0, 4)$, SB: $4x + 9y - 36 = 0$

Y: $5y - 12 = 0$. $y = 2.4$, $x = 3.6$. $Y(3.6, 2.4)$

XY: $2x - 3y = 0$. This passes through $A(0, 0)$ so X, Y and A are collinear.

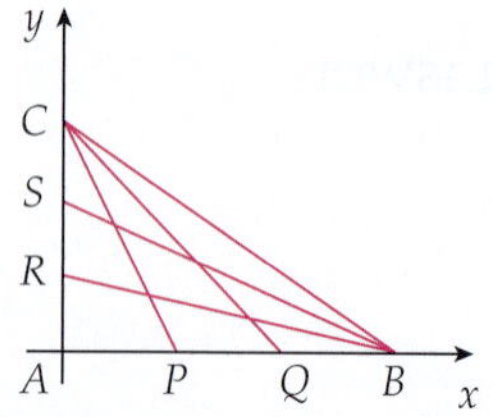

5 (a) -1

(b) CE: $y = -x + 7$ or $x + y - 7 = 0$. $D(0, 7)$

(c) AE: $y - 4 = -3(x - 3)$, $3x + y - 13 = 0$. $A\ 4\frac{1}{3}, 0$. $B(0, 13)$

(d) Shaded area = area ΔODC + area ΔBDE
$= 24.5 + 9$
$= 33.5$ units^2

6 D

$P(z, 5z + 3)$, $Q(3, -2)$. $M\ \frac{z+3}{2}, \frac{5z+1}{2}$. Hence $x = \frac{z+3}{2}$,

$y = \frac{5z+1}{2}$, $z = 2x - 3$, $y = \frac{5(2x-3)+1}{2}$, $y = 5x - 7$

7 B

8 59°

9 $n = 14$, area = 84 cm^2

10 $\cos(\angle MQP) = 0.28$

11 4.58 units

12 13.55 units

13 (a) $AB = 3.2$ m

(b) Length of timber = 16.9 m

(c) $\angle BAX = 22°$, $\angle BYX = 44°$

Chapter 8

Recall 8

1 (a)

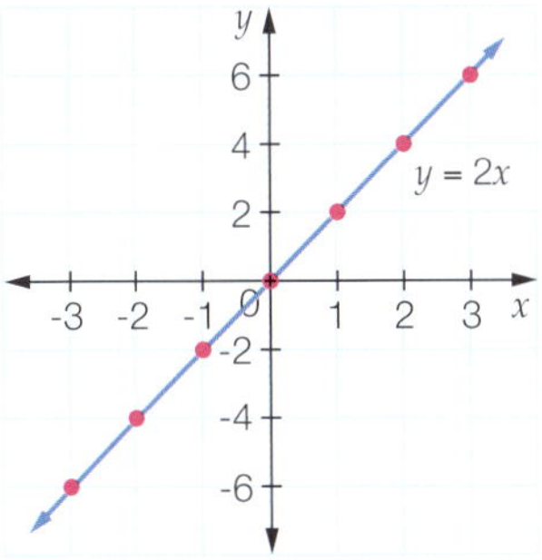

(b)

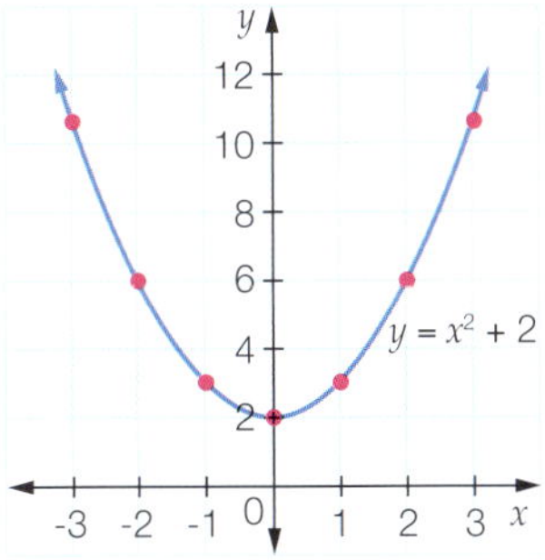

2 (a) neither

(b) linear

(c) quadratic

3 (a)

x	-5	-4	-3	-2	-1	0	1	2	3
y	7	0	-5	-8	-9	-8	-5	0	7

(b)

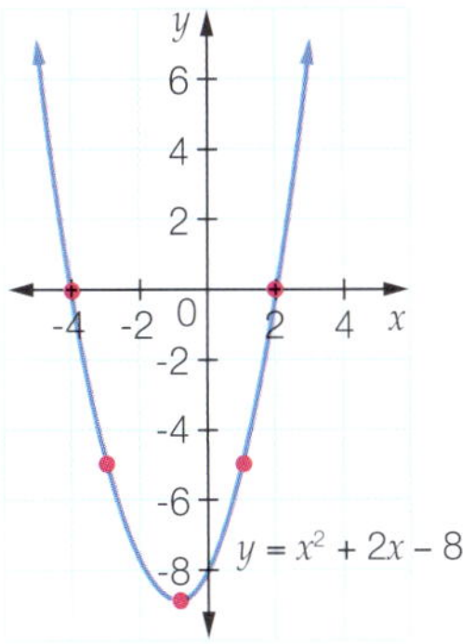

(c) (i) (0, -8) (ii) (-4, 0), (2, 0) (iii) (-1, -9)

(d) minimum (e) $x = -1$

Exercise 8.1

1 (a)

x	-3	-2	-1	0	1	2	3
y	-81	-24	-3	0	3	24	81

(b)

x	-3	-2	-1	0	1	2	3
y	-9	-4	-1	0	-1	-4	-9

(c)

x	-3	-2	-1	0	1	2	3
y	40.5	8	0.5	0	0.5	8	40.5

2 (a)

x	-3	-2	-1	0	1	2	3
y	18	8	2	0	2	8	18

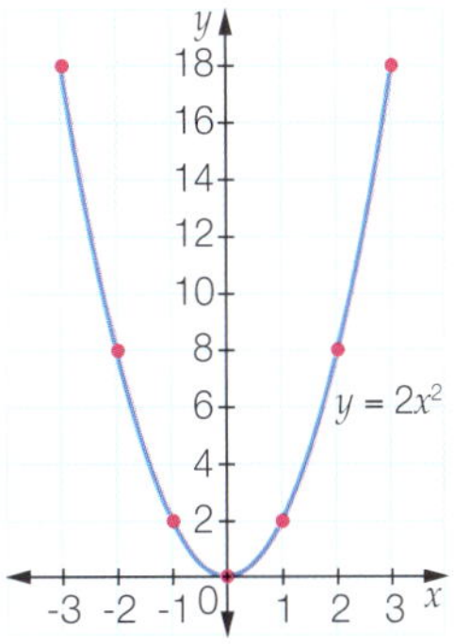

(b)

x	-3	-2	-1	0	1	2	3
y	-54	-16	-2	0	2	16	54

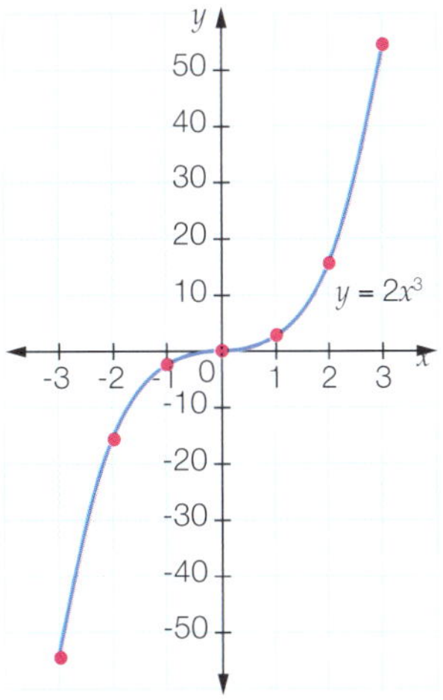

(c)

x	-3	-2	-1	0	1	2	3
y	162	32	2	0	2	32	162

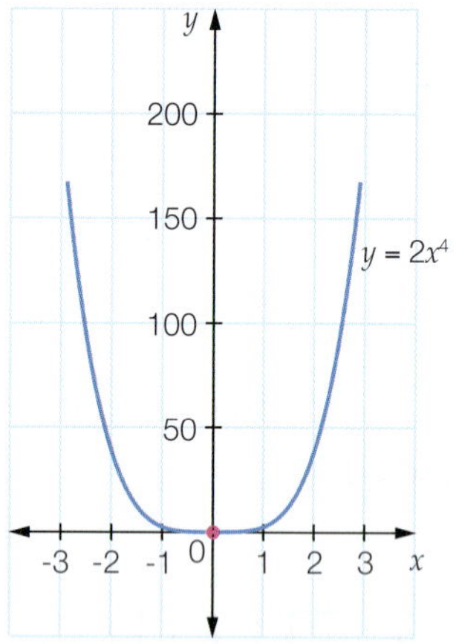

3 D

4 (a) A (b) F (c) C (d) E (e) D (f) B

5 (a)

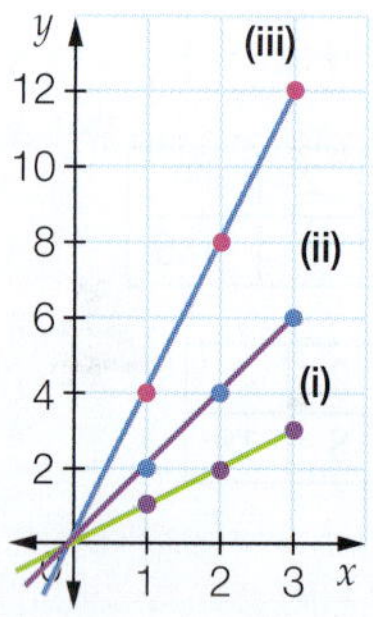

(b) The graph becomes steeper as the coefficient gets larger.

(c) No, graphs with a negative coefficient become steeper as the coefficient of x decreases.

6 (a)

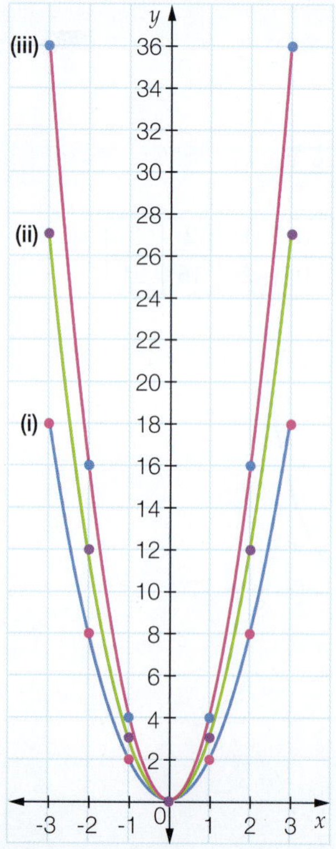

(b) The parabola becomes narrower about the y-axis as the coefficient increases.

(c) No, parabolas with a negative coefficient become narrower as the coefficient of x decreases.

7 (a)

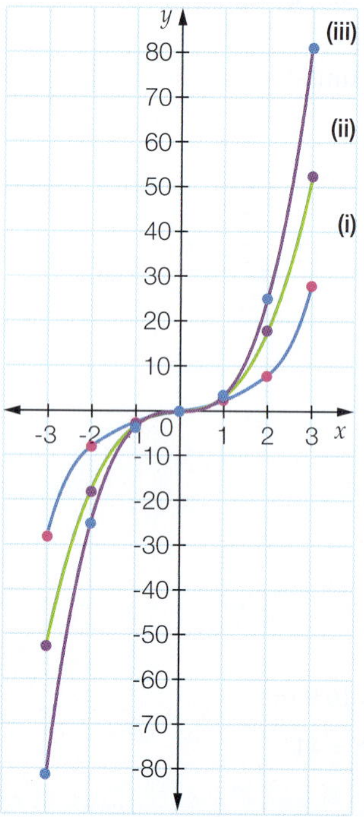

(b) The cubic gets steeper (gets closer to the y-axis) as the coefficient increases.

(c) No, cubics with a negative coefficient are reflected and become narrower as the coefficient of x decreases.

8 (a)

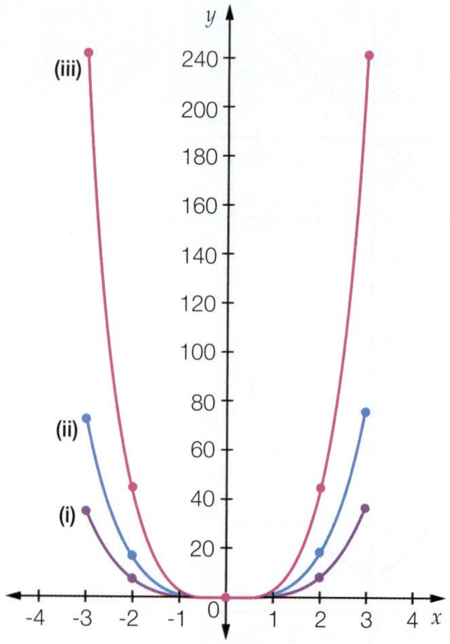

(b) The quartic becomes narrower about the y-axis as the coefficient increases.

(c) No, quartics with a negative coefficient become narrower as the coefficient of x decreases.

9 B

10 (a)

x	0	1	2	3
y	0	0.04	0.64	3.24

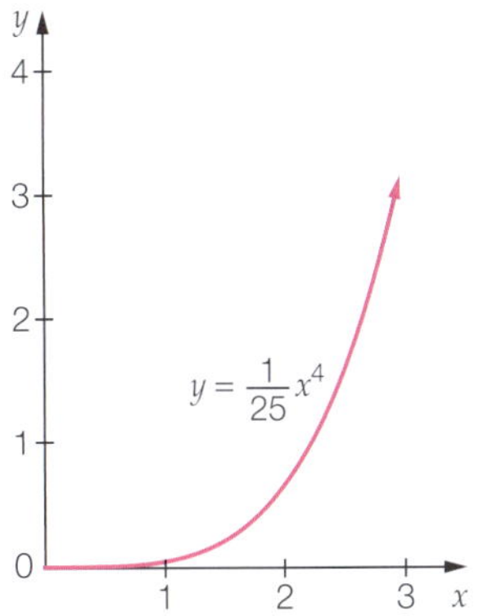

(b)

(c) The new ramp is flatter than the original ramp because the dilation factor a of the new ramp is less than the dilation factor a of the original ramp.

(d)

x	0	1	2	3
y	0	0.02	0.32	1.62

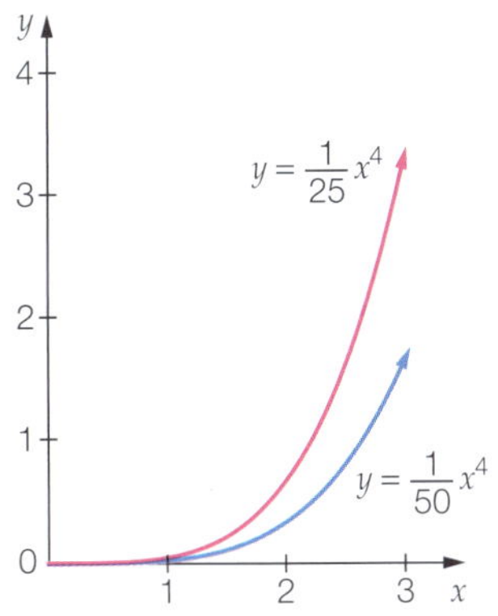

11 (a)

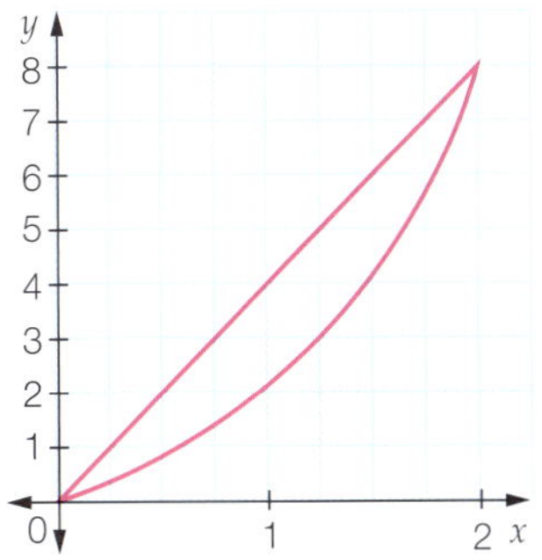

(b) $2\frac{2}{3}$ m^2 (c) $y = 4x$, $y = -2x^2$, $x = -2$ to 0

12 (a), (b)

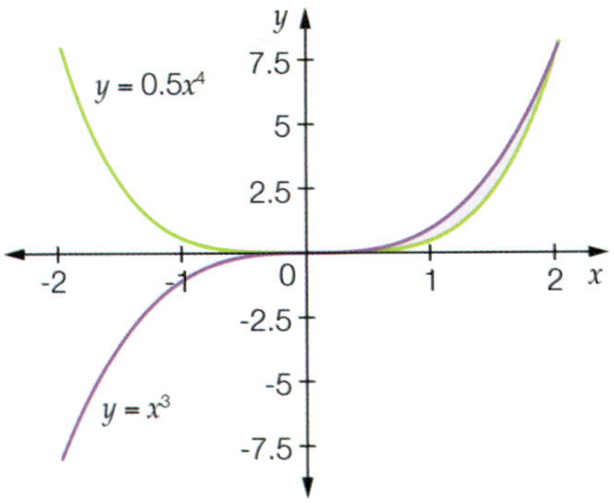

(c)

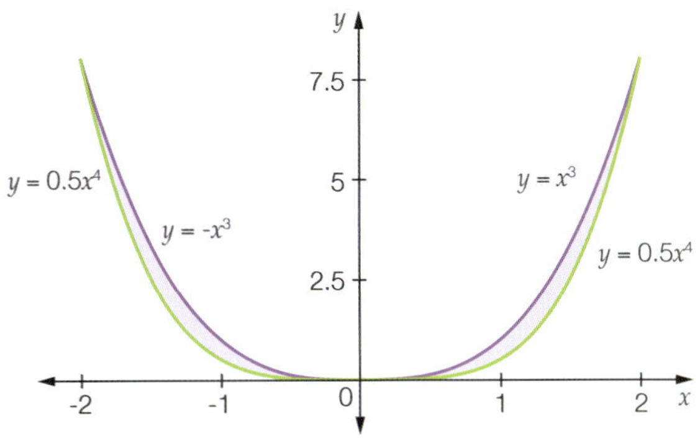

(d) $y = -x^3$ for $-2 \le x \le 0$

Open-ended – Sample answers

13 (a) $c = 10, 11$

(b) Gradient is undefined, the line is vertical.

14 The graphs will intersect at $(-1, a)$ and $(1, a)$. This is true for both positive and negative values of a.

Exercise 8.2

1 (a) The graph of $y = x^2$ has been *dilated* by a factor of 3 parallel to the y-axis. This will cause the graph to be *narrower* than that of $y = x^2$.

(b) The graph of $y = x^2$ has been *reflected* in the x-axis.

(c) The graph of $y = x^2$ has been *translated* 9 units *up*.

(d) The graph of $y = x^2$ has been *translated* 6 units *to the left*.

(e) The graph of $y = x^2$ has been *dilated* by a factor of 2 parallel to the y-axis, *translated* 4 units *to the right* and 7 units *up*.

(f) The graph of $y = x^2$ has been *reflected* in the x-axis, *dilated* by a factor of $\frac{1}{3}$ parallel to the y-axis, *translated* 1 unit *to the left* and 4 units *down*.

2

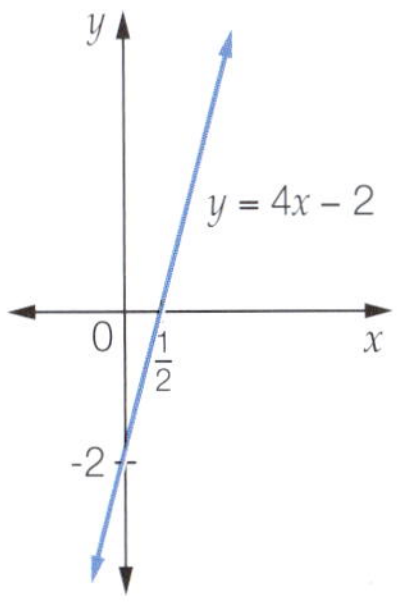

3 (a)

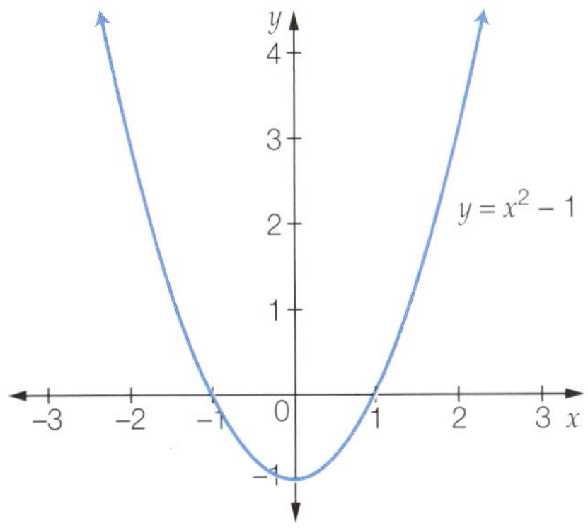

(b)

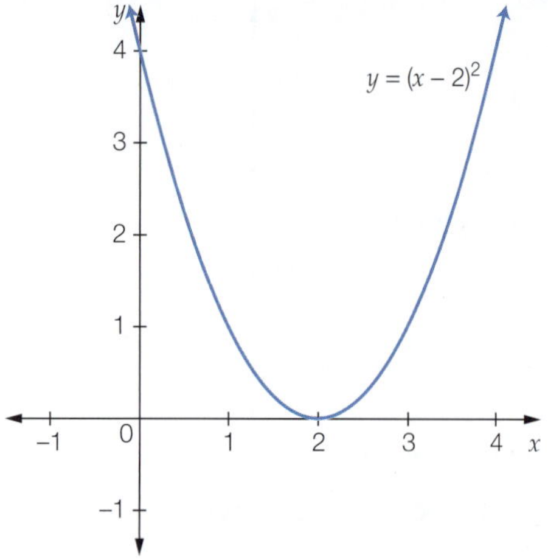

(c)

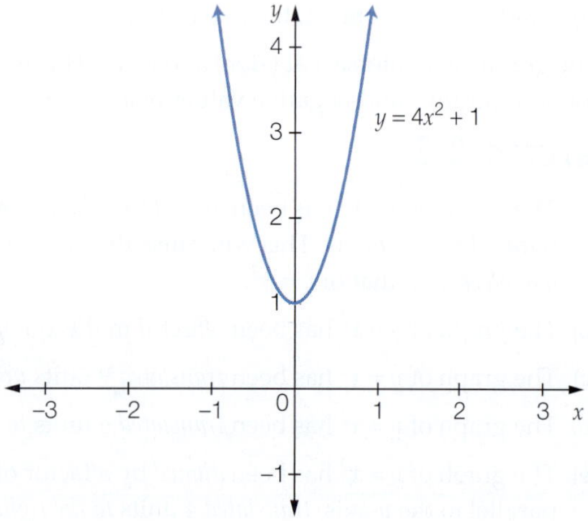

(d)

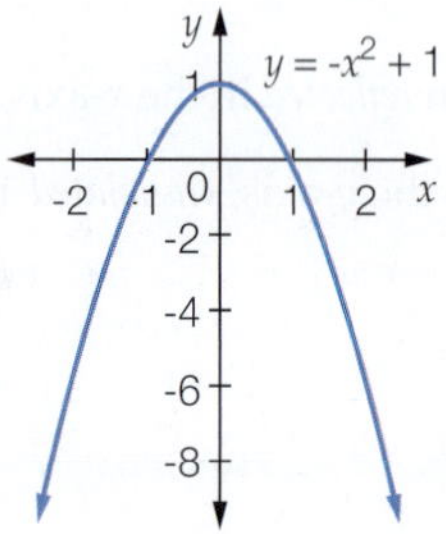

(e)

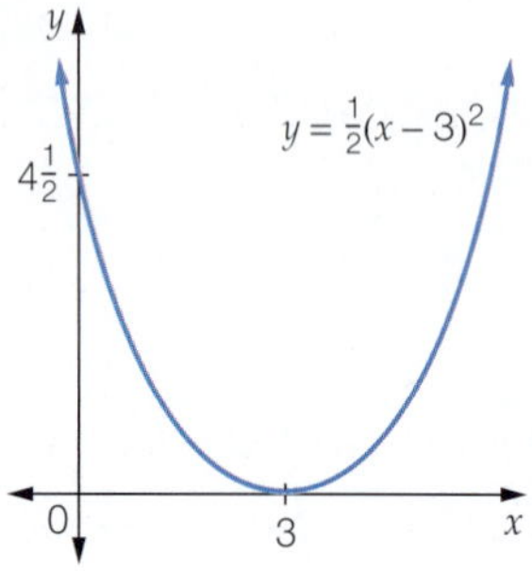

(f)

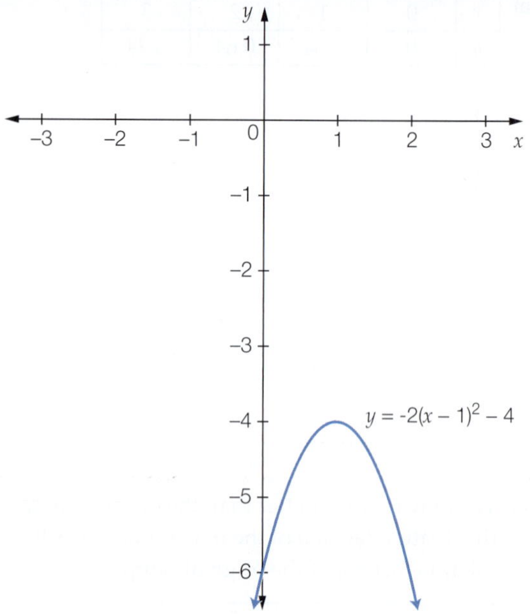

4 (a)

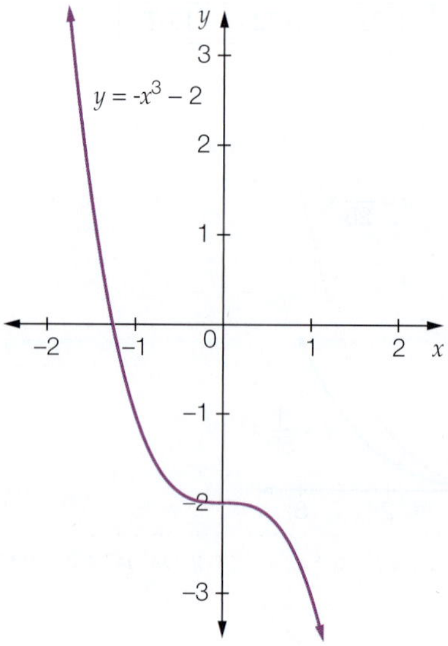

(b)

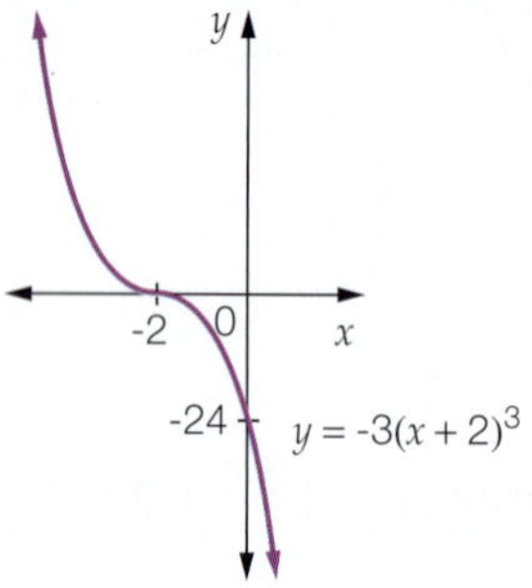

5 (a)

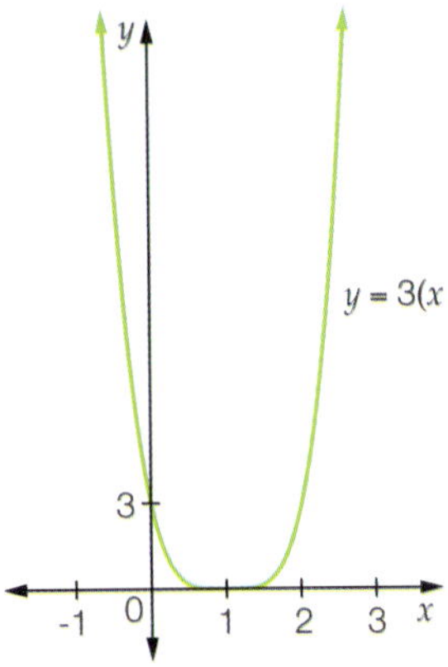

(b)

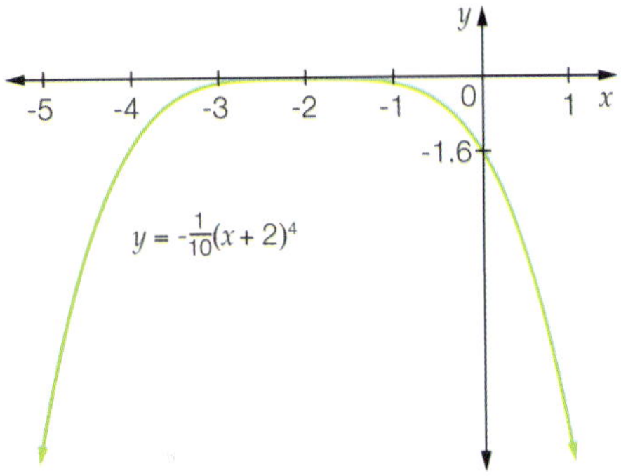

6 D

7 C

8 C

9 (a) $h = 2$ and $k = 4$
$y = a(x-2)^3 + 4$

(b) $a = 3$
$y = 3(x-2)^3 + 4$

10 (a) $h = 1, k = 2$

(b) Negative, the turning point is a maximum.

(c) $a = -3$

11 (a) $y = \frac{x^3}{10}$

(b)

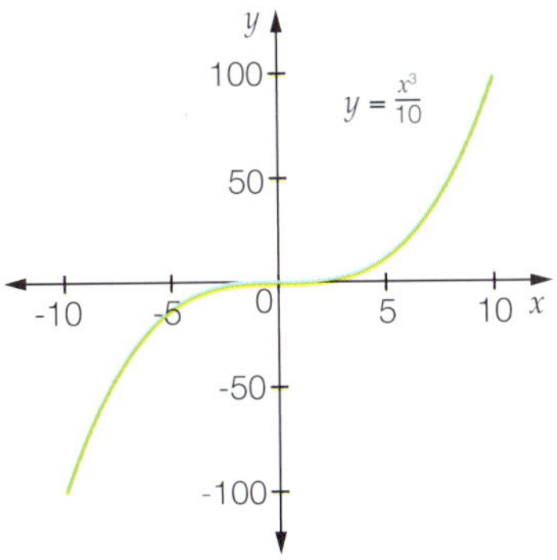

(c)

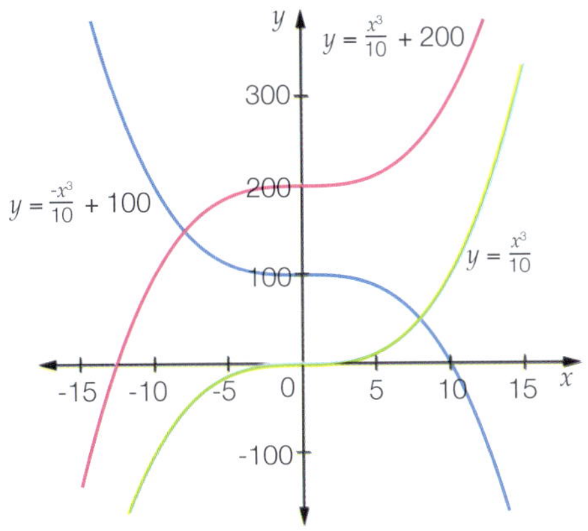

(d) The snake has moved up vertically 100 cm, the shape of the graph is exactly the same, but it is a mirror image of its initial shape. It is reflected in the line $y = 50$.

12 (a) $y = 10 - 0.2(x-20)^2$

(b)

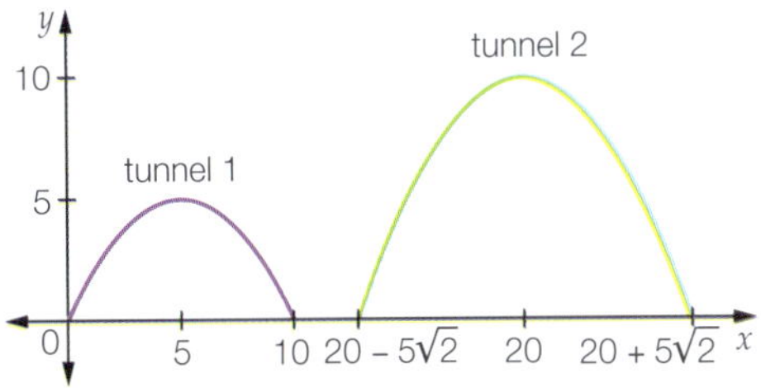

(c) 15.8 m

13 (a–b)

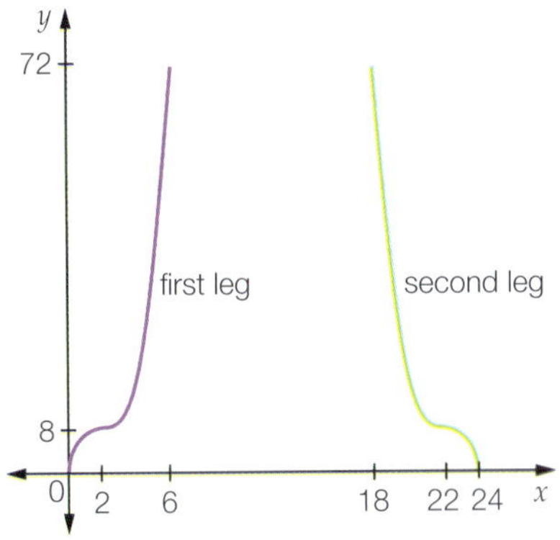

(c) $y = -(x-22)^3 + 8$ (d) 12 cm

Open-ended – Sample answers

14 $y = -3(x+2)^3 + 4$, $y = -5(x+1)^3 - 8$

Exercise 8.3

1 (a) degree 3, leading coefficient 2, constant term 2

(b) degree 4, leading coefficient -4, constant term 5

(c) degree 1, leading coefficient 1, constant term 0

(d) degree 2, leading coefficient -1, constant term -6

2 (a) 62 (b) -142 (c) -28 (d) 8

3 (a) $5x^3 + 11x^2 - 13x - 3$ (b) $10x^4 - 3x^3 - 6x^2 + 29x + 6$

4 $2x^5 - 3x^4 - x^3 + 7x^2 + 15x - 20$

5 (a) $148\frac{15}{25}$ or $148\frac{3}{5}$ (b) $338\frac{5}{12}$

(c) $348\frac{8}{17}$ (d) $106\frac{4}{19}$

(e) $510\frac{7}{29}$ (f) $88\frac{122}{123}$

6 Divisor: $3x + 4$
Dividend: $3x^3 + 31x^2 + 21x - 16$
Quotient: $x^2 + 9x - 5$
Remainder: 4

7 (a) quotient is $x + 3$, remainder is 1
(b) quotient is $x + 5$, remainder is 9
(c) quotient is $5x + 13$, remainder is 30
(d) quotient is $2x - 4$, remainder is 0

8 (a) quotient is $x^2 - 2x + 10$, remainder is -15
(b) quotient is $x^3 + 3x^2 + 6x + 18$, remainder is 55
(c) quotient is $3x^2 + 5x + 18$, remainder is 27
(d) quotient is $x^3 - 6x^2 - x - 2$, remainder is 1

9 $x^3 + 8x^2 - 3x + 8$

10 $12x^3 - 55x^2 + 30x - 20$

11 (a) $x + 4, 1$ (b) $2x + 1, 5$
(c) $2x - 5, 6$ (d) $3x^2 + 6x + 14, 24$

12 A

13 $4x^2 + x - 3 = (4x - 3)(x + 1)$ so remainder is zero as it is a factor.

14 D

15 $x^3 - x^2 - 5x - 25 - \frac{120}{x - 5}$

16 (a) $R(x) = 16x^3 - 6x^2 + 4$; 108
(b) $R(x) = 3x^3 - 16x^2 - 17x + 22$; -152
(c) $R(x) = 4x^3 - 26x^2 - 28x + 36$; 34

17 (a) $x^4 - 3x - 6$ (b) 4
(c) $A(2) = 4$, $B(2) = 1$, $\frac{A(2)}{B(2)} = 4$, which is the same as the answer for (b).

18 (a) $4x^3 + x + 8$
(b) (i) $4x^2 + 1 + \frac{8}{x}$ (ii) if $x = 0$, $\frac{8}{x}$ is undefined.

19 (a) $4x^2 - 10x + 6$
(b) $8x^3 - 30x^2 - 27x$
(c) (i) $8x - 22 - \frac{49}{x - 1}$
(ii) $x = 8$ cm, $x - 1 = 7$, 35 small rectangles

Open-ended – Sample answers

20 $2x^5 - x^4 - 3x^3 - 1$

21 $3x^3 - 14x^2 + 13x + 6$

22 (a) $1, x + \frac{1}{x + 1}, x^2 + 1, x^3 + x + \frac{1}{x + 1}, x^4 + x^2 + 1$

(b) If $P_n = x^n + x^{(n-1)} + x^{(n-2)} + \ldots + x^2 + x + 1$ then for n odd, answers are $1, x^2 + 1, x^4 + x^2 + 1$, so $\frac{P_n(x)}{x + 1} = Q_{n-1}$, where $Q_{n-1}(x) = x^{(n-1)} + x^{(n-3)} + \ldots$ which are even powers of x when n is odd.

(c) If $P_n = x^n + x^{(n-1)} + x^{(n-2)} + \ldots + x^2 + x + 1$ then for n even, answers are $x + \frac{1}{x + 1}$, $x^3 + x + \frac{1}{x + 1}$, so $\frac{P_n(x)}{x + 1} = Q_{n-1} + \frac{1}{x + 1}$, where $Q_{n-1}(x) = x^{(n-1)} + x^{(n-3)} + \ldots$ which are odd powers of x when n is even.

Half-time 8

1 (a)

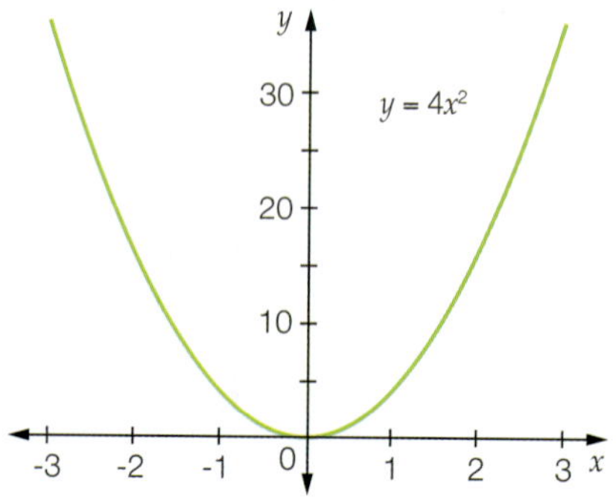

(b)

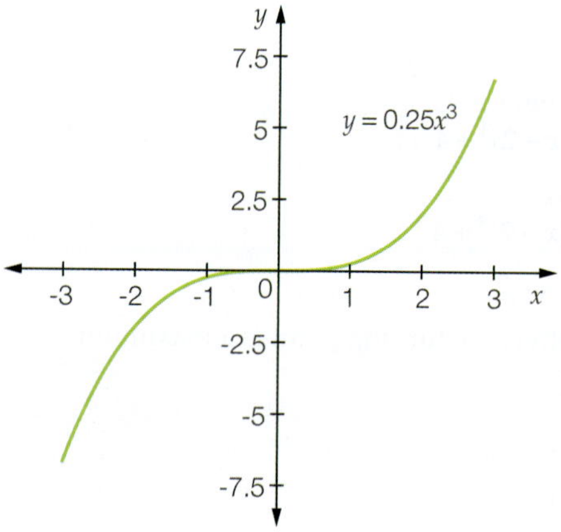

2 $x^2 + x - 7, 2$

3 (a) (i) 8 (ii) 17 (iii) -42
(b) $9x^2 - 5$ (c) $9x^3 - 18x^2 + 11x - 4$ (d) $2x + 1$

4 (a)

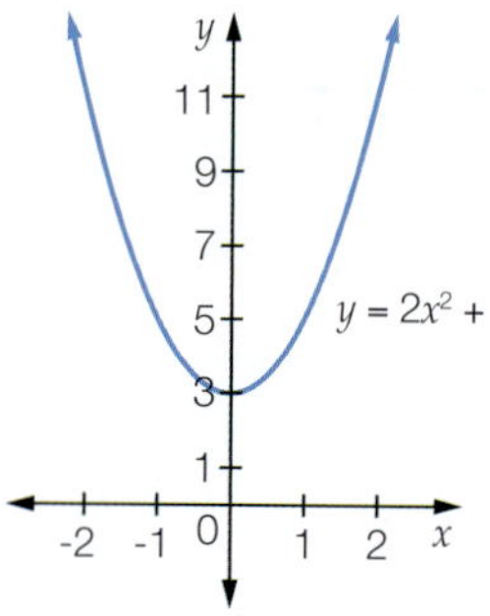

(b)

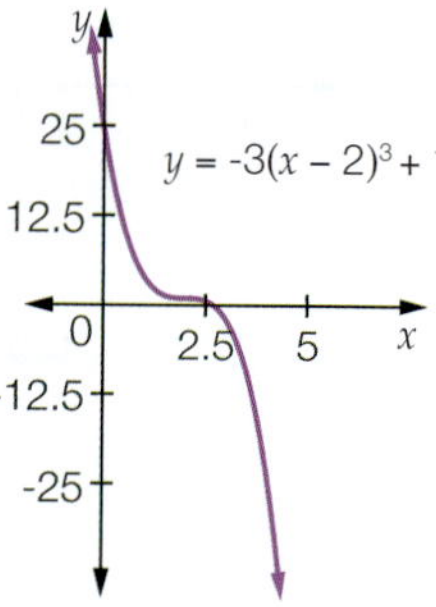

(c)

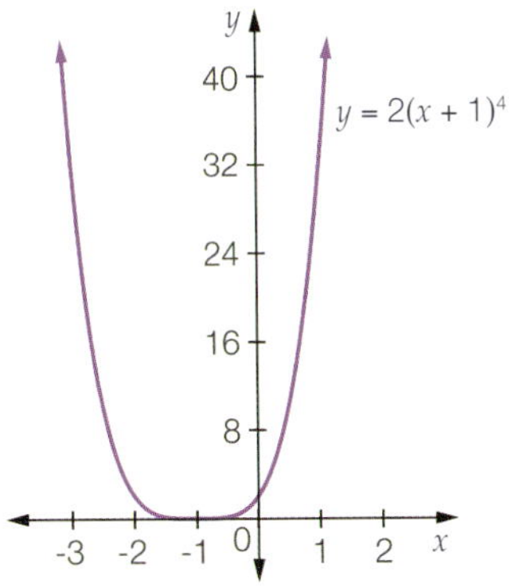

5 (a) 1 (b) $-\frac{47}{81}$

6 (a)

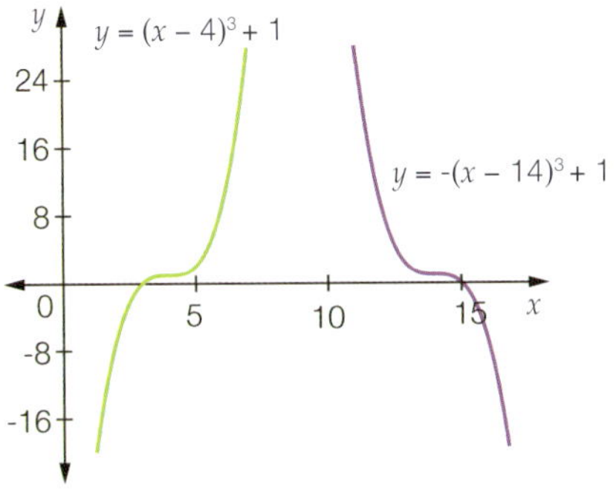

(b) $x = 9$

7 $\frac{x^3 + x - 5}{x - 1} = x^2 + x + 2 - \frac{3}{x - 1}$

8 (a) $(x - 1)(x^2 - 5x - 14)$ (b) $(x - 2)(x + 2)(x - 1)^2$

Exercise 8.4

1 $x^2 + 6x + 7 + \frac{17}{x - 2}$, $P(2) = 17$. The remainder $= P(2)$.

2 (a) $P(2) = 1$ (b) $P(-3) = 104$

(c) $P(\frac{1}{2}) = -4\frac{3}{8}$ (d) $P(-\frac{1}{4}) = 8\frac{3}{8}$

3 (a) $(x - 2)$ (b) $(x + 5)$ (c) $(5x - 2)$

4 $(x - 1)(x - 2)(x - 3)$

5 $(x - 1)(x + 1)(x + 2)(x - 3)$

6 (a) 44 (b) -23 (c) $-\frac{21}{16}$

(d) $-\frac{31}{32}$ (e) 550 (f) -140

7 (a) $(x - 1)(x - 2)(x + 4)$ (b) $(x + 1)(x - 3)^2$

(c) $(x + 1)(x - 2)(x - 5)$ (d) $(x - 1)(x - 2)(x + 6)$

(e) $(x + 2)^3$ (f) $(x + 1)^2(x - 5)$

8 (a) -86 (b) -5 (c) -12 (d) -31

9 (a) $(x - 1)(x + 1)(x + 2)(x + 4)$ (b) $x(x - 1)(x - 2)(x - 4)$

(c) $(x - 1)^2(x - 2)(x + 3)$ (d) $(x - 1)^3(x + 5)$

(e) $(x - 1)(x + 1)(x - 2)^2$ (f) $(x - 1)(x + 2)(x - 3)(x + 4)$

10 B 11 D 12 -6

13 (a) $(x - 1)(x - 2)(x - 3) = 0$, $x = 1, 2$ or 3

(b) $(x + 1)(x - 2)(x + 2) = 0$, $x = -2, -1$ or 2

(c) $(x + 2)(x + 3)(x + 4) = 0$, $x = -2, -3$ or -4

(d) $(x - 1)(x + 1)(x - 2)(x + 3) = 0$, $x = -1, -3, 1$ or 2

(e) $(x - 1)(x + 1)(x - 2)(x + 2) = 0$, $x = \pm 1, \pm 2$

(f) $x(x - 1)(x - \sqrt{6})(x + \sqrt{6}) = 0$, $x = \pm\sqrt{6}, 0$ or 1

14 (a) $(x - 1)^2(x - 2)^2$ (b) 4 cm (c) 1 s, 2 s

(d) The top is moving away from the fixed point.

Open-ended – Sample answers

15 $x^3 - 3x^2 - x + 1$, $x^3 - 2x^2 - 3x - 2$

16 $x^4 - 4x^3 + x^2 + 6x = 0$ and multiples of the LHS.

Exercise 8.5

1 (a) y-intercept: (0, 2)
x-intercepts: (-2, 0), $(-\frac{1}{3}, 0)$

(b) y-intercept: (0, 42)
x-intercepts: $(-\frac{7}{2}, 0)$, (2, 0) and (3, 0)

(c) y-intercept: (0, 6)
x-intercepts: (-3, 0), (-2, 0) and (-1, 0)

(d) y-intercept: (0, -2)
x-intercepts: (-1, 0) and (2, 0)

2 (a)

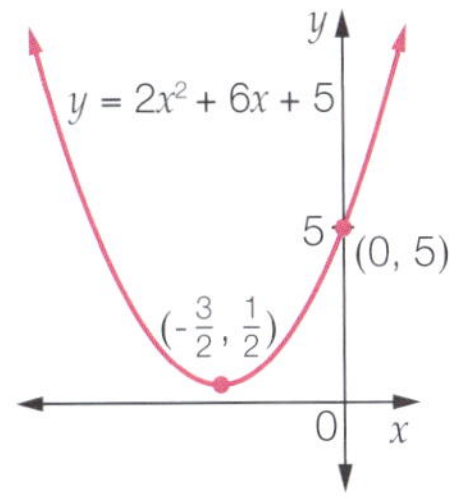

(b)

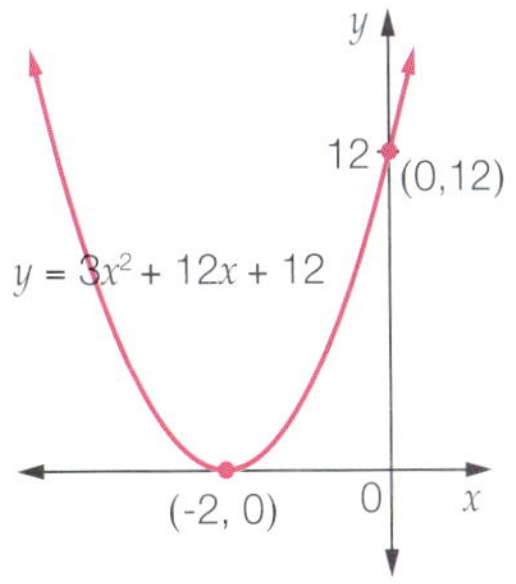

(c)

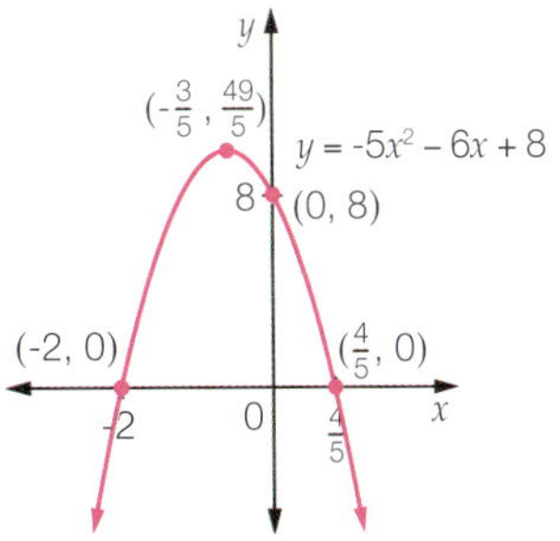

3 (a) (-1, 0)

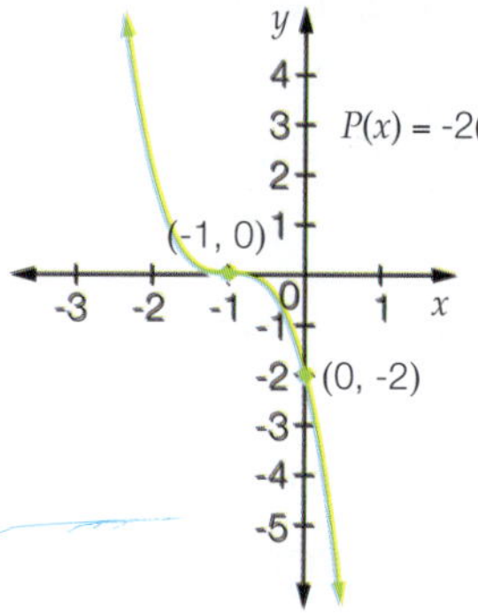

(b) (-1, 0) and (2, 0)

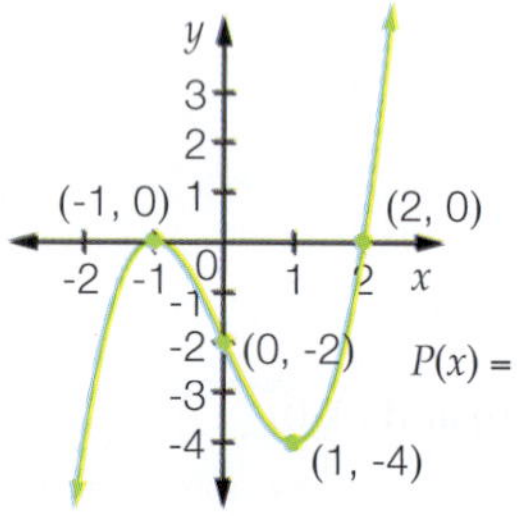

(c) (-3, 0), (-1, 0), (1, 0)

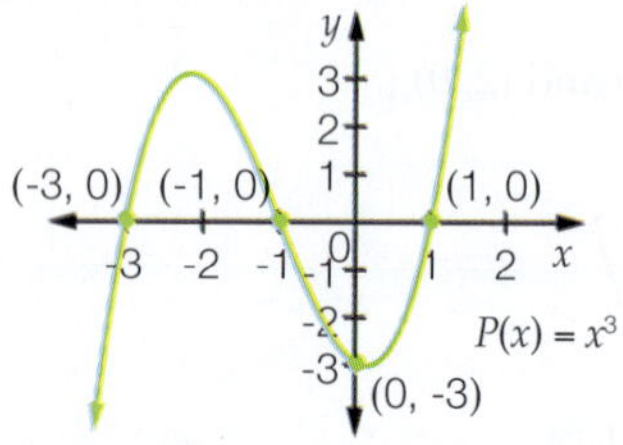

4

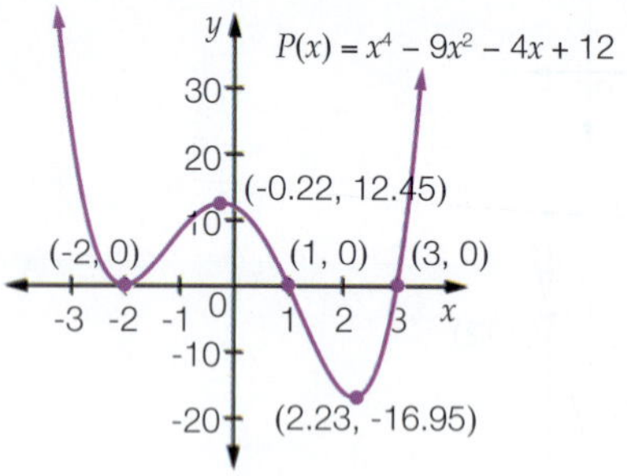

5 (a)

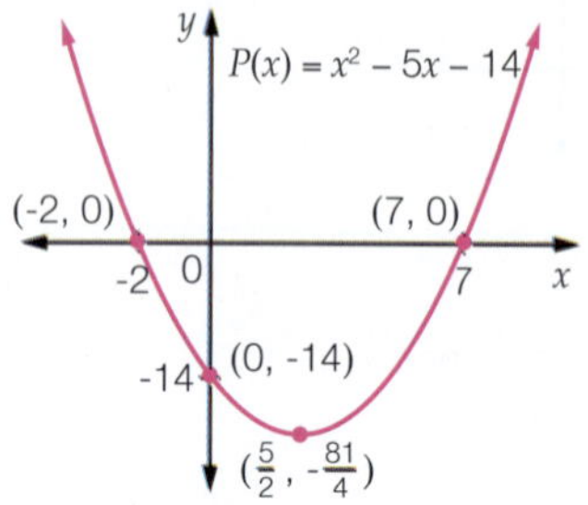

(b)

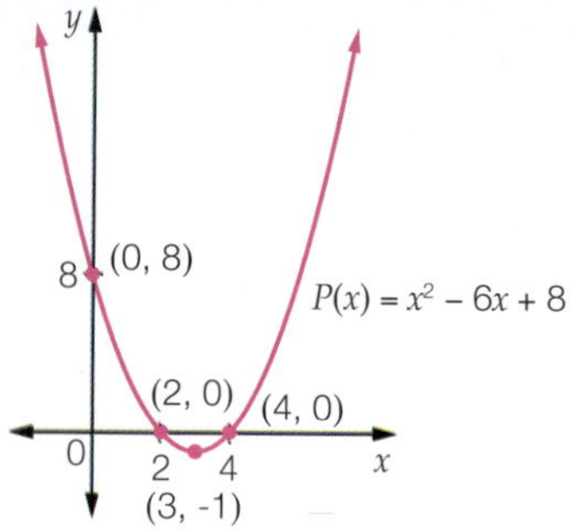

(c)

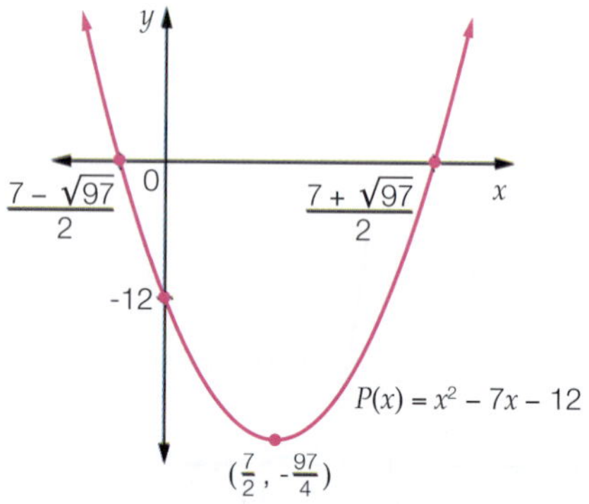

(d)

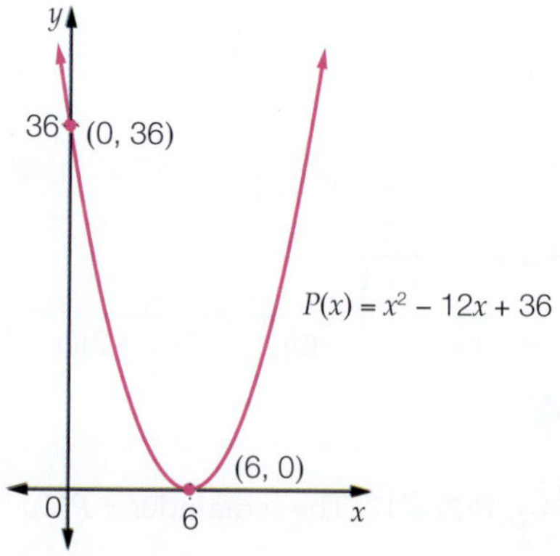

(e)

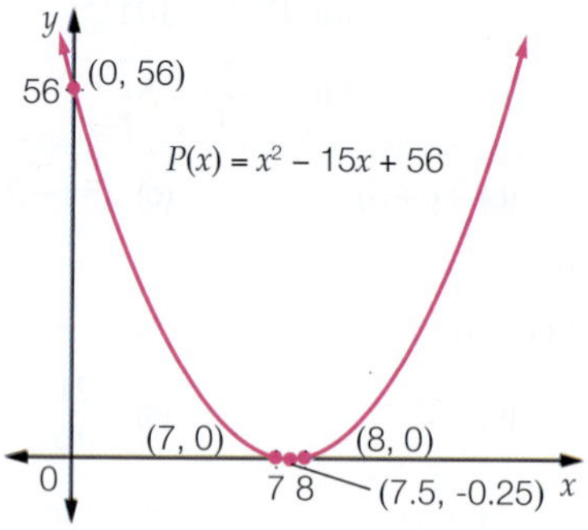

(f)

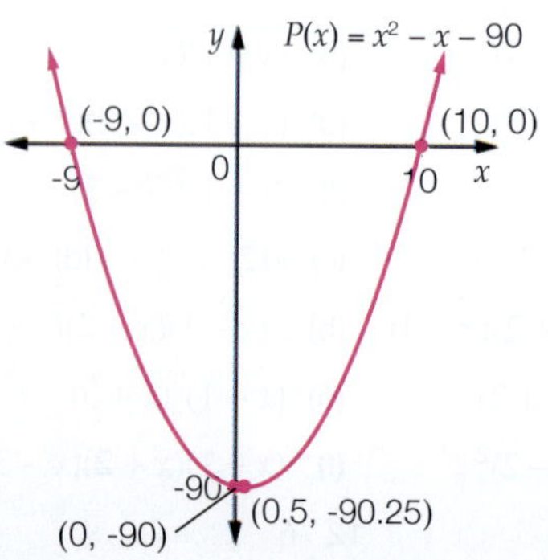

6 (a)

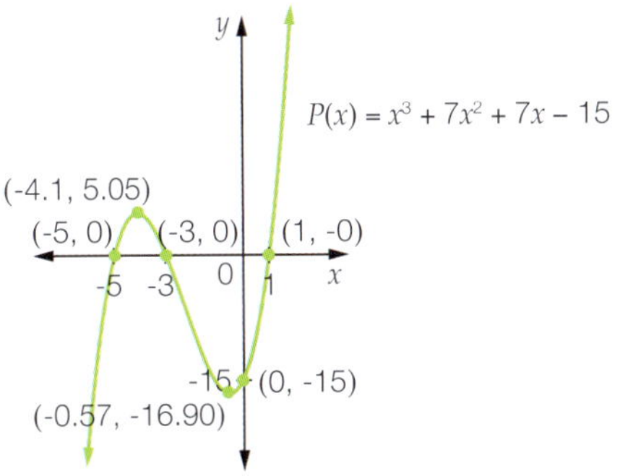

(b)

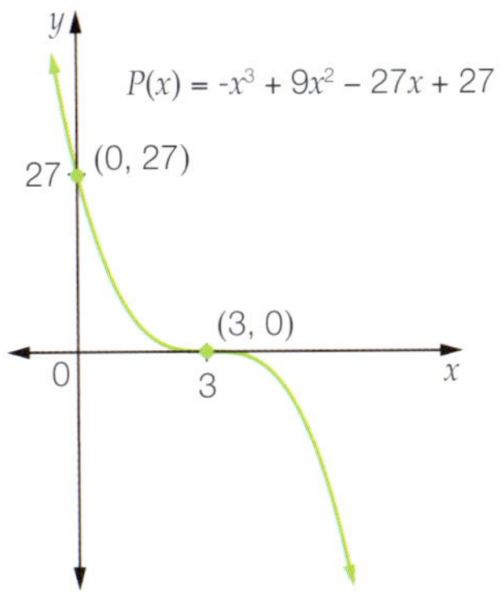

(c)

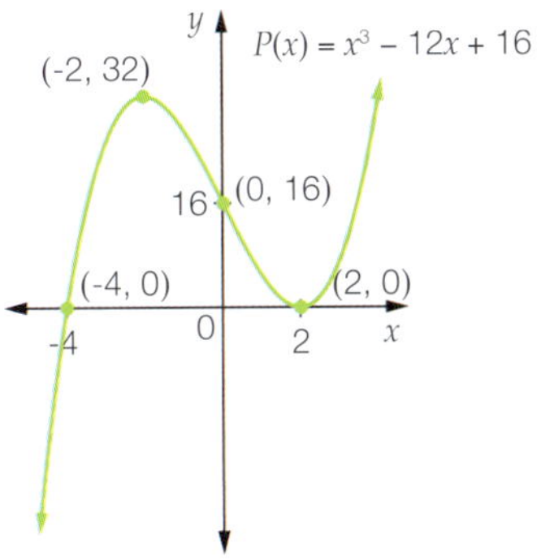

(d)

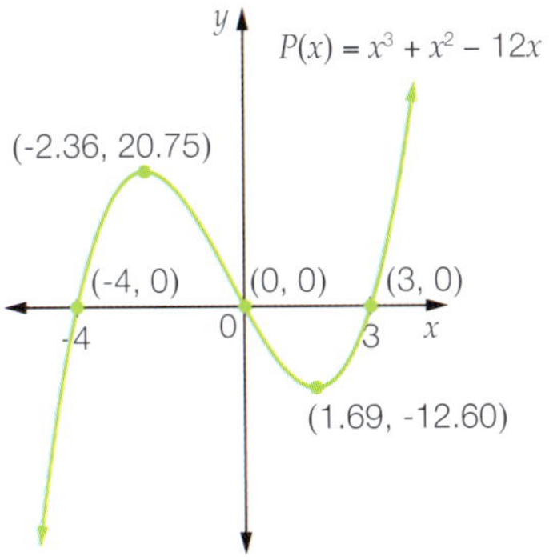

(e)

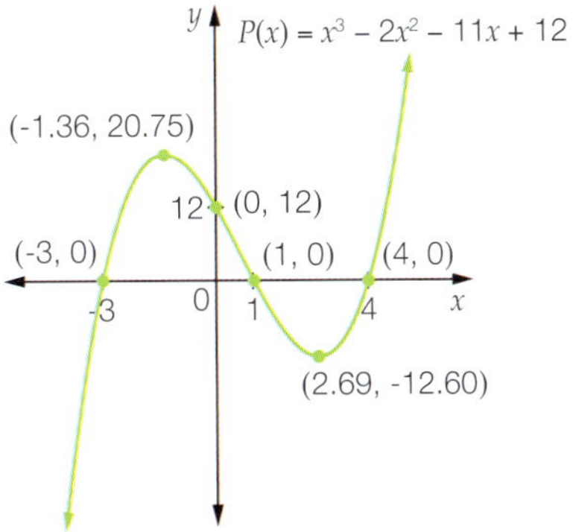

(f)

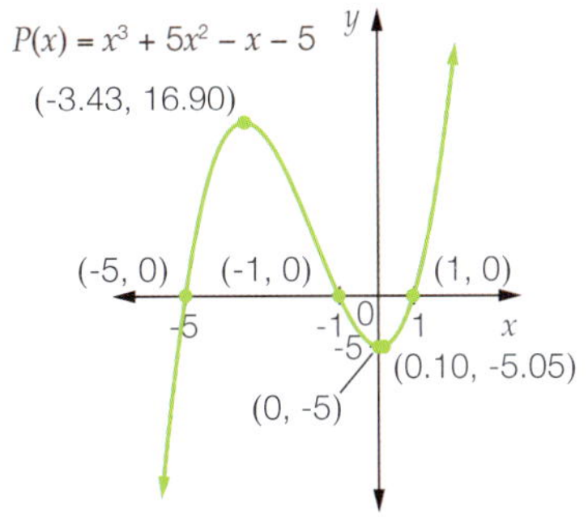

7 (a)

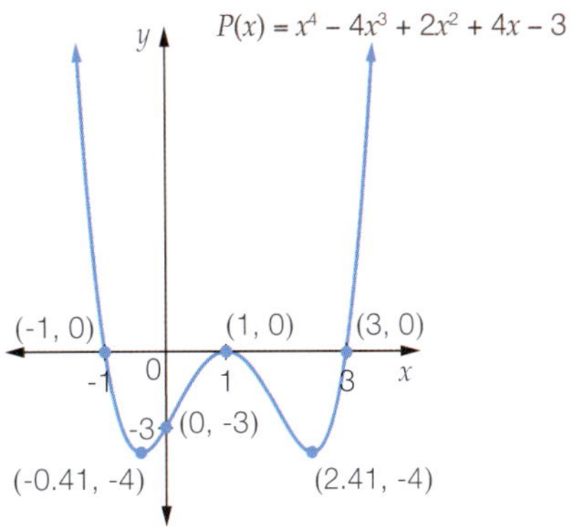

(b)

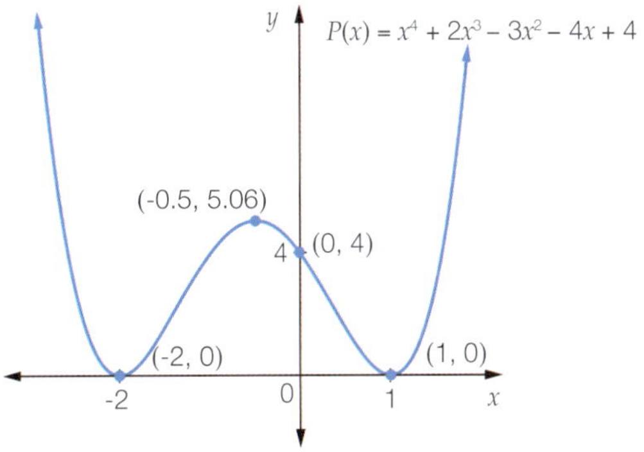

(c)

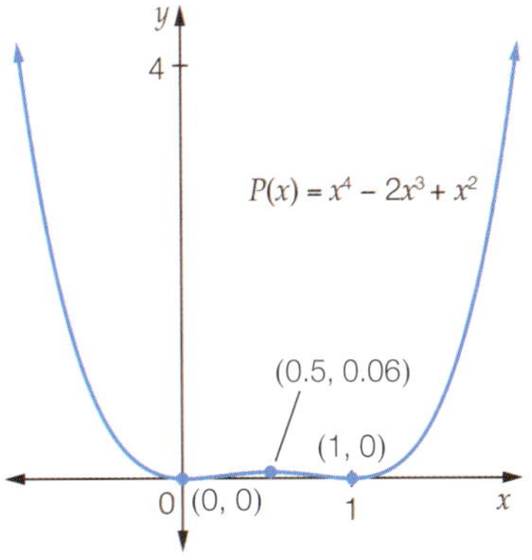

(d)

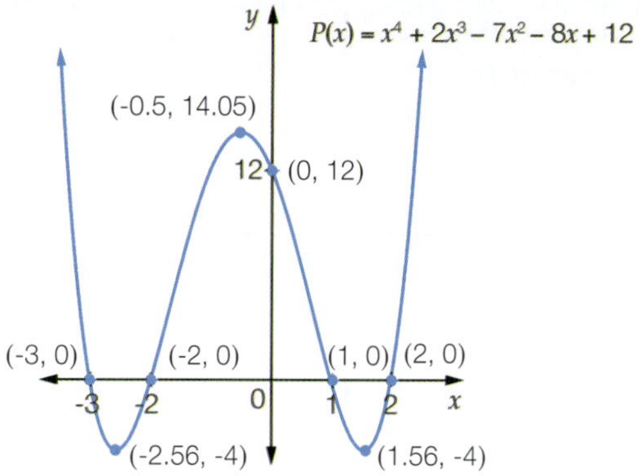

(e)

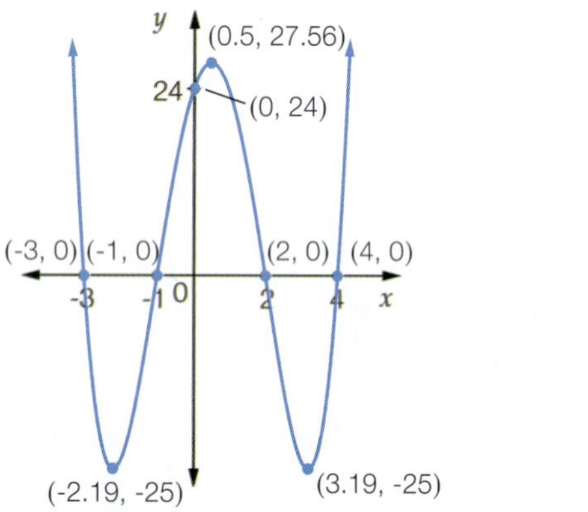

(f)

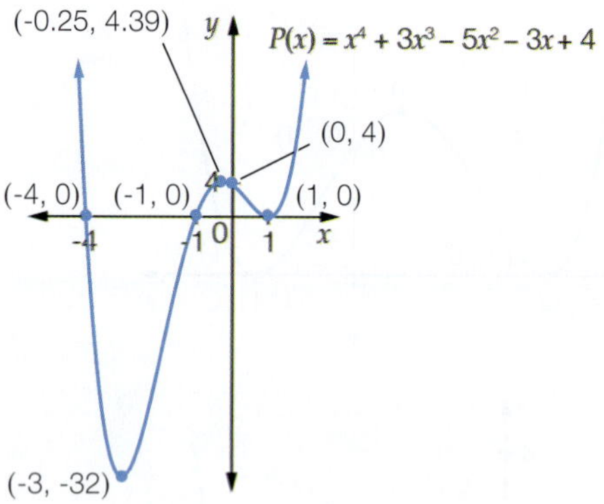

8 (a) B (b) A (c) E (d) C (e) D

9 D

10 D

11 $x = 5, x = -7$

12 $x = 1, x = -2$

13 (a) $P(t) = -t(t-1)(t-2)$

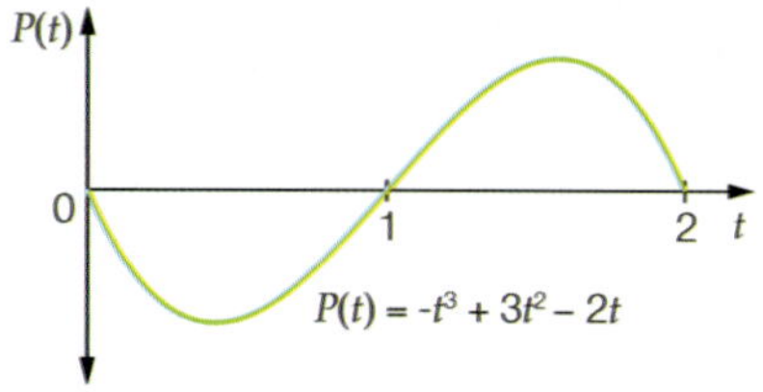

(b) 1 second, 0.39 m (c) 1 second, 0.39 m

14 (a) $y = x(x-3)^2$

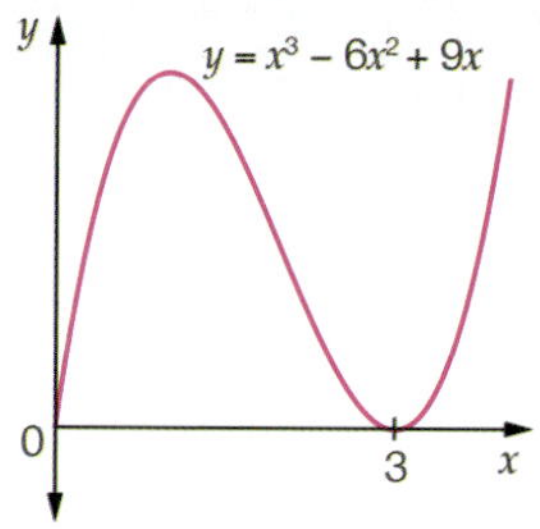

(b) (1, 4), (3, 0) (c) $y = -x(x-3)$

(d) Yes, as (2.5, 1.2) is not on the new parabola.

15 (a) $(x+2)(x-3)(x-4)$

(b)

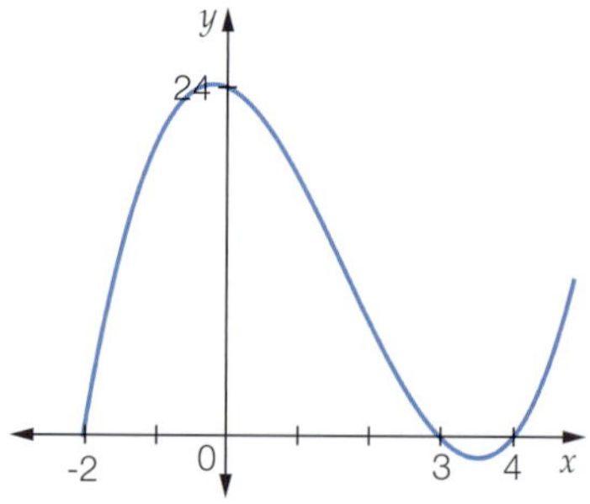

(c) 600 m

Open-ended – Sample answers

16 $P(x) = (x+1)(x-2)(x-2)(x-1)$

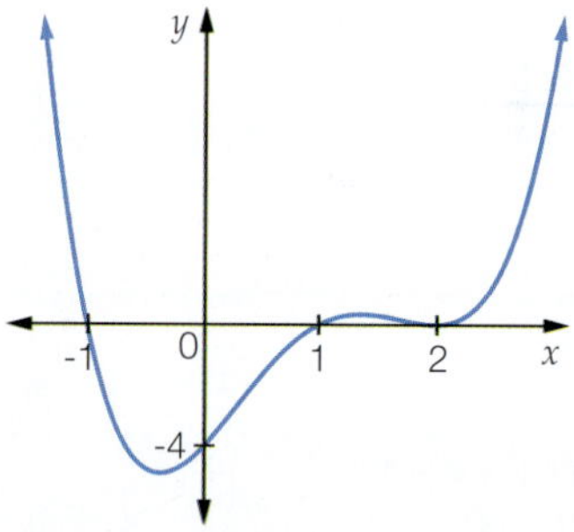

17 (a) $P_1(x) = (x+1)^2(x-3)^2$, $P_2(x) = (x+1)(x-3)^3$

(b) $P_1(x) = (x+1)^2(x-3)(x-4)$, $P_2(x) = (x+1)(x-3)^2(x-5)$

(c) $P_1(x) = x(x+1)(x-3)(x-4)$,
$P_2(x) = (x+1)(x-3)(x-5)(x+6)$

(d) $P_1(x) = 5(x+1)^3(x-3)$, $P_2(x) = 0.5(x+1)(x-3)^3$

Chapter review 8

1 (a)

x	-3	-2	-1	0	1	2	3
y	27	12	3	0	3	12	27

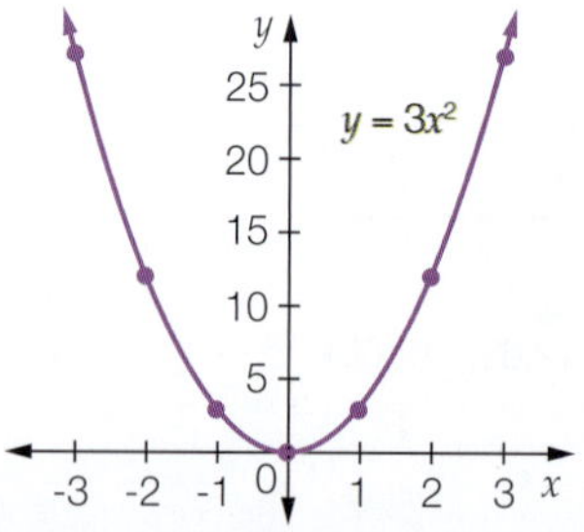

(b)

x	-3	-2	-1	0	1	2	3
y	-13.5	-4	-0.5	0	0.5	4	13.5

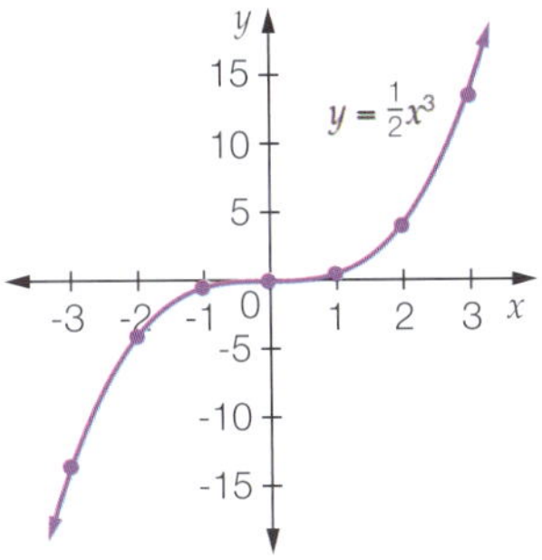

(c)

x	-3	-2	-1	0	1	2	3
y	-81	-16	-1	0	-1	-16	-81

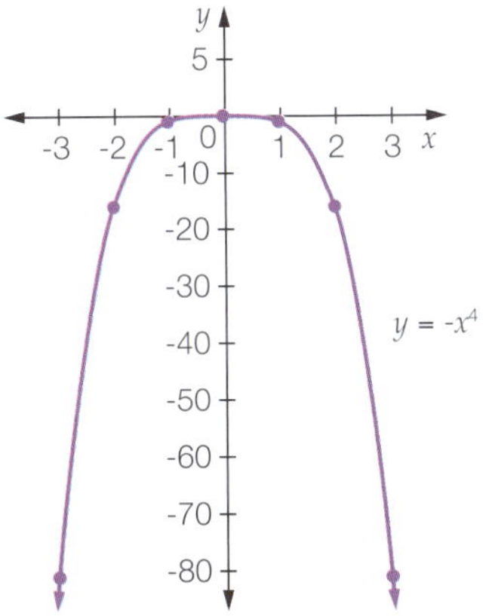

2 (a)

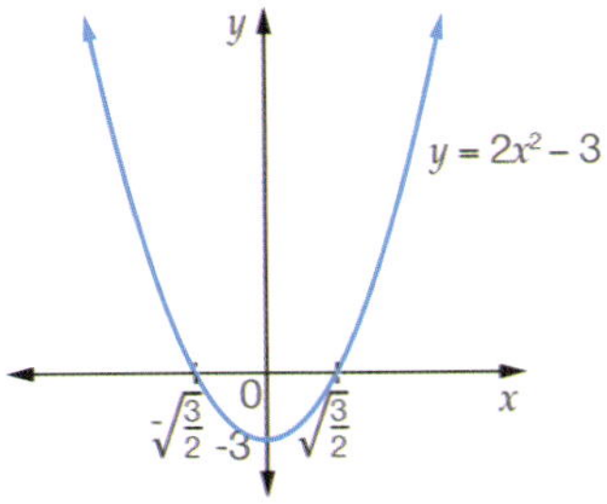

(b)

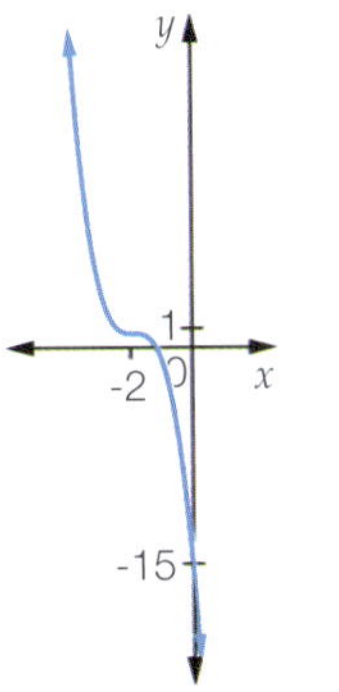

(c)

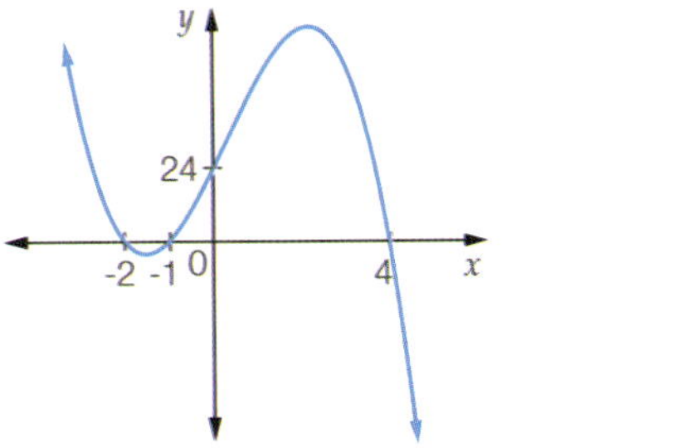

3 (a) (i) 13 (ii) 16 (iii) -42

(b) (i) $9x^2 + 5x$ (ii) $12x^3 - 13x^2 + 11x - 6$

(iii) $3x^2 - 5x + 5$

4 $x^2 + x + 6, 1$

5 (a) -2 (b) $-\frac{5}{16}$

6 (a) $(x + 1)(x - 2)(x + 7)$ (b) $(x + 1)(x - 1)(x + 2)(x - 2)$

7 (a)

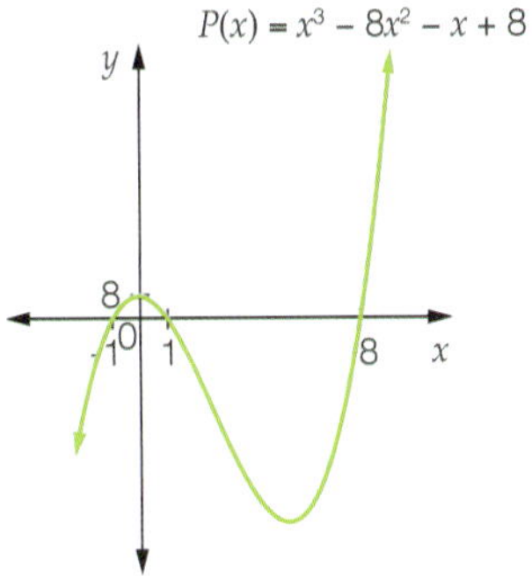

(b)

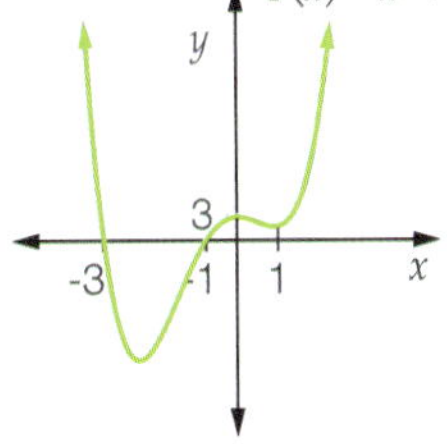

8 (a) $y = -\frac{1}{2}(x + 2)^4 + 3$

(b)

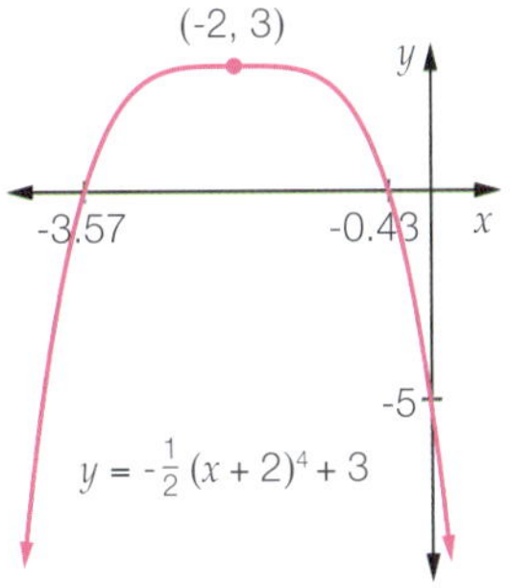

9 remainder = -7

10 (a) $y = 0.001x(x - 60)^2$

(b)

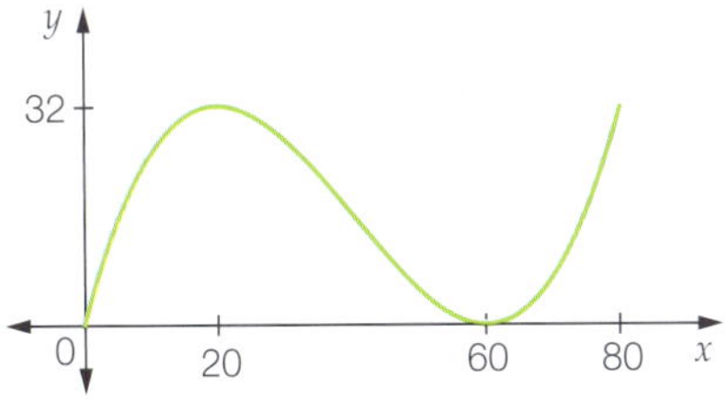

(c) 0.36 min

11 $m^2 - m - 12 = 0, (m - 4)(m + 3) = 0, m = 4$ or $-3, x^2 = 4, x = \pm 2$

12 $\frac{n}{2}(10 + 4(n - 1)) = 860, 2n^2 + 3n - 860 = 0, (2n + 43)(n - 20) = 0.$

$n = 20$

13 (a)

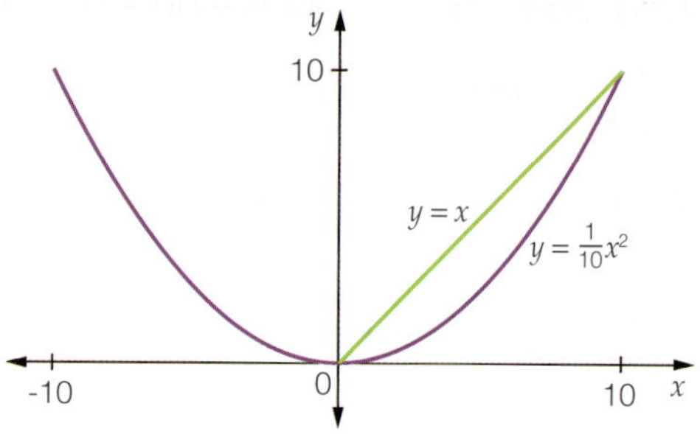

(b) 10 cm

14 (a)

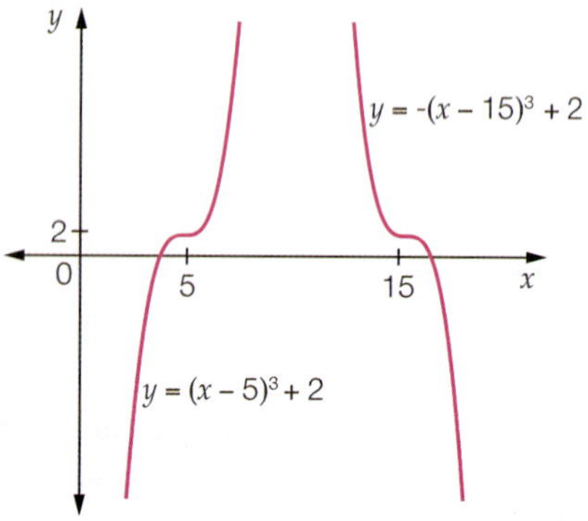

(b) $x = 10$

15 (a) $h = (t - 2)(t - 3)(t - 4)$

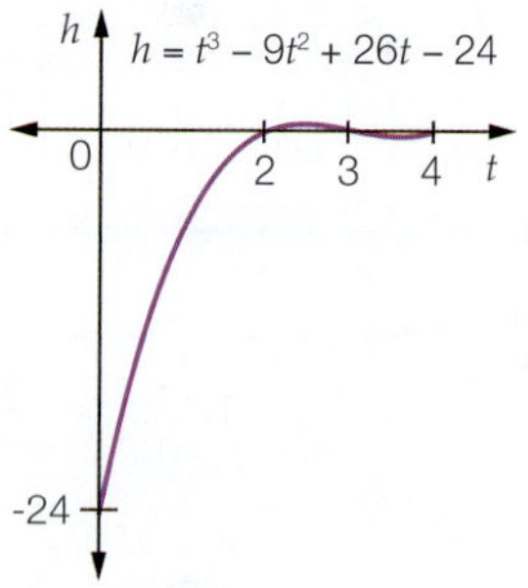

(b) The capsule is released at a depth of 24 m. It breaks the surface after 2 seconds and submerges 1 second later. It stays just below the surface for another second before breaking the surface again.

Chapter 9

Recall 9

1 (a) $a = 23°$ (b) $a = 69°$ (c) $x = 15°$

2 (a) $a = 46°$; $b = 46°$; $c = 46°$

3 (a) equilateral (b) right-angled
(c) scalene (d) isosceles
(e) right-angled and isosceles (f) scalene

4 (a) E (b) B (c) A (d) D (e) C

Exercise 9.1

1 (a) $AB = DE$
$BC = EF$
$AC = DF$

(b) $PQ = HI$
$QR = IJ$
$RP = JH$

(c) $AB = DE$
$BC = EF$
$AC = DF$

(d) $AB = DE$
$CB = CD$
$CA = CE$

(e) $XY = TS = 5$ cm
$ZX = RT = 7$ cm
$ZY = RS = 10$ cm

(f) $AB = CD = 3.86$ units
$AD = CB = 2.79$ units
BD is common to both triangles and is equal to 5.21 units.

2 (a) $XY = PQ$ (both 9 cm), $YZ = QR$ (both 16 cm), $ZX = RP$ (both 8 cm); SSS

(b) $\angle BAC = \angle EFD$ (both 65°), $\angle BCA = \angle FED$ (both 50°), $AC = FE$ (both 10); ASA

(c) $\angle JKL = \angle MPN$ (both 90°), $JL = NM$ (both 5), KJ = PN (both 3); RHS

(d) $XZ = RT$ (both 10), $YZ = ST$ (both 10), $\angle XZY = \angle STR$ (both 40°); SAS

(e) $\angle HGJ = MLK$ (both 56°), $JHG = \angle KML$ (both 44°), $HJ = MK$ (both 6); ASA

(f) $\angle PQR = \angle YXW$ (both 90°), $PR = YW$ (both 8), $RQ = WX$ (both 5); RHS

(g) $\angle CBD = \angle FHG$ (both 28°), $CB = FH$ (both 6), $DB = GH$ (both 10); SAS

(h) $JL = AC$ (both 4), $LK = CB$, $JK = AB$ (both 9); SSS

3 (a) The ratio is 1 : 2. (b) The ratio is 1 : 3.
(c) The ratio is 2 : 1. (d) The ratio is 8 : 5.

4 (a) $x = 14$ (b) $x = 8$
(c) $x = 7.5$ (d) $x = 20$
(e) $x = \frac{50}{3} = 16\frac{2}{3}$ (f) $x = \frac{51}{5} = 10\frac{1}{5}$

5 $A \equiv E$ (SAS), $B \equiv D$ (SSS), $C \equiv F$ (ASA)

6 D 7 C

8 (a) A (b) A

9 (a) A (b) B

10 (a) $x = 10$ (b) $x = 6$ (c) $x = 5\frac{5}{9}$ (d) $x = 26.4$

11 5.25 m 12 316.67 cm 13 7.16 m

14 (a) ΔDFA, ΔBGI, ΔCHJ, ΔDKM, ΔELN, ΔFMO

(b) ΔBKN and ΔCLO; ΔAJG and ΔKNB

15 (a)

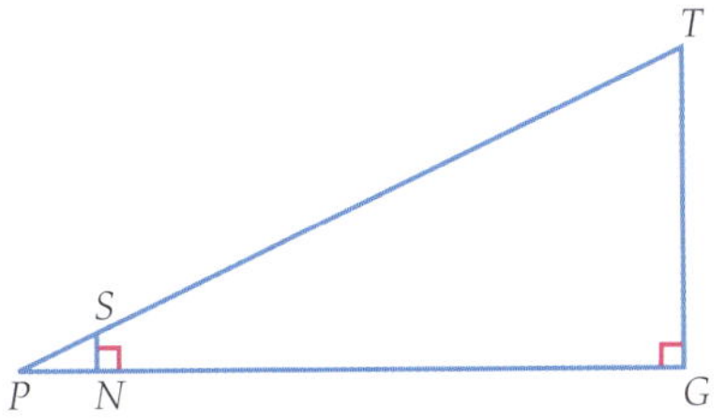

(b) Step 3: Measure SN, PN and NG

$\frac{TG}{PG} = \frac{SN}{PN}$ so $TG = \frac{SN \times PG}{PN}$, where $PG = PN + NG$

(c) Need to measure accurately; can't really get your head level with the ground.

Open-ended – Sample answers

16 AAA does not guarantee any sides are equal in length. If it did, every equilateral triangle would be the same size.

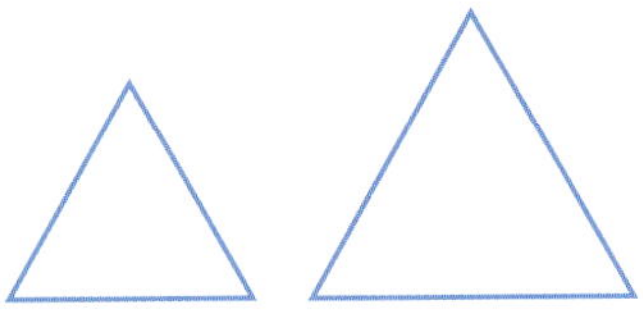

17 Draw any triangle in which $BC = 4$ cm and then measure the other two sides.

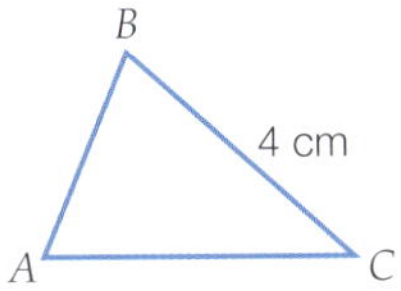

In this triangle $AB = 2.6$ cm and $AC = 5$ cm. So, in ΔDEF, $DE = 2.6 \times 1.5 = 3.9$ cm; $AC = 5 \times 1.5 = 7.5$ cm and $EF = 4 \times 1.5 = 6$ cm

Exercise 9.2

1 (a) $\angle ABC \equiv \angle XYZ$ (SSS) (b) $\angle PQR \equiv \angle FED$ (ASA)

(c) $\angle HIJ \equiv \angle KLM$ (SAS) (d) $\angle XYZ \equiv \angle QPR$ (RHS)

2 (a) $\Delta ABC \equiv \Delta CDA$

$BC = DA$ (given)
$BA = DC$ (given)
$AC = CA$ (same side)
$\Delta ABC \equiv \Delta CDA$ (ASA)
$\angle BAC = \angle DCA$, so $x = 55°$

(b) $\Delta ACD \equiv \Delta ACB$

$\angle ACD = \angle ACB$ (given)
$\angle ADC = \angle ABC$ (given)
$AC = AC$ (same side)
$\Delta ACD \equiv \Delta ACB$ (ASA)
$AD = AB$, so $p = 19$ cm

(c) $\Delta ABE \equiv \Delta CDE$

$AB = CD$ (given)
$\angle ABE = \angle CDE$ (given)
$\angle AEB = \angle CED$ (vertically opposite)
$\Delta ABE \equiv \Delta CDE$ (ASA)
$AE = CE$, so $x = 23$ cm

(d) $\Delta AOB \equiv \Delta COB$

$AO = CO$ (radius of circle)
$\angle OBA = \angle OBC$ (given)
$\angle OAB = \angle OCB$ (isosceles triangle)
$\Delta AOB \equiv \Delta COB$ (ASA)
$\angle OAC = \angle OCB$, so $y = 26°$

(e) $\Delta ABO \equiv \Delta CDO$

$AB = CD$ (given)
$AO = CO$ (radius of circle)
$BO = DO$ (radius of circle)
$\Delta ABO \equiv \Delta CDO$ (SSS)
$AO = CO$, so $y = 9$ m

(f) $\Delta ABE \equiv \Delta DCE$

$\angle AEB = \angle DEC$ (vertically opposite)
$\angle BAE = \angle CDE$ (given)
$AB = DC$ (given)
$\Delta ABE \equiv \Delta CDE$ (ASA)
$AE = ED$, so $p = 6$ cm

3 (a) $\frac{AB}{DE} = \frac{3}{6} = \frac{1}{2}$; $\frac{BC}{EF} = \frac{4}{8} = \frac{1}{2}$; $\frac{AC}{DF} = \frac{5}{10} = \frac{1}{2}$;

$\Delta ABC \sim \Delta DEF$ (SSS)

(b) $\frac{AB}{DB} = \frac{12}{8} = \frac{2}{3}$

$\frac{AC}{DE} = \frac{6}{4} = \frac{2}{3}$

$\angle CAB = \angle EDB$ (corresponding angles equal on parallel lines)
$\Delta ABC \sim \Delta DBE$ (SAS)

(c) $\angle YWX = \angle VWU$ (vertically opposite)
$\angle WYX = \angle UVW$ (alternate angles between parallel lines)
$\angle WXY = \angle WUV$ (alternate angles between parallel lines)
$\Delta WYX \sim \Delta WVU$ (AAA)

(d) $OM = OP$ (given)
$ON = OQ$ (given)
$\angle MON = \angle POQ$ (vertically opposite)
$\Delta MON \sim \Delta POQ$

Congruent triangles are also similar, so $\Delta MON \sim \Delta POQ$.

(e) $\angle RVS = \angle RUT$ (given)
$\angle SRV = \angle TRU$ (same angle)
$\angle RSV = \angle RTU$ (angle sum of the triangle)
$\Delta RSV \sim \Delta RTU$ (AAA)

(f) $\angle ACB = \angle CDB$ (right angles)
$\angle CAB = \angle DCB$ ($\angle CBD$ is common; angle sum of triangles)
$\angle ABC \sim \angle CBD$ (AAA)

4 C

5 B

6 (a) $\angle RUT = \angle RST$ (given—right angle)
$RT = RT$ (same side—hypotenuse)
$UT = ST$ (given)
$\Delta RUT \equiv \Delta RST$ (RHS)

(b) $SP = RQ$ (given)
$\angle SPQ = \angle RQP$ (given)
$PQ = PQ$ (same side)
$\Delta SPQ \sim \Delta RQP$ (SAS)

7 (a)

$\angle QRS = \angle PRT$	(same angle)
$\angle RQS = \angle RPT$	(corresponding angles on parallel lines)
$\angle RSQ = \angle RTP$	(corresponding angles on parallel lines)
$\Delta RQS \sim \Delta RPT$	(AAA)

(b)

$\angle ABC = \angle BDC$	(given—right angle)
$\angle ACB = \angle BCD$	(same angle)
$\angle BAC = \angle DBC$	(third angle in triangle)
$\Delta ABC \sim \Delta BDC$	(AAA)
$\angle ADB = \angle ABC$	(given—right angle)
$\angle BAD = \angle CAB$	(same angle)
$\angle ABD = \angle ACB$	(third angle in triangle)
$\Delta ABC \sim \Delta ADB$	(AAA)
$\therefore, BDC \sim \Delta ADB$	(AAA)

8 (a) (i) not necessarily similar, angles same but sides might not be proportional

(ii) all corresponding sides in same ratio

(iii) not necessarily similar, angles between equal sides might not be equal

(iv) all angles the same (AAA)

(v) all angles the same (AAA)

(vi) radius is only measure so all distances will be in the same ratio

(b) (i) not similar (ii) side lengths the same

(iii) not similar (iv) side lengths the same

(v) one pair of corresponding sides same length

(vi) radius the same

Open-ended – Sample answers

9

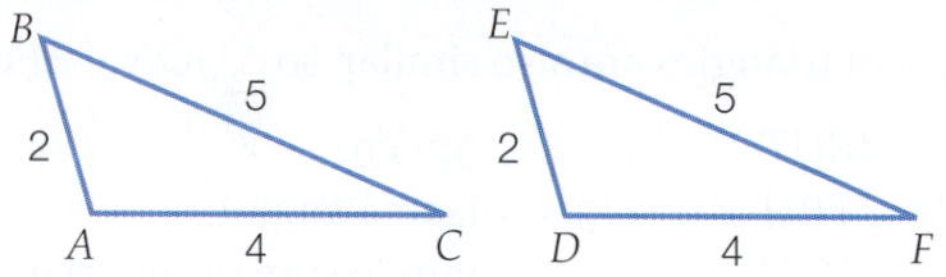

SSS

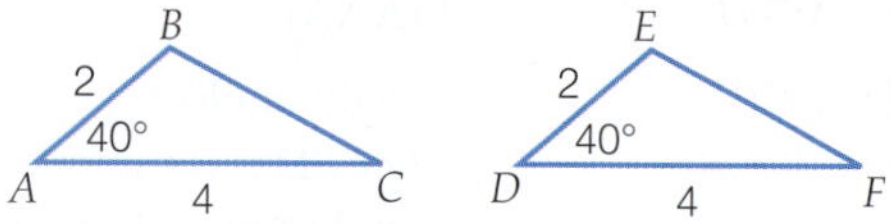

SAS

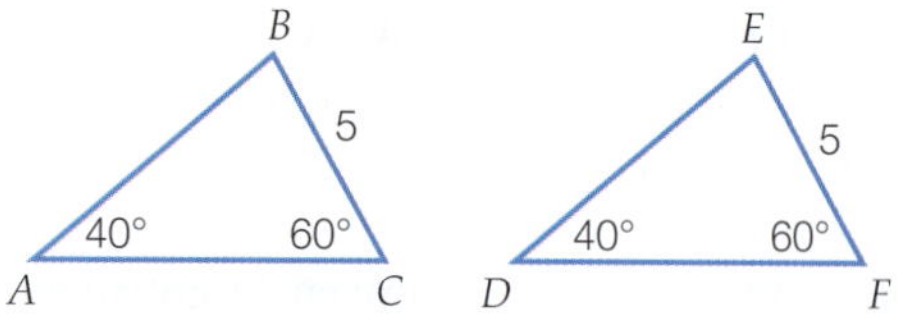

ASA

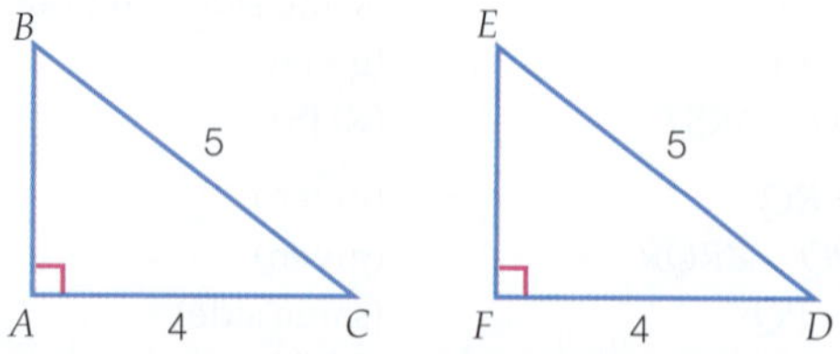

RHS

Exercise 9.3

1 (a)

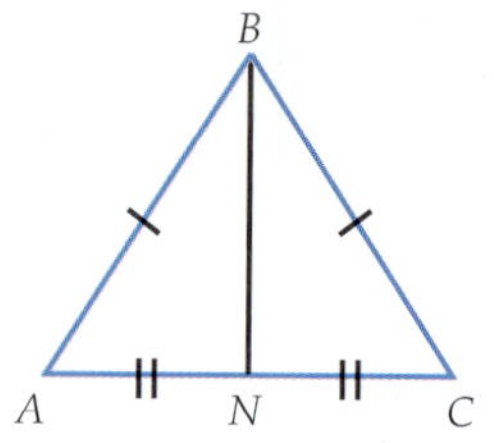

With this construction:

$\Delta ABN \equiv \Delta CBN$	
$AN = CN$	(given by construction)
$BN = BN$	(same side)
$AB = CB$	(given)
$\Delta ABN \equiv \Delta CBN$	(SSS)
So, $\angle BAC = \angle BCA$	

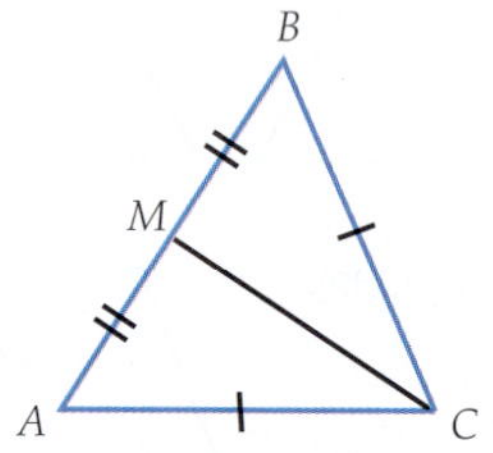

With this construction:

$\Delta BCM \equiv \Delta ACM$	
$AM = BM$	(given by construction)
$CM = CM$	(same side)
$CB = CA$	(given)
$\Delta BCM \equiv \Delta ACM$	(SSS)
So, $\angle CBA = \angle CAB$	

But, $\angle CAB = \angle BCA$ from the first construction, so $\angle CBA = \angle CAB = \angle BCA$

(b) All angles are the same (60°)

2 (a)

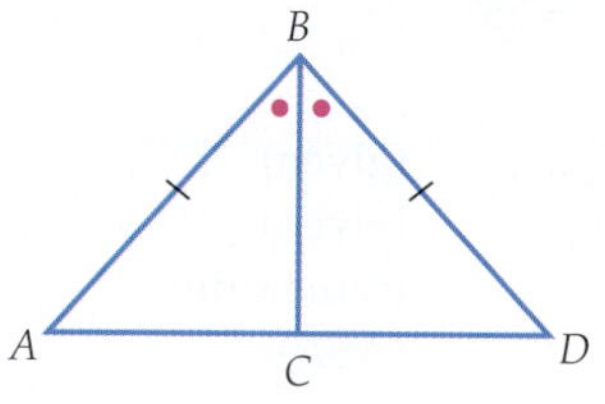

To show: $AC = CD$

$\angle BAC = \angle BDC$	(isosceles triangle)
$\angle ABC = \angle DBC$	(by construction)
$BC = BC$	
So, $\Delta BAC \equiv \Delta BDC$	(ASA)
$\therefore$ so $AC = DC$	(corresponding sides of congruent triangles)

(b)

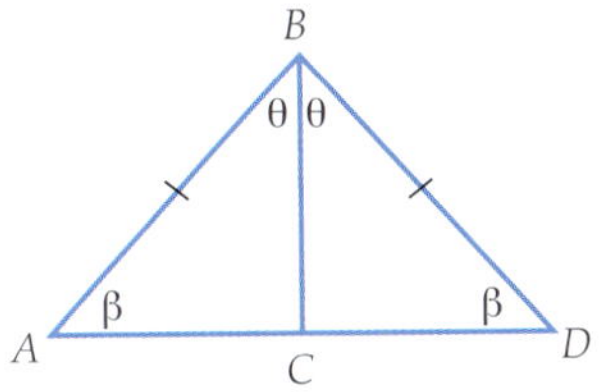

To show: $\angle BCA = \angle BCD = 90°$

$2 \times \theta + 2 \times \beta = 180$ (angle sum of triangle)

$\theta + \beta = 90°$

So, $\angle BCA = 180 - (\theta + \beta) = 90°$

3 $\angle CDE + \angle EDF = 180°$ (straight angle)

$\angle EFG + \angle EFD = 180°$ (straight angle)

$\angle EDF = \angle EFD$ (equal angles of isosceles triangle)

$\angle CDE = 180 - \angle EDF$

$\angle EFG = 180 - \angle EFD$

$= 180 - \angle EDF$

$= \angle CDE$

4 A

5 B

6 $AC^2 = AB^2 + BC^2$

$BC^2 = BD^2 + DC^2$ (Pythagoras)

So, $AC^2 = AB^2 + BD^2 + DC^2$ [1]

Also, $AB^2 = DB^2 + AD^2$ (Pythagoras)

Substituting into [1]:

$AC^2 = DB^2 + AD^2 + BD^2 + DC^2$

$= DC^2 + 2BD^2 + AD^2$

7 $\Delta ABF \equiv \Delta BCD$

$AB = BC$ (given)

$\angle FAB = \angle DBC$ (corresponding angles, $FA \parallel BD$)

$\angle FBA = \angle DCB$ (corresponding angles, $FB \parallel DC$)

$\Delta ABF \equiv \Delta BCD$ (ASA)

8 To show: $\angle RPT = \angle RTP$

$\angle QPU = \angle STU$ as both these angles are the third angle in a triangle where the other two angles are the same.

So in ΔRPT, $RP = RT$ (opposite equal sides)

$\therefore \Delta RPT$ is isosceles.

9

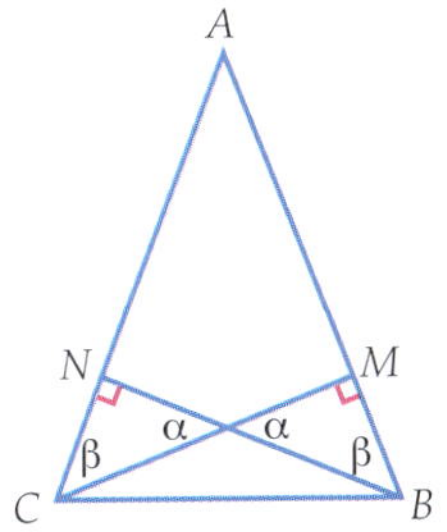

$\angle CNB = \angle BMC$ (90°, given)

$\angle NCB = \angle MBC$ (angles opposite equal sides)

ΔCNB and ΔBMC have a common side, BC

So, $\Delta CNB \equiv \Delta BMC$ (ASA)

So, $NC = MB$ (corresponding sides)

$AC = AN + NC$

$AB = AM + MB$

$AB = AC$ (given)

So $AN = AM$ (equal sides AC and AB; NC and NB)

10 Let $EC = DB = x$

$\therefore ED = x + DC = BC$

In ΔEFD and ΔBAC

$ED = BC$ (above)

$\angle EFD = \angle BAC$ (given)

$\angle FED = \angle ABC$ (given)

$\therefore \Delta ABC \equiv \Delta FED$ (ASA)

11 Let $BF = EC = x$

$\therefore BC = x + FC = FE$

In ΔABC and ΔDEF

$BC = EF$ (above)

$\angle ABC = \angle DEF$ (given)

$\angle ACB = \angle DFE$ (given)

$\therefore \Delta ABC \equiv \Delta DEF$ (ASA)

12 (a) $\Delta BAE \equiv \Delta BCD$

$\angle BAD = \angle BCE$ (given)

$AE = CD$ ($AD = EC$ and DE is common)

$BA = BC$ (given)

$\Delta BAE \equiv \Delta BCD$ (SAS)

(b) $\Delta ABE \equiv \Delta CBD$

$\angle BAE = \angle BCD$ (given)

$\angle ABE = \angle CBD$ ($\angle ABD = \angle CBE$ and $\angle EBD$ is common)

$AB = BC$ (isosceles triangle)

$\Delta ABE \equiv \Delta CBD$ (ASA)

(c) $\Delta ABD \equiv \Delta CBE$

$AD = CE$ ($AD = AE - DE$ and $CE = CD - DE$; DE common)

$\angle BAD = \angle BCE$ (given)

$AB = CB$ (isosceles triangle)

$\Delta ABD \equiv \Delta CBE$ (SAS)

(d) $\Delta ABD \equiv \Delta CBE$

$AB = CB$ (given)

$AD = CE$ ($AD = AE - DE$ and $CE = CD - DE$; DE common)

$\angle BAD = \angle BCE$ (isosceles triangle)

$\Delta ABD \equiv \Delta CBE$ (SAS)

13 $\Delta ADF \equiv \Delta AEF$

$DF = EF$ (given)

$\angle ADF = \angle AEF$ (both 90° as $FD \perp AB$ and $FE \perp AC$)

$AF = AF$ (common side)

$\Delta ADF \equiv \Delta AEF$ (RHS)

14 $\Delta ABC \equiv \Delta CDE$

$\angle ABC = \angle CDE$ (given—right angle)

$AC = CE$ (given C is the midpoint of AE hypotenuse)

$AB = CD$ (given)

$\Delta ABC \equiv \Delta CDE$ (RHS)

15 $\angle OEF = \angle OFE$ (isosceles triangle; call it α)

$\angle OED = \angle ODE$ (isosceles triangle; call it β)

$\angle EDO + \angle OFE + \angle FED = 180°$ (angle sum of triangle)

So, $\beta + \alpha + (\alpha + \beta) = 180°$

$2(\alpha + \beta) = 180°$

$\alpha + \beta = 90°$

As $\angle DFE = \alpha + \beta = 90°$ the angle in a semicircle is a right angle.

Open-ended – Sample answers

16 (a) $\angle ABC = \angle ACB$ (opposite equal angles)

$\angle AXY = \angle ABC$ (corresponding angles $XZ \parallel BC$)

$\angle AZY = \angle ACB$ (corresponding angles $XZ \parallel BC$)

$\therefore \angle AXY = \angle AZY$

$\therefore AX = AZ$ (sides opposite equal angles)

(b) $\Delta AXY \equiv \Delta AZY$

$AX = AZ$ (from **(a)**)

$XY = ZY$ (given)

$\angle AXY = \angle AZX$ (from **(a)**)

$\Delta AXY \equiv \Delta AZY$ (SAS)

(c) $\Delta BXY \equiv \Delta CZY$

$BX = CZ$ ($AB = AC$ and $AX = AZ$)

$XY = ZY$ (given)

$\angle BXY = \angle CZY$ (Each $(180 - b)°$)

$\Delta BXY \equiv \Delta CZY$ (SAS)

(d) $a + b + 90 = 180$ (angle sum in triangle)

$b = 90 - a$

$\angle ACY = c$ and $\angle CAY = a$ but these are equal since ΔACY is isosceles.

So, $c = a$

$c + d = b$ (corresponding angles on parallel lines)

$a + d = 90 - a$

$d = 90 - 2a$

17

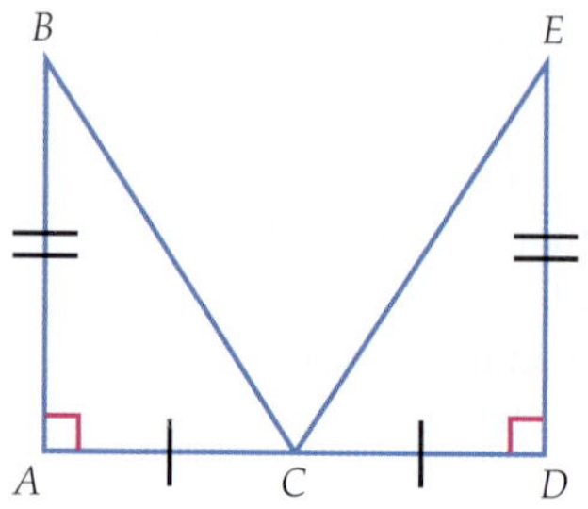

$\Delta ABC \equiv \Delta DEC$

Exercise 9.4

1 (a) To prove: $\Delta ABC \sim \Delta ADE$

$\angle ADE = \angle ABC$ (corresponding angles, $DE \parallel BC$)

$\angle DAE = \angle BAC$ (same angle)

$\angle AED = \angle ACB$ (corresponding angles, $DE \parallel BC$)

$\Delta ABC \sim \Delta ADE$ (AAA)

(b) To prove: $AE = EC$

$\Delta ABC \sim \Delta ADE$ (from **(a)**)

So, $\dfrac{AD}{AB} = \dfrac{AE}{AC}$

So, $\dfrac{AE}{AC} = \dfrac{1}{2}$

So, $2AE = AC$

So, $AE = EC$

2 To prove: $\Delta ADB \sim \Delta DCA$

The orientation of the two triangles must be as shown below as AD cannot be in the same position in both triangles.

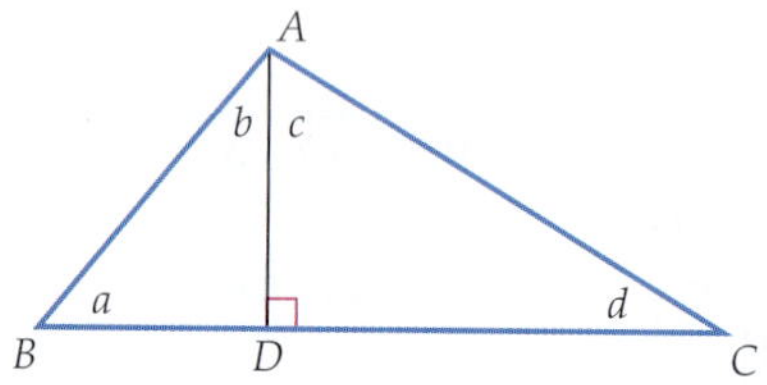

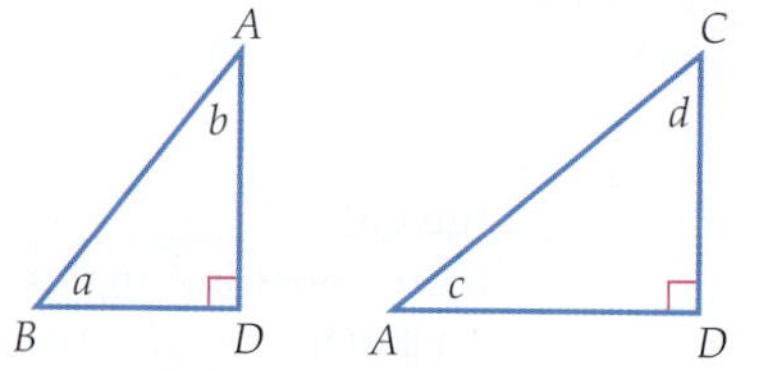

$a + b = 90°$ (rest of angle sum of 180° for triangle)

$c + d = 90°$ (rest of angle sum of 180° for triangle)

$b + c = 90°$ (right angle in original triangle)

$a + b = b + c$, so $a = c$

$b + c = c + d$, so $b = d$

$\Delta ADB \sim \Delta DCA$ (AAA)

To prove: $\Delta DCA \sim \Delta CAB$

$\angle ADC = \angle CAB$ (both 90°)

$\angle ACD = \angle ACB$ (common angle)

$\angle ABD = \angle ABC$ (common angle)

$\Delta DCA \sim \Delta CAB$ (AAA)

3 To prove: $AD^2 = BD \times DC$

$\Delta ADB \sim \Delta DCA$ (from Question **2**)

So, $\dfrac{AD}{DC} = \dfrac{BD}{AD}$ (corresponding sides of similar triangles)

So, $AD \times AD = BD \times DC$

$AD^2 = BD \times DC$

4 To prove: $BA^2 = BD \times BC$ and $CA^2 = CD \times CB$

$\Delta ADB \sim \Delta CAB$ (from Question **2**)

$\frac{BA}{BC} = \frac{BD}{BA}$ (corresponding sides of similar triangles)

So, $BA^2 = BC \times BD$

To prove: $CA^2 = CD \times CB$

$\Delta ADC \sim \Delta BAC$ (from Question **2**)

$\frac{CA}{BC} = \frac{CD}{AC}$ (corresponding sides of similar triangles)

$CA^2 = CD \times BC$

5 $AB^2 = BC \times BD$ (from Question **4**)

$AC^2 = CD \times BC$ (from Question **4**)

$$\begin{aligned} AB^2 + AC^2 &= BC \times BD + CD \times BC \\ &= BC(BD + CD) \\ &= BC \times BC \\ &= BC^2 \end{aligned}$$

6 C

7 (a) To show: $\Delta ABO \sim \Delta DCO$

$\angle ABD = \angle BDC$ (alternate angles, $AB \parallel DC$)

$\angle BAC = \angle ACD$ (alternate angles, $AB \parallel DC$)

$\angle AOB = \angle COD$ (vertically opposite angles)

$\Delta ABO \sim \Delta DCO$ (AAA)

Now, $\frac{OA}{OC} = \frac{OB}{OD} = \frac{AB}{DC}$ (corresponding sides of similar triangles)

So, $OA : OC = OB : OD$

(b) $\frac{AB}{CD} = \frac{1}{2}$ (given)

So, $\frac{AO}{OC} = \frac{1}{2}$ (corresponding sides of similar triangles are in same ratio)

$2AO = OC$

Similarly, $2OB = OD$

$$\begin{aligned} AC &= AO + OC \\ &= AO + 2AO \\ &= 3AO \end{aligned}$$

So, $AO : AC = 1 : 3$ and similarly $BO : BD = 1 : 3$

8 To prove: $\frac{JM}{MK} = \frac{JN}{NL}$

Show that: $\Delta JMN \sim \Delta JKL$

$\angle MJN = \angle KJL$ (common angle)

$\angle JMN = \angle JKL$ (corresponding angles, $MN \parallel KL$)

$\angle JNM = \angle JLK$ (corresponding angles, $MN \parallel KL$)

$\Delta JMN \sim \angle JKL$ (AAA)

So, $\frac{JK}{JM} = \frac{JL}{JN}$ (corresponding sides of similar triangles are proportional)

$\frac{JK}{JM} - 1 = \frac{JL}{JN} - 1$

$\frac{JK - JM}{JM} = \frac{JL - JN}{JN}$

$\frac{MK}{JM} = \frac{NL}{JN}$ ($JK = JM + MK$)

($JL = JN + NL$)

So, $\frac{JM}{MK} = \frac{JN}{NL}$ (reciprocal)

9 To prove: $GH \parallel EF$

Show that: $\Delta DGH \sim \Delta DEF$

$\angle GDH = \angle EDF$ (common angle)

$\frac{DG}{GE} = \frac{DH}{HF}$ (given)

$\frac{GE}{DG} = \frac{HF}{DH}$ (reciprocal)

$\frac{GE}{DG} + 1 = \frac{HF}{DH} + 1$

$\frac{GE + DG}{DG} = \frac{HF + DH}{DH}$

$\frac{DE}{DG} = \frac{DF}{DH}$

Corresponding sides are in the same proportion so

$\Delta DGH \sim \Delta DEF$ (SAS)

So, $\angle DGH = \angle DEF$ (corresponding angles in similar triangles)

And $\angle DHG = \angle DFE$ (corresponding angles in similar triangles)

So, $GH \parallel EF$ (corresponding angles equal on parallel lines)

10 Extend lines AD and BC to point G as shown here.

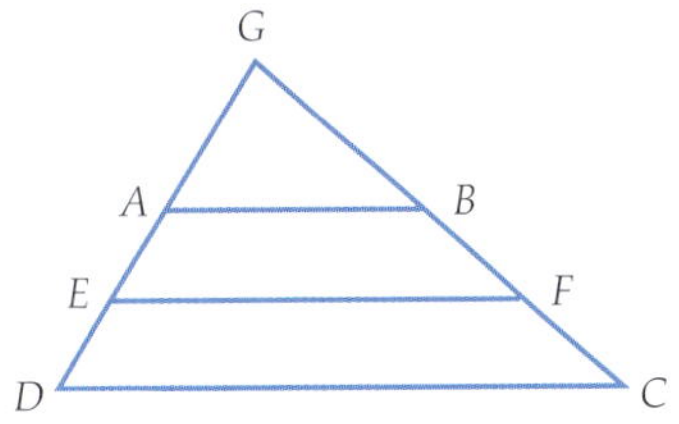

In ΔGAB and ΔGDC

$\angle AGB = \angle DGC$ (common angle)

$\angle GAB = \angle GDC$ (corresponding angles, $AB \parallel DC$)

$\therefore \Delta GAB \sim \Delta GDC$ (AAA)

$\therefore$ corresponding sides are proportional

Let $GA = x$, $AE = y$, $ED = y$

$\therefore GB = mx$, $BC = 2my$, $BF = my$, $FC = my$

$\frac{GA}{GE} = \frac{x}{x + y}$

$\frac{GB}{GF} = \frac{mx}{mx + my} = \frac{x}{x + y}$

So the corresponding sides are in the same ratio.

$\therefore \Delta GAB \sim \Delta GEF$ (SAS)

$\therefore \angle GAB = \angle GEF$ (corresponding angles)

However, these angles are equal corresponding angles

$\therefore AB \parallel EF$

11 To prove: $\Delta BXY \sim \Delta BYC$
Hint: Prove $\Delta BXY \equiv \Delta CZY$

$\Delta BXY \equiv \Delta CZY$
$BX = CZ$
$(AB = AC$ and $AX = AZ)$
$XY = ZY$ (given)
$\angle BXY = \angle CZY$ (Each $(180 - b)°$)
$\Delta BXY \equiv \Delta CZY$ (SAS)
$BY = CY$ (congruent triangles)
$\frac{BX}{BY} = \frac{XY}{YC}$ (as $BX = XY$ and $BY = YC$)
Need to show: $\angle BXY = \angle BYC$
Now, $\angle XBY = \angle XYB$ (isosceles triangle) and call it b
So, $\angle BXY = (180 - 2b)°$
But, $\angle ZYC = b$ (congruent triangle)
And, $\angle XYB + \angle BYC + \angle ZYC = 180°$ (straight angle)
$b + \angle BYC + b = 180°$
$\angle BYC = (180 - 2b)°$
So, $\angle BXY = \angle BYC$
So, $\Delta BXY \sim \Delta BYC$ (SAS)

Open-ended – Sample answers

12

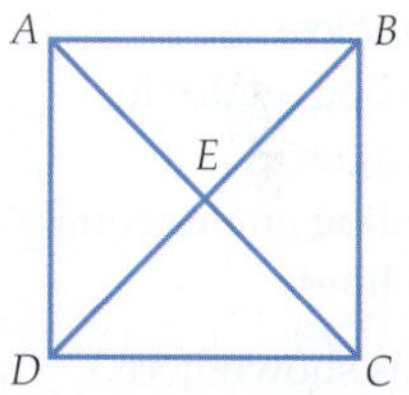

To prove: $\Delta AED \sim \Delta ADC$
The diagonals of a square bisect each other at right angles
So, $\frac{AE}{AD} = \frac{ED}{DC}$ since $AE = ED$ and $AD = DC$
$\angle AED = \angle ADC$ (both right angles)
So, $\Delta AED \sim \Delta ADC$ (SAS)

Half-time 9

1 (a) To prove: $\Delta ABC \equiv \Delta EDC$
$\angle ABC = \angle CDE$ (both right angles)
$\angle BCA = \angle ECD$ (vertically opposite)
So, $\Delta ABC \sim \Delta EDC$ (AAA)
$BC = DC$ (ASA)
$\therefore \Delta ABC \equiv \Delta EDC$

(b) AB and ED are corresponding sides in congruent triangles so must be equal.

2 To prove: $\Delta ABC \sim \Delta APQ$
$\angle ABC = \angle APQ$ (given)
$\angle CAB = \angle QAP$ (common angle)
$\angle AQP = \angle ACB$ (third angle in triangle)
$\Delta ABC \sim \Delta APQ$ (AAA)

3 (a) RHS (b) SAS

4 (a) $OA = OD$ (radii)
$OB = OC$ (radii)
$AB = CD$ (given)
So, $\Delta AOB \equiv \Delta DOC$ (SSS)
$p = 15$

(b) In ΔURT and ΔSTR
$RT = RT$ (common side)
$\angle URT = \angle STR$ (alternate angles, $RU \parallel TS$)
$\angle RTU = \angle SRT$ (alternate angles, $RS \parallel UT$)
$\therefore \Delta URT \equiv \Delta STR$ (ASA)
$\therefore n = 11$ cm (corresponding sides in congruent triangles)

5 (a) $x = 12$ (b) $x = 15$

6 To prove: $\Delta LYZ \sim \Delta MXZ$
$\angle LYZ = \angle MXZ$ (both right angles)
$\angle YZL = \angle XZM$ (vertically opposite)
$\angle YLZ = \angle XMZ$ (third angle in triangle)
$\Delta LYZ \sim \Delta MXZ$ (AAA)

7 In ΔACX and ΔBCX
$AC = BC$ (given)
$AX = BX$ (given)
$CX = CX$ (common side)
$\therefore \Delta ACX \equiv \Delta BCX$ (SSS)

8 2.4 cm

9 To prove: $\Delta ADF \equiv \Delta AEF$
$\angle ADF = \angle AEF$ (both right angles)
$AF = AF$ (common side)
$DF = EF$ (given)
$\Delta ADF \equiv \Delta AEF$ (RHS)

10 To prove: $XY = XZ$
$\angle YAX = \angle ZBX$ (isosceles ΔABC)
$YA = ZB$ (halves of equal sides)
$XA = XB$ (given)
$\Delta YAX \equiv \Delta ZBX$ (SAS)
XY and XZ are corresponding sides in congruent triangles, so $XY = XZ$.

11 To prove: $\Delta ABC \equiv \Delta DEC$
$\angle ACB = \angle ECD$ (vertically opposite)
$\angle CBA = \angle CDE$ (alternate angles, $AB \parallel DE$)
$BC = CD$ (C is the midpoint of DB)
$\therefore \Delta ABC \equiv \Delta DEC$ (ASA)

12 Sample answer:

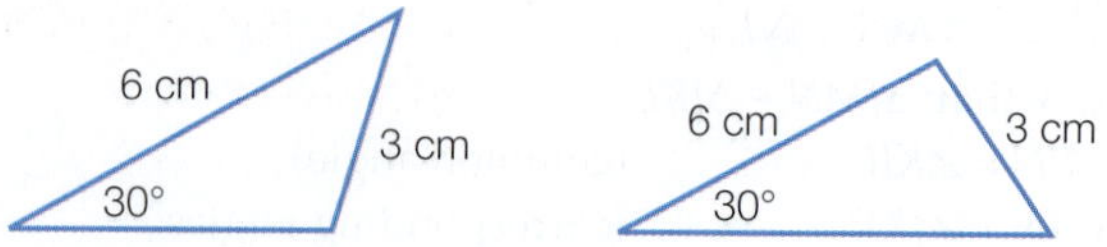

13 5 m

Exercise 9.5

1 (a) (i) Opposite sides are parallel and one pair of adjacent sides is equal
(ii) rhombus

(b) (i) Two pairs of adjacent sides equal.

(ii) kite

(c) (i) Two pairs of opposite sides parallel; one pair of adjacent sides equal; one right angle.

(ii) square

2 (a) To prove: $\Delta DAB \equiv \Delta DCB$

$\angle DBA = \angle BDC$	(alternate angles, $AB \parallel CD$)
$\angle BDA = \angle CBD$	(alternate angles, $AD \parallel BC$)
$DB = DB$	(common side)
$\Delta DAB \equiv \Delta DCB$	(ASA)

(b) $\angle DAB = \angle BCD$ (corresponding angles, congruent triangles from (a))

$\angle ABC$ is supplement of $\angle DAB$ (co-interior angles, $AB \parallel CD$)

$\angle ADC$ is supplement of $\angle DCB$ (co-interior angles, $AD \parallel BC$)

$\therefore \angle ABC = \angle ADC$ (supplements of equal angles)

(c) Opposite angles in a parallelogram are equal

(d) To prove: $\Delta DAE \equiv \Delta BCE$

$AD = BC$	(corresponding sides in congruent triangles)
$\angle AED = \angle BEC$	(vertically opposite)
$\angle ADE = \angle CBE$	(alternate angles, $AD \parallel BC$)
$\Delta DAE \equiv \Delta BCE$	(ASA)

(e) These are corresponding sides in congruent triangles, so $AE = CE$ and $DE = BE$

(f) The diagonals of a parallelogram bisect each other

3 $AC = DB$ and bisect each other

$\therefore AX = XC = DX = XB$

$AC \perp BD$ $\therefore$ all angles at centre are 90°

The four triangles are congruent (SAS)

$\therefore$ all hypotenuses are equal (corresponding sides)

acute angles of each triangle are 45° (isosceles triangle, angle sum of triangle)

$\therefore \angle ABC = \angle BCD = \angle CDA = \angle DAB = 90°$

$\therefore$ quadrilateral is a square (all sides equal, angles are right angles)

4 (a) To prove: $\angle KLM = \angle KNM$

Show that: $\Delta KLM \equiv \Delta KNM$

$KL = KN$	(given)
$LM = NM$	(given)
$KM = KM$	(common side)
$\Delta KLM \equiv \Delta KNM$	(SSS)
$\angle KLM = \angle KNM$	(corresponding angles in congruent triangles)

(b) To prove: $\angle KOL = 90°$

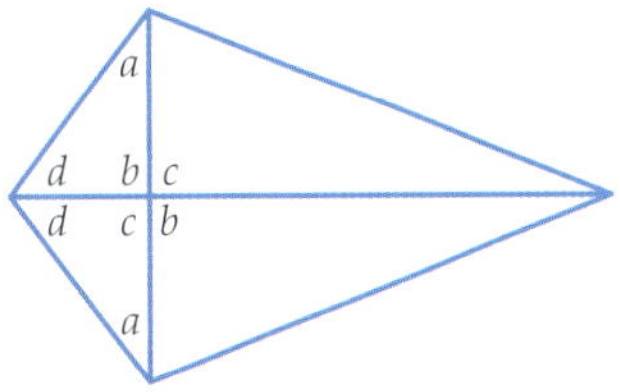

The angles marked a are equal as they are in an isosceles triangle.

The angles marked d are equal because they are corresponding angles in congruent triangles.

The angles marked b and c are pairs of vertically opposite angles.

So, $a + b = a + c$	(supplements of d)
So, $b = c$	
But, $b + c = 180°$	(straight angle)
So, $b = c = 90°$	
So, $\angle KOL = 90°$	

5 B

6 C

7 (a) To prove: $PQ = RS$

Show that: $\Delta PQS \equiv \Delta RSQ$

$SQ = SQ$	(common side)
$\angle PSQ = \angle RQS$	(alternate angles, $PS \parallel RQ$)
$\angle PQS = \angle QSR$	(alternate angles, $PQ \parallel RS$)
$\Delta PQS \equiv \Delta RSQ$	(ASA)
So, $PQ = RS$	(corresponding sides on congruent triangles)

(b) Opposite sides of a rectangle are equal.

8 (a) To show: $\Delta ACD \equiv \Delta ACB$

$AD = BC$	(given)
$DC = AB$	(given)
$AC = AC$	(given)
$\Delta ACD \equiv \Delta ACB$	(SSS)
$\angle BAC = \angle BCA = 45°$	(isosceles triangles)
$\angle DAC = \angle DCA = 45°$	(isosceles triangles)

$\angle BAD = \angle BAC + \angle DAC$

$= 45° + 45°$

$= 90°$

So, $\angle BAD = 90°$, so this is a square.

(b) To show: $\Delta ABD \equiv \Delta CBD$

$AB = BC$	(given)
$AD = DC$	(given)
$DB = DB$	(given)
$\Delta ABD \equiv \Delta CBD$	(SSS)

So, $\angle ABD = \angle BDC$ as these are corresponding angles in congruent triangles. But, these are alternate angles, so the lines must be parallel.

Then, do the same for the other pair of sides.

9 **(a)** To show: $\Delta ABD \equiv \Delta BDC$

$AD = BC$	(given)
$\angle ADB = \angle DBC$	(alternate angles on parallel lines)
$\angle ABD = \angle BDC$	(alternate angles on parallel lines)
$\Delta ABD \equiv \Delta BDC$	(ASA)
So, $AB = DC$	(corresponding sides in congruent triangles)
And, $\angle DAB = \angle DCB$	(corresponding angles in congruent triangles)

So the quadrilateral is a parallelogram.

(b) To show: $\Delta AED \equiv \Delta CEB$

$AE = CE$	(given)
$DE = BE$	(given)
$\angle AED = \angle BEC$	(vertically opposite)
$\Delta AED \equiv \Delta CEB$	(SAS)
So, $AD = BC$	(corresponding sides in congruent triangles)

In a similar way show $\Delta AEB \equiv \Delta DEC$ and so $AB = DC$.

Two pairs of opposite sides are equal and parallel, so the quadrilateral is a parallelogram.

(c) Consider the following diagram:

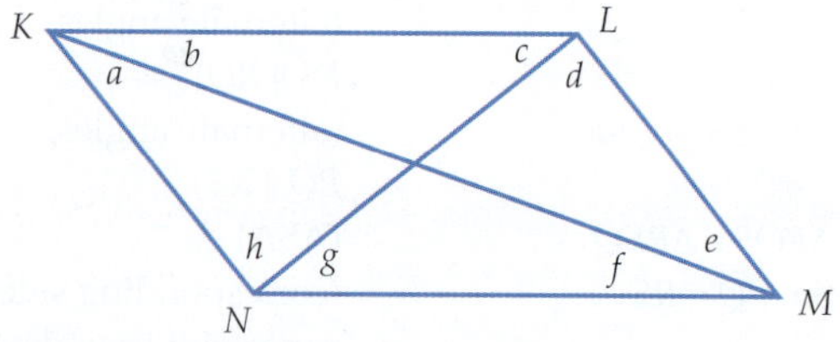

A number of relationships can be established from the diagram.

$a + b + c + h = 180°$	(angles in triangle)
$e + f + d + g = 180°$	(angles in triangle)
but $e + f = a + b$	(given)
So, $a + b + c + h = e + f + d + g$	
becomes $a + b + c + h = a + b + d + g$	
so, $c + h = d + g$	
but $c + d = h + g$	(given)
so, $d = h$ and $g = c$	

These are alternate angles so $KL \parallel NM$.

Then do a similar thing to show $KN \parallel LM$.

(d) Consider this diagram:

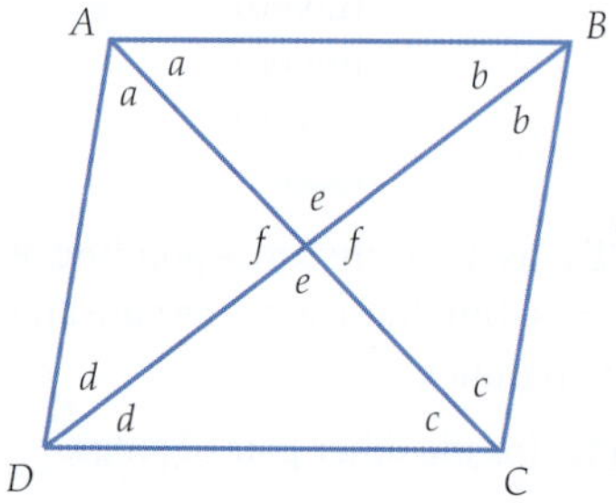

$2a + b + d = 2c + b + d$	(angles in triangle)
so, $a = c$	
$2b + a + c = 2d + a + c$	(angles in triangle)
so, $b = d$	

similarly $a = c$

Redraw the diagram:

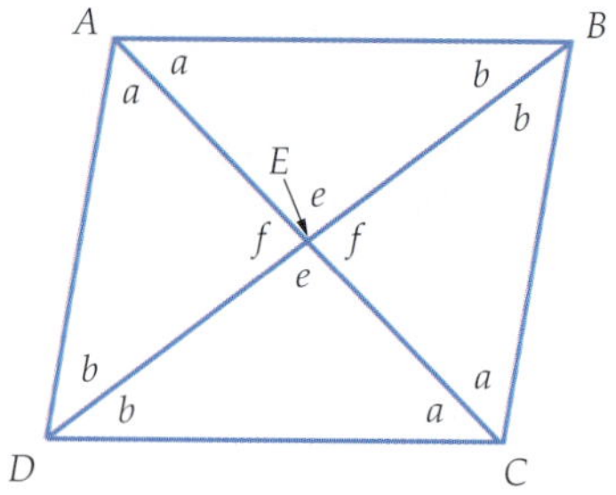

$a + b + e = a + b + f$	(angles in triangle)
so, $e = f$	
$AD = AB$	(opposite equal angles)
$BC = DC$	(opposite equal angles)
$AD = DC$	(opposite equal angles)
$\therefore AB = BC = CD = DA$	
$\therefore ABCD$ is a rhombus	(all sides equal)

10 A

11 To show: $\Delta WMZ \equiv \Delta YNZ$

$ZM = NZ$	(given)
$WZ = YZ$	(given)
$\angle ZWM = \angle ZYN$	(both 90°, given)
$\Delta WMZ \equiv \Delta YNZ$	(RHS)

12 The words link in the following way:

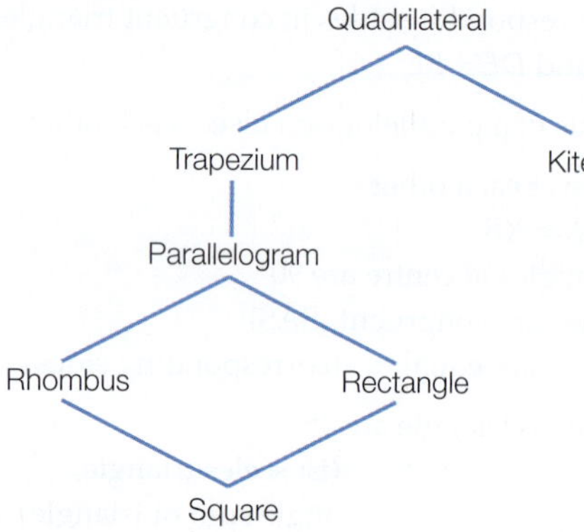

13 To show: $\Delta AKD \equiv \Delta CJB$

As $AJ = CK$ and $AB = DC$ (opposite sides of parallelogram)

then $JB = DK$

So, $JB = DK$

$AD = BC$	(parallelogram)
$\angle ADK = \angle JBC$	(opposite angles in parallelogram)
$\Delta AKD \equiv \Delta CJB$	(SAS)

14 A square has four right angles and two pairs of opposite sides equal and parallel with all sides the same length, while a rhombus has two pairs of opposite sides equal and parallel with all sides the same length. So, the square has an extra condition.

15 A square has four right angles and two pairs of opposite sides equal and parallel with all sides the same length, whereas a rectangle has four right angles and two pairs of opposite sides equal and parallel. So, the square has an extra condition.

Open-ended – Sample answers

16 For the condition *at least one pair of parallel sides* the shape could be: trapezium, square, rectangle, rhombus or parallelogram

For the condition *at least one pair of equal sides* the shape could be: square, rectangle, parallelogram, rhombus or kite

For the condition *at least one pair of equal angles* the shape could be: square, rectangle, rhombus, parallelogram or kite

17 (a) A rhombus has four equal sides but is not a square, because it has no right angle.

(b) Every square has four right angles. Every shape with four right angles is a square. False—it could be a rectangle.

Every kite has two pairs of adjacent sides equal. Every shape with two pairs of adjacent sides equal is a kite. False—it could be a square or a rhombus.

18 (a) Students' own diagrams.

(b) Students to draw and measure from their own diagrams.

(c) The diagonals of a rhombus intersect at right angles.

(d) A demonstration involving measurement is never sufficient as a proof.

19 A square, rhombus and kite all have diagonals that bisect at right angles, so George is correct.

Exercise 9.6

1 Join GO and extend it to H.

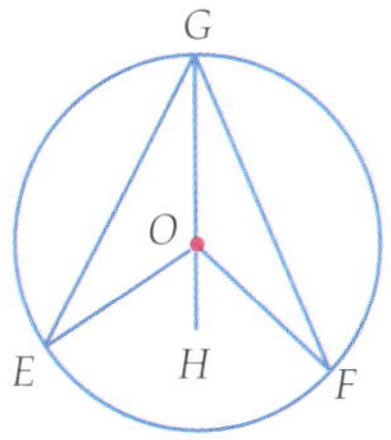

In ΔOGE, $OG = OE$ (radii of a circle)

$\therefore \Delta OGE$ is isosceles

$\therefore \angle OGE = \angle OEG$ (base angles of isosceles triangles)

Hence $\angle HOE = \angle OGE + \angle OEG$ (exterior angle of triangle is sum of interior opposite angles)

$\therefore \angle HOE = 2\angle OGE$

Repeat these steps in ΔOGF to obtain $\angle HOF = 2\angle OGF$.
Now $\angle EOF = \angle HOE + \angle HOF$.

$\therefore \angle EOF = 2\angle OGE + 2\angle OGF$
$= 2(\angle OGE + \angle OGF)$

$\therefore \angle EOF = 2\angle EGF$

2 $\angle AOB = 2\angle ACB$ (angle at the centre is twice the angle at the circumference standing on the same arc)

But $\angle AOB = 180°$ (straight angle)

$\therefore \angle ACB = 90°$

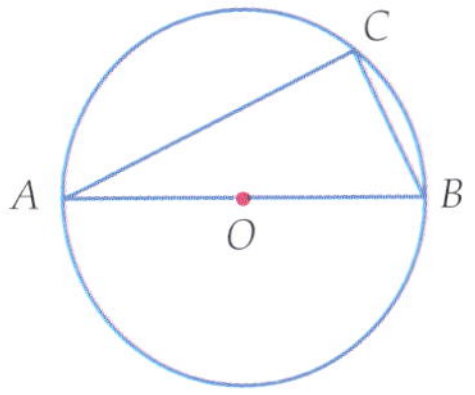

3 Join KO and LO.

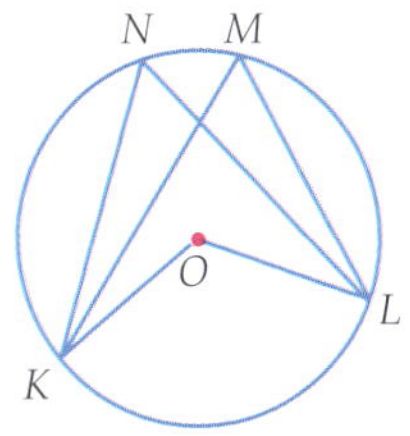

$\angle KOL = 2\angle KNL$ (angle at the centre is twice the angle at the circumference standing on the same arc)

Similarly, $\angle KOL = 2\angle KML$

Hence $2\angle KNL = 2\angle KML$

So $\angle KML = \angle KNL$

4 $\angle EOF = 2\angle EGF$ (angle at the centre is twice the angle at the circumference standing on the same arc)

$\therefore 130° = 2\angle EGF$

$\angle EGF = 65°$

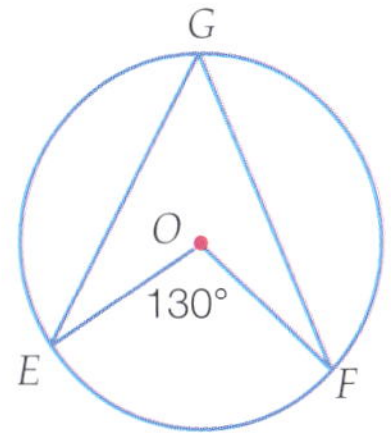

5 $\angle XZY = 90°$ (angle in a semicircle)

But $\angle ZXY = 35°$

So $\angle ZYX = 55°$ (angle sum of triangle is 180°)

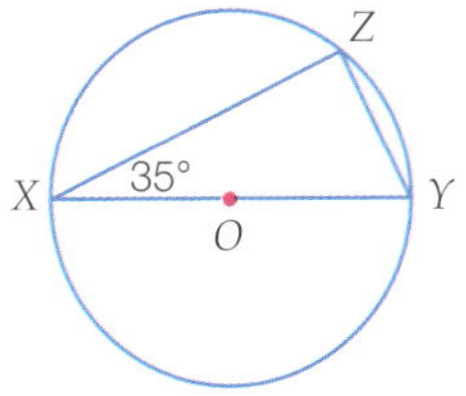

6 $\angle RUS = 46°$ (angles in the same segment)

$\angle UVR = 80°$ (angle sum of ΔUVR)

$\angle TVS = 80°$ (vertically opposite angles)

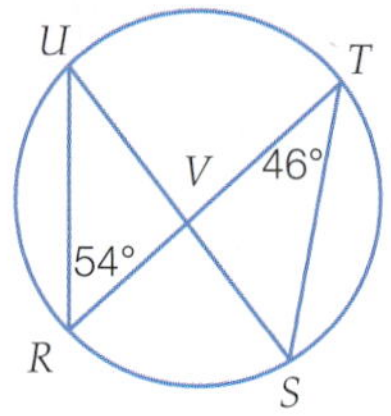

7 $\angle LOK = 50°$ (corresponding angles, $OK \parallel MN$)

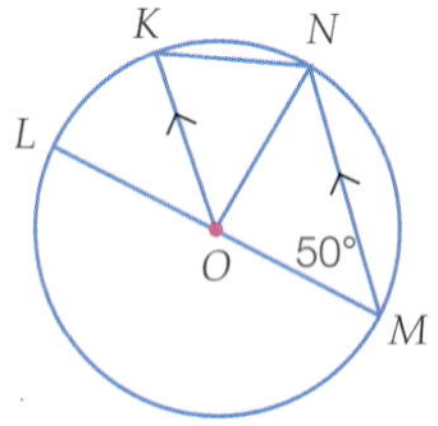

$\angle LON = 2\angle LMN = 100°$ (angle at the centre is twice the angle at the circumference standing on the same arc)

$\angle OMN = \angle ONM$ (opposite equal sides (radii) of ΔONM)

$\angle KON = \angle ONM$ (alternate angles, $OK \parallel MN$)

Hence $\angle KON = \angle LON - \angle LOK = 100° - 50° = 50°$

$OK = ON$ (radii of a circle)

Hence ΔOKN is isosceles and $\angle OKN = \angle ONK$.

$50° + \angle OKN + \angle ONK = 180°$ (angle sum of ΔOKN)

$\therefore 2\angle OKN = 130°$

$\therefore \angle OKN = 65°$

8 (a) $x = 126°$ (angle at the centre is twice the angle at the circumference standing on the same arc)

(b) $y = 70°$ (angle at the centre is twice the angle at the circumference standing on the same arc)

(c) $x = 115°$ (reflex angle at the centre is twice the angle at the circumference standing on the same arc)

9 (a) $\angle ACD = 90°$ (angle in a semicircle), $z = 46°$ (angle sum of ΔACD)

(b) $\angle ECG = 90°$ (angle in a semicircle), $\angle CGE = 40°$ (angle sum of ΔECG), $x + 40° = 95°$ (exterior angle of triangle is sum of interior opposite angles), $x = 55°$

(c) $\angle OGH = y$ (isosceles triangle, $OG = OH$), $2y = 70°$ (exterior angle of triangle is sum of interior opposite angles), $y = 35°$

10 (a) $\angle BDC = 43°$ (angles in the same segment), $x = 107°$ (angle sum of ΔEDC)

(b) $x = 47°$ (angles in the same segment), $\angle ABD = 90°$ (angle in a semi circle), $y = 43°$, $z = 43°$ (angles in the same segment)

(c) $y = 50°$ (angles in the same segment), $ABD = 90°$ (angle in a semi circle), $\angle ABC = 40°$, $z + 40° = 95°$ (exterior angle of triangle is sum of interior opposite angles), $z = 55°$

(d) $n = 32°$, $m = 32°$ (angles in the same segment)

(e) $\angle BDC = 35°$ (angles in the same segment), $e = 51°$ (angle sum of ΔADC)

(f) $\angle DCA = 44°$ (angles in the same segment), $k = 25°$ (exterior angle of triangle is sum of interior opposite angles)

11 (a) $x = 35°$ (angle at the centre is twice the angle at the circumference standing on the same arc), $\angle FOE = 80°$ (angle at the centre is twice the angle at the circumference standing on the same arc), $2y + 80° = 180°$ (angle sum of isosceles triangle FOE), $y = 50°$

(b) $\angle CDB = 49°$ (angle sum of ΔBDC), $\angle ACD = 49°$ (alternate angles, $AC \parallel DB$), $z = 98°$ (angle at the centre is twice the angle at the circumference standing on the same arc)

(c) $\angle MNQ = \angle MQN = 72°$ (base angles of isosceles ΔMNQ), $\angle NQP = 90°$ (angle in a semi circle), $w = 18°$

12 (a) $\angle BOD = 2y$ (angle at centre is twice angle at circumference standing on the same arc)

$OB = OD$ (radii)

$\angle OBD = \angle ODB = x°$ (angles opposite equal sides of a triangle)

$\therefore x + x + 2y = 180$ (angle sum of a triangle)

$\therefore x + y = 90$

(b) $\angle ODB = \angle OBD = 44°$ (base angles of isosceles triangle OBD)

$\angle ODC = 62° - 44° = 18°$

$\angle OCD = 18°$ (base angles of isosceles triangle OCD)

13 ΔBDC is isosceles, so $\angle DBC = \angle DCB = 64°$

$\angle ABC = 90°$ (angle in a semicircle) $\therefore \angle ADE = 38°$

$\angle ABD = 26° \therefore \angle ACB = 38°$

$\angle ACD = 26°$ (angles in the same segment)

14 $\angle DGE = 20°$ (angles in the same segment)

$\angle FDE = 48°$ (angles in the same segment)

$\angle GFD = 42°$ (angles in the same segment)

$\angle GDF = 70°$ (angle sum of triangle)

$\angle GEF = 70°$ (angles in the same segment)

$\angle FGD = \angle FGE + \angle DGE = 68°$

$\angle GDE = \angle GDF + \angle FDE = 118°$

$\angle DEF = \angle GED + \angle GEF = 112°$

$\angle EFG = \angle DFE + \angle GFD = 62°$

$\angle GHF = 90°$ (angle sum of a triangle)

$\therefore$ The diagonals GE and DF are perpendicular.

15 (a) $\angle ODF = 30°$ (isosceles triangle), $\angle FOD = 120°$ (angle sum of triangle), $y = 60°$ (angle at the centre is twice the angle at the circumference standing on the same arc)

(b) $\angle ODF = 33°$ (isosceles triangle), $\angle FOD = 114°$ (angle sum of triangle), reflex $\angle FOD = 246°$ (angles at a point), $\angle FBD = 123°$ (reflex angle at the centre is twice the angle at the circumference standing on the same arc), $x = 22°$ (angle sum of triangle)

(c) $\angle AOD = 2x$ (angle at the centre is twice the angle at the circumference standing on the same arc), reflex $\angle AOD = 4x°$ (reflex angle at the centre is twice the angle at the circumference standing on the same arc), $6x = 360°$ (angles at a point), $x = 60°$

16 Join KO, LO, NO

Let $\angle LNO = x$

$\therefore \angle OLN = \angle ONL = x$ (angles opposite equal sides of a triangle)

$\therefore \angle LON = (180 - 2x)°$ (angles sum of a triangle)

In ΔOKL and ΔONL, $OK = ON$ (radii)

$KL = NL$ (given)

LO is common

$\therefore \Delta OKL \equiv \Delta ONL$

$\therefore \angle OLK = \angle OLN = x$ (corresponding angles in congruent triangles)

$\therefore \angle KLM = (180 - 2x)°$ (MKL is a straight line)

$\therefore \angle KLM = \angle LON$ (both $(180 - 2x)°$)

17 (a) In ΔKPM and ΔQNM,

$\angle KPQ = \angle QNK$ (angles in the same segment)

$\angle PKN = \angle NQP$ (angles in same segment)

$\angle KMP = \angle QMN$ (vertically opposite angles)

$\therefore \Delta KPM \sim \Delta QNM$ (AAA)

(b) $KM = MQ$ (given)

$\angle PKM = \angle NQM$ (corresponding angles in similar triangles)

$\angle MNQ = \angle MPK$ (corresponding angles in similar triangles)

$\therefore \Delta KMP \equiv \Delta QMN$ (ASA)

$\therefore MP = MN$ (corresponding sides in congruent triangles)

18 $\angle CAF = 90°$ (angle in a semicircle)

$\angle DAF = 90°$ (angle in a semicircle)

$\therefore \angle CAF + \angle DAF = 180°$

$\therefore \angle CAD = 180°$, so C, A and D are collinear

19 (a) In ΔVSW and ΔUTW,

$\angle VSU = \angle VTU$ (angles in the same segment)

$\angle SVT = \angle SUT$ (angles in same segment)

But $\angle XWU = \angle SVW$ (opposite equal sides (radii) ΔVSW)

So $\angle VSW = \angle WUT$ (both equal to $\angle VSU$)

But these are a pair of equal alternate angles between VS and UT

$\therefore VS \parallel UT$

(b) $\angle SVT = \angle SUT = 45°$ (angle at centre is twice angle at circumference standing on the same arc)

$\angle UTV = \angle TVS = 45°$ (alternate angles equal, $VS \parallel UT$)

$\therefore \angle TWU = 90°$ (angle sum of ΔTWU)

20 $\angle ACD = \angle ADC$ (base angles of isosceles triangle ACD)

$\angle AFB = \angle ACB$ (angles in the same segment standing on AB)

$\angle AFE = \angle ADE$ (angles in the same segment standing on AE)

$\therefore \angle AFB = \angle AFE$ (from line 1)

$\therefore AF$ bisects $\angle EFB$

21 $\angle KML = 90°$ (angle in a semicircle on diameter KL)

$\angle ONL = 90°$ (angle in a semicircle on diameter OL)

$\therefore \angle KML = \angle ONL = 90°$

$\therefore KM \parallel ON$ (pair of corresponding angles are equal)

22 $\angle COA = 2\angle CBA$ (angle at the centre is twice the angle at the circumference standing on the same arc)

Reflex $\angle COA = 2\angle CDA$ (angle at the centre is twice the angle at the circumference standing on the same arc)

But $\angle COA +$ reflex $\angle COA = 360°$ (angles at a point)

So $2\angle CBA + 2\angle CDA = 360°$

Hence $\angle CBA + \angle CDA = 180°$

Open-ended – Sample answers

23 (a) (i) $\angle DAC = 30°$ (ii) $\angle ADB = 70°$

(iii) $\angle ACB = 70°$

(b) Students' own answer.

(c) (i) ΔADC is isosceles because $\angle DAC = \angle DCA = 30°$

(ii) $\angle ABC = 60°$

$\angle ADC = 120°$

(d) (i) ΔADB is isosceles because $\angle ADB = \angle ABD = 70°$

(ii) $\angle BCD = 140°$

$\angle ADB = 40°$

24 Neither is correct. It is just a quadrilateral with a pair of opposite angles each equal to 90°.

$\angle KJM = \angle KLM = 90°$ (angle in a semicircle)

Nothing is known about the lengths of the sides. To be a kite, you would need $JM = LM$. To be a square, you would need $JM = LM = LK = KJ$.

Exercise 9.7

1 (a) In ΔWOX and ΔYOZ, $WX = YZ$ (given)

$WO = YO$ (radii)

$OX = OZ$ (radii)

$\therefore \Delta WOX \equiv \Delta YOZ$ (SSS)

$\therefore \angle WOX = \angle YOZ$ (corresponding angles in congruent triangles)

(b) $\angle WXO = \angle YZO$ (corresponding angles in congruent triangles)

In ΔOXP and ΔOZQ

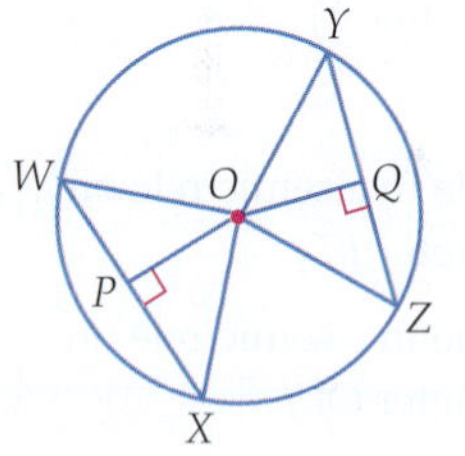

$\angle OPX = \angle OQZ = 90°$ (given)

$\angle PXO = \angle QZO$ (from above)

$OX = OZ$ (radii)

$\therefore \Delta OXP \equiv \Delta OZQ$ (ASA)

$\therefore OP = OQ$ (corresponding sides in congruent triangles)

2 In ΔOGE and ΔOGF, $\angle OGE = \angle OGF = 90°$ (given)

$OE = OF$ (radii)

OG is a common side

$\therefore \Delta OGE \equiv \Delta OGF$ (RHS)

$\therefore EG = GF$ (corresponding sides in congruent triangles)

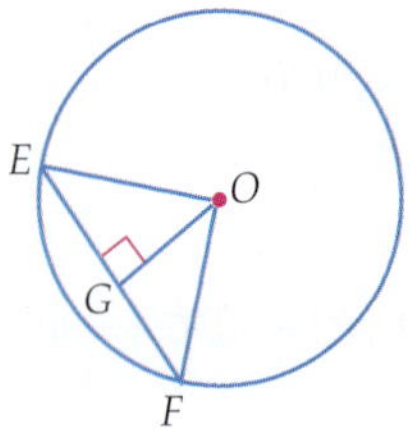

3 In ΔOMK and ΔOML, $KM = ML$ (given)

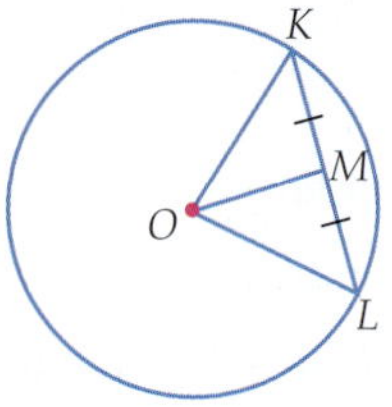

$OK = OL$ (radii)

OM is a common side

$\therefore \Delta OKM \equiv \Delta OLM$ (SSS)

$\therefore \angle OMK = \angle OML$ (corresponding angles in congruent triangles)

But $\angle OMK + \angle OML = 180°$, so $\angle OMK = \angle OML = 90°$. Hence, $OM \perp KL$.

4 $WX = WY$ as all points on the perpendicular bisector of the interval XY are equidistant from the end points of the interval.

$OX = OY$ (radii of the circle centre O on which X and Y lie)

Because O is equidistant from X and Y, it must lie on the perpendicular bisector of XY.

$\therefore O$ lies on WZ.

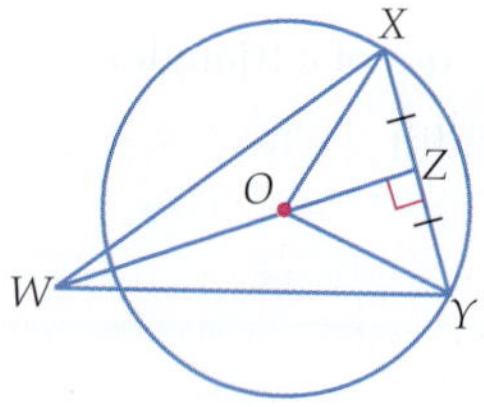

5 $WX = XY$ (given)

Hence, $x = 70$ (equal chords subtend equal angles at the centre)

Because $\angle WOX = \angle YOZ = 70°$, $y = 2$ (equal angles at the centre are subtended by equal chords)

6 Draw OC perpendicular to AB.

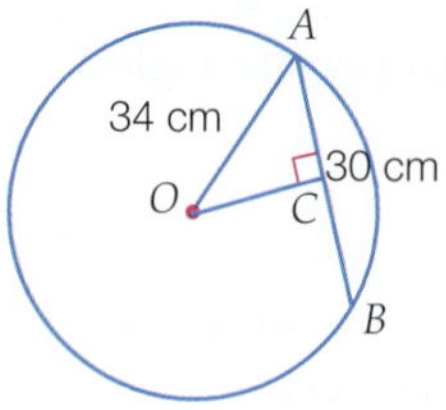

$AC = CB = 15$ cm (perpendicular from the centre of a circle to a chord, bisects the chord)

$34^2 = 15^2 + OC^2$

$OC^2 = 34^2 - 15^2 = 1156 - 225 = 931$

$OC = \sqrt{931} \approx 30.5$ cm

7 (a) $AB = 5$ cm (perpendicular from the centre of a circle to a chord, bisects the chord)

(b) $AB = 2 \times AC = 20$ cm (perpendicular from the centre of a circle to a chord, bisects the chord)

(c) $AB = 15$ cm (Pythagoras' theorem)

8 (a) $\angle AOB = 70°$ (equal chords subtend equal angles at the centre)

(b) $\angle AOB = 80°$ (angle sum of isosceles triangle is 180°)

(c) $\angle OFG = 50°$ (base angles of isosceles triangle), $\angle FOG = 80°$ (angle sum of triangle), $\angle AOB = 80°$ (equal chords subtend equal angles at the centre)

9 (a) $AE = EB = 24$ cm (E is midpoint of AB, given)

$\angle OEA = 90°$ (line joining midpoint of chord to centre of circle $\perp$ chord)

$OA^2 = 7^2 + 24^2$ (Pythagoras' theorem)

$OA = 25$ cm

(b) $OG^2 = 20^2 + 21^2 = 841$ (Pythagoras' theorem), $OG = 29$, $OA = 29$ cm

(c) ΔOGF is equilateral; $OA = OG = 10$ cm

10 (a) $CE^2 = 53^2 - 28^2 = 2025$ (Pythagoras' theorem), $CE = 45$, $CD = 2 \times 45 = 90$ cm (perpendicular from the centre of a circle to a chord, bisects the chord)

(b) $GH^2 = 73^2 - 48^2 = 3025$ (Pythagoras' theorem), $GH = 55$, $GF = 2 \times 55 = 110$ cm (perpendicular from the centre of a circle to a chord, bisects the chord); $FD = GF = 110$ cm (given), $CD = FD = 110$ cm (equal angles at the centre are subtended by equal chords)

(c) $GH^2 = 85^2 - 77^2 = 1296$ (Pythagoras' theorem), $GH = 36$, $GF = 2 \times 36 = 72$ cm (perpendicular from the centre of a circle to a chord, bisects the chord), $CD = GF = 72$ cm (equal angles at the centre are subtended by equal chords)

11 In ΔDOE and ΔBDC, $DO = OB$ (radii)

$DE = EC$ (E is midpoint of CD)

$\therefore \dfrac{DO}{DB} = \dfrac{DE}{DC} = \dfrac{1}{2}$

$\angle ODE = \angle BDC$ (same angle)

$\Delta DOE \sim \Delta DCB$ (common angle with the sides containing it proportional)

$\angle DEO = \angle DCB$ (corresponding angles of similar triangles)

$\therefore OE \parallel BC$ (pair of corresponding angles are equal)

But $\angle DEO = 90°$ (line joining midpoint of chord to centre of circle $\perp$ chord)

$\therefore \angle BCD = 90°$

12 Join OD

$\angle OFC = 90°$ (line joining midpoint of chord to centre of circle $\perp$ chord)

$\therefore \angle OFC + \angle FCD = 180°$ ($\angle FCD = 90°$, given)

But these are a pair of supplementary co-interior angles

$\therefore OF \parallel DC$

$\angle ODC = 90°$ (line joining midpoint of chord to centre of circle $\perp$ chord)

$\therefore \angle ODC + \angle FCD = 180°$ ($\angle FCD = 90°$, given)

But these are a pair of supplementary co-interior angles

$\therefore OD \parallel FC$

$\therefore OFCD$ is a parallelogram

$\therefore OF = DC$ (opposite sides of a parallelogram are equal)

But $BC = 2DC$ (D is midpoint of BC)

$\therefore BC = 2OF$

13 $\angle OKF = 90°$ (line joining midpoint of chord to centre of circle $\perp$ chord)

$\angle QJF = 90°$ (line joining midpoint of chord to centre of circle $\perp$ chord)

$\therefore \angle OKF + \angle QJF = 180°$

$\angle OKJ$ and $\angle QJK$ are a pair of supplementary co-interior angles contained between OK and QJ.

$\therefore OK \parallel QJ$

$\therefore \angle KOQ + \angle OQJ = 180°$ (co-interior angles, $OK \parallel QJ$)

14 Let O be the centre of the circular arch.
By symmetry, the Point C must be in the middle of AB, so $AC = 2$ m.

$OC^2 = 2.5^2 - 2^2$ (Pythagoras' theorem)

$OC = 1.5$ m

$CH = OH - OC = 2.5 - 1.5 = 1$ m

15 Let O be the centre of the circle and P the midpoint of LM.

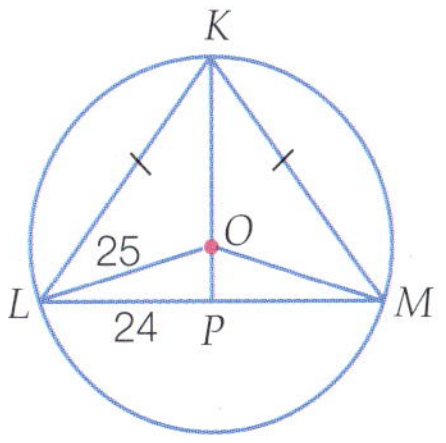

$\angle OPL = 90°$ (line joining midpoint of chord to centre of circle $\perp$ chord)

$OP^2 = 25^2 - 24^2$ (Pythagoras' theorem)

$OP = 7$ mm

$KP = 25 + 7 = 32$ mm ($OK = 25$ mm, radius)

$KL^2 = 32^2 + 24^2$ (Pythagoras' theorem)

$KL = 40$ mm

Open-ended – Sample answers

16 Students' own answers.

17 The centres of all the circles passing through K and L lie on the line PQ.

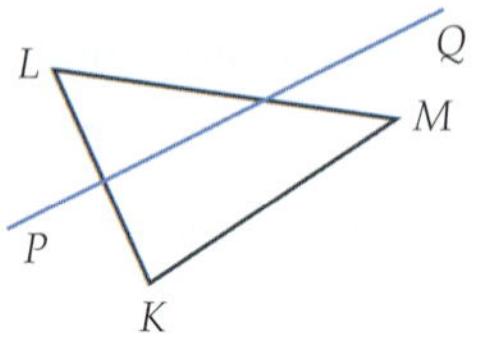

The centres of all the circles passing through L and M lie on the line RS.

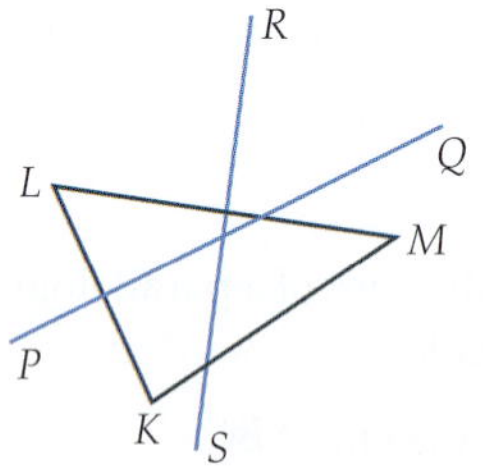

The point of intersection of PQ and RS must be the centre of the only circle that passes through K, L and M.
Label the point of intersection of PQ and RS, X, and draw the circle centre X and radius XL. It will pass through all the points K, L and M.

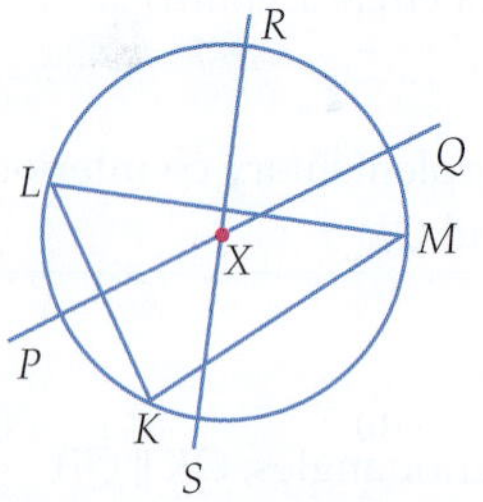

18 **(i)**

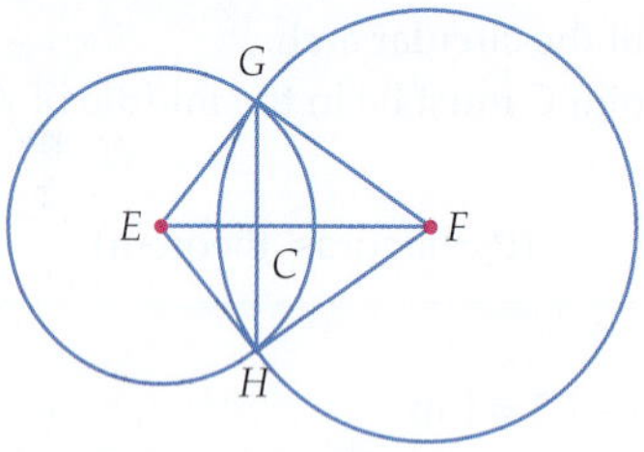

In ΔEGF and ΔEFH,

$EG = EH$	(radii of circle centre E)
$GF = HF$	(radii of circle centre F)
EF is a common side	$\therefore \Delta EGF \equiv \Delta EFH$ (SSS)
$\angle GEF = \angle HEF$	(corresponding angles in congruent triangles)

In ΔEGC and ΔEHC, $EG = EH$ (radii of circle centre E)

$\angle GEC = \angle HEC$	(corresponding angles in congruent triangles)
EC is a common side.	
$\therefore \Delta EGC = \Delta EHC$	(SAS)
$GC = CH$	(corresponding sides in congruent triangles)

$\therefore$ EF biscets GH

(i.e. the line of centres bisects the common chord).

(ii) CE is the line joining the centre of the chord GH to the centre E of the circle.

$\therefore CE \perp GH$

(i.e. the line of centres is perpendicular to the common chord).

Chapter review 9

1 **(a)** SSS **(b)** SAS

2 B

3 **(a)** 5 **(b)** 9

4 **(a)** In ΔAOB and ΔDOC,

$\angle AOB = \angle DOC$	(vertically opposite angles)
$AO = DO$	(radii in the same circle)
$BO = CO$	(radii in the same circle)
$\therefore \Delta AOB = \Delta DOC$	(SAS)
So $AB = DC$	(corresponding sides in congruent triangles)

$\therefore p = 6$ m

(b) In ΔDEH and ΔGEF,

$\angle DEH = \angle GEF$	(vertically opposite angles)
$DH = GF$	(given)
$\angle EDH = \angle EGF$	(alternate angles, $DH \parallel GF$)
$\therefore \Delta DEH \equiv \Delta GEF$	(ASA)
So $HE = FE$	(corresponding sides in congruent triangles)

$\therefore r = 9$ cm

5 **(a)** In ΔABC and ΔADE,

$\angle CAB = \angle EAB$	(common angle)
$\angle ABC = \angle ADE$	(given)
$\therefore \angle ACB = \angle AED$	(third angle in a triangle)
$\therefore \Delta ABC \sim \Delta ADE$	(AAA)

(b) In ΔABC and ΔADE,

$\angle BAC = \angle DAE$	(vertically opposite angles)
$\angle BCA = \angle AED$	(alternate angles and $BC \parallel DE$)
$\angle CBA = \angle ADE$	(alternate angles and $BC \parallel DE$)
$\therefore \Delta ABC \sim \Delta ADE$	(AAA)

6 In ΔAXC and ΔBXC,

$\angle AXC = \angle BXC = 90°$	(given that $CX \perp AB$)
$\angle ACX = \angle BCX = \frac{1}{2} \times \angle ACB$	(given that CX bisects $\angle ACB$)
Now $CX = CX$	(common side in ΔAXC and ΔBXC)

$\therefore \Delta AXC \equiv \Delta BXC$ and so $AX = BX$ (corresponding sides in congruent triangles)

7 In ΔDGH and ΔDEF,

$\angle GDH = \angle EDF$ (common angle)

$\angle DHG = \angle DFE$ (corresponding angles equal, $GH \parallel EF$)

$\angle DGH = \angle DEF$ (corresponding angles equal, $GH \parallel EF$)

$\therefore \Delta DGH \equiv \Delta DEF$ (AAA)

$$\frac{DE}{DG} = \frac{DF}{DH}$$

$$\frac{DG + GE}{DG} = \frac{DH + HF}{DH}$$

$$\frac{DG}{DG} + \frac{GE}{DG} = \frac{DH}{DH} + \frac{HF}{DG}$$

$$1 + \frac{GE}{DG} = 1 + \frac{HF}{DH}$$

$$\therefore \frac{GE}{DG} = \frac{HF}{DH} \text{ so } \frac{DG}{GE} = \frac{DH}{HF}$$

8 In ΔABC and ΔCDA,

$AB = CD$ (given)

$BC = DA$ (given)

$AC = CA$ (common side)

$\therefore \Delta ABC \equiv \Delta CDA$ (SSS)

So $\angle BAC = \angle DCA$ (corresponding angles in congruent triangles)

But $\angle BAC = \angle DCA$ are alternate angles,

$\therefore AB \parallel CD$

And $\angle BCA = \angle DAC$ (corresponding angles in congruent triangles)

But $\angle BCA = \angle DAC$ are alternate angles $\therefore AD \parallel CB$

The quadrilateral is a parallelogram because both pairs of opposite sides are parallel.

Have proved that if both pairs of opposite sides of a quadrilateral are equal, then the quadrilateral is a parallelogram.

9 In ΔPTQ, ΔRTQ, ΔRTS and ΔPTS,

$PT = RT$ (given)

$ST = QT$ (given)

$\angle PTQ = \angle RTQ = \angle RTS = \angle PTS = 90°$ (given)

$\therefore \Delta PTQ \equiv \Delta RTQ \equiv \Delta RTS \equiv \Delta PTS$ (SAS)

So $PQ = RQ$ (corresponding sides in congruent triangles)

ΔPTQ is isosceles so $\angle TPQ = \angle TQP$

$\angle TPQ + \angle TQP + \angle PTQ = 180°$

$2\angle TQP + 90° = 180°$

$2\angle TQP = 90°$

$\angle TQP = 45°$

But $\angle TQP = \angle TQR = \angle TRQ = \angle TRS = \angle TSR = \angle TSP = \angle TPS = \angle TPQ$ (corresponding angles in congruent triangles)

So $\angle TQR = 45°$

$\angle PQR = \angle TQR + \angle TQP = 90°$

$\angle QRS = \angle TRQ + \angle TRS = 90°$

$\angle SPQ = \angle TPQ + \angle TPS = 90°$

$\angle PQR$ and $\angle QRS$ are co-interior angles and they add to 180°.

$\therefore PQ \parallel SR$

$\angle PQR$ and $\angle SPQ$ are co-interior angles and they add to 180°.

$\therefore PS \parallel QR$

Have shown $PQ = RQ$, $\angle PQR = 90°$, $PQ \parallel SR$ and $PS \parallel QR$.

$PQRS$ is a square as it has two adjacent sides, one right angle and both pairs of opposite sides parallel.

Have proved that if the diagonals of a quadrilateral are equal and bisect each other at right angles, then quadrilateral is a square.

10 (a) $y = 59°$ (angle at the centre is twice the angle at the circumference standing on the same arc)

(b) $x = 125°$ (reflex angle at the centre is twice the angle at the circumference standing on the same arc)

(c) $x = 38°$ (vertically opposite angles, angle sum of triangle)

(d) $x = 83°$ (angles in the same segment, angle sum of triangle)

(e) $k = 39°$ (angles in the same segment, exterior angle of a triangle)

(f) $x = 33°$ (angle at the centre is twice the angle at the circumference standing on the same arc)

$y = 90 - 36 = 54°$ (angle in a semicircle, isosceles triangle)

11 (a) $AB = 8$ cm (perpendicular from the centre of a circle to a chord, bisects the chord)

(b) $AB = 16$ cm (perpendicular from the centre of a circle to a chord, bisects the chord)

(c) $AE = 40$ mm (line joining midpoint of chord to centre of circle $\perp$ chord, Pythagoras' theorem), $AB = 80$ mm

12 (a) $\angle AOB = 65°$ (equal chords subtend equal angles at the centre)

(b) $\angle OFG = 55°$ (base angles of isosceles triangle),
$\angle GOF = 70°$ (angle sum of triangle),
$\angle AOB = 70°$ (equal chords subtend equal angles at the centre)

(c) $\angle GOH = 35°$ (perpendicular bisector of chord bisects angle at centre),
$\angle GOF = 70°$, (equal chords subtend equal angles
$\angle AOB = 70°$ at the centre)

13 12.6 m

14 (a) In ΔBAD and ΔBCD,

$\angle BDA = \angle BDC = 90°$ (by construction)

$BA = BC$ (given)

$BD = BD$ (common side)

$\therefore \Delta BAD \equiv \Delta BDC$ (RHS)

So $\angle BAD = \angle BCD$ (corresponding angles in congruent triangles)

$\therefore \angle BAC = \angle BCA$

(b) Have proved that the angles opposite the equal sides in an isosceles triangle are also equal.

15 In ΔABZ and ΔACX,

$AB = AC$ (given)

$AZ = AX$ (given)

$\angle BAZ = \angle CAX$ (common angle)

$\therefore \Delta ABZ \equiv \Delta ACX$ (SAS) (and so $BZ = CX$)

Now consider ΔBYC

$\angle YBC = \angle YCB$ (given)

$BY = CY$ (If two angles in a triangle are equal then the sides opposite these angles are equal—the triangle is isosceles.)

Consider ΔBYX and ΔCYZ

$\angle BYX = \angle CYZ$ (vertically opposite angles)

$BX = CZ$ (as $AB = AC$ and $AZ = AX$)

$\therefore XY = ZY$ (as $BY = CY$ and $BZ = CX$)

16 (a) Let $\angle DFE = x$

$\therefore \angle GEF = (90 - x)$ (angle sum of ΔGEF is 180°)

$\therefore \angle GED = x$ ($\angle DEF = 90°$)

In ΔDEF and ΔDEG,

$\angle DFE = \angle DEG = x$

$\angle DEF = \angle DGE = 90°$

$\therefore \Delta DEF \sim \Delta DGE$ (AAA)

$\frac{EF}{EG} = \frac{DF}{DE} = \frac{DE}{DG}$ [1] (corresponding sides of similar triangles are proportional)

In ΔDEF and ΔEGF,

$\angle DFE = \angle GFE = x$

$\angle DEF = \angle FGE = 90°$

$\therefore \Delta DEF \sim \Delta EGF$ (AAA)

(b) $\frac{EF}{GF} = \frac{DF}{EF} = \frac{DE}{EG}$ [2] (corresponding sides of similar triangles are proportional)

[1] gives $EF = \frac{EG \times DE}{DG}$

[2] gives $EF = \frac{GF \times DE}{EG}$

$\therefore \frac{EG \times DE}{DG} = \frac{GF \times DE}{EG}$

$\therefore EG^2 = DG \times GF$

17 (a) In ΔEAB and ΔECD,

$\angle ABE = \angle CDE$ (alternate angles, $AB \parallel CD$)

$\angle AEB = \angle CED$ (vertically opposite)

$AB = CD$ (given)

$\therefore \Delta EAB = \Delta ECD$ (by ASA)

$AE = CE$ (corresponding sides in congruent triangles)

In ΔEAB and ΔECB,

$AE = CE$ (found earlier)

$AB = CB$ (given)

$BE = BE$ (common side)

$\therefore \Delta EAB = \Delta ECB$ (by SSS)

$\therefore \angle AEB = \angle CEB$ (corresponding angles in congruent triangles)

But $\angle AEB + \angle CEB = 180°$
(angles on a straight line, supplementary)

$\therefore \angle AEB = 90°$ and $\angle CEB = 90°$

$\therefore AC \perp BD$

(b) Have just proved that the diagonals of a rhombus intersect at right angles.

18 (a) $x = 60°, y = 50°$ ($\angle$ at centre, $\angle$s in same segment)

(b) $x = 100°, y = 80°$ (co-interior angles, opposite angles in cyclic quadrilateral)

(c) $x = 73°, y = 13°, z = 107°$
(straight angle, opposite angles in cyclic quadrilateral)

19 $\angle TSR + \angle TSU = 180°$ (angles on a straight line)

$\angle TUS + \angle TUV = 180°$ (angles on a straight line)

$\therefore \angle TSR + \angle TSU + \angle TUS + \angle TUV = 180° + 180° = 360°$

$\angle TSU + \angle TUS + \angle UTS = 180°$
(angles in a triangle)

$\angle TSU + \angle TUS + 90° = 180°$

$\angle TSU + \angle TUS = 90°$

$\angle TSR + \angle TSU + \angle TUS + \angle TUV = 360°$
(found earlier)

$\angle TSR + 90° + \angle TUV = 360°$

$\therefore \angle TSR + \angle TUV = 270°$

20 $a + b = 180°$ (angles on a straight line)

$c + d = 180°$ (angles on a straight line)

$\therefore a + b + c + d = 180° + 180° = 360°$

$\angle GWD = \angle AWB = a$ (vertically opposite angles)

$\angle FXA = \angle DXC = d$ (vertically opposite angles)

$\angle BZE = \angle BWD = b$ (corresponding angles, $AD \parallel HE$)

$\angle CYH = \angle CXA = c$ (corresponding angles, $AD \parallel HE$)

$\angle GWD + \angle BZE + \angle FXA + \angle CYH$

$= a + b + c + d$

$= 360°$

So the interior angles of a trapezium add to 360°.

21 To show: $\Delta LYZ \sim \Delta MXZ$

$\angle LZY = \angle MZX$ (vertically opposite)

$\angle LYZ = \angle MXZ$ (both 90°)

$\angle ZMX = \angle ZLY$ (third angle in a triangle)

$\therefore \Delta LYZ \sim \Delta MXZ$ (AAA)

22 In ΔUXY and ΔVWX,

$YX = XW$ (given)

$UY = VX$ (given)

$\angle UYX = \angle VXW$ (corresponding angles, $UY \parallel VX$)

$\therefore \Delta UXY = \Delta VWX$ (SAS)

23 $RT = \sqrt{36 - 9} = \sqrt{27} = 3\sqrt{3}$ (Pythagoras' theorem in ΔRTQ)

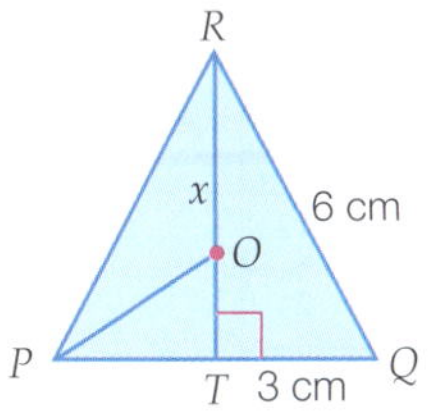

Let radius $= x$. Hence $OP = x$.

Use Pythagoras' theorem in ΔPOT:

$x^2 = 3^2 + (\sqrt{27} - x)^2$

$x^2 = 9 + 27 - 2\sqrt{27}x + x^2$

$2\sqrt{27}x = 36$

$6\sqrt{3}x = 36$

$x = \frac{36}{6\sqrt{3}} = 2\sqrt{3}$

24 $\angle TWU = 90°$ ($SU \perp VT$)

$\angle WUT = 43°$ (angle sum of triangle)

$\angle SVT = \angle SUT = 43°$ (angles in the same segment)

$\angle SOT = 86°$ (angle at the centre twice the angle at the circumference standing on the same arc)

Chapter 10

Recall 10

1 (a) $\frac{1}{5}$ (b) $\frac{4}{13}$

2 (a) $\frac{7}{8}$ (b) $\frac{11}{15}$ (c) $\frac{2}{5}$ (d) $\frac{3}{10}$ (e) 0.4 (f) 0.01

3 (a) boy, girl; boy, boy; girl, boy; girl, girl

(b) heads, tails

4 (a) (i) 25 (ii) 25 (iii) 6 (iv) 9 (v) 15

(b) (i) 1, 2, 3, 4, 5, 6 (ii) 3

Exercise 10.1

1 $\frac{7}{9}$

2 (a) $\frac{2}{7}$ (b) $\frac{5}{7}$

(c) Assume that the probability of the car being hard to start is the same every day.

3 (a) $\frac{1}{8}$ (b) $\frac{1}{2}$ (c) $\frac{3}{8}$ (d) $\frac{1}{4}$ (e) $\frac{7}{8}$

4 (a) $\frac{1}{12}$ (b) $\frac{1}{2}$ (c) $\frac{1}{12}$ (d) $\frac{1}{6}$ (e) $\frac{1}{6}$ (f) $\frac{1}{2}$

5 (a)

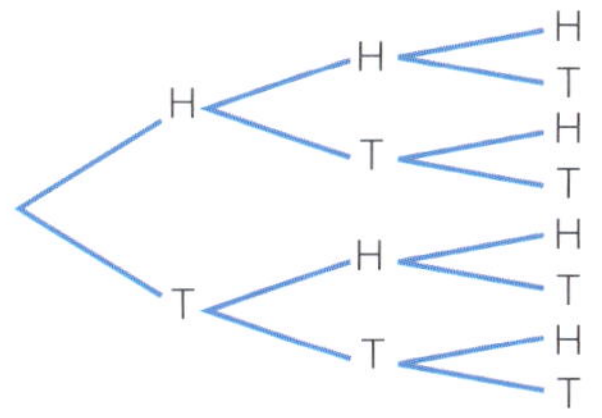

(b) HHH, HHT, HTH, HTT, THH, THT, TTH, TTT

(c) (i) $\frac{1}{8}$ (ii) $\frac{1}{8}$ (iii) $\frac{1}{8}$ (iv) $\frac{1}{4}$ (v) $\frac{1}{2}$

6 (a) $\frac{7}{12}$ (b) $\frac{5}{12}$ (c) $\frac{1}{24}$ (d) $\frac{1}{24}$ (e) $\frac{1}{12}$ (f) $\frac{11}{12}$

7 (a) $\frac{1}{12}$ (b) $\frac{11}{12}$ (c) $\frac{1}{6}$ (d) $\frac{5}{6}$

8 (a) B (b) D (c) C

9 (a) C (b) B (c) B

10 $\frac{4}{9}$

11 (a) A (b) C

12 (a) $\frac{1}{45}$ (b) $\frac{1}{5}$ (c) $\frac{22}{45}$ (d) $\frac{23}{45}$ (e) 0 (f) $\frac{11}{45}$

13 (a) B = black sock, W = white sock, G = green sock

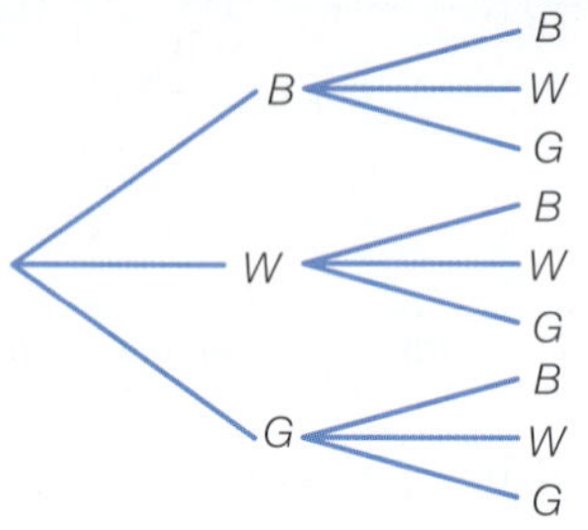

(b) Each sock has an equal probability of being selected from each drawer, so the outcomes for the new event are also equally likely.

(c) (i) $\frac{1}{9}$ (ii) $\frac{2}{9}$ (iii) $\frac{1}{3}$ (iv) $\frac{2}{3}$

14 (a) $\frac{1}{4}$ (b) $\frac{1}{6}$ (c) $\frac{1}{4}$ (d) $\frac{1}{3}$ (e) $\frac{1}{2}$ (f) $\frac{2}{3}$

15 (a)

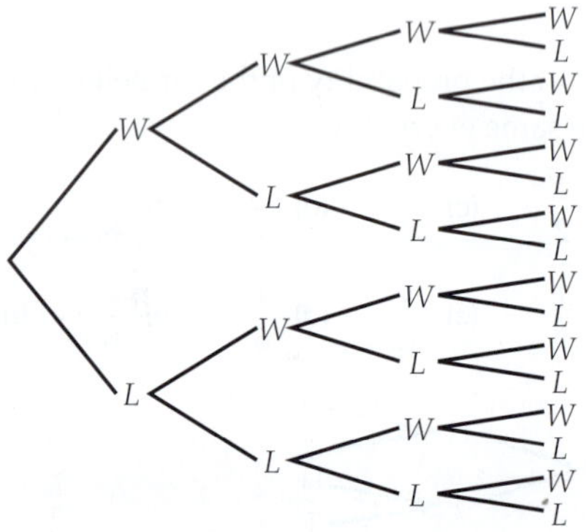

(b) *WWWW, WWWL, WWLW, WWLL, WLWW, WLWL, WLLW, WLLL, LWWW, LWWL, LWLW, LWLL, LLWW, LLWL, LLLW, LLLL*; 16 possible outcomes

(c) (i) $\frac{1}{16}$ (ii) $\frac{3}{8}$ (iii) $\frac{5}{16}$

(iv) $\frac{1}{2}$ (v) $\frac{1}{2}$ (vi) $\frac{1}{4}$

16 (a) Possible totals

		Die 2					
		1	2	3	4	5	6
Die 1	1	2	3	4	5	6	7
	2	3	4	5	6	7	8
	3	4	5	6	7	8	9
	4	5	6	7	8	9	10
	5	6	7	8	9	10	11
	6	7	8	9	10	11	12

(b) 36

(c) Each die individually has an equal chance of producing 1, 2, 3, 4, 5, 6 so together the dice have an equal chance of producing each of the 36 possible outcomes.

(d) $\frac{1}{9}$ (e) $\frac{5}{12}$ (f) $\frac{1}{6}$

17 (a) Sample space

		Die 2					
		2	4	6	8	10	12
Die 1	1	(1, 2)	(1, 4)	(1, 6)	(1, 8)	(1, 10)	(1, 12)
	1	(1, 2)	(1, 4)	(1, 6)	(1, 8)	(1, 10)	(1, 12)
	2	(2, 2)	(2, 4)	(2, 6)	(2, 8)	(2, 10)	(2, 12)
	2	(2, 2)	(2, 4)	(2, 6)	(2, 8)	(2, 10)	(2, 12)
	3	(3, 2)	(3, 4)	(3, 6)	(3, 8)	(3, 10)	(3, 12)
	3	(3, 2)	(3, 4)	(3, 6)	(3, 8)	(3, 10)	(3, 12)

(b) Possible totals

		Die 2					
		2	4	6	8	10	12
Die 1	1	3	5	7	9	11	13
	1	3	5	7	9	11	13
	2	4	6	8	10	12	14
	2	4	6	8	10	12	14
	3	5	7	9	11	13	15
	3	5	7	9	11	13	15

(c) (i) $\frac{1}{3}$ (ii) $\frac{2}{9}$ (iii) $\frac{7}{18}$ (iv) $\frac{7}{18}$

(d) You cannot just look at one diagonal to find the totals; you have to check all the table. However, it is not much more difficult in this case as the numbers are easy to check.

(e) Possible products

		Die 2					
		2	4	6	8	10	12
Die 1	1	2	4	6	8	10	12
	1	2	4	6	8	10	12
	2	4	8	12	16	20	24
	2	4	8	12	16	20	24
	3	6	12	18	24	30	36
	3	6	12	18	24	30	36

(f)

Product	2	4	6	8	10	12	16	18	20	24	30	36
Probability	$\frac{1}{18}$	$\frac{1}{9}$	$\frac{1}{9}$	$\frac{1}{9}$	$\frac{1}{18}$	$\frac{1}{6}$	$\frac{1}{18}$	$\frac{1}{18}$	$\frac{1}{18}$	$\frac{1}{9}$	$\frac{1}{18}$	$\frac{1}{18}$

18 Lesley thinks that the lottery numbers depend on which numbers have come before, but this is not true. Every lottery is independent of all others, so the probability is the same every time.

Open-ended – Sample answers

19 (a) A is the event 'rolling a 1 or 2'

(b) B is the event 'rolling a multiple of 3'

(c) C is the event 'rolling a number greater then 7'

20 4W/1L, 8W/2L, 12W/3L

21 The correct answer is $\frac{7}{8}$. Sharanpreet has calculated the probability of exactly 1 boy, not at least 1 boy.

Exercise 10.2

1 (a)

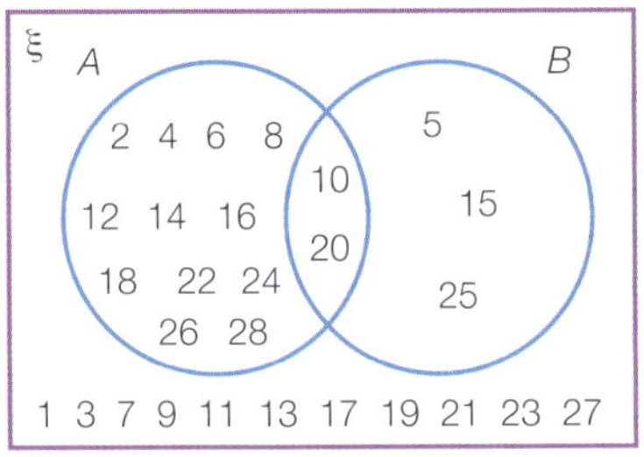

(b) (i) $\frac{1}{2}$ (ii) $\frac{1}{14}$ (iii) $\frac{23}{28}$

(c)

	B	not *B*
A	$\frac{1}{14}$	$\frac{3}{7}$
not *A*	$\frac{3}{28}$	$\frac{11}{28}$

2 (a) $\frac{3}{16}$ (b) $\frac{7}{16}$ (c) $\frac{1}{16}$ (d) $\frac{5}{16}$ (e) $\frac{1}{2}$ (f) $\frac{1}{4}$

3 (a)

	Piano	not Piano	
Singing	4	2	6
not Singing	4	10	14
	8	12	20

(b) (i) $\frac{1}{10}$ (ii) $\frac{1}{5}$ (iii) $\frac{1}{5}$

4 (a)

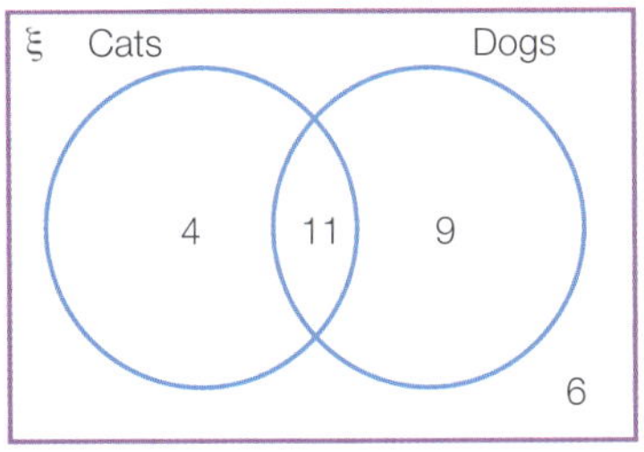

(b) (i) $\frac{2}{15}$ (ii) $\frac{3}{10}$ (iii) $\frac{11}{30}$

5 (a)

	B	not *B*
A	$\frac{1}{8}$	$\frac{5}{16}$
not *A*	$\frac{1}{16}$	$\frac{1}{2}$

(b)

	C	not *C*
D	0.2	0.5
not *D*	0.2	0.1

6 (a) 1, 2, 3, 4, 5, 7, 8, 9, 10, 12, 13, 14, 15, 17, 18, 20

(b) 1, 2, 3, 5, 6, 7, 8, 9, 11, 12, 14, 15, 16, 17, 18, 19, 20

(c) 2, 3, 4, 5, 6, 7, 9, 10, 11, 12, 13, 14, 16, 17, 19, 20

(d) 3, 7, 12, 17, 20

(e) 2, 7, 9, 12

(f) 5, 7, 12, 14

(g) 7, 12

(h) 6, 11, 16, 19

7 (a)

	TV	not TV	
internet	236	1200	1436
not internet	554	10	564
	790	1210	2000

(b) $\frac{59}{500}$

8 (a)

	Twitter	not Twitter	
Facebook	$\frac{1}{4}$	$\frac{1}{2}$	$\frac{3}{4}$
not Facebook	$\frac{1}{8}$	$\frac{1}{8}$	$\frac{1}{4}$
	$\frac{3}{8}$	$\frac{5}{8}$	1

(b) (i) $\frac{1}{8}$ (ii) $\frac{5}{8}$

9 (a)

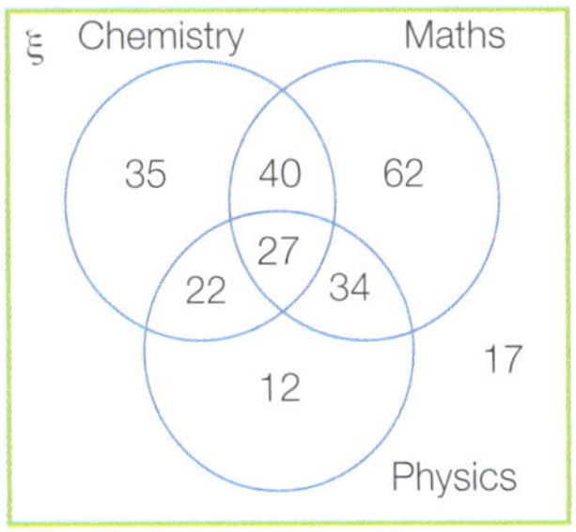

(b) 62 (c) 12 (d) 35 (e) 249

(f) $\frac{163}{249}$, it may be similar but it is unlikely to be exactly the same.

10 (a)

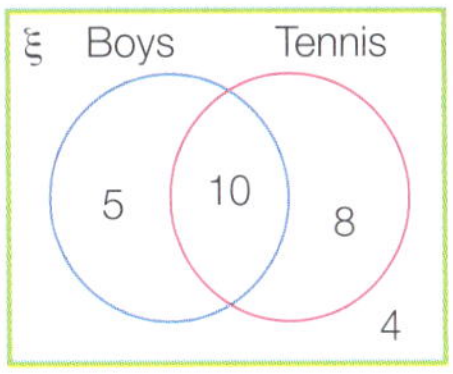

(b) (i) $\frac{8}{27}$ (ii) $\frac{5}{27}$ (iii) $\frac{2}{3}$ (iv) $\frac{1}{3}$

11 (a)

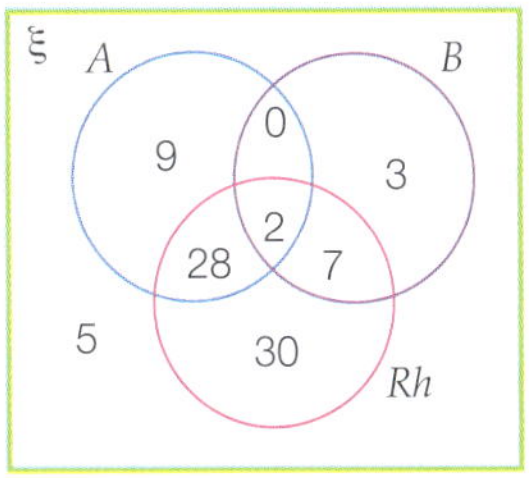

(b) 84 (c) $\frac{1}{2}$ (d) $\frac{5}{12}$

(e) A+: $\frac{1}{3} = 33.3\%$; A–: $\frac{3}{28} = 10.7\%$

B+: $\frac{1}{12} = 8.3\%$; B–: $\frac{1}{28} = 3.6\%$

AB+: $\frac{1}{42} = 2.4\%$; AB–: 0%

O+: $\frac{5}{14} = 35.7\%$; O–: $\frac{5}{84} = 6\%$

(f) The A group is larger than expected and the O group is smaller than expected.

12 (a)

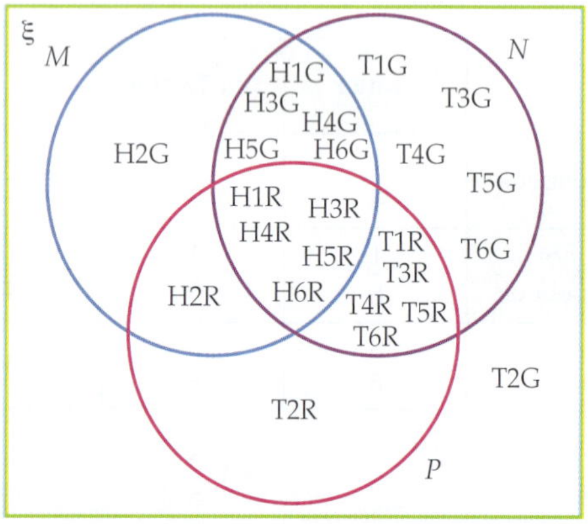

(b) Sample space = {H1R, H1G, H2R, H2G, H3R, H3G, H4R, H4G, H5R, H5G, H6R, H6G, T1R, T1G, T2R, T2G, T3R, T3G, T4R, T4G, T5R, T5G, T6R, T6G}

(c) (i) not M = {T1R, T1G, T2R, T2G, T3R, T3G, T4R, T4G, T5R, T5G, T6R, T6G}

(ii) not N = {H2R, H2G, T2R, T2G}

(iii) not P = {H1G, H2G, H3G, H4G, H5G, H6G, T1G, T2G, T3G, T4G, T5G, T6G}

Open-ended – Sample answers

13 (a) Answers may vary. For example:

	B	not B	
A	9	20	29
not A	10	11	21
	19	31	50

(b) (i) $\frac{9}{50}$ (ii) $\frac{39}{50}$

Exercise 10.3

1 (a) $\frac{1}{4}$ (b) $\frac{5}{8}$ (c) $\frac{3}{8}$

2 (a) $\frac{1}{4}$ (b) $\frac{1}{2}$ (c) $\frac{3}{4}$ (d) $\frac{1}{2}$

3 (a) yes (b) no (c) yes

4 (a) A (b) D

5 (a) $\frac{4}{5}$ (b) $\frac{1}{5}$ (c) $\frac{4}{5}$ (d) $\frac{3}{5}$ (e) $\frac{17}{20}$

6 (a) $\frac{2}{3}$ (b) $\frac{1}{3}$ (c) $\frac{19}{30}$ (d) $\frac{11}{15}$

7 (a) $\frac{1}{2}$ (b) $\frac{1}{4}$ (c) $\frac{1}{13}$ (d) $\frac{1}{52}$ (e) $\frac{27}{52}$

(f) $\frac{4}{13}$ (g) $\frac{1}{26}$ (h) $\frac{3}{26}$ (i) $\frac{9}{13}$ (j) $\frac{19}{26}$

(k) $\frac{11}{13}$ (l) $\frac{1}{4}$

8 (a) Six pairs: AB, AC, AD, BC, BD, CD

(b) AC and BC are mutually exclusive.

(c) $\Pr(A) = \frac{1}{2}$, $\Pr(B) = \frac{1}{3}$, $\Pr(C) = \frac{1}{3}$, $\Pr(D) = \frac{1}{3}$

(d) $\Pr(A \text{ or } B) = \frac{1}{2}$, $\Pr(A \text{ or } C) = \frac{5}{6}$, $\Pr(A \text{ or } D) = \frac{2}{3}$,

$\Pr(B \text{ or } C) = \frac{2}{3}$, $\Pr(B \text{ or } D) = \frac{1}{2}$, $\Pr(C \text{ or } D) = \frac{1}{2}$

(e) $\Pr(A \text{ and } B) = \frac{1}{3}$, $\Pr(A \text{ and } C) = 0$, $\Pr(A \text{ and } D) = \frac{1}{6}$,

$\Pr(B \text{ and } C) = 0$, $\Pr(B \text{ and } D) = \frac{1}{6}$, $\Pr(C \text{ and } D) = \frac{1}{6}$

(f) $\Pr(A) + \Pr(B) = \frac{5}{6}$, $\Pr(A) + \Pr(C) = \frac{5}{6}$,

$\Pr(A) + \Pr(D) = \frac{5}{6}$, $\Pr(B) + \Pr(C) = \frac{2}{3}$,

$\Pr(B) + \Pr(D) = \frac{2}{3}$, $\Pr(C) + \Pr(D) = \frac{2}{3}$

(g) In general, $\Pr(A \text{ or } B) = \Pr(A) + \Pr(B) - \Pr(A \text{ and } B)$. However, if A and B are mutually exclusive $\Pr(A \text{ and } B) = 0$, and so $\Pr(A \text{ or } B) = \Pr(A) + \Pr(B)$.

9 (a) The vehicle types listed are not mutually exclusive, so there could be arguments as to where a particular vehicle should be placed; or a particular vehicle may be registered in more than one category, thereby making later calculations difficult.

(b) The categories have to be clearly mutually exclusive but also cover every type of vehicle that may go past. Including the category 'other' would cover the second requirement.

Open-ended – Sample answers

10 (a) Sample answers:
odd numbers, even numbers
numbers < 3, numbers > 6
multiples of 7, multiples of 8
numbers < 10, multiples of 10

(b) Sample answers:
odd numbers, multiples of 3
numbers < 6, numbers > 3
multiples of 3, multiples of 4
numbers < 10, multiples of 4

(c) Sample answer:
A: even numbers
B: numbers greater than 5

Half-time 10

1 (a)

	smartphone	not smartphone	
laptop	60	15	75
not laptop	150	175	325
	210	190	400

(b) (i) $\frac{3}{8}$ (ii) $\frac{3}{80}$

2 (a) $\frac{2}{10} = \frac{1}{5}$ (b) $\frac{1}{2}$ (c) $\frac{6}{10} = \frac{3}{5}$ (d) $\frac{3}{10}$

3 0.09

4 (a) If W = Suzanne wins and L = Suzanne loses:

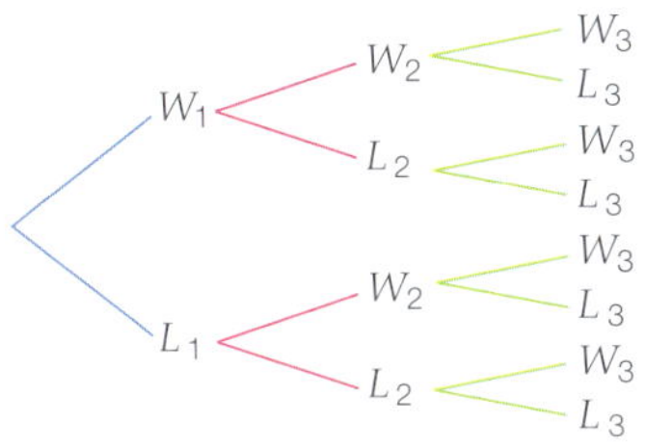

$W_1W_2W_3$, $W_1W_2L_3$, $W_1L_2W_3$, $W_1L_2L_3$, $L_1W_2W_3$, $L_1W_2L_3$, $L_1L_2W_3$, $L_1L_2L_3$

(b) 8

(c) (i) $\frac{1}{8}$ (ii) $\frac{3}{8}$ (iii) $\frac{1}{2}$

(iv) $\frac{1}{2}$ (v) $\frac{1}{2}$ (vi) $\frac{1}{4}$

5 (a) A and B; A and C; B and C

(b) $\Pr(A) = \frac{1}{2}$; $\Pr(B) = \frac{1}{2}$; $\Pr(C) = \frac{1}{8}$

(c) A and C are mutually exclusive.

Exercise 10.4

1 (a)

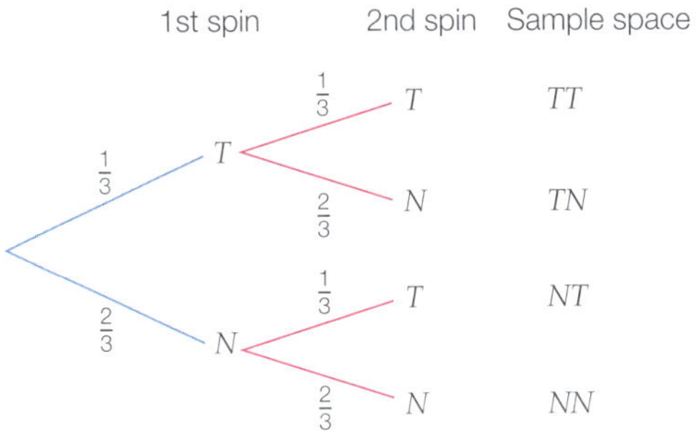

Probability of each outcome

$\Pr(TT) = \frac{1}{3} \times \frac{1}{3} = \frac{1}{9}$

$\Pr(TN) = \frac{1}{3} \times \frac{2}{3} = \frac{2}{9}$

$\Pr(NT) = \frac{2}{3} \times \frac{1}{3} = \frac{2}{9}$

$\Pr(NN) = \frac{2}{3} \times \frac{2}{3} = \frac{4}{9}$

(b) (i) $\frac{1}{9}$ (ii) $\frac{4}{9}$ (iii) $\frac{4}{9}$ (iv) $\frac{5}{9}$

2

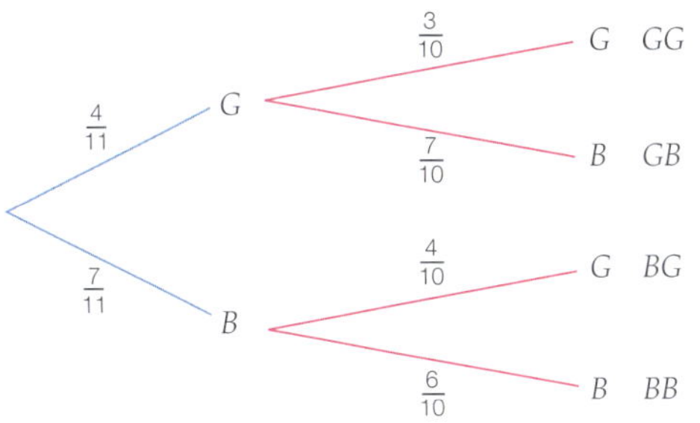

(a) $\frac{42}{110} = \frac{21}{55}$ (b) $\frac{98}{110} = \frac{49}{55}$

3 (a)

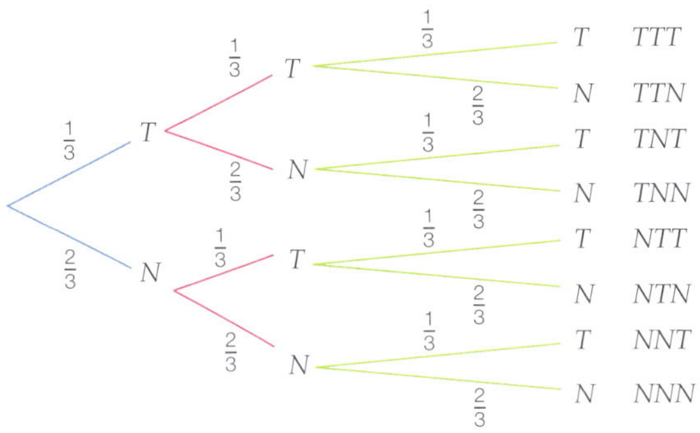

(b) $\frac{1}{27}$ (c) $\frac{8}{27}$

4 (a) B (b) D (c) B (d) C (e) B

5 (a)

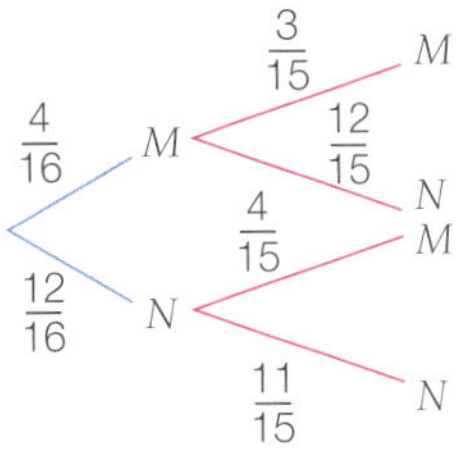

(b) MM, MN, NM, NN

(c) $\frac{1}{20}$ (d) $\frac{2}{5}$ (e) $\frac{11}{20}$

6 (a) 0.2 (b) 0.4

(c)

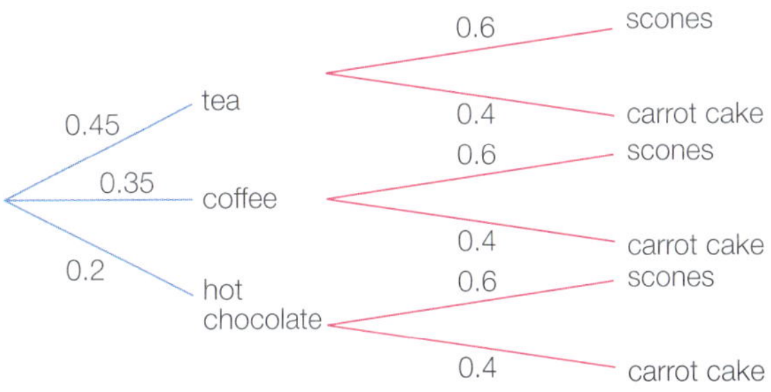

(d) (i) 0.27 (ii) 0.08 (iii) 0.35 (iv) 0.79

7 (a) (i)

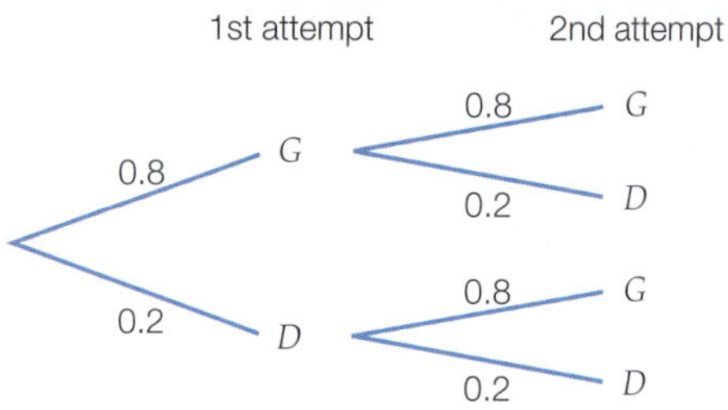

(ii) $\Pr(GG) = 0.64$, $\Pr(GD) = 0.16$, $\Pr(DG) = 0.16$, $\Pr(DD) = 0.04$

(iii) 1

(b) (i)

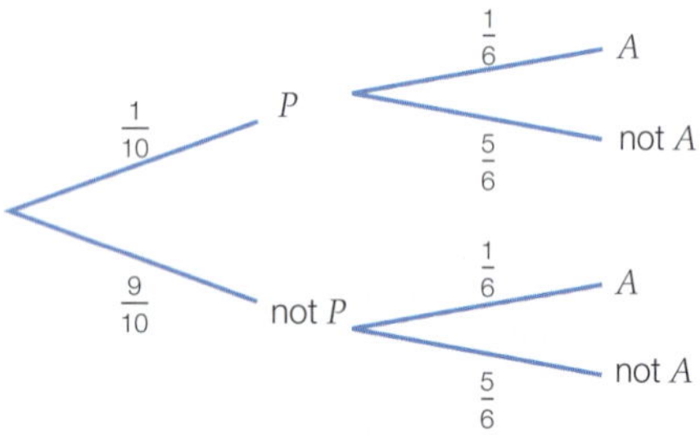

(ii) $\Pr(PA) = \frac{1}{60}$, $\Pr(P, \text{not } A) = \frac{1}{12}$, $\Pr(\text{not } P, A) = \frac{3}{20}$, $\Pr(\text{not } P, \text{not } A) = \frac{3}{4}$

(iii) 1

(c) (i)

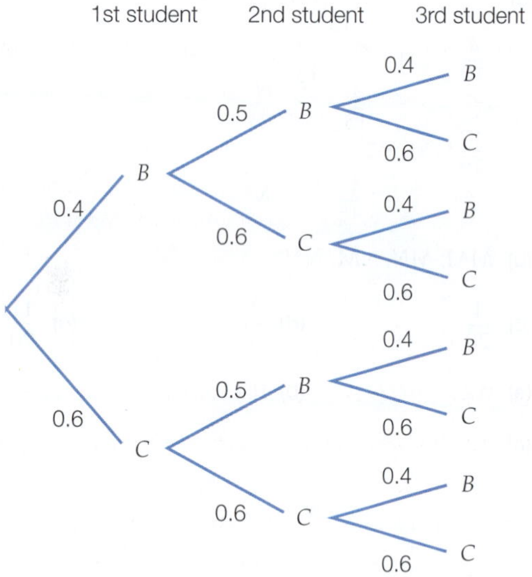

(ii) $\Pr(BBB) = 0.064$, $\Pr(BBC) = 0.096$, $\Pr(BCB) = 0.096$, $\Pr(BCC) = 0.144$, $\Pr(CBB) = 0.096$, $\Pr(CBC) = 0.144$, $\Pr(CCB) = 0.144$, $\Pr(CCC) = 0.216$

(iii) 1

8 (a) 0.008 (b) 0.096 (c) 0.488

9 (a) not T = a 3 is not rolled with a die

(b)

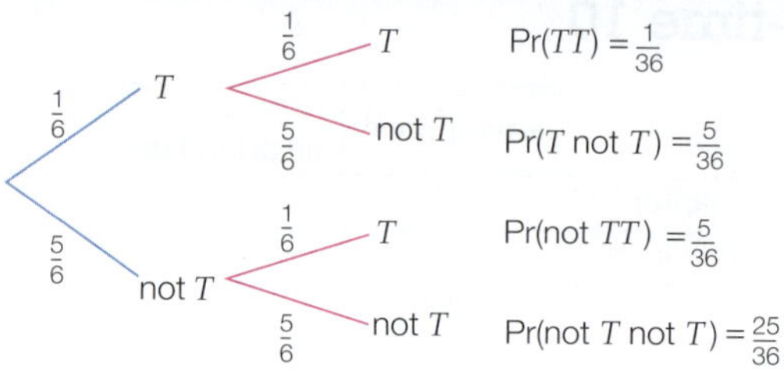

(c) (i) $\frac{1}{36}$ (ii) $\frac{25}{36}$ (iii) $\frac{5}{18}$ (iv) $\frac{11}{36}$ (v) $\frac{35}{36}$

10 (a) $\frac{1}{169}$ (b) $\frac{1}{8}$ (c) $\frac{1}{2}$ (d) $\frac{1}{2}$

11

Total	0	1	2	3	4	5
Probability	0.327 68	0.4096	0.2048	0.0512	0.0064	0.000 32

12 (a)

Number of rolls	1	2	3	4	5	6
Probability	0.167	0.139	0.116	0.096	0.080	0.067

(b) approximately 0.665

(c) This means that it could still take more than six rolls to get the first 6. In fact, it could (theoretically) take an infinite number of rolls.

13 (a) $\frac{7}{29}$ (b) $\frac{3}{11}$

(c) $R^3 = \frac{5}{16}$, $R^4 = \frac{4}{11}$, $R^5 = \frac{3}{7}$, $R^6 = \frac{1}{2}$, $R^7 = \frac{1}{2}$

(d) The probabilities add to more than one. This is because it was assumed that elimination had not taken place before calculating the probability for the next round. This means the events did not all stem from the same tree but a new path was started after each round.

Open-ended – Sample answers

14 Students' own diagrams and explanations.

15 The two events are not mutually exclusive; in fact, playing one of the sports increases the possibility of playing the other, as the person has a demonstrated interest in sport. Armapreet has treated them as mutually exclusive.

Exercise 10.5

1 (a) (i)

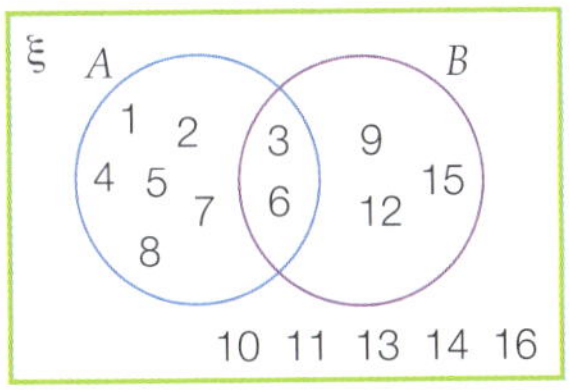

(ii) $\frac{2}{5}$ (iii) $\frac{1}{4}$

(b) (i)

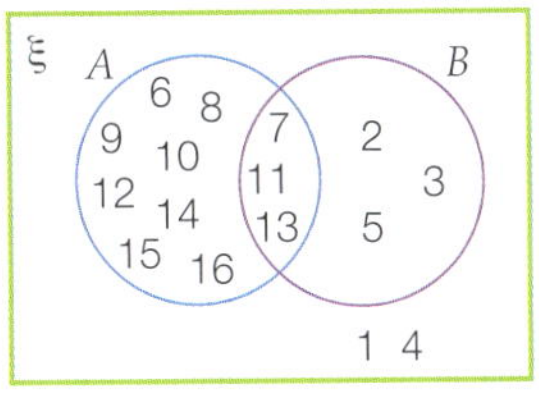

(ii) $\frac{1}{2}$ (iii) $\frac{3}{11}$

2 (a)

	Support PCP	Do not support PCP	
Own property	0.1925	0.1575	0.35
Do not own property	0.26	0.39	0.65
	0.4525	0.5475	

(b) 0.57

3 (a) (i) $\frac{2}{3}$ (ii) $\frac{3}{7}$

(b) (i) $0.1\dot{5}$ (ii) $0.2414 = \frac{7}{29}$

4 (a) C   (b) D

5 (a) $\frac{1}{13}$ (b) $\frac{1}{2}$ (c) $\frac{1}{13}$ (d) $\frac{1}{2}$

6 (a)

	Infected	Not infected	
Positive	0.02375	0.06825	0.092
Negative	0.00125	0.90675	0.908
	0.025	0.975	

(b) 0.26   (c) 0.999

7 (a)

	Defective	Not defective	
A	0.0325	0.6175	0.65
B	0.0315	0.3185	0.35
	0.064	0.936	

(b) 0.51   (c) 0.34

8 (a) 0.9375   (b) 0.7895   (c) 0.8

9 (a) 0.784   (b) 0.91   (c) 0.7

(d) 0.83692   (e) 0.973

10 $0.\dot{5}$

Open-ended – Sample answers

11 (a) Not enough information has been provided to answer the question.

(b) Given Pr(A) = 0.6, Pr(B) = 0.4 and Pr(A and B) = 0.1, find Pr(A given B). Pr(A given B) = 0.25

Exercise 10.6

1 $\frac{1}{10}$

2 (a) $\frac{1}{78}$ (b) $\frac{9}{39} = \frac{3}{13}$

3 (a) $\frac{10}{27}$ (b) $\frac{1}{9}$ (c) $\frac{2}{9}$ (d) $\frac{16}{27}$

4 (a) $\frac{3}{40}$ (b) $\frac{3}{40}$ (c) $\frac{27}{40}$

5 (a) I   (b) D   (c) I   (d) D   (e) D   (f) D

6 (a) $\frac{1}{35}$ (b) $\frac{2}{35}$ (c) $\frac{4}{15}$ (d) $\frac{2}{35}$

(e) $\frac{16}{105}$ (f) $\frac{16}{105}$ (g) $\frac{37}{105}$ (h) $\frac{68}{105}$

7 (a) A   (b) D   (c) A

8 (a) $\frac{3}{7}$ (b) $\frac{4}{35}$ (c) $\frac{2}{7}$

9 (a) $\frac{5}{51}$ (b) $\frac{28}{153}$ (c) $\frac{2}{51}$ (d) $\frac{49}{153}$

(e) $\frac{104}{153}$ (f) $\frac{8}{17}$ (g) $\frac{29}{51}$ (h) $\frac{91}{153}$

10 (a) $\frac{1}{8}$ (b) $\frac{1}{32}$ (c) $\frac{3}{64}$ (d) $\frac{1}{4}$

11 (a) 0.35   (b) 0.5

12 (a) 0.3   (b) 0.49   (c) 0.027   (d) 0.0081

13 (a) $\frac{4}{91}$ (b) $\frac{24}{91}$

14 (a) (1, 1), (1, 2), (1, 3), (1, 4), (1, 5), (1, 6), (2, 1), (2, 2), (2, 3), (2, 4), (2, 5), (2, 6), (3, 1), (3, 2), (3, 3), (3, 4), (3, 5), (3, 6), (4, 1), (4, 2), (4, 3), (4, 4), (4, 5), (4, 6), (5, 1), (5, 2), (5, 3), (5, 4), (5, 5), (5, 6), (6, 1), (6, 2), (6, 3), (6, 4), (6, 5), (6, 6)

(b) Pr(A) = $\frac{1}{6}$; Pr(B) = $\frac{1}{6}$; Pr(C) = $\frac{5}{36}$; Pr(D) = $\frac{1}{12}$

(c) 6 pairs: AB, AC, AD, BC, BD, CD

(d) See if Pr(A) × Pr(B) = Pr(A and B). If it is then the events are independent.

(e) Only AB are independent.

15 (a) 0.5

(b) Use Pr(A or B) = Pr(A) + Pr(B) − Pr(A and B) = 0.65

16 If independent Pr(B) × Pr(M) = Pr(B and M), where B stands for 'black hair' and M stands for 'mathematics'.

Pr(B) = $\frac{475}{900}$, Pr(M) = $\frac{870}{900}$, Pr(B and M) = $\frac{470}{900}$,

Pr(B) × Pr(M) = $\frac{551}{1080}$.

So, Pr(B) × Pr(M) ≠ Pr(B and M).

17 (a) 0   (b) 0.1575

Open-ended – Sample answers

18 (a) Possible solutions:
1, 2, 3, … 12 or 1, 2, 3, … 24 or 2, 3, 4, … 13 etc.

(b) Possible solutions:
1, 2, 3, … 10 or 1, 2, 3, … 20 etc.

19 (a) Possible solution:
$x = y = z = 4$

(b) Possible solution:
$x = 4, y = 10, z = 4$

(c) Possible solution:
$x = 4, y = 20, z = 4$

Chapter review 10

1 (a) $\frac{1}{5}$ (b) $\frac{4}{5}$ (c) $\frac{3}{5}$ (d) $\frac{2}{5}$ (e) $\frac{3}{5}$ (f) $\frac{4}{5}$

2 (a) $\frac{3}{15} = \frac{1}{5}$ (b) $\frac{8}{15}$ (c) $\frac{10}{15} = \frac{2}{3}$

3 $\frac{4}{6} = \frac{2}{3}$

4 (a) Possible outcomes

		Spin 2			
		1	2	3	4
Spin 1	1	(1, 1)	(2, 1)	(3, 1)	(4, 1)
	2	(1, 2)	(2, 2)	(3, 2)	(4, 2)
	3	(1, 3)	(2, 3)	(3, 3)	(4, 3)
	4	(1, 4)	(2, 4)	(3, 4)	(4, 4)

(b) (i) $\frac{1}{16}$ (ii) $\frac{3}{8}$ (iii) $\frac{9}{16}$ (iv) $\frac{1}{4}$

5 $\frac{1}{4}$

6 (a) 18 (b) 8 (c) 13 (d) 4 (e) 10 (f) 3

7 (a) D (b) B

8 (a)

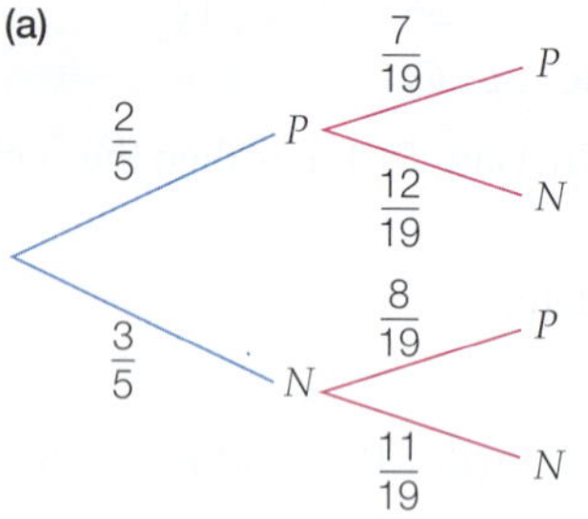

$Pr(PP) = \frac{14}{95}$

$Pr(PN) = \frac{24}{95}$

$Pr(NP) = \frac{24}{95}$

$Pr(NN) = \frac{33}{95}$

(b) $\frac{14}{95}$ (c) $\frac{33}{95}$ (d) $\frac{62}{95}$

9 (a) (i) 0.4286 (ii) 0.3333
(b) (i) 0.4706 (ii) 0.2759

10 $\frac{6}{99} = \frac{2}{33}$

11 (a) $\frac{1}{7}$ (b) $\frac{3}{7}$ (c) $\frac{3}{7}$ (d) $\frac{6}{7}$ (e) $\frac{2}{7}$ (f) $\frac{3}{7}$

12 (a) 0.25

(b) If H = hit and M = miss:

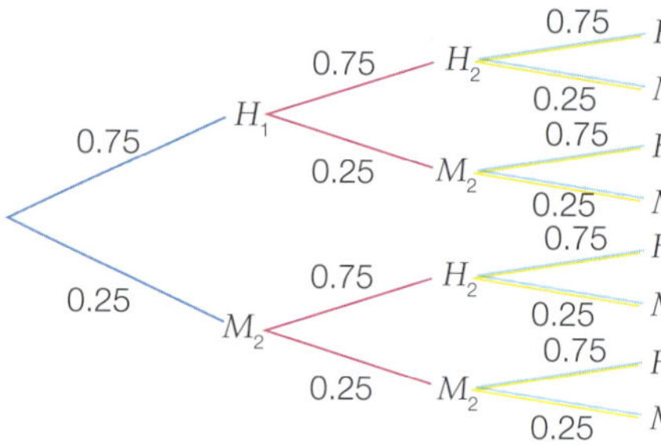

$Pr(HHH) = 0.422$
$Pr(HHM) = 0.141$
$Pr(HMH) = 0.141$
$Pr(HMM) = 0.047$
$Pr(MHH) = 0.141$
$Pr(MHM) = 0.047$
$Pr(MMH) = 0.047$
$Pr(MMM) = 0.016$

(c) (i) 0.422 (ii) 0.141 (iii) 0.984

13 (a)

	lemonade	not lemonade	
cola	12	18	30
not cola	28	2	30
	40	20	60

(b) (i) $\frac{2}{3}$ (ii) $\frac{7}{15}$ (iii) $\frac{1}{2}$ (iv) $\frac{3}{10}$
(v) $\frac{1}{5}$ (vi) $\frac{1}{3}$ (vii) $\frac{29}{30}$ (viii) $\frac{1}{30}$

14 (a) $\frac{7}{16}$ (b) $\frac{9}{16}$ (c) $\frac{13}{16}$ (d) $\frac{3}{8}$

15 (a) 0.35 (b) 0.275
(c) 0.015 (d) 0.012

16 (a) (i) $\frac{1}{8}$ (ii) $\frac{3}{8}$ (iii) $\frac{3}{8}$ (iv) $\frac{1}{8}$

(b) 1. Yes, this is an expected result as these outcomes cover the sample space.

(c) Pr(0 tails) = Pr(3 heads); Pr(1 tails) = Pr(2 heads); etc.

(d) (i) $\frac{1}{8}$ (ii) $\frac{1}{8}$ (iii) $\frac{1}{8}$

(e) Any outcome with T exactly once has the same probability as any other outcome with T exactly once. The order does not matter.

17 (a) $\frac{16}{81}$ (b) $\frac{344}{1215}$ (c) $\frac{296}{1215}$

18 (a) 0.015 (b) 0.16 (c) 0.032 (d) 0.12

(e) Sydney and Canberra are relatively close, so if it rains in one city then the probability of rain in both cities will be higher. The two events are not independent.

Mixed review C

1 (a) $(a - 9)(a + 4)$ (b) $(a - 6)(a - 2)$
(c) $(2a + 5)^2$

2 (a) 104 cm^2 (b) 37 m^2 (c) 103.5 cm^2

3 (a) -6 (b) $-\frac{15}{16}$ (c) $\frac{11}{6}$

4 (a) 42.7° (b) 23.9° (c) 50.5°

5 4.85 cm

6 The proof makes use of the following diagram, and its named vertices.

To show: $\Delta ACD \equiv \Delta ABD$

$AC = AB$ (given)

$CD = BD$ (given)

$AD = AD$ (shared)

So, $\Delta ACD \equiv \Delta ABD$

So, $\angle BAE = \angle CAE$

(corresponding angles in congruent triangles)

Now to show: $\Delta ABE \equiv \Delta ACE$

$\angle BAE = \angle CAE$ (from above)

$AC = AB$ (given)

$AE = AE$ (shared)

So, $\Delta ABE \equiv \Delta ACE$

So, $\angle AEB = \angle AEC$

(corresponding angles in congruent triangles)

But, $\angle AEB + \angle AEC = 180°$ (straight angle)

So, $\angle AEB = \angle AEC = 90°$

So, the diagonals of a kite intersect at right angles.

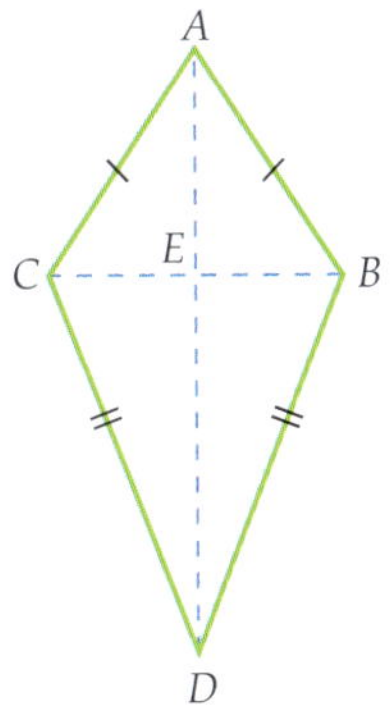

7 (a) 33.36 (b) 125.24

8 (a) 2786.89 cm^2 (b) 256.31 m^2

9 (a) 301.59 cm^3 (b) 164.64 cm^3

10 (a)

x	-3	-2	-1	0	1	2	3
y	15	6	1	0	3	10	21

(b)

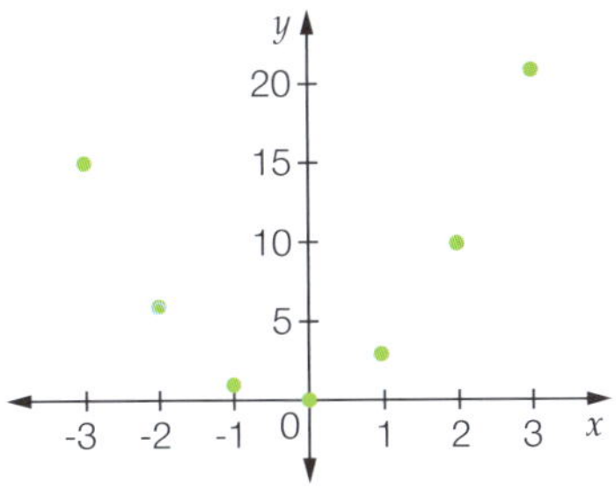

(c) (-0.25, -0.125)

(d) x-intercepts: -0.5 and 0; y-intercept: 0

11 (a)

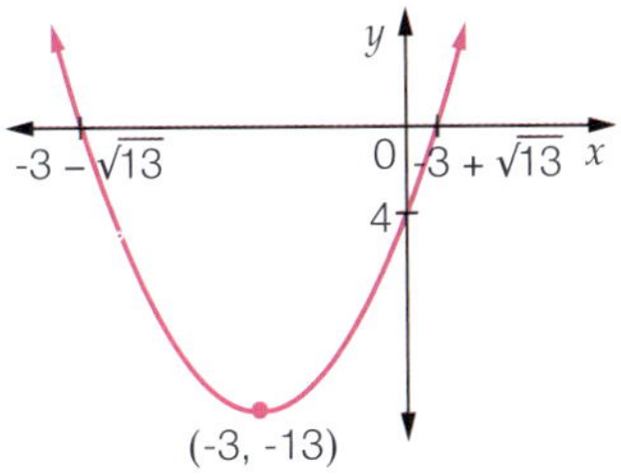

(b)

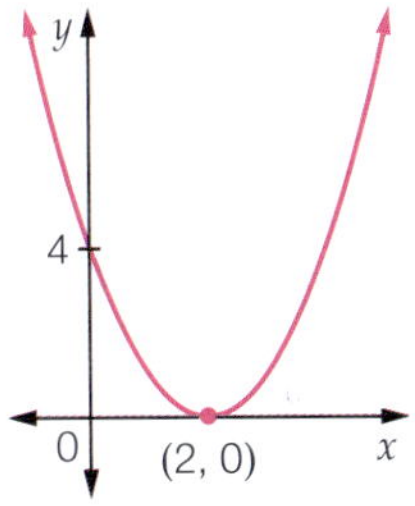

(c)

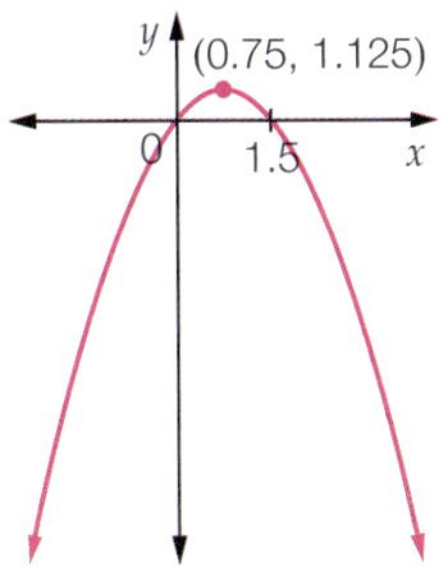

12 (a) $x = \pm 5$ (b) $x = 0$ and -0.5 (c) $x = 2$ and 3

13 (a) (2, 3) (b) (-4, -1) (c) (7, -6)

14 (a) $A = w(w + 2.5)$ cm^2

(b)

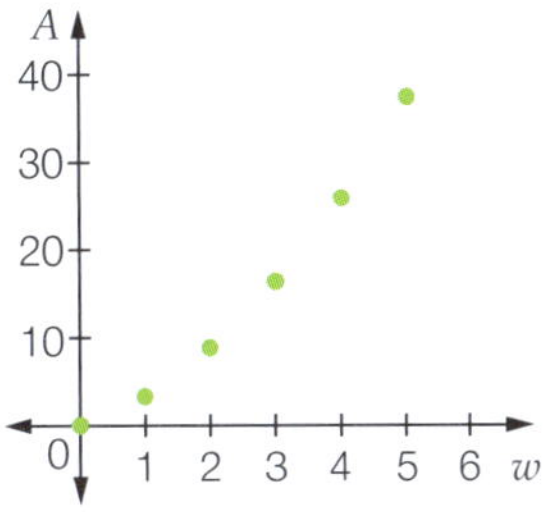

(c) $w = 5$ from the graph so the dimensions are 5 cm by 7.5 cm

15 (a) $x = 5$

(b) $y = (x - 4)(x - 6)$

$y = x^2 - 10x + 24$

16 384 cm^2

17 (a) 81 cm^2 (b) 66.21 cm^2 (c) 37.12 cm^2

18 (a) 24.5 m^2 (b) 34.1 m

Challenge C

1 C 2 B

3 ΔADC and ΔDBC are equal in area. They have the same height, the perpendicular distance from C to AB.

Hence their bases must be the same, so $AD = DB$.

ΔADC is isosceles.

ΔDBC is equilateral so $\angle BDC = 60°$

Hence, $\angle ADC = 120°$ and $\angle DCA = \angle DAC = 30°$

$\therefore x = 150°$. (You can now say that ΔABC is right angled at C.)

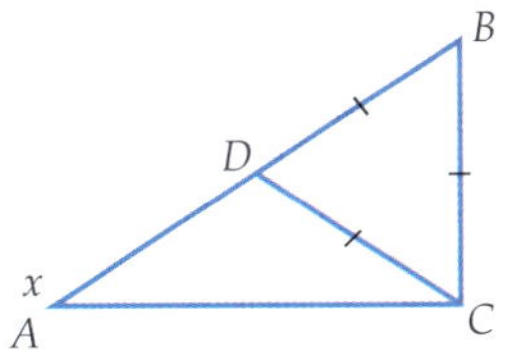

4 B $AE = 25, AD = 24, ED = 7$, by Pythagoras' theorem

$\angle AFD = 90°$ (angle in a semicircle)

$AB = DF, BC = FE$ (corresponding matching sides of similar triangles)

$AC = ED = 7, CB = EF$
$FC = 25 - (EF + 7)$
$DB = 24 - AB$

Perimeter of $CBDF = DB + BC + CF + FD$
$= 24 - FD + EF + 25 - (EF + 7) + FD = 49 - 7 = 42$ units

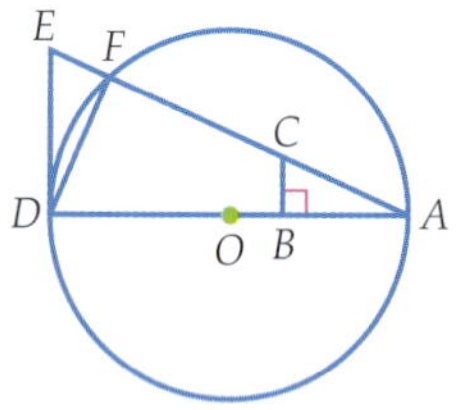

5 45°

6 (a) $x^2 - 3x - 18 = -14, x^2 - 3x - 4 = 0,$
$(x - 4)(x + 1) = 0, x = -1, 4$

(b) $(x^2 - 3x)^2 - 3(x^2 - 3x) - 4 = 0, x^2 - 3x = -1, (x^2 - 3x) = 4$
$x^2 - 3x + 1 = 0, x^2 - 3x - 4 = 0, x = \frac{3 \pm \sqrt{9 - 4}}{2},$
$(x - 4)(x + 1) = 0, x = \frac{3 + \sqrt{5}}{2}, \frac{3 - \sqrt{5}}{2}, -1, 4$

7 (a) $(x - 7)(x + 1) = 0, x = -1, 7$ Sum = 6

(b) $(x - 2)(x + 2) = 0, x = -2, 2$ Product = -4

(c) $x(x^2 - 6x + 5) = 0, x(x - 1)(x - 5) = 0, x = 0, 1, 5$ Average = 2

8 (a) $7 = (c - 2)(c - 8) + 7$
$(c - 2)(c - 8) = 0$
$c = 2$ or $c = 8$

(b) vertex (5, -2)

9 (a) $y = -3, 2$ (b) $(-\sqrt{5}, 2), (0, -3), (\sqrt{5}, 2)$

Chapter 11

Recall 11

1 (a) $\pm\sqrt{3}$ (b) $\sqrt[3]{5}$ (c) $\pm\sqrt{7}$ (d) $\sqrt[3]{6}$ (e) $\pm\sqrt[4]{10}$

2 (a) 2 (b) 6 (c) -3 (d) 1 (e) 2 (f) 4 (g) -3 (h) 1 (i) 2 (j) -1

3 (a) 3.32 (b) -2.65 (c) 2.47 (d) 1156 (e) 19 683

4 (a) $3x + 12$ (b) $-8y + 12$

5 (a) $x^2 - y^2$ (b) $c^2 - 81$ (c) $4p^2 - 9$

Exercise 11.1

1 (a) irrational (b) rational (c) rational (d) irrational (e) rational (f) irrational (g) irrational (h) rational (i) irrational (j) irrational (k) irrational (l) rational

2 (a) 8 (b) -5 (c) -6 (d) 9 (e) 42 (f) -28 (g) 13 (h) 7 (i) 69 (j) 30 (k) 16 (l) 73

3 (a) 3.146 (b) 0.592 (c) -0.960 (d) 5.916 (e) -1.732 (f) 2.571 (g) 2.154 (h) 2.449 (i) -4.160

4 (a) C (b) D (c) D (d) D

5 (a) C (b) A

6 (a) B (b) D

7 (a) B (b) C

8 (a) F (b) T (c) T (d) F (e) T

9 (a) 5 m (b) $\sqrt{17}$ m

10 (a) $\sqrt{60}$ m (b) 31 m

11 (a) 2 m (b) $\sqrt[3]{15}$ cm

12 (a) $\sqrt[3]{7}$ cm (b) 19 mm

Open-ended – Sample answers

13 (a) $\sqrt{2}, \pi, \sqrt[3]{5}$ (b) $-\sqrt{3}, -2\sqrt{5}, -2\sqrt[3]{6}$

(c)

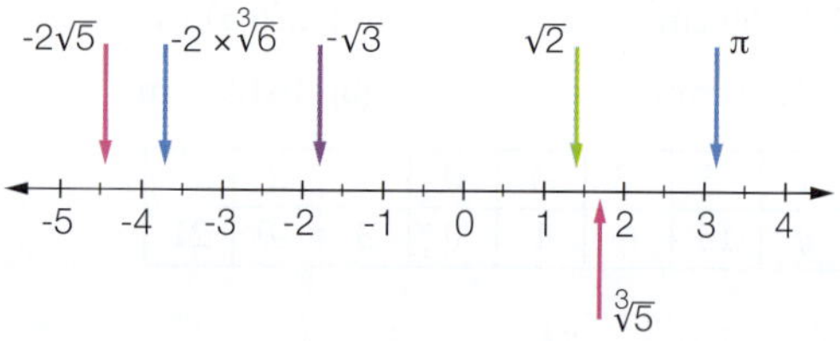

14 Lie the metre ruler on a flat surface and rule a 1 m line with the pencil. Place an end of the ruler on the line and rule another metre line at right angles to the first. Run the string from one end of a metre line to the other (as if forming a triangle) this length is $\sqrt{2}$ metres.

Exercise 11.2

1 (a) $\sqrt{6}$ (b) $\sqrt{55}$ (c) $\sqrt{70}$ (d) 7 (e) 4 (f) 8 (g) $6\sqrt{2}$ (h) $5\sqrt{7}$ (i) $3\sqrt{2}$ (j) $10\sqrt{6}$ (k) $4\sqrt{33}$ (l) $40\sqrt{2}$ (m) 35 (n) 24 (o) 40 (p) 42 (q) $15\sqrt{6}$ (r) $144\sqrt{2}$

2 (a) 3 (b) 4 (c) 5 (d) 2 (e) 6 (f) 3 (g) 1 (h) 2 (i) $\frac{5}{3}$ (j) $\frac{10}{7}$ (k) $\frac{2}{9}$ (l) $\frac{6}{11}$ (m) $\frac{1}{4}$ (n) $\frac{1}{3}$ (o) 4 (p) 5 (q) 1 (r) $\frac{1}{3}$ (s) 30 (t) 10 (u) $6\sqrt{6}$ (v) $\frac{\sqrt{5}}{5}$ (w) $\frac{2\sqrt{7}}{5}$ (x) $\frac{2\sqrt{3}}{3}$

3 (a) $3\sqrt{3}$ (b) $2\sqrt{5}$ (c) $9\sqrt{2}$ (d) $4\sqrt{6}$ (e) $3\sqrt{11}$ (f) $10\sqrt{5}$ (g) $9\sqrt{6}$ (h) $9\sqrt{5}$ (i) $10\sqrt{2}$ (j) $-8\sqrt{6}$ (k) $-15\sqrt{7}$ (l) $36\sqrt{3}$

4 (a) $\sqrt{12}$ (b) $\sqrt{63}$ (c) $\sqrt{72}$ (d) $\sqrt{27}$
(e) $-\sqrt{600}$ (f) $-\sqrt{40}$ (g) $-\sqrt{192}$ (h) $\sqrt{175}$
(i) $\sqrt{128}$ (j) $\sqrt{245}$ (k) $-\sqrt{108}$ (l) $-\sqrt{275}$

5 (a) C (b) B (c) C (d) D (e) B

6 (a) $45\sqrt{2}$ (b) $16\sqrt{7}$ (c) $-28\sqrt{15}$ (d) $-32\sqrt{5}$
(e) $-15\sqrt{5}$ (f) 120 (g) 48 (h) $36\sqrt{6}$
(i) $168\sqrt{21}$ (j) $100\sqrt{15}$ (k) $216\sqrt{14}$ (l) $54\sqrt{3}$

7 (a) $2\sqrt{2}$ (b) $3\sqrt{3}$ (c) $9\sqrt{5}$ (d) $-12\sqrt{6}$
(e) $\frac{3\sqrt{3}}{2}$ (f) $\frac{4\sqrt{2}}{5}$ (g) $6\sqrt{2}$ (h) $3\sqrt{7}$
(i) $\frac{5\sqrt{5}}{6}$ (j) $-\frac{2\sqrt{11}}{3}$ (k) $-\frac{3\sqrt{3}}{5}$ (l) $\frac{7\sqrt{3}}{10}$

8 (a) 75 m^2 (b) 16 m^2 (c) 12 km^2 (d) $9\sqrt{6}\text{ cm}^2$

9 (a) $2\sqrt{10}$ mm (b) $6\sqrt{5}\pi$ m

10 (a) $\sqrt{0.2825}$ m or $\sqrt{2825}$ cm (b) 6 windows

Open-ended – Sample answers

11 $b = 9, c = 2$

12 $\sqrt{24}, \frac{\sqrt{96}}{2}, \frac{\sqrt{216}}{3}$

13 (a) $\sqrt{5} \times \sqrt{5}$ (b) $\frac{8\sqrt{2}}{\sqrt{8}}$

Exercise 11.3

1 (a) $10\sqrt{2}$ (b) $3\sqrt{5}$ (c) $5\sqrt{3}$
(d) $8\sqrt{7}$ (e) $-4\sqrt{3}$ (f) $-23\sqrt{11}$
(g) $11\sqrt{2}$ (h) 0 (i) $-3\sqrt{5}$
(j) $5\sqrt{6} - 5\sqrt{3}$ (k) $7\sqrt{2} - 5\sqrt{5}$ (l) $-\sqrt{3} - 4\sqrt{11}$

2 (a) $4\sqrt{2}$ (b) $\sqrt{7}$ (c) $3\sqrt{5}$
(d) $36\sqrt{2}$ (e) $\sqrt{7}$ (f) $-4\sqrt{10}$
(g) $9\sqrt{10}$ (h) $27\sqrt{2}$ (i) $8\sqrt{6}$
(j) $49\sqrt{3}$ (k) $-30\sqrt{5}$ (l) $19\sqrt{2}$
(m) $2\sqrt{7}$ (n) 0

3 (a) B (b) A (c) D

4 (a) C (b) D

5 (a) $8\sqrt{2}$ (b) $20\sqrt{3}$
(c) $11\sqrt{2}$ (d) $8\sqrt{10} + 12\sqrt{7}$

6 (a) $x = 4\sqrt{5}$ (b) $y = \sqrt{6}$

7 $\frac{3\pi}{4}\text{ cm}^2$

8 (a) $6\sqrt{5} + 4\sqrt{3}$ m (b) $6\sqrt{15}\text{ m}^2$ (c) no

Open-ended – Sample answers

9 length = $2\sqrt{6}$ cm, width = 2 cm, perimeter = $4\sqrt{6} + 4$ cm;
length = $2\sqrt{3}$ cm, width = $2\sqrt{2}$ cm,
perimeter = $4\sqrt{3} + 4\sqrt{2}$ cm

10 (a) (i) Ruby has added the numbers underneath the square root.
(ii) When subtracting one surd from another, Ruby has forgotten to include the surd part of the solution.
(b) (i) $7\sqrt{3}$ (ii) $3\sqrt{6}$
(c) (i)
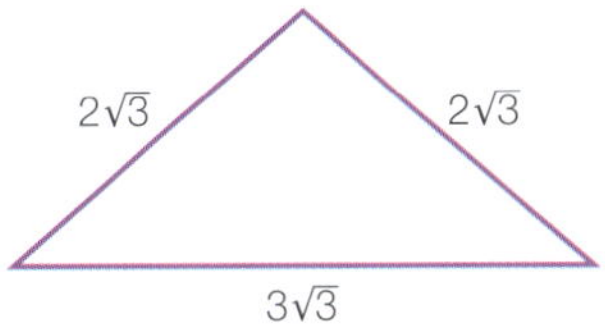

(ii)
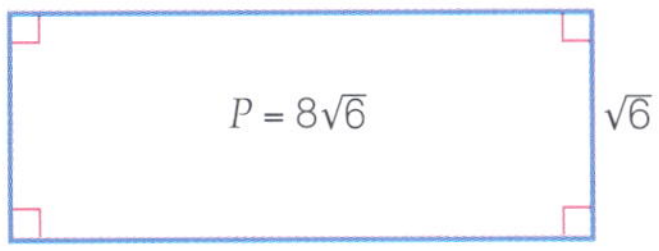

Half-time 11

1 (a) 6 (b) 88 (c) $\frac{9}{4}$

2 $15\sqrt{2}\text{ m}^2$

3 $3\sqrt{5}$ m

4 (a) rational (b) rational (c) irrational (d) rational

5 $(18 + 12\sqrt{15})\text{ m}^2$

6 (a) $6\sqrt{2}$ (b) 360 (c) $\frac{3}{5}$

7 (a) -15 (b) 17 (c) 34

Exercise 11.4

1 (a) $3\sqrt{7} + 6$ (b) $-4\sqrt{11} - 4\sqrt{3}$
(c) $18\sqrt{5} - 9\sqrt{2}$ (d) $3\sqrt{35} - 5\sqrt{14}$
(e) $8\sqrt{10} + 24\sqrt{6}$ (f) $3\sqrt{10} - 21\sqrt{15} + 9\sqrt{35}$
(g) $3 + 2\sqrt{3}$ (h) $15\sqrt{14} - 70$
(i) $24\sqrt{15} - 6\sqrt{5} - 15$ (j) $6\sqrt{7} - 12$
(k) $12 - 12\sqrt{2} + 20\sqrt{3}$ (l) $40 - 4\sqrt{5} - 20\sqrt{3}$
(m) $\sqrt{10} + 3\sqrt{5} + 4\sqrt{2} + 12$ (n) $3 + \sqrt{21} - \sqrt{15} - \sqrt{35}$
(o) $5 + 2\sqrt{6}$ (p) $88 - 16\sqrt{10}$
(q) 25 (r) -53

2 (a) B (b) D (c) C (d) C

3 (a) $\pi(19 - 6\sqrt{10})\text{ m}^2$ (b) $(2\sqrt{35} + 12\sqrt{7})\text{ cm}^2$

4 $\pi(21 + 8\sqrt{5})\text{ cm}^2$

5 (a) 21 km^2 (b) $(14 + 3\sqrt{2} + \sqrt{10})$ km
(c) 21 405 m

6 (a) (i) $(18\sqrt{2} - 12)$ m (ii) $(12\sqrt{2} - 8)$ m
(iii) $(15\sqrt{2} - 10)$ m (iv) $(21\sqrt{2} - 14)$ m
(b) (i) 13.46 m (ii) 8.97 m
(iii) 11.21 m (iv) 15.70 m

7 (a) $7\pi\text{ cm}^2$ (b) $(5 + 2\sqrt{7})$ cm by $(3 + 2\sqrt{7})$ cm
(c) $(43 + 16\sqrt{7})\text{ cm}^2$ (d) $(43 + 16\sqrt{7} - 7\pi)\text{ cm}^2$
(e) 63 cm^2

8 (a) 47.74 cm (b) $22\,791$ cm^2

(c) $12\,110 + 1440\sqrt{55}$ cm^2 (d) $22\,789.3$ cm^2

(e) (i) 0%

(ii) The difference is negligible. It is slightly higher due to rounding up.

9 (a) $\sqrt{18}$ or $3\sqrt{2}$ cm (b) $150\sqrt{2}$ cm

(c) 212.1 cm (d) 4.2 cm

(e) 210 cm (f) 1% lower from rounding down.

Open-ended – Sample answers

10 base 4 cm, height $2\sqrt{3}$ cm

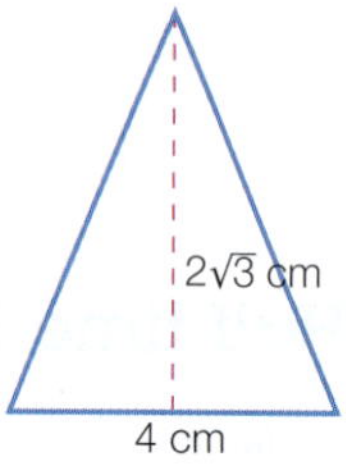

base $\frac{4\sqrt{3}}{3}$ cm, height 6 cm

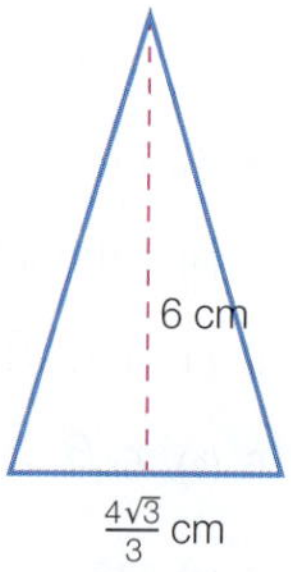

11 (a) $2m\sqrt{n} + 2p$ or $2(m\sqrt{n} + p)$

(b)

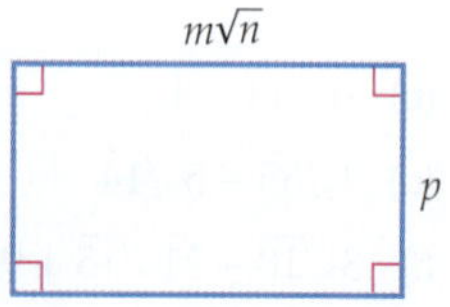

$P = 2(m\sqrt{n} + p)$

If $m = 10$, $n = 3$ and $p = 4$

Then $P = 2(10\sqrt{3} + 4) = 42.641$ cm

(c) $A = m\sqrt{n} \times p = mp\sqrt{n}$

(d) If $m = 2$, $n = 2$ and $p = 3$ then $A = 8.485$ m^2

Exercise 11.5

1 (a) $\frac{\sqrt{7}}{7}$ (b) $\frac{4\sqrt{3}}{3}$ (c) $-2\sqrt{6}$ (d) $\frac{\sqrt{6}}{2}$

(e) $\sqrt{55}$ (f) $\frac{\sqrt{30}}{2}$ (g) $-\frac{3\sqrt{7}}{28}$ (h) $\frac{2\sqrt{15}}{7}$

2 (a) $\frac{3 + \sqrt{7}}{2}$ (b) $8 - 4\sqrt{3}$

(c) $\frac{\sqrt{15} + \sqrt{6}}{3}$ (d) $\frac{4\sqrt{7} - 10}{3}$

(e) $\frac{24 + 15\sqrt{3}}{3} = 8 + 5\sqrt{3}$ (f) $\frac{22 + 10\sqrt{5}}{4} = \frac{11 + 5\sqrt{5}}{2}$

(g) $\frac{6\sqrt{3} - \sqrt{42} - 3\sqrt{2} + \sqrt{7}}{11}$ (h) $\frac{5\sqrt{6} - 8\sqrt{3}}{6}$

3 (a) D (b) A (c) B

4 (a) B (b) D

5 (a) $\frac{2 + 9\sqrt{2}}{6}$ (b) $\frac{12\sqrt{5} - 5\sqrt{3}}{15}$

(c) $\frac{4\sqrt{30} - 3\sqrt{5}}{10}$ (d) $\frac{3\sqrt{2} + \sqrt{10}}{4}$

6 (a) $\frac{2\sqrt{3} + \sqrt{5} + 1}{4}$ (b) $\frac{\sqrt{7} + 2\sqrt{6} + 5}{6}$

(c) $\frac{5\sqrt{6} - 5\sqrt{3} - 12 - 3\sqrt{11}}{15}$ (d) $\frac{3\sqrt{5} - 3\sqrt{3} - 2\sqrt{7} - 4}{6}$

7 (a) $15\,000\sqrt{6}$ cm^2 (b) $50\sqrt{6}$ tiles

(c) As $122 < 50\sqrt{6} < 123$, the minimum number of tiles is 123, but maybe buy 130 tiles to allow for wastage.

(d) $15\,000\sqrt{6} + 6000\sqrt{15} + 600\sqrt{35} + 1500\sqrt{14}$ cm^2

8 $\frac{112\sqrt{3} - 294}{249}$ or $\frac{14(8\sqrt{3} - 21)}{249}$

Open-ended – Sample answers

9 $\frac{4\sqrt{6}}{3}, \frac{8}{\sqrt{6}}, \frac{4\sqrt{10}}{\sqrt{15}}$

10 (a) $\frac{\sqrt{2}}{3}, \frac{2}{3}$ (b) $\frac{2\sqrt{7}}{7}, \frac{4\sqrt{7}}{7}$

(c) $\frac{\sqrt{15}}{5}, \frac{3\sqrt{15}}{5}$ (d) $\frac{\sqrt{2}}{6}, \frac{\sqrt{2}}{2}$

11 $2 - \sqrt{3}$ and $4 - 2\sqrt{3}$

12 When two surds that are the same are multiplied, the result is a whole number, e.g. $\sqrt{6} \times \sqrt{6} = 6$. Multiplying by the conjugate will remove the square roots from the denominator, resulting in a rationalised denominator. It is also like the difference of two squares.

Chapter review 11

1 (a) 4.123 (b) 2.924 (c) 34.641 (d) 2.280

2 (a) $12\sqrt{30}$ (b) 70 (c) $\frac{8}{5}$ or $1\frac{3}{5}$ (d) $9\sqrt{2}$

(e) $2\sqrt{10}$ (f) $10\sqrt{5}$ (g) $30\sqrt{5}$ (h) $5\sqrt{3}$

3 (a) $\sqrt{72}$ (b) $-\sqrt{63}$

4 (a) $\sqrt{2}$ (b) $4\sqrt{7} - \sqrt{3}$

(c) $29\sqrt{3}$ (d) $46\sqrt{2} - \sqrt{6}$

5 (a) $4\sqrt{15} - 10 + \sqrt{10}$ (b) $12 - 18\sqrt{14} - 6\sqrt{2}$

(c) $5\sqrt{5} - 8$ (d) $95 + 24\sqrt{14}$

6 (a) $\frac{5\sqrt{2}}{2}$ (b) $2\sqrt{2}$ (c) $\frac{7\sqrt{5}}{10}$

7 (a) $\frac{21 + 7\sqrt{5}}{4}$ (b) $\frac{2\sqrt{6} - \sqrt{14} - 4\sqrt{3} + 2\sqrt{7}}{5}$

8 (a) C (b) D

9 (a) $2\sqrt{58}$ cm (b) 15.23 cm

10 (a) length 25.3 cm, height 20.2 cm

(b) 1749.3 cm^2

(c) $(180\sqrt{6} + 108\sqrt{7} + 160\sqrt{15} + 48\sqrt{70})\text{ cm}^2$

(d) 1747.9 cm^2 (e) (b) was larger by 1.4 cm^2.

11 (a) $5\sqrt{2}$ cm (b) $\sqrt{\dfrac{160}{\sqrt{3}}}$ cm (c) $10\pi\sqrt{3}$ m

12 $-\dfrac{17 + 136\sqrt{5}}{319}$

Chapter 12

Recall 12

1 (a) $(x+5)(x+2)$ (b) $(x-3)(x-2)$ (c) $(x-14)(x+1)$

(d) $(x+8)(x-3)$ (e) $(x+9)(x-4)$ (f) $2(x+1)(x+6)$

(g) $-(x+2)(x+1)$ (h) $-(x-5)(x-2)$ (i) $-2(x+3)(x-4)$

2 (a) 27 (b) $\dfrac{1}{4}$ (c) $3\dfrac{1}{25}$

3 (a) $x = -3, 1$ (b) $x = -2, -4$

(c) $x = 1$ (d) $x = 0, 5$

4 (a) $x = 3$ (b) $x = \pm\dfrac{5}{2}$ (c) $x = -4$ (d) $x = \pm\dfrac{6}{11}$

5 (a) $(x + 2 - \sqrt{2})(x + 2 + \sqrt{2})$

(b) $(x - 3 + \sqrt{3})(x - 3 - \sqrt{3})$

(c) $(x - \dfrac{3}{2} + \dfrac{\sqrt{29}}{2})(x - \dfrac{3}{2} - \dfrac{\sqrt{29}}{2})$

6 (a) m^5 (b) $12k^3p^9$ (c) x^7

(d) 3 (e) $m^{12}n^4$ (f) $\dfrac{x^6}{y^4}$

Exercise 12.1

1 (a)

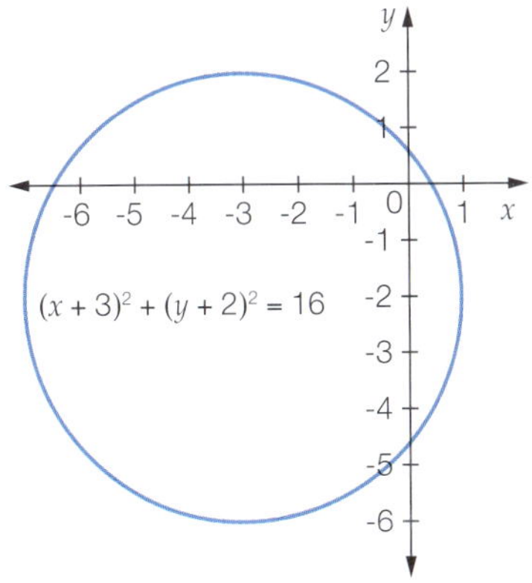

(b)

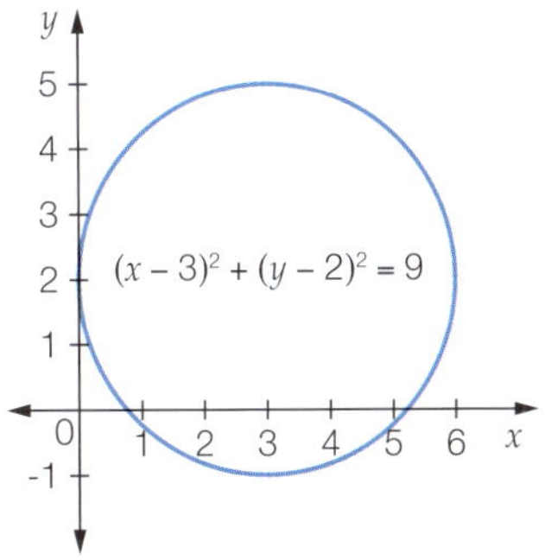

(c)

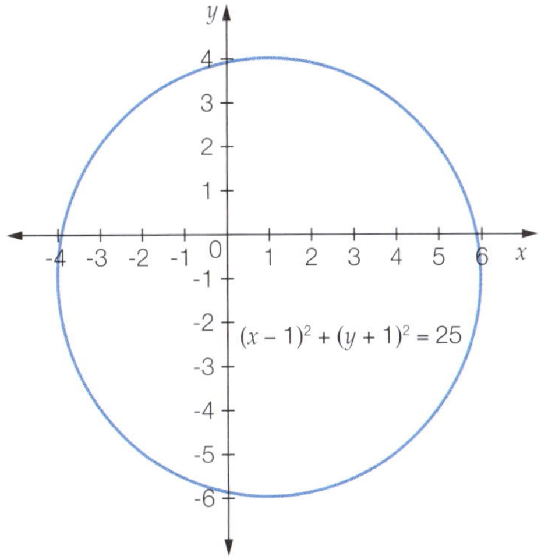

(d)

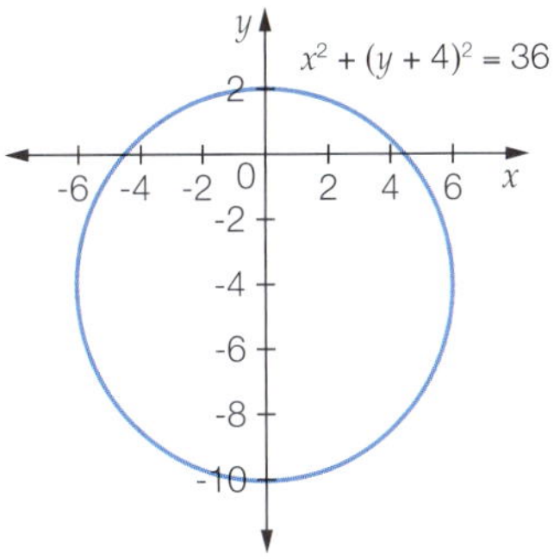

(e)

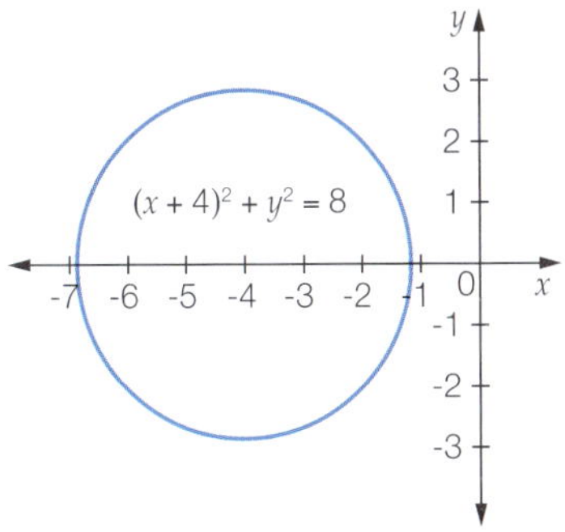

(f)

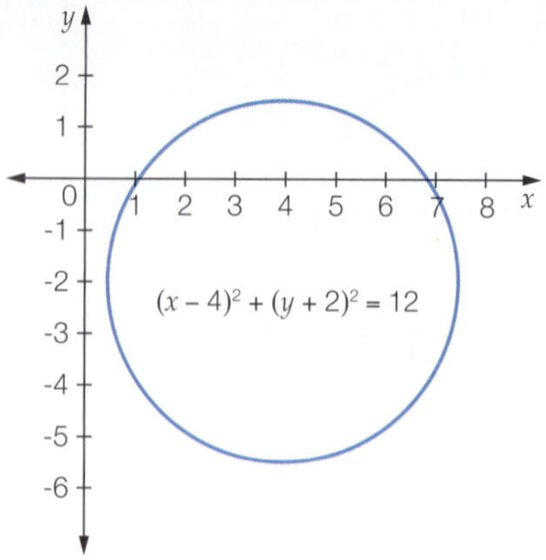

2 (a)

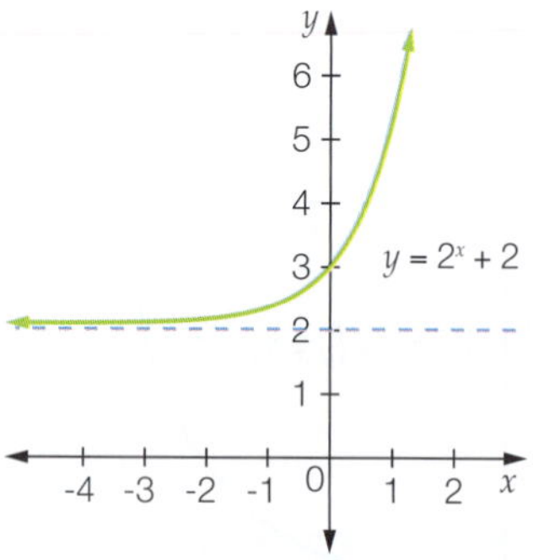

(b)

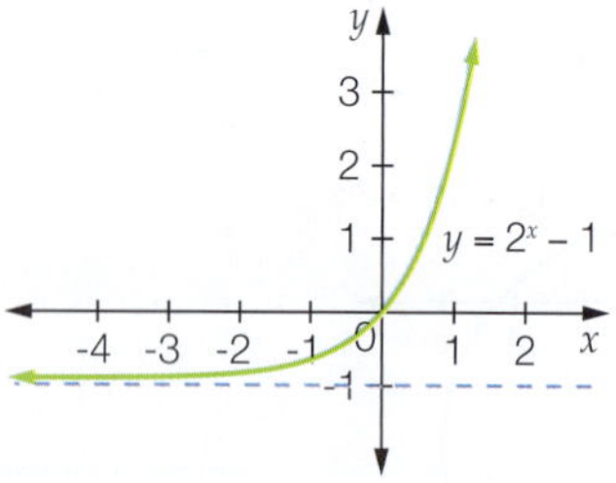

(c)

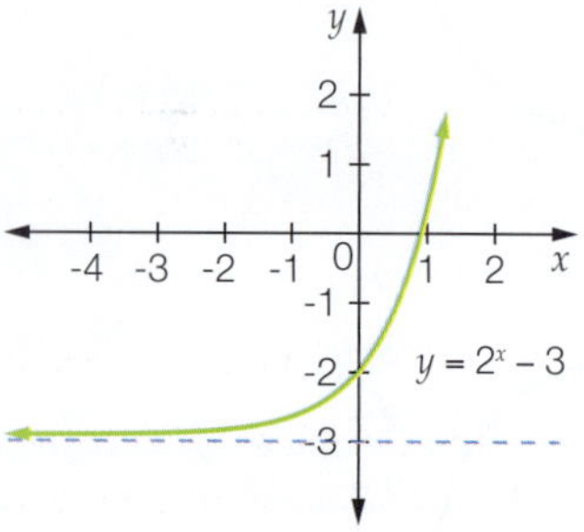

(d)

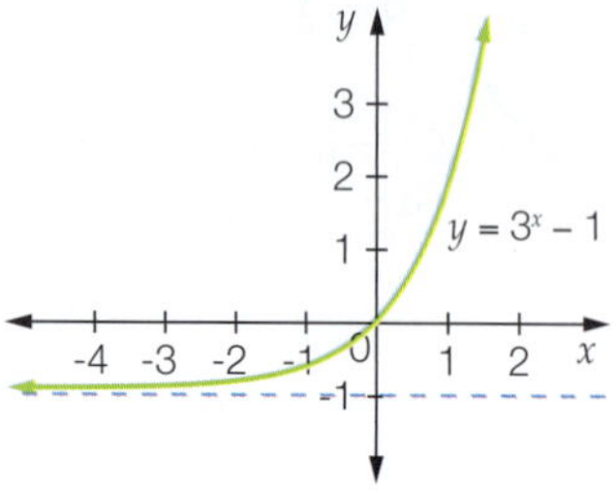

(e)

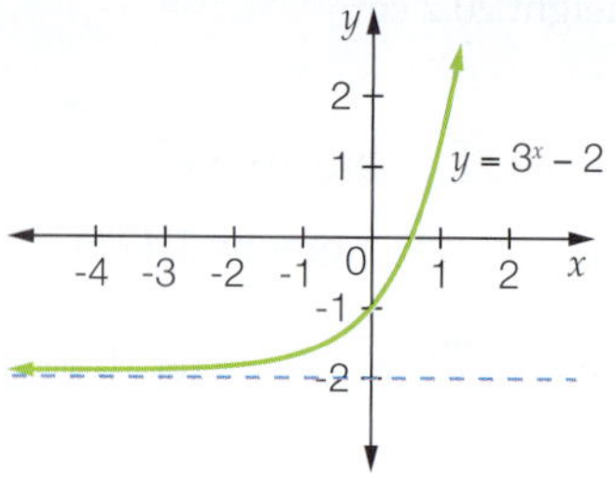

(f)

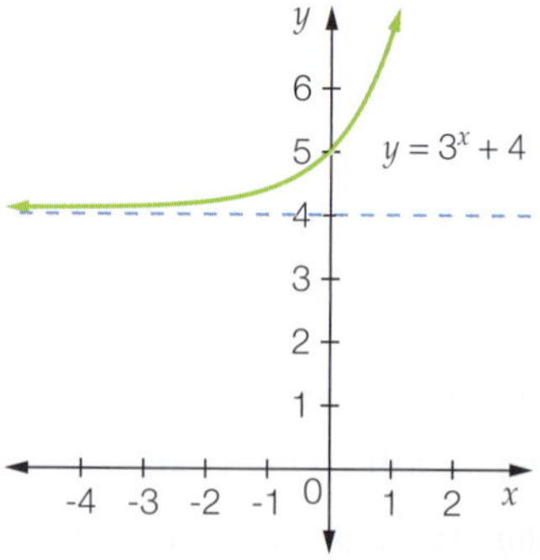

3 (a)

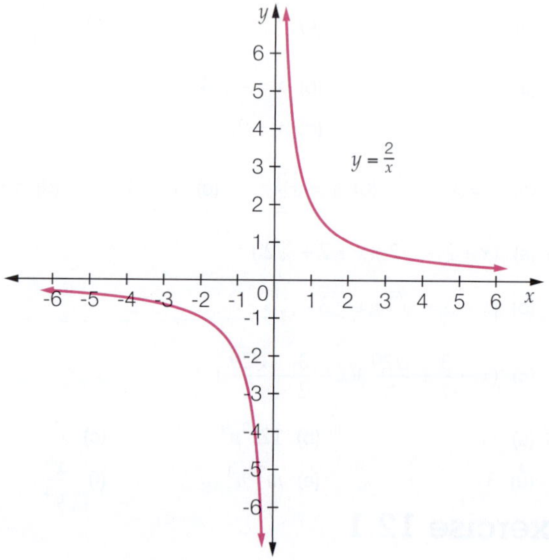

(b)

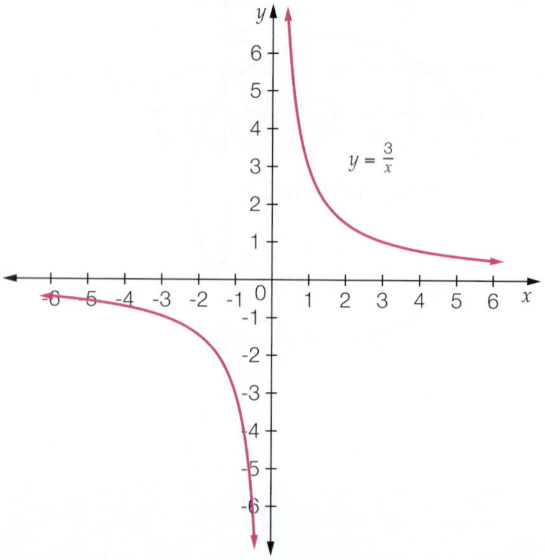

(c)

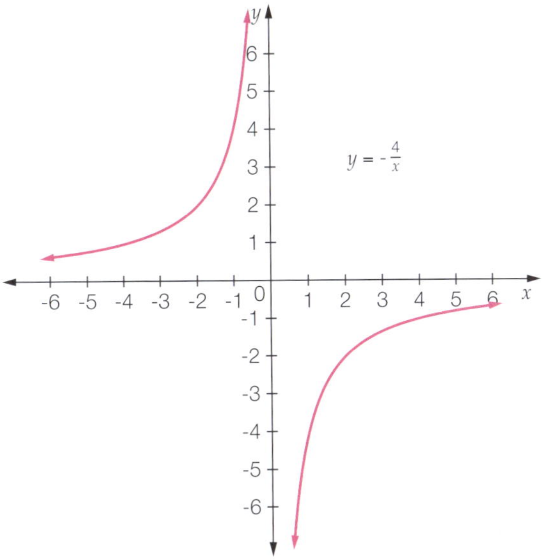

(d)

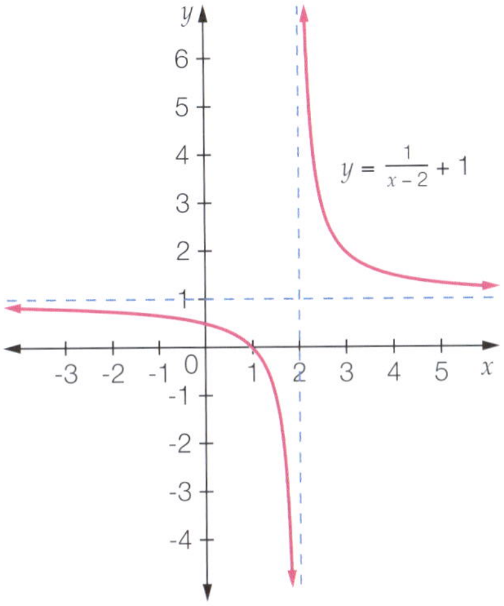

(e)

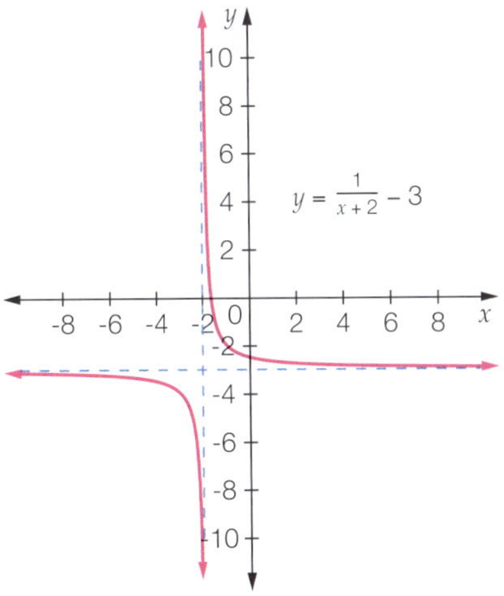

(f)

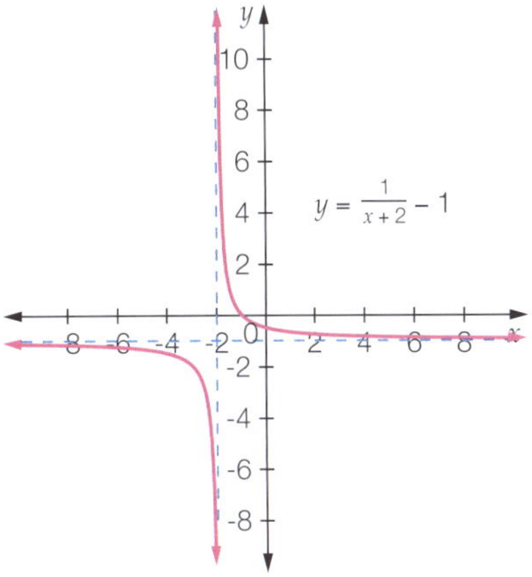

4 A　　**5** B　　**6** D

7 **(a)**

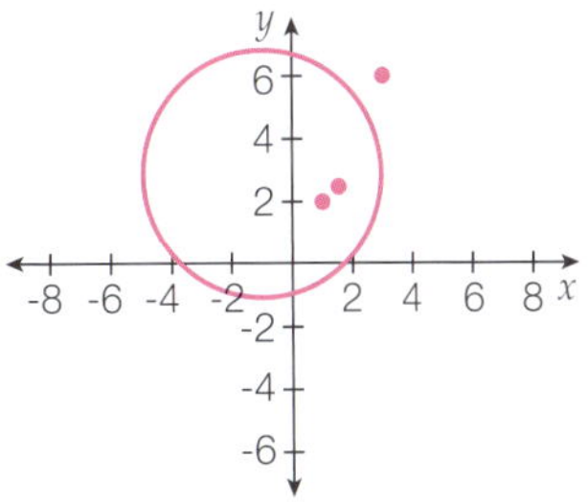

(b) (1, 2) and (1.5, 2.5) are inside and (3, 6) is outside.

8 **(a)**

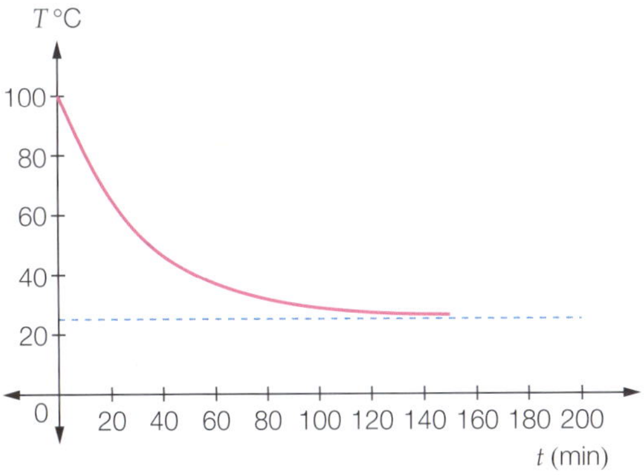

(b) 100 °C

(c) **(i)** 79 °C　　**(ii)** 65 °C　　**(iii)** 46 °C　　**(iv)** 40 °C

(d) The first 10 minutes. The coffee lost 21 °C in the first 10 minutes, which is the most in any 10-minute interval. This is the steepest part of the graph.

(e) After 107 minutes.

(f) After 3 hours, the temperature of the coffee will be approximately 25 °C.

9 **(a)** **(i)** $(x-3)^2 + (y-1)^2 = 49$

(ii) $(x+2)^2 + (y+4)^2 = 9$

(iii) $(x-2)^2 + (y+5)^2 = 4$

(b) (i)

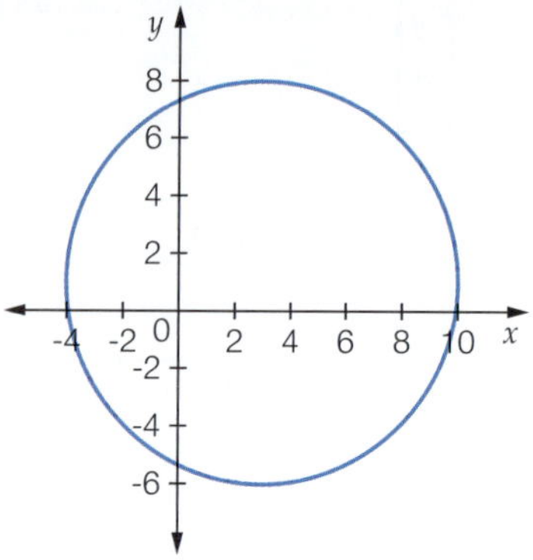

(ii)

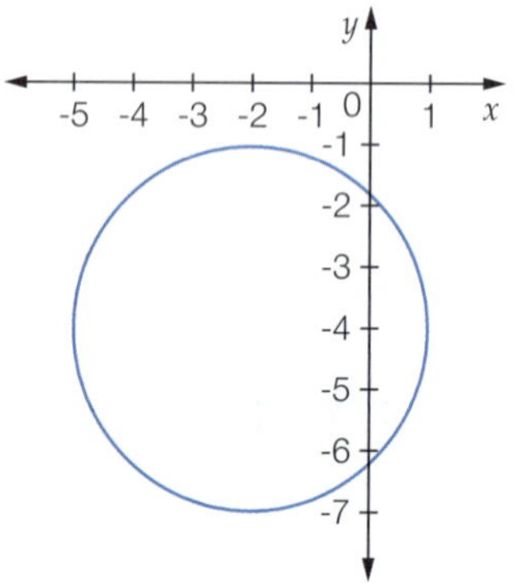

(iii)

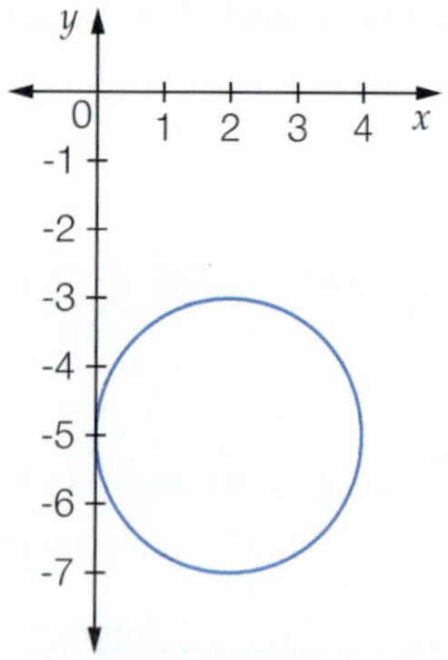

10 B 11 D

12 (a)

x	-2	-1	0	1	2	3	4
y	0.25	0.5	1	2	4	8	16

(b)

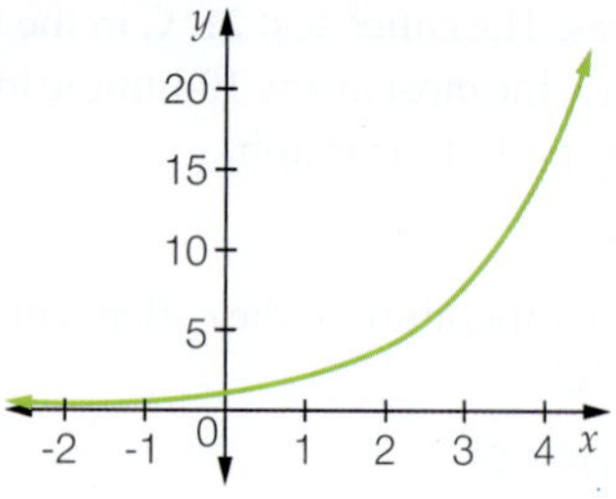

(c) All of these answers are approximate.

(i) 2.8 (ii) 1.7 (iii) 0.7

(iv) 0.4 (v) 1.4 (vi) 2.8

(d) Students should check (c) using a calculator.

(e) All of these answers are approximate.

(i) 2.3 (ii) 3.6 (iii) 3.9

13 (a) 4240

(b)

n	0	1	2	3	4	5
P	4000	4240	4494	4764	5050	5353

(c) $P = 4000(1.06)^n$

(d)

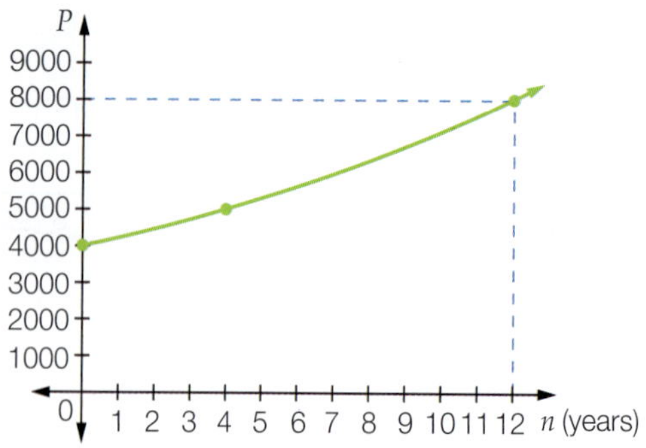

(e) 12 years

(f) $P = 4000(1.06)^{12}$

$= 8048.78$

(g) Future time and population can only be positive numbers.

14 $(x-3)^2 + (y+2)^2 = 1$

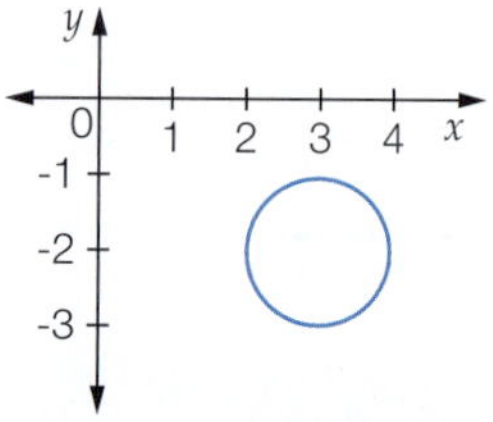

The point (2.4, -2.7) is inside the circle, so a tree could not be planted there. Substituting (2.4, -2.7) into the LHS of the equation gives 0.85. Any value less than 1 is inside the circle, exactly 1 is on the perimeter, and greater than 1 is outside the circle.

Open-ended – Sample answers

15 Equations should be in the form $(x+h)^2 + (y+k)^2 = r^2$, where h and k are both non-zero.

16 Equations should be in the form $y = a^x - 1$.

17 Equations should be in the form $y = \frac{1}{(x+2)} + 6$.

18 $(x+3)^2 + (y+1)^2 = 25$

$(x-4)^2 + (y+2)^2 = 25$

Exercise 12.2

1 (a) $x^2 + (y+3)^2 = 64$

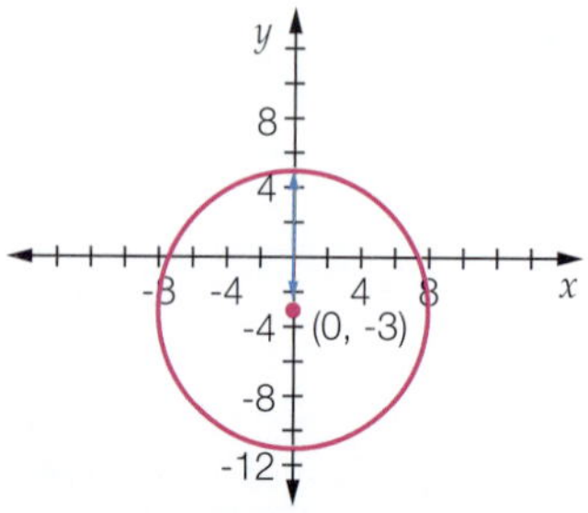

(b) $(x-1)^2 + y^2 = 36$

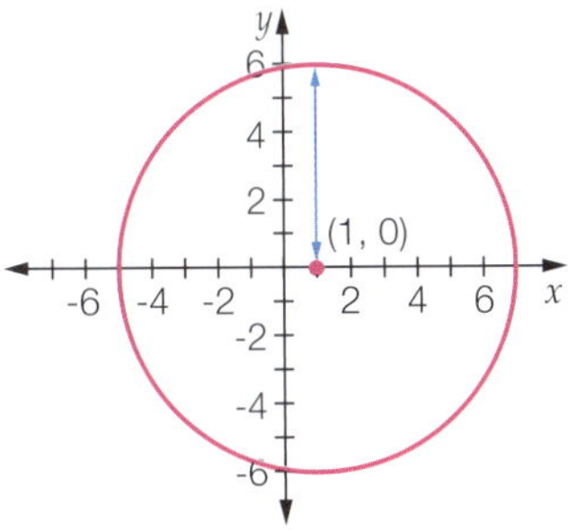

(c) $(x-2)^2 + (y+1)^2 = 16$

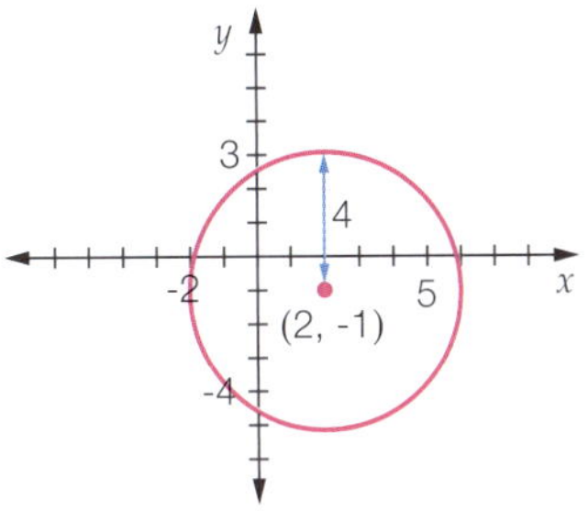

(d) $(x+3)^2 + (y-4)^2 = 25$

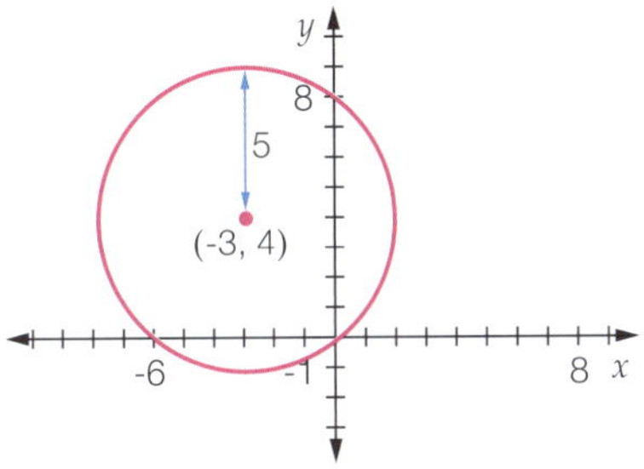

(e) $(x-2)^2 + (y-2)^2 = 4$

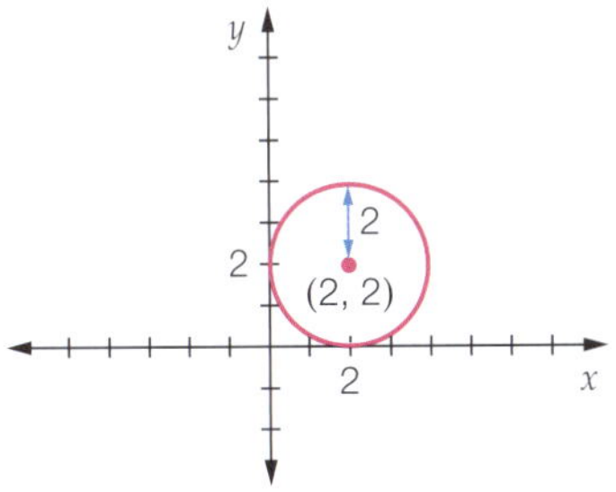

(f) $(x+4)^2 + (y+1)^2 = 16$

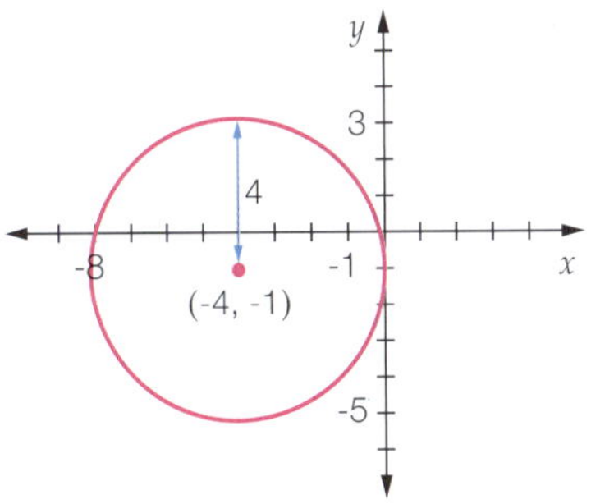

2 (a) $y = 2^{(x+3)}$

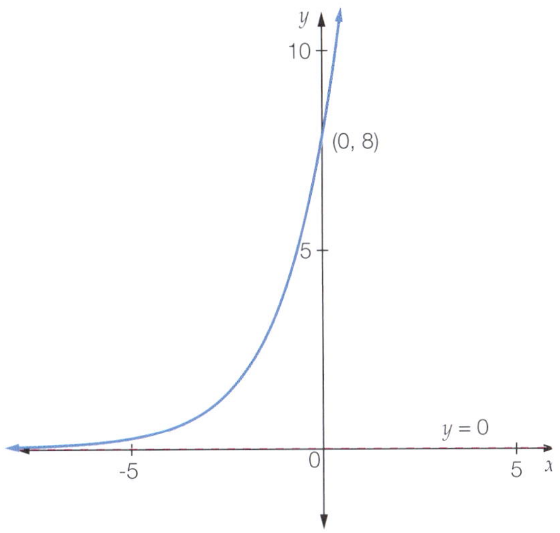

(b) $y = 3^x + 2$

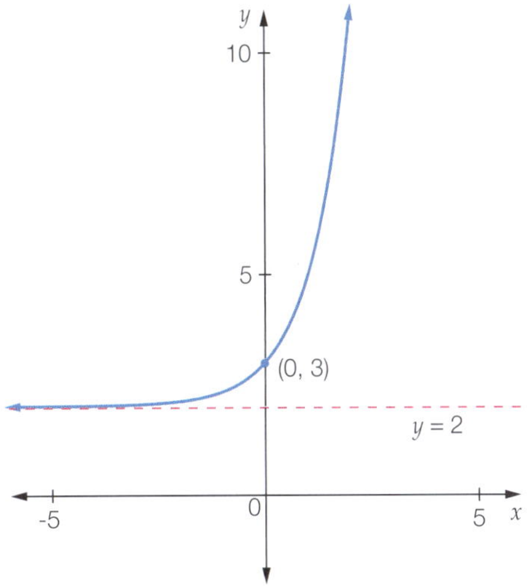

(c) $y = 2^{(x-2)} + 1$

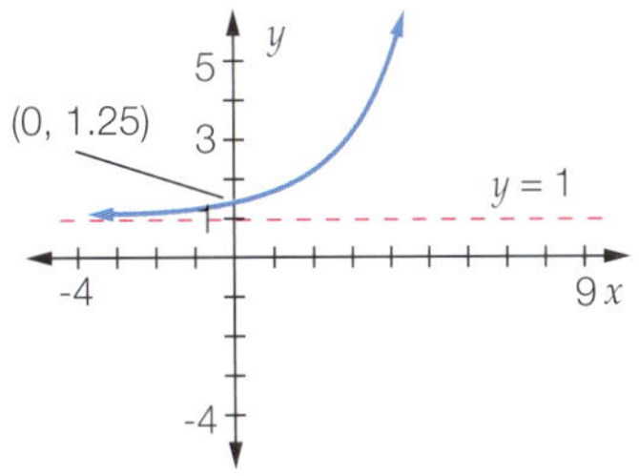

(d) $y = 2^{(x+1)} + 3$

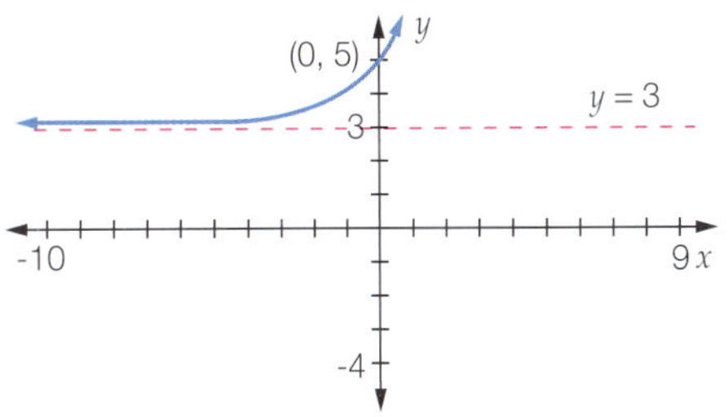

(e) $y = 2^x - 3$

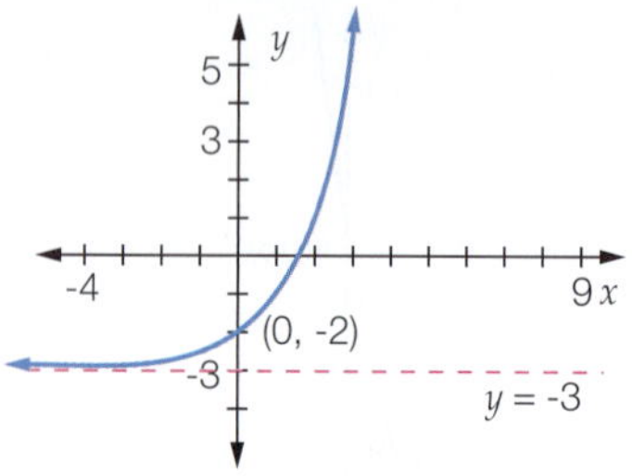

(f) $y = 3^{(x-4)} - 1$

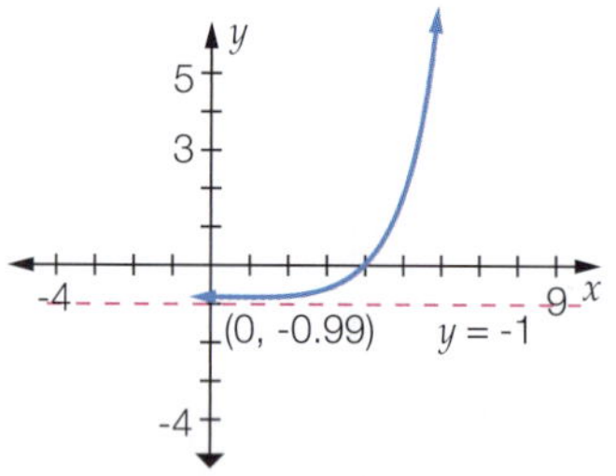

3 (a) $x^2 + (y+1)^2 = 4$ (b) $(x+1)^2 + (y+1)^2 = 9$

4 (a) $(x-2)^2 + (y+1)^2 = 49$

(b)

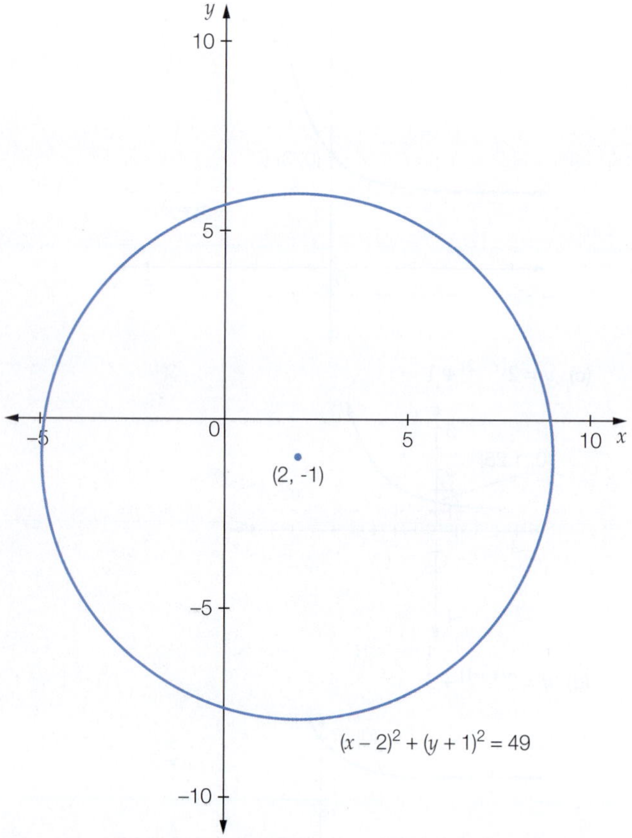

5 (a) $y = 2^{(x+1)} - 2$

(b)

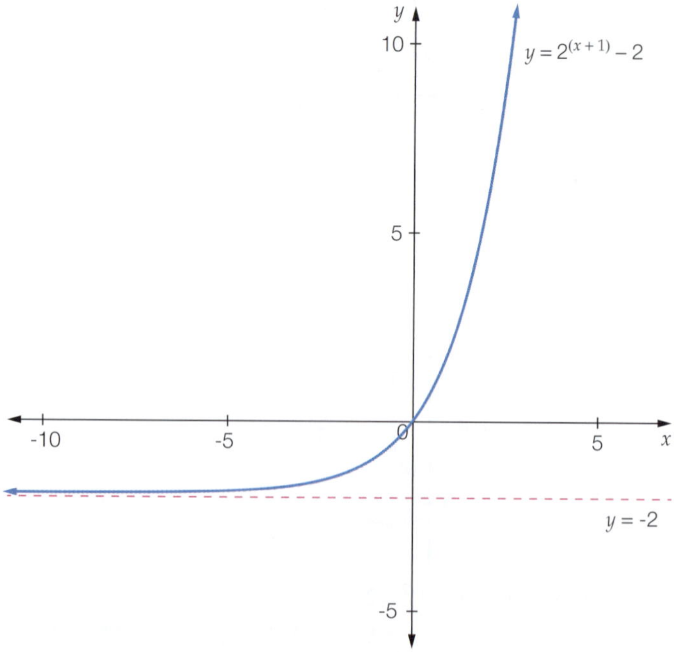

6 (a) $(x-1)^2 + (y+7)^2 = 25$

(b) $(x-3)^2 + (y+4)^2 = 64$

7 (a) $y = \dfrac{1}{x+3} - 2$

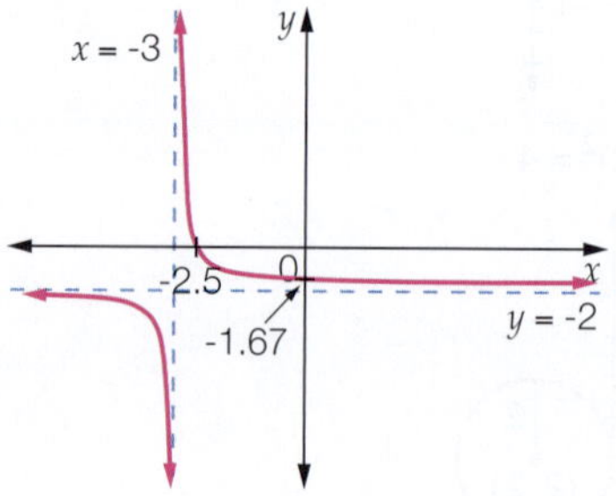

(b) $y = \dfrac{1}{x-5} + 1$

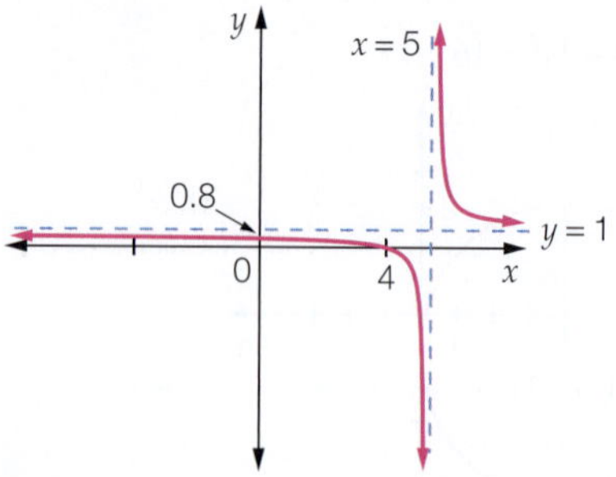

(c) $y = \frac{1}{x+2} + 3$

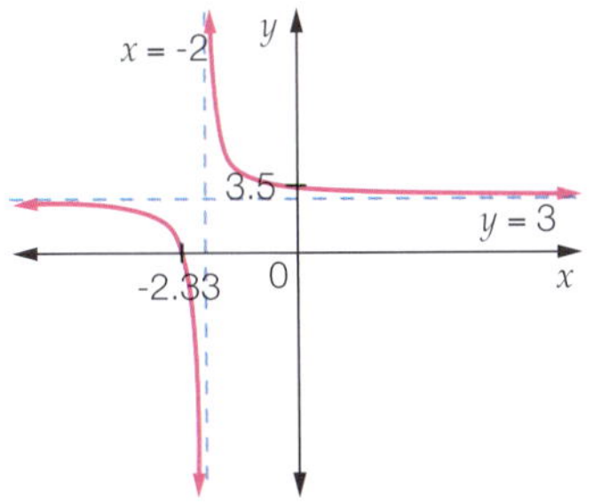

8 C

9 (a) $\frac{1}{x-1} + 3 = \frac{1}{x-1} + \frac{3}{x-1} = \frac{1+3(x-1)}{x-1} = \frac{1+3x-3}{x-1}$

$= \frac{3x-2}{x-1} = \text{RHS}$

(b)

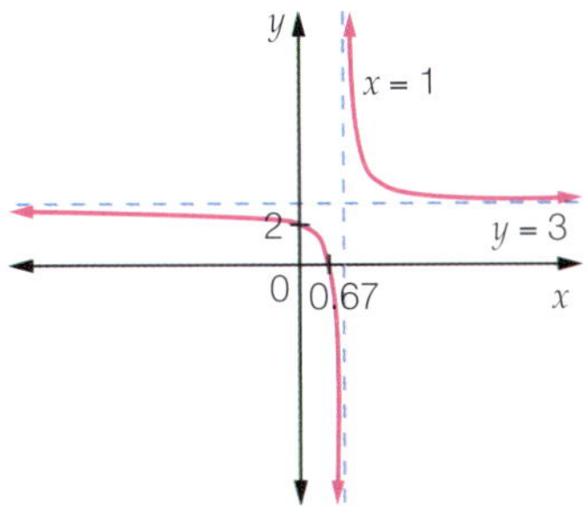

10 (a) (12.5, 7.5)

(b) $(x-12.5)^2 + (y-7.5)^2 = 9$

(c)

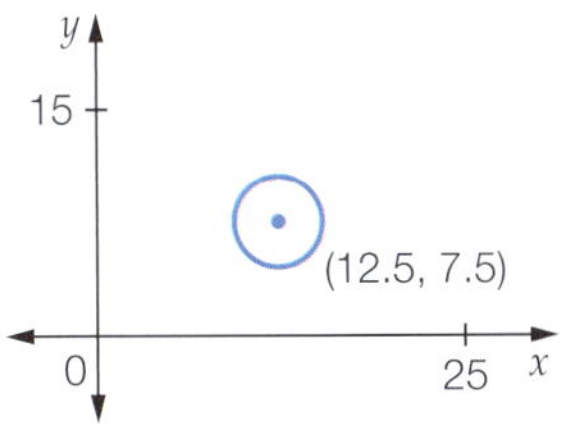

(d) (i) $(x-6)^2 + (y-7)^2 = 9$

(ii)

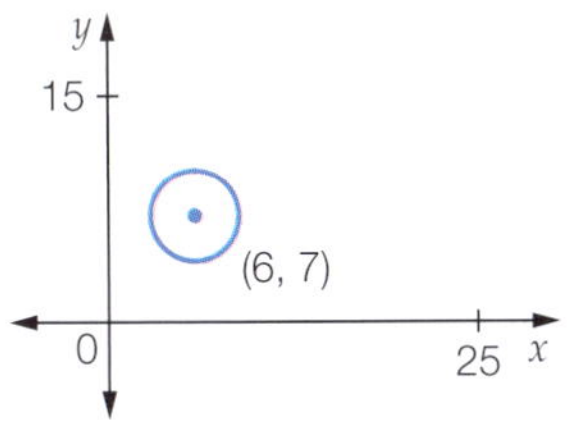

(e) (i) $(x-6)^2 + (y-7)^2 = 4$

(ii)

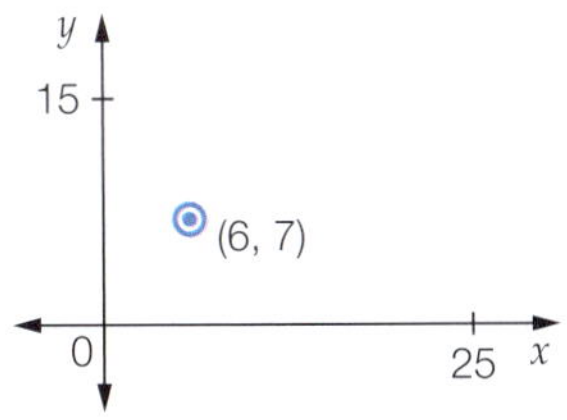

Open-ended – Sample answers

11 (a) Something in the form: $y = \frac{1}{x-h} + k$

(b) $x = h$ and $y = k$

(c) Students' own graph.

12 $(x-2)^2 + (y-1)^2 = 1$

Exercise 12.3

1 (a) x^8 (b) $12a^5$ (c) x^4x^7 (d) a^4

(e) $3x^4y$ (f) $\frac{p}{q^3}$ (g) m^8 (h) p^5q^{10}

(i) $a^{20}b^9$ (j) $\frac{a^5}{b^3}$ (k) $\frac{6q^4}{p^2}$ (l) x^2

(m) $\frac{1}{q^7}$ (n) $\frac{1}{c^9d^5}$ (o) 2 (p) q

(q) $\frac{3k^3}{10h}$ (r) $\frac{2a^6b^4}{3}$

2 (a) k^7p^6 (b) $\frac{6m^6n^5}{5}$ (c) $\frac{7y^{10}}{x^4}$ (d) $\frac{1}{k^7p^4}$

(e) $\frac{2a^5}{3b}$ (f) $\frac{5fg^4}{2}$ (g) x^2y^4 (h) $\frac{2m^3}{3}$

(i) $\frac{q^2}{5}$

3 (a) B (b) A (c) C

4 (a) x^9y^4 (b) $\frac{a^{12}}{b^{11}}$ (c) $\frac{4}{k^{18}p^2}$ (d) $9m^{20}n^2$

(e) $\frac{x^{27}}{y^{11}}$ (f) $\frac{m^6}{n^{20}}$ (g) $\frac{a}{b^7}$ (h) $\frac{m^{14}}{n^{11}}$

(i) $\frac{4x^{13}}{y^{18}}$ (j) $\frac{12d^{15}}{c^3}$

5 (a) x^2 (b) $\frac{a^3}{b^5}$ (c) pq (d) d^4f^2

(e) $\frac{1}{y^2}$ (f) $\frac{1}{a^3}$ (g) $\frac{n}{m^2}$ (h) $\frac{q^2}{p^7}$

(i) $\frac{b^9}{a^{16}}$ (j) $\frac{x^{13}}{y^7}$ (k) $\frac{12q^4}{p^{17}}$ (l) $\frac{9y^7}{8x^{14}}$

6 (a) $\frac{m^9}{n^2}$ (b) $\frac{b^3}{a}$ (c) $\frac{x^5}{y^4}$ (d) $\frac{y^{19}}{x^6}$ (e) $\frac{9}{a^2b^7}$

(f) $\frac{8}{d^5f^7}$ (g) $\frac{10b}{a^9}$ (h) $\frac{n^3}{m^8}$ (i) $\frac{1}{x^2}$

7 (a) 1 divided by the original number. (b) $192a$

8 (a) Murray just said $-2 \times -4 = 8$ and possibly put it as the power in the denominator because he thought the negatives needed to be accounted for. This is not how indices work.

(b) Raphael knows that indices are added when numbers are multiplied. $\frac{1}{a^6}$ may be Raphael's answer, which would be correct.

(c) When numbers with indices and the same base are multiplied, keep the base the same and add the indices.

$a^{-2} \times a^{-4} = a^{-6} = \frac{1}{a^6}$.

Open-ended – Sample answers

9 (a) $(6m^2n^3)^2 \times n^2$; $9m^3n^4 \times 4mn^4$

(b) $\frac{12m^7n^3}{15m^3n^9}$; $24m^4 \div 30n^6$

10 $a = 2, b = 3$; $a = 3, b = 2$; $a = 5, b = 1$

These give the fractions: $\frac{1}{8}, \frac{1}{9}$ and $\frac{1}{5}$

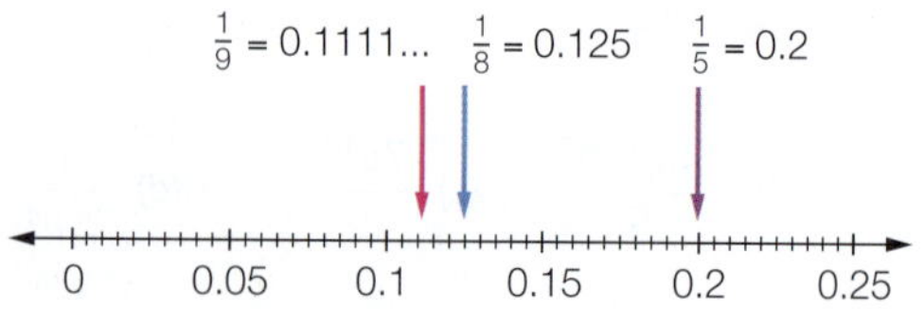

Half-time 12

1 radius 2 and centre (-1, 2)

2 (a)

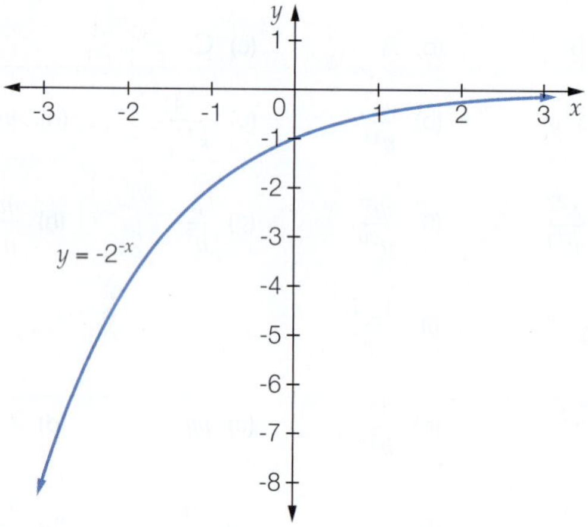

(b)

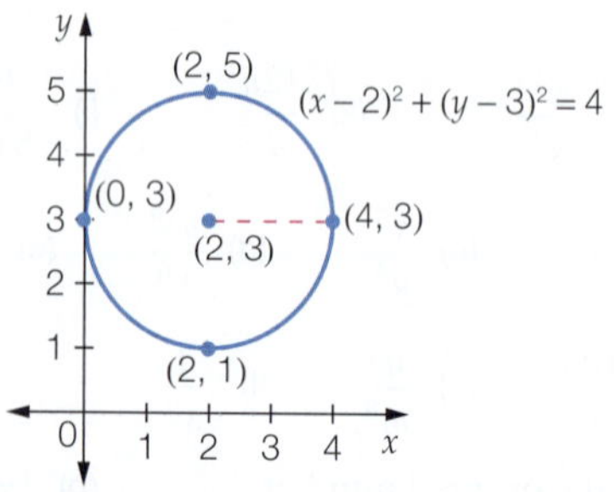

3 (a) rectangular hyperbola

(b) $x = 3$ and $y = 4$

4 (a)

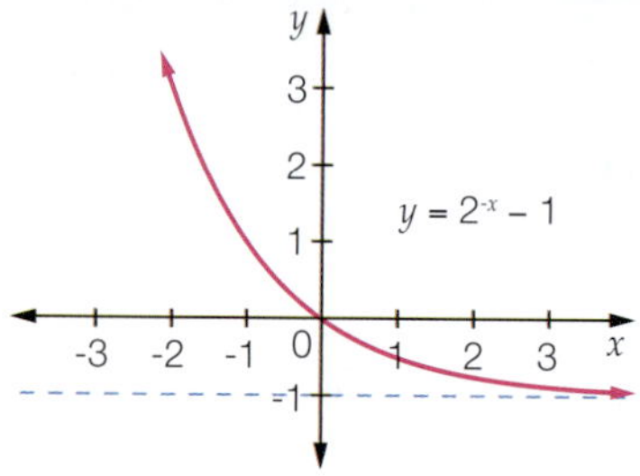

(b) $y = -1$ (c) $(0, 0)$

5 (a) $(x-2)^2 + (y-3)^2 = 16$ (b) $(x+5)^2 + (y+3)^2 = 4$

6 (a) $12x^5$ (b) $6m^{10}n$ (c) $\frac{a^2}{b^2}$

7 (a) $\frac{1}{q^{22}}$ (b) f^6g^{11}

8 (a) $y = \frac{1}{x+3} + 4$

(b)

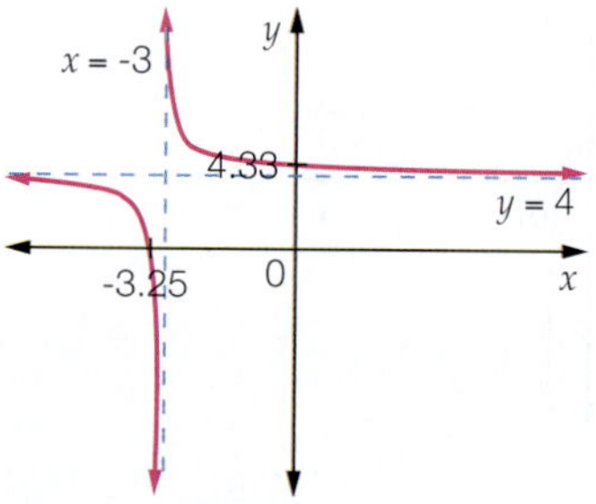

Exercise 12.4

1 (a) $x^{\frac{2}{3}}$ (b) $a^{\frac{5}{6}}$ (c) $k^{\frac{8}{15}}$ (d) $m^{\frac{1}{4}}$

(e) $p^{\frac{1}{6}}$ (f) $y^{\frac{1}{4}}$ (g) $5p^2$ (h) $7x^3$

(i) $2a^{\frac{1}{3}}b$ (j) $x^2y^{\frac{1}{2}}$ (k) $k^{\frac{1}{2}}n^4$ (l) $\frac{p^2}{q^5}$

2 (a) 14 (b) 19 (c) 6 (d) 8 (e) 5

(f) 7.94 (g) 54.77 (h) 2.15 (i) 2.71

3 $4^{-1}, 6^{-\frac{1}{5}}, 2^{\frac{1}{2}}, 3^{\frac{1}{3}}$

4 (a) B (b) C

5 (a) a (b) d^3 (c) $\frac{x}{y}$

(d) $\frac{1}{m^2n^4}$ (e) $\frac{1}{q^2}$ (f) $\frac{y^{\frac{7}{6}}}{5x^{\frac{1}{3}}}$

6 (a) $\sqrt[3]{36}$ (b) $\frac{1}{\sqrt[4]{x^3}}$

(c) $p^2\sqrt{p}$ or $\sqrt{p^5}$ (d) $\frac{1}{5\sqrt[4]{5}}$ or $\frac{1}{\sqrt[4]{3125}}$

7 (a) $r^{\frac{3}{2}}$ (b) $m^{-\frac{2}{3}}$ (c) $7^{\frac{3}{2}}$ (d) $a^{-\frac{5}{2}}$

8 (a) B (b) A

9 (a) 0.500 (b) 0.164 (c) 0.369 (d) 0.398

10 (a) $x = \sqrt[3]{3}$ (b) $x = 4$ (c) $x = \frac{1}{2}$

11 (a) x^3 (b) a^3 (c) $2m^2n$

12 (a) $x = \pm 2$ (b) $x = 3$ (c) $x = -2$ (d) $x = 2$

13 (a) 100 000 (b) 32 (c) 4 (d) $\frac{1}{36}$

14 (a) $12g^{\frac{11}{3}}$ (b) $\frac{10}{a^{\frac{1}{2}}}$ (c) $12p^3q$

Open-ended – Sample answers

15 $1476.106^{\frac{4}{5}}$, 7^3, 1.793^{10}

16 $\left(n^{\frac{1}{4}}m^{\frac{1}{5}}\right)^3$, $(\sqrt[4]{n} \times \sqrt[5]{m})^3$; $(\sqrt[4]{n^3})(\sqrt[5]{m^3})$

Exercise 12.5

1 (a) $\log_2(32) = 5$ (b) $\log_2(4096) = 12$

(c) $\log_3(27) = 3$ (d) $\log_3(243) = 5$

(e) $\log_3(6561) = 8$ (f) $\log_{10}(100) = 2$

(g) $\log_{10}(1\,000\,000) = 6$ (h) $\log_2\left(\frac{1}{4}\right) = -2$

(i) $\log_2\left(\frac{1}{8}\right) = -3$ (j) $\log_2\left(\frac{1}{16}\right) = -4$

(k) $\log_{10}(0.0001) = -4$ (l) $\log_3\left(\frac{1}{9}\right) = -2$

2 (a) $10^2 = 100$ (b) $10^5 = 100\,000$

(c) $2^4 = 16$ (d) $2^7 = 128$

(e) $2^9 = 512$ (f) $5^6 = 15\,625$

(g) $10^{-3} = \frac{1}{1000}$ (h) $7^{-2} = \frac{1}{49}$

(i) $10^{-6} = \frac{1}{1\,000\,000}$ (j) $2^{-3} = \frac{1}{8}$

(k) $2^{-5} = \frac{1}{32}$ (l) $2^{-7} = \frac{1}{128}$

3 (a) C (b) D

4 (a) 4 (b) 8 (c) 9 (d) -2 (e) -3

(f) -6 (g) -3 (h) -5 (i) 4

5 (a) 16 (b) 64 (c) 128 (d) -4 (e) -5

(f) -8 (g) 5 (h) 7 (i) 5

6 (a) (i) 10 (ii) 5 (iii) 7

(b) $\log_a(a) = 1$

7 (a) (i) $x = 1$ (ii) $x = 1$ (iii) $x = 1$

(b) $\log_a 1 = 0$

8 The square root of a negative number is not a real number.

9 (a) Students to verify that both equations are true.

(b) 6

(c) $2^{12} = 4096$ is the index equivalent of the first statement and $4^y = 4096$ is the second. But, this could be written as $(2^2)^y = 4096$ or $2^{2y} = 4096$, so $2y = 12$ and $y = 6$.

Open-ended – Sample answers

10 $\log_9(531\,441) = 6$ and $\log_{27}(531\,441) = 4$

11 Effectively Lee is saying $\log_6(-36) = -2$, which would mean that $6^{-2} = -36$, which is clearly incorrect. Lee should have divided both sides by -1 to get $(6)^2 = 36$, which has the equivalent form $\log_6(36) = 2$.

Exercise 12.6

1 (a) $\log_2(32)$ (b) $\log_2(64)$ (c) $\log_2(256)$

(d) $\log_{10}(30)$ (e) $\log_{10}(21)$ (f) $\log_{10}(36)$

2 (a) $\log_2(2)$ (b) $\log_2(5)$ (c) $\log_2(5)$

(d) $\log_{10}(14)$ (e) $\log_{10}(5)$ (f) $\log_{10}(6)$

3 (a) $4\log_2(5)$ (b) $3\log_2(6)$ (c) $4\log_2(7)$

(d) $3\log_{10}(7)$ (e) $4\log_{10}(8)$ (f) $5\log_{10}(11)$

4 (a) 5 (b) 7 (c) 4 (d) 3

5 (a) A (b) C

6 (a) 5 (b) 11 (c) 1 (d) 2 (e) 8 (f) -9

7 (a) 3 (b) 4 (c) 3 (d) 2 (e) 1 (f) -1

8 (a) 2 (b) $\frac{5}{6}$ (c) 3 (d) 2 (e) $\frac{3}{2}$ (f) $\frac{4}{3}$

9 (a) $\frac{1}{4}$ (b) 2 (c) $\frac{2}{3}$

10 (a) $y = 4x^3$ (b) $y = 11x^5$

11 $5\log_2(y) = \log_2(x-1) + 2$

$\log_2\left(\frac{y^5}{x-1}\right) = 2$

$2^2 = \frac{y^5}{x-1}$

$4(x-1) = y^5$

Open-ended – Sample answers

12 The log term with the power of $\frac{1}{2}$ is on the denominator, so the fraction cancels to:

$1 + \frac{1}{2} = 2$

The correct answer is 2.

Chapter review 12

1 (a) exponential; y-intercept: (0, 3.6); asymptote: $y = 3$

(b) circle; centre (-1, 3) and $r = 3$

2 (a) $(x+2)^2+(y-3)^2=16$

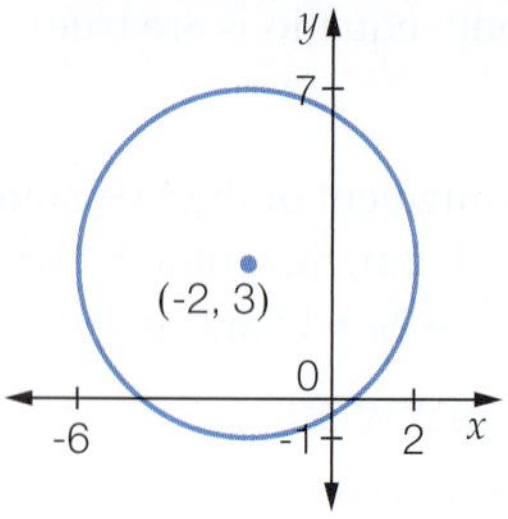

(b) $x^2+(y-3)^2=4$

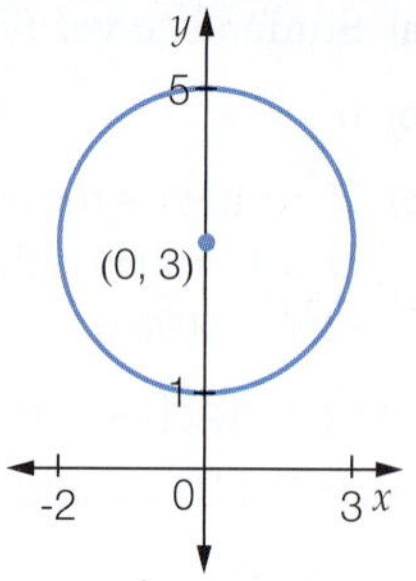

(c) $y=2^{(x-2)}+1$

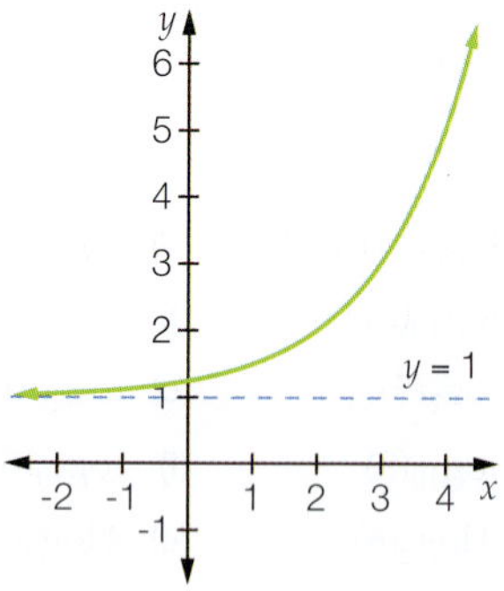

(d) $y=2^{(x+2)}-4$

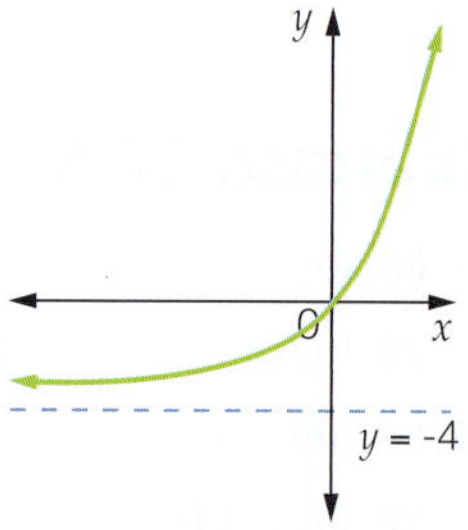

3 (a) $y=\dfrac{1}{x-4}+1$

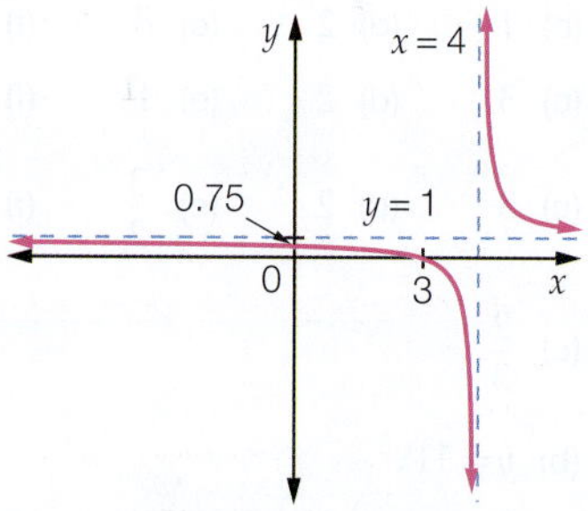

(b) $y=\dfrac{1}{x+2}-1$

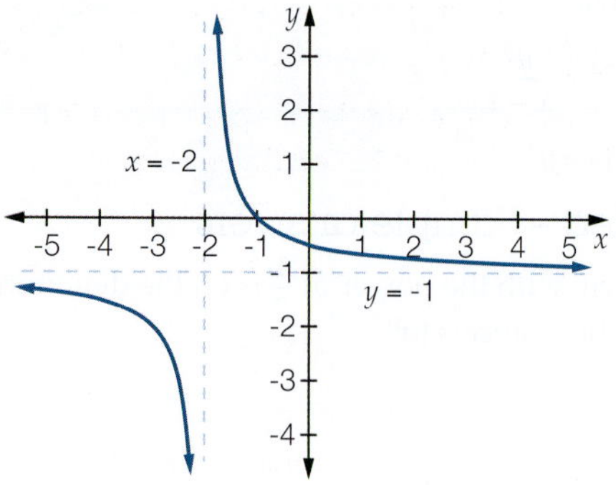

4 (a) $9a^2b^{10}$ (b) c^2d^{14} (c) u^2t

(d) $2xy^{10}$ (e) $m^{12}n^9$ (f) $a^{12}b$

5 (a) $\dfrac{6}{a^6}$ (b) $\dfrac{27x^4}{16y}$ (c) $\dfrac{n^6}{2}$

6 (a) $4mn^{\frac{1}{3}}$ (b) $p^3q^{\frac{1}{2}}$ (c) $a^{\frac{1}{2}}$

(d) $\dfrac{b}{a^9}$ (e) $\dfrac{2n^{\frac{4}{3}}}{m^{\frac{1}{3}}}$ (f) $\dfrac{729n^7}{m^{19}}$

7 (a) $k^{\frac{5}{2}}$ (b) $x^{\frac{3}{4}}$ (c) $m^{-\frac{2}{3}}$

8 (a) $\log_4(16\,384)=7$ (b) $\log_2(1024)=10$

(c) $\log_{10}\left(\dfrac{1}{1000}\right)=-3$

9 (a) $2^{14}=16\,384$ (b) $3^5=243$ (c) $10^{-5}=\dfrac{1}{100\,000}$

10 (a) 9 (b) -2 (c) $\log_3(50)$ (d) $\log_5(10)$

11 (a) A and D

(b)

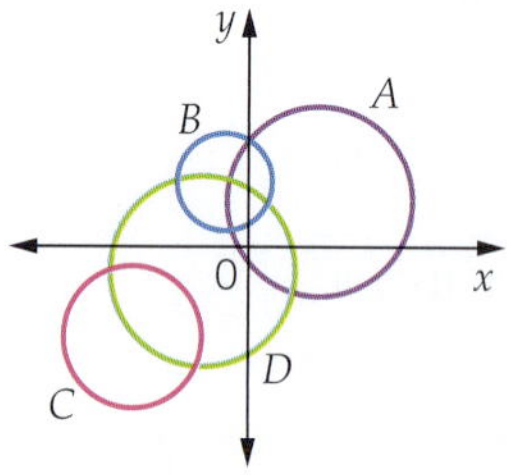

D intersects the other three circles.

12 (a) 5 (b) 7 (c) -6 (d) $\dfrac{1}{16}$ (e) 8 (f) -4

13 (a) 3 (b) 3 (c) 2 (d) $\dfrac{3}{4}$

14 (a) $y=7x^3$ (b) $y=\dfrac{5}{x^2}$

Chapter 13

Recall 13

1 (a) 0.17 (b) 0.06 (c) 0.192

(d) 0.048 (e) 0.0201 (f) 0.0025

2 (a) $3000 (b) $600 (c) $33 250 (d) $3375

3 (a) (i) 80 (ii) 60 (iii) 50 (iv) 120

(b) (i) 30 (ii) 42 (iii) 54 (iv) 15

4 (a) 49 (b) 8 (c) 1 000 000

(d) 0.36 (e) $\dfrac{1}{32}$ (f) 1.4641

5 (a) 0.5 (b) 0.25 (c) 8.75

(d) 0.5 (e) 0.25 (f) 0.4

Exercise 13.1

1 (a) $I=PrT$ (b) $I=\$80$ (c) $I=\$700$

2 (a) $1000 (b) 5.1% p.a. (c) 0.051 (d) 5 years

3 (a) $P=\$1041.67$ (b) $P=\$3857.14$ (c) $r=9.2\%$

(d) $r=2.5\%$ (e) $T=9$ years (f) $T=16$ years

4 (a) $P = \$500$ (b) $T = 10$ years
(c) $r = 3.2\%$ p.a. (d) $I = \$160$

5 (a) (i) \$3200 (ii) \$43 200
(b) (i) \$10 500 (ii) \$40 500
(c) (i) \$24 000 (ii) \$64 000
(d) (i) \$281 250 (ii) \$431 250
(e) (i) \$22 275 (ii) \$322 275
(f) (i) \$27 912.50 (ii) \$377 912.50
(g) (i) \$6175 (ii) \$19 175
(h) (i) \$28 125 (ii) \$73 125
(i) (i) \$3375 (ii) \$13 375

6 (a) I is missing. (b) P is missing. (c) T is missing.

7 (a) \$4000 (b) \$8000 (c) 14% p.a.
(d) 11% p.a. (e) 10 years (f) 12 years

8 (a) (i) $A = P(1 + r)^n$ (ii) $A = \$13\,785.11$
(iii) $A = \$187\,789.83$
(b) (i) $I = A - P$ (ii) $I = \$10\,000$
(iii) $I = \$5780$

9 (a) $P = \$594\,500$ (b) $A = \$58\,750$

10 (a) (i) \$43 200 (ii) \$3200
(b) (i) \$42 076.55 (ii) \$12 076.55
(c) (i) \$71 834.25 (ii) \$31 834.25
(d) (i) \$914 750.94 (ii) \$764 750.94
(e) (i) \$322 348.70 (ii) \$22 348.70
(f) (i) \$378 011.55 (ii) \$28 011.55
(g) (i) \$20 465.10 (ii) \$7465.10
(h) (i) \$82 809.23 (ii) \$37 809.23
(i) (i) \$13 695.26 (ii) \$3695.26

11 Lily (a) (i) \$5830.96 (ii) \$5250
(b) Compound interest is \$580.96 higher.
Jarrah (a) (i) \$39 939.82 (ii) \$29 925
(b) Compound interest is \$10 014.82 higher.

12 (a) \$9419.20 (b) \$9240

13 C 14 \$129.17

15 (a) (i) simple interest (ii) 40% p.a., \$12
(b) (i) compound interest (ii) \$994
(c) (i) compound interest (ii) \$25 729

16 A

17 (a) 100 years (b) 47 years

18 5% p.a.

Open-ended – Sample answers

19 \$12 000 invested at 11.8% p.a. would give a final value of \$14 999 (clearly within the \$100 buffer!).

20 $r \times T = 0.8$, e.g. 16% p.a. for 5 years, 15% p.a. for 5 years and 4 months, 12% p.a. for 6 years and 8 months.

Exercise 13.2

1 (a) 4 periods (b) 2 periods (c) 365 periods
(d) 12 periods (e) 1 period

2 (a) $\frac{9}{4}\%$ (b) $\frac{9}{2}\%$ (c) $\frac{9}{365}\%$
(d) $\frac{3}{4}\%$ (e) 9%

3 (a) $P = \$25\,000$, $n = 26$, $r = 0.0125$ (b) $P = \$9000$, $n = 40$, $r = 0.018$ (c) $P = \$200\,000$, $n = 12\,775$, $r = \frac{36}{365\,000}$

4 (a) $A = P(1 + r)^n$
(b) (i) \$10 056.78 (ii) \$112 523.74 (iii) \$69 712.70

5 (a) \$20 793.60 (b) \$20 972.74 (c) \$21 068.94

6 (a) \$9528.13 (b) \$9552.42

7 B

8 C

9 (a) \$64 656.50 (b) \$65 731.65 (c) \$66 472.15

10 (a) \$647.54 (b) \$13 499.14

11 \$5468.12 and \$5459.91; Martina is better off with the 4.5% p.a. compounded quarterly. The difference is \$8.21.

12 (a) (i) \$3215.09 (ii) \$2879.21
(iii) \$2545.44 (i) gives the best return
(b) no
(c) (i) $I \approx 16\%$ of P (ii) $I \approx 14\%$ of P (iii) $I \approx 13\%$ of P

Open-ended – Sample answers

13 The statement means that if compound interest is calculated more frequently (e.g. every month instead of every quarter), it will result in a greater amount of interest.
Quarterly compounding gives interest of \$4859.47. (48.6%)
Monthly compounding gives interest of \$4898.46. (49.0%)
Weekly compounding gives interest of \$4913.66. (49.1%)

14 Using the suggested example, simple interest is \$600 and interest compounded annually is \$630.50. The principal gets bigger after each time period.

Exercise 13.3

1 (a) $P = \$900$ (b) $P = \$2000$

2 (a) P (b) n (c) r

3 (a) $P = \$16\,000$, $n = 72$, $r = \frac{0.031}{12}$, $A = \$19\,266.13$
(b) $P = \$46\,000$, $n = 1460$, $r = \frac{0.084}{365}$, $A = \$64\,367.11$
(c) $P = \$4000$, $n = 60$, $r = \frac{1}{600}$, $A = \$4420.32$
(d) $P = \$2500$, $n = 16$, $r = \frac{0.063}{4}$, $A = \$4122.10$

4 (a) \$5996 (b) \$5437 (c) \$5456.76 (d) B

5 (a) $r = 1.97\%$ (b) $r = 1.15\%$ (c) $r = 1.64\%$ (d) $r = 2.66\%$

6 (a) $r = 1.7\%$ p.a. (b) $r = 3.4\%$ p.a.

(c) $r = 6.8\%$ p.a. (d) $r = 7.3\%$ p.a.

7 (a) 10.2% p.a. (b) 3.24% p.a. (c) 2.6% p.a. (d) B

8 (a) $n = 17$ (b) $n = 24$ (c) $n = 8$

9 (a) 12 (b) 7 (c) 3 (d) 19

10 (a) 12 years (b) 10 years (c) 4.2 years

11 (a) \$5956.42 (b) \$956.42

(c) $I = \$956.46$, only marginally better

(d) \$956.42 (e) 4.47% p.a.

(f) At least 4 years, as it would be worth \$5788.13 at the end of the third year and \$6077.53 at the end of the fourth year.

12 2 periods

13 (a) every 4 months (b) 9.3% p.a.

(c) monthly compounding

14 At least every 4 months.

15 Rounding error: $\frac{0.065}{4} = 0.01625$, $P = \$18\,751.14$

Open-ended – Sample answers

16 11% p.a. with annual compounding, 10.5% p.a. compounding half-yearly, 10.2% p.a. compounding daily

Half-time 13

1 B 2 \$8673.33 3 \$8889.26 4 200% p.a.

5 (a) \$50 649.49 (b) \$51 824.82 (c) \$52 059.90

6 9

7 (a) \$10 866 (b) \$9000

(c) Compound interest is \$1866 more than simple interest.

8 17.3% p.a.

Exercise 13.4

1 (a) 8.5% (b) 12.0% (c) 1.8% (d) 3.1%

2 (a) 7% p.a. (b) 9.4% p.a. (c) 13.52% p.a.

(d) 21.73% p.a. (e) 17.88% p.a. (f) 6.45% p.a.

3 (a) 13.1% (b) 6.9% (c) 9.8%

(d) 19.4% (e) 24.4% (f) 9.9%

4 B

5 (a) (i) $r = 7.3\%$ p.a., $n = 5$ (ii) 7.52% p.a.

(b) (i) $r = 15.5\%$ p.a., $n = 12$ (ii) 16.65% p.a.

(c) (i) $r = 21\%$ p.a., $n = 4$ (ii) 22.71% p.a.

(d) (i) $r = 16\frac{1}{4}\%$ p.a., $n = 3$ (ii) 17.15% p.a.

(e) (i) $r = 20\frac{1}{3}\%$ p.a., $n = 365$ (ii) 22.54% p.a.

(f) (i) $r = 5.25\%$ p.a., $n = 2$ (ii) 5.32% p.a.

6 (a) C Bank: 15.87% p.a., D Bank: 16.06% p.a.

(b) D Bank

7 (a) 22.13% p.a. (b) 22.26% p.a.

(c) Personal loan because the effective rate is lower.

8 (a) 25.85% p.a. (b) online

9 (a) 15% p.a. compounded daily—higher rate and more frequent compounding

(b) 10.5% p.a. compounded weekly—rate is the same so more frequent compounding is better

(c) 12% p.a. compounded daily—compounding frequency is the same so higher rate is better

(d) 15% p.a. compounded annually for 2 years—compounded beats flat rate over 2 years

10 Yes. In each case choose the other option because for loans you want the lowest effective rate.

11 (a) car company loan

(b) flat rate

Open-ended – Sample answers

12 1-year loan of \$10 000 at 6% p.a. compounded daily. Nominal rate is 6% p.a. The interest is compounded, not simple. However, the same amount of interest will be paid if \$10 000 is borrowed for a year at a simple interest rate of 6.18% p.a. Either way nominal isn't the same as simple.

13 42% p.a. compounded half-yearly, 40% p.a. compounded quarterly

Exercise 13.5

1 (a) \$6750 (b) \$1122

2 (a) \$18 250 (b) \$2278

3 (a) (i) \$10 500 (ii) \$19 500

(b) (i) \$28 500 (ii) \$19 000

(c) (i) \$100 000 (ii) No value

(d) (i) \$772 225 (ii) \$17 775

(e) (i) \$60 900 (ii) \$84 100

(f) (i) \$107 250 (ii) \$192 750

(g) (i) \$4928.13 (ii) \$42 571.87

(h) (i) \$76 159.13 (ii) \$76 540.87

4 (a) \$4305.92 (b) \$17 075.91

5 (a) (i) \$20 870.65 (ii) \$9129.35

(b) (i) \$25 667.10 (ii) \$21 832.90

(c) (i) \$35 848.59 (ii) \$64 151.41

(d) (i) \$290 950.49 (ii) \$499 049.51

(e) (i) \$93 037.95 (ii) \$51 962.05

(f) (i) \$207 293.23 (ii) \$92 706.77

(g) (i) \$42 285.67 (ii) \$5214.33

(h) (i) \$83 275.25 (ii) \$69 424.75

6 (a) \$2466 (b) \$13 422 (c) \$2980

7 (a) (i) \$6000 (ii) \$2/hour (iii) \$20 (iv) \$6130

(b) (i) \$1000 (ii) 50c/hour (iii) \$500 (iv) \$500

8 \$526.50 9 B 10 \$2240

11 \$2923.24. Reducing balance depreciation is less so written-down value after 4 years is higher than under straight-line depreciation.

12 Depreciation 1st year = \$9000
Total depreciation 1st year = \$9000

Written-down value 1st year = \$36 000

Depreciation 2nd year = \$9000
Total depreciation 1st year = \$9000 2nd year = \$18 000

Written-down value 2nd year = \$27 000

Depreciation 3rd year = \$9000
Total depreciation 1st year = \$9000 3rd year = \$27 000

Written-down value 3rd year = \$18 000

13 Written-down value 1st year = \$36 000
Total depreciation 1st year = \$9000
Depreciation 1st year = \$9000

Written-down value 2nd year = \$28 800
Total depreciation 2nd year = \$16 200
Depreciation 2nd year = \$7200

Written-down value 3rd year = \$23 040
Total depreciation 3rd year = \$21 960
Depreciation 3rd year = \$5760

1st year is the same as it was for linear. After that, the depreciation slows down and depreciation for each year is less than the year before. After the 1st year, written-down value is also better than it was for linear.

14 C

15 (a) D (b) A

16 \$9943.11 17 \$2415

18 (a) \$3353.75 (b) \$2768.75

19 (a) (i) \$50 000 (ii) \$36 000 (iii) \$25 800 (iv) \$18 390

(b) after 21 years

Open-ended – Sample answers

20 5% p.a. and 14 years, 10% p.a. and 7 years, 20% p.a. and 3.5 years, 70% p.a. and 1 year. Must have $r \times T = 0.7$

21 Any values that satisfy $0.3 = (1 - r)^T$; e.g. 70% p.a. and 1 year, 45.23% p.a. and 2 years, 21.4% p.a. and 5 years

Exercise 13.6

1 (a) growth (b) decay (c) decay

2 (a) $A = 2503(1.126)^n$ (b) $A = 125\,000(1.052)^n$

(c) $A = 2750(1.02)^n$

3 (a) 20 524 (b) 15 200 (c) 1171

(d) \$875 855 (e) 242 057 (f) 1065

4 (a) $A = 25\,590(0.83)^n$ (b) $A = 3623(0.976)^n$

(c) $A = 2400(0.888)^n$

5 (a) 42 (b) \$896 (c) 6218

(d) 164 870 (e) 1533 (f) 4 566 189

6 (a) (i) $k = 4$ (ii) growth (iii) $a = 2$ (iv) $N = 4(2)^T$

(b) (i) $k = 1$ (ii) growth (iii) $a = 5$ (iv) $N = (5)^T$

(c) (i) $k = 324$ (ii) decay

(iii) $a = \frac{1}{3}$ (iv) $N = 324\left(\frac{1}{3}\right)^T$

(d) (i) $k = 320$ (ii) decay

(iii) $a = \frac{1}{2}$ (iv) $N = 320\left(\frac{1}{2}\right)^T$

7 (a)

4	5	6
405	1 215	3 645

(b) $S = 5 \times 3^T$ (c) 388 564

(d)

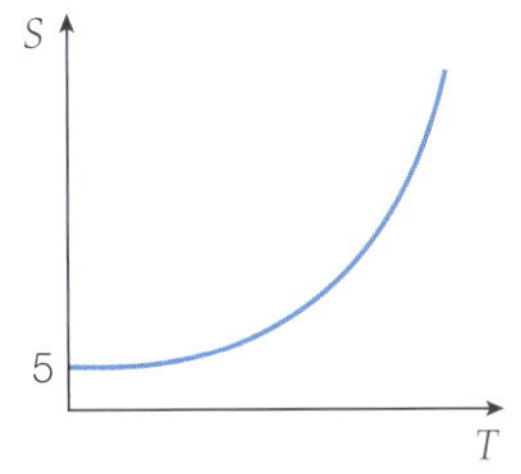

(e) after 2.5 years

8 (a) C (b) B

9 (a) \$5500 (b) 110% (c) 1.1 (d) \$6050 (e) 1.1

(f)

Year	Balance (\$)
0	5 000
1	5 500
2	6 050
3	6 655
4	7 320.50
5	8 052.55
6	8 857.51
7	9 743.59
8	10 717.94
9	11 789.74
10	12 968.71

(g) $A = 5000 \times 1.1^n$

(h)

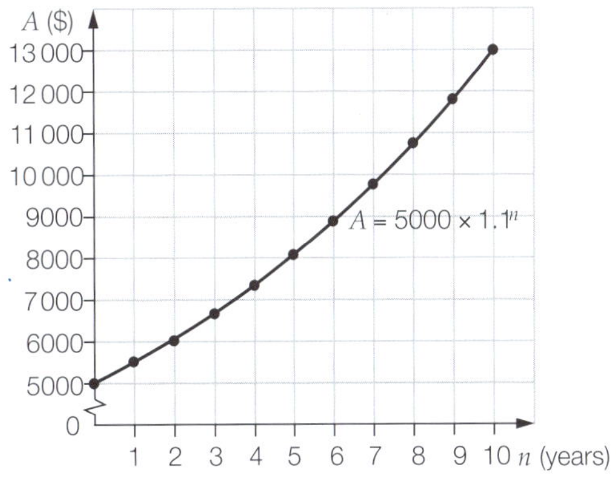

(i) \$6979.82 (j) after 4 years, 11 months

(k) after 7.27 years (or 7 years 3 months 1 week)

10 (a) 95% (b) 0.95 (c) $A = 45\,000 \times 0.95^n$

(d)

Year	Balance ($)
0	45 000
1	42 750
2	40 612.50
3	38 581.88
4	36 652.78
5	34 820.14

(e)

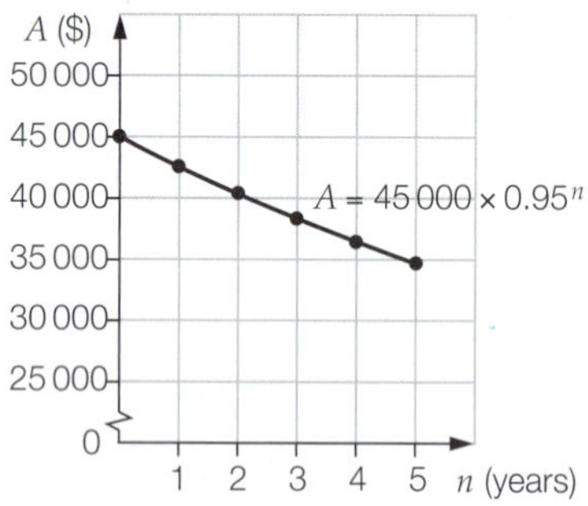

(f) $39 584.17

(g) after 4 years, 4 months

11 (a) $I = 10^6 \times 0.5^T$ (b) 95c

(c) When she's 7 years old.

12 (a) $A = 5\left(\frac{2}{3}\right)^n$ (b) 7 g

Open-ended – Sample answers

13 $T = 6$ years, $r = 200\%$ p.a., $T = 3$ years, $r = 800\%$ p.a., $T = 20$ years, $r \approx 39\%$ p.a.

14 There is no error in the report. Janet has assumed a linear model of decay. Radioactive material follows an exponential decay model. Need to show:

$$0.1 = \left(1 - \frac{0.15}{60}\right)^{850}$$

Chapter review 13

1 (a) D (b) D

2 (a) $25 913.02 (b) $26 221.74 (c) $26 438.68

3 $30 000

4 (a) 8.7% p.a. (b) 9.5% p.a.

(c) 4% p.a. (d) 18.7% p.a.

5 (a) $39 000 (b) $41 741.30

6 417 7 11.5% p.a. 8 $59 193

9 B 10 8.1% p.a. 11 B

12 (a) If the prime cost of a photocopier is $10 800 and it has a scrap value of $500, then the total possible depreciation is $10 300. If it has an expected life of 206 000 pages, then the unit cost depreciation is $0.05 per page. After it copies 125 000, the depreciation is $6250 and the written-down value is $4550.

(b) If the prime cost of an item is $52 500 and it has scrap value of $750 then the total possible depreciation is $51 750. If it has an expected life of 5000 days, then the unit cost depreciation is $10.35 per day. After it does a year's work, the depreciation is $3777.75 and the written-down value is $48 722.25.

13 63 680 14 20.4 °C 15 4.2 years

16 (a) 11% p.a. compounded quarterly—lower nominal rate and less frequent compounding.

(b) 10% p.a. compounded monthly—same nominal rate but less frequent compounding.

(c) 17% p.a. compounded daily—same compounding period but lower rate.

(d) 12% p.a. flat rate for 2 years—after 1 year, flat rate is cheaper than compounded.

17 77.8% per month

18 10% per month

Mixed review D

1 $20.23

2 area = 36.2 cm^2, perimeter = 24.7 cm

3 (a) $2\sqrt{6}$ (b) 18 (c) 6

(d) $40\sqrt{2}$ (e) $12\sqrt{15}$ (f) $126\sqrt{2}$

4 (a) $4\sqrt{5}$ (b) $12\sqrt{15}$ (c) $35\sqrt{21}$

5 (a) $a = -6$ (b) $a = -\frac{15}{16}$ (c) $a = \frac{11}{6}$

6 (a) $\frac{4\sqrt{15}}{5}$ (b) $\frac{5 + \sqrt{3}}{11}$ (c) $\frac{\sqrt{10} - \sqrt{6}}{2}$

7 (a) $(k - 1)(k - 3)$ (b) $(3t - 5)(t + 2)$

8 (a) $\frac{2a^6b^4}{3}$ (b) $a^{10}b^7$ (c) $\frac{1}{n^3k^{17}}$

9 (a) 82.45° (b) 69.12° (c) 31.54°

10 (a) $\frac{7}{3}$ (b) -2

11 (a) $788.20 (b) $796.82 (c) $801.22

12 4.4% 13 $x = -1$ and $y = -3$

14 (a)

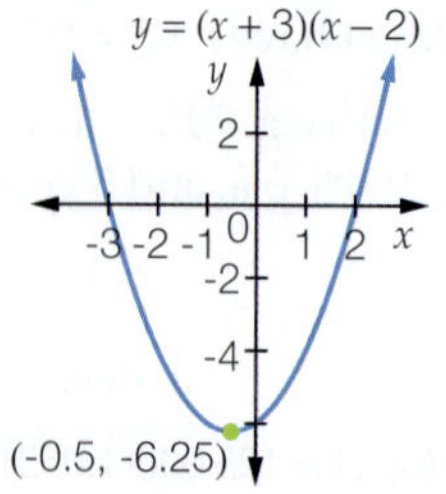

(b)

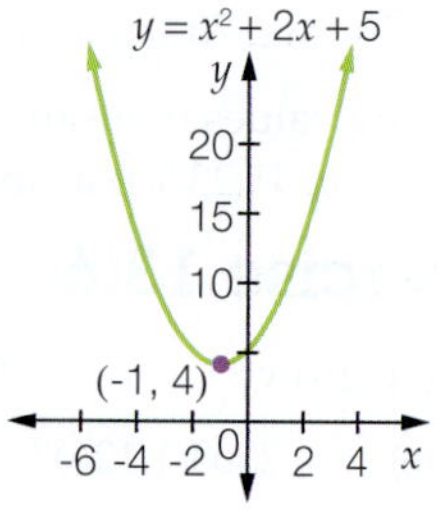

(c)

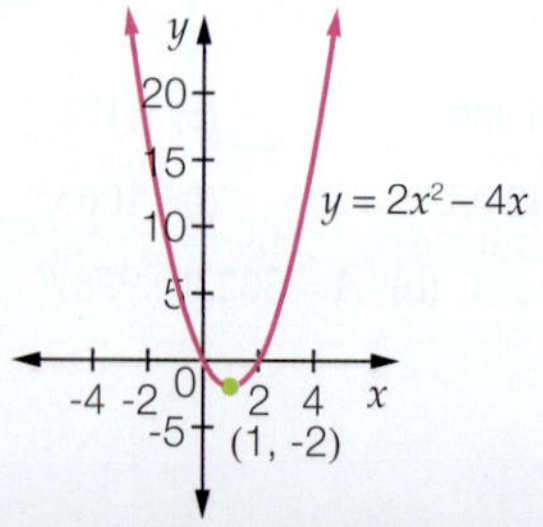

15 (a) $(8\sqrt{5} + 2\sqrt{3})$ cm (b) $15 + \sqrt{15}\ \text{cm}^2$

16 (a) A rectangle is a quadrilateral in which all angles are right angles.

(b) A rhombus is a quadrilateral with all sides equal in length.

(c) A kite is a quadrilateral with two pairs of adjacent sides equal in length.

17 (a) $\frac{1}{12}$ (b) $\frac{5}{36}$ (c) $\frac{1}{6}$

18 (a) $3^{-5} < 5^{-3}$ (b) $9^{\frac{1}{2}} > 9^{-1}$ (c) $8^{-\frac{1}{3}} = \frac{2^{-3}}{2^{-2}}$

19

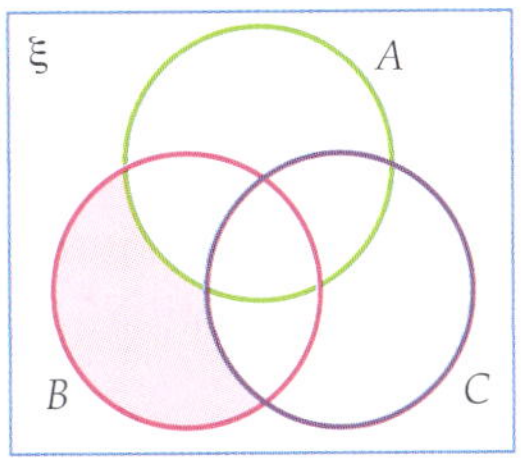

20 For an equation to be a quadratic equation, the highest powers of the variable must be a 2, and all powers of the variable must be positive integers.

21 (a) $10 - x$ (b) $A = x(10 - x) = -x^2 + 10x$

(c)

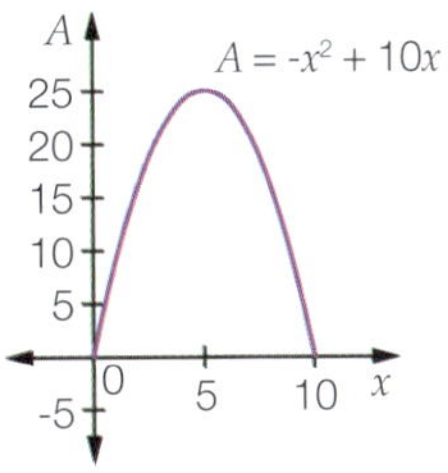

(d) 5 m by 5 m (e) 25 m^2

22 $8\sqrt{2}$ cm

23 (a) Isosceles triangle with base and height of 10 cm, and equal sides of about 11.2 cm.

(b) Draw a line from the top of the triangle to the midpoint of the base to create two congruent right-angled triangles. You can now use Pythagoras' theorem to calculate the equal side lengths of the isosceles triangle:

$l^2 = 5^2 + 10^2$
$l^2 = 125$
$l = \sqrt{125} = 5\sqrt{5}$ cm

24 $(x-4)^2 + (y-2)^2 = 25$ 25 C 26 C

Challenge D

1 D

There are 36 possible combinations from two rolls, 9 of which score a 3 or less: (1, 1), (1, 2), (1, 3), (2, 1), (2, 2), (2, 3), (3, 1), (3, 2), (3, 3)

2 (a) $\frac{1}{6}$; the second spin is an independent event from the first.

(b) (i)

Prize	Frequency	Relative frequency
\$0	8	0.16
\$5	7	0.14
\$10	9	0.18
\$20	10	0.20
\$30	8	0.16
\$60	8	0.16
Total	50	1

(ii) \$1045; $5 \times 7 + 10 \times 9 + 20 \times 10 + 30 \times 8 + 60 \times 8 = 1045$

(c) (i) You would make a profit of \$40:

$5 \times 8 + 10 \times 8 + 20 \times 8 + 30 \times 8 + 60 \times 8 = 1000$
$48 \times 20 = 960$
$1000 - 960 = 40$

(ii) You would make a \$45 profit.

3 B

Side length of base $= 12\sqrt{10}$, so each triangular face a height of $\frac{1680}{12\sqrt{10}} = 14\sqrt{10}$. Using Pythagoras' theorem, the height of the pyramid, h, is given by $h^2 = (14\sqrt{10})^2 - (6\sqrt{10})^2$, so the height $= \sqrt{1600} = 40$.

4 A

Area of four semicircles $= 2\pi$, area of inner square $= 4$, radius of inner circle $= \sqrt{2}$, area of inner circle $= 2\pi$.

Area of shaded region $= 4 + 2\pi - 2\pi = 4$.

5 (a) $\frac{1}{x} = \frac{1}{3} + \frac{1}{4}, x = \frac{12}{7}$

(b) $x + 3 = 5, x = 2$

(c) $2^x = 2 \times 2^{48} + 2 \times 2^{48} = 2 \times 2^{49} = 2^{50}, x = 50$

(d) $2^{2x-4} = 2^3, x = 3.5$

(e) $2^{4x} = 2^{x+4}(2-1)$
$2^{4x} = 2^{x+4}$
$4x = x + 4$
$x = \frac{4}{3}$

Glossary and index

addition rule for mutually exclusive events if A and B are mutually exclusive events (so that they cannot occur at the same time), then: $\Pr(A \text{ or } B) = \Pr(A) + \Pr(B)$ (p. 591)

adjacent side the side of a right-angled triangle next to the reference angle that is not the hypotenuse (p. 322)

adjusted value *see* written-down value (p. 740)

algebraic fractions expressions written in fraction form that contain at least one pronumeral (p. 196)

amplitude half the distance between the minimum and maximum y-values of the graph of a trigonometric function (p. 392)

angle at the centre an angle that has its vertex at the centre of a circle (p. 541)

angle at the circumference an angle formed by the intersection of two chords at the circumference of a circle (p. 541)

angle in a semicircle an angle formed by joining the ends of the diameter of a circle to a point on the circumference (p. 541)

angle of depression the angle between the horizontal and the line of sight when looking down (p. 343)

angle of elevation the angle between the horizontal and the line of sight when you looking up (p. 343)

angles in the same segment angles at the circumference of a circle, drawn from the same two points on the circumference and on the same side of the chord (p. 541)

appreciation the increase in the cost or value of something over a period of time (p. 712)

arc a section of the circumference of a circle (p. 540)

asymptote a line that a graph approaches but does not touch or cross (pp. 394, 669)

axis of symmetry a line that divides a figure into two identical but flipped (mirror-image) parts (p. 162)

bimodal two data values have the same highest frequency (p.65)

binomial product the product of two expressions that each have two terms, e.g. $(x + 1)(x + 2)$ (p. 150)

bivariate data a data set that contains values for two variables (p. 100)

bisect to cut in half; when a point is placed on a line interval such that the distance from one end to the point is the same as the distance from the other end to the point (p. 518)

box plot a visual representation of the statistical five-number summary (p. 74)

capacity the volume of gas or liquid that a container can hold, usually measured in litres (p. 283)

chord a line segment connecting any two points on the circumference of a circle (p. 540)

coincident lines lines that lie on top of each other (p. 46)

compass bearing a direction measured as an angle either side of north or south (p. 351)

complementary event two outcomes of an event (often called success and failure) whose probabilities add to 1 (p. 573)

completing the square factorising quadratic expressions by adding a term to form a perfect square, then subtracting the same term to keep the expression equivalent, then factorising the complete expression as the difference of two squares (p. 217)

compound interest interest that is calculated each time on the current total amount (including principal as well as previous interest), not just on the original principal (p. 710)

concentration the amount of a substance dissolved in a solution, per unit quantity of the solution (p. 312)

conditional statement extra information (conditions) given in a probability situation that changes the sample space being considered (p. 611)

cone a solid with a circular base that tapers through the height to a single vertex (p. 289)

congruent exactly the same in size and shape (p. 494)

conjugate surds two surd expressions of the form $(\sqrt{a} + \sqrt{b})$ and $(\sqrt{a} - \sqrt{b})$ that differ only in that one has an addition sign between the two terms and the other has a subtraction sign (p. 659)

constant a term with no pronumeral part; a fixed number value

conventional bearings *see* compass bearings (p. 351)

cosine (cos) the ratio $\dfrac{\text{adjacent}}{\text{hypotenuse}}$ for a reference angle in a right-angled triangle (p. 322)

cosine rule in any triangle, $a^2 = b^2 + c^2 - 2bc\cos(A)$, where a, b and c are the side lengths and A is the angle opposite side a; may also be stated as $c^2 = a^2 + b^2 - 2ab\cos(C)$ (p. 406)

coterminal angles angles that differ by multiples of 360°, so that they have the same position when shown on the unit circle (p. 399)

cubic an expression or equation where the highest power of a pronumeral is 3 (p. 441)

cumulative frequency curve a graph that shows the frequencies of all data values less than each particular data value (p. 67)

cylinder a solid with a circular base and a uniform cross-section (p. 275)

definition the minimum amount of information necessary to describe a figure exactly (p. 509)

degree

1. a unit with symbol ° used to measure the size of an angle's turn, so that a full turn is equal to 360°
2. a unit with symbol ° used to measure temperature, usually either as degrees Celsius (°C) or as degrees Fahrenheit (°F)
3. in a polynomial expression, the highest power of the variable (p. 457)

density the mass per unit volume of a substance (p. 310)

dependent events in probability, when the outcome of one event affects the outcome of another event (p. 618)

dependent variable a variable whose value depends on the value of another variable (the independent variable) (pp. 4, 109)

depreciation the decrease over time of the value of an item (p. 740)

depreciation value the loss in dollar value of an item (p. 740)

difference of two squares the algebraic rule: $a^2 - b^2 = (a + b)(a + b)$ (p. 151)

dilation a change in the size of a graph or shape; for quadratic or quartic equations, a transformation that makes the graph wider or narrower; for linear or cubic equations, a transformation that makes the graph more or less steep (pp. 167, 441)

distributive law the algebraic rule to expand brackets: $a(b + c) = ab + ac$ (p. 653)

dividend in a division calculation, the number or expression being divided; e.g. in 300 ÷ 15, the dividend is 300 (p. 460)

divisor in a division calculation, the number or expression being divided by; e.g. in 300 ÷ 15, the divisor is 15 (p. 460)

effective interest rate the annual interest rate that includes the effect of the number of compounding periods per year, expressed as the percentage of the principal that is earned in the first year; *see also* nominal interest rate (p. 733)

entire surd a surd expression where the entire number is under the root sign, e.g. $\sqrt{50}$ rather than $5\sqrt{2}$ (p. 644)

equidistant at equal distance from a point (p. 551)

equivalent angle an angle in the first quadrant that has the same sin, cos or tan value as an angle in another quadrant (p. 385)

expected life the period of time for which an item is expected to operate effectively, expressed as units of time or units of work output (p. 743)

exponential a relationship between variables where one of the variables is an exponent, such as $N = 2^t$ (p. 669)

exponential decay a relationship in which the value of one variable decreases by a fixed percentage for each unit change in the other variable (p. 749)

exponential growth a relationship in which the value of one variable increases by a fixed percentage for each unit change in the other variable (p. 748)

extrapolation making predictions from a graph outside the range of known values (p. 117)

factor theorem if, after dividing a polynomial by an expression, the remainder is zero, then the expression is a factor of the polynomial (p. 469)

factorising writing an expression as a product of factors (p. 158)

five-number summary the five statistical values used to draw a box plot: minimum, lower quartile, median, upper quartile and maximum (p. 74)

flat rate a rate of change calculated using the simple interest formula, so that the amount of change is a regular fixed amount (p. 708)

general addition rule for events if A and B are events, then: $\Pr(A \text{ or } B) = \Pr(A) + \Pr(B) - \Pr(A \text{ and } B)$ (p. 592)

gradient the measure of a line's steepness, also called slope (p. 15)

grouping in pairs factorising an expression with four terms by grouping the terms as two pairs, with a common factor in each pair, then taking out the common factor to write the expression as a binomial product (p. 159)

highest common factor (HCF) the highest factor that is common to all parts of an expression (p. 158)

hypotenuse the longest side of a right-angled triangle, opposite the right angle (p. 322)

included angle the angle between two lines (p. 495)

independent events in probability, when the outcome of one event does not affect the other event (p. 617)

independent variable a variable that determines the value of another variable (the dependent variable) (pp. 4, 109)

inflation rate the rate of increase of prices in an economy, usually expressed as a percentage per year (p. 712)

initial cost *see* prime cost (p. 740)

interest additional money that is paid or earned when money is borrowed or invested, calculated at regular times as a proportion of the total amount (p. 708)

interpolation making predictions from a graph within the range of known values (p. 117)

interquartile range (IQR) the difference between the upper quartile and the lower quartile of a data set; the range that includes the middle 50% of data values lie (p. 73)

irrational number a number that cannot be expressed in the form $\frac{a}{b}$, where a and b are whole numbers and $b \neq 0$; when written as a decimal number, the digits are non-terminating and non-recurring (p. 637)

leading coefficient the coefficient of the leading term in a polynomial (p. 457)

leading term the first term of a polynomial expression, which includes the highest power of the variable (p. 457)

line of best fit a line drawn to approximate the relationship between variables, where data has been plotted as a scatter plot (p. 115)

linear equation an equation in which the highest power of the pronumeral(s) is 1 (p. 4)

linear graph a straight-line graph (p. 4)

linear inequality a statement that relates two expressions by using one of the four inequality symbols, $>, \geq, <, \leq$; e.g. the statement $x \geq 7$ or $3 < x < 8$ (p. 40)

linear relationship a relationship between two variables that gives a straight line when graphed (p. 4)

logarithm the index or exponent written so that e.g. $2^3 = 8$ is written as $\log_2(8) = 3$ (p. 695)

major arc an arc that covers more than half the circumference of a circle (p. 540)

major segment a segment that covers more than half the area of a circle (p. 540)

mean the sum of all the data divided by the number of data values (p.65)

measures of centre represents the 'centre' of a set of numerical data (p.65)

median the middle value when the data set is written in order (p.65)

minor arc an arc that covers less than half the circumference of a circle (p. 540)

minor segment a segment that covers less than half the area of a circle (p. 540)

minute
1. one sixtieth of an hour (time measurement)
2. one sixtieth of a degree (angle measurement) (p. 325)

mode the most common data value (p. 65)

monic quadratic a quadratic expression where the coefficient of the squared value is 1 (p. 180)

mutually exclusive events impossible for two or more events to occur at the same time (p. 591)

net a series of connected shapes drawn on paper that can be folded together to form a solid (p. 274)

nominal interest rate a stated annual interest rate that does not include the number of compounding periods per year; *see also* effective interest rate (p. 733)

non-monic quadratic a quadratic expression where the coefficient of the squared pronumeral is not 1 (p. 241)

null factor law if the product of two factors is zero, than at least one of the factors is zero; in algebra, if $ab = 0$, then $a = 0$ and/or $b = 0$ (p. 208)

opposite side the side of a right-angled triangle opposite the reference angle (p. 322)

outlier a data value in a data set that is more than $1.5 \times$ IQR below the lower quartile, or more than $1.5 \times$ IQR above the upper quartile (p. 79)

parabola the shape of the graph of a quadratic relationship (p. 162)

parallel box plot two or more box plots drawn on the same scale (p. 87)

parallel dot plot two or more dot plots drawn on the same scale (p. 87)

per annum literally 'for each year' (p. 708)

percentage cumulative frequency curve a cumulative frequency curve where the vertical axis (cumulative frequency) has been converted to percentages (p. 77)

percentile the value below which a stated percentage of results lie (p. 77)

perfect square
1. the result of multiplying an expression by itself
2. the algebraic rules: $(a + b)^2 = a^2 + 2ab + b^2$ and $(a - b)^2 = a^2 - 2ab + b^2$ (p. 151)

period the horizontal distance covered by one full cycle of the graph of a trigonometric function (p. 392)

perpendicular bisector a line that intersects another line at a right angle and cuts the other line in half (p. 524)

perpendicular lines two lines that intersect or meet at right angles; the product of the gradients of two perpendicular lines is -1 (p. 524)

point of inflection a point on a curve where the shape of the curve changes direction (e.g. from concave to convex, or vice versa); points of inflection in cubics and quartics have a zero gradient (p. 442)

polynomial an expression involving a sum of terms with pronumerals raised to different powers (p. 457)

primary data data collected directly by an observer (p. 100)

prime cost the original purchase price of an item (p. 740)

principal the original amount of money borrowed or invested (p. 708)

prism a three-dimensional shape with a uniform cross-section, where the cross-section is a polygon and the edges are straight (p. 274)

probability the likelihood of a specific event, expressed as a number from 0 to 1 (p. 572)

probability tree diagram a tree diagram showing the probabilities of successive events on each of the branches (p. 601)

property any characteristic of a particular geometrical figure (p. 509)

prove to demonstrate that a theorem is true in all cases (p. 503)

pyramid a three-dimensional shape with a polygon base and triangular sides that meet at a vertex (p. 289)

quadrant one quarter of a circle or one of the four parts of a Cartesian plane, when the plane is divided vertically and horizontally by the y-axis and x-axis (p. 383)

quadratic an expression or equation where the highest power of a pronumeral is 2 (p. 180)

quadratic trinomial a quadratic expression with three terms (p. 180)

quantile a percentile expressed in decimal form (p. 77)

quartic an expression or equation where the highest power of a pronumeral is 4 (p. 441)

quartile each quarter of an ordered data set; the lower quartile or first quartile is equivalent to the 25th percentile, while the upper quartile or third quartile is equivalent to the 75th percentile (p. 73)

quotient the result of a division calculation (p. 460)

range the difference between the largest value and smallest value in a data set (p. 73)

rational number a number that can be expressed in the form $\frac{a}{b}$, where a and b are integers and $b \neq 0$ (p. 637)

rationalising the denominator the process of changing a fractional expression with a surd component in the denominator into an equivalent form with no surd component in the denominator (p. 658)

real numbers all rational and irrational numbers (p. 637)

rectangular hyperbola a non-linear graph of equation $y = \frac{a}{x-h} + k$ (p. 671)

reducing balance depreciation when an item depreciates by smaller and smaller amounts each year (p. 741)

reference angle the angle in a right-angled triangle that determines the opposite and adjacent sides used in trigonometric ratios (p. 322)

reflection the transformation of a graph or figure by finding its mirror image on the other side of a line of reflection (p. 167)

regression line a line of best fit created by analysing how well the data fits a particular mathematical relationship (p. 115)

remainder the number or expression 'left over' at the end of a division calculation (p. 460)

remainder theorem if the polynomial $P(x)$ is divided by $(x - a)$, then the remainder is $P(a)$; if $P(x)$ is divided by $(ax - b)$, then the remainder is $P\left(\frac{b}{a}\right)$ (p. 469)

residual value the value of an item (such as a piece of machinery) at the end of its useful life (p. 743)

sample space a list of all the possible outcomes for a particular situation (p. 572)

scatter plot a graph of bivariate data represented as points on a number plane (p. 101)

scrap value *see* residual value (p. 743)

second one sixtieth of a minute, as a time measurement or as an angle measurement (p. 325)

secondary data data not collected directly, instead taken from another source (p. 100)

sector of a circle the area formed between an arc and the two radii drawn from its end points (p. 540)

segment the area formed between a chord and the circumference of a circle (p. 540)

similar having the same shape but not necessarily the same size (p. 496)

simple interest interest calculated on the initial principal for the entire term of a loan/investment (p. 708)

simultaneous equations two or more equations that are true for the same pronumeral values (p. 46)

sine (sin) the ratio $\frac{\text{opposite}}{\text{hypotenuse}}$ for a reference angle in a right-angled triangle (p. 322)

sine rule in any triangle, $\frac{\sin(A)}{a} = \frac{\sin(B)}{b} = \frac{\sin(C)}{c}$, where A, B and C are the three angles and a, b, and c are the sides opposite the angles, respectively (p. 407)

sphere a ball-shaped solid, where every point on the surface is the same distance from the centre (p. 291)

standard deviation a measure of the spread of a data set, calculated by finding the square root of the variance (p. 133)

straight-line depreciation decrease in value by a fixed amount each year (p. 740)

subtended when an angle is formed by lines that come directly from the endpoints of an arc (or other curve or line) (p. 540)

surd an irrational number that can only be expressed exactly using the radical ($\sqrt{\ }$) symbol (p. 637)

surface area the total area of all the surfaces of a solid (p. 274)

tangent (tan) the ratio $\frac{\text{opposite}}{\text{adjacent}}$ for a reference angle in a right-angled triangle (p. 322)

tapered solid a solid for which the cross-section gets smaller at a constant rate ('tapers') from the base to the top (p. 289)

term

1. in algebra, a single item that is added or subtracted in an expression (p. 5)
2. in financial mathematics, the amount of time set for a loan or investment (p. 711)

theorem a mathematical statement that is proved to be always true (p. 517)

total possible depreciation the prime cost of an item, less any resale value or scrap value (p. 743)

transformation any change that can be made to a graph or figure using rotation, dilation, reflection or translation (p. 167)

translation a transformation of a graph or figure that keeps the shape, size and orientation exactly the same, but changes the location (p. 167)

tree diagram a diagram, consisting of lines branching out from other lines, used to represent a sample space; probabilities can be assigned to each branch and then used to calculate the probability of a particular outcome (p. 578)

trigonometric ratio a ratio of sides of a right-angled triangle, for a given reference angle; *see* cosine, *see* sine, *see* tangent (p. 322)

true bearing a direction measured as an angle clockwise from north (p. 351)

turning point a point where the gradient of a curve changes from positive to negative or from negative to positive (p. 162)

turning point form the form of a quadratic equation written as $y = (x - h)^2 + k$, which has the turning point (h, k) (p. 228)

two-way table a table that categorises information in rows and columns according to two different criteria (p. 585)

uniform cross-section when an object's cross-section is identical to its base, no matter where the cross-section is cut parallel to the base (p. 274)

unique equations equations that have non-parallel and non-coincident graphs, with only one point of intersection and one unique solution (p. 47)

unit circle a circle with a radius of one unit, used to define trigonometric functions (p. 383)

unit cost depreciation the amount of depreciation, in dollars, per unit of an item's expected life (p. 743)

universal set the set of all elements (all possible outcomes) in a situation; it is often labelled by a symbol ξ or U (p. 585)

variables in algebra, an unknown amount that may vary, represented by a pronumeral

variance a measure of the spread of a data set, calculated by first finding the difference of each data value from the mean, then squaring each of these differences, then finding the mean of these squared differences (p. 133)

Venn diagram a diagram used to show the relationships between sets, using areas to show the elements (or number of elements) in different sets; areas that represent sets may be disjoint (separate, no overlap), intersecting (partial overlap), or a subset (complete overlap) (p. 585)

vertex (*plural* **vertices**)

1. a point where lines or edges meet
2. *see* turning point (p. 162)

written-down value the value of an item after depreciation (p. 740)